MW01502510

CALCULUS
WITH
FINITE
MATHEMATICS

CALCULUS
WITH
FINITE
MATHEMATICS

Geoffrey C. Berresford
Long Island University

Andrew M. Rockett
Long Island University

Houghton Mifflin Company
Boston   New York

Cover designer: Harold Burch Designs, NYC

Cover image: H. Kuwajima

Senior Sponsoring Editor: Maureen O'Connor
Senior Associate Editor: Dawn Nuttall
Senior Project Editor: Maria A. Morelli
Editorial Assistant: Lauren M. Gagliardi
Senior Production/Design Coordinator: Carol Merrigan
Senior Manufacturing Coordinator: Sally Culler

Photo credits:
Chapter 0: The Harold E. Edgerton 1992 Trust, courtesy of Palm Press, Inc.; Shroud of Turin: The Bettmann Archives; Chapter 1: Alan Schein/ The Stock Market; Ben Franklin: Pennsylvania Academy of the Fine Arts; Chapter 2: Steve Leonard/Tony Stone Images; Chapter 3: Stock quotron: Jim Pickerell/Tony Stone Images; Chapter 4: Hank de Lespinasse/The Image Bank; Chapter 5 © The Photo Works/Photo Researchers, Inc.; IBM probability machine: Courtesy IBM Inc.; Chapter 6: The Harold E. Edgerton 1992 Trust, courtesy of Palm Press, Inc.; Gottfried Leibniz: Stock Montage, Inc.; Chapter 7: St. Louis arch: Andrea Pistolesi/The Image Bank; Chapter 8: Owen Franken/Stock, Boston; Chapter 9: Joseph Nettis/Stock, Boston; Chapter 10: David W. Hamilton/The Image Bank; Chapter 11: Topographic map: © Dynamic Graphics Inc., available from Raven Maps and Images.

Copyright © 1999 by Houghton Mifflin Company. All rights reserved.

No part of this work may be reproduced or transmitted in any form or by any means, electronic or mechanical, including photocopying and recording, or by any information storage or retrieval system without the prior written permission of Houghton Mifflin Company unless such copying is expressly permitted by federal copyright law. Address inquiries to College Permissions, Houghton Mifflin Company, 222 Berkeley Street, Boston, MA 02116-3764.

Printed in the U.S.A.

ISBN
Text: 0-395-70825-7
123456789-DC-03 02 01 00 99

PREFACE

A recent scientific study of yawning found that more yawns occurred in calculus class than anywhere else.* This book hopes to remedy that situation. Rather than being another dry recitation of standard results, our presentation endeavors to exhibit some of the many fascinating and useful applications of mathematics in business, the sciences, and everyday life. Even beyond its utility, however, there is a beauty to mathematics, and we hope that this book will convey some of its elegance and simplicity.

This book is an introduction to finite mathematics and calculus and their applications to the management, social, behavioral, and biomedical sciences, and other fields. Chapter 0 consists of a brief review of functions. Part I: Finite Mathematics covers compound interest and related topics, matrices, linear programming, finite probability, and statistics. Part II: Calculus covers derivatives and integrals, and concludes with introductory material on differential equations and calculus of several variables.

Features

Realistic Applications The basic nature of courses using this book is very "applied"; therefore, this book contains an unusually large number of applications, many appearing in no other textbook. In Part I we calculate the real cost of a car loan, maximize a factory's profit, find the fair price of a lottery ticket, and the probability of a jury returning the correct verdict. In Part II we use calculus to predict the national debt, maximize longevity, estimate the dangers of cigarette smoking, study global warming, judge the authenticity of the Shroud of Turin, and evaluate strategies for controlling heroin, marijuana, and liquor sales. These applications show that calculus is not just the manipulation of abstract symbols but is deeply connected to everyday life.

Graphing Calculators (Optional) Calculators with capabilities previously available only on computers have changed the way mathe-

---

* Baenninger, Ronald. "Some Comparative Aspects of Yawning in Betta splendens, Homo sapiens, Panthera leo, and Papoi spinx". *Journal of Comparative Psychology*, 1987, vol. 101, No. 4, 349–354.

matics can be taught and used. Reading this book does *not* require a graphing calculator, but having one will simplify the calculations in many problems, and may at the same time deepen understanding of mathematics by permitting students to concentrate on *concepts*. Throughout the book are **Graphing Calculator Explorations** and **Exercises,** which explore new topics, carry out otherwise "messy" calculations, or show the limitations and pitfalls of technology. While any graphing calculator (or a computer) may be used, the displays shown in the text are from the Texas Instruments TI-83. A discussion of the essentials of graphing calculators follows this preface. For those not using a graphing calculator, the Graphing Calculator Explorations are boxed so that they can be easily omitted or simply read for enrichment.

Enhanced Readability For the sake of continuity, references to earlier material have been minimized by restating results whenever they are used. Where references are necessary, explicit page numbers are given.

Application Previews Most sections begin with an Application Preview that presents an interesting application or historical development of the mathematics developed in the section. They are self-contained (although some exercises are based on them), and serve to motivate interest in the coming section. Topics include world records in the mile run, retirement planning, airline scheduling, political polling, inflation, AIDS, and predicting personal wealth.

Practice Problems Learning requires active participation—"mathematics is not a spectator sport." Throughout the reading are short pencil and paper "Practice Problems" designed to consolidate understanding of one topic before another is introduced. Complete solutions to all practice problems are given at the end of the book.

Annotations Notes to the right of many mathematical formulas and manipulations state the results in words, emphasizing the important skill of "reading mathematics." They also provide explanation and justification for the steps in calculations, and interpretation of the results.

Extensive Exercises Anyone who has ever learned any mathematics did so by solving many problems, and the exercises are the most essential part of the learning process. The exercises are graded from routine drill to significant applications, including some with Explorations and Excursions that extend and augment the material presented in the text. Most **Applied Exercises** have both general and specific titles, such as "Environmental Science: Pollution Control." Exercises marked with the symbol ▦ require a business or scientific calculator with keys like ⟨ln x⟩ and ⟨yˣ⟩ for natural logarithms and powers. Exercises marked by ▦ require a graphing calculator. At the end of the book are answers to the odd-numbered exercises, and answers to *all* Chapter Review Exercises and Cumulative Review Exercises.

Constant Reinforcement Because of the many new ideas and techniques in this book, frequent summaries and reviews are provided at several different levels. **Section Summaries** briefly state essential formulas and key concepts. **Chapter Summaries** review the major developments of the chapter (keyed to particular review exercises) and offer **Hints and Suggestions** that unify the chapter, give specific reminders of basic ideas that are sometimes overlooked or forgotten, and list a selection of **Review Exercises** for a **Practice Test** of the chapter material. **Cumulative Reviews** contain exercises from groups of related chapters.

Projects and Essays Concluding each chapter is a collection of small research projects and topics for student essays that ask the student (or a group of students) to research a relevant person or an idea, to compare several different mathematical ideas, or to relate a concept to their lives. This feature is in keeping with recent recommendations from the Mathematical Association of America and the American Mathematical Association of Two-Year Colleges. Other more challenging projects can be found in the highly recommended *MAA Notes* Numbers 27–30, available from the Mathematical Association of America in Washington, D.C.

Explorations and Excursions At the end of some exercise sets are optional problems of a more advanced nature that carry the development of certain topics beyond the level of the text.

Accuracy and Proofs All of the answers and other mathematics have been checked carefully by several mathematicians. The statements of definitions and theorems are mathematically accurate. Because the treatment is applied rather than theoretical, intuitive and geometric justifications have often been preferred to formal proofs. When proofs are given, however, they are correct and "honest."

Philosophy We wrote this book with several principles in mind. One is that to learn something, it is best to begin doing it as soon as possible.

Therefore the preliminary material is brief, so that students begin doing useful mathematics as soon as possible. An early start allows more time during the course for interesting applications and necessary review. Another principle is that the mathematics should be done together with the applications. Consequently every section contains applications (there are no "pure math" sections).

Prerequisites The only prerequisite for this course is some knowledge of algebra, functions, and graphing, which are reviewed in Chapter 0. Other review material has been placed in relevant locations throughout the book.

How to Obtain Graphing Calculator Programs

The optional graphing calculator programs used in the text have been written for a variety of Texas Instruments Graphing Calculators (including the TI-82, TI-83, TI-85, TI-86, TI-89, and TI-92), and may be obtained for free, by any of the following ways:

- If you know someone who already has the programs on a Texas Instruments graphing calculator of the same model as yours, you can easily transfer the programs from their calculator to yours using the black cable that came with the calculator and the LINK button. (Even if you use one of the following methods to get a "first" copy of the programs, you can use this method to transfer the programs to others in your class.)

- If you have access to the Internet, you may download the programs from the Houghton Mifflin website at **www.hmco.com** onto a computer. Then use the TI-GRAPH LINK™ software and grey cable (available for purchase in most stores that sell graphing calculators) to transfer the programs to your calculator. Descriptive materials for the programs are also available at this website.

- You may send a formatted 3-1/2 "floppy" disk, carefully packed, to the authors at the following address, specifying your type of computer (IBM-compatible or Macintosh) and specifying your type of Texas Instruments calculator (TI-82, TI-83, TI-85, TI-86, TI-89, or TI-92). We will return a disk containing the appropriate programs and a packet of descriptive information. After you use the disk to load the programs into your computer, you will need the TI-GRAPH LINK™ software and grey cable (available for purchase in most stores that sell graphing calculators) to transfer the programs to your calculator.

- You may send your calculator (TI-82, TI-83, TI-85, TI-86, TI-89, or TI-92), carefully packed, to the authors at the following address, and we will return it loaded with the appropriate programs, along with a packet of descriptive information.

Authors' Address: Dr. G.C. Berresford and Dr. A.M. Rockett
Department of Mathematics
C.W. Post Campus of Long Island University
720 Northern Boulevard
Brookville, New York 11548-1300

Supplements for Instructors

Instructor's Manual This booklet contains full solutions for all exercises in the book.

Computerized Test Bank (IBM and Macintosh) The test bank contains more than 2000 test questions arranged by chapter and section, allowing instructors to create customized tests efficiently. Many of these test questions are applied problems. Test questions can be selected by section number as well as other criteria. Both versions have full editing capabilities and high-quality graph reproduction. They produce scrambled and multiple test versions in multiple-choice or free-response format, and provide answer keys. Both versions also provide **on-line testing** and **gradebook** functions.

Printed Test Bank with Chapter Tests This is a printed version of the Computerized Test Bank for instructors who do not use computers. Also included are two comprehensive Chapter Tests for each chapter (one multiple choice and one free response). Answers to all test questions are included.

Supplement for Students

Student Solutions Manual This booklet, available from your bookstore, contains worked-out solutions to selected exercises.

Acknowledgments

We are indebted to many people for their useful suggestions, conversations, and correspondence during the writing of this book. We thank Chris Berresford, Anne Burns, Richard Cavaliere, Ruth Enoch, Theodore Faticoni, Jeff Goodman, Susan Halter, Brita and Ed Immergut, Ethel Matin, Gary Patric, Shelly Rothman, Charlene Russert, Stuart Saal, Bob Sickles, John Stevenson, and all of our "Math 5 and 6" students at C. W. Post over the past several years for serving as proofreaders and critics.

We had the good fortune to have had supportive and expert editors at Houghton Mifflin: Maureen O'Connor (Senior Sponsoring Editor), Dawn Nuttall (Senior Associate Editor), and Maria Morelli (Senior Project Editor). They made the difficult tasks seem easy, and helped beyond words. We also express our gratitude to the many others at Houghton Mifflin who made important contributions too numerous to mention.

The following reviewers have contributed greatly to the progression from first draft to finished book:

Paul Allen, *University of Alabama*
Bob Bradshaw, *Ohlone College,* California
James Brasel, *Philips Community College,* Arkansas
Linda Buchanan, *Howard College,* Texas
Charles E. Cleaver, *The Citadel,* South Carolina
Barbara Cohen, *West Los Angeles College,* California
Catherine Cron, *Fairfield University,* Connecticut
R. D. Derderian, *Providence College,* Rhode Island
Chris Edwards, *University of Wisconsin—Oshkosh*
Dauhrice K. Gibson, *Gulf Coast Community College,* Florida
Lee H. LaRue, *Paris Junior College,* Texas
Marsha May, *Midwestern State University,* Texas
Theodore F. Moore, *Mohawk Valley Community College,* New York
Sharon Wilson, *University of Tulsa,* Oklahoma

Dedication

We dedicate this book to our wives, Barbara and Kathryn, and our children, Lee, Chris, Justin and Joshua, for their understanding and patience, without which this book would not exist.

Comments Welcomed

With the knowledge that any book can always be improved, we welcome corrections, constructive criticisms and suggestions from every reader.

Graphing Calculator Terminology

The graphing calculator applications have been kept as generic as possible for use with any of the popular graphing calculators. It is assumed that either the instructor or the student (or both) is familiar with the sequence of buttons necessary to accomplish various operations on the calculator being used. Certain standard calculator terms are capitalized in this book and are described below. Your calculator may use slightly different terminology.

The viewing or graphing **WINDOW** is the part of the Cartesian plane shown in the display screen of your graphing calculator. **XMIN** and **XMAX** are the smallest and largest x-values shown, and **YMIN** and **YMAX** are the smallest and largest y-values shown. These values can be set by using the **WINDOW** or **RANGE** command and are changed automatically by using any of the **ZOOM** operations. **XSCALE** and **YSCALE** define the distance between tick marks on the x- and y-axes.

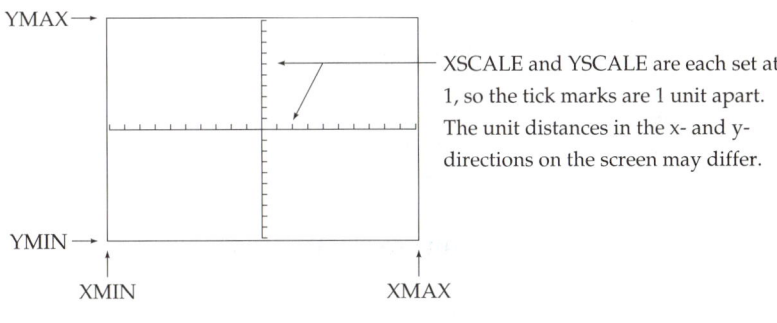

XSCALE and YSCALE are each set at 1, so the tick marks are 1 unit apart. The unit distances in the x- and y-directions on the screen may differ.

Viewing Window [−10, 10] by [−10, 10]

The viewing window is always [XMIN, XMAX] by [YMIN, YMAX]. We will set XSCALE and YSCALE so that there are a reasonable number of tick marks (generally 2 to 20) on each axis. The x- and y-axes will not be visible if the viewing window does not include the origin.

Pixel, an abbreviation for *pic*ture *el*ement, refers to a tiny rectangle on the screen that can be darkened to represent a dot on a graph. Pixels are arranged in a rectangular array on the screen. In the above window, the axes and tick marks are formed by darkened pixels. The size of the screen and number of pixels varies with different calculators.

TRACE allows you to move a flashing pixel, or *cursor*, along a curve in the viewing window with the *x*- and *y*-coordinates shown at the bottom of the screen.

> *Useful Hint:* To make the *x*-values in TRACE take simple values like .1, .2, and .3, choose XMIN and XMAX to be multiples of one less than the number of pixels across the screen. For example, on the TI-82 and TI-83, which have 95 pixels across the screen, using an *x*-window like [−9.4, 9.4] or [−4.7, 4.7] or [940, −940] will TRACE with simpler *x*-values than the standard windows stated in this book.

ZOOM IN allows you to magnify any part of the viewing window to see finer detail around a chosen point. **ZOOM OUT** does the opposite, like stepping back to see a larger portion of the plane but with less detail. These and other **ZOOM** commands change the viewing window.

VALUE or **EVALUATE** finds the value of a previously entered expression at a specified *x*-value.

SOLVE or **ROOT** finds the *x*-value that solves $f(x) = 0$, or equivalently, the *x*-intercepts of a curve. When applied to a difference $f(x) - g(x)$, it finds the *x*-value where the two curves meet (also done by the **INTERSECT** command).

MAX and **MIN** find the maximum and minimum values of a previously entered curve between specified *x*-values.

NDERIV or **DERIV** or *dy/dx* approximates the *derivative* of a function at a point. **FnInt** or $\int f(x)\ dx$ approximates the definite integral of a function on an interval.

In **CONNECTED MODE** your calculator will darken pixels to connect calculated points on a graph to show it as a continuous or "unbroken" curve. However, this may lead to "false lines" in a graph that should have breaks or "jumps." False lines can be eliminated by using **DOT MODE.**

The **TABLE** command lists in table form the values of a function, just as you have probably done when graphing a curve. The *x*-values may be chosen by you or by the calculator.

The **Order of Operations** used by most calculators evaluates operations in the following order: first powers and roots, then operations like **LN** and **LOG,** then multiplication and division, then addition and subtraction—left to right within each level. For example, $5 \wedge 2x$ means $(5 \wedge 2)x$, *not* $5 \wedge (2x)$. Also, $1/x+1$ means $(1/x)+1$, *not* $1/(x+1)$. See your calculator's instuction manual for further information. *Be careful:* Some calculators evaluate $1/2x$ as $(1/2)x$ and some as $1/(2x)$. When in doubt, use parentheses to clarify the expression.

Much more information can be found in the manual for your graphing calculator. Other features will be discussed later as needed.

CONTENTS

CHAPTER 2

MATRICES AND SYSTEMS OF EQUATIONS   *167*

CHAPTER 3

LINEAR PROGRAMMING   *280*

CHAPTER 11   **CALCULUS OF SEVERAL VARIABLES**   *911*

CALCULUS
WITH
FINITE
MATHEMATICS

0

FUNCTIONS

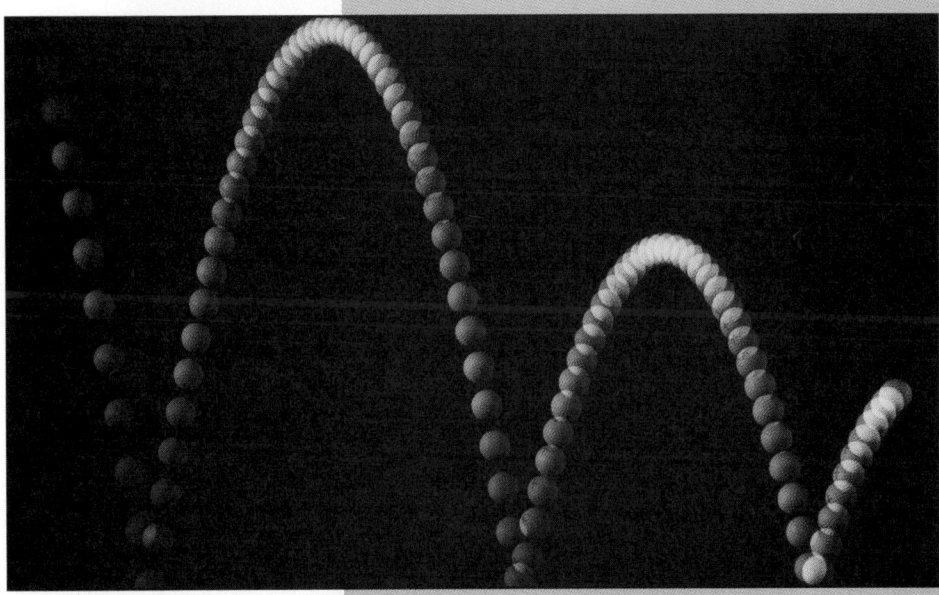

Parabolas described by a bouncing ball

0.1 Real Numbers, Inequalities, and Lines

APPLICATION PREVIEW

World Record Mile Runs

The dots on the graph below show the world record times for the mile run from 1865 to the 1993 world record of 3 minutes 44.39 seconds, set by the Algerian runner Noureddine Morceli. These points fall roughly along a line, called the **regression line.** The regression line is easily found using a graphing calculator, based on a method called **least squares,** which is explained in Chapters 2 and 11. Several exercises in this chapter involve using a graphing calculator to find the regression line for a collection of points.

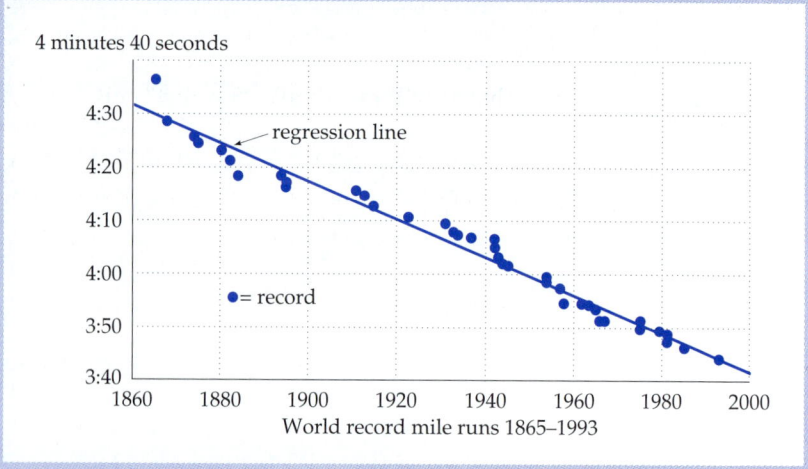

HISTORY OF THE RECORD FOR THE MILE RUN

Time	Year	Athlete	Time	Year	Athlete
4:36.5	1865	Richard Webster	4:15.6	1895	Thomas Conneff
4:29.0	1868	William Chinnery	4:15.4	1911	John Paul Jones
4:28.8	1868	Walter Gibbs	4:14.4	1913	John Paul Jones
4:26.0	1874	Walter Slade	4:12.6	1915	Norman Taber
4:24.5	1875	Walter Slade	4:10.4	1923	Paavo Nurmi
4:23.2	1880	Walter George	4:09.2	1931	Jules Ladoumegue
4:21.4	1882	Walter George	4:07.6	1933	Jack Lovelock
4:18.4	1884	Walter George	4:06.8	1934	Glenn Cunningham
4:18.2	1894	Fred Bacon	4:06.4	1937	Sydney Wooderson
4:17.0	1895	Fred Bacon	4:06.2	1942	Gunder Hägg

Time	Year	Athlete	Time	Year	Athlete
4:06.2	1942	Arne Andersson	3:51.3	1966	Jim Ryun
4:04.6	1942	Gunder Hägg	3:51.1	1967	Jim Ryun
4:02.6	1943	Arne Andersson	3:51.0	1975	Filbert Bayi
4:01.6	1944	Arne Andersson	3:49.4	1975	John Walker
4:01.4	1945	Gunder Hägg	3:49.0	1979	Sebastian Coe
3:59.4	1954	Roger Bannister	3:48.8	1980	Steve Ovett
3:58.0	1954	John Landy	3:48.53	1981	Sebastian Coe
3:57.2	1957	Derek Ibbotson	3:48.40	1981	Steve Ovett
3:54.5	1958	Herb Elliott	3:47.33	1981	Sebastian Coe
3:54.4	1962	Peter Snell	3:46.31	1985	Steve Cram
3:54.1	1964	Peter Snell	3:44.39	1993	Noureddine Morceli
3:53.6	1965	Michel Jazy			

Source: USA Track & Field

The equation of the regression line shown in the graph is $y = -0.357x + 257.46$, where x represents years after 1900 and y is in seconds. The regression line can be used to predict the world mile record in future years. Notice that the most recent world record would have been predicted quite accurately by this line, since the rightmost dot falls almost exactly on the line. Linear trends, however, must not be extended too far. The downward slope of this line means that it will eventually "predict" mile runs in a fraction of a second, or in *negative* time. Moral: In the real world, linear trends do not continue indefinitely. This and other topics in "linear" mathematics will be developed in Section 0.1 and thereafter.

Introduction

In this section we will study *linear* relationships between two quantities, that is, relationships that can be represented by *lines*. Later we will turn to *nonlinear* relationships, such as learning curves and population growth rates.

When reading this book, it will be helpful (but not necessary) to have a graphing calculator. The **Graphing Calculator Explorations** show how to use a graphing calculator to explore a concept more deeply or to analyze an application in more detail. The parts of the book that require graphing calculators are marked by the symbol 🖩.

Exercises that can be done with a graphing *or* scientific *or* business calculator (with keys like [ln x] and [y^x]) are marked by 🖩.

Real Numbers and Inequalities

In this book the word "number" means *real* number, a number that can be represented by a point on the number line (also called the *real line*).

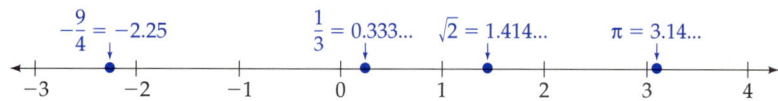

The *order* of the real numbers is expressed by inequalities, with $a < b$ meaning "*a* is to the *left* of *b*."

Inequalities

Inequality	In words
$a < b$	*a* is less than (smaller than) *b*
$a \le b$	*a* is less than or equal to *b*
$a > b$	*a* is greater than (larger than) *b*
$a \ge b$	*a* is greater than or equal to *b*

The inequalities $a < b$ and $a > b$ are called "strict" inequalities, and $a \le b$ and $a \ge b$ are called "nonstrict" inequalities.

EXAMPLE 1   Inequalities Between Numbers

a. $3 \le 5$ **b.** $6 > -2$ **c.** $-10 < -5$

-10 is less than (smaller than) -5

Throughout this book are many **Practice Problems,** short questions designed to check your understanding of a topic before moving on to new material. Full solutions are given at the back of the book. Solve the following Practice Problem and then check your answer.

PRACTICE PROBLEM 1   Which number is smaller: $\frac{1}{100}$ or $-1,000,000$?

Solution at the back of the book

A *double* inequality, such as $a < x < b$, means that *both* the inequalities $a < x$ and $x < b$ hold. The inequality $a < x < b$ can be interpreted graphically as "*x* is between *a* and *b*."

Sets and Intervals

Braces { } are read "the set of all" and a vertical bar | is read "such that."

EXAMPLE 2   **Interpreting Sets**

The set of all

a. $\{x \mid x > 3\}$ means "the set of all x such that x is greater than 3."

Such that

b. $\{x \mid -2 < x < 5\}$ means "the set of all x such that x is between -2 and 5." ∎

PRACTICE PROBLEM 2   **a.**   Write in set notation "the set of all x such that x is greater than or equal to -7."

b.   Express in words: $\{x \mid x < -1\}$.   *Solutions at the back of the book*

The set $\{x \mid 2 \leq x \leq 5\}$ can be expressed in *interval* notation by enclosing the endpoints 2 and 5 in square brackets, [2, 5]. The *square* brackets indicate that the endpoints are *in*cluded. The set $\{x \mid 2 < x < 5\}$ can be written (2, 5). The *parentheses* indicate that the endpoints 2 and 5 are *ex*cluded. An interval is *closed* if it includes both endpoints, and *open* if it includes neither endpoint. The four types of intervals are shown below: a *solid* dot ● on the graph indicates that the point is *in*cluded in the interval; a *hollow* dot ○ indicates that the point is *ex*cluded.

Finite Intervals

Interval Notation	Set Notation	Graph	Type
[a, b]	$\{x \mid a \leq x \leq b\}$		Closed
(a, b)	$\{x \mid a < x < b\}$		Open
[a, b)	$\{x \mid a \leq x < b\}$		Half-open
(a, b]	$\{x \mid a < x \leq b\}$		or half-closed

An interval may extend infinitely far to the right (indicated by the symbol ∞ for "infinity") or infinitely far to the left (indicated by $-\infty$ for "negative infinity"). Note that ∞ and $-\infty$ are not numbers, but are merely symbols to indicate that the interval extends endlessly in one direction or the other. The infinite intervals in the next box are said to be *closed* or *open* depending on whether they *include* or *exclude* their single endpoint.

Infinite Intervals

Interval Notation	Set Notation	Graph	Type
$[a, \infty)$	$\{x \mid x \geq a\}$		Closed
(a, ∞)	$\{x \mid x > a\}$		Open
$(-\infty, a]$	$\{x \mid x \leq a\}$		Closed
$(-\infty, a)$	$\{x \mid x < a\}$		Open

The interval extending infinitely far in *both* directions (meaning the entire real line, or the set of all real numbers) is denoted by the symbol $\mathbb{R}$, or by $(-\infty, \infty)$.

$$\mathbb{R} = (-\infty, \infty)$$

Cartesian Plane

Two real lines or *axes*, one horizontal and one vertical, intersecting at their zero points, define the *Cartesian plane*.* The axes divide the plane into four *quadrants*, I through IV. Any point in the Cartesian plane can be specified uniquely by an ordered pair of numbers (x, y); x, called the *abscissa* or *x-coordinate*, is the number on the horizontal axis corresponding to the point; y, called the *ordinate* or *y-coordinate*, is the number on the vertical axis corresponding to the point.

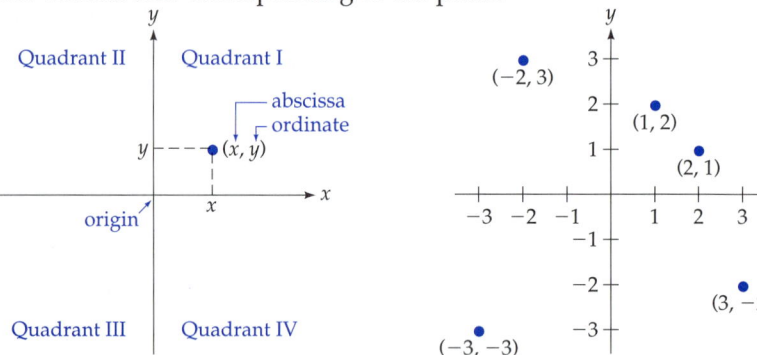

The Cartesian plane

The Cartesian plane with several points. Order matters: $(1, 2)$ is not the same as $(2, 1)$

Slope of a Line

The symbol Δ (read "delta," the Greek letter D) means "the change in." For any two points (x_1, y_1) and (x_2, y_2) we define

* So named because it was originated by the French philosopher and mathematician René Descartes (1596–1650). Following the custom of the day, Descartes signed his scholarly papers with his Latin name Cartesius, hence "Cartesian" plane.

$$\Delta x = x_2 - x_1 \qquad \text{The change in } x \text{ is the difference in the } x\text{-coordinates}$$
$$\Delta y = y_2 - y_1 \qquad \text{(and similarly for } y\text{)}$$

The *slope* of a nonvertical line measures the steepness of the line, and is defined as *the change in y divided by the change in x* for any two points on the line.

Slope of Line Through (x_1, y_1) and (x_2, y_2)

$$m = \frac{\Delta y}{\Delta x} = \frac{y_2 - y_1}{x_2 - x_1} \qquad \text{Slope is the change in } y \text{ over the}$$
$$\text{change in } x \ (x_2 \neq x_1)$$

The changes Δy and Δx are often called, respectively, the "rise" and the "run," with the understanding that a negative "rise" means a "fall." Slope is then "rise over run."

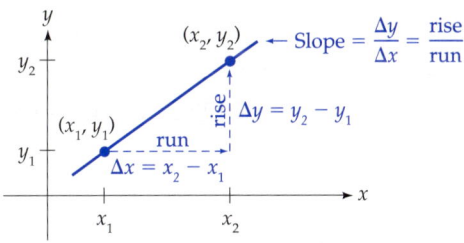

EXAMPLE 3 Finding Slopes and Graphing Lines

Find the slope of the line through each pair of points, and graph the line.

a. $(1, 3), (2, 5)$ **b.** $(2, 4), (3, 1)$ **c.** $(-1, 3), (2, 3)$ **d.** $(2, -1), (2, 3)$

Solution

We use the slope formula $m = \dfrac{y_2 - y_1}{x_2 - x_1}$ for each pair $(x_1, y_1), (x_2, y_2)$.

a. For $(1, 3)$ and $(2, 5)$ the slope is

$$\frac{5 - 3}{2 - 1} = \frac{2}{1} = 2$$

b. For $(2, 4)$ and $(3, 1)$ the slope is

$$\frac{1 - 4}{3 - 2} = \frac{-3}{1} = -3$$

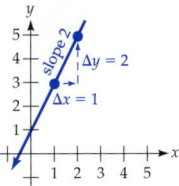

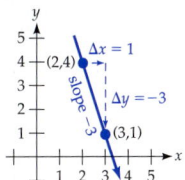

c. For $(-1, 3)$ and $(2, 3)$ the slope is $\dfrac{3 - 3}{2 - (-1)} = \dfrac{0}{3} = 0$

d. For $(2, -1)$ and $(2, 3)$ the slope is $\dfrac{3 - (-1)}{2 - 2} = \dfrac{4}{0}$ Undefined!

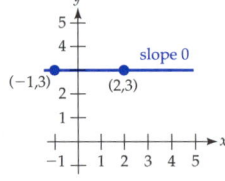

If $\Delta x = 1$, as in Examples 3a and 3b, then the slope is just the "rise," giving

$$\text{Slope} = \left(\begin{array}{c} \text{the amount that the line rises} \\ \text{when } x \text{ increases by 1} \end{array} \right)$$

PRACTICE PROBLEM 3

A company president is considering four different business strategies, called S_1, S_2, S_3, and S_4, each with different projected future profits. The graph on the right shows the annual projected profit for the first few years for each of the strategies. Which strategy yields:

a. The highest projected profit in year 1?

b. The highest projected profit in the long run?   *Solutions at the back of the book*

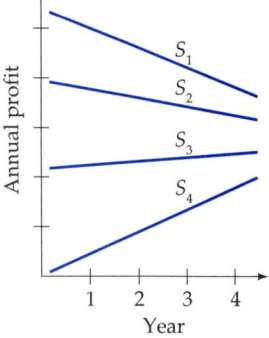

Equations of Lines

The point where a nonvertical line crosses the y-axis is called the *y-intercept* of the line. The y-intercept can be given either as the y-coordinate b or as the point $(0, b)$. Such a line can be expressed very simply in terms of its slope and y-intercept.

Slope-Intercept Form of a Line

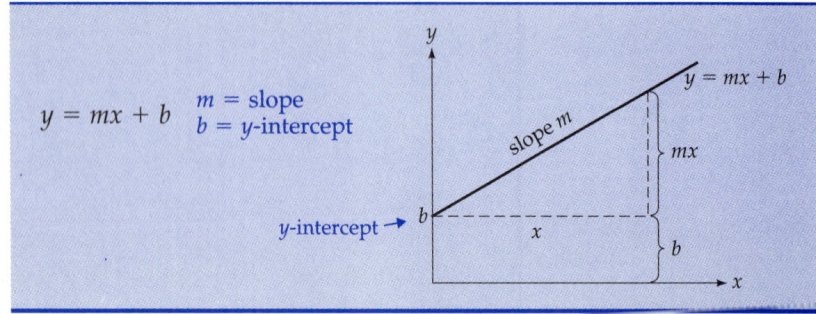

EXAMPLE 4   **Using the Slope-Intercept Form**

Find an equation of the line with slope -2 and y-intercept 4.

Solution

$$y = -2x + 4 \qquad\qquad y = mx + b \text{ with } m = -2 \text{ and } b = 4$$

We graph the line by first plotting the y-intercept $(0, 4)$. Using the slope $m = -2$, we plot another point 1 unit over and 2 units *down* from the y-intercept. We then draw the line through these two points, as shown below.

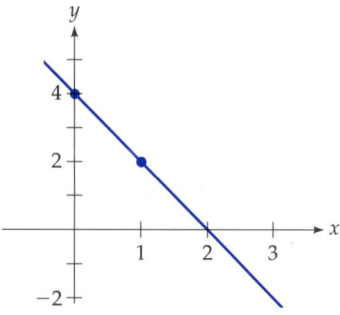

Graphing Calculator Exploration

a. Use a graphing calculator to graph the lines $y_1 = x$, $y_2 = 2x$, and $y_3 = 3x$ simultaneously on the viewing window $[-10, 10]$ by $[-10, 10]$. How do the graphs change as the coefficient of x increases from 1 to 2 to 3?

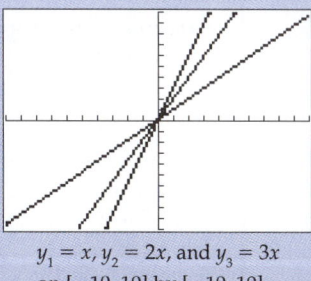

$y_1 = x, y_2 = 2x,$ and $y_3 = 3x$
on $[-10, 10]$ by $[-10, 10]$

b. Predict what the graph of $y = 0.5x$ would look like. What about $y = -2x$? Check your predictions by graphing them.

c. Describe the graph of the line $y = mx$ for any number m.

Point-Slope Form of a Line

$$y - y_1 = m(x - x_1)$$

(x_1, y_1) = point on the line
m = slope

This form comes directly from the slope formula $m = \dfrac{y_2 - y_1}{x_2 - x_1}$, dropping the subscript 2 and multiplying each side by $(x - x_1)$.

EXAMPLE 5   Using the Point-Slope Form

Find an equation of the line through $(6, -2)$ with slope $-\frac{1}{2}$.

Solution

$$y - (-2) = -\tfrac{1}{2}(x - 6)$$

$y - y_1 = m(x - x_1)$ with
$y_1 = -2$, $m = -\frac{1}{2}$, and $x_1 = 6$

$$y + 2 = -\tfrac{1}{2}x + 3$$

Eliminating parentheses

$$y = -\tfrac{1}{2}x + 1$$

Subtracting 2 from each side

■

Alternatively, we could have found this equation using $y = mx + b$, replacing m by the given slope $-\frac{1}{2}$, and then substituting the given $x = 6$, $y = -2$ to evaluate b.

EXAMPLE 6   Finding an Equation for a Line Through Two Points

Find an equation for the line through the points $(4, 1)$ and $(7, -2)$.

Solution   The slope is not given, so we calculate it from the two points.

$$m = \frac{-2 - 1}{7 - 4} = \frac{-3}{3} = -1 \qquad m = \frac{y_2 - y_1}{x_2 - x_1} \text{ with } (4, 1) \text{ and } (7, -2)$$

Then we use the point-slope formula with this slope and either of the two points.

$$y - 1 = -1(x - 4)$$

$y - y_1 = m(x - x_1)$ with
slope -1 and point $(4, 1)$

$$y - 1 = -x + 4$$

Eliminating parentheses

$$y = -x + 5$$

Adding 1 to each side

■

PRACTICE PROBLEM 4

Find the slope-intercept form of the line through the points (2, 1) and (4, 7).   *Solution at the back of the book*

Vertical and horizontal lines have particularly simple equations: a variable equaling a constant.

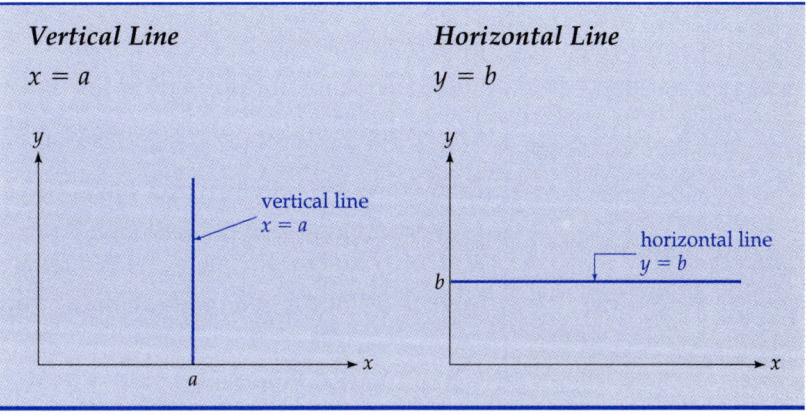

EXAMPLE 7 Graphing Vertical and Horizontal Lines

Graph the lines $x = 2$ and $y = 6$.

Solution

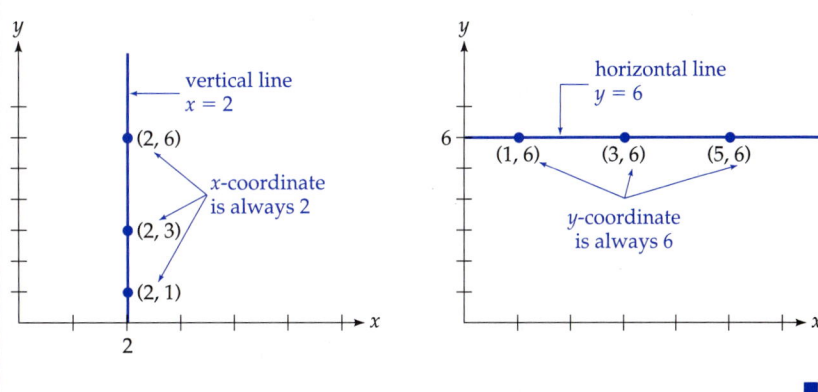

EXAMPLE 8 Finding Equations of Vertical and Horizontal Lines

a. Find an equation for the *vertical* line through the point (3, 2).

b. Find an equation for the *horizontal* line through the point (3, 2).

Solution

a. Vertical line $x = 3$

$x = a$, with a being the
x-coordinate from $(3, 2)$

b. Horizontal line $y = 2$

$y = b$, with b being the
y-coordinate from $(3, 2)$

■

PRACTICE PROBLEM 5   Find an equation for the vertical line through the point $(-2, 10)$.

Solution at the back of the book

Distinguish carefully between slopes of vertical and horizontal lines:
Vertical line: slope is *undefined*.
Horizontal line: slope *is* defined, and is zero.
There is one form that covers *all* lines, vertical and nonvertical.

General Linear Equation

$$ax + by = c$$

For constants a, b, c, with a
and b not both zero

Any equation that can be written in this form is called a *linear equation*, and the variables are said to *depend linearly* on each other.

EXAMPLE 9   **Finding the Slope and the *y*-intercept from a Linear Equation**

Find the slope and y-intercept of the line $2x + 3y = 12$.

Solution   We write the line in slope-intercept form. Solving for y:

$$3y = -2x + 12$$

$2x + 3y = 12$ after subtracting $2x$
from both sides

$$y = -\tfrac{2}{3}x + 4$$

Dividing each side by 3 gives the
slope-intercept form $y = mx + b$

Therefore, the slope is $-\tfrac{2}{3}$ and the y-intercept is $(0, 4)$.

■

PRACTICE PROBLEM 6   Find the slope and y-intercept of the line $x - \dfrac{y}{3} = 2$.

Solution at the back of the book

SUMMARY

An *interval* is a set of real numbers corresponding to a section of the real line. The interval is *closed* if it contains all of its endpoints, and *open* if it contains none of its endpoints.

The nonvertical line through two points (x_1, y_1) and (x_2, y_2) has slope

$$m = \frac{\Delta y}{\Delta x} = \frac{y_2 - y_1}{x_2 - x_1} \qquad x_1 \neq x_2$$

There are five equations or "forms" for lines:

$$y = mx + b$$

Slope-intercept form
m = slope, b = y-intercept

$$y - y_1 = m(x - x_1)$$

Point-slope form
(x_1, y_1) = point, m = slope

$$x = a$$

Vertical line (slope undefined)
a = x-intercept

$$y = b$$

Horizontal line (slope zero)
b = y-intercept

$$ax + by = c$$

General linear equation

EXERCISES 0.1

Write each interval in set notation and graph it on the real line.

1. $[0, 6)$ 2. $(-3, 5]$ 3. $(-\infty, 2]$ 4. $[7, \infty)$

5. Given the equation $y = 5x - 12$, how will y change if x:
   a. Increases by 3 units?
   b. Decreases by 2 units?

6. Given the equation $y = -2x + 7$, how will y change if x:
   a. Increases by 5 units?
   b. Decreases by 4 units?

Find the slope (if it is defined) of the line through each pair of points.

7. $(2, 3)$ and $(4, -1)$ 8. $(3, -1)$ and $(5, 7)$

9. $(-4, 0)$ and $(2, 2)$ 10. $(-1, 4)$ and $(5, 1)$

11. $(0, -1)$ and $(4, -1)$ 12. $(-2, \frac{1}{2})$ and $(5, \frac{1}{2})$

13. $(2, -1)$ and $(2, 5)$ 14. $(6, -4)$ and $(6, -3)$

For each equation, find the slope m and y-intercept $(0, b)$ (if they exist) and draw the graph.

15. $y = 3x - 4$ 16. $y = 2x$

17. $y = -\frac{1}{2}x$ 18. $y = -\frac{1}{3}x + 2$

19. $y = 4$ 20. $y = -3$

21. $x = 4$ 22. $x = -3$

23. $2x - 3y = 12$ 24. $3x + 2y = 18$

25. $x + y = 0$ 26. $x = 2y + 4$

27. $x - y = 0$ 28. $y = \frac{2}{3}(x - 3)$

29. $y = \frac{x + 2}{3}$ 30. $\frac{x}{2} + \frac{y}{3} = 1$

31. $\frac{2x}{3} - y = 1$ 32. $\frac{x + 1}{2} + \frac{y + 1}{2} = 1$

 Use a graphing calculator to graph each line. [*Note:* Your graph will depend on the viewing window you choose. Begin with a "standard" window like $[-10, 10]$ by $[-10, 10]$, choosing a larger window (or "zooming out") if the line does not appear.]

33. $y = 2x - 8$ **34.** $y = 3x - 6$

35. $y = 7 - 3x$ **36.** $y = 5 - 2x$

37. $y = 50 - x$ **38.** $y = x - 40$

Find an equation of the line satisfying the following conditions. If possible, write your answer in the form $y = mx + b$.

39. Slope -2.25 and y-intercept 3

40. Slope $\frac{2}{3}$ and y-intercept -8

41. Slope 5 and passing through the point $(-1, -2)$

42. Slope -1 and passing through the point $(4, 3)$

43. Horizontal and passing through the point $(1.5, -4)$

44. Horizontal and passing through the point $(\frac{1}{2}, \frac{3}{4})$

45. Vertical and passing through the point $(1.5, -4)$

46. Vertical and passing through the point $(\frac{1}{2}, \frac{3}{4})$

47. Passing through the points $(5, 3)$ and $(7, -1)$

48. Passing through the points $(3, -1)$ and $(6, 0)$

49. Passing through the points $(1, -1)$ and $(5, -1)$

50. Passing through the points $(2, 0)$ and $(2, -4)$

Write an equation of the form $y = mx + b$ for each line graphed below. (*Hint:* Either find the slope and y-intercept or use any two points on the line.)

51.

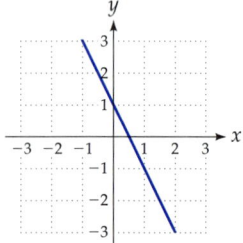

52.

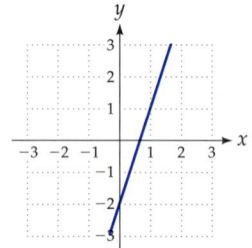

53.

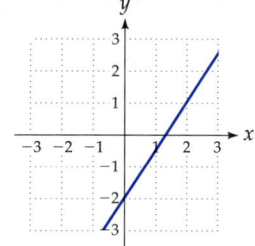

54.

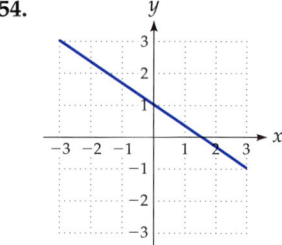

Find equations for the lines that make up the four-sided figures shown below.

55.

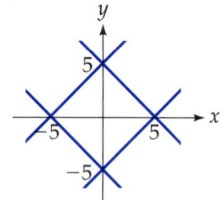

56.

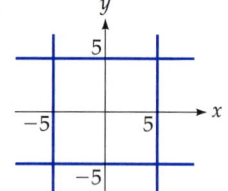

57. Show that $y - y_1 = m(x - x_1)$ simplifies to $y = mx + b$ if the point (x_1, y_1) is the y-intercept $(0, b)$.

58. Show that the linear equation $\frac{x}{a} + \frac{y}{b} = 1$ has x-intercept $(a, 0)$ and y-intercept $(0, b)$. (The x-intercept is the point where the line crosses the x-axis.)

59. Find the x-intercept $(a, 0)$ where the line $y = mx + b$ crosses the x-axis. Under what condition on m will a single x-intercept exist?

60. i. Show that the general linear equation $ax + by = c$ with $b \neq 0$ can be written as $y = -\dfrac{a}{b} x + \dfrac{c}{b}$, which is the equation of a line in slope-intercept form.

ii. Show that the general linear equation $ax + by = c$ with $b = 0$ but $a \neq 0$ can be written as $x = \dfrac{c}{a}$, which is the equation of a vertical line.

[*Note:* Since these steps are reversible, parts (i) and (ii) together show that the general linear equation $ax + by = c$ (for a and b not both zero) includes vertical and nonvertical lines.]

61. a. Graph the lines $y_1 = -x$, $y_2 = -2x$, and $y_3 = -3x$ on the window $[-5, 5]$ by $[-5, 5]$ (using the *negation* key $(-)$, not the *subtraction* key $-$). Observe how the coefficient of x changes the slope of the line.

b. Predict what the line $y = -9x$ would look like, then check your prediction by graphing the line.

62. a. Graph the lines $y_1 = x + 2$, $y_2 = x + 1$, $y_3 = x$, $y_4 = x - 1$, and $y_5 = x - 2$ on the window $[-5, 5]$ by $[-5, 5]$. Observe how the constant changes the position of the line.

b. Predict what the lines $y = x + 4$ and $y = x - 4$ would look like, then check your prediction by graphing them.

APPLIED EXERCISES

63. Business: Energy Usage A utility considers demand for electricity "low" if it is below 8 mkW (million kilowatts), "average" if it is at least 8 mkW but below 20 mkW, "high" if it is at least 20 mkW but below 40 mkW, and "critical" if it is 40 mkW or more. Express these demand levels in interval notation. *Hint:* The interval for "low" is $[0, 8)$.

64. General: Grades If a grade of 90 through 100 is an A, at least 80 but less than 90 is a B, at least 70 but less than 80 a C, at least 60 but less than 70 a D, and below 60 an F, write these grade levels in interval form (ignoring rounding). *Hint:* F would be $[0, 60)$.

65. General: Mile Run Read the Application Preview on pages 2–3.

a. Use the regression line $y - 0.357x + 257.46$ to predict the world record in the year 2010. (*Hint:* If x represents years after 1900, what value of x corresponds to the year 2010? The result will be in seconds, and should be converted to minutes and seconds.)

b. According to this formula, when will the record be 3 minutes 30 seconds? (*Hint:* Set the formula equal to 210 seconds and solve. What year corresponds to this x-value?)

66. General: Mile Run Read the Application Preview on pages 2–3. Evaluate the regression line $y = -0.357x + 257.46$ at $x = 720$ and at $x = 722$ (corresponding to the years 2620 and 2622). Does the formula give reasonable times for the mile record in these years?

67. Business: Corporate Profit A company's profit increased linearly from $6 million at the end of year 1 to $14 million at the end of year 3.

a. Use the two (year, profit) data points $(1, 6)$ and $(3, 14)$ to find the linear relationship $y = mx + b$ between $x =$ year and $y =$ profit.

b. Find the company's profit at the end of 2 years.

c. Predict the company's profit at the end of 5 years.

68. Economics: Per Capita Personal Income In the short run, per capita personal income (PCPI) in the United States grows approximately linearly. In 1993 PCPI was 21.2, and in 1996 it had grown to 24.2 (both in thousands of dollars).

a. Use the two (year, PCPI) data points $(0, 21.2)$ and $(3, 24.2)$ to find the linear relationship $y = mx + b$ between $x =$ years since 1993 and $y =$ PCPI.

b. Use your linear relationship to predict PCPI in 2010.

69. General: Temperature   On the Fahrenheit temperature scale, water freezes at 32° and boils at 212°. On the Celsius (centigrade) scale, water freezes at 0° and boils at 100°.

     a. Use the two (Celsius, Fahrenheit) data points (0, 32) and (100, 212) to find the linear relationship $y = mx + b$ between $x =$ Celsius temperature and $y =$ Fahrenheit temperature.

     b. Find the Fahrenheit temperature that corresponds to 20° Celsius.

70. Ecology: Waste Disposal   The amount of solid waste generated per person annually in the United States has increased approximately linearly, from 200 pounds in 1960 to 308 pounds in 1990.

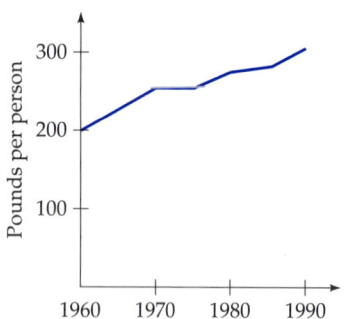

U.S. Solid Waste Generated Per Capita
(Source: U.S. Environmental Protection Agency)

     a. Use the two (year, pounds) data points (0, 200) and (30, 308) to find the linear relationship $y = mx + b$ between $x =$ years since 1960 and $y =$ per capita waste.

     b. Use your formula to predict the amount in the year 2010.

71–72: Business: Straight-Line Depreciation
Straight-line depreciation is a method for estimating the value of an asset (such as a piece of machinery) as it loses value ("depreciates") through use. Given the original *price* of an asset, its *useful lifetime*, and its *scrap value* (its value at the end of its useful lifetime), the value of the asset after t years

is given by the formula:

$$\text{Value} = (\text{price}) - \left(\frac{(\text{price}) - (\text{scrap value})}{(\text{useful lifetime})}\right) \cdot t$$

$$0 \le t \le (\text{useful lifetime})$$

71. a. A farmer buys a harvester for $50,000 and estimates its useful life to be 20 years, at the end of which its scrap value will be $6000. Use the above formula to find a formula for the value V of the harvester after t years, for $0 \le t \le 20$.

     b. Use your formula to find the value of the harvester after 5 years.

     c. Graph the function found in part (a) on a graphing calculator on the window [0, 20] by [0, 50,000]. (*Hint:* Use x instead of t.)

72. a. A newspaper buys a printing press for $800,000 and estimates its useful life to be 20 years, at the end of which its scrap value will be $60,000. Use the above formula to find a formula for the value V of the press after t years, for $0 \le t \le 20$.

     b. Use your formula to find the value of the press after 10 years.

     c. Graph the function found in part (a) on a graphing calculator on the window [0, 20] by [0, 800,000]. (*Hint:* Use x instead of t.)

73–74: General: Life Expectancy   The following tables give the life expectancy (years of life expected) for a newborn child born in the indicated year (Exercise 73 is for males, Exercise 74 for females). For each exercise:

a. Enter the data into a graphing calculator and make a plot of the resulting points, with Years Since 1950 on the x-axis and Life Expectancy on the y-axis.

b. Use the graphing calculator to find the linear regression line for these points. Enter the resulting function as y_1, which then estimates life expectancy based on the year of birth. Graph the points together with the regression line.

c. Use your line y_1 to estimate the life expectancy of a child born in the year 2025. (This might be your child or grandchild. *Hint:* What x-value corresponds to 2025?)

73.

Birth Year (Years Since 1950)	Life Expectancy (Male)
1950	0	65.6
1960	10	66.6
1970	20	67.1
1980	30	70.0
1990	40	71.8

74.

Birth Year (Years Since 1950)	Life Expectancy (Female)
1950	0	71.1
1960	10	73.1
1970	20	74.7
1980	30	77.5
1990	40	78.8

0.2   Exponents

A P P L I C A T I O N   P R E V I E W

Size, Shape, and Exponents

The study of shape and size is called "allometry," and many allometric relationships involve exponents that are fractions or decimals. For example, among all four-legged animals, from mice to elephants, (average) leg width and body length are governed (approximately) by the law

$$\left(\begin{array}{c}\text{Leg} \\ \text{width}\end{array}\right) = (\text{constant}) \cdot \left(\begin{array}{c}\text{body} \\ \text{length}\end{array}\right)^{3/2}$$

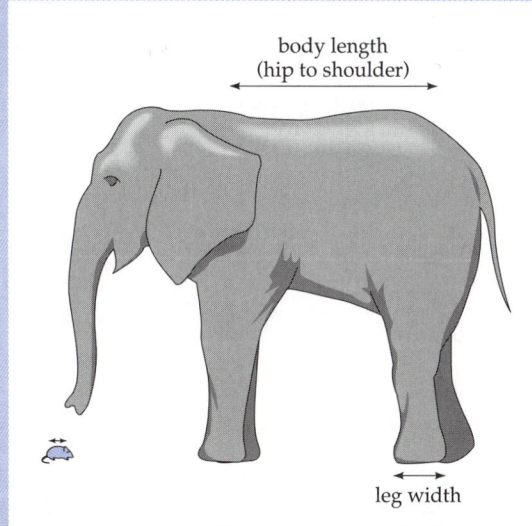

body length
(hip to shoulder)

leg width

The constant depends on the units (inches, centimeters, etc.), but the exponent is always $\frac{3}{2}$. The precise meaning of the exponent $\frac{3}{2}$ will be explained later in this section, but the fact that it is greater than 1 means that for larger animals, leg width increases *faster* than body length. This is the reason why nature cannot build a land animal very much larger than an elephant—its oversized legs would get in each other's way. The above formula is an example of what is called the *law of simple allometry*, which states that two measurements x and y will be related by a *power law* of the form $y = a \cdot x^b$ for constants a and b.*

Another application of fractional exponents concerns maps and illustrations. Studies have shown that when people are asked to estimate the size of objects, the *perceived* size and the *actual* size are related by a power law

$$\left(\begin{array}{c} \text{Perceived} \\ \text{size} \end{array} \right) = (\text{constant}) \cdot \left(\begin{array}{c} \text{actual} \\ \text{size} \end{array} \right)^{a}$$

with exponent $a < 1$. The value of the exponent a depends on whether the object is one-, two-, or three-dimensional, but the exponent is always less than 1. This means that if you were using a geometric symbol on a map to indicate rainfall, to suggest that one country's rainfall was twice as great as another's, you should use a symbol *more* than twice as large. Many other uses of fractional exponents will be discussed in this section.

Introduction

Not all variables are related linearly. In this section we will discuss exponents, which will enable us to express many nonlinear relationships.

Positive Integer Exponents

Numbers may be expressed with exponents, as in $2^3 = 2 \cdot 2 \cdot 2 = 8$. More generally, for any positive integer n, x^n means the product of n x's.

$$x^n = \overbrace{x \cdot x \cdots x}^{n}$$

The number being raised to the power is called the *base:*

* For further information on allometry, see D'Arcy Wentworth Thompson, *On Growth and Form* (Cambridge University Press, 1942; Dover Publications, 1992); Stephen Jay Gould, "Allometry and Size in Ontogeny and Philogeny," *Biol. Rev.*, **41**:587–640, 1966; Stefan Hildebrandt and Anthony Tromba, *Mathematics and Optimal Form* (Scientific American Books, 1985).

There are several *laws of exponents* for simplifying expressions. The first three are known, respectively, as the addition, subtraction, and multiplication laws of exponents.

Laws of Exponents

$x^m \cdot x^n = x^{m+n}$	To *multiply* powers of the same base, *add* the exponents
$\dfrac{x^m}{x^n} = x^{m-n}$	To *divide* powers of the same base, *subtract* the exponents (top exponent minus bottom exponent)
$(x^m)^n = x^{m \cdot n}$	To raise a power to a power, *multiply* the powers
$(xy)^n = x^n \cdot y^n$	To raise a product to a power, raise *each factor* to the power
$\left(\dfrac{x}{y}\right)^n = \dfrac{x^n}{y^n}$	To raise a fraction to a power, raise the numerator *and* denominator to the power

EXAMPLE 1 Simplifying Exponents

a. $x^2 \cdot x^3 = x^5$ ⟵ $2 + 3$

b. $\dfrac{x^5}{x^3} = x^2$ ⟵ $5 - 3$

c. $(x^2)^3 = x^6$ ⟵ $2 \cdot 3$

d. $\dfrac{[(x^2)^3]^4}{x^5 \cdot x^7 \cdot x} = \dfrac{x^{24}}{x^{13}} = x^{11}$ ⟵ $2 \cdot 3 \cdot 4$, $24 - 13$, $5 + 7 + 1$

e. $(2w)^3 = 2^3 w^3 = 8w^3$

f. $\left(\dfrac{x}{3}\right)^4 = \dfrac{x^4}{3^4} = \dfrac{x^4}{81}$

■

PRACTICE PROBLEM 1   Simplify: **a.** $\dfrac{x^5 \cdot x}{x^2}$ **b.** $[(x^3)^2]^2$ *Solutions at the back of the book*

Remember: For exponents in the form $x^2 \cdot x^3 = x^5$, *add* exponents.
For exponents in the form $(x^2)^3 = x^6$, *multiply* exponents.

Graphing Calculator Exploration

a. Use a graphing calculator to graph $y_1 = x$, $y_2 = x^2$, $y_3 = x^3$, and $y_4 = x^4$ on the viewing window [0, 2] by [0, 2]. Use TRACE to identify which curve goes with which power.

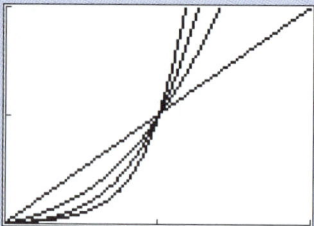

b. Which curve is highest for values of x between 0 and 1? Which is lowest?

c. Which curve is highest for values of x greater than 1? Which is lowest?

d. Predict what the curve $y = x^5$ would look like. Check your prediction by graphing it.

e. Predict which of these curves will be positive when x is negative. Check your prediction by changing the viewing window to [−2, 2] by [−2, 2].

Zero and Negative Exponents

For any number x other than zero, we define

$x^0 = 1$	x to the power zero is one
$x^{-1} = \dfrac{1}{x}$	x to the power -1 is one over x
$x^{-2} = \dfrac{1}{x^2}$	x to the power -2 is one over x squared
$x^{-n} = \dfrac{1}{x^n}$	x to a negative power is one over x to the positive power

EXAMPLE 2   **Simplifying Zero and Negative Exponents**

a. $5^0 = 1$ **b.** $7^{-1} = \dfrac{1}{7}$

c. $3^{-2} = \dfrac{1}{3^2} = \dfrac{1}{9}$ **d.** $(-2)^{-3} = \dfrac{1}{(-2)^3} = \dfrac{1}{-8} = -\dfrac{1}{8}$

e. 0^0 and 0^{-3} are undefined.

■

PRACTICE PROBLEM 2   Evaluate: **a.** 2^0 **b.** 2^{-4} *Solutions at the back of the book*

The definitions of x^0 and x^{-n} are motivated by the following calculations.

$$1 = \frac{x^2}{x^2} = x^{2-2} = x^0$$

The subtraction law of exponents leads to $x^0 = 1$

$$\frac{1}{x^n} = \frac{x^0}{x^n} = x^{0-n} = x^{-n}$$

$x^0 = 1$ and the subtraction law of exponents lead to $x^{-n} = \dfrac{1}{x^n}$

A fraction to a negative power means *division* by the fraction, so we "invert and multiply."

$$\left(\frac{x}{y}\right)^{-1} = \frac{1}{\frac{x}{y}} = 1 \cdot \frac{y}{x} = \frac{y}{x}$$

Reciprocal of the original fraction

Therefore, for $x \neq 0$ and $y \neq 0$,

$$\left(\frac{x}{y}\right)^{-1} = \frac{y}{x}$$

A fraction to the power -1 is the reciprocal of the fraction

$$\left(\frac{x}{y}\right)^{-n} = \left(\frac{y}{x}\right)^{n}$$

A fraction to the negative power is the reciprocal of the fraction to the positive power

EXAMPLE 3   **Simplifying Fractions to Negative Exponents**

a. $\left(\dfrac{3}{2}\right)^{-1} = \dfrac{2}{3}$ **b.** $\left(\dfrac{1}{2}\right)^{-3} = \left(\dfrac{2}{1}\right)^{3} = \dfrac{2^3}{1^3} = 8$

Reciprocal of $\dfrac{3}{2}$

■

PRACTICE PROBLEM 3   Simplify: $\left(\dfrac{2}{3}\right)^{-2}$ *Solution at the back of the book*

Roots and Fractional Exponents

We may take the square root of any *nonnegative* number, and the cube root of *any* number.

EXAMPLE 4   Evaluating Roots

a. $\sqrt{9} = 3$ **b.** $\sqrt{-9}$ is undefined.   Square roots of negative numbers are not defined

c. $\sqrt[3]{8} = 2$ **d.** $\sqrt[3]{-8} = -2$ Cube roots of negative numbers *are* defined

e. $\sqrt[3]{\dfrac{27}{8}} = \dfrac{\sqrt[3]{27}}{\sqrt[3]{8}} = \dfrac{3}{2}$

■

There are *two* square roots of 9, namely 3 and -3, but the radical sign $\sqrt{}$ means just the *positive* one (the "principal" square root).

$\sqrt[n]{a}$ means the principal *n*th root of *a*.   Principal means the positive root if there are two

In general, we may take *odd* roots of *any* number, but *even* roots only if the number is positive or zero.

EXAMPLE 5   Evaluating Roots of Positive and Negative Numbers

Odd roots of negative numbers are defined

a. $\sqrt[4]{81} = 3$ **b.** $\sqrt[5]{-32} = -2$ Since $(-2)^5 = -32$

■

Graphing Calculator Exploration

a. Use a graphing calculator to graph $y_1 = x$, $y_2 = \sqrt{x}$, $y_3 = \sqrt[3]{x}$, and $y_4 = \sqrt[4]{x}$ simultaneously on the viewing window $[0, 3]$ by $[0, 2]$. Use TRACE to identify which curve goes with which root.

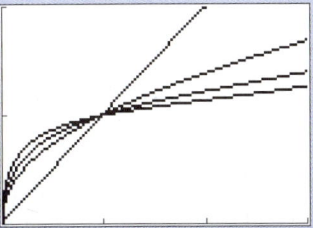

b. Which curve is highest for values of x between 0 and 1? Which is lowest?

c. Which curve is highest for values of x greater than 1? Which is lowest?

d. Predict what the curve $y = \sqrt[7]{x}$ would look like. Check your prediction by graphing it.

e. Which of these roots are defined for *negative* values of x? Check your answer by changing the window to $[-3, 3]$ by $[-2, 2]$ and using TRACE where x is negative.

Fractional Exponents

Fractional exponents are defined as follows:

$x^{\frac{1}{2}} = \sqrt{x}$	Power $\frac{1}{2}$ means the principal square root
$x^{\frac{1}{3}} = \sqrt[3]{x}$	Power $\frac{1}{3}$ means the cube root
$x^{\frac{1}{n}} = \sqrt[n]{x}$	Power $\frac{1}{n}$ means the principal nth root (for a positive integer n)

EXAMPLE 6   Evaluating Fractional Exponents

a. $9^{\frac{1}{2}} = \sqrt{9} = 3$ $\qquad\qquad$ **b.** $125^{\frac{1}{3}} = \sqrt[3]{125} = 5$

c. $81^{\frac{1}{4}} = \sqrt[4]{81} = 3$ $\qquad\qquad$ **d.** $(-32)^{\frac{1}{5}} = \sqrt[5]{-32} = -2$

e. $\left(-\dfrac{27}{8}\right)^{\frac{1}{3}} = \sqrt[3]{-\dfrac{27}{8}} = -\dfrac{\sqrt[3]{27}}{\sqrt[3]{8}} = -\dfrac{3}{2}$

■

PRACTICE PROBLEM 4   Evaluate: **a.** $(-27)^{\frac{1}{3}}$ **b.** $\left(\dfrac{16}{81}\right)^{\frac{1}{4}}$ *Solutions at the back of the book*

The definition of $x^{1/2}$ is motivated by the multiplication law of exponents:

$$\left(x^{\frac{1}{2}}\right)^2 = x^{\frac{1}{2}\cdot 2} = x^1 = x$$

Taking square roots of each side of $\left(x^{\frac{1}{2}}\right)^2 = x$ gives

$$x^{\frac{1}{2}} = \sqrt{x}$$ *x to the half power means the square root of x*

To define $x^{\frac{m}{n}}$ for positive integers m and n, the exponent $\frac{m}{n}$ must be fully reduced (for example, $\frac{4}{6}$ must be reduced to $\frac{2}{3}$). Then

$$x^{\frac{m}{n}} = \left(x^{\frac{1}{n}}\right)^m = \left(x^m\right)^{\frac{1}{n}}$$ *Since in both cases the exponents multiply to $\frac{m}{n}$*

Therefore we define:

Fractional Exponents

$$x^{\frac{m}{n}} = (\sqrt[n]{x})^m = \sqrt[n]{x^m}$$ $x^{m/n}$ means the *m*th power of the *n*th root, or equivalently, the *n*th root of the *m*th power

Both expressions, $(\sqrt[n]{x})^m$ and $\sqrt[n]{x^m}$, will give the same answer. In either case the numerator determines the power and the denominator determines the root.

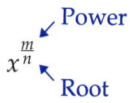
Power
$x^{\frac{m}{n}}$
Root

Power over root

EXAMPLE 7 Evaluating Fractional Exponents

a. $8^{2/3} = \sqrt[3]{8^2} = \sqrt[3]{64} = 4$ First the power, then the root

b. $8^{2/3} = (\sqrt[3]{8})^2 = (2)^2 = 4$ First the root, then the power

c. $25^{3/2} = (\sqrt{25})^3 = (5)^3 = 125$

d. $\left(\dfrac{-27}{8}\right)^{2/3} = \left(\sqrt[3]{\dfrac{-27}{8}}\right)^2 = \left(\dfrac{-3}{2}\right)^2 = \dfrac{9}{4}$

PRACTICE PROBLEM 5   Evaluate: **a.** $16^{3/2}$ **b.** $(-8)^{2/3}$ *Solutions at the back of the book*

Graphing Calculator Exploration

a. Use a graphing calculator to evaluate $25^{3/2}$. [On some calculators, press 25^(3 ÷ 2).] Your answer should agree with Exercise 7c above.

b. Evaluate $(-8)^{2/3}$. Use the ⎢(–)⎥ key for negation, and parentheses around the exponent. Your answer should be 4. If you get an "error," try evaluating the expression as $[(-8)^{1/3}]^2$ or $[(-8)^2]^{1/3}$. Whichever way works, remember it for evaluating negative numbers to fractional powers in the future.

EXAMPLE 8 Evaluating Negative Fractional Exponents

a. $8^{-2/3} = \dfrac{1}{8^{2/3}} = \dfrac{1}{(\sqrt[3]{8})^2} = \dfrac{1}{2^2} = \dfrac{1}{4}$ A negative exponent means the reciprocal of the number to the positive exponent, which is then evaluated as before

b. $\left(\dfrac{9}{4}\right)^{-3/2} = \left(\dfrac{4}{9}\right)^{3/2} = \left(\sqrt{\dfrac{4}{9}}\right)^3 = \left(\dfrac{2}{3}\right)^3 = \dfrac{8}{27}$

    Interpreting the power 3/2
    Reciprocal to the positive exponent
    Negative exponent

PRACTICE PROBLEM 6   Evaluate: **a.** $25^{-3/2}$ **b.** $\left(\dfrac{1}{4}\right)^{-1/2}$ **c.** $5^{1.3}$ (*Hint:* Use 🖩 .)

Solutions at the back of the book

Avoiding Pitfalls in Simplifying

The square root of a product is equal to the product of the square roots:

$$\sqrt{a \cdot b} = \sqrt{a} \cdot \sqrt{b}$$

However, the corresponding statement for *sums* is *not* true:

$$\sqrt{a + b} \quad \text{is } not \text{ equal to} \quad \sqrt{a} + \sqrt{b}$$

For example,

$$\underbrace{\sqrt{9 + 16}}_{\sqrt{25}} \neq \underbrace{\sqrt{9}}_{3} + \underbrace{\sqrt{16}}_{4}$$

The two sides are not equal:
one is 5 and the other is 7

Therefore, do not "simplify" $\sqrt{x^2 + 9}$ into $x + 3$. The expression $\sqrt{x^2 + 9}$ *cannot be simplified*. Similarly,

$$(x + y)^2 \quad \text{is } not \text{ equal to} \quad x^2 + y^2$$

The expression $(x + y)^2$ means $(x + y)$ times itself:

$$(x + y)^2 = (x + y)(x + y) = x^2 + xy + yx + y^2 = x^2 + 2xy + y^2$$

This result will be very useful in Chapter 6.

$(x + y)^2 = x^2 + 2xy + y^2$	$(x + y)^2$ is the first number squared plus twice the product of the numbers plus the second number squared

Learning Curves in Airplane Production

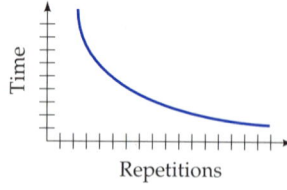

Time

Repetitions

It is a truism that the more you practice a task, the faster you can do it. Successive repetitions generally take less time, following a "learning curve" like that on the left. Learning curves are used in industrial production. For example, it took 150,000 work-hours to build the first Boeing 707 airliner, while later planes ($n = 2, 3, \ldots, 300$) took less time.*

---

* A work-hour is the amount of work that a person can do in one hour. For further information on learning curves in industrial production, see J. M. Dutton et al., "The History of Progress Functions as a Managerial Technology," *Bus. Hist. Rev.,* 58:204–233, 1984.

$$\left(\begin{array}{c}\text{Time to build}\\ \text{plane number } n\end{array}\right) = 150\, n^{-0.322} \qquad \text{thousand work-hours}$$

The time for the tenth Boeing 707 is found by substituting $n = 10$:

$$\left(\begin{array}{c}\text{Time to build}\\ \text{plane 10}\end{array}\right) = 150(10)^{-0.322}$$
$$\approx 71.46 \text{ thousand work-hours}$$

$150n^{-0.322}$ with
$n = 10$; using
a calculator

This shows that building the tenth Boeing 707 took about 71,460 work-hours, which is less than half of the 150,000 work-hours needed for the first. For the 100th Boeing 707:

$$\left(\begin{array}{c}\text{Time to build}\\ \text{plane 100}\end{array}\right) = 150(100)^{-0.322}$$
$$\approx 34.05 \text{ thousand work-hours}$$

$150n^{-0.322}$ with
$n = 100$

or about 34,050 work-hours, which is less than half the time needed to build the tenth. Such learning curves are used for determining the cost of a contract to build several planes.

Notice that the learning curve graphed on the previous page decreases *less steeply* as the number of repetitions increases. This means that while construction time continues to decrease, it does so more slowly for later planes. This behavior, called "diminishing returns," is typical of learning curves.

SUMMARY

We defined zero, negative, and fractional exponents as follows:

$$x^0 = 1 \qquad\qquad \text{for } x \neq 0$$

$$x^{-n} = \frac{1}{x^n} \qquad\quad \text{for } x \neq 0$$

$$x^{\frac{m}{n}} = (\sqrt[n]{x})^m = \sqrt[n]{x^m} \qquad m > 0, \quad n > 0, \quad \frac{m}{n} \text{ fully reduced}$$

With these definitions, the following laws of exponents hold for *all* exponents, whether integral or fractional, positive or negative.

$$x^m \cdot x^n = x^{m+n} \qquad (x^m)^n = x^{m \cdot n} \qquad \left(\frac{x}{y}\right)^n = \frac{x^n}{y^n}$$

$$\frac{x^m}{x^n} = x^{m-n} \qquad (xy)^n = x^n \cdot y^n$$

EXERCISES 0.2

Evaluate each expression *without* using a calculator:

1. $(2^2 \cdot 2)^2$ **2.** $(5^2 \cdot 4)^2$ **3.** 2^{-4}

4. 3^{-3} **5.** $\left(\dfrac{1}{2}\right)^{-3}$ **6.** $\left(\dfrac{1}{3}\right)^{-2}$

7. $\left(\dfrac{5}{8}\right)^{-1}$ **8.** $\left(\dfrac{3}{4}\right)^{-1}$ **9.** $4^{-2} \cdot 2^{-1}$

10. $3^{-2} \cdot 9^{-1}$ **11.** $\left(\dfrac{3}{2}\right)^{-3}$ **12.** $\left(\dfrac{2}{3}\right)^{-3}$

13. $\left(\dfrac{1}{3}\right)^{-2} - \left(\dfrac{1}{2}\right)^{-3}$ **14.** $\left(\dfrac{1}{3}\right)^{-2} - \left(\dfrac{1}{2}\right)^{-2}$

15. $\left[\left(\dfrac{2}{3}\right)^{-2}\right]^{-1}$ **16.** $\left[\left(\dfrac{2}{5}\right)^{-2}\right]^{-1}$

17. $25^{1/2}$ **18.** $36^{1/2}$ **19.** $25^{3/2}$

20. $16^{3/2}$ **21.** $16^{3/4}$ **22.** $27^{2/3}$

23. $(-8)^{2/3}$ **24.** $(-27)^{2/3}$ **25.** $(-8)^{5/3}$

26. $(-27)^{5/3}$ **27.** $\left(\dfrac{25}{36}\right)^{3/2}$ **28.** $\left(\dfrac{16}{25}\right)^{3/2}$

29. $\left(\dfrac{27}{125}\right)^{2/3}$ **30.** $\left(\dfrac{125}{8}\right)^{2/3}$ **31.** $\left(\dfrac{1}{32}\right)^{2/5}$

32. $\left(\dfrac{1}{32}\right)^{3/5}$ **33.** $4^{-1/2}$ **34.** $9^{-1/2}$

35. $4^{-3/2}$ **36.** $9^{-3/2}$ **37.** $8^{-2/3}$

38. $16^{-3/4}$ **39.** $(-8)^{-1/3}$ **40.** $(-27)^{-1/3}$

41. $(-8)^{-2/3}$ **42.** $(-27)^{-2/3}$ **43.** $\left(\dfrac{25}{16}\right)^{-1/2}$

44. $\left(\dfrac{16}{9}\right)^{-1/2}$ **45.** $\left(\dfrac{25}{16}\right)^{-3/2}$ **46.** $\left(\dfrac{16}{9}\right)^{-3/2}$

47. $\left(-\dfrac{1}{27}\right)^{-5/3}$ **48.** $\left(-\dfrac{1}{8}\right)^{-5/3}$

 Use a calculator to evaluate each expression. Round answers to 2 decimal places.

49. $7^{0.39}$ **50.** $5^{0.47}$ **51.** $8^{2.7}$ **52.** $5^{3.9}$

 Use a graphing calculator to evaluate each expression.

53. $(-8)^{7/3}$ **54.** $(-8)^{5/3}$ **55.** $[(5/2)^{-1}]^{-2}$

56. $[(3/2)^{-2}]^{-1}$ **57.** $[(4)^{-1}]^{0.5}$ **58.** $[(0.25)^{-1}]^{0.5}$

59. $(0.4^{-7})^{-1/7}$ **60.** $[(0.5^{-1})^{-2}]^{-3}$

61. $[(0.1)^{0.1}]^{0.1}$ **62.** $\left(1 + \dfrac{1}{1000}\right)^{1000}$

63. $\left(1 - \dfrac{1}{1000}\right)^{-1000}$ **64.** $(1 + 10^{-6})^{10^6}$

Simplify.

65. $(x^3 \cdot x^2)^2$ **66.** $(x^4 \cdot x^3)^2$ **67.** $[z^2(z \cdot z^2)^2 z]^3$

68. $[z(z^3 \cdot z)z^2]^2$ **69.** $[(x^2)^2]^2$ **70.** $[(x^3)^3]^3$

71. $\dfrac{(ww^2)^3}{w^3 w}$ **72.** $\dfrac{(ww^3)^2}{w^3 w^2}$ **73.** $\dfrac{(5xy^4)^2}{25x^3y^3}$

74. $\dfrac{(4x^3y)^2}{8x^2y^3}$ **75.** $\dfrac{(9xy^3z)^2}{3(xyz)^2}$ **76.** $\dfrac{(5x^2y^3z)^2}{5(xyz)^2}$

77. $\dfrac{(2u^3vw^3)^2}{4(uw^2)^2}$ **78.** $\dfrac{(u^3vw^2)^2}{9(u^2w)^2}$

APPLIED EXERCISES

 79–80: Allometry: Dinosaurs For most four-legged animals, from mice to elephants, their body measurements obey (approximately) the following power law:

$$\left(\begin{array}{c}\text{Average body}\\\text{thickness}\end{array}\right) = 0.4 \ (\text{hip-to-shoulder length})^{3/2}$$

where body thickness is measured vertically and all measurements are in feet. Assuming that this same relationship held for dinosaurs, find the average body thickness of the following dinosaurs, whose hip-to-shoulder length can be measured from their skeletons:

79. Diplodocus, whose hip-to-shoulder length was 16 ft.

80. Triceratops, whose hip-to-shoulder length was 14 ft.

 81–82: Business: The Rule of 0.6 in Industrial Production Many chemical and refining companies use "The Rule of Point Six" to estimate the cost of new equipment. According to this rule, if a piece of equipment (such as a storage tank) originally cost C dollars, then the cost of similar equipment that is x times as large will be approximately $x^{0.6}C$ dollars. For example, if the original equipment cost C dollars, then new equipment with twice the capacity of the old equipment ($x = 2$) would cost $2^{0.6}C = 1.516C$ dollars, that is, about 1.5 times as much. Therefore, to increase capacity by 100% costs only about 50% more. While the rule of 0.6 is only a rough "rule of thumb," it can be somewhat justified on the basis that the equipment of such industries consists mainly of containers, and the cost of a container depends on its surface area (square units), which increases more slowly than its capacity (cubic units).

81. Use the Rule of 0.6 to find how costs change if a company wants to quadruple ($x = 4$) its capacity.

82. Use the Rule of 0.6 to find how costs change if a company wants to triple ($x = 3$) its capacity.

 83–84: Business: Continuation Use a graphing calculator to graph $y = x^{0.6}$, expressing y, the cost multiple for larger equipment, in terms of x, the size multiple. Use the viewing window $[0, 5]$ by $[0, 3]$.

83. By how much can a company multiply its capacity for twice the money? That is, find the value of x that satisfies $x^{0.6} = 2$. (*Hint:* Either use TRACE or find where $y_1 = x^{0.6}$ INTERSECTs $y_2 = 2$.)

84. Does the curve rise more steeply or less steeply as x increases? What does this mean about how rapidly cost increases as equipment size increases?

85–86: Biomedical: Heart Rate It is well known that the hearts of smaller animals beat faster than the hearts of larger animals. The actual relationship is approximately

$$\text{(Heart rate)} = 250(\text{weight})^{-1/4}$$

where the heart rate is in beats per minute and the weight is in pounds. Use this relationship to estimate the heart rate of:

85. A 16-pound dog

86. A 625-pound grizzly bear

 87–88: Biomedical: Continuation Use a graphing calculator to graph $y = 250x^{-0.25}$, which expresses y, heartbeats per minute, in terms of x, the animal's weight. Use the viewing window $[0, 200]$ by $[0, 150]$.

87. Notice that the curve decreases less steeply for larger values of x. Explain what this means about how rapidly heart rate decreases as body weight increases.

88. Evaluate this formula at your own weight, x, to find your predicted heart rate. Then take your pulse and see if the numbers (roughly) agree.

 89–90: Business and Psychology: Learning Curves in Airplane Production Recall (pages 26–27) that the learning curve for the production of Boeing 707 airplanes is $150x^{-0.322}$ (thousand work-hours). Find how many work-hours it took to build:

89. The 50th Boeing 707

90. The 250th Boeing 707

 91. General: Richter Scale The Richter scale (developed by Charles Richter in 1935) is widely used to measure the strength of earthquakes. Every increase of one on the Richter scale corresponds to a tenfold increase in ground motion. Therefore, an increase on the Richter scale from A to B means that ground motion increases by a factor of 10^{B-A} (for $B > A$). Find the increase in ground motion between the following earthquakes:

a. The 1994 Northridge, California, earthquake, measuring 6.8 on the Richter scale, and the 1906 San Francisco earthquake, measuring 8.3. (The San Francisco earthquake resulted in 500 deaths and a 3-day fire that destroyed 4 square miles of San Francisco.)

b. The 1995 Kobe (Japan) earthquake, measuring 7.2 on the Richter scale, and the 1933 Miyagi earthquake, measuring 8.1. (The Miyagi earthquake caused a 90-foot-high tsunami, or "tidal wave," that killed 3064 people. The death toll in the Kobe earthquake was more than 5000.)

92. Continuation   Every increase of one on the Richter scale corresponds to an approximately *thirtyfold* increase in *energy released*. Therefore, an increase on the Richter scale from A to B means that the energy released increases by a factor of 30^{B-A} (for $B > A$).

   a. Find the increase in *energy released* between the earthquakes in Exercise 91a.
   b. Find the increase in *energy released* between the earthquakes in Exercise 91b.

93–94: General: Waterfalls   Water falling from a waterfall that is x feet high will hit the ground with speed $\frac{60}{11}x^{0.5}$ miles per hour (neglecting air resistance).

93. Find the speed of the water at the bottom of the highest waterfall in the world, Angel Falls in Venezuela (3281 feet high).

94. Find the speed of the water at the bottom of the highest waterfall in the United States, Ribbon Falls in Yosemite, California (1650 feet high).

95–96: Environmental Science: Biodiversity   It is well known that larger land areas can support a larger number of species. According to one study,* multiplying the land area by a factor of x multiplies the number of species by a factor of $x^{0.239}$. Use a graphing calculator to graph $y = x^{0.239}$. Use the viewing window [0, 100] by [0, 4].

95. Find the multiple x for the land area that leads to *double* the number of species. That is, find the value of x such that $x^{0.239} = 2$. (*Hint:* Either use TRACE or find where $y_1 = x^{0.239}$ INTERSECTs $y_2 = 2$.)

96. Find the multiple x for the land area that leads to triple the number of species. That is, find the value of x such that $x^{0.239} = 3$. (*Hint:* Either use TRACE or find where $y_1 = x^{0.239}$ INTERSECTs $y_2 = 3$.)

* *The Theory of Island Biogeography,* by Robert H. MacArthur and Edward O. Wilson (Princeton University Press, 1967).

97. Business: Learning Curves   A manufacturer of supercomputers finds that the number of work-hours required to build the first, the tenth, the twentieth, and the thirtieth supercomputer are as follows:

Supercomputer Number	Work-Hours Required
1	3200
10	1900
20	1400
30	1300

   a. Enter these numbers into a graphing calculator and make a plot of the resulting points (Supercomputer Number on the x axis and Work-Hours Required on the y axis).
   b. Have the calculator find the power regression formula for these data, fitting a curve of the form $y = ax^b$ to the points. Enter the results as y_1. Plot the points together with the regression line. Observe that the line fits the points rather well.
   c. Evaluate y_1 at $x = 50$ to predict the number of work-hours required to build the fiftieth supercomputer.

98. General: Paper Stacking   Suppose that you take an ordinary piece of paper (about $\frac{1}{250}$ of an inch thick), cut it in half, stack the two halves, and repeat this cutting and stacking operation many times. Each operation doubles the height of the stack, so that after a total of 25 such operations, the stack will be $2^{25} \cdot \dfrac{1}{250} \cdot \dfrac{1}{12} \cdot \dfrac{1}{5280}$ miles high.

   a. Evaluate this height.
   b. Use a graphing calculator to make a TABLE showing the height of the stack when the number of operations is 15 or more. (*Hint:* Use $y_1 = 2^x/(250 \cdot 12 \cdot 5280)$ for values of x beginning at 15.) For what number of operations is the stack over a mile high? over 10 miles high? over 100 miles high?

0.3 FUNCTIONS

Functional Drunkenness

When you drink liquor, the alcohol begins to enter your blood-stream almost immediately and your blood-alcohol level rises rapidly. After you stop drinking, your natural metabolic processes slowly eliminate the alcohol and your blood-alcohol level begins to fall. How long must you wait after drinking before you can drive safely?

Experiments show that after drinking 6 beers, 4 glasses of wine, or 4 shots of liquor in an hour, your blood-alcohol level typically rises to 0.11 gram per deciliter of blood. Thereafter, alcohol is eliminated at the rate of 0.02 gram per hour. (Many substances are eliminated from the bloodstream *non*linearly, but alcohol is eliminated *linearly*.) The following graph shows the resulting blood-alcohol level over time, first rising rapidly and then falling linearly. By calculating where the graph falls below the legal limit (0.08 in most states), you can find how long you must wait—about one and a half hours. Incidentally, coffee and exercise have no effect; only time helps.

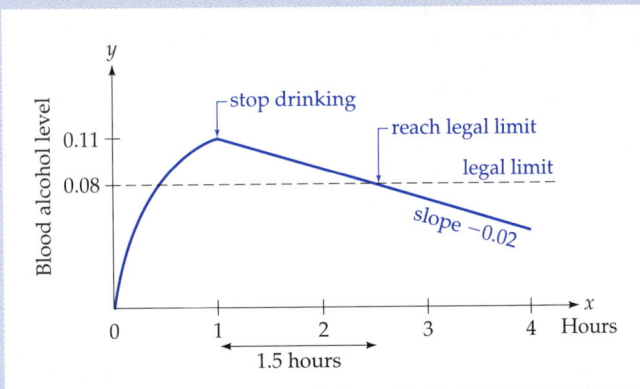

Mathematical relationships like this, in which each *x* (time, in hours) gives exactly one *y* (blood-alcohol level) are called *functions*.

Introduction

In the previous section we saw that the time required to build a Boeing 707 airliner depends upon the number that have already been built. Mathematical relationships like this, in which one number depends upon another, are called *functions*, and are central to the study of mathematics. In this section we define and give some applications of functions.

Functions

A *function** is a rule or procedure for finding, from a given number, a new number. If the function is denoted by f and the given number by x, then the resulting number is written $f(x)$ (read "f of x") and is called *the value of the function f at x.* The set of numbers x for which a function f is defined is called the *domain* of f, and the set of all function values $f(x)$ is called the *range* of f. For any x in the domain, $f(x)$ must be a *single* number. Formally,

Function

> A *function f* is a rule that assigns to each number x in a set a number $f(x)$. The set of all values of x is called the domain, and the set of all values $f(x)$ for x in the domain is called the *range*.

For example, recording the temperature at a given location throughout a particular day would define a *temperature* function:

$$f(x) = \begin{pmatrix} \text{Temperature at} \\ \text{time } x \text{ hours} \end{pmatrix} \qquad \text{Domain would be [0, 24]}$$

A function f may be thought of as a numerical procedure or "machine" that takes an "input" number x (in its domain) and produces an "output" number $f(x)$.

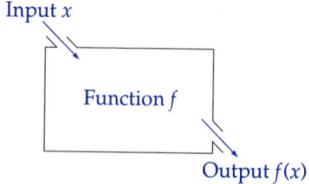

We will be mostly concerned with functions that are defined by *formulas* for calculating $f(x)$ from x. If the domain of such a function is

---

* In this chapter the word "function" will mean *function of one variable.* In Chapters 3 and 11 we will discuss functions of more than one variable.

not stated, then it is always taken to be the *largest* set of numbers for which the function is defined, called the *natural domain* of the function. To *graph* a function *f*, we plot all points (x, y) such that *x* is in the domain and $y = f(x)$. We call *x* the *independent variable* and *y* the *dependent variable*, since *y depends on* (is calculated from) *x*. The domain and range can be illustrated graphically.

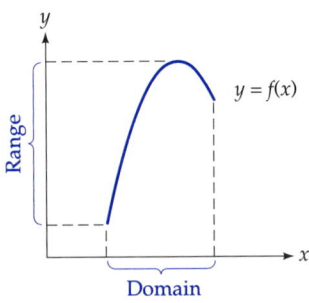

The domain of a function $y = f(x)$ is the set of all possible *x*-values, and the range is the set of all corresponding *y*-values.

PRACTICE PROBLEM 1

Find the domain and range of the function graphed on the right.

Solution at the back of the book

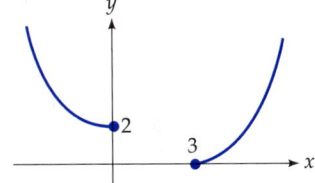

EXAMPLE 1   **Finding the Domain and Range of a Rational Function**

For the function $f(x) = \dfrac{1}{x - 1}$, find:

a. $f(5)$,   **b.** the domain,   **c.** the range.

Solution

a. $f(5) = \dfrac{1}{5 - 1} = \dfrac{1}{4}$ $f(x) = \dfrac{1}{x - 1}$ with $x = 5$

b. Domain $= \{x \mid x \neq 1\}$ $f(x) = \dfrac{1}{x - 1}$ is defined for all
                                            x except x = 1.

c. The graph of the function (from a graphing calculator) is shown on the right. From it, and realizing that the curve continues upward and downward (as may be verified by zooming out), it is clear that every y-value is taken on except for $y = 0$ (since the curve does not touch the x-axis). Therefore:

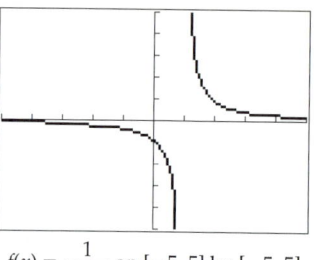

$f(x) = \dfrac{1}{x - 1}$ on $[-5, 5]$ by $[-5, 5]$

$$\text{Range} = \{y \mid y \neq 0\}$$

May also be written $\{z \mid z \neq 0\}$ or with any other letter

---

EXAMPLE 2   Finding the Domain and Range of a Polynomial

For $f(x) = 2x^2 + 4x - 5$, find:

a. $f(-3)$,   **b.** the domain,   **c.** the range.

Solution

a. $f(-3) = 2(-3)^2 + 4 \cdot (-3) - 5$

$\qquad = 18 - 12 - 5 = 1$

$f(x) = 2x^2 + 4x - 5$ with each x replaced by -3

b. Domain $= \mathbb{R}$

$2x^2 + 4x - 5$ is defined for *all* real numbers

 c. From the graph of $f(x) = 2x^2 + 4x - 5$ on the right, the lowest y-value is -7 (as can be found from TRACE or MIN), and all higher y-values are taken on (since the curve is a parabola opening upward). Therefore:

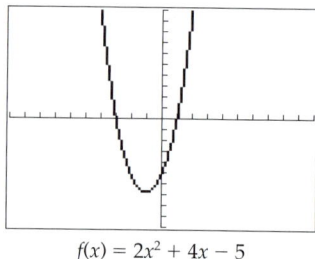

$f(x) = 2x^2 + 4x - 5$
on $[-10, 10]$ by $[-10, 10]$

$$\text{Range} = \{y \mid y \geq -7\}$$

Any letters may be used for defining a function.

PRACTICE PROBLEM 2   For $g(z) = \sqrt{z - 2}$, find: **a.** $g(27)$,   **b.** the domain,   **c.** the range.

Solutions at the back of the book

For each x in the domain of a function, there must be a *single* number $y = f(x)$, and so the graph of a function cannot have two points (x, y) with the same x-value but different y-values. This leads to the following *graphical* test for functions.

Vertical Line Test for Functions

A curve in the Cartesian plane is the graph of a *function* if and only if no vertical line intersects the curve at more than one point.

EXAMPLE 3 Using the Vertical Line Test

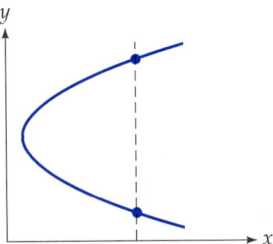

This is *not* the graph of a function of x because there is a vertical line (shown dashed) that intersects the curve at more than one point.

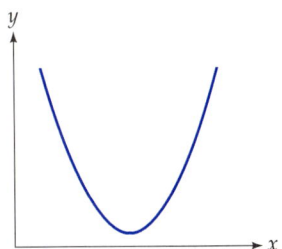

This *is* the graph of a function of x because no vertical line intersects the curve at more than one point.

A graph that may have two or more points (x, y) with the same x-value but different y-values, like the one on the left above, defines a *relation* rather than a function. We will be concerned exclusively with *functions*, and so we will use the terms "function," "graph," and "curve" interchangeably.

Linear Function

A *linear function* is a function that can be expressed in the form

$$f(x) = mx + b$$

with constants m and b. Its graph is a line with slope m and y-intercept b.

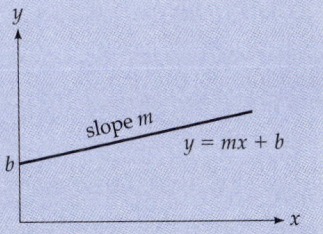

EXAMPLE 4 Finding a Company's Cost Function

An electronics company manufactures pocket calculators at a cost of $9 each, and the company's fixed costs (such as rent) amount to $400 per day. Find a function $C(x)$ that gives the total cost of producing x pocket calculators in a day.

Solution

Each calculator costs $9 to produce, and so x calculators will cost $9x$ dollars, to which we must add the fixed costs of $400.

$$C(x) \quad = \quad 9x \quad + \quad 400$$

Total	Unit	Number	Fixed
cost	cost	of units	cost

The graph of $C(x) = 9x + 400$ is a line with slope 9 and y-intercept 400, as shown below. Notice that the *slope* is the same as the *rate of change* of cost (costs increase at the rate of $9 per additional calculator), which is also the company's *marginal cost* (the cost of producing one more calculator is $9). The *slope*, the *rate of change*, and the *marginal* cost are always the same, as we will see in Chapter 6.

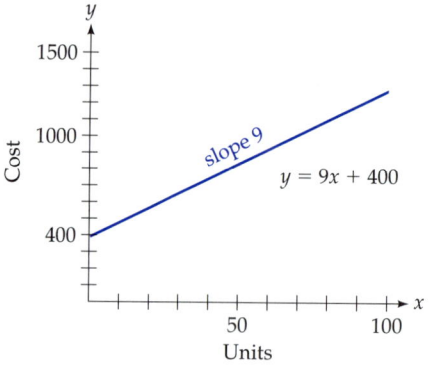

Cost function $C(x) = 9x + 400$

PRACTICE PROBLEM 3   A trucking company will deliver furniture for a charge of $25 plus 5% of the purchase price of the furniture. Find a function $D(x)$ that gives the delivery charge for a piece of furniture that cost x dollars.

Solution at the back of the book

A mathematical description of a real-world situation is called a *mathematical model*. For example, the cost function $C(x) = 9x + 400$ is a mathematical model for the cost of manufacturing calculators. In this model, x, the number of calculators, should take only integer values (0, 1, 2, 3, . . .), and the graph should consist of discrete dots rather than a continuous curve. Instead, we will find it easier to let x take *continuous* values, and round up or down as necessary at the end.

Quadratic Function

> A *quadratic function* is a function that can be expressed in the form
>
> $$f(x) = ax^2 + bx + c$$
>
> with constants ("coefficients") $a \neq 0$, b, and c. Its graph is called a *parabola*.

The condition $a \neq 0$ keeps the function from becoming $f(x) = bx + c$, which would be linear. Many familiar curves are parabolas.

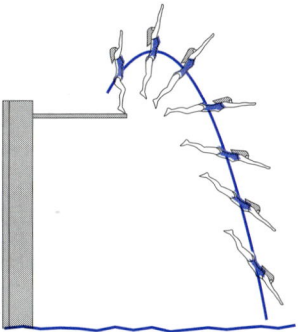

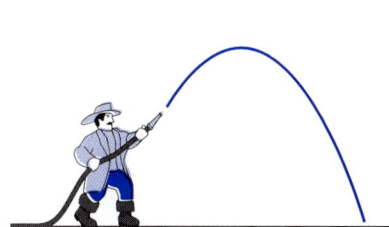

The center of gravity of a diver describes a parabola.

A stream of water from a hose takes the shape of a parabola.

The parabola $f(x) = ax^2 + bx + c$ opens *upward* if the constant a is *positive* and *downward* if the constant a is *negative*. The *vertex* of a parabola is its "central" point. The vertex is the *lowest* point on the parabola if it opens *up,* and the *highest* point if it opens *down.*

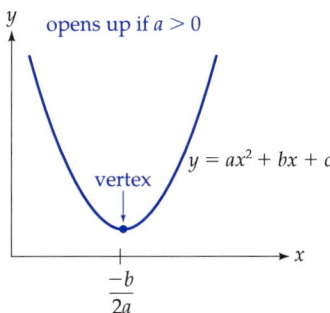

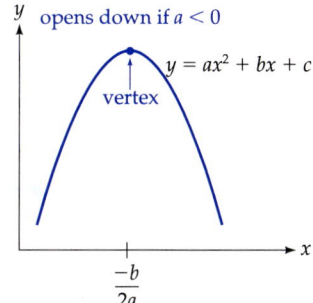

Graphing Calculator Exploration

a. Graph the parabolas $y_1 = x^2$, $y_2 = 2x^2$, and $y_3 = 4x^2$ on the graphing window $[-10, 10]$ by $[-10, 10]$. Use TRACE to identify which curve goes with which formula. How does the shape of the parabola change when the coefficient of x^2 increases?

b. Graph $y_4 = -x^2$. What did the negative sign do to the parabola?

c. Predict the shape of the parabolas $y_5 = -2x^2$ and $y_6 = \frac{1}{3}x^2$.

Then check your predictions by graphing the functions.

The x coordinate of the vertex of a parabola may be found by a formula, which will be derived in Exercise 57 of Section 7.1.

Vertex Formula for a Parabola

The x-coordinate of the vertex of the parabola $f(x) = ax^2 + bx + c$ is

$$x = \frac{-b}{2a}$$

EXAMPLE 5   Graphing a Quadratic Function

Graph the quadratic function $f(x) = 2x^2 - 40x + 104$.

Solution

Graphing using a graphing calculator is largely a matter of finding an appropriate viewing window, as the following three unsatisfactory windows show.

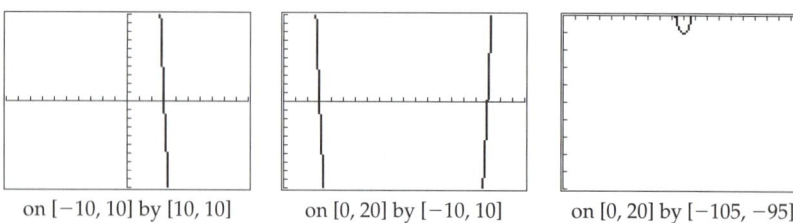

on $[-10, 10]$ by $[10, 10]$ on $[0, 20]$ by $[-10, 10]$ on $[0, 20]$ by $[-105, -95]$

To find an appropriate viewing window, we use the vertex formula:

$$x = \frac{-b}{2a} = \frac{-(-40)}{2 \cdot 2} = \frac{40}{4} = 10 \qquad \begin{array}{l} x \text{ coordinate of the vertex, from} \\ x = -b/2a \text{ with } a = 2 \text{ and } b = -40 \end{array}$$

We move a few units, say 5, to either side of $x = 10$, making the x window [5, 15]. Using the calculator to EVALUATE the given function at $x = 10$ (or evaluating by hand) gives $y(10) = -96$. Since the parabola opens upward (the coefficient of x^2 is positive), the curve rises up from its vertex, so we select a y interval from -96 upward, say $[-96, -70]$. Graphing the function on the window [5, 15] by $[-96, -70]$ gives the result shown below. (Some other graphing windows are just as good.)

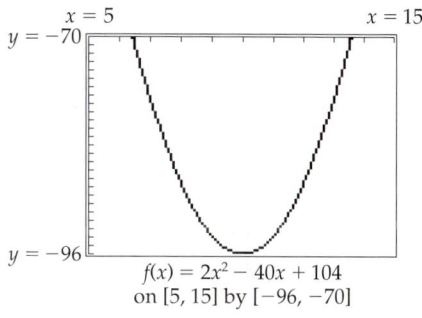

$f(x) = 2x^2 - 40x + 104$
on [5, 15] by [−96, −70]

Solving Quadratic Equations

A value of x that solves an equation $f(x) = 0$ is called a *root* of the equation, or a *zero* of the function. The (real) roots of a quadratic equation can often be found by factoring.

EXAMPLE 6   Solving a Quadratic Equation by Factoring

Solve $2x^2 - 4x = 6$.

Solution

$$2x^2 - 4x - 6 = 0 \qquad \text{Subtracting 6 from each side to get zero on the right}$$

$$2(x^2 - 2x - 3) = 0 \qquad \text{Factoring out a 2}$$

$$2\underbrace{(x - 3)}_{\substack{\text{Equals 0} \\ \text{at } x = 3}} \cdot \underbrace{(x + 1)}_{\substack{\text{Equals 0} \\ \text{at } x = -1}} = 0 \qquad \text{Factoring } x^2 - 2x - 3$$

Finding x-values that make each factor zero

$$x = 3, x = -1 \qquad \text{Solutions}$$

Graphing Calculator Exploration

Find the solutions to the equation in Example 6 by graphing the function $f(x) = 2x^2 - 4x - 6$ and using ZERO or TRACE to find where the curve crosses the x-axis. Your answers should agree with those found in Example 6.

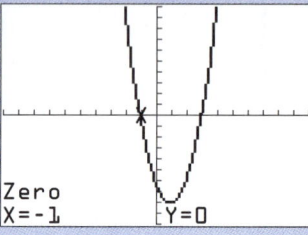

PRACTICE PROBLEM 4

Solve by factoring or graphically: $9x - 3x^2 = -30$

Solution at the back of the book

Quadratic equations can often be solved by the "quadratic formula." A derivation of this formula is given on pages 43–44.

Quadratic Formula

The solutions to $ax^2 + bx + c = 0$ are

$$x = \frac{-b \pm \sqrt{b^2 - 4ac}}{2a}$$

The "plus or minus" sign $\pm$ means calculate both ways, first using the $+$ sign and then using the $-$ sign

In a business, it is often important to find a company's break-even points, the numbers of units of production where a company's costs are equal to its revenue.

EXAMPLE 7 **Finding Break-Even Points**

A company that installs automobile compact disc (CD) players finds that if it installs x CD players per day, then its costs will be $C(x) = 120x + 4800$ and its revenue will be $R(x) = -2x^2 + 400x$ (both in dollars). Find the company's break-even points. (*Note*: In Chapter 7 we will see how such cost and revenue functions are found.)

Solution

$$120x + 4800 = -2x^2 + 400x$$ Setting $C(x) = R(x)$

$$2x^2 - 280x + 4800 = 0$$ Combining all terms on one side

$$x = \frac{280 \pm \sqrt{(-280)^2 - 4 \cdot 2 \cdot 4800}}{2 \cdot 2}$$ Quadratic formula with $a = 2, b = -280$, and $c = 4800$

$$= \frac{280 \pm \sqrt{40{,}000}}{4} = \frac{280 \pm 200}{4}$$

$$= \frac{480}{4} \text{ or } \frac{80}{4} = 120 \text{ or } 20$$ Working out the formula on a calculator

The company will break even when it makes either 20 units or 120 units. ∎

Although it is important for a company to know where its break-even points are, most companies want to do better than break even—they want to maximize their profits. Profit is defined as *revenue minus cost* (since profit is what is left over after subtracting expenses from income).

Profit = Revenue − Cost

EXAMPLE 8 **Maximizing Profit**

For the CD installer whose daily revenue and cost functions were given in Example 7, find the number of units that maximizes profit, and the maximum profit.

Solution

The profit function is the revenue function minus the cost function.

$$P(x) = \underbrace{-2x^2 + 400x}_{R(x)} - \underbrace{(120x + 4800)}_{C(x)}$$ $P(x) = R(x) - C(x)$ with $R(x) = -2x^2 + 400x$ and $C(x) = 120x + 4800$

$$= -2x^2 + 280x - 4800$$ Simplifying

Since this function represents a parabola opening downward (because of the -2), it is maximized at its vertex, which is found using the vertex formula.

$$x = \frac{-280}{2(-2)} = \frac{-280}{-4} = 70 \qquad \begin{array}{l} x = \dfrac{-b}{2a} \text{ with} \\ a = -2 \text{ and } b = 280 \end{array}$$

Thus, profit is maximized when 70 units are installed. For the maximum profit, we substitute $x = 70$ into the profit function:

$$P(70) = -2(70)^2 + 280 \cdot 70 - 4800$$
$$= 5000$$

$P(70) = -2x^2 + 280x - 4800$
with $x = 70$

Multiplying and combining

Therefore, the company will maximize its profit when it installs 70 CD players per day. Its maximum profit will be $5000 per day.

■

Why doesn't a company make more profit the more it sells? Because to increase its sales it must lower its prices, which eventually leads to lower profits. The relationship among the cost, revenue, and profit functions can be seen graphically as follows.

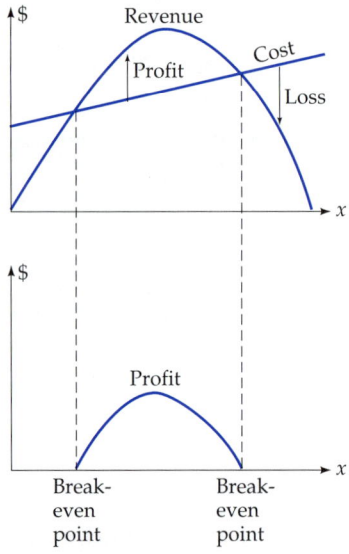

Not all quadratic equations have (real) solutions.

EXAMPLE 9   Using the Quadratic Formula

Solve $\frac{1}{2}x^2 - 3x + 5 = 0$.

Solution The quadratic formula with $a = \frac{1}{2}$, $b = -3$, and $c = 5$ gives

$$x = \frac{3 \pm \sqrt{9 - 4(\frac{1}{2})(5)}}{2(\frac{1}{2})} = \frac{3 \pm \sqrt{9 - 10}}{1} = 3 \pm \sqrt{-1} \qquad \text{Undefined!}$$

Therefore, the equation $\frac{1}{2}x^2 - 3x + 5 = 0$ has *no real solutions* (because of the undefined $\sqrt{-1}$). The geometrical reason that there are no solutions can be seen in the graph below: The curve never reaches the *x*-axis, so the function never equals zero.

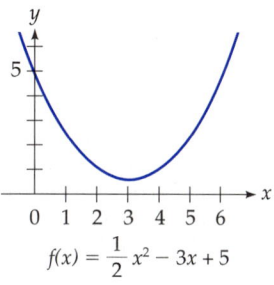

$$f(x) = \frac{1}{2}x^2 - 3x + 5$$

The quantity $b^2 - 4ac$, whose square root appears in the quadratic formula, is called the *discriminant*. If the discriminant is *positive* (as in Example 7), the equation $ax^2 + bx + c = 0$ has *two* solutions (since the square root is added and subtracted). If the discriminant is *zero*, there is only *one* root (since adding and subtracting zero gives the same answer). If the discriminant is *negative* (as in Example 9), then the equation has *no* real roots. Therefore, the discriminant being positive, zero, or negative corresponds to the parabola meeting the *x*-axis at 2, 1, or 0 points, as shown below.

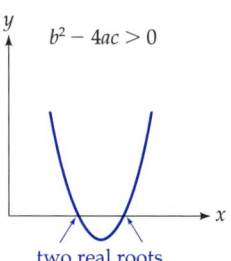

two real roots

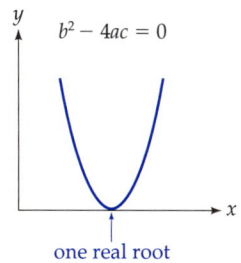

one real root

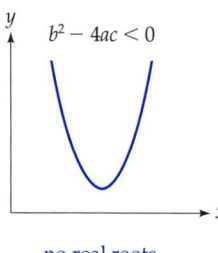

no real roots

Derivation of the Quadratic Formula

$$ax^2 + bx + c = 0 \qquad \text{The quadratic set equal to zero}$$

$$ax^2 + bx = -c \qquad \text{Subtracting } c$$

$$4a^2x^2 + 4abx = -4ac \qquad \text{Multiplying by } 4a$$

$$4a^2x^2 + 4abx + b^2 = b^2 - 4ac \qquad \text{Adding } b^2$$

$$(2ax + b)^2 = b^2 - 4ac \qquad \text{Since } 4a^2x^2 + 4abx + b^2 = (2ax + b)^2$$

$$2ax + b = \pm\sqrt{b^2 - 4ac} \qquad \text{Taking square roots}$$

$$2ax = -b \pm \sqrt{b^2 - 4ac} \qquad \text{Subtracting } b$$

$$x = \frac{-b \pm \sqrt{b^2 - 4ac}}{2a} \qquad \begin{array}{l}\text{Dividing by } 2a \text{ gives} \\ \text{the quadratic formula}\end{array}$$

SUMMARY

In this section we defined and gave examples of *functions*, and saw how to find their domains and ranges. The most important characteristic of a function *f* is that for any given "input" number *x* in the domain, there is exactly one "output" number *f(x)*. This requirement is stated geometrically in the *vertical line test*, that no vertical line can intersect the graph of a function at more than one point. We then defined *linear functions* (whose graphs are lines) and *quadratic functions* (whose graphs are parabolas), and solved quadratic equations by factoring, graphing, and using the quadratic formula.

EXERCISES 0.3

Determine whether each graph defines a function of *x*.

1.

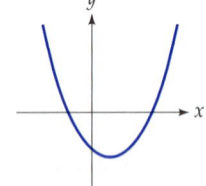

2.

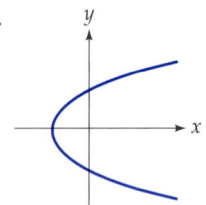

5.

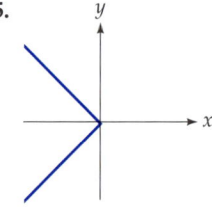

6.

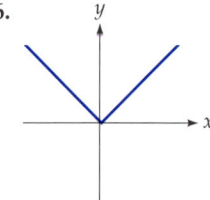

3.

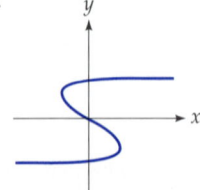

4.

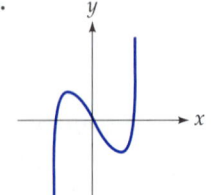

7.

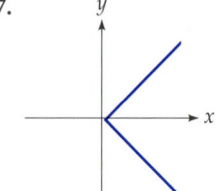

8.

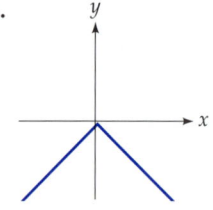

Find the domain and range of each function graphed below.

9.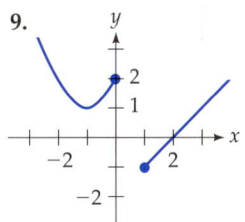

10.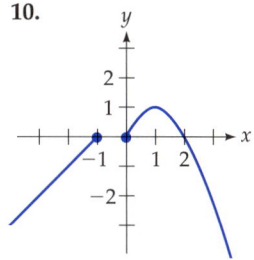

11–22: For each function:

a. Evaluate the given expression.
b. Find the domain of the function.
c. Find the range.
(*Hint:* Use a graphing calculator.)

11. $f(x) = \sqrt{x - 1}$, find $f(10)$

12. $f(x) = \sqrt{x - 4}$, find $f(40)$

13. $h(z) = \dfrac{1}{z + 4}$, find $h(-5)$

14. $h(z) = \dfrac{1}{z + 7}$, find $h(-8)$

15. $h(x) = x^{1/4}$, find $h(81)$

16. $h(x) = x^{1/6}$, find $h(64)$

17. $f(x) = x^{2/3}$, find $f(-8)$

(*Hint for Exercises 17 and 18:* You may need to enter $x^{m/n}$ as $(x^m)^{1/n}$ or as $(x^{1/n})^m$, as discussed on page 25.)

18. $f(x) = x^{4/5}$, find $f(-32)$

19. $f(x) = \sqrt{4 - x^2}$, find $f(0)$

20. $f(x) = \dfrac{1}{\sqrt{x}}$, find $f(4)$

21. $f(x) = \sqrt{-x}$, find $f(-25)$

22. $f(x) = -\sqrt{-x}$, find $f(-100)$

23–30: Graph each function "by hand." (*Note:* Even if you have a graphing calculator, it is important to

be able to sketch simple curves by finding a few important points.)

23. $f(x) = 3x - 2$

24. $f(x) = 2x - 3$

25. $f(x) = -x + 1$

26. $f(x) = -3x + 5$

27. $f(x) = 2x^2 + 4x - 16$

28. $f(x) = 3x^2 - 6x - 9$

29. $f(x) = -3x^2 + 6x + 9$

30. $f(x) = -2x^2 + 4x + 16$

 31–34: For each quadratic function:

a. Find the vertex using the vertex formula.
b. Graph the function on an appropriate viewing window. (Answers may differ.)

31. $f(x) = x^2 - 40x + 500$

32. $f(x) = x^2 + 40x + 500$

33. $f(x) = -x^2 - 80x - 1800$

34. $f(x) = -x^2 + 80x - 1800$

35–52: Solve each equation by factoring or the quadratic formula.

35. $x^2 - 6x - 7 = 0$ **36.** $x^2 - x - 20 = 0$

37. $x^2 + 2x = 15$ **38.** $x^2 - 3x = 54$

39. $2x^2 + 40 = 18x$ **40.** $3x^2 + 18 = 15x$

41. $5x^2 - 50x = 0$ **42.** $3x^2 - 36x = 0$

43. $2x^2 - 50 = 0$ **44.** $3x^2 - 27 = 0$

45. $4x^2 + 24x + 40 = 4$ **46.** $3x^2 - 6x + 9 = 6$

47. $-4x^2 + 12x = 8$ **48.** $-3x^2 + 6x = -24$

49. $2x^2 - 12x + 20 = 0$ **50.** $2x^2 - 8x + 10 = 0$

51. $3x^2 + 12 = 0$ **52.** $5x^2 + 20 = 0$

53–62: Solve each equation using a graphing calculator.
[*Hint:* Begin with the viewing window $[-10, 10]$ by $[-10, 10]$ or another of your choice (see Useful Hint in Graphing Calculator Terminology fol-

lowing the Preface) and use ZERO, SOLVE, or TRACE and ZOOM IN.] (In Exercises 61 and 62, round answers to two decimal places.)

53. $x^2 - x - 20 = 0$ **54.** $x^2 + 2x = 15$

55. $2x^2 + 40 = 18x$ **56.** $3x^2 + 18 = 15x$

57. $4x^2 + 24x + 45 = 9$ **58.** $3x^2 - 6x + 5 = 2$

59. $3x^2 + 7x + 12 = 0$ **60.** $5x^2 + 14x + 20 = 0$

61. $2x^2 + 3x - 6 = 0$ **62.** $3x^2 + 5x - 7 = 0$

 63. Use a graphing calculator to graph the following four equations simultaneously on the viewing window $[-10, 10]$ by $[-10, 10]$:

$$y_1 = 2x + 6$$
$$y_2 = 2x + 2$$
$$y_3 = 2x - 2$$
$$y_4 = 2x - 6$$

a. What do the lines have in common and how do they differ?

b. Write the equation of another line with the same slope that lies two units below the lowest line. Then check your answer by graphing it with the others.

 64. Use a graphing calculator to graph the following four equations simultaneously on the viewing window $[-10, 10]$ by $[-10, 10]$:

$$y_1 = 3x + 4$$
$$y_2 = 1x + 4$$
$$y_3 = -1x + 4 \quad \text{(Use } (-) \text{ to get } -1x.)$$
$$y_4 = -3x + 4$$

a. What do the lines have in common and how do they differ?

b. Write the equation of a line through this y-intercept with slope $\frac{1}{2}$. Then check your answer by graphing it with the others.

 65–68: Shifts of Parabolas Set your graphing window to $[-5, 5]$ by $[-5, 5]$ and graph the parabola $y_1 = x^2$. Then do the following:

65. a. Horizontal Shifts of Parabolas Graph $y_2 = (x - 4)^2$ (a parabola in factored form)

together with the original parabola. How is the new parabola shifted compared to the original one? What are the coordinates of the vertex of the new parabola?

b. Return to y_2, change it to $y_2 = (x + 3)^2$, and graph it together with the original parabola $y_1 = x^2$. Now how is the new parabola shifted compared to the original one? What are the coordinates of its vertex?

c. In general, for any number a, how would you describe the parabola $y = (x - a)^2$ compared to the original parabola? What are the coordinates of its vertex? What if the sign is plus instead of minus?

66. a. Vertical Shifts of Parabolas See instructions above Exercise 65. Graph $y_2 = x^2 + 2$ together with the original parabola. How is the new parabola shifted compared to the original one? What are the coordinates of the vertex of the new parabola?

b. Return to y_2, change it to $y_2 = x^2 - 3$, and graph it together with the original parabola $y_1 = x^2$. Now how is the new parabola shifted compared to the original one? What are the coordinates of its vertex?

c. In general, for any number b, how would you describe the parabola $y = x^2 + b$ compared to the original parabola? What if the sign is minus instead of plus?

67. a. Continuation of Exercises 65 and 66 Graph $y_2 = (x - 3)^2 + 2$ together with the original parabola $y = x^2$. How is the new parabola shifted compared to the original one? What are the coordinates of the vertex of the new parabola?

b. Return to y_2, change it to $y_2 = (x + 2)^2 - 5$, and graph it together with the original parabola $y_1 = x^2$. Now how is the new parabola shifted compared to the original one? What are the coordinates of its vertex?

c. In general, for any numbers a and b, how would you describe the parabola $y = (x - a)^2 + b$ compared to the original parabola?

68. a. Continuation of Exercise 67 In the Graphing Calculator Exploration on page 38 you investigated the effect of the coefficient of x^2 on the shape of the parabola. From that, together with the results of the previous exercise, describe the parabola $y = 5(x - 2)^2 - 1$ as compared to the original parabola. Then check your description by graphing the new parabola together with the original one.

b. Describe the parabola $y = -5(x - 2)^2 - 1$ as compared to the original parabola. Check your description by graphing it.
c. Describe the parabola $y = c(x - a)^2 + b$ (for any numbers a, b, and c) as compared with the original parabola $y_1 = x^2$.

APPLIED EXERCISES

69. Business: Cost Functions   A lumberyard will deliver wood for $4 per board foot plus a delivery charge of $20. Find a function $C(x)$ for the cost of having x board feet of lumber delivered.

70. Business: Cost Functions   A company manufactures bicycles at a cost of $55 each. If the company's fixed costs are $900, express the company's costs as a linear function of x, the number of bicycles produced.

71. Business: Salary   An employee's weekly salary is $500 plus $15 per hour of overtime. Find a function $P(x)$ giving his pay for a week in which he worked x hours of overtime.

72. Business: Salary   A sales clerk's weekly salary is $300 plus 2% of her total week's sales. Find a function $P(x)$ for her pay for a week in which she sold x dollars of merchandise.

73. General: Water Pressure   At a depth of d feet underwater, the water pressure is $p(d) = 0.45d + 15$ pounds per square inch. Find the pressure at

a. The bottom of a 6-foot-deep swimming pool.
b. The maximum ocean depth of 35,000 feet.

74. General: Boiling Point   At higher altitudes water boils at lower temperatures. This is why at high altitudes foods must be boiled for longer times—the lower boiling point imparts less heat to the food. At an altitude of h thousand feet above sea level, water boils at a temperature of $B(h) = -1.8h + 212$ degrees Fahrenheit. Find the altitude at which water boils at 98.6 degrees Fahrenheit. (Your answer will show that at a high enough altitude, water boils at

normal body temperature. This is why airplane cabins must be pressurized—at high enough altitudes one's blood would boil.)

75–76: General: Stopping Distance   According to data from the National Transportation Safety Board, a car traveling at speed v miles per hour should be able to come to a full stop in a distance of

$$D(v) = 0.055v^2 + 1.1v \qquad \text{feet}$$

Find the stopping distance required for a car traveling at:

75. 40 mph

76. 60 mph

77. Biomedical: Cell Growth   The number of cells in a culture after t days is given by $N(t) = 200 + 50t^2$. Find the size of the culture after

a. 2 days,
b. 10 days.

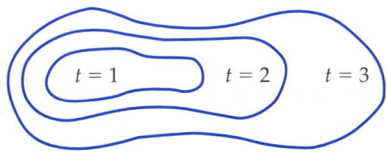

78. General: Juggling   If you toss a ball h feet straight up, it will return to your hand after $T(h) = 0.5\sqrt{h}$ seconds. This leads to the "jugglers dilemma." Juggling more balls means tossing them higher. However, the square root in the above formula means that tossing them twice as high does not gain twice as much time, but only $\sqrt{2} \approx 1.4$ times as much time. Because of this, there is a limit to the number of balls that a person can juggle, which seems to be about ten. Use this formula to find:

a. How long will a ball spend in the air if it is tossed to a height of 4 ft? 8 ft?
b. How high must it be tossed to spend 2 seconds in the air? 3 seconds in the air?

 79. General: Impact Velocity   If a marble is dropped from a height of x feet, it will hit the ground with velocity $v(x) = \frac{60}{11}\sqrt{x}$ miles per hour (neglecting air resistance). Use this formula to find the velocity with which a marble will strike the ground if it is dropped from the top of the tallest building in the United States, the 1454-foot Sears Tower in Chicago.

 80. General: Tsunamis   The speed of a tsunami (popularly known as a tidal wave, although it has nothing whatever to do with tides) depends on the depth of the water through which it is traveling. At a depth of d feet, the speed of a tsunami will be $s(d) = 3.86\sqrt{d}$ miles per hour. Find the speed of a tsunami in the Pacific basin, where the average depth is 15,000 feet.

 81–82: Impact Time of a Projectile   If an object is thrown upward so that its height (in feet) above the ground t seconds after it is thrown is given by the function $h(t)$ below, find when the object hits the ground. That is, find the positive value of t such that $h(t) = 0$. Give the answer correct to two decimal places.
(*Hint:* Enter the function in terms of x rather than t. Use the ZERO operation, or TRACE and ZOOM IN, or use similar operations.)

81. $h(t) = -16t^2 + 45t + 5$

82. $h(t) = -16t^2 + 40t + 4$

83. A company that produces devices for computer disk drives finds that if it produces x devices per week, then its costs will be $C(x) = 180x + 16{,}000$ and its revenue will be $R(x) = -2x^2 + 660x$ (both in dollars).

a. Find the company's break-even points.
b. Find the number of devices that maximizes profit, and the maximum profit.

84. A bicycle store finds that if it sells x racing bicycles per month, then its costs will be $C(x) = 420x + 72{,}000$ and its revenue will be $R(x) = -3x^2 + 1800x$ (both in dollars).

a. Find the store's break-even points.
b. Find the number of bicycles that maximizes profit, and the maximum profit.

85. A sporting goods store finds that if it sells x exercise machines per day, then its costs will be $C(x) = 100x + 3200$ and its revenue will be $R(x) = -2x^2 + 300x$ (both in dollars).

a. Find the store's break-even points.
b. Find the number of sales that will maximize profit, and the maximum profit.

86. A company that installs car alarm systems finds that if it installs x systems per week, then its costs will be $C(x) = 210x + 72{,}000$ and its revenue will be $R(x) = -3x^2 + 1230x$ (both in dollars).

a. Find the company's break-even points.
b. Find the number of installations that will maximize profit, and the maximum profit.

87. Business: Sales   The following table gives a company's annual sales (in millions of units) at the ends of its first through fourth years.

Year	Sales (millions)
1	3.8
2	3.6
3	3.7
4	4.0

a. Enter the numbers from the table into a graphing calculator and make a plot of the resulting points (Year on the x-axis and Sales on the y-axis).
b. Have your calculator find the quadratic (parabolic) regression formula for these data. Then enter the result as y_1, which gives a formula for sales for each year. Plot the points together with the regression line.
c. Predict the sales at the end of year 5 by evaluating $y_1(5)$.

0.4 Functions, continued

Automobile Efficiency and Rational Functions

Automobile efficiency is measured by miles per gallon (mpg), the number of miles that a car can drive on 1 gallon of gas (under specified conditions). How much money will you actually save if you change from a car getting 10 mpg to one getting 20 mpg? Suppose that you drive 12,000 miles in a typical year, and that gas costs $1.50 per gallon. Your savings will be

$$\underbrace{\frac{12{,}000}{10}\,(1.50)}_{\substack{\text{Cost of driving} \\ \text{12,000 miles} \\ \text{at 10 mpg}}} - \underbrace{\frac{12{,}000}{20}\,(1.50)}_{\substack{\text{Cost of driving} \\ \text{12,000 miles} \\ \text{at 20 mpg}}} = \underbrace{1800 - 900 = \$900}_{\substack{\text{Annual} \\ \text{savings}}}$$

Instead of asking about improving efficiency from 10 to 20 mpg, let us ask the more general question: How much is saved by improving efficiency from x to $(x + 10)$ mpg? Based on the preceding calculation, the savings $f(x)$ in changing from x to $(x + 10)$ mpg is

$$f(x) = \frac{12{,}000}{x}\,(1.50) - \frac{12{,}000}{x + 10}\,(1.50)$$

After some algebra, this simplifies to

$$f(x) = \frac{180{,}000}{x(x + 10)}$$

Such functions are called *rational functions*, and will be discussed in this section. The graph of this rational function is shown on the following page, although for our purposes the graph is meaningful only for positive values of x. Note that the graph becomes arbitrarily high and low near the values $x = 0$ and $x = -10$, where the denominator becomes zero.

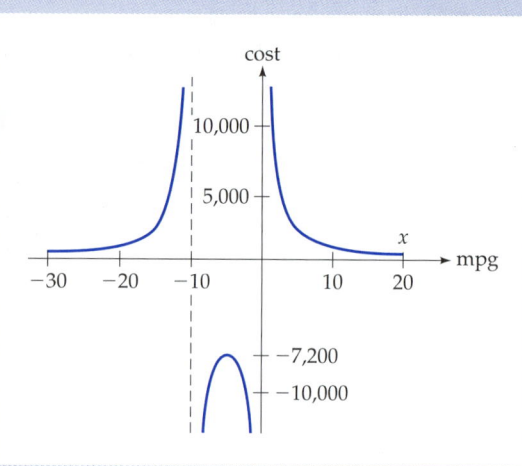

The rightmost "branch" of the curve shows that as x increases, the function takes values closer and closer to zero. This means that at high mpg efficiencies, another 10 mpg yields only a very small saving. For example, improving from 10 mpg to 20 mpg saves $900, but improving from 50 mpg to 60 mpg would save only $60 for the year, with negligible savings thereafter.

An ultralight four-passenger car achieving 66 mpg, called the Prius, is now available in Japan, and hybrid gas–electric cars achieving up to 300 mpg now seem possible. (*Source: Rocky Mountain Institute*)

An ultralight concept car

Introduction

In this section we will define other useful types of functions and an important operation, the *composition* of two functions.

Polynomial Functions

A *polynomial function* (or simply a *polynomial*) is a function that can be written in the form

$$f(x) = a_n x^n + a_{n-1} x^{n-1} + \cdots + a_2 x^2 + a_1 x + a_0$$

where n is a nonnegative integer and $a_0, a_1, \ldots, a_n$ are (real) numbers, called *coefficients*. The *domain* of a polynomial is $\mathbb{R}$, the set of all (real) numbers. The *degree* of a polynomial is the highest power of the variable. The following are polynomials.

$$f(x) = 2x^8 - 3x^7 + 4x^5 - 5$$
A polynomial of degree 8 (since the highest power of x is 8)

$$f(x) = -4x^2 - \tfrac{1}{3}x + 19$$
A polynomial of degree 2 (a quadratic function)

$$f(x) = x - 1$$
A polynomial of degree 1 (a linear function)

$$f(x) = 6$$
A polynomial of degree 0 (a constant function)

Polynomials are used to model many situations in which change occurs at different rates. For example, the polynomial in the following graph might represent the total cost of manufacturing x units of a product. At first costs rise quite steeply, as a result of high start-up expenses,

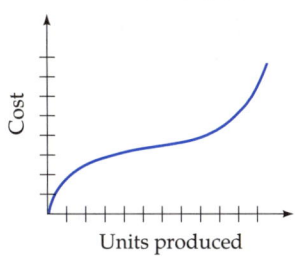

A cost function may increase at different rates at different production levels.

then they rise more slowly as the economies of mass production come into play, and finally they rise more steeply as new production facilities need to be built.

Polynomial equations can often be solved by factoring (just as with quadratic equations).

EXAMPLE 1 Solving a Polynomial Equation

Solve $3x^4 - 6x^3 = 24x^2$

Solution

$$3x^4 - 6x^3 - 24x^2 = 0$$ Written with all the terms on the left side

$$3x^2(x^2 - 2x - 8) = 0$$ Factoring out $3x^2$

$$3x^2 \ (x - 4) \ (x + 2) = 0$$ Factoring further

Equals zero at $x = 0$ Equals zero at $x = 4$ Equals zero at $x = -2$ Finding the zeros of each factor

$$x = 0, \quad x = 4, \quad x = -2$$ Solutions

∎

As in this example, if a positive power of x can be factored out of a polynomial, then $x = 0$ is one of the roots.

PRACTICE PROBLEM 1   Solve $2x^3 - 4x^2 = 48x$.   *Solutions at the back of the book*

Rational Functions

The word "ratio" means fraction or quotient, and the quotient of two polynomials is called a *rational function*. The following are rational functions.

$$f(x) = \frac{4x^3 + 3x^2}{x^2 - 2x + 1} \qquad g(x) = \frac{1}{x^2 + 1}$$ A rational function is a polynomial over a polynomial.

The domain of a rational function is the set of all numbers for which the denominator is not zero. For example, the domain of the function on the left above is $\{x \mid x \neq 1\}$ (since $x = 1$ makes the denominator zero), and the domain of the function on the right is $\mathbb{R}$ (since $x^2 + 1$ is never zero).

PRACTICE PROBLEM 2   What are the domain and range of the rational function graphed on page 50?   *Solution at the back of the book*

Simplifying a rational function by canceling a common factor from the numerator and the denominator can change the domain of the function, so that the "simplified" and "original" versions may not be equal (since they have different domains). For example, the rational function on the next page is not defined at $x = 1$, whereas the simplified version on its right *is* defined at $x = 1$, so that the two functions are technically not equal.

$$\frac{x^2 - 1}{x - 1} = \frac{(x + 1)(x - 1)}{x - 1} \neq x + 1$$

Not defined at $x = 1$,
so the domain is $\{x \mid x \neq 1\}$

Is defined at $x = 1$,
so the domain is $\mathbb{R}$

However, the functions *are* equal at every x-value *except* $x = 1$, and the graphs (shown below) are the same except that the rational function omits the point at $x = 1$. We will return to this technical issue when we discuss limits in Chapter 6.

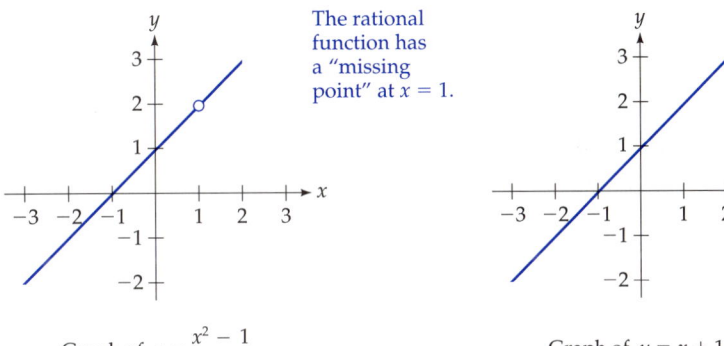

The rational function has a "missing point" at $x = 1$.

Graph of $y = \dfrac{x^2 - 1}{x - 1}$

Graph of $y = x + 1$

Piecewise Linear Functions

The rule for calculating the values of a function may be given in several parts. If each part is linear, the function is called a *piecewise linear function*, and its graph consists of "pieces" of straight lines.

EXAMPLE 2   **Graphing a Piecewise Linear Function**

Graph

$$f(x) = \begin{cases} 5 - 2x & \text{if } x \geq 2 \\ x + 3 & \text{if } x < 2 \end{cases}$$

This notation means: Use the top formula for $x \geq 2$ and the bottom formula for $x < 2$

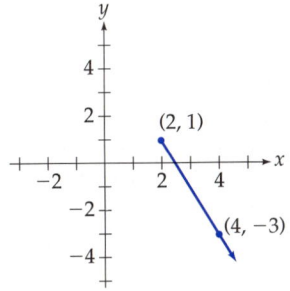

Solution   We graph one "piece" at a time.

Step 1: To graph the first part, $f(x) = 5 - 2x$ if $x \geq 2$, we use the "end-point" $x = 2$ and also $x = 4$ (or any other x-value satisfying $x \geq 2$). The points are $(2, 1)$ and $(4, -3)$, with the y-coordinates calculated from $f(x) = 5 - 2x$. Draw the line through these two points, but only for $x \geq 2$ (from $x = 2$ to the *right*).

Step 2: For the second part, $f(x) = x + 3$ if $x < 2$, the restriction $x < 2$ means that the line ends just *before* $x = 2$. We mark this "missing point" (2, 5) by an "open circle" (○) to indicate that it is *not* included in the graph (the y-coordinate comes from $f(x) = x + 3$). For a second point, choose $x = 0$ (or any other $x < 2$), giving (0, 3). Draw the line through these two points, but only for $x < 2$ (to the *left* of $x = 2$), completing the graph of the function.

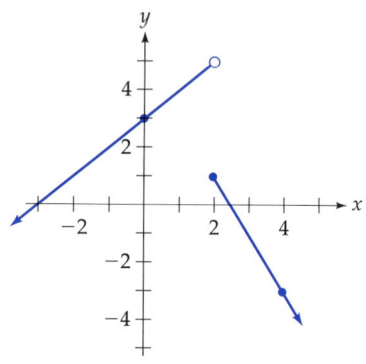

■

EXAMPLE 3   Graphing the Absolute Value Function

The absolute value function is defined as

$$f(x) = \begin{cases} x & \text{if } x \geq 0 \\ -x & \text{if } x < 0 \end{cases}$$

The second line, for *negative x*, attaches a *second* negative sign to make the result *positive*

For example, when applied to either 3 or -3, the function gives *positive* 3:

$$f(3) = 3$$

Using the top formula (since $3 \geq 0$)

$$f(-3) = -(-3) = 3$$

Using the bottom formula (since $-3 < 0$)

Its graph, drawn as in Example 2, is shown on the following page.

■

Absolute Value Function

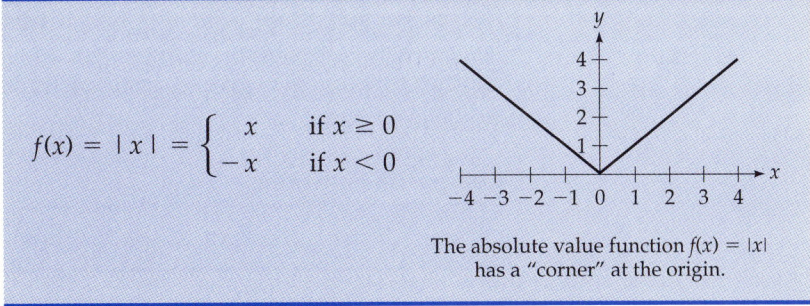

$$f(x) = |x| = \begin{cases} x & \text{if } x \geq 0 \\ -x & \text{if } x < 0 \end{cases}$$

The absolute value function $f(x) = |x|$ has a "corner" at the origin.

Examples 2 and 3 show that the "pieces" of a piecewise linear function may or may not be connected.

EXAMPLE 4 Graphing an Income Tax Function

Federal income taxes are "progressive," meaning that they take a higher percentage of higher incomes. For example, the 1997 federal income tax for a single taxpayer whose taxable income is less than $124,650 is determined by a three-part rule: 15% of income up to $24,650, plus 28% of any amount over $24,650 up to $59,750, plus 31% of any amount over $59,750 up to $124,650. For an income of x dollars, the tax $f(x)$ may then be expressed as follows:

$$f(x) = \begin{cases} 0.15x & \text{if } 0 \leq x \leq 24{,}650 \\ 3697.50 + 0.28(x - 24{,}650) & \text{if } 24{,}650 < x \leq 59{,}750 \\ 13{,}525.50 + 0.31(x - 59{,}750) & \text{if } 59{,}750 < x \leq 124{,}650 \end{cases}$$

Graphing this by the same technique as before leads to the graph shown below. The slopes 0.15, 0.28, and 0.31 are called the *marginal tax rates*.

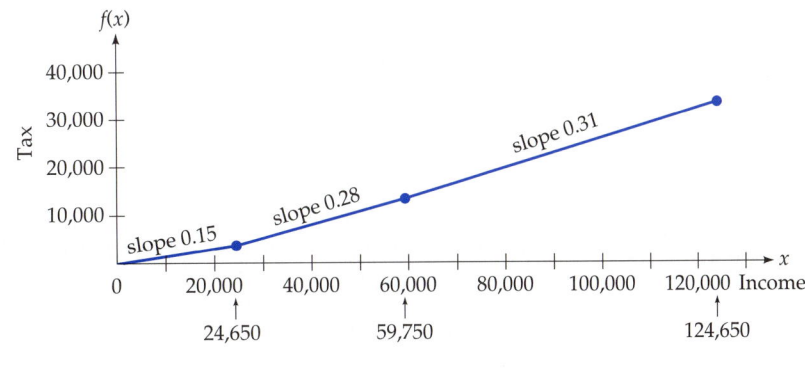

Composite Functions

Just as we substitute a *number* into a function, we may substitute a *function* into a function. Taking a function $f(x)$ and replacing each x by another function $g(x)$ gives a *composite* function $f(g(x))$, called the *composition* of f and g.*

Composite Functions

The composition of the functions f and g is $f(g(x))$.

The *domain* of the composite function $f(g(x))$ is the set of all numbers x in the domain of g such that $g(x)$ is in the domain of f. If we think of $f(x)$ and $g(x)$ as "numerical machines," then the composite function $f(g(x))$ may be thought of as a combined machine in which the output of $g(x)$ is connected to the input of $f(x)$.

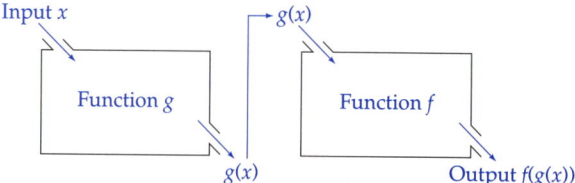

A "machine" for generating the composite function $f(g(x))$. A number x is fed into the function g, and the output $g(x)$ is then fed into the function f, resulting in $f(g(x))$.

EXAMPLE 5   **Finding a Composite Function**

If $f(x) = x^7$ and $g(x) = x^3 - 2x$, find the composition $f(g(x))$.

Solution

$$f(g(x)) \quad = \quad [\,g(x)\,]^7 \quad = \quad (x^3 - 2x)^7$$

$\underbrace{f(x) = x^7 \text{ with } x}$ $\underbrace{\text{Using}}$
replaced by $g(x)$ $g(x) = x^3 - 2x$

EXAMPLE 6   **Finding Both Composite Functions**

If $f(x) = \dfrac{x + 8}{x - 1}$ and $g(x) = \sqrt{x}$, find $f(g(x))$ and $g(f(x))$.

---

* The composite function $f(g(x))$ may also be written $(f \circ g)(x)$, although we will not use this notation.

Solution

$$f(g(x)) = \frac{g(x) + 8}{g(x) - 1} = \frac{\sqrt{x} + 8}{\sqrt{x} - 1}$$

$f(x) = \dfrac{x + 8}{x - 1}$ with x replaced by $g(x) = \sqrt{x}$

$$g(f(x)) = \sqrt{f(x)} = \sqrt{\frac{x + 8}{x - 1}}$$

$g(x) = \sqrt{x}$ with x replaced by $f(x) = \dfrac{x + 8}{x - 1}$

The order of composition makes a difference: $f(g(x))$ is *not* the same as $g(f(x))$. To show this, we evaluate the above $f(g(x))$ and $g(f(x))$ at $x = 4$:

$$f(g(4)) = \frac{\sqrt{4} + 8}{\sqrt{4} - 1} = \frac{2 + 8}{2 - 1} = \frac{10}{1} = 10$$

$f(g(x)) = \dfrac{\sqrt{x} + 8}{\sqrt{x} - 1}$ at $x = 4$

Different answers

$$g(f(4)) = \sqrt{\frac{4 + 8}{4 - 1}} = \sqrt{\frac{12}{3}} = \sqrt{4} = 2$$

$g(f(x)) = \sqrt{\dfrac{x + 8}{x - 1}}$ at $x = 4$

PRACTICE PROBLEM 3   If $f(x) = x^2 + 1$ and $g(x) = \sqrt[3]{x}$, find: **a.** $f(g(x))$,   **b.** $g(f(x))$.

Solutions at the back of the book

EXAMPLE 7   Predicting Water Usage

A planning commission estimates that if a city's population is p thousand people, its daily water usage will be $W(p) = 30p^{1.2}$ thousand gallons. The commission further predicts that the population in t years will be $p(t) = 60 + 2t$ thousand people. Express the water usage W as a function of t, the number of years from now, and find the water usage 10 years from now.

Solution

Water usage W as a function of t is the *composition* of $W(p)$ and $p(t)$:

$$W(p(t)) = 30[p(t)]^{1.2} = 30(60 + 2t)^{1.2}$$

$W = 30p^{1.2}$ with p replaced by $p(t) = 60 + 2t$

To find water usage in 10 years, we evaluate $W(p(t))$ at $t = 10$:

$$W(p(10)) = 30(60 + 2 \cdot 10)^{1.2}$$

$30(60 + 2t)^{1.2}$ with $t = 10$

$$= 30(80)^{1.2} \approx 5765$$

Using a calculator

Thousand gallons

Therefore, in 10 years the city will need about 5,765,000 gallons of water per day.

Shifts of Graphs

In certain cases, the graph of a composite function is simply a "shift" of an original graph. That is, adding a number to or subtracting it from a variable or a function results in a horizontal or vertical shift of the graph. The diagram below shows the graph of $y = x^2$ in the center, along with *horizontal* shifts (adding to or subtracting from x) and *vertical* shifts (adding to or subtracting from the *function*).

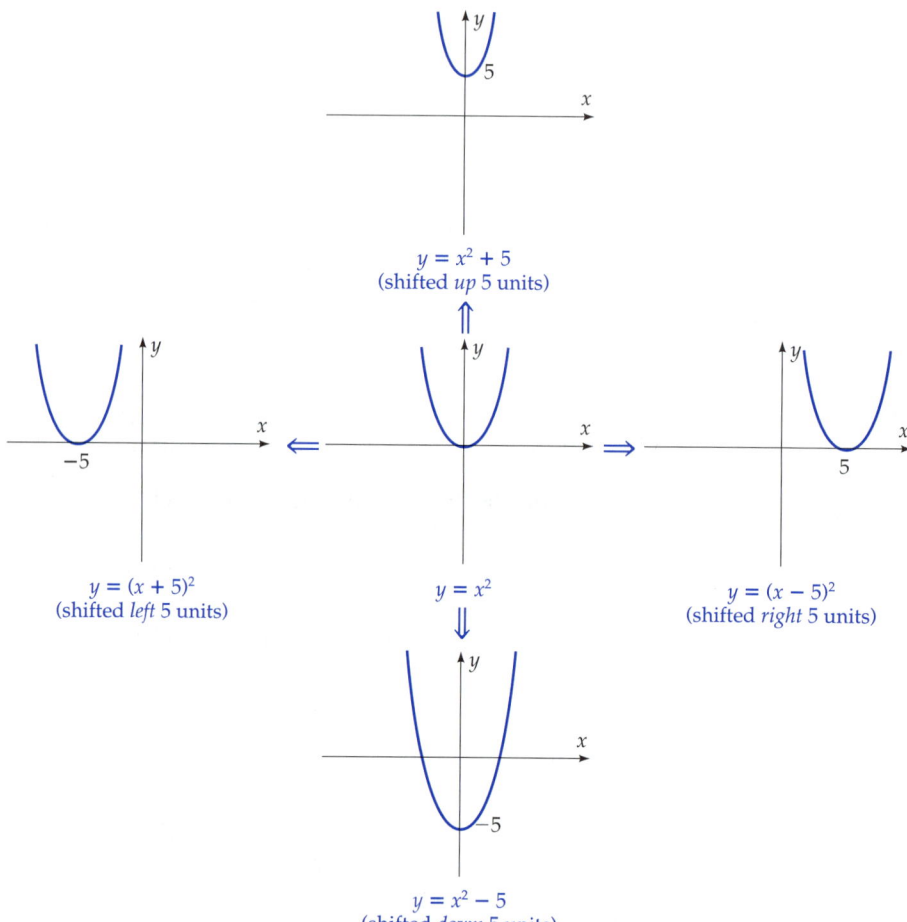

$y = x^2 + 5$
(shifted *up* 5 units)

$y = (x + 5)^2$
(shifted *left* 5 units)

$y = x^2$

$y = (x - 5)^2$
(shifted *right* 5 units)

$y = x^2 - 5$
(shifted *down* 5 units)

In general, for any function $y = f(x)$ and positive numbers a and b:

The graph of	is the graph of $y = f(x)$ shifted
$y = f(x + a)$	*left* by a units
$y = f(x - a)$	*right* by a units
$y = f(x) + b$	*up* by b units
$y = f(x) - b$	*down* by b units

Of course, a graph can be shifted both horizontally and vertically:

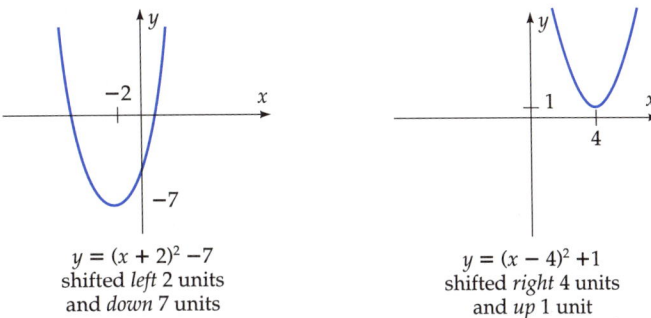

$$y = (x + 2)^2 - 7$$
shifted *left* 2 units
and *down* 7 units

$$y = (x - 4)^2 + 1$$
shifted *right* 4 units
and *up* 1 unit

Such shifts apply to *any* function $y = f(x)$: the graph of $y = f(x + a) + b$ is shifted *left a* units and *up b* units (with the understanding that a *negative a* or *b* means that the direction is reversed).

Be careful: Remember that adding a *positive* number to x means a *left* shift.

Graphing Calculator Exploration

The absolute value function $y = |x|$ may be graphed on a graphing calculator as $y_1 = abs(x)$.

a.   Graph $y_1 = abs(x - 2) - 6$ and observe that the absolute value function is shifted *right* 2 units and *down* 6 units. (The graph shown is drawn using ZOOM ZSquare.)

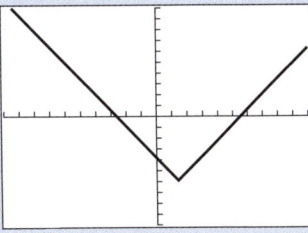

b.   Predict the shift of $y_1 = abs(x + 4) + 2$ and then verify your prediction by graphing the function on your calculator.

Difference Quotients

Given $f(x)$, to find an algebraic expression for the "shifted" function $f(x + h)$ we simply replace each occurrence of x by $x + h$.

EXAMPLE 8   Finding $f(x + h)$ from $f(x)$

If $f(x) = x^2 - 5x$, find $f(x + h)$.

Solution

$f(x + h) = \underbrace{(x + h)^2} - \underbrace{5(x + h)}$ $f(x) = x^2 - 5x$ with each x replaced by $x + h$

$= x^2 + 2xh + h^2 - 5x - 5h$ Expanding

◼

The quantity $\dfrac{f(x + h) - f(x)}{h}$ is called the *difference quotient*, since it is a quotient whose numerator is a difference. Finding difference quotients for given functions will be important in Chapter 6 when we begin calculus.

EXAMPLE 9   Finding $\dfrac{f(x + h) - f(x)}{h}$

If $f(x) = x^2 - 4x + 1$, find and simplify $\dfrac{f(x + h) - f(x)}{h}$ $(h \neq 0)$

Solution

$$\frac{f(x + h) - f(x)}{h} = \frac{\overbrace{(x + h)^2 - 4(x + h) + 1}^{f(x+h)} - \overbrace{(x^2 - 4x + 1)}^{f(x)}}{h}$$

$$= \frac{x^2 + 2xh + h^2 - 4x - 4h + 1 - x^2 + 4x - 1}{h} \qquad \text{Expanding}$$

$$= \frac{\cancel{x^2} + 2xh + h^2 - \cancel{4x} - 4h + \cancel{1} - \cancel{x^2} + \cancel{4x} - \cancel{1}}{h} \qquad \text{Canceling}$$

$$= \frac{2xh + h^2 - 4h}{h} = \frac{h(2x + h - 4)}{h} \qquad \begin{array}{l}\text{Factoring an } h \\ \text{from the top}\end{array}$$

$$= \frac{\cancel{h}(2x + h - 4)}{\cancel{h}} = 2x + h - 4 \qquad \begin{array}{l}\text{Canceling } h \text{ from} \\ \text{top and bottom} \\ \text{(since } h \neq 0)\end{array}$$

◼

PRACTICE PROBLEM 4   If $f(x) = 3x^2 - 2x + 1$, find and simplify $\dfrac{f(x + h) - f(x)}{h}$.

Solution at the back of the book

SUMMARY

We have introduced a variety of functions: polynomials (which include linear and quadratic functions), rational functions, and piecewise linear functions. Examples of these are shown below and on the following page. You should be able to identify these basic types of functions from their algebraic forms. We also added constants to perform horizontal and vertical *shifts* of graphs of functions, and combined functions by using the "output" of one as the "input" of the other, resulting in *composite* functions, such as $f(g(x))$ and $g(f(x))$.

A Gallery of Functions

POLYNOMIALS

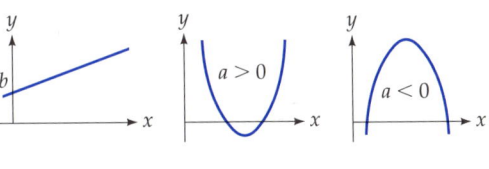

Linear function
$f(x) = mx + b$

Quadratic functions
$f(x) = ax^2 + bx + c$

$f(x) = ax^4 + bx^3 + cx^2 + dx + e$

RATIONAL FUNCTIONS

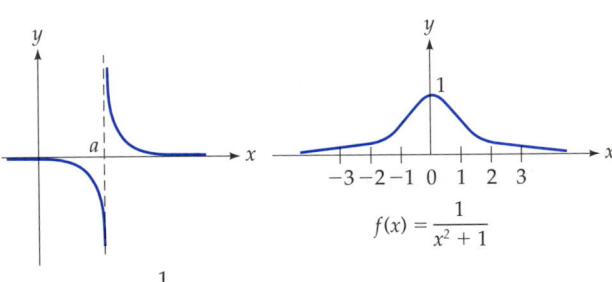

$f(x) = \dfrac{1}{x - a}$

$f(x) = \dfrac{1}{x^2 + 1}$

PIECEWISE LINEAR FUNCTIONS

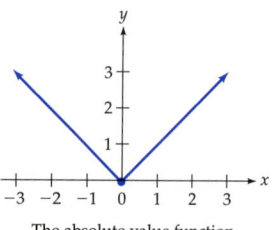

The absolute value function
$f(x) = |x|$

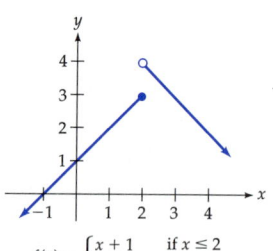

$$f(x) = \begin{cases} x + 1 & \text{if } x \le 2 \\ 6 - x & \text{if } x > 2 \end{cases}$$

EXERCISES 0.4

Find the domain and range of each function graphed below.

1.

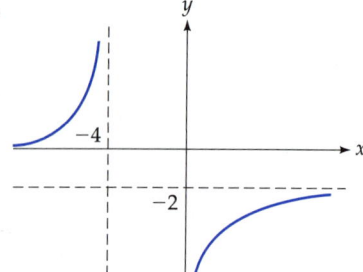

2.

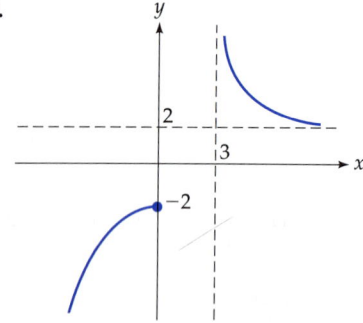

For each function in Exercises 3–10:

a. Evaluate the given expression.
b. Find the domain of the function.
 c. Find the range. (*Hint*: Use a graphing calculator. You may have to ignore some false lines on

the graph. Graphing in "dot mode" will also eliminate false lines.)

3. $f(x) = \dfrac{1}{x + 4}$, find $f(-3)$

4. $f(x) = \dfrac{1}{(x - 1)^2}$, find $f(-1)$

5. $f(x) = \dfrac{x^2}{x - 1}$, find $f(-1)$

6. $f(x) = \dfrac{x^2}{x + 2}$, find $f(2)$

7. $f(x) = \dfrac{12}{x(x + 4)}$, find $f(2)$

8. $f(x) = \dfrac{16}{x(x - 4)}$, find $f(-4)$

9. $g(x) = |x + 2|$, find $g(-5)$
10. $g(x) = |x| + 2$, find $g(-5)$

Solve each equation by factoring.

11. $x^5 + 2x^4 - 3x^3 = 0$ **12.** $x^6 - x^5 - 6x^4 = 0$

13. $5x^3 - 20x = 0$ **14.** $2x^5 - 50x^3 = 0$

15. $2x^3 + 18x = 12x^2$ **16.** $3x^4 + 12x^2 = 12x^3$

17. $6x^5 = 30x^4$ **18.** $5x^4 = 20x^3$

19. $3x^{5/2} - 6x^{3/2} = 9x^{1/2}$ **20.** $2x^{7/2} + 8x^{5/2} = 24x^{3/2}$
(*Hint for Exercises 19 and 20:* First factor out a fractional power.)

Solve each equation using a graphing calculator.

21. $x^3 - 2x^2 - 8x = 0$ **22.** $x^3 + 2x^2 - 8x = 0$

23. $x^5 - 2x^4 - 3x^3 = 0$ **24.** $x^6 + x^5 - 6x^4 = 0$

25. $2x^3 = 12x^2 - 18x$

26. $3x^4 + 12x^3 + 12x^2 = 0$

27. $6x^5 + 30x^4 = 0$ **28.** $5x^4 + 20x^3 = 0$

29. $2x^{5/2} + 4x^{3/2} = 6x^{1/2}$

30. $3x^{7/2} - 12x^{5/2} = 36x^{3/2}$

31. $x^5 - x^4 - 5x^3 = 0$

(For Exercises 31 and 32, round answers to two decimal places.)

Graph each function.

32. $x^6 + 2x^5 - 5x^4 = 0$

33. $f(x) = |x - 3| - 3$

34. $f(x) = |x + 2| - 2$

35. $f(x) = \begin{cases} 2x - 7 & \text{if } x \geq 4 \\ 2 - x & \text{if } x < 4 \end{cases}$

36. $f(x) = \begin{cases} 2 - x & \text{if } x \geq 3 \\ 2x - 4 & \text{if } x < 3 \end{cases}$

37. $f(x) = \begin{cases} 8 - 2x & \text{if } x \geq 2 \\ x + 2 & \text{if } x < 2 \end{cases}$

38. $f(x) = \begin{cases} 2x - 4 & \text{if } x > 3 \\ 5 - x & \text{if } x \leq 3 \end{cases}$

Identify each function below as a polynomial, a rational function, a piecewise linear function, or none of these. (Do not graph the functions, just identify their types.)

39. $f(x) = x^5$ **40.** $f(x) = 3|x|$

41. $f(x) = 1 - |x|$ **42.** $f(x) = x^4$

43. $f(x) = x + 2$

44. $f(x) = \begin{cases} 3x - 1 & \text{if } x \geq 2 \\ 1 - x & \text{if } x < 2 \end{cases}$

45. $f(x) = \dfrac{1}{x + 2}$ **46.** $f(x) = x^2 + 9$

47. $f(x) = \begin{cases} x - 2 & \text{if } x < 3 \\ 7 - 4x & \text{if } x \geq 3 \end{cases}$

 48. $f(x) = \dfrac{x}{x^2 + 9}$ **49.** $f(x) = 3x^2 - 2x$

50. $f(x) = x^3 - x^{2/3}$ **51.** $f(x) = x^2 + x^{1/2}$

52. $f(x) = 5$

 53. For the functions $y_1 = 100x$, $y_2 = 10x^2$, $y_3 = x^3$:

    a. Predict which curve will be the highest for large values of x.

    b. Predict which curve will be the lowest for large values of x.

    c. Check your prediction by graphing the functions on the window [0, 12] by [0, 2000].

    d. From your graph, where do all these curves meet?

 54. Graph the parabola $y_1 = 1 - x^2$ and the semicircle $y_2 = \sqrt{1 - x^2}$ on the graphing window [-1, 1] by [0, 1]. (You may want to adjust the window to make the semicircle look more like a semicircle.) Use TRACE to determine which is the "inside" curve (the parabola or the semicircle) and which is the "outside" curve. These graphs show that when you graph a parabola, you should draw the curve near the vertex to be slightly more "pointed" than a circular curve.

 55. For any x, the function INT(x) is defined as the greatest integer less than or equal to x. For example, INT(3.7) = 3 and INT(-4.2) = -5.

    a. Use a graphing calculator to graph the function $y_1 = $ INT(x). (You may need to graph it in DOT mode to eliminate false connecting lines.)

    b. From your graph, what are the domain and range of this function?

56. a. Use a graphing calculator to graph the function $y_1 = 2$ INT(x). (See the previous exercise for a definition of INT(x).)

    b. From your graph, what are the domain and range of this function?

For each pair $f(x)$ and $g(x)$, find **a.** $f(g(x))$, **b.** $g(f(x))$.

57. $f(x) = x^5$, $g(x) = 7x - 1$

58. $f(x) = x^8$, $g(x) = 2x + 5$

59. $f(x) = \frac{1}{x}$, $g(x) = x^2 + 1$

60. $f(x) = \sqrt{x}$, $g(x) = x^3 - 1$

61. $f(x) = x^3 - x^2$, $g(x) = \sqrt{x} - 1$

62. $f(x) = x - \sqrt{x}$, $g(x) = x^2 + 1$

63. $f(x) = \dfrac{x^3 - 1}{x^3 + 1}$, $g(x) = x^2 - x$

64. $f(x) = \dfrac{x^4 + 1}{x^4 - 1}$, $g(x) = x^3 + x$

65. a. Find the composition $f(g(x))$ of the two linear functions $f(x) = ax + b$ and $g(x) = cx + d$ (for constants a, b, c, and d).

   b. Is the composition of two linear functions always a linear function?

66. a. Is the composition of two quadratic functions always a quadratic function? (*Hint:* Find the composition of $f(x) = x^2$ and $g(x) = x^2$.)

   b. Is the composition of two polynomials always a polynomial?

For each function in Exercises 67–70, find and simplify $f(x + h)$.

67. $f(x) = 5x^2$

68. $f(x) = 3x^2$

69. $f(x) = 2x^2 - 5x + 1$

70. $f(x) = 3x^2 - 5x + 2$

For each function in Exercises 71–82, find and simplify $\dfrac{f(x + h) - f(x)}{h}$. (Assume $h \neq 0$.)

71. $f(x) = 5x^2$ **72.** $f(x) = 3x^2$

73. $f(x) = 2x^2 - 5x + 1$ **74.** $f(x) = 3x^2 - 5x + 2$

75. $f(x) = 7x^2 - 3x + 2$ **76.** $f(x) = 4x^2 - 5x + 3$

77. $f(x) = x^3$
(*Hint:* Use $(x + h)^3 = x^3 + 3x^2h + 3xh^2 + h^3$.)

78. $f(x) = x^4$
(*Hint:* Use $(x + h)^4 = x^4 + 4x^3h + 6x^2h^2 + 4xh^3 + h^4$.)

79. $f(x) = \frac{2}{x}$ **80.** $f(x) = \frac{3}{x}$

81. $f(x) = \frac{1}{x^2}$ **82.** $f(x) = \sqrt{x}$
[*Hint:* Multiply the top and bottom of the fraction by $(\sqrt{x + h} + \sqrt{x})$.]

 83. Find, rounding to five decimal places:

   a. $\left(1 + \dfrac{1}{100}\right)^{100}$

   b. $\left(1 + \dfrac{1}{10,000}\right)^{10,000}$

   c. $\left(1 + \dfrac{1}{1,000,000}\right)^{1,000,000}$

   d. Do the resulting numbers seem to be approaching a limiting value? Estimate the limiting value to five decimal places. The number that you have approximated is denoted e, and will be used in the next section and throughout calculus.

84. Use the TABLE feature of a graphing calculator to evaluate $\left(1 + \dfrac{1}{x}\right)^{x}$ for values of x like 100, 10,000, and 1,000,000 and higher. Do the resulting numbers seem to be approaching a limiting value? Estimate the limiting value to five decimal places. The number that you have approximated is denoted e, and will be used in the next section and throughout calculus.

APPLIED EXERCISES

85. Economics: Income Tax The function below expresses an income tax that is 10% for incomes below $5000, and otherwise is $500 plus 30% of income in excess of $5000.

$$f(x) = \begin{cases} 0.10x & \text{if } x \le 5000 \\ 500 + 0.30(x - 5000) & \text{if } x \ge 5000 \end{cases}$$

   a. Calculate the tax on an income of $3000.
   b. Calculate the tax on an income of $5000.
   c. Calculate the tax on an income of $10,000.
   d. Graph the function.

86. Economics: Income Tax The following function expresses an income tax that is 15% for incomes below $6000, and otherwise is $900 plus 40% of incomes in excess of $6000.

$$f(x) = \begin{cases} 0.15x & \text{if } x \leq 6000 \\ 900 + 0.40(x - 6000) & \text{if } x \geq 6000 \end{cases}$$

a. Calculate the tax on an income of $3000.
b. Calculate the tax on an income of $6000.
c. Calculate the tax on an income of $10,000.
d. Graph the function.

 87. Business: Insurance Reserves An insurance company keeps reserves (money to pay claims) of $R(v) = 2v^{0.3}$, where v is the value of all of its policies, and the value of its policies is predicted to be $v(t) = 60 + 3t$, where t is the number of years from now. (Both R and v are in millions of dollars.) Express the reserves R as a function of t, and evaluate the function at $t = 10$.

 88. Business: Research Expenditures An electronics company's research budget is $R(p) = 3p^{0.25}$, where p is the company's profit, and the profit is predicted to be $p(t) = 55 + 4t$, where t is the number of years from now. (Both R and p are in millions of dollars.) Express the research expenditure R as a function of t, and evaluate the function at $t = 5$.

 89. General: Oil Imports and Fuel Efficiency The average fuel efficiency of all cars in America, called the *fleet mpg*, is 21.6 mpg. The amount of crude oil that would be saved if the fleet mpg were increased to a value of x is

$$S(x) = 3208 - \frac{69{,}300}{x} \quad \begin{array}{l}\text{million barrels} \\ \text{annually}\end{array}$$

a. Graph the function $S(x)$ on the viewing window [21.6, 40] by [0, 2000].
b. For what fleet mpg x will the savings reach 720 million barrels, which is the amount annually imported from the Middle East OPEC nations? (*Hint*: Either ZOOM IN around the point at $y = 720$ or, better, use INTERSECT with $y_2 = 720$.) (*Note*: New cars average 29 mpg, but fleet mpg always lags. Ten manufacturers have already built

and tested prototype cars getting from 67 to 138 mpg.) (*Source: Information Please Almanac and the Rocky Mountain Institute.*)

 90. General: Long-Term Trends in Cigarette Smoking The table below gives the annual cigarette consumption in the United States per adult from 1910 to 1990.

Years since 1900	Cigarettes per Adult
1910	10	200
1930	30	1500
1950	50	3400
1970	70	3900
1990	90	3300

a. Enter these data into a graphing calculator and make a plot of the resulting points (Years since 1900 on the x-axis and Cigarettes per Adult on the y-axis).
b. Have your calculator find the cubic (third-order polynomial) regression formula for these data. Then enter the result as y_1, which gives a formula for cigarette consumption for each year. Plot the points together with the regression line.
c. Use the regression formula to predict the cigarette consumption in the years 2000 ($x = 100$) and 2005 ($x = 105$).
d. Use the regression formula to predict the cigarette consumption in the year 2020 ($x = 120$). The answer shows that polynomial predictions, if extended too far, give nonsensical results. (*Source: Worldwatch Institute.*)

 91. How will the graph of $y = (x + 3)^3 + 6$ differ from the graph of $y = x^3$? Check by graphing both functions together.

 92. How will the graph of $y = -(x - 4)^2 + 8$ differ from the graph of $y = -x^2$? Check by graphing both functions together.

0.5 Exponential Functions

APPLICATION PREVIEW

Exponential Functions and the World's Worst Currency

Most economies of the world suffer from some degree of inflation, which drives up prices over the years. For example, if the inflation rate is 5%, prices rise by 5% each year, so that a $100 item rises in a year to $105. This is equivalent to multiplying the original price by 1.05, and so another year of 5% inflation increases the already inflated prices by multiplying them by 1.05 again, to $100(1.05)^2$. After x years, the price would rise to $100(1.05)^x$. This sort of function, with the variable in the exponent, is called an *exponential function,* and is the subject of this section. Exponential functions are used extensively for calculating bank interest and the effects of inflation.

What country suffers the worst inflation? At the end of 1993, the inflation rate in Yugoslavia exceeded *1 million percent per year.* (Yugoslavia was under United Nations sanctions for supporting a war in neighboring Bosnia.) This rate of inflation rendered its currency, the Yugoslavian dinar, almost worthless within days of issue. Near the end of 1993 the government issued a 500 billion dinar note. At the time of issue it was worth about $5. Within a month it was worth one-thousandth of a cent. Such is the power of exponential functions.

From $5 to 0.001¢ in a month

Introduction

In this section we introduce *exponential* functions, in which the variable is in the exponent. Exponential functions will be used extensively in

Chapter 1 for compound interest, and throughout calculus. We will also introduce the very important mathematical constant e.

Exponential functions are often used to model the processes of growth and decay.

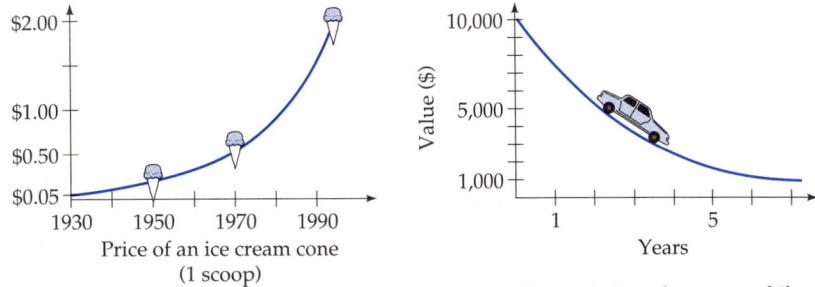

Price of an ice cream cone
(1 scoop)

Depreciation of an automobile

Exponential Functions

A function that has a variable in an exponent, such as $f(x) = 2^x$, is called an *exponential function*.

$$f(x) = 2^x \quad \text{Exponent} \quad \text{Base}$$

The table below shows some values of the exponential function $f(x) = 2^x$, and its graph (based on these points) is shown on the right.

x	$y = 2^x$
-3	$2^{-3} = \frac{1}{8}$
-2	$2^{-2} = \frac{1}{4}$
-1	$2^{-1} = \frac{1}{2}$
0	$2^0 = 1$
1	$2^1 = 2$
2	$2^2 = 4$
3	$2^3 = 8$

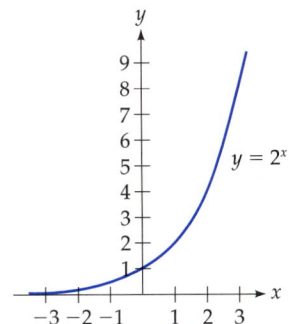

Domain of 2^x is $\mathbb{R} = (-\infty, \infty)$
and range is $(0, \infty)$.

Clearly, the exponential function 2^x is quite different from the parabola x^2.

The exponential function $f(x) = \left(\frac{1}{2}\right)^x$ has base $\frac{1}{2}$. The following table shows some of its values, and its graph is shown on the right. Notice that it is the mirror image of the curve $y = 2^x$.

x	$y = \left(\frac{1}{2}\right)^x$
-3	$\left(\frac{1}{2}\right)^{-3} = 8$
-2	$\left(\frac{1}{2}\right)^{-2} = 4$
-1	$\left(\frac{1}{2}\right)^{-1} = 2$
0	$\left(\frac{1}{2}\right)^{0} = 1$
1	$\left(\frac{1}{2}\right)^{1} = \frac{1}{2}$
2	$\left(\frac{1}{2}\right)^{2} = \frac{1}{4}$

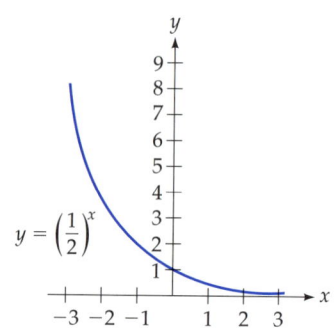

Domain of $\left(\frac{1}{2}\right)^x$ is $\mathbb{R} = (-\infty, \infty)$ and range is $(0, \infty)$.

We can define an exponential function $f(x) = a^x$ for any positive base a. We always take the base to be positive, so for the rest of this section *the letter a will stand for a positive constant.*

Exponential functions with bases $a > 1$ are used to model *growth,* as in populations or savings accounts, and exponential functions with bases $a < 1$ are used to model *decay,* as in depreciation. (For base $a = 1$ the graph is a horizontal line, since $1^x = 1$ for any x.)

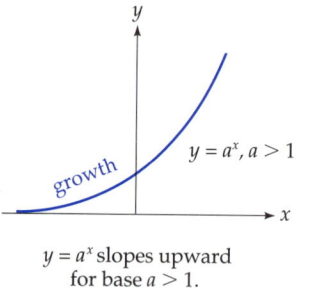

$y = a^x$ slopes upward
for base $a > 1$.

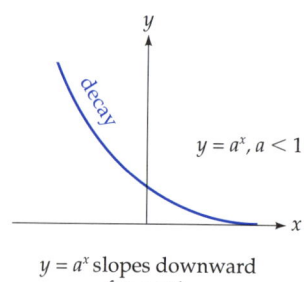

$y = a^x$ slopes downward
for $a < 1$.

Value Appreciation

Suppose that you make an investment of P dollars that increases in value ("appreciates") by 8% each year. Then after 1 year the value will be

$$\left(\begin{array}{c}\text{Value after}\\ \text{1 year}\end{array}\right) = P + 0.08P = P(1 + 0.08)$$

Notice that increasing a quantity by 8% is the same as multiplying it by $(1 + 0.08)$. Therefore, to find the value of the investment after a *second* year we would multiply the above answer by $(1 + 0.08)$, giving $P(1 + 0.08)(1 + 0.08)$ or $P(1 + 0.08)^2$. For the value after t years we simply multiply P by $(1 + 0.08)$ t times:

$$\begin{pmatrix} \text{Value after} \\ t \text{ years} \end{pmatrix} = P\overbrace{(1 + 0.08) \cdot (1 + 0.08) \ldots (1 + 0.08)}^{t \text{ times}}$$

$$= P(1 + 0.08)^t$$

Clearly, the 8% can be replaced by *any* growth rate r (written in decimal form).

For an investment of P dollars that increases at rate r each year,

$$\begin{pmatrix} \text{Value after} \\ t \text{ years} \end{pmatrix} = P(1 + r)^t$$

EXAMPLE 1   Finding the Value of an Investment

A coin collection recently appraised at \$25,000 is expected to appreciate in value by 8.5% each year. Predict its value 6 years from now.

Solution

$$25,000(1 + 0.085)^6 = \qquad \begin{array}{l} P(1 + r)^t \text{ with } P = 25,000, \\ r = 0.085, \text{ and } t = 6 \end{array}$$

$$25,000 \cdot 1.085^6 \approx 40,786.69 \qquad \text{Using a calculator}$$

The coin collection will be worth approximately \$40,787.

PRACTICE PROBLEM 1

Find the value of a 30-year treasury bond (called a "long bond") at the end of its 30-year term if it is now selling for \$10,000 and if the interest rate is 7%.   *Solution at the back of the book*

Depreciation by a Fixed Percentage

Depreciation by a fixed percentage means that a piece of equipment loses a fixed percentage (say 30%) of its value each year. Losing a percentage of value is like compound interest but with a *negative* interest rate. Therefore, we use the compound interest formula, $P(1 + r)^t$, but for depreciation r is *negative*.

EXAMPLE 2   Depreciating an Asset

A car worth $10,000 depreciates in value by 40% each year. How much is it worth after 3 years?

Solution

The car loses 40% of its value each year, which is equivalent to an interest rate of *negative* 40%. The compound interest formula with $P = 10,000$, $r = -0.40$, and $t = 3$ gives

$$P(1 + r)^t = 10,000(1 - 0.40)^3 = 10,000(0.60)^3 = \$2160$$

Using a
calculator

The exponential function $f(x) = 10,000(0.60)^x$, giving the value of the car after x years of depreciation, is graphed below.

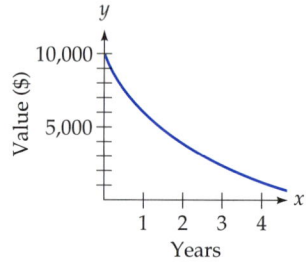

PRACTICE PROBLEM 2

A printing press, originally worth $50,000, loses 20% of its value each year. What is its value after 4 years?   *Solution at the back of the book*

The graph above shows that depreciation by a fixed percentage is quite different from "straight-line" depreciation. Under straight-line depreciation the same *dollar* value is lost each year, whereas under fixed percentage depreciation the same *percentage* of value is lost each year, resulting in larger dollar losses in the early years and smaller dollar losses in later years. Depreciation by a fixed percentage (also called the "declining balance" method) is one type of "accelerated" depreciation. The method of depreciation that one uses depends on how one chooses to estimate value, and in practice is often determined by the tax laws.

The Number *e*

Which is better: a one-time 100% increase or two successive 50% increases? You might think that they are the same, but let's apply both to $1 and compare the answers. If $1 is increased by 100%, the result is obviously $2. But if $1 is first increased by 50% to $1.50, and then this $1.50 is increased by 50% (adding $0.75), the total is $2.25. Therefore, the single 100% increase gave $2.00 while two successive 50% increases gave $2.25, or 25¢ more. (While an additional 25¢ may seem insignificant, it is a significant increase based on $1, and if instead we had begun with $1000, the difference would be $250.)

There is a faster way to carry out this calculation: Increasing something by 50% means keeping the original and adding half more, which is equivalent to multiplying it by $\left(1 + \dfrac{1}{2}\right)$. Increasing by 50% again means multiplying a *second* time, that is, finding $\left(1 + \dfrac{1}{2}\right)^2 = 2.25$, the same answer as before.

Even better is to have four successive 25% increases, since $\left(1 + \dfrac{1}{4}\right)^4 \approx 2.44$, which is better than $2.25 by 19¢.

How about 100 successive increases of 1% each, or 200 increases of $\frac{1}{2}$%, and so on? The general formula for the result of performing n successive equal increases that add up to 100% is $\left(1 + \dfrac{1}{n}\right)^n$. The following graphing calculator display evaluates this expression for values up to a million (replacing n by x).

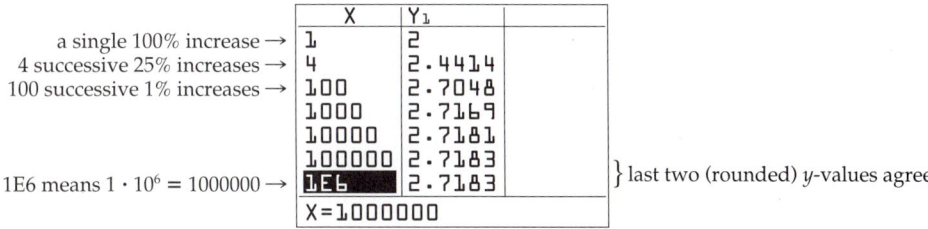

	X	Y₁
a single 100% increase →	1	2
4 successive 25% increases →	4	2.4414
100 successive 1% increases →	100	2.7048
	1000	2.7169
	10000	2.7181
	100000	2.7183
1E6 means 1·10⁶ = 1000000 →	1E6	2.7183

$\}$ last two (rounded) y-values agree

X=1000000

Notice that the y-values do not become arbitrarily large but are "settling down" to a result approximately equal to 2.7183. That is, breaking 100% up into smaller and smaller successive increases does *not* result in an arbitrarily large increase, but results in a multiple of approximately 2.7183 (which is still much better than the multiple of 2 resulting from a single 100% increase). This number is so important in mathematics that it is given its own symbol (like π)—it is called *e*.

More formally, using the notation $\lim\limits_{n\to\infty}$ to mean the value *approached* by a quantity as n becomes arbitrarily large, we can represent e as follows:

$$e = \lim_{n\to\infty}\left(1 + \frac{1}{n}\right)^n = 2.71828\ldots$$

$\lim\limits_{n\to\infty}\left(1 + \dfrac{1}{n}\right)^n$ is read: the limit as n approaches infinity of 1 plus $1/n$ to the nth power

The number e has an infinitely long decimal representation, and has been calculated to several million decimal places. This same e appears on the e^x key of most calculators (including business calculators), and is used in statistics to define the famous "bell-shaped" or "normal" curve (see Chapter 5). The value of e may be found on most calculators by pressing ⌈2nd⌉ ⌈LN⌉ ⌈1⌉ or simply the "e" key to give $e \approx 2.718281828$ (rounded to nine decimal places), as you should check.

The Function $y = e^x$

The number e gives us a new exponential function $f(x) = e^x$. This function is used extensively in business, economics, and all areas of science. The table below shows the values of e^x for various values of x. These values lead to the graph of $f(x) = e^x$ shown on the right.

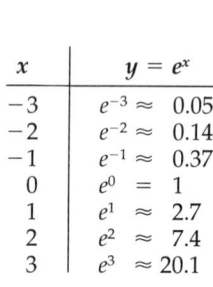

x	$y = e^x$
-3	$e^{-3} \approx 0.05$
-2	$e^{-2} \approx 0.14$
-1	$e^{-1} \approx 0.37$
0	$e^0 = 1$
1	$e^1 \approx 2.7$
2	$e^2 \approx 7.4$
3	$e^3 \approx 20.1$

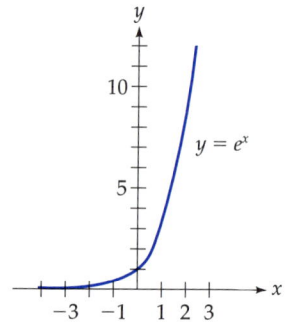

Domain of e^x is $\mathbb{R} = (-\infty, \infty)$ and range is $(0, \infty)$.

Notice that e^x is never zero, and is positive for all values of x, even when x is negative. We restate this important observation as follows:

e to any power is positive.

The function $f(x) = e^{kx}$ for various values of the constant k is graphed on the following page. For positive values of k the curve rises, and for negative values of k the function falls (as you move to the right). For higher values of k the curve rises more steeply.

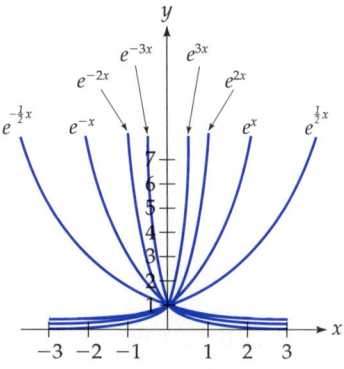

$f(x) = e^{kx}$ for various values of k.

EXAMPLE 3 Predicting the Population of the United States

As the following graph shows, the population of the United States since 1900 has grown approximately exponentially, closely matching the curve $y = 80.1e^{.013x}$, where x is the number of years since 1900. Use this exponential function to predict the U.S. population in the year 2010.

Solution

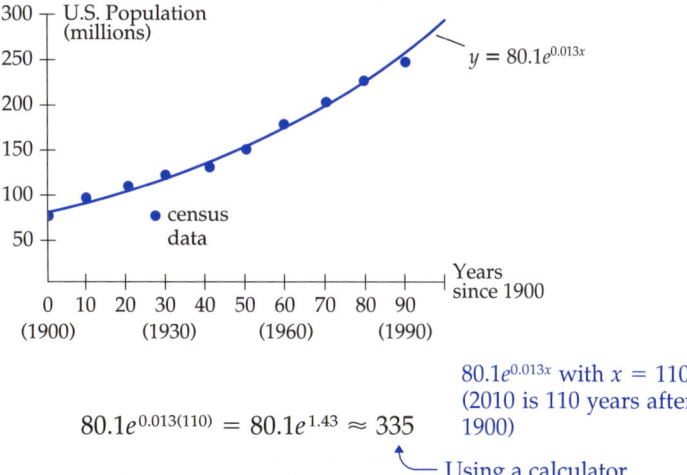

$$80.1e^{0.013(110)} = 80.1e^{1.43} \approx 335$$

$80.1e^{0.013x}$ with $x = 110$ (2010 is 110 years after 1900)

Using a calculator

In the year 2010 the population of the United States will be approximately 335 million (based on the last nine censuses).

Graphing Calculator Exploration

The most populous states are California and Texas, with New York third and Florida fourth but gaining. According to data from the Census Bureau, x years after 1990 the population of New York will be $17.99e^{0.0025x}$ and the population of Florida will be $12.94e^{0.0283x}$ (both in millions).

a. Graph these two functions on a calculator on the window $[0, 20]$ by $[0, 25]$. (Use the $\boxed{e^x}$ key above the $\boxed{\text{LN}}$ key for entering the power of e.)

b. Use INTERSECT to find the x-value where the curves intersect.

c. From your answer to (b), in which year is Florida projected to overtake New York as the third largest state? (*Hint:* x is years after 1990.)

Exponential Growth

All exponential growth has one common characteristic: The amount of growth is proportional to the size of the quantity. For example, the interest that a bank account earns is proportional to the size of the account, and the growth of a population is proportional to the size of the population. This is in contrast, for example, to a person's height, which does not increase exponentially. That is, exponential growth occurs in those situations when a quantity grows *in proportion to its size.*

SUMMARY

Exponential functions have exponents that involve variables. The exponential functions $f(x) = a^x$ slope *upward* or *downward* depending on whether the base a (which must be positive) satisfies $a > 1$ or $a < 1$.

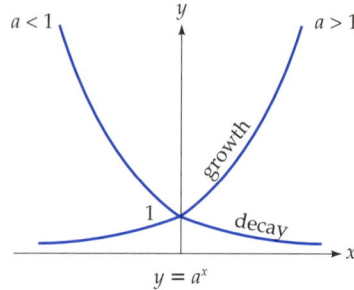

The formula $P(1 + r)^t$ gives the value after t years of an investment originally worth P dollars that increases at rate r each year. This same formula, with a *negative* growth rate r, governs depreciation by a fixed percentage.

By considering a 100% increase not given all at once but divided up into smaller and smaller successive increases, we defined a new constant e:

$$e = \lim_{n \to \infty} \left(1 + \frac{1}{n} \right)^n \approx 2.71828$$

Exponential functions with base e are used extensively in modeling many types of growth, such as the growth of populations and interest that is compounded continuously, and will be very important in both finance and calculus.

EXERCISES 0.5 (Most need .)

Graph each function. If you are using a graphing calculator, make a hand-drawn sketch from the screen.

1. $y = 3^x$ **2.** $y = 5^x$

3. $y = \left(\frac{1}{3}\right)^x$ **4.** $y = \left(\frac{1}{5}\right)^x$

Calculate each value of e^x using a calculator.

5. $e^{1.74}$ **6.** $e^{-0.09}$

 7. e^x versus x^n   Which curve is eventually higher, x to a power or e^x?

a. Graph x^2 and e^x on the window $[0,5]$ by $[0,20]$. Which curve is higher?

b. Graph x^3 and e^x on the window $[0,6]$ by $[0,200]$. Which curve is higher for large values of x?

c. Graph x^4 and e^x on the window $[0,10]$ by $[0,10{,}000]$. Which curve is higher for large values of x?

d. Graph x^5 and e^x on the window $[0,15]$ by $[0,1{,}000{,}000]$. Which curve is higher for large values of x?

e. Do you think that e^x will exceed x^5 for large values of x? Based on these observations, can you make a conjecture about e^x and *any* power of x?

8. Geometric versus Exponential Growth

a. Graph $y_1 = x$ and $y_2 = e^{0.01x}$ on the window $[0, 10]$ by $[0, 10]$. Which curve is higher for x near 10?

b. Then graph the same curves on the window $[0, 1000]$ by $[0, 1000]$. Which curve is higher for x near 1000?

A linear function like y_1 represents *geometric* growth, and y_2 represents *exponential* growth, and the result here is true in general: Exponential growth always beats geometric growth (eventually, no matter what the constants).*

---

* The realization that populations grow exponentially while food supplies grow only geometrically caused the great nineteenth-century essayist Thomas Carlyle to dub economics the "dismal science." He was commenting not on how interesting economics is, but on the grim conclusions that follow from populations outstripping their food supplies.

9. **General: Art Appreciation**   A Picasso etching, *The Frugal Repast*, bought for $250 in 1944, has been appreciating in value by 15% per year. Predict its value in the year 2004.

10. **General: Art Appreciation**   A Van Gogh painting, *Intérieur d'un Restaurant*, was auctioned in 1996 for $10,342,500. Assuming that it continues to appreciate in value by 12% per year, predict its value in the year 2016.

11. **General: Asset Appreciation**   A mint condition 1910 Honus Wagner* baseball card was sold at an auction in 1996 for $640,500, a record price for an item of sports memorabilia. Assuming that it will continue to appreciate in value by 12% per year, predict its value in the year 2006.

12. **General: Asset Appreciation**   In 1626, Peter Minuit purchased Manhattan Island from the Shinnecock Indians for trinkets and beads worth 60 guilders, or approximately $24. If this $24 had been invested to grow by 5% annually, what would it be worth now?

13. **General: Retirement Funds**   At the birth of your child you put $1000 into an investment that will grow by 8% per year. Find its value if your child holds it until retirement, 60 years later.

14. **General: College Costs**   In an effort to defray the cost of a college education, at the birth of your child you put $15,000 in an investment that will increase by 6% per year. What will be the investment's value when your child is ready for college at age 18?

15. **General: Depreciation**   A $25,000 automobile depreciates by 35% per year. Find its value after:

   a. 10 years   **b.** 6 months

*Honus Wagner, regarded as the greatest player of his time, played for the Pittsburgh Pirates. Baseball cards were then sold with cigarettes, and Wagner became even more famous when he refused to accept payment for the use of his picture as a protest against the use of tobacco.

16. **Business: Depreciation**   A $22 million corporate jet depreciates by 18% per year. Find its value after:

   a. 10 years   **b.** 6 months

17. **General: Population**   According to the United Nations Fund for Population Activities, the population of the world x years after the year 2000 will be $P(x) = 5.89e^{0.0125x}$ billion people (for $0 \leq x \leq 20$). Use this formula to predict the world population in the year 2015.

18. **General: Population**   The most populous country is China, with a (1997) population of 1.22 billion, growing at the rate of 0.7% per year. The 1997 population of India was 968 million, growing at 1.5% per year. Assuming that these growth rates continue, which population will be larger in the year 2100?

19. **General: Nuclear Meltdown**   According to the Nuclear Regulatory Commission, the probability of a "severe core meltdown accident" at a nuclear reactor in the United States within the next n years is

$$P(n) = 1 - (0.9997)^{100n}$$

   Find the probability of a meltdown

   a. within 25 years.
   b. within 40 years.

   (The 1986 core meltdown in the Chernobyl reactor in the Soviet Union spread radiation over much of Eastern Europe, leading to an undetermined number of fatalities.)

20. **General: Mosquitoes**   Female mosquitoes (*Culex pipiens*) feed on blood (only the females drink blood) and then lay several hundred eggs. In this way each mosquito can, on the average, breed another 300 mosquitoes in about 9 days. Find the number of great-grandchildren mosquitoes that will be descended from one female mosquito, assuming that all eggs hatch and mature.

21. **Environmental Science: Light**   According to the Bouguer–Lambert law, the proportion of

light that penetrates ordinary seawater to a depth of x feet is $e^{-0.44x}$. Find the proportion of light that penetrates to a depth of

a. 3 feet   **b.** 10 feet

22–23: Biomedical: Drug Dosage   If a dosage d of a drug is administered to a patient, the amount of the drug remaining in the tissues t hours later will be $f(t) = d \cdot e^{-kt}$, where k (the "absorption constant") depends on the drug.*

22. For the immunosuppressant cyclosporine, the absorption constant is $k = 0.012$. For a dose of $d = 400$ mg, use the above formula to find the amount of cyclosporine remaining in the tissues after

a. 24 hours   **b.** 48 hours

23. For the cardioregulator Digoxin, the absorption constant is $k = 0.018$. For a dose of $d = 2$ mg, use the above formula to find the amount remaining in the tissues after

a. 24 hours   **b.** 48 hours

24. Biomedical: Bacterial Growth   A colony of bacteria in a Petri dish doubles in size every hour. At noon the Petri dish is just covered with bacteria. At what time was the Petri dish

a. 50% covered (*Hint*: No calculation needed.)
b. 25% covered

25. Business: Advertising   A company finds that x days after the conclusion of an advertising campaign the daily sales of a new product are $S(x) = 100 + 800e^{-0.2x}$. Find the daily sales 10 days after the advertising campaign.

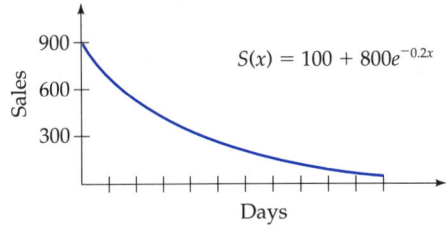

26. Business: Quality Control   A company finds that the proportion of its light bulbs that will burn continuously for longer than t weeks is $p(t) = e^{-0.01t}$. Find the proportion of bulbs that burn for longer than 10 weeks.

27. General: Temperature   A covered mug of coffee originally at 200 degrees, if left for t hours in a room whose temperature is 70 degrees, will cool to a temperature of $T(t) = 70 + 130e^{-1.8t}$ degrees. Find the temperature of the coffee after

a. 15 minutes   **b.** half an hour

28. Behavioral Science: Learning   In certain experiments the percentage of items that are remembered after t time units is

$$p(t) = 100 \frac{1 + e}{1 + e^{t+1}}$$

Such curves are called "forgetting" curves. Find the percentage remembered after

a. 0 time units   **b.** 2 time units

29. Biomedical: Epidemics   The Reed–Frost model for the spread of an epidemic predicts that the number I of newly infected people is $I = S(1 - e^{-rx})$, where S is the number of susceptible people, r is the effective contact rate, and x is the number of infectious people. Suppose that a school reports an outbreak of measles with $x = 10$ cases, and that the effective contact rate is $r = 0.01$. If the number of susceptibles is $S = 400$, use the Reed–Frost model to estimate how many students will be newly infected during this stage of the epidemic.

30. Social Science: Election Cost   The cost of winning a seat in the House of Representatives in recent years has been approximately $C(x) = 655e^{0.0728x}$ thousand dollars, where x is the number of years since 1996. Estimate the cost of winning a House seat in the year 2004. (*Source: Center for Responsive Politics*)

31. General: Rate of Return   An investment of $8000 grows to $10,291.73 in 4 years. Find the annual rate of return. [*Hint:* Use $P(1 + r)^t$ and solve for r (rounded).]

32. General: Rate of Return   An investment of $9000 grows to $10,380.65 in 2 years. Find the

*For further details, see T. R. Harrison (ed), *Principles of Internal Medicine*, 11th ed. (McGraw-Hill, New York, 1987), pp. 342–352.

annual rate of return. [*Hint:* Use $P(1 + r)^t$ and solve for r (rounded).]

 33. **General: Population** As stated earlier, the most populous states are California and Texas, with New York third and Florida fourth but gaining. According to the Census Bureau, x years after 1990 the population of New York will be $17.99e^{0.0025x}$, the population of Texas will be $16.99e^{0.0177x}$, and the population of Florida will be $12.94e^{0.0283x}$ (all in millions).

a. Graph these three functions on a calculator on the window $[0, 30]$ by $[0, 30]$.

b. Note that the order of these states (largest to smallest) in 1995 was Texas, New York, Florida. What will be the (projected) order in 2005?

c. What will be the (projected) order in 2020?

d. In which year is Florida projected to overtake Texas as the third largest state? (*Hint:* Use INTERSECT.)

 34. **General: St. Louis Arch** The Gateway Arch in St. Louis (see page 597) is built around a mathematical curve called a "catenary." The height of this catenary above the ground at a point x feet from the center line is

$$y = 688 - 31.5(e^{0.01033x} + e^{-0.01033x})$$

a. Graph this curve on a calculator on the window $[-400, 400]$ by $[0, 700]$.

b. Find the height of the Gateway Arch at its highest point, using the fact that the top of the arch is 5 feet higher than the top of the central catenary.

0.6   Logarithmic Functions

APPLICATION   PREVIEW

Carbon 14 Dating and the Shroud of Turin

Carbon 14 dating is a method for estimating the age of ancient plants and animals. While a plant is alive, it absorbs carbon dioxide, and with it both "ordinary" carbon and carbon 14, a radioactive form of carbon produced by cosmic rays in the upper atmosphere. When the plant dies, it stops absorbing both types of carbon, and the carbon 14 decays exponentially at a known rate, as shown in the following graph.

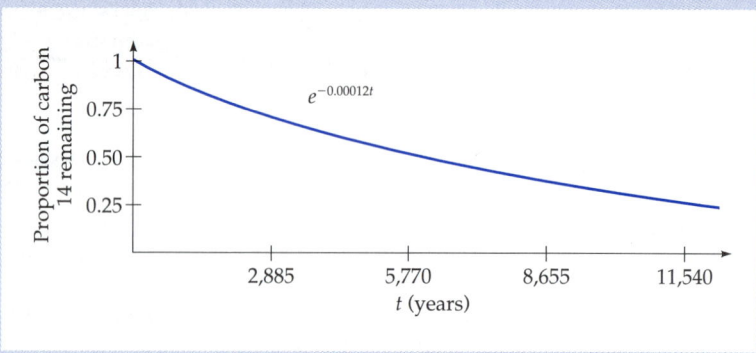

The time since the plant died can then be estimated by comparing the amount of carbon 14 that remains to the amount of ordinary carbon remaining, since they were originally in a known proportion. The comparison technique involves using *logarithms,* as explained in this section. Animal remains can also be dated in this way, since their diets contain plant matter.

In 1988, the Roman Catholic Church used carbon 14 dating to determine the authenticity of the Shroud of Turin, believed by many to be the burial cloth of Jesus. This was done by estimating the age of the flax plants from which the linen shroud was woven. In Exercise 32 you will find that the shroud is only a few hundred years old, and therefore cannot be the burial cloth of Christ.

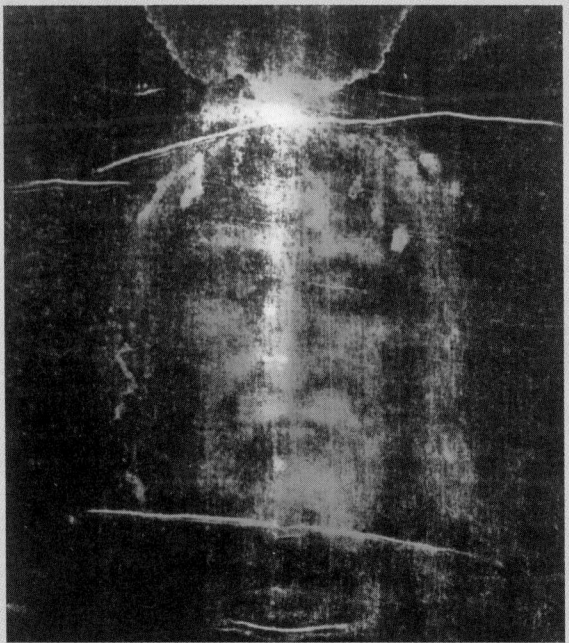

A few of the technical details of carbon 14 dating are as follows: The amount of carbon 14 remaining in a sample can be found by burning the sample near a Geiger counter to measure the radioactivity. For the shroud, this would have required a handkerchief-sized sample, and so instead a more advanced method was employed, using a tandem accelerator mass spectrometer and requiring only a postage-stamp-sized sample.

Carbon 14 dating assumes, however, that the ratio of carbon 14 to ordinary carbon in the atmosphere has remained steady over the centuries. This assumption seemed difficult to verify until the discovery in 1955 of a bristlecone pine tree in the White Mountains of California that was over 2000 years old (according to its growth rings). Each annual ring had absorbed carbon and carbon 14 from the air, and so provided a year-by-year record of their ratio. The analysis showed that the ratio has remained relatively steady over the centuries, and where variations did occur, scientists were able to construct a table for making corrections in the original carbon 14 dates. (Incidentally, bristlecone pine trees that are 4900 years old have been found, making them the oldest living things on earth.) For extremely old remains, such as dinosaur fossils, archeologists use longer-lasting radioactive elements, like potassium 40.

Carbon 14 dating, for which its inventor Willard Libby received a Nobel Prize in 1960, has become an invaluable tool in the social and biological sciences.

Introduction

In this section we introduce logarithmic functions, concentrating on *common* (base 10) and *natural* (base e) logarithms, and then apply them to the doubling of investments and carbon 14 dating. Both types of logarithms will be used in Chapter 1, and the natural logarithm function will be used throughout calculus.

Common Logarithms

The word "logarithm" (abbreviated "log") means *power* or *exponent*. The number being raised to the power is called the *base* and is written as a subscript. For example, the expression

$$\log_{10} 1000 \qquad\qquad \text{Read: log (base 10) of 1000}$$

base

means the *exponent* of 10 that gives 1000. Since $10^3 = 1000$, the exponent is 3, so the *logarithm* is 3:

$$\log_{10} 1000 = 3 \qquad\qquad \text{Since } 10^3 = 1000$$

Logarithms with base 10 are called *common logarithms*. For common logarithms we often omit the subscript, with base 10 understood.

$$\log x = y \quad \text{is equivalent to} \quad 10^y = x \qquad \log x \text{ means } \log_{10} x$$

Common logarithms can often be found by expressing them as exponents of 10 and then finding the exponent.

EXAMPLE 1   Finding a Common Logarithm

Find log 100.

Solution

$$\log 100 = y \qquad \text{is equivalent to} \qquad 10^y = 100 \qquad \begin{array}{l} y = 2 \text{ works, so 2} \\ \text{is the logarithm} \end{array}$$

The logarithm y is the exponent that solves

Therefore

$$\log 100 = 2 \qquad\qquad \text{Since } 10^2 = 100$$

■

EXAMPLE 2   Finding a Common Logarithm

Find $\log \dfrac{1}{10}$.

Solution

$$\log \frac{1}{10} = y \qquad \text{is equivalent to} \qquad 10^y = \frac{1}{10} \qquad \begin{array}{l} y = -1 \text{ works, so } -1 \\ \text{is the logarithm} \end{array}$$

The logarithm y is the exponent that solves

Therefore

$$\log \frac{1}{10} = -1 \qquad\qquad \text{Since } 10^{-1} = \tfrac{1}{10}$$

■

PRACTICE PROBLEM 1   Find log 10,000.   *Solution at the back of the book*

Graphing Calculator Exploration

On a graphing calculator, common logarithms are found using the LOG key. Verify that the last three common logarithms can be found in this way.

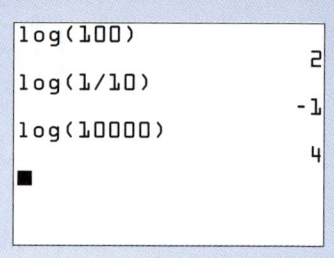

Properties of Common Logarithms

Since logarithms are exponents, the properties of exponents can be restated as properties of logarithms.

Properties of Common Logarithms

1. $\log 1 = 0$	The log of 1 is 0 (since $10^0 = 1$)
2. $\log 10 = 1$	The log of 10 is 1 (since $10^1 = 10$)
3. $\log 10^x = x$	The log of 10 to a power is just the power (since $10^x = 10^x$)
4. $10^{\log x} = x$	10 raised to the log of a number is just the number
5. $\log (M \cdot N) = \log M + \log N$	The log of a product is the sum of the logs
6. $\log \left(\dfrac{1}{N}\right) = -\log N$	The log of 1 over a number is minus the log of the number
7. $\log \left(\dfrac{M}{N}\right) = \log M - \log N$	The log of a quotient is the difference of the logs
8. $\log (M^N) = N \cdot \log M$	The log of a number to a power is the power times the log

The first two properties are simply special cases of the third (with $x = 0$ and $x = 1$). Since logs are exponents, the third property simply says that the exponent of 10 that gives 10^x is x, which is obvious when you think about it. Since $\log x$ is the power of 10 that gives x, raising 10 to that power must give x, which is the fourth property. Justifications for properties 5–8 are given on page 89. Property 8 will be particularly useful in applications, and can be summarized: *logarithms bring down exponents.*

EXAMPLE 3   Using the Properties of Common Logarithms

a. $\log 10^7 = 7$ Property 3

b. $10^{\log 13} = 13$ Property 4

c. $\log (3 \cdot 4) = \log 3 + \log 4$ Property 5

d. $\log \left(\dfrac{1}{4}\right) = -\log 4$ Property 6

e. $\log \left(\dfrac{3}{4}\right) = \log 3 - \log 4$ Property 7

f. $\log (5^3) = 3 \log 5$ Property 8: $\ln(5^3) = 3 \ln 5$

EXAMPLE 4   Finding the Doubling Time of an Investment

An investment grows in value by 7% per year. How many years will it take to double?

Solution

Let P stand for the size of the investment (since we are not told the amount). As we saw on page 69, an investment grows according to the formula $P(1 + r)^t$, and for the investment to double it must reach $2P$. This gives the following equation, which we solve for t:

$$P(1 + 0.07)^t = 2P \qquad \text{$P(1 + r)^t$ with $r = 0.07$ set equal to $2P$}$$

$$1.07^t = 2 \qquad \text{Canceling Ps and simplifying}$$

$$\log 1.07^t = \log 2 \qquad \text{Taking log of each side}$$

$$t \cdot \log 1.07 = \log 2 \qquad \text{Bringing down t (property 8)}$$

$$t = \frac{\log 2}{\log 1.07} \qquad \text{Dividing by log 1.07}$$

$$t \approx 10.24 \qquad \text{Using a calculator}$$

The investment will have doubled in value in 11 years.

Note that we round *up* to the next whole year to ensure that the value actually doubles, by which time the value will have somewhat *more* than doubled. This same technique can be used to find the doubling time for a population or for any other quantity that grows exponentially.

Graphs of Logarithmic and Exponential Functions

If a point (x, y) lies on the graph of $y = \log x$, or equivalently $x = 10^y$, then, reversing x and y, the point (y, x) lies on the graph of $y = 10^x$. That is, the curves $y = \log x$ and $y = 10^x$ are related by having their x and y coordinates reversed, so the curves are mirror images of each other in the line $y = x$.

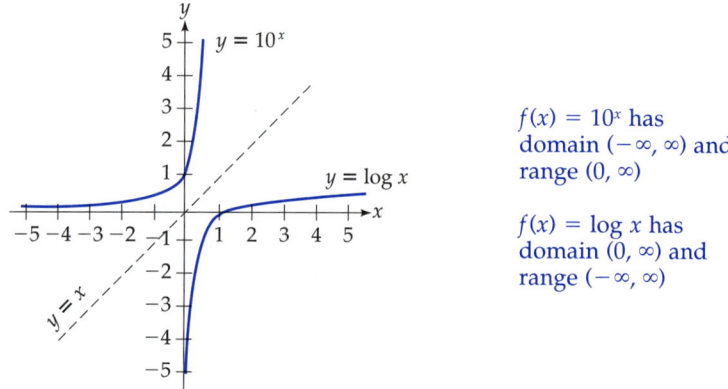

$f(x) = 10^x$ has domain $(-\infty, \infty)$ and range $(0, \infty)$

$f(x) = \log x$ has domain $(0, \infty)$ and range $(-\infty, \infty)$

This graphical relationship is equivalent to the fact that the functions $\log x$ and 10^x "undo" or "reverse" each other, as is shown by properties 3 and 4 on page 82. Such functions are called *inverse functions*:

$$\log x \quad \text{and} \quad 10^x \quad \text{are inverse functions}$$

Logarithms to Other Bases

We may calculate logarithms to bases other than 10. In fact, *any* positive number other than 1 may be used as a base for logs, using the following definition.

$$\log_a x = y \qquad \text{is equivalent to} \qquad a^y = x$$

For example,

$$\log_9 3 = \frac{1}{2} \qquad \qquad \text{since } 9^{1/2} = \sqrt{9} = 3$$

Natural Logarithms

We will use only one other base, the number e (approximately 2.718) that we defined on page 72. Logarithms to the base e are called *natural* or *Napierian* logarithms.* The natural logarithm of x is written $\ln x$ ("n" for "natural"), and is found using the $\boxed{\text{LN}}$ key on a calculator.

$$\ln x = y \qquad \text{is equivalent to} \qquad e^y = x \qquad \ln x \text{ means } \log_e x$$

PRACTICE PROBLEM 2   Use a calculator to find $\ln 8.34$.   *Solution at the back of the book*

The properties of natural logarithms are similar to those of common logarithms but with "ln" instead of "log" and "e" instead of "10".

Properties of Natural Logarithms

1. $\ln 1 = 0$	The natural log of 1 is 0 (since $e^0 = 1$)
2. $\ln e = 1$	The natural log of e is 1 (since $e^1 = e$)
3. $\ln e^x = x$	The natural log of e to a power is just the power (since $e^x = e^x$)
4. $e^{\ln x} = x$	e raised to the natural log of a number is just the number
5. $\ln (M \cdot N) = \ln M + \ln N$	The natural log of a product is the sum of the logs
6. $\ln \left(\dfrac{1}{N} \right) = -\ln N$	The natural log of 1 over a number is minus the log of the number
7. $\ln \left(\dfrac{M}{N} \right) = \ln M - \ln N$	The natural log of a quotient is the difference of the logs
8. $\ln (M^N) = N \cdot \ln M$	The natural log of a number to a power is the power times the log

As with common logs, the first two properties are special cases of the third (with $x = 0$ and $x = 1$). The third and fourth properties have interpretations analogous to their "common" counterparts (for example, the third says that the exponent of e that gives e^x is x). As before, Property 8 can be summarized: *logarithms bring down exponents.*

---

* After John Napier, a seventeenth-century Scottish mathematician and, incidentally, the inventor of the decimal point.

EXAMPLE 5   **Using the Properties of Natural Logarithms**

a. $\ln e^7 = 7$ Property 3

b. $e^{\ln 13} = 13$ Property 4

c. $\ln (3 \cdot 4) = \ln 3 + \ln 4$ Property 5

d. $\ln \left(\dfrac{1}{4}\right) = -\ln 4$ Property 6

e. $\ln \left(\dfrac{3}{4}\right) = \ln 3 - \ln 4$ Property 7

f. $\ln (5^3) = 3 \ln 5$ Property 8: $\ln(5^3) = 3 \ln 5$

Properties 3 and 4 show that $y = \ln x$ and $y = e^x$ are *inverse functions*, so their graphs are reflections of each other in the diagonal line $y = x$.

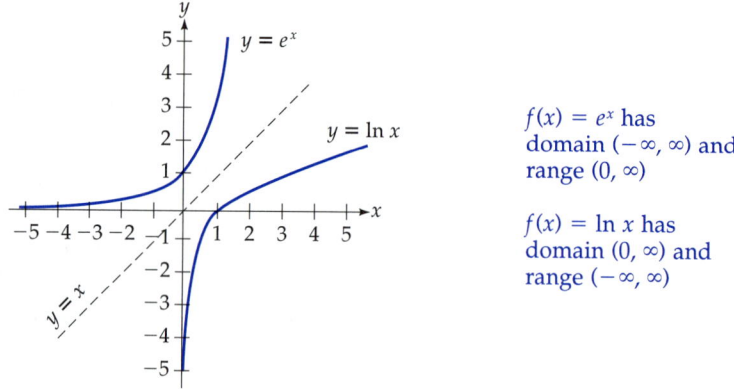

$f(x) = e^x$ has domain $(-\infty, \infty)$ and range $(0, \infty)$

$f(x) = \ln x$ has domain $(0, \infty)$ and range $(-\infty, \infty)$

The properties of natural logarithms are helpful for simplifying functions.

EXAMPLE 6   **Simplifying a Function**

$f(x) = \ln (2x) - \ln 2$

$= \ln 2 + \ln x - \ln 2$ Since $\ln (2x) = \ln 2 + \ln x$ by property 5

$= \ln x$ Canceling

EXAMPLE 7 **Simplifying a Function**

$f(x) = \ln\left(\dfrac{x}{e}\right) + 1$

$= \ln x - \ln e + 1$ Since $\ln (x/e) = \ln x - \ln e$ by property 7

$= \ln x - 1 + 1$ Since $\ln e = 1$ by property 2

$= \ln x$ Canceling

∎

EXAMPLE 8 **Simplifying a Function**

$f(x) = \ln (x^5) - \ln (x^3)$

$= 5 \ln x - 3 \ln x$ Bringing down exponents by property 8

$= 2 \ln x$ Combining

∎

Graphing Calculator Exploration

a. Graph the function $y = \ln x^2$ on the window $[-5, 5]$ by $[-5, 5]$.

b. Change the function to $y = 2 \ln x$ and explain why the two graphs are different. (Doesn't property 8 say that the two functions should be the same?)

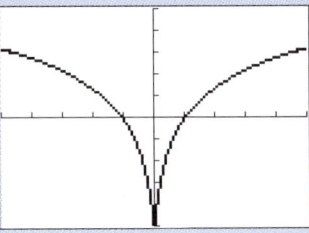
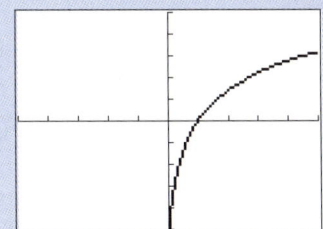

Carbon 14 Dating

All living things absorb small amounts of radioactive carbon 14 from the atmosphere. When they die, the carbon 14 stops being absorbed and decays exponentially into ordinary carbon. Therefore, the proportion of carbon 14 still present in a fossil or other ancient remain can be used to estimate how old it is. The proportion of the original carbon 14 that will be present after t years is

$$\begin{pmatrix} \text{Proportion of carbon 14} \\ \text{remaining after } t \text{ years} \end{pmatrix} = e^{-0.00012t}$$

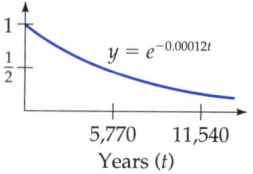

$y = e^{-0.00012t}$

5,770   11,540

Years (t)

EXAMPLE 9   Dating by Carbon 14

The Dead Sea Scrolls, discovered in a cave near the Dead Sea in Jordan, are among the earliest documents of Western civilization. Estimate the age of the Dead Sea Scrolls if the animal skins on which some were written contain 78% of their original carbon 14.

Solution   The proportion of carbon 14 remaining after t years is $e^{-0.00012t}$. We equate this formula to the actual proportion (expressed as a decimal):

$e^{-0.00012t} = 0.78$ Equating the proportions

$\ln e^{-0.00012t} = \ln 0.78$ Taking natural logs

$-0.00012t = \ln 0.78$ $\ln e^{-0.00012t} = -0.00012t$ by property 3

$t = \dfrac{\ln 0.78}{-0.00012} \approx \dfrac{-0.24846}{-0.00012} \approx 2071$ Solving for t using a calculator

Therefore, the Dead Sea Scrolls are approximately 2070 years old.

Graphing Calculator Exploration

Solve Example 9 by graphing $y_1 = e^{-0.00012x}$ and $y_2 = 0.78$ on the window [0, 10000] by [0, 1] and using INTERSECT to find where they meet.

Both here and in Example 4 on page 83 we solved for the variable in the exponent by taking logarithms, but here we used *natural* logarithms and there we used *common* logarithms. Is there a difference? Not really—you can use logarithms to *any* base to solve for exponents, and the final answer will be the same (see Exercise 41). However, when *e* is involved, it is usually easier to use *natural* logs.

Justification of Properties 5–8 of Logarithms

Properties 1–4 of logarithms (both common and natural) were justified earlier. Properties 5–8 of common logarithms are justified below, with the "natural" justifications obtained simply by replacing 10 by *e*.

The addition law of exponents, $10^x \cdot 10^y = 10^{x+y}$ can be stated in words: "the exponent of a product is the sum of the exponents." Since logs are exponents, this can be restated "the log of a product is the sum of the logs," which is just property 5.

The subtraction law of exponents, $10^x/10^y = 10^{x-y}$ can be stated: "the exponent of a quotient is the difference of the exponents." This translates into "the log of a quotient is the difference of the logs," which is just property 7. Property 6 is simply a special case of property 7 with $M = 1$ (and using property 1).

The law of exponents $(10^x)^y = 10^{x \cdot y}$ says that the exponent y can be "brought down" and multiplied by the x. Since logs are exponents, this says that in $\log(M^N)$ the exponent N can be brought down and multiplied by the logarithm $\log M$, giving $N \cdot \log M$, which is just property 8.

SUMMARY

Logarithms are exponents: common logs are exponents of 10, and natural logs are exponents of *e*. That is,

$y = \log x$ is equivalent to $x = 10^y$ $\log x$ means $\log_{10} x$ (base 10)

$y = \ln x$ is equivalent to $x = e^y$ $\ln x$ means $\log_e x$ (base e)

Each property of logarithms is equivalent to a property of exponents.

Logarithmic Property	*Exponential Property*
$\log 1 = 0$	$10^0 = 1$
$\log (M \cdot N) = \log M + \log N$	$10^x \cdot 10^y = 10^{x+y}$
$\log \left(\dfrac{1}{N} \right) = -\log N$	$\dfrac{1}{10^y} = 10^{-y}$
$\log \left(\dfrac{M}{N} \right) = \log M - \log N$	$\dfrac{10^x}{10^y} = 10^{x-y}$

In practice, logarithms are found using the $\boxed{\text{LOG}}$ and $\boxed{\text{LN}}$ keys on a calculator. The properties of exponents lead to properties of logarithms, which are listed on pages 82 and 85. Property 8, that *logs bring down exponents,* is particularly useful in applications that require solving for a variable in the exponent.

EXERCISES 0.6 (Most need $\boxed{}$ or $\boxed{}$.)

Find each logarithm *without* using a calculator.

1. a. $\log 100{,}000$ **b.** $\log \frac{1}{100}$ **c.** $\log \sqrt{10}$

2. a. $\log 1000$ **b.** $\log \frac{1}{1000}$ **c.** $\log \sqrt[3]{10}$

3. a. $\ln e^5$ **b.** $\ln \frac{1}{e}$ **c.** $\ln \sqrt[3]{e}$

4. a. $\ln e^3$ **b.** $\ln \frac{1}{e^2}$ **c.** $\ln \sqrt{e}$

5. a. $\ln 1$ **b.** $\ln (\ln e^e)$ **c.** $\ln \sqrt[3]{e^2}$

6. a. $\ln e$ **b.** $\ln (\ln e)$ **c.** $\ln \sqrt{e^3}$

7. a. $\log_4 16$ **b.** $\log_4 \frac{1}{4}$ **c.** $\log_4 2$

8. a. $\log_9 81$ **b.** $\log_9 3$ **c.** $\log_9 \frac{1}{9}$

For each logarithm:

i. Find the logarithm using a calculator, rounding your answer to three decimal places.

ii. Raise the appropriate base (either 10 or e) to the power you found in part (i) and check that the result agrees with the number in the original problem. (*Note:* For the power, include *all* of the digits that your calculator showed for part (i) to minimize the error.)

9. a. $\log 22.3$ **b.** $\ln 22.3$

10. a. $\log 44.9$ **b.** $\ln 44.9$

Use the properties of natural logarithms to simplify each function.

11. $f(x) = \ln (9x) - \ln 9$ **12.** $f(x) = \ln \left(\frac{x}{2}\right) + \ln 2$

13. $f(x) = \ln (x^3) - \ln x$ **14.** $f(x) = \ln (4x) - \ln 4$

15. $f(x) = \ln \left(\frac{x}{4}\right) + \ln 4$

16. $f(x) = \ln (x^5) - 3 \ln x$

17. $f(x) = \ln (e^{5x}) - 2x - \ln 1$

18. $f(x) = \ln (e^{-2x}) + 3x + \ln 1$

19. $f(x) = 8x - e^{\ln x}$

20. $f(x) = e^{\ln x} + \ln (e^{-x})$

21. Without using a calculator, sketch the graph of $f(x) = \log (x + 1)$.

22. Without using a calculator, sketch the graph of $f(x) = \ln (x + e)$.

23. Find the domain and range and graph the function $f(x) = \ln (x^2 - 1)$.

24. Find the domain and range and graph the function $f(x) = \ln (1 - x^2)$.

APPLIED EXERCISES

25. Business: Investments A real estate investment increases in value by 10% per year. How soon will it double in value?

26. Business: Investments An art collection increases in value by 4% each year. How soon will it double in value?

27. General: Population During the years 1990–1995, the population of Georgia increased by 2.1% annually. At this rate, how soon will the population double?

28. General: Population During the years 1990–1995, the population of Nevada increased by 4.9% annually. At this rate, how soon will the population double?

29. Economics: Energy Output The world's output of primary energy (petroleum, natural gas, coal, hydroelectricity, and nuclear electricity) is increasing by 2.4% annually. How soon will it double? (*Source: Worldwatch Institute.*)

30. Environmental Science: Carbon Emissions
In 1996, the global carbon emissions increased by 2.8%. At this rate, how soon will carbon emissions double? (*Note:* Carbon dioxide is the most prevalent "greenhouse gas" that many scientists believe is destabilizing the Earth's climate.) (*Source: Worldwatch Institute.*)

31–32: General: Carbon 14 Dating The proportion of carbon 14 still present in a sample after t years is $e^{-0.00012t}$.

31. Estimate the age of the cave paintings discovered in 1994 in the Ardèche region of France if the carbon with which they were drawn contains only 2.3% of its original carbon 14. They are the oldest known paintings in the world.

32. Estimate the age of the Shroud of Turin, believed by many to be the burial cloth of Christ (see the Application Preview on pages 78–80), from the fact that its linen fibers contained only 92.3% of their original carbon 14.

33–34: General: Potassium 40 Dating The radioactive isotope potassium 40 is used to date very old remains. The proportion of potassium 40 that remains after t million years is $e^{-0.00054t}$. Use this function to estimate the age of the following fossils.

33. The most complete skeleton of an early human ancestor ever found was discovered in Kenya in 1984. Use the above formula to estimate the age of the remains if they contained 99.91% of their original potassium 40.

34. Dating Older Women Use the above formula to estimate the age of the partial skeleton of *Australopithecus afarensis* (known as "Lucy") that was found in Ethiopia in 1977 if it had 99.82% of its original potassium 40.

35. According to a 1992 study,* each additional year of education increases one's income by

16%. How many additional years of schooling would then be required to double one's income?

 Solve the following exercises on a graphing calculator by graphing an appropriate exponential function (using x for ease of entry) together with a constant function and using INTERSECT to find where they meet. You will have to choose an appropriate window.

36. General: Value of an Automobile An antique automobile increases in value by 7.5% annually. How soon will it double in value?

37. General: Value of a Home A vacation home increases in value by 5.5% annually. How soon will it double in value?

38. General: Carbon 14 Dating In 1991 two hikers in the Italian Alps found the frozen but well-preserved body of the most ancient human ever found, dubbed "Iceman." Estimate the age of Iceman if his grass cape contained 53% of its original carbon 14. (Use the carbon 14 decay function stated in Exercises 31–32.)

39. General: Potassium 40 Dating Estimate the age of the oldest known dinosaur, a dog-sized creature called Herrerasaurus, found in Argentina in 1988, if volcanic material found with it contained 88.4% of its original potassium 40. (Use the potassium 40 decay function given in Exercises 33–34.)

40. Environmental Science: Nuclear Waste Half a century after the beginning of the nuclear age, not a single country has found a safe or permanent way of disposing of long-lived radioactive waste. Among the most hazardous radioactive waste is irradiated fuel from nuclear power plants, totaling 84,000 tons in 1990 and growing by 11.3% annually. At this rate, how long will it take for this amount to double? (*Source: Worldwatch Institute.*)

41. Change of Base Formula for Logarithms Let a and b be any two bases, and let x be any number.

a. Give a justification for each numbered equals sign.

$$\log_a x \overset{1}{=} \log_a b^{\log_b x} \overset{2}{=} (\log_b x) \cdot (\log_a b)$$

b. Show that the result can be written as:

$$\log_a x = (\log_a b) \cdot (\log_b x)$$

This is the "change of base" formula for logarithms, enabling one to express $\log_b x$ in terms of $\log_a x$.

c. Use the change of base formula in the numerator and denominator of the following fraction to justify each numbered equals sign.

$$\frac{\log_a x}{\log_a y} \overset{3}{=} \frac{(\log_a b) \cdot (\log_b x)}{(\log_a b) \cdot (\log_b y)} \overset{4}{=} \frac{\log_b x}{\log_b y}$$

The above equation shows that when finding a *ratio* of logarithms, using one base gives the same result as using any other base. Since solving for a variable in the exponent involves calculating *ratios* of logarithms, you may do so using logarithms to *any* base.

Chapter Summary with Hints and Suggestions

The reading and exercises of this chapter have helped you to learn the following skills. Each skill indicates the section from which it came (in case you need to review it) and some exercises in this section that use it. Answers to all exercises are at the end of the book; full solutions to all exercises are in the Student Solutions Manual.

0.1   Real Numbers, Inequalities, and Lines

● Translate an interval into set notation and graph it on the real line. *(Review Exercises 1–4.)*

$$[a, b] \quad (a, b) \quad [a, b) \quad (a, b] \quad (-\infty, b]$$
$$(-\infty, b) \quad [a, \infty) \quad (a, \infty) \quad (-\infty, \infty)$$

● Express given information in interval form. *(Review Exercises 5–6.)*

● Find an equation for a line that satisfies certain conditions. *(Review Exercises 7–12.)*

$$m = \frac{y_2 - y_1}{x_2 - x_1} \qquad y = mx + b$$

$$y - y_1 = m(x - x_1) \qquad x = a \qquad y = b$$

$$ax + by = c$$

● Find an equation of a line from its graph. *(Review Exercises 13–14.)*

● Use straight-line depreciation to find the value of an asset. *(Review Exercises 15–16.)*

● Use 📊 on real-world data to find a regression line and make a prediction. *(Review Exercises 17–18.)*

0.2   Exponents

● Evaluate negative and fractional exponents without using a calculator. *(Review Exercises 19–26.)*

$$x^0 = 1 \qquad x^{-n} = \frac{1}{x^n} \qquad x^{m/n} = \sqrt[n]{x^m} = (\sqrt[n]{x})^m$$

● Evaluate an exponential expression using a calculator. *(Review Exercises 27–28.)*

0.3   Functions

● Evaluate and find the domain and range of a function. *(Review Exercises 29–32.)*

A function f is a rule that assigns to each number x in a set (the domain) a (single) number $f(x)$. The range is the set of all resulting values $f(x)$.

● Use the vertical line test to see if a graph defines a function. *(Review Exercises 33–34.)*

- Graph a linear function.
  (*Review Exercises 35–36.*)

$$f(x) = mx + b$$

- Graph a quadratic function.
  (*Review Exercises 37–38.*)

$$f(x) = ax^2 + bx + c$$

- Solve a quadratic equation by factoring and by the quadratic formula.
  (*Review Exercises 39–42.*)

$$\text{vertex:} \qquad x = \frac{-b}{2a}$$

$$x\text{-intercepts:} \; x = \frac{-b \pm \sqrt{b^2 - 4ac}}{2a}$$

- Use $\boxed{\sim}$ to graph a quadratic function.

  (*Review Exercises 43–44.*)

- Construct a linear function from an applied problem or from real-life data, and then use the function in an application.
  (*Review Exercises 45–48.*)

- For given cost and revenue functions, find the break-even points and maximum profit.
  (*Review Exercises 49–50.*)

0.4   Functions, continued

- Evaluate and find the domain and range of a more complicated function.
  (*Review Exercises 51–54.*)

- Solve a polynomial equation by factoring.
  (*Review Exercises 55–58.*)

- Graph a "shifted" function.
  (*Review Exercises 59–60.*)

- Graph a piecewise linear function.
  (*Review Exercises 61–62.*)

- Given two functions $f(x)$ and $g(x)$, find their composition.   (*Review Exercises 63–66.*)

$$f(g(x)) \qquad g(f(x))$$

- For a given function $f(x)$, find and simplify $\dfrac{f(x + h) - f(x)}{h}$.   (*Review Exercises 67–68.*)

- Solve an appllied problem involving the composition of functions.   (*Review Exercise 69.*)

- Use factoring or $\boxed{\sim}$ to solve a polynomial equation.   (*Review Exercises 70–71.*)

- Use $\boxed{\sim}$ to fit a curve to real-life data and make a prediction.   (*Review Exercise 72.*)

0.5   Exponential Functions

- Sketch the graph of an exponential function.
  (*Review Exercises 73–74.*)

- Find the appreciated value of an investment.
  (*Review Exercises 75–76.*)

$$P(1 + r)^t$$

- Depreciate an asset by a fixed percentage.
  (*Review Exercises 77–78.*)

$$P(1 + r)^t \qquad \text{(for depreciation, } r \text{ is negative)}$$

- Use exponential functions to estimate the growth of populations or other quantities.
  (*Review Exercises 79–80.*)

0.6   Logarithmic Functions

- Evaluate common and natural logarithms without using a calculator.
  (*Review Exercises 81–82.*)

$$\log x = y \quad \text{is equivalent to} \quad 10^y = x$$
$$\ln x = y \quad \text{is equivalent to} \quad e^y = x$$

- Use the properties of natural logarithms to simplify a function.   (*Review Exercises 83–84.*)

$$\ln 1 = 0 \qquad \ln e = 1 \qquad \ln e^x = x$$
$$e^{\ln x} = x \qquad \ln (M \cdot N) = \ln M + \ln N$$
$$\ln \left(\frac{1}{N}\right) = -\ln N \qquad \ln \left(\frac{M}{N}\right) = \ln M - \ln N$$
$$\ln (M^N) = N \cdot \ln M$$

- Find the doubling time for an investment, a population, or some other quantity. (*Review Exercises 85–88.*)

$$P(1 + r)^t = 2P$$

- Estimate the age of a fossil. (*Review Exercises 89–90.*)

Hints and Suggestions

- **Overview:** In reviewing this chapter, try to understand the difference between two different kinds of mathematical objects, *geometric* objects (points, curves, etc.) and *analytic* objects (numbers, functions, etc.), and the connections between them. Descartes first made this connection: By drawing axes in the plane, he saw that points could be specified by numerical coordinates, and so *curves* could be specified by *equations* governing their coordinates. This idea connected geometry to algebra, previously distinct subjects. You should be able to express geometric objects analytically, and vice versa. For example, given a *graph* of a line, you should

be able to find an *equation* for it, and given a quadratic *function*, you should be able to *graph* it. As you do the review problems, try to be aware of the difference between geometric and analytic objects and their interrelations.

- A graphing calculator or a computer with appropriate software can help you to *explore* a concept more fully (for example, seeing how a curve changes as a coefficient or exponent changes), and also to *solve* a problem (for example, eliminating the point-plotting aspect of graphing, or finding a regression line).

- If you don't have a graphing calculator, you should have a scientific or business calculator to carry out calculations, especially in later chapters.

- The Practice Problems help you to check your mastery of the skills presented. Complete solutions are given at the back of the book.

- The Student Solutions Manual, available separately from your bookstore, provides fully worked-out solutions to selected exercises.

- **Practice for Test:** Review Exercises 1, 9, 11, 13, 15, 17, 19, 31, 33, 35, 37, 43, 47, 49, 55, 61, 65, 67, 71, 75, 77, 81, 83, and 85.

Review Exercises   *Practice test exercises are in blue.*

0.1   Real Numbers, Inequalities, and Lines

For each interval, write it in set notation and graph it on the real line.

1. (2, 5]   **2.** [−2, 0)   **3.** [100, ∞)   **4.** (−∞, 6]

5. General: Wind Speed The United States Coast Guard defines a "hurricane" as winds of at least 74 mph, a "storm" as winds of at least 55 mph but less than 74 mph, a "gale" as winds of at least 38 mph but less than 55 mph, and a "small craft warning" as winds of at least 21 mph but less than 38 mph. Express each of these wind conditions in interval form. (*Hint:* A small craft warning is [21, 38).)

6. State in interval form:

   a. The set of all positive numbers
   b. The set of all negative numbers
   c. The set of all nonnegative numbers
   d. The set of all nonpositive numbers

Find an equation of the line satisfying the following conditions. If possible, write your answer in the form $y = mx + b$.

7. Slope 2 and passing through the point (1, −3)

8. Slope −3 and passing through the point (−1, 6)

9. Vertical and passing through the point (2, 3)

10. Horizontal and passing through the point (2, 3)

11. Passing through the points $(-1, 3)$ and $(2, -3)$

12. Passing through the points $(1, -2)$ and $(3, 4)$

Write an equation of the form $y = mx + b$ for each line graphed below.

13.

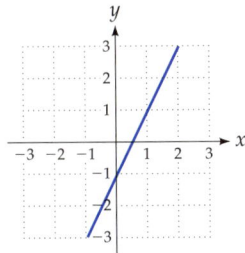

14.

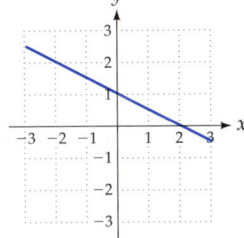

15. **Business: Straight-line Depreciation**   A contractor buys a backhoe for $25,000 and estimates its useful life to be 8 years, at the end of which its scrap value will be $1,000.

  a. Use straight-line depreciation to find a formula for the value V of the backhoe after t years, for $0 \leq t \leq 8$.
  b. Use your formula to find the value of the backhoe after 4 years.

16. **Business: Straight-line   Depreciation**   A trucking company buys a satellite communication system for $78,000 and estimates its useful life to be 15 years, at the end of which its scrap value will be $3,000.

  a. Use straight-line depreciation to find a formula for the value V of the system after t years, for $0 \leq t \leq 15$.
  b. Use your formula to find the value of the system after 8 years.

17. **Ecology: Sulfur Oxide Pollution**   Sulfur oxide pollution has decreased significantly in the

United States during the last two decades, mostly because of antipollution devices on automobiles and on coal- and oil-fired power plants. The following table shows sulfur oxide emissions (in millions of tons) in the United States from 1975 to 1995. To avoid large numbers, years are listed in the table as years since 1975.

	Years Since 1975	Sulfur Oxide Emissions
1975	0	26.0
1980	5	23.5
1985	10	21.6
1990	15	19.3
1995	20	18.2

Source: Worldwatch Institute and U.S. Environmental Protection Agency.

  a. Enter the table numbers into a graphing calculator and make a plot of the resulting points (Years Since 1975 on the x-axis and Sulfur Oxide Emissions on the y-axis).
  b. Have your calculator find the linear regression formula for this data. Then enter the result as y_1, which gives a formula for Sulfur Oxide Emissions in each year. Plot the points together with the regression line. How well does the line fit the data?
  c. Use your formula to predict the sulfur oxide pollution in the years 2005 and 2010 (assuming that the current trend continues).

18. **Social Science: Gap Between Rich and Poor**   During the last few decades, the richest 20% of the world's people have been growing richer, while the poorest 20% have been growing poorer. The probable consequences of this growing gap are not only social instability but also environmental decline, since the richest consume more wastefully while the poorest must cut down rainforest and overgraze land just to survive. The table on the next page shows the ratio of income of the richest 20% to the poorest 20% from 1960 to 1990. To avoid large numbers, years are listed in the table as years since 1960.

Years Since 1960	Ratio of Richest to Poorest
1960	0	30 to 1
1970	10	32 to 1
1980	20	45 to 1
1990	30	60 to 1

Source: United Nations Development Programme.

a. Enter the table numbers into a graphing calculator and make a plot of the resulting points [Years Since 1960 on the x-axis and the larger number in the Ratio column (the 30, 32, etc.) on the y-axis].

b. Have your calculator find the linear regression formula for these data. Then enter the result as y_1, which gives a formula for the ratio of richest to poorest in each year. Plot the points together with the regression line. How well does the line fit the data?

c. Use your formula to predict the ratio in the years 2005 and 2010 (assuming that the current trend continues).

0.2   Exponents

Evaluate each expression *without* using a calculator.

19. $\left(\frac{1}{6}\right)^{-2}$ **20.** $\left(\frac{4}{3}\right)^{-1}$ **21.** $64^{1/2}$

22. $1000^{1/3}$ **23.** $81^{-3/4}$ **24.** $100^{-3/2}$

25. $\left(-\frac{8}{27}\right)^{-2/3}$ **26.** $\left(\frac{9}{16}\right)^{-3/2}$

 Use a calculator to evaluate each expression. Round answers to 2 decimal places.

27. $3^{2.4}$ **28.** $12^{1.9}$

0.3   Functions

For each function in Exercises 29–32:

a. Evaluate the given expression.
b. Find the domain.
 c. Find the range.

29. $f(x) = \sqrt{x - 7}, f(11)$ **30.** $g(t) = \dfrac{1}{t + 3}, g(-1)$

31. $h(w) = w^{-3/4}, h(16)$

32. $w(z) = z^{-4/3}, w(8)$

Determine whether each graph defines a function of x.

33. y

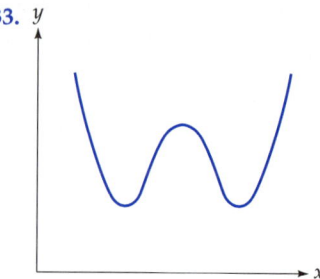

34. y

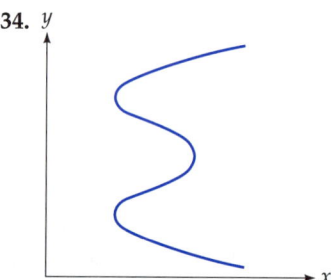

Graph each function.

35. $f(x) = 4x - 8$ **36.** $f(x) = 6 - 2x$

37. $f(x) = -2x^2 - 4x + 6$

38. $f(x) = 3x^2 - 6x$

Solve each equation by: **a.** factoring, **b.** the quadratic formula.

39. $3x^2 + 9x = 0$ **40.** $2x^2 - 8x - 10 = 0$

41. $3x^2 + 3x + 5 = 11$ **42.** $4x^2 - 2 = 2$

 For each quadratic function in Exercises 43–44:

a. Find the vertex using the vertex formula.
b. Graph the function on an appropriate viewing window. (Answers may vary.)

43. $f(x) = x^2 - 10x - 25$

44. $f(x) = x^2 + 14x - 15$

45. Business: Car Rentals   A rental company rents cars for $45 per day and $0.12 per mile. Find a function $C(x)$ for the cost of a rented car driven for x miles in a day.

46. Business: Simple Interest   If money is borrowed for a short period of time, generally less than a year, the interest is often calculated as *simple* interest, according to the formula Interest $= P \cdot r \cdot t$, where P is the principal, r is the rate (expressed as a decimal), and t is the time (in years). Find a function $I(t)$ for the interest charged on a loan of \$10,000 at an interest rate of 8% for t years. Simplify your answer.

47. General: Air Temperature The air temperature decreases by about 1 degree Fahrenheit for each 300 ft of altitude. Find a function $T(x)$ for the temperature at an altitude of x feet if the sea level temperature is 70°.

48. Ecology: Carbon Dioxide Pollution The burning of fossil fuels (such as oil and coal) added 24 billion tons of carbon dioxide to the atmosphere during 1995, and this annual amount grows by 0.46 billion tons per year. Find a function $C(t)$ for the amount of carbon dioxide added during the year t years after 1995, and use the formula to find how soon this annual amount will reach 30 billion tons. (*Note:* Carbon dioxide traps solar heat, increasing the earth's temperature, and may lead to flooding of lowland areas by melting the polar ice.) *Source: Oak Ridge National Laboratory.*

49. A store that installs satellite TV receivers finds that if it installs x receivers per week, then its costs will be $C(x) = 80x + 1950$ and its revenue will be $R(x) = -2x^2 + 240x$ (both in dollars).

   a. Find the store's break-even points.
   b. Find the number of receivers it should install to maximize profit, and the maximum profit.

50. An air conditioner outlet finds that if it sells x air conditioners per month, then its costs will be $C(x) = 220x + 202,500$ and its revenue will be $R(x) = -3x^2 + 2020x$ (both in dollars).

   a. Find the break-even points.
   b. Find the number of air conditioners it should sell to maximize profit, and the maximum profit.

0.4 Functions, continued

For each function in Exercises 51–54:

a. Evaluate the given expression.

b. Find the domain.
c. Find the range.

51. $f(x) = \dfrac{3}{x(x-2)}$, find $f(-1)$

52. $f(x) = \dfrac{16}{x(x+4)}$, find $f(-8)$

53. $g(x) = |x + 2| - 2$, find $g(-4)$

54. $g(x) = x - |x|$, find $g(-5)$

Solve each equation by factoring.

55. $5x^4 + 10x^3 = 15x^2$ **56.** $4x^5 + 8x^4 = 32x^3$

57. $2x^{5/2} - 8x^{3/2} = 10x^{1/2}$

58. $3x^{5/2} + 3x^{3/2} = 18x^{1/2}$

Graph each function.

59. $f(x) = (x + 1)^2 - 1$ **60.** $f(x) = (x - 2)^2 - 4$

61. $f(x) = \begin{cases} 3x - 7 & \text{if } x \geq 2 \\ -x - 1 & \text{if } x < 2 \end{cases}$

If you use a graphing calculator for Exercises 61 and 62, be sure to indicate any missing points.

62. $f(x) = \begin{cases} 6 - 2x & \text{if } x > 2 \\ 2x - 1 & \text{if } x \leq 2 \end{cases}$

For each pair $f(x)$ and $g(x)$, find **a.** $f(g(x))$, **b.** $g(f(x))$.

63. $f(x) = x^2 + 1$, $g(x) = \dfrac{1}{x}$

64. $f(x) = \sqrt{x}$, $g(x) = 5x - 4$

65. $f(x) = \dfrac{x + 1}{x - 1}$, $g(x) = x^3$

66. $f(x) = |x|$, $g(x) = x + 2$

For each function, find and simplify

$$\dfrac{f(x + h) - f(x)}{h}$$

67. $f(x) = 2x^2 - 3x + 1$ **68.** $f(x) = \dfrac{5}{x}$

69. Business: Advertising Budget A company's advertising budget is $A(p) = 2p^{0.15}$, where p is the company's profit, and the profit is predicted to be $p(t) = 18 + 2t$, where t is the number of years from now. (Both A and p are in millions of dollars.) Express the advertising budget A as a function of t, and evaluate the function at $t = 4$.

 70. a. Solve the equation $x^4 - 2x^3 - 3x^2 = 0$ by factoring.
   b. Use a graphing calculator to graph $y = x^4 - 2x^3 - 3x^2$ and find the x-intercepts of the graph. Be sure that you understand why your answers to parts (a) and (b) agree.

 71. a. Solve the equation $x^3 + 2x^2 - 3x = 0$ by factoring.
   b. Use a graphing calculator to graph $y = x^3 + 2x^2 - 3x$ and find the x-intercepts of the graph. Be sure that you understand why your answers to parts (a) and (b) agree.

 72. Business: Revenue The following table gives a company's annual revenue (in millions of dollars) from its overseas operations during its first 5 years.

Year	Revenue
1	2.0
2	1.8
3	1.9
4	2.1
5	2.8

a. Enter the table numbers into a graphing calculator and make a plot of the resulting points (Years on the x-axis and Revenue on the y-axis). What kind of curve do the points suggest?
b. Have your calculator fit such a curve to the data. Then enter the result as y_1, which gives a formula for the annual revenue each year. Plot the points together with the regression curve.
c. Use your formula to predict the revenue in years 6 and 7.

0.5   Exponential Functions

Graph each function.

73. $f(x) = 4^x$

74. $f(x) = (\tfrac{1}{4})^x$

75. General: Appreciation A fifteenth-century chalk drawing by Raphael, *Study for the Head and Hand of an Apostle*, was sold in 1996 for $8.7 million. Assuming that it continues to appreciate in value by 6.5% per year, estimate its value in the year 2005.

76. General: Appreciation A 1949 Scuderia Ferrari (166MM Barchetta) automobile was auctioned in 1996 for $1.65 million. If it continues to appreciate in value by 5.5% per year, find its value in the year 2010.

77. Business: Depreciation An $800,000 computer depreciates by 20% each year.
  a. Give a formula for its value after t years.
  b. Find its value after 4 years.

78. General: Depreciation A $5.4 million offshore oil drilling platform depreciates by 12% each year.
  a. Give a formula for its value after t years.
  b. Find its value after 10 years.

79. General: Population The largest city in the world is Tokyo, followed by Mexico City and São Paulo (Brazil). According to the Census Bureau, x years after 1995 the population of Tokyo will be $27e^{0.0155x}$, the population of Mexico City will be $16.9e^{0.0194x}$, and the population of São Paulo will be $16.1e^{0.0249x}$ (all in millions). Graph these three functions on a calculator on the window [0, 100] by [0, 100]. Assuming that this growth continues:
  a. When will São Paulo overtake Mexico City as the second largest city?
  b. When will São Paulo overtake Tokyo as the largest city?

80. Computers: Moore's Law The amount of information that can be stored on a computer chip can be measured in megabits (a "bit" is a binary digit, 0 or 1, and a "megabit" is a million bits). The first 1-megabit chips became available in 1987, and 4-megabit chips became available in 1990. This quadrupling of capacity every three years is expected to continue, so that t years after 1987 chip capacity will be $C(t) = 4^{t/3}$ megabits. Use this formula (known as Moore's law, after Gordon Moore, a founder of the Intel Corporation) to predict chip capacity in the year

2002. (*Hint:* What value of t corresponds to 2002?)

0.6   Logarithmic Functions

81. Find each logarithm *without* using a calculator.

   a. $\log 1000$ **b.** $\log \frac{1}{1000}$ **c.** $\ln e^3$ **d.** $\ln \sqrt[4]{e}$

82. Find each logarithm *without* using a calculator.

   a. $\log \sqrt{10}$ **b.** $\log 10^8$ **c.** $\ln \frac{1}{e}$ **d.** $\ln e^{3/2}$

Use the properties of natural logarithms to simplify each function.

83. $f(x) = \ln (x^4) - \ln (x^3) - \ln 1$

84. $f(x) = \ln (e^{7x}) - 5x - \ln e$

85. **Business: Demand** Demand for a computer memory chip is increasing by 18% per year. How soon will demand double?

86. **Economics: Geothermal Power** Geothermal electricity generating power (using heat from the center of the Earth) has been increasing worldwide by 5.5% annually. How soon will it double? (*Source: Worldwatch.*)

87. **General:' Population** According to the Census Bureau, the population of Arizona is increasing by 3.1% annually. At this rate, how soon will Arizona double in population?

88. **General: Population** According to the Census Bureau, the population of Colorado is increasing by 2.4% annually. At this rate, how soon will Colorado double in population?

89–90: General: Fossils In the following exercises, use the fact that the proportion of potassium 40 remaining after t million years is $e^{-0.00054t}$.

89. In 1984 in the Wind River Basin of Wyoming, scientists discovered a fossil of a small, three-toed horse, an ancestor of the modern horse. Estimate the age of this fossil if it contained 97.3% of its original potassium 40.

90. Estimate the age of a skull found in 1959 in Tanzania (dubbed "Nutcracker Man" because of its huge jawbone) that contained 99.9% of its original potassium 40.

Projects and Essays

The following projects and essays are based on Chapter 0. There are no right and wrong answers—the results depend only on your imagination and resourcefulness.

1. Write a page about the mile record example on pages 2–3, discussing the following questions: What factors might make modern runners faster than earlier runners? Would you expect this approximately linear trend to continue indefinitely? For how long do you think it might continue? What would you expect the trend to look like *eventually*, and why?

2. Discuss the relationships between the various forms of straight lines.
For example:

   a. Explain how the slope-intercept form comes from the diagram beside it on page 8.

   b. Show how the point-slope form leads to the slope-intercept form if the point is an intercept.

   c. Show how the point-slope and slope-intercept forms lead to the horizontal line form.

   d. Show how the general linear equation $ax + by = c$ includes both vertical and non-vertical lines.

3. Look up René Descartes in a book on the history of mathematics and write an essay about his discovery of Cartesian geometry and his other contributions to mathematics.

4. Give two examples of real-life situations that can be modeled by *linear* functions, and explain why they are linear. Then give and discuss two examples of situations that can be modeled by *quadratic* functions. Finally, give and discuss two examples of situations that are more complicated than linear or quadratic functions.

5. For each of the following two stories, find the graph that best matches it, and explain (briefly) why it matches. Then make up a story to match the remaining graph. Finally, make up two new stories and graphs that match them. (Be imaginative.)

   Story a. I drove away from home in a hurry, but got a speeding ticket, after which I continued on more slowly.

   Story b. I went out for a jog around town, returning home too exhausted to go out again.

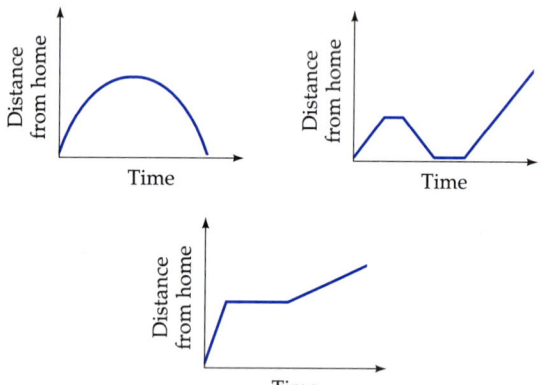

6. Define a "function" on people as follows: for any person x, let

   $$f(x) = \text{the } father \text{ of person } x$$

   $$m(x) = \text{the } mother \text{ of person } x$$

   Then the maternal grandmother of x would be $m(m(x))$. Define all grandparents of x, and state how many of them there are. Do the same for great-grandparents. Can you generalize even further?

7. The constant e is defined as $e = \lim\limits_{n \to \infty} \left(1 + \frac{1}{n}\right)^n \approx 2.71828$. Use a calculator to evaluate $\left(1 + \frac{1}{n}\right)^n$ for large values of n. Compare e to $\lim\limits_{n \to \infty} \left(1 + \frac{1}{100}\right)^n$ and $\lim\limits_{n \to \infty} \left(1 + \frac{1}{n}\right)^{100}$ where, in each case, one of the n's is replaced by a constant. Estimate $\lim\limits_{n \to \infty} \left(1 + \frac{1}{n}\right)^{2n}$ and $\lim\limits_{n \to \infty} \left(1 + \frac{1}{2n}\right)^n$ and try to determine how their values are related to e. What if 2 is replaced by another number?

8. The "Money Angles" column in *Time* magazine (May 17, 1993) recommended buying wine by the case, saving an estimated 10% every 12 weeks. The column then suggests that repeating this purchase every 12 weeks will lead to a saving of "more than 40% per year." Questions: If you did this for two years, would you save 80%? For three years, would you save 120% (so the wine would be free, or even better)? Explain what is wrong with *Time*'s reasoning.

9. Find out how logarithms (base 10) were used for calculating products and quotients before the invention of pocket calculators. Which properties of logs enable multiplication problems to be changed into addition problems, and division problems into subtraction problems? Look up "slide rule" and explain how slide rules were used for calculations, and how they are based on the addition and subtraction of logarithms.

10. Write about the difference between straight-line depreciation and depreciation by a fixed percentage, describing the advantages of each. For example, if the line (from straight-line depreciation) and the curve (from fixed percentage depreciation) both begin at the same point and end at the same point, which method of depreciation provides the greater dollar drop in value in the first year? In the last year? If you had to pay income tax on the value each year, which method would you prefer? If you had to

sell the asset midway, which method would you prefer? If you had to buy the asset midway?

11. Write about logarithms and their connection with exponents. Include discussions of the following questions: Why is the logarithm of 1 always zero (for any base)? Why can't 1 be a base for logarithms? Why can't a negative number be a base? Why can't you find logarithms of zero and negative numbers? What does the graph of the logarithm function look like for different bases (including bases between zero and one)? Look up John Napier and include information about the history of logarithms. Why do we use the letter e to represent the constant defined by the limit on page 72. (*Hint:* Look up Leonhard Euler.)

12. Reread the first item under "Hints and Suggestions" on page 94. Give other examples of geometric and analytic objects, and discuss the relationship between them. Be sure to include a discussion of curves and functions and the relationships between them, including the vertical line test.

13. Look over your notes, your homework, and the text, and write a page about how a graphing calculator has helped you in this chapter. Include examples of how it has helped you to *explore concepts* and how it has helped to *simplify your work*. What was its *most* helpful or interesting use? What was its *least* helpful or interesting use? Are there problems that can be done on a graphing calculator but that are easier to do "by hand"?

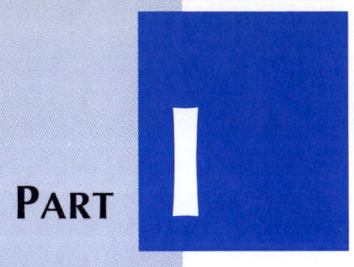

PART **I**

FINITE MATHEMATICS

1 MATHEMATICS OF FINANCE

Lenders offer borrowers many choices. This chapter explains how loans work and how to compare them.

Simple Interest

APPLICATION PREVIEW

Benjamin Franklin's Will

In 1789, the eighty-three-year-old Benjamin Franklin added a codicil to his will creating a long-term bequest of £1000 (about $4500) to the Town of Boston. For the first hundred years, the money was to start a loan fund to help young married tradesmen start their own businesses. Each year the borrowers would repay one tenth of the principal together with 5% interest, and this money would then be lent to fresh borrowers. Franklin hoped ". . . that no part of the money will at any time lie dead or be diverted to other purposes . . ." and calculated that after one hundred years the fund would grow to £131,000 (about $500,000). This money would then be divided, with £100,000 for public works in Boston and the rest to continue the loan fund for another hundred years. At the end of the second hundred years, Franklin estimated the fund would total over £4,061,000 (more than $18 million) and he directed that it then be divided between Boston and the state government, ". . . not presuming to carry my views farther."

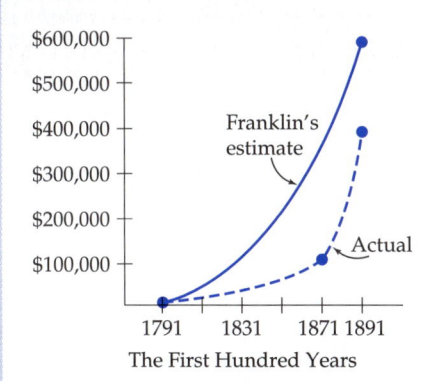

The First Hundred Years

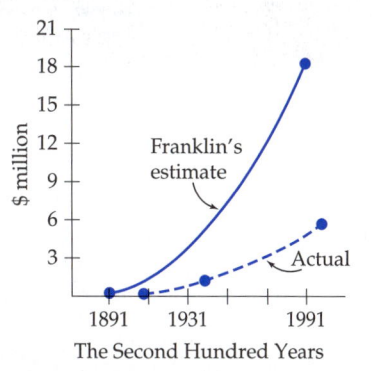

The Second Hundred Years

After Franklin's death in 1790, the Town of Boston accepted his offer, and the money arrived in March 1791. Despite the fund's early popularity, by the 1830s the changes brought about by the Industrial Revolution forced many workers into factories instead of starting up businesses on their own, and demand for loans all but disappeared. The managers of the fund placed some of the money in other investments, and by the hundredth anniversary in July 1891, its value reached $391,000. After various legal squabbles over the division of this considerable sum, a technical training school was established that has now become the Franklin Institute of Boston. In 1994, the second part of the Franklin Fund was ended and its $4.6 million assets were turned over to the school. The 200-year history of Franklin's £1000 gift provides an example of the relation between money and time, a subject that will be explored further in this chapter.

Introduction

In the modern credit world, the old adage "time is money" has become a basic fact of economic life. When you open a savings account or take out a car loan, you directly experience the "time value of money." This chapter covers the basic financial properties of a loan between a lender and a borrower and the calculation of the interest and payments involved. We begin with simple interest.

Simple Interest Formula

The *principal* of a loan is the amount of money borrowed from the lender, the *term* is the time the borrower has the money, and the *interest* is the additional money paid by the borrower for the use of the lender's money. The interest is called *simple interest* if it is calculated as a fixed percentage of the principal and is paid at the end of the term. The *interest rate* of the loan is the dollars of interest per $100 of principal per year (or "per annum") and is usually stated as a percentage but is always written as a decimal in calculations. Simple interest is calculated as follows.

Simple Interest Formula

The interest I on a loan of P dollars at simple interest rate r for t years is

$$I = Prt$$

EXAMPLE 1   Finding Simple Interest

Find the interest on a loan of two million dollars at 7.2% simple interest for 4 months.

Solution

Writing the interest rate in decimal form and changing the term to years,

$$I = (2,000,000)(0.072)\left(\frac{4}{12}\right) = 48,000$$

$I = Prt$ with
$P = \$2,000,000,$
$r = 0.072,$ and $t = 4/12$

The interest is \$48,000.

◼

Be Careful! When using a calculator, *round off only your final answer.* For example, if you use $t = 0.33$ in the previous example instead of $\frac{4}{12}$ or $\frac{1}{3}$, you would get the value $(2,000,000)(0.072)(0.33) = \$47,520$, which is wrong by \$480.

Banker's Rule

Loan agreements for short terms (that is, less than a year) often add the phrase *using the Banker's rule* after stating the interest rate. This means that the term is calculated as the number of days divided by 360. At first glance, this seems merely to simplify the banker's arithmetic by making each year into 12 months of 30 days each, but in fact it sometimes gives a slight advantage to the lender.

EXAMPLE 2 Using the Banker's Rule

Find the interest on a loan of \$2,000,000 at 7.2% simple interest from June 15 to October 15 using the Banker's rule.

Solution

Since the term is 4 months, this appears to be the same as Example 1. But the particular 4 months from June 15 to October 15 total 122 days, and by the Banker's rule the interest is

$$I = (2,000,000)(0.072)\left(\frac{122}{360}\right) = 48,800 \qquad I = Prt$$

So in this case the Banker's rule resulted in an \$800 advantage to the lender.

◼

PRACTICE PROBLEM 1

Find the interest on a loan of \$50,000 at 19.8% simple interest from March 3 to June 3 using the Banker's rule. How does this compare to a loan of \$50,000 at 19.8% simple interest for 3 months?

Solution at the back of the book

If you know any three unknowns in the interest formula $I = Prt$, you can solve for the fourth one.

EXAMPLE 3   Finding the Simple Interest Rate

What is the interest rate of a loan charging $18 simple interest on a principal of $150 after 2 years?

Solution

We solve $I = Prt$ for interest rate r:

$$r = \frac{I}{Pt}$$

$I = Prt$ divided by Pt

Then

$$r = \frac{18}{(150)(2)} = \frac{18}{300} = 0.06$$

Substituting $I = 18$, $P = 150$, and $t = 2$

The interest rate is 6%.   ■

Total Amount Due on a Loan

When the term of a loan is over, the borrower repays the principal and interest. So the total amount due is

$$\underbrace{P}_{\text{Principal}} + \underbrace{Prt}_{\text{Interest}} = \underbrace{P(1 + rt)}_{\text{Total amount}}$$

Factoring out the common term

This gives the following formula:

Total Amount Due for Simple Interest

The total amount A due at the end of a loan of P dollars at simple interest rate r for t years is

$$A = P(1 + rt)$$

As before, this formula may be used as it is or it may be solved for any one of the other variables, as the next few examples will show.

EXAMPLE 4   Finding the Total Amount Due

What it the total amount due on a loan of $3000 at 6% simple interest for 4 years?

Solution

$$A = 3000(1 + (0.06)(4))$$
$$= 3000(1.24) = 3720$$

```
3000→P
              3000
.06→R
               .06
4→T
                 4
P(1+RT)
              3720
```

The total due on the loan is $3720.

The amount of the loan in Example 4 grew from $3000 to $3720 in 4 years. The $3720 is sometimes called the *future value* of the original $3000. Reversing our viewpoint, we say that the principal of $3000 is the *present value* of the later $3720. To find a formula for the present value, we solve the "total amount due" formula (on page 108) for P by dividing by $1 + rt$ and rename the result PV:

$$PV = \frac{A}{1 + rt}$$

PV means **P**resent **V**alue of the future amount A

EXAMPLE 5   Finding a Present Value

How much should be invested now at 8.6% simple interest if $10,000 is needed in 6 years?

Solution

This is the same as finding the present value of $10,000 in 6 years:

$$PV = \frac{10,000}{1 + (0.086)(6)} = \frac{10,000}{1.516} \approx 6596.31$$

$PV = A/(1 + rt)$ with $A = 10,000$, $r = 0.086$, and $t = 6$

The amount required is $6596.31.

PRACTICE PROBLEM 2

Find the present value of a "promissory note" that will pay $5000 in 4 years at 12% simple interest.

Solution at the back of the book

We found the present value by deriving a formula for it. We could instead have substituted the given numbers into the "total amount due" formula and then solved for P. We will do the next example in this way, substituting numbers and then solving for the remaining variable.

EXAMPLE 6   Finding the Term of a Simple Interest Loan

What is the term of a loan of $2000 at 4% simple interest if the amount due is $2400?

Solution

Substituting the given numbers for the appropriate variables in the "total amount due" formula (page 108) gives:

$$2400 = 2000(1 + 0.04t)$$ $A = P(1 + rt)$ with $A = 2400$, $P = 2000$, and $r = 0.04$

$$1.2 = 1 + 0.04t$$ Dividing by 2000

$$0.2 = 0.04t$$ Subtracting 1

$$t = \frac{0.2}{0.04} = 5$$ Dividing by 0.04 and reversing sides

The term is 5 years.

 ## Graphing Calculator Exploration

The formula $A = P(1 + rt)$ may be written as $A = (Pr)t + P$ so that it has the familiar $y = mx + b$ form of a straight line with slope Pr and y-intercept P (but with t and A instead of x and y).

a.   Using the values $P = 2000$ and $P \cdot r = (2000)(0.04) = 80$ from Example 6, graph the line $y_1 = 80x + 2000$ on the window [0, 8] by [0, 4000], so that y gives the amount of the loan for any term x.

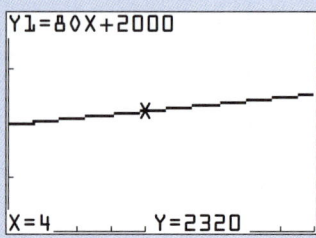

b. Use TRACE to estimate the term x that gives $y = 2400$. How does your graphical answer compare to the answer for Example 6? Try ZOOMing IN to improve your estimate.

c. Reset the window to [0, 8] by [0, 4000] and add the horizontal line $y_2 = 2400$ to your graph. Use INTERSECT to find the intersection point. Does the x-value of this point exactly match the answer for Example 6?

Discounted Loans and Effective Interest Rates

A *discounted* loan is a loan in which the lender deducts the interest from the amount the borrower receives at the start. How different is this from a simple interest loan? It must be better for the lender because getting the money early (so it can earn interest somewhere else) is always better than later. Since the borrower actually is receiving a smaller amount, we may recalculate the interest rate as a "standard" loan on this smaller amount. The resulting rate is called the *effective* simple interest rate of the loan.

EXAMPLE 7   Finding the Effective Simple Interest Rate

Find the effective simple interest rate on a discounted loan of $1000 at 6% simple interest for 2 years.

Solution

The interest is ($1000)(0.06)(2) = $120, so as a simple interest loan, the borrower receives only $P = 1000 - 120 = \$880$ and agrees to pay back $1000 at the end of 2 years. If we write r_s for the simple interest rate of this loan, we see that

$$1000 = 880(1 + r_s 2)$$

$A = P(1 + rt)$ with $A = 1000$, $P = 880$, and $t = 2$

Solving for r_s,

$$r_s = \frac{1}{2}\left(\frac{1000}{880} - 1\right) \approx 0.0682$$

Dividing by 880, subtract 1, and divide by 2

The effective simple interest rate on this discounted loan is 6.82%, which is significantly larger than the stated rate of 6%.

Using these same steps, we can find a formula for the effective rate. For a discounted loan of amount A at interest rate r for term t, the borrower receives $A - Art = A(1 - rt)$ dollars. Therefore, the simple interest equation with effective simple interest rate r_s gives:

$$A = \underbrace{A(1 - rt)}_{\text{Principal}}(1 + r_s t)$$

$A = P(1 + rt)$ with $A(1 - rt)$ for P

Solving for r_s:

$$\frac{1}{1 - rt} = 1 + r_s t$$

Canceling the As and dividing by $(1 - rt)$

$$r_s t = \frac{1}{1 - rt} - 1 = \frac{1 - (1 - rt)}{1 - rt} = \frac{rt}{1 - rt}$$

Reversing sides, subtracting 1, and simplifying

$$r_s = \frac{1}{t} \frac{rt}{1 - rt} = \frac{r}{1 - rt}$$

Dividing by t

This gives the following formula for r_s.

Effective Simple Interest Rate for a Discounted Loan

For a discounted loan at interest rate r for t years, the effective simple interest rate r_s is

$$r_s = \frac{r}{1 - rt}$$

Notice that the effective rate r_s will be larger than r (since the denominator is less than 1) and that it depends only on the *rate* and *term* of the loan, and not on its amount. For the loan in Example 7, our formula gives the answer that we found:

$$r_s = \frac{0.06}{1 - (0.06)(2)} = \frac{0.06}{0.88} \approx 0.0682$$

Using $r = 0.06$ and $t = 2$

PRACTICE PROBLEM 3

What is the effective simple interest rate of a discounted loan at 6% interest for 3 years? for 5 years?   *Solution at the back of the book*

SUMMARY

The simple interest formula is

$$I = Prt$$

$I =$ simple interest
$P =$ principal
$r =$ interest rate
$t =$ term in years

We can solve for any one of the variables if the others are known. Using the Banker's rule, the term is calculated as the number of days divided by 360. The total amount due at the end of the loan (principal plus interest) is

$$A = P(1 + rt) \qquad A = \text{total amount due}$$

This equation may be solved for any one of the variables, keeping in mind that the amount A at the end of the loan is sometimes called the *future value* of the principal P, and P is sometimes called the *present value* of the future amount A.

For a discounted loan, the interest is subtracted from the principal at the beginning of the loan. The effective simple interest rate r_s for a discounted loan at rate r for t years is

$$r_s = \frac{r}{1 - rt}$$

EXERCISES 1.1 (Most require ⊞ .)

Find the simple interest on each loan.

1. $1500 at 7% for 10 years.

2. $2000 at 9% for 7 years.

3. $6000 at 6.5% for 8 years.

4. $4500 at 4.25% for 9 years.

5. $825 at 6.58% for 5 years 6 months.

6. $950 at 5.87% for 6 years 3 months.

7. $1280 at 4.8% for 3 months.

8. $5275 at 5.3% for 2 months.

9. $1280 at 4.8% from March 8 to June 8 using the Banker's rule.

10. $5275 at 5.3% from October 14 to December 14 using the Banker's rule.

Find the total amount due for each simple interest loan.

11. $1500 at 7% for 10 years.

12. $2000 at 9% for 7 years.

13. $6100 at 5.7% for 4 years 9 months.

14. $4500 at 6.3% for 3 years 6 months.

15. $3125 at 4.81% for 10 months.

16. $8775 at 13.11% for 7 months.

APPLIED EXERCISES

17. Interest Rate Find the interest rate on a loan charging $704 simple interest on a principal of $2750 after 4 years.

18. Interest Rate Find the interest rate on a loan charging $1127 simple interest on a principal of $4900 after 5 years.

19. Principal Find the principal of a loan at 8.4% if the simple interest after 5 years 6 months is $1155.

20. Principal Find the principal of a loan at 7.6% if the simple interest after 9 years 3 months is $2109.

21. Term Find the term of a loan of $175 at 9% if the simple interest is $63.

22. Term Find the term of a loan of $225 at 7% if the simple interest is $94.50.

23. Present Value How much should be invested now at 5.2% simple interest if $8670 is needed in 3 years?

24. Present Value How much should be invested now at 4.8% simple interest if $4530 is needed in 4 years 4 months?

Zero Coupon Bonds A *zero coupon bond* pays only its *face value* on maturity (getting its name because it has no "coupons" for interest payments before that date). The *fair market price* is the present value of the face value at the current interest rate.

25. What would be the fair market price of a $10,000 zero coupon bond due in 1 year if today's long-term simple interest rate is 5.81%?

26. What would be the fair market price of a $15,000 zero coupon bond due in 1 year if today's long-term simple interest rate is 7.23%?

27. What would be the fair market price of a $5000 zero coupon bond due in 2 years if today's long-term simple interest rate is 3.54%?

 28. How Interest Rates Affect Present Value The present value formula $PV = A/(1 + rt)$ can be viewed on your graphing calculator as a function y of the simple interest rate x by rewriting it as $y_1 = A/(1 + xt)$. Using $A = \$10,000$ and $t = 1$ year, graph this expression with window $[0, 1]$ by $[0, 12000]$ to see the present value of a $10,000 zero coupon bond due in 1 year as a function of the interest rate. Use TRACE or VALUE to find the present value of the bond for interest rates of 4%, 5%, 6%, 7%, 8%, and 9%. Does a 1% increase in the interest rate always cause the same decrease in the present value of the bond?

29. Term What should be the term for a loan of $6500 at 7.3% simple interest if the lender wants to receive $9347 when the loan is paid off?

30. Term What should be the term for a loan of $5400 at 5.8% simple interest if the lender wants to receive $6966 when the loan is paid off?

31. Term How long will it take an investment at 8% simple interest to increase by 70%?

32. Term How long will it take an investment at 4.2% simple interest to increase by 26.6%?

Doubling Time The *doubling time* of an investment is the number of years it takes for the value to double. This is the same as the number of years for the value to increase by 100%.

33. What is the doubling time of a 5% simple interest investment?

34. Show that the doubling time of a simple interest investment is $1/r$ where r is the simple interest rate.

35. Effective Rate What is the effective simple interest rate of a discounted loan at 4.6% interest for 3 years 6 months?

36. Effective Rate What is the effective simple interest rate of a discounted loan at 7.2% interest for 2 years 10 months?

37. Discounted Loan Would you agree to borrow $1000 as a discounted loan at 5% for 20 years? Explain.

38. How the Term Affects the Discounted Rate The effective rate formula $r_s = r/(1 - rt)$ can be viewed on your graphing calculator as a function of the term t by rewriting it as $y_1 = r/(1 - rx)$. Using $r = 0.05$, graph this expression with window $[0, 20]$ by $[0, 1]$ to see the effective rate y of the discounted loan as a function of the term x. Use TRACE or VALUE to find the effective rate of the loan for terms of 5, 10, and 15 years. Does a 1-year increase in the term always cause the same increase in the effective rate?

39. Bridge Loan You have a buyer for your condominium but the seller of your dream house wants to close now. In order to get enough money to go to the closing, you take out a 2-month bridge loan for $72,000 at 7.92% simple interest. Hopefully, you will close on the sale of your condominium before the 2 months are up. How much interest will you have to pay on the loan?

40. Banker's Rule Suppose you wanted a loan for the month of February. Would the Banker's rule work in your favor or the bank's favor? (*Hint:* Calculate *t* both ways. Which will give a larger value of *I*?) What about a 3-month loan starting in January and ending in March?

41. **Lawyer's Fees** After successfully defending a client against a driving while intoxicated (DWI) charge, a lawyer reluctantly accepted as payment a promissory note for $5000 plus 20% simple interest due in 270 days. Learning subsequently of the client's "deadbeat" reputation, she sold the note 90 days later to an experienced bill collector for $4600. Using the Banker's rule and assuming the collector will actually obtain the full value of the note, what simple interest rate will the collector earn for his troubles?

42. **Home Improvements** Since the contractor was short of cash when the job was finished, a plumber accepted as payment a promissory note for $2000 plus 18% simple interest due in 60 days. Needing cash himself, the plumber

sold the note 15 days later for $2000 to a local loan agency. Using the Banker's rule, what simple interest rate will the agency earn on its investment?

Brokerage Commissions An Internet discount brokerage firm charges a commission of 10% on the first $20,000 plus 5% of the excess over $20,000 on each buy or sell transaction. Find the simple interest rate earned by each of the following investments after including the commissions paid to the brokerage firm.

43. Purchase 900 shares of American WebWide Education at $18.50 per share and sell them 4 months later at $26.75 per share.

44. Purchase 1500 shares of Well Care Deluxe at $11.90 per share and sell them 10 months later at $14.80 per share.

45. Purchase 800 shares of DucoFood Services at $28.70 per share and sell them 7 months later at $37.80 per share.

1.2 Compound Interest

APPLICATION PREVIEW

Musical Instruments as Investments*

No two guitars sound alike—good ones, anyway. Over the last 30 years, prices for high-quality acoustic instruments have soared. In 1993, a John D'Angelico New Yorker 18-inch-wide cutaway archtop ("jazz") guitar sold for $150,000, setting the record for the price of an acoustic fretted instrument not previously owned by a deceased superstar. Many pre-World War II Gibson and Martin instruments are valued not only for their craftsmanship but also for their distinctive sound. A mint condition Martin D-45 made in 1932–1942 that might have sold for $3500 in 1971 could sell for $125,000 in 1996, while a Gibson 1932–1939 Mastertone Granada 5-string (replacement neck) banjo selling for $18,000 in 1996 might have sold for just $1200 in 1971.

*Much of this material was generously provided by Stanley M. Jay and Larry Wexer of *mandolin bros. Ltd.* (Staten Island, New York), dealers in fine mandolins, guitars, and banjos.

How can these price increases be compared to other investments such as stocks, bonds, or bank savings accounts? The dots on the graph show the typical selling prices over the 25-year period from 1971 to 1996 for a Gibson F-5 mandolin made between 1922 and 1924 and signed by Gibson acoustic engineer Lloyd A. Loar, together with a curve showing the increasing value of a $2000 investment earning 12% compound interest.

Selling Prices for a Gibson Mandolin

The dots closely match the curve, showing that this mandolin has increased in value by about 12% annually, better than many bond and stock investments. The values of fine musical instruments and other investments can be found by the methods of this section.

Introduction

When a simple interest loan is not paid off but "rolled over" into another loan, the new principal includes the unpaid interest from the old loan. The interest on this new loan is called *compound interest* because it combines interest on both the original principal and on the unpaid interest. Naturally, this "interest on interest" is to the benefit of the lender. For example, if you lend $1000 at 6% simple interest for 3 years, your return will be $A = \$1000(1 + (0.06)(3)) = \$1180$. Now suppose you lend it for the first year only—the amount due will be $\$1000(1 + 0.06) = \$1060$. You then lend this $1060 for the next year, at the end of which the amount due is $\$1060(1 + 0.06) = \$1123.60$. For the third and final year you lend this $1123.60 and receive $\$1123.60(1 + 0.06) = \$1191.02$. This amount exceeds the simple interest amount by $\$1191.02 - \$1180 = \$11.02$, the difference coming from *interest on interest*. For a term longer than 3 years you would continue multiplying the debt by $(1 + 0.06)$—that is, by 1 plus the interest rate—each year.

Compound Interest Formula

In general, the amount A due on a compound interest loan of P dollars at interest rate r per year compounded yearly for t years is found by repeatedly multiplying the principal P by $(1 + r)$, once for each year:

$$A = P\underbrace{(1 + r)(1 + r) \cdots (1 + r)}_{\substack{t \text{ multiplications} \\ \text{by } (1 + r)}} = P(1 + r)^t$$

Graphing Calculator Exploration

To see that compound interest eventually surpasses simple interest (even with a higher rate and principal), compare $500 invested at 4% compounded annually with $2500 invested at 8% simple interest.

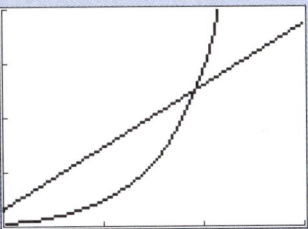

a. Graph the simple interest amount $A = P(1 + rt)$ as $y_1 = 2500(1 + .08x)$ on the window $[0, 150]$ by $[0, 35000]$ so that y_1 is the amount of the investment after x years.

b. Graph the compound interest amount $A = P(1 + r)^t$ as $y_2 = 500(1 + .04)^x$ on the same window.

c. Use TRACE or INTERSECT to find when the compound interest amount equals the simple interest amount. What happens after this intersection point? Try your graphs with the larger windows $[0, 300]$ by $[0, 70000]$ and by $[0, 500000]$.

For compounding done more frequently than once a year, we must adjust both the interest rate and the number of compoundings that will be done over the term of the loan. Some standard compounding periods are described in the following table.

Compounding Frequency	Periods per Year
Annually	1
Semiannually	2
Quarterly	4
Bimonthly	6
Monthly	12
Biweekly	26
Weekly	52
Daily	365

With m compounding periods per year, the interest rate for each period is decreased to r/m while the number of compoundings is increased to mt. Compound interest is then calculated as follows.

Compound Interest Formula

The amount A due on a loan of P dollars at yearly interest rate r compounded m times per year for t years is

$$A = P\left(1 + \frac{r}{m}\right)^{mt}$$

The yearly interest rate r is also called the *nominal rate* of the loan.

EXAMPLE 1   Finding the Amount Due on a Compound Interest Loan

Find the amount due on a loan of $1500 at 4.8% compounded monthly for 5 years.

Solution

The nominal rate of 4.8% expressed as a decimal is 0.048, and since monthly compounding means that $m = 12$, we have:

$$A = 1500\left(1 + \frac{0.048}{12}\right)^{(12)(5)}$$

$A = P(1 + r/m)^{mt}$
with $P = 1500$, $r = 0.048$, $m = 12$, and $t = 5$

$$= 1500(1.004)^{60} \approx 1905.96$$

The amount due is $1905.96.

PRACTICE PROBLEM 1   Find the amount that is due on a loan of $3500 at 5.1% compounded bimonthly for 8 years.   *Solution at the back of the book*

In Example 1, the amount $1905.96 may be called the *future value* of the original $1500 and, conversely, the $1500 is the *present value* of the later amount $1905.96. To find a formula for the present value, we solve the compound interest formula (page 118) for *P*, using the familiar $1/x^n = x^{-n}$ rule of exponents and renaming the result *PV*.

$$PV = \frac{A}{\left(1 + \dfrac{r}{m}\right)^{mt}} = A\left(1 + \frac{r}{m}\right)^{-mt} \qquad PV \text{ means } \textbf{Present Value}$$

EXAMPLE 2   Finding the Present Value

How much should be invested now at 8.6% compounded weekly if $10,000 is needed in 6 years?

Solution

$$PV = 10,000 \left(1 + \frac{0.086}{52}\right)^{-(52)(6)} \approx 5971.58$$

$PV = A(1 + r/m)^{-mt}$
with $A = 10,000$, $r = 0.086$, $m = 52$, and $t = 6$

The amount required is $5971.58.   ■

Growth Times

We solved Example 2 by using the formula that we had derived for present value. Instead, we could have substituted the given numbers into the compound interest formula and *then* solved for the remaining variable. We will do the next example in this way, finding a "growth time," the time for a loan to reach a given value. We will also use the rule of logarithms $\log(x^n) = n \log(x)$ to "bring down the power."

EXAMPLE 3   Finding the Term of a Compound Interest Loan

What is the term of a loan of $2000 at 6% compounded monthly that will have an amount due of $2400?

Solution

Substituting the given numbers into the compound interest formula:

$$2400 = 2000 \left(1 + \frac{0.06}{12}\right)^{12 \cdot t}$$

$A = P(1 + r/m)^{mt}$ with $A = 2400$, $P = 2000$, $r = 0.06$, and $m = 12$

$$1.2 = 1.005^{12t}$$ Dividing by 2000 and simplifying

$$\log 1.2 = \log 1.005^{12t}$$ Taking logarithms

$$\log 1.2 = 12t \log 1.005$$ Using $\log (x^n) = n \log (x)$ to bring down the power

$$12t = \frac{\log 1.2}{\log 1.005} \approx 36.555$$ Dividing by log 1.005, reversing sides, and calculating logarithms

Months

This amount of time, $12t = 36.555$, is in *months* (t is years, but $12t$ is *months*). We round *up* to the nearest month (since a shorter term will not reach the needed amount), so the term needed is 37 months, or 3 years 1 month.   ■

Hint: Instead of solving for t, if the compounding is monthly stop when you find $12t$ (the number of months), if quarterly stop when you find $4t$ (the number of quarters), if weekly stop when you find $52t$ (the number of weeks), and so on. Then round *up* to the next whole number of periods to ensure that the needed amount will actually be reached.

Notice that the actual amounts $2400 and $2000 did not matter in this calculation, only their *ratio* $\frac{2400}{2000} = 1.2$. Thus, a loan of $2000 grows to $2400 in the same amount of time that a loan of 20 *million* dollars would grow to 24 million dollars (at the stated interest rate) since the ratios are the same. This is true in general because in the compound interest formula (page 118), the total amount and the principal are *proportional*. In Example 3 we could have omitted the dollar amounts and simply asked for the term of a loan that would *multiply the principal by 1.2*. Furthermore, multiplying by 1.2 means *increasing by 0.2 or 20%*, so we could have asked for the term of a loan that would *increase the principal by 20%*—all three formulations are equivalent.

Be careful! Distinguish carefully between the *increase* and the *multiplier. Increasing by 20%* means *multiplying by 1.2* (the "1" keeps the original amount and the "0.2" increases it by 20%). For example, to *increase* an amount by 35% you would *multiply* it by 1.35.

EXAMPLE 4   Finding the Term to Increase the Principal

What is the term of a loan at 6% compounded weekly that will increase the principal by 40%?

Solution

Increasing by 40% means multiplying the amount by 1.4. Using P for the unknown principal, we must solve:

$$1.4P = P\left(1 + \frac{0.06}{52}\right)^{52 \cdot t}$$ *P (for principal) on both sides*

$$1.4 = (1 + 0.06/52)^{52t}$$ *Canceling the Ps and simplifying*

$$\log 1.4 = 52t \log (1 + 0.06/52)$$ *Taking logarithms and bringing down the power*

$$52t = \frac{\log (1.4)}{\log (1 + 0.06/52)} \approx 291.8$$ *Dividing by log (1 + 0.06/52), reversing sides, and calculating logarithms*

                Weeks

Rounding *up*, the term needed is 292 weeks, or 5 years 32 weeks.

■

PRACTICE PROBLEM 2

What is the term of a loan at 6% compounded quarterly that multiplies the principal by 2?   *Solution at the back of the book*

Rule of 72

The *doubling time* of an investment is the number of years it takes for the value to double. Doubling times are often estimated by using the *rule of 72*:

$$\left(\begin{matrix}\text{Doubling} \\ \text{time}\end{matrix}\right) \approx \frac{72}{r \times 100}$$ *Divide 72 by the rate times 100*

The rule of 72, however, gives only an *approximation* for the doubling time, but it is often quite accurate. For example, for a loan at 6% compounded quarterly the rule of 72 gives:

$$\left(\begin{matrix}\text{Doubling} \\ \text{time}\end{matrix}\right) \approx \frac{72}{0.06 \times 100} = \frac{72}{6} = 12 \text{ years}$$ *Rule of 72 estimate*

In Practice Problem 2 we found the correct doubling time to be $11\frac{3}{4}$ years, so the rule of 72 is not far off. The following Graphing Calculator Exploration shows that the rule of 72 is reasonably accurate for many "everyday" interest rates (but see also Exercise 100 on page 132).

 ## Graphing Calculator Exploration

Compare the "rule of 72" with the exact formula for the doubling time by graphing both as functions of the interest rate.

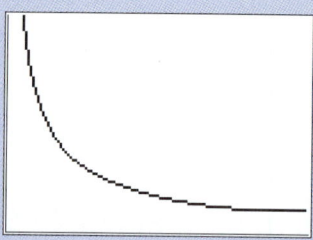

a. Graph $y_1 = 72/(x \times 100)$ on the window [0, .5] by [0, 20] to show the "rule of 72" estimate for the number of years y at interest rate x.

b. Graph the curve $y_2 = \log(2)/(12 \log(1 + x/12))$ on the same window to show the doubling time of a compounded monthly loan with interest rate x.

c. Verify that the graph shows *both* curves.

Effective Rates

How different is a loan charging 6.2% compounded *quarterly* from a loan charging 6.2% compounded *monthly*? In general, how can we compare different compound interest rates? We can compare them by calculating the *actual percentage increase* that each will generate during 1 year. More formally, we calculate the *effective rate r_e*, which is the simple interest rate that will return the same amount on a 1-year loan. The effective rate is also called the *annual percentage rate* or *APR* and by law must be stated in consumer loan agreements. The effective rate r_e is found by solving:

$$P(1 + r_e) = P\left(1 + \frac{r}{m}\right)^m$$

Finding the "simple" rate that gives the "compound" rate for $t = 1$

Canceling the P's and subtracting the 1 on the left gives:

Effective Rate for a Compound Interest Loan

For a compound interest loan at interest rate r compounded m times per year, the effective rate r_e is

$$r_e = \left(1 + \frac{r}{m}\right)^m - 1$$

The effective rate can be interpreted as the interest earned by $1 for 1 year.

EXAMPLE 5 **Finding the Effective Rate**

What is the effective rate of a loan at 6.2% compounded quarterly?

Solution

$$r_e = \left(1 + \frac{0.062}{4}\right)^4 - 1 \approx 0.0635 \qquad \begin{array}{l} r_e = (1 + r/m)^m - 1 \\ \text{with } r = 0.062 \text{ and } m = 4 \end{array}$$

The effective rate is 6.35%. That is, 6.2% compounded quarterly results in an annual gain of 6.35%.

■

PRACTICE PROBLEM 3   What is the effective rate of a loan at 6.2% compounded monthly?

Solution at the back of the book

Graphing Calculator Exploration

The effective rate should be larger for more frequent compounding since the sooner the interest is credited, the sooner it begins to earn more interest.

a. Using the value $r = 0.062$ from Example 5, graph the curve $y_1 = (1 + .062/x)^x - 1$ on the window [0, 100] by [.061, .065] so that y is the effective rate of a 6.2% loan compounded x times a year.

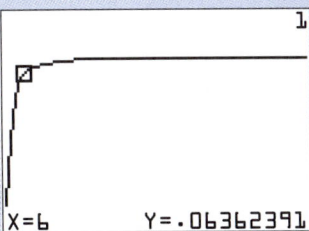

b. Use TRACE or VALUE to find y when $x = 4$ and when $x = 12$. How do your answers compare to the answers to Example 5 and Practice Problem 3?

c. What happens for large values of x? Does the effective rate become arbitrarily large or does it "settle down"?

d. Can the effective rate be as large as 6.395674%? (You may need a larger window!) Can it be as large as 6.3963%?

Choices between different loans or investments are often made on the basis of their effective rates.

EXAMPLE 6   Using Effective Rates to Compare Investments

A self-employed carpenter is setting up a Keogh retirement plan (which defers taxes on interest until withdrawals are made) and is considering a certificate of deposit (CD) with either the First & Federal Bank at 6.2% compounded semiannually or the Chicago Nationswide Bank at 6.15% compounded monthly. Which bank offers the greater effective rate?

Solution

The effective rate for 6.2% compounded semiannually (First & Federal) is

$$r_e = \left(1 + \frac{0.062}{2}\right)^2 - 1 \approx 0.0630 \qquad r_e \approx 6.30\%$$

whereas for 6.15% compounded monthly (Chicago Nationswide) the effective rate is

$$r_e = \left(1 + \frac{0.0615}{12}\right)^{12} - 1 \approx 0.0633 \qquad r_e \approx 6.33\%$$

The CD at the Chicago Nationswide Bank is better.

◼

The effective rate of return for an investment of P dollars that returns an amount A after t years can be found from the compound interest formula (page 118) with $m = 1$:

$$A = P(1 + r_e)^t$$

$A = P(1 + r/m)^{mt}$
with $m = 1$ and $r = r_e$

An example of this is a *zero coupon bond*, which pays its *face value* on maturity and sells now for a lower price. This kind of bond makes no payments before maturity and so has no interest "coupons."

EXAMPLE 7   Effective Rate for a Zero Coupon Bond

What is the effective rate of return of a $10,000 zero coupon bond maturing in 6 years and offered now for sale at $5500?

Solution

We solve:

$$10{,}000 = 5500(1 + r_e)^6 \qquad \begin{array}{l} A = P(1 + r_e)^t \text{ with} \\ A = 10{,}000, P = 5500, \\ \text{and } t = 6 \end{array}$$

$$\frac{10{,}000}{5500} = (1 + r_e)^6 \qquad \text{Dividing by 5500}$$

$$(100/55)^{1/6} = 1 + r_e \qquad \text{Raising to power } \tfrac{1}{6}$$

$$r_e = (100/55)^{1/6} - 1 \approx 0.1048 \qquad \text{Solving for } r_e$$

The effective rate is 10.48%.

Continuous Compounding of Interest

How much does the amount due on a loan change as the compounding changes? Clearly, if the term and interest rate are fixed, the amount due will increase if the compounding is more frequent, since the sooner the interest is deposited the sooner it can begin earning more interest. But will more frequent compounding increase the amount arbitrarily or is there a limit? To answer these questions we look at the compound interest formula and let the number of compounding periods m approach infinity, $m \to \infty$. We will use the fact from page 72 that as $n \to \infty$, the quantity $\left(1 + \dfrac{1}{n}\right)^n$ approaches the number $e \approx 2.71828$.

$$A = P\left(1 + \frac{r}{m}\right)^{mt} = P\left(1 + \frac{r}{m}\right)^{\frac{m}{r}\, rt}$$

Compound   Multiplying and
interest formula   dividing by r in
the exponent

$$= P\left(1 + \frac{1}{n}\right)^{nrt} = P\left[\left(1 + \frac{1}{n}\right)^n\right]^{rt} \xrightarrow[\text{As } n \to \infty]{} Pe^{rt}$$

Letting $n = \frac{m}{r}$ $\to e$ as $n \to \infty$
so $\frac{1}{n} = \frac{r}{m}$

The arrow on the right means that the amount A approaches the quantity Pe^{rt} as n (or equivalently m) approaches infinity. That is, as the number m of compounding periods per year gets larger, the amount A approaches a limiting value of Pe^{rt}. This "infinitely frequent" compounding is called *continuous compounding*.

Interest Compounded Continuously

> The amount A due on a loan of P dollars at annual interest rate r compounded *continuously* for t years is
>
> $$A = Pe^{rt}$$

To evaluate e to a power, use the [2nd] and [LN] keys on your calculator.

EXAMPLE 8   **Finding Amount Due with Continuous Compounding**

Find the amount due on a loan of $1500 at 4.8% compounded continuously for 5 years.

Solution

$$A = 1500e^{0.048 \cdot 5} = 1500e^{0.24} \approx 1906.87$$

```
1,500→P
              1,500
.048→R
               .048
5→T
                  5
Pe^(RT)
        1906.873725
```

The amount is $1906.87.

■

Intuitive Meaning of Continuous Compounding

In Example 1 on page 118, we found that the amount due on a similar loan but compounded *monthly* was $1905.96. Notice that the amount found here for *continuous* compounding is slightly greater. Why does continuous compounding grow faster? With monthly compounding, the interest is not added until the *end* of the month. Under *continuous* compounding the interest is added to the balance *continuously, without delay*, so that the interest starts earning interest immediately. The extra growth in continuous compounding comes from this "instant crediting" of interest, generating more interest from the start.

For "present value under continuous compounding," we simply solve the continuous compounding formula $A = Pe^{rt}$ for P and rename it PV.

$$PV = Ae^{-rt}$$ Present value with continuous compounding

EXAMPLE 9   Finding Present Value with Continuous Compounding

How much should be invested now at $6\frac{1}{2}\%$ compounded continuously if $5000 is needed in 4 years?

Solution

$$PV = 5000e^{-0.065 \cdot 4} = 5000e^{-0.26} \approx 3855.26$$

$PV = Ae^{-rt}$
with $A = 5000$,
$r = 0.065$, and $t = 4$

The amount needed now is $3855.26.

Growth Times under Continuous Compounding

To find the time needed to reach a particular amount, we solve the continuous compounding formula $A = Pe^{rt}$ for t. It is easiest to use *natural* logarithms, which have the property that $\ln e^x = x$ (see page 85).

EXAMPLE 10   Finding the Doubling Time with Continuous Compounding

What is the doubling time of a loan at 6% compounded continuously?

Solution

We solve:

$$2P = Pe^{0.06t}$$ $A = Pe^{rt}$ with A equal to $2P$

$$2 = e^{0.06t}$$ Canceling Ps

$$\ln 2 = \ln e^{0.06t}$$ Taking natural logs of both sides

$$\ln 2 = 0.06t$$ Since $\ln e^{0.06t} = 0.06t$

$$t = \frac{\ln 2}{0.06} \approx 11.6$$ Dividing by 0.06, reversing sides, and calculating using the $\boxed{\text{LN}}$ key

The term is about 11.6 years.

In Practice Problem 2 (page 121), you found that the doubling time for a similar loan but with *quarterly* compounding was $11\frac{3}{4}$ years. Do you understand why continuous compounding should give a slightly shorter doubling time?

Hint: Remember that for continuous compounding there is no "rounding to the next period," since there is no "period"—the term is simply an amount of years. How do you know when to use which formula? Use $A = Pe^{rt}$ when the compounding is *continuous* and $A = P(1 + r/m)^{mt}$ (the "discrete" formula) in all other cases.

Other types of continuous growth problems can be done just as Examples 3 and 4 (pages 119–121), but using the "continuous" formula $A = Pe^{rt}$ and then taking *natural* logarithms. That is, if a loan with continuous compounding is to increase by 75%, then we would solve $1.75P = Pe^{rt}$ for t. If specific dollar amounts are given, we would substitute them for P and A and then solve for t.

Effective Rates for Continuous Compounding

For *continuous* compounding, the effective rate is found from:

$$P(1 + r_e) = Pe^r$$

r_e is the "simple" rate that gives the continuous compound rate for $t = 1$

Canceling the Ps and subtracting 1 gives:

$$r_e = e^r - 1$$

Effective rate for continuous compounding at rate r

EXAMPLE 11   Finding the Effective Rate with Continuous Compounding

What is the effective rate of a loan at 6.2% compounded continuously?

Solution

$$r_e = e^{0.062} - 1 \approx 0.0640$$

$r_e = e^r - 1$ with $r = 0.062$

The effective rate is 6.40%. That is, 6.2% compounded continuously results in an annual gain of 6.4%. ∎

SUMMARY

The *compound interest formula* is

$$A = P\left(1 + \frac{r}{m}\right)^{mt}$$

A = amount due
P = principal
r = interest rate
m = compoundings per year
t = term in years

A is the *future value* of P while P is the *present value* of A. Given values for four of the variables, we can solve for the fifth. Remember, however, to round the term *up* to the next whole number of compounding periods. The *doubling time,* the amount of time for an amount to double ($A = 2P$), can be approximated by the *rule of 72*: $t \approx 72/(r \times 100)$ years.

The *effective rate* or *APR* of a loan is

$$r_e = \left(1 + \frac{r}{m}\right)^m - 1$$

$r_e =$ effective rate of interest at rate r compounded m times each year

The effective rate of return for *any* investment of P dollars that returns A dollars after t years (such as a zero coupon bond) is found from the compound interest formula with $m = 1$, $A = P(1 + r_e)^t$.

For continuous compounding, the formula is

$$A = Pe^{rt}$$

$A =$ amount
$r =$ rate
$t =$ years

For the present value under continuous compounding we solve this formula for P, whereas for the term we solve for t using natural logarithms. The effective rate for nominal rate r compounded continuously is $r_e = e^r - 1$.

EXERCISES 1.2 (Most require 🖩 or 🖩.)

Find the amount due on each compound interest loan.

1. $1500 at 7% compounded annually for 10 years.

2. $2000 at 9% compounded annually for 7 years.

3. $6000 at 6.5% compounded semiannually for 8 years.

4. $4500 at 4.25% compounded semiannually for 9 years.

5. $825 at 6.58% compounded quarterly for 5 years 6 months.

6. $950 at 5.87% compounded quarterly for 6 years 3 months.

7. $1280 at 4.8% compounded monthly for 3 years.

8. $5275 at 5.3% compounded monthly for 2 years.

9. $1280 at 4.8% compounded weekly for 6 years.

10. $5275 at 5.3% compounded weekly for 2 years.

Find the present value of each compound interest loan.

11. $15,000 after 10 years at 7% compounded annually.

12. $20,000 after 7 years at 9% compounded annually.

13. $3000 after 8 years at 6% compounded semiannually.

14. $4000 after 9 years at 7% compounded semiannually.

15. $17,500 after 6 years 9 months at 8.5% compounded quarterly.

16. $19,200 after 5 years 3 months at 7.9% compounded quarterly.

17. $179,000 after 25 years at 11.2% compounded monthly.

18. $153,000 after 30 years at 10.9% compounded monthly.

19. $21,500 after 4 years at 12.3% compounded weekly.

20. $20,700 after 3 years at 11.8% compounded weekly.

Find the term of each compound interest loan.

21. 8.2% compounded quarterly to obtain $8400 from a principal of $2000.

22. 4.7% compounded annually to obtain $18,750 from a principal of $5000.

23. 10.28% compounded monthly to obtain $32,130 from a principal of $6300.

24. 5.48% compounded monthly to multiply the principal by 1.50.

25. 6.8% compounded quarterly to multiply the principal by 1.80.

26. 3.58% compounded monthly to multiply the principal by 1.75.

27. 8.5% compounded monthly to increase the principal by 65%.

28. 7.2% compounded quarterly to increase the principal by 40%.

29. 9.3% compounded annually to increase the principal by 55%.

30. 7.65% compounded monthly to triple the principal.

Use the "rule of 72" to estimate the doubling time (in years) for each interest rate and then calculate it exactly.

31. 9% compounded annually.

32. 8% compounded annually.

33. 6% compounded quarterly.

34. 12% compounded quarterly.

35. 7.9% compounded semiannually.

36. 8.1% compounded semiannually.

37. 2.05% compounded monthly.

38. 2.9% compounded monthly.

39. 5.9% compounded weekly.

40. 4.1% compounded weekly.

Find the effective rate of each compound interest rate or investment.

41. 4.3% compounded weekly.

42. 7.9% compounded biweekly.

43. 8.57% compounded monthly.

44. 11.2% compounded semiannually.

45. 9.8% compounded quarterly.

46. 1.5% compounded monthly (the typical credit card interest rate!).

47. A $50,000 zero coupon bond maturing in 8 years and selling now for $23,500.

48. A $20,000 zero coupon bond maturing in 19 years and selling now for $8600.

49. A $10,000 zero coupon bond maturing in 12 years and selling now for $6400.

50. A $5,000,000 zero coupon bond maturing in 27 years and selling now for $1,675,000.

Find the amount due on each continuously compounded loan. If you did the corresponding problem in Exercises 1–10, compare your answers.

51. $1500 at 7% compounded continuously for 10 years.

52. $2000 at 9% compounded continuously for 7 years.

53. $6000 at 6.5% compounded continuously for 8 years.

54. $4500 at 4.25% compounded continuously for 9 years.

55. $825 at 6.58% compounded continuously for 5 years 6 months.

56. $950 at 5.87% compounded continuously for 6 years 3 months.

57. $1280 at 4.8% compounded continuously for 3 years.

58. $5275 at 5.3% compounded continuously for 2 years.

59. $1280 at 4.8% compounded continuously for 6 years.

60. $5275 at 5.3% compounded continuously for 2 years.

Find the present value of each continuously compounded loan. If you did the corresponding problem in Exercises 11–20, compare your answers.

61. $15,000 after 10 years at 7% compounded continuously.

62. $20,000 after 7 years at 9% compounded continuously.

63. $3000 after 8 years at 6% compounded continuously.

64. $4000 after 9 years at 7% compounded continuously.

65. $17,500 after 6 years 9 months at 8.5% compounded continuously.

66. $19,200 after 5 years 3 months at 7.9% compounded continuously.

67. $179,000 after 25 years at 11.2% compounded continuously.

68. $153,000 after 30 years at 10.9% compounded continuously.

69. $21,500 after 4 years at 12.3% compounded continuously.

70. $20,700 after 3 years at 11.8% compounded continuously.

Find the term of each continuously compounded loan. If you did the corresponding problem in Exercises 21–30, compare your answers.

71. 8.2% compounded continuously to obtain $8400 from a principal of $2000.

72. 4.7% compounded continuously to obtain $18,750 from a principal of $5000.

73. 10.28% compounded continuously to obtain $32,130 from a principal of $6300.

74. 5.48% compounded continuously to multiply the principal by 1.50.

75. 6.8% compounded continuously to multiply the principal by 1.80.

76. 3.58% compounded continuously to multiply the principal by 1.75.

77. 8.5% compounded continuously to increase the principal by 65%.

78. 7.2% compounded continuously to increase the principal by 40%.

79. 9.3% compounded continuously to increase the principal by 55%.

80. 7.65% compounded continuously to triple the principal.

Find the effective rate of each continuously compounded rate. If you did the corresponding problem in Exercises 41–46, compare your answers.

81. 4.3% compounded continuously.

82. 7.9% compounded continuously.

83. 8.57% compounded continuously.

84. 11.2% compounded continuously.

85. 9.8% compounded continuously.

86. 18% compounded continuously (the typical credit card nominal rate!).

87. 28.8% compounded continuously.

88. 69.2% compounded continuously.

89. 69.5% compounded continuously.

90. 100% compounded continuously.

91. Show that after two years the (yearly) compound interest amount $A = P(1 + r)^2$ exceeds the simple interest amount $A = P(1 + 2r)$ by $P(r^2)$.

92. Show that after three years the (yearly) compound interest amount $A = P(1 + r)^3$ exceeds the simple interest amount $A = P(1 + 3r)$ by $P(3r^2 + r^3)$.

93. Solve any five of Exercises 1–10 as follows. Enter the compound interest formula as $y = P(1 + r/m)^{mt}$. Then for each problem, STORE the values for P, r, m, and t, and evaluate y to find the amount due.

94. Solve any five of Exercises 11–20 as follows. Enter the present value formula as $y = A(1 + r/m)^{-mt}$. Then for each problem, STORE the values for A, r, m, and t, and evaluate y to find the present value.

95. Solve the compound interest formula (page 118) for the number of compounding periods mt and show that the result is $mt = \log (A/P)/\log (1 + r/m)$. Then show that if this number is not an integer, rounding it *up* is the same as using the formula $n = \lfloor \log (A/P)/\log (1 + r/m) \rfloor + 1$ where $\lfloor x \rfloor$ denotes the greatest integer less than or equal to the number x.

96. The formula for the number y of compounding periods from Exercise 95 can be viewed on your graphing calculator as a function of the interest rate x by entering it as $y = \text{int}(\log (A/P)/\log (1 + x/m)) + 1$. STORE the values 2400 for A, 2000 for P, and 12 for m (these are the numbers from Example 3 on page 119), and then graph y on the window $[0, .1]$ by $[0, 100]$. Does this curve have the same shape as the curve in the "rule of 72" Graphing Calculator Exploration on pages 121–122? Use TRACE to explore the y values for various values of x. Is y always a whole number? Use VALUE to find the number of compounding periods when the interest rate is 6%. Does this agree with the answer to Example 3?

97. Solve Exercises 21–23 as follows. Enter the formula for the number of compounding periods from Exercise 95 as $y = \text{int}(\log (A/P)/\log (1 + x/m)) + 1$. Then for each problem, STORE the values for A, P, r, and m, and evaluate y to find the number of compounding periods. Can you modify the expression for y to solve Exercises 24–30 in a similar manner?

98. Show that $P(1 + r/m)^{mt} = P(1 + r_e)^t$.

99. Show that the *effective m-compound rate* r_m is $r_m = m((A/P)^{1/mt} - 1)$ for an investment compounded m times per year and returning A dollars from P dollars after t years. Then check this formula by showing that the rate r_m compounded m times for 1 year gives the effective rate $r_e = (A/P)^{1/t} - 1$.

100. This problem continues the "rule of 72" Graphing Calculator Exploration on pages 121–122. Use the same $[0, .5]$ by $[0, 20]$ window for all your graphs.

a. For interest compounded quarterly instead of monthly, graph the curve $y_1 = \log (2)/(4 \log (1 + x/4))$ with the curve $y_2 = 72/(x \times 100)$. Do these two curves still closely match?

b. For interest compounded weekly instead of monthly, graph the curve $y_1 = \log (2)/(52 \log (1 + x/52))$ with the curve $y_2 = 72/(x \times 100)$. Do these two curves still closely match?

c. Using your monthly, quarterly, or weekly compounding curve, replace the curve $y_2 = 72/(x \times 100)$ with the curve $y_2 = 70/(x \times 100)$. Try it again with the curve $y_2 = 71/(x \times 100)$. Do these curves still closely match the compound interest curve? Which "rule" is easiest to use in mental calculations: 70, 71, or 72?

d. Using your monthly, quarterly, or weekly compounding curve, replace the curve $y_2 = 72/(x \times 100)$ with the curves $y_2 = 60/(x \times 100)$ and $y_3 = 80/(x \times 100)$. Is 60 much too small and 80 much too big?

101. Suppose that you found a bank that offered 100% interest compounded continuously. Being duly cautious, you decided to "test" it with a $1 deposit for 1 year. Show that your deposit would be worth exactly e dollars after 1 year.

102. Find the value of $100 at 20% compounded quarterly for the following terms: 1 year, 5 years, 25 years. Repeat these calculations for the same amount, rate, and terms but with *continuous* compounding. Is continuous compounding significantly better over 1 year? Over 25 years?

103. Show that the doubling time for interest rate r compounded continuously is $\frac{\ln 2}{r}$.

104. Show that the time it takes for an amount at interest rate r compounded continuously to be multiplied by a factor of m is $\frac{\ln m}{r}$.

105. Solve any five of Exercises 51–60 as follows. Enter the continuous compound interest formula as $y = Pe^{rt}$. Then for each problem, STORE the values for P, r, and t, and evaluate y to find the amount due.

106. Mutual Funds   During the first half of the 1990s, the Twentieth Century Ultra aggressive growth mutual fund returned 19.83% compounded quarterly. How much would a $10,000 investment in this fund have been worth after 5 years?

107. Bond Funds   During the mid 1990s, the T. Rowe Price International Bond fund returned 10.43% compounded monthly. How much would a $5000 investment in this fund have been worth after 3 years?

108. College Savings   How much would your grandparents have needed to set aside 18 years ago at 5.6% compounded quarterly to give you $45,000 for college expenses today?

109. College Savings   How much would your parents have needed to set aside 17 years ago at 7.3% compounded weekly to give you $50,000 for college expenses today?

110. College Savings   How much would your parents have needed to set aside 16 years ago at 6.7% compounded continuously to give you $60,000 for college expenses today?

111. Bond Funds   You have just received $125,000 from the estate of a long-lost rich uncle. If you invest all of your inheritance in a tax-free bond fund earning 6.9% compounded quarterly, how long do you have to wait to become a millionaire?

112. Home Buying   You and your new spouse have decided to use all $6500 of your wedding present monies as a nest egg for a house down payment. Investing at 11.47% compounded monthly, how long must you wait to have enough to put 10% down on a $140,000 house in the suburbs?

113. Mutual Funds   A $10,000 investment in the Fidelity Blue Chip Growth mutual fund in 1987 would have been worth $30,832 seven years later. What was the effective rate of this investment?

114. Rate Comparisons   The First Federal Bank offers 4.7% compounded weekly passbook savings accounts while Consolidated Nationwide Savings offers 4.73% compounded quarterly. Which bank offers the higher rate?

115. Rate Comparisons   The Second Peoples National Bank offers a long-term certificate of deposit earning 6.43% compounded monthly. Your broker locates a $20,000 zero coupon bond rated AA by Standard & Poor's for $7965 and maturing in 14 years. Which investment will give the greater rate of return?

116. Horse Trading   In March 1995, a descendant of an early Texas settler sent $100 to Sam Houston IV to make good on a $100 debt (possibly from the sale of a horse) owed for 160 years to Sam Houston, the hero of Texas independence. The check was donated to the Sam Houston Museum and the debt considered as settled. However, newspapers reported that a banker had estimated the accumulated interest on the loan to be $420 million. What yearly interest rate did the banker use in her calculations?

117. Lottery Winnings   You have just won $100,000 from a lottery. If you invest all this amount in a tax-free money market fund earning 7% compounded continuously, how long do you have to wait to become a millionaire?

118. Becoming a Millionaire   How much would you have to invest now at 6.5% interest compounded continuously to have a million dollars in 40 years?

119. Rate Comparisons   The People's State Bank offers 4.2% compounded quarterly while Statewide Federal offers 4.1% compounded continuously. Which bank offers the better rate?

120. Rate Comparisons   The Southwestern Savings and Loan Bank offers a 10-year certificate of deposit earning 6.4% compounded continuously. Your broker offers you a $10,000 zero coupon bond costing $5325 and maturing in 10 years. Which gives the greater rate of return?

1.3 Annuities

Retirement Planning

Although buying your first car and your first house will give you the most immediate satisfaction when you take possession, planning and saving for your retirement will extend over more years and involve more money than perhaps anything else. Most financial planners will tell you that the amount you set aside is less important than the fact that you are saving *something* regularly, and the earlier you start, the greater the results. An individual retirement account or IRA allows almost anyone to set aside up to $2000 each year to earn interest tax free until retirement age. The power of compound interest is so great that if you fund an IRA *only* during your twenties, you can end up with more money than if you wait until your thirties and then invest every year until retirement.

But where should you invest your IRA money? From 1926 to 1995, the average annual return was 3.72% on treasury bills, 5.69% on long-term bonds, and 10.54% on common stocks. Although stocks are the most volatile, they do have the most growth potential over time, and investment in a mutual fund can spread your money over a wide variety of companies. Some mutual funds charge annual fees for IRA accounts while others do not. Can a $30 annual fee make much difference over the 40 years before you retire? The graph shows the difference between the values of two IRA accounts earning 10% with annual $2000 deposits when one charges a $30 annual fee and the other does not—after 40 years, the difference is $13,278.

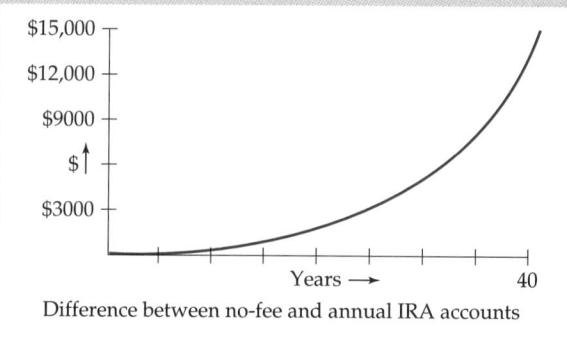

Difference between no-fee and annual IRA accounts

Such accounts are called *annuities*, and are the subject of this section.

Introduction

An *annuity* is a scheduled sequence of payments. Some annuities, such as retirement pensions, provide money to you while others, such as car loans and home mortgages, require that you make regular payments. In this section we will examine the relation between the payments and the value of an *ordinary annuity*, which is an annuity with equal payments at regular intervals, a fixed interest rate, and the interest compounded at the end of each payment period. For example, a car loan for 4 years with monthly payments of $200 at 6% compounded monthly is an ordinary annuity.

While having a new car is nice, when the first payment comes due at the end of the first month you might wish that instead you were saving that $200 for yourself. If you were, then 4 years later your first $200 would have grown to $200(1 + 0.06/12)^{47} (the exponent is 47 rather than 48 because the first payment was made at the *end* of the first month, and so earns interest for only 47 months). Similarly, your second $200 would grow to $200(1 + 0.06/12)^{46}, your third to $200(1 + 0.06/12)^{45}, and so on. The total amount you could have saved would be the sum of the 48 payments with the interest earned on each one:

$$200 \left(1 + \frac{0.06}{12} \right)^{47} + 200 \left(1 + \frac{0.06}{12} \right)^{46} + 200 \left(1 + \frac{0.06}{12} \right)^{45}$$

$$+ \cdots + 200 \left(1 + \frac{0.06}{12} \right)^{2} + 200 \left(1 + \frac{0.06}{12} \right)^{1} + 200$$

Notice that your last $200 earns no interest since you pay it at the end of the 4 years. With patience and careful attention to detail, you could add up the 48 values to find this sum. However, there is an easier way.

Geometric Series

The above sum is an example of a *geometric series*, where each value in the sum is a fixed multiple of the previous value. The simplest geometric series is of the form:

$$x^{n-1} + x^{n-2} + x^{n-3} + \cdots + x^2 + x + 1$$

If we multiply this sum by $x - 1$ we obtain:

$$(x^{n-1} + x^{n-2} + x^{n-3} + \cdots + x^2 + x + 1)(x - 1)$$
$$= x^{n-1}(x - 1) + x^{n-2}(x - 1) + x^{n-3}(x - 1)$$
$$+ \cdots + x^2(x - 1) + x^1(x - 1) + 1(x - 1)$$

Multiplying each by $x - 1$

$$= x^n - x^{n-1} + x^{n-1} - x^{n-2} + x^{n-2} - x^{n-3} + \cdots$$

Cancel   Cancel   Cancels with
next term

All cancel
except $x^n - 1$

$$\cdots + x^3 - x^2 + x^2 - x + x - 1$$

Cancels   Cancel   Cancel
with
previous
term

$$= x^n - 1$$

Setting the first and last expressions equal to each other, we have:

$$(x^{n-1} + x^{n-2} + x^{n-3} + \cdots + x^2 + x + 1)(x - 1) = x^n - 1$$

or

$$x^{n-1} + x^{n-2} + x^{n-3} + \cdots + x^2 + x + 1 = \frac{x^n - 1}{x - 1} \qquad \text{Dividing by } x - 1$$

Multiplying both sides by a gives a formula for the sum of a geometric series:

$$ax^{n-1} + ax^{n-2} + ax^{n-3} + \cdots + ax^2 + ax + a = a\,\frac{x^n - 1}{x - 1}$$

For instance, the geometric series $48 + 24 + 12 + 6 + 3$, in which $a = 3$, $x = 2$, and $n = 5$, has the value $3(2^5 - 1)/(2 - 1) = 93$, as you should check by adding up the five numbers. Applying this formula to our 4-year savings account, we find that:

$$200\left(1 + \frac{0.06}{12}\right)^{47} + 200\left(1 + \frac{0.06}{12}\right)^{46} + \cdots$$

$$+ 200\left(1 + \frac{0.06}{12}\right)^{2} + 200\left(1 + \frac{0.06}{12}\right)^{1} + 200$$

$$= 200\,\frac{\left(1 + \dfrac{0.06}{12}\right)^{48} - 1}{\left(1 + \dfrac{0.06}{12}\right) - 1} \approx 10{,}819.57 \qquad a\,\frac{x^n - 1}{x - 1}\text{ with } a = 200,\; x = (1 + 0.06/12),\text{ and } n = 48$$

Therefore, regular $200 deposits at the end of each month for 4 years into a savings account earning 6% interest compounded monthly will total $10,819.57. (This is significantly more than simply adding the 48 payments of $200, which total only $9600.)

Accumulated Amount Formula

Applying the same method to any annuity with regular payments of P dollars repeated m times per year for t years with the interest rate r compounded at each payment, the final accumulated amount A will be:

$$A = P\left(1 + \frac{r}{m}\right)^{mt-1} + P\left(1 + \frac{r}{m}\right)^{mt-2} + \cdots$$

$$+ P\left(1 + \frac{r}{m}\right)^{1} + P = P\frac{\left(1 + \frac{r}{m}\right)^{mt} - 1}{\left(1 + \frac{r}{m}\right) - 1}$$

Canceling the 1 and the -1 in the denominator, we have the following formula.

Accumulated Amount of an Annuity

> The accumulated amount A of an ordinary annuity with payments of P dollars m times per year for t years at interest rate r compounded at the end of each payment period is
>
> $$A = P\frac{\left(1 + \frac{r}{m}\right)^{mt} - 1}{\frac{r}{m}}$$

While this formula is simple to evaluate using a calculator, compound interest and annuity calculations have been common features of finance for many centuries. Writing i for r/m and n for mt, the archaic symbol $s_{n|i}$ (pronounced "s sub n angle i") is sometimes still used to denote the value of $((1 + i)^n - 1)/i$ in business textbooks and tables. Using this notation, the above formula takes the form $A = Ps_{n|i}$ with $n = mt$ and $i = r/m$.

EXAMPLE 1 Finding the Accumulated Amount of an Annuity

What is the final balance of a retirement account earning 4% interest compounded weekly if $50 is deposited at the end of every week for 40 years? How much of this final balance is interest?

Solution

$$A = 50 \, \frac{\left(1 + \dfrac{0.04}{52}\right)^{(52)(40)} - 1}{\dfrac{0.04}{52}} \approx 256{,}749.15$$

$A = P \dfrac{(1 + r/m)^{mt} - 1}{r/m}$
with $P = 50$, $r = 0.04$,
$m = 52$, and $t = 40$

The final balance is \$256,749.15. The deposits total only \$50 × 52 × 40 = \$104,000, so the account has earned \$152,749.15 in interest during the 40 years. ∎

PRACTICE PROBLEM 1

What is the final balance of a retirement account earning 5% interest compounded weekly if \$40 is deposited at the end of every week for 35 years? How much of this final balance is interest?

Solution at the back of the book

Sinking Funds

A *sinking fund* is a regular savings plan designed to provide a given amount after a certain number of years. For example, suppose you decide to save up for a \$18,000 car instead of taking out a loan. What amount should you save each month to have the \$18,000 in 3 years? If you just put your money in a shoe box, you will need to set aside \$18,000 ÷ (12 × 3) = \$500 each month. It would make more sense to put your money each month into a savings account and let it earn compound interest. This is the same as setting up an annuity with a given accumulated amount and asking what the regular payment should be. Solving the Accumulated Amount formula (page 137) for P by dividing gives the following formula.

Sinking Fund Payment

The payment P to make m times per year for t years at interest rate r compounded at each payment to accumulate amount A is

$$P = A \, \frac{\dfrac{r}{m}}{\left(1 + \dfrac{r}{m}\right)^{mt} - 1}$$

This formula may also be written $P = A/s_{\overline{n}|i}$ with $n = mt$ and $i = r/m$.

EXAMPLE 2 Finding a Sinking Fund Payment

What amount should be deposited at the end of each month for 3 years in a savings account earning 4.5% interest compounded monthly to accumulate $18,000 to buy a new car?

Solution

$$P = 18{,}000 \, \frac{\dfrac{0.045}{12}}{\left(1 + \dfrac{0.045}{12}\right)^{(12)(3)} - 1} \approx 467.945$$

$P = A \dfrac{r/m}{(1 + r/m)^{mt} - 1}$
with $A = 18{,}000$,
$r = 0.045$, $m = 12$,
and $t = 3$

Rounding up to the next penny, $467.95 should be saved each month. (If we rounded down to $467.94, the savings account would accumulate slightly less than the amount needed. Notice that this amount is significantly lower than the $500 that would be needed *without* compound interest.) ∎

This example could also have been done using the Accumulated Amount formula (page 137) by substituting $A = 18{,}000$, $r = 0.045$, $m = 12$, and $t = 3$, and then solving for the remaining variable to find $P = \$467.95$.

PRACTICE PROBLEM 2

For wage earners paid every other week instead of monthly, Example 2 becomes: "What amount should be deposited biweekly for 3 years in a savings account earning 4.5% interest compounded biweekly to accumulate $18,000 for a new car?" Find this amount.

Solution at the back of the book

How Long Will It Take?

Continuing with our sinking fund example, suppose you could save only $200 each month. How long would it take to accumulate the $18,000?

We solve the Accumulated Amount formula (page 137) for the number of years t.

$$\frac{A}{P}\frac{r}{m} + 1 = \left(1 + \frac{r}{m}\right)^{mt}$$

$A = P \dfrac{(1 + r/m)^{mt} - 1}{r/m}$
Dividing by P, multiplying by r/m, and adding 1

$$mt \log\left(1 + \frac{r}{m}\right) = \log\left(\frac{A}{P}\frac{r}{m} + 1\right)$$

Switching sides and taking logs of both to bring down the exponent

$$mt = \frac{\log \left(\dfrac{A}{P} \dfrac{r}{m} + 1 \right)}{\log \left(1 + \dfrac{r}{m} \right)} \text{ periods} \qquad \begin{array}{l}\text{Dividing by} \\ \log (1 + r/m)\end{array}$$

As usual, we round *up* to the next whole number of compounding periods and then express the answer as years plus any extra periods. Notice that the time t depends only on the ratio $\frac{A}{P}$ between the accumulated amount A and the regular payment P.

EXAMPLE 3   Finding the Term for a Sinking Fund

How long will it take to accumulate $18,000 by depositing $200 at the end of each month in a savings account earning 4.5% interest compounded monthly?

Solution

$$12t = \frac{\log \left(\dfrac{18{,}000}{200} \dfrac{0.045}{12} + 1 \right)}{\log \left(1 + \dfrac{0.045}{12} \right)} \approx 77.7 \qquad \begin{array}{l} mt = \dfrac{\log \left(\dfrac{A}{P} \dfrac{r}{m} + 1 \right)}{\log \left(1 + \dfrac{r}{m} \right)} \\[4pt] \text{with } A = 18{,}000, P = 200, \\ r = 0.045, \text{ and } m = 12 \end{array}$$

The time is 77.7 months, which we round *up* to 78 months. Therefore, it will take 6 years 6 months to accumulate the $18,000.   ∎

This example could also have been done using the Accumulated Amount formula from page 137 by substituting $A = 18{,}000$, $P = 200$, $r = 0.045$, and $m = 12$, and then solving (using logarithms) for the remaining variable to find $12t = 78$ months.

PRACTICE PROBLEM 3   How long will it take to accumulate $18,000 by depositing $250 each month in a savings account earning 4.5% interest compounded monthly?   *Solution at the back of the book*

Graphing Calculator Exploration

Depending on the size of the monthly deposit, the time needed to accumulate a given amount may vary greatly.

a. Using the values $r = 0.045$, $m = 12$, and $A = 18,000$ from Example 3, and replacing the payment amount P by x, graph the curve $y_1 = \log((18000/x)(.045/12) + 1)/\log(1 + .045/12)$ on the window $[0, 1000]$ by $[0, 240]$. The graph shows how y, the number of months needed to accumulate \$18,000, decreases as the monthly payment x increases.

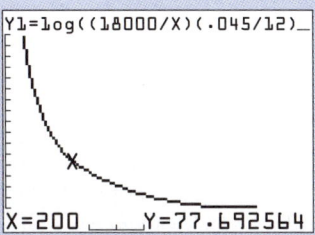

b. Use TRACE or VALUE to find y when $x = 200$ and when $x = 250$. How do your answers compare to the answers to Example 3 and Practice Problem 3?

c. Use TRACE or VALUE to find the years y needed when the monthly deposits x are \$100, \$150, \$200, \$250, and \$300. Does each additional \$50 reduce the time needed by the same amount? How does the graph show this?

Internal Rate of Return

Suppose you make regular investments into a stock market fund, and after several years you want to find the "rate of return" on your money. One measure of this is the *internal rate of return*, which is the interest rate r of the annuity having the same payments, payment frequency, accumulated amount, and term. That is, in the Accumulated Amount formula from page 137:

$$A = P \frac{\left(1 + \dfrac{r}{m}\right)^{mt} - 1}{\dfrac{r}{m}}$$

we know the values of P, m, t, and A, and we want to solve for the interest rate r. Unfortunately, there is no simple algebraic solution of this problem. However, the solution can be approximated using a graphing calculator by replacing r by x and graphing $y = P((1 + x/m)^{mt} - 1)/(x/m)$ using the known values of P, m, and t. You can then search the graph for an interest rate x giving a y-value near the given amount A. (Other more tedious methods include searching a $s_{\overline{n}|i}$ table for values near $\frac{A}{P}$ or by trial and error using guesses for the interest rate in the annuity formula.)

 Graphing Calculator Exploration

What is the internal rate of return of $500 quarterly investments in a mutual fund account valued at $27,660 after 8 years?

Solution

We have the values $P = 500$, $m = 4$, $t = 8$, and $A = 27,660$. The exponent mt is $(4)(8) = 32$.

a. Graph the curve $y = 500((1 + x/4)^{32} - 1)/(x/4)$ on the window [0, .2] by [0, 40000] to show the relation between the amount y and the internal rate of return x.

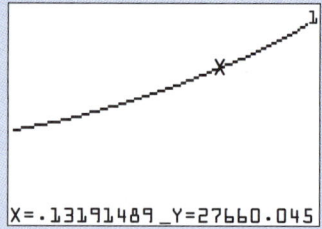

X=.13191489_Y=27660.045

b. Use TRACE or INTERSECT to estimate the value for x that corresponds to a y-value of 27,660.

The internal rate of return of this mutual fund is 13.19% over the last 8 years. You can verify this estimate by checking that this value for r with the given values for P, m, and t in the annuity formula gives a value for A near the actual value of $27,660.

Overview

In the previous section we found the eventual value of an investment, the required initial amount to reach that later amount, how much time it takes, and the internal rate of return, just as we did in this section. What's the difference between the two sections? *There* we were considering investments of a *single* payment of P dollars, whereas *here* we are considering *multiple and regular payments* of P dollars.

SUMMARY

For an *ordinary annuity* with regular payments of P dollars m times per year for t years at interest rate r compounded at the end of each payment, the accumulated amount A is:

$$A = P\,\dfrac{\left(1 + \dfrac{r}{m}\right)^{mt} - 1}{\dfrac{r}{m}}$$

A = accumulated amount,
P = payment, r = interest rate,
m = times per year, t = years

This formula may be solved for the other variables. Solving for P finds the regular payment to make into a *sinking fund* to accumulate the amount A after t years:

$$P = A\,\dfrac{\dfrac{r}{m}}{\left(1 + \dfrac{r}{m}\right)^{mt} - 1}$$

Solving for t finds the number of years required for the payments into a sinking fund to accumulate to a final amount A:

$$mt = \dfrac{\log\left(\dfrac{A}{P}\dfrac{r}{m} + 1\right)}{\log\left(1 + \dfrac{r}{m}\right)}$$

mt = number of periods
(other variables as above)

Round *up* to find the number of payments and then express your answer as a number of years plus any additional periods.

There is no simple algebraic way to solve for the interest rate r. However, the *internal rate of return* of an investment with known regular payments and accumulated amount can be approximated using a graphing calculator.

EXERCISES 1.3 (Most exercises need  or .)

In the following ordinary annuities, the interest is compounded with each payment and the payment is made at the end of the compounding period.

Find the accumulated amount of each annuity.

1. $1500 annually at 7% for 10 years.

2. $1750 annually at 8% for 7 years.

3. $2000 semiannually at 6% for 12 years.

4. $2500 semiannually at 7% for 10 years.

5. $950 quarterly at 10.7% for 6 years.

6. $750 quarterly at 8.7% for 11 years.

7. $1000 monthly at 6.9% for 20 years.

8. $1100 monthly at 7% for 10 years.

9. $200 weekly at 11.3% for 5 years.

10. $175 weekly at 4.5% for 18 years.

Find the required payment for each sinking fund.

11. Monthly deposits earning 5% to accumulate $5000 after 10 years.

12. Monthly deposits earning 4% to accumulate $6000 after 15 years.

13. Monthly deposits earning 11.7% to accumulate $14,000 after 5 years.

14. Monthly deposits earning 13.9% to accumulate $9500 after 8 years.

15. Quarterly deposits earning 6.4% to accumulate $50,000 after 20 years.

16. Quarterly deposits earning 5.9% to accumulate $45,000 after 25 years.

17. Yearly deposits earning 12.3% to accumulate $8500 after 12 years.

18. Yearly deposits earning 10.2% to accumulate $9750 after 11 years.

19. Weekly deposits earning 9.8% to accumulate $15,000 after 6 years.

20. Weekly deposits earning 7.7% to accumulate $11,500 after 7 years.

Find the amount of time needed for each sinking fund to reach the given accumulated amount.

21. $1500 yearly at 8% to accumulate $100,000.

22. $2000 yearly at 9% to accumulate $125,000.

23. $800 semiannually at 6.6% to accumulate $80,000.

24. $1000 semiannually at 6.4% to accumulate $75,000.

25. $500 quarterly at 8.2% to accumulate $19,500.

26. $350 monthly at 7.9% to accumulate $21,750.

27. $235 monthly at 5.9% to accumulate $25,000.

28. $190 monthly at 7.2% to accumulate $20,000.

29. $70 weekly at 10.9% to accumulate $7500.

30. $65 weekly at 10.8% to accumulate $18,100.

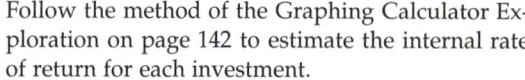 Follow the method of the Graphing Calculator Exploration on page 142 to estimate the internal rate of return for each investment.

31. $350 quarterly to accumulate $21,600 after 10 years.

32. $300 quarterly to accumulate $16,940 after 9 years.

33. $50 monthly to accumulate $5270 after 6 years.

34. $100 monthly to accumulate $8080 after 5 years.

35. $180 monthly to accumulate $52,800 after 14 years.

36. $225 monthly to accumulate $134,525 after 20 years.

37. $10 weekly to accumulate $9572 after 12 years.

38. $25 weekly to accumulate $13,000 after 8 years.

39. $1.50 daily to accumulate $106,795 after 50 years.

40. $0.75 daily to accumulate $52,500 after 45 years.

41. Solve any five of Exercises 1–10 as follows. Enter the accumulated amount formula as $y = P((1 + r/m)^{mt} - 1)/(r/m)$. Then for each exercise, STORE the values for P, r, m, and t, and evaluate y to find the amount due.

42. Solve any five of Exercises 11–20 as follows. Enter the sinking fund payment formula as $y = A(r/m)/((1 + r/m)^{mt} - 1)$. Then for each problem, STORE the values for A, r, m, and t, and evaluate y to find the payment.

43. Show that if the number of deposits (or equivalently, the number of periods) $mt = \log ((A/P)(r/m) + 1))/(\log (1 + r/m))$ from page 140 is not a whole number, then rounding up is the same as using the formula $n = \lfloor \log ((A/P)(r/m) + 1)/\log (1 + r/m) \rfloor + 1$ where $\lfloor x \rfloor$ denotes the greatest integer less than or equal to the number x.

44. The formula for the number y of deposits from Exercise 43 can be viewed on your graphing calculator as a function of the deposit x by entering it as $y = \text{int}(\log ((A/x)(r/m) + 1)/\log (1 + r/m)) + 1$. STORE the values 18,000 for A, 0.045 for r, and 12 for m (these are the numbers from Example 3 on page 140), and then graph y on the window [0, 1000] by [0, 500]. Does this curve have the same shape as the curve in the Graphing Calculator Exploration on pages 140–141 after Practice Problem 3? Use TRACE to explore the y values for various values of x. Is y always a whole number? Use VALUE to find the number y of deposits when the deposit x is 200. Does this agree with the answer to Example 3?

 45. Solve any five of Exercises 21–30 as follows. Enter the formula for the number of deposits from Exercise 43 as $y = \text{int}(\log((A/P)(r/m)+1)/\log(1+r/m)) + 1$. For each exercise, STORE the values for A, P, r, and m, and evaluate y to find the number of deposits needed.

The following exercises provide further information about geometric series.

46. For each choice of values for a, x, and n, write out the geometric series $ax^{n-1} + ax^{n-2} + ax^{n-3} + \cdots + ax^2 + ax + a$, find the sum, and then evaluate the corresponding $a\dfrac{x^n - 1}{x - 1}$ expression to check that both have the same value.

  a. $a = 2$, $x = 3$, $n = 5$
  b. $a = 162$, $x = \frac{1}{3}$, $n = 5$. How do (a) and (b) differ?
  c. $a = 3$, $x = 10$, $n = 6$

47. Check the formula $x^{n-1} + x^{n-2} + x^{n-3} + \cdots + x^2 + x + 1 = \dfrac{x^n - 1}{x - 1}$ for the case $n = 5$ by carrying out the long division $x - 1 \overline{)x^5 - 1}$.

48. Show that

$$x^{n-1} + x^{n-2} + x^{n-3} + \cdots + x^2 + x + 1 = \frac{x^n - 1}{x - 1}$$

by *induction* as follows.

  a. Check that $(x + 1)(x - 1) = x^2 - 1$ by multiplying out the left side. Then divide by $x - 1$ to show that $x + 1 = \dfrac{x^2 - 1}{x - 1}$.

  b. Factor out an x to check that $x^2 + x + 1 = x(x + 1) + 1$. Using (a), substitute $\dfrac{x^2 - 1}{x - 1}$ for $x + 1$ on the right side, get a common denominator, and collect like terms to show that $x^2 + x + 1 = \dfrac{x^3 - 1}{x - 1}$.

  c. Factor out an x to check that $x^3 + x^2 + x + 1 = x(x^2 + x + 1) + 1$. Using (b), substitute $\dfrac{x^3 - 1}{x - 1}$ for $x^2 + x + 1$ on the right side, get a common denominator, and collect

like terms to show that $x^3 + x^2 + x + 1 = \dfrac{x^4 - 1}{x - 1}$. Do you now see how to use this method to show that $x^4 + x^3 + x^2 + x + 1 = \dfrac{x^5 - 1}{x - 1}$?

  d. We can now show that $x^{n-1} + x^{n-2} + x^{n-3} + \cdots + x^2 + x + 1 = \dfrac{x^n - 1}{x - 1}$ for every value of $n \geq 2$. Part (a) showed that this formula is true for $n = 2$, part (b) showed it true for $n = 3$ because it is true for $n = 2$, and part (c) showed that this formula is true for $n = 4$ because it is true for $n = 3$. A *proof by induction* concludes that the formula is true for every $n \geq 2$ if we know that it is true for $n = 2$ (the "starting" value) and that the formula is true for $n + 1$ if it is true for n (that is, we can always prove the "next" version of the formula from the one we just proved). To check the *induction step* to complete the proof, show that

$$\begin{aligned} x^n &+ x^{n-1} + x^{n-2} + \cdots + x^2 + x + 1 \\ &= x(x^{n-1} + x^{n-2} + x^{n-3} + \cdots \\ &\qquad\qquad + x^2 + x + 1) + 1 \end{aligned}$$

Substitute $\dfrac{x^n - 1}{x - 1}$ for $(x^{n-1} + x^{n-2} + x^{n-3} + \cdots + x^2 + x + 1)$, get a common denominator, and collect like terms to show that

$$\begin{aligned} x^n &+ x^{n-1} + x^{n-2} + \cdots \\ &+ x^2 + x + 1 = \frac{x^{n+1} - 1}{x - 1} \end{aligned}$$

  e. By the way, is the formula true for $n = 1$?

 49. *Requires a calculator with series operations.* Solve Exercise 46 as follows. Your calculator can LIST the numbers making up a geometric series as a *sequence* of values by using SEQUENCE. For the values of a, x, and n in Exercise 46a, list the numbers as the SEQUENCE $2(3^n)$ for n starting at 0, ending at 4, and increasing by 1. This list of numbers can be added using SUM. In this way, check each of the SUMs in Exercise 46. Then check that the geometric SEQUENCE $200(1 + 0.06/12)^n$ on page 136 for n from 0 to 47 SUMs to \$10,819.57.

50. *Requires a calculator with series operations.* With the values $a = 1$ and $x = \frac{1}{2}$, the geometric series formula becomes

$$\left(\frac{1}{2}\right)^{n-1} + \left(\frac{1}{2}\right)^{n-2} + \left(\frac{1}{2}\right)^{n-3} + \cdots$$

$$+ \left(\frac{1}{2}\right)^2 + \frac{1}{2} + 1 = \frac{\left(\frac{1}{2}\right)^n - 1}{\frac{1}{2} - 1}$$

a. Check this formula by evaluating both sides for $n = 5$, $n = 10$, and $n = 15$. As n gets larger, do these values get closer to 2?

b. Graph the SUM y of the first x terms of this geometric SEQUENCE on the window [0, 25] by [0, 3]. Use TRACE to explore the graph. As x gets larger, do these values get closer to 2?

An *infinite geometric series* is a sum of the form $a + ax + ax^2 + ax^3 + ax^4 + \cdots$ where the $+ \cdots$ indicates that the sum continues on to include every power of x. For $a = 1$ and $x = \frac{1}{2}$, we suspect from (a) and (b) that

$$1 + \frac{1}{2} + \left(\frac{1}{2}\right)^2 + \left(\frac{1}{2}\right)^3 + \left(\frac{1}{2}\right)^4 + \cdots = 2$$

c. Check this arithmetic: Since $\left(\frac{1}{2}\right)^n$ gets

closer to 0 as n gets larger, the geometric

series formula $\dfrac{\left(\frac{1}{2}\right)^n - 1}{\frac{1}{2} - 1}$ is close to

$$\frac{0 - 1}{\frac{1}{2} - 1} = 2 \text{ as } n \text{ gets larger.}$$

d. If $\quad 1 + \frac{1}{2} + \left(\frac{1}{2}\right)^2 + \left(\frac{1}{2}\right)^3 + \left(\frac{1}{2}\right)^4 + \cdots$
really is 2, it should behave just like 2.
Show that $\frac{1}{2}\left(1 + \frac{1}{2} + \left(\frac{1}{2}\right)^2 + \left(\frac{1}{2}\right)^3 + \right.$

$\left. \left(\frac{1}{2}\right)^4 + \cdots\right) = 1$ by multiplying out the

left side and using the fact that $1 + \frac{1}{2} +$

$\left(\frac{1}{2}\right)^2 + \left(\frac{1}{2}\right)^3 + \left(\frac{1}{2}\right)^4 + \cdots = 2.$

e. Find the value of $1 + \frac{1}{3} + \left(\frac{1}{3}\right)^2 + \left(\frac{1}{3}\right)^3 +$

$\left(\frac{1}{3}\right)^4 + \cdots.$

f. Find the value of $9 + 9\left(\frac{1}{10}\right) + 9\left(\frac{1}{10}\right)^2 +$

$9\left(\frac{1}{10}\right)^3 + 9\left(\frac{1}{10}\right)^4 + \cdots.$ Can you write
this geometric series as a decimal?

APPLIED EXERCISES

51. Retirement Savings   An individual retirement account or IRA earns tax–deferred interest and allows the owner to invest up to $2000 each year. Joe and Jill both will make IRA deposits for 30 years (from age 35 to 65) into stock mutual funds yielding 9.8%. Joe deposits $2000 once each year while Jill has $38.46 (which is 2000/52) withheld from her weekly paycheck and deposited automatically. How much will each have at age 65?

52. Retirement Savings

  a. Steve opens a retirement account yielding 9% and deposits $150 each month for the

next 30 years (from age 35 to 65). How much will he have at age 65?

  b. Sue opens her own retirement account yielding 8% and deposits $150 each month for the 10 years from age 25 to 35. She then makes no further deposits but lets the accumulated amount earn compound interest for the next 30 years. How much will she have at age 65?

  c. Who has the most money for retirement?

53. Mutual Funds   How much must you invest each month in a mutual fund yielding 13.4% compounded monthly to become a millionaire in 10 years?

54. Bond Funds How much must you invest each quarter in a bond fund yielding 9.7% compounded quarterly to become a millionaire in 20 years?

55. Lifetime Savings The Oseola McCarty Scholarship Fund at the University of Southern Mississippi was established by a $150,000 gift from an 87-year-old black woman who had dropped out of sixth grade and worked for most of her life as a washerwoman. How much must be saved each week in a bank account earning 3.9% compounded weekly to have $150,000 after 75 years?

56. Lifetime Savings If each day your fairy godmother put $1 into a savings account at 8.37% compounded daily, how much would the account be worth when you are 65 years old?

57. Home Buying You and your new spouse each bring home $1500 each month after taxes and other payroll deductions. By living frugally, you intend to live on just one paycheck and save the other in a bond fund yielding 7.86% compounded monthly. How long will it take to have enough for a 20% down payment on a $165,000 condo in the city?

58. Boat Buying How long will it take to save $16,000 for a new motorboat if you deposit $375 each month into a money market fund yielding 5.73% compounded monthly?

59. Stock Market According to *The Wall Street Journal*, the $16\frac{1}{2}$ years from February 1966 to August 1982 was perhaps the most brutal period for stock market investors since the Great Depression in the 1930s. Ibbotson Associates estimates that if you had invested $100 each month into the stocks that make up the S&P 500 during this period, by the market low at mid-1982 your portfolio would have been worth almost $32,600 (with all dividends reinvested). What would the internal rate of return have been on such an investment?

60. Technology Stocks What is the internal rate of return for $500 quarterly investments in a science and technology specialty stock fund valued at $17,375 after 6 years?

1.4 Amortization

APPLICATION PREVIEW

Loan Repayment and the Rule of 78

Many car loans include the clause that early repayment of the loan will be calculated using the Rule of 78. While most buyers expect to satisfy their loans by making the agreed monthly payments, unforeseen circumstances could make it necessary to end the loan early. The Rule of 78 is a method of dividing the remaining payments between the interest owed and the principal borrowed in order to find the amount yet to be repaid. Like the Banker's Rule, the simplicity of the Rule of 78 hides the advantage it gives the lender. For a 1-year loan with monthly payments, the Rule of 78 states that the last payment contains one part of the total interest, the second to last contains two parts of the total interest, and so on back to the first monthly payment, which then contains 12 parts of the total interest. So the interest for the year has been divided into $1 + 2 + 3 + \cdots + 11 + 12 = 78$ equal parts, which gives this rule its name. This means that the first payment pays $\frac{12}{78}$ of the interest and that $\frac{66}{78}$ remain. For a longer term, the Rule of 78 similarly sums the number of payments.

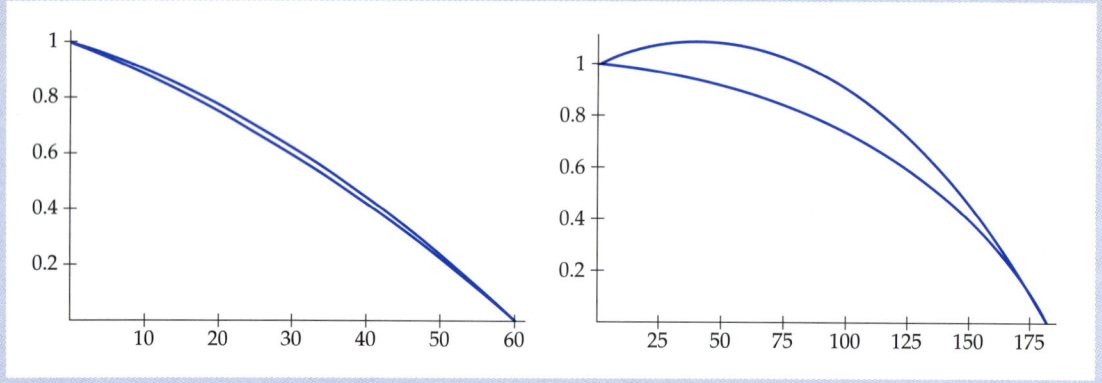

Is this fair to the borrower? The first graph shows the amount still owed on a five-year car loan at 18% calculated by both the Rule of 78 and as an annuity formed by the remaining payments. The Rule of 78 curve is significantly higher for the first two-thirds of the loan period. The second graph shows the amount still owed on a fifteen-year home improvement loan at 18%. Not only is the Rule of 78 amount larger, for the first seven years it is even larger than the amount of the loan! This "negative amortization" provides an extreme yet possible demonstration of the advantage the Rule of 78 gives the lender.

Introduction

In this section we will find the *present value of an annuity*—the total value now of all the future payments. Such calculations are important for buying and selling annuities and for using them to pay off debts (amortization). For example, on page 136 we found the accumulated amount for monthly payments of $200 for 4 years at 6% compounded monthly. But how much would a loan company give right now for that promise of future payments?

Present Value of an Annuity

We know that for an ordinary annuity with payments of P dollars m times per year for t years at interest rate r compounded at each payment, the accumulated amount A is

$$A = P \frac{\left(1 + \dfrac{r}{m}\right)^{mt} - 1}{\dfrac{r}{m}}$$

From page 137

Since the present value PV of this annuity must grow to match the accumulated amount of the annuity at the end of the t years, we must also have by the compound interest formula:

$$A = PV \left(1 + \frac{r}{m}\right)^{mt}$$

From page 118

To find the present value PV, we set these two expressions for A equal to each other and divide by $(1 + r/m)^{mt}$:

$$PV = P \frac{\left(1 + \dfrac{r}{m}\right)^{mt} - 1}{\dfrac{r}{m} \left(1 + \dfrac{r}{m}\right)^{mt}}$$

Equating As and dividing by $(1 + r/m)^{mt}$

$$= P \frac{1 - \left(1 + \dfrac{r}{m}\right)^{-mt}}{\dfrac{r}{m}}$$

Dividing numerator and denominator by $(1 + r/m)^{mt}$

Therefore:

Present Value of an Annuity

The present value PV of an ordinary annuity with payments of P dollars m times per year for t years at interest rate r compounded at each payment is

$$PV = P \frac{1 - \left(1 + \dfrac{r}{m}\right)^{-mt}}{\dfrac{r}{m}}$$

Writing i for r/m and n for mt, the archaic symbol $a_{\overline{n}|i}$ (pronounced "a sub n angle i") is sometimes still used to denote the value of $(1 - (1 + i)^{-n})/i$ in business textbooks and tables. Using this notation, the above formula takes the form $PV = Pa_{\overline{n}|i}$ with $n = mt$ and $i = r/m$.

EXAMPLE 1   Finding the Present Value of an Annuity

What is the present value of a 6% car loan for 4 years with monthly payments of $200?

Solution

$$PV = 200 \; \frac{1 - \left(1 + \dfrac{0.06}{12}\right)^{-(12)(4)}}{\dfrac{0.06}{12}} \approx 8516.06 \qquad PV = P\,\frac{1 - (1 + r/m)^{-mt}}{r/m}$$

with $P = 200$, $r = 0.06$, $m = 12$, and $t = 4$

The present value is $8516.06.

■

Graphing Calculator Exploration

We know from page 136 that the accumulated amount after 4 years of monthly deposits of $200 into a savings account earning 6% compounded monthly is $10,819.57. In Example 1, we found that the present value of this annuity is $8516.06.

a. To find the future value y_1 of the $8516.06 after x years, use the values $r = 0.06$, $m = 12$, and $PV = 8516.06$ and graph the compound interest formula $y_1 = 8516.06(1 + .06/12)^{12x}$ on the window $[0, 9.4]$ by $[0, 30000]$.

b. To find the accumulated amount y_2 after x years, use the values $r = 0.06$, $m = 12$, and $P = 200$ and graph the annuity formula $y_2 = 200((1 + .06/12)^{12x} - 1)/(.06/12)$.

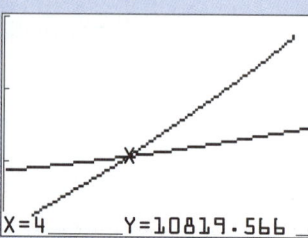

c.   Use TRACE or INTERSECT to find the intersection point of these two curves.

Notice that the loan company will get the same final return on its $8516.06 investment from either putting the money into a compound interest savings account or "buying an annuity" in the form of loan payments from you for the same time period.

PRACTICE PROBLEM 1   What is the present value of a 20-year retirement annuity paying $850 per month if the current long-term interest rate is 7.53%?

Solution at the back of the book

Amortization

A debt is *amortized* (or "killed off") if it is repaid by a regular sequence of payments. These payments form an annuity with present value *PV* equal to the debt *D*. To find the payments *P* to amortize debt *D*, we solve for *P* in the "present value of an annuity" formula (page 149).

Amortization Payment

The payment *P* to make *m* times per year for *t* years at interest rate *r* compounded at each payment to amortize a debt of *D* dollars is

$$P = D \, \frac{\dfrac{r}{m}}{1 - \left(1 + \dfrac{r}{m}\right)^{-mt}}$$

This formula may also be written $P = D/a_{\overline{n}|i}$ with $n = mt$ and $i = r/m$.

EXAMPLE 2   **Finding the Payment to Amortize a Debt**

What monthly payment will amortize a $150,000 home mortgage at 8.6% in 30 years?

Solution

$$P = 150{,}000 \; \frac{\dfrac{0.086}{12}}{1 - \left(1 + \dfrac{0.086}{12}\right)^{-(12)(30)}} \approx 1164.02$$

$P = D \dfrac{r/m}{1 - (1 + r/m)^{-mt}}$
with $D = 150{,}000$,
$r = 0.086$, $m = 12$,
and $t = 30$

The required payment is \$1164.02 each month. Notice that the borrower will pay a total of \$1164.02 × 12 × 30 = \$419,047.20 during the 30 years of the loan.

PRACTICE PROBLEM 2

What monthly payment will amortize a \$150,000 home mortgage at 8.6% in 25 years? What is the total amount the borrower will pay?

Solution at the back of the book

Graphing Calculator Exploration

When amortizing a loan, a longer term means smaller payments but the total amount the borrower pays is much larger.

a. To find the monthly payment y to amortize a debt over x years using the values $D = 150{,}000$, $r = 0.086$, and $m = 12$ from Practice Problem 2, graph the curve $y_1 = 150000(.086/12)/(1 - (1 + .086/12)^{-12x})$ on the window [0, 50] by [0, 5000].

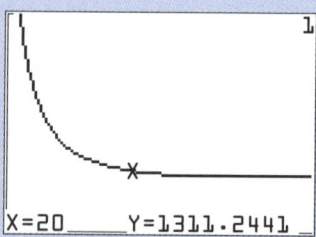

X=20   Y=1311.2441

b. Use TRACE or VALUE to find the payments for mortgage terms of 15, 20, 25, and 30 years. How do your answers compare to the answers to Example 2 and Practice Problem 2?

c. To find the total amount the borrower pays, multiply the payment by m and t (here replaced by x). Graph the new curve $y_1 = 12x(150000)(.086/12)/(1 - (1 + .086/12)^{-12x})$ on the window [0, 50] by [0, 600000] to find the total amount paid y on a loan for x years.

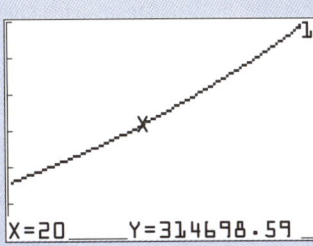

X=20   Y=314698.59

d. Use TRACE or VALUE to find the total amount paid for mortgage terms of 15, 20, 25, and 30 years. Does a small change in the payment make a large difference in both the term and the total amount paid on a debt?

Unpaid Balance

How much does the borrower still owe partway through an agreed amortization schedule? Suppose you have made mortgage payments for 10 years on a 30-year loan and now you must move. How much should you pay to settle your debt? Your remaining payments form another annuity and are worth precisely the present value of this new annuity.

EXAMPLE 3 Finding the Unpaid Balance

What is the unpaid balance after 10 years of monthly payments on a 30-year mortgage of $150,000 at 8.6%?

Solution

First we must calculate the monthly payments. But for this problem, we know from Example 2 (pages 151–152) that the monthly payment to amortize the original mortgage is $1164.02. The remaining payments thus form an annuity with $P = \$1164.02$, $r = 0.086$, $m = 12$, and $t = 20$ years. The amount still owed on the mortgage is the present value of this annuity, which we can calculate by the formula on page 149.

$$PV = 1164.02 \, \frac{1 - \left(1 + \dfrac{0.086}{12}\right)^{-(12)(20)}}{\dfrac{0.086}{12}} \qquad \begin{array}{l} PV = P\dfrac{1 - (1 + r/m)^{-mt}}{r/m} \\ \text{with } P = 1164.02, \, r = 0.086, \\ m = 12, \text{ and } t = 20 \end{array}$$

$$\approx 133{,}158.27$$

The unpaid balance after 10 years of payments on this $150,000 mortgage is $133,158.27.

Notice that after the first 10 of 30 years, much less than one third of the mortgage has been paid. The early payments of any loan pay mostly interest and only a little of the principal.

PRACTICE PROBLEM 3

How much is still owed on a 4-year car loan of $12,000 at 4.7% after 1 year of monthly payments?　　*Solution at the back of the book*

Graphing Calculator Exploration

The program AMORTABL* constructs an *amortization table* showing the application of the payments on a loan toward the interest due each period and the corresponding reductions of the outstanding debt by the remaining parts of the payments. To explore the table for the situation used in Examples 2 and 3, run this program by selecting it from the program menu and proceed as follows:

a.　Enter 150000 for the debt, .086 for the interest rate, 30 for the number of years, and 12 for the number of compoundings per year (since the payments are monthly).

```
DEBT (DOLLARS)
  150000
RATE (DECIMAL)
  .086
NUMBER OF YEARS
  30
COMPOUNDINGS/YR
  12
```

b.　After the table of values has been calculated by finding the payment rounded to the nearest penny and then applying each payment to the rounded interest due on the debt for that period and reducing the debt by the excess, you may choose which part of the table you wish to see.

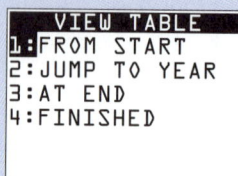

```
   VIEW TABLE
1:FROM START
2:JUMP TO YEAR
3:AT END
4:FINISHED
```

*See the Preface for information on how to obtain this program.

c. The amortization table contains five columns: X is the number of the payment (negative X's or X's larger than the number of payments display zeros in the other columns), Y_1 is the payment, Y_2 is the interest part of the payment, Y_3 is the remaining part used to reduce the debt, and Y_4 is the debt remaining after this reduction. Use the arrow keys to scroll right and left through these columns.

X	Y_1	Y_2
0	0	0
1	1164	1075
2	1164	1074.4
3	1164	1073.7
4	1164	1073.1
5	1164	1072.4
6	1164	1071.8
X=0		

X	Y_3	Y_4
0	0	150000
1	89.02	149911
2	89.66	149821
3	90.3	149731
4	90.95	149640
5	91.6	149548
6	92.26	149456
Y_4=150000		

X	Y_1	Y_2
356	1164	40.8
357	1164	32.75
358	1164	24.64
359	1164	16.48
360	1160	8.25
361	0	0
362	419043	269043
X=361		

X	Y_3	Y_4
356	1123.2	4570
357	1131.3	3438.7
358	1139.4	2299.3
359	1147.5	1151.8
360	1151.8	0
361	0	0
362	150000	0
Y_4=0		

The second line after the end of the amortization contains the total payments, interest, and debt reduction. Notice that the final payment is adjusted to correct for rounding errors and the debt is reduced to zero.

d. To use the table to find the remaining debt after a given number of years, "jump" to that position in the table and scroll to the Y_4 column. After ten years of payments, the table shows \$133,157.56 remaining.

X	Y_3	Y_4
118	205.28	133573
119	206.75	133366
120	208.23	133158
121	209.72	132948
122	211.23	132737
123	212.74	132524
124	214.27	132310
Y_4=133157.56		

SUMMARY

The *present value* of an ordinary annuity compounded at each payment is

$$PV = P \, \frac{1 - \left(1 + \dfrac{r}{m}\right)^{-mt}}{\dfrac{r}{m}}$$

PV = present value,
P = payment, r = rate,
m = times per year, and
t = years

The payment to *amortize a debt* of D dollars is found by solving this formula for P:

$$P = D \, \frac{\dfrac{r}{m}}{1 - \left(1 + \dfrac{r}{m}\right)^{-mt}}$$

D = debt
(other variables
as above)

The amount still owed after making amortization payments for part of the term is the present value of a new annuity formed by the remaining payments, and is found by the first formula.

EXERCISES 1.4 (Most need ▦ or ▧ .)

Find the present value of each annuity.

1. $15,000 annually at 7% for 10 years.

2. $18,000 annually at 6% for 15 years.

3. $10,000 semiannually at 11% for 7 years.

4. $8000 semiannually at 10.3% for 12 years.

5. $3000 quarterly at 8% for 8 years.

6. $4500 quarterly at 6% for 15 years.

7. $1400 monthly at 6.9% for 30 years.

8. $1997 monthly at 5.3% for 25 years.

9. $25 weekly at 7.7% for 6 years.

10. $75 weekly at 9.2% for 9 years.

Find the payment to amortize each debt.

11. Monthly payments on $100,000 at 5% for 25 years.

12. Monthly payments on $125,000 at 5.4% for 20 years.

13. Monthly payments on $6000 at 6% for 5 years.

14. Monthly payments on $8000 at 7% for 4 years.

15. Annual payments on $50,000 at 8.2% for 10 years.

16. Semiannual payments on $45,000 at 7.9% for 8 years.

17. Quarterly payments on $14,500 at 12.7% for 6 years.

18. Quarterly payments on $13,900 at 13.4% for 5 years.

19. Weekly payments on $2500 at 18% for 3 years.

20. Weekly payments on $2000 at 19.5% for 4 years.

Find the unpaid balance on each debt (you should already know the payments from Exercises 11–20).

21. After 6 years of monthly payments on $100,000 at 5% for 25 years.

22. After 15 years of monthly payments on $125,000 at 5.4% for 20 years.

23. After 2 years 11 months of monthly payments on $6000 at 6% for 5 years.

24. After 1 year 5 months of monthly payments on $8000 at 7% for 4 years.

25. After 3 years of annual payments on $50,000 at 8.2% for 10 years.

26. After 4 years 6 months of semiannual payments on $45,000 at 7.9% for 8 years.

27. After 2 years 3 months of quarterly payments on $14,500 at 12.7% for 6 years.

28. After 1 year 9 months of quarterly payments on $13,900 at 13.4% for 5 years.

29. After 2 years of weekly payments on $2500 at 18% for 3 years.

30. After 2 years 30 weeks of weekly payments on $2000 at 19.5% for 4 years.

31. Solve any five of Exercises 1–10 as follows. Enter the present value of an annuity formula as $y = P(1 - (1 + r/m)^{-mt})/(r/m)$. Then for each exercise, STORE the values for P, r, m, and t, and evaluate y to find the present value.

32. The unpaid balance y of the mortgage in Example 3 on page 153 can be viewed as a function of the number of years x that payments have been made by graphing the function $y = (1164.02)(1 - (1 + .086/12)^{-12(30-x)})/(.086/12)$ on the window $[0, 35]$ by $[0, 160000]$. The "t" in the annuity present value formula has been replaced by "$30 - x$" since the number of years remaining is 30 minus the number of years already paid. Use TRACE to explore the curve. Does it start at $(0, 150000)$ and end at $(30, 0)$? Notice that these points correspond to the initial value of the mortgage and the length of the loan. Does the unpaid balance decrease by the same amount each year? How long does it take to reduce the unpaid balance to half the initial amount?

33. Solve any five of Exercises 11–20 as follows. Enter the amortization payment formula as $y = D(r/m)/(1 - (1 + r/m)^{-mt}$. Then for each exercise, STORE the values for m, D, r, and t, and evaluate y to find the required payment.

34. **Unpaid Balance Formula**   Combine the present value of an annuity formula with the amortization payment formula to show that the unpaid balance after x years of payments made m times each year for t years on a debt of D dollars at interest rate r is

$$D \left(1 + \frac{r}{m}\right)^{mx} \frac{\left(1 + \frac{r}{m}\right)^{m(t-x)} - 1}{\left(1 + \frac{r}{m}\right)^{mt} - 1}$$

35. Solve any five of Exercises 21–30 as follows. Enter the unpaid balance formula from Exercise 34 as $y = D(1 + r/m)^{mx}((1 + r/m)^{m(t-x)} - 1)/((1 + r/m)^{mt} - 1)$. Then for each exercise, STORE the values for x, m, t, D, and r, and evaluate y to find the unpaid balance after x years of payments. Why does this unpaid balance formula give slightly different answers for these exercises?

36. Show that the present value of an annuity formula may be rewritten in the form:

$$PV = P \frac{1 - (1 + i)^{-n}}{i}$$

where $i = r/m$ and $n = mt$.

37. Show that the present value of an annuity is the sum of the present values of the payments. (*Hint:* Use the notation of Exercise 36 and show that the present value of the first payment is $P(1 + i)^{-1}$, that of the second is $P(1 + i)^{-2}$, and so on for the n payments. Factor out the common terms from the sum and use the geometric series formula on page 136.)

38. Use the program AMORTABL to solve Exercise 21. How does the amortization table solution differ from your previous solution to Exercise 21? How much does the final payment in the table differ from the payments made during the rest of the table?

39. Use the program AMORTABL to solve Exercise 25. Be sure to enter "1" for the "compoundings per year." How does the amortization table solution differ from your previous solution to Exercise 25? How much does the final payment in the table differ from the payments made during the rest of the table?

40. Use the program AMORTABL to solve Exercise 29. Be sure to enter "52" for the "compoundings per year." How does the amortization table solution differ from your previous solution to Exercise 29? How much does the final payment in the table differ from the payments made during the rest of the table?

41. Contest Prizes   The super prize in a contest is $10 million. This prize will be paid out in equal yearly payments over the next 25 years. If the prize money is guaranteed by U.S. Treasury bonds yielding 7.9% and is placed into an escrow account when the contest is announced 1 year before the first payment, how much do the contest sponsors have to deposit in the escrow account?

42. Grant Funding   In September 1994, the New York City public schools system was awarded a grant of $50 million over 5 years from the Annenberg Foundation. If the grant paid equal yearly amounts for the next 5 years and was financed at 6.8%, how much did this grant cost the Foundation when it was announced?

43. NFL Contracts   When Michael Westbrook signed a seven-year, $18 million contract with the Washington Redskins, it included a $6.5 million signing bonus that was the largest for a wide receiver in NFL history. If the $18 million was paid out in equal quarterly payments for the seven years and the current long-term interest rate was 6.85%, how much was the contract worth when it was signed?

44. Escrow Accounts   On September 1, 1994, Judge Sam C. Pointer, Jr., of the Federal District Court in Birmingham approved a $4.25 billion settlement against the manufacturers of silicon breast implants to be paid out over the next 30 years. If this money was to be paid out in equal semiannual amounts from an escrow account earning 8.7% interest, how much would the manufacturers have to deposit into the account to meet the terms of the settlement?

45. College Savings   When I graduated from high school last spring, my dear Aunt Sallie gave me a savings account passbook for an account earning 5.61%. She told me that starting next fall, there would be enough money for me to take out $10,000 every six months for the next four years to pay for my college expenses. I thought "Wow! Thanks! That's $80,000!" but she just smiled, shook her head "No," and told me to look in the passbook. How much was in the account when she gave it to me?

46. Car Buying   A Cadillac dealer offers the following terms on a two-year lease of a new Eldorado: either $1995 down and $339 per month for 24 months or $9599 down and nothing more to pay for 24 months. (a) If the current interest rate were 5.2%, which option would be cheaper? (b) Since the car dealer is an expert money manager, what must the current interest rate be if the offers are equally good for the dealer?

47. Life Insurance   Just before his first attempt at bungee jumping, John decides to buy a life insurance policy. Since his annual income at age 30 is $35,000, he figures he should get enough insurance to provide his wife and new baby with that amount each year for the next 35 years. If the long-term interest rate is 6.7%, what is the present value of John's future annual earnings? Rounding up to the next $50,000, how much life insurance should he buy?

48. Life Insurance   A 60-year-old grandmother wants a life insurance policy that could replace her annual $55,000 earnings for the next 10 years. If the long-term interest rate is now 6.2%, how much life insurance (rounded up to the next $10,000) should she buy?

49. Mortgages   Real estate prices are so high in Tokyo that some mortgages are written for 99 years in an attempt to keep down the monthly payments. (a) What is the monthly payment on a $500,000 mortgage at 8% for 99 years? (b) How much does this save each month compared to a 30-year mortgage at 8% for the same amount? (c) Which term results in the higher total payment?

50. Apartment Rents   *The New York Times* reported that the Associates for International Research had estimated the following monthly rents for a two-bedroom apartment in major cities around the world.

Two-Bedroom Apartment Monthly Rents Around the World			
Sydney	$1100	London	$2950
Cairo	$1200	Moscow	$5000
Buenos Aires	$1500	Shanghai	$6300
Paris	$2150	Tokyo	$7100
New York	$2300	Hong Kong	$7200

Assuming that the monthly rent represents the payment on the current market value of the apartment as a 30-year investment and that the long-term interest rate is 6.52%, estimate the current market value for an apartment in (a) Sydney, (b) Paris, (c) Moscow, and (d) Tokyo.

51. **Credit Cards**   A MasterCard statement shows a balance of $560 at 13.9% compounded monthly. What monthly payment will pay off this debt in 1 year 8 months?

 52. **Credit Cards**   The MasterCard statement in Exercise 51 also states that the "minimum payment" is $15. How long will it take to pay off this debt by making this minimum payment each month?

   a. To estimate the number of years required, STORE the values 15 for P, 0.139 for r, and 12 for m before graphing the present value formula $y = P(1 - (1 + r/m)^{-mx})/(r/m)$ on the window [0, 10] by [0, 1000]. Use TRACE or INTERSECT to estimate the number of years x that corresponds to the debt's present value $y = \$560$. Be sure to change this number of years into the number of months the payments must be made.

   b. To estimate the number of years required, STORE the values 560 for D, 0.139 for r, and 12 for m before graphing the amortization payment formula $y = D(r/m)/(1 - (1 + r/m)^{-mx})$ on the window [0, 10] by [0, 20]. Use TRACE or INTERSECT to estimate the number of years x that corresponds to the payment $y = \$15$. Be sure to change this number of years into the number of months the payments must be made.

   c. Why are the answers from (a) and (b) exactly the same?

53. **Credit Cards**   What monthly payment should you make on your Visa card to pay off a new $1575 stereo in 15 months if the interest rate is 14.8% compounded monthly?

54. **Mortgages**   The *Town Gossip* local newspaper lost the libel suit filed by the village's former mayor. What quarterly payment will pay off the $150,000 judgment in 5 years if the editor takes out a second mortgage on her business office at 11.6% compounded quarterly?

55. **College Costs**   When Jill graduated from college, her loans with interest totaled $58,720. Since her last co-op employer invited her to stay on as a full-time employee, she was able to handle the monthly payments even though the interest rate was 9.4% and the bank expected everything to be paid back in 8 years. How much debt remained after 3 years of payments?

56. **Car Buying**   Tom bought a new turbo Ford Thunderbird and financed it with a new car loan of $18,000 at 13.78% for 4 years. After making the monthly payments for 7 months, he decided he couldn't study for his business courses at college and put in enough hours at his part-time job to keep up his grades, so he offered to give the car to his sister if she would take over the payments. What purchase price would she be paying for the car?

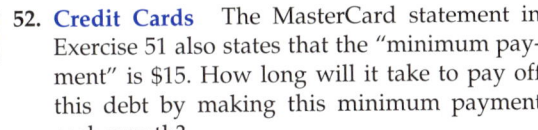 57. **Car Buying**   A car dealer's newspaper ad offers "0% financing with guaranteed credit approval" and includes an example: "finance $10,000 for 48 months—normal payment is $262.84 while 0% financing is $208.33—you save $2616." What is the "normal" finance rate? To estimate the answer, graph the payment formula $y = 10000(x/12)/(1 - (1 + x/12)^{-48})$ on the window [0, .2] by [0, 500] to see the monthly payment y for a 4-year loan of $10,000 at interest rate x compounded monthly. Use TRACE or INTERSECT to find the rate x that gives a y-value near $262.84. (By the way, the small print at the bottom of the ad contains the statement "0% financing many affect selling price of car.")

58. Computer Buying   A Radio Shack back-to-school sale ad offers a "complete PC system for $39 per month after no payments for six months." The ad explains that "if you do not pay the full amount of purchase [$1270] by the end of the deferred period," then "finance charges will accrue from the date of purchase and will be added to the purchase balance" and states that the APR is 22.55%.

a. Show that the nominal interest rate for this loan is 20.508%.

b. How much will the balance be after six months if you make no payments?

c. How long will it then take to pay off the debt by paying the $39 each month? To estimate the number of years required, STORE the values 39 for P, 0.20508 for r, and 12 for m before graphing the present value formula $y = P(1 - (1 + r/m)^{-mx})/(r/m)$ on the window [0, 20] by [0, 2000]. Use TRACE or INTERSECT to estimate the number of years x that corresponds to the present value y from (b). Be sure to change this number of years

into the number of months the payments must be made.

d. Using the advertised financing, what is the buyer's total cost?

59. Yearly Interest   A home owner finances a new luxury car using a "home equity" loan so that the interest can be deducted each year when calculating her taxable income. If she plans to repay the $40,000 in equal monthly payments over 4 years and the interest rate is fixed at 9.35%, how much interest will she pay in the second year of the loan?

60. Home Refinancing   A widow needs cash and decides to refinance the home she and her husband bought 20 years ago for $35,000 with a monthly payment, 30-year mortgage at 6.8% on 80% of the purchase price. The Home Sweet Home Mortgage Corporation has appraised her house at $125,000 and agreed to loan her 70% of this value. How much cash will she have after paying off the old mortgage?

Chapter Summary with Hints and Suggestions

The reading and exercises of this chapter have helped you to learn the following skills. For each skill, the section from which it came (in case you need to review it) and some exercises in this section that use it are indicated. Answers to all exercises are at the end of the book, and full solutions to all exercises are in the Student Solutions Manual.

1.1   Simple Interest

- Find the simple interest due on a simple interest loan.   *(Review Exercises 1–4.)*

$$I = Prt$$

- Use the Banker's rule in a simple interest calculation.   *(Review Exercises 5–8.)*

- Find the total amount due on a simple interest loan.   *(Review Exercises 9–12.)*

$$A = P(1 + rt)$$

- Solve a simple interest situation for the interest,

the simple interest rate, the principal, or the term.   *(Review Exercises 13–17.)*

- Find a future or present value. *(Review Exercises 18–20.)*

$$A = P(1 + rt) \quad \text{and} \quad PV = \frac{A}{1 + rt}$$

- Find the effective simple interest rate of a discounted loan.   *(Review Exercises 21–22.)*

$$r_s = \frac{r}{1 - rt}$$

- Solve an applied problem using these skills. (*Review Exercises 23–25.*)

1.2   Compound Interest

- Find the amount due on a compound interest loan.   (*Review Exercises 26–30.*)

$$A = P \left(1 + \frac{r}{m}\right)^{mt} \quad \text{and} \quad A = Pe^{rt}$$

- Find a future or present value. (*Review Exercises 31–35.*)

$$A = P\left(1 + \frac{r}{m}\right)^{mt} \quad \text{and} \quad PV = \frac{A}{\left(1 + \frac{r}{m}\right)^{mt}}$$

- Find the term needed for a given principal to grow to a future value. (*Review Exercises 36–40.*)
- Use the "rule of 72" to estimate a doubling time.   (*Review Exercises 41–45.*)

$$\left(\begin{array}{c}\text{Doubling} \\ \text{time}\end{array}\right) \approx \frac{72}{r \times 100}$$

- Find the effective rate of a loan or investment. (*Review Exercises 46–50.*)

$$r_e = \left(1 + \frac{r}{m}\right)^m - 1 \quad \text{and} \quad r_e = e^r - 1$$

1.3   Annuities

- Find the accumulated amount of an ordinary annuity.   (*Review Exercises 51–55.*)

$$A = P \frac{\left(1 + \frac{r}{m}\right)^{mt} - 1}{\frac{r}{m}}$$

- Find the regular payment to make into a sinking fund to accumulate a given amount. (*Review Exercises 56–60.*)

$$P = A \frac{\frac{r}{m}}{\left(1 + \frac{r}{m}\right)^{mt} - 1}$$

- Find the number of years to make payments into a sinking fund to accumulate a given amount.   (*Review Exercises 61–65.*)

$$t = \frac{\log \left(\frac{A}{P} \frac{r}{m} + 1\right)}{m \log \left(1 + \frac{r}{m}\right)}$$

- Estimate the internal rate of return of regular investments using a graphing calculator. (*Review Exercises 66–70.*)

1.4   Amortization

- Find the present value of an ordinary annuity. (*Review Exercises 71–75.*)

$$PV = P \frac{1 - \left(1 + \frac{r}{m}\right)^{-mt}}{\frac{r}{m}}$$

- Find the regular payment to amortize a debt. (*Review Exercises 76–80.*)

$$P = D \frac{\frac{r}{m}}{1 - \left(1 + \frac{r}{m}\right)^{-mt}}$$

- Find the amount still owed after making amortization payments for part of the term. (*Review Exercises 81–85.*)

Hints and Suggestions

- (*Overview*) Compound interest is repeated simple interest with the interest added to the principal in each successive period. The *future value* is the amount the principal will become after all the interest is included. Reversing the point of view, the *present value* is the principal needed now that will grow to the final amount. The three basic amount formulas are: $P(1 + rt)$ for simple interest, $P(1 + r/m)^{mt}$ for compound interest, and $P((1 + r/m)^{mt} - 1)/(r/m)$ for an annuity (multiple payments). The other formulas all follow from these basic formulas.

- The rate is usually stated as a percent but is always used in decimal form for calculations: 6.7% in decimal form is 0.067.

- Round off only your final answer when using your calculator. Don't use the decimal 0.33 for the fraction $\frac{1}{3}$.

- Don't confuse a *percent increase* with the *multiplier. Increasing* an amount by 25% means *multiplying* it by 1.25.

- Round *up* to find the whole number of compoundings needed to reach a future value because a smaller number won't reach the stated goal.

- The "rule of 72" is only an approximation but is a helpful check that can catch "button pushing" errors when using your calculator.

- Practice for test: Review Exercises 1, 5, 9, 13, 17, 22, 28, 31, 36, 38, 44, 47, 51, 54, 58, 62, 66, 72, 77, 82.

Review Exercises for Chapter 1   *Practice test exercises are in blue.*

Round all dollar amounts to the nearest penny, all times in years to two decimal places, and all terms to the appropriate number of compounding periods. Most exercises need ▦ or ▦.

1.1   Simple Interest

Find the simple interest on each loan.

1. $1875 at 5.8% for 2 years.

2. $1150 at 9.2% for 6 months.

3. $8000 at $7\frac{1}{2}$% for 3 years 9 months.

4. $2385 at 11.3% for 1 year 3 months.

Use the Banker's rule to find the simple interest on each loan.

5. $1575 at 8.6% from January 10 to April 10 (not a leap year).

6. $2835 at 4.7% from March 15 to August 15.

7. $800 at 12% from July 21 to September 21.

8. $10,000 at 11.1% from September 9 to November 9.

Find the total amount due on each simple interest loan.

9. $8900 at 5.9% for 2 years 6 months.

10. $1375 at 11.3% for 9 months.

11. $1795 at 6.38% for 3 years.

12. $3700 at 8.3% for 1 year 11 months.

Solve each problem.

13. Find the interest rate on a loan charging $272 simple interest on a principal of $2000 after 2 years.

14. Find the principal of a loan at 9.3% if the simple interest after 1 year 8 months is $279.

15. Find the term of a loan of $7600 at 9% if the simple interest is $1026.

16. What should be the term for a loan of $5500 at 8.7% simple interest if the lender wants to receive $7414 when the loan is paid off?

17. How long will it take an investment at 10% simple interest to increase by 50%?

18. How much should be invested now at 5.9% simple interest if $5208 is needed in 2 years 8 months?

19. What would be the fair market price of a $10,000 zero coupon bond due in 1 year if today's long-term simple interest rate is 6.45%?

20. What would be the fair market price of a $50,000 zero coupon bond due in 1 year if today's long-term simple interest rate is 5.78%?

21. What is the effective simple interest rate of a discounted loan at 12% interest for 6 years?

22. What is the effective simple interest rate of a discounted loan at 15% interest for 2 years?

23. **Treasury Bonds**   *Forbes* magazine ranked John W. Kluge as the richest American, with a net worth of $5.2 billion, at the time of his divorce from Patricia Kluge. A newspaper reported that besides a 45-room Georgian mansion, Mrs. Kluge ". . . received $1 billion that invested conservatively in 30-year Treasury bonds would throw off $66 million a year." What was the interest rate when the article was written?

24. **Insurance Settlements**   How large an insurance settlement does an accident victim need for a yearly income of $29,000 from long-term bond investments paying 5.8%?

25. **Furniture Buying**   A furniture store offers a complete living room suite for $999 and will finance the entire price as a discounted loan at 10% for 1 year. What is the effective simple interest rate for this loan and how much will you need to pay at the end of the year?

1.2   Compound Interest

Find the amount due on each compound interest loan.

26. $15,000 at 7.5% compounded quarterly for 10 years.

27. $65,000 at 8.25% compounded continuously for 17 years.

28. **Mutual Funds**   From 1970 to 1995, the AIM Charter Fund posted an average annual return of 15.55% including sales charges. How much would a $10,000 investment in this fund have been worth after 25 years?

29. **Mutual Funds**   During the mid 1990s, the Brandywine long-term growth fund returned 19.6% compounded quarterly. How much would a $5000 investment in this fund have been worth after 4 years?

30. **National Debt**   From 1835 to 1837, the United States not only was free of debt but actually had a surplus in the Treasury. On January 1, 1837, after $5 million had been set aside as a reserve fund, there remained $37,468,859. If that money had not been distributed to the 26 states but instead had been invested at 5% compounded continuously, would there be enough by 1997 to pay off a $4.3 trillion national debt?

Find the present value of each compound interest loan.

31. $25,000 after 10 years at 9% compounded monthly.

32. $30,000 after 8 years at 6% compounded continuously.

33. $175,000 after 16 years 6 months at 5.5% compounded quarterly.

34. **Certificates of Deposit**   From 1968 to 1995, the average annual return on 6-month CDs was 7.79% compounded semiannually. How much would have been invested in 1968 to have $100,000 in 1995?

35. **Bond Funds**   During the mid-1990s, the Northeast Investors Trust general-term bond fund returned 15.7% compounded monthly. How much would you need to invest in this fund to have $15,000 after 3 years 8 months?

Find the term of each compound interest loan.

36. 8.9% compounded quarterly to obtain $10,000 from a principal of $2000.

37. 11.5% compounded monthly to multiply the principal by 1.60.

38. How long will it take the value of an apartment building to increase by 80% if real estate prices are increasing 8% continuously?

39. **Treasury Bonds**   How long will you have to wait to become a millionaire if you invest all of your $850,000 lottery winnings in treasury bonds paying 5.6% compounded semiannually?

40. **Art Appreciation**   How long will it take for the value of an early Picasso pencil sketch to triple if its market value is increasing 6.5% annually?

Use the "Rule of 72" to estimate the doubling time (in years) for each interest rate and then calculate it exactly.

41. 8% compounded annually.

42. 6% compounded monthly.

43. 11.9% compounded weekly.

44. Real Estate How long will it take for the value of a house to double if real estate prices are increasing 12% each year?

45. Index Funds How long will it take an investment in a stock market index fund yielding 10% annually to double?

46. Find the effective rate of 13.25% compounded quarterly and compounded continuously.

47. Baseball Cards A mint condition 1955 Topps "Sandy Koufax" baseball card selling for $500 in 1989 was worth $1350 in 1993. What is the effective rate of this price increase?

48. Stocks One of Wall Street's most cherished buys in the 1950s was Texas Instruments stock, which rose spectacularly from $72\frac{1}{8}$ to $214\frac{1}{4}$ in 18 months. What is the effective rate of this price increase?

49. Baseball Teams The Haas family bought the Oakland Athletics from Charles O. Finley for $12.7 million in 1980 and sold it in 1995 for $85 million. Neglecting all other expenses, what is the effective rate of return on their investment?

50. Pianos A properly maintained 90-year-old Steinway grand piano is now worth 13.6 times its initial purchase price. What is the effective rate of this increase?

1.3   Annuities

Find the accumulated amount of each annuity.

51. $1500 annually at 11% for 20 years.

52. $500 semiannually at 8% for 12 years.

53. Home Buying How much will you have for a vacation home if you save $25 each week for 16 years in a 5.5% passbook savings account?

54. Retirement Savings How much will an IRA stock fund earning 11.2% be worth if you deposit $180 each month for 35 years?

55. Apartments When André-François Raffray died at age 77 in Arles, France, he had been making $500 monthly payments to Jeanne Calment for 30 years for the right to take over her apartment when she died. Mrs. Calment, who at 120 was then the world's oldest person with the records to prove it, remarked that "In life, one sometimes makes bad deals." How much would Mr. Raffray's money be worth if instead he had deposited it into a 7.8% bond fund?

Find the required payment for each sinking fund.

56. Yearly deposits earning 7% to accumulate $16,000 after 8 years.

57. Quarterly deposits earning 4.9% to accumulate $40,000 after 25 years.

58. How much must you pay biweekly to pay off a 13.2% car loan of $19,500 in 5 years?

59. Stock Funds How much must you deposit each month into a stock fund earning 11.2% to accumulate $150,000 in 20 years?

60. Home Buying How much must you save each week in a 5.5% passbook savings account to have $26,000 for a vacation home in 16 years?

Find the number of years needed for each sinking fund to reach the given accumulated amount.

61. $1000 yearly at 10% to accumulate $100,000.

62. $250 quarterly at 8.1% to accumulate $75,000.

63. Money Market Funds How long must you save $275 each month in a 6.2% money market fund to accumulate $19,500 for a new car?

64. Stock Funds How long must you deposit $300 each month into a stock fund earning 11.2% to accumulate $150,000?

65. Home Buying How long must you save $25 each week in a 5.5% passbook savings account to have $26,000 for a vacation home?

Follow the method of the Graphing Calculator Exploration on page 142 to estimate the internal rate of return for each investment.

66. $150 quarterly to accumulate $10,000 after 10 years.

67. $225 monthly to accumulate $100,000 after 25 years.

68. $500 monthly to accumulate $1,000,000 after 30 years.

69. $1.25 weekly to accumulate $50,000 after 60 years.

70. $10 daily to accumulate $100,000 after 20 years.

1.4   Amortization

71. Find the present value of an annuity of $25,000 annually at 6% for 25 years.

72. Football Contracts   Deion Sanders became the highest-paid defensive player in football when he signed a $25 million, five-year contract with the Dallas Cowboys. If the $25 million was paid out in equal weekly payments over the five years and the current interest rate was 7.32%, how much was the contract worth when it was signed?

73. Grant Funding   A foundation is funding a new program that will award twenty $250,000 grants each year for the next 8 years to promote art and music instruction in elementary schools. How much must be placed in a money market account paying 6.75% to fund this initiative?

74. Contest Prizes   The grand prize in a lottery drawing is $14 million to be paid out in equal quarterly payments over the next 20 years. If the prize money is guaranteed by U.S. Treasury bonds yielding 7.6% and is placed into an escrow account when the lottery is announced 1 year before the first payment, how much do the lottery sponsors have to deposit in the escrow account?

75. Home Buying   Before looking for their first house, the Jones family calculates that they could afford a $1250 monthly mortgage pay-

ment. How large a mortgage could they afford at 8.75% for 30 years?

76. Find the weekly payments to amortize a debt of $45,000 at 9.15% over 15 years.

77. Credit Cards   A Discover card statement shows a balance of $945 at 15.7% compounded monthly. What monthly payment will pay off this debt in 2 years?

78. Credit Cards   What monthly payment should you make on your Optima card to pay off $2784 of spring break vacation charges in 7 months if the interest rate is 18.9% compounded monthly?

79. Home Buying   The Jones family has found a nice house in the suburbs. What is the monthly payment on their $150,000 mortgage at 8.75% for 30 years?

80. Stock Funds   How much must you save each week in a stock fund yielding 10.1% to become a millionaire in 15 years?

Find the unpaid balance on each debt.

81. After 5 years of annual payments on $85,000 at 9.1% for 20 years.

82. After 3 years 8 months of monthly payments on $150,000 at 8.7% for 25 years.

83. Car Buying   After 1 year 5 months of $297.92 monthly car payments on a 3-year loan at 4.6%, Karen wants to get rid of her sedan and get a red sports car. How much should she pay to settle her debt?

84. Personnel Negotiations   The Middleville Central School Board made a mistake when they hired the new superintendent and agreed to a 6-year contract paying $125,000 per year. With current interest rates at 6.2%, how much should they offer to break the contract after the first year?

85. Home Buying   After 4 years 9 months of monthly mortgage payments on their 30-year, $150,000 mortgage at 8.75%, the Jones family has to relocate to Atlanta and must sell their nice house in the suburbs. How much do they still owe on their mortgage?

Projects and Essays

The following projects and essays are based on Chapter 1. Most have no right or wrong answers—the results depend only on your imagination and resourcefulness.

1. Discuss the differences between two $1000 loans for 8 years: one at 5% simple interest and the other at 5% compounded annually.

2. Make a list of the basic financial concepts from Section 1.1 that were used in the later sections of this chapter and discuss how your understanding of each changed as you mastered the subsequent material.

3. Visit your local bank and copy down the nominal and effective interest rates on several CDs or loans. How accurately did the bank do its calculations?

4. Ask your parents, grandparents, or other relatives for stories about how much things cost when they were your age. Use the rule of 72 or your calculator to estimate the yearly inflation rate since the time of these stories. Are these rates approximately the same for all the stories you can collect?

5. Find a car loan contract and read the "fine print." Does it contain a reference to the Rule of 78 in the early repayment section? If there is an example, compare it to the actual amount using the formulas from Section 1.4.

6. Look up the article "A Hidden Case of Negative Amortization" by Bert K. Waits and Franklin Demana in the March 1990 issue of *The College Mathematics Journal* and write a page about their

example. Would you ever agree to use the Rule of 78 with a loan longer than 5 years?

7. Ask your parents or other relatives if you may read their home mortgage contract. Check the bank's calculations and write a page about what you find and how much of the total mortgage payments is interest.

8. Ask your grandparents or other relatives to show you their monthly Social Security checks. Look up the current long-term interest rate in today's newspaper and calculate the present value of an annuity with the same monthly payment for the next 20 years.

9. If your state runs a lottery game, read the rules and regulations to find out how the payments are made. Work out an example showing when and how much a $10 million winner would actually receive and find the present value an equivalent annuity using the current long-term interest rates from today's newspaper.

10. Look over your notes, your homework, and the text, and write a page about how your graphing calculator has helped you in this chapter. Include examples of how it has helped you to explore concepts and how it has helped to simplify your work. What was the *most* helpful or interesting use? What was the *least* helpful or interesting use? Are there problems that are easier to do "by hand" that you tried to solve on your graphing calculator?

2 MATRICES AND SYSTEMS OF EQUATIONS

Each number in a grid has a special meaning given by its row and column position. This chapter uses matrices to represent and solve problems that depend on many numbers.

2.1   Systems of Two Linear Equations in Two Variables

A "Taxing" Problem

To be acceptable, a tax law must be perceived as fair by the taxpayers. When several government divisions each attempt to tax the same resource, abhorrence of "double jeopardy" suggests that the part paid as tax to one government authority should not be subjected to further taxation by another. Consider a simplified situation in which the federal income tax is 21% of the taxable income after the state income tax has first been deducted from the taxable income, and the state income tax is 4% of the taxable income after the federal income tax has first been deducted. How could a taxpayer with a taxable income of $29,748 find the correct taxes due? There is a widespread misconception that the resulting federal and state taxes can only be approximated but not found exactly. This confusion starts with the observation that since neither tax amount is obvious, it is reasonable to estimate the state tax, calculate the corresponding federal tax, and then readjust the state tax estimate and try again. Of course, this "solution method" will be even worse for an example involving federal, state, and city taxes. Perhaps as a result of this general confusion, state income tax codes do not allow for the deduction of federal taxes, although the federal income tax Schedule A form allows the deduction of state income taxes. This chapter develops the mathematical tools that directly and simply solve the kinds of problems represented by this tax situation.

Form **W-4**	**Employee's Withholding Allowance Certificate**	OMB No. 1545-0010
Department of the Treasury Internal Revenue Service	▶ For Privacy Act and Paperwork Reduction Act Notice, see reverse.	**1995**

1	Type or print your first name and middle initial	Last name		2	Your social security number

Home address (number and street or rural route)	3	☐ Single ☐ Married ☐ Married, but withhold at higher Single rate. Note: If married, but legally separated, or spouse is a nonresident alien, check the Single box.
City or town, state, and ZIP code	4	If your last name differs from that on your social security card, check here and call 1-800-772-1213 for a new card . . . . ▶ ☐

5	Total number of allowances you are claiming (from line G above or from the worksheets on page 2 if they apply) .	5	
6	Additional amount, if any, you want withheld from each paycheck	6	$
7	I claim exemption from withholding for 1995 and I certify that I meet **BOTH** of the following conditions for exemption:

Introduction

Several different pieces of information about a situation are often sufficient to determine the exact nature of each of the components. Neither of the statements "Bob and Sue together have $100" and "Bob has $20 less than Sue" tells us how much either has. But taken together, they force the conclusion that Bob has $40 and Sue has $60. Many methods of solving such simple problems have been used since antiquity, and perhaps even more ways are known to justify the solution once it has been found. Unfortunately, as the problem situations become more complicated and the statements more numerous, many methods that succeed for simple problems become unwieldy and often fail. It is the purpose of this chapter to explore a solution method that solves many kinds of interesting problems and has the pleasing property that the method for large situations is an easy extension of the method for the simplest.

Systems of Equations

We begin with the simplest form of these problems.

Systems of Two Linear Equations in Two Variables

A system of two linear equations in two variables is any problem expressed in the form

$$\begin{cases} ax + by = c \\ Ax + By = C \end{cases}$$

where x and y are the variables and the constants a, b, c, A, B, C are such that at least one of a, b, A, B is not zero.

EXAMPLE 1   **System of Two Linear Equations in Two Variables**

Express the statements "Bob and Sue together have $100" and "Bob has $20 less than Sue" as a system of two linear equations in two variables.

Solution

Let x represent the amount of money that Bob has and y represent the amount that Sue has. The first statement may be written "$x + y = 100$"

and the second as "$x = y - 20$". Rearranging this second equation by subtracting y from both sides, we obtain the system of equations:

$$\begin{cases} x + y = 100 \\ x - y = -20 \end{cases}$$

$ax + by = c$ with
$a = 1, b = 1, c = 100$

$Ax + By = C$ with
$A = 1, B = -1, C = -20$

This is not the only possible representation of this situation. Other possible representations are:

$$\begin{cases} x - y = -20 \\ x + y = 100 \end{cases}$$

Switch the order
of the equations

or

$$\begin{cases} x + y = 100 \\ -x + y = 20 \end{cases}$$

Multiply $x - y = -20$
through by -1

or

$$\begin{cases} 2x + 2y = 200 \\ x - y = -20 \end{cases}$$

Multiply $x + y = 100$
through by 2

■

While we have already noted that $x = 40$, $y = 60$ is the solution of this example, it is important to notice that this solution is itself a system of two linear equations in two variables:

$$\begin{cases} 1x + 0y = 40 \\ 0x + 1y = 60 \end{cases}$$

$x = 40, y = 60$ is a
system of equations

PRACTICE PROBLEM 1   Express the statements "a jar of pennies and nickels contains 80 coins" and "the coins in the jar are worth $1.60" as a system of two linear equations in two variables.   *Solution at the back of the book*

A *solution* of a system of two linear equations in two variables is a pair of values for the variables that satisfy all the equations (such as $x = 40$, $y = 60$ for Example 1). The *solution set* is the collection of all solutions. *Solving* the system of equations means finding this solution set. If the two equations represent different information about the situation, the system is *independent,* and if these two pieces of information are not contradictory, the system is *consistent.* If the two equations represent the *same* information about the situation, the system is *dependent.* The two statements "Bob has $20 less than Sue" and "Sue has $20 more than Bob" may appear at first to be two different pieces of information, but they really express the same relationship between Bob's and Sue's wealth ("$x = y - 20$" is the same as "$y = x + 20$"). If the two equations represent contradictory statements, the system is *inconsistent.* The two statements "Bob has $20 less than Sue" and "Sue has $50 more than Bob" are inconsistent because if one were true, the other necessarily would be false. If the equations are inconsistent, there can be no solution; if they are consistent, there will be at least one solution.

Graphical Representations

An equation of the form $ax + by = c$ (with at least one of a and b not zero) is the equation of a line written in *general form*. If both a and b are not zero, the line $ax + by = c$ has x-intercept $(\frac{c}{a}, 0)$ and y-intercept $(0, \frac{c}{b})$. If $a = 0$, the line is the horizontal line $y = \frac{c}{b}$. If $b = 0$, the line is the vertical line $x = \frac{c}{a}$. If $c = 0$, the line passes through the origin $(0, 0)$ and the points $(b, -a)$ and $(-b, a)$.

EXAMPLE 2 **Sketching a Linear Equation in Two Variables**

Sketch each two-variable linear equation.

a. $2x + 3y = 12$

b. $0x + 3y = 12$

c. $2x + 0y = 12$

d. $2x + 3y = 0$

Solution

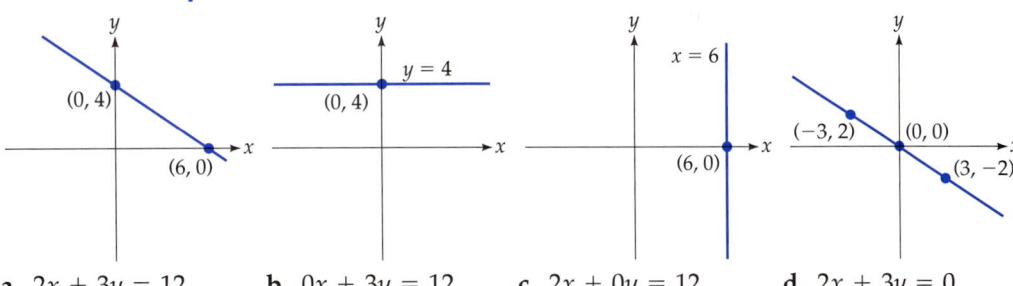

a. $2x + 3y = 12$ **b.** $0x + 3y = 12$ **c.** $2x + 0y = 12$ **d.** $2x + 3y = 0$

If the system $\begin{cases} ax + by = c \\ Ax + By = C \end{cases}$ is *independent*, it represents two *distinct* lines, whereas if it is *dependent*, the equations represent the *same* line. If it does represent two distinct lines, they are *parallel* if the system is *inconsistent* and *intersecting* if it is *consistent*.

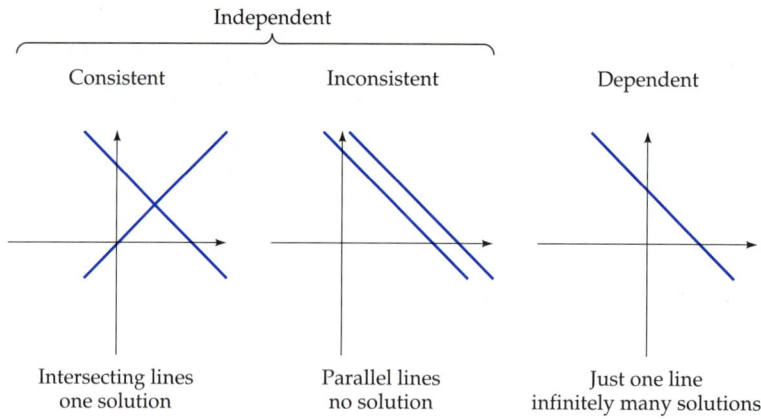

Intersecting lines
one solution

Parallel lines
no solution

Just one line
infinitely many solutions

The nature of the solution of a system of equations (and any values) can be determined from the graph if it is drawn with sufficient accuracy.

EXAMPLE 3 Solving by Graphing

Solve each system of equations by sketching the graphs of the lines.

a. $\begin{cases} 2x + 3y = 12 \\ x - y = 1 \end{cases}$ **b.** $\begin{cases} 2x + 3y = 12 \\ 4x + 6y = 12 \end{cases}$ **c.** $\begin{cases} 2x + 3y = 12 \\ 4x + 6y = 24 \end{cases}$

Solution

The line $2x + 3y = 12$ was sketched in Example 2a. For part (a), the second line has intercepts $(1, 0)$ and $(0, -1)$; for (b), the second line has intercepts $(3, 0)$ and $(0, 2)$; and for (c), the second line has intercepts $(6, 0)$ and $(0, 4)$.

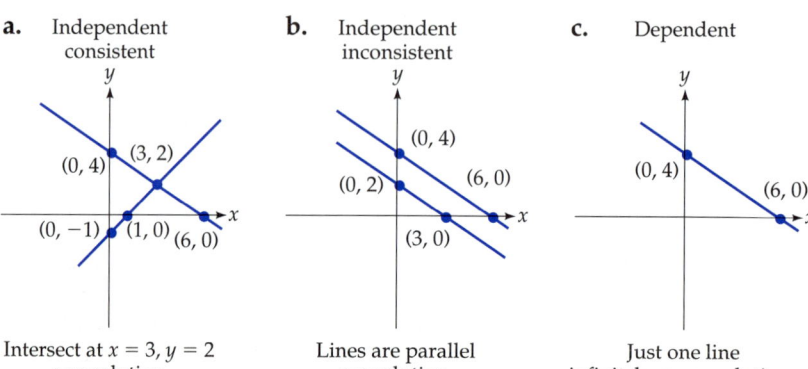

a. Independent consistent

Intersect at $x = 3, y = 2$
one solution

b. Independent inconsistent

Lines are parallel
no solution

c. Dependent

Just one line
infinitely many solutions

Notice that the solution to (a) checks because $2(3) + 3(2) = 6 + 6 = 12$ and $(3) - (2) = 1$ as required by the equations. In (c), *every* point on the line $2x + 3y = 12$ is a solution. Rewriting $2x + 3y = 12$ as $x = (12 - 3y)/2 = 6 - \frac{3}{2}y$, we can *parameterize* these solutions as $x = 6 - \frac{3}{2}t, y = t$ where t is any real number. Choosing $t = 8$, we have the particular solution $x = -6, y = 8$ and similarly for any other value of t. This parameterized solution for (c) checks because $2(6 - \frac{3}{2}t) + 3(t) = 12 - 3t + 3t = 12$ and $4(6 - \frac{3}{2}t) + 6(t) = 24 - 6t + 6t = 24$ as required by the equations.

■

Graphing Calculator Exploration

You can view a nonvertical line $ax + by = c$ by entering it as $y = (c - ax)/b$. To see the system of equations $\begin{cases} 2x + 3y = 12 \\ x - y = 1 \end{cases}$ from Example 3a,

a. Enter the first equation as $y_1 = (12 - 2x)/3$ and the second as $y_2 = (1 - 1x)/(-1)$. Graph them on the window $[-3, 7]$ by $[-3, 5]$.

b. Use TRACE or EVALUATE to check the x- and y-intercepts.

c. Use TRACE or INTERSECT to find the intersection point of these two lines.

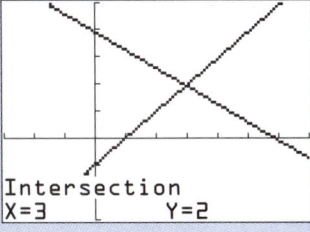

```
Intersection
X=3          Y=2
```

Substitution Method

The *substitution method* quickly finds the solution of an independent and consistent system when one of the equations represents a vertical or horizontal line.

EXAMPLE 4   Solving by the Substitution Method

Solve $\begin{cases} x + 3y = 15 \\ x = 9 \end{cases}$ by the substitution method.

Solution

Since the second equation states that $x = 9$, we may substitute 9 for x in the first equation:

$$(9) + 3y = 15 \qquad \text{$x + 3y = 15$ with $x = 9$}$$

$$3y = 6 \qquad \text{Subtract 9 from both sides}$$

$$y = 2 \qquad \text{Divide by 3}$$

The solution is $x = 9$, $y = 2$. The point $(9, 2)$ is the intersection of the line $x + 3y = 15$ with the vertical line $x = 9$.

■

The substitution method also can be used when one of the equations can be solved for a variable appearing in the other equation.

EXAMPLE 5   Solving by the Substitution Method, Continued

Solve $\begin{cases} x + 3y = 15 \\ 2x - 5y = 8 \end{cases}$ by the substitution method.

Solution

Since the first equation can easily be solved for x as $x = 15 - 3y$, we may substitute $15 - 3y$ for x in the second equation:

$$2(15 - 3y) - 5y = 8 \qquad \text{Replace x with $15 - 3y$}$$

$$30 - 6y - 5y = 8 \qquad \text{Multiply out}$$

$$-11y = -22 \qquad \text{Collect like terms}$$

$$y = 2 \qquad \text{Divide by -11}$$

and then

$$x = 15 - 3(2) = 15 - 6 = 9 \qquad \text{$x = 15 - 3y$ with $y = 2$}$$

The solution is $x = 9$, $y = 2$ and the original system is independent and consistent.

■

PRACTICE PROBLEM 2   Solve $\begin{cases} 2x + y = 10 \\ x + 2y = 8 \end{cases}$ by the substitution method.

Solution at back of the book

An *inconsistent* system of equations always leads to an impossible equation stating that two different numbers are equal. For instance, if we attempt to solve $\begin{cases} x + y = 10 \\ x + y = 20 \end{cases}$ by using the first equation to find that $x = 10 - y$ and then substituting this for x in the second equation, we find $(10 - y) + y = 20$ so that $10 = 20$. But this is contradictory, so there is no solution and the equations are inconsistent.

A *dependent* system of equations always results in a useless equation stating that a number equals itself. For instance, if we attempt to solve $\begin{cases} x + y = 10 \\ 2x + 2y = 20 \end{cases}$ by using the first equation to find that $x = 10 - y$ and then substituting this for x in the second equation, we find $2(10 - y) + 2y = 20$ so that $20 - 2y + 2y = 20$. But $20 = 20$ is just uninformative but not contradictory, giving as solutions *all* the points on the line $x + y = 10$, so the system is dependent.

Equivalent Systems of Equations

Two systems of equations are *equivalent* if they have the same solution. Returning to Example 1 (page 169), we have already remarked that the following systems are all equivalent:

$$\begin{cases} x + y = 100 \\ x - y = -20 \end{cases} \qquad \begin{cases} x - y = -20 \\ x + y = 100 \end{cases} \qquad \begin{cases} x + y = 100 \\ -x + y = 20 \end{cases}$$

$$\begin{cases} 2x + 2y = 200 \\ x - y = -20 \end{cases} \quad \text{and} \quad \begin{cases} 1x + 0y = 40 \\ 0x + 1y = 60 \end{cases}$$

An equivalent system can be obtained from a given system of equations by doing one or more of the following:

1. Switch the order of the equations.
$$\begin{cases} x + y = 100 \\ x - y = -20 \end{cases} \xrightarrow[\text{equations}]{\text{Switch}} \begin{cases} x - y = -20 \\ x + y = 100 \end{cases}$$

2. Multiply or divide one of the equations by a nonzero number.
$$\begin{cases} x + y = 100 \\ x - y = -20 \end{cases} \xrightarrow[\text{equation by } -1]{\text{Multiply second}} \begin{cases} x + y = 100 \\ -x + y = 20 \end{cases}$$

3. Replace an equation with its sum or difference with the other equation.
$$\begin{cases} x + y = 100 \\ x - y = -20 \end{cases} \xrightarrow[\text{to the second}]{\text{Add first equation}} \begin{cases} x + y = 100 \\ 2x + 0y = 80 \end{cases}$$

Elimination Method

The *elimination method* solves $\begin{cases} ax + by = c \\ Ax + By = C \end{cases}$ by attempting to change it into an equivalent system of the form $\begin{cases} 1x + 0y = P \\ 0x + 1y = Q \end{cases}$ so that the y variable has been removed from the first equation and the x variable has been removed from the second. If this can be done, the solution is $x = P$, $y = Q$ and the system of equations is independent and consistent. Even if this cannot be done, the attempt will still identify the system as dependent or inconsistent.

EXAMPLE 6   Solving by the Elimination Method

Solve $\begin{cases} x + 3y = 15 \\ 2x - 5y = 8 \end{cases}$ by the elimination method.

Solution

We hope to remove the x variable from the second equation and the y variable from the first. The following sequence of equivalent systems is one possible way of reaching the solution. First, we eliminate the x variable from the second equation by subtracting an appropriate multiple of the first equation:

$$\begin{cases} 2x + 6y = 30 \\ 2x - 5y = 8 \end{cases}$$
Multiply the first equation by 2 to get a second $2x$

$$\begin{cases} 2x + 6y = 30 \\ 0x - 11y = -22 \end{cases}$$
Subtract the first equation from the second to get a zero

Then we eliminate the y variable from the first equation by adding an appropriate multiple of the second equation:

$$\begin{cases} 2x + 6y = 30 \\ 0x - 6y = -12 \end{cases}$$
Multiply the second equation by $\frac{6}{11}$ to get a second $6y$

$$\begin{cases} 2x + 0y = 18 \\ 0x - 6y = -12 \end{cases}$$
Add the second equation to the first to get a zero

We finish by dividing to get just one x in the first equation and one y in the second equation:

$$\begin{cases} 1x + 0y = 9 \\ 0x + 1y = 2 \end{cases}$$
Divide the first equation by 2
Divide the second equation by -6

The solution is $x = 9$, $y = 2$ and the original system is independent and consistent. ■

An inconsistent system of equations always leads to the impossible equation that zero equals a nonzero number. For instance, solving Example 3b by the elimination method could be done as:

$$\begin{cases} 2x + 3y = 12 \\ 4x + 6y = 12 \end{cases} \xrightarrow[\text{equation by 2}]{\text{Divide second}} \begin{cases} 2x + 3y = 12 \\ 2x + 3y = 6 \end{cases} \xrightarrow[\text{from second}]{\substack{\text{Subtract first} \\ \text{equation}}} \begin{cases} 2x + 3y = 12 \\ 0x + 0y = -6 \end{cases}$$

Since the second equation says $0 = -6$, which is contradictory, there is no solution and the equations are inconsistent.

A dependent system of equations always results in one equation becoming $0x + 0y = 0$. For instance, solving Example 3c by the elimination method could be done as follows:

$$\begin{cases} 2x + 3y = 12 \\ 4x + 6y = 24 \end{cases} \xrightarrow[\text{equation by 2}]{\text{Divide second}} \begin{cases} 2x + 3y = 12 \\ 2x + 3y = 12 \end{cases} \xrightarrow[\text{from second}]{\substack{\text{Subtract first} \\ \text{equation}}} \begin{cases} 2x + 3y = 12 \\ 0x + 0y = 0 \end{cases}$$

Since the second equation says $0 = 0$, which is uninformative but not contradictory, the solutions are all the points on the line $2x + 3y = 12$ and the system is dependent.

PRACTICE PROBLEM 3   Solve $\begin{cases} x + y = 100 \\ x - y = -20 \end{cases}$ by the elimination method.

Solution at the back of the book

We conclude this section with the solution of the tax problem posed in the Application Preview.

EXAMPLE 7   A Taxing Problem

What are the federal and state taxes on an income of $29,748 if the federal tax is 21% of the income after first deducting the state tax and the state tax is 4% of the income after first deducting the federal tax?

Solution

Let x represent the federal tax and y represent the state tax. The statements:

"the federal tax is 21% of ($29,748 less the state tax)"
"the state tax is 4% of ($29,748 less the federal tax)"

become the equations:

$$x = 0.21(29{,}748 - y) \quad \text{and} \quad y = 0.04(29{,}748 - x)$$

Multiplying out to remove the parentheses:

$$x = 6247.08 - 0.21y \quad \text{and} \quad y = 1189.92 - 0.04x$$

Moving the variables to the left sides of the equals signs, we can represent the given situation as a system of two linear equations in two variables:

$$\begin{cases} x + 0.21y = 6247.08 \\ 0.04x + y = 1189.92 \end{cases}$$

Solving by the elimination method, we create the following sequence of equivalent systems of equations. First, we eliminate the x variable from the second equation by subtracting an appropriate multiple of the first equation:

$$\begin{cases} 0.04x + 0.0084y = 249.8832 \\ 0.04x + y = 1189.92 \end{cases}$$

Multiply first equation by 0.04

$$\begin{cases} 0.04x + 0.0084y = 249.8832 \\ 0x + 0.9916y = 940.0368 \end{cases}$$

Subtract first equation from the second

Simplifying the numbers by dividing, we obtain

$$\begin{cases} 1x + 0.21y = 6247.08 \\ 0x + 1y = 948 \end{cases}$$

Divide first equation by 0.04
Divide second equation by 0.9916

Next we eliminate the y variable from the first equation by subtracting an appropriate multiple of the second equation:

$$\begin{cases} 1x + 0.21y = 6247.08 \\ 0x + 0.21y = 199.08 \end{cases}$$

Multiply second equation by 0.21

$$\begin{cases} 1x + 0y = 6048 \\ 0x + 0.21y = 199.08 \end{cases}$$

Subtract second equation from first

We finish by dividing to get just one y in the second equation:

$$\begin{cases} 1x + 0y = 6048 \\ 0x + 1y = 948 \end{cases}$$

Divide second equation by 0.21

The solution is $x = 6048$, $y = 948$. The federal tax is \$6048 and the state tax is \$948. You should check that each of these tax amounts is the correct percentage of the original income once the other tax is subtracted.

SUMMARY

A *system of two linear equations in two variables* can be written in the form:

$$\begin{cases} ax + by = c \\ Ax + By = C \end{cases}$$

Since the graphs of these linear equations are lines, the system of equations can be solved by sketching the lines.

$\begin{cases} ax + by = c \\ Ax + By = C \end{cases}$	Graph	Number of Solutions
Independent and consistent	Lines intersect	One
Independent and inconsistent	Parallel lines	None
Dependent	Just one line	Infinitely many

The *substitution method* solves one of the equations for a variable appearing in the other and uses this new expression to rewrite the other equation with just one variable. Solving this new equation and then finding the first variable solves the system of equations.

Equivalent systems of equations have the same solution. Systems of equations equivalent to a given system can be found by:

1. Switching the order of the equations,

2. Multiplying or dividing one equation by a nonzero number, or

3. Replacing an equation with its sum or difference with the other.

The *elimination method* solves the system by attempting to change it into an equivalent system of the form $\begin{cases} 1x + 0y = P \\ 0x + 1y = Q \end{cases}$. If this can be done, the solution is $x = P$, $y = Q$ and the system of equations is independent and consistent. If an equation of the form $0x + 0y = 1$ is obtained, there is *no solution* and the system is *independent* and *inconsistent*. If an equation of the form $0x + 0y = 0$ is obtained, there are *infinitely many solutions* and the system is *dependent*.

EXERCISES 2.1

Represent each pair of statements as a system of two linear equations in two variables using the given x and y definitions. Verify that the values given for x and y are a solution for the system of equations.

1. "The sum of two numbers is eighteen" and "the first number is two more than the second number." Let x be the first number and y be the second number. $x = 10, y = 8$

2. "The sum of two numbers is twenty-five" and "twice the first number added to the second number totals thirty-two." Let x be the first number and y be the second number. $x = 7, y = 18$

3. "Tom has $6 more than Alice" and "together, they have $40." Let x be the amount of money that Tom has and y be the amount Alice has. $x = 23, y = 17$

4. "Bill and Jessica together have $25" and "Jessica has $12." Let x be the amount of money that Bill has and y be the amount Jessica has. $x = 13, y = 12$

5. "An envelope of $1 and $5 bills contains thirty bills" and "the money in the envelope is worth $70." Let x be the number of $1 bills and y be the number of $5 bills. $x = 20, y = 10$

6. "An envelope of $10 and $20 bills contains eight bills" and "the money in the envelope is worth $110." Let x be the number of $10 bills and y be the number of $20 bills. $x = 5, y = 3$

7. "A small theater sold tickets for all one hundred seats" and "the box office receipts of $650 came from adult tickets at $10 and child tickets at $5." Let x be the number of adult tickets sold and y be the number of child tickets sold. $x = 30, y = 70$

8. "A movie theater sold tickets for three hundred seats" and "the box office receipts of $2400 came from adult tickets at $9 and child tickets at $6." Let x be the number of adult tickets sold and y be the number of child tickets sold. $x = 200, y = 100$

9. "A corn and beet farmer planted 225 acres of crops" and "he planted twice as many acres of corn as acres of beets." Let x be the number of acres of corn and y be the number of acres of beets. $x = 150, y = 75$

10. "A stock and bond speculator invested $10,000 in the market" and "she invested three times as much in stocks as in bonds." Let x be the amount in stocks and y be the amount in bonds. $x = 7500, y = 2500$

Solve each of the following systems of equations by graphing. Identify each system as "independent and consistent," "independent and inconsistent," or "dependent."

You may use a graphing calculator if permitted by your instructor.

11. $\begin{cases} x + y = 6 \\ x - y = 2 \end{cases}$

12. $\begin{cases} -x + y = 2 \\ x + y = 4 \end{cases}$

13. $\begin{cases} 2x + y = 8 \\ x = 3 \end{cases}$

14. $\begin{cases} x + 2y = 10 \\ y = 4 \end{cases}$

15. $\begin{cases} x - y = 4 \\ -x + 2y = -6 \end{cases}$

16. $\begin{cases} 2x - y = 2 \\ x + 2y = 6 \end{cases}$

17. $\begin{cases} x + y = 10 \\ -x - y = 10 \end{cases}$

18. $\begin{cases} -2x + 4y = -16 \\ x - 2y = 4 \end{cases}$

19. $\begin{cases} x + y = 10 \\ -x - y = -10 \end{cases}$

20. $\begin{cases} 2x - 4y = 16 \\ -x + 2y = -8 \end{cases}$

Solve each of the following systems of equations by the substitution method. Identify each system as "independent and consistent," "independent and inconsistent," or "dependent."

21. $\begin{cases} x + 2y = 10 \\ y = 3 \end{cases}$

22. $\begin{cases} 2x + y = 8 \\ x = 2 \end{cases}$

23. $\begin{cases} 2x + y = 20 \\ x + y = 15 \end{cases}$

24. $\begin{cases} x + y = 12 \\ x + 2y = 14 \end{cases}$

25. $\begin{cases} 5x + 2y = 30 \\ 2x + y = 10 \end{cases}$

26. $\begin{cases} 2x + 3y = 25 \\ x - y = 5 \end{cases}$

27. $\begin{cases} 3x + 2y = 30 \\ x - y = -5 \end{cases}$

28. $\begin{cases} 2x + y = 20 \\ x + 3y = 15 \end{cases}$

29. $\begin{cases} -2x + 2y = -20 \\ x - y = 10 \end{cases}$

30. $\begin{cases} 3x - 2y = 30 \\ -6x + 4y = 30 \end{cases}$

Solve each of the following systems of equations by the elimination method. Identify each system as "independent and consistent," "independent and inconsistent," or "dependent."

31. $\begin{cases} x + y = 11 \\ 2x + 3y = 30 \end{cases}$

32. $\begin{cases} 3x + 2y = 30 \\ x + y = 13 \end{cases}$

33. $\begin{cases} 3x + y = 15 \\ x + 2y = 10 \end{cases}$

34. $\begin{cases} x + 3y = 30 \\ 2x + y = 10 \end{cases}$

35. $\begin{cases} x + 2y = 14 \\ 3x + 4y = 36 \end{cases}$

36. $\begin{cases} 2x + 3y = 30 \\ x - y = 10 \end{cases}$

37. $\begin{cases} 2x + 5y = 60 \\ 2x + 3y = 48 \end{cases}$

38. $\begin{cases} 4x + 3y = 36 \\ x + 3y = 18 \end{cases}$

39. $\begin{cases} 3x + 4y = -24 \\ 6x + 8y = 24 \end{cases}$

40. $\begin{cases} x - 4y = 20 \\ -2x + 8y = -40 \end{cases}$

APPLIED EXERCISES

Formulate each situation as a system of two linear equations in two variables. Be sure to state clearly the meaning of your x and y variables. Solve the system of equations by the elimination method. Be sure to state your final answer in terms of the original question.

41. Coins in a Jar A jar contains 60 nickels and dimes worth $4.30. How many of each are in the jar?

42. Bills in an Envelope An envelope found in a safe deposit box contains ninety $5 and $20 bills worth $1200. How many of each are in the envelope?

43. Financial Planning A retired couple wish to invest their nest egg of $10,000 in a money market account paying 6% and in a stock mutual fund returning 11%. If their income tax and Social Security situation requires that they earn $1000 from these investments, how much should they invest in each?

44. Home Financing A young couple needs to borrow $168,000 to finance their first house. They could borrow the whole amount at 12% from their bank, but her father offers to lend them enough of the money at 5% (with the same terms as the bank loan) to reduce the overall interest rate to just 8%. How much does he lend them and how much do they borrow from the bank?

45. Ice Hockey Concession Receipts The concession stand at an ice hockey rink had receipts of $7200 from selling 3000 sodas and hot dogs. If each soda sold for $2 and each hot dog sold for $3, how many of each did it sell?

46. Baseball Tickets A college baseball game generated box office receipts of $4800 from 600 ticket sales. If general admission tickets were $12 and student tickets were half-price, how many of each were sold?

47. Income Taxes Find the federal and state taxes on a taxable income of $49,900 if the federal tax is 10% of the taxable income after first deducting the state tax and the state tax is 2% of the taxable income after first deducting the federal tax.

48. Estate Division A will specifies that each of two sons receive one-half of the $3 million estate after first deducting the other's share, and that any remainder is then to be given to their sister. How much does each son receive? How much is left for the sister?

49. Sports Nutrition The dietician at a sports training facility has determined that one of her athletes needs an additional 600 mg each of calcium and phosphorus daily. Two supplements are available containing the milligrams of calcium and phosphorus per tablet as given by the table. How many tablets of each supplement will provide the required calcium and phosphorus?

	Calcium	Phosphorus
Supplement A	150	100
Supplement B	120	120

50. Bicycle Shop Management A bicycle shop has $10,500 to spend on new bikes and 390 hours of assembly time to put them together. Each mountain bike costs $50 wholesale and takes two hours to assemble. Each racing bike costs $70 wholesale and takes two-and-one-half hours to assemble. How many of each can the shop buy and assemble to use all the available money and time?

Explorations and Excursions

The following problems extend and augment the material presented in the text.

Systems of Three Linear Equations in Two Variables. If we have a system that represents three lines instead of just two, there are more possibilities. The second line could be parallel to the first and the third might be the same as the first, so the system could be both inconsistent and dependent.

Sketch each of the following systems and identify them as "consistent" or "inconsistent" and as "dependent" or "independent." If the system is consistent, find the solution.

 You may use a graphing calculator if permitted by your instructor.

51. $\begin{cases} x + y = 10 \\ x + y = 5 \\ 2x + 2y = 20 \end{cases}$ **52.** $\begin{cases} x + y = 10 \\ x - y = 0 \\ 2x + 2y = 20 \end{cases}$

53. $\begin{cases} x + 3y = 27 \\ 2x + 5y = 50 \\ x + y = 19 \end{cases}$ **54.** $\begin{cases} 2x + 3y = 12 \\ x - y = 1 \\ 4x + y = 14 \end{cases}$

55. $\begin{cases} 2x + y = 24 \\ x - y = 6 \\ x + y = 16 \end{cases}$

 More About Parameterizations

56. In Example 2a, we sketched the line $2x + 3y = 12$ and in Example 3c, we parameterized this line as $x = 6 - \frac{3}{2}t$, $y = t$. To graph these *parametric equations* on your calculator, change the MODE from FUNCtion to PARametric and enter the pair of equations $x_{1T} = 6 - (3/2)T$, $y_{1T} = T$. Set the WINDOW with Tmin $= -2$, Tmax $= 2$, Tstep $= 0.1$, Xmin $= -10$, Xmax $= 10$, Ymin $= -10$, and Ymax $= 10$ and watch as your calculator draws the graph. Change Tmin to -3, Tmax to 3, and GRAPH it again. Experiment with different val-

ues for Tmin and Tmax until you see how to get a "complete" picture of this line in this window.

57. Check that the line $2x - 5y = 8$ can be parameterized as $x = 4 + \frac{5}{2}t$, $y = t$. Set the WINDOW with Tmin $= -2$, Tmax $= 2$, Tstep $= 0.1$, Xmin $= -10$, Xmax $= 10$, Ymin $= -10$, and Ymax $= 10$. GRAPH these equations (be sure to first change the MODE from FUNCtion to PARametric) and use TRACE to explore this line segment. Experiment with different values for Tmin and Tmax until you see how to get a "complete" picture of this line in this window.

58. Find a parameterization for the line $x + 3y = 15$ and use your calculator to verify that your parametric equations determine a line with x-intercept (15, 0) and y-intercept (0, 5).

59. Use the parametric equations from Exercises 57 and 58 to GRAPH Example 5 on the WINDOW with Tmin $= -5$, Tmax $= 10$, Tstep $= 0.1$, Xmin $= -5$, Xmax $= 20$, Ymin $= -10$, and Ymax $= 10$. Use TRACE to explore both lines and verify that the intersection point is (9, 2).

60. Using the WINDOW with Tmin $= -5$, Tmax $= 10$, Tstep $= 0.1$, Xmin $= -5$, Xmax $= 5$, Ymin $= -5$, and Ymax $= 5$, explore each of the following pairs of parametric equations. Can you tell in advance which give lines and which give curves? Why does (c) give just part of the line?

a. $\begin{cases} x = T - 1 \\ y = T \end{cases}$ **b.** $\begin{cases} x = T^2 - 1 \\ y = T \end{cases}$

c. $\begin{cases} x = T^2 - 1 \\ y = T^2 \end{cases}$ **d.** $\begin{cases} x = T^2 - 1 \\ y = T^3 \end{cases}$

e. $\begin{cases} x = T^3 - 1 \\ y = T^2 \end{cases}$ **f.** $\begin{cases} x = T^3 - 1 \\ y = T^3 \end{cases}$

Change Tstep from 0.1 to 0.05 and redo (a) and (f). Watch while your calculator draws each line. How do these parameterizations of the line $x - y = -1$ differ?

Round-off errors can completely misrepresent the true nature of a system of equations. (If you should wish to remove fractions from a problem, multiply each equation through by the least common denominator instead of rounding off.)

61. Use the elimination method to show that the system $\begin{cases} x + \frac{1}{3}y = 39 \\ 2x + \frac{2}{3}y = 84 \end{cases}$ is inconsistent.

62. Rounding to one decimal place, the system in Exercise 61 becomes $\begin{cases} x + 0.3y = 39 \\ 2x + 0.7y = 84 \end{cases}$. Use the elimination method to show that this new system is consistent and independent with solution $x = 21$, $y = 60$.

63. Use the elimination method to show that the system $\begin{cases} x + \frac{2}{9}y = 9.79 \\ 4x + \frac{8}{9}y = 39.16 \end{cases}$ is dependent.

64. Rounding to two decimal places, the system in Exercise 63 becomes $\begin{cases} x + 0.22y = 9.79 \\ 4x + 0.89y = 39.16 \end{cases}$. Use the elimination method to show that this new system is consistent and independent with solution $x = 9.79$, $y = 0$.

65. Use the elimination method to show that the system $\begin{cases} 2.1x + \frac{1}{7}y = 157 \\ 3x + \frac{1}{5}y = 224 \end{cases}$ is consistent and independent with solution $x = 70$, $y = 70$. Rounding to two decimal places, this system becomes $\begin{cases} 2.10x + 0.14y = 157 \\ 3.00x + 0.20y = 224 \end{cases}$. Use the elimination method to show that this new system is inconsistent.

2.2 | Matrices and Linear Equations in Two Variables

APPLICATION PREVIEW

Spreadsheets

For hundreds of years the business spreadsheet has been a standard accounting tool. Originally written on slates or paper, it organizes data and calculated values into rows and columns containing similar quantities. Since even simple business situations may lead to large collections of numbers linked by complicated formulas, carrying out the required calculations can be a formidable, time-consuming task. Even worse, should one of the data values change, the effect on the rest of the spreadsheet can only be found by recalculating the entire grid of numbers.

When Dan Bricklin was a graduate student at the Harvard Business School, he had the idea that business calculations could be greatly simplified if the spreadsheet was really a computer program with the formulas embedded in the positions where the calculated values should appear. Together with Robert Frankston, a friend from their undergraduate days at the Massachusetts Institute of Technology, he created the first personal computer spreadsheet program in 1979. Written for the Apple II computer with 32K memory, VisiCalc had only 254 rows and 63 columns and a crude user interface. But it worked and became the original "killer application" of the revolution that changed the personal computer from an interesting toy to an essential business tool.

C1 = A1 + B1		
A	**B**	**C**
1 300	800	1100
2 200	700	900

The formula in cell C1 is A1 + B1, so C1 displays the sum of the current values in A1 and B1. Can you find a similar formula for the value in cell C2?

Introduction

The solutions of many problems are greatly simplified when expressed in appropriate notations. In this section, we will use matrix notation to streamline the elimination method used in the previous section to solve systems of two linear equations in two variables. In the next section, we shall extend this simplification to the solutions of systems of many linear equations in many variables.

Matrices

A *matrix* is a rectangular array of numbers called *elements.* This rectangular array may be viewed as consisting of *rows* (with the first at the top, the second below the first, and so on) or *columns* (with the first on the left, the second to the right of the first, and so on). The *dimension* of a matrix is the size expressed as "rows by columns" and we will write $m \times n$ for the dimension of a matrix with m rows and n columns. Thus a 5×2 matrix is "tall and thin" while a 2×5 matrix is "short and wide." A *square matrix* has the same number of rows as columns. A *row matrix* has just one row, and a *column matrix* has just one column.

$$(1 \quad 2 \quad 3) \qquad \begin{pmatrix} 1 \\ 2 \\ 3 \end{pmatrix}$$

Row matrix   Column matrix

Each element of a matrix has both a *value* and a *position* given by its row and column address. We name matrices with capital letters ($A, B, C, \ldots$) and then the elements are named by the corresponding lowercase letter together with the appropriate row and column address. For instance, if

$$A = \begin{pmatrix} 1 & 2 & 3 & 4 \\ 5 & 6 & 7 & 8 \\ 9 & 10 & 11 & 12 \end{pmatrix}$$

then the dimension of A is 3×4 and $a_{2,3} = 7$ because 7 is the value of the element of A in the second row and third column. This *double subscript* notation for the elements of the matrix is sometimes used without the comma between the row and column addresses, so that a_{23} means $a_{2,3}$. The "L" shape of the letter and subscripts suggests the "follow the L" nickname sometimes given this notation: just as you make the letter L by a downstroke then a right-stroke, the element $a_{2,3}$ is found by moving *down* to row 2 and then *right* to column 3. The elements on the *main diagonal* are those with the same row address as column address. For the matrix A above, the main diagonal consists of the elements $a_{1,1} = 1$, $a_{2,2} = 6$, and $a_{3,3} = 11$.

Augmented Matrices from Systems of Equations

An *augmented matrix* is a matrix created from two "smaller" matrices having the same number of rows by placing them beside each other and joining them into one "larger" matrix. The system of equations $\begin{cases} ax + by = c \\ Ax + By = C \end{cases}$ naturally gives rise to a *coefficient matrix* $\begin{pmatrix} a & b \\ A & B \end{pmatrix}$ and a *constant term matrix* $\begin{pmatrix} c \\ C \end{pmatrix}$. Taken together, these form the augmented matrix $\begin{pmatrix} a & b & c \\ A & B & C \end{pmatrix}$, which represents the system of equations. The first column gives the coefficients of the x-variable, the second gives those of the y-variable, and the last column contains the constant terms after the equals signs.

Augmented Matrix of a System of Equations

The augmented matrix $\begin{pmatrix} a & b & c \\ A & B & C \end{pmatrix}$ represents the system of equations $\begin{cases} ax + by = c \\ Ax + By = C \end{cases}$.

EXAMPLE 1 Augmented Matrices and Systems of Equations

a. Find the augmented matrix representing the system
$$\begin{cases} 2x + 3y = 24 \\ 4x + 5y = 60 \end{cases}.$$

b. Find the system of equations represented by the augmented matrix $\begin{pmatrix} 6 & 8 & 84 \\ 4 & 5 & 60 \end{pmatrix}$.

Solution

a. The system of equations $\begin{cases} 2x + 3y = 24 \\ 4x + 5y = 60 \end{cases}$ is represented by $\begin{pmatrix} 2 & 3 & 24 \\ 4 & 5 & 60 \end{pmatrix}$.

b. The augmented matrix $\begin{pmatrix} 6 & 8 & 84 \\ 4 & 5 & 60 \end{pmatrix}$ represents $\begin{cases} 6x + 8y = 84 \\ 4x + 5y = 60 \end{cases}$.

■

PRACTICE PROBLEM 1

a. Find the augmented matrix representing the system of equations $\begin{cases} 2x - y = 14 \\ x + 3y = 21 \end{cases}$.

b. Find the system of equations represented by the augmented matrix $\begin{pmatrix} 3 & 2 & 35 \\ 1 & 3 & 21 \end{pmatrix}$.

Solutions at the back of the book

Row Operations

In order to solve the system of equations represented by an augmented matrix, we transform it into *equivalent matrices* using *row operations* that correspond to the manipulations we used to solve the system by the elimination method (see page 175).

Matrix Row Operations

1. Switch any two rows.
2. Multiply or divide one of the rows by a nonzero number.
3. Replace a row by its sum or difference with another row.

We have phrased the above definition to apply to matrices of any size. We shall indicate the row operation used by writing next to the changed row a short formula explaining what was done. We shall write *R* for "row" and *R'* (read "R prime") for the "new row."

EXAMPLE 2 **Performing Matrix Row Operations**

Carry out the indicated row operation on the given matrix.

a. $R'1 = R2$ and $R'2 = R1$ on $\begin{pmatrix} 2 & 3 & 25 \\ 3 & 4 & 36 \end{pmatrix}$

b. $R'2 = 3R2$ on $\begin{pmatrix} 2 & 3 & 30 \\ 1 & 1 & 13 \end{pmatrix}$

c. $R'1 = R1 - R2$ on $\begin{pmatrix} 2 & 3 & 30 \\ 2 & -2 & 20 \end{pmatrix}$

Solution

a. $R'1 = R2$ says that the "new row 1 is the (old) row 2" and $R'2 = R1$ says that the "new row 2 is the (old) row 1," so this row operation switches rows 1 and 2:

$$\begin{pmatrix} 3 & 4 & 36 \\ 2 & 3 & 25 \end{pmatrix} \begin{matrix} R'1 = R2 \\ R'2 = R1 \end{matrix}$$

b. $R'2 = 3R2$ says that the "new row 2 is 3 times the (old) row 2," so the second row is multiplied by 3:

$$\begin{pmatrix} 2 & 3 & 30 \\ 3 & 3 & 39 \end{pmatrix} R'2 = 3R2$$

c. $R'1 = R1 - R2$ says that the "new row 1 is the (old) row 1 minus the (old) row 2," so the first row is replaced by its difference with the second row:

$$\begin{pmatrix} 0 & 5 & 10 \\ 2 & -2 & 20 \end{pmatrix} R'1 = R1 - R2$$

■

PRACTICE PROBLEM 2

a. Carry out $R'1 = 3R1$ on $\begin{pmatrix} 2 & 1 & 14 \\ 1 & -3 & 21 \end{pmatrix}$.

b. Carry out $R'1 = R1 + R2$ on $\begin{pmatrix} 6 & 3 & 42 \\ 1 & -3 & 21 \end{pmatrix}$.

c. Carry out $R'1 = \frac{1}{7}R1$ on $\begin{pmatrix} 7 & 0 & 63 \\ 1 & -3 & 21 \end{pmatrix}$.

d. Is $R'1 = 0R1$ a valid row operation?

e. Is $R'2 = 5R1$ a valid row operation? *Solutions at the back of the book*

Solving Equations by Row Reduction

Two matrices are *equivalent* if there is a sequence of row operations that transforms one into the other. Since the augmented matrix represents the system of equations and the row operations correspond to the steps used to solve the system, we can "solve" the augmented matrix by *row reducing* it to an equivalent matrix that displays the solution. If we can row reduce an augmented matrix so that it takes the form:

$$\begin{pmatrix} 1 & 0 & P \\ 0 & 1 & Q \end{pmatrix} \qquad \begin{cases} 1x + 0y = P \\ 0x + 1y = Q \end{cases} \text{ or } \begin{cases} x = P \\ y = Q \end{cases}$$

then the solution of the corresponding system of equations is $x = P$, $y = Q$, and the system is independent and consistent. However, if we obtain a row of zeros ending with a nonzero number:

$$0 \quad 0 \quad N \qquad\qquad N \neq 0$$

then the system is *inconsistent* and has *no solution* since the equation $0x + 0y = N$ makes the impossible claim that zero equals a nonzero number. On the other hand, if there is no such "inconsistent" row but there is a row consisting entirely of zeros:

$$0 \quad 0 \quad 0$$

then the system is *dependent* and there are *infinitely many solutions* since the equation $0x + 0y = 0$ is *always* true and represents no restriction at all. These important observations will be extended in the next section.

EXAMPLE 3   Solving Equations by Row Reduction

Solve $\begin{cases} x + 3y = 15 \\ 2x - 5y = 8 \end{cases}$ by row reducing the corresponding augmented matrix.

Solution

The augmented matrix for this system of equations is

$$\begin{pmatrix} 1 & 3 & 15 \\ 2 & -5 & 8 \end{pmatrix}$$

We hope to change the first column from $\begin{matrix} 1 \\ 2 \end{matrix}$ to $\begin{matrix} 1 \\ 0 \end{matrix}$ (which is the same as removing the x variable from the second equation) and then the second column to $\begin{matrix} 0 \\ 1 \end{matrix}$ (to remove the y variable from the first equation). If we are successful, we will have the matrix $\begin{pmatrix} 1 & 0 & P \\ 0 & 1 & Q \end{pmatrix}$ and the solution

$x = P$, $y = Q$. If not, we will find that the system is inconsistent or dependent. The following sequence of row operations and equivalent matrices is one possible way of reaching the solution. First, we get a zero at the bottom of the first column by using the row 1, column 1 element:

$$\begin{pmatrix} 2 & 6 & 30 \\ 2 & -5 & 8 \end{pmatrix} R'1 = 2R1 \qquad \text{To get a 2 above the other 2}$$

$$\begin{pmatrix} 2 & 6 & 30 \\ 0 & -11 & -22 \end{pmatrix} R'2 = R2 - R1 \qquad \text{Subtracting to get the zero}$$

Next, we get a zero at the top of the second column by using the row 2, column 2 element:

$$\begin{pmatrix} 2 & 6 & 30 \\ 0 & -6 & -12 \end{pmatrix} R'2 = \frac{6}{11}R2 \qquad \text{To get a } -6 \text{ below the 6}$$

$$\begin{pmatrix} 2 & 0 & 18 \\ 0 & -6 & -12 \end{pmatrix} R'1 = R1 + R2 \qquad \text{Adding to get another zero}$$

Since we now have zeros in the desired locations, we conclude by changing the 2 and -6 into ones by multiplying the rows by the appropriate numbers:

$$\begin{pmatrix} 1 & 0 & 9 \\ 0 & -6 & -12 \end{pmatrix} R'1 = \frac{1}{2}R1$$

$$\begin{pmatrix} 1 & 0 & 9 \\ 0 & 1 & 2 \end{pmatrix} R'2 = -\frac{1}{6}R2$$

The solution is $x = 9$, $y = 2$, and the original system is consistent and independent. [We solved this same system in the previous section by the elimination method in Example 6 (page 176) and found the same solution.] ∎

PRACTICE PROBLEM 3   Solve $\begin{cases} 2x + y = 14 \\ x - 3y = 21 \end{cases}$ by row reducing the corresponding augmented matrix. (*Hint:* Use your work from Practice Problem 2.)

Solution at the back of the book

 Graphing Calculator Exploration

Some graphing calculators can row reduce matrices. If your calculator has a RREF(command (for "reduced row-echelon form," as explained in the next section), you can easily check your row reduction. For the system of equations in Example 3,

a.   Enter $\begin{pmatrix} 1 & 3 & 15 \\ 2 & -5 & 8 \end{pmatrix}$ as matrix [A] using MATRX EDIT.

```
MATRIX[A] 2 X3
[1      3      15  ]
[2     -5      8   ]



2,3=8
```

b.   Quit and select the RREF(command from the MATRX MATH menu.

```
NAMES  MATH  EDIT
0↑cumSum(
A:ref(
B:rref(
C:rowSwap(
D:row+(
E:*row(
F:*row+(
```

c.   Apply RREF(to the matrix [A].

```
rref([A])
          [[1  0  9]
           [0  1  2]]
```

While this serves as a useful check of our answer, do not rely on it completely because the calculator sometimes returns an answer with round-off errors. Furthermore, this calculator command may not work for matrices with more rows than columns, and we will be interested in such matrices in the next section.

EXAMPLE 4   Solving Equations by Row Reduction, Continued

Solve $\begin{cases} 6x - 3y = 30 \\ -8x + 4y = -40 \end{cases}$ by row reducing the corresponding augmented matrix.

Solution

The augmented matrix is

$$\begin{pmatrix} 6 & -3 & 30 \\ -8 & 4 & -40 \end{pmatrix}$$

When the numbers in a row have an obvious common factor, removing that factor can sometimes simplify the reduction. Since the first row is divisible by 3, we begin with:

$$\begin{pmatrix} 2 & -1 & 10 \\ -8 & 4 & -40 \end{pmatrix} R'1 = \tfrac{1}{3}R1$$

Similarly for the second row:

$$\begin{pmatrix} 2 & -1 & 10 \\ -2 & 1 & -10 \end{pmatrix} R'2 = \tfrac{1}{4}R2$$

Adding the first row to the second, we get a zero at the bottom of the first column:

$$\begin{pmatrix} 2 & -1 & 10 \\ 0 & 0 & 0 \end{pmatrix} R'2 = R2 + R1$$

To finish, we divide the first row by 2 to make the row begin with a one on the left:

$$\begin{pmatrix} 1 & -\tfrac{1}{2} & 5 \\ 0 & 0 & 0 \end{pmatrix} R'1 = \tfrac{1}{2}R1$$

The zero row means that the system is *dependent*, so there are infinitely many solutions. The first row of this final matrix says that

$$x - \tfrac{1}{2}y = 5$$

or, solving for x,

$$x = 5 + \tfrac{1}{2}y$$

We may let y be *any* number t and then determine x from the above equation. That is, the solutions may be parameterized as $x = 5 + \tfrac{1}{2}t$, $y = t$, where t is *any* number. The following table lists some of these solutions for various values of the parameter t.

t	$x = 5 + \tfrac{1}{2}t$	$y = t$
-20	-5	-20
-10	0	-10
0	5	0
10	10	10
20	15	20

There are many other sequences of row operations to reduce this augmented matrix and all reach the same conclusion.

■

PRACTICE PROBLEM 4

Verify that $x = -5$, $y = -20$, and $x = 15$, $y = 20$ from the above table solve $\begin{cases} 6x - 3y = 30 \\ -8x + 4y = -40 \end{cases}$.

Solution at the back of the book

EXAMPLE 5   **Production Management**

A worker in a plastics factory breaks apart sheets of component A and strips of component B and then snaps one of each together to make a finished item. If the worker can break off 20 A's or 30 B's per minute and snap together 10 pairs of A's and B's per minute, how should the worker's 440-minute workday be divided so as to complete as many items as possible with no unused pieces left over?

Solution

Let x be the minutes spent breaking apart sheets of component A and y be the minutes breaking apart strips of component B. The remainder of the worker's time, $440 - (x + y)$, will then be spent snapping A's and B's together. Since the worker needs as many A's as B's,

$$20x = 30y$$ Rate $\times$ time gives number finished

As many finished items as A's will be completed:

$$20x = 10(440 - (x + y))$$

Rewriting these as a system of equations, the problem becomes:

$$\begin{cases} 20x - 30y = 0 \\ 30x + 10y = 4400 \end{cases}$$

The augmented matrix is

$$\begin{pmatrix} 20 & -30 & 0 \\ 30 & 10 & 4400 \end{pmatrix}$$

We begin the row reduction by removing from both rows the common factor of 10:

$$\begin{pmatrix} 2 & -3 & 0 \\ 3 & 1 & 440 \end{pmatrix} \begin{matrix} R'1 = \frac{1}{10}R1 \\ R'2 = \frac{1}{10}R2 \end{matrix}$$

To get a one in the upper lefthand corner:

$$\begin{pmatrix} 1 & 4 & 440 \\ 3 & 1 & 440 \end{pmatrix} R'1 = R2 - R1$$

To get a zero below the one in the upper lefthand corner:

$$\begin{pmatrix} 3 & 12 & 1320 \\ 3 & 1 & 440 \end{pmatrix} R'1 = 3R1$$

$$\begin{pmatrix} 3 & 12 & 1320 \\ 0 & 11 & 880 \end{pmatrix} R'2 = R1 - R2$$

Removing common factors again:

$$\begin{pmatrix} 1 & 4 & 440 \\ 0 & 1 & 80 \end{pmatrix} \begin{matrix} R'1 = \frac{1}{3}R1 \\ R'2 = \frac{1}{11}R2 \end{matrix}$$

To get a zero above the one at the bottom of the second column:

$$\begin{pmatrix} 1 & 4 & 440 \\ 0 & 4 & 320 \end{pmatrix} R'2 = 4R2$$

$$\begin{pmatrix} 1 & 0 & 120 \\ 0 & 4 & 320 \end{pmatrix} R'1 = R1 - R2$$

Now that we have zeros in the proper positions, we finish by returning the column 2, row 2 entry back to one:

$$\begin{pmatrix} 1 & 0 & 120 \\ 0 & 1 & 80 \end{pmatrix} R'2 = \frac{1}{4}R2$$

There are many other sequences of row operations to reduce this augmented matrix and all reach the same conclusion. The solution is $x = 120$, $y = 80$. In terms of the original question, the worker should break apart sheets of component A for 120 minutes, break apart strips of component B for 80 minutes, and snap A's and B's together for the remaining 240 minutes. ∎

 If you have a graphing calculator with a RREF(command, use it to check the result of the row reduction in Example 5.

SUMMARY

Each element of a matrix has both a *value* and a *position*. The augmented matrix $\begin{pmatrix} a & b & c \\ A & B & C \end{pmatrix}$ represents the system of equations $\begin{cases} ax + by = c \\ Ax + By = C \end{cases}$. The rows correspond to the equations, while the first column corresponds to the x-variable, the second column to the y-variable, and the third column to the constant terms of the equations.

The three basic row operations are:

1. Switch any two rows.

2. Multiply or divide one of the rows by a nonzero number.

3. Replace a row by its sum or difference with another row.

Any 2×3 augmented matrix corresponding to a system of equations is equivalent to exactly one of the matrices in the following table, and these special matrices display the solutions of the original systems of equations.

The augmented matrix representation $\begin{pmatrix} a & b & c \\ A & B & C \end{pmatrix}$ of the system $\begin{cases} ax + by = c \\ Ax + By = C \end{cases}$ is equivalent to exactly one of the following matrices (where P and Q stand for any numbers).

Matrix	Equations	Solution
$\begin{pmatrix} 1 & 0 & P \\ 0 & 1 & Q \end{pmatrix}$	Independent and consistent	One solution: $x = P, y = Q$
$\begin{pmatrix} 1 & P & 0 \\ 0 & 0 & 1 \end{pmatrix}$	Independent and inconsistent	No solution
$\begin{pmatrix} 0 & 1 & 0 \\ 0 & 0 & 1 \end{pmatrix}$	Independent and inconsistent	No solution
$\begin{pmatrix} 1 & P & Q \\ 0 & 0 & 0 \end{pmatrix}$	Dependent	Infinitely many solutions: $x = Q - Pt, y = t$
$\begin{pmatrix} 0 & 1 & P \\ 0 & 0 & 0 \end{pmatrix}$	Dependent	Infinitely many solutions: $x = t, y = P$

EXERCISES 2.2

Find the dimension of each matrix and the value(s) of the specified element(s).

1. $\begin{pmatrix} 1 & 4 \\ 2 & 5 \\ 3 & 6 \end{pmatrix}$; $a_{1,1}, a_{3,2}$

2. $\begin{pmatrix} 6 & 3 \\ 5 & 2 \\ 4 & 1 \end{pmatrix}$; $a_{1,1}, a_{3,2}$

3. $\begin{pmatrix} 1 & -1 & 2 & -2 \\ -3 & 3 & -4 & 4 \\ 5 & -5 & 6 & -6 \end{pmatrix}$; $a_{2,2}, a_{3,4}$

4. $\begin{pmatrix} 1 & -2 & -1 \\ -3 & 4 & 3 \\ 5 & -6 & -5 \\ -7 & 8 & 7 \end{pmatrix}$; $a_{2,2}, a_{4,3}$

5. $\begin{pmatrix} 1 & 0 & 0 & 0 \\ 0 & 1 & 0 & 0 \\ 0 & 0 & 1 & 0 \\ 0 & 0 & 0 & 1 \end{pmatrix}$; $a_{1,1}, a_{2,2}, a_{3,3}, a_{4,4}, a_{1,4}$

6. $\begin{pmatrix} 1 & 0 & 0 \\ 0 & 1 & 0 \\ 0 & 0 & 1 \end{pmatrix}$; $a_{1,1}, a_{2,2}, a_{3,3}, a_{2,1}$

7. $(4 \quad 5 \quad 6 \quad 7)$; $a_{1,3}$

8. $(2 \quad 3 \quad 4 \quad 5 \quad 6)$; $a_{1,4}$

9. $\begin{pmatrix} 9 \\ 8 \\ 7 \\ 6 \\ 5 \end{pmatrix}$; $a_{2,1}, a_{4,1}$

10. $\begin{pmatrix} 2 \\ 8 \\ 3 \\ 7 \end{pmatrix}$; $a_{2,1}, a_{3,1}$

Find the augmented matrix representing the system of equations.

11. $\begin{cases} x + 2y = 2 \\ 3x + 4y = 12 \end{cases}$

12. $\begin{cases} 2x - y = 10 \\ -3x + 4y = 60 \end{cases}$

13. $\begin{cases} -4x + 3y = 84 \\ 5x - 2y = 70 \end{cases}$

14. $\begin{cases} -x + 2y = 2 \\ 2x - 3y = 6 \end{cases}$

15. $\begin{cases} 3x - 2y = 24 \\ x = 6 \end{cases}$

16. $\begin{cases} 4x - 3y = 24 \\ y = 8 \end{cases}$

17. $\begin{cases} 5x - 15y = 30 \\ -4x + 12y = 24 \end{cases}$

18. $\begin{cases} 12x - 4y = 36 \\ -15x + 5y = 45 \end{cases}$

19. $\begin{cases} x = 20 \\ y = 30 \end{cases}$

20. $\begin{cases} y = 18 \\ x = 12 \end{cases}$

Find the system of equations represented by the augmented matrix.

21. $\begin{pmatrix} 1 & 1 & 9 \\ 0 & 1 & 4 \end{pmatrix}$ **22.** $\begin{pmatrix} 1 & 0 & -3 \\ 1 & 1 & 5 \end{pmatrix}$

23. $\begin{pmatrix} -4 & 3 & -60 \\ 1 & -2 & 20 \end{pmatrix}$ **24.** $\begin{pmatrix} 3 & 4 & 24 \\ 1 & 2 & 6 \end{pmatrix}$

25. $\begin{pmatrix} 1 & -3 & -70 \\ 1 & 1 & 10 \end{pmatrix}$ **26.** $\begin{pmatrix} 1 & 1 & 5 \\ 2 & 3 & 7 \end{pmatrix}$

27. $\begin{pmatrix} 2 & 1 & 6 \\ 1 & 2 & -6 \end{pmatrix}$ **28.** $\begin{pmatrix} 3 & 1 & -24 \\ 1 & 3 & 24 \end{pmatrix}$

29. $\begin{pmatrix} 20 & -15 & 60 \\ -16 & 12 & -48 \end{pmatrix}$

30. $\begin{pmatrix} 20 & -15 & 60 \\ -16 & 12 & 48 \end{pmatrix}$

Carry out the row operation on the matrix.

31. $R'1 = R2$ and $R'2 = R1$ on $\begin{pmatrix} 3 & 4 & 24 \\ 5 & 6 & 30 \end{pmatrix}$

32. $R'1 = R2$ and $R'2 = R1$ on $\begin{pmatrix} 8 & 7 & 56 \\ 6 & 5 & 60 \end{pmatrix}$

33. $R'1 = R1 - R2$ on $\begin{pmatrix} 8 & 7 & 56 \\ 6 & 5 & 60 \end{pmatrix}$

34. $R'1 = 2R1$ on $\begin{pmatrix} 3 & 4 & 24 \\ 5 & 6 & 30 \end{pmatrix}$

35. $R'2 = 5R2$ on $\begin{pmatrix} 5 & 6 & 30 \\ 1 & 2 & 18 \end{pmatrix}$

36. $R'2 = 3R2$ on $\begin{pmatrix} 6 & 5 & 60 \\ 2 & 2 & -4 \end{pmatrix}$

37. $R'2 = R1 - R2$ on $\begin{pmatrix} 6 & 6 & -12 \\ 6 & 5 & 60 \end{pmatrix}$

38. $R'1 = R1 - R2$ on $\begin{pmatrix} 5 & 6 & 30 \\ 4 & 8 & 72 \end{pmatrix}$

39. $R'2 = \frac{1}{8}R2$ on $\begin{pmatrix} 1 & -2 & -42 \\ 0 & 8 & 120 \end{pmatrix}$

40. $R'1 = \frac{1}{6}R1$ on $\begin{pmatrix} 6 & 0 & 420 \\ 0 & 1 & -72 \end{pmatrix}$

Interpret each reduced row-echelon form matrix as the solution of a system of equations. Identify each system as "independent and consistent," "independent and inconsistent," or "dependent."

41. $\begin{pmatrix} 1 & 0 & 7 \\ 0 & 1 & -3 \end{pmatrix}$ **42.** $\begin{pmatrix} 1 & 0 & -5 \\ 0 & 1 & 8 \end{pmatrix}$

43. $\begin{pmatrix} 1 & 1 & 0 \\ 0 & 0 & 1 \end{pmatrix}$ **44.** $\begin{pmatrix} 1 & -1 & 0 \\ 0 & 0 & 1 \end{pmatrix}$

45. $\begin{pmatrix} 0 & 1 & 0 \\ 0 & 0 & 1 \end{pmatrix}$ **46.** $\begin{pmatrix} 1 & 0 & 0 \\ 0 & 1 & 0 \end{pmatrix}$

47. $\begin{pmatrix} 1 & 2 & 3 \\ 0 & 0 & 0 \end{pmatrix}$ **48.** $\begin{pmatrix} 1 & -4 & 6 \\ 0 & 0 & 0 \end{pmatrix}$

49. $\begin{pmatrix} 0 & 1 & -3 \\ 0 & 0 & 0 \end{pmatrix}$ **50.** $\begin{pmatrix} 0 & 1 & 9 \\ 0 & 0 & 0 \end{pmatrix}$

Solve each system of equations by row reducing the corresponding augmented matrix. Identify each system as "independent and consistent," "independent and inconsistent," or "dependent."

If you have a graphing calculator with a RREF(command, use it to check your row reduction.

51. $\begin{cases} x + y = 5 \\ x = 3 \end{cases}$ **52.** $\begin{cases} x + y = 8 \\ y = 6 \end{cases}$

53. $\begin{cases} x + y = 4 \\ x - y = 2 \end{cases}$ **54.** $\begin{cases} x - 2y = 6 \\ x + y = 3 \end{cases}$

55. $\begin{cases} 2x + y = 4 \\ x + y = 3 \end{cases}$ **56.** $\begin{cases} x + y = 4 \\ 3x + y = 6 \end{cases}$

57. $\begin{cases} x + y = 5 \\ 2x + 3y = 12 \end{cases}$ **58.** $\begin{cases} 3x + y = 9 \\ 2x + y = 4 \end{cases}$

59. $\begin{cases} 2x + y = 20 \\ x + 3y = 15 \end{cases}$ **60.** $\begin{cases} x + y = 8 \\ 3x + 5y = 30 \end{cases}$

61. $\begin{cases} -x + 2y = 4 \\ x - 2y = 6 \end{cases}$ **62.** $\begin{cases} x - 3y = 12 \\ -x + 3y = 9 \end{cases}$

63. $\begin{cases} 6x + 2y = 18 \\ 5x + 2y = 10 \end{cases}$ **64.** $\begin{cases} 5x + 3y = 15 \\ 4x + 2y = 8 \end{cases}$

65. $\begin{cases} -3x + y = 3 \\ 5x - 2y = -10 \end{cases}$ **66.** $\begin{cases} 3x - y = 3 \\ -2x + y = 2 \end{cases}$

67. $\begin{cases} 2x - 6y = 18 \\ -3x + 9y = -27 \end{cases}$ **68.** $\begin{cases} 3x - 6y = 24 \\ -5x + 10y = -40 \end{cases}$

69. $\begin{cases} 4x + 7y = 56 \\ 2x + 3y = 30 \end{cases}$ **70.** $\begin{cases} 4x + 3y = 24 \\ 6x + 5y = 30 \end{cases}$

APPLIED EXERCISES

Formulate each situation as a system of two linear equations in two variables. Be sure to state clearly the meaning of your x- and y-variables. Solve the system of equations by row reducing the corresponding augmented matrix. Be sure to state your final answer in terms of the original question.

 If you have a graphing calculator with a RREF(command, use it to check your row reduction.

71. Commodities   A corn and soybean commodities speculator invested $15,000 yesterday with twice as much in soybean futures as in corn futures. How much did she invest in each?

72. Stamps and Coins   A stamp and coin dealer spent $8000 at a numismatics auction last weekend. If he spent three times as much on coins as on stamps, how much did he spend on each?

73. Estate Division   A will specifies that the older brother is to receive one-half of the $12 million estate after first deducting the younger brother's share, the younger brother is to receive one-third of the estate after first deducting the older brother's share, and that the remainder is to be given to their sister. How much does each brother receive? How much is left over for their sister?

74. Real Estate Taxes   Find the state and city property taxes on an apartment building assessed at $833,000 if the state tax is 4% of the assessed value after first deducting the city tax and the city tax is 1% of the assessed value after first deducting the state tax.

75. Coins in a Jar   A jar contains nickels and quarters worth $36.75. If there were twice as many nickels and half as many quarters, they would be worth $31.50 instead. How many of each are in the jar?

76. Financial Planning   Last year, a retired couple's investments in a money market account yielding 5% and in a stock mutual fund yielding 14% paid a total of $16,800. If the stock mutual fund can return 17% this year and the money market account can continue to yield 5%, the couple will receive $17,850. How much is invested in each?

77. Geriatrics Nutrition   The dietician at a senior care facility has decided to supplement the weekly menu with "Fountain of Youth" NutraDrink and "Get Up & Go" VitaPills, which contain calcium and vitamin C as listed in the table. If each resident needs an additional 300 mg of calcium and 240 units of vitamin C per week, how many cans of NutraDrink and tablets of VitaPills should each resident be given per week?

	Calcium (mg)	Units of Vitamin C
NutraDrink	25	20
VitaPills	20	16

78. Plant Fertilizer   A garden field needs 308 pounds of potash and 330 pounds of nitrogen. Two brands of fertilizer, GrowRite and GreatGreen, are available and contain the amounts of potash and nitrogen per bag listed in the table. How many bags of each brand should be used to provide the required potash and nitrogen?

(Pounds per Bag)	GrowRite	GreatGreen
Potash	4	7
Nitrogen	5	6

79. Political Advertising   For the final days before the election, the campaign manager has set aside $36,000 to spend on TV and radio campaign advertisements. Each TV ad costs $3000 and is seen by 10,000 voters while each radio ad costs $500 and is heard by 2000 voters. Ignoring repeated exposures to the same voter, how many TV and radio ads will contact 130,000 voters using the allocated funds?

80. Consumer Preferences   A marketing company wants to gather information on consumer preferences for laundry detergents using telephone interviews and direct mail questionnaires. Each attempted telephone interview costs $2.25 whether successful or not and each mailed questionnaire costs $0.75 whether returned or not. The research director estimates that 21% of the telephone calls will result in usable interviews while only 7% of the questionnaires will be returned in usable condition. If the director's budget is $15,750 and she needs 1470 usable responses, how many telephone calls should she attempt and how many questionnaires should she mail out?

Explorations and Excursions

The following problems extend and augment the material presented in the text.

More about 2 × 3 Row Reduced Matrices The table on page 194 lists exactly five possible final matrices for the solution of a system of equations by reducing the corresponding augmented matrix. The following problems provide examples of each possibility.

Sketch each system of equations and then solve it by row reducing the corresponding augmented matrix. Identify each system as "independent and consistent," "independent and inconsistent," or "dependent."

If permitted by your instructor, graph each system on the window $[-30, 50]$ by $[-30, 50]$, use the RREF(command to row reduce the augmented matrix, and graph the parametric solution if the system is dependent.

81. $\begin{cases} x - 2y = -14 \\ 2x + 3y = 84 \end{cases}$ **82.** $\begin{cases} x + 2y = 56 \\ 2x - 3y = 42 \end{cases}$

83. $\begin{cases} 3x - 6y = 18 \\ -2x + 4y = 16 \end{cases}$ **84.** $\begin{cases} 4x - 6y = 36 \\ -2x + 3y = 30 \end{cases}$

85. $\begin{cases} 2y = 8 \\ 3y = 18 \end{cases}$ **86.** $\begin{cases} 5y = 20 \\ 3y = 30 \end{cases}$

87. $\begin{cases} 4x + 14y = 56 \\ 6x + 21y = 84 \end{cases}$ **88.** $\begin{cases} 6x + 14y = 42 \\ 9x + 21y = 63 \end{cases}$

89. $\begin{cases} 5y = 30 \\ 3y = 18 \end{cases}$ **90.** $\begin{cases} 2y = 18 \\ 3y = 27 \end{cases}$

Row Operations are Reversible   Carry out each row operation on the augmented matrix $\begin{pmatrix} 3 & -10 & -65 \\ -4 & 13 & 84 \end{pmatrix}$ and then find a row operation to perform on your new matrix that will return it back to $\begin{pmatrix} 3 & -10 & -65 \\ -4 & 13 & 84 \end{pmatrix}$.

91. $R'1 = R2$ and $R'2 = R1$

92. $R'2 = R1$ and $R'1 = R2$

93. $R'1 = 5R1$ **94.** $R'2 = 3R2$

95. $R'2 = \frac{1}{4}R2$ **96.** $R'1 = \frac{1}{3}R1$

97. $R'1 = R1 + R2$ **98.** $R'2 = R2 + R1$

99. $R'1 = R1 - R2$ **100.** $R'2 = R2 - R1$

2.3 Matrix Row Reduction and Systems of Linear Equations

APPLICATION PREVIEW

Ancient Chinese Mathematics*

The earliest known instance of matrix notation used for the systematic solution of linear equations appears in the eighth section of the ancient Chinese mathematics treatise *Arithmetic in Nine Sections*, written about 200 B.C. The method is explained by solving example problems, the first of which may be rephrased as follows.

* Based on material in Yoshio Mikami, *The Development of Mathematics in China and Japan*, 2nd ed. (Chelsea Publishing Company: New York, 1974).

A farmer grows three kinds of corn and each kind has been harvested and gathered into bundles. Three bundles of the first kind, two of the second, and one of the third make 39 bushels. Two of the first, three of the second, and one of the third make 34 bushels. And one of the first, two of the second, and three of the third make 26 bushels. How many bushels of corn are contained in one bundle of each kind of corn?

The rule for finding the solution is to write the 3, 2, 1 bundles of the three kinds and the 39 bushels as a column on the right, and then write the other conditions in the middle and on the left. The problem leads to the following table, whose columns correspond to our rows:

1	2	3	First kind
2	3	2	Second kind
3	1	1	Third kind
26	34	39	Total bushels

The calculation begins by multiplying the middle column by the top number of the right column:

1	6	3	First kind
2	9	2	Second kind
3	3	1	Third kind
26	102	39	Total bushels

The right column of numbers is then taken away as many times as possible from the middle column. In this case, it can be subtracted two times to leave:

1	0	3	First kind
2	5	2	Second kind
3	1	1	Third kind
26	24	39	Total bushels

The process then continues by multiplying columns by numbers and subtracting other columns as many times as possible until each column describes only one kind of corn and gives a number of bushels at the bottom:

0	0	540	First kind
0	180	0	Second kind
36	0	0	Third kind
99	765	4995	Total bushels

The solution is finished by dividing each lower number by the number of the kind to find the number of bushels: $\frac{4995}{540} = 9\frac{1}{4}$ for the first kind, $\frac{765}{180} = 4\frac{1}{4}$ for the second, and $\frac{99}{36} = 2\frac{3}{4}$ for the third.

Although this method was used in China over 2000 years ago, it is interesting how similar the problem statement and the column manipulations are to the augmented matrices and row operations used in the last section.

Introduction

In this section, we use augmented matrices to solve systems of many linear equations in many variables. We shall simply enlarge the size of the augmented matrix to allow for more equations (rows) and more variables (columns), and then apply row operations to find an equivalent matrix in reduced row-echelon form that will display the solution.

Names for Many Variables

To deal with many variables, we shall now distinguish them by subscripts instead of different letters. Our first variable will be named x_1 ("x sub one" or "the first x"), the second x_2 ("x sub two" or "the second x"), and so on for as many as we need (possibly x_{10}, x_{20}, or even x_{100}). With this new notation, we can rewrite the problem from the Application Preview as:

$$\begin{cases} 3x_1 + 2x_2 + x_3 = 39 \\ 2x_1 + 3x_2 + x_3 = 34 \\ x_1 + 2x_2 + 3x_3 = 26 \end{cases}$$

We form the augmented matrix exactly as before, so for this system of equations we have:

$$
\begin{array}{cccc}
x_1 & x_2 & x_3 & \\
\begin{pmatrix} 3 & 2 & 1 & 39 \\ 2 & 3 & 1 & 34 \\ 1 & 2 & 3 & 26 \end{pmatrix}
\end{array}
$$

← Variables corresponding to columns (last column represents constant terms)

← Each row represents an equation

Reduced Row-Echelon Form

We shall continue to use row operations to "solve" the augmented matrix by finding an equivalent matrix that displays the solution. With our list of reduced row-echelon form 2×3 matrices (page 194) as a guide, we make the following definition for matrices of any dimension. In this definition, a *zero row* is a row containing only zeros, and a *nonzero row* is a row with at least one nonzero element.

Reduced Row-Echelon Form

A matrix is in *reduced row-echelon form* if it satisfies the following four conditions.

1. All zero rows are below every nonzero row.

2. The first nonzero element of every nonzero row is a one (we will call these special ones "leftmost ones").

3. The column of each leftmost one [from (2)] contains only zeros in the other positions.

4. Each leftmost one [from (2)] appears to the right of the leftmost ones in the rows above it.

The following matrices are in reduced row-echelon form:

$$
\begin{pmatrix} 0 & 1 & 2 & 0 & 1 \\ 0 & 0 & 0 & 1 & 4 \end{pmatrix}, \begin{pmatrix} 1 & 2 & 3 & 4 & 0 \\ 0 & 0 & 0 & 0 & 1 \\ 0 & 0 & 0 & 0 & 0 \end{pmatrix}, \text{ and } \begin{pmatrix} 1 & 0 & 0 & 0 & 1 \\ 0 & 1 & 0 & 0 & 2 \\ 0 & 0 & 1 & 0 & 3 \\ 0 & 0 & 0 & 1 & 4 \end{pmatrix}
$$

Leftmost ones   Zero row

Notice that the ones in the upper right corners of the first and last matrices above are not *leftmost* ones and do not need to have zeros below them. The following matrix is *not* in reduced row-echelon form because although conditions (1), (2), and (3) are satisfied, condition (4) fails for the third row.

$$
\begin{pmatrix} 1 & 0 & 0 & 2 & 3 \\ 0 & 0 & 1 & 4 & 5 \\ 0 & 1 & 0 & 6 & 7 \\ 0 & 0 & 0 & 0 & 0 \end{pmatrix}
$$

The 1 in the third row is not to the right of the 1 in the second row

Can you think of a row operation that will correct this defect and result in a matrix in reduced row-echelon form?

Solutions from Augmented Matrices

If a reduced row-echelon form matrix has a zero row, then the system of equations from which it comes is *dependent*; otherwise, the system is *independent*. If a reduced row-echelon form matrix has a row of zeros ending in a 1 (such as "0 0 0 1"), then the system of equations is *inconsistent*; otherwise, the system is *consistent*. The following are examples of each possibility.

	Independent	Dependent
Consistent	$\begin{pmatrix} 1 & 0 & 0 & 2 \\ 0 & 1 & 0 & 3 \\ 0 & 0 & 1 & 4 \end{pmatrix}$	$\begin{pmatrix} 1 & 0 & 0 & 2 \\ 0 & 1 & 0 & 3 \\ 0 & 0 & 0 & 0 \end{pmatrix}$
Inconsistent	$\begin{pmatrix} 1 & 0 & 0 & 0 \\ 0 & 1 & 0 & 0 \\ 0 & 0 & 0 & 1 \end{pmatrix}$	$\begin{pmatrix} 1 & 0 & 0 & 0 \\ 0 & 0 & 0 & 1 \\ 0 & 0 & 0 & 0 \end{pmatrix}$

In a consistent system, a variable is *determined* or *dependent* if its column in the reduced row-echelon form matrix has a leftmost one, and otherwise it is *free* or *independent*. The free variables may take any values, and then the values of the determined variables follow from the equations represented by their rows. The free variables give a *parameterization* of the solution by determining the values of the other variables. For instance, for the matrix:

$$\begin{pmatrix} 1 & 2 & 3 \\ 0 & 0 & 0 \end{pmatrix}$$

Three columns means two variables x_1, x_2

the x_1 is *determined* (its column has a leftmost one) while the x_2 is *free*. Since x_2 is "free" to take any value t and then $x_1 = 3 - 2t$ is "determined," the solution is $x_1 = 3 - 2t$, $x_2 = t$, where t may take any value.

EXAMPLE 1 Finding a Parameterized Solution

Find the solution of a system of equations with augmented matrix equivalent to the reduced row-echelon form matrix:

$$\begin{pmatrix} 1 & 0 & 2 & 3 & 0 & 4 \\ 0 & 1 & 5 & 6 & 0 & 7 \\ 0 & 0 & 0 & 0 & 1 & 8 \end{pmatrix}$$

Solution

There were five variables in the system of equations because the augmented matrix has six columns. The variables x_1, x_2, and x_5 are determined while x_3 and x_4 are free. Letting the free variables take the values $x_3 = t_1$ and $x_4 = t_2$, we can use the rows to find the determined variables:

$$1x_1 + 0x_2 + 2x_3 + 3x_4 + 0x_5 = 4 \quad \text{so} \quad x_1 = 4 - 2t_1 - 3t_2$$
$$0x_1 + 1x_2 + 5x_3 + 6x_4 + 0x_5 = 7 \quad \text{so} \quad x_2 = 7 - 5t_1 - 6t_2$$
$$0x_1 + 0x_2 + 0x_3 + 0x_4 + 1x_5 = 8 \quad \text{so} \quad x_5 = 8$$

For instance, if we choose $t_1 = 10, t_2 = -10$, we obtain the solution

$$x_1 = 14$$
$$x_2 = 17$$
$$x_3 = 10$$
$$x_4 = -10$$
$$x_5 = 8$$

PRACTICE PROBLEM 1

Find the solution of a system of equations with augmented matrix equivalent to the reduced row-echelon form matrix:

$$\begin{pmatrix} 1 & 2 & 0 & 0 & 0 \\ 0 & 0 & 1 & 0 & 0 \\ 0 & 0 & 0 & 1 & 0 \\ 0 & 0 & 0 & 0 & 1 \end{pmatrix}$$

Solution at the back of the book

Row Reducing a Large Matrix

Row reducing a large matrix is essentially the same process as reducing a 2×3 matrix. Before an example, three suggestions.

1. To make the numbers more manageable, you may wish to often check for and remove common factors in a row.

2. You will often repeat sequences of row operations similar to the following from Example 5 in the last section (page 193):

$$\begin{pmatrix} 1 & 4 & 440 \\ 0 & 1 & 80 \end{pmatrix}$$
Need a 0 where the 4 now appears

$$\begin{pmatrix} 1 & 4 & 440 \\ 0 & 4 & 320 \end{pmatrix} R'2 = 4R2$$
Multiply to duplicate the 4 (or get a -4)

$$\begin{pmatrix} 1 & 0 & 120 \\ 0 & 4 & 320 \end{pmatrix} R'1 = R1 - R2$$

Subtract (or add)
to get zero

$$\begin{pmatrix} 1 & 0 & 120 \\ 0 & 1 & 80 \end{pmatrix} R'2 = \frac{1}{4}R2$$

Remove the
common factor

This method of removing some other nonzero element from the column of a leftmost one is used so often that you should do it in just one row operation:

$$\begin{pmatrix} 1 & 4 & 440 \\ 0 & 4 & 320 \\ 0 & 1 & 80 \end{pmatrix}$$

Write in multiples of $R2$
so you can see them while
doing your arithmetic

$$\begin{pmatrix} 1 & 0 & 120 \\ 0 & 1 & 80 \end{pmatrix} R'1 = R1 - 4R2$$

3. While it is best to work systematically through the matrix from the top left moving down and over, if you see a quick way of getting many zeros, then that should be done first.

Remember that there are "good" choices for your next row operation, "better" ones, and "bad" choices that ruin entries that you want to keep (usually a zero or a one). Try to use the fewest number of steps when you row reduce a matrix.

EXAMPLE 2 **Row Reducing a Large Matrix**

Row reduce the matrix $\begin{pmatrix} 5 & 5 & 0 & 5 & 50 \\ 2 & 3 & 1 & 0 & 17 \\ 2 & 2 & 1 & -1 & 9 \\ 2 & 3 & 1 & 1 & 22 \end{pmatrix}$.

Solution

Before setting to work, always look for simple row operations that may make the problem easier. Since the elements of the first row have 5 as a common factor, we begin by simplifying that row:

$$\begin{pmatrix} 1 & 1 & 0 & 1 & 10 \\ 2 & 3 & 1 & 0 & 17 \\ 2 & 2 & 1 & -1 & 9 \\ 2 & 3 & 1 & 1 & 22 \end{pmatrix} R'1 = \frac{1}{5}R1$$

Remove
common
factor

Next, notice that the second and fourth rows have the same first three elements, so we can get "many zeros" by subtracting:

$$\begin{pmatrix} 1 & 1 & 0 & 1 & 10 \\ 2 & 3 & 1 & 0 & 17 \\ 2 & 2 & 1 & -1 & 9 \\ 0 & 0 & 0 & 1 & 5 \end{pmatrix} R'4 = R4 - R2$$

Get "easy" zeros

In the same way, the second and third rows can give us zeros and a one where we would wish in the second row:

$$\begin{pmatrix} 1 & 1 & 0 & 1 & 10 \\ 0 & 1 & 0 & 1 & 8 \\ 2 & 2 & 1 & -1 & 9 \\ 0 & 0 & 0 & 1 & 5 \end{pmatrix} R'2 = R2 - R3$$

Get easy "0 1 0" pattern

The third column is now as we would like, as is the bottom row. Using the one in the bottom row, we can "zero out" the rest of the fourth column. To save space, we write the results of the next three row operations all at once:

$$\begin{pmatrix} 1 & 1 & 0 & 0 & 5 \\ 0 & 1 & 0 & 0 & 3 \\ 2 & 2 & 1 & 0 & 14 \\ 0 & 0 & 0 & 1 & 5 \end{pmatrix} \begin{matrix} R'1 = R1 - R4 \\ R'2 = R2 - R4 \\ R'3 = R3 + R4 \end{matrix}$$

"Zero out" rest of column using the leftmost one

We finish by zeroing out the rest of the first and second columns:

$$\begin{pmatrix} 1 & 1 & 0 & 0 & 5 \\ 0 & 1 & 0 & 0 & 3 \\ 0 & 0 & 1 & 0 & 4 \\ 0 & 0 & 0 & 1 & 5 \end{pmatrix} R'3 = R3 - 2R1$$

$$\begin{pmatrix} 1 & 0 & 0 & 0 & 2 \\ 0 & 1 & 0 & 0 & 3 \\ 0 & 0 & 1 & 0 & 4 \\ 0 & 0 & 0 & 1 & 5 \end{pmatrix} R'1 = R1 - R2$$

Reduced row echelon form

■

PRACTICE PROBLEM 2   Solve the system of equations $\begin{cases} 5x_1 + 5x_2 + 5x_4 = 50 \\ 2x_1 + 3x_2 + 1x_3 = 17 \\ 2x_1 + 2x_2 + x_3 - x_4 = 9 \\ 2x_1 + 3x_2 + x_3 + x_4 = 22 \end{cases}$. Be sure to

check your answer in the equations. (*Hint:* No work necessary—just use the reduced row-echelon form matrix from Example 2.)

Solution at the back of the book

Graphing Calculator Exploration

The program ROWOPS* carries out the arithmetic for the type of row operation you select from a menu. To carry out the row operations used in Example 2 on pages 203–204 on your calculator, proceed as follows.

a. Enter the augmented matrix as matrix [A]. Since this matrix is too large to fit on the calculator screen, it will scroll from left to right and back again as you enter your numbers.

b. Run the program ROWOPS. It will display the current values in matrix [A] and you can use the arrows to scroll the screen to see the rest of it. Press ENTER to select the type of row operation you would like to perform.

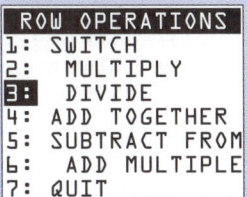

c. Choose the type of row operation you want by using the arrows to move to its number and pressing ENTER, or just press the number. Enter the specific details for the particular operation you want and the program will carry out your request.

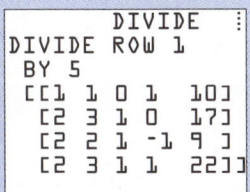

---

* See the Preface for information on how to obtain this and other programs.

d. Press [ENTER] to select another row operation or to choose 7 and QUIT the program.

The program ROWOPS allows you to multiply rows by fractions and displays fractions in the usual 3/5 notation. The reduction of this matrix took eight row operations to complete in Example 2. Can you reduce it in just seven?

EXAMPLE 3   Managing Production

A hand-thrown pottery shop manufactures plates, cups, and vases. Each plate requires 4 ounces of clay, 6 minutes of shaping, and 5 minutes of painting; each cup requires 4 ounces of clay, 5 minutes of shaping, and 3 minutes of painting; and each vase requires 3 ounces of clay, 4 minutes of shaping, and 4 minutes of painting. This week the shop has 165 pounds of clay, 59 hours of skilled labor for shaping, and 46 hours of skilled labor for painting. If the shop manager wishes to fully use all these resources, how many of each product should the shop produce?

Solution

Let x_1, x_2, and x_3 be the numbers of plates, cups and vases produced. The ounces of clay required is then $4x_1 + 4x_2 + 3x_3$ and this must match the 165 pounds available:

$$4x_1 + 4x_2 + 3x_3 = 2640$$

Use ounces on both sides of the equation

Similarly, for the time in minutes required for shaping and painting:

$$6x_1 + 5x_2 + 4x_3 = 3540$$
$$5x_1 + 3x_2 + 4x_3 = 2760$$

Use minutes on both sides of the equation

Therefore, the augmented matrix is

$$\begin{pmatrix} 4 & 4 & 3 & 2640 \\ 6 & 5 & 4 & 3540 \\ 5 & 3 & 4 & 2760 \end{pmatrix}$$

There are many different sequences of row operations to reduce this augmented matrix, and all reach the same conclusion. One possible way is as follows.

$$\begin{pmatrix} 4 & 4 & 3 & 2640 \\ 2 & 1 & 1 & 900 \\ 1 & -1 & 1 & 120 \end{pmatrix} \begin{matrix} \\ R'2 = R2 - R1 \\ R'3 = R3 - R1 \end{matrix}$$

Use a row to make numbers in other rows closer to zero

$$\begin{pmatrix} 0 & 8 & -1 & 2160 \\ 0 & 3 & -1 & 660 \\ 1 & -1 & 1 & 120 \end{pmatrix} \begin{matrix} R'1 = R1 - 4R3 \\ R'2 = R2 - 2R3 \\ \\ \end{matrix}$$

Use 1 in first column to "zero out" the rest of that column

$$\begin{pmatrix} 0 & 5 & 0 & 1500 \\ 0 & 3 & -1 & 660 \\ 1 & -1 & 1 & 120 \end{pmatrix} \begin{matrix} R'1 = R1 - R2 \\ \\ \\ \end{matrix}$$

Use same number in two rows to get a zero

$$\begin{pmatrix} 0 & 1 & 0 & 300 \\ 0 & 3 & -1 & 660 \\ 1 & -1 & 1 & 120 \end{pmatrix} \begin{matrix} R'1 = \frac{1}{5}R1 \\ \\ \\ \end{matrix}$$

Remove common factor

$$\begin{pmatrix} 0 & 1 & 0 & 300 \\ 0 & 0 & 1 & 240 \\ 1 & 0 & 1 & 420 \end{pmatrix} \begin{matrix} \\ R'2 = 3R1 - R2 \\ R'3 = R3 + R1 \end{matrix}$$

Use 1 in second column to "zero out" the rest of that column

$$\begin{pmatrix} 0 & 1 & 0 & 300 \\ 0 & 0 & 1 & 240 \\ 1 & 0 & 0 & 180 \end{pmatrix} \begin{matrix} \\ \\ R'3 = R3 - R2 \end{matrix}$$

Use 1 in third column to "zero out" the rest of that column

$$\begin{pmatrix} 1 & 0 & 0 & 180 \\ 0 & 1 & 0 & 300 \\ 0 & 0 & 1 & 240 \end{pmatrix} \begin{matrix} R'1 = R3 \\ R'2 = R1 \\ R'3 = R2 \end{matrix}$$

Switch rows to achieve correct order

The system of equations is independent and consistent with solution $x_1 = 180$, $x_2 = 300$, $x_3 = 240$. You should check that these values satisfy the original equations and that the clay and skilled labor resources are fully used. In terms of the original question, the shop should produce 180 plates, 300 cups, and 240 vases this week.

■

EXAMPLE 4   Modeling a Computer Network

The office manager for the accounting division of a large company is writing a recommendation report to her supervisor about the demands placed on the computers serving her division. Besides meeting the needs of the accounting division, her four "file servers" are expected to accept and relay messages to and from other areas. The diagram shows the four file servers (I, II, III, IV) and the arrows and numbers indicate the data packets per minute passing from the senders to the receivers. Computers A through H are outside the accounting division. Connections passing an unknown number of data packets per minute are marked with variables (for instance, x_1 is the number of data packets per minute sent from server I to server II, while x_5 is the number leaving server IV to outside computer G). In order for the network to function, the number of data packets per minute arriving at each computer must match the number leaving it. How many data packets per minute must leave IV for G? Is it possible for each connection within the accounting division to carry no more than 1500 data packets per minute?

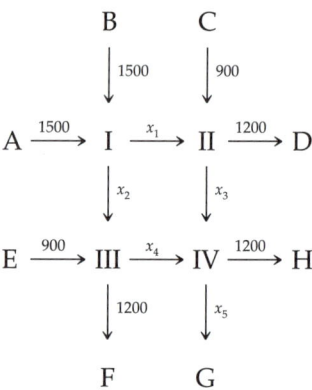

Solution

Using the notations from the diagram, the requirements that the data packets per minute arriving at each file server match the number leaving are the equations:

		"In" equals "out"
$1500 + 1500 = x_1 + x_2$		Server I
$x_1 + 900 = x_3 + 1200$		Server II
$900 + x_2 = x_4 + 1200$		Server III
$x_3 + x_4 = 1200 + x_5$		Server IV

Rewriting these to put the variables on the left, we have the system:

$$\begin{cases} x_1 + x_2 = 3000 \\ x_1 - x_3 = 300 \\ x_2 - x_4 = 300 \\ x_3 + x_4 - x_5 = 1200 \end{cases}$$

Since there are fewer equations than variables, the solution will not consist of unique values for all the variables. If the system is consistent, we will find a parameterized solution, whereas if it is inconsistent, there will be no solution. The augmented matrix representing these equations is shown on the left below, with its reduced row-echelon form shown on the right. (You may wish to carry out the reduction—it can be done easily because of the many zeros and ones—or use your calculator's RREF(command to check the reduction.)

$$\begin{pmatrix} 1 & 1 & 0 & 0 & 0 & 3000 \\ 1 & 0 & -1 & 0 & 0 & 300 \\ 0 & 1 & 0 & -1 & 0 & 300 \\ 0 & 0 & 1 & 1 & -1 & 1200 \end{pmatrix} \xrightarrow{\text{Leads to}} \begin{pmatrix} 1 & 0 & 0 & 1 & 0 & 2700 \\ 0 & 1 & 0 & -1 & 0 & 300 \\ 0 & 0 & 1 & 1 & 0 & 2400 \\ 0 & 0 & 0 & 0 & 1 & 1200 \end{pmatrix}$$

The system is consistent and x_4 is a free variable (since its column does *not* have a leftmost one). We may parameterize the solution as follows:

$$x_1 = 2700 - t$$

$$x_2 = 300 + t$$

$$x_3 = 2400 - t$$

$$x_4 = t \qquad\qquad \text{The free variable}$$

$$x_5 = 1200$$

To prevent a "back flow" along the connections, each variable must stay nonnegative, which means that $t \geq 0$ (since $x_4 = t$) and $t \leq 2400$ (since $x_3 = 2400 - t$). For example, the choice $t = 1200$ gives the solution:

$$x_1 = 1500, x_2 = 1500, x_3 = 1200, x_4 = 1200, x_5 = 1200$$

In terms of the original questions, 1200 data packets per minute must leave computer IV for G, and the example with $t = 1200$ shows that it *is* possible to arrange the transmissions so that each connection within the accounting division carries no more than 1500 data packets per minute. ■

SUMMARY

The solution of any system of linear equations can be found by representing it as an augmented matrix, applying row operations (page 186) to find the equivalent reduced row-echelon form matrix (page 200), and interpreting this final matrix as a statement about the original system of equations (page 201). Each step in this method is a direct extension of the method for solving systems of two linear equations in two variables presented in the previous section.

EXERCISES 2.3

Find the augmented matrix representing the system of equations.

1. $\begin{cases} x_1 + x_2 + x_3 = 4 \\ x_1 + 2x_2 + x_3 = 3 \\ x_1 + 2x_2 + 2x_3 = 5 \end{cases}$

2. $\begin{cases} x_1 + 5x_2 + 4x_3 = 6 \\ x_1 + x_2 + x_3 = 4 \\ 2x_1 + 3x_2 + 3x_3 = 9 \end{cases}$

3. $\begin{cases} 2x_1 - x_2 + 2x_3 = 11 \\ -x_1 + x_2 - 3x_3 = -12 \\ 2x_1 - 2x_2 + 7x_3 = 27 \end{cases}$

4. $\begin{cases} 4x_1 + 3x_2 - x_3 = 2 \\ 3x_1 + 3x_2 + 2x_3 = 9 \\ 2x_1 + x_2 - 3x_3 = -6 \end{cases}$

5. $\begin{cases} 2x_1 + x_2 + 5x_3 + 4x_4 + 5x_5 = 2 \\ x_1 + x_2 + 3x_3 + 3x_4 + 3x_5 = -1 \end{cases}$

6. $\begin{cases} 5x_1 + 2x_2 - 4x_3 + x_4 + 5x_5 = 7 \\ 3x_1 + x_2 - 3x_3 + x_4 + 3x_5 = 5 \end{cases}$

7. $\begin{cases} 6x_1 + 3x_2 + 5x_3 = 8 \\ x_1 + 2x_2 + 2x_3 = 1 \\ 4x_1 + 3x_2 + 4x_3 = 5 \\ 5x_1 + x_2 + 3x_3 = 7 \end{cases}$

8. $\begin{cases} 5x_1 + 9x_2 + 9x_3 = 11 \\ 4x_1 + 7x_2 + 6x_3 = 9 \\ 3x_1 + 5x_2 + 3x_3 = 8 \\ 4x_1 + 7x_2 + 5x_3 = 10 \end{cases}$

9. $\begin{cases} 3x_1 + 4x_2 + 2x_3 + 4x_4 = 12 \\ x_1 + 2x_2 + x_3 + x_4 = 4 \\ 4x_1 + 5x_2 + 2x_3 + 5x_4 = 14 \\ 6x_1 + 6x_2 + x_3 + 6x_4 = 15 \end{cases}$

10. $\begin{cases} 5x_1 + 4x_2 + 7x_3 + 6x_4 = 18 \\ 2x_1 + 2x_2 + 3x_3 + 3x_4 = 9 \\ 4x_1 + 3x_2 + 5x_3 + 5x_4 = 16 \\ 3x_1 + 2x_2 + 3x_3 + 3x_4 = 11 \end{cases}$

Find the system of equations represented by the augmented matrix.

11. $\begin{pmatrix} 4 & 3 & 2 & 11 \\ 3 & 3 & 1 & 6 \\ 1 & -2 & 3 & 13 \end{pmatrix}$

12. $\begin{pmatrix} 3 & -2 & 5 & 23 \\ -1 & 1 & -3 & -12 \\ 2 & -2 & 7 & 27 \end{pmatrix}$

13. $\begin{pmatrix} 2 & 1 & 1 & 7 \\ 2 & 2 & 1 & 6 \\ 3 & 3 & 2 & 10 \end{pmatrix}$

14. $\begin{pmatrix} 5 & 1 & 3 & 20 \\ 1 & 1 & 2 & 6 \\ 4 & 1 & 3 & 17 \end{pmatrix}$

15. $\begin{pmatrix} 8 & 3 & -2 & 19 & 15 \\ 3 & 1 & -1 & 7 & 6 \end{pmatrix}$

16. $\begin{pmatrix} 6 & -2 & -4 & -2 & 36 \\ 2 & -1 & -10 & 5 & 6 \end{pmatrix}$

17. $\begin{pmatrix} 2 & 3 & 2 & 5 \\ 3 & 5 & 3 & 8 \\ 1 & 2 & 2 & 2 \\ 4 & 7 & 5 & 9 \end{pmatrix}$

18. $\begin{pmatrix} 3 & 1 & 3 & 5 \\ 2 & 2 & 1 & 1 \\ 3 & 2 & 2 & 3 \\ 5 & 3 & 4 & 6 \end{pmatrix}$

19. $\begin{pmatrix} 3 & 3 & 5 & 4 & 11 \\ 2 & 2 & 3 & 3 & 9 \\ 2 & 1 & 2 & 2 & 7 \\ 3 & 2 & 3 & 3 & 11 \end{pmatrix}$

20. $\begin{pmatrix} 2 & 1 & -1 & 1 & 1 \\ 1 & 1 & 0 & 1 & 2 \\ -2 & 1 & 4 & 1 & 8 \\ 1 & 2 & 1 & 1 & 4 \end{pmatrix}$

Interpret each reduced row-echelon form matrix as the solution of a system of equations. Identify each system as "independent" or "dependent" and as "consistent" or "inconsistent."

21. $\begin{pmatrix} 1 & 0 & 0 & 4 \\ 0 & 1 & 0 & 5 \\ 0 & 0 & 1 & -4 \end{pmatrix}$

22. $\begin{pmatrix} 1 & 0 & 0 & 0 \\ 0 & 1 & 0 & 0 \\ 0 & 0 & 1 & 1 \end{pmatrix}$

23. $\begin{pmatrix} 1 & 0 & 0 & 0 & 2 \\ 0 & 1 & 0 & 0 & -1 \\ 0 & 0 & 1 & 0 & 3 \\ 0 & 0 & 0 & 1 & 1 \end{pmatrix}$

24. $\begin{pmatrix} 1 & 0 & 7 \\ 0 & 1 & 3 \\ 0 & 0 & 0 \end{pmatrix}$

25. $\begin{pmatrix} 1 & 0 & 1 & 0 \\ 0 & 1 & 0 & 0 \\ 0 & 0 & 0 & 1 \end{pmatrix}$

26. $\begin{pmatrix} 1 & 1 & 0 \\ 0 & 0 & 1 \end{pmatrix}$

27. $\begin{pmatrix} 1 & 0 & -1 & -5 \\ 0 & 1 & 1 & 5 \\ 0 & 0 & 0 & 0 \end{pmatrix}$

28. $\begin{pmatrix} 1 & 1 & 0 & 0 & 2 \\ 0 & 0 & 1 & 0 & -1 \\ 0 & 0 & 0 & 1 & 3 \\ 0 & 0 & 0 & 0 & 0 \end{pmatrix}$

29. $\begin{pmatrix} 1 & -1 & 0 & 1 & 8 \\ 0 & 0 & 1 & -1 & 4 \\ 0 & 0 & 0 & 0 & 0 \end{pmatrix}$

30. $\begin{pmatrix} 1 & 0 & 0 & 6 \\ 0 & 0 & 1 & 3 \\ 0 & 0 & 0 & 0 \end{pmatrix}$

Use an appropriate row operation or sequence of row operations to find the equivalent reduced row-echelon form matrix.

 If your instructor permits, you may carry out the calculations on a graphing calculator using the ROWOPS program.

31. $\begin{pmatrix} 0 & 1 & 0 & 2 \\ 1 & 0 & 0 & 1 \\ 0 & 0 & 1 & 3 \end{pmatrix}$
32. $\begin{pmatrix} 1 & 0 & 0 & 1 \\ 0 & 0 & 1 & 3 \\ 0 & 1 & 0 & 2 \end{pmatrix}$

33. $\begin{pmatrix} 1 & 0 & 1 & 4 \\ 0 & 1 & 0 & 2 \\ 0 & 0 & 1 & 3 \end{pmatrix}$
34. $\begin{pmatrix} 1 & 0 & 0 & 1 \\ 1 & 1 & 0 & 3 \\ 0 & 0 & 1 & 3 \end{pmatrix}$

35. $\begin{pmatrix} 2 & 4 & 0 & 6 \\ 0 & 0 & 1 & -3 \\ 0 & 0 & 0 & 0 \end{pmatrix}$
36. $\begin{pmatrix} 1 & 0 & 2 & 5 \\ 0 & 3 & -6 & 3 \\ 0 & 0 & 0 & 0 \end{pmatrix}$

37. $\begin{pmatrix} 2 & 0 & 4 & 0 & 6 \\ 0 & 1 & 0 & -1 & 1 \\ 0 & 0 & 1 & 1 & 1 \\ 0 & 0 & 1 & 1 & 2 \end{pmatrix}$

38. $\begin{pmatrix} 1 & 0 & -1 & 0 & 1 \\ 0 & 1 & 2 & 1 & 3 \\ 0 & 2 & 4 & 2 & 8 \\ 0 & 0 & 0 & 1 & -2 \end{pmatrix}$

39. $\begin{pmatrix} 1 & 1 & 0 & 0 & 4 \\ 0 & 1 & 0 & 0 & 2 \\ 0 & 0 & 0 & 1 & 1 \\ 0 & 0 & 1 & 0 & 3 \end{pmatrix}$

40. $\begin{pmatrix} 0 & 1 & 0 & 1 & 4 \\ 1 & 1 & 0 & 1 & 2 \\ 0 & 0 & 1 & 1 & 3 \\ 0 & 0 & 1 & 0 & 2 \end{pmatrix}$

Solve each system of equations by row reducing the corresponding augmented matrix. Identify each system as "independent" or "dependent" and as "consistent" or "inconsistent."

If you have a graphing calculator with a RREF(command, use it to check your row reduction.

41. $\begin{cases} x_1 + x_2 + x_3 = 2 \\ x_1 + 2x_2 + 2x_3 = 3 \\ x_1 + 3x_2 + 2x_3 = 1 \end{cases}$

42. $\begin{cases} 2x_1 + 3x_2 + x_3 = 4 \\ 3x_1 + 3x_2 + 2x_3 = 12 \\ x_1 + 2x_2 + x_3 = 4 \end{cases}$

43. $\begin{cases} 2x_1 + 2x_2 - x_3 = -5 \\ -2x_1 - x_2 + x_3 = 3 \\ 3x_1 + 4x_2 - x_3 = -8 \end{cases}$

44. $\begin{cases} 3x_1 - 4x_2 + 2x_3 = -15 \\ -x_1 + 2x_2 - x_3 = 6 \\ 4x_1 - 3x_2 + 2x_3 = -16 \end{cases}$

45. $\begin{cases} 2x_1 + x_2 - 2x_3 + x_4 = 2 \\ x_1 + x_2 + 2x_3 + x_4 = 5 \\ x_1 + x_2 + x_3 + x_4 = 4 \\ 2x_1 + 2x_2 + 3x_3 + x_4 = 8 \end{cases}$

46. $\begin{cases} x_1 - x_2 + x_3 + 2x_4 = 7 \\ x_1 - 2x_2 + x_3 + 2x_4 = 8 \\ 2x_1 + 2x_2 + x_3 + 3x_4 = 7 \\ x_1 - x_2 + x_3 + x_4 = 6 \end{cases}$

47. $\begin{cases} 2x_1 + 3x_2 + x_3 = 4 \\ 3x_1 + 5x_2 + 2x_3 = 12 \\ x_1 + 2x_2 + x_3 = 3 \end{cases}$

48. $\begin{cases} x_1 + 3x_2 + 2x_3 = 6 \\ x_1 + 2x_2 + 2x_3 = 3 \\ 2x_1 + 5x_2 + 4x_3 = 8 \end{cases}$

49. $\begin{cases} 4x_1 + 3x_2 + 2x_3 = 24 \\ x_1 + x_2 + 3x_3 = 7 \\ 5x_1 + 4x_2 + 5x_3 = 31 \end{cases}$

50. $\begin{cases} 5x_1 - 7x_2 + 3x_3 = -9 \\ -x_1 + x_2 - x_3 = 1 \\ 4x_1 - 5x_2 + 3x_3 = -6 \end{cases}$

51. $\begin{cases} x_1 + x_2 + 2x_3 + x_4 = 2 \\ 2x_1 + 2x_2 + 3x_3 + 3x_4 = 9 \\ 2x_1 + x_2 + 2x_3 + 2x_4 = 7 \\ x_1 + x_2 + x_3 + x_4 = 4 \end{cases}$

52. $\begin{cases} 3x_1 + 7x_2 + 4x_3 + 4x_4 = -7 \\ 7x_1 + 7x_2 + 4x_3 + 7x_4 = 3 \\ 4x_1 + 3x_2 + 2x_3 + 3x_4 = 2 \\ 3x_1 + 2x_2 + x_3 + 3x_4 = 4 \end{cases}$

53. $\begin{cases} 2x_1 + 2x_2 - 2x_3 + x_4 = -6 \\ -3x_1 - x_2 + x_3 - 2x_4 = 8 \\ 2x_1 + x_2 - x_3 + x_4 = -5 \\ -x_1 - x_2 + 2x_3 - x_4 = 6 \end{cases}$

54. $\begin{cases} 2x_1 - 5x_2 + 3x_3 + 6x_4 = -16 \\ 5x_1 - 11x_2 + 5x_3 + 14x_4 = -41 \\ -4x_1 + 9x_2 - 3x_3 - 13x_4 = 37 \\ -3x_1 + 7x_2 - 4x_3 - 8x_4 = 23 \end{cases}$

55. $\begin{cases} 4x_1 - x_2 + 3x_3 - x_4 + x_5 = -2 \\ 2x_1 + 2x_2 + x_3 + 4x_4 = 0 \\ 4x_1 - x_2 + 2x_3 - 2x_4 + x_5 = 1 \\ 2x_1 + x_2 + x_3 + 2x_4 = 0 \\ 3x_1 - x_2 + x_3 - 2x_4 + x_5 = 2 \end{cases}$

56. $\begin{cases} x_1 + x_3 + x_5 = 4 \\ -x_1 + 2x_2 - x_3 + x_4 - x_5 = -9 \\ 4x_1 + 2x_2 + 4x_3 + 2x_4 + 3x_5 = 9 \\ -x_1 + 4x_2 - 2x_3 + 2x_4 - 2x_5 = -17 \\ 3x_1 + x_2 + 2x_3 + x_4 + x_5 = 5 \end{cases}$

57. $\begin{cases} x_1 - x_2 + x_4 = 1 \\ 2x_1 - 2x_2 + 2x_4 = 2 \\ x_1 - x_2 - x_3 - x_4 = 1 \\ 2x_1 - 2x_2 - x_3 = 1 \end{cases}$

58. $\begin{cases} x_1 - x_3 + x_4 = 2 \\ x_2 + 2x_3 + x_4 = 0 \\ 2x_1 - 2x_2 - 6x_3 = 5 \\ x_1 + x_2 + x_3 + 2x_4 = 2 \end{cases}$

59. $\begin{cases} x_1 + x_2 + x_3 + 2x_4 = 3 \\ x_1 - x_3 + x_4 = 2 \\ x_1 + 2x_2 + 3x_3 + 3x_4 = 4 \\ x_2 + 2x_3 + x_4 = 1 \end{cases}$

60. $\begin{cases} x_1 + 2x_2 + x_3 - 2x_4 = -3 \\ 2x_1 + 4x_2 + 2x_3 - x_4 = 0 \\ x_1 + 2x_2 + x_3 + x_4 = 3 \\ x_1 + 2x_2 + x_3 = 1 \end{cases}$

APPLIED EXERCISES

Formulate each situation as a system of linear equations in an appropriate number of variables and be sure to state clearly the meaning of each. Solve the system of equations by row reducing the corresponding augmented matrix. Be sure to state your final answer in terms of the original question.

 If you have a graphing calculator with a RREF(command, use it to check your row reduction.

(Pounds per Bag)	GrowRite	MiracleMix	GreatGreen
Potash	4	6	7
Nitrogen	5	10	16
Phosphoric acid	3	4	5

61. Plant Fertilizer   A backyard garden needs 35 pounds of potash, 68 pounds of nitrogen, and 25 pounds of phosphoric acid. Three brands of fertilizer, GrowRite, MiracleMix, and Great-Green, are available and contain the amounts of potash, nitrogen, and phosphoric acid per bag listed in the table. How many bags of each brand should be used to provide the required potash, nitrogen, and phosphoric acid?

62. Nutrition   A student athlete has decided to "bulk up" before the wrestling season by supplementing his weekly diet with an additional 325 grams of protein, 185 grams of fiber, and 110 grams of fat. If a hamburger contains 20 grams of protein, 10 grams of fiber, and 5 grams of fat, a cheeseburger contains 25 grams of protein, 10 grams of fiber, and 5 grams of fat, and a "sloppy-joe" contains 20 grams of protein, 15 grams of fiber, and 10 grams of fat, how many of each should he eat this week to meet his goal?

63. Coins in a Jar A jar contains 700 nickels, dimes, and quarters worth $60. (a) How many of each are in the jar? (b) What is the greatest possible number of quarters in the jar?

64. Financial Planning An international investment banker wishes to invest $150,000 in U.S. and German stocks and bonds. Since stocks are not as secure as bonds, he plans to spread his investments so that he has three times as much in U.S. bonds as in U.S. stocks and twice as much in German bonds as in German stocks. Furthermore, he intends to invest just $45,000 in stocks altogether. How much does he invest in each?

65. Income Taxes (a) Find the federal, state, and city income taxes on a taxable income of $180,000 if the federal tax is 50% of the taxable income after first deducting the state and city taxes, the state tax is 17% of the taxable income after first deducting the federal and city taxes, and the city tax is 3% of the taxable income after first deducting the federal and state taxes. (b) Although it appears that the taxpayer is facing a 70% nominal tax rate, use the actual taxes to determine the effective combined tax rate for this situation.

66. Estate Division As the old patriarch lay dying, he spoke to his four sons: "I leave you my fortune of 11,100 pieces of gold, which you must divide as I command. The oldest is to take one half, the second oldest is to take one third, the third oldest is to take one quarter, and the youngest is to take one fifth. But each of you must take your part from what remains after the others have taken theirs and then you must give any remainder to your mother and sister." The old woman was frantic: "How can there be any left for our daughter when you have given away more than everything?" "Don't worry, Mother," said the daughter, who had gotten an MBA while her brothers had been off fighting at Troy. How much did each son receive and how much was left over for the mother and daughter?

67. Apparel Production (a) A "limited edition" ladies fashion shop has a 240-yard supply of a silk fabric suitable for scarves, dresses, blouses, and skirts. Each scarf requires 1 yard of material, 3 minutes of cutting, and 5 minutes of sewing; each dress requires 3 yards of material, 14 minutes of cutting, and 40 minutes of sewing; each blouse requires 1.5 yards of material, 9 minutes of cutting, and 30 minutes of sewing; and each skirt requires 2 yards of material, 8 minutes of cutting, and 20 minutes of sewing. If the shop has 17 hours of skilled pattern cutter labor and 45 hours of skilled seamstress labor available, how many of each can be made? (b) If 20 blouses and 10 skirts are made, how many scarves and dresses can be made?

68. Agriculture Management A farmer grows wheat, barley, and oats on his 360-acre farm. The labor (for planting, tending, and harvesting) and capital (for seeds, fertilizers, and pesticides) requirements for each crop are given in the table. If the farmer can contract for 1050 days of migrant worker labor and has $18,600 set aside for capital expenses, how many acres of each crop can he grow?

(Per Acre)	Wheat	Barley	Oats
Days of labor	2	4	3
Capital expenses	$50	$40	$60

69. Advertising The promotional director for a new movie has $110,000 to spend on TV, radio, and newspaper advertisements in the metropolitan area. Each TV ad costs $2500 and is seen by 10,000 moviegoers, each radio ad costs $500 and is heard by 2000 moviegoers, and each newspaper ad costs $1000 and is read by 5000 moviegoers. Ignoring repeated exposures to the same person, how many of each ad will contact 500,000 moviegoers using the allocated funds?

70. Mass Transit   Part of a subway system is shown in the diagram. The numbers of subway cars per hour arriving and leaving stations I, II, III, and IV are indicated by arrows with numbers or variables. In order for the subway to function, the numbers of cars arriving and leaving per hour must match at each station. To prevent collisions, the subway cars must travel in the directions indicated by the arrows and so none of the variables may be negative. A group of concerned citizens has petitioned the city council to build an express service from station I to IV (indicated on the diagram as x_5) capable of handling 40 cars per hour. Is this a good idea or is it a waste of money?

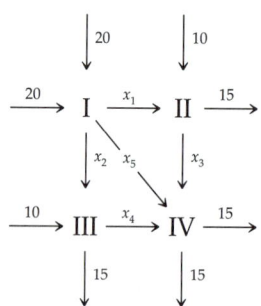

Explorations and Excursions

The following problems extend and augment the material presented in the text.

Row-echelon form has the same requirements as reduced row-echelon form (page 200) except that condition (3) is relaxed to require only that *the column of each leftmost one contains zeros below it.* Such a matrix is also called a *triangular matrix,* since the nonzero elements can only be on and above the main diagonal.

Find the system of equations represented by each row-echelon form matrix:

71. $\begin{pmatrix} 1 & 2 & 3 \\ 0 & 1 & 1 \end{pmatrix}$ **72.** $\begin{pmatrix} 1 & 3 & 9 \\ 0 & 1 & 2 \end{pmatrix}$

73. $\begin{pmatrix} 1 & 1 & 1 & 4 \\ 0 & 1 & 1 & 3 \\ 0 & 0 & 1 & 2 \end{pmatrix}$ **74.** $\begin{pmatrix} 1 & 0 & 1 & 0 & 7 \\ 0 & 1 & 0 & 1 & 3 \\ 0 & 0 & 1 & 0 & 4 \\ 0 & 0 & 0 & 1 & 1 \end{pmatrix}$

Unlike reduced row-echelon form matrices, there are many row-echelon form matrices equivalent to a given augmented matrix.

Row reduce each group of matrices to show that they are all equivalent because they are all equivalent to the same reduced row-echelon form matrix.

75. $\begin{pmatrix} 1 & 2 & 3 \\ 2 & 3 & 5 \end{pmatrix}, \begin{pmatrix} 1 & 2 & 3 \\ 0 & 1 & 1 \end{pmatrix}, \begin{pmatrix} 1 & 3 & 4 \\ 0 & 1 & 1 \end{pmatrix}$

76. $\begin{pmatrix} 2 & 5 & 16 \\ 1 & 4 & 11 \end{pmatrix}, \begin{pmatrix} 1 & 3 & 9 \\ 0 & 1 & 2 \end{pmatrix}, \begin{pmatrix} 1 & -1 & 1 \\ 0 & 1 & 2 \end{pmatrix}$

77. $\begin{pmatrix} 1 & 2 & 2 & 7 \\ 1 & 1 & -1 & 0 \\ 2 & 2 & 1 & 6 \end{pmatrix}, \begin{pmatrix} 1 & 1 & 1 & 4 \\ 0 & 1 & 1 & 3 \\ 0 & 0 & 1 & 2 \end{pmatrix},$
$\begin{pmatrix} 1 & 2 & 2 & 7 \\ 0 & 1 & 2 & 5 \\ 0 & 0 & 1 & 2 \end{pmatrix}$

78. $\begin{pmatrix} 1 & 1 & 1 & 1 & 10 \\ 1 & 0 & 2 & 0 & 11 \\ 1 & 1 & 0 & 1 & 6 \\ 1 & 0 & 1 & 1 & 8 \end{pmatrix}, \begin{pmatrix} 1 & 0 & 1 & 0 & 7 \\ 0 & 1 & 0 & 1 & 3 \\ 0 & 0 & 1 & 0 & 4 \\ 0 & 0 & 0 & 1 & 1 \end{pmatrix},$
$\begin{pmatrix} 1 & -1 & 1 & -1 & 4 \\ 0 & 1 & -1 & 1 & -1 \\ 0 & 0 & 1 & -1 & 3 \\ 0 & 0 & 0 & 1 & 1 \end{pmatrix}$

Find an equivalent matrix in row-echelon form using the REF(matrix command.

79. $\begin{pmatrix} 1 & 2 & 3 \\ 2 & 3 & 5 \end{pmatrix}$ **80.** $\begin{pmatrix} 1 & 2 & 2 & 7 \\ 1 & 1 & -1 & 0 \\ 2 & 2 & 1 & 6 \end{pmatrix}$

Back Substitution   If the augmented matrix for a system of equations is in row-echelon form, the last equation gives the value for the last variable. Substituting this value into the second-to-last equation, the value of the second-to-last variable is easy to find. Continuing upward through the equations, the values of all the remaining variables are easily found.

Solve each system of equations by back substitution.

81. $\begin{cases} x_1 + 2x_2 = 3 \\ x_2 = 1 \end{cases}$ **82.** $\begin{cases} x_1 + 3x_2 = 9 \\ x_2 = 2 \end{cases}$

83. $\begin{cases} x_1 + 3x_2 = 4 \\ x_2 = 1 \end{cases}$

84. $\begin{cases} x_1 - x_2 = 1 \\ x_2 = 2 \end{cases}$

85. $\begin{cases} x_1 + x_2 + x_3 = 4 \\ x_2 + x_3 = 3 \\ x_3 = 2 \end{cases}$

86. $\begin{cases} x_1 + 2x_2 + 2x_3 = 7 \\ x_2 + 2x_3 = 5 \\ x_3 = 2 \end{cases}$

87. $\begin{cases} x_1 + x_3 = 7 \\ x_2 + x_4 = 3 \\ x_3 = 4 \\ x_4 = 1 \end{cases}$

88. $\begin{cases} x_1 - x_2 + x_3 - x_4 = 4 \\ x_2 - x_3 + x_4 = -1 \\ x_3 - x_4 = 3 \\ x_4 = 1 \end{cases}$

The Gauss–Jordan Method Named after Carl F. Gauss (1777–1855), the "prince of mathematicians,"

and Wilhelm Jordan (1842–1899), the German geodesist, the Gauss–Jordan method of row reducing an augmented matrix consists of two steps: first use row operations to find an equivalent row-echelon form matrix, and then use more row operations to back-substitute and complete the solution. Of course, if the row-echelon form matrix contains a row of zeros ending in a one, the corresponding system of equations is inconsistent, there can be no solution, and the method stops without attempting to carry out the back substitution.

Solve each system of equations by the Gauss–Jordan method.

89. $\begin{cases} x_1 + 3x_2 + x_3 = 2 \\ 2x_1 + 5x_2 + 2x_3 = 5 \\ x_1 + 2x_2 + 2x_3 = 5 \end{cases}$

90. $\begin{cases} 2x_1 + 2x_2 + x_3 + 3x_4 = 8 \\ x_1 + 2x_2 + x_3 + x_4 = 4 \\ 2x_1 + 3x_2 + x_3 + 2x_4 = 6 \\ 3x_1 + 2x_2 - x_3 + 2x_4 = 3 \end{cases}$

2.4 Matrix Arithmetic

APPLICATION PREVIEW

Matrices and Computers

Since matrices and their arithmetic are naturally suited to representing and manipulating vast quantities of numerical information, it is hardly surprising that even some of the earliest programming languages included commands to define and process matrices. BASIC, perhaps the most widely used computer programming language, was created at Dartmouth College in May 1964 by John G. Kemeny and Thomas E. Kurtz. This very first version contained an array data structure so that whole grids of values could be conveniently represented. A variety of BASIC introduced during the summer of 1964 called CARD-BASIC included the now-standard MAT instructions, so that single commands could specify arithmetic operations on entire matrices. For instance, if the manufacturing costs of a company's entire product line are entered in a matrix named Costs and the corresponding product revenues are stored in a matrix named Revenues, the single program statement

$$\text{MAT Profits} = \text{Revenues} - \text{Costs}$$

will subtract each cost from the corresponding revenue and place the result at the correct position in a matrix named Profits. From the beginning, BASIC has stored arrays in row-major order, in which the values of the first column are placed in adjacent computer memory locations, with those of the next column immediately following, to permit the fastest possible access to the matrix elements during MAT calculations. The inclusion of matrix commands in virtually all computer languages has stimulated many useful applications of matrix methods.

Introduction

Our focus thus far has been on the augmented matrix representation of a system of equations. The augmented matrix representation is "unnatural" in the sense that it somehow "contains" the equals signs from the equations. We shall now define the arithmetic of matrices so that we may write an entire system of equations as a single *matrix equation*. The solution in the next section of such an equation of matrices will deepen our understanding of what it means to solve a system of equations.

Some Basic Terms

We write $A = (a_{i,j})$ to indicate that the matrix A is composed of the elements $a_{i,j}$ and $A^{m \times n}$ to indicate that the dimension of A is $m \times n$. Two matrices are *equal* if they have the same dimension and if elements in corresponding locations are equal; that is, $A^{m \times n} = B^{m \times n}$ if $a_{i,j} = b_{i,j}$ for every row i ($1 \leq i \leq m$) and column j ($1 \leq j \leq n$). For instance,

$$\begin{pmatrix} 1 & 2 & 3 \\ 4 & 5 & 6 \end{pmatrix} = \begin{pmatrix} 1 & 1+1 & 4-1 \\ 2^2 & 10/2 & 2 \cdot 3 \end{pmatrix}$$

but

$$\begin{pmatrix} 1 & 2 \\ 3 & 4 \end{pmatrix} \neq \begin{pmatrix} 1 \\ 2 \\ 3 \\ 4 \end{pmatrix} \quad \text{and} \quad \begin{pmatrix} 1 & 2 \\ 3 & 4 \end{pmatrix} \neq \begin{pmatrix} 1 & 2 \\ 4 & 3 \end{pmatrix}$$

The *transpose* of a matrix is formed by turning each row into a column (or equivalently, each column into a row). More precisely, the transpose (denoted by a superscript t) of the $m \times n$ matrix $A = (a_{i,j})$ is the $n \times m$ matrix $A^t = (a_{j,i})$. For instance,

$$\begin{pmatrix} 1 & 2 & 3 \\ 4 & 5 & 6 \end{pmatrix}^t = \begin{pmatrix} 1 & 4 \\ 2 & 5 \\ 3 & 6 \end{pmatrix}$$

First row becomes first column
Second row becomes second column

A matrix is *symmetric* if it is equal to its transpose. Notice that a symmetric matrix must be square and that the elements "above" the main diagonal must mirror those "below" it. For instance,

$$\begin{pmatrix} 1 & 5 & 6 \\ 5 & 2 & 7 \\ 6 & 7 & 3 \end{pmatrix} \text{ is symmetric but } \begin{pmatrix} 1 & 2 & 3 \\ 2 & 1 & 2 \\ 2 & 3 & 1 \end{pmatrix} \text{ is not}$$

An *identity matrix* is a square matrix with ones on the main diagonal and zeros elsewhere. We write $I = I^{n \times n}$ for the $n \times n$ identity matrix. Notice that identity matrices are symmetric. The identity matrices $I^{1 \times 1}$, $I^{2 \times 2}$, $I^{3 \times 3}$, and $I^{4 \times 4}$ are written below.

$$(1), \begin{pmatrix} 1 & 0 \\ 0 & 1 \end{pmatrix}, \begin{pmatrix} 1 & 0 & 0 \\ 0 & 1 & 0 \\ 0 & 0 & 1 \end{pmatrix}, \begin{pmatrix} 1 & 0 & 0 & 0 \\ 0 & 1 & 0 & 0 \\ 0 & 0 & 1 & 0 \\ 0 & 0 & 0 & 1 \end{pmatrix}$$

Scalar multiplication of a matrix simply means multiplying each element of a matrix by the same number. More precisely, a number (or "scalar") s times a matrix A is defined as $sA = (sa_{i,j})$. In particular, the *negative* of a matrix is $-A = (-a_{i,j})$.

$$\text{For } A = \begin{pmatrix} 1 & 2 & 3 \\ 4 & 5 & 6 \end{pmatrix}, \quad 3A = \begin{pmatrix} 3 & 6 & 9 \\ 12 & 15 & 18 \end{pmatrix}$$

$$\text{and} \quad -A = \begin{pmatrix} -1 & -2 & -3 \\ -4 & -5 & -6 \end{pmatrix}$$

Matrix Addition

For two matrices with the same dimensions, *matrix addition* means adding elements in corresponding locations. That is, the *sum* of $A^{m \times n}$ and $B^{m \times n}$ is $A + B = (a_{i,j} + b_{i,j})$. Matrices with *different* dimensions cannot be added.

$$\begin{pmatrix} 1 & 4 \\ 2 & 5 \\ 3 & 6 \end{pmatrix} + \begin{pmatrix} 11 & 12 \\ 10 & 9 \\ 7 & 8 \end{pmatrix} = \begin{pmatrix} 12 & 16 \\ 12 & 14 \\ 10 & 14 \end{pmatrix}$$

but
$$\begin{pmatrix} 1 & 4 \\ 2 & 5 \\ 3 & 6 \end{pmatrix} + \begin{pmatrix} 7 & 8 & 9 \\ 10 & 11 & 12 \end{pmatrix}$$
is not possible.

Similarly, two matrices with the same dimensions can be *subtracted* by subtracting corresponding elements (or equivalently, adding the negative of the second matrix).

$$\begin{pmatrix} 11 & 12 \\ 10 & 9 \\ 7 & 8 \end{pmatrix} - \begin{pmatrix} 1 & 4 \\ 2 & 5 \\ 3 & 6 \end{pmatrix} = \begin{pmatrix} 10 & 8 \\ 8 & 4 \\ 4 & 2 \end{pmatrix}$$

A *zero matrix* has all elements equal to zero. If two matrices are equal, their difference is a zero matrix. We write 0 for the zero matrix of whatever size is appropriate for the situation. That is, if $A^{m \times n} = B^{m \times n}$, then $A - B = 0$ means that 0 is the $m \times n$ zero matrix.

$$\begin{pmatrix} 1 & 2 \\ 3 & 4 \\ 5 & 6 \end{pmatrix} - \begin{pmatrix} 1 & 2 \\ 3 & 4 \\ 5 & 6 \end{pmatrix} = 0 \text{ means } \begin{pmatrix} 1 & 2 \\ 3 & 4 \\ 5 & 6 \end{pmatrix} - \begin{pmatrix} 1 & 2 \\ 3 & 4 \\ 5 & 6 \end{pmatrix} = \begin{pmatrix} 0 & 0 \\ 0 & 0 \\ 0 & 0 \end{pmatrix}$$

Graphing Calculator Exploration

The basic matrix operations are found in the MATRX MATH menu. To have the calculator demonstrate the preceding concepts, proceed as follows.

a. Use MATRX EDIT to enter $\begin{pmatrix} 1 & 2 & 3 \\ 4 & 5 & 6 \end{pmatrix}$ in [A].

```
MATRIX[A]  2 x3
[1      2      3   ]
[4      5      6   ]




2,3=6
```

b. Find the transpose of A.

```
[A]
          [[1 2 3]
           [4 5 6]]
[A]ᵀ
          [[1 4]
           [2 5]
           [3 6]]
```

c. Find the scalar multiples $3A$ and $(1/2)A$. Check that attempting the division $[A]/2$ gives a "data type" error.

```
3[A]
      [[3  6  9 ]
       [12 15 18]]
(1/2)[A]
      [[.5 1    1.5]
       [2  2.5 3    ]]
```

d. Find $A + A$ and $A - A$.

```
[A]+[A]
      [[2 4  6 ]
       [8 10 12]]

[A]-[A]
      [[0 0 0]
       [0 0 0]]
```

e. Find the 4×4 identity matrix.

```
identity(4)
      [[1 0 0 0]
       [0 1 0 0]
       [0 0 1 0]
       [0 0 0 1]]
```

EXAMPLE 1   Using Matrix Arithmetic

A McBurger restaurant sells hamburgers for $1, cheeseburgers for $1.50, and fries for 75¢ while the BurgerQueen across the street sells hamburgers for $1.25, cheeseburgers for $1.75, and fries for 50¢. At McBurger, the preparation costs are 30¢ per hamburger, 40¢ per cheeseburger, and 25¢ per order of fries, while at BurgerQueen the costs are 35¢ per hamburger, 45¢ per cheeseburger, and 20¢ per order of fries.

a. Represent the selling prices and preparation costs as matrices.

b. Using the matrices from (a), find the profit margin matrix for the items at the restaurants.

c. Using the matrix from (b), find the franchise fee charged on each item at each restaurant if this fee is 30% of the profit margin.

Solution

a. Let R be the revenue matrix of selling prices and let C be the cost matrix of preparation costs. Since we have 3 items from 2 restaurants, we can choose the matrices to be either 2×3 or 3×2. We choose the matrices to be 2 × 3, with the rows corresponding to the different restaurants and the columns to the different menu items:

$$
\begin{array}{ccc}
\text{Hamburger} & \text{Cheeseburger} & \text{Fries} \\
\downarrow & \downarrow & \downarrow
\end{array}
$$

$$
\begin{array}{c}
\text{McBurger} \rightarrow \\
\text{BurgerQueen} \rightarrow
\end{array}
\begin{pmatrix}
\underline{} & \underline{} & \underline{} \\
\underline{} & \underline{} & \underline{}
\end{pmatrix}
$$

From the given information, we obtain the matrices:

$$
R = \begin{pmatrix} 1.00 & 1.50 & 0.75 \\ 1.25 & 1.75 & 0.50 \end{pmatrix} \quad \text{and} \quad C = \begin{pmatrix} 0.30 & 0.40 & 0.25 \\ 0.35 & 0.45 & 0.20 \end{pmatrix}
$$

b. Let P be the profit margin matrix, which is revenue R minus cost C:

$$
P = R - C = \begin{pmatrix} 1.00 & 1.50 & 0.75 \\ 1.25 & 1.75 & 0.50 \end{pmatrix} - \begin{pmatrix} 0.30 & 0.40 & 0.25 \\ 0.35 & 0.45 & 0.20 \end{pmatrix}
$$

$$
= \begin{pmatrix} 0.70 & 1.10 & 0.50 \\ 0.90 & 1.30 & 0.30 \end{pmatrix} \qquad \text{Using matrix subtraction}
$$

c. Let F be the franchise fee matrix, which is 30% of the profit matrix P:

$$
F = 0.30P = 0.30 \cdot \begin{pmatrix} 0.70 & 1.10 & 0.50 \\ 0.90 & 1.30 & 0.30 \end{pmatrix}
$$

$$
= \begin{pmatrix} 0.21 & 0.33 & 0.15 \\ 0.27 & 0.39 & 0.09 \end{pmatrix} \qquad \text{Using scalar multiplication}
$$

■

Matrix Multiplication as Evaluation

To evaluate the expression $5x_1 + 6x_2 + 7x_3$ at $x_1 = 2, x_2 = 3$, and $x_3 = 4$, we substitute the values for the variables, multiply them out, and add them up: $5 \cdot 2 + 6 \cdot 3 + 7 \cdot 4 = 56$. With this kind of evaluation in mind, we define *matrix multiplication* as multiplying in order the numbers from a row by the numbers from a column and adding the results. Thus, *we link matrix arithmetic to systems of equations by defining matrix multiplication as "evaluation" of the row "expression" on the left by the column "list of values" on the right.* The matrix product of a row matrix by a column matrix is a 1 × 1 matrix:

$$
\begin{pmatrix} 5 & 6 & 7 \end{pmatrix} \cdot \begin{pmatrix} 2 \\ 3 \\ 4 \end{pmatrix} = (5 \cdot 2 + 6 \cdot 3 + 7 \cdot 4) = (56)
$$

Of course there must be exactly as many elements in the row on the left as in the column on the right.

PRACTICE PROBLEM 1 Find $(1 \quad 2)\begin{pmatrix} 3 \\ 4 \end{pmatrix}$. *Solution at the back of the book*

If there are several rows on the left, and several columns on the right, then we multiply *each row* times *each column*, with each answer placed in the product matrix at the row and column position from whence it came:

First row times second column goes here, in first row, second column

$$\begin{pmatrix} 3 & 2 & 1 \\ 2 & 0 & -2 \end{pmatrix}\begin{pmatrix} 2 & 7 \\ 3 & 6 \\ 4 & 5 \end{pmatrix} = \begin{pmatrix} (3 \ 2 \ 1)\begin{pmatrix} 2 \\ 3 \\ 4 \end{pmatrix} & (3 \ 2 \ 1)\begin{pmatrix} 7 \\ 6 \\ 5 \end{pmatrix} \\ (2 \ 0 \ -2)\begin{pmatrix} 2 \\ 3 \\ 4 \end{pmatrix} & (2 \ 0 \ -2)\begin{pmatrix} 7 \\ 6 \\ 5 \end{pmatrix} \end{pmatrix} = \begin{pmatrix} 16 & 38 \\ -4 & 4 \end{pmatrix}$$

Go from here

Multiply each row times column "in your head"

Directly to here

In general, to multiply two matrices, the row length of the first must match the column length of the second, with the other numbers giving the dimension of the product:

$$A^{m \times p} \ B^{p \times n} = C^{m \times n} \qquad \text{Inside numbers get "absorbed"}$$

Formally, matrix multiplication is defined as follows:

Matrix Multiplication

The product of $A^{m \times p}$ with $B^{q \times n}$ is defined only if $p = q$, in which case $A \cdot B$ is the matrix $C^{m \times n} = (c_{i,j})$ with elements $c_{i,j} = a_{i,1}b_{1,j} + a_{i,2}b_{2,j} + \cdots + a_{i,p}b_{p,j}$.

Not every matrix product can be found; for instance, the product:

$$\begin{pmatrix} 2 & 7 \\ 3 & 6 \\ 4 & 5 \end{pmatrix}\begin{pmatrix} 5 & 6 & 7 \\ 1 & -1 & 1 \\ 2 & 0 & -2 \\ 3 & 2 & 1 \end{pmatrix} \text{ is not defined.}$$

The row length (2) of the first does not match the column length (4) of the second

Notice that while $(5 \quad 6 \quad 7)\begin{pmatrix} 2 \\ 3 \\ 4 \end{pmatrix} = (56)$ is a 1×1 matrix, the product in the reverse order is a very different 3×3 matrix:

$$\begin{pmatrix} 2 \\ 3 \\ 4 \end{pmatrix}(5 \quad 6 \quad 7) = \begin{pmatrix} 2 \cdot 5 & 2 \cdot 6 & 2 \cdot 7 \\ 3 \cdot 5 & 3 \cdot 6 & 3 \cdot 7 \\ 4 \cdot 5 & 4 \cdot 6 & 4 \cdot 7 \end{pmatrix} = \begin{pmatrix} 10 & 12 & 14 \\ 15 & 18 & 21 \\ 20 & 24 & 28 \end{pmatrix}$$

 Graphing Calculator Exploration

Calculator matrix multiplication is easy to do. After entering your matrices as [A] and [B], use the usual multiplication key.

```
[A]
           [[5 6 7]]
[B]
           [[2]
            [3]
            [4]]
```

```
[A]*[B]
                [[56]]
[B]*[A]
           [[10 12 14]
            [15 18 21]
            [20 24 28]]
```

Unlike number multiplication, *matrix multiplication is not commutative.* That is, $A \cdot B$ may not be the same as $B \cdot A$. However, if we change rows into columns and columns into rows (that is, transpose each matrix) and then reverse the order, we will be multiplying and adding the same collections of numbers, with only their positions changed. That is, *the transpose of a product is the product of the transposes in the opposite order.*

If $A \cdot B$ is possible, then $(A \cdot B)^t = B^t \cdot A^t$.

EXAMPLE 2 Matrix Multiplication

Use the selling price ("revenue") matrix from Example 1 (page 219) to find the total selling price for a meal of three hamburgers, a cheeseburger, and two fries at each of the restaurants as a matrix product.

Solution

The revenue matrix R from Example 1 gives the selling prices in rows, one row for each restaurant. If we write the three hamburgers, one cheeseburger, and two fries as a *column*, we can use matrix multiplication to find the total price at each restaurant.

$$\begin{pmatrix} 1.00 & 1.50 & 0.75 \\ 1.25 & 1.75 & 0.50 \end{pmatrix} \begin{pmatrix} 3 \\ 1 \\ 2 \end{pmatrix} = \begin{pmatrix} 1.00 \cdot 3 + 1.50 \cdot 1 + 0.75 \cdot 2 \\ 1.25 \cdot 3 + 1.75 \cdot 1 + 0.50 \cdot 2 \end{pmatrix}$$

$\qquad\qquad \uparrow \qquad\qquad\qquad \uparrow$

R (from Example 1)   Order (3 hamburgers, 1 cheeseburger, 2 fries)

$$= \begin{pmatrix} 6.00 \\ 6.50 \end{pmatrix} \begin{matrix} \leftarrow \text{Price at McBurger} \\ \leftarrow \text{Price at BurgerQueen} \end{matrix}$$

Notice that we could have found the same answers by multiplying the transposes in the opposite order:

$$(3 \quad 1 \quad 2) \begin{pmatrix} 1.00 & 1.25 \\ 1.50 & 1.75 \\ 0.75 & 0.50 \end{pmatrix} = (6.00 \quad 6.50)$$

■

PRACTICE PROBLEM 2

Use the selling price ("revenue") matrix R from Example 1 or 2 to find the total selling price for a meal of four hamburgers, two cheeseburgers, and five fries at each of the restaurants as a matrix product.

Solution at the back of the book

Identity Matrices

Multiplication by an identity matrix (of the proper size) does not change the matrix:

$$\begin{pmatrix} 1 & 0 & 0 \\ 0 & 1 & 0 \\ 0 & 0 & 1 \end{pmatrix} \begin{pmatrix} 2 & 7 \\ 3 & 6 \\ 4 & 5 \end{pmatrix} = \begin{pmatrix} 2 & 7 \\ 3 & 6 \\ 4 & 5 \end{pmatrix} \qquad \begin{matrix} \text{Same matrix} \\ \text{on the right} \\ \text{as on the left} \end{matrix}$$

The first row $(1 \quad 0 \quad 0)$ picks out just the first value, the second $(0 \quad 1 \quad 0)$ picks out the second value, and so on. (You should carefully verify this multiplication to see how the identity works.) To multiply by the identity matrix on the other side, we need in this case to use an identity matrix of a different size:

$$\begin{pmatrix} 2 & 7 \\ 3 & 6 \\ 4 & 5 \end{pmatrix} \begin{pmatrix} 1 & 0 \\ 0 & 1 \end{pmatrix} = \begin{pmatrix} 2 & 7 \\ 3 & 6 \\ 4 & 5 \end{pmatrix} \qquad \begin{matrix} \text{Again, the} \\ \text{matrix is} \\ \text{duplicated} \end{matrix}$$

Writing I for an identity matrix of the appropriate size, we have for any matrix A:

$$I \cdot A = A \cdot I = A$$

That is, the matrix I plays the role in matrix arithmetic that the number 1 plays in ordinary arithmetic—multiplying by it gives back exactly what you started with.

Matrix Multiplication and Systems of Equations

Since matrix multiplication was defined in order to evaluate linear expressions (page 220), we may use it to write an entire system of equations as a single matrix equation. Consider the system of equations:

$$\begin{cases} 2x_1 + 3x_2 + 3x_3 = 15 \\ 3x_1 + 4x_2 + 4x_3 = 22 \\ 3x_1 + 4x_2 + 5x_3 = 30 \end{cases}$$

The "coefficient" matrix A and the "constant term" matrix B are:

$$A = \begin{pmatrix} 2 & 3 & 3 \\ 3 & 4 & 4 \\ 3 & 4 & 5 \end{pmatrix} \quad \text{and} \quad B = \begin{pmatrix} 15 \\ 22 \\ 30 \end{pmatrix}$$

Let X be the column matrix of the variables:

$$X = \begin{pmatrix} x_1 \\ x_2 \\ x_3 \end{pmatrix}$$

Then the matrix product $A \cdot X$ is precisely the left sides of the equations (as you should check):

$$\underbrace{\begin{pmatrix} 2 & 3 & 3 \\ 3 & 4 & 4 \\ 3 & 4 & 5 \end{pmatrix}}_{A} \underbrace{\begin{pmatrix} x_1 \\ x_2 \\ x_3 \end{pmatrix}}_{X} = \begin{pmatrix} 2x_1 + 3x_2 + 3x_3 \\ 3x_1 + 4x_2 + 4x_3 \\ 3x_1 + 4x_2 + 5x_3 \end{pmatrix}$$

The system of equations is the same as the matrix equation $AX = B$:

$$\begin{cases} 2x_1 + 3x_2 + 3x_3 = 15 \\ 3x_1 + 4x_2 + 4x_3 = 22 \\ 3x_1 + 4x_2 + 5x_3 = 30 \end{cases} \quad \text{is the same as} \quad \underbrace{\begin{pmatrix} 2 & 3 & 3 \\ 3 & 4 & 4 \\ 3 & 4 & 5 \end{pmatrix}}_{A} \underbrace{\begin{pmatrix} x_1 \\ x_2 \\ x_3 \end{pmatrix}}_{X} = \underbrace{\begin{pmatrix} 15 \\ 22 \\ 30 \end{pmatrix}}_{B}$$

That is, the system of equations on the left can be written as one matrix equation $A \cdot X = B$. Such matrix representations will be particularly useful for even larger systems of equations.

Matrix Multiplication and Row Operations

Since matrix multiplication of A by I picks out the elements of A in the correct order, switching the rows of I before multiplying will pick out the elements of A in a different order:

$$\text{I with rows 1 and 2 switched} \longrightarrow \begin{pmatrix} 0 & 1 & 0 \\ 1 & 0 & 0 \\ 0 & 0 & 1 \end{pmatrix} \begin{pmatrix} 2 & 7 \\ 3 & 6 \\ 4 & 5 \end{pmatrix} = \begin{pmatrix} 3 & 6 \\ 2 & 7 \\ 4 & 5 \end{pmatrix} \longleftarrow \text{A with rows 1 and 2 switched}$$

Replacing one of the 1s in I with a different value before multiplying will pick out the corresponding element of A that many times:

$$\text{I with row 3 multiplied by 8} \longrightarrow \begin{pmatrix} 1 & 0 & 0 \\ 0 & 1 & 0 \\ 0 & 0 & 8 \end{pmatrix} \begin{pmatrix} 2 & 7 \\ 3 & 6 \\ 4 & 5 \end{pmatrix} = \begin{pmatrix} 2 & 7 \\ 3 & 6 \\ 32 & 40 \end{pmatrix} \longleftarrow \text{A with row 3 multiplied by 8}$$

Replacing one of the 0s in I with a 1 before multiplying will pick out the elements of another row and add them to the current row of A:

$$\text{I with row 3 added to row 1} \longrightarrow \begin{pmatrix} 1 & 0 & 1 \\ 0 & 1 & 0 \\ 0 & 0 & 1 \end{pmatrix} \begin{pmatrix} 2 & 7 \\ 3 & 6 \\ 4 & 5 \end{pmatrix} = \begin{pmatrix} 6 & 12 \\ 3 & 6 \\ 4 & 5 \end{pmatrix} \longleftarrow \text{A with row 3 added to row 1}$$

Thus the row operations of switching two rows, multiplying a row by a constant, and adding a row to another can be accomplished by matrix multiplications. Furthermore, to find the matrix that performs a given row operation, we need only apply the same row operation to the identity matrix (of the appropriate size).

> Any row operation can be accomplished by a matrix multiplication.

This observation provides a second reason why matrix multiplication is not commutative: changing the order of a sequence of row operations usually changes the result, so changing the order of the corresponding matrix multiplications must similarly change the result.

Any sequence of row operations can be carried out by multiplying by a single matrix. For example, let M_1 be the matrix that switches rows 1 and 2, and let M_2 be the matrix that multiplies row 1 by 8, when applied to a matrix A. Then the two row operations are accomplished in order by multiplying on the left first by M_1 and then by M_2:

$$M_2 \cdot M_1 \cdot A$$

Since M_1 and M_2 could first be multiplied together to obtain a new matrix, M_3, the two row operations can also be accomplished by just multiplying by M_3:

$$M_3 \cdot A \text{ is the same as } M_2 \cdot M_1 \cdot A \text{ provided } M_3 = M_2 \cdot M_1$$

By repeating this process of applying successive row operations to the identity matrix I and building the product of these matrices, we can produce a single matrix that will perform any desired sequence of row operations. Thus if A is equivalent to the identity matrix, a matrix R representing a sequence of row operations that reduce A to I will satisfy $R \cdot A = I$. This observation will be very important in the next section.

SUMMARY

The elements of a matrix have both values and positions. Matrix arithmetic uses the following terms and operations.

Equal matrices	$A^{m \times n} = B^{m \times n}$	Same values in same positions
Transpose of a matrix	A^t	Row and column positions reversed
Symmetric matrix	$A^t = A$	"Above" diagonal mirrors "below" diagonal
Identity matrix	I	Square, zeros except for ones on diagonal
Scalar multiple	sA	Multiply every element by s
Matrix addition	$A^{m \times n} + B^{m \times n}$	Add elements in same positions
Matrix subtraction	$A^{m \times n} - B^{m \times n}$	Subtract elements in same positions
Zero matrix	0	All elements are zeros
Matrix multiplication	$A^{m \times p} \cdot B^{p \times n}$	Rows on left times columns on right

Any system of linear equations may be written as a matrix equation $A \cdot X = B$ where A is the coefficient matrix, X is the column matrix of variables, and B is the constant term matrix.

Any row operation can be accomplished by a matrix multiplication, and this "row operation" matrix can be found by applying the row operation to the identity matrix. Furthermore, a succession of such row operations can be carried out by multiplication on the left by a *single* matrix, the product of all of the matrices corresponding to the row operations.

EXERCISES 2.4

Use the given matrices to find each matrix expression.

$$A = \begin{pmatrix} 1 & 2 & 3 \\ 4 & 5 & 6 \\ 7 & 8 & 9 \end{pmatrix} \quad B = \begin{pmatrix} 9 & 8 & 7 \\ 6 & 5 & 4 \\ 3 & 2 & 1 \end{pmatrix} \quad C = \begin{pmatrix} 1 & 6 & 8 \\ 4 & 2 & 7 \\ 9 & 5 & 3 \end{pmatrix}$$

1. A^t
2. B^t
3. $3C$
4. $2A$
5. $-B$
6. $-C$
7. $A + C$
8. $B + C$
9. $C - (A + I)$
10. $(A + I) - B$

Find each matrix product.

11. $(1 \quad -1 \quad 1) \begin{pmatrix} 4 \\ 3 \\ 5 \end{pmatrix}$

12. $(1 \quad 1 \quad -1 \quad -1) \begin{pmatrix} 3 \\ 2 \\ 4 \\ 1 \end{pmatrix}$

13. $(1 \quad 0 \quad 1 \quad 1) \begin{pmatrix} 4 \\ 5 \\ -6 \\ 2 \end{pmatrix}$

14. $(1 \quad 0 \quad 1) \begin{pmatrix} 4 \\ 5 \\ 2 \end{pmatrix}$

15. $\begin{pmatrix} 1 & 2 & 1 \\ 2 & 1 & 2 \end{pmatrix} \begin{pmatrix} 2 \\ -3 \\ 2 \end{pmatrix}$

16. $\begin{pmatrix} 1 & 3 & 1 \\ 3 & 1 & 3 \end{pmatrix} \begin{pmatrix} 4 \\ -2 \\ 4 \end{pmatrix}$

17. $\begin{pmatrix} 1 & 3 & 1 \\ 2 & 1 & 2 \end{pmatrix} \begin{pmatrix} 1 & 2 \\ 1 & 1 \\ 2 & 1 \end{pmatrix}$

18. $\begin{pmatrix} 1 & 2 & 1 \\ 3 & 1 & 3 \end{pmatrix} \begin{pmatrix} 2 & 3 \\ 3 & 1 \\ 2 & -1 \end{pmatrix}$

19. $\begin{pmatrix} 2 & 3 \\ 3 & 1 \\ 2 & -1 \end{pmatrix} \begin{pmatrix} 1 & 2 & 1 \\ 3 & 1 & 3 \end{pmatrix}$

20. $\begin{pmatrix} 1 & 2 \\ 1 & 1 \\ 2 & 1 \end{pmatrix} \begin{pmatrix} 1 & 3 & 1 \\ 2 & 1 & 2 \end{pmatrix}$

Use the given matrices to find each matrix expression.

 If your instructor permits, you may carry out the calculations on a graphing calculator.

$A = \begin{pmatrix} 2 & -1 & 1 \\ 1 & 0 & 1 \end{pmatrix} \qquad B = \begin{pmatrix} 1 & 2 & 1 \\ 3 & -2 & 0 \end{pmatrix}$

$C = \begin{pmatrix} 2 & -1 & 2 \\ 1 & 1 & 1 \\ 0 & 2 & 1 \end{pmatrix}$

21. $A \cdot C$ **22.** $B \cdot C$

23. $C \cdot B^t$ **24.** $C \cdot A^t$

25. $(A - B) \cdot C$ **26.** $(B - A) \cdot C$

27. $A^t \cdot B + C$ **28.** $B^t \cdot A - C$

29. $B \cdot (C + I)$ **30.** $A \cdot (C - I)$

Rewrite each system of linear equations as a matrix equation $A \cdot X = B$.

31. $\begin{cases} x_1 + 5x_2 + 4x_3 = 6 \\ x_1 + x_2 + x_3 = 4 \\ 2x_1 + 3x_2 + 3x_3 = 9 \end{cases}$

32. $\begin{cases} x_1 + x_2 + x_3 = 4 \\ x_1 + 2x_2 + x_3 = 3 \\ x_1 + 2x_2 + 2x_3 = 5 \end{cases}$

33. $\begin{cases} 4x_1 + 3x_2 - x_3 = 2 \\ 3x_1 + 3x_2 + 2x_3 = 9 \\ 2x_1 + x_2 - 3x_3 = -6 \end{cases}$

34. $\begin{cases} 2x_1 - x_2 + 2x_3 = 11 \\ -x_1 + x_2 - 3x_3 = -12 \\ 2x_1 - 2x_2 + 7x_3 = 27 \end{cases}$

35. $\begin{cases} 5x_1 + 2x_2 - 4x_3 + x_4 + 5x_5 = 7 \\ 3x_1 + x_2 - 3x_3 + x_4 + 3x_5 = 5 \end{cases}$

36. $\begin{cases} 2x_1 + x_2 + 5x_3 + 4x_4 + 5x_5 = 2 \\ x_1 + x_2 + 3x_3 + 3x_4 + 3x_5 = -1 \end{cases}$

Rewrite each matrix equation $A \cdot X = B$ as a system of linear equations.

37. $\begin{pmatrix} 5 & 9 & 9 \\ 4 & 7 & 6 \\ 3 & 5 & 3 \\ 4 & 7 & 5 \end{pmatrix} \cdot \begin{pmatrix} x_1 \\ x_2 \\ x_3 \end{pmatrix} = \begin{pmatrix} 11 \\ 9 \\ 8 \\ 10 \end{pmatrix}$

38. $\begin{pmatrix} 6 & 3 & 5 \\ 1 & 2 & 2 \\ 4 & 3 & 4 \\ 5 & 1 & 3 \end{pmatrix} \cdot \begin{pmatrix} x_1 \\ x_2 \\ x_3 \end{pmatrix} = \begin{pmatrix} 8 \\ 1 \\ 5 \\ 7 \end{pmatrix}$

39. $\begin{pmatrix} 5 & 4 & 7 & 6 \\ 2 & 2 & 3 & 3 \\ 4 & 3 & 5 & 5 \\ 3 & 2 & 3 & 3 \end{pmatrix} \cdot \begin{pmatrix} x_1 \\ x_2 \\ x_3 \\ x_4 \end{pmatrix} = \begin{pmatrix} 18 \\ 9 \\ 16 \\ 11 \end{pmatrix}$

40. $\begin{pmatrix} 3 & 4 & 2 & 4 \\ 1 & 2 & 1 & 1 \\ 4 & 5 & 2 & 5 \\ 6 & 6 & 1 & 6 \end{pmatrix} \cdot \begin{pmatrix} x_1 \\ x_2 \\ x_3 \\ x_4 \end{pmatrix} = \begin{pmatrix} 12 \\ 4 \\ 14 \\ 15 \end{pmatrix}$

For each row operation (or sequence of row operations), find a 4×4 matrix R such that the matrix product $R \cdot A$ is the same as the result of carrying out the row operation(s) on the matrix:

$$A = \begin{pmatrix} 3 & 4 & 3 & 2 \\ 1 & 6 & 2 & 5 \\ 2 & -3 & 1 & -4 \\ 1 & 3 & 1 & 2 \end{pmatrix}$$

41. $R'2 = R4$ and $R'4 = R2$

42. $R'3 = R1$ and $R'1 = R3$

43. $R'4 = 3R4$

44. $R'2 = 2R2$

45. $R'1 = R1 - R3$

46. $R'4 = R2 - R4$

47. $R'3 = R3 - 2R4$

48. $R'1 = R1 - 3R2$

49. $R'1 = R1 - 3R3,$
$R'2 = R2 - 2R3,$ and
$R'4 = R4 - R3$

50. $R'1 = R1 - 3R2,$
$R'3 = R3 - 2R2,$ and
$R'4 = R4 - R2$

APPLIED EXERCISES

Formulate each situation in matrix form. Be sure to indicate the meaning of your rows and columns. Find the requested quantities using the appropriate matrix arithmetic.

51. Sales Commissions   A salesman at a furniture store sells bed mattresses manufactured by SlumberKing, DreamOn, and RestEasy. The selling prices for the three models of SlumberKing mattresses are $300 for the economy, $350 for the best, and $500 for the deluxe; DreamOn mattresses are $350 for the economy, $400 for the best, and $550 for the deluxe; and RestEasy mattresses are $400 for the economy, $500 for the best, and $700 for the deluxe. Represent these selling prices as a price matrix. Use this matrix to find the salesperson's commission matrix for these mattresses from these manufacturers if the commission is 15% of the selling price.

52. Sales Taxes   The ToysForYou stores in Rockland and Martinville sell "Little Tykes" baseballs, bats, gloves, and caps. At the Rockland store, the retail prices are $1 per baseball, $8 per bat, $7 per glove, and $10 per cap while at the Martinville store, the prices are $1 per baseball, $9 per bat, $8 per glove, and $11 per cap. Represent these retail prices as a price matrix. Use this matrix to find the sales tax matrix for these items at these stores if the state sales tax is 5% of the retail price.

53. Car Sales   A car dealer sells sedans, station wagons, vans, and pickup trucks at sales lots in Oakdale and Roanoke. The "dealer markup" is the difference between the sticker price and the dealer invoice price. The dealer invoice prices at both locations are the same: $15,000 per sedan, $19,000 per wagon, $23,000 per van, and $25,000 per pickup. The sticker prices at the Oakdale lot are $18,900 per sedan, $22,900 per wagon, $26,900 per van, and $29,900 per pickup while at the Roanoke lot the sticker prices are $19,900 per sedan, $21,900 per wagon, $27,900 per van, and $28,900 per pickup. Represent these prices as a dealer invoice matrix and a sticker price matrix. Use these matrices to find the dealer markup matrix for these vehicles at these sales lots.

54. Fuel Prices   An oil refinery in Louisiana produces gasoline, kerosene, and diesel fuel for sale at service stations in Tennessee, Alabama, and Florida. In Tennessee, the pump prices per gallon are $1.18 for gasoline, $0.87 for kerosene, and $1.09 for diesel fuel while the combined federal and state taxes per gallon are 52¢ for gasoline, 35¢ for kerosene, and 46¢ for diesel fuel. In Alabama, the pump prices per gallon are $1.15 for gasoline, $0.88 for kerosene, and $1.04 for diesel fuel while the combined federal and state taxes per gallon are 48¢ for gasoline, 33¢ for kerosene, and 41¢ for diesel fuel. And in Florida, the pump prices per gallon are $1.27 for gasoline, $0.93 for kerosene, and $1.16 for diesel fuel while the combined federal and state taxes per gallon are 56¢ for gasoline, 41¢ for kerosene, and 48¢ for diesel fuel. Represent these prices and taxes as matrices. Use these matrices to find the pretax price matrix for these fuels in these states.

55. **Overseas Manufacturing** A sports apparel company manufactures shorts, tee shirts, and caps in Costa Rica and Honduras for importation and sale in the United States. In Costa Rica, the labor costs per item are 75¢ per pair of shorts, 25¢ per tee shirt, and 45¢ per cap while the costs of the necessary materials are $1.60 per pair of shorts, 95¢ per tee shirt, and $1.15 per cap. In Honduras, the labor costs per item are 80¢ per pair of shorts, 20¢ per tee shirt, and 55¢ per cap while the costs of the necessary materials are $1.50 per pair of shorts, 80¢ per tee shirt, and $1.10 per cap. Represent these costs as a labor cost matrix and a materials cost matrix. Use these matrices to find the total cost matrix for these products in these countries.

56. **Retirement Income** A study of retired Chicago municipal workers now living in the Ozark Plateau found that in Missouri the average monthly pension benefits were $2700 for former police officers, $2500 for former mass transit workers, and $2800 for former firefighters, while in Arkansas the averages were $2750 for former police officers, $2300 for former mass transit workers, and $2900 for former firefighters. The average monthly Social Security benefits received by the same individuals in Missouri were $1100 for former police officers, $800 for former mass transit workers, and $1300 for former firefighters, while in Arkansas the averages were $1000 for former police officers, $900 for former mass transit workers, and $1200 for former firefighters. Represent these average monthly incomes as a Chicago pension matrix and a Social Security matrix. Use these matrices to find the average monthly retirement income matrix for these groups of retired employees in these states.

57. **Picnic Supplies** A supermarket advertises a "summer picnic sale" with 2 liters of soda for 89¢, a large bottle of pickles for $1.29, packages of hot dogs for $2.39, and large bags of chips for $1.69. The Culbert family wants 12 sodas, 2 bottles of pickles, 3 packages of hot dogs, and 4 bags of chips. Represent the sale prices as a row matrix and the Culbert's shopping list as a

column matrix. Use these matrices to find the total cost of these items at these prices.

58. **Part-Time Jobs** A college student makes money during the semester shelving books in the library at $5.50 per hour, tutoring freshmen at $20 per hour, and pumping gas on weekends at $4.75 per hour. Last week this student shelved books for 10 hours, tutored for 3 hours, and pumped gas for 16 hours. Represent the hourly pay rates as a row matrix and the hours worked as a column matrix. Use these matrices to find the total amount this student earned last week.

59. **Furniture Production** A furniture company manufactures pine tables, chairs, and desks at factories in Wytheville and Andersen. Each table requires 2 hours of cutting and milling, 1 hour of assembly, and 2 hours of finishing, each chair requires 1.5 hours of cutting and milling, 1 hour of assembly, and 0.5 hours of finishing, and each desk requires 3 hours of cutting and milling, 2 hours of assembly, and 3 hours of finishing. At the Wytheville factory, the per-hour labor costs are $9 for cutting and milling work, $14 for assembly work, and $13 for finishing work, while at the Andersen factory, the per-hour labor costs are $10 for cutting and milling work, $13 for assembly work, and $12 for finishing work. Represent the table, chair, and desk labor requirements as a time matrix and the factory per-hour labor costs as a cost matrix. Use these matrices to find the production costs of these pieces of furniture at these factories.

60. **Municipal Management** The Hollins County business manager has received a request from the police department for 3 new cars, 5 new motorcycles, and 1 van as well as a request from the rescue squad for 1 new car, 2 new vans, and 4 new ambulances. The local Ford dealer's prices are $20,000 per car, $8000 per motorcycle, $27,000 per van, and $52,000 per ambulance, while the local GM dealer's prices are $19,000 per car, $6000 per motorcycle, $29,000 per van, and $58,000 per ambulance. Represent the police and rescue requests as a vehicle matrix and the Ford and GM bids as a price matrix. Use these matrices to find the costs of these requests at these dealers.

Explorations and Excursions

The following problems extend and augment the material presented in the text.

Symmetric Matrices   Find $A^t \cdot A$ and $A \cdot A^t$ for each matrix A and identify these matrix products as "symmetric" or "not symmetric."

 If your instructor permits, you may carry out the calculations on a graphing calculator.

61. $\begin{pmatrix} 1 & 2 & 3 \\ 4 & 5 & 6 \end{pmatrix}$ **62.** $\begin{pmatrix} 1 & 0 & 1 & 0 \\ 0 & -1 & 0 & -1 \end{pmatrix}$

63. $\begin{pmatrix} 1 & -1 & 2 \\ 0 & 2 & -1 \\ 0 & 0 & 1 \end{pmatrix}$ **64.** $\begin{pmatrix} 1 & 0 & 1 & -1 \\ 0 & 2 & -1 & 1 \\ 0 & 1 & 2 & 0 \\ 1 & 0 & 0 & 1 \end{pmatrix}$

65. Show that the matrix products $A^t \cdot A$ and $A \cdot A^t$ are symmetric for any matrix A by showing that $(A^t \cdot A)^t = A^t \cdot A$ and $(A \cdot A^t)^t = A \cdot A^t$. (*Hint:* Use the fact that $(A \cdot B)^t = B^t \cdot A^t$.)

Matrix Multiplication and Function Composition   A *fractional linear transformation* is a function $f(x) = \dfrac{ax + b}{cx + d}$ with at least one of c and d not zero. The composition of two fractional linear transformations $f(x)$ and $g(x)$ is the fractional linear transformation $f(g(x))$. Each fractional linear transformation $f(x) = \dfrac{ax + b}{cx + d}$ corresponds to a 2×2 matrix $F = \begin{pmatrix} a & b \\ c & d \end{pmatrix}$.

Find the composition $f(g(x))$ and matrix product $F \cdot G$ for each pair of fractional linear transformations $f(x)$ and $g(x)$ and corresponding matrices F and G.

66. $f(x) = \dfrac{2x + 3}{x + 2}$ and $g(x) = \dfrac{x - 2}{-x + 1}$

$F = \begin{pmatrix} 2 & 3 \\ 1 & 2 \end{pmatrix}$ and $G = \begin{pmatrix} 1 & -2 \\ -1 & 1 \end{pmatrix}$

67. $f(x) = \dfrac{3x + 2}{2x + 1}$ and $g(x) = \dfrac{2x + 1}{x - 2}$

$F = \begin{pmatrix} 3 & 2 \\ 2 & 1 \end{pmatrix}$ and $G = \begin{pmatrix} 2 & 1 \\ 1 & -2 \end{pmatrix}$

68. $f(x) = \dfrac{x + 1}{2x - 1}$ and $g(x) = \dfrac{1}{x}$

$F = \begin{pmatrix} 1 & 1 \\ 2 & -1 \end{pmatrix}$ and $G = \begin{pmatrix} 0 & 1 \\ 1 & 0 \end{pmatrix}$

69. $f(x) = \dfrac{2x - 1}{x + 1}$ and $g(x) = 2x$

$F = \begin{pmatrix} 2 & -1 \\ 1 & 1 \end{pmatrix}$ and $G = \begin{pmatrix} 2 & 0 \\ 0 & 1 \end{pmatrix}$

70. Show that the composition $f(g(x))$ of $f(x) = \dfrac{ax + b}{cx + d}$ and $g(x) = \dfrac{Ax + B}{Cx + D}$ corresponds to the matrix product $F \cdot G$ of $F = \begin{pmatrix} a & b \\ c & d \end{pmatrix}$ and $G = \begin{pmatrix} A & B \\ C & D \end{pmatrix}$.

Zero Divisors   A familiar property of the real numbers is that if the product of two numbers is zero, then at least one of the numbers must have been zero as well. This is *not* true for matrix multiplication. A *zero divisor* is a nonzero matrix A such that $A \cdot B = 0$ for some nonzero matrix B.

Show that the first of each pair of matrices is a zero divisor by finding the product of the two matrices in the order they are given.

71. $\begin{pmatrix} 1 & 0 \\ 1 & 0 \end{pmatrix}, \begin{pmatrix} 0 & 0 \\ 1 & 1 \end{pmatrix}$

72. $\begin{pmatrix} 1 & 1 & 0 \\ 1 & 1 & 0 \end{pmatrix}, \begin{pmatrix} 0 & 1 & -1 \\ 0 & -1 & 1 \\ 1 & 1 & 1 \end{pmatrix}$

73. $\begin{pmatrix} 1 & 0 & 1 \\ 0 & 1 & 0 \\ 1 & 0 & 1 \end{pmatrix}, \begin{pmatrix} 1 & -2 & 1 \\ 0 & 0 & 0 \\ -1 & 2 & -1 \end{pmatrix}$

74. $\begin{pmatrix} 0 & 1 & 1 & 0 \\ 1 & 1 & 1 & 1 \\ 1 & 1 & 1 & 1 \\ 0 & 1 & 1 & 0 \end{pmatrix}, \begin{pmatrix} 1 & 0 & 0 & -1 \\ 0 & -1 & 1 & 0 \\ 0 & 1 & -1 & 0 \\ -1 & 0 & 0 & 1 \end{pmatrix}$

75. $\begin{pmatrix} 2 & 4 \\ 3 & 6 \end{pmatrix}, \begin{pmatrix} 2 & -2 \\ -1 & 1 \end{pmatrix}$

76. $\begin{pmatrix} 1 & 2 & 3 \\ 4 & 5 & 6 \end{pmatrix}, \begin{pmatrix} 7 & -8 \\ -14 & 16 \\ 7 & -8 \end{pmatrix}$

77. $\begin{pmatrix} 1 & 2 & 3 \\ 4 & 5 & 6 \\ 7 & 8 & 9 \end{pmatrix}$, $\begin{pmatrix} 1 & -1 & 1 \\ -2 & 2 & -2 \\ 1 & -1 & 1 \end{pmatrix}$

78. $\begin{pmatrix} 1 & 2 & 3 & 4 \\ 5 & 6 & 7 & 8 \\ 9 & 10 & 11 & 12 \\ 13 & 14 & 15 & 16 \end{pmatrix}$, $\begin{pmatrix} -1 & -1 & 1 & 1 \\ 1 & 3 & -3 & -1 \\ 1 & -3 & 3 & -1 \\ -1 & 1 & -1 & 1 \end{pmatrix}$

79. $\begin{pmatrix} 1 & 5 & 9 & 13 \\ 2 & 6 & 10 & 14 \\ 3 & 7 & 11 & 15 \\ 4 & 8 & 12 & 16 \end{pmatrix}$, $\begin{pmatrix} 12 & -11 & 10 & -9 \\ -16 & 15 & -14 & 13 \\ -4 & 3 & -2 & 1 \\ 8 & -7 & 6 & -5 \end{pmatrix}$

80. $\begin{pmatrix} 2 & -1 & 2 & -1 & 2 \\ -1 & 3 & -1 & 3 & -1 \\ 3 & -2 & 3 & -2 & 3 \end{pmatrix}$, $\begin{pmatrix} 1 & 2 & 1 \\ -2 & -2 & 1 \\ 1 & -1 & -2 \\ 2 & 2 & -1 \\ -2 & -1 & 1 \end{pmatrix}$

The Pivot Operation The *pivot operation* is a sequence of row operations that changes a nonzero element of the matrix into a one and then "clears out" the column above and below this "pivot element." Let *PE* be the pivot element, R_{pivot} be the pivot row, R_{other} be any other row, and *PCE* be the entry in this other row in the same column as the pivot element. The pivot operation is $R'_{pivot} = \frac{1}{PE}R_{pivot}$ ("divide the pivot row by the pivot element") followed by $R'_{other} = R_{other} - PCE \cdot R_{pivot}^{new}$ for every other row ("subtract multiples of the new pivot row from the other rows to get zeros in the rest of the pivot element's column").

For each matrix, carry out the given sequence of row operations to pivot on the given pivot element.

81. Pivot on the 2 in row 1 and column 1 of
$\begin{pmatrix} 2 & 1 & 1 \\ 1 & 1 & 2 \\ 1 & 1 & 1 \end{pmatrix}$

$R'1 = \frac{1}{2}R1$, $R'2 = R2 - R1$, and $R'3 = R3 - R1$

82. Pivot on the 3 in row 1 and column 1 of
$\begin{pmatrix} 3 & 2 & 2 \\ 1 & 1 & 2 \\ 2 & 2 & 3 \end{pmatrix}$

$R'1 = \frac{1}{3}R1$, $R'2 = R2 - R1$, and $R'3 = R3 - 2R1$

83. Pivot on the $\frac{1}{3}$ in row 2 and column 2 of
$\begin{pmatrix} 1 & 2/3 & 2/3 \\ 0 & 1/3 & 4/3 \\ 0 & 2/3 & 5/3 \end{pmatrix}$

$R'2 = 3R2$, $R'1 = R1 - \frac{2}{3}R2$, and $R'3 = R3 - \frac{2}{3}R2$

84. Pivot on the -1 in row 3 and column 3 of
$\begin{pmatrix} 1 & 0 & -2 \\ 0 & 1 & 4 \\ 0 & 0 & -1 \end{pmatrix}$

$R'3 = -R3$, $R'1 = R1 + 2R3$, and $R'2 = R2 - 4R3$

Determinants The *determinant* of a square matrix that is row equivalent to the identity matrix is the product of any sequence of pivot elements that reduce the matrix to the identity matrix. If the square matrix is not row equivalent to the identity matrix, the determinant is zero. The determinant of a nonsquare matrix is not defined. The basic properties of determinants were first published in 1750 by Gabriel Cramer (1704–1752) in an investigation of the solutions of systems of linear equations. By the mid-nineteenth century, James Joseph Sylvester (1814–1897) and Arthur Cayley (1821–1895) had realized that determinants were just part of a larger "theory of invariants" and Sylvester's invention of the term "matrix" (based on the Latin *mater* for "mother") was meant to indicate the larger context within which determinants should be understood. The following century saw determinants completely supplanted by the matrix methods presented in this chapter.

Find the determinant of each matrix by pivoting at row 1 and column 1, row 2 and column 2, and then row 3 and column 3.

If your calculator has a DETERMINANT(command, use it to check your answers.

85. $\begin{pmatrix} 2 & 1 & 1 \\ 1 & 1 & 2 \\ 1 & 1 & 1 \end{pmatrix}$

86. $\begin{pmatrix} 3 & 2 & 2 \\ 1 & 1 & 2 \\ 2 & 2 & 3 \end{pmatrix}$

87. $\begin{pmatrix} 3 & -2 & 4 \\ 1 & -7 & -2 \\ -3 & -7 & -9 \end{pmatrix}$

88. $\begin{pmatrix} 6 & 5 & -3 \\ -5 & 5 & -4 \\ 0 & -4 & 3 \end{pmatrix}$

Properties of Determinants   The determinant is multiplied by -1 if two rows are switched; is multiplied by a number if a row is multiplied by that same number; and is unchanged if one row is added to (or subtracted from) another row. Furthermore, the determinant of the transpose of a matrix is the same as the determinant of the original matrix.

Use the properties of the determinant to predict the determinant of the second matrix of each pair given that the determinant of the first is -1.

Verify your prediction using the DET(calculator matrix math command.

89. $\begin{pmatrix} 3 & 2 & 2 \\ 1 & 1 & 2 \\ 2 & 2 & 3 \end{pmatrix}$ and $\begin{pmatrix} 1 & 1 & 2 \\ 3 & 2 & 2 \\ 2 & 2 & 3 \end{pmatrix}$

90. $\begin{pmatrix} 2 & 1 & 1 \\ 1 & 1 & 2 \\ 1 & 1 & 1 \end{pmatrix}$ and $\begin{pmatrix} 2 & 1 & 1 \\ 1 & 1 & 1 \\ 1 & 1 & 2 \end{pmatrix}$

91. $\begin{pmatrix} 3 & 2 & 2 \\ 1 & 1 & 2 \\ 2 & 2 & 3 \end{pmatrix}$ and $\begin{pmatrix} 3 & 2 & 2 \\ 5 & 5 & 10 \\ 2 & 2 & 3 \end{pmatrix}$

92. $\begin{pmatrix} 2 & 1 & 1 \\ 1 & 1 & 2 \\ 1 & 1 & 1 \end{pmatrix}$ and $\begin{pmatrix} 2 & 1 & 1 \\ 1 & 1 & 2 \\ 6 & 6 & 6 \end{pmatrix}$

93. $\begin{pmatrix} 3 & 2 & 2 \\ 1 & 1 & 2 \\ 2 & 2 & 3 \end{pmatrix}$ and $\begin{pmatrix} 3 & 2 & 2 \\ 16 & 11 & 12 \\ 2 & 2 & 3 \end{pmatrix}$

94. $\begin{pmatrix} 2 & 1 & 1 \\ 1 & 1 & 2 \\ 1 & 1 & 1 \end{pmatrix}$ and $\begin{pmatrix} 2 & 1 & 1 \\ 6 & 6 & 7 \\ 1 & 1 & 1 \end{pmatrix}$

95. $\begin{pmatrix} 3 & 2 & 2 \\ 1 & 1 & 2 \\ 2 & 2 & 3 \end{pmatrix}$ and $\begin{pmatrix} 3 & 1 & 2 \\ 2 & 1 & 2 \\ 2 & 2 & 3 \end{pmatrix}$

96. $\begin{pmatrix} 2 & 1 & 1 \\ 1 & 1 & 2 \\ 1 & 1 & 1 \end{pmatrix}$ and $\begin{pmatrix} 2 & 1 & 1 \\ 1 & 1 & 1 \\ 1 & 2 & 1 \end{pmatrix}$

97. A square matrix is a *diagonal matrix* if every element not on the main diagonal is zero. Use the definition and properties of the determinant to show that the determinant of any diagonal matrix is the product of the elements on the main diagonal.

98. A square matrix is an *upper triangular matrix* if every element "below" the main diagonal is zero. Use the definition and properties of the determinant to show that the determinant of any upper triangular matrix is the product of the elements on the main diagonal.

99. A square matrix is a *lower triangular matrix* if every element "above" the main diagonal is zero. Use the definition and properties of the determinant to show that the determinant of any lower triangular matrix is the product of the elements on the main diagonal.

100. Show that the determinant of any 2×2 matrix $\begin{pmatrix} a & b \\ c & d \end{pmatrix}$ is $ad - bc$ by *symbolically* pivoting at row 1 and column 1 and then again at row 2 and column 2 or by *symbolically* pivoting at row 1 and column 1 and using the result stated in Exercise 98.

2.5   Inverse Matrices and Systems of Linear Equations

APPLICATION PREVIEW

Sensitivity Analysis

While every dawn does greet a new day, one day's business problems are usually not entirely new but rather just variations on the problems of the previous day. But if yesterday's problems were understood and solved yesterday, there is little reason to begin today's work by solving familiar problems all over again. For example, a factory that manufactures products in response to orders received each day should require only minor changes in its production lines to fulfill the new day's orders. *Sensitivity analysis* is the study of how small changes in a problem affect the solution. Consider the following system of equations and the solution:

$$\begin{cases} x_1 + x_2 + x_3 = 15 \\ 2x_1 + 2x_2 + x_3 = 23 \\ x_1 + 2x_2 + x_3 = 18 \end{cases} \longrightarrow \begin{array}{l} x_1 = 5 \\ x_2 = 3 \\ x_3 = 7 \end{array}$$

Problem   Solution

If we change the 15 in the first equation to 16, the solution becomes $x_1 = 5$, $x_2 = 2$, $x_3 = 9$. Notice that this change to the first equation did not change the value of the first variable. If we change the 15 to 17, the solution becomes $x_1 = 5$, $x_2 = 1$, $x_3 = 11$. Is it possible that every increase to the 15 results in a decrease of the x_2 value by the same amount, together with an increase of the x_3 value by twice that same amount? The answers to this and other similar questions are simply found using the inverse matrix discussed in this section.

Introduction

Since we can write a system of linear equations as a single matrix equation, we will solve the system by solving a matrix equation of the form $AX = B$. Just as the simple equation $2x = 10$ is easily solved by multiplying each side by $\frac{1}{2}$ (the inverse of 2) to obtain $1x = 5$, we will solve the matrix equation $AX = B$ by multiplying by an inverse matrix to transform the left side from AX to IX or simply X (since the identity matrix I plays the role of 1 in matrix arithmetic).

Inverse Matrices

The inverse of the number 2 is $\frac{1}{2}$ because $\frac{1}{2} \cdot 2 = 1$ and $2 \cdot \frac{1}{2} = 1$. Similarly, the *inverse* of a matrix is defined as another matrix whose product with the first (in either order) gives the identity matrix.

Inverse Matrix

The square matrix A has an *inverse matrix* A^{-1} such that

$$A^{-1} \cdot A = I \quad \text{and} \quad A \cdot A^{-1} = I$$

if A is row equivalent to the identity matrix.

The inverse of the matrix A is denoted A^{-1} just as the inverse of the number 2 is $2^{-1} = \frac{1}{2}$. Not all matrices have inverses. A square matrix is *invertible* if it has an inverse, and it is *singular* if it does not. We saw on page 226 that if a matrix A is equivalent to the identity matrix I, then the row reduction can be accomplished by multiplying A by the product R of the row operation matrices: $R \cdot A = I$. Thus a matrix is invertible if it is equivalent to the identity matrix, and any sequence of row operations demonstrating this equivalence determines the inverse matrix R. (See Exercise 69 on page 247.)

EXAMPLE 1 Checking Inverse Matrices

For $A = \begin{pmatrix} \frac{1}{2} & 3 \\ 1 & 5 \end{pmatrix}$ and $A^{-1} = \begin{pmatrix} -10 & 6 \\ 2 & -1 \end{pmatrix}$, verify that A^{-1} is indeed the inverse of A by showing that $A^{-1} \cdot A = I$ and $A \cdot A^{-1} = I$.

Solution

$$A^{-1} \cdot A = \begin{pmatrix} -10 & 6 \\ 2 & -1 \end{pmatrix} \begin{pmatrix} \frac{1}{2} & 3 \\ 1 & 5 \end{pmatrix} = \begin{pmatrix} 1 & 0 \\ 0 & 1 \end{pmatrix} = I$$

and

$$A \cdot A^{-1} = \begin{pmatrix} \frac{1}{2} & 3 \\ 1 & 5 \end{pmatrix} \begin{pmatrix} -10 & 6 \\ 2 & -1 \end{pmatrix} = \begin{pmatrix} 1 & 0 \\ 0 & 1 \end{pmatrix} = I$$

as required. (You should check the arithmetic in both multiplications.) ∎

How to Find Inverse Matrices

We could calculate A^{-1} as the product of the row operation matrices corresponding to a row reduction of A to I. But since $A^{-1} \cdot A = I$ and $A^{-1} \cdot I = A^{-1}$, the augmented matrix $(A \mid I)$ formed by adjoining an identity matrix to the right side of A has the property that $A^{-1} \cdot (A \mid I) = (A^{-1} \cdot A \mid A^{-1} \cdot I) = (I \mid A^{-1})$. That is, if we row reduce $(A \mid I)$ to get I on the left side, then the result is $(I \mid A^{-1})$, with the inverse A^{-1} on the right.

Calculating the Inverse Matrix

If the square matrix A is equivalent to the identity matrix, then the augmented matrix $(A \mid I)$ is equivalent to $(I \mid A^{-1})$ and A^{-1} appears in the right half of this reduced row-echelon form matrix.

Of course, if A cannot be row reduced to the identity matrix, this will be discovered during the attempt to row reduce $(A \mid I)$ and the reduction process can be stopped immediately with the conclusion that A is singular.

EXAMPLE 2   Calculating an Inverse Matrix

Find the inverse of $A = \begin{pmatrix} 1 & 0 & 2 \\ 1 & 1 & 1 \\ 1 & 1 & 2 \end{pmatrix}$.

Solution

A is 3×3, so we augment it by $I^{3 \times 3}$ and row reduce $(A \mid I)$:

$$\left(\begin{array}{ccc|ccc} 1 & 0 & 2 & 1 & 0 & 0 \\ 1 & 1 & 1 & 0 & 1 & 0 \\ 1 & 1 & 2 & 0 & 0 & 1 \end{array} \right) \qquad (A \mid I)$$

$$\underbrace{}_{A} \quad \underbrace{}_{I}$$

There are many different sequences of row operations to reduce this augmented matrix, and all reach the same conclusion. One way is as follows.

$$\left(\begin{array}{ccc|ccc} 1 & 0 & 2 & 1 & 0 & 0 \\ 1 & 1 & 1 & 0 & 1 & 0 \\ 0 & 0 & 1 & 0 & -1 & 1 \end{array} \right) \begin{array}{l} \\ \\ R'3 = R3 - R2 \end{array}$$

$$\left(\begin{array}{ccc|ccc} 1 & 0 & 0 & 1 & 2 & -2 \\ 1 & 1 & 0 & 0 & 2 & -1 \\ 0 & 0 & 1 & 0 & -1 & 1 \end{array} \right) \begin{array}{l} R'1 = R1 - 2R3 \\ R'2 = R2 - R3 \\ \end{array}$$

$$\left(\begin{array}{ccc|ccc} 1 & 0 & 0 & 1 & 2 & -2 \\ 0 & 1 & 0 & -1 & 0 & 1 \\ 0 & 0 & 1 & 0 & -1 & 1 \end{array} \right) \begin{array}{l} \\ R'2 = R2 - R1 \\ \end{array} \qquad \begin{array}{l} \text{Since } I \text{ is on} \\ \text{the left, } A^{-1} \text{ is} \\ \text{on the right} \end{array}$$

$$\underbrace{}_{I} \quad \underbrace{}_{A^{-1}}$$

Therefore, $A^{-1} = \begin{pmatrix} 1 & 2 & -2 \\ -1 & 0 & 1 \\ 0 & -1 & 1 \end{pmatrix}$ is the inverse of the original matrix

$A = \begin{pmatrix} 1 & 0 & 2 \\ 1 & 1 & 1 \\ 1 & 1 & 2 \end{pmatrix}$.

∎

PRACTICE PROBLEM 1   For the matrices A and A^{-1} in Example 2 on the previous page, verify that A^{-1} *is* the inverse of A by showing that $A^{-1} \cdot A = I$ and $A \cdot A^{-1} = I$.

Solution at the back of the book

 Graphing Calculator Exploration

Inverse matrices can be found directly using the $\boxed{x^{-1}}$ calculator button.

a.   Enter $\begin{pmatrix} 1 & 0 & 2 \\ 1 & 1 & 1 \\ 1 & 1 & 2 \end{pmatrix}$ from Example 2 as matrix [A] using MATRX EDIT.

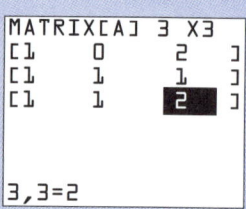

```
MATRIX[A] 3 X3
[1      0      2   ]
[1      1      1   ]
[1      1      2   ]



3,3=2
```

b.   Find A^{-1} using [A] and the $\boxed{x^{-1}}$ button.

```
[A]⁻¹
         [[1    2   -2]
          [-1   0    1 ]
          [0   -1    1 ]]
```

The method used by the calculator to find A^{-1} sometimes produces round-off errors (such as the small number 2.9×10^{-13} instead of zero), which can usually be corrected using the ROUND(command.

```
[A]
    [[1  -6  2  6 ]
     [2   5  4  0 ]
     [2   2  3  2 ]
     [1   4  2 -1]]
```

```
[A]⁻¹
[[-5        14   2...
 [2         -7   0...
 [2.9ɛ-13   2    -...
 [3        -10   0...
```

```
round([A]⁻¹,6)
[[-5  14   2   -26...
 [2   -7   0   12 ...
 [0    2  -1   -2 ...
 [3  -10   0   17 ...
```

Be Careful! Since apparent round-off errors might also indicate that A is invertible when it is in fact singular, always verify that the calculator answer does indeed satisfy $A^{-1} \cdot A = I$.

Solving $A \cdot X = B$ Using A^{-1}

Just as the equation $2x = 10$ is solved simply by multiplying by 2^{-1} or $\frac{1}{2}$, we can solve $AX = B$ by multiplying both sides (on the left) by A^{-1}.

$$AX = B \qquad \text{Original equation}$$

$$A^{-1} \cdot A \cdot X = A^{-1} \cdot B \qquad \text{Left-multiplying by } A^{-1}$$

$$I \cdot X = A^{-1} \cdot B \qquad \text{Since } A^{-1} \cdot A = I$$

$$X = A^{-1}B \qquad \text{Since } I \cdot X = X$$

Notice that since matrix multiplication is *not* commutative, we *must* multiply both sides of $A \cdot X = B$ on the *left* by A^{-1}. (And besides, the product $B \cdot A^{-1}$ is not defined because B is a column matrix and A^{-1} is a square matrix.)

Solving $A \cdot X = B$ Using A^{-1}

The solution of $A \cdot X = B$ is $X = A^{-1} \cdot B$, provided the square matrix A is invertible.

If A is singular, the system of equations is *dependent* and we must use regular row reduction (as in Section 2.3) to find the solution if the equations are consistent or to establish that there is no solution because they are inconsistent.

EXAMPLE 3   Solving a System of Equations Using the Inverse Matrix

Use the inverse matrix to solve $\begin{cases} x_1 + 2x_3 = 22 \\ x_1 + x_2 + x_3 = 11 \\ x_1 + x_2 + 2x_3 = 20 \end{cases}$.

Solution

Writing this system of equations as the matrix equation $A \cdot X = B$, we have:

$$\underbrace{\begin{pmatrix} 1 & 0 & 2 \\ 1 & 1 & 1 \\ 1 & 1 & 2 \end{pmatrix}}_{A} \cdot \underbrace{\begin{pmatrix} x_1 \\ x_2 \\ x_3 \end{pmatrix}}_{X} = \underbrace{\begin{pmatrix} 22 \\ 11 \\ 20 \end{pmatrix}}_{B} \qquad A \cdot X = B$$

Ordinarily, we would now find the inverse of matrix A by row reducing $(A \mid I)$. However, this is exactly the matrix that we just row reduced in Example 2 (page 235), so we will simply use the A^{-1} found in Example 2 and write the solution as $X = A^{-1} \cdot B$:

$$\underbrace{\begin{pmatrix} x_1 \\ x_2 \\ x_3 \end{pmatrix}}_{X} = \underbrace{\begin{pmatrix} 1 & 2 & -2 \\ -1 & 0 & 1 \\ 0 & -1 & 1 \end{pmatrix}}_{A^{-1}} \cdot \underbrace{\begin{pmatrix} 22 \\ 11 \\ 20 \end{pmatrix}}_{B} = \begin{pmatrix} 4 \\ -2 \\ 9 \end{pmatrix}$$

Using A^{-1} from Example 2

That is, the solution is $x_1 = 4$, $x_2 = -2$, $x_3 = 9$. We check this solution by substituting into the original system of equations:

$$4 + 2 \cdot 9 \qquad = 22$$
$$4 - 2 + 9 \qquad = 11$$
$$4 - 2 + 2 \cdot 9 = 20$$

It checks!   ∎

Graphing Calculator Exploration

If A is invertible, you can solve $A \cdot X = B$ by entering the matrices A and B and then calculating $A^{-1} \cdot B$. To solve the problem from Example 3, enter $\begin{pmatrix} 1 & 0 & 2 \\ 1 & 1 & 1 \\ 1 & 1 & 2 \end{pmatrix}$ in [A], $\begin{pmatrix} 22 \\ 11 \\ 20 \end{pmatrix}$ in [B], and calculate $[A]^{-1}[B]$:

```
[A]
        [[1 0 2]
         [1 1 1]
         [1 1 2]]
```

```
[B]
        [[22]
         [11]
         [20]]
```

```
[A]⁻¹[B]
        [[4 ]
         [-2]
         [9 ]]
```

Now enter your choices for any three integers in [B] and find $[A]^{-1}[B]$. Do these values satisfy your new equations?

PRACTICE PROBLEM 2   Solve $\begin{cases} x_1 + 2x_3 = -5 \\ x_1 + x_2 + x_3 = 10 \\ x_1 + x_2 + 2x_3 = 0 \end{cases}$.   *Solution at the back of the book*

(*Hint*: Write this system as $AX = B$ and notice that A is the same as in Example 3 but B is different and so the solution can be found by matrix multiplication using the inverse found in Example 2.)

Solving $A \cdot X = B$ by finding the inverse and then using $X = A^{-1}B$ is only slightly more difficult than row reducing the augmented matrix for the system of equations. However, if you need to solve $A \cdot X = B$ with the same A but different values in B, using the inverse A^{-1} means that you only do the row reductions *once* and then the solution for any new matrix B is given by a simple matrix multiplication. Such problems, where A remains the same but B changes, occur frequently in applications.

Solving $AX = B$ for Many Different B's

> If A is invertible, the solutions of the matrix equations
>
> $$A \cdot X = B_1, \quad A \cdot X = B_2, \ldots, A \cdot X = B_k$$
>
> may all be found by calculating A^{-1} once and then finding the solutions as the products
>
> $$X = A^{-1} \cdot B_1, \quad X = A^{-1} \cdot B_2, \ldots, X = A^{-1} \cdot B_k$$

EXAMPLE 4   Jewelry Production

An employee-owned jewelry company fabricates enameled gold rings, pendants, and bracelets. Each ring requires 3 grams of gold, 1 gram of enameling compound, and 2 hours of labor; each pendant requires 6 grams of gold, 2 grams of enameling compound, and 3 hours of labor; and each bracelet requires 8 grams of gold, 3 grams of enameling compound, and 2 hours of labor. Each of the five employee-owners works 160 hours each month and the company has contracts guaranteeing the delivery of the grams of gold and enameling compound shown in the table on the first day of the months of March, April, May, and June. How many rings, pendants, and bracelets should the company fabricate each month to use all the available materials and time?

	March	April	May	June
Gold	1720	2620	2460	2220
Enamel	600	960	900	800

Solution

Let x_1, x_2, and x_3 be the numbers of rings, pendants, and bracelets produced in one month. Then the required grams of gold, grams of enameling compound, and hours of labor are:

Gold	$3x_1 + 6x_2 + 8x_3$
Enamel	$x_1 + 2x_2 + 3x_3$
Labor	$2x_1 + 3x_2 + 2x_3$

For the month of March, these quantities must match the amounts available:

$$\begin{cases} 3x_1 + 6x_2 + 8x_3 = 1720 \\ x_1 + 2x_2 + 3x_3 = 600 \\ 2x_1 + 3x_2 + 2x_3 = 800 \end{cases}$$

From the table and
5 workers @ 160 hours each

The systems of equations for the other months follow in a similar manner and we have four problems to solve:

March	April
$\begin{cases} 3x_1 + 6x_2 + 8x_3 = 1720 \\ x_1 + 2x_2 + 3x_3 = 600 \\ 2x_1 + 3x_2 + 2x_3 = 800 \end{cases}$	$\begin{cases} 3x_1 + 6x_2 + 8x_3 = 2620 \\ x_1 + 2x_2 + 3x_3 = 960 \\ 2x_1 + 3x_2 + 2x_3 = 800 \end{cases}$
May	**June**
$\begin{cases} 3x_1 + 6x_2 + 8x_3 = 2460 \\ x_1 + 2x_2 + 3x_3 = 900 \\ 2x_1 + 3x_2 + 2x_3 = 800 \end{cases}$	$\begin{cases} 3x_1 + 6x_2 + 8x_3 = 2220 \\ x_1 + 2x_2 + 3x_3 = 800 \\ 2x_1 + 3x_2 + 2x_3 = 800 \end{cases}$

We could solve these four problems by row reducing the corresponding augmented matrix for the March problem and then repeating the same sequence of row operations to solve the augmented matrices for the April, May, and June problems. Since the coefficient matrices for these four problems are the same, if we find the inverse matrix for the March problem by row reducing $(A \mid I)$, we will only have to carry out the row operations to reduce A once and then the four problems can be solved with just four matrix multiplications.

We first find A^{-1} by row reducing $(A \mid I)$. There are many different sequences of row operations to reduce this augmented matrix and all reach the same conclusion. One possible way is as follows.

$$(A \mid I) = \begin{pmatrix} 3 & 6 & 8 & 1 & 0 & 0 \\ 1 & 2 & 3 & 0 & 1 & 0 \\ 2 & 3 & 2 & 0 & 0 & 1 \end{pmatrix}$$

$$\begin{pmatrix} 0 & 0 & 1 & -1 & 3 & 0 \\ 1 & 2 & 3 & 0 & 1 & 0 \\ 0 & 1 & 4 & 0 & 2 & -1 \end{pmatrix} \begin{matrix} R'1 = 3R2 - R1 \\ \\ R'3 = 2R2 - R3 \end{matrix}$$

Use 1 in first column
to zero out the rest
of that column

$$\begin{pmatrix} 0 & 0 & 1 & -1 & 3 & 0 \\ 1 & 0 & -5 & 0 & -3 & 2 \\ 0 & 1 & 4 & 0 & 2 & -1 \end{pmatrix}$$
$R'2 = R2 - 2R3$

Use 1 in second column
to zero out the rest
of that column

$$\begin{pmatrix} 0 & 0 & 1 & -1 & 3 & 0 \\ 1 & 0 & 0 & -5 & 12 & 2 \\ 0 & 1 & 0 & 4 & -10 & -1 \end{pmatrix}$$
$R'2 = R2 + 5R1$
$R'3 = R2 - 4R1$

Use 1 in third column
to zero out the rest
of that column

$$\begin{pmatrix} 1 & 0 & 0 & -5 & 12 & 2 \\ 0 & 1 & 0 & 4 & -10 & -1 \\ 0 & 0 & 1 & -1 & 3 & 0 \end{pmatrix}$$
$R'1 = R2$
$R'2 = R3$
$R'3 = R1$

Switch rows
to achieve the
correct order

Since the left side is I, the right side is the inverse matrix:

$$A^{-1} = \begin{pmatrix} -5 & 12 & 2 \\ 4 & -10 & -1 \\ -1 & 3 & 0 \end{pmatrix}$$

The solution for each month is simply the product of this inverse matrix with the column matrix of the constant terms for that month.

Month	*X = A^{-1} · B*	*Solution*
March	$\begin{pmatrix} x_1 \\ x_2 \\ x_3 \end{pmatrix} = \begin{pmatrix} -5 & 12 & 2 \\ 4 & -10 & -1 \\ -1 & 3 & 0 \end{pmatrix} \cdot \begin{pmatrix} 1720 \\ 600 \\ 800 \end{pmatrix} = \begin{pmatrix} 200 \\ 80 \\ 80 \end{pmatrix}$	200 rings 80 pendants 80 bracelets
April	$\begin{pmatrix} x_1 \\ x_2 \\ x_3 \end{pmatrix} = \begin{pmatrix} -5 & 12 & 2 \\ 4 & -10 & -1 \\ -1 & 3 & 0 \end{pmatrix} \cdot \begin{pmatrix} 2620 \\ 960 \\ 800 \end{pmatrix} = \begin{pmatrix} 20 \\ 80 \\ 260 \end{pmatrix}$	20 rings 80 pendants 260 bracelets
May	$\begin{pmatrix} x_1 \\ x_2 \\ x_3 \end{pmatrix} = \begin{pmatrix} -5 & 12 & 2 \\ 4 & -10 & -1 \\ -1 & 3 & 0 \end{pmatrix} \cdot \begin{pmatrix} 2460 \\ 900 \\ 800 \end{pmatrix} = \begin{pmatrix} 100 \\ 40 \\ 240 \end{pmatrix}$	100 rings 40 pendants 240 bracelets
June	$\begin{pmatrix} x_1 \\ x_2 \\ x_3 \end{pmatrix} = \begin{pmatrix} -5 & 12 & 2 \\ 4 & -10 & -1 \\ -1 & 3 & 0 \end{pmatrix} \cdot \begin{pmatrix} 2220 \\ 800 \\ 800 \end{pmatrix} = \begin{pmatrix} 100 \\ 80 \\ 180 \end{pmatrix}$	100 rings 80 pendants 180 bracelets

■

SUMMARY

A square matrix A is *invertible* if there is an *inverse matrix* A^{-1} such that $A^{-1} \cdot A = I$ and $A \cdot A^{-1} = I$. A square matrix is *singular* if it is not invertible.

The inverse of an invertible matrix A may be found by row reducing the augmented matrix $(A | I)$ to obtain $(I | A^{-1})$.

The solution of the matrix equation $A \cdot X = B$ is $X = A^{-1} \cdot B$, provided the matrix A is invertible.

EXERCISES 2.5

Find each matrix product. Identify each pair of matrices as "a matrix and its inverse" or "not a matrix and its inverse."

1. $\begin{pmatrix} 1 & 2 \\ -1 & -1 \end{pmatrix}$ and $\begin{pmatrix} -1 & -2 \\ 1 & 1 \end{pmatrix}$

2. $\begin{pmatrix} 5 & 3 \\ -3 & -2 \end{pmatrix}$ and $\begin{pmatrix} 2 & 3 \\ -3 & -5 \end{pmatrix}$

3. $\begin{pmatrix} 1 & 1 & 0 \\ 2 & 1 & 1 \\ 1 & 0 & 0 \end{pmatrix}$ and $\begin{pmatrix} 0 & 0 & 1 \\ 1 & 0 & -1 \\ -1 & 1 & -1 \end{pmatrix}$

4. $\begin{pmatrix} 2 & 1 & -1 \\ -2 & 0 & 1 \\ -3 & -1 & 2 \end{pmatrix}$ and $\begin{pmatrix} 1 & -1 & 1 \\ 1 & 1 & 0 \\ 2 & -1 & 2 \end{pmatrix}$

5. $\begin{pmatrix} 4 & 6 & 3 \\ 3 & 4 & 1 \\ 5 & 7 & 3 \end{pmatrix}$ and $\begin{pmatrix} -5 & -3 & 6 \\ 4 & 3 & -5 \\ -1 & -2 & 2 \end{pmatrix}$

6. $\begin{pmatrix} 3 & 2 & 3 \\ 5 & 2 & 6 \\ 2 & 3 & 1 \end{pmatrix}$ and $\begin{pmatrix} 16 & -7 & -6 \\ -7 & 3 & 3 \\ -11 & 5 & 4 \end{pmatrix}$

7. $\begin{pmatrix} 10 & -4 & -7 \\ -7 & 3 & 5 \\ 4 & -1 & -3 \end{pmatrix}$ and $\begin{pmatrix} 4 & 5 & -1 \\ 1 & 2 & 1 \\ 5 & 6 & 2 \end{pmatrix}$

8. $\begin{pmatrix} 4 & 1 & 5 \\ -1 & 1 & -2 \\ 3 & 4 & 2 \end{pmatrix}$ and $\begin{pmatrix} 10 & 18 & -7 \\ -4 & -7 & -3 \\ -7 & -13 & 5 \end{pmatrix}$

9. $\begin{pmatrix} 2 & 0 & 1 & 0 \\ 1 & 1 & 1 & 0 \\ -2 & 0 & -1 & 1 \\ 1 & 0 & 0 & 1 \end{pmatrix}$ and $\begin{pmatrix} -1 & 0 & -1 & 1 \\ -2 & 1 & -1 & 1 \\ 3 & 0 & 2 & -2 \\ 1 & 0 & 1 & 0 \end{pmatrix}$

10. $\begin{pmatrix} 1 & -1 & 0 & 0 \\ -1 & 1 & 0 & 1 \\ -3 & 2 & 1 & 2 \\ 1 & 0 & 0 & -1 \end{pmatrix}$ and $\begin{pmatrix} 1 & 1 & 0 & 1 \\ 0 & 1 & 0 & 1 \\ 1 & -1 & 1 & 1 \\ 1 & 1 & 0 & 0 \end{pmatrix}$

Row reduce $(A \mid I)$ for each matrix A to find the inverse matrix A^{-1} or to identify A as a singular matrix.

11. $\begin{pmatrix} 1 & 3 \\ 0 & 1 \end{pmatrix}$

12. $\begin{pmatrix} 1 & 4 \\ 0 & 1 \end{pmatrix}$

13. $\begin{pmatrix} 11 & 2 \\ 6 & 1 \end{pmatrix}$

14. $\begin{pmatrix} 5 & 7 \\ 2 & 3 \end{pmatrix}$

15. $\begin{pmatrix} 1 & 1 & 0 \\ 3 & 0 & 2 \\ 1 & 0 & 1 \end{pmatrix}$

16. $\begin{pmatrix} 1 & 0 & 1 \\ 1 & 1 & 0 \\ 5 & 0 & 4 \end{pmatrix}$

17. $\begin{pmatrix} 1 & 1 & 0 \\ 0 & 1 & 1 \\ 1 & 2 & 1 \end{pmatrix}$

18. $\begin{pmatrix} 1 & 0 & 1 \\ 2 & 1 & 3 \\ 0 & 1 & 1 \end{pmatrix}$

19. $\begin{pmatrix} 1 & 1 & 0 & 1 \\ 0 & 1 & 0 & 0 \\ 1 & 0 & 1 & 0 \\ 0 & 1 & 0 & 1 \end{pmatrix}$

20. $\begin{pmatrix} 1 & 0 & 0 & 1 \\ 0 & 1 & 1 & 1 \\ 2 & 0 & 0 & 1 \\ 0 & 1 & 0 & 1 \end{pmatrix}$

21. $\begin{pmatrix} 2 & 3 & 1 \\ 1 & 2 & 1 \\ 2 & 3 & 2 \end{pmatrix}$

22. $\begin{pmatrix} 1 & 2 & 2 \\ 1 & 1 & 1 \\ 1 & 3 & 2 \end{pmatrix}$

23. $\begin{pmatrix} 3 & -4 & 2 \\ -1 & 2 & -1 \\ 4 & -3 & 2 \end{pmatrix}$

24. $\begin{pmatrix} 2 & 2 & -1 \\ -2 & -1 & 1 \\ 3 & 4 & -1 \end{pmatrix}$

25. $\begin{pmatrix} 1 & -2 & 1 & 2 \\ 1 & -1 & 1 & 2 \\ 2 & 2 & 1 & 3 \\ 1 & -1 & 1 & 1 \end{pmatrix}$

26. $\begin{pmatrix} 2 & 1 & -2 & 1 \\ 1 & 1 & 2 & 1 \\ 1 & 1 & 1 & 1 \\ 2 & 2 & 3 & 1 \end{pmatrix}$

27. $\begin{pmatrix} 1 & 0 & -1 & 1 \\ 1 & 1 & 1 & 2 \\ 2 & -2 & -6 & 0 \\ 1 & 0 & 2 & 1 \end{pmatrix}$

28. $\begin{pmatrix} 1 & -1 & 0 & 1 \\ 2 & -2 & -1 & 0 \\ 1 & -1 & -1 & -1 \\ 2 & -2 & 0 & 2 \end{pmatrix}$

29. $\begin{pmatrix} 1 & 0 & 1 & 0 & 1 \\ -1 & 2 & -1 & 1 & -1 \\ 4 & 2 & 4 & 2 & 3 \\ -1 & 4 & -2 & 2 & -2 \\ 3 & 1 & 2 & 1 & 1 \end{pmatrix}$

30. $\begin{pmatrix} 4 & -1 & 3 & -1 & 1 \\ 2 & 2 & 1 & 4 & 0 \\ 4 & -1 & 2 & -2 & 1 \\ 2 & 1 & 1 & 2 & 0 \\ 3 & -1 & 1 & -2 & 1 \end{pmatrix}$

Rewrite each system of equations as a matrix equation $A \cdot X = B$ and use the inverse of A to find the solution. Be sure to check your solution in the original system of equations.

31. $\begin{cases} 11x_1 + 2x_2 = 9 \\ 6x_1 + x_2 = 5 \end{cases}$

32. $\begin{cases} 5x_1 + 7x_2 = 2 \\ 2x_1 + 3x_2 = 1 \end{cases}$

33. $\begin{cases} x_1 + x_2 = 2 \\ 3x_1 + 2x_3 = 5 \\ x_1 + x_3 = 2 \end{cases}$

34. $\begin{cases} x_1 + x_3 = 1 \\ x_1 + x_2 = 3 \\ 5x_1 + 4x_3 = 6 \end{cases}$

35. $\begin{cases} 2x_1 + 3x_2 + x_3 = 6 \\ x_1 + 2x_2 + x_3 = 4 \\ 2x_1 + 3x_2 + 2x_3 = 7 \end{cases}$

36. $\begin{cases} x_1 + 2x_2 + 2x_3 = 5 \\ x_1 + x_2 + x_3 = 3 \\ x_1 + 3x_2 + 2x_3 = 6 \end{cases}$

37. $\begin{cases} 3x_1 - 4x_2 + 2x_3 = 12 \\ -x_1 + 2x_2 - x_3 = -4 \\ 4x_1 - 3x_2 + 2x_3 = 15 \end{cases}$

38. $\begin{cases} 2x_1 + 2x_2 - x_3 = 1 \\ -2x_1 - x_2 + x_3 = 0 \\ 3x_1 + 4x_2 - x_3 = 2 \end{cases}$

39. $\begin{cases} x_1 - 2x_2 + x_3 + 2x_4 = 8 \\ x_1 - x_2 + x_3 + 2x_4 = 7 \\ 2x_1 + 2x_2 + x_3 + 3x_4 = 7 \\ x_1 - x_2 + x_3 + x_4 = 6 \end{cases}$

40. $\begin{cases} 2x_1 + x_2 - 2x_3 + x_4 = 2 \\ x_1 + x_2 + 2x_3 + x_4 = 5 \\ x_1 + x_2 + x_3 + x_4 = 4 \\ 2x_1 + 2x_2 + 3x_3 + x_4 = 8 \end{cases}$

41. $\begin{cases} x_1 - 2x_2 + x_3 + 2x_4 = 10 \\ x_1 - x_2 + x_3 + 2x_4 = 8 \\ 2x_1 + 2x_2 + x_3 + 3x_4 = 5 \\ x_1 - x_2 + x_3 + x_4 = 6 \end{cases}$

42. $\begin{cases} 2x_1 + x_2 - 2x_3 + x_4 = 8 \\ x_1 + x_2 + 2x_3 + x_4 = 3 \\ x_1 + x_2 + x_3 + x_4 = 4 \\ 2x_1 + 2x_2 + 3x_3 + x_4 = 5 \end{cases}$

43. $\begin{cases} x_1 + x_3 + x_5 = 4 \\ -x_1 + 2x_2 - x_3 + x_4 - x_5 = -9 \\ 4x_1 + 2x_2 + 4x_3 + 2x_4 + 3x_5 = 9 \\ -x_1 + 4x_2 - 2x_3 + 2x_4 - 2x_5 = -17 \\ 3x_1 + x_2 + 2x_3 + x_4 + x_5 = 5 \end{cases}$

44. $\begin{cases} 4x_1 - x_2 + 3x_3 - x_4 + x_5 = -2 \\ 2x_1 + 2x_2 + x_3 + 4x_4 = 0 \\ 4x_1 - x_2 + 2x_3 - 2x_4 + x_5 = 1 \\ 2x_1 + x_2 + x_3 + 2x_4 = 0 \\ 3x_1 - x_2 + x_3 - 2x_4 + x_5 = 2 \end{cases}$

45. $\begin{cases} x_1 + x_3 + x_5 = 3 \\ -x_1 + 2x_2 - x_3 + x_4 - x_5 = -9 \\ 4x_1 + 2x_2 + 4x_3 + 2x_4 + 3x_5 = 3 \\ -x_1 + 4x_2 - 2x_3 + 2x_4 - 2x_5 = -17 \\ 3x_1 + x_2 + 2x_3 + x_4 + x_5 = 2 \end{cases}$

46. $\begin{cases} 4x_1 - x_2 + 3x_3 - x_4 + x_5 = 12 \\ 2x_1 + 2x_2 + x_3 + 4x_4 = -9 \\ 4x_1 - x_2 + 2x_3 - 2x_4 + x_5 = 13 \\ 2x_1 + x_2 + x_3 + 2x_4 = -3 \\ 3x_1 - x_2 + x_3 - 2x_4 + x_5 = 11 \end{cases}$

47. $\begin{cases} x_1 + 2x_2 + x_3 = 11 \\ x_1 + 4x_2 + x_3 = 19 \\ 2x_1 + 2x_2 + x_3 = 13 \end{cases}$

48. $\begin{cases} 2x_1 + 3x_2 + x_3 = 4 \\ 3x_1 + 3x_2 + 2x_3 = 12 \\ x_1 + 2x_2 + x_3 = 4 \end{cases}$

49. $\begin{cases} 4x_1 + 6x_2 + 5x_3 + 9x_4 = 75 \\ 4x_1 + x_2 + 2x_3 + 4x_4 = 10 \\ x_1 + 4x_2 + 3x_3 + 6x_4 = 65 \\ 4x_1 + 3x_2 + 3x_3 + 4x_4 = 10 \end{cases}$

50. $\begin{cases} 3x_1 + 2x_2 + x_3 + 3x_4 = 10 \\ 4x_1 + 3x_2 + 3x_3 + 7x_4 = 15 \\ 6x_1 + 7x_2 + 3x_3 + 8x_4 = 50 \\ 5x_1 + 3x_2 + 3x_3 + 7x_4 = 10 \end{cases}$

APPLIED EXERCISES

Formulate each situation as a collection of systems of linear equations. Be sure to state clearly the meaning of each variable. Solve each collection by finding the inverse of the coefficient matrix and then using matrix multiplications. Be sure to state your final answers in terms of the original questions.

 If permitted by your instructor, you may use a graphing calculator to find the necessary inverse matrices and matrix products.

51. Movie Tickets   A five-screen multiplex cinema charges $10 for adults and $5 for children under twelve. The number of tickets sold for each of today's shows and the corresponding gross receipts are given in the table. How many tickets of each kind were sold for each film?

	Film No. 1	Film No. 2	Film No. 3	Film No. 4	Film No. 5
Tickets sold	500	400	450	500	600
Gross receipts	$3250	$3000	$3500	$4500	$6000

52. Summer Day Care   An inner city antipoverty foundation staffs summer day care sites serving children aged 6 to 12 at Hollis Avenue, Beaverton Boulevard, Gramson Park, and Riverside Street. Certified instructors earn $350 per week and supervise 8 children and college student group leaders earn $250 per week and supervise 6 children. The number of children and the weekly payroll at each site are given in the table. How many instructors and group leaders work at each site?

	Hollis Avenue	Beaverton Boulevard
Children	92	130
Payroll	$3900	$5500

	Gramson Park	Riverside Street
Children	124	152
Payroll	$5300	$6500

53. Coins in Jars   Three glass jars (red, green, and blue) each contain pennies, nickels, and dimes. The value and number of coins in each jar are given in the table, together with the "altered value" of the coins when the pennies are replaced by the same number of nickels. How many of each coin are in each jar?

	Red Jar	Green Jar	Blue Jar
Value	$20	$39	$25
Number of coins	500	700	600
"Altered" value	$30	$45	$38

54. Plant Fertilizer   The manager of a garden supply store has received the soil test results for the gardens of Mr. Smith, Mrs. Jones, Miss Roberts, and Mr. Wheeler. Using the size of each garden plot, the manager calculated the ounces of nutrients needed for each garden as listed in the first table. The store sells three brands of fertilizer, GreatGreen, MiracleMix, and GrowRite, which contain the amounts of potash, nitrogen, and phosphoric acid per box listed in the second table. How many boxes of each fertilizer should be sold to each customer?

Ounces Needed	Mr. Smith	Mrs. Jones	Miss Roberts	Mr. Wheeler
Potash	26	31	25	25
Nitrogen	31	33	31	29
Phosphoric acid	19	20	19	18

Ounces per Box	GreatGreen	MiracleMix	GrowRite
Potash	4	5	2
Nitrogen	5	5	3
Phosphoric acid	3	3	2

55. Infant Nutrition A pediatric dietician at an inner-city foundling hospital needs to supplement each bottle of baby formula given to three infants in her ward with the units of vitamin A, vitamin D, calcium, and iron given in the first table. If four diet supplements are available with the nutrient content per drop given in the second table, how many drops of each supplement per bottle of formula should each infant receive?

Units Needed	Vitamin A	Vitamin D	Calcium	Iron
Billy	48	26	19	49
Susie	46	26	27	65
Jimmy	47	26	19	49

Units per Drop	Vitamin A	Vitamin D	Calcium	Iron
Supplement No. 1	5	3	0	1
Supplement No. 2	0	1	3	6
Supplement No. 3	4	2	3	7
Supplement No. 4	4	2	2	5

56. International Investments An international investment advisor recommends industrial stocks in Nigeria, Bolivia, Thailand, and Hungary, with four times as much in Thailand as in

Bolivia and three times as much in Hungary as in Nigeria. Furthermore, he suggests that more than 20% but less than 25% of the investor's portfolio be Bolivian and Nigerian stocks. Acting on this advice, four investors commit the amounts given in the table. How much does each invest in each country?

	Mr. Croft	Mrs. Fredericks
Total investment	$100,000	$150,000
Bolivia and Nigeria	$23,000	$33,000
	Mr. Spencer	Ms. Winpeace
Total investment	$90,000	$135,000
Bolivia and Nigeria	$22,000	$28,000

57. Financial Planning A retirement planning councelor recommends investing in a stock fund yielding 18%, a money market fund returning 6%, and a bond fund paying 8%, with twice as much in the bond and money market funds together as in the stock fund. Mr. and Mrs. Jordan have $300,000 to invest and need an annual return of $31,000; Mr. and Mrs. French have $234,900 to invest and need an annual return of $25,600; and Mr. and Mrs. Daimen have $270,000 to invest and need an annual return of $28,500. How much should each elderly couple place in each investment to receive their desired income?

58. Year-End Bonuses At the end of each year, the owner of a small company gives himself, his salesman, and his secretary a bonus by dividing up whatever remains in the "office supplies" account. He takes one third for himself, gives one quarter to his salesman, and one fifth to his secretary, but each share is taken after the others have been given out. Any remaining money is then spent on the company's New Years Eve dinner at the best restaurant in town. The ending balances in the office supplies account for four years in the 1990s are given in the table. How much did each person receive each year and how much was left over each year for the dinner party?

1994	1995	1996	1997
$2500	$2000	$3000	$2800

59. **Mass Transit**   The metropolitan area mass transit manager is revising the subway, bus, and jitney service to the suburbs of Brighton, Conway, Longwood, and Oakley. To meet federal clean air mandates, the mayor's office demands that twice as many electric subway cars as diesel buses and jitneys combined be used for each suburb. The transit workers' union contract requires that like the buses and jitneys, each subway car must have a driver/ticket taker whether or not that subway car is part of a longer train. The number of commuters from each suburb using mass transit together with the number of transit workers assigned to each area are given in the table. If each subway car carries 70 commuters, each bus carries 60 commuters, and each jitney carries 10 commuters, how many of each vehicle must be assigned to each suburb?

	Brighton	**Conway**
Commuters	11,500	9000
Transit workers	180	150
	Longwood	**Oakley**
Commuters	9500	10,250
Transit workers	150	165

60. **Community Food Pantry**   A town-wide food drive to aid the interfaith ministries' food pantry for the needy was supported by the Boy Scouts, Girl Scouts, and the Lions Club. Both dry food (in 10-ounce boxes) and canned food, in small (15-ounce) and large (40-ounce) sizes,

were collected. The Boy Scouts collected 240 items weighing 355 pounds, of which 315 pounds were canned goods; the Girl Scouts collected 240 items weighing 320 pounds, of which 260 pounds were canned goods; and the Lions Club collected 416 items weighing 565 pounds, of which 465 pounds were canned goods. How many boxes, small cans, and large cans did each group collect?

Explorations and Excursions

The following problems extend and augment the material presented in the text.

Solve each matrix equation *symbolically* for the unknown matrix X. Evaluate this expression for X using the given matrices and verify that it satisfies the original equation.

If permitted by your instructor, store the matrices A, B, C, and D in your calculator and use your calculator to evaluate each expression for X.

$$A = \begin{pmatrix} 3 & 1 & 2 \\ 1 & 2 & 1 \\ 4 & 2 & 3 \end{pmatrix} \quad B = \begin{pmatrix} 1 & -1 & 2 \\ 4 & 1 & 1 \\ 6 & 1 & 2 \end{pmatrix}$$

$$C = \begin{pmatrix} 280 \\ 240 \\ 320 \end{pmatrix} \quad D = \begin{pmatrix} 200 \\ 120 \\ 280 \end{pmatrix}$$

61. $A \cdot X + X = C$

62. $A \cdot X - X = D$

63. $A \cdot X + B \cdot X = C + D$

64. $A \cdot X - B \cdot X = C - D$

65. Show that the inverse of an inverse matrix is the original matrix; that is, show that $(A^{-1})^{-1} = A$ for any invertible matrix A.

66. Show that the inverse of a transposed matrix is the transpose of the inverse matrix; that is, show that $(A^t)^{-1} = (A^{-1})^t$ for any invertible matrix A.

67. Show that the inverse of a product is the product of the inverses in the reverse order; that is, show that $(A \cdot B)^{-1} = (B^{-1}) \cdot (A^{-1})$ for any invertible matrices $A^{n \times n}$ and $B^{n \times n}$.

68. Show that the $n \times n$ identity matrix is unique; that is, given that $I \cdot A^{n \times n} = A$ and $J \cdot A^{n \times n} = A$, establish that $I = J$.

69. Show that if $B^{n \times n} \cdot A^{n \times n} = I$, then $A \cdot B = I$ and $B = A^{-1}$.

70. Show that if the determinant $ad - bc$ of the matrix $\begin{pmatrix} a & b \\ c & d \end{pmatrix}$ is 1, then the inverse of the matrix $\begin{pmatrix} a & b \\ c & d \end{pmatrix}$ is the matrix $\begin{pmatrix} d & -b \\ -c & a \end{pmatrix}$.

2.6 Three Applications

Introduction

We conclude this chapter with three applications of the matrix methods: the Leontief input-output model of an economy, Markov chains, and least squares estimation. Because of the computational nature of these topics, the use of a graphing calculator or a computer for matrix calculations is appropriate throughout this section.

Leontief "Open" Input-Output Models

Input-output analysis was developed by Wassily Leontief (who received the 1973 Noble Memorial Prize for economics) to study the flow of goods and services among different sectors of an economy. In a "closed" model, all the goods produced are used by the producers, while the economy of an "open" model produces more than is needed by the producers, with the extra output available to consumers. In the global economy of today, each country may be considered an open system with the gross national product measuring the extra production. Given the relationships among the different sectors of an economy, it should be possible to calculate the extra production from the amount of economic activity within each sector. And, conversely, it should be possible to determine the level of economic activity necessary within each sector to achieve a desired level of excess production. The Leontief "open" input-output economic model solves these two computational tasks by expressing the relation between the activity levels of the sectors and the extra production as a matrix equation.

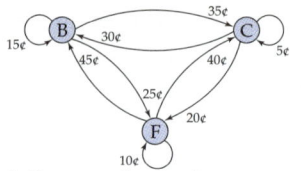

A diagram of a simple economy.

We begin by considering the simple economy shown in the diagram, consisting of a blacksmith (B), who makes nails, plows, and other tools, a carpenter (C), who builds barns and other useful buildings, and a farmer (F), who grows food. Each arrow connects a producer with a consumer and indicates that goods and services from the producer are a part of the *input* needed by the consumer to manufacture *output*, which in turn becomes the input used by the others. Since each producer also takes as input some of his or her own output, each has a circular arrow indicating his or her consumption of a part of his or her own production. The number near the head of each arrow indicates the value of that input needed to produce one dollar of output from the target. Thus the topmost arrow from B to C labeled 35¢ means that each dollar of value produced by the carpenter requires the input of 35¢ worth of the blacksmith's output. The connections between the sectors of the economy and the values of materials needed to produce each one dollar of output from each are the basic data needed to construct the Leontief model of the economy.

Let x_1, x_2, and x_3 be the values of the goods and services produced by the blacksmith, the carpenter, and the farmer, and let y_1, y_2, and y_3 be the values of the *excess* production from each that can be sold outside this economic system. The value produced by the blacksmith is the same as the values of the blacksmith's products used by the blacksmith, the carpenter, and the farmer together with the amount sold outside this economy. Using the numbers from the diagram:

$$x_1 = (\quad 0.15x_1 \quad + \quad 0.35x_2 \quad + \quad 0.25x_3 \quad) + \quad y_1$$

Value of blacksmith's products used by the blacksmith	Value of blacksmith's products used by the carpenter	Value of blacksmith's products used by the farmer	Value of blacksmith's products sold to the "outside"

Similarly, for the carpenter and the blacksmith:

$$x_2 = (\quad 0.30x_1 \quad + \quad 0.05x_2 \quad + \quad 0.20x_3 \quad) + \quad y_2$$

Value of carpenter's products used by the blacksmith	Value of carpenter's products used by the carpenter	Value of carpenter's products used by the farmer	Value of carpenter's products sold to the "outside"

$$x_3 = (\quad 0.45x_1 \quad + \quad 0.40x_2 \quad + \quad 0.10x_3 \quad) + \quad y_3$$

Value of farmer's products used by the blacksmith	Value of farmer's products used by the carpenter	Value of farmer's products used by the farmer	Value of farmer's products sold to the "outside"

In matrix form, this system of equations becomes:

$$\begin{pmatrix} x_1 \\ x_2 \\ x_3 \end{pmatrix} = \begin{pmatrix} 0.15 & 0.35 & 0.25 \\ 0.30 & 0.05 & 0.20 \\ 0.45 & 0.40 & 0.10 \end{pmatrix} \begin{pmatrix} x_1 \\ x_2 \\ x_3 \end{pmatrix} + \begin{pmatrix} y_1 \\ y_2 \\ y_3 \end{pmatrix}$$

The numerical matrix in this equation was called the "interindustry matrix of technical coefficients" by Leontief and is now usually referred to as the "technology matrix" of the economy. The columns of this matrix are the input values taken from each sector as shown in the economy diagram. In general,

Leonteiff "Open" Input–Output Model

A Leontief "open" input-output model is a matrix equation

$$X = A \cdot X + Y$$

where the column matrix $X^{n \times 1}$ lists the values produced by each of the n economic sectors, the column matrix $Y^{n \times 1}$ lists the values of the excess production of these same sectors, and the element $a_{i,j}$ of the technology matrix $A^{n \times n}$ represents the value of sector i goods and services needed to produce one dollar of sector j output.

Solving this matrix equation for the excess productions Y in terms of the sector production values X, we have $X - A \cdot X = Y$ so that $Y = (I - A) \cdot X$. Alternatively, solving for the sector production values X required for given excess productions Y, we have $X = (I - A)^{-1} \cdot Y$. That is,

Excess and Sector Productions

In a Leontief "open" input-output model economy, the excess productions Y from given sector production values X is

$$Y = (I - A) \cdot X$$

while the sector production values X necessary to provide required excess productions Y is

$$X = (I - A)^{-1} \cdot Y$$

Returning to our example economy, if a government survey of economic activity establishes that both the blacksmith and the farmer each produce $200 of value each year and the carpenter produces $160, then the amount of extra production is given by $Y = (I - A) \cdot X$:

$$\begin{pmatrix} x_1 \\ x_2 \\ x_3 \end{pmatrix} = \begin{pmatrix} 200 \\ 160 \\ 200 \end{pmatrix}$$

means that

$$\begin{pmatrix} y_1 \\ y_2 \\ y_3 \end{pmatrix} = \left(\begin{pmatrix} 1 & 0 & 0 \\ 0 & 1 & 0 \\ 0 & 0 & 1 \end{pmatrix} - \begin{pmatrix} 0.15 & 0.35 & 0.25 \\ 0.30 & 0.05 & 0.20 \\ 0.45 & 0.40 & 0.10 \end{pmatrix} \right) \cdot \begin{pmatrix} 200 \\ 160 \\ 200 \end{pmatrix} = \begin{pmatrix} 64 \\ 52 \\ 26 \end{pmatrix}$$

Therefore, the amount of extra production will be $64 from the black-smith, $52 from the carpenter, and $26 from the farmer, for a total of $142.

How much will the production of each sector have to change in order to provide $151 of excess production, specifically $59 from the blacksmith, $48 from the carpenter, and $44 from the farmer? We use the formula $X = (I - A)^{-1} \cdot Y$:

$$\begin{pmatrix} y_1 \\ y_2 \\ y_3 \end{pmatrix} = \begin{pmatrix} 59 \\ 48 \\ 44 \end{pmatrix}$$

means that

$$\begin{pmatrix} x_1 \\ x_2 \\ x_3 \end{pmatrix} = \left(\begin{pmatrix} 1 & 0 & 0 \\ 0 & 1 & 0 \\ 0 & 0 & 1 \end{pmatrix} - \begin{pmatrix} 0.15 & 0.35 & 0.25 \\ 0.30 & 0.05 & 0.20 \\ 0.45 & 0.40 & 0.10 \end{pmatrix} \right)^{-1} \cdot \begin{pmatrix} 59 \\ 48 \\ 44 \end{pmatrix} = \begin{pmatrix} 200 \\ 160 \\ 220 \end{pmatrix}$$

Thus if the blacksmith and the carpenter continue to produce the same value as before but the farmer increases production by $20 to $220, the economy can generate an additional $9 in excess production (but notice that the respective sources of this excess change drastically).

Graphing Calculator Exploration

Leontief "open" input-output model calculations are simple to do on your calculator. To verify the above results for the blacksmith-carpenter-farmer economy example:

a.   Store the technology matrix A in [A].

```
[A]
[[.15 .35 .25]
 [.3  .05 .2 ]
 [.45 .4   .1 ]]
```

b.   To find the excess productions Y for productions $X = \begin{pmatrix} 200 \\ 160 \\ 200 \end{pmatrix}$, store these values in [B].

```
[B]
       [[200]
        [160]
        [200]]
```

c. Then use the MATRX MATH IDENTITY(3) command to find $Y = (I - A) \cdot X$.

```
(identity(3)-[A]
)*[B]
                [[64]
                 [52]
                 [26]]
```

d. To find the sector productions X necessary to provide $Y = \begin{pmatrix} 59 \\ 48 \\ 44 \end{pmatrix}$ excess productions, store these values in [C] and use the $\boxed{x^{-1}}$ button to find $X = (I - A)^{-1} \cdot Y$.

```
(identity(3)-[A]
)⁻¹*[C]
                [[200]
                 [160]
                 [220]]
```

There are many reasons to alter the production levels or the excess production levels of the sectors, such as raising the standard of living, reducing social debts, or supporting government programs through taxes. The Leontief "open" input-output model provides a tool to evaluate the effects of such possible changes.

Markov Chains

Markov processes were first investigated by the Russian mathematician A. A. Markov (1856–1922) to analyze the long-term consequences of repeated short-term changes. A *Markov chain* is a collection of "states" to which the members of a population belong together with "transition percentages" of each state's population that will have moved to each other state the next time a count is taken. The transition percentages are sometimes called "transition probabilities" when the focus of attention is a particular member of the population rather than the population as a whole. A *state-transition diagram* of the Markov chain indicates the transition from one state to another with an arrow from the starting state to the new state labeled with the percentage of the starting state's population that will make this change.

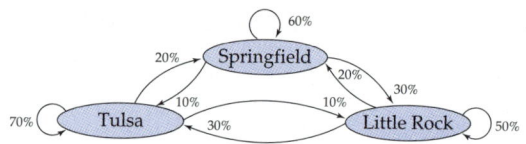

A state-transition diagram with three states.

For instance, consider a truck rental company that has offices in Springfield, Tulsa, and Little Rock (the three "states"), and each rents trucks by the week for local use or to be turned in at one of the other offices. The "population" is the collection of rental trucks, and the above state-transition diagram shows that 60% of the trucks rented in Springfield are returned at the end of the week to Springfield while 10% are turned in at Tulsa and 30% are turned in at Little Rock. Notice that the diagram accounts for 100% of the trucks at each of the rental offices. As the weeks pass, what happens to the distribution of trucks among the rental offices—do they bunch up in one city or do they mix around in some predictable fashion?

Suppose we start with x_1 trucks in Springfield, x_2 trucks in Tulsa, and x_3 trucks in Little Rock. At the end of the first week, the number of trucks turned in to the Springfield office is

$$\underbrace{0.60x_1}_{\substack{\text{Springfield trucks} \\ \text{staying in Springfield}}} + \underbrace{0.20x_2}_{\substack{\text{Tulsa trucks} \\ \text{taken to Springfield}}} + \underbrace{0.20x_3}_{\substack{\text{Little Rock trucks} \\ \text{taken to Springfield}}} \quad \substack{\text{Number of trucks} \\ \text{in Springfield} \\ \text{after 1 week}}$$

Similarly for the number of trucks turned in to Tulsa and Little Rock:

$$\underbrace{0.10x_1}_{\substack{\text{Springfield trucks} \\ \text{taken to Tulsa}}} + \underbrace{0.70x_2}_{\substack{\text{Tulsa trucks} \\ \text{staying in Tulsa}}} + \underbrace{0.30x_3}_{\substack{\text{Little Rock trucks} \\ \text{taken to Tulsa}}} \quad \substack{\text{Number of trucks} \\ \text{in Tulsa} \\ \text{after 1 week}}$$

$$\underbrace{0.30x_1}_{\substack{\text{Springfield trucks} \\ \text{taken to Little Rock}}} + \underbrace{0.10x_2}_{\substack{\text{Tulsa trucks} \\ \text{taken to Little Rock}}} + \underbrace{0.50x_3}_{\substack{\text{Little Rock trucks} \\ \text{staying in Little Rock}}} \quad \substack{\text{Number of trucks} \\ \text{in Little Rock} \\ \text{after 1 week}}$$

These expressions may be written as the matrix product:

$$(x_1 \quad x_2 \quad x_3) \begin{pmatrix} 0.60 & 0.10 & 0.30 \\ 0.20 & 0.70 & 0.10 \\ 0.20 & 0.30 & 0.50 \end{pmatrix} \quad \substack{\text{Number of trucks} \\ \text{in each city} \\ \text{after 1 week}}$$

This numerical matrix is the *transition matrix* of the Markov chain and the element in row i, column j is the percentage of the population in state i that changes to state j in one week (the transition time period). Each row consists of the transition percentages from that state and totals 100% (that is, the row decimals add up to 1).

What happens after two weeks? Multiplying by the transition matrix changes the distribution at one time to the distribution one week later, so to find the distribution after *two* weeks we simply multiply twice, or equivalently, by the *square* of the transition matrix.

Initial distribution
↓

$$(x_1 \quad x_2 \quad x_3) \underbrace{\begin{pmatrix} 0.60 & 0.10 & 0.30 \\ 0.20 & 0.70 & 0.10 \\ 0.20 & 0.30 & 0.50 \end{pmatrix}}_{\substack{\text{Distribution} \\ \text{after one week}}} \cdot \underbrace{\begin{pmatrix} 0.60 & 0.10 & 0.30 \\ 0.20 & 0.70 & 0.10 \\ 0.20 & 0.30 & 0.50 \end{pmatrix}}_{\substack{\text{Multiply by transition} \\ \text{matrix to get to} \\ \text{the next week}}} = (x_1 \quad x_2 \quad x_3) \underbrace{\begin{pmatrix} 0.60 & 0.10 & 0.30 \\ 0.20 & 0.70 & 0.10 \\ 0.20 & 0.30 & 0.50 \end{pmatrix}^2}_{\substack{\text{Distribution} \\ \text{after two weeks}}}$$

For future weeks, we simply multiply by a higher power of the transition matrix:

Markov Chain

> Let A be the transition matrix of a Markov chain with n states and let $X = (x_1 \quad x_2 \quad \cdots \quad x_n)$ be the initial populations of these states. Then the populations in the states after k transitions is $X \cdot A^k$.

Graphing Calculator Exploration

```
[A]
    [[.6 .1 .3]
     [.2 .7 .1]
     [.2 .3 .5]]
[B]
    [[90 45 45]]
```

Suppose that the initial distribution of trucks when the company started was 90 in Springfield and 45 each in Tulsa and in Little Rock. What happens after one year of business?

a. Store the transition matrix in matrix [A] and the initial distribution in matrix [B].

b. Find [B]*[A]^k for k = 4 (one month), k = 12 (three months), k = 26 (six months), and k = 52 (one year). To see all of the values in these row matrices on one screen, use the transpose command to see the answer as a column.

```
([B]*[A]^4)ᵀ
      [[60.768]
       [67.824]
       [51.408]]
```

```
([B]*[A]^12)ᵀ
      [[60.00050332]
       [69.99656067]
       [50.00293601]]
```

```
((CBJ*CAJ^26)ᵀ                  ((BJ*CAJ^52)ᵀ
  CC60          ]                 CC60J
  C69.999999983                   C70J
  C50.0000000233                  C50JJ
```

Thus at the end of a year, there will be 60 trucks in Springfield, 70 trucks in Tulsa, and 50 trucks in Little Rock. Repeat this experiment with [B] = [[120   20   40]]. Do you come to the same final distribution of trucks at the end of the year?

For this particular Markov chain, an initial distribution of 180 trucks among the three states results in a long-term or *steady-state distribution* of $\frac{1}{3}$ of them in Springfield, $\frac{7}{18}$ of them in Tulsa, and $\frac{5}{18}$ of them in Little Rock. These fractions add up to 1 (since they explain the entire group of trucks) and they are unchanged by multiplication by the transition matrix:

$$(\tfrac{1}{3} \quad \tfrac{7}{18} \quad \tfrac{5}{18}) \begin{pmatrix} 0.60 & 0.10 & 0.30 \\ 0.20 & 0.70 & 0.10 \\ 0.20 & 0.30 & 0.50 \end{pmatrix} = (\tfrac{1}{3} \quad \tfrac{7}{18} \quad \tfrac{5}{18})$$

Steady-State Distribution

A steady-state distribution for a Markov chain satisfies the matrix equation $X \cdot A = X$.

Not every Markov chain has a steady-state distribution. For instance, the state-transition diagram below depicts a Markov chain in which the entire population switches back and forth between two states so that no part remains in the same state.

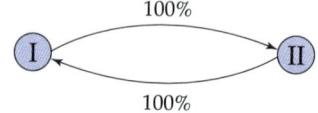

A Markov chain with no steady state distribution.

If a Markov chain *does* have a steady-state distribution, then the higher powers A^k of the transition matrix A must similarly settle down so that the expressions $X \cdot A^k$ become more and more the same for every high power k.

Graphing Calculator Exploration

To see if the higher powers of the transition matrix from the truck rental example have this property,

a. Check that [A] still has the correct values.

```
[A]
      [[.6 .1 .3]
       [.2 .7 .1]
       [.2 .3 .5]]
```

b. Find [A]^k for k = 4, 6, 8, and 52. Display these values in fraction form if possible.

```
[A]^4▶Frac
[[219/625 431/1...
 [203/625 527/1...
 [203/625 99/25...
```

```
[A]^6▶Frac
[[.336064 .3791...
 [.331968 .3954...
 [.331968 .3913...
```

```
[A]^8▶Frac
[[.33377024 .38...
 [.33311488 .39...
 [.33311488 .38...
```

```
[A]^52▶Frac
[[1/3 7/18 5/18...
 [1/3 7/18 5/18...
 [1/3 7/18 5/18...
```

Why does each column settle down to a single value? Consider the product of one row of A with one column of A^k. If all the numbers in the column are the same, their product with the row will also be this same number since the row numbers add up to 1:

$$(0.2 \quad 0.7 \quad 0.1) \cdot \begin{pmatrix} a \\ a \\ a \end{pmatrix} = 0.2a + 0.7a + 0.1a = a(0.2 + 0.7 + 0.1) = a$$

Now suppose that the numbers in the column are different: then there is a biggest value b and a smallest value s. Since some of the numbers in the column are smaller than b, the product of the row (which adds up to 1) with the column must be smaller than the product of this same row with a column consisting of all b's:

$$(0.2 \quad 0.7 \quad 0.1) \cdot \begin{pmatrix} b \\ \vdots \\ s \end{pmatrix} < (0.2 \quad 0.7 \quad 0.1) \cdot \begin{pmatrix} b \\ b \\ b \end{pmatrix} = b$$

Similarly, the product of a row and column must be larger than the product of that row with a column consisting of all s's. Hence *every multiplication by A reduces the size of the largest entry in each column and increases the size of the smallest entry.* This means that as we raise A to higher and higher powers, the biggest and smallest values in each column will become closer and closer to each other, tending toward a single value, exactly as we saw in our calculator experiment. Of course, this reasoning would fail if any of the elements of the transition matrix were one (that could keep the biggest value) or zero (that could lower the smallest value).

Ergodic Matrix

A square matrix A is *ergodic* or *mixing* if every entry is positive and the sum of the entries in each row is 1.

The powers of any ergodic matrix A approximate arbitrarily closely an ergodic matrix consisting of identical rows, with each row solving the matrix equation $X \cdot A = X$.

We conclude by finding the steady-state distribution of the truck rental example by solving $X \cdot A = X$. Since X is a row matrix with the elements summing to 1, we need to solve the equations:

$$\begin{cases} (x_1 \quad x_2 \quad x_3) \begin{pmatrix} 0.6 & 0.1 & 0.3 \\ 0.2 & 0.7 & 0.1 \\ 0.2 & 0.3 & 0.5 \end{pmatrix} = (x_1 \quad x_2 \quad x_3) \\[2em] (x_1 \quad x_2 \quad x_3) \begin{pmatrix} 1 \\ 1 \\ 1 \end{pmatrix} = (1) \end{cases}$$

Taking transposes and recalling that $(A \cdot B)^t = B^t \cdot A^t$, we can rewrite this problem with the variables in columns:

$$\begin{cases} \begin{pmatrix} 0.6 & 0.2 & 0.2 \\ 0.1 & 0.7 & 0.3 \\ 0.3 & 0.1 & 0.5 \end{pmatrix} \begin{pmatrix} x_1 \\ x_2 \\ x_3 \end{pmatrix} = \begin{pmatrix} x_1 \\ x_2 \\ x_3 \end{pmatrix} \\[2em] (1 \quad 1 \quad 1) \begin{pmatrix} x_1 \\ x_2 \\ x_3 \end{pmatrix} = (1) \end{cases}$$

Subtracting the variables from both sides of the first matrix equation, $A^t \cdot X^t = X^t$ becomes $(A^t - I) \cdot X^t = 0$, and we need to solve the large matrix equation:

$$\begin{pmatrix} -0.4 & 0.2 & 0.2 \\ 0.1 & -0.3 & 0.3 \\ 0.3 & 0.1 & -0.5 \\ 1 & 1 & 1 \end{pmatrix} \begin{pmatrix} x_1 \\ x_2 \\ x_3 \end{pmatrix} = \begin{pmatrix} 0 \\ 0 \\ 0 \\ 1 \end{pmatrix}$$

Since the coefficient matrix of this problem is not square, we must solve it by row reduction.

Graphing Calculator Exploration

To find the steady-state distribution for the truck rental example, we need to solve the system of equations $\begin{cases} (A^t - I) \cdot X^t = 0 \\ (1 \quad 1 \cdots 1)X^t = 1 \end{cases}$:

a. Enter the augmented matrix for this system of equations in matrix [C].

```
MATRIX[C]  4 ×4
-.2      .2     0    ]
-.3      .3     0    ]
-.1     -.5     0    ]
 1       1      1    ]

4,4=1
```

b. Use the RREF(command to row reduce this matrix.

```
rref([C])▶Frac
  [[1 0 0 1/3 ]
   [0 1 0 7/18]
   [0 0 1 5/18]
   [0 0 0 0   ]]
```

Notice that the steady-state distribution found here, 1/3, 7/18, 5/18 (the fractions of the trucks in the three cities) could also have been guessed from the high powers of the transition matrix that we calculated (see page 255).

This calculation could be done starting from the original transition matrix A stored as matrix [A] by first finding $A^t - I$ as $[A]^T - \text{IDENTITY}(N) \to [C]$ (with the correct value for n), using MATRX EDIT to increase the dimensions of [C] by one row and one column, and then filling in the new bottom row with 1's. The steady-state distribution is then found by the same RREF([C]) ▶ FRAC command used in (b) above.

Not all transition matrices are ergodic (some have zeros). However, if some *power* of the transition matrix is ergodic (that is, if multiplying the transition matrix by itself many times gives a matrix with all positive elements), then the matrix is said to be *regular*. Since all higher powers are then also ergodic, any regular Markov chain has a steady-state distribution and the rows of the powers of the transition matrix approximate this distribution arbitrarily closely. By our previous remarks, this means that the long-term behavior of a regular Markov chain with transition matrix A can be found by solving the system of linear equations $\begin{cases} (A^t - I) \cdot X^t = 0 \\ (1 \quad 1 \cdots 1)X^t = 1 \end{cases}$.

Least Squares

The method of least squares was invented in 1794 by Carl Friedrich Gauss (1777–1855) to find the best compromise solution to an inconsistent system of linear equations. He became world famous as a scientist in 1801 when he used his method to predict when and where the asteroid Ceres could next be seen after it had been lost in the sun's glare on February 11, 1801 shortly after its discovery on January 1, 1801 by the Italian astronomer Piazzi. The sensation created by the accuracy of Gauss's prediction ultimately led to his appointment as astronomer at the Gottingen Observatory, assuring him the financial security that allowed him to pursue his many other ideas.

We have seen that a system of linear equations written in the matrix form $A \cdot X = B$ may be solved as $X = A^{-1} \cdot B$ only in the special case that A is invertible (and hence square). But for any matrix A, the product $A^t \cdot A$ is square and symmetric. Let us now attempt to solve any matrix equation $A \cdot X = B$ by multiplying both sides by A^t and then trying to solve $A^t A \cdot X = A^t B$ as $X = (A^t A)^{-1} \cdot (A^t B)$. We begin with two examples in two variables.

The three lines $2x + y = 12$, $x + y = 8$, and $x + 2y = 12$ intersect in the common point $x = 4$, $y = 4$. Writing these equations as a single matrix equation $A \cdot X = B$ and multiplying by A^t, we obtain $A^t A \cdot X = A^t B$:

$$\begin{pmatrix} 2 & 1 \\ 1 & 1 \\ 1 & 2 \end{pmatrix} \begin{pmatrix} x \\ y \end{pmatrix} = \begin{pmatrix} 12 \\ 8 \\ 12 \end{pmatrix}$$

becomes

$$\begin{pmatrix} 2 & 1 & 1 \\ 1 & 1 & 2 \end{pmatrix} \begin{pmatrix} 2 & 1 \\ 1 & 1 \\ 1 & 2 \end{pmatrix} \begin{pmatrix} x \\ y \end{pmatrix} = \begin{pmatrix} 2 & 1 & 1 \\ 1 & 1 & 2 \end{pmatrix} \begin{pmatrix} 12 \\ 8 \\ 12 \end{pmatrix} \qquad A^t A \cdot X = A^t B$$

Multiplying out:

$$\begin{pmatrix} 6 & 5 \\ 5 & 6 \end{pmatrix} \begin{pmatrix} x \\ y \end{pmatrix} = \begin{pmatrix} 44 \\ 44 \end{pmatrix}$$

so $$\begin{pmatrix} x \\ y \end{pmatrix} = \begin{pmatrix} 6 & 5 \\ 5 & 6 \end{pmatrix}^{-1} \begin{pmatrix} 44 \\ 44 \end{pmatrix}$$

$$= \begin{pmatrix} 6/11 & -5/11 \\ -5/11 & 6/11 \end{pmatrix} \begin{pmatrix} 44 \\ 44 \end{pmatrix} = \begin{pmatrix} 4 \\ 4 \end{pmatrix} \quad \text{The known solution}$$

Thus in this particular example, we were able to solve a consistent system of equations with a unique solution using an inverse matrix after multiplying both sides by the transpose of the coefficient matrix.

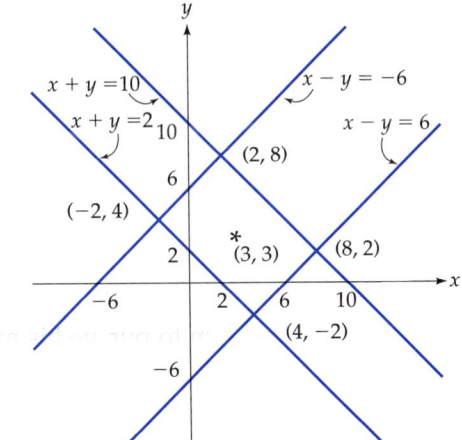

An inconsistent system of lines

What happens if the equations are *inconsistent*? The above diagram shows that the four lines $x + y = 10$, $x + y = 2$, $x - y = 6$, and $x - y = -6$ have no point in common. If again we multiply the matrix equation $A \cdot X = B$ by A^t, we obtain:

$$\begin{pmatrix} 1 & 1 \\ 1 & 1 \\ 1 & -1 \\ 1 & -1 \end{pmatrix} \begin{pmatrix} x \\ y \end{pmatrix} = \begin{pmatrix} 10 \\ 2 \\ 6 \\ -6 \end{pmatrix}$$

becomes

$$\begin{pmatrix} 1 & 1 & 1 & 1 \\ 1 & 1 & -1 & -1 \end{pmatrix} \begin{pmatrix} 1 & 1 \\ 1 & 1 \\ 1 & -1 \\ 1 & -1 \end{pmatrix} \begin{pmatrix} x \\ y \end{pmatrix} = \begin{pmatrix} 1 & 1 & 1 & 1 \\ 1 & 1 & -1 & -1 \end{pmatrix} \begin{pmatrix} 10 \\ 2 \\ 6 \\ -6 \end{pmatrix}$$

Multiplying out:

$$\begin{pmatrix} 4 & 0 \\ 0 & 4 \end{pmatrix}\begin{pmatrix} x \\ y \end{pmatrix} = \begin{pmatrix} 12 \\ 12 \end{pmatrix}$$

so

$$\begin{pmatrix} x \\ y \end{pmatrix} = \begin{pmatrix} 4 & 0 \\ 0 & 4 \end{pmatrix}^{-1}\begin{pmatrix} 12 \\ 12 \end{pmatrix}$$

$$= \begin{pmatrix} 1/4 & 0 \\ 0 & 1/4 \end{pmatrix}\begin{pmatrix} 12 \\ 12 \end{pmatrix} = \begin{pmatrix} 3 \\ 3 \end{pmatrix} \qquad \begin{matrix} x = 3 \\ y = 3 \end{matrix}$$

However, $x = 3$, $y = 3$ fails to satisfy any of the original equations. But looking back at the above diagram, we see that the point $(3, 3)$ is "in the middle" of the four lines. In a clear geometrical sense, this "solution" $(3, 3)$ is the "best compromise" for an answer in that it is as close as possible to each of the lines at the same time. A common measure of the error in an estimate is the sum of the squares of the vertical distances. For this example, the sum of the squares of the vertical distances to the lines is $(4)^2 + (-4)^2 + (-6)^2 + (6)^2 = 104$ (the minus signs mean the lines were below the point). Any other point will give a larger sum of squares; for instance, repeating the previous calculation for the point $(3, 4)$ gives the larger value $(3)^2 + (-5)^2 + (-5)^2 + (7)^2 = 108$. Thus in this special case, if $A \cdot X = B$ is inconsistent but $A^t \cdot A$ is invertible, the "solution" $X = (A^tA)^{-1} \cdot (A^tB)$ is *the best compromise answer* in that it minimizes the sum of the squares of the vertical errors. (On pages 961–962 we will give a proof of this fact.)

We now turn to prediction problems similar to those solved by Gauss with such spectacular results. Suppose we know that some quantity y should depend on some other quantity x in such a way that y is a linear function of x: $y = mx + b$. If we could take *perfect* measurements of x and y, we might obtain the data shown below.

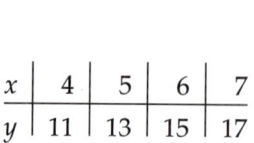

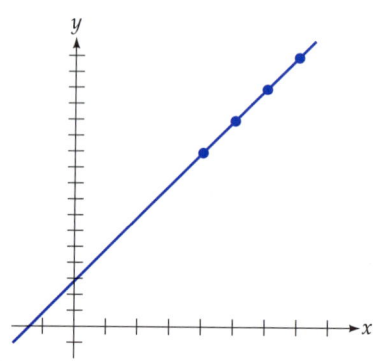

The above graph shows the points together with the line $y = 2x + 3$ passing through them (the values $m = 2$ and $b = 3$ are easily found by the usual slope and intercept methods).

However, in the "real world," measurements are not exact, and data often have errors. Suppose that the data were as follows.

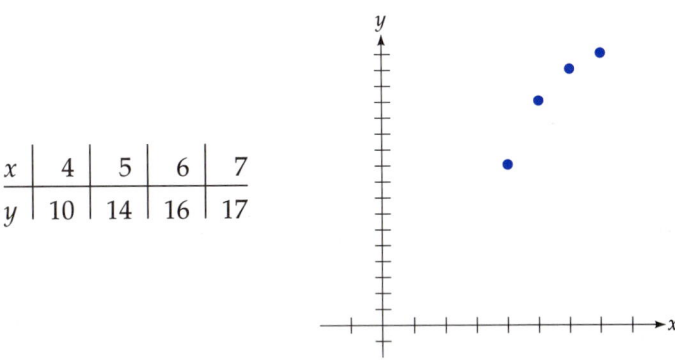

x	4	5	6	7
y	10	14	16	17

Clearly, these four data points are *not* collinear, so it is impossible to find a line $y = mx + b$ passing through all four of them. If we were to write the problems of finding m and b as a system of linear equations:

$$\begin{cases} 4m + b = 10 \\ 5m + b = 14 \\ 6m + b = 16 \\ 7m + b = 17 \end{cases} \rightarrow \begin{pmatrix} 4 & 1 \\ 5 & 1 \\ 6 & 1 \\ 7 & 1 \end{pmatrix} \begin{pmatrix} m \\ b \end{pmatrix} = \begin{pmatrix} 10 \\ 14 \\ 16 \\ 17 \end{pmatrix} \qquad AX = B$$

then the system would be *inconsistent* since there is no solution. However, instead of an *exact* solution we can look for the best *compromise* line passing *closest* to the points. As before, we multiply both sides of the above matrix equation by A^t:

$$\begin{pmatrix} 4 & 5 & 6 & 7 \\ 1 & 1 & 1 & 1 \end{pmatrix} \begin{pmatrix} 4 & 1 \\ 5 & 1 \\ 6 & 1 \\ 7 & 1 \end{pmatrix} \begin{pmatrix} m \\ b \end{pmatrix} = \begin{pmatrix} 4 & 5 & 6 & 7 \\ 1 & 1 & 1 & 1 \end{pmatrix} \begin{pmatrix} 10 \\ 14 \\ 16 \\ 17 \end{pmatrix} \qquad A^tAX = A^tB$$

Multiplying out:

$$\begin{pmatrix} 126 & 22 \\ 22 & 4 \end{pmatrix} \begin{pmatrix} x \\ y \end{pmatrix} = \begin{pmatrix} 325 \\ 57 \end{pmatrix}$$

so

$$\begin{pmatrix} x \\ y \end{pmatrix} = \begin{pmatrix} 126 & 22 \\ 22 & 4 \end{pmatrix}^{-1} \begin{pmatrix} 325 \\ 57 \end{pmatrix} = \begin{pmatrix} 1/5 & -11/10 \\ -11/10 & 63/10 \end{pmatrix} \begin{pmatrix} 325 \\ 57 \end{pmatrix} = \begin{pmatrix} 2.3 \\ 1.6 \end{pmatrix}$$

These values, $m = 2.3$ and $b = 1.6$, give the line $y = 2.3x + 1.6$, which is called the "least squares" line, providing the best fit to the four points in that it minimizes the sum of the squared vertical distances. The graph below shows the four points together with the least squares line.

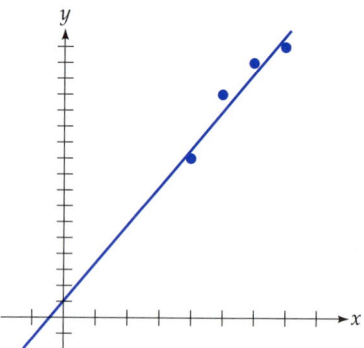

If the four points *did* lie on a line, then the *m* and *b* given by this procedure would be the *exact* values for the correct line. The least squares line is used widely in business and the sciences to predict future trends from current but imperfect data.

 ## Graphing Calculator Exploration

The least squares line calculation is simple to do on your calculator. To verify the above results for the data with errors,

a.   Store the *x* data values and the 1's in [A].

```
[A]
           [[4 1]
            [5 1]
            [6 1]
            [7 1]]
```

b.   Store the *y* data values in [B].

```
[B]
           [[10]
            [14]
            [16]
            [17]]
```

c.   Then use the **TRANSPOSE** command and the $\boxed{x^{-1}}$ button to find $X = (A^tA)^{-1} \cdot (A^tB)$.

```
([A]^T[A])^-1([A]^T[
B])
           [[2.3]
            [1.6]]
```

The method of least squares can be summarized as follows.

Least Squares Approximation

The slope m and y-intercept b of the least squares best approximation line $y = mx + b$ for the data points (x_1, y_1), (x_2, y_2), . . . , (x_n, y_n) are the unique solution of the system of linear equations

$$\begin{pmatrix} x_1 & x_2 \cdots x_n \\ 1 & 1 \cdots 1 \end{pmatrix} \begin{pmatrix} x_1 & 1 \\ x_2 & 1 \\ \vdots & \vdots \\ x_n & 1 \end{pmatrix} \begin{pmatrix} m \\ b \end{pmatrix} = \begin{pmatrix} x_1 & x_2 \cdots x_n \\ 1 & 1 \cdots 1 \end{pmatrix} \begin{pmatrix} y_1 \\ y_2 \\ \vdots \\ y_n \end{pmatrix}$$

provided the values $x_1, x_2, \ldots, x_n$ are distinct.

SUMMARY

A *Leontief "open" input-output model* is a matrix equation $X = AX + Y$, where the column matrix X lists the values produced by each of the n economic sectors, the column matrix Y lists the values of the excess production of these same sectors, and the element $a_{i,j}$ of the $n \times n$ technology matrix A represents the value of sector i goods and services needed to produce one dollar of sector j output. Solving this model for the excess productions Y in terms of the sector production values X finds that $Y = (I - A) \cdot X$, while the sector production values X necessary to provide excess productions Y is $X = (I - A)^{-1} \cdot Y$.

A state-transition diagram of a *Markov chain* indicates the transition from one state to another with an arrow from the starting state to the new state labeled with the percentage of the starting state's population that will make this change. The element in row i, column j of the transition matrix $A^{n \times n}$ of a Markov chain with n states is the decimal form of the percentage of the population in state i that changes to state j. Each element in A is at least zero and no more than 1 and the sum of the elements in each row equals 1. A transition matrix is *ergodic* if every entry is positive and *regular* if some power is ergodic. If the initial populations of the states are $X^{1 \times n} = (x_1 \quad x_2 \cdots x_n)$, then the populations in the states after k transitions is $X \cdot A^k$. The steady-state distribution of a regular Markov chain can be found exactly by solving the system of linear equations

$$\begin{cases} (A^t - I) \cdot X^t = 0 \\ (1 \quad 1 \cdots 1)X^t = 1 \end{cases}$$

and each row of the powers of the transition matrix approximates this steady-state distribution.

If A is not invertible but $A^t \cdot A$ is, the "solution" $X = (A^tA)^{-1} \cdot (A^tB)$ of the system of linear equations $A \cdot X = B$ has the smallest possible sum of the squares of the errors $B - A \cdot X$. For data points $(x_1, y_1), (x_2, y_2), \ldots , (x_n, y_n)$ with distinct $x_1, x_2, \ldots , x_n$ values, the slope m and y-intercept b of the *least squares best approximation line* $y = mx + b$ are the unique solution of the system of linear equations

$$\begin{pmatrix} x_1 & x_2 \cdots x_n \\ 1 & 1 \cdots 1 \end{pmatrix} \begin{pmatrix} x_1 & 1 \\ x_2 & 1 \\ \vdots & \vdots \\ x_n & 1 \end{pmatrix} \begin{pmatrix} m \\ b \end{pmatrix} = \begin{pmatrix} x_1 & x_2 \cdots x_n \\ 1 & 1 \cdots 1 \end{pmatrix} \begin{pmatrix} y_1 \\ y_2 \\ \vdots \\ y_n \end{pmatrix}$$

and this least squares line can be used to predict values for y from values of x.

EXERCISES 2.6

 A graphing calculator will be helpful for most exercises.

Leontief "Open" Input-Output Models

Let A denote agriculture, C denote construction, E denote electronics, F denote fishing, H denote heavy industry, L denote light industry, M denote mining, R denote railroads, and T denote tourism.

Find the technology matrix for each economy diagram.

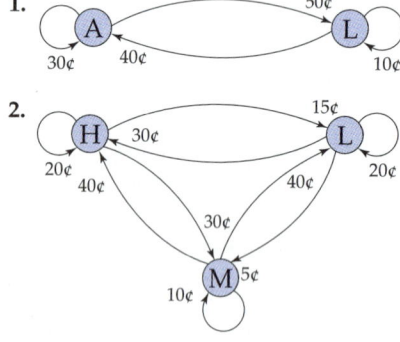

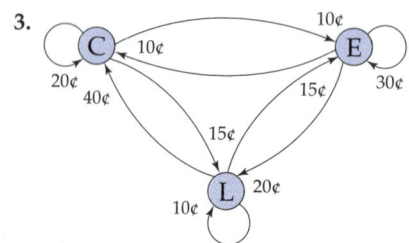

4.

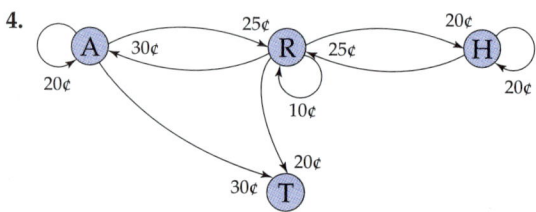

Draw an economy diagram for each technology matrix.

5. $\begin{pmatrix} 0.15 & 0.10 \\ 0.25 & 0.30 \end{pmatrix}$ for sectors M and H

6. $\begin{pmatrix} 0.10 & 0.30 & 0.10 \\ 0.20 & 0.40 & 0.20 \\ 0.10 & 0.30 & 0.10 \end{pmatrix}$ for sectors A, R, and M

7. $\begin{pmatrix} 0.20 & 0.30 & 0.10 \\ 0.50 & 0.20 & 0.20 \\ 0.30 & 0.10 & 0.20 \end{pmatrix}$ for sectors C, A, and L

8. $\begin{pmatrix} 0.10 & 0 & 0.20 \\ 0 & 0.20 & 0.10 \\ 0.10 & 0.40 & 0 \end{pmatrix}$ for sectors A, C, and R

Find the excess production Y of each economy with technology matrix A and economic activity level X.

9. $A = \begin{pmatrix} 0.20 & 0.30 \\ 0.35 & 0.25 \end{pmatrix}$ and $X = \begin{pmatrix} 130 \\ 110 \end{pmatrix}$

10. $A = \begin{pmatrix} 0.45 & 0.20 \\ 0.30 & 0.35 \end{pmatrix}$ and $X = \begin{pmatrix} 140 \\ 100 \end{pmatrix}$

11. $A = \begin{pmatrix} 0.05 & 0.15 & 0.20 \\ 0.15 & 0.05 & 0.15 \\ 0.10 & 0.10 & 0.05 \end{pmatrix}$ and $X = \begin{pmatrix} 150 \\ 170 \\ 140 \end{pmatrix}$

12. $A = \begin{pmatrix} 0.10 & 0.05 & 0.10 & 0.15 \\ 0.15 & 0.10 & 0.10 & 0.05 \\ 0 & 0.05 & 0.10 & 0 \\ 0.10 & 0 & 0.10 & 0.05 \end{pmatrix}$ and $X = \begin{pmatrix} 120 \\ 100 \\ 110 \\ 80 \end{pmatrix}$

Find the economic activity level X for each economy with technology matrix A necessary to generate excess production Y.

13. $A = \begin{pmatrix} 0.20 & 0.30 \\ 0.30 & 0.20 \end{pmatrix}$ and $Y = \begin{pmatrix} 84 \\ 51 \end{pmatrix}$

14. $A = \begin{pmatrix} 0.25 & 0.35 \\ 0.45 & 0.15 \end{pmatrix}$ and $Y = \begin{pmatrix} 33 \\ 57 \end{pmatrix}$

15. $A = \begin{pmatrix} 0.10 & 0.20 & 0 \\ 0 & 0.15 & 0.20 \\ 0.30 & 0.10 & 0.20 \end{pmatrix}$ and $Y = \begin{pmatrix} 60 \\ 31 \\ 50 \end{pmatrix}$

16. $A = \begin{pmatrix} 0.10 & 0.15 & 0.10 & 0.05 \\ 0.15 & 0.15 & 0.05 & 0.10 \\ 0 & 0.20 & 0.10 & 0.10 \\ 0.10 & 0 & 0.15 & 0.05 \end{pmatrix}$ and $Y = \begin{pmatrix} 60 \\ 55 \\ 60 \\ 70 \end{pmatrix}$

Represent each situation as a Leontief "open" input-output model by constructing an economy diagram and the corresponding technology matrix. Find the required excess production or level of economic activity and be sure to state your final answer in terms of the original question.

17. Industrial Production The heavy and light industry sectors of the Birmingham economy depend on each other in the following way: each dollar of production from the heavy industry sector requires $0.25 of heavy industry produce and $0.15 of light industry produce, while each dollar of production from the light industry sector requires $0.35 of heavy industry produce and $0.05 of light industry produce. How much must each type of industry produce to yield an excess production of $127 million of heavy industry produce and $221 million of light industry produce?

18. County Production Planning An analysis of the mining, railroad, construction, and light industry sectors of the Hanover County economy revealed that each dollar produced by the mining sector requires $0.20 of mining products,

$0.10 of railroad services, $0.10 of construction, and $0.10 of light industry products. Each dollar produced by the railroad sector requires $0.10 of mining products, $0.20 of railroad services, and $0.10 of light industry products. Each dollar produced by the construction sector requires $0.30 of mining products, $0.20 of construction, and $0.10 of light industry products. Each dollar produced by the light industry sector requires $0.10 of mining products, $0.20 of railroad services, and $0.10 of light industry products. If these industries in Hanover County presently produce excess productions valued at $8 million from mining, $30 million from railroads, $32 million from construction, and $31 million from light industry, what is the current production level of each of these economic sectors?

19. Third-World GNP The new government of a third-world country wants to increase its gross national product by stimulating the heavy industry sector of the national economy. An analysis of the relationships between its heavy industry, light industry, and railroad sectors found that the current production levels for these sectors are $100 million from heavy industry, $150 million from light industry, and $100 million from the railroads, with technology matrix as given in the table. Find the current GNP from these sectors of the economy. If the light industry and railroad productions remain the same, how much does this GNP increase with each $10 million increase in heavy industry production? What is the greatest heavy industry production level that this part of the national economy can tolerate?

	Heavy industry	Light industry	Railroads
Heavy industry	0.20	0.20	0.30
Light industry	0.50	0.20	0.20
Railroads	0.20	0.10	0.20

20. Island Economy An economist is studying the agriculture-, fishing-, and tourism-based economy of a small Pacific island nation. Although the country's gross national product was $158

million last year, she has found that the economy actually produced $300 million, of which $142 million was consumed in the course of production. Her breakdown of each sector's production is given in the table. (The first line shows that the total value of agriculture production was $100 million, with $20 million consumed by the agriculture sector, $8 million consumed by the fishing sector, and $36 million consumed by the tourism sector, leaving an excess of $36 million that became the agriculture sector's part of the national GNP.) If the same relations continue among the sectors of the economy, what will happen to the national GNP next year if the fishing sector declines to $60 million while the other sectors produce the same amounts as before?

in Millions of $	Agriculture	Fishing	Tourism	GNP
Agriculture	$20	$8	$36	$36
Fishing	$10	$8	$36	$26
Tourism	$0	$0	$24	$96

MARKOV CHAINS

Find the transition matrix of each Markov chain represented by the state-transition diagram.

21.

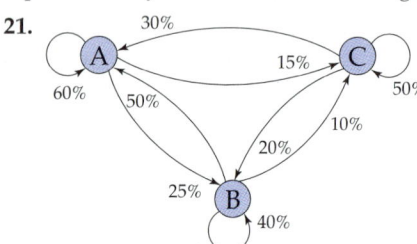

22.

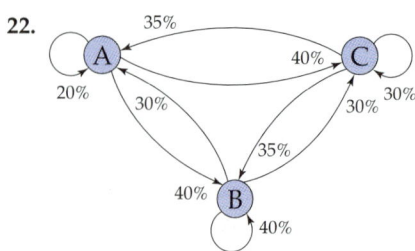

23.

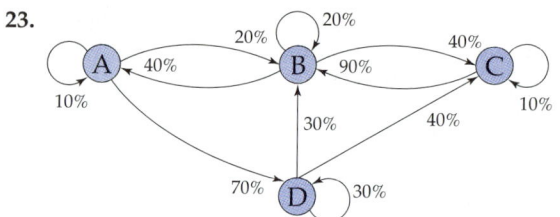

24.

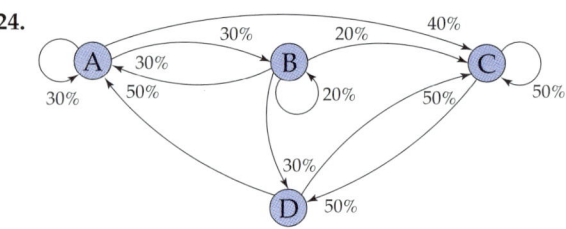

Identify each transition matrix as "ergodic," "regular," or "neither."

25. $\begin{pmatrix} 0.20 & 0.80 \\ 0.70 & 0.30 \end{pmatrix}$

26. $\begin{pmatrix} 0.50 & 0.50 \\ 0.40 & 0.60 \end{pmatrix}$

27. $\begin{pmatrix} 0.30 & 0.40 & 0.30 \\ 0.20 & 0.30 & 0.50 \\ 0.40 & 0.30 & 0.30 \end{pmatrix}$

28. $\begin{pmatrix} 0 & 0.40 & 0.60 \\ 0 & 0.60 & 0.40 \\ 1 & 0 & 0 \end{pmatrix}$

29. $\begin{pmatrix} 0.70 & 0.30 & 0 \\ 0 & 0.30 & 0.70 \\ 0.70 & 0 & 0.30 \end{pmatrix}$

30. $\begin{pmatrix} 1 & 0 & 0 \\ 0 & 0 & 1 \\ 0 & 1 & 0 \end{pmatrix}$

Find the steady-state distribution for each Markov chain with the given transition matrix.

31. $\begin{pmatrix} 0.25 & 0.75 \\ 0.75 & 0.25 \end{pmatrix}$

32. $\begin{pmatrix} 0.40 & 0.60 \\ 0.40 & 0.60 \end{pmatrix}$

33. $\begin{pmatrix} 0 & 0.50 & 0.50 \\ 0.50 & 0 & 0.50 \\ 0.50 & 0.50 & 0 \end{pmatrix}$

34. $\begin{pmatrix} 0 & 1 & 0 \\ 0.40 & 0.20 & 0.40 \\ 0 & 1 & 0 \end{pmatrix}$

35. $\begin{pmatrix} 0.50 & 0.50 & 0 & 0 \\ 0 & 0.50 & 0.50 & 0 \\ 0 & 0 & 0.50 & 0.50 \\ 0.50 & 0 & 0 & 0.50 \end{pmatrix}$

36. $\begin{pmatrix} 0 & 0.25 & 0.25 & 0.25 & 0.25 \\ 0 & 0 & 1 & 0 & 0 \\ 0 & 0 & 0 & 1 & 0 \\ 0 & 0 & 0 & 0 & 1 \\ 1 & 0 & 0 & 0 & 0 \end{pmatrix}$

Represent each situation as a Markov chain by constructing a state-transition diagram and the corresponding transition matrix. Find the steady-state distribution and interpret it in terms of the original situation. Be sure to state your final answer in terms of the original question.

37. Population Dynamics   Life is so good in Lucas, Marion, and Warren counties in upstate Maine that no one ever moves away, although each year 3% of the Lucas residents move to Marion and 2% move to Warren; 4% of the Marion residents move to Lucas and 1% move to Warren; and 2% of the Warren residents move to Lucas and 3% move to Marion. If the combined population of these three counties is 11,200, how many reside in each county?

38. Mass Transit   Commuters in the Pittsburgh metropolitan area travel alone in their cars, join a carpool so they may drive in the High Occupancy Vehicle (HOV) lane, or take the bus. Each month, of those who drive by themselves, 20% join a carpool, 30% switch to the bus, and the rest continue driving alone; of those who are in a carpool, 30% switch to driving alone, 20% switch to the bus, and the rest stay in a carpool; and of those who take the bus, 20% switch to driving alone, 30% join a carpool, and the rest continue taking the bus. While each month many people may change the way they get to work, how many of the 3 million commuters drive alone each month?

39. Art Gallery Shows   The Harmon Gallery's "by invitation only" showing last night of new paintings by Clyberg, Stevensen, and Georgan was a huge success. One thousand fifty-five art dealers and collectors jammed the exhibition rooms from 8:00 P.M. to well after midnight. The security staff hired to protect both the artwork and the guests reported the following pattern of crowd movement every five minutes throughout the evening: of the crowd in the Clyberg display room, 10% stayed, 30% moved on to the Stevensen exhibit, and 60% went for more hors d'oeuvres; of those in the Stevensen display room, 10% stayed, 30% moved on to the Clyberg exhibit, 10% moved on to the Georgan exhibit, and 50% went for more hors d'oeuvres; of those in the Georgan display room, 10% stayed,

80% moved on to the Stevensen exhibit, and 10% went for more hors d'oeuvres; and of those in the central refreshments room, 20% stayed for more hors d'oeuvres, 20% went to the Clyberg exhibit, 20% went to the Stevensen exhibit, and 40% went to the Georgan exhibit. After several hours of milling about in this fashion, how many people were in each room at the gallery?

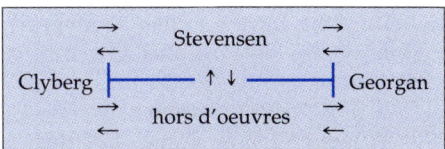

40. Market Share   The Peerless Products Corporation has decided to market a new toothpaste named HappyFace designed to compete successfully with the market leaders: GreatGrin, SuperSmile, and WhataWhite. Test marketing results from several cities indicate that each week, of those who used GreatGrin the previous week, 40% will buy it again, 20% will switch to SuperSmile, 30% will switch to WhataWhite, and 10% will switch to HappyFace; of those who used SuperSmile the previous week, 40% will buy it again, 20% will switch to GreatGrin, 20% will switch to WhataWhite, and 20% will switch to HappyFace; of those who used WhataWhite the previous week, 40% will buy it again, 10% will switch to GreatGrin, 30% will switch to SuperSmile, and 20% will switch to HappyFace; and of those who used HappyFace the previous week, 40% will buy it again, 20% will switch to GreatGrin, 10% will switch to SuperSmile, and 30% will switch to WhataWhite. If these buying patterns continue, what will the long-term market share be for HappyFace toothpaste?

Least Squares

Solve each system of equations by row reducing the corresponding augmented matrix. Rewrite the system of equations in the matrix form $A \cdot X = B$, calculate $X = (A^t A)^{-1} \cdot (A^t B)$, and verify that this is the same solution.

41. $\begin{cases} x + y = 1 \\ y = -4 \\ x - y = 9 \end{cases}$

42. $\begin{cases} x - y = 1 \\ x = 4 \\ x + y = 7 \end{cases}$

43. $\begin{cases} x_1 + x_2 = 5 \\ x_2 + x_3 = 7 \\ x_1 + x_3 = 6 \\ x_1 - x_2 + x_3 = 3 \end{cases}$

44. $\begin{cases} x_1 - x_2 = 2 \\ -x_2 + x_3 = 1 \\ x_1 - x_3 = 1 \\ x_1 - x_2 + x_3 = 5 \end{cases}$

Row reduce the corresponding augmented matrix for each system of equations to verify that the system is inconsistent. Rewrite the system of equations in the matrix form $A \cdot X = B$, calculate $X = (A^tA)^{-1} \cdot (A^tB)$, and verify that this compromise solution "almost" satisfies the original equations.

45. $\begin{cases} x + y = 12 \\ x = 3 \\ y = 3 \end{cases}$

46. $\begin{cases} 2x - y = 6 \\ x + y = 6 \\ -x + 2y = 6 \end{cases}$

47. $\begin{cases} x_1 + x_2 = 5 \\ x_2 + x_3 = 7 \\ x_1 + x_3 = 1 \\ x_1 - x_2 + x_3 = 3 \end{cases}$

48. $\begin{cases} x_1 - x_2 = 2 \\ -x_2 + x_3 = 4 \\ x_1 - x_3 = 4 \\ x_1 - x_2 + x_3 = 5 \end{cases}$

49. $\begin{cases} x_1 + x_3 = 6 \\ x_2 + x_4 = 3 \\ x_2 = 4 \\ x_3 = 3 \\ x_1 + x_4 = 7 \end{cases}$

50. $\begin{cases} x_1 = 4 \\ x_2 + x_3 = 6 \\ x_2 - x_3 = 2 \\ x_4 = 3 \\ x_1 + x_4 = 4 \end{cases}$

Find the least squares best approximation line $y = mx + b$ for each collection of x and y data pairs.

51.

x	−1	0	1
y	24	36	42

52.

x	−1	0	1
y	8	8	14

53.

x	−1	0	1	2
y	120	90	70	30

54.

x	−1	0	1	3
y	761	656	691	551

55.

x	1	4	5	8	12
y	185	195	195	245	275

56.

x	1	4	5	7	8	11
y	50	80	110	140	170	200

Use the least squares best approximation line to make each prediction. Be sure to state your final answer in terms of the original question.

57. Real Estate   The new salesman at Abbott Associates, Real Estate Brokers, had an impressive first three months with sales of $300,000, $480,000, and $600,000. If he can keep improving this much every month, how much will he sell in his fifth month?

58. Commodity Futures   A soybean speculator has been nervously watching the December delivery price per bushel drop from $6.59 on Monday to $6.41 on Tuesday and then to $6.29 on Wednesday. If this trend continues, what will the price be on Friday?

59. Price-Demand   The market research division of a large candy manufacturer has test marketed the new treat YummieCrunchies in five different markets at five different prices per eight-ounce bag to determine the relation between the selling price and the sales volume. The results of its study are presented in the table and are normalized to give the weekly sales per 20,000 consumers. How many weekly sales per 20,000 consumers can the manufacturer expect when it begins national distribution next week with an introductory price of 79¢ per eight-ounce bag?

Selling price	Sales volume
70¢	2050
55¢	2675
75¢	1975
90¢	1250
85¢	1425

60. TV Advertising The accounts manager at television station WXXB claims that every dollar spent on commercials by local retailers generates $23 of sales. As proof, she shows prospective advertisers the following table of advertising budgets and gross receipts for five area companies that aired commercials last month. Is her claim justified?

Dollars spent on WXXB commercials	Gross receipts
$8000	$245,000
$5000	$215,000
$2000	$125,000
$9000	$305,000
$6000	$230,000

Chapter Summary with Hints and Suggestions

The reading and exercises of this chapter have helped you learn the following skills. For each skill, the section from which it came (in case you need to review it) and some exercises in this section that use it are indicated. Answers to all exercises are at the end of the book, and full solutions to all exercises are in the Student Solutions Manual.

2.1   Systems of Two Linear Equations in Two Variables

- Represent a pair of statements as a system of two linear equations in two variables after making an appropriate choice for the meaning of the x- and y-variables. (*Review Exercises 1–2.*)

$$\begin{cases} ax + by = c \\ Ax + By = C \end{cases}$$

- Solve a system of two linear equations in two variables by graphing and identify the system as "independent and consistent," "independent and inconsistent" or "dependent." (*Review Exercises 3–4.*)

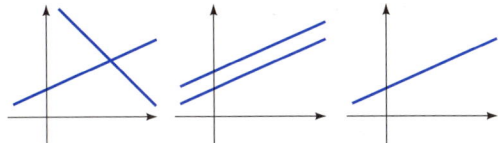

- Solve a system of two linear equations in two variables by the substitution method and identify the system as "independent and consistent," "independent and inconsistent" or "dependent." (*Review Exercises 5–6.*)

- Solve a system of two linear equations in two variables by the elimination method and identify the system as "independent and consistent," "independent and inconsistent" or "dependent." (*Review Exercises 7–8.*)

- Formulate an application as a system of two linear equations in two variables after making an appropriate choice for the meaning of the x- and y-variables, solve the system of equations by the elimination method, and then state the final answer in terms of the original question. (*Review Exercises 9–10.*)

2.2   Matrices and Linear Equations in Two Variables

- Find the dimension of a matrix and identify a particular element using double subscript notation. (*Review Exercises 11–12.*)

$$A = \begin{pmatrix} a_{1,1} & \cdots & a_{1,n} \\ \vdots & & \\ a_{m,1} & \cdots & a_{m,n} \end{pmatrix}$$

- Carry out a given row operation on an augmented matrix. (*Review Exercises 13–14.*)

$$\begin{cases} ax + by = c \\ Ax + By = C \end{cases} \rightarrow \begin{pmatrix} a & b & c \\ A & B & C \end{pmatrix}$$

- Solve a system of two linear equations in two variables by row reducing the corresponding augmented matrix, interpreting the final reduced row-echelon form matrix as the solution of the system of equations, and identifying the system as "independent and consistent," "independent and inconsistent," or "dependent." See table on page 194.
  (Review Exercises 15–18.)

- Formulate an application as a system of two linear equations in two variables after making an appropriate choice for the meaning of the x- and y-variables, solve the system of equations by row reducing the corresponding augmented matrix, and then state the final answer in terms of the original question.
  (Review Exercises 19–20.)

2.3   Matrix Row Reduction and Systems of Linear Equations

- Interpret a reduced row-echelon form matrix as the solution of a system of equations and identify the system as "independent" or "dependent" and as "consistent" or "inconsistent."
  (Review Exercises 21–22.)

  $0 \ 0 \ \cdots \ 0 \ 0$ means dependent.
  $0 \ 0 \ \cdots \ 0 \ 1$ means inconsistent.

- Use an appropriate row operation or sequence of row operations to find the equivalent reduced row-echelon form matrix.
  (Review Exercises 23–24.)

- Solve a system of linear equations by row reducing the corresponding augmented matrix, interpreting the final reduced row-echelon form matrix as the solution of the system of equations, and identifying the system as "independent" or "dependent" and as "consistent" or "inconsistent."
  (Review Exercises 25–28.)

- Formulate an application as a system of linear equations in an appropriate number of variables after making an appropriate choice for the meaning of each, solve the system of equations

by row reducing the corresponding augmented matrix, and then state the final answer in terms of the original question.
(Review Exercises 29–30.)

2.4   Matrix Arithmetic

- Find a matrix product or determine that the product is not defined.
  (Review Exercises 31–32.)

  $$A^{m \times p} \cdot B^{p \times n} = C^{m \times n}$$

- Find the value of a matrix expression involving scalar multiplication, matrix addition, subtraction, transposition, or multiplication, and the zero and identity matrices.
  (Review Exercises 33–34.)

  $$A \cdot I = I \cdot A = A \qquad A + 0 = 0 + A = A$$

- Rewrite a system of linear equations as a matrix equation $A \cdot X = B$.   *(Review Exercises 35–36.)*

- Find a matrix R so that the matrix product $R \cdot A$ is the same as the result of carrying out a given row operation or sequence of row operations on the matrix A.
  (Review Exercises 37–38.)

- Formulate an application in matrix form after making an appropriate choice for the meaning of each row and column and then find the requested quantity using the appropriate matrix arithmetic.   *(Review Exercises 39–40.)*

2.5   Inverse Matrices and Systems of Linear Equations

- Find the product of a pair of matrices to identify the pair as "a matrix and its inverse" or "not a matrix and its inverse."
  (Review Exercises 41–42.)

  $$A \cdot A^{-1} = A^{-1} \cdot A = I$$

- Row reduce $(A \mid I)$ for a square matrix A to find the inverse matrix A^{-1} or to identify A as a singular matrix.   *(Review Exercises 43–44.)*

  $$(A \mid I) \rightarrow (I \mid A^{-1})$$

- Rewrite a system of equations as a matrix equation $A \cdot X = B$ and use the inverse of A to find the solution as $X = A^{-1} \cdot B$. *(Review Exercises 45–48.)*

- Formulate an application as a collection of systems of linear equations in an appropriate number of variables after making an appropriate choice for the meaning of each, solve the collection of systems of equations by finding the inverse of the common coefficient matrix and using matrix multiplications of this inverse times the various constant term matrices, and then state the final answer in terms of the original question. *(Review Exercises 49–50.)*

2.6   Three Applications

Leontief "Open" Input-Output Models

- Find the technology matrix from an economy diagram. *(Review Exercises 51–52.)*

- Draw an economy diagram from a technology matrix. *(Review Exercises 53–54.)*

- Find the excess production of an economy with a given technology matrix and economic activity level. *(Review Exercises 55–56.)*

$$Y = (I - A) \cdot X$$

- Find the economic activity level of an economy with a given technology matrix necessary to generate a specified excess production. *(Review Exercises 57–58.)*

$$X = (I - A)^{-1} \cdot Y$$

- Represent an application as a Leontief "open" input-output model by constructing an economy diagram and the corresponding technology matrix, find the required excess production or level of economic activity, and state the final answer in terms of the original question. *(Review Exercises 59–60.)*

Markov Chains

- Find the transition matrix for a Markov chain represented by a state-transition diagram. *(Review Exercises 61–62.)*

- Identify a transition matrix as "ergodic," "regular," or "neither." *(Review Exercises 63–66.)*

- Find the steady-state distribution for a Markov chain from the transition matrix. *(Review Exercises 67–68.)*

$$X \cdot A = X$$

- Represent an application as a Markov chain by constructing a state-transition diagram and the corresponding transition matrix, find the steady-state distribution, and then interpret it in terms of the original situation. *(Review Exercises 69–70.)*

Least Squares

- Solve a consistent and independent system of linear equations by row reducing the corresponding augmented matrix, rewrite the system of equations in the matrix form $A \cdot X = B$, calculate $X = (A^t A)^{-1} \cdot (A^t B)$, and verify that this is the same solution. *(Review Exercises 71–72.)*

- Row reduce the corresponding augmented matrix for an inconsistent system of linear equations to verify that the system is inconsistent, rewrite the system of equations in the matrix form $A \cdot X = B$, calculate $X = (A^t A)^{-1} \cdot (A^t B)$, and verify that this compromise solution "almost" satisfies the original equations. *(Review Exercises 73–74.)*

- Find the least squares best approximation line $y = mx + b$ for a collection of x and y data pairs such that all the x values are distinct. *(Review Exercises 75–78.)*

$$\begin{pmatrix} x_1 & x_2 \cdots x_n \\ 1 & 1 \cdots 1 \end{pmatrix} \begin{pmatrix} x_1 & 1 \\ x_2 & 1 \\ \vdots & \vdots \\ x_n & 1 \end{pmatrix} \begin{pmatrix} m \\ b \end{pmatrix}$$

$$= \begin{pmatrix} x_1 & x_2 \cdots x_n \\ 1 & 1 \cdots 1 \end{pmatrix} \begin{pmatrix} y_1 \\ y_2 \\ \vdots \\ y_n \end{pmatrix}$$

- Use the least squares best approximation line to make a prediction from information presented in an application. *(Review Exercises 79–80.)*

Hints and Suggestions

- (*Overview*) Row operations on matrices are a generalization of the elimination method of finding the intersection of two lines, extending the technique to problems with many equations in many variables. Matrix arithmetic is similar to real number operations except that matrix multiplication is not commutative and many matrices do not have inverses. However, if a square matrix A *does* have an inverse, then the equation $AX = B$ can be solved as $X = A^{-1}B$, just as a real number linear equation can be solved by dividing. Matrices are used to represent and solve many large and complicated problems important to both science and society.

- While not every system of linear equations has a solution, they can all be identified as "consistent" or "inconsistent" and as "dependent" or "independent" by row reducing the corresponding augmented matrix. Only "consistent" equations have a solution, which may be a single collection of values for the variables or many values for the variables given in terms of one or more parameters.

- When setting up a word problem, look first for the questions "how many" or "how much" to help identify the variables. Be sure that finding values for your variables will answer the question stated in the problem. Use the rest of the given information to build equations describing facts about your variables.

- Many row reduction problems require many steps to solve, so don't give up: keep improving the matrix until it meets all the requirements to be row reduced. Make sure that each row operation you choose will move you toward your goal without undoing the parts you already have gotten the way you need.

- Do not confuse the way the technology matrix for a Leontief model is constructed (the *columns* are the input costs taken from each sector in the diagram) with the transition matrix for a Markov chain (the *rows* are the transition percentages from the states).

- Practice for test: Review Exercises 4, 6, 8, 11, 17, 20, 25, 30, 33, 35, 40, 43, 49, 51, 53, 57, 59, 61, 65, 68, 69, 75, and 79.

Review Exercises for Chapter 2   *Practice test exercises are in blue.*

2.1   Systems of Two Linear Equations in Two Variables

Represent each pair of statements as a system of two linear equations in two variables. Be sure to state clearly the meaning of your x- and y-variables.

1. "A small commuter airplane has thirty passengers" and "the ticket receipts of $3970 come from 30-day advance sale tickets at $79 and full fare tickets at $159."

2. "A cow and horse rancher has four hundred twenty animals" and "there are twice as many cows as horses."

Solve each system of equations by graphing. Identify each system as "independent and consistent," "independent and inconsistent," or "dependent."

3. $\begin{cases} x + y = 18 \\ x - y = 8 \end{cases}$

4. $\begin{cases} 2x - 4y = 36 \\ -3x + 6y = -54 \end{cases}$

Solve each system of equations by the substitution method. Identify each system as "independent and consistent," "independent and inconsistent," or "dependent."

5. $\begin{cases} 5x - 2y = 10 \\ -2x + y = 2 \end{cases}$

6. $\begin{cases} -4x + 2y = 12 \\ 2x - y = 12 \end{cases}$

Solve each system of equations by the elimination method. Identify each system as "independent and consistent," "independent and inconsistent," or "dependent."

7. $\begin{cases} x + y = 12 \\ x + 3y = 18 \end{cases}$

8. $\begin{cases} 4x + 5y = 60 \\ 2x + 3y = 42 \end{cases}$

Formulate each situation as a system of two linear equations in two variables. Be sure to state clearly the meaning of your x- and y-variables. Solve the system of equations by the elimination method. Be sure to state your final answer in terms of the original question.

9. Garden Plants A retired investment broker has rosebushes in his flower garden and tomato plants in his vegetable garden. He spends one hour each day tending his twenty-five plants. If each rosebush takes three minutes of care and each tomato plant takes two minutes, how many of each does he have?

10. Fraternity Convention Twenty-six members of the Alpha Alpha Alpha fraternity want to go to the national convention in Orlando and each has put in $10 for gas. If each car holds 5 people and uses $45 worth of gas and each van holds 8 people and uses $85 worth of gas, how many of each vehicle do they need for the trip?

2.2 Matrices and Linear Equations in Two Variables

Find the dimension of each matrix and the values of the specified elements.

11. $\begin{pmatrix} 8 & 3 & 4 \\ 1 & 5 & 9 \\ 6 & 7 & 2 \end{pmatrix}$; $a_{2,2}, a_{3,1}, a_{1,3}$

12. $\begin{pmatrix} 1 & 15 & 14 & 4 \\ 12 & 6 & 7 & 9 \\ 8 & 10 & 11 & 5 \\ 13 & 3 & 2 & 16 \end{pmatrix}$; $a_{2,3}, a_{3,2}, a_{4,1}$

Carry out the row operation on the matrix.

13. $R'2 = 3R2$ on $\begin{pmatrix} 3 & 4 & 12 \\ 1 & 2 & 2 \end{pmatrix}$

14. $R'1 = R1 + R2$ on $\begin{pmatrix} -1 & 2 & 2 \\ 2 & -3 & 6 \end{pmatrix}$

Solve each system of equations by row reducing the corresponding augmented matrix and interpreting the final reduced row-echelon form matrix as the solution. Identify each system as "independent and consistent," "independent and inconsistent," or "dependent."

15. $\begin{cases} x + 3y = 63 \\ 4x + 5y = 140 \end{cases}$ **16.** $\begin{cases} -3x + 4y = 60 \\ 2x - y = 10 \end{cases}$

17. $\begin{cases} 12x - 4y = 36 \\ -15x + 5y = 45 \end{cases}$

18. $\begin{cases} -16x + 12y = -48 \\ 20x - 15y = 60 \end{cases}$

Formulate each situation as a system of two linear equations in two variables. Be sure to state clearly the meaning of your x- and y-variables. Solve the system of equations by row reducing the corresponding augmented matrix. Be sure to state your final answer in terms of the original question.

19. Pharmaceuticals The pharmacist at the Charter Drug Shop filled ninety-two prescriptions today for antibiotics and cough suppressants. If there were thirty-four more prescriptions for antibiotics than for cough suppressants, how many prescriptions for each were filled?

20. Classic Magazines A used book store offers grab bag packages of old *Life* and *The New Yorker* magazines from the 1940s containing 4 copies of *Life* and 3 copies of *The New Yorker* for $39 and larger bags of 12 copies of *Life* and 10 copies of *The New Yorker* for $122. What is the price of one copy of each old magazine?

2.3 Matrix Row Reduction and Systems of Linear Equations

Interpret each reduced row-echelon form matrix as the solution of a system of equations. Identify each system as "independent" or "dependent" and as "consistent" or "inconsistent."

21. $\begin{pmatrix} 1 & 0 & 0 & 3 \\ 0 & 1 & 0 & -3 \\ 0 & 0 & 1 & 6 \end{pmatrix}$ **22.** $\begin{pmatrix} 1 & 1 & 0 & 4 \\ 0 & 0 & 1 & 2 \\ 0 & 0 & 0 & 0 \end{pmatrix}$

Use an appropriate row operation or sequence of row operations to find the equivalent reduced row-echelon form matrix.

23. $\begin{pmatrix} 0 & 1 & 1 & -1 \\ 1 & 0 & 1 & 6 \\ 2 & 0 & 1 & 10 \end{pmatrix}$ **24.** $\begin{pmatrix} 1 & 0 & 1 & 1 & 4 \\ 5 & 1 & 1 & 4 & 12 \\ 2 & 1 & 0 & 1 & 5 \\ 3 & 0 & 1 & 3 & 8 \end{pmatrix}$

Solve each system of equations by row reducing the corresponding augmented matrix. Identify each system as "independent" or "dependent" and as "consistent" or "inconsistent."

 If permitted by your instructor, you may use a graphing calculator.

25. $\begin{cases} x_1 - 2x_3 = 2 \\ -x_1 + x_2 + 2x_3 = 1 \\ -x_1 + 2x_2 + 3x_3 + x_4 = 7 \\ x_1 - 2x_3 + x_4 = 4 \end{cases}$

26. $\begin{cases} x_1 - x_2 + x_4 = 3 \\ x_2 + x_3 = 3 \\ 2x_1 - x_2 + x_3 + 2x_4 = 9 \\ 2x_1 - x_2 + x_3 + x_4 = 5 \end{cases}$

27. $\begin{cases} x_1 + 2x_3 + x_4 = 3 \\ x_1 + x_2 + 3x_3 + x_4 = 2 \\ 3x_1 + 3x_2 + 9x_3 + 4x_4 = 7 \\ 2x_1 + 4x_3 + 2x_4 = 7 \end{cases}$

28. $\begin{cases} x_1 + x_2 + x_3 + x_4 = 2 \\ x_1 + x_2 + x_3 + x_4 = 1 \\ x_1 + x_2 + x_3 + x_4 = 0 \\ x_1 + x_2 + x_3 + x_4 = 3 \end{cases}$

Formulate each situation as a system of linear equations in an appropriate number of variables and be sure to state clearly the meaning of each. Solve the system of equations by row reducing the corresponding augmented matrix. Be sure to state your final answer in terms of the original question.

 If permitted by your instructor, you may use a graphing calculator.

29. Nursery Management   The Nyack Nursery starts plants from seeds and sells potted plants to garden stores for resale. Each dahlia costs 16¢ to start and needs a 10¢ flowerpot and 5 ounces of soil; each chrysanthemum costs 11¢ to start and needs a 12¢ flowerpot and 6 ounces of soil; and each daisy costs 13¢ to start and needs an 8¢ flowerpot and 5 ounces of soil. If the nursery has $50 to spend on starting the plants, $48 to spend on flowerpots, and 153 pounds of potting soil, how many of each plant can it raise using all of the available resources?

30. Family Entertainment   The Family Fun Center in Asheville offers the package specials listed in the table for an afternoon of fun at its go-kart track, miniature golf course, house of funny mirrors, and snack stand serving hot dogs and sodas. How much does one hot dog cost? How much does a go-kart ride cost if the mirror house is $1.50 and sodas are $1?

Go-kart	Golf	Mirror House	Hot Dog	Soda	Package Price
2	1	1	1	1	$15
3	2	1	2	1	$23
4	2	2	3	2	$31

2.4   Matrix Arithmetic

Find each matrix product.

31. $\begin{pmatrix} 1 & 2 & 3 & 4 \\ 4 & 3 & 2 & 1 \end{pmatrix} \begin{pmatrix} 1 & 0 \\ 0 & -1 \\ 1 & 0 \\ 0 & -1 \end{pmatrix}$

32. $\begin{pmatrix} 1 & 0 \\ 0 & -1 \\ 1 & 0 \\ 0 & -1 \end{pmatrix} \begin{pmatrix} 1 & 2 & 3 & 4 \\ 4 & 3 & 2 & 1 \end{pmatrix}$

Use the given matrices to find each matrix expression.

$$A = \begin{pmatrix} -1 & 2 & -1 \\ 2 & -1 & 2 \end{pmatrix} \quad B = \begin{pmatrix} 3 & 6 & 8 \\ 7 & 5 & 4 \end{pmatrix}$$

$$C = \begin{pmatrix} 2 & 1 & 2 \\ 1 & -2 & 1 \\ 2 & 1 & 2 \end{pmatrix} \quad D = \begin{pmatrix} 5 & 8 \\ 7 & 6 \end{pmatrix}$$

33. $3D - A \cdot B^t + I$ **34.** $3I + A^t \cdot B - C$

Rewrite each system of linear equations as a matrix equation $A \cdot X = B$.

35. $\begin{cases} x_1 + 4x_2 + x_3 = 15 \\ 2x_1 + 8x_2 + 3x_3 = 26 \\ x_1 + 5x_2 + 2x_3 = 17 \end{cases}$

36. $\begin{cases} 2x_1 + 3x_2 - x_3 + x_4 = 20 \\ 5x_1 + 4x_2 + x_3 + 2x_4 = 35 \\ 2x_1 + x_2 + x_3 + x_4 = 12 \end{cases}$

For each row operation (or sequence of row operations), find a 3×3 matrix R so that the matrix product $R \cdot A$ is the same as the result of carrying out the row operation(s) on the matrix:

$$A = \begin{pmatrix} 2 & 5 & 23 \\ 1 & 3 & 13 \\ 2 & 4 & 20 \end{pmatrix}$$

37. $R'1 = R1 - R2$

38. $R'1 = R2$ and $R'2 = R1$,
$R'2 = R2 - R3$, and
$R'3 = R3 - 2R1$

Formulate each situation in matrix form. Be sure to indicate the meaning of your rows and columns. Find the requested quantities using the appropriate matrix arithmetic.

39. Growing Grandchildren   A proud grandmother's record book show that her grandson Thomas is now 61 inches tall and weighs 90 pounds while last year he was 58 inches tall and weighed 80 pounds; her grandson Richard is now 54 inches tall and weighs 75 pounds while last year he was 52 inches tall and weighed 70 pounds; and her granddaughter Harriet is now 47 inches tall and weighs 60 pounds while last year she was 46 inches tall and weighed 55 pounds. Represent these facts as a "this year" matrix and a "last year" matrix. Use these matrices to find how much each grandchild grew.

40. Spring Fashions   The buyer for the ladies sportswear division of a large department store needs 200 jackets, 300 blouses, 250 skirts, and 175 pairs of slacks. The spring lines shown by both an East Coast designer and an Italian team are acceptable to her but the East Coast designer wants $195 for each jacket, $85 for each blouse, $145 for each skirt, and $130 for each pair of slacks while the Italian company wants $190 for each jacket, $90 for each blouse, $150 for each skirt, and $125 for each pair of slacks. Represent her needs as a column matrix and the prices as a price matrix. Use these matrices to find the cost of her order from each source.

2.5   *Inverse Matrices and Systems of Linear Equations*

Find each matrix product. Identify each pair of matrices as "a matrix and its inverse" or "not a matrix and its inverse."

41. $\begin{pmatrix} 1 & 2 & 3 \\ 1 & 1 & 1 \\ 0 & 1 & 3 \end{pmatrix}$ and $\begin{pmatrix} -2 & 3 & 1 \\ 3 & -3 & -2 \\ -1 & 1 & 1 \end{pmatrix}$

42. $\begin{pmatrix} -3 & 0 & 1 \\ 1 & 3 & 1 \\ -3 & 2 & 2 \end{pmatrix}$ and $\begin{pmatrix} -4 & -2 & 3 \\ 5 & 3 & -4 \\ -11 & -6 & 8 \end{pmatrix}$

Row reduce $(A \mid I)$ for each matrix A to find the inverse matrix A^{-1} or to identify A as a singular matrix.

43. $\begin{pmatrix} 1 & 1 & 0 \\ 0 & -3 & 1 \\ 2 & 3 & 0 \end{pmatrix}$ **44.** $\begin{pmatrix} 1 & 0 & 1 \\ 1 & 1 & 0 \\ 2 & 1 & 1 \end{pmatrix}$

Rewrite each system of equations as a matrix equation $A \cdot X = B$ and use the inverse of A to find the solution. Be sure to check the solution in the original system of equations.

45. $\begin{cases} 13x_1 + 4x_2 = 33 \\ 3x_1 + x_2 = 8 \end{cases}$ **46.** $\begin{cases} 3x_1 + 8x_2 = 25 \\ 2x_1 + 5x_2 = 16 \end{cases}$

47. $\begin{cases} 5x_1 + x_2 + 2x_3 = 11 \\ 2x_1 + 2x_2 + x_3 = 7 \\ 2x_1 + x_2 + x_3 = 5 \end{cases}$

48. $\begin{cases} 3x_1 + 2x_2 + x_3 + 2x_4 = 7 \\ 2x_1 + 5x_2 + 2x_3 + 2x_4 = 10 \\ x_1 + 2x_2 + x_3 + x_4 = 4 \\ 2x_1 + 2x_2 + x_3 + 2x_4 = 6 \end{cases}$

Formulate each situation as a collection of systems of linear equations. Be sure to state clearly the meaning of each variable. Solve each collection of systems of equations by finding the inverse of the common coefficient matrix and using matrix multiplications of this inverse times the various constant term matrices. Be sure to state your final answer in terms of the original question.

If permitted by your instructor, you may use a graphing calculator.

49. Retail Displays   A chain of furniture stores sells living room and bedroom suites and displays them in the store windows and inside on the showroom floor. A living room suite window display requires 3 square yards of window space and then the same furniture group is allowed 7 square yards of showroom space, while a bedroom suite window display requires 4 square yards of window space and then the same furniture group is allowed 9 square yards of showroom space. For furniture styles not given window space but shown only on the showroom floor, each living room suite is allowed 8 square yards and each bedroom suite is allowed 10 square yards. The setup costs per square yard are $50 for furniture shown in the window and on the showroom floor, while for furniture shown only on the showroom floor, it is $60 for living room suites and $70 for bedroom suites. The table lists the available space (in square yards) at the chain's stores in Kingman, Prescott, and Holbrook, together with the approved setup budgets and the amount of the total showroom space that must be given to furniture not displayed in the window. How many different living room and bedroom suites can be shown at each location?

	Window Space	Showroom Floor	Floor Space for Nonwindow Suites	Setup Budget
Kingman	18	131	90	$8850
Prescott	21	182	134	$12,190
Holbrook	17	187	148	$12,680

50. **Income Taxes**   Find the federal, state, and city income taxes on the taxable incomes of the individuals listed in the table if the federal income tax is 20% of the taxable income after first deducting the state and city taxes; the state income tax is 10% of the taxable income after first deducting the federal and city taxes; and the city income tax is 5% of the taxable income after first deducting the federal and state taxes.

Taxpayer	Mr. Dahlman	Mrs. Farrell	Ms. Mazlin	Mr. Seidner
Taxable income	$96,700	$48,350	$77,360	$145,050

2.6   Three Applications

Leontief "Open" Input–Output Models
Find the technology matrix for each economy diagram.

51.

52.
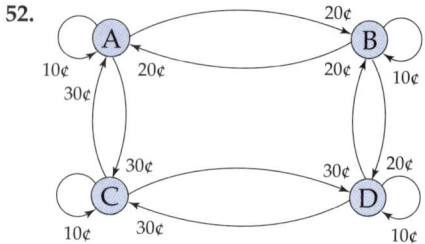

Draw an economy diagram for each technology matrix.

53. $\begin{pmatrix} 0.10 & 0 & 0.20 \\ 0.10 & 0.15 & 0 \\ 0 & 0.15 & 0.20 \end{pmatrix}$

54. $\begin{pmatrix} 0.15 & 0.10 & 0 & 0.15 \\ 0 & 0.10 & 0.20 & 0 \\ 0.15 & 0 & 0.10 & 0 \\ 0 & 0.10 & 0 & 0.15 \end{pmatrix}$

Find the excess production Y of each economy with technology matrix A and economic activity level X.

55. $A = \begin{pmatrix} 0.10 & 0.05 & 0.15 \\ 0.15 & 0.10 & 0.05 \\ 0.10 & 0.05 & 0.10 \end{pmatrix}$, $X = \begin{pmatrix} 540 \\ 620 \\ 560 \end{pmatrix}$

56. $A = \begin{pmatrix} 0.15 & 0.15 & 0.10 & 0.05 \\ 0.10 & 0.10 & 0.05 & 0.10 \\ 0.10 & 0.05 & 0.10 & 0.15 \\ 0.05 & 0.10 & 0.15 & 0.10 \end{pmatrix}$, $X = \begin{pmatrix} 280 \\ 200 \\ 260 \\ 240 \end{pmatrix}$

Find the economic activity level X for each economy with technology matrix A necessary to generate excess production Y.

57. $A = \begin{pmatrix} 0.10 & 0.35 \\ 0.40 & 0.15 \end{pmatrix}$, $Y = \begin{pmatrix} 175 \\ 225 \end{pmatrix}$

58. $A = \begin{pmatrix} 0.10 & 0.10 & 0.20 \\ 0.30 & 0.20 & 0.10 \\ 0.10 & 0.30 & 0.20 \end{pmatrix}$, $Y = \begin{pmatrix} 245 \\ 294 \\ 196 \end{pmatrix}$

Represent each situation as a Leontief "open" input-output model by constructing an economy diagram and the corresponding technology matrix. Find the required excess production or level of economic activity and be sure to state your final answer in terms of the original question.

59. A 5&10 Problem   Each of the four divisions of the Woolworth Corporation depends on the others in the following way: each dollar of production from each division requires 10¢ of production from that division and 5¢ of production from each of the others. How much must each division produce to yield an excess production of $3 million from each?

60. Energy Dependence   As a result of treaty obligations, the domestic and foreign oil consumption of an industrialized nation is linked to the protection provided by its military forces. Each dollar of domestic oil produced requires 5¢ of domestic oil and 10¢ of military protection; each dollar of foreign oil requires 20¢ of military protection; and each dollar of military protection consumes 10¢ of domestic oil, 10¢ of foreign oil, and 5¢ of military protection. If the domestic oil production level is $940 million, the foreign production is $2000 million, and the military protection budget is $520 million, how much of each can be used elsewhere in the nation's economy?

Markov Chains
Find the transition matrix of each Markov chain represented by the state-transition diagram.

61.

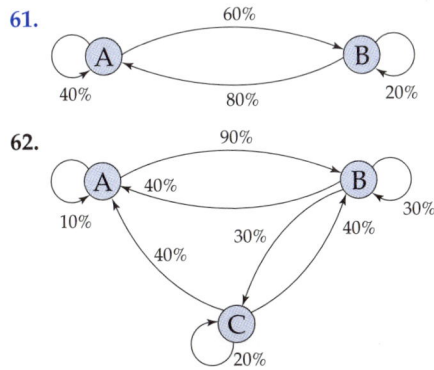

62.

For Exercises 63–66, identify each transition matrix as "ergodic," "regular," or "neither."

63. $\begin{pmatrix} 0.30 & 0.70 \\ 0.70 & 0.30 \end{pmatrix}$

64. $\begin{pmatrix} 0.30 & 0.70 \\ 0 & 1.00 \end{pmatrix}$

65. $\begin{pmatrix} 0.30 & 0.20 & 0.50 \\ 0.50 & 0.50 & 0 \\ 0.20 & 0.50 & 0.30 \end{pmatrix}$

66. $\begin{pmatrix} 0.10 & 0.10 & 0.80 \\ 0.70 & 0.20 & 0.10 \\ 0.20 & 0.60 & 0.20 \end{pmatrix}$

 Find the steady state distribution for each Markov chain with the given transition matrix.

67. $\begin{pmatrix} 0.10 & 0.20 & 0.70 \\ 0.20 & 0.30 & 0.50 \\ 0.30 & 0.40 & 0.30 \end{pmatrix}$

68. $\begin{pmatrix} 0 & 0.20 & 0.20 & 0.60 \\ 0.20 & 0 & 0.60 & 0.20 \\ 0.20 & 0.60 & 0 & 0.20 \\ 0.60 & 0.20 & 0.20 & 0 \end{pmatrix}$

Represent each situation as a Markov chain by constructing a state-transition diagram and the corresponding transition matrix. Find the steady-state distribution and interpret it in terms of the original situation. Be sure to state your final answer in terms of the original question.

69. Customer Satisfaction   Surveys of customer satisfaction with the repair service departments of dealers for a major car manufacturer show that of those rated "below average" one year, the next year 10% will remain that way, 70% will improve to "satisfactory," and the remaining 20% will improve to "excellent"; of those rated "satisfactory" one year, the next year 60% will remain that way, 30% will improve to "excellent," and the remaining 10% will slip to "below average"; and of those rated "excellent" one year, the next year 70% will remain that way and the remaining 30% will slip to "satisfactory." Assuming this pattern has repeated for many years, how many of the manufacturer's 2655 dealers nationwide have service departments rated "excellent" by their customers?

70. Weather   The old-timers at Edna's Cafe in downtown Nora Springs will tell you that the weather just keeps getting better and better for growing corn. In fact, if it was bad last year, there is a 20% chance it will now be terrific and a 60% chance it will be great this year; if last year was great, there is a 50% chance it will now be terrific and a 40% chance it will be great again this year; and if it was terrific last year, there is a 30% chance it will be that way again this year and a 60% chance it will be great this year. Of course, they don't bother to mention that it might be bad this year because that might change their luck. If they are right, how many of every 18 years will have terrific weather for growing corn?

Least Squares

 Solve each system of equations by row reducing the corresponding augmented matrix. Rewrite the system of equations in the matrix form $A \cdot X = B$, calculate $X = (A^t A)^{-1} \cdot (A^t B)$, and verify that this is the same solution.

71. $\begin{cases} x - y = 1 \\ 3x + y = 15 \\ x + 2y = 10 \end{cases}$ **72.** $\begin{cases} 4x + 3y = 12 \\ x + y = 5 \\ -2x + y = 14 \end{cases}$

 Row reduce the corresponding augmented matrix for each system of equations to verify that the system is inconsistent. Rewrite the system of equations in the matrix form $A \cdot X = B$, calculate $X = (A^t A)^{-1} \cdot (A^t B)$, and verify that this compromise solution "almost" satisfies the original equations.

73. $\begin{cases} x - 3y = -9 \\ 2x - y = 12 \\ x + 2y = 6 \end{cases}$ **74.** $\begin{cases} x + 3y = 3 \\ 3x - y = 9 \\ x + y = -9 \end{cases}$

 Find the least squares best approximation line $y = mx + b$ for each collection of x and y data pairs.

75.

x	2	3	4
y	22	32	36

76.

x	3	5	6
y	238	364	462

77.

x	2	3	5	6
y	40	50	80	90

78.

x	10	11	12	13	14
y	140	150	180	200	210

 Use the least squares best approximation line to make each prediction. Be sure to state your final answer in terms of the original question.

79. Store Hours   The owners of a "Mom and Pop" corner store have had their store open various numbers of hours on recent Mondays, the usual day off. The table shows the number of hours the store was open and the sales receipts for the day. How much could they expect to sell next Monday if they keep their store open for 12 hours?

Hours open	6	8	10	14
Total sales	$2230	$3035	$3770	$5065

80. Production Accidents   The supervisor of a factory assembly line gathered the data in the table over the last year to compare the number of minor accidents each month with the number of extra five-minute mini-breaks she allows during the day. How many minor accidents could she expect next month if she allows four mini-breaks during the day?

Mini-breaks	0	2	6	8
Minor accidents	21	16	8	3

Projects and Essays

The following projects and essays are based on Chapter 2. Most have no right and wrong answers—the results depend only on your imagination and resourcefulness.

1. Go to a library (especially a business or technical library) and browse through the last several issues of the journals *Management Science* or *Operations Research*. Write a one-page report on the use of matrices to analyze a real-world problem.

2. Read the essay "Gauss, the Prince of Mathematicians" by Eric Temple Bell in Volume One of *The World of Mathematics* by James R. Newman (Simon and Schuster, New York: 1956), pages 295–339, and write a one-page report on what you learn.

3. Read the essay "Invariant Twins, Cayley and Sylvester" by Eric Temple Bell in Volume One of *The World of Mathematics* by James R. Newman (Simon and Schuster, New York: 1956), pages 341–365, and write a one-page report on what you learn.

4. Read the article "The Structure of the U.S. Economy" by Wassily W. Leontief on pages 25–35 of the April 1965 issue of *Scientific American* (Volume 212, Number 4), and write a one-page report on what you find.

5. Read the article "The World Economy of the Year 2000" by Wassily W. Leontief on pages 206–231 of the September 1980 issue of *Scientific American* (Volume 243, Number 3), and write a one-page report on the progress made since this article was written on reducing the gap between the rich and poor of the world.

6. Find examples of matrices representing "independent" and "dependent" systems of equations and representing "consistent" and "inconsistent" systems of equations that are different from those given on page 201.

7. Make up your own example of a Leontief "open" input-output economic model and explore how each one-dollar change in the production level of the first product changes the total excess production.

8. Find an example of an inconsistent system of equations $A \cdot X = B$ such that the square matrix $A^t \cdot A$ is not invertible.

9. Show by example that if the x and y data for a least squares best approximation line contains two pairs with the same x-value, then the matrix $A^t \cdot A$ is not invertible.

10. If the statistical commands for your graphing calculator will automatically find the least squares best approximation line, compare the calculator answer to your matrix solution of Exercise 56 on page 268.

3 LINEAR PROGRAMMING

An assembly line can work only if enough pieces and workers are always ready at the correct times and places. Such management problems can be solved by the method of linear programming, as explained in this chapter.

3.1 Linear Inequalities

Northwest Airlines and Crew Schedules

Northwest Airlines spends over $1 billion each year on pilot and flight attendant salaries, benefits, overtime, and other expenses. As part of the airline's direct operating costs, these crew costs are exceeded only by airplane fuel expenditures. But unlike fuel costs, the airline chooses how it uses its employees. If the available personnel could be used more effectively, a minor improvement of just 1% would represent a $10 million savings that year. For this reason, Northwest Airlines employs mathematicians and computer analysts in its schedule development department and provides them with the latest supercomputers. Since the airline's flight schedule changes monthly, the scheduling department is never out of work. Its objective is to minimize the crew costs while adequately staffing the scheduled flights and satisfying Federal Aviation Agency regulations and union contract requirements. These constraints may be expressed as inequalities: each flight must have at least a pilot and copilot in the cockpit, each crew member can fly for no more than a given number of hours and must have at least a corresponding number of rest hours each day, each crew member's flight plans must form a round trip back to the starting airport, and so on. The solution of each month's problem is found in two stages: finding a "feasible solution" that provides a crew for each flight while satisfying all of the constraints, and then searching for ways to alter this initial solution to lower the costs while still respecting all of the restrictions. While the large size of Northwest Airline's scheduling problem makes it impossible to find the "best" solution even using today's computers, the scheduling department's solutions (found using the methods described in this chapter) are sufficiently close to the best to provide the company with significant savings.

Introduction

This chapter describes and solves a large class of problems known as *linear programming problems*. These problems and their solutions by the *simplex method* are the cornerstone of the science of *operations research*. Many problems in modern business may be addressed by these methods. Although the simplex method is an algebraic procedure, it is based on the geometry of linear inequalities, where we now begin.

Inequalities

On pages 4–5 we discussed inequalities such as $x < y$ ("x is less than y") and $x \geq y$ ("x is greater than or equal to y" or equivalently "x is at least y"). Many everyday relationships are easily expressed using inequalities.

EXAMPLE 1   Writing an Inequality in Algebraic Notation

Express the statement "Joe is richer than Fred" as an inequality.

Solution

Since this means that "Joe has more money than Fred," we could write $J > F$ where J stands for Joe's money and F represents Fred's money. The statement could also be rephrased "Fred has less money than Joe" and then written as $F < J$. Since a number minus a smaller number is positive, we might even write that $J - F > 0$, or, looking at it from the other viewpoint, $F - J < 0$. From this we see that every inequality may be written in several different ways.

∎

Since inequalities with two variables can be understood in the same manner as one-variable inequalities, we start with the simpler situation.

Inequalities in One Variable

To *sketch* an inequality means to graph the points that satisfy the inequality. The *boundary* of an inequality is the corresponding *equality*. For example, the boundary of the inequality $x \geq 2$ is the equality $x = 2$. On the number line, $x = 2$ represents just a single point, and the numbers satisfying $x \geq 2$ are all on the same side of $x = 2$. That is, we sketch $x \geq 2$ on the number line in three steps:

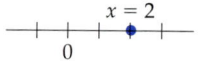

1. Draw the boundary

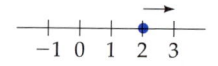

2. $x > 2$ is on the *right* side of the boundary

3. Shade the correct side of the boundary

You can also find the correct side of the boundary by checking the inequality at a "test point," substituting a number for the variable. For example, given the inequality $x \geq 2$, substituting 5 for x gives $5 \geq 2$, which is *true*, so the point $x = 5$ is on the *correct* side of $x \geq 2$. On the other hand, substituting -3 for x gives $-3 \geq 2$, which is *false*, so $x = -3$ is on the *wrong* side of $x \geq 2$. Any one point on either side will do, since finding the "wrong" side just means that the other side is "right."

The set of all the points satisfying the inequality is called the *feasible region* of the inequality. To graph the feasible region, sketch the inequality using the following three steps.

How to Sketch a Linear Inequality

1. Draw the boundary.
2. Choose the side of the boundary corresponding to the inequality (and verify it with a "test point").
3. Shade the correct side to show the feasible region.

A *system* of linear inequalities is two or more linear inequalities joined by a brace, {, meaning that *each* of the inequalities holds. The feasible region of a system is the set of points where *each* of the inequalities holds, and may be graphed using the same three steps. For example, the numbers x on the real number line satisfying the system $\begin{cases} -1 \leq x \\ x \leq 2 \end{cases}$ are found as follows.

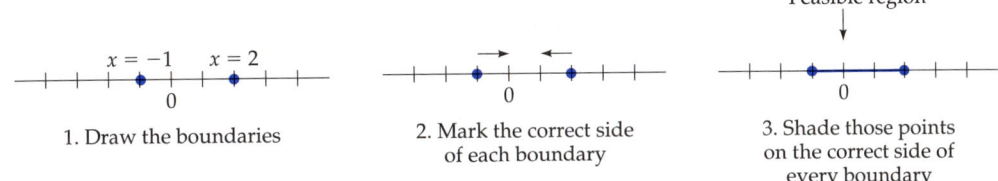

1. Draw the boundaries

2. Mark the correct side of each boundary

3. Shade those points on the correct side of every boundary

Since there *are* points between $x = -1$ and $x = 2$, the system of linear inequalities $-1 \leq x$ and $x \leq 2$ is said to be *feasible*. If there are *no* feasible points, then the system is *infeasible*. For example, there are no points such that $x \geq 3$ and $x \leq 1$:

1. Draw the boundaries

2. Mark the correct side of each boundary

$x \geq 3$ and $x \leq 1$ is INFEASIBLE

3. Because there are no points that are both to the right of $x = 3$ and also to the left of $x = 1$

Two-Variable Inequalities

Inequalities in two variables can be sketched by the same method, but now in the *plane* rather than on the number line.

EXAMPLE 2　Sketching a Linear Inequality in the Plane

Sketch the linear inequality $2x + 3y \geq 12$.

Solution

The boundary is the line $2x + 3y = 12$. Such a line can be sketched by plotting the x- and y-intercepts. Since the x-intercept has $y = 0$, the equation becomes $2x + 0 = 12$, so $x = 12/2 = 6$ and the x-intercept is $(6, 0)$. Similarly, since the y-intercept has $x = 0$, the equation becomes $0 + 3y = 12$, so $y = 12/3 = 4$ and the y-intercept is $(0, 4)$. Continuing as before:

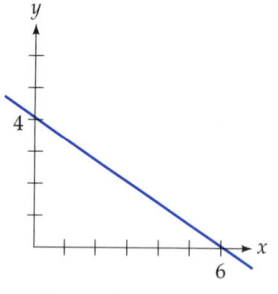

1. Draw the boundary

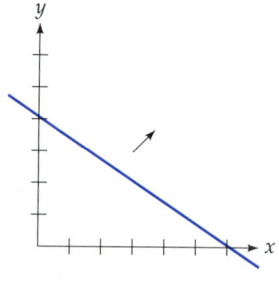

2. Since $2 \cdot 0 + 3 \cdot 0$ is not ≥ 12, the origin $(0, 0)$ is *not* on the correct side of the boundary

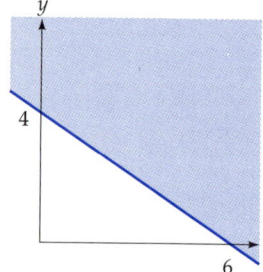

3. Shade the correct side of the boundary

In general:

> The boundary of the linear inequality $ax + by \leq c$ (or $\geq c$) is the line $ax + by = c$ with x-intercept at $\dfrac{c}{a}$ and y-intercept at $\dfrac{c}{b}$.

Notice that if the ratio c/a or c/b does not exist, then neither does the corresponding intercept. For example, the boundary of the inequality $1x + 0y \geq 2$ is the line $x = 2$. This vertical line does not have a y-intercept and the ratio "2/0" does not exist (since division by zero is not defined).

If $c = 0$, then the boundary line $ax + by = 0$ passes through the origin $(0, 0)$ and the points $(b, -a)$ and $(-b, a)$.

For example, the boundary of $2x - 3y \geq 0$ is the line $2x - 3y = 0$, which passes through the origin $(0, 0)$ and the points $(-3, -2)$ and $(3, 2)$. Any two of these points will suffice to sketch the boundary line.

Graphing Calculator Exploration

You can view any nonvertical boundary line $ax + by = c$ by entering it as $y = (c - ax)/b$. For example, to see the boundary line $2x + 3y = 12$ from Example 2:

a. Enter the boundary line as $y_1 = (12 - 2x)/3$ and graph it on the window $[-5, 10]$ by $[-5, 10]$.

b. Use TRACE or EVALUATE to check the x- and y-intercepts.

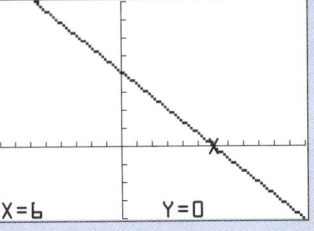

PRACTICE PROBLEM 1   Sketch the linear inequality $3x - 5y \leq 60$.   *Solution at the back of the book*

EXAMPLE 3   Sketching a System of Linear Inequalities

a. Sketch the system $\begin{cases} x + y \geq 3 \\ x - y \geq 3 \\ x \leq 6 \end{cases}$.

b. Is this system feasible?

c. What happens if the last inequality ($x \leq 6$) is replaced by $x \leq 2$?

Solution

a. The boundaries are the lines $x + y = 3$ (with intercepts $(3, 0)$ and $(0, 3)$), $x - y = 3$ (with intercepts $(3, 0)$ and $(0, -3)$), and $x = 6$ (a vertical line with x-intercept $(6, 0)$ and no y-intercept). The origin $(0, 0)$ does not satisfy either $x + y \geq 3$ or $x - y \geq 3$ but it does satisfy $x \leq 6$.

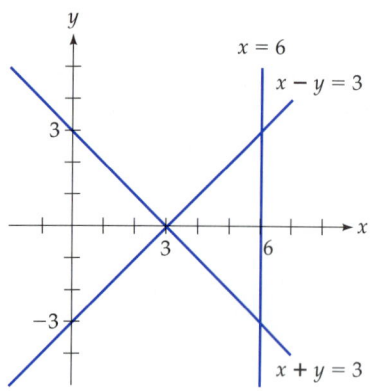

1. Draw the boundary lines

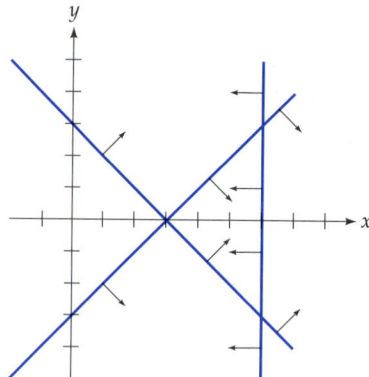

2. Mark the correct sides

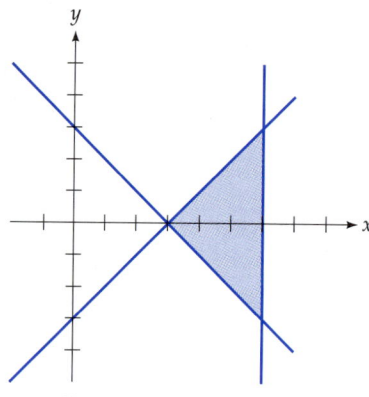

3. Shade the feasible region

b. Since there are points that lie on the correct sides of all the boundaries, this system of linear inequalities *is* feasible.

c. If we replace the third inequality ($x \leq 6$) by $x \leq 2$, the vertical
boundary line $x = 6$ is replaced by the vertical boundary line
$x = 2$:

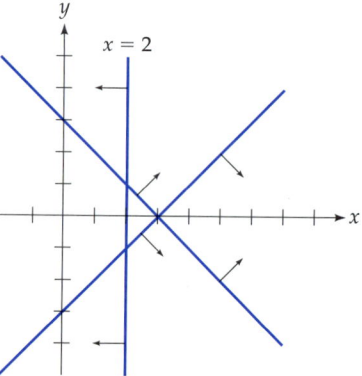

This new system of linear inequalities is *infeasible* because there are no
points that lie on the correct sides of all the boundaries at once (as you
can see from the sketch).

◼

PRACTICE PROBLEM 2   Sketch the system $\begin{cases} x + 2y \leq 20 \\ x + y \geq 10 \\ x \leq 10 \end{cases}$. Is this system feasible?

Solution at the back of the book

Convex Regions and Vertices

A region is *convex* if it contains the line segment joining any two points in the region. Since any linear inequality represents just one side of the boundary line, the feasible region of any linear inequality is convex. Thus *the feasible region of any system of linear inequalities is convex* because given any two points in the region, those two points and the line segment between them are on the correct side of each of the boundary lines, and so the line segment is also in the region.

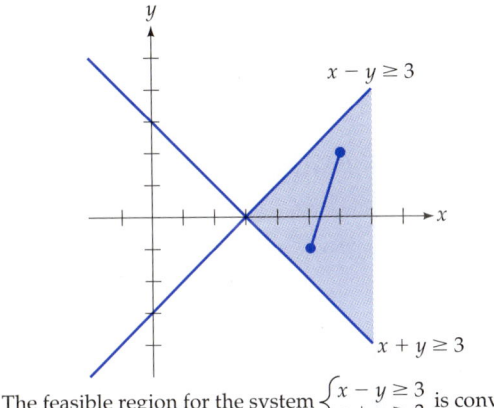

The feasible region for the system $\begin{cases} x - y \geq 3 \\ x + y \geq 3 \end{cases}$ is convex.

On the other hand, the following region is *not* convex because the line segment between (1, 1) and (5, 1) does *not* lie completely inside the region.

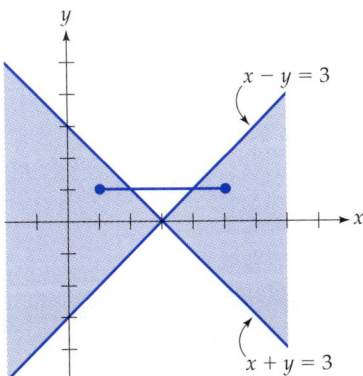

This region is *not* convex and so *cannot* be a feasible region.

The "corners" of the feasible region are called *vertices*. More precisely:

> A *vertex* of a system of linear inequalities is an intersection point of two (or more) of the boundaries that satisfies all of the inequalities.

Each vertex may be found by solving one of the boundaries for x or y in terms of the other variable, substituting this expression into the other boundary, solving for the value of the remaining variable, and then finding the value of the first variable. If this point satisfies all of the inequalities, then it is a vertex of the region.

EXAMPLE 4   Finding the Vertices of a Feasible Region

Sketch the system $\begin{cases} x + 2y \geq 12 \\ x - y \geq 0 \\ x \leq 8 \end{cases}$ and find the vertices of the feasible region.

Solution

The boundaries are the lines $x + 2y = 12$ (with intercepts (12, 0) and (0, 6)), $x - y = 0$ (with intercept (0, 0) and passing through (1, 1)), and $x = 8$ (a vertical line with x-intercept (8, 0) and no y-intercept). The origin does not satisfy $x + 2y \geq 12$, the point (12, 0) satisfies $x - y \geq 0$, and the origin satisfies $x \leq 8$.

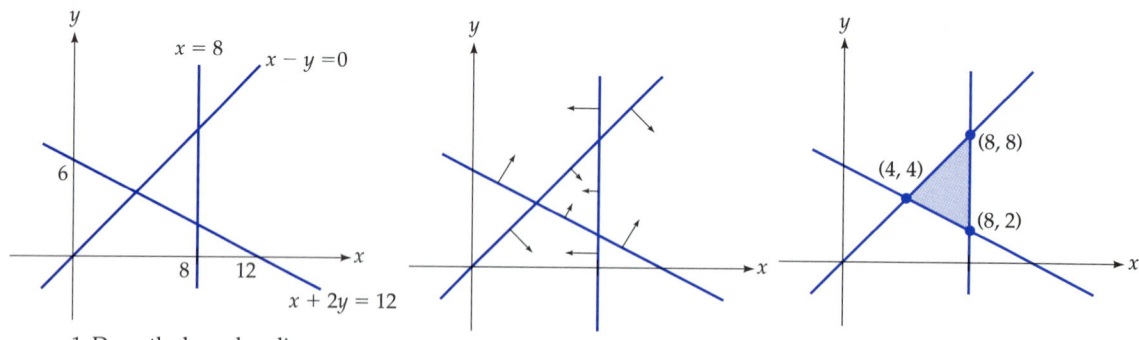

1. Draw the boundary lines   2. Mark the correct sides   3. Shade the feasible region

Two vertices of this region lie on the vertical line $x = 8$. Substituting $x = 8$ into $x + 2y = 12$ gives $(8) + 2y = 12$ so $y = 2$ and the intersection point is (8, 2). Substituting $x = 8$ into $x - y = 0$ gives $(8) - y = 0$

so $y = 8$ and the intersection point is (8, 8). The third vertex is the intersection of $x + 2y = 12$ and $x - y = 0$:

$x = 12 - 2y$	Solve $x + 2y = 12$ for x
$(12 - 2y) - y = 0$	Substitute into $x - y = 0$
$12 - 3y = 0$ means $12 = 3y$ so $y = 12/3 = 4$	Simplify and solve for y
$x = 12 - 2(4) = 12 - 8 = 4$	$x = 12 - 2y$ with $y = 4$
(4, 4)	Vertex at $x = 4$, $y = 4$

The vertices of this region are the points (8, 2), (8, 8), and (4, 4), as shown on page 288.

■

A region is *bounded* if it can be completely contained inside a rectangular region of the form $L \le x \le R$ and $B \le y \le T$ for some values of L, R, B, and T (for "left," "right," "bottom," and "top," respectively). A region is *unbounded* if it is not bounded. Of course, for a bounded region there are many possible choices for the values of L, R, B, and T. The region in Example 4 is bounded because all of the x-values are between 2 and 10 and all the y-values are between 1 and 10.

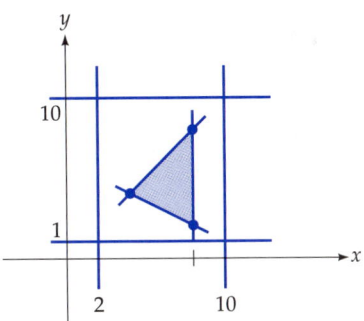

PRACTICE PROBLEM 3

Find the vertices of the feasible region for the system $\begin{cases} x + 2y \le 20 \\ x + y \ge 10 \\ x \le 10 \end{cases}$

from Practice Problem 2. Is this region bounded?

Solution at the back of the book

Two Typical Regions

In many applications, the variables represent quantities of materials and thus may not take negative values. The conditions $x \ge 0$ and $y \ge 0$ are *nonnegativity constraints* and force the feasible region to be in the first quadrant. The next two examples display typical feasible regions.

EXAMPLE 5　Finding the Feasible Region for a Word Problem

A small jewelry company prepares and mounts semiprecious stones. There are 10 lapidaries (who cut and polish the stones) and 12 jewelers (who mount the stones in gold settings). Each employee works 7 hours each day. Each tray of agates requires 5 hours of cutting and polishing and 4 hours of mounting, while each tray of onyxes requires 2 hours of cutting and polishing and 3 hours of mounting. How many trays of each stone can be processed each day?

Formulate this situation as a system of linear inequalities, sketch the feasible region, and find the vertices.

Solution

Clearly, there are many different quantities of stones that this company could process each day: it could choose to do nothing and give the employees the day off, it could process just one kind of stone, or it could process some combination. But it could not exceed the amount of time the lapidaries could work (10 workers at 7 hours each is 70 work-hours) nor the amount of time the jewelers could work (12 workers at 7 hours each is 84 work-hours). Nor could it process a negative number of trays.

Let

$$x = \left(\begin{array}{c}\text{Number of trays} \\ \text{of agates}\end{array}\right) \text{ and } y = \left(\begin{array}{c}\text{Number of trays} \\ \text{of onyxes}\end{array}\right)$$

For the lapidaries,

$$\underbrace{5x}_{\substack{x \text{ trays} \\ @ \text{ 5 hours} \\ \text{each}}} + \underbrace{2y}_{\substack{y \text{ trays} \\ @ \text{ 2 hours} \\ \text{each}}} \underbrace{\leq}_{\substack{\text{No} \\ \text{more} \\ \text{than}}} \underbrace{70}_{\substack{10 \text{ workers} \\ @ \text{ 7 hours} \\ \text{each}}} \qquad \left(\begin{array}{c}\text{Time} \\ \text{for} \\ \text{agates}\end{array}\right) + \left(\begin{array}{c}\text{Time} \\ \text{for} \\ \text{onyxes}\end{array}\right) \leq \left(\begin{array}{c}\text{Total} \\ \text{time for} \\ \text{lapidaries}\end{array}\right)$$

For the jewelers,

$$\underbrace{4x}_{\substack{x \text{ trays} \\ @ \text{ 4 hours} \\ \text{each}}} + \underbrace{3y}_{\substack{y \text{ trays} \\ @ \text{ 3 hours} \\ \text{each}}} \underbrace{\leq}_{\substack{\text{No} \\ \text{more} \\ \text{than}}} \underbrace{84}_{\substack{12 \text{ workers} \\ @ \text{ 7 hours} \\ \text{each}}} \qquad \left(\begin{array}{c}\text{Time} \\ \text{for} \\ \text{agates}\end{array}\right) + \left(\begin{array}{c}\text{Time} \\ \text{for} \\ \text{onyxes}\end{array}\right) \leq \left(\begin{array}{c}\text{Total} \\ \text{time for} \\ \text{jewelers}\end{array}\right)$$

Combining these constraints with the nonnegativity conditions $x \geq 0$ and $y \geq 0$, the problem may be represented by the system of linear inequalities

$$\begin{cases} 5x + 2y \leq 70 \\ 4x + 3y \leq 84 \\ x \geq 0 \\ y \geq 0 \end{cases}$$

We can now proceed as usual. The intercepts of the boundary line $5x + 2y = 70$ are $(70/5, 0) = (14, 0)$ and $(0, 70/2) = (0, 35)$. The intercepts of the boundary line $4x + 3y = 84$ are $(84/4, 0) = (21, 0)$ and $(0, 84/3) = (0, 28)$. The nonnegativity conditions, $x \geq 0$ and $y \geq 0$, place the feasible region in the first quadrant. Since the test point $(0, 0)$ satisfies the constraints, this region has the origin for one of its vertices.

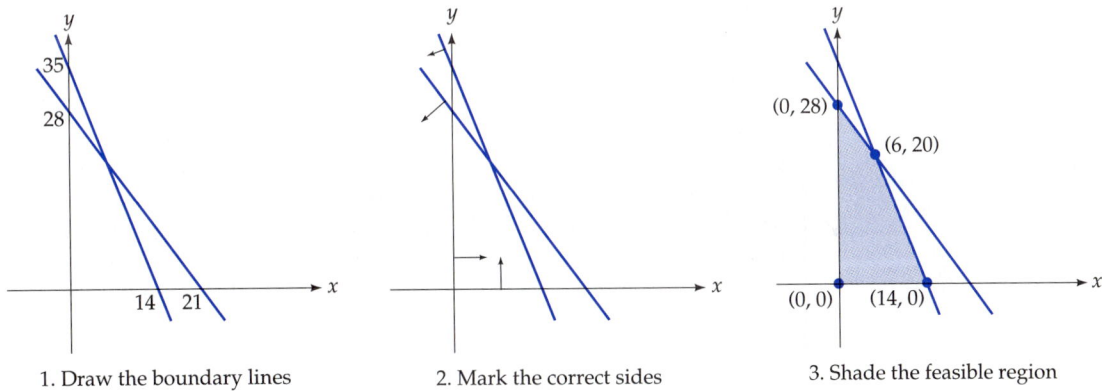

1. Draw the boundary lines   2. Mark the correct sides   3. Shade the feasible region

Three of the vertices of this bounded region are already known from the x- and y-intercepts of the boundary lines. The fourth is the intersection of $5x + 2y = 70$ and $4x + 3y = 84$:

$$5x = 70 - 2y \quad \text{so} \quad x = 14 - \tfrac{2}{5}y \qquad \text{Solve } 5x + 2y = 70 \text{ for } x$$

$$4\left(14 - \tfrac{2}{5}y\right) + 3y = 84 \qquad \text{Substitute into } 4x + 3y = 84$$

$$56 - \tfrac{8}{5}y + 3y = 84 \quad \text{so} \quad \tfrac{7}{5}y = 28 \quad \text{and then} \quad y = \tfrac{5}{7} \cdot 28 = 20 \qquad \text{Simplify and solve for } y$$

$$x = 14 - \tfrac{2}{5}(20) = 14 - 8 = 6 \qquad x = 14 - \tfrac{2}{5}y \text{ with } y = 20$$

$$(6, 20) \qquad \text{Vertex at } x = 6, y = 20$$

The four vertices of this region are $(0, 0)$, $(14, 0)$, $(6, 20)$, and $(0, 28)$, as shown above.

■

Graphing Calculator Exploration

You can sketch the boundary lines from Example 5 on your graphing calculator by entering them as $y_1 = (70 - 5x)/2$ and $y_2 = (84 - 4x)/3$.

a.   Set the window to $[-10, 60]$ by $[-10, 40]$ and graph your lines. Remember that the non-negativity conditions $x \geq 0$ and $y \geq 0$ put this region in the first quadrant.

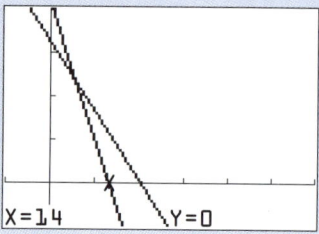

b.   Use TRACE or EVALUATE to check the x- and y-intercepts.

c.   You can use TRACE or INTERSECT to find the intersection point of the lines y_1 and y_2.

EXAMPLE 6   An Unbounded Feasible Region

The Marshall County trash incinerator in Norton burns 10 tons of trash per hour and co-generates 6 kilowatts (kW) of electricity, while the Wiseburg incinerator burns 5 tons per hour and co-generates 4 kilowatts. If the county needs to burn at least 70 tons of trash and co-generate at least 48 kilowatts of electricity each day, how many hours should each plant operate?

Formulate this situation as a system of linear inequalities, sketch the feasible region, and find the vertices.

Solution

Again, there are many different schedules of operating times that could burn all the trash and co-generate enough electricity. But there must be at least enough time to get the job done and neither plant can operate a negative number of hours. Let

$$x = \begin{pmatrix} \text{Number of hours} \\ \text{Norton operates} \end{pmatrix} \text{ and } y = \begin{pmatrix} \text{Number of hours} \\ \text{Wiseburg operates} \end{pmatrix}$$

For the amount of trash to be burned, we have the inequality

$$10x \quad + \quad 5y \quad \geq \quad 70$$

$\underbrace{}$ $\underbrace{}$ $\underbrace{}$ $\underbrace{}$
x hours   y hours   at   Tons of
@ 10 tons   @ 5 tons   least   trash
each   each   to burn

$$\begin{pmatrix} \text{Tons} \\ \text{burned} \\ \text{at Norton} \end{pmatrix} + \begin{pmatrix} \text{Tons} \\ \text{burned at} \\ \text{Wiseburg} \end{pmatrix} \geq \begin{pmatrix} \text{Tons} \\ \text{to} \\ \text{burn} \end{pmatrix}$$

| For the electricity to be produced,

$$6x \quad + \quad 4y \quad \geq \quad 48$$

$\underbrace{}$ $\underbrace{}$ $\underbrace{}$ $\underbrace{}$
x hours   y hours   at   kWs
@ 6 kWs   @ 4 kWs   least   needed
each   each

$$\begin{pmatrix} \text{Norton} \\ \text{electricity} \end{pmatrix} + \begin{pmatrix} \text{Wiseburg} \\ \text{electricity} \end{pmatrix} \geq \begin{pmatrix} \text{Electricity} \\ \text{needed} \end{pmatrix}$$

Combining these constraints with the nonnegativity conditions $x \geq 0$ and $y \geq 0$, the problem may be represented by the system of linear inequalities

$$\begin{cases} 10x + 5y \geq 70 \\ 6x + 4y \geq 48 \\ x \geq 0 \\ y \geq 0 \end{cases}$$

The intercepts of the boundary line $10x + 5y = 70$ are $(70/10, 0) = (7, 0)$ and $(0, 70/5) = (0, 14)$. The intercepts of the boundary line $6x + 4y = 48$ are $(48/6, 0) = (8, 0)$ and $(0, 48/4) = (0, 12)$. The non-negativity conditions place the feasible region in the first quadrant. Since the origin does not satisfy either of the first two inequalities, this region does *not* have the origin for one of its vertices.

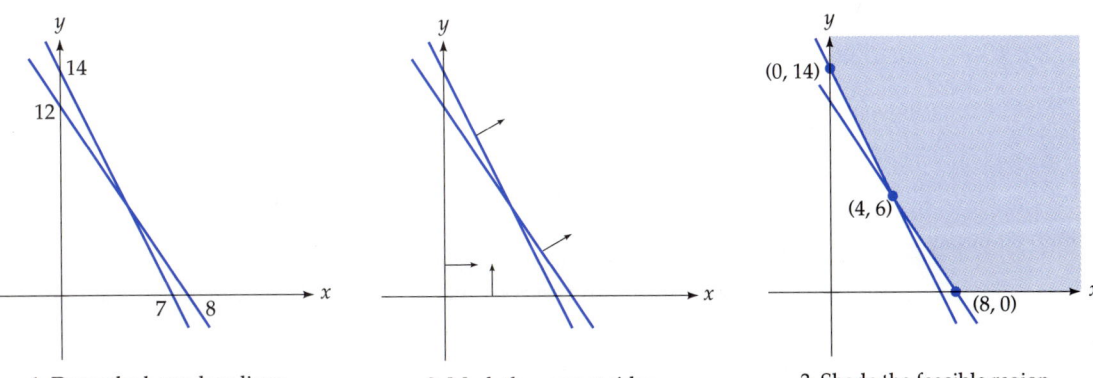

1. Draw the boundary lines   2. Mark the correct sides   3. Shade the feasible region

Two of the vertices of this unbounded region are already known from the x- and y-intercepts of the boundary lines. The third is the intersection of $10x + 5y = 70$ and $6x + 4y = 48$:

$$5y = 70 - 10x \quad \text{so} \quad y = 14 - 2x \qquad \text{Solve } 10x + 5y = 70 \text{ for } y$$

$$6x + 4(14 - 2x) = 48 \qquad \text{Substitute into } 6x + 4y = 48$$

$$6x + 56 - 8x = 48 \quad \text{means} \quad -2x = -8 \quad \text{so} \quad x = 4 \quad \text{Simplify and solve for } x$$

$$y = 14 - 2 \cdot 4 = 14 - 8 = 6 \qquad y = 14 - 2x \text{ with } x = 4$$

$$(4, 6) \qquad \text{Vertex at } x = 4, y = 6$$

The three vertices of this region are $(8, 0)$, $(4, 6)$, and $(0, 14)$, as shown on page 293.

■

SUMMARY

The *feasible region* of a *system of linear inequalities* $\begin{cases} ax + by \leq c \\ \cdot \quad \cdot \quad \cdot \end{cases}$ consists of all the (x, y) points on the correct sides of the *boundary lines* $\begin{cases} ax + by = c \\ \cdot \quad \cdot \quad \cdot \end{cases}$. Each boundary line $ax + by = c$ has x-intercept $(\frac{c}{a}, 0)$ and y-intercept $(0, \frac{c}{b})$ (if $c = 0$, the boundary passes through the points $(0, 0)$, $(-b, a)$, and $(b, -a)$). The correct side of the boundary can be found by trying a "test point" in the inequality (the origin $(0, 0)$ is usually the easiest to use).

The *vertices* are the "corners" of the region and are the intersections of two (or more) boundary lines that satisfy the inequalities. These intersections can be found by solving the equations representing the lines.

A feasible region is *bounded* if it is contained within a rectangular box, and *unbounded* if it is not. The feasible region of any system of linear inequalities is *convex* because it contains the line segment joining any two points in the region.

Two special inequalities are the *nonnegativity conditions* $x \geq 0$ and $y \geq 0$, which place the feasible region in the first quadrant and frequently appear in word problems when the variables represent amounts of materials or other real objects.

EXERCISES 3.1

For each region, select the linear inequality it represents.

1. **a.** $5x + 8y \geq 40$
b. $8x + 5y \geq 40$
c. $5x + 8y \leq 40$
d. $8x + 5y \leq 40$
e. No such inequality because the region is not convex

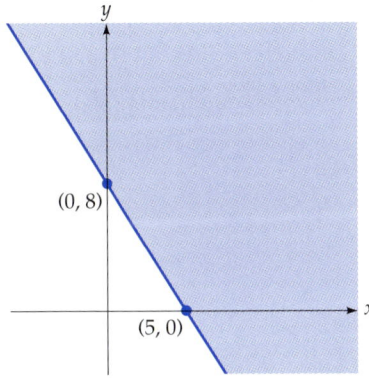

2. **a.** $7x + 6y \geq 42$
b. $6x + 7y \geq 42$
c. $7x + 6y \leq 42$
d. $6x + 7y \leq 42$
e. No such inequality because the region is not convex

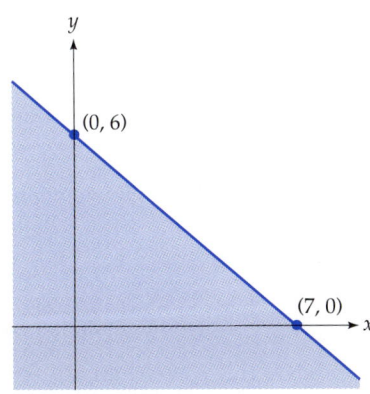

3. **a.** $-12x + 5y \geq -60$
b. $5x - 12y \geq -60$
c. $5x - 12y \leq 60$
d. $10x + 24y \leq 60$
e. No such inequality because the region is not convex

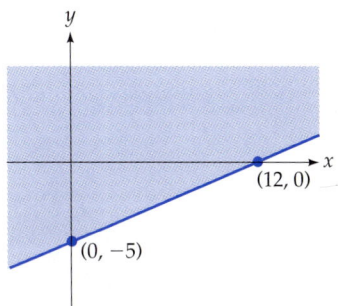

4. **a.** $5x - 3y \leq -15$
b. $5x - 3y \geq -15$
c. $-3x + 5y \leq 15$
d. $-5x + 3y \leq 15$
e. No such inequality because the region is not convex

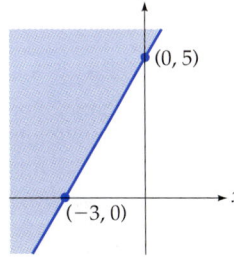

For each region, select the system of linear inequalities it represents.

5. **a.** $5x + 2y \leq 20$ and $y \geq 0$
b. $4x + 10y \leq 40$ and $x \leq 0$
c. $10x - 4y \leq 40$ and $y \leq 0$
d. $-2x + 5y \geq -20$ and $y \geq 0$
e. No such inequalities because the region is not convex

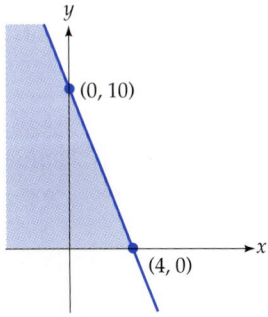

6. a. $5x - 3y \leq -30$ and $x \geq 0$
   b. $-6x + 10y \leq -60$ and $x \geq 0$
   c. $-5x + 3y \leq 30$ and $x \leq 0$
   d. $5x + 3y \leq -30$ and $x \geq 0$
   e. No such inequalities because the region is not convex

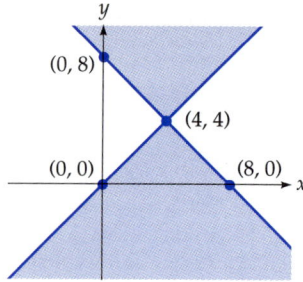

7. a. $-x + 2y \leq 2$ and $2x - y \leq 2$
   b. $x - 2y \leq 2$ and $x - 2y \leq -2$
   c. $x + 2y \leq 2$ and $x + 2y \geq -2$
   d. $x - 2y \leq 2$ and $-x + 2y \leq 2$
   e. No such inequalities because the region is not convex

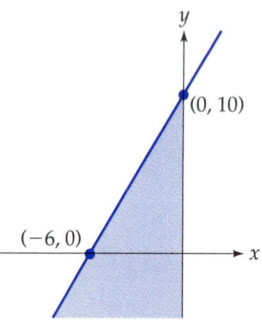

9. a. $x - 3y \leq -60$, $x \geq 0$ and $y \geq 0$
   b. $2x - 6y \geq -120$, $x \geq 0$ and $y \geq 0$
   c. $-60x + 20y \leq 240$ and $y \geq 0$
   d. $3x - y \geq -60$, $x \leq 0$ and $y \geq 0$
   e. No such inequalities because the region is not convex

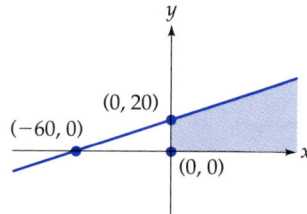

10. a. $2x + y \geq 16$ and $x \geq 0$
   b. $8x + 16y \leq 32$, $y \leq 0$ and $x \geq 0$
   c. $x + 2y \geq 16$ and $x \geq 0$
   d. $2x + y \geq -16$ and $y \geq 0$
   e. No such inequalities because the region is not convex

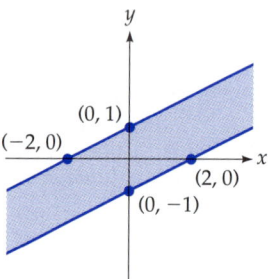

8. a. $x + y \leq 8$ and $x - y \leq 0$
   b. $x + y \geq 8$ and $x + y \geq 0$
   c. $x + 8y \leq 8$ and $x - y \geq 0$
   d. $x - y \leq -8$ and $-x + y \leq 0$
   e. No such inequalities because the region is not convex

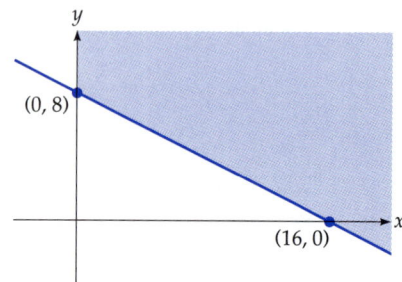

Sketch each system of linear inequalities. List all vertices and identify the region as "bounded" or "unbounded."

11. $\begin{cases} x + 2y \leq 40 \\ x \geq 0, y \geq 0 \end{cases}$ **12.** $\begin{cases} 3x + y \leq 90 \\ x \geq 0, y \geq 0 \end{cases}$

13. $\begin{cases} -2x + y \leq 10 \\ x \leq 10 \\ x \geq 0, y \geq 0 \end{cases}$ **14.** $\begin{cases} 2x - y \leq 20 \\ y \leq 40 \\ x \geq 0, y \geq 0 \end{cases}$

15. $\begin{cases} x + 2y \leq 8 \\ x + y \leq 6 \\ x \geq 0, y \geq 0 \end{cases}$ **16.** $\begin{cases} 2x + y \leq 10 \\ x + y \leq 8 \\ x \geq 0, y \geq 0 \end{cases}$

17. $\begin{cases} 5x + 2y \geq 20 \\ x \geq 0, y \geq 0 \end{cases}$ **18.** $\begin{cases} 4x + 5y \geq 20 \\ x \geq 0, y \geq 0 \end{cases}$

19. $\begin{cases} 4x + 3y \geq 24 \\ y \geq 4 \\ x \geq 0, y \geq 0 \end{cases}$ **20.** $\begin{cases} 3x + 4y \geq 12 \\ x \leq 8 \\ x \geq 0, y \geq 0 \end{cases}$

21. $\begin{cases} 3x + y \geq 12 \\ x + y \geq 8 \\ x \geq 0, y \geq 0 \end{cases}$ **22.** $\begin{cases} x + 3y \geq 15 \\ x + y \geq 9 \\ x \geq 0, y \geq 0 \end{cases}$

23. $\begin{cases} x + y \leq 20 \\ x \leq 15 \\ y \leq 10 \\ x \geq 0, y \geq 0 \end{cases}$ **24.** $\begin{cases} 2x + y \leq 20 \\ x \leq 8 \\ y \leq 10 \\ x \geq 0, y \geq 0 \end{cases}$

25. $\begin{cases} 2x + y \leq 80 \\ x + 3y \geq 30 \\ x \geq 0, y \geq 0 \end{cases}$ **26.** $\begin{cases} 3x + y \leq 90 \\ x + 2y \geq 20 \\ x \geq 0, y \geq 0 \end{cases}$

27. $\begin{cases} x - 3y \leq 15 \\ 2x + y \geq 30 \\ x \geq 0, y \geq 0 \end{cases}$ **28.** $\begin{cases} 3x + y \geq 30 \\ x - 2y \geq -60 \\ x \geq 0, y \geq 0 \end{cases}$

29. $\begin{cases} 2x + y \leq 18 \\ x + y \leq 10 \\ x + 3y \leq 24 \\ x \geq 0, y \geq 0 \end{cases}$ **30.** $\begin{cases} 3x + y \geq 24 \\ 3x + 2y \geq 42 \\ x + 2y \geq 18 \\ x \geq 0, y \geq 0 \end{cases}$

APPLIED EXERCISES

Formulate each situation as a system of linear inequalities, sketch the feasible region, and find the vertices.

31. Livestock Management   A rancher raises goats and llamas on his 400-acre ranch. Each goat needs 2 acres of land and requires $100 of veterinary care per year while each llama needs 5 acres of land and requires $80 of veterinary care per year. If the rancher can afford no more than $13,200 for veterinary care this year, how many of each animal can he raise?

32. Agriculture Management   A farmer grows wheat and barley on her 500-acre farm. Each acre of wheat requires 3 days of labor to plant, tend, and harvest while each acre of barley requires 2 days of labor. If the farmer and her hired field hands can provide no more than

1200 days of labor this year, how many acres of each crop can she grow?

33. Production Planning   A boat company manufactures aluminum dinghies and rowboats. The amount of metal work and painting needed for each is shown in the table, together with the number of hours of skilled labor available for each task. How many of each kind of boat can the company manufacture?

	Dinghy	Rowboat	Labor Available
Metal work	2 hours	3 hours	120 hours
Painting	2 hours	2 hours	100 hours

34. Resource Allocation   A sailboat company manufactures fiberglass prams and yawls. The amount of molding, painting, and finishing needed for each is shown in the table, together with the number of hours of skilled labor available for each task. How many of each kind of sailboat can the company manufacture?

	Pram	Yawl	Labor Available
Molding	3 hours	6 hours	150 hours
Painting	3 hours	2 hours	114 hours
Finishing	2 hours	6 hours	132 hours

35. Nutrition   Joshua loves "junk food" but wants to stay within the recommended daily limits of 80 grams of fat and 2250 calories. If each serving of SugarSnaks contains 5 grams of fat and 125 calories and each bag of Gobbl'Ems contains 8 grams of fat and 250 calories, how much of each food may he eat today?

36. Diet Planning   Justin plays a lot of sports and wants to make sure he gets at least 70 grams of protein and 20 milligrams of iron each day. If each serving of Pro-Team Power Bars has 10 grams of protein and 2 milligrams of iron and each glass of Bulk-Up-Delight has 5 grams of protein and 2 milligrams of iron, how much of each food should he eat each day?

37. Pollution Control   A smelting company refines metals at two factories located in Ohio and in Pennsylvania. The smokestacks release both sulfur dioxide (which combines with water vapor to form "acid rain") and particulates (solid matter such as soot which can cause respiratory problems) at the rates shown in the table. The EPA has obtained an injunction against the company preventing it from releasing more than 64 pounds of sulfur dioxide and 60 pounds of particulates into the atmosphere each day.

How many hours can the company operate these factories each day?

(Pounds Per Hour)	Sulfur Dioxide	Particulates
Ohio factory	4	5
Pennsylvania factory	4	3

38. Pollution Control   A chemical company manufactures batteries at two factories located in Connecticut and in Alabama. The factories discharge both heavy metals (such as mercury and cadmium, which are very toxic) and nitric acid into the local river systems at the rates shown in the table. An environmental organization has obtained an injunction against the company preventing it from discharging more than 54 pounds of heavy metals and 60 pounds of nitric acid each day. How many hours can the company operate these factories each day?

(Pounds Per Hour)	Heavy Metals	Nitric Acid
Connecticut factory	6	4
Alabama factory	3	4

39. Investment Strategy   An investment portfolio manager has $8 million to invest in stock and bond funds. If the amount invested in stocks can be no more than the amount invested in bonds, how much can be invested in each type of fund?

40. Financial Planning   A retired couple want to invest their $20,000 life savings in bank certificates of deposit and treasury bonds. If they want at least $5000 in each type of investment, how much can they invest in each?

3.2   Two-Variable Linear Programming Problems

Managing an Investment Portfolio

The adage "never put all your eggs in one basket" particularly applies to investing money. Since there is no such thing as a low-risk yet high-return opportunity, each investor must reconcile the desire for high returns with the possibility that even the principal might be lost. An "investment strategy" balances greed against safety and thus determines both the types of investments made and the amounts of each. Given the many different opportunities in today's global economy, the actual solution of a particular instance of this portfolio management problem can be very difficult.

Let us imagine a simplified situation in which there are only two possible investments: one with low risk and low return and the other with high risk and corresponding high return. Suppose you have $10,000 to invest and wish to protect yourself by investing at least $1000 in the low-risk investment and no more in the high-risk investment than in the low-risk one. Writing these restrictions as inequalities, where x represents the money in the low-risk investment and y represents the money in the high-risk investment, we have that $x + y \leq 10,000$ ("the amounts invested cannot exceed the available funds"), $x \geq 1000$ ("at least 1000 in the low-risk investment"), $x \geq y$ ("at least as much in the low-risk investment as in the high-risk one") and, of course, $x \geq 0$ and $y \geq 0$ ("each investment must be at least nothing"). As in the previous section, we can sketch the feasible region of possible investments. The portfolio management problem is to find the point (x, y) of amounts that represents the investment choice with the greatest total return.

It should be immediately clear that in this simple situation, the greatest return is gotten by investing as much as possible in the high-return investment and all the rest in the low-return investment. In this section, we shall see that the solution of any problem of this general type is solved by "pushing" the values of x and y to an extreme point of the constraint region.

Introduction

In this section we will explain what a linear programming problem is and then use the geometry of feasible regions from the previous section to show that the solution of a linear programming problem occurs at a vertex of the region.

Linear Programming Problems

A *linear programming problem* asks for the greatest or smallest value of a linear *objective function* subject to *constraints* in the form of a system of linear inequalities.

EXAMPLE 1   A Linear Programming Problem

A farmer grows corn and soybeans on his 200-acre farm. To maintain soil fertility, the farmer rotates the crops and always plants at least as many acres of soybeans as acres of corn. If each acre of corn yields a profit of $150 and each acre of soybeans yields a profit of $100, how many acres of each crop should the farmer plant to obtain the greatest possible profit?

Formulate this situation as a linear programming problem by identifying the variables, the objective function, and the constraints.

Solution

Since the question asks "how many acres of each crop," we let:

$$x = \left(\begin{matrix} \text{Number of} \\ \text{acres of corn} \end{matrix} \right) \text{ and } y = \left(\begin{matrix} \text{Number of} \\ \text{acres of soybeans} \end{matrix} \right)$$

The objective is to maximize the farmer's profit, and this profit is $P = 150x + 100y$ because the profits per acre are $150 for corn and $100 for soybeans. The 200-acre size of the farm leads to the constraint $x + y \leq 200$ while the crop rotation requirement that $y \geq x$ can be written as $x - y \leq 0$. Since x and y cannot be negative, we also have the nonnegativity constraints $x \geq 0$ and $y \geq 0$. This maximum linear programming problem may be written as:

Objective function

$$\text{Maximize } P = 150x + 100y \qquad \text{Corn and soybean profits}$$

$$\text{Subject to } \begin{cases} x + y \leq 200 & \text{Size of farm} \\ x - y \leq 0 & \text{Crop rotation} \\ x \geq 0 \text{ and } y \geq 0 & \text{Nonnegativity} \end{cases}$$

Constraints

■

Every linear programming problem in two variables may be written in one of the forms:

$$\text{Maximize } P = Mx + Ny$$
$$\text{Subject to } \begin{cases} ax + by \leq c \\ \cdots \end{cases}$$

or

$$\text{Minimize } C = Mx + Ny$$
$$\text{Subject to } \begin{cases} ax + by \geq c \\ \cdots \end{cases}$$

We have named the objective function P for profit in the maximum problem and C for cost in the minimum problem, but problems with different goals are perfectly acceptable. Nonnegativity conditions should be included only when appropriate.

Fundamental Theorem of Linear Programming

A *solution* of a *maximum* linear programming problem is a feasible point for the system of linear inequalities that gives the *largest* possible value the objective function can take on the region. Similarly, a solution of a *minimum* linear programming problem is a feasible point for the system of linear inequalities that gives the *smallest* possible value the objective function can take on the region. With either problem, if the constraints are infeasible, then the problem has no solution.

How can the solution be found? As with systems of linear inequalities, we begin with a simpler problem in just one variable. Omitting the y-variables from the maximum linear programming problem above, we are left with the task of finding the largest value of Mx subject to constraints of the form $x \leq R$. It is immediate that the solution of this simple problem must occur at one of the extreme values allowed for x. (For instance, the largest value of $5x$ where $2 \leq x$ and $x \leq 4$ is 20 when $x = 4$.)

Turning back to two-variable problems, let us first suppose that the feasible region is bounded. For any interior point and any vertex (see the sketch below), the line segment from the vertex to the interior point can be extended to meet the boundary (because the region is bounded) and this line segment lies completely within the feasible region (because it is convex). On just this line segment, the largest (or smallest) value must occur at one end or the other of the segment (just as in the one-variable case).

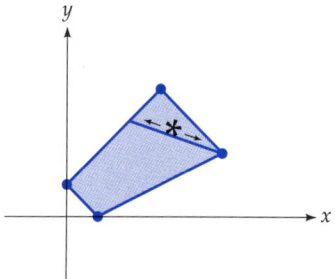

Thus the solution must occur on the boundary of the region. Since the boundary consists of line segments joining pairs of vertices, the same reasoning shows that the solution must occur at an endpoint of these segments. That is, *the solution of the linear programming problem with a bounded region must occur at a vertex of the region.*

On an unbounded region, it is possible that a solution does not exist because the line segment from a vertex to an interior point may extend "forever" and we may wish to move in this unbounded direction to improve the value of the objective function. However, if improving the value of the objective function always means moving toward a vertex, then the solution will exist and it must occur at a vertex.

Combining these observations, we have established the following result about the solution of a linear programming problem.

The Fundamental Theorem of Linear Programming

If a linear programming problem has a solution, then it occurs at a *vertex* of the region determined by the constraints.

Graphing Calculator Exploration

You can use the program ViewLP* to explore the values of an objective function on a feasible region. The following demonstration uses the problem from Example 1.

a. Enter the boundary line $x + y = 200$ as $y_1 = (200 - 1x)/1$ (or just $y_1 = 200 - x$). Then enter the boundary line $x - y = 0$ as $y_2 = (0 - 1x)/-1$ (or just $y_2 = x$).

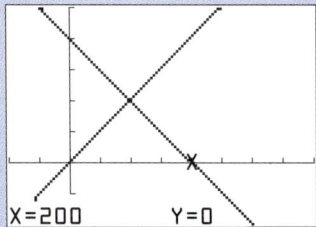

b. Graph them on the window $[-80, 390]$ by $[-80, 230]$ and check that you see the expected region.

c. Run the program ViewLP and enter the values for the objective function: 150 for M and 100 for N (press ENTER after each value).

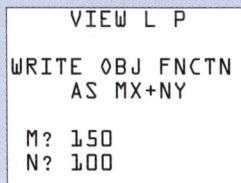

d. Use the arrow buttons to move the dot into the region. At each (x, y) point, the bottom half of the screen shows the value of the objective function MX + NY at that X and Y.

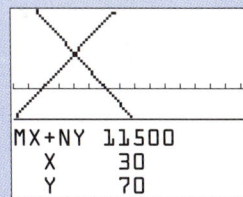

e. Move around in the region and watch how the values of the objective function change. Be careful not to move outside the region. Move to each of the vertices and note the value of the objective function. Does the greatest value occur at a vertex?

f. When you are finished exploring the region, press ENTER to reset the viewing screen and exit the program.

ViewLP can be used in this way to explore and verify the solution of any linear programming problem in two variables provided you can find an appropriate viewing window.

* See Preface for information on how to obtain this program.

The Fundamental Theorem of Linear Programming gives us the following procedure to find a solution.

How to Solve a Linear Programming Problem

On a bounded region, list the vertices, calculate the value of the objective function at each, and select the vertex giving the largest (or smallest) value.

On an unbounded region, first check whether the objective function "improves" in an unbounded direction: if it does, there is no solution; if it does not, list the vertices, calculate the value of the objective function at each, and select the vertex giving the largest (or smallest) value.

EXAMPLE 2   **Solution of Example 1**

Solve the linear programming problem:

$$\text{Maximize } P = 150x + 100y \qquad \text{Objective function}$$

$$\text{Subject to } \begin{cases} x + y \le 200 \\ x - y \le 0 \\ x \ge 0 \text{ and } y \ge 0 \end{cases} \qquad \text{Constraints}$$

Solution

We begin by sketching the region determined by the constraints. The nonnegativity conditions $x \ge 0$ and $y \ge 0$ place the region in the first quadrant. The boundary line $x + y = 200$ has intercepts $(200, 0)$ and $(0, 200)$ while the boundary line $x - y = 0$ passes through the origin $(0, 0)$ and the point $(1, 1)$. The origin satisfies the inequality $x + y \le 200$ and the point $(0, 200)$ satisfies the inequality $x - y \le 0$.

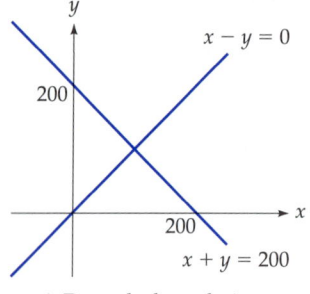

1. Draw the boundaries

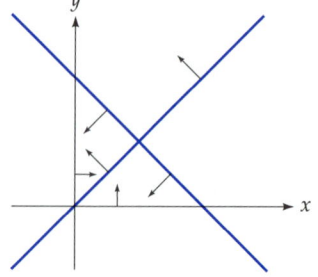

2. Mark the correct sides

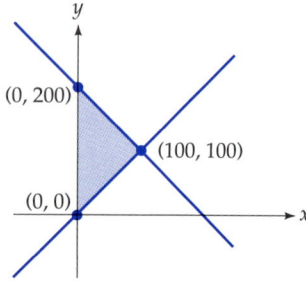

3. Feasible region

The region has three vertices and two are already known from the x- and y-intercepts of the boundary lines. The third is the intersection of $x - y = 0$ and $x + y = 200$:

$x - y = 0$ so $x = y$	Solve $x - y = 0$ for x
$(y) + y = 200$	Substitute into $x + y = 200$
$2y = 200$ so $y = 200/2 = 100$	Simplify and solve for y
$x = 100$	$x = y$ with $y = 100$
$(100, 100)$	Vertex at $x = 100$, $y = 100$

The vertices are $(0, 0)$, $(0, 200)$, and $(100, 100)$, as shown on page 304.

Since the region is bounded, this problem *does* have a solution and it occurs at a vertex. Evaluating the objective function at the vertices, we find the following values.

Vertex	Value of $P = 150x + 100y$	
$(0, 0)$	0	$0 = 150 \cdot 0 + 100 \cdot 0$
$(0, 200)$	$20{,}000$	$20{,}000 = 150 \cdot 0 + 100 \cdot 200$
$(100, 100)$	$25{,}000 \leftarrow$ Largest	$25{,}000 = 150 \cdot 100 + 100 \cdot 100$

Since the largest value of the objective function occurs at the vertex $(100, 100)$, the solution of this problem is: The maximum value of P is 25,000 at the vertex $(100, 100)$. In terms of the original word problem in Example 1, the maximum profit is $25,000 when the farmer plants 100 acres of corn and 100 acres of soybeans.

■

PRACTICE PROBLEM 1   Solve the linear programming problem:

$$\text{Maximize } P = 5x + 2y$$

$$\text{Subject to } \begin{cases} 3x + y \le 60 \\ y \le 36 \\ x \ge 0 \text{ and } y \ge 0 \end{cases}$$

Solution at the back of the book

You may also want to use the program ViewLP (page 303) to explore this problem on your graphing calculator.

EXAMPLE 3   A Minimum Problem on an Unbounded Region

Solve the linear programming problem:

$$\text{Minimize } C = 3x + 5y$$

$$\text{Subject to } \begin{cases} 2x + y \geq 12 \\ x + y \geq 8 \\ x \geq 0 \text{ and } y \geq 0 \end{cases}$$

Solution

We begin by sketching the region determined by the constraints. The nonnegativity conditions $x \geq 0$ and $y \geq 0$ place the region in the first quadrant. The boundary line $2x + y = 12$ has intercepts $(12/2, 0) = (6, 0)$ and $(0, 12)$ while the boundary line $x + y = 8$ has intercepts $(8, 0)$ and $(0, 8)$. Since the test point $(0, 0)$ does not satisfy $2x + y \geq 12$ and $x + y \geq 8$, the region is on the sides of these boundaries *away* from the origin.

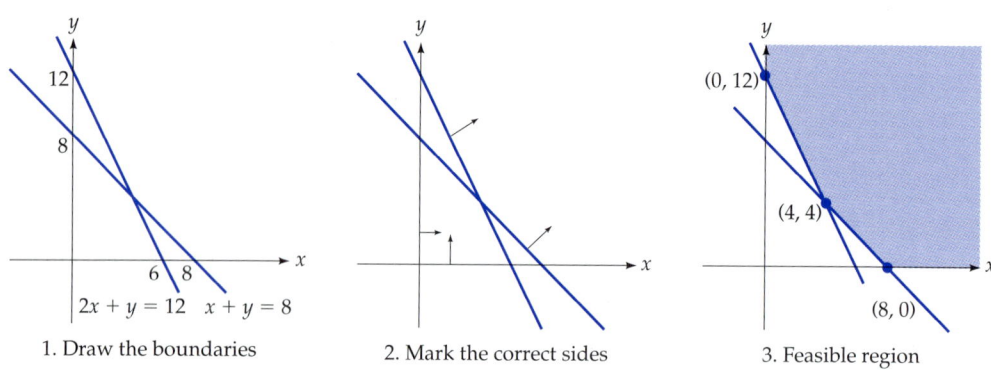

1. Draw the boundaries   2. Mark the correct sides   3. Feasible region

This unbounded region has three vertices and two are already known from the x- and y-intercepts of the boundary lines. The third is the intersection of $2x + y = 12$ and $x + y = 8$. Solving the second as $x = 8 - y$ and substituting into the first equation, $2(8 - y) + y = 12$ means $16 - 2y + y = 12$ so $y = 4$ and then $x = 8 - 4 = 4$. The vertices are $(8, 0)$, $(4, 4)$, and $(12, 0)$, as shown above.

Since the region is unbounded, before looking at the vertices we must find whether a solution exits. The value of the objective function C at the interior point $(10, 10)$, for example, is $C = 3 \cdot 10 + 5 \cdot 10 = 80$. If we shift this point upward to $(10, 15)$ or right to $(15, 10)$, the larger values for x or y will *increase* the value of $C = 3x + 5y$, which is the *opposite* of what we should do to minimize C. (You may want to use the program ViewLP (page 303) to explore this further.) Thus the objective function does not "improve" in an unbounded direction and the solution does exist. Evaluating the objective function at the vertices, we find the following values.

Vertex	$C = 3x + 5y$		
(8, 0)	24	← Smallest	$24 = 3 \cdot 8 + 5 \cdot 0$
(4, 4)	32		$32 = 3 \cdot 4 + 5 \cdot 4$
(0, 12)	60		$60 = 3 \cdot 0 + 5 \cdot 12$

Since the smallest value of the objective function occurs at the vertex (8, 0), the solution of this problem is: The minimum value of C is 24 at the vertex (8, 0). ∎

PRACTICE PROBLEM 2

Solve the linear programming problem:

$$\text{Minimize } C = 5x + 11y$$

$$\text{Subject to } \begin{cases} x + 3y \geq 60 \\ x + 2y \geq 50 \\ x \geq 0 \text{ and } y \geq 0 \end{cases}$$

Solution at the back of the book

 You may also want to use the program ViewLP (page 303) to explore this problem on your graphing calculator.

EXAMPLE 4   **A Manufacturing Problem**

A fully-automated plastics factory produces two toys, a racing car and a jet airplane, in three stages: molding, painting, and packaging. After allowing for routine maintenance, the equipment for each stage can operate no more than 150 hours per week. Each batch of racing car toys requires 6 hours of molding, 2.5 hours of painting, and 5 hours of packaging, while each batch of jet airplane toys requires 3 hours of molding, 7.5 hours of painting, and 5 hours of packaging. If the profit per batch of toys is $120 for cars and $100 for airplanes, how many batches of each toy should be produced each week to obtain the greatest possible profit?

Solution

Let:

$$x = \begin{pmatrix} \text{Number of batches} \\ \text{of racing car toys} \end{pmatrix} \text{ and } y = \begin{pmatrix} \text{Number of batches} \\ \text{of jet airplane toys} \end{pmatrix}$$

made during the week. The profit is $P = 120x + 100y$ and the problem is to maximize P subject to the constraints that the molding, painting, and packaging processes can each take no more than 150 hours:

$$6x + 3y \leq 150 \qquad \text{Molding time}$$

$$2.5x + 7.5y \leq 150 \qquad \text{Painting time}$$

$$5x + 5y \leq 150 \qquad \text{Packaging time}$$

and the nonnegativity conditions $x \geq 0$ and $y \geq 0$. Simplifying the constraints by removing common factors (that is, by dividing the molding constraint by 3, the painting constraint by 2.5, and the packaging constraint by 5), this problem may be rewritten as the linear programming problem:

$$\text{Maximize } P = 120x + 100y \qquad \text{Objective function}$$

$$\text{Subject to } \begin{cases} 2x + y \leq 50 \\ x + 3y \leq 60 \\ x + y \leq 30 \\ x \geq 0 \text{ and } y \geq 0 \end{cases} \qquad \text{Constraints}$$

The nonnegativity conditions $x \geq 0$ and $y \geq 0$ place the region in the first quadrant. The origin satisfies all the constraints, the region is bounded, and one vertex is the origin. For the boundary line intercepts, $2x + y = 50$ has intercepts $(50/2, 0) = (25, 0)$ and $(0, 50)$, $x + 3y = 60$ has intercepts $(60, 0)$ and $(0, 60/3) = (0, 20)$, and $x + y = 30$ has intercepts $(30, 0)$ and $(0, 30)$.

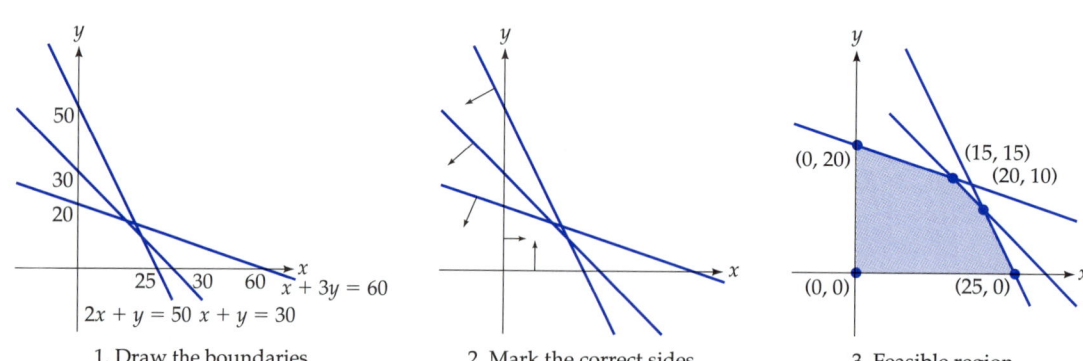

1. Draw the boundaries   2. Mark the correct sides   3. Feasible region

The region has five vertices and three are already known from the x- and y-intercepts of the boundary lines. The other two vertices are intersections of the lines $2x + y = 50$, $x + 3y = 60$, and $x + y = 30$. Since these three lines have three intersection points, one of these three points is not a vertex of the region. The intersection of $2x + y = 50$

and $x + 3y = 60$ is the point $(18, 14)$, but this is *not* a vertex of the region because it does not also satisfy $x + y \leq 30$: $18 + 14$ is not ≤ 30. The intersection of $2x + y = 50$ and $x + y = 30$ is $(20, 10)$, and this point *is* a vertex of the region because it also satisfies the inequality $x + 3y \leq 60$. The intersection of $x + 3y = 60$ and $x + y = 30$ is $(15, 15)$, and this point *is* a vertex because it also satisfies the inequality $2x + y \leq 50$. The vertices are $(0, 0)$, $(25, 0)$, $(20, 10)$, $(15, 15)$, and $(0, 20)$.

Since the region is bounded, this problem *has* a solution and it occurs at a vertex. Evaluating the objective function at the vertices, we find the following values.

Vertex	$P = 120x + 100y$
$(0, 0)$	0
$(25, 0)$	3000
$(20, 10)$	3400 ← Largest
$(15, 15)$	3300
$(0, 20)$	2000

Since the largest value of the objective function occurs at the vertex $(20, 10)$, the solution is: The maximum value is 3400 at the vertex $(20, 10)$. In terms of the original question, the maximum profit is \$3400 when the factory produces 20 batches of racing car toys and 10 batches of jet airplane toys each week. ∎

Be careful! Except in very simple problems, some boundary line intersections are *not* vertices of the feasible region because they violate at least one of the other constraints. Particularly when sketching regions by hand, it is best to find all the intersection points anyway and then verify which are feasible and which are not.

Extensions to Larger Problems

There are two difficulties with our geometric solution of linear programming problems. First, as the number of constraints increases, the number of intersection points rapidly increases, with relatively fewer and fewer being vertices. For a problem with many constraints, this will cause an enormous amount of wasted effort in calculating and discarding these irrelevant points. Second, just two variables is not realistic: a manufacturing problem such as Example 4 should really be concerned with *many* different toys instead of just two. But three or more variables make the feasible regions difficult if not impossible to sketch.

One solution to the first difficulty would be to find only the vertices needed for the solution of the problem and omit the calculation of all the other intersection points. In a region such as in Example 4, it is clear that the origin is a vertex (the company could choose to manufacture nothing). It might then be possible to start at this first vertex (which we have found with no effort) and move to the solution by finding a "path of vertices" giving larger and larger profits until there is nowhere to go that could make the profit greater. For the region of Example 4, this might result in the path shown below using only the vertices (0, 0), (25, 0), and (20, 10) to arrive at the solution.

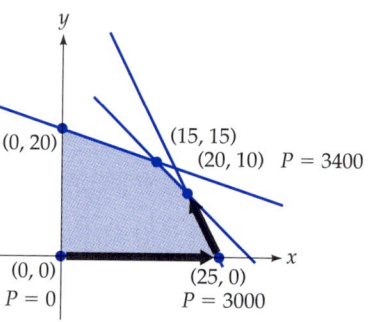

A solution method based on this idea is the subject of the next section (Section 3.3).

The second difficulty of how to deal with more variables suggests that we must leave the geometric method and search for an *algebraic* method of calculating the solution. One possibility would be to use the constraints to make upper estimates on the size of the objective function. If these upper estimates could be made small enough, then perhaps the solution might be found. For the problem from Example 4, if we multiply the first constraint by 20 and add it to 80 times the third, we find:

Maximize $P = 120x + 100y$

Subject to $\begin{cases} 2x + y \le 50 \\ x + 3y \le 60 \\ x + y \le 30 \\ x \ge 0 \text{ and } y \ge 0 \end{cases}$

$\times 20$ is $\qquad 40x + 20y \le 1000$

$\times 80$ is $\qquad \dfrac{80x + 80y \le 2400}{}$

and add to get: $\quad 120x + 100y \le 3400$

The resulting inequality, $120x + 100y \le 3400$, not only contains the exact objective function, but also gives an estimate that is the best possible since there actually is a vertex where the objective function takes the value 3400. While this may seem too good to be true in general, we will see in Section 3.4 that this idea leads to a powerful method for solving linear programming problems.

SUMMARY

A *linear programming problem* consists of a linear *objective function* to be *maximized* or *minimized* subject to *constraints*, written as a system of linear inequalities. Every such problem can be written in the form:

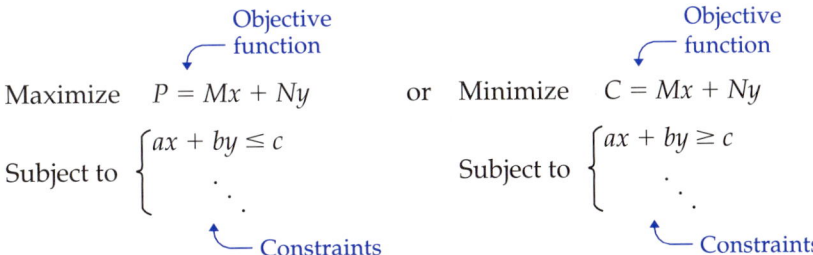

A *solution* is a feasible point giving the largest (or smallest) possible value for the objective function. The *fundamental theorem of linear programming* says that if there is a solution, it occurs at a *vertex* of the feasible region. So the problem can be solved by sketching the feasible region of the constraints, evaluating the objective function at the vertices, and selecting the vertex with the optimal value. If the region is unbounded, you must check that a solution exists before selecting the "best" vertex.

EXERCISES 3.2

For Exercises 1–5: Find each maximum or minimum value for the given region. If such a value does not exist, explain why not.

4. Minimum of $C = 4x + 3y$

5. Maximum of $P = x - y$

For Exercises 6–10: Find each maximum or minimum value for the given region. If such a value does not exist, explain why not.

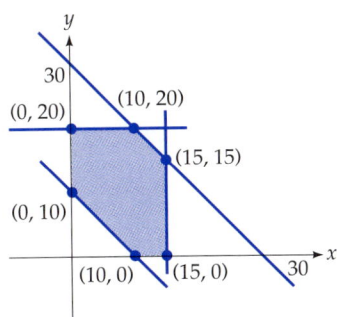

1. Maximum of $P = 2x + y$

2. Maximum of $P = x + 2y$

3. Minimum of $C = 3x + 4y$

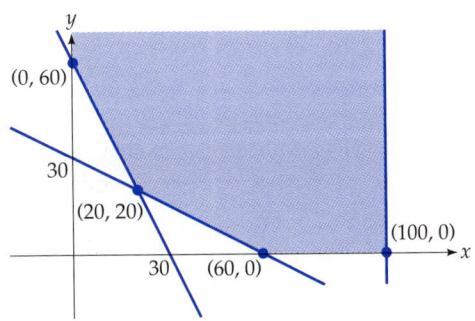

6. Minimum of $C = x + y$

7. Minimum of $C = 3x + y$

8. Maximum of $P = x + 3y$

9. Maximum of $P = 5x + 8y$

10. Maximum of $P = -x - y$

Solve each linear programming problem by sketching the region and labeling the vertices, deciding whether a solution exists, and then finding it if it does exist.

11. Maximize $P = 30x + 40y$

Subject to $\begin{cases} 2x + y \le 16 \\ x + y \le 10 \\ x \ge 0, y \ge 0 \end{cases}$

12. Maximize $P = 80x + 70y$

Subject to $\begin{cases} x + 2y \le 18 \\ x + y \le 10 \\ x \ge 0, y \ge 0 \end{cases}$

13. Minimize $C = 15x + 45y$

Subject to $\begin{cases} 2x + 5y \ge 20 \\ x \ge 0, y \ge 0 \end{cases}$

14. Minimize $C = 35x + 25y$

Subject to $\begin{cases} 5x + 3y \ge 60 \\ x \ge 0, y \ge 0 \end{cases}$

15. Maximize $P = 4x + 5y$

Subject to $\begin{cases} 2x + y \le 50 \\ x + 3y \le 75 \\ x \ge 0, y \ge 0 \end{cases}$

16. Maximize $P = 7x + 8y$

Subject to $\begin{cases} 3x + y \le 90 \\ x + 2y \le 60 \\ x \ge 0, y \ge 0 \end{cases}$

17. Minimize $C = 12x + 10y$

Subject to $\begin{cases} 4x + y \ge 40 \\ 2x + 3y \ge 60 \\ x \ge 0, y \ge 0 \end{cases}$

18. Minimize $C = 20x + 30y$

Subject to $\begin{cases} 3x + 2y \ge 120 \\ x + 4y \ge 80 \\ x \ge 0, y \ge 0 \end{cases}$

19. Maximize $P = 5x + 7y$

Subject to $\begin{cases} 4x + 3y \le 60 \\ x - y \le 8 \\ x \ge 0, y \ge 0 \end{cases}$

20. Maximize $P = 2x + y$

Subject to $\begin{cases} 3x - 4y \le 60 \\ x + y \le 48 \\ x \ge 0, y \ge 0 \end{cases}$

21. Minimize $C = 6x + 10y$

Subject to $\begin{cases} -x + 4y \le 60 \\ 2x + y \ge 60 \\ x \ge 0, y \ge 0 \end{cases}$

22. Minimize $C = 5x - 10y$

Subject to $\begin{cases} -3x + y \le 60 \\ y \le 120 \\ x \ge 0, y \ge 0 \end{cases}$

23. Maximize $P = 5x + 3y$

Subject to $\begin{cases} 2x + y \le 90 \\ x + y \le 50 \\ x + 2y \le 90 \\ x \ge 0, y \ge 0 \end{cases}$

24. Maximize $P = 6x + 5y$

Subject to $\begin{cases} x + 2y \le 96 \\ x + y \le 54 \\ 2x + y \le 96 \\ x \ge 0, y \ge 0 \end{cases}$

25. Maximize $P = 10x + 12y$

Subject to $\begin{cases} 3x + 2y \le 180 \\ 4x + y \le 120 \\ 3x + y \le 105 \\ x \ge 0, y \ge 0 \end{cases}$

26. Maximize $P = 20x + 15y$

Subject to $\begin{cases} 2x + 3y \le 60 \\ x + 4y \le 40 \\ x + 3y \le 33 \\ x \ge 0, y \ge 0 \end{cases}$

27. Minimize $C = 20x + 25y$

Subject to $\begin{cases} 3x + y \geq 60 \\ x + y \geq 42 \\ x + 3y \geq 60 \\ x \geq 0, y \geq 0 \end{cases}$

28. Minimize $C = 50x + 35y$

Subject to $\begin{cases} x + 3y \geq 72 \\ x + y \geq 48 \\ 3x + y \geq 72 \\ x \geq 0, y \geq 0 \end{cases}$

29. Minimize $C = 5x + 6y$

Subject to $\begin{cases} 4x + 3y \geq 840 \\ 2x + 5y \geq 700 \\ 4x + 5y \geq 1280 \\ x \geq 0, y \geq 0 \end{cases}$

30. Minimize $C = 15x + 10y$

Subject to $\begin{cases} 3x + 4y \geq 336 \\ 5x + 2y \geq 280 \\ 5x + 4y \geq 520 \\ x \geq 0, y \geq 0 \end{cases}$

APPLIED EXERCISES

Formulate each situation as a linear programming problem by identifying the variables, the objective function, and the constraints. Determine if a solution exists and, if it does, find it. Be sure to state your final answer in terms of the original question.

31. Livestock Management A rancher raises goats and llamas on his 400-acre ranch. Each goat needs 2 acres of land and requires $100 of veterinary care per year, while each llama needs 5 acres of land and requires $80 of veterinary care per year. The rancher can afford no more than $13,200 for veterinary care this year. If the expected profit is $60 for each goat and $90 for each llama, how many of each animal should he raise to obtain the greatest possible profit?

32. Agriculture Management A farmer grows wheat and barley on her 500-acre farm. Each acre of wheat requires 3 days of labor to plant, tend, and harvest, while each acre of barley requires 2 days of labor. The farmer and her hired field hands can provide no more than 1200 days of labor this year. If the expected profit is $50 for each acre of wheat and $40 for each acre of barley, how many acres of each crop should she grow to obtain the greatest possible profit?

33. Resource Allocation A sailboat company manufactures fiberglass prams and yawls. The amount of molding, painting, and finishing needed for each is shown in the table, together with the number of hours of skilled labor available for each task. If the expected profit is $150 for each pram and $180 for each yawl, how many of each kind of sailboat should the company manufacture to obtain the greatest possible profit?

	Pram	Yawl	Labor Available
Molding	3 hours	6 hours	150 hours
Painting	3 hours	2 hours	114 hours
Finishing	2 hours	6 hours	132 hours

34. Production Planning A small jewelry company prepares and mounts semiprecious stones. There are 10 lapidaries (who cut and polish the stones) and 12 jewelers (who mount the stones in gold settings). Each employee works 7 hours each day. Each tray of agates requires 5 hours of cutting and polishing and 4 hours of mounting, while each tray of onyxes requires 2 hours of cutting and polishing and 3 hours of mounting. If the profit is $15 for each tray of agates and $10 for each tray of onyxes, how many trays of each stone should be processed each day to obtain the greatest possible profit?

35. Waste Management The Marshall County trash incinerator in Norton burns 10 tons of trash per hour and co-generates 6 kilowatts of electricity, while the Wiseburg incinerator burns 5 tons per hour and co-generates 4 kilowatts. The county needs to burn at least 70 tons of trash and co-generate at least 48 kilowatts of electricity every day. If the Norton incinerator costs $80 per hour to operate and the Wiseburg incinerator costs $50, how many hours should each incinerator operate each day with the least cost to the county?

36. **Disaster Relief**   An international relief agency has been asked to provide medical support to a Caribbean island devastated by a recent hurricane. The agency estimates that it must be able to perform at least 50 major surgeries, 78 minor surgeries, and 130 outpatient services each day. The daily capacities of its portable field hospitals and clinics are given in the table, along with the daily operating costs. How many field hospitals and clinics should the agency airlift to the island to provide the needed help at the least daily cost?

	Major Surgeries	Minor Surgeries	Outpatient Services	Daily Cost
Field hospitals	5	3	2	$2000
Clinics	0	2	10	$500

37. **Nutrition**   A pet store owner raises baby bunnies and feeds them leftover salad greens from a nearby restaurant. The weekly nutritional needs of each bunny are given in the table, as well as the nutrition provided by the salad greens and nutritional supplement drops. If the salad greens cost 2¢ per handful and the nutritional supplement costs 1¢ per drop, how many handfuls of greens and drops of supplement should each bunny receive weekly to minimize the owner's costs?

	Salad Greens (Per Handful)	Nutritional Supplement (Per Drop)	Minimum Weekly Requirement
Fiber (grams):	3	0	90
Vitamins (mg):	2	5	140
Minerals (mg):	1	6	84

38. **Diet Planning**   The director of a school district's hot lunch program estimates that each lunch should contain at least 1000 calories and 15 grams of protein and no more than 30 grams of fat. The nutritional content per ounce of the cafeteria's famous mystery foods X and Y are given in the table. If each ounce of X costs 5¢ and each ounce of Y costs 6¢, how much of each food should be served to meet the director's nutritional goals at the least cost?

	Calories	Protein	Fat
Mystery food X	200	1	3
Mystery food Y	100	3	3

39. **Land Reclamation**   A state government included $1.2 million in its current appropriations bill to reclaim land at a 3000-acre strip mine site. It will cost $800 per acre to return the land to productive grassland suitable for livestock and $500 per acre to return the land to forest suitable for commercial timber production. The appropriations bill also requires that at least 1800 acres be reclaimed immediately and then the income from leasing it to ranchers and/or paper companies be used to reclaim the rest of the site in the coming years. If long-term-use agreements yield $200 per acre of grassland and $150 per acre of forest, how many acres of each should be reclaimed this year to raise the greatest amount for next year's reclamation efforts?

40. **Urban Renewal**   A nonprofit urban development corporation has agreed to rebuild at least 24 city blocks in the south side of Megatropolis with at least 15 blocks of semidetached single family homes and at least 3 but not more than 10 blocks of commercial buildings (retail stores and/or light industry). If it will cost $6 million to rebuild 1 block with homes and $7 million to rebuild 1 block for commercial use, how many blocks of each should be rebuilt to meet the goals with the least cost?

3.3 The Simplex Method for Standard Maximum Problems

George Dantzig and the Origins of the Simplex Method*

The military term "program" refers to a training, supply, and troop deployment schedule. During World War II, such programs became critically important and took enormous effort to develop. George Dantzig* became an expert at solving such problems while serving as the mathematical advisor to the U.S. Air Force Comptroller. Since there were no computers, finding a solution meant managing large rooms of clerks carrying out their assigned computations on mechanical calculators (with numbers handed them on slips of paper), writing their results on more pieces of paper, and passing these to other clerks for further calculations.

After the war, Dantzig sought a way to compute solutions in less time. He was familiar with the matrix input-output model of the American economy proposed by Wassily Leontief in 1932 and he saw the need to generalize it to include alternative activities and make it more easily computable. He wrote years later that "initially there was no objective function: explicit goals did not exist because practical planners had no way to implement such a concept," but by mid-1947 he "decided that the objective had to be made explicit." He formulated a "linear programming problem" as the optimization of "a linear form subject to linear equations and inequalities." Assuming that economists were familiar with this type of problem, he visited T. C. Koopmans (who had worked on transportation models for the Allied Shipping Board during the war) at the University of Chicago. Koopmans immediately saw the implications of Dantzig's work for general economic planning, but could not offer a method of solution. Dantzig at first rejected on intuitive grounds the obvious idea of moving along the edges from vertex to vertex as being too inefficient. However, when he tried it, "by good luck it worked!" This "simplex" method was named for the geometric objects from which the convex feasible region can be built, just as in two variables the region can be viewed as a collection of adjacent triangles whose vertices form the corners of the region.

In a 1981 address to the American Mathematical Society, Dantzig concluded by remarking that "The ability to state general objectives and then find optimal policy solutions to practical decision problems

* George B. Dantzig, "Reminiscences about the Origins of Linear Programming," *Essays in the History of Mathematics,* Memoirs of the American Mathematical Society, Number 298, March 1984, pp. 1–11.

of great complexity is a revolutionary development. . . . [But in] areas such as modeling the dynamics of growing populations of the world against a diminishing resource base, its potential for raising the standard of living has scarcely been realized."

Introduction

Using the geometric method from the previous section as a guide, we shall explain and demonstrate the *simplex method* for solving linear programming problems. This method finds a "path of vertices" leading from an initial vertex to a solution without finding all the intersection points of the boundary lines or even all the vertices of the feasible region.

Standard Maximum Problems

A linear programming problem is a *standard maximum problem* if the objective function is to be maximized, the constraints include non-negativity conditions for all the variables, and the origin is a vertex of the feasible region. Writing $x_1, x_2, x_3, \ldots, x_n$ for the variables instead of just x and y, a standard maximum problem can be written in the following form.

Standard Maximum Problem

Maximize $P = c_1 x_1 + c_2 x_2 + \cdots + c_n x_n$

Subject to
$$\begin{cases} a_{1,1} x_1 + a_{1,2} x_2 + \cdots + a_{1,n} x_n \leq b_1 \\ a_{2,1} x_1 + a_{2,2} x_2 + \cdots + a_{2,n} x_n \leq b_2 \\ \quad \vdots \\ a_{m,1} x_1 + a_{m,2} x_2 + \cdots + a_{m,n} x_n \leq b_m \\ x_1 \geq 0,\, x_2 \geq 0,\, \ldots,\, x_n \geq 0 \end{cases}$$

A "larger" version of Maximize
$P = Mx + Ny$
Subject to $\begin{cases} ax + by \leq c \\ \quad \vdots \end{cases}$

where $b_1 \geq 0,\, b_2 \geq 0,\, \ldots,\, b_m \geq 0$

Here n is the number of variables and m is the number of constraints other than the nonnegativity conditions $x_1 \geq 0,\, x_2 \geq 0,\, \ldots,\, x_n \geq 0$.

Since the numbers b_1, b_2, . . . , b_m after the "at most" inequalities are all nonnegative, $x_1 = 0, x_2 = 0, \ldots, x_n = 0$ satisfies all the constraints and so the origin *is* a vertex of the region.

If we define the matrices c^t, A, X, and b to be:

$$c^t = (c_1, c_2, \ldots, c_n), \quad A = \begin{pmatrix} a_{1,1} & a_{1,2} & \cdots & a_{1,n} \\ \vdots & \vdots & \vdots & \vdots \\ a_{m,1} & a_{m,2} & \cdots & a_{m,n} \end{pmatrix},$$

From the objective function   From constraints before $\leq$

$$X = \begin{pmatrix} x_1 \\ \vdots \\ x_n \end{pmatrix}, \quad \text{and} \quad b = \begin{pmatrix} b_1 \\ \vdots \\ b_m \end{pmatrix},$$

Variables   From constraints after $\leq$

then we may write this standard maximum problem in the compact form:

Standard Maximum Problem

Maximize $P = c^t X$

Subject to $\begin{cases} AX \leq b \\ X \geq 0 \end{cases}$

where $b \geq 0$

The Initial Simplex Tableau

The constraints in the standard maximum problem are written as inequalities. Since equations are easier to solve than inequalities, we first simplify the problem by changing the inequalities into equations by adding to each a *slack variable* that represents the "amount not used." For example, the two sides of the inequality $4 \leq 7$ differ by 3, so adding 3 to the left side gives an *equation* $4 + 3 = 7$. Similarly, the inequality $x \leq 7$ can be rewritten as the equation $x + s = 7$ by adding a slack variable $s \geq 0$ to the left side to make up the difference between the sides. We do this for each inequality, using a different slack variable $s_1, s_2, s_3, \ldots, s_m$ for each inequality. Using zeros to indicate the slacks that do not appear in a constraint, we can rewrite the constraints as *equations:*

$$a_{1,1}x_1 + a_{1,2}x_2 + \cdots + a_{1,n}x_n + 1s_1 + 0s_2 + \cdots + 0s_m = b_1$$

$$a_{2,1}x_1 + a_{2,2}x_2 + \cdots + a_{2,n}x_n + 0s_1 + 1s_2 + \cdots + 0s_m = b_2$$

$$a_{m,1}x_1 + a_{m,2}x_2 + \cdots + a_{m,n}x_n + 0s_1 + 0s_2 + \cdots + 1s_m = b_m$$

Constraints with variables arranged in columns

To express the objective function $P = c_1x_1 + c_2x_2 + \cdots + c_nx_n$ in the same form, we include the slacks by adding on $0s_1 + 0s_2 + \cdots + 0s_m$ and then rewriting it as the equation:

$$P - c_1x_1 - c_2x_2 - \cdots - c_nx_n + 0s_1 + 0s_2 + \cdots + 0s_m = 0$$

Since both the objective function and the constraints are now written as equations with the variables appearing in the same order, this standard maximum problem can be represented by just a rectangular table of numbers. Dantzig named this representation the "simplex tableau" for the problem.

Initial Simplex Tableau

The *initial simplex tableau* of the standard maximum linear programming problem is the rectangular table of variables and coefficients

	x_1	x_2	$\cdots$	x_n	s_1	s_2	$\cdots$		s_m		
s_1	$a_{1,1}$	$a_{1,2}$	$\cdots$	$a_{1,n}$	1	0	0	$\cdots$	0	0	b_1
s_2	$a_{2,1}$	$a_{2,2}$	$\cdots$	$a_{2,n}$	0	1	0	$\cdots$	0	0	b_2
$\vdots$	$\vdots$	$\vdots$	$\ddots$	$\vdots$	$\vdots$	$\vdots$		$\ddots$	$\vdots$		$\vdots$
s_m	$a_{m,1}$	$a_{m,2}$	$\cdots$	$a_{m,n}$	0	0	0	$\cdots$	0	1	b_m
P	$-c_1$	$-c_2$	$\cdots$	$-c_m$	0	0	0	$\cdots$	0	0	0

Column variables

Constraints

Objective function

Represents an equal sign

Using the matrices defined on page 317, the initial simplex tableau takes the form:

Initial Simplex Tableau

	X	S	
S	A	I	b
P	$-c^t$	0	0

$$I = \begin{pmatrix} 1 & 0 & 0 & \cdots & 0 \\ 0 & 1 & 0 & \ddots & \vdots \\ 0 & 0 & 1 & \ddots & 0 \\ 0 & \ddots & \ddots & \ddots & 0 \\ 0 & 0 & \cdots & 0 & 1 \end{pmatrix}$$

The matrix I is the "m by m identity matrix" consisting of all zeros except for 1's along the diagonal from top left to bottom right.

EXAMPLE 1 Constructing an Initial Simplex Tableau

Rewrite the following problem in matrix form, determine if it is a standard maximum problem and, if it is, construct the initial simplex tableau.

$$\text{Maximize } P = 9x_1 - 2x_2 + 8x_3$$

$$\text{Subject to } \begin{cases} -16x_1 + 4x_2 + 11x_3 \le 176 \\ 15x_1 + 3x_2 + 9x_3 \le 135 \\ x_1 \ge 0,\ x_2 \ge 0,\ \text{and } x_3 \ge 0 \end{cases}$$

Solution

The matrix form of the problem is

$$\text{Maximize } P = \overset{c^t}{(\ 9 \quad -2 \quad 8\ )} \overset{X}{\begin{pmatrix} x_1 \\ x_2 \\ x_3 \end{pmatrix}} \qquad \text{Maximize } P = c^t\, X$$

$$\text{Subject to } \left\{ \begin{pmatrix} -16 & 4 & 11 \\ 15 & 3 & 9 \end{pmatrix} \begin{pmatrix} x_1 \\ x_2 \\ x_3 \end{pmatrix} \le \begin{pmatrix} 176 \\ 135 \end{pmatrix} \right. \qquad \text{Subject to } \begin{cases} A\,X \le b \\ X \ge 0 \end{cases}$$

$$\text{and } \begin{pmatrix} x_1 \\ x_2 \\ x_3 \end{pmatrix} \ge 0$$

None of the numbers in the matrix $b = \begin{pmatrix} 176 \\ 135 \end{pmatrix}$ is negative, so this *is* a standard maximum problem. Since there are two constraint inequalities besides the nonnegativity conditions, there will be two slack variables. The initial simplex tableau is

	x_1	x_2	x_3	s_1	s_2	
s_1	-16	4	11	1	0	176
s_2	15	3	9	0	1	135
P	-9	2	-8	0	0	0

	X	S	
S	A	I	b
P	$-c^t$	0	0

where $I = \begin{pmatrix} 1 & 0 \\ 0 & 1 \end{pmatrix}$

> The block of numbers $\begin{matrix} 1 & 0 \\ 0 & 1 \end{matrix}$ under the slack variables represents the
> 2×2 identity matrix consisting of all zeros except for 1's along the
> diagonal from top left to bottom right.
>
> ■

PRACTICE PROBLEM 1   Rewrite the following problem in matrix form, determine if it is a standard maximum problem, and, if it is, construct the initial simplex tableau.

$$\text{Maximize } P = 9x_1 - x_2 + 10x_3 + 12x_4$$

$$\text{Subject to } \begin{cases} x_1 + x_2 + 2x_3 + 2x_4 \le 16 \\ 2x_1 + 2x_2 + x_3 + x_4 \le 20 \\ x_1 + 2x_2 + 2x_3 + x_4 \le 18 \\ x_1 \ge 0, \, x_2 \ge 0, \, x_3 \ge 0, \text{ and } x_4 \ge 0 \end{cases}$$

Solution at the back of the book

Basic and Nonbasic Variables

In an initial simplex tableau such as:

	x_1	x_2	x_3	s_1	s_2	
s_1	−16	4	11	1	0	176
s_2	15	3	9	0	1	135
P	−9	2	−8	0	0	0

The slack variables s_1 and s_2 have two special properties:

1. The columns for these variables each contain all zeros except for exactly one 1,

2. Each row above the bottom row has exactly one of these special 1's.

These special variables determine a feasible point in the region without any further calculation: just let each take the value at the right end of its row and set all the remaining variables equal to zero. For the tableau above, this means:

	x_1	x_2	x_3	s_1	s_2		
s_1	-16	4	11	1	0	176	$\leftarrow s_1 = 176$
s_2	15	3	9	0	1	135	$\leftarrow s_2 = 135$
P	-9	2	-8	0	0	0	and $x_1 = x_2 = x_3 = 0$

Other variables
are zero

In general, any m variables in a simplex tableau satisfying properties (1) and (2) above are called *basic* variables and form a *basis*. The other variables are the *nonbasic* variables. Because of their importance, the basic variables are listed on the left side of the simplex tableau, each next to the row containing its special 1. The tableau is to be interpreted as specifying a *basic feasible point* (a vertex) of the region: just let each basic variable take the value at the right end of its row and set the nonbasic variables equal to zero. Moreover, the tableau also displays the value of the objective function at this point in the bottom right corner. For example, the simplex tableau below has basic variables s_1, $s_2, \ldots, s_m$, nonbasic variables $x_1, x_2, \ldots, x_n$, and displays the basic feasible point shown on the right.

Basic variables $\left\{ \begin{array}{c} \\ \\ \\ \end{array} \right.$

	x_1	x_2	$\cdots$	x_n	s_1	s_2	$\cdots$	s_m	
s_1	$a_{1,1}$	$a_{1,2}$	$\cdots$	$a_{1,n}$	1	0	0	0	b_1
$\vdots$	$\vdots$	$\vdots$		$\vdots$	$\vdots$			$\vdots$	$\vdots$
s_m	$a_{m,1}$	$a_{m,2}$	$\cdots$	$a_{m,n}$	0	0	0	1	b_m
P	$-c_1$	$-c_2$	$\cdots$	$-c_m$	0	0	0	0	0

Values for the
basic variables

$$s_1 = b_1$$
$$\vdots$$
$$s_m = b_m$$

$$x_1 = 0, \ldots, x_n = 0$$

Value of P at this vertex

The simplex method begins by taking the slack variables as the basic variables, and then finding successively better basic variables (ones that increase the value of the objective function) until the solution is found. This procedure mimics the geometric procedure that we used earlier for two-variable problems: calculating ratios to find intercepts (an operation now called *finding the pivot element*), and then using the substitution method to find intersection points (now called *performing the pivot operation*).

The Pivot Element

To find the pivot element, we first find its column and then its row.

The best variable to use to increase the objective function is the variable with the largest positive coefficient, because increasing it can have the greatest effect. Since the bottom row of the tableau contains the *negatives* of these coefficients, this means choosing the variable with the *smallest negative entry in the bottom row.*

Pivot Column

> The *pivot column* is the column with the smallest negative entry in the bottom row of the tableau. If there is a tie, the pivot column is the leftmost such column.

Having chosen the variable, how much can it be increased without violating a constraint? Recall from the previous section that in a constraint like $3x + 4y \le 24$, the x cannot be increased further than the intercept $\frac{24}{3} = 8$ (the ratio of the rightmost number divided by the coefficient of x). Analogously, the chosen variable cannot exceed the ratio of the rightmost number of any constraint divided by its pivot column entry. Not exceeding *each* ratio means not exceeding the *smallest* of them. Of course, ratios involving division by zero or by negative numbers should not even be considered since they correspond to boundary lines parallel to an axis or intercepts not in the first quadrant. Therefore:

Pivot Row

> For each row except the bottom row, divide the rightmost entry by the pivot column entry (omitting any row with a zero or negative pivot column entry). The *pivot row* is the row with the *smallest* such ratio. If there is a tie, the pivot row is the uppermost such row.

Pivot Element

> The *pivot element* is the entry in the pivot column and the pivot row.

If the tableau does not have a pivot column, then the objective function cannot be increased, so the solution has been found: The maximum value of the objective function appears in the bottom right corner of the tableau and it occurs at the basic feasible point given by the tableau (see pages 320–321).

If the tableau has a pivot column but no pivot row, then there is *no solution* because there is a direction that will increase the objective function but there is no boundary to prevent the variable from increasing in this direction without bound.

The Pivot Operation

Once we have identified the pivot element, we need to increase the pivot column variable x_c until we reach the boundary of the feasible region. The *pivot operation* rewrites the simplex tableau to display the basic feasible point, with x_c equal to the ratio that determined the pivot row. The pivot operation moves x_c into the basis by making the pivot element into a 1 and then making the rest of the pivot column into zeros. This rewriting process is the tableau version of the substitution method: solving the equation represented by the pivot row to find x_c and then using the resulting expression to eliminate x_c from the other equations.

The Pivot Operation

Let PE be the **pivot element**, R_{pivot} be the **pivot row**, R_{other} be any **other** row, and PCE be the **pivot column entry** in that other row. The pivot operation builds the **new rows** R^{new} of the next tableau in two steps:

1. Divide every entry in the pivot row by the pivot element:

$$R_{pivot}^{new} = \frac{R_{pivot}}{PE}$$

   The variable at the far left of the pivot row is replaced by the variable of the pivot column to show the new basis.

2. Subtract from each of the *other* rows the pivot column entry of that row times the new pivot row:

$$R_{other}^{new} = R_{other} - (PCE)\, R_{pivot}^{new}$$

EXAMPLE 2   The Pivot Element and the Pivot Operation

Find the pivot element in the following simplex tableau and carry out the pivot operation.

	x_1	x_2	x_3	x_4	s_1	s_2	s_3	
s_1	1	1	-2	1	1	0	0	8
s_2	6	-2	2	4	0	1	0	10
s_3	5	3	1	3	0	0	1	6
P	-8	12	-10	-6	0	0	0	0

Solution

The smallest negative entry in the bottom row is -10, so the pivot column is column 3. To find the pivot row, we divide the last entry of each row by the pivot column entry for that row (except the bottom row and any row having a zero or negative pivot column entry) and choose the row with the smallest nonnegative ratio. The calculations on the right below show that the pivot row is row 2. The pivot element is the 2 at the intersection of the pivot column and the pivot row.

Pivot element

	x_1	x_2	x_3	x_4	s_1	s_2	s_3		
s_1	1	1	-2	1	1	0	0	8	(Can't use since -2 is < 0)
s_2	6	-2	2	4	0	1	0	10	$\frac{10}{2} = 5$ ← Smallest ratio ← Pivot row
s_3	5	3	1	3	0	0	1	6	$\frac{6}{1} = 6$
P	-8	12	-10	-6	0	0	0	0	

Smallest negative

Pivot column

Having found the pivot element, we carry out the pivot operation. The first step is to divide the pivot row by the pivot element, so we divide the second row by 2. We also update the basis on the left (the pivot column variable replaces the pivot row variable).

x_3 replaces s_2 in basis →

	x_1	x_2	x_3	x_4	s_1	s_2	s_3	
s_1	1	1	-2	1	1	0	0	8
x_3	3	-1	1	2	0	1/2	0	5
s_3	5	3	1	3	0	0	1	6
P	-8	12	-10	-6	0	0	0	0

$R^{new}_{pivot} = \dfrac{R_{pivot}}{2}$ Other rows remain the same

The second step of the pivot operation is to subtract multiples of the new pivot row from the other rows. For the first row, the pivot column entry (PCE) is -2. So the formula $R^{new}_{other} = R_{other} - (PCE)\, R^{new}_{pivot}$ (on page 323) says to multiply the pivot row by -2 and subtract it from the first row: $R^{new}_1 = R_1 - (-2)R^{new}_{pivot}$. Since this is the same as $R^{new}_1 =$

$R_1 + (2)R_{pivot}^{new}$, we multiply the pivot row by 2 and add it to the first row:

1	1	−2	1	1	0	0	8	R_1	Row 1 (above)
+ 6	−2	2	4	0	1	0	10	$(2)R_{pivot}^{new}$	Twice pivot row
7	−1	0	5	1	1	0	18	$R_1^{new} = R_1 + (2)R_{pivot}^{new}$	New row 1

For the third row, the pivot column entry (*PCE*) is 1, so the formula on page 323 gives $R_3^{new} = R_3 - (1)R_{pivot}^{new}$ or simply $R_3^{new} = R_3 - R_{pivot}^{new}$. For the bottom row, the PCE is −10, so the formula gives $R_3^{new} = R_3 - (-10)R_{pivot}^{new}$ or simply $R_3^{new} = R_3 + 10R_{pivot}^{new}$. Carrying out these two calculations on scratch paper (as we did above) and writing each of the four new rows in its proper place gives the new simplex tableau:

<table>
<tr><td></td><td>x_1</td><td>x_2</td><td>x_3</td><td>x_4</td><td>s_1</td><td>s_2</td><td>s_3</td><td></td><td colspan="2"></td></tr>
<tr><td>s_1</td><td>7</td><td>−1</td><td>0</td><td>5</td><td>1</td><td>1</td><td>0</td><td>18</td><td>$R_1^{new} = R_1 + (2)R_{pivot}^{new}$</td><td></td></tr>
<tr><td>x_3</td><td>3</td><td>−1</td><td>1</td><td>2</td><td>0</td><td>1/2</td><td>0</td><td>5</td><td>$R_{pivot}^{new} = R_{pivot}/2$</td><td>← Do pivot</td></tr>
<tr><td>s_3</td><td>2</td><td>4</td><td>0</td><td>1</td><td>0</td><td>−1/2</td><td>1</td><td>1</td><td>$R_3^{new} = R_3 - R_{pivot}^{new}$</td><td>row first</td></tr>
<tr><td>P</td><td>22</td><td>2</td><td>0</td><td>14</td><td>0</td><td>5</td><td>0</td><td>50</td><td>$R_4^{new} = R_4 + (10)R_{pivot}^{new}$</td><td></td></tr>
</table>

The new basis takes the values at the ends of the rows

All of these steps—finding the pivot element, updating the basic variables on the left of the tableau, and changing each row according to the formulas in the box on page 323 (always beginning with the pivot row)—comprise *one* pivot operation. You will have to practice with scrap paper before you fully master this process. Be sure to begin by finding the new pivot row, writing the new basic variable on the left, and then working on the other rows from the top down. From now on we will just state the resulting tableau, omitting the separate calculations.

Graphing Calculator Exploration

The program PIVOT* carries out the pivot operation after you specify the pivot column and pivot row. To carry out the pivot operation on your calculator with the simplex tableau from Example 2 on page 323, proceed as follows.

---

* See Preface for information on how to obtain this program.

a.   Enter the simplex tableau as one large matrix [A]. The tableau from Example 2 has four rows of numbers and eight columns. Enter the numbers one by one until the tableau is complete. Since this tableau is too large to fit on the calculator screen, it will scroll from left to right and back again as you enter your numbers.

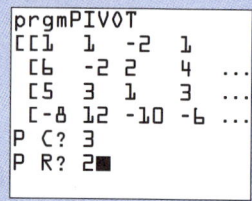

b.   Run the program PIVOT. It will display the current tableau in matrix [A] and you can use the arrows to scroll the screen to see the rest of it. Press ENTER to enter the Pivot Column and then again to enter the Pivot Row.

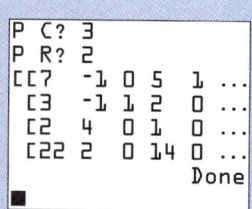

c.   After you ENTER the pivot row, the calculator will perform the pivot operation and display the new tableau. Check this result for Example 2 with the new simplex tableau on page 325. (Of course, if you select a zero for your pivot element, the program will respond with an error message.)

Press ENTER to exit the program.

If your problem requires several pivot operations, you can rerun the program with the new tableau by pressing the ENTER key again.

PRACTICE PROBLEM 2   In each of the following simplex tableaux, find the pivot element and carry out the pivot operation.

a.

	x_1	x_2	x_3	x_4	s_1	s_2	
s_1	-1	2	1	1	1	0	4
s_2	3	6	3	2	0	1	9
P	-5	-3	-6	-2	0	0	0

b.

	x_1	x_2	x_3	x_4	s_1	s_2	
s_1	3	6	-1	2	1	0	18
s_2	2	-8	0	3	0	1	24
P	-3	-1	-4	-2	0	0	0

Solution at the back of the book

The Simplex Method

The *simplex method* solves a standard maximum problem by repeating the pivot operation until the tableau does not have a pivot element and this final tableau gives the solution of the problem.

The Simplex Method

To solve a standard maximum problem by the simplex method:

1. Construct the initial simplex tableau (page 318).

2. Locate the pivot element (page 322) and go to Step 3. If the tableau does not have a pivot element, go to Step 4.

3. Perform the pivot operation (page 323) on the pivot element and return to Step 2.

4. If the final tableau does not have a pivot *column*, then the solution occurs at the vertex given by the basic variables (whose values appear at the right ends of their rows, with the nonbasic variables equal to zero), and the maximum value of the objective function appears in the bottom right corner of the tableau. If the final tableau *does* have a pivot *column* but no pivot *row*, then there is *no solution* to the problem.

The simplex method uses the pivot operation to make the bottom row "less negative" while keeping the right column nonnegative. That is, *the simplex method makes the tableau "more optimal" while keeping it feasible.* When the bottom row contains no negative entries, the tableau is *optimal* since the largest value for the objective function has been found.

EXAMPLE 3   The Simplex Method

Solve the following linear programming problem by the simplex method.

$$\text{Maximize } P = 3x_1 + 5x_2$$

$$\text{Subject to } \begin{cases} 2x_1 + x_2 \le 8 \\ x_1 - x_2 \le 1 \\ x_2 \le 4 \\ x_1 \ge 0 \text{ and } x_2 \ge 0 \end{cases}$$

Solution

This problem in matrix form is

Maximize $P = \begin{pmatrix} 3 & 5 \end{pmatrix} \begin{pmatrix} x_1 \\ x_2 \end{pmatrix}$ Maximize $P = c^t X$

Subject to $\left\{ \begin{pmatrix} 2 & 1 \\ 1 & -1 \\ 0 & 1 \end{pmatrix} \begin{pmatrix} x_1 \\ x_2 \end{pmatrix} \leq \begin{pmatrix} 8 \\ 1 \\ 4 \end{pmatrix} \right.$ Subject to $\begin{cases} A X \leq b \\ X \geq 0 \end{cases}$

and $\begin{pmatrix} x_1 \\ x_2 \end{pmatrix} \geq 0$

None of the numbers in the matrix $b = \begin{pmatrix} 8 \\ 1 \\ 4 \end{pmatrix}$ is negative, so this *is* a standard maximum problem. Besides the nonnegativity conditions, there are three constraint inequalities so there will be three slack variables. In the initial simplex tableau, these slacks will be the basic variables:

	x_1	x_2	s_1	s_2	s_3	
s_1	2	1	1	0	0	8
s_2	1	−1	0	1	0	1
s_3	0	1	0	0	1	4
P	−3	−5	0	0	0	0

Basic variables $\left\{ \vphantom{\begin{matrix} s_1 \\ s_2 \\ s_3 \end{matrix}} \right.$

	X	S	
S	A	I	b
P	$-c^t$	0	0

where $I = \begin{pmatrix} 1 & 0 & 0 \\ 0 & 1 & 0 \\ 0 & 0 & 1 \end{pmatrix}$

The pivot column is column 2 (with the smallest negative number in the bottom row). The pivot row is row 3 (with the smallest ratio of the rightmost number divided by the pivot column entry). The pivot element is the 1 in column 2 and row 3:

Pivot element

	x_1	x_2	s_1	s_2	s_3		
s_1	2	1	1	0	0	8	$\frac{8}{1} = 8$
s_2	1	−1	0	1	0	1	(Omit since $-1 < 0$)
s_3	0	1	0	0	1	4	$\frac{4}{1} = 4$ ← Pivot row
P	−3	−5	0	0	0	0	

↑
Smallest
negative

Pivot column

Carrying out the pivot operation (applying the formulas on page 323 to each row, showing the row operations on the right) gives the following tableau.

	x_1	x_2	s_1	s_2	s_3	
s_1	2	0	1	0	-1	4
s_2	1	0	0	1	1	5
x_2	0	1	0	0	1	4
P	-3	0	0	0	5	20

x_2 replaces s_3 in basis →

$R_1^{new} = R_1 - (1)R_{pivot}^{new}$

$R_2^{new} = R_2 - (-1)R_{pivot}^{new}$ Or just $R_2 + R_{pivot}^{new}$

$R_{pivot}^{new} = R_{pivot}$ Unchanged since divided by 1

$R_4^{new} = R_4 - (-5)R_{pivot}^{new}$ Or just $R_4 + 5R_{pivot}^{new}$

The -3 at the bottom of the first column means that there is yet another pivot column. The smallest ratio from the rows is $\frac{4}{2} = 2$ for the first row, so row 1 is the pivot row. Carrying out the pivot operation on the 2 in the first column and first row gives the tableau shown below (as you should verify).

	x_1	x_2	s_1	s_2	s_3	
x_1	1	0	$1/2$	0	$-1/2$	2
s_2	0	0	$-1/2$	1	$3/2$	3
x_2	0	1	0	0	1	4
P	0	0	$3/2$	0	$7/2$	26

x_1 replaces s_1 in basis →

$R_{pivot}^{new} = R_{pivot}/2$ ← Do this first

$R_2^{new} = R_2 - (1)R_{pivot}^{new}$

$R_3^{new} = R_3 - (0)R_{pivot}^{new}$ Unchanged since subtract nothing

$R_4^{new} = R_4 - (-3)R_{pivot}^{new}$ Or just $R_4 + 3R_{pivot}^{new}$

Since the bottom row of this tableau does not contain any negative entries, *this is the final tableau.* The basic variables take the values at the ends of their rows, so $x_1 = 2$, $s_2 = 3$, and $x_2 = 4$, the nonbasic variables are zero, so $s_1 = 0$ and $s_3 = 0$, and the objective function takes the value at the end of the bottom row, so $P = 26$. (You should check that these values satisfy the original constraints and give the claimed value in the objective function.)

Answer: The maximum value of P is 26 at the vertex $x_1 = 2$, $x_2 = 4$. ∎

We can compare this solution with the graphical method of the previous section by graphing the feasible region with x_1 on the horizontal axis and x_2 on the vertical axis, as shown at the top of page 330. The five vertices and the values of the objective function are given in the table and the "path of vertices" visited by the simplex method is marked in bold.

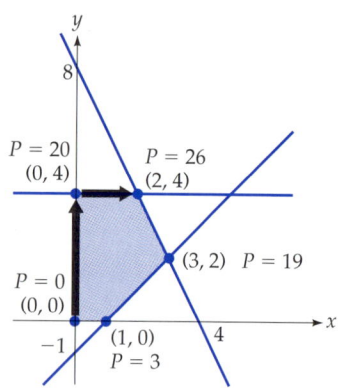

Vertex (x_1, x_2)	$P = 3x_1 + 5x_2$
$(0, 0)$	0
$(1, 0)$	3
$(0, 4)$	20
$(3, 2)$	19
$(2, 4)$	26

Initial tableau → $(0, 0)$

Second tableau → $(0, 4)$

Final tableau → $(2, 4)$

 Graphing Calculator Exploration

Use the program PIVOT with the simplex tableau from Example 3 on page 327 to carry out the pivot operation calculations on the pivot elements in [column 2, row 3] and in [column 1, row 1], thus verifying that the final tableau given on page 329 is correct.

PRACTICE PROBLEM 3   Find the solution of the following linear programming problem from its final tableau shown on the right. (*Hint:* No calculations are necessary.)

Maximize $P = 5x_1 + 7x_2$

Subject to $\begin{cases} x_1 + 2x_2 \leq 16 \\ x_1 - x_2 \leq 4 \\ 2x_1 + x_2 \leq 14 \\ x_1 \geq 0 \text{ and } x_2 \geq 0 \end{cases}$

$\xrightarrow[\text{leads to}]{\text{Simplex method}}$

	x_1	x_2	s_1	s_2	s_3	
x_2	0	1	3/2	0	$-1/3$	6
s_2	0	0	1	1	-1	6
x_1	1	0	$-1/3$	0	2/3	4
P	0	0	3	0	1	62

Solution at the back of the book

 *EXAMPLE 4* **A Manufacturing Problem with Many Variables**

A pottery shop manufactures dinnerware in four different patterns by shaping the clay, decorating it, and then kiln firing it. The numbers of hours required per place setting for the four designs are given in the following table, together with the expected profits. The shop employs two skilled workers to do the initial shaping and three artists to do the decorating, none of whom will work more than 40 hours each week. The kiln can be used no more than 55 hours per week. How many place settings of each pattern should be made this week to obtain the greatest possible profit?

	Classic	Modern	Art Deco	Floral
Shaping	2 hours	1 hour	4 hours	2 hours
Decorating	3 hours	1 hour	6 hours	4 hours
Kiln firing	1 hour	1 hour	1 hour	1 hour
Expected profit	$10	$6	$9	$8

Solution

Let

$$x_1 = \left(\begin{array}{c}\text{Number of Classic}\\ \text{place settings}\end{array}\right), x_2 = \left(\begin{array}{c}\text{Number of Modern}\\ \text{place settings}\end{array}\right),$$

$$x_3 = \left(\begin{array}{c}\text{Number of Art Deco}\\ \text{place settings}\end{array}\right), \text{ and } x_4 = \left(\begin{array}{c}\text{Number of Floral}\\ \text{place settings}\end{array}\right)$$

made during the week, so $x_1 \geq 0$, $x_2 \geq 0$, $x_3 \geq 0$, and $x_4 \geq 0$. From the last line in the table, the profit to be maximized is $P = 10x_1 + 6x_2 + 9x_3 + 8x_4$. The constraints on the time spent shaping, decorating, and firing come from the other numbers in the table, along with the time available for each:

$$2x_1 + x_2 + 4x_3 + 2x_4 \leq 80 \qquad \text{2 shapers @ 40 hours each}$$

$$3x_1 + x_2 + 6x_3 + 4x_4 \leq 120 \qquad \text{3 decorators @ 40 hours each}$$

$$x + x_2 + x_3 + x_4 \leq 55 \qquad \text{Time kiln can be used}$$

In matrix form this problem is

$$\text{Maximize } P = (10 \quad 6 \quad 9 \quad 8) \begin{pmatrix} x_1 \\ x_2 \\ x_3 \\ x_4 \end{pmatrix}$$

$$\text{Subject to } \begin{cases} \begin{pmatrix} 2 & 1 & 4 & 2 \\ 3 & 1 & 6 & 4 \\ 1 & 1 & 1 & 1 \end{pmatrix} \begin{pmatrix} x_1 \\ x_2 \\ x_3 \\ x_4 \end{pmatrix} \leq \begin{pmatrix} 80 \\ 120 \\ 55 \end{pmatrix} \\ \\ \text{and } \begin{pmatrix} x_1 \\ x_2 \\ x_3 \\ x_4 \end{pmatrix} \geq 0 \end{cases}$$

Standard since $b = \begin{pmatrix} 80 \\ 120 \\ 55 \end{pmatrix} \geq 0$

The initial simplex tableau is

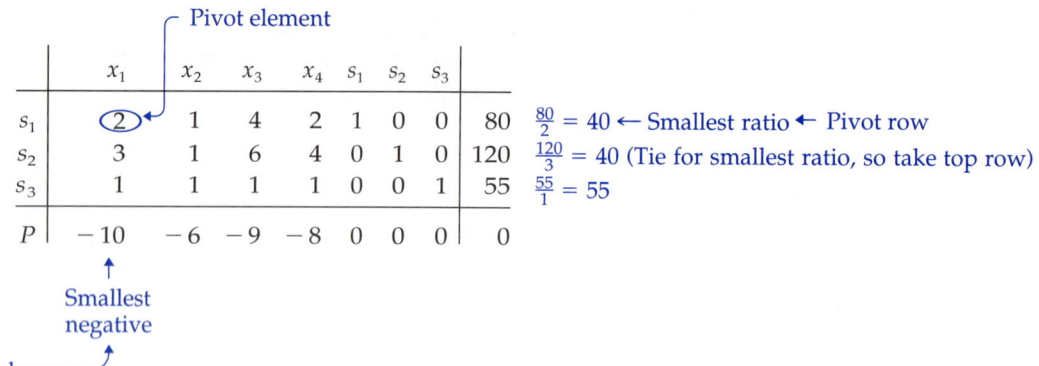

Pivot element

	x_1	x_2	x_3	x_4	s_1	s_2	s_3	
s_1	②	1	4	2	1	0	0	80	$\frac{80}{2} = 40$ ← Smallest ratio ← Pivot row
s_2	3	1	6	4	0	1	0	120	$\frac{120}{3} = 40$ (Tie for smallest ratio, so take top row)
s_3	1	1	1	1	0	0	1	55	$\frac{55}{1} = 55$
P	−10	−6	−9	−8	0	0	0	0

Smallest
negative

Pivot column

Pivoting (on the 2 in column 1, row 1), the tableau becomes:

Pivot element

	x_1	x_2	x_3	x_4	s_1	s_2	s_3	
x_1	1	1/2	2	1	1/2	0	0	40	$\frac{40}{1/2} = 80$
s_2	0	−1/2	0	1	−3/2	1	0	0	(Cannot use since $-\frac{1}{2}$ is < 0)
s_3	0	⑴/2	−1	0	−1/2	0	1	15	$\frac{15}{1/2} = 30$ ← Smallest ratio
P	0	−1	11	2	5	0	0	400

x_1 replaces x_1
s_1 in basis

Smallest
negative

Pivot column

Pivot row

Pivoting again (on the 1/2 in column 2, row 3), we reach the final tableau:

	x_1	x_2	x_3	x_4	s_1	s_2	s_3	
x_1	1	0	3	1	1	0	−1	25
s_2	0	0	−1	1	−2	1	1	15
x_2	0	1	−2	0	−1	0	2	30
P	0	0	9	2	4	0	2	430

x_2 replaces s_3 in basis →

No negatives in bottom row
so final tableau

The basic variables take the values at the right ends of their rows, the nonbasic variables are zero, and the objective function value appears in the bottom right corner. Therefore, the maximum value of P is 430 when $x_1 = 25$, $x_2 = 30$, $x_3 = 0$, and $x_4 = 0$. The fact that $s_2 = 15$ means that 15 hours of the available decorating time will not be needed.

Answer: The pottery shop should manufacture 25 place settings of Classic, 30 of Modern, and none of Art Deco and Floral.

How Good Is the Simplex Method?

Will the simplex method always find the solution? Even if it does find the solution, will it get to the answer after a reasonable number of tableaux?

In 1955, E. M. I. Beale published a linear programming problem with the awful property that the simplex method found the same sequence of tableaux over and over again. This repetition of tableaux is known as "cycling" because the tableaux repeat in a cyclic pattern. If a computer was solving this problem, it would never stop. Several examples of cycling and changes to the simplex method that prevent cycling are given in Exercises 56–60.

In 1972, V. Klee and G. J. Minty described a class of problems that take many more tableaux than their small size would suggest: with each new variable, the number of tableaux doubles. This means that a Klee–Minty problem with 2 variables takes 4 tableaux to solve, one with 3 variables takes 8 tableaux, one with 4 variables takes 16 tableaux, and so on. The 10-variable problem in this sequence will take 1024 tableaux. An example of these kinds of problems is given in Exercises 61–65.

Most "real world" computer programs for solving linear programming problems include subroutines to protect against cycling and Klee–Minty situations. S. Smale has shown that ". . . the number of pivots required to solve a linear programming problem grows in proportion to the number of variables on average." Several other methods (by L. Khachiyan and N. Karmarkar) to solve linear programming problems guarantee that the number of steps required does not grow too fast for every problem. However, the simplex method continues to provide the fastest solutions to commercial problems.

SUMMARY

The maximum linear programming problem:

$$\text{Maximize } P = c^t X$$

$$\text{Subject to } \begin{cases} AX \leq b \\ X \geq 0 \end{cases}$$

is a *standard maximum problem* if $b \geq 0$. The *initial simplex tableau* is

	X	S	
S	A	I	b
P	$-c^t$	0	0

The variables listed on the far left form a *basis*, and setting them equal to the values on the far right of their rows and setting the nonbasic variables equal to zero gives a *basic feasible point* that is a vertex of the region. The pivot column is the column with the smallest negative entry in the bottom row (omitting the right column), and the pivot row is the row with the smallest ratio between the last entry and the pivot column entry (omitting any row with a zero or negative pivot column entry, and the bottom row). The *pivot element* is the entry in the pivot column and pivot row.

The *pivot operation* divides the pivot row by the pivot element and then subtracts from each of the other rows the pivot column entry of that row times the new pivot row. The pivot column variable enters the basis by replacing the variable of the pivot row.

$$R_{pivot}^{new} = R_{pivot} / (Pivot\ Element)$$

$$R_{other}^{new} = R_{other} - \begin{pmatrix} Pivot \\ Column \\ Entry \end{pmatrix} R_{pivot}^{new}$$

The *simplex method* solves a standard maximum problem by pivoting until the tableau does not have a pivot element. If there is a pivot *column* but no pivot *row*, then there is no solution. If there is no pivot *column*, the solution occurs at the vertex given by the basic variables and the maximum value of the objective function appears in the bottom right corner.

EXERCISES 3.3

Construct the initial simplex tableau for each standard maximum problem.

1. Maximize $P = 8x_1 + 9x_2 + 7x_3$

Subject to $\begin{cases} 3x_1 + 2x_2 + 4x_3 \leq 12 \\ 6x_1 + x_2 + 5x_3 \leq 15 \\ x_1 \geq 0, x_2 \geq 0, x_3 \geq 0 \end{cases}$

2. Maximize $P = 10x_1 + 15x_2 + 12x_3$

Subject to $\begin{cases} 5x_1 + 2x_2 + 6x_3 \leq 30 \\ 3x_1 + 4x_2 + x_3 \leq 36 \\ x_1 \geq 0, x_2 \geq 0, x_3 \geq 0 \end{cases}$

3. Maximize $P = 13x_1 + 7x_2$

Subject to $\begin{cases} 4x_1 + 3x_2 \leq 12 \\ 5x_1 + 2x_2 \leq 20 \\ x_1 + 6x_2 \leq 12 \\ x_1 \geq 0, x_2 \geq 0 \end{cases}$

4. Maximize $P = 8x_1 + 33x_2$

Subject to $\begin{cases} 3x_1 + 2x_2 \le 30 \\ x_1 + 4x_2 \le 20 \\ 5x_1 + 6x_2 \le 60 \\ x_1 \ge 0, x_2 \ge 0 \end{cases}$

5. Maximize $P = 5x_1 - 2x_2 + 10x_3 - 5x_4$

Subject to $\begin{cases} 2x_1 + x_2 + x_3 + 3x_4 \le 6 \\ x_1 + 4x_2 - 2x_3 + x_4 \le 8 \\ x_1 \ge 0, x_2 \ge 0, x_3 \ge 0, x_4 \ge 0 \end{cases}$

6. Maximize $P = 5x_1 + 10x_2 - 30x_3 + 5x_4$

Subject to $\begin{cases} 3x_1 + x_2 - 2x_3 + 4x_4 \le 12 \\ 2x_1 - x_2 + 3x_3 + x_4 \le 12 \\ x_1 \ge 0, x_2 \ge 0, x_3 \ge 0, x_4 \ge 0 \end{cases}$

7. Maximize $P = 10x_1 + 20x_2 + 15x_3$

Subject to $\begin{cases} 8x_1 + x_2 + 4x_3 \le 32 \\ 3x_1 + 5x_2 + 7x_3 \le 30 \\ 6x_1 + 2x_2 + 9x_3 \le 28 \\ x_1 \ge 0, x_2 \ge 0, x_3 \ge 0 \end{cases}$

8. Maximize $P = 10x_1 + 15x_2 + 5x_3$

Subject to $\begin{cases} 7x_1 + 3x_2 + x_3 \le 21 \\ 6x_1 + 5x_2 + 4x_3 \le 20 \\ 8x_1 + 9x_2 + 2x_3 \le 45 \\ x_1 \ge 0, x_2 \ge 0, x_3 \ge 0 \end{cases}$

9. Maximize $P = 90x_1 + 80x_2 + 100x_3$

Subject to $\begin{cases} x_1 + 2x_2 + 3x_3 \le 45 \\ 6x_1 + 5x_2 + 4x_3 \le 40 \\ 7x_1 + 8x_2 + 9x_3 \le 63 \\ 12x_1 + 11x_2 + 10x_3 \le 60 \\ x_1 \ge 0, x_2 \ge 0, x_3 \ge 0 \end{cases}$

10. Maximize $P = 10x_1 + 20x_2 + 5x_3$

Subject to $\begin{cases} 9x_1 + x_2 + 8x_3 \le 20 \\ 10x_1 - 2x_2 + 7x_3 \le 40 \\ -11x_1 + 3x_2 - 6x_3 \le 80 \\ 12x_1 - 4x_2 + 5x_3 \le 60 \\ x_1 \ge 0, x_2 \ge 0, x_3 \ge 0 \end{cases}$

For each simplex tableau, find the pivot element and carry out one complete pivot operation. If there is no pivot element, explain what the tableau shows about the solution of the original standard maximum problem.

11.

	x_1	x_2	x_3	s_1	s_2	s_3	
s_1	2	0	1	1	0	0	4
s_2	1	1	1	0	1	0	5
s_3	1	-1	3	0	0	1	6
P	-7	-8	-9	0	0	0	0

12.

	x_1	x_2	x_3	s_1	s_2	s_3	
s_1	1	-2	1	1	0	0	5
s_2	3	1	2	0	1	0	6
s_3	2	0	3	0	0	1	7
P	-4	-5	-3	0	0	0	0

13.

	x_1	x_2	x_3	s_1	s_2	s_3	
s_1	1	0	1	1	0	0	3
s_2	0	1	1	0	1	0	4
s_3	1	1	0	0	0	1	5
P	-6	-7	-8	0	0	0	0

14.

	x_1	x_2	x_3	s_1	s_2	s_3	
s_1	0	1	1	1	0	0	5
s_2	1	0	1	0	1	0	3
s_3	1	1	0	0	0	1	4
P	-7	-8	-6	0	0	0	0

15.

	x_1	x_2	x_3	x_4	s_1	s_2	
s_1	4	2	6	2	1	0	12
s_2	3	1	2	1	0	1	8
P	-4	-5	6	-3	0	0	0

16.

	x_1	x_2	x_3	x_4	s_1	s_2	
s_1	5	3	1	3	1	0	30
s_2	6	2	2	4	0	1	16
P	-4	-6	-5	7	0	0	0

17.

	x_1	x_2	x_3	x_4	s_1	s_2	s_3	
s_1	0	0	2	1	1	1	-1	10
x_2	0	1	1	1	0	1	0	15
x_1	1	0	-2	0	0	-1	1	10
P	0	0	0	2	0	1	3	90

	x_1	x_2	x_3	x_4	s_1	s_2	s_3	
x_2	0	1	-1	0	1	-1	0	20
x_4	1	0	1	1	0	1	0	40
s_3	1	0	2	0	-1	1	1	30
P	2	0	8	0	3	3	0	300

18.

	x_1	x_2	s_1	s_2	s_3	s_4	
s_1	0	0	1	-1	0	0	10
x_1	1	0	0	1	-1	0	10
x_2	0	1	0	1	0	0	20
s_4	0	0	0	-1	0	1	5
P	0	0	0	25	-10	0	400

19.

	x_1	x_2	s_1	s_2	s_3	s_4	
s_1	0	0	1	0	-1	0	5
x_2	0	1	0	1	-1	0	10
x_1	1	0	0	1	0	0	20
s_4	0	0	0	-1	0	1	5
P	0	0	0	35	-15	0	550

20.

Solve each problem by the simplex method. (Exercises 21, 22, 31, and 32 may also be solved by the graphical method.)

You may use the PIVOT program if permitted by your instructor.

21. Maximize $P = x_1 + 2x_2$

Subject to $\begin{cases} 3x_1 + x_2 \le 24 \\ x_1 + x_2 \le 14 \\ x_1 \ge 0, x_2 \ge 0 \end{cases}$

22. Maximize $P = x_1 + 3x_2$

Subject to $\begin{cases} 2x_1 + x_2 \le 24 \\ x_1 + x_2 \le 15 \\ x_1 \ge 0, x_2 \ge 0 \end{cases}$

23. Maximize $P = 3x_1 + 4x_2 + 5x_3 + 2x_4$

Subject to $\begin{cases} x_1 + 2x_2 + x_3 + x_4 \le 8 \\ x_1 + x_2 + 2x_4 \le 7 \\ 2x_1 + x_2 + x_3 + x_4 \le 6 \\ x_1 \ge 0, x_2 \ge 0, x_3 \ge 0, x_4 \ge 0 \end{cases}$

24. Maximize $P = 3x_1 + 6x_2 + 5x_3 + 4x_4$

Subject to $\begin{cases} 3x_1 + x_2 + x_3 + 2x_4 \le 10 \\ x_1 + 2x_3 + x_4 \le 2 \\ x_1 + x_2 + 3x_3 + x_4 \le 8 \\ x_1 \ge 0, x_2 \ge 0, x_3 \ge 0, x_4 \ge 0 \end{cases}$

25. Maximize $P = 30x_1 + 40x_2 + 15x_3$

Subject to $\begin{cases} 2x_1 + x_2 + 3x_3 \le 150 \\ 3x_1 + 2x_2 + x_3 \le 100 \\ x_1 \ge 0, x_2 \ge 0, x_3 \ge 0 \end{cases}$

26. Maximize $P = 30x_1 + 60x_2 + 15x_3$

Subject to $\begin{cases} 2x_1 + 3x_2 + x_3 \le 75 \\ 3x_1 + x_2 + 2x_3 \le 50 \\ x_1 \ge 0, x_2 \ge 0, x_3 \ge 0 \end{cases}$

27. Maximize $P = 6x_1 + 4x_2 + 5x_3$

Subject to $\begin{cases} x_1 - x_2 + x_3 \le 20 \\ x_1 + 2x_3 \le 10 \\ x_1 \ge 0, x_2 \ge 0, x_3 \ge 0 \end{cases}$

28. Maximize $P = 6x_1 + 5x_2 + 8x_3$

Subject to $\begin{cases} 2x_1 + x_3 \le 40 \\ x_1 - x_2 + x_3 \le 30 \\ x_1 \ge 0, x_2 \ge 0, x_3 \ge 0 \end{cases}$

29. Maximize $P = 4x_1 + 3x_2 - 5x_3 + 6x_4$

Subject to $\begin{cases} x_1 + x_2 + x_4 \le 60 \\ x_1 + x_3 + x_4 \le 40 \\ x_1 + x_2 + x_3 \le 50 \\ x_1 \ge 0, x_2 \ge 0, x_3 \ge 0, x_4 \ge 0 \end{cases}$

30. Maximize $P = 2x_1 + 3x_2 + 4x_3 + x_4$

Subject to $\begin{cases} x_1 + x_2 + x_4 \le 20 \\ x_1 + x_3 + x_4 \le 15 \\ x_1 + x_2 + x_3 \le 25 \\ x_1 \ge 0, x_2 \ge 0, x_3 \ge 0, x_4 \ge 0 \end{cases}$

31. Maximize $P = 4x_1 + 2x_2$

Subject to $\begin{cases} x_1 - x_2 \le 1 \\ -x_1 + x_2 \le 3 \\ x_1 + x_2 \le 5 \\ x_1 \ge 0, x_2 \ge 0 \end{cases}$

32. Maximize $P = 12x_1 + 6x_2$

Subject to $\begin{cases} x_1 + x_2 \le 8 \\ x_1 - x_2 \le 2 \\ -x_1 + 2x_2 \le 10 \\ x_1 \ge 0, x_2 \ge 0 \end{cases}$

33. Maximize $P = 40x_1 + 60x_2 + 50x_3$

Subject to $\begin{cases} 2x_1 + x_2 + 4x_3 \le 400 \\ 4x_1 + 3x_2 + 2x_3 \le 600 \\ x_1 \ge 0, x_2 \ge 0, x_3 \ge 0 \end{cases}$

34. Maximize $P = 140x_1 + 80x_2 + 100x_3$

Subject to $\begin{cases} 4x_1 + 2x_2 + x_3 \le 100 \\ 2x_1 + 3x_2 + 3x_3 \le 150 \\ x_1 \ge 0,\, x_2 \ge 0,\, x_3 \ge 0 \end{cases}$

35. Maximize $P = x_1 + 2x_2 + 3x_3$

Subject to $\begin{cases} x_1 + x_2 + x_3 \le 15 \\ x_2 + x_3 \le 10 \\ x_3 \le 5 \\ x_1 \ge 0,\, x_2 \ge 0,\, x_3 \ge 0 \end{cases}$

36. Maximize $P = 15x_1 + 10x_2 + 5x_3$

Subject to $\begin{cases} x_1 + x_2 + x_3 \le 3 \\ x_1 + x_2 \le 2 \\ x_1 \le 1 \\ x_1 \ge 0,\, x_2 \ge 0,\, x_3 \ge 0 \end{cases}$

37. Maximize $P = 50x_1 + 30x_2 + 40x_3$

Subject to $\begin{cases} 4x_1 + 2x_2 + x_3 \le 80 \\ 2x_1 + x_2 + 3x_3 \le 120 \\ x_1 \ge 0,\, x_2 \ge 0,\, x_3 \ge 0 \end{cases}$

38. Maximize $P = 64x_1 + 56x_2 + 80x_3$

Subject to $\begin{cases} 3x_1 + x_2 + 2x_3 \le 240 \\ x_1 + 3x_2 + 4x_3 \le 160 \\ x_1 \ge 0,\, x_2 \ge 0,\, x_3 \ge 0 \end{cases}$

39. Maximize $P = 14x_1 + 5x_2 + 10x_3$

Subject to $\begin{cases} 4x_1 + x_2 + 3x_3 \le 130 \\ 3x_1 + x_2 + 2x_3 \le 95 \\ 4x_1 + 2x_2 + 3x_3 \le 140 \\ x_1 \ge 0,\, x_2 \ge 0,\, x_3 \ge 0 \end{cases}$

40. Maximize $P = 8x_1 + 12x_2 + 7x_3$

Subject to $\begin{cases} 4x_1 + 4x_2 + 3x_3 \le 440 \\ x_1 + 2x_2 + x_3 \le 150 \\ 3x_1 + 3x_2 + 2x_3 \le 320 \\ x_1 \ge 0,\, x_2 \ge 0,\, x_3 \ge 0 \end{cases}$

APPLIED EXERCISES

Formulate each situation as a linear programming problem by identifying the variables, the objective function, and the constraints. Check that the problem is a standard maximum problem and then solve it by the simplex method. Be sure to state your final answer in terms of the original question.

 You may use the PIVOT program if permitted by your instructor.

41. Production Planning   An automotive parts shop rebuilds carburetors, fuel pumps, and alternators. The number of hours to rebuild and then inspect and pack each part is shown in the table, together with the number of hours of skilled labor available for each task. If the profit is $12 for each carburetor, $14 for each fuel pump, and $10 for each alternator, how many of each should the shop rebuild to obtain the greatest possible profit?

	Carburetor	Fuel Pump
Rebuilding	5 hours	4 hours
Final inspection and packaging	1 hour	1 hour

	Alternator	Labor Available
Rebuilding	3 hours	200 hours
Final inspection and packaging	0.5 hour	45 hours

42. Pollution Control   An empty storage yard at a coal burning electric power plant can hold no more than 100,000 tons of coal. Two grades of coal are available: low sulfur (1%) with an energy content of 20 million British thermal units (BTU) per ton and high sulfur (2%) with an energy content of 30 million BTU per ton. If the next coal purchase may contain no more than 1400 tons of sulfur, how many tons of each type of coal should be purchased to obtain the most energy?

43. Recycling Management   A volunteer recycle center accepts both used paper and empty glass bottles, which are then sorted and sold to a reprocessing company. The center has room to accept 800 crates of paper and glass each week and 50 hours of volunteer help to do the sorting.

Each crate of paper products takes 5 minutes to sort and sells for 8¢ while each crate of bottles takes 3 minutes to sort and sells for 7¢. How many crates of each should the center accept each week to raise the most money for its ecology scholarship fund?

44. Production Planning   A small jewelry company prepares and mounts semiprecious stones. There are twenty lapidaries (who cut and polish the stones) and twenty-four jewelers (who mount the stones in gold settings). Each employee works seven hours each day. Each tray of agates requires five hours of cutting and polishing and four hours to mount, each tray of onyxes requires two hours of cutting and polishing and three hours to mount, and each tray of garnets requires six hours of cutting and polishing and three hours to mount. If the profit is $15 for each tray of agates, $10 for each tray of onyxes, and $12 for each tray of garnets, how many trays of each stone should be processed each day to obtain the greatest possible profit?

45. Production Planning   Repeat the previous problem but with a profit of $13 per tray of garnets.

46. Resource Allocation   A furniture shop manufactures wooden desks, tables, and chairs. The numbers of hours to assemble and finish each piece are shown in the table, together with the numbers of hours of skilled labor available for each task. If the profit is $80 for each desk, $84 for each table, and $68 for each chair, how many of each should the shop manufacture to obtain the greatest possible profit?

	Desk	Table
Assembly	2 hours	1 hour
Finishing	2 hours	3 hours

	Chair	Labor Available
Assembly	2 hours	200 hours
Finishing	1 hour	150 hours

47. Advertising   In the last few days before the election, a politician can afford to spend no more than $27,000 on TV advertisements and can arrange for no more than 10 ads. Each daytime ad costs $2000 and reaches 4000 viewers, each prime-time ad costs $3000 and reaches 5000 viewers, and each late-night ad costs $1000 and reaches 2000 viewers. Ignoring repeated viewings by the same person and assuming every viewer can vote, how many ads in each of the time periods will reach the most voters?

48. Advertising   The manager of a new mall may spend up to $18,000 on grand opening announcements in newspapers, on radio, and on TV. Each newspaper ad costs $300 and reaches 6000 readers, each one-minute radio commercial costs $800 and is heard by 10,000 listeners, and each 15-second TV spot costs $900 and is seen by 11,000 viewers. If there is time to arrange for no more than 5 newspaper ads and no more than 20 radio commercials and TV spots combined, how many of each should be placed to reach the largest number of potential customers? (Ignore multiple exposures to the same consumer.)

49. Agriculture   A farmer grows corn, peanuts, and soybeans on his two-hundred-forty-acre farm. To maintain soil fertility, the farmer rotates the crops and always plants at least as many acres of soybeans as the total acres of the other crops. Each acre of corn requires 2 days of labor and yields a profit of $150, each acre of peanuts requires 5 days of labor and yields a profit of $300, and each acre of soybeans requires 1 day of labor and yields a profit of $100. If the farmer and his family can put in at most 630 days of labor, how many acres of each crop should the farmer plant to obtain the greatest possible profit?

50. Agriculture   A farmer grows wheat, barley, and oats on her five-hundred-acre farm. Each acre of wheat requires 3 days of labor (to plant, tend, and harvest) and costs $21 (for seed, fertilizer, and pesticides), each acre of barley requires 2 days of labor and costs $27, and each acre of oats requires 3 days of labor and costs

$24. The farmer and her hired field hands can provide no more than 1200 days of labor this year and she can afford to spend no more than $15,120. If the expected profit is $50 for each acre of wheat, $40 for each acre of barley, and $45 for each acre of oats, how many acres of each crop should she grow to obtain the greatest possible profit?

Explorations and Excursions

The following problems extend and augment the material presented in the text.

 The PIVOT program may be helpful (if permitted by your instructor).

More About the Pivot Operation

51. a. Find the intersection of the lines $5x + 2y = 70$ and $4x + 3y = 84$ by solving the first equation for x and substituting this expression for x into the second equation. Then solve for y and find the value of x. (See Section 3.1, Example 5 on page 291.)

    b. Carry out one complete pivot operation using the 5 for the pivot element in the

$$\text{mini-tableau} \quad \begin{array}{c|cc|c} & x & y & \\ \hline & 5 & 2 & 70 \\ & 4 & 3 & 84 \end{array}. \quad \text{How do the}$$

rows of your new mini-tableau compare with the two equations you created for the first sentence of part (a)?

52. Show that pivoting on a tableau obeys the diagram:

$$
\begin{array}{ccc}
\vdots \quad \vdots & & \vdots \quad \vdots \\
\cdots \; c \;\; q \;\; \cdots & & \cdots \; 0 \;\; q - \dfrac{rc}{p} \;\; \cdots \\
\cdots \; p \;\; r \;\; \cdots & \longrightarrow & \cdots \; 1 \;\; r/c \;\; \cdots \\
\vdots \quad \vdots & & \vdots \quad \vdots
\end{array}
$$

where p is the pivot element, c is any other entry in the pivot column, r is any other entry in the pivot row, and q is any entry in the tableau not in the pivot row or pivot column. In words, this diagram defines the pivot operation in four steps: "(1) the pivot element becomes 1, (2) the rest of the pivot column becomes 0, (3) the rest of the pivot row is divided by the pivot element, and (4) every entry not in the pivot row or pivot column is decreased by the product of the pivot column entry in the same row with the pivot row entry in the same column divided by the pivot element." This form of the pivot operation makes it easy to check any particular number in a new tableau without repeating the row operations.

53. In Example 2 (pages 323–325), we carried out the pivot operation on the 2 in column 3 and row 2:

	x_1	x_2	x_3	x_4	s_1	s_2	s_3	
s_1	1	1	−2	1	1	0	0	8
s_2	6	−2	2	4	0	1	0	10
s_3	5	3	1	3	0	0	1	6
P	−8	12	−10	−6	0	0	0	0

	x_1	x_2	x_3	x_4	s_1	s_2	s_3	
s_1	7	−1	0	5	1	1	0	18
$\rightarrow\; x_3$	3	−1	1	2	0	1/2	0	5
s_3	2	4	0	1	0	−1/2	1	1
P	22	2	0	14	0	5	0	50

To undo this pivot operation, pivot on the $\frac{1}{2}$ in column 6 and row 2 of the final tableau. Does this return you to the initial simplex tableau? (This shows how you can undo a mistaken pivot operation and return to your previous tableau.)

54. The final tableaux in Exercises 17 and 18 both came from initial simplex tableaux having the form:

	x_1	x_2	x_3	x_4	s_1	s_2	s_3	
s_1	?	?	?	?	1	0	0	?
s_2	?	?	?	?	0	1	0	?
s_3	?	?	?	?	0	0	1	?
P	?	?	?	?	0	0	0	0

To find the initial simplex tableaux for these problems, begin with the actual final tableaux and pivot where necessary to return the columns of the slack variables to their initial form. Reconstruct the original problems from these initial tableaux and check that the answers you found in Exercises 17 and 18 satisfy these original constraints and objective functions.

55. The final tableaux in Exercises 19 and 20 both came from initial simplex tableaux having the form:

	x_1	x_2	s_1	s_2	s_3	s_4	
s_1	?	?	1	0	0	0	?
s_2	?	?	0	1	0	0	?
s_3	?	?	0	0	1	0	?
s_4	?	?	0	0	0	1	?
P	?	?	0	0	0	0	0

To find the initial simplex tableaux for these problems, begin with the actual final tableaux and pivot where necessary to return the columns of the slack variables to their initial form. Reconstruct the original problems from these initial tableaux and sketch the regions determined by their constraints. Do these regions confirm your answers to Exercises 19 and 20?

More About Cycling

56. Beale's Cycling Example Attempt to solve the following problem by the simplex method:

$$\text{Maximize } P = (3/4 \quad -20 \quad 1/2 \quad -6)\begin{pmatrix} x_1 \\ x_2 \\ x_3 \\ x_4 \end{pmatrix}$$

$$\text{Subject to } \begin{cases} \begin{pmatrix} 1/4 & -8 & -1 & 9 \\ 1/2 & -12 & -1/2 & 3 \\ 0 & 0 & 1 & 0 \end{pmatrix}\begin{pmatrix} x_1 \\ x_2 \\ x_3 \\ x_4 \end{pmatrix} \le \begin{pmatrix} 0 \\ 0 \\ 1 \end{pmatrix} \\ \\ \text{and } \begin{pmatrix} x_1 \\ x_2 \\ x_3 \\ x_4 \end{pmatrix} \ge 0 \end{cases}$$

You should return to the initial simplex tableau after pivoting at [column 1, row 1], [column 2, row 2], [column 3, row 1], [column 4, row 2], [column 5, row 1], and [column 6, row 2].

57. Rescaling Changing the size of the numbers in a problem without altering the feasible region can prevent cycling.

a. Multiply the first constraint in Beale's cycling example (Exercise 56) by 4 to get the equivalent problem:

$$\text{Maximize } P = (3/4 \quad -20 \quad 1/2 \quad -6)\begin{pmatrix} x_1 \\ x_2 \\ x_3 \\ x_4 \end{pmatrix}$$

$$\text{Subject to } \begin{cases} \begin{pmatrix} 1 & -32 & -4 & 36 \\ 1/2 & -12 & -1/2 & 3 \\ 0 & 0 & 1 & 0 \end{pmatrix}\begin{pmatrix} x_1 \\ x_2 \\ x_3 \\ x_4 \end{pmatrix} \le \begin{pmatrix} 0 \\ 0 \\ 1 \end{pmatrix} \\ \\ \text{and } \begin{pmatrix} x_1 \\ x_2 \\ x_3 \\ x_4 \end{pmatrix} \ge 0 \end{cases}$$

You should reach the final simplex tableau after pivoting at [column 1, row 1], [column 2, row 2], [column 3, row 1], [column 4, row 2], [column 1, row 3], and [column 5, row 2].

b. Divide every coefficient of x_2 in Beale's cycling example (Exercise 56) by 4 to get the equivalent problem:

$$\text{Maximize } P = (3/4 \quad -5 \quad 1/2 \quad -6)\begin{pmatrix} x_1 \\ x_2 \\ x_3 \\ x_4 \end{pmatrix}$$

$$\text{Subject to } \begin{cases} \begin{pmatrix} 1/4 & -2 & -1 & 9 \\ 1/2 & -3 & -1/2 & 3 \\ 0 & 0 & 1 & 0 \end{pmatrix}\begin{pmatrix} x_1 \\ x_2 \\ x_3 \\ x_4 \end{pmatrix} \le \begin{pmatrix} 0 \\ 0 \\ 1 \end{pmatrix} \\ \\ \text{and } \begin{pmatrix} x_1 \\ x_2 \\ x_3 \\ x_4 \end{pmatrix} \ge 0 \end{cases}$$

You should reach the final simplex tableau after pivoting at [column 1, row 1], [column 3, row 2], [column 4, row 3] and [column 5, row 3].

58. Bland's Rule Eliminates cycling by changing the choice of the pivot column to be *the leftmost column with a negative entry in the bottom row* (that is, ignore the size of the numbers and just look for the first negative). Use Bland's rule to solve Beale's cycling example (Exercise 56). You should reach the final simplex tableau after pivoting at [column 1, row 1], [column 2, row 2], [column 3, row 1], [column 4, row 2], [column 1, row 3], and [column 5, row 2].

59. A Nondeterministic Simplex Algorithm The simplex method (page 327) is a *deterministic* procedure to solve standard maximum problems because once the process is started, the rules *always* select the *same* sequence of pivot elements. If we allow *chance* to play a role in the selection of the pivot column or pivot row, the resulting solution method is nondeterministic in that the sequence of pivot elements in the solution is no longer determined but rather may change from one solution attempt to another. Let us change the rule on page 322 for the selection of the pivot row to state that *if there is a tie for the smallest ratio between two of the rows, the pivot row is to be selected by a coin flip*. Solve Beale's cycling example (Exercise 56) several times using this modified procedure. Do your solutions all use the same number of pivot operations?

60. More Cycling Examples A large class of cyclic examples was found by K. T. Marshall and J. W. Suurballe in 1969.

a. Try to solve this typical Marshall–Suurballe problem by the simplex method.

$$\text{Maximize } P = (1 \quad -28 \quad -2 \quad -2)\begin{pmatrix} x_1 \\ x_2 \\ x_3 \\ x_4 \end{pmatrix}$$

$$\text{Subject to } \begin{cases} \begin{pmatrix} 1/2 & -18 & -2 & 4 \\ 1/4 & -5 & -1/2 & 1/2 \end{pmatrix}\begin{pmatrix} x_1 \\ x_2 \\ x_3 \\ x_4 \end{pmatrix} \le \begin{pmatrix} 0 \\ 0 \end{pmatrix} \\ \\ \text{and } \begin{pmatrix} x_1 \\ x_2 \\ x_3 \\ x_4 \end{pmatrix} \ge 0 \end{cases}$$

You should return to the initial simplex tableau after pivoting at [column 1, row 1], [column 2, row 2], [column 3, row 1], [column 4, row 2], [column 5, row 1], and [column 6, row 2].

b. Rescale the constraints of the problem in part (a) by multiplying the first by 2 and the second by 4. How many pivots does it now take to solve this rescaled problem?

c. Solve the problem in part (a) using Bland's rule (see Exercise 58).

d. Solve the problem in part (a) using the nondeterministic simplex algorithm given in Exercise 59.

Klee–Minty Problems

61. Solve the following two-variable problem by the simplex method. (Your fourth tableau should be the final tableau.)

$$\text{Maximize } P = (1 \quad 10)\begin{pmatrix} x_1 \\ x_2 \end{pmatrix}$$

$$\text{Subject to } \begin{cases} \begin{pmatrix} 0 & 1 \\ 1 & 20 \end{pmatrix}\begin{pmatrix} x_1 \\ x_2 \end{pmatrix} \le \begin{pmatrix} 1 \\ 100 \end{pmatrix} \\ \\ \text{and } \begin{pmatrix} x_1 \\ x_2 \end{pmatrix} \ge 0 \end{cases}$$

62. Solve the following three-variable problem by the simplex method. (Your eighth tableau should be the final tableau.)

$$\text{Maximize } P = (1 \quad 10 \quad 100)\begin{pmatrix} x_1 \\ x_2 \\ x_3 \end{pmatrix}$$

$$\text{Subject to } \begin{cases} \begin{pmatrix} 0 & 0 & 1 \\ 0 & 1 & 20 \\ 1 & 20 & 200 \end{pmatrix}\begin{pmatrix} x_1 \\ x_2 \\ x_3 \end{pmatrix} \le \begin{pmatrix} 1 \\ 100 \\ 10{,}000 \end{pmatrix} \\ \\ \text{and } \begin{pmatrix} x_1 \\ x_2 \\ x_3 \end{pmatrix} \ge 0 \end{cases}$$

63. Following the form of Exercises 61 and 62, write down a problem using four variables that will continue the pattern and take sixteen tableaux to solve.

64. Following the form of Exercises 61–63, write down a problem using five variables that will continue the pattern and take thirty-two tableaux to solve.

65. Using the problems from Exercises 61–64, check that each of this family of Klee–Minty problems can be solved in one pivot by using Bland's rule to choose the pivot column.

3.4 Duality and Standard Minimum Problems

APPLICATION PREVIEW

Diet Problems and "The Cost of Subsistence"

In 1945 the University of Minnesota economist George J. Stigler* published "The Cost of Subsistence," in which he pointed out that although "elaborate investigations have been made of the adequacy of diets at various income levels, . . . no one has determined the minimum cost of obtaining the amounts of calories, protein, minerals, and vitamins which these studies accept as adequate or optimum. This will be done in the present paper, not only for its own interest but because it sheds much light on the meaning of conventional 'low-cost' diets." He examined 77 foods whose nutritive values and 1939 prices were known. This preliminary food list was first reduced down to 15 foods by excluding those whose nutritive values per dollar of expenditure were less than some others. "For example, white bread has less than half the nutrients (per dollar) of white flour, except for calcium, for which neither is an economical source." Stigler then remarked that ". . . *there does not appear to be any direct method of finding the minimum of a linear function subject to linear conditions*. . . . There is no reason to believe that the cheapest combination was found, for only a handful of the 510 possible combinations . . . were examined."

STIGLER'S MINIMUM-COST ($39.93) ANNUAL DIET (1939 PRICES)

Commodity	Quantity	Cost
Wheat flour	370 pounds	$13.33
Evaporated milk	57 cans	3.84
Cabbage	111 pounds	4.11
Spinach	23 pounds	1.85
Dried navy beans	285 pounds	16.80

In the fall of 1947, Laderman of the Mathematical Tables Project in New York computed the optimal solution of Stigler's problem in a

* George J. Stigler, "The Cost of Subsistence," *Journal of Farm Economics* **27**:303–314, 1945.

test of Dantzig's new simplex method. Interestingly enough, this minimum solution gave a yearly cost of $36.64, while Stigler's trial-and-error solution (using the intuition of an economist) cost just $3.29 more. Comparing his minimal diet to those recommended by professional dieticians, Stigler found that they cost two to three times as much as his. He ended his paper with the following remarks.

> Why do these conventional diets cost so much? The answer is evident from their composition. The dieticians take account of the palatability of foods, variety of diet, prestige of various foods, and other cultural facets of consumption. . . . No one can now say with any certainty what the cultural requirements of a particular person may be. . . . If the dieticians persist in presenting minimum diets, they should at least report separately the physical and cultural components. . . .

Introduction

The simplex method for standard maximum problems begins at the origin, with an objective function value of zero, and moves along a path of vertices to arrive at a solution vertex, where the objective function has the largest possible value. On page 310 we remarked that it might also be possible to find the solution by making smaller and smaller upper estimates on the objective function using multiples of the constraints. In this section, we shall see that this idea leads to a pairing between maximum and minimum problems in such a way that the solution of one gives the solution of the other. Since we can already solve maximum problems using the simplex method, this observation will allow us to solve minimum problems by solving the corresponding maximum problems.

The Dual of a Standard Maximum Problem

While we could solve the standard maximum problem:

$$\text{Maximize } P = 24x_1 + 16x_2 + 42x_3$$

$$\text{Subject to } \begin{cases} 3x_1 + x_2 + 3x_3 \leq 9 \\ x_1 + x_2 + 2x_3 \leq 8 \\ x_1 \geq 0, x_2 \geq 0, \text{ and } x_3 \geq 0 \end{cases}$$

by the simplex method, let us instead attempt to estimate P in terms of the constraints. If we multiply the first constraint by 11 and add it to 5 times the second, we have that P can be no more than 139:

Maximize $P = 24x_1 + 16x_2 + 42x_3$

Subject to $\begin{cases} 3x_1 + x_2 + 3x_3 \leq 9 \\ x_1 + x_2 + 2x_3 \leq 8 \\ x_1 \geq 0, x_2 \geq 0, x_3 \geq 0 \end{cases}$

$\times$ 11 is
$\times$ 5 is
Add to give:

$\begin{array}{r} 33x_1 + 11x_2 + 33x_3 \leq 99 \\ 5x_1 + 5x_2 + 10x_3 \leq 40 \\ \hline 38x_1 + 16x_2 + 43x_3 \leq 139 \end{array}$

so $P = 24x_1 + 16x_2 + 42x_3$ is ≤ 139

As we want the best possible estimate, we should make more estimates using other multipliers. But we should (1) never multiply an inequality by a negative number (since a negative multiple would become a lower estimate), (2) always get at least 24 of the x_1's, at least 16 of the x_2's, and at least 42 of the x_3's, and (3) make the upper estimate for P as small as possible. If we multiply the first constraint by a number y_1 and add it to y_2 times the second constraint, we get

$3x_1 + x_2 + 3x_3 \leq 9$
$x_1 + x_2 + 2x_3 \leq 8$

$\times y_1$ is
$\times y_2$ is
Add to give:

$\begin{array}{l} 3y_1 x_1 + y_1 x_2 + 3y_1 x_3 \leq 9y_1 \\ y_2 x_1 + y_2 x_2 + 2y_2 x_3 \leq 8y_2 \\ \hline (3y_1 + y_2)x_1 + (y_1 + y_2)x_2 + (3y_1 + 2y_2)x_3 \leq 9y_1 + 8y_2 \end{array}$

Rule (1) now means that $y_1 \geq 0$ and $y_2 \geq 0$, rule (2) becomes $3y_1 + y_2 \geq 24$, $y_1 + y_2 \geq 16$, and $3y_1 + 2y_2 \geq 42$, and rule (3) means we want to minimize $9y_1 + 8y_2$. This new linear programming problem:

Minimize $C = 9y_1 + 8y_2$

Subject to $\begin{cases} 3y_1 + y_2 \geq 24 \\ y_1 + y_2 \geq 16 \\ 3y_1 + 2y_2 \geq 42 \\ y_1 \geq 0 \text{ and } y_2 \geq 0 \end{cases}$

is called the *dual* of the original problem. These two problems have the property that $P \leq C$ for any x_1, x_2, x_3 and y_1, y_2 values in the feasible regions. Furthermore, if we could find feasible points such that $P = C$, then we would have the solutions to both problems because the value of P could be made no larger and the value of C could be made no smaller.

The relation between our maximum problem and its dual mini-mum problem becomes clearer when we write both problems in matrix form.

$$\text{Maximize } P = \begin{pmatrix} 24 & 16 & 42 \end{pmatrix} \begin{pmatrix} x_1 \\ x_2 \\ x_3 \end{pmatrix} \qquad \text{Minimize } C = \begin{pmatrix} 9 & 8 \end{pmatrix} \begin{pmatrix} y_1 \\ y_2 \end{pmatrix}$$

and

$$\text{Subject to } \begin{cases} \begin{pmatrix} 3 & 1 & 3 \\ 1 & 1 & 2 \end{pmatrix} \begin{pmatrix} x_1 \\ x_2 \\ x_3 \end{pmatrix} \leq \begin{pmatrix} 9 \\ 8 \end{pmatrix} \\[2em] \text{and } \begin{pmatrix} x_1 \\ x_2 \\ x_3 \end{pmatrix} \geq 0 \end{cases} \qquad \text{Subject to } \begin{cases} \begin{pmatrix} 3 & 1 \\ 1 & 1 \\ 3 & 2 \end{pmatrix} \begin{pmatrix} y_1 \\ y_2 \end{pmatrix} \geq \begin{pmatrix} 24 \\ 16 \\ 42 \end{pmatrix} \\[2em] \text{and } \begin{pmatrix} y_1 \\ y_2 \end{pmatrix} \geq 0 \end{cases}$$

Notice that rows in one problem become columns of the same numbers in the other problem.

Two matrices with the property that the columns of one are the rows of the other in the same order are *transposes* of each other. The *transpose* of a matrix M is written M^t and is found by turning each row into a column. For example, the transpose of:

$$M = \begin{pmatrix} 1 & 2 & 3 & 4 \\ 5 & 6 & 7 & 8 \\ 9 & 10 & 11 & 12 \end{pmatrix} \text{ is } M^t = \begin{pmatrix} 1 & 5 & 9 \\ 2 & 6 & 10 \\ 3 & 7 & 11 \\ 4 & 8 & 12 \end{pmatrix}$$

It makes no difference whether you change rows into columns or columns into rows. Furthermore, the transpose of a transpose is the original matrix again: $(M^t)^t = M$.

Dual Problems and the Duality Theorem

Giving the matrices in the maximum problem the usual names, we have the following definition:

Dual Problems

Each of the following linear programming problems is the *dual* of the other:

$$\text{Maximize } P = c^t X \qquad \text{Minimize } C = b^t Y$$

and

$$\text{Subject to } \begin{cases} AX \leq b \\ X \geq 0 \end{cases} \qquad \text{Subject to } \begin{cases} A^t Y \geq c \\ Y \geq 0 \end{cases}$$

The dual of a maximum problem is a minimum problem and the dual of a minimum problem is a maximum problem. The numbers in the objective function of one problem are the numbers after the inequalities in the constraints of the other. The maximum problem has $\leq$ constraints while the minimum problem has $\geq$ constraints. There are as many slack variables (one for each constraint) in the *maximum* problem as variables $(y_1, y_2, \ldots, y_m)$ in the *minimum* problem. There are as many slack variables (one for each constraint) in the *minimum* problem as variables $(x_1, x_2, \ldots, x_n)$ in the *maximum* problem. We shall write $t_1, t_2, \ldots, t_n$ for the slack variables in the minimum problem constraints $A^t Y \geq c$, and $t_1 \geq 0, t_2 \geq 0, \ldots, t_n \geq 0$ because we *subtract* them to lower $A^t Y$ down to equal c.

EXAMPLE 1   The Dual of a Minimum Problem

Find the dual of the following minimum problem and then construct the initial simplex tableau for this maximum problem.

$$\text{Minimize } C = 440y_1 + 300y_2 + 200y_3$$

$$\text{Subject to } \begin{cases} 3y_1 + 2y_2 + 2y_3 \geq 156 \\ 2y_1 + y_2 + 4y_3 \geq 120 \\ 2y_1 + 2y_2 + 3y_3 \geq 132 \\ 4y_1 + 3y_2 - y_3 \geq 180 \\ y_1 \geq 0, y_2 \geq 0, \text{ and } y_3 \geq 0 \end{cases}$$

Solution

In matrix form, this problem and its dual maximum problem are:

$$\text{Minimize } C = \begin{pmatrix} 440 & 300 & 200 \end{pmatrix} \begin{pmatrix} y_1 \\ y_2 \\ y_3 \end{pmatrix}$$

$$\text{Maximize } P = \begin{pmatrix} 156 & 120 & 132 & 180 \end{pmatrix} \begin{pmatrix} x_1 \\ x_2 \\ x_3 \\ x_4 \end{pmatrix}$$

and

$$\text{Subject to } \begin{cases} \begin{pmatrix} 3 & 2 & 2 \\ 2 & 1 & 4 \\ 2 & 2 & 3 \\ 4 & 3 & -1 \end{pmatrix} \begin{pmatrix} y_1 \\ y_2 \\ y_3 \end{pmatrix} \geq \begin{pmatrix} 156 \\ 120 \\ 132 \\ 180 \end{pmatrix} \\ \text{and } \begin{pmatrix} y_1 \\ y_2 \\ y_3 \end{pmatrix} \geq 0 \end{cases}$$

$$\text{Subject to } \begin{cases} \begin{pmatrix} 3 & 2 & 2 & 4 \\ 2 & 1 & 2 & 3 \\ 2 & 4 & 3 & -1 \end{pmatrix} \begin{pmatrix} x_1 \\ x_2 \\ x_3 \\ x_4 \end{pmatrix} \leq \begin{pmatrix} 440 \\ 300 \\ 200 \end{pmatrix} \\ \text{and } \begin{pmatrix} x_1 \\ x_2 \\ x_3 \\ x_4 \end{pmatrix} \geq 0 \end{cases}$$

The minimum problem has 3 variables y_1, y_2, y_3 and 4 slacks t_1, t_2, t_3, t_4 while the dual maximum problem has 4 variables x_1, x_2, x_3, x_4 and 3 slacks s_1, s_2, s_3. The initial simplex tableau for this dual maximum problem is

	x_1	x_2	x_3	x_4	s_1	s_2	s_3	
s_1	3	2	2	4	1	0	0	440
s_2	2	1	2	3	0	1	0	300
s_3	2	4	3	−1	0	0	1	200
P	−156	−120	−132	−180	0	0	0	0

PRACTICE PROBLEM 1

Find the dual of the following minimum problem and then construct the initial simplex tableau for this maximum problem.

$$\text{Minimize } C = 11y_1 + 9y_2 + 7y_3$$

$$\text{Subject to } \begin{cases} 2y_1 + y_2 + y_3 \geq 4 \\ y_1 + y_2 \geq 7 \\ y_2 + y_3 \geq 5 \\ y_1 + y_2 + 3y_3 \geq 6 \\ y_1 \geq 0,\ y_2 \geq 0,\ \text{and } y_3 \geq 0 \end{cases}$$

Solution at the back of the book

How can we solve a minimum problem? The following theorem explains that any tableau for a maximum problem displays *at the same time* information about the dual minimum problem.

Duality Theorem for the Simplex Method

The bottom row of any simplex tableau for a maximum linear programming problem displays values for the dual minimum problem's slack variables, variables, and objective function:

	x_1	$\cdots$	x_n	s_1	$\cdots$	s_m	
P	t_1	$\cdots$	t_n	y_1	$\cdots$	y_m	C

$\underbrace{\qquad\qquad}_{\text{Slack variables}}$ $\underbrace{\qquad\qquad}_{\text{Variables}}$ Objective function

These values for the minimum problem satisfy its constraints and yield the given value in its objective function even if some are negative.

Because dual problems are related by transposition, it is not surprising to find that the solution of the *minimum* problem appears in the bottom *row*, just as the solution of the *maximum* problem appears in the last *column*. Since the choice of the pivot element makes the pivot operation keep the right column of the tableau nonnegative while making the bottom row "less negative," *the simplex method makes the maximum problem optimal while keeping it feasible and makes the dual minimum problem feasible while keeping it optimal.* If the tableau represents a feasible point for both the maximum and the dual minimum problems, the tableau is the final tableau for both problems.

Solving Standard Minimum Problems

Using the duality theorem, we can solve a minimum problem by the simplex method provided its dual maximum problem is a *standard* problem.

Standard Minimum Problem

$$\text{Minimize } C = b^t Y$$

$$\text{Subject to } \begin{cases} A^t Y \geq c \\ Y \geq 0 \end{cases}$$

$$\text{where} \quad b^t \geq 0$$

The solution of a standard minimum problem can be found by solving the dual maximum problem by the simplex method and interpreting the final tableau as a statement about the original minimum problem.

Solution of a Standard Minimum Problem

To solve a standard minimum problem:

1. Construct the dual maximum problem (page 345).

2. Solve the dual maximum problem by the simplex method (page 327).

3. If there is a solution to the dual maximum problem, then there is a solution to the minimum problem and the values of the slacks, the variables, and the objective function appear in the bottom row of the final tableau of the dual maximum problem. If there is no solution to the dual maximum problem, then the minimum problem has no solution.

EXAMPLE 2   Solution of a Standard Minimum Problem

Solve the following standard minimum problem by finding the dual maximum problem and using the simplex method.

$$\text{Minimize } C = 50y_1 + 60y_2$$

$$\text{Subject to } \begin{cases} y_1 + 4y_2 \geq 20 \\ 2y_1 + 3y_2 \geq 30 \\ y_1 - y_2 \geq 5 \\ y_1 \geq 0 \text{ and } y_2 \geq 0 \end{cases}$$

Solution

In matrix form, this problem and its dual maximum problem are:

$$\text{Minimize } C = (50 \quad 60)\begin{pmatrix} y_1 \\ y_2 \end{pmatrix}$$

$$\text{Subject to } \begin{cases} \begin{pmatrix} 1 & 4 \\ 2 & 3 \\ 1 & -1 \end{pmatrix}\begin{pmatrix} y_1 \\ y_2 \end{pmatrix} \geq \begin{pmatrix} 20 \\ 30 \\ 5 \end{pmatrix} \\ \text{and } \begin{pmatrix} y_1 \\ y_2 \end{pmatrix} \geq 0 \end{cases}$$

and

$$\text{Maximize } P = (20 \quad 30 \quad 5)\begin{pmatrix} x_1 \\ x_2 \\ x_3 \end{pmatrix}$$

$$\text{Subject to } \begin{cases} \begin{pmatrix} 1 & 2 & 1 \\ 4 & 3 & -1 \end{pmatrix}\begin{pmatrix} x_1 \\ x_2 \\ x_3 \end{pmatrix} \leq \begin{pmatrix} 50 \\ 60 \end{pmatrix} \\ \text{and } \begin{pmatrix} x_1 \\ x_2 \\ x_3 \end{pmatrix} \geq 0 \end{cases}$$

The initial simplex tableau for the dual maximum problem is

	x_1	x_2	x_3	s_1	s_2	
s_1	1	2	1	1	0	50
s_2	4	3	-1	0	1	60
P	-20	-30	-5	0	0	0

Bottom row states that
$t_1 = -20$, $t_2 = -30$, $t_3 = -5$,
$y_1 = 0$, $y_2 = 0$, and $C = 0$

To solve the dual maximum problem, we first pivot on the 3 in column 2 and row 2 to find the tableau:

	x_1	x_2	x_3	s_1	s_2	
s_1	-5/3	0	5/3	1	-2/3	10
x_2	4/3	1	-1/3	0	1/3	20
P	20	0	-15	0	10	600

Bottom row states that
$t_1 = 20$, $t_2 = 0$, $t_3 = -15$,
$y_1 = 0$, $y_2 = 10$,
and $C = 600$

Then we pivot again on the 5/3 in column 3 and row 1 to reach the final tableau:

	x_1	x_2	x_3	s_1	s_2	
x_3	-1	0	1	3/5	$-2/5$	6
x_2	1	1	0	1/5	1/5	22
P	5	0	0	9	4	690

Bottom row states that
$t_1 = 5, t_2 = 0, t_3 = 0,$
$y_1 = 9, y_2 = 4,$
and $C = 690$

Answer: The minimum value is 690 when $y_1 = 9$ and $y_2 = 4$.

■

To compare this solution with the graphical method, we graph the feasible region of the minimum problem with y_1 on the horizontal axis and y_2 on the vertical axis. The three vertices and the values of the objective function are given in the table. The path of nonfeasible points visited by the simplex tableaux of the dual maximum problem and leading to a feasible vertex is marked in bold.

Vertex (y_1, y_2)	$C = 50y_1 + 60y_2$
(20, 0)	1000
(12, 2)	720
(9, 4)	690

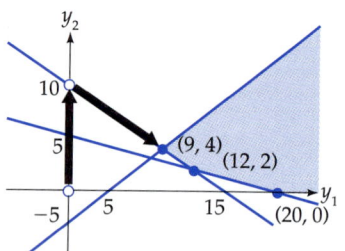

Since *the dual variables are multipliers for the constraints of the other problem,* let us check this with both the values for y_1, y_2 and for x_1, x_2, x_3. Using $y_1 = 9, y_2 = 4$ as multipliers for the dual maximum problem, we find the estimate $P \leq 690$:

Maximize $P = 20x_1 + 30x_2 + 5x_3$

Subject to $\begin{cases} x_1 + 2x_2 + x_3 \leq 50 \\ 4x_1 + 3x_2 - x_3 \leq 60 \\ x_1 \geq 0, x_2 \geq 0, x_3 \geq 0 \end{cases}$

$\times 9$ is
$\times 4$ is
Add to give:

$9x_1 + 18x_2 + 9x_3 \leq 450$
$\underline{16x_1 + 12x_2 - 4x_3 \leq 240}$
$25x_1 + 30x_2 + 5x_3 \leq 690$

so $P = 20x_1 + 30x_2 + 5x_3$ is ≤ 690

If P is the profit from goods manufactured by processes limited by the constraints, the multipliers $y_1, y_2, \ldots, y_m$ are called the *marginal values* for these processes because they show how much value each contributes to the final profit. For example, if the above problem represented the profit of a pottery shop as in Example 4 on page 330, but now with just two constraints ("shaping" and "decorating"), y_1 and y_2 show that the shaping process contributes \$9 per hour to the profit and the decorating process contributes \$4 per hour.

Using $x_1 = 0$, $x_2 = 22$, and $x_3 = 6$ as multipliers for the minimum problem, we similarly find the estimate $C \geq 690$:

Minimize $C = 50y_1 + 60y_2$

Subject to $\begin{cases} y_1 + 4y_2 \geq 20 \\ 2y_1 + 3y_2 \geq 30 \\ y_1 - y_2 \geq 5 \\ y_1 \geq 0, y_2 \geq 0 \end{cases}$

$\times 0$ is
$\times 22$ is
$\times 6$ is
Add to give:

$\begin{array}{r} 0y_1 + 0y_2 \geq 0 \\ 44y_1 + 66y_2 \geq 660 \\ 6y_1 - 6y_2 \geq 30 \\ \hline 50y_1 + 60y_2 \geq 690 \end{array}$

so $C = 50y_1 + 60y_2$ is ≥ 690

If C is the cost of meeting the requirements represented by the constraints, the multipliers $x_1, x_2, \ldots, x_n$ are called the *shadow prices* of these requirements because they show how much each contributes to the total cost. For example, if the above problem represented the cost of meeting nutritional needs as in Exercise 38 on page 314 with the constraints representing "calories," "protein," and "fat," then x_1, x_2, and x_3 show that the calorie requirement does not contribute to the cost but the price of the protein requirement is 22¢ per gram and the price of the fat requirement is 6¢ per gram.

PRACTICE PROBLEM 2

Find the solution of the linear programming problem:

Minimize $C = 255y_1 + 435y_2 + 300y_3 + 465y_4$

Subject to $\begin{cases} 2y_1 + 3y_2 + 2y_3 + 4y_4 \geq 25 \\ y_1 + 3y_2 + 2y_3 + 2y_4 \geq 18 \\ 3y_1 + 4y_2 + 3y_3 + 2y_4 \geq 36 \\ y_1 \geq 0, y_2 \geq 0, y_3 \geq 0, \text{ and } y_4 \geq 0 \end{cases}$

from the fact that the final tableau of its dual maximum problem is

	x_1	x_2	x_3	s_1	s_2	s_3	s_4	
x_3	0	0	1	0	-2	3	0	30
x_1	1	0	0	1	3	-5	0	60
x_2	0	1	0	-1	0	1	0	45
s_4	0	0	0	-2	-8	12	1	75
P	0	0	0	7	3	1	0	3390

Solution at the back of the book

Graphing Calculator Exploration

The initial simplex tableau of the dual maximum problem in Practice Problem 2 (above) is

	x_1	x_2	x_3	s_1	s_2	s_3	s_4	
s_1	2	1	3	1	0	0	0	255
s_2	3	3	4	0	1	0	0	435
s_3	2	2	3	0	0	1	0	300
s_4	4	2	2	0	0	0	1	465
P	-25	-18	-36	0	0	0	0	0

Use the program PIVOT to pivot at [column 3, row 1], [column 2, row 3], and [column 1, row 2] and thus verify that the final tableau is as given.

Mixed Constraints: A Transportation Problem

Although we have been careful to write our standard minimum problem constraints as $A^t Y \geq c$, we do not mean to exclude the possibility of *mixed constraints*, some $\leq$ and some $\geq$, from the initial statement of the problem. However, we must write them all as $\geq$ inequalities (by multiplying by -1 as necessary) before forming the matrix inequality $A^t Y \geq c$. (The requirement $b \geq 0$ refers to the objective function of the minimum problem and places no restrictions on the inequalities.) The following example belongs to a general type of minimization problem with mixed constraints having the nice property that the pivot elements will all be 1's.

EXAMPLE 3   A Transportation Problem

A retail store chain has cartons of goods stored at warehouses in Maryland and Washington that must be distributed to its stores in Ohio and Louisiana. The cost to ship each carton from Maryland to Ohio is $6 and from Maryland to Louisiana is $7, while the cost to ship each carton from Washington to Ohio is $8 and from Washington to Louisiana is $9. There are 300 cartons at the Maryland warehouse and 300 at the warehouse in Washington. If the Ohio stores need 200 cartons and the Louisiana stores need 300 cartons, how many cartons should be shipped from each warehouse to each state to incur the smallest shipping costs?

Solution

This is a minimization problem that requires four variables: one for the amount shipped from each warehouse to each state. Let:

$$y_1 = \begin{pmatrix} \text{Cartons shipped} \\ \text{from MD to OH} \end{pmatrix}, \; y_2 = \begin{pmatrix} \text{Cartons shipped} \\ \text{from MD to LA} \end{pmatrix},$$

$$y_3 = \begin{pmatrix} \text{Cartons shipped} \\ \text{from WA to OH} \end{pmatrix}, \; \text{and } y_4 = \begin{pmatrix} \text{Cartons shipped} \\ \text{from WA to LA} \end{pmatrix}$$

Since each warehouse can ship no more than the number of cartons known to be stored there and each state must receive at least the required number of cartons, this problem is the linear programming problem:

Minimize $C = 6y_1 + 7y_2 + 8y_3 + 9y_4$ Total shipping costs

$$\text{Subject to} \begin{cases} y_1 + y_2 & \leq 300 & \text{Have 300 at MD} \\ y_3 + y_4 & \leq 300 & \text{Have 300 at WA} \\ y_1 + y_3 & \geq 200 & \text{Need 200 in OH} \\ y_2 + y_4 & \geq 300 & \text{Need 300 in LA} \\ y_1 \geq 0, \, y_2 \geq 0, \, y_3 \geq 0, \text{ and } y_4 \geq 0 & & \text{Nonnegativity} \end{cases}$$

Multiplying the first two constraints by -1 to put them into $\geq$ form, this problem and its dual maximum problem in matrix form are:

$$\text{Minimize } C = \begin{pmatrix} 6 & 7 & 8 & 9 \end{pmatrix} \begin{pmatrix} y_1 \\ y_2 \\ y_3 \\ y_4 \end{pmatrix}$$

$$\text{Subject to} \begin{cases} \begin{pmatrix} -1 & -1 & 0 & 0 \\ 0 & 0 & -1 & -1 \\ 1 & 0 & 1 & 0 \\ 0 & 1 & 0 & 1 \end{pmatrix} \begin{pmatrix} y_1 \\ y_2 \\ y_3 \\ y_4 \end{pmatrix} \geq \begin{pmatrix} -300 \\ -300 \\ 200 \\ 300 \end{pmatrix} \\ \\ \text{and } \begin{pmatrix} y_1 \\ y_2 \\ y_3 \\ y_4 \end{pmatrix} \geq 0 \end{cases}$$

and

$$\text{Maximize } P = \begin{pmatrix} -300 & -300 & 200 & 300 \end{pmatrix} \begin{pmatrix} x_1 \\ x_2 \\ x_3 \\ x_4 \end{pmatrix}$$

$$\text{Subject to } \begin{cases} \begin{pmatrix} -1 & 0 & 1 & 0 \\ -1 & 0 & 0 & 1 \\ 0 & -1 & 1 & 0 \\ 0 & -1 & 0 & 1 \end{pmatrix} \begin{pmatrix} x_1 \\ x_2 \\ x_3 \\ x_4 \end{pmatrix} \le \begin{pmatrix} 6 \\ 7 \\ 8 \\ 9 \end{pmatrix} \\ \\ \text{and } \begin{pmatrix} x_1 \\ x_2 \\ x_3 \\ x_4 \end{pmatrix} \ge 0 \end{cases}$$

The initial simplex tableau is

	x_1	x_2	x_3	x_4	s_1	s_2	s_3	s_4	
s_1	−1	0	1	0	1	0	0	0	6
s_2	−1	0	0	1	0	1	0	0	7
s_3	0	−1	1	0	0	0	1	0	8
s_4	0	−1	0	1	0	0	0	1	9
P	300	300	−200	−300	0	0	0	0	0

We pivot at [column 4, row 2], [column 3, row 1], and [column 1, row 3] to reach the final tableau:

	x_1	x_2	x_3	x_4	s_1	s_2	s_3	s_4	
x_3	0	−1	1	0	0	0	1	0	8
x_4	0	−1	0	1	−1	1	1	0	9
x_1	1	−1	0	0	−1	0	1	0	2
s_4	0	0	0	0	1	−1	−1	1	0
P	0	100	0	0	0	300	200	0	3700
					y_1	y_2	y_3	y_4	C

Bottom row states that $t_1 = 0, t_2 = 100$, $t_3 = 0, t_4 = 0$, $y_1 = 0, y_2 = 300$, $y_3 = 200, y_4 = 0$, and $C = 3700$

The Maryland warehouse should send nothing to Ohio and 300 cartons to Louisiana while the Washington warehouse should send 200 cartons to Ohio and nothing to Louisiana to achieve the least shipping cost of $3700. The second slack shows that there will be 100 cartons unused in the Washington warehouse.

Graphing Calculator Exploration

Use the program PIVOT with the above initial simplex tableau to pivot at [column 4, row 2], [column 3, row 1], and [column 1, row 3] and thus verify that the final tableau is as given.

SUMMARY

The dual of the minimum problem:

$$\text{Minimize } C = b^t Y$$

$$\text{Subject to } \begin{cases} A^t Y \geq c \\ Y \geq 0 \end{cases}$$

is the maximum problem:

$$\text{Maximize } P = c^t X$$

$$\text{Subject to } \begin{cases} AX \leq b \\ X \geq 0 \end{cases}$$

Both problems are standard problems if $b \geq 0$. The rows of one problem form the columns of the other. The minimum constraints are all written as $\geq$ inequalities and the maximum constraints are all written as $\leq$ inequalities. The maximum of P is the same as the minimum of C. The simplex method solution of the maximum problem displays in the bottom row the solution of the dual minimum problem (slack variables, variables, and objective function). If one problem has no solution, then neither does its dual.

EXERCISES 3.4

Write each of the following standard minimum problems in the matrix form "minimize $C = b^t Y$ subject to $A^t Y \geq c$ and $Y \geq 0$ where $b^t \geq 0$," and construct the dual maximum problem and its initial simplex tableau.

1. Minimize $C = 60y_1 + 100y_2 + 300y_3$

$$\text{Subject to } \begin{cases} y_1 + 2y_2 + 3y_3 \geq 180 \\ 4y_1 + 5y_2 + 6y_3 \geq 120 \\ y_1 \geq 0, \, y_2 \geq 0, \, y_3 \geq 0 \end{cases}$$

2. Minimize $C = 100y_1 + 60y_2 + 280y_3$

$$\text{Subject to } \begin{cases} 5y_1 + 9y_2 + 7y_3 \geq 315 \\ 2y_1 + 6y_2 + 4y_3 \geq 480 \\ y_1 \geq 0, \, y_2 \geq 0, \, y_3 \geq 0 \end{cases}$$

3. Minimize $C = 3y_1 + 20y_2$

$$\text{Subject to } \begin{cases} 3y_1 + 2y_2 \geq 150 \\ y_1 + 4y_2 \geq 100 \\ 3y_1 + 4y_2 \geq 228 \\ y_1 \geq 0, \, y_2 \geq 0 \end{cases}$$

4. Minimize $C = 60y_1 + 16y_2$

$$\text{Subject to } \begin{cases} y_1 + 4y_2 \geq 20 \\ 3y_1 + 2y_2 \geq 30 \\ 3y_1 + 4y_2 \geq 48 \\ y_1 \geq 0, \, y_2 \geq 0 \end{cases}$$

5. Minimize $C = 84y_1 + 21y_2$

$$\text{Subject to } \begin{cases} 3y_1 + y_2 \geq 21 \\ 4y_1 - y_2 \geq 0 \\ y_1 \geq 0, \, y_2 \geq 0 \end{cases}$$

6. Minimize $C = 7y_1 + 7y_2$

$$\text{Subject to } \begin{cases} 2y_1 + 3y_2 \geq 42 \\ 3y_1 + y_2 \geq 21 \\ y_1 \geq 0, \, y_2 \geq 0 \end{cases}$$

7. Minimize $C = 15y_1 + 20y_2 + 5y_3$

$$\text{Subject to } \begin{cases} y_1 - y_2 - 2y_3 \leq 30 \\ y_1 + 2y_2 + y_3 \geq 30 \\ y_1 \geq 0, \, y_2 \geq 0, \, y_3 \geq 0 \end{cases}$$

8. Minimize $C = 30y_1 + 40y_2 + 80y_3$

Subject to $\begin{cases} -3y_1 + y_2 - y_3 \le 60 \\ y_1 + y_2 + 2y_3 \ge 60 \\ y_1 \ge 0,\, y_2 \ge 0,\, y_3 \ge 0 \end{cases}$

9. Minimize $C = 105y_1 + 40y_2$

Subject to $\begin{cases} 7y_1 + 5y_2 \ge 70 \\ 3y_1 + y_2 \ge 45 \\ 5y_1 + 2y_2 \ge 80 \\ y_1 \ge 0,\, y_2 \ge 0 \end{cases}$

10. Minimize $C = 40y_1 + 105y_2$

Subject to $\begin{cases} 3y_1 + 5y_2 \ge 75 \\ 2y_1 + 5y_2 \ge 80 \\ y_1 + 3y_2 \ge 45 \\ y_1 \ge 0,\, y_2 \ge 0 \end{cases}$

Solve each of the following standard minimum problems by finding the dual maximum problem and using the simplex method. (Exercises 13–22 may also be solved by the graphical method.)

 You may use the PIVOT program if permitted by your instructor.

11. Minimize $C = 10y_1 + 20y_2 + 10y_3$

Subject to $\begin{cases} -y_1 + y_2 + y_3 \ge 50 \\ y_1 + y_2 - y_3 \ge 30 \\ y_1 \ge 0,\, y_2 \ge 0,\, y_3 \ge 0 \end{cases}$

12. Minimize $C = 30y_1 + 50y_2 + 30y_3$

Subject to $\begin{cases} -y_1 + y_2 + y_3 \ge 10 \\ y_1 + y_2 - y_3 \ge 20 \\ y_1 \ge 0,\, y_2 \ge 0,\, y_3 \ge 0 \end{cases}$

13. Minimize $C = 4y_1 + 5y_2$

Subject to $\begin{cases} y_1 + y_2 \ge 10 \\ y_1 \ge 2 \\ y_2 \ge 3 \\ y_1 \ge 0,\, y_2 \ge 0 \end{cases}$

14. Minimize $C = 3y_1 + 2y_2$

Subject to $\begin{cases} y_1 + y_2 \ge 20 \\ y_1 \ge 4 \\ y_2 \ge 5 \\ y_1 \ge 0,\, y_2 \ge 0 \end{cases}$

15. Minimize $C = 15y_1 + 10y_2$

Subject to $\begin{cases} y_1 + 2y_2 \ge 20 \\ 3y_1 - y_2 \ge 60 \\ y_1 \ge 0,\, y_2 \ge 0 \end{cases}$

16. Minimize $C = 40y_1 + y_2$

Subject to $\begin{cases} 5y_1 - y_2 \ge 5 \\ 4y_1 + y_2 \ge 4 \\ y_1 \ge 0,\, y_2 \ge 0 \end{cases}$

17. Minimize $C = 4y_1 + 3y_2$

Subject to $\begin{cases} y_1 + y_2 \ge 15 \\ 3y_1 + y_2 \le 60 \\ y_1 \ge 0,\, y_2 \ge 0 \end{cases}$

18. Minimize $C = 4y_1 + 5y_2$

Subject to $\begin{cases} y_1 + y_2 \ge 20 \\ 2y_1 + y_2 \le 50 \\ y_1 \ge 0,\, y_2 \ge 0 \end{cases}$

19. Minimize $C = 2y_1 + y_2$

Subject to $\begin{cases} -y_1 + y_2 \le 20 \\ y_1 - y_2 \le 20 \\ y_1 + y_2 \ge 10 \\ y_1 \ge 0,\, y_2 \ge 0 \end{cases}$

20. Minimize $C = 2y_1 + 3y_2$

Subject to $\begin{cases} -y_1 + y_2 \le 30 \\ y_1 - y_2 \le 30 \\ y_1 + y_2 \ge 10 \\ y_1 \ge 0,\, y_2 \ge 0 \end{cases}$

21. Minimize $C = 30y_1 + 90y_2$

Subject to $\begin{cases} -y_1 + 2y_2 \ge 60 \\ 3y_1 - y_2 \ge 45 \\ y_1 \ge 0,\, y_2 \ge 0 \end{cases}$

22. Minimize $C = 60y_1 + 45y_2$

Subject to $\begin{cases} 2y_1 - y_2 \ge 90 \\ -y_1 + 3y_2 \ge 30 \\ y_1 \ge 0,\, y_2 \ge 0 \end{cases}$

23. Minimize $C = 30y_1 + 19y_2 + 30y_3$

Subject to $\begin{cases} 2y_1 + y_2 + y_3 \geq 6 \\ y_1 + y_2 + 2y_3 \geq 4 \\ y_1 \geq 0,\, y_2 \geq 0,\, y_3 \geq 0 \end{cases}$

24. Minimize $C = 60y_1 + 39y_2 + 60y_3$

Subject to $\begin{cases} 2y_1 + y_2 + y_3 \geq 6 \\ y_1 + y_2 + 2y_3 \geq 8 \\ y_1 \geq 0,\, y_2 \geq 0,\, y_3 \geq 0 \end{cases}$

25. Minimize $C = 20y_1 + 30y_2 + 40y_3$

Subject to $\begin{cases} y_1 - y_2 + y_3 \geq 15 \\ y_1 + y_2 + y_3 \geq 20 \\ y_1 - y_2 + y_3 \leq 10 \\ y_1 \geq 0,\, y_2 \geq 0,\, y_3 \geq 0 \end{cases}$

26. Minimize $C = 20y_1 + 50y_2 + 30y_3$

Subject to $\begin{cases} 2y_1 - y_2 + y_3 \leq 10 \\ y_1 + y_2 + y_3 \geq 30 \\ 2y_1 - y_2 + y_3 \geq 20 \\ y_1 \geq 0,\, y_2 \geq 0,\, y_3 \geq 0 \end{cases}$

27. Minimize $C = 5y_1 + 3y_2 + 2y_3 + 4y_4$

Subject to $\begin{cases} y_1 + y_2 \leq 20 \\ y_3 + y_4 \leq 30 \\ y_1 + y_2 + y_3 + y_4 \geq 40 \\ y_1 \geq 0,\, y_2 \geq 0,\, y_3 \geq 0,\, y_4 \geq 0 \end{cases}$

28. Minimize $C = 7y_1 + 6y_2 + 9y_3 + 8y_4$

Subject to $\begin{cases} y_1 + y_3 \leq 40 \\ y_2 + y_4 \leq 50 \\ y_1 + y_2 + y_3 + y_4 \geq 60 \\ y_1 \geq 0,\, y_2 \geq 0,\, y_3 \geq 0,\, y_4 \geq 0 \end{cases}$

29. Minimize $C = 132y_1 + 102y_2 + 60y_3$

Subject to $\begin{cases} 3y_1 + 2y_2 + y_3 \geq 48 \\ 4y_1 + 3y_2 + 2y_3 \geq 72 \\ 2y_1 + 2y_2 + y_3 \geq 42 \\ y_1 \geq 0,\, y_2 \geq 0,\, y_3 \geq 0 \end{cases}$

30. Minimize $C = 96y_1 + 144y_2 + 84y_3$

Subject to $\begin{cases} 2y_1 + 3y_2 + 2y_3 \geq 102 \\ 3y_1 + 4y_2 + 2y_3 \geq 132 \\ y_1 + 2y_2 + y_3 \geq 60 \\ y_1 \geq 0,\, y_2 \geq 0,\, y_3 \geq 0 \end{cases}$

APPLIED EXERCISES

Formulate each situation as a standard minimum linear programming problem by identifying the variables, the objective function, and the constraints. Solve it by finding the dual maximum problem and using the simplex method. Be sure to state your final answer in terms of the original question.

 You may use the PIVOT program if permitted by your instructor.

31. Nutrition An athlete's training diet needs at least 44 more grams of carbohydrates, 12 more grams of fat, and 16 more grams of protein each day. A dietician recommends two food supplements, Bulk-Up Bars (costing 48¢ each) and Power Drink (costing 45¢ per can), with nutritional contents (in grams) as given in the table. How much of each food supplement will provide the extra needed nutrition at the least cost?

	Bulk-Up Bar	Power Drink
Carbohydrates	4	3
Fat	1	1
Protein	2	1

32. Diet Planning The residents of an elder care facility need at least 156 more milligrams of calcium, 180 more micrograms of folate, and 66 more grams of protein in their weekly diets. The staff chef has created a new fish entree and a salad that the residents should find appealing, and the nutritional contents of the new menu items are given in the accompanying table. If the fish entree costs 21¢ per ounce and the salad costs 15¢ per ounce, how many ounces of each should be added to every resident's weekly menu to provide the additional nutrition at the least cost?

	Fish Entree (per Ounce)	Salad (per Ounce)
Calcium (milligrams)	6	4
Folate (micrograms)	6	5
Protein (grams)	2	2

33. **Purchasing**   The office manager of a large accounting firm needs at least 315 more boxes of pens and 120 more boxes of pencils but has only $1500 left in his budget. Jack's Office Supplies has packages of 5 boxes of pens with 3 boxes of pencils on sale for $20, while John's Discount offers packages of 2 boxes of pens and 1 box of pencils for $11. How many packages from each store should he buy to restock the store room at the least cost?

34. **Highway Construction**   The project engineer for a highway construction company needs to cut through a small hill to make way for a new road. She estimates that at least 3500 cubic yards of dirt and at least 2400 cubic yards of crushed rock will have to be hauled away. A heavy-duty dump truck can haul either 10 cubic yards of dirt or 6 cubic yards of crushed rock, while a regular dump truck can haul either 5 cubic yards of dirt or 4 cubic yards of crushed rock. The dirt and crushed rock cannot be mixed since they go to different dumping sites. The union contract demands that this job use at least 400 loads carried in heavy-duty dump trucks. If each heavy-duty dump truck load costs $90 and each regular dump truck load costs $50, how many truckloads with each type of truck and cargo will be needed to complete the job at the least cost?

35. **Highway Construction**   Repeat Exercise 34 with the additional requirement that there can

be no more than 800 truckloads hauled from the job site in order to get the project finished on time.

36. **Purchasing**   Repeat Problem 33 with the additional requirement that the office manager feels he must purchase at least $1200 worth of pens and pencils from Jack's Office Supplies because it gave him such a good deal on some filing cabinets last month.

37. **Agriculture**   A soil analysis of a farmer's field showed that he needs to apply at least 3000 pounds of nitrogen, 2400 pounds of phosphoric acid, and 2100 pounds of potash. Plant fertilizer is labeled with three numbers giving the percentages of nitrogen, phosphoric acid, and potash. The local farm supply store sells 15-30-15 Miracle Mix for 15¢ per pound and a 10-5-5 store brand for 8¢ per pound. How many pounds of each fertilizer should the farmer buy to meet the needs of the field at the least cost?

38. **Gardening**   A weekend gardener's vegetable patch needs at least 10.2 pounds of nitrogen, 7.8 pounds of phosphoric acid, and 6.6 pounds of potash. Plant fertilizer is labeled with three numbers giving the percentages of nitrogen, phosphoric acid, and potash. The local garden center sells 15-30-15 Miracle Mix for 16¢ per pound, 15-10-10 Grow Great for 13¢ per pound, and a 10-5-5 store brand for 8¢ per pound. How many pounds of each fertilizer should the gardener buy to meet the needs of the vegetable patch at the least cost?

39. **Transportation**   A retail store chain has cartons of goods stored at warehouses in Kentucky and Utah that must be distributed to its stores in Kansas, Texas, and Oregon. Each carton shipped from the Utah warehouse costs $2 whether it goes to Kansas, Texas, or Oregon. However, the cost to ship one carton from the Kentucky warehouse to Kansas is $2, to Texas is $4, and to Oregon is $5. There are 200 cartons at the Utah warehouse and 400 at the warehouse in Kentucky. If the Kansas stores need 200 cartons, the Texas stores need 300 cartons, and the Oregon stores need 100 cartons, how many cartons should be shipped from each

warehouse to each state to incur the smallest shipping costs?

40. Transportation   A soda distributor has warehouses in Seaford and Centerville and needs to supply stores in Huntington and Towson. The costs of shipping one case of sodas are: 8¢ from Seaford to Huntington, 5¢ from Seaford to Towson, 6¢ from Centerville to Huntington, and 4¢ from Centerville to Towson. There are 800 cases in the Seaford warehouse and 1200 in the Centerville warehouse. If the Huntington stores need at least 1000 cases and the Towson stores need at least 600 cases, how many cases should be shipped from each warehouse to each city to incur the smallest shipping costs?

Explorations and Excursions

The following problems extend and augment the material presented in the text.

41. Why $P \leq C$ for Dual Problems   *(Requires matrix algebra)* For the pair of dual problems

Maximize $P = c^t X$ and   Minimize $C = b^t Y$

$$\text{Subject to } \begin{cases} AX \leq b \\ X \geq 0 \end{cases} \qquad \text{Subject to } \begin{cases} A^t Y \geq c \\ Y \geq 0 \end{cases}$$

show that $P \leq C$ for any feasible X and Y by justifying each $=$ and $\leq$ in the chain of statements

$$P = c^t \cdot X = X^t \cdot c \leq X^t \cdot A^t Y$$
$$= (X^t A^t) \cdot Y = (AX)^t \cdot Y \leq b^t \cdot Y = C$$

42. *(Requires matrix algebra)* Check the chain of matrix statements in Exercise 41 for the matrices

$$A = \begin{pmatrix} 1 & 2 & 1 \\ 4 & 3 & -1 \end{pmatrix}, \ b = \begin{pmatrix} 50 \\ 60 \end{pmatrix}, \ c = \begin{pmatrix} 20 \\ 30 \\ 5 \end{pmatrix}, \ X = \begin{pmatrix} x_1 \\ x_2 \\ x_3 \end{pmatrix}, \text{ and } Y = \begin{pmatrix} y_1 \\ y_2 \end{pmatrix}$$ from Example 2 on page 349.

43. A Proof of the Duality Theorem   *(Requires matrix algebra)* The following sequence of statements provides a proof of the Duality Theorem

on page 347. Justify each statement to verify this proof.

a. If $mx + b$ is the same as $0x + v$ for every value of x, then $m = 0$ and $b = v$.

b. The dual problems in Exercise 41 may be rewritten as:

Maximize $P = c^t X$ and   Minimize $C = b^t Y$

$$\text{Subject to } \begin{cases} AX + S = b \\ X, S \geq 0 \end{cases} \qquad \text{Subject to } \begin{cases} A^t Y - T = c \\ Y, T \geq 0 \end{cases}$$

to clearly show both the variables (X and Y) and the slack variables (S and T).

c. The simplex tableau for the maximum problem (both the initial tableau and after *any* sequence of pivot operations) has numbers in the bottom row:

	x_1	$\cdots$	x_n	s_1	$\cdots$	s_m	
P	w_1	$\cdots$	w_n	z_1	$\cdots$	z_m	V

To prove the Duality Theorem, we must show that these values can be used for the values of Y and T in the minimum problem. That is, if we set $y_1 = z_1, \ldots, y_m = z_m$, and $t_1 = w_1, \ldots, t_n = w_n$, then the number V is the same as the value $C = b^t Y$ and the constraint $A^t Y - T = c$ is satisfied.

d. If we use the notations $W = \begin{pmatrix} w_1 \\ \vdots \\ w_n \end{pmatrix}$ and $Z = \begin{pmatrix} z_1 \\ \vdots \\ z_m \end{pmatrix}$, the bottom row of the tableau is the same as the matrix equation $P + W^t X + Z^t S = V$ and this equation holds for *any* values of X and S such that $AX + S = b$.

e. Since $P = c^t X$ and $S = b - AX$, we may rewrite the matrix equation $P + W^t X + Z^t S = V$ as $c^t X + W^t X + Z^t(b - AX) = V$.

f. That is, $X^t c + X^t W + b^t Z - (AX)^t Z = V$ and so $X^t \cdot (c + W - A^t Z) + b^t Z = V$. Thus

$(c + W - A^tZ) = 0$ and $b^tZ = V$. So the number V is the value of the minimum problem objective function $C = b^tY$ when Y takes the value Z and this value for Y satisfies the constraint $A^tY - T = c$ when the slack T takes the value W.

44. **Complementary Slackness** Use the Duality Theorem for the simplex method (see page 347) to show that *at least* one of *every* pair of variables

x_1 and t_1, x_2 and t_2, . . . , x_n and t_n, y_1 and s_1, y_2 and s_2, . . . , y_m and s_m is zero for the solution points of a linear programming problem and its dual. (*Hint:* What can you say about the bottoms of the basic variable columns in the simplex tableau of a maximum problem?)

45. Verify the Complementary Slackness Theorem (Exercise 44) for the final tableau given in Practice Problem 2 (see page 351).

3.5 Nonstandard Problems

APPLICATION PREVIEW

Kantorovich and Production Planning

The earliest discussion and solution of what we now call "linear programming problems" was published in 1939 by L. V. Kantorovich, a young professor at Leningrad State University in the USSR. In his introduction, he pointed out that "[t]here are two ways of increasing the efficiency of the work of a shop, an enterprise, or a whole branch of industry. One way is by various improvements in technology . . . . The other way—thus far much less used—is improvement in the organization of planning and production. Here are included, for instance, such questions as the distribution of work among individual machines . . . , the correct distribution of orders among enterprises, the correct distribution of different kinds of raw materials, fuel, and other factors." He then discussed many examples and described a method of solution.

Kantorovich was careful to note the differences between a capitalist society and the Soviet system of economic planning and management. In particular, he included requirements that the machines be in use for all the available time and that the workers be fully occupied. He thus considered maximum linear programming problems that included "at least" inequalities as well as the familiar "at most" restrictions. For some reason, probably based on an ideological assertion that an abstract subject like mathematics could have no possible use in the economic problems of the Soviet Union, Kantorovich was ignored. Dantzig discovered his article while collecting references to linear programming problems and arranged for its translation and publication* in 1960. In 1975, Kantorovich and Koopmans shared the Nobel Prize in Economics "for their contributions to the theory of optimum allocation of resources."

Introduction

In this section we will extend the simplex method to maximum problems that include constraints with $\geq$ inequalities as well as $\leq$ inequalities. Since a $\geq$ inequality can be changed into a $\leq$ inequality by multiplying through by -1, this extension is equivalent to dropping the requirement that $b \geq 0$ in the constraint $AX \leq b$.

The Dual Pivot Element

The origin is not a vertex for a nonstandard maximum problem because an inequality like $x_1 \geq 3$ separates the feasible region from the origin. Since the initial tableau begins at the origin, it is not feasible as well as not optimal. The question becomes: How can we move to a feasible point so that we can then work on making it optimal? On page 348 we saw that the pivot operations for a standard maximum problem make the dual minimum problem feasible. This gives the clue to the answer to our question: pivoting on "dual pivot elements" would make the problem feasible and then pivoting on (regular) pivot elements would make it optimal.

Since dual problems interchange the roles of rows and columns, the following definition is just the definition of the pivot element with the roles of the rows and columns reversed, and we must take the *largest* ratio instead of the smallest because we are now dividing by negative numbers.

Dual Pivot Element

> The *dual pivot row* is the row with the smallest negative entry in the rightmost column of the tableau (omitting the bottom row). If there is a tie, the dual pivot row is the uppermost such row.
>   The *dual pivot column* is the column with the largest ratio found by dividing the bottom entry by the dual pivot row entry, omitting any column with a zero or positive dual pivot row entry

* L. V. Kantorovich, "Mathematical Methods of Organizing and Planning Production," *Management Science* 6:366–422, 1960.

and the rightmost column. If there is a tie, the dual pivot column is the leftmost such column.

The *dual pivot element* is the entry in the dual pivot row and the dual pivot column.

EXAMPLE 1   The Dual Pivot Element

Find the dual pivot element in the following simplex tableau or explain why the tableau does not have a dual pivot element.

	x_1	x_2	x_3	x_4	s_1	s_2	s_3	
s_1	2	1	1	2	1	0	0	30
s_2	−1	−2	−1	−1	0	1	0	−5
s_3	−1	−1	−1	−2	0	0	1	−10
P	−8	12	−10	−14	0	0	0	0

Solution

The dual pivot row is row 3 because -10 is the smallest negative entry on the right (omitting the bottom row). The dual pivot column is column 3 because the ratio $\frac{-10}{-1} = 10$ is greater than the others (the remaining columns are not considered because their dual pivot row entries are zero or positive and the rightmost column is never considered).

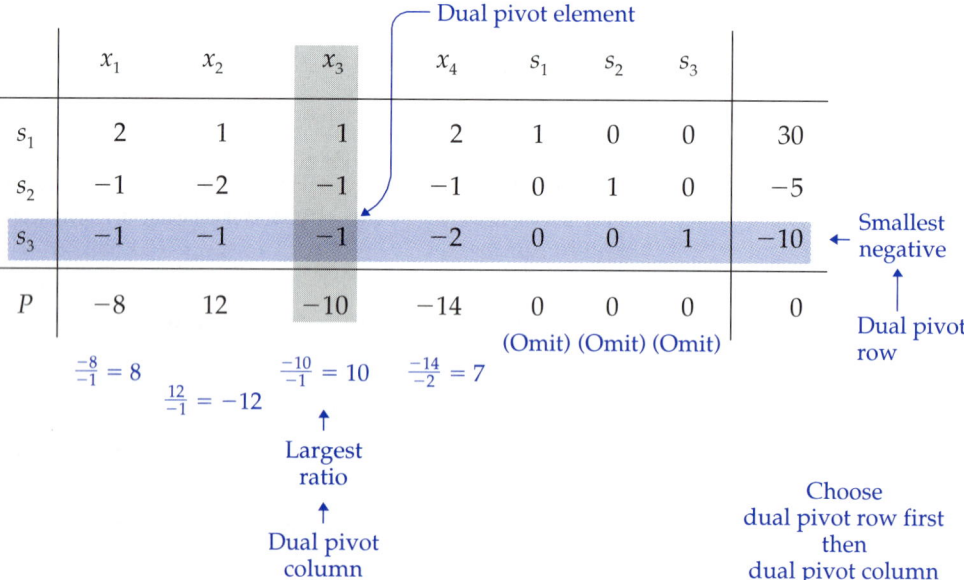

The dual pivot element is the -1 in row 3 and column 3.

PRACTICE PROBLEM 1

For each of the following simplex tableaux, find the dual pivot element or explain why the tableau does not have a dual pivot element.

a.

	x_1	x_2	x_3	x_4	s_1	s_2	s_3	
s_1	-1	-1	2	-2	1	0	0	-10
s_2	1	1	-1	-1	0	1	0	20
s_3	0	-1	3	2	0	0	1	-5
P	-5	-6	14	8	0	0	0	0

b.

	x_1	x_2	x_3	x_4	s_1	s_2	s_3	s_4	
s_1	1	1	5	1	1	0	0	0	25
s_2	-1	-1	-2	-1	0	1	0	0	-15
s_3	3	1	1	2	0	0	1	0	30
s_4	2	1	1	4	0	0	0	1	-20
P	-6	-8	10	-10	0	0	0	0	0

Solution at the back of the book

The Two-Stage Simplex Method

Pivoting on dual pivot elements will make the tableau feasible. After the tableau becomes feasible, pivoting on *regular* pivot elements will lead to the solution of the problem.

Two-Stage Simplex Method

To solve any maximum problem by the simplex method:

1. Write the constraints in the form $AX \leq b$ with $X \geq 0$ and construct the initial simplex tableau.

2. If the tableau is not feasible (at least one basic variable has a negative value in the rightmost column), go to Step 3. If the tableau is feasible (all the basic variables have nonnegative values in the rightmost column), go to Step 4.

3. Locate the dual pivot element (page 361), perform the pivot operation (page 323), and return to Step 2. If the tableau has a dual pivot row but no dual pivot column, then there is *no solution* to the problem (the region is infeasible).

4. Locate the pivot element (page 322), perform the pivot operation (page 323), and return to Step 2. If the tableau does not have a pivot column, then the solution occurs at the vertex given by the basic variables and the maximum value of the objective function appears in the bottom right corner of the tableau. If the tableau has a pivot column but no pivot row, then there is *no solution* to the problem (the region is unbounded).

EXAMPLE 2   **The Two-Stage Simplex Method**

Solve the following nonstandard linear programming problem by the two-stage simplex method.

$$\text{Maximize } P = 5x_1 + 7x_2$$

$$\text{Subject to } \begin{cases} x_1 + x_2 \le 8 \\ x_1 + x_2 \ge 4 \\ 2x_1 + x_2 \ge 6 \\ x_1 \ge 0 \text{ and } x_2 \ge 0 \end{cases}$$

"Mixed constraints" because they have $\le$ and $\ge$ inequalities

Solution

We first change the inequalities from $\ge$ to $\le$ by multiplying by -1 as needed: the second inequality becomes $-x_1 - x_2 \le -4$ and the third becomes $-2x_1 - x_2 \le -6$. This problem in matrix form is

$$\text{Maximize } P = (5 \quad 7)\begin{pmatrix} x_1 \\ x_2 \end{pmatrix}$$

$$\text{Subject to } \begin{cases} \begin{pmatrix} 1 & 1 \\ -1 & -1 \\ -2 & -1 \end{pmatrix}\begin{pmatrix} x_1 \\ x_2 \end{pmatrix} \le \begin{pmatrix} 8 \\ -4 \\ -6 \end{pmatrix} \\ \text{and } \begin{pmatrix} x_1 \\ x_2 \end{pmatrix} \ge 0 \end{cases}$$

The initial simplex tableau is

	x_1	x_2	s_1	s_2	s_3	
s_1	1	1	1	0	0	8
s_2	-1	-1	0	1	0	-4
s_3	-2	-1	0	0	1	-6
P	-5	-7	0	0	0	0

$\left.\begin{array}{l} s_1 = 8 \\ s_2 = -4 \\ s_3 = -6 \end{array}\right\}$ Not feasible because some basic variables are negative

The tableau is not feasible because s_2 and s_3 are negative. The dual pivot row is row 3 (-6 is the smallest negative entry in the right column, omitting the bottom row). The dual pivot column is column 2 (the ratio $\frac{-7}{-1} = 7$ is greater than $\frac{-5}{-2} = 2.5$ and none of the other columns may be considered). The dual pivot element is the -1 in row 3 and column 2. Pivoting on this dual pivot element, the tableau becomes feasible:

	x_1	x_2	s_1	s_2	s_3	
s_1	-1	0	1	0	1	2
s_2	1	0	0	1	-1	2
x_2	2	1	0	0	-1	6
P	9	0	0	0	-7	42

$\left.\begin{array}{l} s_1 = 2 \\ s_2 = 2 \\ x_2 = 6 \end{array}\right\}$ Feasible because all basic variables are nonnegative

Not optimal because bottom row has a negative entry

If it were still not feasible, we would look for another dual pivot element. Since the tableau is now feasible but not optimal, we look for a (regular) pivot element. The pivot column is column 5 (-7 is the only negative entry in the bottom row). The pivot row is row 1 (the other rows cannot be considered since their pivot column entries are negative). The pivot element is the 1 in column 5 and row 1. Pivoting on this pivot element, the tableau becomes optimal:

	x_1	x_2	s_1	s_2	s_3	
s_3	-1	0	1	0	1	2
s_2	0	0	1	1	0	4
x_2	1	1	1	0	0	8
P	2	0	7	0	0	56

$\left.\begin{array}{l} \\ \\ \end{array}\right\}$ Feasible: all are nonnegative

Optimal: all are nonnegative

This is the final tableau since it is both feasible and optimal. The maximum is $P = 56$ when $x_1 = 0$ and $x_2 = 8$.

■

To compare this solution with the graphical method, we graph the feasible region with x_1 on the horizontal axis and x_2 on the vertical axis. The five vertices and the values of the objective function are given in the table. The path of vertices visited by the two-stage simplex method is marked in bold.

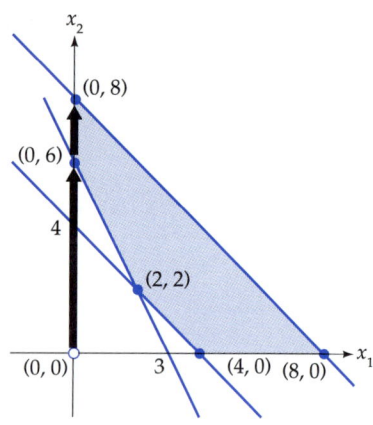

Vertex (x_1, x_2)	$P = 5x_1 + 7x_2$
$(8, 0)$	40
$(4, 0)$	20
$(2, 2)$	24
$(0, 6)$	42
$(0, 8)$	56

PRACTICE PROBLEM 2

Solve the following nonstandard linear programming problem by the two-stage simplex method.

$$\text{Maximize} \quad P = 4x_1 + x_2 + 3x_3$$

$$\text{Subject to} \begin{cases} x_1 + 2x_2 + x_3 \leq 50 \\ 2x_1 + x_2 + 2x_3 \geq 10 \\ x_1 \geq 0, x_2 \geq 0, \text{ and } x_3 \geq 0 \end{cases}$$

Solution at the back of the book

EXAMPLE 3   Managing an Investment Portfolio

The manager of an $80 million mutual fund has three investment possibilities: secure government bonds yielding 4%, a blue chip growth stock paying 6%, and a biotechnology start-up company returning 12%. If the manager wants to invest at least $50 million in a combination of safe government bonds and blue chip stock, and at least as much in the bonds as the total in blue chip and biotechnology stocks, how much should be invested in each to obtain the greatest possible return?

Solution

Let:

$$x_1 = \begin{pmatrix} \text{Money invested in} \\ \text{government bonds} \end{pmatrix},$$

$$x_2 = \begin{pmatrix} \text{Money invested in} \\ \text{the blue chip stock} \end{pmatrix} \text{ and}$$

$$x_3 = \begin{pmatrix} \text{Money invested in the} \\ \text{biotechnology company} \end{pmatrix}$$

be given in millions of dollars. The objective is to maximize the return:

$$P = 0.04x_1 + 0.06x_2 + 0.12x_3$$

Rate × amount
for each investment

subject to the restrictions:

$$x_1 + x_2 + x_3 \leq 80$$ $80 million available to invest

$$x_1 + x_2 \geq 50$$ $50 million in safe investments

$$x_1 \geq x_2 + x_3$$ Bonds ≥ total in stocks

$$x_1 \geq 0, x_2 \geq 0, x_3 \geq 0$$ Nonnegativity

Multiplying the second constraint by -1 and rewriting the third as $-x_1 + x_2 + x_3 \leq 0$, this problem in matrix form is

$$\text{Maximize } P = (0.04 \quad 0.06 \quad 0.12)\begin{pmatrix} x_1 \\ x_2 \\ x_3 \end{pmatrix}$$

$$\text{Subject to } \begin{cases} \begin{pmatrix} 1 & 1 & 1 \\ -1 & -1 & 0 \\ -1 & 1 & 1 \end{pmatrix}\begin{pmatrix} x_1 \\ x_2 \\ x_3 \end{pmatrix} \leq \begin{pmatrix} 80 \\ -50 \\ 0 \end{pmatrix} \\ \\ \text{and } \begin{bmatrix} x_1 \\ x_2 \\ x_3 \end{bmatrix} \geq 0 \end{cases}$$

The initial simplex tableau is not feasible:

	x_1	x_2	x_3	s_1	s_2	s_3	
s_1	1	1	1	1	0	0	80
s_2	-1	-1	0	0	1	0	-50
s_3	-1	1	1	0	0	1	0
P	-0.04	-0.06	-0.12	0	0	0	0

$s_2 < 0$ ← Dual pivot row

So we look for a dual pivot element. The dual pivot row is row 2 and the dual pivot column is column 2 (since the ratio $\frac{-0.06}{-1} = 0.06$ is greater than $\frac{-0.04}{-1} = 0.04$ and none of the other columns may be considered). Pivoting on the -1 in row 2 and column 2, the tableau is still not feasible:

	x_1	x_2	x_3	s_1	s_2	s_3	
s_1	0	0	1	1	1	0	30
x_2	1	1	0	0	-1	0	50
s_3	-2	0	1	0	1	1	-50
P	0.02	0	-0.12	0	-0.06	0	3

$s_3 < 0$ ← Dual pivot row

Pivoting on the dual pivot element in row 3 and column 1, the tableau becomes feasible but not optimal:

	x_1	x_2	x_3	s_1	s_2	s_3	
s_1	0	0	1	1	1	0	30
x_2	0	1	1/2	0	$-1/2$	1/2	25
x_1	1	0	$-1/2$	0	1/2	$-1/2$	25
P	0	0	-0.11	0	-0.05	0.01	2.5

Feasible: all are ≥ 0

Not optimal: some are < 0

Pivoting on the (regular) pivot element in column 3 and row 1, the tableau becomes both feasible and optimal:

	x_1	x_2	x_3	s_1	s_2	s_3	
x_3	0	0	1	1	1	0	30
x_2	0	1	0	$-1/2$	-1	1/2	10
x_1	1	0	0	1/2	0	$-1/2$	40
P	0	0	0	0.11	0.06	0.01	5.8

Feasible: all are ≥ 0

Optimal: all are ≥ 0

This is the final tableau because it is both feasible and optimal. The maximum is $P = 5.8$ when $x_1 = 40$, $x_2 = 10$, and $x_3 = 30$. In terms of the original situation, the manager should invest $40 million in the government bonds, $10 million in the blue chip stock, and $30 million in the biotechnology start-up company for a maximum return of $5.8 million.

Graphing Calculator Exploration

Formulate the following situation as a linear programming problem. Use the program PIVOT to solve it by the two-stage simplex method.

The mutual fund has grown and now the manager has $100 million to invest among three possibilities: secure government bonds still yielding 4%, a blue chip growth stock now paying 10% (and looking not so secure), and a biotechnology start-up company returning 14%. If the manager now decides to invest at least $60 million in safe government bonds and at least as much in the bonds as the total in blue chip and biotechnology stocks, how much should be invested in each to obtain the greatest possible return?

(*Answer:* The manager should invest $60 million in the government bonds, nothing in the blue chip stock, and $40 million in the biotechnology start-up company for a maximum return of $8 million.)

Can Any Linear Programming Problem Now Be Solved?

The two-stage simplex method allows us to solve *any* maximum linear programming problem, no matter what mixture of $\leq$ and $\geq$ inequalities are present in the constraints, provided that the variables are nonnegative. A direct way to change a minimum problem into a maximum problem is described in Exercises 41–42. Should the initial formulation of a problem include equality constraints, there are several ways to change the problem into a form that we can solve. Exercises 43–45 explore equality constraints by discussing several solutions of an example of Kantorovich, and additional equality constraint problems are given in Exercises 46–47. Exercises 48–50 explain how to reformulate a problem so that all the variables are nonnegative. Taken together, these techniques allow us to write *any* linear programming problem as a maximum problem with inequality constraints and nonnegative variables. Thus we may conclude that we are now able to solve any linear programming problem.

SUMMARY

A nonstandard maximum problem has both $\geq$ and $\leq$ constraints. Change the $\geq$ inequalities into $\leq$ form by multiplying by -1, and then construct the initial simplex tableau. In general, this tableau will not be feasible (it will have negative numbers in the rightmost column) as well as not optimal (it will have negative numbers in the bottom row).

The dual pivot *row* is the row with the smallest negative entry in the rightmost column of the tableau (omitting the bottom row). The dual pivot *column* is the column with the largest ratio of the bottom entry divided by the dual pivot row entry, omitting any column with a zero or positive dual pivot row entry and the rightmost column.

The two-stage simplex method pivots on dual pivot elements until the tableau is feasible, and then continues to pivot on (regular) pivot elements until the tableau is optimal. The problem has no solution if the dual pivot column or the (regular) pivot row does not exist.

EXERCISES 3.5

Write each of the following nonstandard linear programming problems in the matrix form "maximize $P = c^t X$ subject to $AX \leq b$ and $X \geq 0$," construct the initial simplex tableau, and locate the dual pivot element.

1. Maximize $P = 15x_1 + 20x_2 + 18x_3$

Subject to $\begin{cases} 3x_1 + 2x_2 + 8x_3 \leq 96 \\ 5x_1 + x_2 + 6x_3 \geq 30 \\ x_1 \geq 0, x_2 \geq 0, x_3 \geq 0 \end{cases}$

2. Maximize $P = 60x_1 + 90x_2 + 30x_3$

Subject to $\begin{cases} 4x_1 + 5x_2 + x_3 \geq 40 \\ 10x_1 + 18x_2 + 3x_3 \leq 150 \\ x_1 \geq 0, x_2 \geq 0, x_3 \geq 0 \end{cases}$

3. Maximize $P = 6x_1 + 4x_2 + 6x_3$

Subject to $\begin{cases} 2x_1 + x_2 + 3x_3 \geq 30 \\ x_1 + x_2 + 2x_3 \geq 20 \\ x_1 \geq 0, x_2 \geq 0, x_3 \geq 0 \end{cases}$

4. Maximize $P = 9x_1 + 5x_2 + 4x_3$

Subject to $\begin{cases} 3x_1 + x_2 + x_3 \geq 40 \\ 2x_1 + 2x_2 + x_3 \geq 30 \\ x_1 \geq 0, x_2 \geq 0, x_3 \geq 0 \end{cases}$

5. Maximize $P = 20x_1 + 30x_2 + 10x_3$

Subject to $\begin{cases} x_1 + x_2 + x_3 \geq 8 \\ x_1 + 2x_2 + 3x_3 \leq 30 \\ x_1 + 2x_2 + x_3 \leq 18 \\ x_1 \geq 0, x_2 \geq 0, x_3 \geq 0 \end{cases}$

6. Maximize $P = 24x_1 + 18x_2 + 30x_3$

Subject to $\begin{cases} x_1 + x_2 + x_3 \geq 4 \\ x_1 + 2x_2 + 3x_3 \leq 15 \\ x_1 + 2x_2 + x_3 \leq 9 \\ x_1 \geq 0, x_2 \geq 0, x_3 \geq 0 \end{cases}$

7. Maximize $P = 3x_1 + 2x_2 + 5x_3 + 4x_4$

Subject to $\begin{cases} x_1 + x_2 + x_3 + x_4 \geq 30 \\ 2x_1 + 3x_2 + 2x_3 + x_4 \geq 20 \\ 4x_1 + 2x_2 + x_3 + 2x_4 \leq 80 \\ x_1 \geq 0, x_2 \geq 0, x_3 \geq 0, x_4 \geq 0 \end{cases}$

8. Maximize $P = 6x_1 + 8x_2 + 3x_3 + 5x_4$

Subject to $\begin{cases} x_1 + x_2 + x_3 + x_4 \geq 70 \\ 3x_1 + 2x_2 + 4x_3 + 3x_4 \geq 60 \\ 4x_1 + x_2 + 2x_3 + 5x_4 \leq 100 \\ x_1 \geq 0, x_2 \geq 0, x_3 \geq 0, x_4 \geq 0 \end{cases}$

9. Maximize $P = 2x_1 + 2x_2 + x_3$

Subject to $\begin{cases} x_1 + 2x_2 + x_3 \geq 40 \\ 2x_1 + x_2 + x_3 \geq 50 \\ 5x_1 + 3x_2 + 2x_3 \leq 120 \\ 3x_1 + x_2 + 2x_3 \leq 150 \\ x_1 \geq 0, x_2 \geq 0, x_3 \geq 0 \end{cases}$

10. Maximize $P = x_1 + 2x_2 + 3x_3$

Subject to $\begin{cases} x_1 + x_2 + 2x_3 \geq 30 \\ x_1 + 2x_2 + x_3 \geq 60 \\ 2x_1 + 5x_2 + 3x_3 \leq 150 \\ 2x_1 + 3x_2 + x_3 \leq 120 \\ x_1 \geq 0, x_2 \geq 0, x_3 \geq 0 \end{cases}$

Solve each of the following nonstandard linear programming problems by the two-stage simplex method. (Exercises 11, 12, 15, 16, 19, and 20 may also be solved by the graphical method.)

 You may use the PIVOT program if permitted by your instructor.

11. Maximize $P = 4x_1 + x_2$

Subject to $\begin{cases} 3x_1 + 2x_2 \leq 120 \\ x_1 + x_2 \geq 50 \\ 2x_1 + x_2 \leq 60 \\ x_1 \geq 0, x_2 \geq 0 \end{cases}$

12. Maximize $P = x_1 + 4x_2$

Subject to $\begin{cases} 3x_1 + 2x_2 \leq 120 \\ x_1 + x_2 \leq 50 \\ 2x_1 + x_2 \geq 60 \\ x_1 \geq 0, x_2 \geq 0 \end{cases}$

13. Maximize $P = 30x_1 + 20x_2 + 28x_3$

Subject to $\begin{cases} 3x_1 + x_2 + 2x_3 \geq 30 \\ 4x_1 + x_2 + 3x_3 \leq 60 \\ x_1 \geq 0, x_2 \geq 0, x_3 \geq 0 \end{cases}$

14. Maximize $P = 18x_1 + 10x_2 + 20x_3$

Subject to $\begin{cases} 3x_1 + x_2 + 4x_3 \geq 30 \\ 2x_1 + x_2 + 5x_3 \leq 50 \\ x_1 \geq 0, x_2 \geq 0, x_3 \geq 0 \end{cases}$

15. Maximize $P = 4x_1 + 5x_2$

Subject to $\begin{cases} x_1 + 2x_2 \leq 12 \\ x_1 + x_2 \geq 15 \\ 2x_1 + x_2 \leq 12 \\ x_1 \geq 0, x_2 \geq 0 \end{cases}$

16. Maximize $P = 5x_1 + 4x_2$

Subject to $\begin{cases} 2x_1 + 3x_2 \leq 12 \\ x_1 + x_2 \geq 10 \\ 3x_1 + 2x_2 \leq 12 \\ x_1 \geq 0, x_2 \geq 0 \end{cases}$

17. Maximize $P = 15x_1 + 12x_2 + 18x_3$

Subject to $\begin{cases} 5x_1 + x_2 + 2x_3 \geq 30 \\ 2x_1 + x_2 + 3x_3 \leq 24 \\ x_1 \geq 0, x_2 \geq 0, x_3 \geq 0 \end{cases}$

18. Maximize $P = 4x_1 + 5x_2 + 6x_3$

Subject to $\begin{cases} 5x_1 + x_2 + 3x_3 \leq 30 \\ 2x_1 + x_2 + 2x_3 \geq 24 \\ x_1 \geq 0, x_2 \geq 0, x_3 \geq 0 \end{cases}$

19. Maximize $P = x_1 + x_2$

Subject to $\begin{cases} x_1 \leq 8 \\ x_2 \leq 5 \\ x_1 + 2x_2 \geq 6 \\ x_1 \geq 0, x_2 \geq 0 \end{cases}$

20. Maximize $P = x_1 + x_2$

Subject to $\begin{cases} x_1 \leq 6 \\ x_2 \leq 9 \\ 3x_1 + x_2 \geq 6 \\ x_1 \geq 0, x_2 \geq 0 \end{cases}$

21. Maximize $P = 2x_1 + 3x_2 + 2x_3$

Subject to $\begin{cases} 2x_1 + x_2 + x_3 \geq 20 \\ x_1 - x_2 + x_3 \leq 10 \\ x_1 \geq 0, x_2 \geq 0, x_3 \geq 0 \end{cases}$

22. Maximize $P = 6x_1 + 5x_2 + 3x_3$

Subject to $\begin{cases} 2x_1 + x_2 - x_3 \leq 20 \\ 3x_1 + 2x_2 + x_3 \geq 30 \\ x_1 \geq 0, x_2 \geq 0, x_3 \geq 0 \end{cases}$

23. Maximize $P = 4x_1 + 6x_2 + 12x_3 + 10x_4$

Subject to $\begin{cases} x_1 + x_2 + x_3 + x_4 \leq 60 \\ 2x_1 + x_2 + x_3 + 2x_4 \geq 10 \\ 2x_1 + 2x_2 + x_3 + x_4 \leq 100 \\ x_1 \geq 0, x_2 \geq 0, x_3 \geq 0, x_4 \geq 0 \end{cases}$

24. Maximize $P = 8x_1 - x_2 + 5x_3 + 14x_4$

Subject to $\begin{cases} x_1 + 2x_2 + 3x_4 \leq 55 \\ x_1 + x_2 + x_3 + 2x_4 \geq 25 \\ x_1 + 3x_3 + 3x_4 \leq 45 \\ x_1 \geq 0, x_2 \geq 0, x_3 \geq 0, x_4 \geq 0 \end{cases}$

25. Maximize $P = 5x_1 + 32x_2 + 3x_3 + 15x_4$

Subject to $\begin{cases} 5x_1 + 8x_2 + x_3 + 3x_4 \geq 18 \\ 2x_1 - x_2 + x_3 + x_4 \leq 10 \\ 3x_1 - 3x_2 + x_3 + 2x_4 \geq 6 \\ x_1 \geq 0, x_2 \geq 0, x_3 \geq 0, x_4 \geq 0 \end{cases}$

26. Maximize $P = 70x_1 + 12x_2 + 60x_3 + 20x_4$

Subject to $\begin{cases} 2x_1 + x_2 + x_3 + x_4 \leq 14 \\ 3x_1 + 3x_2 + x_3 + 2x_4 \leq 24 \\ 5x_1 + 3x_2 + 4x_3 + 2x_4 \geq 44 \\ x_1 \geq 0, x_2 \geq 0, x_3 \geq 0, x_4 \geq 0 \end{cases}$

27. Maximize $P = -x_1 + x_2 - 2x_3$

Subject to $\begin{cases} x_1 + x_2 + x_3 \leq 50 \\ x_1 + x_3 \geq 10 \\ x_2 + x_3 \geq 20 \\ x_1 \geq 0, x_2 \geq 0, x_3 \geq 0 \end{cases}$

28. Maximize $P = x_1 + 2x_2 + x_3$

Subject to $\begin{cases} x_1 + x_2 + x_3 \leq 30 \\ x_1 + x_2 \leq 25 \\ x_2 + x_3 \geq 15 \\ x_1 \geq 0, x_2 \geq 0, x_3 \geq 0 \end{cases}$

29. Maximize $P = 12x_1 + 2x_2 + 14x_3 + 8x_4$

Subject to $\begin{cases} x_1 + x_2 + x_3 + x_4 \geq 20 \\ 2x_1 + x_2 + 2x_3 + x_4 \leq 30 \\ x_3 + x_4 \geq 10 \\ x_1 \geq 0, x_2 \geq 0, x_3 \geq 0, x_4 \geq 0 \end{cases}$

30. Maximize $P = 8x_1 + 10x_2 + 6x_3 + 8x_4$

Subject to $\begin{cases} x_1 + x_2 + x_4 \leq 30 \\ 2x_1 + 2x_2 + x_3 + x_4 \leq 80 \\ x_1 + x_2 + x_3 + x_4 \geq 20 \\ x_1 \geq 0, x_2 \geq 0, x_3 \geq 0, x_4 \geq 0 \end{cases}$

APPLIED EXERCISES

Formulate each situation as a linear programming problem by identifying the variables, the objective function, and the constraints. Check that the problem is a nonstandard maximum problem and then solve it by the two-stage simplex method. Be sure to state your final answer in terms of the original question.

 You may use the PIVOT program if permitted by your instructor.

31. Financial Planning   A retired couple want to invest their $20,000 life savings in bank certificates of deposit yielding 6% and treasury bonds yielding 5%. If they want at least $5000 in each type of investment, how much should they invest in each to receive the greatest possible income?

32. Recycling Management   A volunteer recycling center accepts both used paper and empty glass bottles, which it then sorts and sells to a reprocessing company. The center has room to accept a total of 800 crates of paper and glass each week and has 50 hours of volunteer help to do the sorting. Each crate of paper products takes 5 minutes to sort and sells for 8¢, while each crate of bottles takes 3 minutes to sort and sells for 7¢. To support the city's "grab that glass" recycle theme this week, the center wants to accept at least 600 crates of glass. How many crates of each should the center accept this week to raise the most money for its ecology scholarship fund?

33. Advertising   The manager of a new mall may spend up to $18,000 on "grand opening" announcements in newspapers, on radio, and on TV. Each newspaper ad costs $300 and reaches 5000 readers, each one-minute radio commercial costs $800 and is heard by 13,000 listeners, and each 15-second TV spot costs $900 and is seen by 15,000 viewers. If the manager wants at least 5 newspaper ads and at least 20 radio commercials and TV spots combined, how many of each should be placed to reach the largest number of potential customers? (Ignore multiple exposures to the same customer.)

34. Advertising   In the last few days before the election, a politician can afford to spend no more than $27,000 on TV advertisements and can arrange for no more than 10 ads. Each daytime ad costs $2000 and reaches 4000 viewers, each prime time ad costs $3000 and reaches 5000 viewers, and each late night ad costs $1000 and reaches 2000 viewers. To be sure to reach the widest variety of voters, the politician's advisors insist on at least 5 ads scheduled during daytime and late night. Ignoring repeated viewings by the same person and assuming every viewer can vote, how many ads in each of the time periods will reach the most voters?

35. Resource Allocation   A furniture shop manufactures wooden desks, tables, and chairs. The numbers of hours to assemble and finish each piece are shown in the table, together with the numbers of hours of skilled labor available for each task. To meet expected demand, a total of at least 30 desks and tables combined must be made. If the profit is $75 for each desk, $84 for each table, and $66 for each chair, how many of each should the company manufacture to obtain the greatest possible profit?

	Desk	Table	Chair	Labor Available
Assembly	2 hours	1 hour	2 hours	210 hours
Finishing	2 hours	3 hours	1 hour	150 hours

36. Production Planning   An automotive parts shop rebuilds carburetors, fuel pumps, and alternators. The number of hours to rebuild and then inspect and pack each part is shown in the table, together with the number of hours of skilled labor available for each task. At least 12 carburetors must be rebuilt. If the profit is $12 for each carburetor, $14 for each fuel pump, and $10 for each alternator, how many of each should the shop rebuild to obtain the greatest possible profit?

	Carburetor	Fuel Pump
Rebuilding	5 hours	4 hours
Final inspection and packaging	1 hour	1 hour

	Alternator	Labor Available
Rebuilding	3 hours	200 hours
Final inspection and packaging	0.5 hour	45 hours

37. Agriculture A farmer grows wheat, barley, and oats on her 500-acre farm. Each acre of wheat requires 3 days of labor (to plant, tend, and harvest) and costs $21 (for seed, fertilizer, and pesticides), each acre of barley requires 2 days of labor and costs $27, and each acre of oats requires 3 days of labor and costs $24. The farmer and her hired field hands can provide no more than 1200 days of labor this year. She wants to grow at least 100 acres of oats and can afford to spend no more than $15,120. If the profit is $50 for each acre of wheat, $40 for each acre of barley, and $45 for each acre of oats, how many acres of each crop should she grow to obtain the greatest possible profit?

38. Agriculture A farmer grows corn, peanuts, and soybeans on his two-hundred-forty-acre farm. To maintain soil fertility, the farmer rotates the crops and always plants at least as many acres of soybeans as the total acres of the other crops. Since he has promised to sell some of his corn to a neighbor who raises cattle, he must plant at least 42 acres of corn. Each acre of corn requires 2 days of labor and yields a profit of $150, each acre of peanuts requires 5 days of labor and yields a profit of $300, and each acre of soybeans requires 1 day of labor and yields a profit of $100. If the farmer and his children can put in at most 630 days of labor, how many acres of each crop should the farmer plant to obtain the greatest possible profit?

39. Pollution Control An empty storage yard at a coal burning electric power plant can hold no more than 100,000 tons of coal. Two grades of coal are available: low sulfur (1%) with an energy content of 20 million BTU per ton and high sulfur (2%) with an energy content of 30 million BTU per ton. If existing contracts with the high-sulfur coal mine operator require that at least 50,000 tons of high sulfur coal be purchased, and if the next coal purchase may contain no more than 1400 tons of sulfur, how many tons of each type of coal should be purchased to obtain the most energy?

40. Production Planning A small jewelry company prepares and mounts semiprecious stones. There are 20 lapidaries (who cut and polish the stones) and 24 jewelers (who mount the stones in gold settings). Each employee works 7 hours each day, 5 days each week. Each tray of agates requires 5 hours of cutting and polishing and 4 hours to mount, each tray of onyxes requires 2 hours of cutting and polishing and 3 hours to mount, and each tray of garnets requires 6 hours of cutting and polishing and 3 hours to mount. Furthermore, the company's owner has decided that the company will process at least 12 trays of agates each week. If the profit is $15 for each tray of agates, $10 for each tray of onyxes, and $13 for each tray of garnets, how many trays of each stone should be processed each week to obtain the greatest possible profit?

Explorations and Excursions

The following problems extend and augment the material presented in the text.

 The PIVOT program may be helpful (if permitted by your instructor).

More About the Dual Pivot Element

41. a. Solve both of the following problems by the graphical method from Section 3.2. Is it true that the minimum of C is the same as -1 times the maximum of $P = -C$?

Minimize $C = 3x + 4y$

Subject to $\begin{cases} x + y \le 10 \\ 2x + y \ge 8 \\ x \ge 0 \text{ and } y \ge 0 \end{cases}$

Maximize $P = -3x - 4y$

Subject to $\begin{cases} x + y \le 10 \\ 2x + y \ge 8 \\ x \ge 0 \text{ and } y \ge 0 \end{cases}$

b. Solve the following problem by finding the dual maximum problem and using the (regular) simplex method as we did in Section 3.4.

Minimize $C = 3y_1 + 4y_2$

Subject to $\begin{cases} -y_1 - y_2 \ge -10 \\ 2y_1 + y_2 \ge 8 \\ y_1 \ge 0 \text{ and } y_2 \ge 0 \end{cases}$

c. Solve the following nonstandard problem by the two-stage simplex method.

Maximize $P = -3x_1 - 4x_2$

Subject to $\begin{cases} x_1 + x_2 \le 10 \\ -2x_1 - x_2 \le -8 \\ x_1 \ge 0 \text{ and } x_2 \ge 0 \end{cases}$

d. Compare the tableaux and pivot elements in parts (b) and (c). Are the pivot elements the same numbers? Is the arithmetic to choose each pair of pivot elements the same? Is the "dual pivot element" just the "dual" of the (regular) pivot element?

42. Repeat the process of Exercise 41 to find "dual" solutions to the following problems.

Minimize $C = 20y_1 + 18y_2$

Subject to $\begin{cases} 2y_1 + y_2 \ge 18 \\ y_1 + y_2 \ge 14 \\ 4y_1 + 3y_2 \ge 48 \\ y_1 \ge 0 \text{ and } y_2 \ge 0 \end{cases}$

Maximize $P = -20x_1 - 18x_2$

Subject to $\begin{cases} -2x_1 - x_2 \le -18 \\ -x_1 - x_2 \le -14 \\ -4x_1 - 3x_2 \le -48 \\ x_1 \ge 0 \text{ and } x_2 \ge 0 \end{cases}$

Kantorovich's First Example   The first example from the 1939 paper "Mathematical Methods for Organizing and Planning Production" by L. V. Kantorovich may be phrased as follows.

A machine shop has three types of equipment that can produce either or both of two kinds of parts (see the table). These parts are used to make finished items containing one of each, so the shop must produce the same number of both parts. Furthermore, the equipment must never be idle. How should the production be divided so that the greatest number of finished items are produced?

Productivity of the Machines for Two Parts

Type of Machine	Number of Machines	Output per Machine	
		First Part	Second Part
Milling machines	3	10	20
Turret lathes	3	20	30
Automatic lathe	1	30	80

Type of Machine	Number of Machines	Total Output	
		First Part	Second Part
Milling machines	3	30	60
Turret lathes	3	60	90
Automatic lathe	1	30	80

Let x_1 represent the fraction of the work day that the milling machines make the first part, so that they make the second part for the remaining $1 - x_1$ fraction of the work day. Similarly, let x_2 and x_3 be the corresponding fractions for the turret lathes and the automatic lathe. We must have that $0 \le x_1 \le 1$, $0 \le x_2 \le 1$, and $0 \le x_3 \le 1$ for these fractions. The number of first parts produced is then $30x_1 + 60x_2 + 30x_3$ and, since the remaining time is spent making the second part, the number of second parts produced is $60(1 - x_1) + 90(1 - x_2) + 80(1 - x_3)$. In order to make the finished items without leftover parts, these must be equal:

$$30x_1 + 60x_2 + 30x_3 =$$
$$60(1 - x_1) + 90(1 - x_2) + 80(1 - x_3)$$

That is:

$$9x_1 + 15x_2 + 11x_3 = 23 \qquad \begin{array}{l}\text{Divide by 10 and}\\ \text{combine terms}\end{array}$$

Since the number of finished parts will be $30x_1 + 60x_2 + 30x_3$, we have the linear programming problem:

Maximize $P = 30x_1 + 60x_2 + 30x_3$

Subject to $\begin{cases} x_1 \le 1 \\ x_2 \le 1 \\ x_3 \le 1 \\ 9x_1 + 15x_2 + 11x_3 = 23 \quad \leftarrow \begin{array}{l}\text{Equality}\\ \text{constraint}\end{array} \\ x_1 \ge 0,\, x_2 \ge 0,\text{ and } x_3 \ge 0 \end{cases}$

Exercises 43–45 explore three possible ways to solve this problem with an equality constraint.

43. One way to solve Kantorovich's First Example is to just go ahead and write down the initial simplex tableau for the constraints just as they are:

	x_1	x_2	x_3	s_1	s_2	s_3	
s_1	1	0	0	1	0	0	1
s_2	0	1	0	0	1	0	1
s_3	0	0	1	0	0	1	1
??	9	15	11	0	0	0	23
P	-30	-60	-30	0	0	0	0

  a. Since this tableau does not display a basis, we must select one of the non-basic variables to replace the "??" in the basis listed on the left of the tableau. The "best" variable to enter the basis is the one with the smallest negative value in the bottom row. Show that pivoting on the 15 in column 2 and row 4 of the tableau will enter x_2 into the basis and result in an optimal but not feasible tableau.

  b. Find the dual pivot element in the tableau from part (a) and pivot to show that the maximum occurs when $x_1 = 8/9$, $x_2 = 1$, and $x_3 = 0$. (In terms of Kantorovich's original problem, the milling machines should produce the first part $\frac{8}{9}$ of the time and the second part the remaining $\frac{1}{9}$, the turret lathes should produce the first part all of the time, and the automatic lathe should produce the second part all of the time.)

44. Since the linear equation $9x_1 + 15x_2 + 11x_3 = 23$ is the same as the two linear inequalities $9x_1 + 15x_2 + 11x_3 \le 23$ and $9x_1 + 15x_2 + 11x_3 \ge 23$, another way of finding the answer to Kantorovich's First Example is to use the two-stage simplex method to solve the nonstandard problem:

Maximize $P = 30x_1 + 60x_2 + 30x_3$

Subject to $\begin{cases} x_1 \le 1 \\ x_2 \le 1 \\ x_3 \le 1 \\ 9x_1 + 15x_2 + 11x_3 \le 23 \\ 9x_1 + 15x_2 + 11x_3 \ge 23 \\ x_1 \ge 0,\, x_2 \ge 0,\text{ and } x_3 \ge 0 \end{cases}$

Carry out the solution using the two-stage simplex method and check that it agrees with the solution found in Exercise 43.

45. Another approach to Kantorovich's First Example is to solve the equation $9x_1 + 15x_2 + 11x_3 = 23$ for x_3 and use this information to rewrite the problem in just x_1 and x_2.

  a. Show that $x_3 = \frac{23}{11} - \frac{9}{11}x_1 - \frac{15}{11}x_2$ and that Kantorovich's First Example may be rewritten (remember that $x_3 \ge 0$) as:

Maximize $P = \frac{1}{11}(60x_1 + 210x_2 + 690)$

Subject to $\begin{cases} x_1 \le 1 \\ x_2 \le 1 \\ -9x_1 - 15x_2 \le -12 \\ 9x_1 + 15x_2 \le 23 \\ x_1 \ge 0 \text{ and } x_2 \ge 0 \end{cases}$

  b. Explain how the solution to the problem in part (a) can be found from the solution of the problem:

Maximize $P = 2x_1 + 7x_2$

Subject to $\begin{cases} x_1 \le 1 \\ x_2 \le 1 \\ -3x_1 - 5x_2 \le -4 \\ 9x_1 + 15x_2 \le 23 \\ x_1 \ge 0 \text{ and } x_2 \ge 0 \end{cases}$

c. Solve the problem in part (b) by the two-stage simplex method.

d. Use the solution found in part (c) to find the solution of the problem in part (a) and then deduce the solution to Kantorovich's First Example. Does this solution agree with the solutions found in Exercises 43 and 44?

46. Use the methods discussed in Exercises 43–45 to solve the following equality constraint linear programming problem. Check your answer by solving this problem by the graphical method from Section 3.2.

Maximize $P = 5x_1 + 3x_2$

Subject to $\begin{cases} 2x_1 + x_2 = 6 \\ x_1 + x_2 \le 4 \\ x_1 \ge 0 \text{ and } x_2 \ge 0 \end{cases}$

47. Use the methods discussed in Exercises 43–45 to solve the following equality constraint linear programming problem.

Maximize $P = 5x_1 + 4x_2 + 6x_3 + 3x_4$

Subject to $\begin{cases} 2x_1 + x_2 + x_3 \le 40 \\ 2x_1 + 2x_3 + x_4 \le 60 \\ x_1 + x_2 = 20 \\ x_3 + x_4 = 30 \\ x_1 \ge 0, x_2 \ge 0, x_3 \ge 0, \text{ and } x_4 \ge 0 \end{cases}$

More About Nonnegativity Conditions   The following problems show how to reformulate linear programming problems to have nonnegativity conditions on *all* the variables when the original versions do not.

48. Show that the first problem may be rewritten as the second by replacing the variable x_1 by a new variable u_1 where $u_1 = x_1 + 8$ and $u_1 \ge 0$. Solve the first problem by the graphical method from Section 3.2, solve the second by the two-stage simplex method, and check that your answers agree.

Maximize $P = 5x_1 + 3x_2$

Subject to $\begin{cases} 2x_1 + x_2 \le 6 \\ x_1 + x_2 \le 4 \\ x_1 \ge -8 \text{ and } x_2 \ge 0 \end{cases}$

Maximize $P = 5u_1 + 3x_2 - 40$

Subject to $\begin{cases} 2u_1 + x_2 \le 22 \\ u_1 + x_2 \le 12 \\ u_1 \ge 0 \text{ and } x_2 \ge 0 \end{cases}$

49. Show that the first problem may be rewritten as the second by replacing the variable x_1 by a new variable u_1 where $u_1 = -x_1$ and $u_1 \ge 0$. Solve the first problem by the graphical method from Section 3.2, solve the second by the two-stage simplex method, and check that your answers agree.

Maximize $P = 5x_1 + 3x_2$

Subject to $\begin{cases} 2x_1 + x_2 \le 10 \\ -x_1 + x_2 \le 4 \\ x_1 \le 0 \text{ and } x_2 \ge 0 \end{cases}$

Maximize $P = -5u_1 + 3x_2$

Subject to $\begin{cases} -2u_1 + x_2 \le 10 \\ u_1 + x_2 \le 4 \\ u_1 \ge 0 \text{ and } x_2 \ge 0 \end{cases}$

50. A variable is *unrestricted* if it may take *any* value (positive, negative, or zero). Show that the first problem may be rewritten as the second by replacing the variable x_1 by two new variables u_1 and u_2 where $u_1 - u_2 = x_1$ and $u_1, u_2 \ge 0$. Solve the first problem by the graphical method from Section 3.2, solve the second by the two-stage simplex method, and check that your answers agree.

Maximize $P = 3x_1 + 5x_2$

Subject to $\begin{cases} 2x_1 + x_2 \le 6 \\ x_2 \le 4 \\ x_1 \text{ unrestricted and } x_2 \ge 0 \end{cases}$

Maximize $P = 3u_1 - 3u_2 + 5x_2$

Subject to $\begin{cases} 2u_1 - 2u_2 + x_2 \le 6 \\ x_2 \le 4 \\ u_1 \ge 0, u_2 \ge 0, \text{ and } x_2 \ge 0 \end{cases}$

(*Hint:* Can any real number be written as the difference of two nonnegative numbers?)

Chapter Summary with Hints and Suggestions

The reading and exercises of this chapter have helped you learn the following skills. For each skill, the section from which it came (in case you need to review it) and some exercises in this section that use it are indicated. Answers to all exercises are at the end of the book, and full solutions to all exercises are in the Student Solutions Manual.

3.1 Linear Inequalities

- Sketch a linear inequality $ax + by \leq c$ by drawing the boundary line $ax + by = c$ from its intercepts, using a test point to choose the correct side, and shading in the feasible region. (*Review Exercises 1–2.*)

$$\left(\frac{c}{a}, 0\right) \quad \text{and} \quad \left(0, \frac{c}{b}\right)$$

- Sketch a system of linear inequalities $\begin{cases} ax + by \leq c \\ \vdots \end{cases}$ by drawing the boundary lines, finding the correct sides of the boundaries, and shading in the feasible region. List the vertices or "corners" of the region by finding the intersections of the boundary lines that are feasible. Identify the region as bounded or unbounded. (*Review Exercises 3–8.*)

- Formulate an applied situation as a system of linear inequalities and sketch the feasible region with the vertices. Many situations include nonnegativity conditions for the variables because of what they represent. (*Review Exercises 9–10.*)

3.2 Two-Variable Linear Programming Problems

- Find the maximum or minimum of a linear function on a given region or explain why such a value does not exist. (*Review Exercises 11–14.*)

- Solve a two-variable linear programming problem by sketching the feasible region, determining if a solution exists, and then finding it from the values of the objective function at the vertices. (*Review Exercises 15–18.*)

- Formulate an application as a linear programming problem by identifying the variables, the objective function, and the constraints, and then solve it using the vertices of the feasible region. (*Review Exercises 19–20.*)

3.3 The Simplex Method for Standard Maximum Problems

- Rewrite a maximum linear programming problem in matrix form, check that it is a standard problem, and construct the initial simplex tableau. (*Review Exercises 21–22.*)

$$\text{Maximize } P = c^t X$$

$$\text{Subject to } \begin{cases} AX \leq b \\ X \geq 0 \end{cases}$$

$$\text{where } b \geq 0$$

	X	S	
S	A	I	b
P	$-c^t$	0	0

- Find the pivot element in a given simplex tableau and carry out one complete pivot operation. If there is no pivot element, explain what the tableau shows about the original standard maximum problem. (*Review Exercises 23–24.*)

- Solve a standard maximum problem by the simplex method (construct the initial simplex tableau, pivot until the tableau does not have a pivot element, and interpret this final tableau). (*Review Exercises 25–28.*)

- Formulate an application as a linear programming problem and check that the problem is a standard maximum problem. Solve it by the simplex method. (*Review Exercises 29–30.*)

3.4   Duality and Standard Minimum Problems

- Rewrite a minimum linear programming problem in matrix form, check that it is a standard problem, and construct the dual maximum problem and the initial simplex tableau. (*Review Exercises 31–34.*)

$$\text{Minimize } C = b^t Y$$

$$\text{Subject to } \begin{cases} A^t Y \geq c \\ Y \geq 0 \end{cases}$$

$$\text{where } b^t \geq 0$$

$$\text{Maximize } P = c^t X$$

$$\text{Subject to } \begin{cases} AX \leq b \\ X \geq 0 \end{cases}$$

$$\text{where } b \geq 0$$

- Solve a standard minimum problem by finding the dual maximum problem and using the simplex method.   (*Review Exercises 35–38.*)

- Formulate an application as a standard minimum problem and solve it by finding the dual maximum problem and using the simplex method.   (*Review Exercises 39–40.*)

3.5   Nonstandard Problems

- Rewrite a nonstandard maximum problem in matrix form, construct the initial simplex tableau, and locate the dual pivot element. (*Review Exercises 41–42.*)

- Solve a nonstandard maximum problem by the two-stage simplex method. (*Review Exercises 43–48.*)

- Formulate an application as a nonstandard maximum problem and solve it by the two-stage simplex method. (*Review Exercises 49–50.*)

Hints and Suggestions

- (*Overview*) A linear programming problem asks for the maximum or minimum of a linear objective function subject to constraints in the form of linear inequalities. If there is a solution to the problem, it occurs at a *vertex* of the feasible region determined by the constraints. A problem with more than two variables must be solved algebraically using the simplex method, duality (for a standard minimum problem), or the two-stage simplex method (for a nonstandard problem).

- While a problem with a *bounded* region always has a solution, a problem with an *unbounded* region may or may not have a solution (if it *does*, then it occurs at a vertex).

- When setting up a word problem, look first for the questions "how many" or "how much" to help identify the variables. Be sure that finding values for your variables will answer the question stated in the problem. Use the rest of the given information to find the objective function and constraints. Include nonnegativity conditions for your variables as appropriate (for example, a farmer *cannot* raise a negative number of cows).

- A simplex tableau is *feasible* if it has no negative numbers in the rightmost *column* and is *optimal* if it has no negative numbers in the bottom *row*. If a simplex tableau is both feasible and optimal, it displays the solution of the problem.

- The pivot operation is not completed until *all* the rows of the tableau have been recalculated.

- If you are solving a standard maximum problem by the simplex method and you have a negative number in the rightmost column after pivoting, you've made an error in your choice of the pivot element and/or your arithmetic.

- Many simplex tableaux require several pivots to solve, so don't give up: keep pivoting until the tableau does not have a pivot element.

- Practice for test: Review Exercises 1, 7, 9, 11, 13, 16, 17, 20, 25, 29, 35, 39, 41, 45, 50.

Review Exercises for Chapter 3   *Practice test exercises are in blue.*

3.1 Linear Inequalities

Sketch each linear inequality.

1. $3x - 2y \leq 24$ **2.** $x + y \geq -6$

Sketch each system of linear inequalities. List all vertices and identify the region as "bounded" or "unbounded."

3. $\begin{cases} x + y \leq 20 \\ x \geq 0, y \geq 0 \end{cases}$ **4.** $\begin{cases} x + 2y \geq 10 \\ x \geq 0, y \geq 0 \end{cases}$

5. $\begin{cases} 2x + y \leq 20 \\ x + y \leq 15 \\ x \geq 0, y \geq 0 \end{cases}$ **6.** $\begin{cases} 2x + y \geq 12 \\ x + y \geq 8 \\ x \geq 0, y \geq 0 \end{cases}$

7. $\begin{cases} 3x + 2y \leq 24 \\ x + y \leq 10 \\ x \geq 2 \\ x \geq 0, y \geq 0 \end{cases}$ **8.** $\begin{cases} x + y \leq 10 \\ -x + y \leq 4 \\ x \leq 9 \\ x \geq 0, y \geq 0 \end{cases}$

Formulate each application as a system of linear inequalities, sketch the feasible region, and find the vertices.

9. Dog Training   A dog trainer works with Irish setters and Labrador retrievers for no more than six hours each day. Each training session with a "setter" takes one half hour and requires eight dog treats, while each session with a "lab" takes three quarters of an hour and also requires eight dog treats. If the trainer has only eighty dog treats left, how many of each breed can he train today?

10. Blending Bird Seed   A pet store owner blends custom bird seed by mixing SongBird and MeadowMix brands together. Each pound of SongBird contains 2 ounces of sunflower hearts and 3 ounces of crushed peanuts (along with a lot of cheap filler seed) while each pound of MeadowMix contains 4 ounces of sunflower hearts and 2 ounces of crushed peanuts. If each bag of the custom blend is labeled "contains at least 104 ounces of sunflower hearts and 84 ounces of crushed peanuts," how many pounds of each brand must be put in each bag?

3.2 Two-Variable Linear Programming Problems

Find each maximum or minimum value for the given region. If such a value does not exist, explain why not.

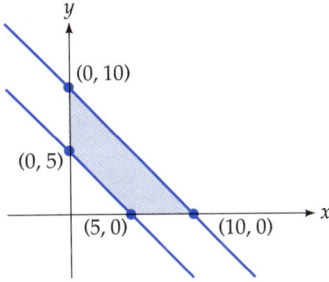

11. Maximum of $P = 3x + 4y$

12. Minimum of $C = 3x + 2y$

Find each maximum or minimum value for the given region. If such a value does not exist, explain why not.

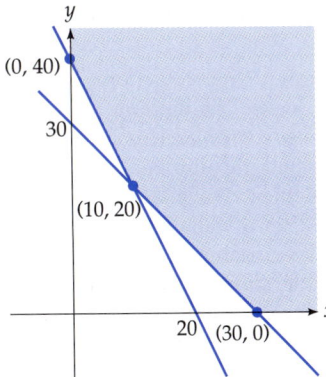

13. Maximum of $P = 4x + 3y$

14. Minimum of $C = 2x + 5y$

Solve each linear programming problem by sketching the region, labeling the vertices, checking whether a solution exists, and then finding it if it does.

15. Maximize $P = 20x + 15y$

Subject to $\begin{cases} x + 3y \le 18 \\ x + y \le 12 \\ x \ge 0, y \ge 0 \end{cases}$

16. Minimize $C = 30x + 40y$

Subject to $\begin{cases} 2x + y \ge 120 \\ x + 3y \ge 120 \\ x \ge 0, y \ge 0 \end{cases}$

17. Maximize $P = 20x + 30y$

Subject to $\begin{cases} 2x + y \le 24 \\ x + 4y \le 40 \\ x + y \le 13 \\ x \ge 0, y \ge 0 \end{cases}$

18. Minimize $C = 7x + 11y$

Subject to $\begin{cases} x + y \ge 20 \\ x - y \ge -10 \\ x \ge 0, y \ge 0 \end{cases}$

Formulate each application as a linear programming problem by identifying the variables, the objective function, and the constraints. Determine if a solution exists and, if it does, find it using the vertices of the feasible region. Be sure to state your final answer in terms of the original question.

19. Production Planning   A wood working shop makes antique reproduction cases for wall and mantel clocks. The amount of cutting and finishing needed for each is shown in the table, together with the number of hours of skilled labor available for each task. If the profit for each case is $250 for wall clocks and $200 for mantel clocks, how many of each kind of clock case should the shop make to obtain the greatest possible profit?

	Wall Clock Case	Mantel Clock Case	Labor Available
Cutting	5 hours	3 hours	165 hours
Finishing	3 hours	4 hours	132 hours

20. Nutrition   A large and active cat needs at least 70 grams of fat and 134.4 grams of protein each week. Each ounce of canned cat food contains 2 grams of fat and 3.2 grams of protein, while each ounce of dry food contains 2.5 grams of fat and 9.6 grams of protein. If canned cat food costs 5¢ per ounce and dry food costs 8¢ per ounce, how many ounces of each type of food will meet the cat's nutritional needs for the week at the least cost?

3.3   The Simplex Method for Standard Maximum Problems

Construct the initial simplex tableau for each standard maximum problem.

21. Maximize $P = 4x_1 + 10x_2 + 9x_3$

Subject to $\begin{cases} 5x_1 + 2x_2 + 3x_3 \le 30 \\ 2x_1 + 3x_2 + 4x_3 \le 24 \\ x_1 \ge 0, x_2 \ge 0, x_3 \ge 0 \end{cases}$

22. Maximize $P = 5x_1 + 7x_2$

Subject to $\begin{cases} x_1 + x_2 \le 10 \\ 2x_1 + x_2 \le 14 \\ x_1 - x_2 \le 4 \\ x_1 \ge 0, x_2 \ge 0 \end{cases}$

For each simplex tableau, find the pivot element and carry out one complete pivot operation. If there is no pivot element, explain what the tableau shows about the solution of the original standard maximum problem.

23.

	x_1	x_2	x_3	s_1	s_2	s_3	
s_1	1	1	1	1	0	0	6
s_2	3	1	2	0	1	0	8
s_3	1	1	-3	0	0	1	6
P	-6	4	-8	0	0	0	0

24.

	x_1	x_2	s_1	s_2	s_3	
s_1	2	0	1	0	-2	8
s_2	6	0	0	1	3	42
x_2	1	1	0	0	1	10
P	5	0	0	0	15	150

Solve each problem by the simplex method.

25. Maximize $P = 7x_1 + 8x_2 + 7x_3$

Subject to $\begin{cases} 2x_1 + x_2 + 4x_3 \le 16 \\ x_1 + 2x_2 - x_3 \le 30 \\ x_1 + x_2 + 3x_3 \le 12 \\ x_1 \ge 0, x_2 \ge 0, x_3 \ge 0 \end{cases}$

26. Maximize $P = 8x_1 + 10x_2 + 9x_3$

Subject to $\begin{cases} 3x_1 + 2x_2 + 3x_3 \le 18 \\ 2x_1 + x_2 + 4x_3 \le 14 \\ x_1 \ge 0, x_2 \ge 0, x_3 \ge 0 \end{cases}$

27. Maximize $P = 8x_1 + 7x_2 + 5x_3 + 7x_4$

Subject to $\begin{cases} 3x_1 + 2x_2 + 4x_3 + 3x_4 \le 12 \\ x_1 + x_2 + x_3 + x_4 \le 5 \\ 3x_1 + x_2 + 5x_3 + 3x_4 \le 15 \\ x_1 \ge 0, x_2 \ge 0, x_3 \ge 0, x_4 \ge 0 \end{cases}$

28. Maximize $P = 5x_1 + 4x_2 + 3x_3$

Subject to $\begin{cases} x_1 + x_2 \le 8 \\ x_2 + x_3 \le 7 \\ x_1 + x_2 - x_3 \le 6 \\ x_1 \ge 0, x_2 \ge 0, x_3 \ge 0 \end{cases}$

Formulate each application as a linear programming problem by identifying the variables, the objective function, and the constraints. Check that the problem is a standard maximum problem and then solve it by the simplex method. Be sure to state your final answer in terms of the original question.

29. Book Publishing A book publisher needs more copies of its latest best seller. By rescheduling several other projects over the next two weeks, it can arrange up to 300 hours of printing and 250 hours of binding. For each thousand copies, the trade edition requires 3 hours of printing and 2 hours of binding, the book club edition requires 2 hours of printing and 2 hours of binding, and the paperback edition requires 1 hour of printing and 1 hour of binding. If the profit per book is $7 for the trade edition, $6 for the book club edition, and $4 for the paperback, how many of each edition should be printed to obtain the greatest possible profit?

30. Production Planning A fully automated plastics factory produces three toys, a racing car, a jet airplane, and a speed boat, in three stages: molding, painting, and packaging. After allowing for routine maintenance, the equipment can operate no more than 150 hours per week. Each batch of racing car toys requires 6 hours of molding, 2.5 hours of painting, and 5 hours of packaging, each batch of jet airplane toys requires 3 hours of molding, 7.5 hours of painting, and 5 hours of packaging, and each batch of speed boat toys requires 3 hours of molding, 5 hours of painting, and 5 hours of packaging. If the profits per batch of toys are $120 for cars, $100 for airplanes, and $110 for boats, how many batches of each toy should be produced each week to obtain the greatest possible profit?

3.4 Duality and Standard Minimum Problems

Rewrite each minimum linear programming problem in matrix form, check that it is a standard problem, and then construct the dual maximum problem and the initial simplex tableau. (Do not solve the problem any further.)

31. Minimize $C = 10y_1 + 15y_2 + 20y_3$

Subject to $\begin{cases} 2y_1 - y_2 + y_3 \ge 40 \\ y_1 + y_2 + 3y_3 \ge 30 \\ y_1 \ge 0, y_2 \ge 0, y_3 \ge 0 \end{cases}$

32. Minimize $C = 30y_1 + 20y_2$

Subject to $\begin{cases} 5y_1 + 2y_2 \ge 210 \\ 3y_1 + 4y_2 \ge 252 \\ 5y_1 + 4y_2 \le 380 \\ y_1 \ge 0, y_2 \ge 0 \end{cases}$

33. Minimize $C = 42y_1 + 36y_2$

Subject to $\begin{cases} 7y_1 + 4y_2 \ge 84 \\ y_1 + y_2 \ge 18 \\ y_1 \ge 0, y_2 \ge 0 \end{cases}$

34. Minimize $C = 130y_1 + 40y_2 + 98y_3$

Subject to $\begin{cases} 3y_1 + y_2 + 2y_3 \ge 51 \\ 4y_1 + y_2 + 2y_3 \ge 60 \\ 3y_1 + y_2 + 3y_3 \ge 57 \\ y_1 \ge 0, y_2 \ge 0, y_3 \ge 0 \end{cases}$

Solve each standard minimum problem by finding the dual maximum problem and using the simplex method.

35. Minimize $C = 9y_1 + 24y_2 + 6y_3$

Subject to $\begin{cases} y_1 + 2y_2 \geq 18 \\ 3y_2 + y_3 \geq 15 \\ y_1 \geq 0, y_2 \geq 0, y_3 \geq 0 \end{cases}$

36. Minimize $C = 30y_1 + 36y_2$

Subject to $\begin{cases} 5y_1 + 3y_2 \geq 60 \\ y_1 + y_2 \geq 16 \\ y_2 \geq 5 \\ y_1 \geq 0, y_2 \geq 0 \end{cases}$

37. Minimize $C = 3y_1 + 4y_2 + 10y_3 + 8y_4$

Subject to $\begin{cases} -y_1 + y_2 + y_3 - 2y_4 \geq 10 \\ y_1 - y_2 - 2y_3 + y_4 \geq 20 \\ y_1 \geq 0, y_2 \geq 0, y_3 \geq 0, y_4 \geq 0 \end{cases}$

38. Minimize $C = 12y_1 + 5y_2 + 6y_3$

Subject to $\begin{cases} 3y_1 + y_2 + 3y_3 \geq 9 \\ 2y_1 + y_2 - 3y_3 \geq 12 \\ 4y_1 + 2y_2 + 2y_3 \geq 8 \\ y_1 \geq 0, y_2 \geq 0, y_3 \geq 0 \end{cases}$

Formulate each application as a standard minimum problem and then solve it by finding the dual maximum problem and using the simplex method. Be sure to state your final answer in terms of the original question.

39. Pens and Ink   A student needs at least 3 new pens and 2 dozen ink cartridges. The bookstore sells single pens for 99¢, packages of 6 ink cartridges for 89¢, and a $1.19 "writer's combo" package containing 1 pen and 2 ink cartridges. How many of each will meet the student's needs at the least cost?

40. Production Planning   A pasta company is expanding its linguini production facility. Two machines are available: a small-capacity machine costing $5000 that produces 20 pounds per minute and needs 1 operator and a large-capacity machine costing $6000 that produces 30 pounds per minute and needs 2 operators. The company wants to hire no more than 34 additional employees yet increase produc-

tion by at least 600 pounds per minute. How many of each machine should the company buy to expand its production at the least cost?

3.5   Nonstandard Problems

Rewrite each nonstandard maximum problem in matrix form, construct the initial simplex tableau, and locate the dual pivot element. (Do not carry out the pivot operation.)

41. Maximize $P = 80x_1 + 30x_2$

Subject to $\begin{cases} 4x_1 + x_2 \leq 40 \\ 2x_1 + 3x_2 \leq 60 \\ x_1 + x_2 \geq 10 \\ x_1 \geq 0, x_2 \geq 0 \end{cases}$

42. Maximize $P = 4x_1 + 8x_2 + 6x_3 + 10x_4$

Subject to $\begin{cases} 2x_1 + 3x_2 - 3x_3 + x_4 \leq 30 \\ 7x_1 - 5x_2 + x_3 + 5x_4 \geq 35 \\ x_1 \geq 0, x_2 \geq 0, x_3 \geq 0, x_4 \geq 0 \end{cases}$

Solve each nonstandard maximum problem by the two-stage simplex method.

43. Maximize $P = 3x_1 + 4x_2$

Subject to $\begin{cases} x_1 + x_2 \leq 10 \\ x_1 + x_2 \geq 5 \\ x_1 \geq 0, x_2 \geq 0 \end{cases}$

44. Maximize $P = 8x_1 - 20x_2 - 18x_3$

Subject to $\begin{cases} x_1 + 3x_2 + 2x_3 \leq 55 \\ x_1 + x_2 + x_3 \geq 25 \\ 2x_1 + 3x_2 + 3x_3 \leq 70 \\ x_1 \geq 0, x_2 \geq 0, x_3 \geq 0 \end{cases}$

45. Maximize $P = 6x_1 - 9x_2$

Subject to $\begin{cases} 3x_1 + x_2 \leq 24 \\ x_1 + 3x_2 \leq 24 \\ x_1 + x_2 \geq 6 \\ x_1 \geq 0, x_2 \geq 0 \end{cases}$

46. Maximize $P = 12x_1 - 21x_2 + 2x_3$

Subject to $\begin{cases} 2x_1 - 3x_2 + x_3 \geq 20 \\ x_1 - x_2 + x_3 \leq 25 \\ x_1 - 3x_2 \geq 5 \\ x_1 \geq 0, x_2 \geq 0, x_3 \geq 0 \end{cases}$

47. Maximize $P = 6x_1 + 7x_2$

Subject to $\begin{cases} x_1 \geq 2 \\ x_2 \geq 3 \\ x_1 \leq 4 \\ x_2 \leq 5 \\ x_1 \geq 0, x_2 \geq 0 \end{cases}$

48. Maximize $P = 8x_1 + 2x_2 + 10x_3$

Subject to $\begin{cases} 2x_1 + x_2 + x_3 \leq 30 \\ x_1 + 2x_2 + x_3 \geq 5 \\ x_1 - x_2 + x_3 \geq 10 \\ x_1 \geq 0, x_2 \geq 0, x_3 \geq 0 \end{cases}$

Formulate each application as a nonstandard maximum problem and solve it by the two-stage simplex method. Be sure to state your final answer in terms of the original question.

49. Lumber Production   A saw mill rough-cuts lumber and planes some of it to make finished-grade boards. The saw cuts 1 thousand board-feet per hour and can operate up to 10 hours each day, while the plane finishes 1 thousand board-feet per hour and can operate up to 8 hours each day. The profit per thousand board-feet is $120 for rough-cut lumber and $100 for finished-grade boards. If at least 4 thousand board-feet of finished-grade boards must be produced each day, how many board-feet of each type of lumber should be produced daily to obtain the greatest possible profit?

50. Financial Planning   An international money manager wishes to invest no more than $150,000 in United States and Canadian stocks and bonds, with at least twice as much in bonds as in stocks. The current market yields are: 10% on U.S. stocks, 5% on U.S. bonds, 11% on Canadian stocks, and 4% on Canadian bonds. If at least $30,000 must be invested in Canadian stocks, how much should be invested in each to obtain the greatest possible return?

Projects and Essays

The following projects and essays are based on Chapter 3. Most have no right and wrong answers—the results depend only on your imagination and resourcefulness.

1. Visit your business school's library and browse through the last several issues of the journals *Management Science* or *Operations Research.* Write a one-page report on the application of linear programming to a "real world" problem.

2. Use an internet connection to visit the web site of the Operations Research Department of Stanford University. Write a report about that department, linear programming problems, and George Dantzig.

3. Improve the program PIVOT so that it will find the pivot column and row for you. Be careful that your new program checks that the tableau is feasible before looking for a pivot column. What should the program do if the tableau is not feasible?

4. Another method of solving linear programming problems is the "ellipsoid method" or "Khachiyan's algorithm." Look up the articles "Some References for the Ellipsoid Algorithm" by Philip Wolfe in *Management Science* **26**(8):747–749, August 1980, and "Khachiyan's Algorithm" by G. C. Berresford, A. M. Rockett, and J. C. Stevenson in *Byte* **5**(8):198–208, August 1980, and *Byte* **5**(9):242–255, September 1980, and write a one-page report on what you find.

5. Another method of solving linear programming problems is N. Karmarkar's "interior point" method. Look up the article "Karmarkar's Algorithm" by A. M. Rockett and J. C. Stevenson in *Byte* **12**(10):146–160, September 1987, and write a one-page report on what you find.

4 PROBABILITY

4.1 Counting Techniques

4.2 Probability Spaces

4.3 Conditional Probability and Independence

4.4 Random Variables and Distributions

Probability, which was first developed to answer questions arising from gambling in the seventeenth century, has now become an essential tool for dealing with the randomness and uncertainty that are a part of our world.

4.1 Counting Techniques

APPLICATION PREVIEW

The Chevalier de Méré

Probability theory as a mathematical discipline began with the analysis by Blaise Pascal (1623–1662) and Pierre de Fermat (1601–1665) of gambling problems posed by a French nobleman, Antoine Gombaude, Chevalier de Méré. The Chevalier's concerns were practical: after winning so often that he had no more takers for his wager that he could throw at least one six in 4 rolls of a die, he had changed his bet to be that he could throw at least one double six in 24 rolls of two dice, and he was now losing his bets. Since 4 rolls with chance $\frac{1}{6}$ on each gives the same ratio as 24 rolls with chance $\frac{1}{36}$, the Chevalier was at a loss to understand his change of luck. Rather than give up on this paradox, the Chevalier had the good sense to ask Blaise Pascal, one of the foremost mathematicians of his era. Pascal's analysis of the Chevalier's problems and his correspondence with Pierre de Fermat (the other great French mathematician of this period) showed that reasonable explanations followed from careful counting of all the possible throws of the dice. (We will analyze the Chevalier's bets on pages 406 and 409.)

Introduction

Although the word "probability" has various meanings in everyday speech, it also has a precise mathematical meaning. Before defining it mathematically, we must clarify some basic counting principles upon which the definition will be based. We begin by reviewing sets and how to count their members.

Sets and Set Operations

A *set* is any well-defined collection of objects (also called *elements* or *members*). Sets can be illustrated by *Venn* diagrams*, which represent sets by ellipses.

---

* After the English logician John Venn (1834–1923), author of *Logic of Chance.*

The *intersection* of sets A and B, denoted $A \cap B$, is the set of all elements that are in *both A and B.*

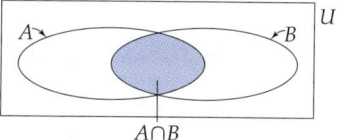

Two sets A and B are *disjoint* if their intersection is empty (that is, if they have no elements in common).

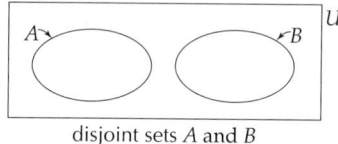

disjoint sets A and B

The *union* of sets A and B, denoted $A \cup B$, is the set of all elements that are in *either A or B* (or both).

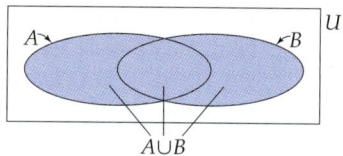

The *complement* of a set A, denoted A^c, is the set of all elements *not* in A.

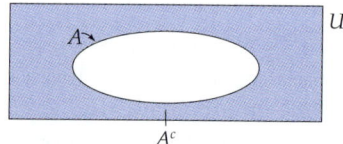

In general, the words "and," "or," and "not" translate into "intersection," "union," and "complement." Throughout this chapter, we will use the word "or" in the *inclusive* sense to mean one possibility or the other or both. We always draw sets inside a larger rectangle, labeled U for *universal* set, which represents the set of all possible elements being considered. The complement of the universal set is the *empty* set, also called the *null* set, denoted $\varnothing$, which has no members. A *subset* of a set is a set (possibly empty) of elements from the original set. We use the symbol $n(A)$ to mean the number of elements in a set A:

$$n(A) = \left(\begin{array}{c} \text{Number of} \\ \text{elements in } A \end{array} \right)$$

For example, $n(\emptyset) = 0$. From the last Venn diagram it is clear that the elements of A^c are precisely the elements of the universal set U that are *not* in A. Therefore:

Complementary Principle for Counting

$$n(A^c) = n(U) - n(A)$$

In words, the number of elements in the complement of a set is the number of elements in the universal set minus the number of elements in the original set. We will find this principle useful when it is easier to count the elements in the complement of a set rather than those in the set itself.

Addition Principle for Counting

Notice that in the following Venn diagram there are $n(A) = 4 + 3 = 7$ elements (dots) in A and $n(B) = 3 + 2 = 5$ elements in B.

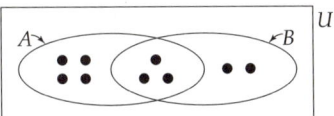

Adding $n(A) = 7$ and $n(B) = 5$ gives 12, but this is *not* the number of elements in $A \cup B$, since $n(A \cup B) = 4 + 3 + 2 = 9$. Why doesn't adding $n(A)$ and $n(B)$ give $n(A \cup B)$? The problem is that the elements in the middle of the diagram, in $A \cap B$, get counted *twice*, once in A and once in B. Therefore, to get the correct number of elements in $A \cup B$ we must add $n(A)$ to $n(B)$ and then *subtract* the number of elements in $A \cap B$ (since they were double-counted). This gives the general rule:

Addition Principle for Counting

$$n(A \cup B) = n(A) + n(B) - n(A \cap B)$$

In words, the number of elements in a union of two sets is the number of elements in one plus the number of elements in the other minus the number of elements in both. Of course, if A and B are disjoint, then $A \cap B$ is empty, so $n(A \cap B) = 0$, giving a simpler addition principle for disjoint sets:

$$n(A \cup B) = n(A) + n(B) \qquad \text{for } A, B \text{ disjoint}$$

EXAMPLE 1   Counting Cars in a Parking Lot

A mall parking lot contains 150 convertibles, 200 cars with sound systems, and 90 convertibles with sound systems. How many cars in the lot are convertibles or have sound systems?

Solution

Let C be the set of convertibles and S be the set of cars with sound systems. Since "or" means "union" we want $n(C \cup S)$. We are given that $n(C) = 150$, $n(S) = 200$, and $n(C \cap S) = 90$, and using the addition principle:

$$n(C \cup S) = \underbrace{n(C)}_{150} + \underbrace{n(S)}_{200} - \underbrace{n(C \cap S)}_{90} = 150 + 200 - 90 = 260$$

There are 260 cars that are convertibles or have sound systems.

PRACTICE PROBLEM 1

A survey of insurance coverage in 300 metropolitan businesses revealed that 150 offer their employees dental insurance, 150 offer vision coverage, and 100 offer both dental and vision coverage. How many of these businesses offer their employees dental or vision insurance?

Solution at the back of the book

We may identify all parts of a Venn diagram in terms of sets and complements.

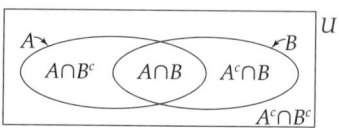

$A \cap B$: in A and B

$A \cap B^c$: in A but outside B

$A^c \cap B$: outside A but in B

$A^c \cap B^c$: outside A and outside B

EXAMPLE 2   Counting Customers

A regional telephone company with 200,000 customers provides 60,000 with call waiting and 70,000 with call forwarding. If 20,000 customers have both, how many have neither?

Solution

We could solve this problem by the addition principle, as before, but instead we use Venn diagrams. Let W be the set of customers with call waiting and let F be the set of customers with call forwarding. We want to find $n(W^c \cap F^c)$, and we are given that $n(U) = 200,000$, $n(W) = 60,000$, $n(F) = 70,000$, and $n(W \cap F) = 20,000$. We fill in the "pieces" of a Venn diagram in stages, beginning with the only piece that we are given, $n(W \cap F)$. For simplicity, we enter the numbers in thousands.

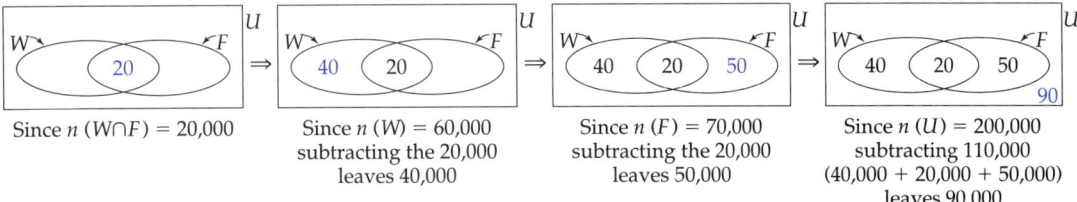

Since $n (W \cap F) = 20,000$

Since $n (W) = 60,000$
subtracting the 20,000
leaves 40,000

Since $n (F) = 70,000$
subtracting the 20,000
leaves 50,000

Since $n (U) = 200,000$
subtracting 110,000
$(40,000 + 20,000 + 50,000)$
leaves 90,000

This last number, $n(W^c \cap F^c) = 90,000$, is the answer we wanted: 90,000 people have neither call waiting nor call forwarding.

■

Notice that in this example we used the complementary principle for counting, finding the "outside" by finding the inside parts and subtracting.

The Multiplication Principle for Counting

The following multiplication principle will be very useful throughout this chapter.

Multiplication Principle for Counting

> If two choices are to be made, and there are m possibilities for the first choice and n possibilities for the second choice, and if any first choice can be combined with any second choice, then the *combination* of the two choices can be made in $m \cdot n$ ways.

The multiplication principle can be proved by enumerating all of the possibilities, as in the following example.

EXAMPLE 3 **Counting Different Products**

A toy company makes red, green, blue, and yellow plastic cars, trucks, and planes. How many different kinds of toys does it make?

Solution

Let the set of colors be C = {red, green, blue, yellow}, and the set of shapes be S = {car, truck, plane}. Each possible toy is described by its color and shape:

	Red	Green	Blue	Yellow
Car	(red, car)	(green, car)	(blue, car)	(yellow, car)
Truck	(red, truck)	(green, truck)	(blue, truck)	(yellow, truck)
Plane	(red, plane)	(green, plane)	(blue, plane)	(yellow, plane)

This table contains all possible combinations of colors and shapes, and since the table is a rectangle, the number of boxes is found by multiplying length times width: There are $4 \cdot 3 = 12$ possible toys that can be made.   ■

Another way of thinking about the multiplication principle is to imagine making up all possible pairs of elements, one from A followed by one from B:

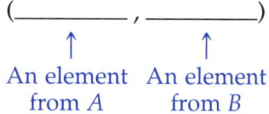

Since there are $n(A)$ choices for the first space and $n(B)$ choices for the second, the total number of possible pairs is $n(A) \cdot n(B)$. Applying this to Example 3 gives:

$4 \cdot 3 = 12$ possible toys

EXAMPLE 4   **Counting Parking Permits**

A parking lot permit displays an identification code consisting of a letter (A to Z) followed by two digits (0 to 9). How many different permits can be issued?

Solution

Since each identification code consists of three symbols, we need to fill three blanks:

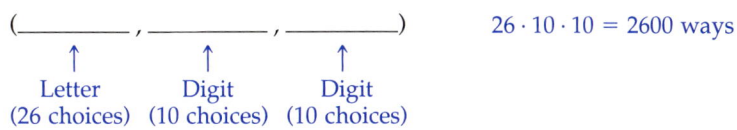

$$26 \cdot 10 \cdot 10 = 2600 \text{ ways}$$

Letter   Digit   Digit
(26 choices) (10 choices) (10 choices)

Therefore, 2600 different permits can be issued.

PRACTICE PROBLEM 2   A computer network password consists of four letters (A–Z) followed by four digits (0–9). How many different passwords are there?

Solution at the back of the book

Permutations

How many different orderings are there for the letters *a*, *b*, and *c*? We may list them all as *abc*, *acb*, *bac*, *bca*, *cab*, and *cba*, so there are 6. Each of these orderings is called a *permutation*. Another way to count them is to observe that there are 3 ways of choosing the first letter, then 2 ways of choosing the second (because one letter was "used up" in the first choice), and 1 way of choosing the last (whichever is left), so by the multiplication principle there are $3 \cdot 2 \cdot 1 = 6$ possible orderings, just as we found before.

How many such permutations are there if we have *n* distinct objects and want to arrange some *r* of them in all possible orders? Clearly there are *n* choices for the first, then $n - 1$ choices for the second, $n - 2$ choices for the third, down to $n - r + 1$ choices for the *r*th. Notice that you always subtract one *less* than the number of the choice, so for the *r*th choice the number of ways is $n - (r - 1) = n - r + 1$.

Permutations

The number of permutations (ordered arrangements) of *n* distinct objects taken *r* at a time is

$$_nP_r = n \cdot (n - 1) \cdot \ldots \cdot (n - r + 1)$$

Product of *r* numbers beginning with *n*

In the case $r = 0$, we define $_nP_0 = 1$. The quantity $_nP_r$ is sometimes written as $P_{n,r}$, P_r^n, or $P(n, r)$.

EXAMPLE 5 Counting Nonsense Words

How many five-letter "nonsense" words (that is, without regard to meaning) can be made from the letters A to Z with no letter repeated?

Solution

We use the permutation formula with $n = 26$ and $r = 5$:

$$_{26}P_5 = 26 \cdot 25 \cdot 24 \cdot 23 \cdot 22 = 7{,}893{,}600 \qquad \text{Product of 5 numbers beginning with 26}$$

There are 7,893,600 five-letter nonsense words with distinct letters.   ■

We obtain the same answer if we think of building each "word" by filling in five blanks with letters, without repetitions:

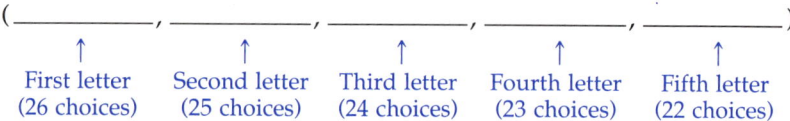

(——————, ——————, ——————, ——————, ——————)

↑	↑	↑	↑	↑
First letter	Second letter	Third letter	Fourth letter	Fifth letter
(26 choices)	(25 choices)	(24 choices)	(23 choices)	(22 choices)

By the multiplication principle, there are $26 \cdot 25 \cdot 24 \cdot 23 \cdot 22 = 7{,}893{,}600$ such nonsense words, just as we found above.

Graphing Calculator Exploration

Values of $_nP_r$ are easy to find if your calculator includes this command. The following screens show several values of $_nP_r$ for $n = 26$. Notice that $_nP_r$ gets large quickly as r increases.

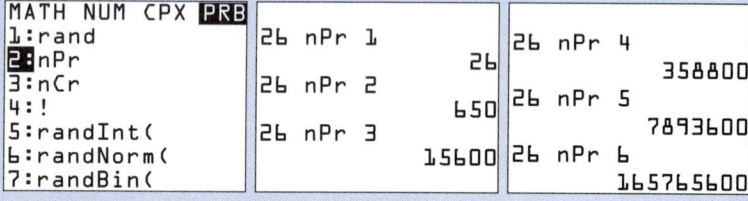

```
MATH NUM CPX PRB
1:rand
2:nPr
3:nCr
4:!
5:randInt(
6:randNorm(
7:randBin(
```

```
26 nPr 1
                26
26 nPr 2
                650
26 nPr 3
             15600
```

```
26 nPr 4
             358800
26 nPr 5
            7893600
26 nPr 6
          165765600
```

REAL
WORLD

EXAMPLE 6   **Counting License Plates**

A car license plate consists of three distinct letters followed by three distinct digits. How many different license plates are possible?

Solution

The number of three-letter patterns with no repeats is $_{26}P_3$, while the number of three-digit patterns with no repeats is $_{10}P_3$. By the multiplication principle, the number of possible license plates is

$$(_{26}P_3) \cdot (_{10}P_3) = (26 \cdot 25 \cdot 24) \cdot (10 \cdot 9 \cdot 8) = 11{,}232{,}000$$

There are 11,232,000 possible license plates.

■

PRACTICE PROBLEM 3
The jurors at an art exhibition must select the first-, second-, and third-place paintings. If there are 35 paintings, in how many different ways can the winners be chosen?   *Solution at the back of the book*

Combinations

How many different pairs of letters can be made from the letters a, b, c, and d without repetitions and *without regard to order*? We may write out the pairs as ab, ac, ad, bc, bd, and cd, so there are six, each called a *combination*. Notice that combinations are counted *without regard to order*, so that ba is considered to be the same object as ab (as distinct from *permutations*, in which they would be considered different objects). In the same way, each of our six pairs would generate a new pair, in reverse, if we were counting order. This observation leads to a way of counting combinations: we can count the number of *permutations* and then divide by the number of possible *reorderings* of each to eliminate duplicates. In the case of the four letters taken two at a time, we would first count the number of *permutations*, $_4P_2 = 4 \cdot 3 = 12$, and then divide by the number of possible reorderings of each pair, $_2P_2 = 2 \cdot 1 = 2$, obtaining $12/2 = 6$, which is the same answer that we obtained from listing the combinations.

In general, let $_nC_r$ be the number of combinations of n things taken r at a time. To find $_nC_r$ in terms of permutations, we simply take $_nP_r$ and divide by $_rP_r$ to eliminate multiple countings of the same combination in its different orders. That is:

Combinations

The number $_nC_r$ of combinations (unordered arrangements) of n distinct objects taken r at a time is

$$_nC_r = \frac{_nP_r}{_rP_r} = \frac{n \cdot (n-1) \cdot \ldots \cdot (n-r+1)}{r \cdot (r-1) \cdot \ldots \cdot (1)}$$

r numbers beginning with n

r numbers beginning with r

In the case of $r = 0$ we define $_nC_0 = 1$. The quantity $_nC_r$ is sometimes written as $C_{n,r}$, C_r^n, $C(n, r)$ or $\binom{n}{r}$. The denominator may be written as $r!$ (read: r factorial).

EXAMPLE 7   Counting Permutations and Combinations

A student club has 15 members. **a.** How many ways can a president, vice-president, and treasurer be chosen? **b.** How many ways can a committee of 3 members be chosen?

Solution

Each question involves choosing three members from the club, but for the officers we want *ordered* arrangements (the order determines who is president, vice-president, and treasurer), while for the committee we want *unordered* arrangements. Thus (a) asks for permutations while (b) asks for combinations.

For the officers:

$$_{15}P_3 = 15 \cdot 14 \cdot 13 = 2730$$

For the committee:

$$_{15}C_3 = \frac{_{15}P_3}{_3P_3} = \frac{15 \cdot 14 \cdot 13}{3 \cdot 2 \cdot 1} = 455 \qquad \begin{array}{l}\leftarrow \text{Three numbers beginning with 15} \\ \leftarrow \text{Three numbers beginning with 3}\end{array}$$

There are 2730 different ways of choosing the president, vice-president, and treasurer, and 455 ways of choosing the committee.   ∎

PRACTICE PROBLEM 4

A college business major can also minor in computer science by taking any six courses from an approved list of 10 courses. How many different collections of courses will satisfy the computer science minor requirements?   *Solution at the back of the book*

Graphing Calculator Exploration

Values of $_nC_r$ are easy to find if your calculator includes this command. The following screens show several values of $_nC_r$ for $n = 15$. Notice that $_nC_r$ gets larger and then smaller as r increases.

```
MATH NUM CPX PRB
1:rand
2:nPr
3:nCr
4:!
5:randInt(
6:randNorm(
7:randBin(
```

```
15 nCr 2
             105
15 nCr 3
             455
15 nCr 4
            1365
```

```
15 nCr 12
             455
15 nCr 13
             105
15 nCr 14
              15
```

A *standard deck* of 52 playing cards contains four *suits* (clubs, diamonds, hearts, and spades), each containing three *face cards* (jack, queen, and king), cards numbered 2 through 10, and an *ace*, which may be counted as the lowest card ("1") for some games or the highest (above the king) for others. A *hand* of cards is a selection of cards from the deck.

EXAMPLE 8   Counting 5-Card Hands

How many 5-card hands can be dealt?

Solution

Since the order in which the cards are dealt makes no difference, we want the number of *combinations* of 52 cards taken 5 at a time:

$$_{52}C_5 = \frac{52 \cdot 51 \cdot 50 \cdot 49 \cdot 48}{5 \cdot 4 \cdot 3 \cdot 2 \cdot 1} = 2{,}598{,}960$$

There are 2,598,960 possible hands.

EXAMPLE 9   Counting Specified Hands

How many 5-card hands will have two 8's and three jacks?

Solution

We can choose the two 8's (from the four 8's) in $_4C_2$ different ways, and the three jacks (from the four jacks) in $_4C_3$ different ways. Then, by the multiplication principle, the number of ways that the combined choice can be made is

$$(_4C_2) \cdot (_4C_3) = \frac{4 \cdot 3}{2 \cdot 1} \cdot \frac{4 \cdot 3 \cdot 2}{3 \cdot 2 \cdot 1} = 6 \cdot 4 = 24$$

There are 24 possible hands with two 8's and three jacks.

SUMMARY

A prerequisite to probability is mastery of several counting techniques for accurately enumerating large collections of possibilities without having to count "one by one." These techniques are summarized as follows:

Complementary Principle	$n(A^c) = n(U) - n(A)$
Addition Principle	$n(A \cup B) = n(A) + n(B) - n(A \cap B)$
Multiplication Principle	$\left(\begin{array}{l}\text{Numbers of pairs } (a, b)\\ \text{with } a \text{ in } A \text{ and } b \text{ in } B\end{array}\right) = n(A) \cdot n(B)$
Permutations (ordered arrangements)	$_nP_r = n \cdot (n - 1) \cdot \ldots \cdot (n - r + 1)$
Combinations (unordered arrangements)	$_nC_r = \dfrac{_nP_r}{_rP_r} = \dfrac{n \cdot (n - 1) \cdot \ldots \cdot (n - r + 1)}{r \cdot (r - 1) \cdot \ldots \cdot (1)}$

EXERCISES 4.1

Find each quantity using the following Venn diagram.

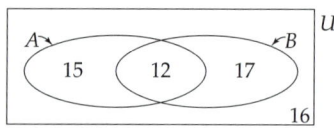

1. $n(A)$ **2.** $n(B)$

3. $n(U)$ **4.** $n(A \cap B)$

5. $n(A \cup B)$ **6.** $n(A^c)$

7. $n(B^c)$ **8.** $n(A \cap B^c)$

9. $n(A^c \cap B)$ **10.** $n(A \cup B^c)$

Identify each numbered part in the following Venn diagram in terms of intersections of sets A, B, and C, and their complements.

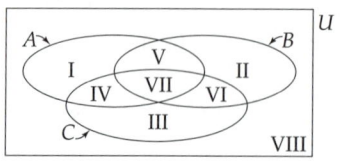

11. Part I   **12.** Part II

13. Part III   **14.** Part IV

15. Part V   **16.** Part VI

17. Part VII   **18.** Part VIII

Find each number of permutations.

19. $_5P_3$ **20.** $_6P_2$

21. $_8P_1$ **22.** $_9P_8$

 23. $_{13}P_4$, $_{13}P_5$, and $_{13}P_6$ **24.** $_{11}P_5$, $_{11}P_6$, and $_{11}P_7$

25. $_{10}P_1$, $_{10}P_5$, and $_{10}P_{10}$ **26.** $_{12}P_1$, $_{12}P_6$, and $_{12}P_{12}$

Find each number of combinations.

27. $_6C_2$ **28.** $_5C_3$

29. $_7C_3$ **30.** $_8C_1$

 31. $_{11}C_4$, $_{11}C_5$, $_{11}C_6$, and $_{11}C_7$

32. $_{13}C_5$, $_{13}C_6$, $_{13}C_7$, and $_{13}C_8$

33. $_{12}C_1$, $_{12}C_2$, $_{12}C_6$, $_{12}C_{11}$, and $_{12}C_{12}$

34. $_{10}C_1$, $_{10}C_2$, $_{10}C_5$, $_{10}C_9$, and $_{10}C_{10}$

APPLIED EXERCISES

(*helpful.*)

35. Lacrosse The sophomore lacrosse team has 24 players, of whom 10 played defense last year, 12 played offense, and 5 played both defense and offense, while the rest of the players did not play last year. How many members of the team played last year?

36. Political Fund Raising If 12,300 individuals contributed to the governor's first election campaign, 15,200 contributed to her second, and 7800 contributed to both, how many individuals made contributions?

37. Current Events A survey of 1200 residents of New Orleans revealed that 760 get information on international events by watching TV news programs, 530 from listening to radio news shows, 290 from both TV and radio, while the rest have no interest in international events. How many have no interest in international events?

38. International Marketing Of the 30 freshmen majoring in international marketing, 12 speak French, 8 speak German, and 5 speak both languages. How many do not speak either French or German?

39. Parking Permits Look back at Example 4 on pages 390–391 and find the number of parking permits that can be issued if the letter O and the number 0 are omitted to avoid confusion.

40. Computer Security A computer network password is made up of letters A–Z, and may be either four or five letters long. How many such passwords are there? Find the answer in two different ways:

  a. Find the number of four-letter passwords and add it to the number of five-letter passwords.

  b. Think of a four-letter password as really having length five, where in the fifth position is a blank, which can be thought of as one more member of an "extended alphabet."

41. Computer Security How many eight-symbol computer passwords can be formed using the letters A to J and the digits 2 to 6?

42. Bagels A bagel shop offers 12 different kinds of bagels and 18 flavors of cream cheese. How many different orders for a bagel and cream cheese can a customer place?

43. Combination Locks A suitcase combination lock has four wheels, with each labeled with the digits 0–9. How many lock combinations are possible?

44. Mix & Match Fashions A designer has created a completely mixable line of five shirts, nine ties, and four pairs of slacks, and advertises that "the possibilities are endless." Exactly how "endless" are the different outfits consisting of a shirt, a tie, and a pair of slacks?

45. Commercial Art A bank lobby wall has space to display 4 paintings of local scenes by local artists. If 12 paintings are available, how many different ways can they be displayed on the wall?

46. Postal Codes How many five-digit ZIP codes have no repeated digits?

47. Election Ballots How many different ways can the eight candidates for the school board election be listed on the ballot?

48. Horse Racing How many different ways can the 14 horses in a race finish in first, second, and third place?

49. Basketball Teams A junior high girls basketball team is to consist of five players. How many different teams can the manager select from a roster of 12 girls?

50. Playing Cards How many five-card hands from a standard deck will contain only face cards?

51. Baseball Teams A junior league boys baseball team is to consist of nine players. How many different teams can the coach select from a roster of 15 boys?

52. Lines and Circles Ten distinct points have been marked on the circumference of a circle. How many lines can be drawn through these points?

53. Telephone Numbers How many seven-digit telephone numbers start with three even digits and end with four odd digits? How many such numbers have no repeated digits?

54. Social Security Numbers How many nine-digit Social Security numbers start with three odd digits, then have two even digits, and end with four distinct digits from 0 to 9?

55. Labor Negotiations The union leadership and the management strike team have agreed to each select 4 representatives to enter into a "closed room, around the clock" marathon in a final attempt to reach a settlement. If there are 10 members of the union leadership and 12 members of the management team, how many ways can the marathon negotiators be chosen?

56. Homework Grading   A sociology professor assigned 5 problems from Unit One, 10 problems from Unit Two, and 8 problems from Unit Three for next Wednesday and announced that she will grade only two of the assigned problems from Unit One, four of those from Unit Two, and three from Unit Three. How many different ways can she choose the problems that she will grade?

Explorations and Excursions

The following problems extend and augment the material presented in the text.

De Morgan's Laws explain the relation between the complement of a set and the operations of union and intersection.

57. Show that $(A \cup B)^c = A^c \cap B^c$ by carefully sketching Venn diagrams for $(A \cup B)^c$ and for $A^c \cap B^c$.

58. Show that $(A \cap B)^c = A^c \cup B^c$ by carefully sketching Venn diagrams for $(A \cap B)^c$ and for $A^c \cup B^c$.

59. Verify that $n((A \cup B)^c) = n(A^c \cap B^c)$ for the sets $A = W$ and $B = F$ discussed in Example 2 (pages 388–389).

60. Verify that $n((A \cap B)^c) = n(A^c \cup B^c)$ for the sets $A = W$ and $B = F$ discussed in Example 2 (pages 388–389).

Factorials   The factorial function for positive integers n is defined by $n! = n \cdot (n-1) \cdot \ldots \cdot 1$, along with $0! = 1$. (For example, $4! = 4 \cdot 3 \cdot 2 \cdot 1 = 24$.)

61. Show that $_nP_n = n!$.

62. Find the values of $n!$ for the first 10 positive integers.

63. Show that $_nP_r = \dfrac{n!}{(n-r)!}$.

64. Show that $_nC_r = \dfrac{n!}{(n-r)!r!}$.

65. Explain why the following calculator screens show that 70! is larger than 10^{100}, so this particular calculator cannot evaluate 70!.

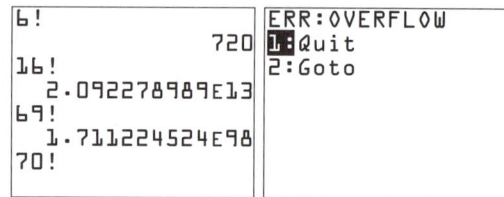

66. Explain why the following calculator screens from the same calculator used in Exercise 65 show that this particular calculator does not use the factorial formulas in Exercises 63 and 64 to evaluate permutations and combinations.

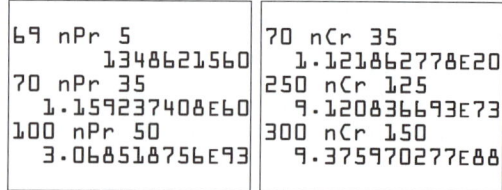

Tree-diagrams   are another way of counting all the possible pairs found by the multiplication principle. A tree diagram lists all the possible choices for the first part of the pairs and each branches to a list of all the possible choices for the second part. For instance, a tree diagram for Example 3 (pages 389–390) would list the toy colors with branches to each toy shape:

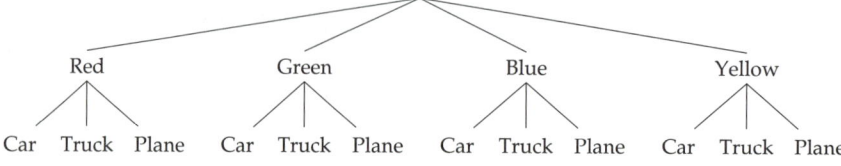

By counting all the "leaves" at the ends of the branches, we can find the number of pairs. A tree diagram may have several layers of branches if more than two different things are being paired together, as in Example 4 (pages 390–391).

Draw a tree diagram for each situation and count the number of leaves to find the number of possible pairs.

67. A parking lot permit sticker displays an identification code consisting of a letter (from A, B, and C) followed by a digit (from 1, 2, 3, and 4). How many different permits can be issued?

68. A computer network password consists of two letters, with the first being either X, Y, or Z and the second being either A, B, C, D, or E. How many different passwords are there?

69. A clothing store sells windbreakers, ski jackets, and overcoats, and each is available in red or blue. How many different kinds of coats does it sell?

70. Ted, Bob, Fred, and Jim share a student dorm suite. How many ways can they choose who takes out the garbage and who sweeps the floor if they can choose the same person to do both? What if different men must do the chores?

71. Is it practical to solve Example 4 (pages 390–391) using a tree diagram? Sketch at least part of the diagram to see how large it might be.

4.2 Probability Spaces

APPLICATION PREVIEW

Coincidences

We often hear of "amazing" coincidences. For example, a woman named Evelyn Marie Adams won the New Jersey Lottery twice, in 1985 and in 1986, an event that was widely reported to have a probability of 1 in 17 trillion.

Actually, the figure 1 in 17 trillion is misleading; such events are not all that unlikely. In fact, given enough tries, the most outrageous things are virtually certain to happen. For example, if a coincidence is defined as an event with a one-in-a-million chance of happening to you today, then in the United States, with over 250 million people, we should expect more than 250 coincidences each day, and almost 100,000 in a year.

Returning to the supposed 1-in-17-trillion double lottery winning, that figure is the right answer to the wrong question: What is the probability that a *preselected* person who buys just *two* tickets for separate lotteries will win on both? The more relevant question is: What is the probability that *some* person, among the many millions who buy lottery tickets (most buying multiple tickets), will win twice in a lifetime? It has been calculated* that such a double winning is likely to occur once in 7 years, with the likelihood approaching certainty for longer periods.

Some knowledge of probability is necessary for an understanding of the world, if only to cast doubt on the misleading statements that one often hears. For further information on probability, see the two books listed below.†

Introduction

Some experiments always achieve the same result from the same conditions, while others involve "chance" or "random" effects that produce a variety of outcomes. In this section we will use such "chance" experiments with a finite number of possible outcomes as models for probability spaces, which provide the mathematical framework for our discussion of probability.

Random Experiments and Sample Spaces

A *random* experiment is one that, when repeated under identical conditions, may produce different outcomes. Each repetition of the experiment is called a *trial,* and each result is an *outcome.* The set of all possible outcomes is called the *sample space* for the experiment. Of course,

---

* See Persi Diaconis and Frederick Mosteller, "Methods for Studying Coincidences," *Journal of the American Statistical Association,* **84**(408):853–861, December 1989.

† For a readable introduction to probability, see Warren Weaver, *Lady Luck* (Dover Publications, 1982). For a more complete exposition, see William Feller, *An Introduction to Probability Theory and Its Applications,* Vol. 1, 3rd ed. (John Wiley, 1957).

the same random experiment can be described by several different sample spaces, depending on the interest of the observer. The only requirement is that exactly *one* of the possible outcomes occurs whenever the experiment is performed. Some simple random experiments and possible sample spaces are listed below.

Random Experiment	Sample Space
a. Flip a coin to obtain "heads" (H) or "tails" (T).	{H, T}
b. Flip a coin to obtain H or T and roll a die to obtain a number 1, 2, 3, 4, 5, or 6.	{(H, 1), (H, 2), (H, 3), (H, 4), (H, 5), (H, 6), (T, 1), (T, 2), (T, 3), (T, 4), (T, 5), (T, 6)}
c. Flip a coin twice to obtain H or T followed by another H or T.	{(H, H), (H, T), (T, H), (T, T)}
d. Flip a coin twice to see if the outcomes match (M) or differ (D).	{M, D}
e. Flip a coin twice and count the number of heads.	{0, 1, 2}

We will denote the possible outcomes by $e_1, e_2, \ldots, e_n$ (read: "e sub 1," "e sub 2," and so on), so the sample space S is

$$S = \{e_1, e_2, \ldots, e_n\}$$

n is the number of possible outcomes

For example, flipping a coin twice [experiment (c) above] involves $n = 4$ possible outcomes, which may be listed $e_1 = (H, H)$, $e_2 = (H, T)$, $e_3 = (T, H)$, and $e_4 = (T, T)$.

Events

An *event* is a subset of the sample space. An event *occurs* if the outcome of the experiment is an element of the event. We will usually pick a sample space consisting of the *most basic* possible outcomes of the experiment, since more complicated events can then be discussed in terms of these. For instance, when tossing a coin twice, we will generally use the sample space $S = \{(H, H), (H, T), (T, H), (T, T)\}$ [experiment (c) above], since it can be used for questions like the "match or differ" experiment (d) by defining the events M and D to be:

$$M = \{(H, H), (T, T)\} \qquad \text{Coins match}$$

$$D = \{(H, T), (T, H)\} \qquad \text{Coins differ}$$

PRACTICE PROBLEM 1   In the experiment of tossing a coin twice and recording the number of heads [experiment (e) on page 401], represent the events 0, 1, and 2 using events from the sample space $S = \{(H, H), (H, T), (T, H), (T, T)\}$.

Solution at the back of the book

Among all events there is one *certain* event, the sample space S consisting of *all* possible outcomes. There is also one *impossible* event, the *empty* set (or *null* set) $\emptyset$. The empty set $\emptyset$ is a subset of every set, and so it is a subset of the sample space, and therefore an *event*. However, since $\emptyset$ is empty, it contains *no* possible outcomes and so cannot occur. It is a good practice to choose the sample space so that $\emptyset$ is the *only* event that cannot occur.

Assigning Probabilities to Possible Outcomes

Having identified the possible outcomes of an experiment, we now assign to each of them a probability from 0 (impossible) to 1 (certain). How do we assign probabilities? This depends upon our knowledge of the experiment. For example, if there are three possible outcomes, and each seems equally likely, then we assign probability $\frac{1}{3}$ to each; if there are 10 outcomes that seem equally likely, then we assign probability $\frac{1}{10}$ to each. In general,

Equally Likely Outcomes

If each of n possible outcomes in the sample space S is equally likely to occur, then the probability of each is $\frac{1}{n}$.

EXAMPLE 1   Assigning Equal Probabilities

a. A coin is said to be *fair* if "heads" and "tails" are equally likely. Therefore, for a fair coin we assign: $P(H) = \frac{1}{2}$ and $P(T) = \frac{1}{2}$.

b. A die is said to be *fair* if each of its six faces is equally likely. Therefore, for a fair die we assign: $P(1) = P(2) = P(3) = P(4) = P(5) = P(6) = \frac{1}{6}$.

c. A card randomly drawn from a standard deck is equally likely to be any one of the 52 cards. Therefore, the probability of drawing any particular card (such as the queen of hearts) is $\frac{1}{52}$.

From now on when we speak of coins or dice, we will assume that they are fair, unless otherwise stated.

EXAMPLE 2 Assigning Equal Probabilities Using Combinations

A student club has 15 members. If a 3-member fund raising committee is selected at random, what is the probability that the committee will consist of Bob, Sue, and Tim?

Solution

Since there are ${}_{15}C_3 = \frac{15 \cdot 14 \cdot 13}{3 \cdot 2 \cdot 1} = 455$ different ways of selecting a 3-member committee from the 15 club members, and each committee is equally likely, the probability that any one particular committee is selected is $\frac{1}{455}$.

Not all outcomes are equally likely. For instance, when tossing a coin twice and counting the number of heads [experiment (e) on page 401], there are three possible outcomes (0, 1, and 2), but assigning probability $\frac{1}{3}$ to each would be unrealistic since there is only one way to get 0 (namely T, T) but two ways to get 1 (H, T or T, H). Probabilities must be assigned to the possible outcomes on the basis of experience and knowledge of the basic experiment, and subject to two conditions: each probability must be between 0 and 1 (inclusive), and they must add to 1.

EXAMPLE 3 Assigning Unequal Probabilities

The arrow of the spinner shown here can point to one of three regions labeled A, B, and C. If the probability of pointing to a region is proportional to its area, find the probability of pointing to each of the areas A, B, and C.

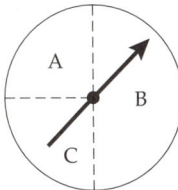

Solution

Let the sample space be $S = \{A, B, C\}$, representing the events that the spinner lands in regions A, B, or C. Clearly, regions A and C are each one quarter of the circle, and region B is one half. Therefore:

$$P(A) = \tfrac{1}{4}$$
$$P(B) = \tfrac{1}{2}$$
$$P(C) = \tfrac{1}{4}$$

PRACTICE PROBLEM 2   Suppose that the spinner on the right is spun. What are the probabilities of the three outcomes?

Solution at the back of the book

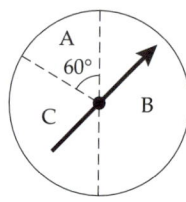

Probabilities of Events

How do we assign probabilities to events other than the individual outcomes? First, consider two extreme cases. Since the sample space S includes all possible outcomes, and one of them *must* occur, the probability of the event S is 1.

$$P(S) = 1 \qquad \text{Something \textit{must} happen}$$

Similarly, since the null set $\varnothing$ is empty, it contains *no* possible events and so cannot occur, and therefore has probability zero.

$$P(\varnothing) = 0 \qquad \text{``Nothing'' cannot happen}$$

All probabilities must lie somewhere between these extreme probabilities, so for any event E we must have $0 \le P(E) \le 1$. If E consists of just one possible outcome, then $P(E)$ is just the probability of that outcome. If E consists of two outcomes, then $P(E)$ must be the *sum* of the probabilities of those two outcomes, since E occurs when *either* of those outcomes occurs. More generally, we define the probability of an event to be the *sum* of the probabilities of the possible outcomes forming the event:

Probability Summation Formula

> The probability of an event E is
>
> $$P(E) = \sum_{\text{All } e_i \text{ in } E} P(e_i) \qquad e_i \text{ are possible events}$$

Σ is the Greek capital letter "sigma," equivalent to our capital S, and stands for "sum." This formula may be read: "the probability of an event E is the sum of the probabilities of outcomes e_i for all e_i's in the event E."

EXAMPLE 4   Finding the Probability of an Event

A fair die is rolled. What is the probability of rolling an even number?

Solution

Each of the six outcomes in the sample space $S = \{1, 2, 3, 4, 5, 6\}$ has probability $\frac{1}{6}$. The event E that an even number occurs is $\{2, 4, 6\}$. Therefore, according to the summation formula, we add the probabilities of these three outcomes:

$$P(E) = \sum_{\text{All } e_i \text{ in } \{2,4,6\}} P(e_i) = P(2) + P(4) + P(6) = \frac{1}{6} + \frac{1}{6} + \frac{1}{6} = \frac{1}{2}$$

The probability of rolling an even number is $\frac{1}{2}$.

■

For the above example, with equally likely outcomes, we added up the equal probabilities as many times as there were outcomes in the event, and obtained $\frac{3}{6} = \frac{1}{2}$. We may interpret the $\frac{3}{6}$ as taking the number of *favorable* outcomes divided by the *total* number of outcomes, although it is important to remember that this only holds for *equally likely* outcomes. In general,

$$P(E) = \frac{\left(\begin{array}{c}\text{Number of outcomes} \\ \text{favorable to } E\end{array}\right)}{\left(\begin{array}{c}\text{Total number} \\ \text{of outcomes}\end{array}\right)} \qquad \text{For equally likely outcomes}$$

Probability That an Event Does *Not* Occur

Since an event E is a subset of the sample space S, the event that E does *not* occur consists of those outcomes that are *not* in the subset E, which is the complement, E^c. Since E and E^c are disjoint and between them contain all outcomes, we must have $P(E) + P(E^c) = 1$. Solving this equation for $P(E^c)$ gives:

Complementary Probability

The probability that an event E does *not* occur is

$$P(E^c) = 1 - P(E)$$

In words: the probability that an event does *not* occur is 1 minus the probability that it *does* occur.

EXAMPLE 5 Using Complementary Probabilities

For a student club consisting of 15 members, find the probability that a randomly chosen committee of 3 does *not* consist of Bob, Sue, and Tim.

Solution

Let B be the event that the committee *does* consist of Bob, Sue, and Tim. In Example 2 we found that $P(B) = \frac{1}{455}$. The probability that the committee does *not* consist of Bob, Sue, and Tim is then the probability of the complement B^c. The complementary probability formula gives:

$$P(B^c) = 1 - P(B) = 1 - \tfrac{1}{455} = \tfrac{454}{455}$$

Therefore, the probability that the committee does *not* consist of Bob, Sue, and Tim is $\frac{454}{455}$, or about 99.8%.

EXAMPLE 6   Why the Chevalier Won His First Bet

Recall from the Application Preview on page 385 that the Chevalier de Méré's first bet was that he could roll at least one six in four rolls of a die. Find the probability that he won his bet.

Solution

In one roll of a die there are 6 possible outcomes, so by the multiplication principle, for *four* rolls there are $6 \cdot 6 \cdot 6 \cdot 6 = 6^4 = 1296$ possible outcomes, each being equally likely, and each with probability $\frac{1}{1296}$. For these four rolls, let W (for "win") be the event that he rolls at least one six, so that the complement W^c is the event of *no* sixes. How many ways are there to roll *no* sixes in four rolls? Each roll must come up 1 through 5, and by the same reasoning as before, four rolls can be done in $5 \cdot 5 \cdot 5 \cdot 5 = 5^4 = 625$ ways. According to the summation formula, we add up the probabilities of these 625 events, each of which is $\frac{1}{1296}$:

$$P(W^c) = \sum_{\text{All } e_i \text{ in } W^c} P(e_i) = \underbrace{\tfrac{1}{1296} + \cdots + \tfrac{1}{1296}}_{625 \text{ times}} = \tfrac{625}{1296} \approx 0.482 \qquad \text{Probability of no sixes}$$

Using the complementary probability principle:

$$P(W) = 1 - P(W^c) \approx 1 - 0.482 = 0.518 \qquad \text{Probability of at least one six}$$

Therefore, the Chevalier's guess was correct—betting on rolling at least one six in four rolls will win about 52% of the time.

PRACTICE PROBLEM 3

Why didn't the Chevalier choose *three* rolls? That is, find the probability of rolling at least one six in *three* rolls of a fair die.

Solution at the back of the book

Probability Space

The sample space together with the probabilities of the elementary events make up the *probability space*. To summarize:

Probability Space

A *probability space* is a sample space $S = \{e_1, e_2, \ldots, e_n\}$ of possible outcomes together with probabilities

$$P(e_1), P(e_2), \ldots, P(e_n)$$

satisfying two conditions:

1. $0 \leq P(e_i) \leq 1$ for each outcome e_i in S
2. $P(e_1) + P(e_2) + \cdots + P(e_n) = 1$

An *event* E is a subset of S, with probability

$$P(E) = \sum_{\text{All } e_i \text{ in } E} P(e_i)$$

In particular, $P(S) = 1$, $P(\varnothing) = 0$, and $P(E^c) = 1 - P(E)$.

Addition Rule for Probability

On page 387 we found that the number of elements in the union of two sets is given by the formula $n(A \cup B) = n(A) + n(B) - n(A \cap B)$, with the subtraction at the end done to avoid double-counting the elements that are in both A and B. For exactly the same reason, to find the probability of a union $A \cup B$ we add the probabilities of events A and B but then must subtract the probability of the intersection (to avoid double-counting outcomes in both events). Therefore:

Addition Rule for Probability

$$P(A \cup B) = P(A) + P(B) - P(A \cap B)$$

For disjoint events (also called *mutually exclusive events*) the intersection is empty and so has probability zero, leading to a simpler additive rule:

$$P(A \cup B) = P(A) + P(B) \qquad \text{For } A, B \text{ disjoint}$$

EXAMPLE 7   **Probability of a Union**

If you flip a coin and roll a die, what is the probability that the coin comes up heads or the die rolls a four?

Solution

Let A be the event that the coin comes up heads, and let B be the event that the die rolls a four. Then the sample space can be diagrammed as follows [see experiment (b) on page 401]:

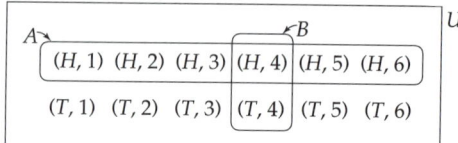

Since each of the 12 possible outcomes is equally likely, each has probability $\frac{1}{12}$. To find the probability of an event we add up $\frac{1}{12}$ as many times as there are elements in it:

$$P(A) = 6 \cdot \frac{1}{12} = \frac{6}{12} \qquad A \text{ has six elements}$$

$$P(B) = 2 \cdot \frac{1}{12} = \frac{2}{12} \qquad B \text{ has two elements}$$

$$P(A \cap B) = \frac{1}{12} \qquad \begin{array}{l} A \text{ and } B \text{ intersect at} \\ \text{one element, } (H, 4) \end{array}$$

The addition rule gives:

$$P(A \cup B) = P(A) + P(B) - P(A \cap B) = \frac{6}{12} + \frac{2}{12} - \frac{1}{12} = \frac{7}{12}$$

The probability of tossing a head or rolling a four is $\frac{7}{12}$.

EXAMPLE 8   **Finding the Probability of a Disjoint Union**

In the same probability space as in Example 7 (flipping a coin and rolling a die) what is the probability that the die rolls a three or a five?

Solution

We want the union of the events A and B shown in the following diagram.

$$
\begin{array}{cc|c|cc|c|c}
 & A\rightarrow & & & \llcorner B & & u \\
(H, 1) & (H, 2) & (H, 3) & (H, 4) & (H, 5) & (H, 6) \\
(T, 1) & (T, 2) & (T, 3) & (T, 4) & (T, 5) & (T, 6) \\
\end{array}
$$

These events are *disjoint*, so we use the simpler version of the addition rule with $P(A) = 2 \cdot \frac{1}{12} = \frac{2}{12}$ and $P(B) = 2 \cdot \frac{1}{12} = \frac{2}{12}$:

$$
P(A \cup B) = P(A) + P(B) = \frac{2}{12} + \frac{2}{12} = \frac{4}{12} = \frac{1}{3}
$$

The probability of the die coming up three or five is $\frac{1}{3}$.

◼

EXAMPLE 9   Why the Chevalier Lost His Second Bet

The Chevalier de Méré's second bet was that he could roll at least one *double* six in 24 rolls of two dice. Find the probability that he won this bet.

Solution

Rolling two dice has $6 \cdot 6 = 6^2 = 36$ possible outcomes. Repeating this 24 times gives 36^{24} possible outcomes, each being equally likely, so each with probability $\frac{1}{36^{24}}$. For these 24 rolls, let D be the event of rolling at least one double six, so that the complement D^c represents 24 rolls with *no* double sixes. How many outcomes are in D^c? *One roll of two dice gives 36 possible outcomes, of which one is a double six and 35 are not. Repeating this 24 times gives, by the multiplication principle, $\underbrace{35 \cdot 35 \cdot \ldots \cdot 35}_{24 \text{ terms}} = 35^{24}$ ways of *not* rolling a double six. Therefore, D^c contains 35^{24} elements, each having probability $\frac{1}{36^{24}}$. By the probability summation formula, we add up $\frac{1}{36^{24}}$ a total of 35^{24} times:

$$
P(D^c) = 35^{24} \cdot \frac{1}{36^{24}} = \frac{35^{24}}{36^{24}} = \left(\frac{35}{36}\right)^{24} \approx 0.509 \qquad \text{\textcolor{blue}{Using a calculator}}
$$

Therefore:

$$
P(D) = 1 - P(D^c) \approx 1 - 0.509 = 0.491 \qquad \text{\textcolor{blue}{Complementary probability principle}}
$$

The probability of throwing at least one double six in 24 rolls of a pair of dice is (approximately) 0.49.

◼

While both of the Chevalier's bets had probabilities close to $\frac{1}{2}$, his observation that, in the long run, the first was a winning bet and the second was a losing bet is borne out by these calculations.

For a random experiment with possible outcomes $e_1, e_2, \ldots, e_n$, the *sample space S* is the *set* of these outcomes, $S = \{e_1, e_2, \ldots, e_n\}$. We assign probabilities to the possible outcomes satisfying two conditions:

1. $0 \le P(e_i) \le 1$ for each outcome e_i in S *All probabilities between 0 and 1*

2. $P(e_1) + P(e_2) + \cdots + P(e_n) = 1$ *Probabilities add to 1*

If the n possible events are equally likely, we assign probability $\frac{1}{n}$ to each. A *probability space* is a sample space S together with an assignment of probabilities. An *event E* is a subset of S, and the probability of the event E is defined as:

$$P(E) = \sum_{\text{All } e_i \text{ in } E} P(e_i)$$

Adding the probabilities for each outcome in the event

In particular, $P(S) = 1$, $P(\varnothing) = 0$, and $P(E^c) = 1 - P(E)$. The probability of a *union* $A \cup B$ is found by the *addition rule*:

$$P(A \cup B) = P(A) + P(B) - P(A \cap B)$$

If events A and B are *mutually exclusive* (so that $A \cap B = \varnothing$), then this rule simplifies to $P(A \cup B) = P(A) + P(B)$.

EXERCISES 4.2

1. **Committees** Find the sample space for a committee of two chosen from Alice, Bill, Carol, and Dan. Then find the sample space if both sexes must be represented. (*Use initials.*)

2. **Committees** Find the sample space for a committee of two formed from Alice, Bill, Carol, Dan, and Edgar. Then find the sample space if the committee must include at least one woman. (*Use initials.*)

3. **Wardrobes** Find the sample space for choosing an outfit consisting of one of four shirts (S_1, S_2, S_3, or S_4) and one of two jackets (J_1 or J_2).

4. **Wardrobes** Find the sample space for choosing an outfit consisting of one of two shirts (S_1 or S_2), one of two jackets (J_1 or J_2), and one of two pairs of pants (P_1 or P_2).

5. **Marbles** A box contains three marbles, one red, one green, and one blue. A first marble is chosen, its color recorded, and then it is replaced in the box and a second marble is chosen, and its color is recorded. Find the sample space. Then find the sample space if the first marble is *not* replaced before the second is chosen.

6. Coin Tossing You toss a coin until you get the first head or until you have tossed it five times, whichever comes first. Find the sample space.

Dice A die is rolled, and $A = \{1, 3, 5\}$ (rolling an odd number), and $B = \{3, 6\}$ (rolling a three or a six). Specify each event as a subset of the sample space $S = \{1, 2, 3, 4, 5, 6\}$.

7. a. $A \cap B$ **b.** $A^c \cup B$
8. a. $A \cup B$ **b.** $A \cap B^c$

Dice Two dice are rolled. E is the event that the sum is even, F is the event of rolling at least one six, and G is the event that the sum is eight. List the outcomes for the following events.

9. a. $E \cap F$ **b.** $E^c \cap G$
10. a. $F \cap G$ **b.** $E^c \cap F \cap G$

Probabilities For the sample space $S = \{e_1, e_2, e_3, e_4, e_5\}$ with probabilities $P(e_1) = 0.10$, $P(e_2) = 0.30$, $P(e_3) = 0.40$, $P(e_4) = 0.05$, and $P(e_5) = 0.15$, and events $A = \{e_1, e_2, e_3\}$, $B = \{e_1, e_3, e_5\}$, and $C = \{e_2, e_4\}$, find each probability.

11. a. $P(A)$ **b.** $P(B \cup C)$
12. a. $P(B)$ **b.** $P(A \cup C)$
13. a. $P(A^c)$ **b.** $P(A^c \cap B)$
14. a. $P(B^c)$ **b.** $P(A^c \cap C)$

15. Committees If a committee of 3 is to be chosen at random from a class of 12 students, what is the probability of any particular committee being selected? What if the committee is to consist of a president, a vice-president, and a treasurer?

16. Committees A committee of 3 is to be chosen at random from a class of 20 students. What is the probability that a particular committee will be selected? What if the 3 are to be a president, a vice-president, and a treasurer?

17. Marbles One marble is selected at random from a box containing 6 red and 4 blue marbles, and its color noted, so the sample space is $\{R, B\}$. What probability should be assigned to each outcome?

Spinners Find the probability of the arrow landing in each numbered region. Assume that the probability of any region is proportional to its area and that areas that look the same size are the same size.

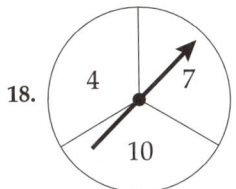

18.

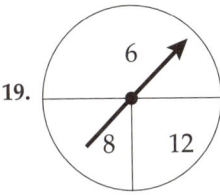

19.

20. Dartboards A circular dartboard has a radius of 12 inches with a circular bullseye of radius 2 inches at the center. You throw a dart and hit the dartboard. Assume that the probability of hitting any region is proportional to the area of the region.

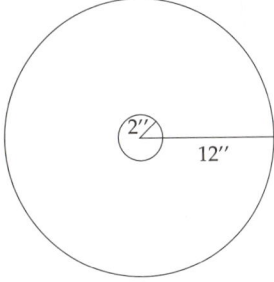

a. What is the probability of hitting the bullseye?

b. What is the probability of hitting the rest of the dartboard?

21. Baseball A baseball hits the 12-foot-by-10-foot wall of your house. Assume that the probability of its hitting any region is proportional to the area of the region.

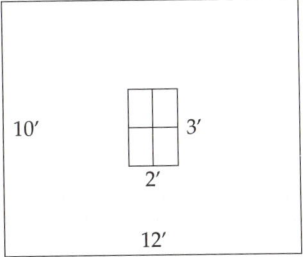

a. What is the probability that it hits the 3-foot-by-2-foot window?

b. What is the probability that it misses the window?

22. Choosing Colors One ball is to be chosen from a box containing red, blue, and green balls. If the probability that the chosen ball is red is $\frac{1}{5}$, and that it is blue is $\frac{1}{3}$, what is the probability that it is green?

23. Political Contributions In a town, 38% of the citizens contributed to the Republicans, 42% contributed to the Democrats, and 12% contributed to both. What percentage contributed to neither party?

24. Credit Cards A store accepts both Visa cards and Mastercards. If 61% of its customers carry Visa cards, 52% carry Mastercards, and 28% carry both, what proportion carry a card that the store will accept?

25. Surveys A college survey claimed that 63% of students had taken English Composition, 48% took calculus, 15% took both, and 10% took neither. Show that these figures cannot be correct.

26. Dice One die is rolled. Find the probability of:
a. rolling at least 3.
b. rolling an odd number.

27. Marbles A box contains 4 red and 8 green marbles. You reach in and remove three marbles all at once. Find the probability that the three marbles:
a. are all red.
b. are all of the same color.

28. Lottery In a lottery you choose 5 numbers out of 40. Then 6 numbers are announced, and you win something if you have 5 of the 6 numbers. What is the probability that you win something?

29. Guessing Numbers Someone picks a number between 1 and 10 (inclusive) and you have three guesses. What is the probability that you will get it?

30. Words Among all 5-letter nonsense words (that is, without regard to meaning), what is the probability that a word has:

a. no vowels?
b. at least one vowel?

31. Cards If you are dealt 5 cards at random from an ordinary deck, what is the probability that your hand contains all four aces?

32. Cards If you are dealt 5 cards at random from an ordinary deck, what is the probability of being dealt a flush (all 5 cards of the same suit)?

33. Elevator Stops An elevator has 5 people and makes 7 stops. What is the probability that no two people get off on the same floor?

34. Defective Products A carton of 24 CD players includes four that are missing a part. If you choose four at random, what is the probability that you get those four?

35. Defective Products A box of 100 screws contains 10 that are defective. If you choose 10 at random, what is the probability that none are defective?

36. Committees A committee of 12 is to be formed from your class of 100 students. What is the probability that you and your best friend are on the committee?

37. Senate The United States Senate consists of 100 members, 2 from each state. A committee of 8 senators is formed. What is the probability that it contains at least one senator from your state?

38. Light Bulbs Your house uses fifteen light bulbs, five in each of three different wattages, and you keep one spare bulb of each wattage. Two bulbs burn out. What is the probability that you have spares for both?

39. Keys You carry six keys in your pocket, two of which are for the two locks on your front door. You lose one key. What is the probability that you can get into your house through the front door?

40. Shoes You are rushing to leave on a trip, and you randomly grab four shoes from the five pairs in your closet. What is the probability that you take at least one pair?

Explorations and Excursions

The following problems extend and augment the material presented in the text.

Odds If E is an event with probability $P(E)$ such that $0 < P(E) < 1$, then the *odds for* the event E is the ratio $P(E):P(E^c)$ [read: "$P(E)$ to $P(E^c)$"] while the *odds against* the event E is the ratio $P(E^c):P(E)$ [read: "$P(E^c)$ to $P(E)$"]. Since odds are ratios, we multiply to clear the fractions, so that, for example, $\frac{2}{3}:\frac{1}{3}$ becomes $2:1$. For equally likely outcomes, the odds for an event can be interpreted as the ratio of the number of *favorable* outcomes to the number of *unfavorable* outcomes.

41. Show that the odds for heads in a flip of a coin is $1:1$.

42. Show that the odds for a two on a roll of a six-sided die is $1:5$.

43. Show that if $P(E) = \frac{n}{m}$ then the odds for E are $n:(m - n)$.

44. Show that if $P(E) = \frac{n}{m}$ then the odds against E are $(m - n):n$.

45. Show that if the odds for E are $n:m$ then $P(E) = \frac{n}{n+m}$.

46. Show that if the odds against E are $n:m$ then $P(E) = \frac{m}{n+m}$.

47. **Baseball** A radio announcer says the odds that the Yankees will beat the Orioles are $7:4$. What does the announcer believe is the probability that the Yankees will beat the Orioles?

48. **Horse Racing** A bookie has changed his odds for Win-By-A-Neck from $3:14$ to $5:24$. Does he think the probability that this three-year-old will win has gone up or down?

49. **Stocks** A stock exchange announces that gainers beat loosers by 7 to 2. What does this mean about the probability that a stock gained value?

4.3 Conditional Probability and Independence

Introduction

We often ask about the probability of one event *given another*. For example, a player being dealt cards may want to know the probability that a third card will be an ace given that the first two were aces. A teenager might ask about the probability of developing cancer given that one continues to smoke. Such questions involve *conditional probability*, which is the subject of this section. We will also discuss Bayes' Formula and the very important concept of independence.

Conditional Probability

When rolling two dice (always fair unless otherwise stated), there are $6 \cdot 6 = 36$ possible outcomes in the sample space

$$S = \{(1, 1), (1, 2), \ldots, (2, 1), (2, 2), \ldots, (6, 6)\}$$

EXAMPLE 1   **Finding a Conditional Probability**

Suppose that you roll a die twice. What is the probability that the sum is 5 given the additional information that the first roll is a 3?

Solution

Let B be the event that the first die comes up 3, so $B =$ {(3, 1), (3, 2), (3, 3), (3, 4), (3, 5), (3, 6)}. B contains six possible outcomes, each being equally likely, and only one has sum 5, namely (3, 2). Therefore, the probability that event A (the sum is 5) occurs given that B occurred is one out of six, which we write as:

$$P(A \text{ given } B) = \tfrac{1}{6}$$

In the above example, we restricted ourselves to a new, smaller sample space corresponding to the "given" event B. The conditional probability was then the probability of A in B, that is, $P(A \text{ and } B)$ *relative* to this new sample space B. This leads to the following general definition.

Conditional Probability

> Let A and B be events with $P(B) > 0$. Then the *conditional probability* that A occurs given that B occurs is
>
> $$P(A \text{ given } B) = \frac{P(A \text{ and } B)}{P(B)}$$

In words, the conditional probability of A *given* B is the probability of A and B divided by the probability of B. (We assume $P(B) > 0$ to avoid zero denominators.) Note that for *events* we prefer to say A *and* B rather than $A \cap B$, although they are equivalent. The conditional probability $P(A \text{ given } B)$ may also be written $P(A \mid B)$. If we represent events by a Venn diagram, and probability is represented by area, then the conditional probability is just the ratio of two areas.

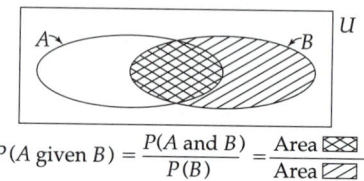

We may check the result of Example 1 using this formula.

$$P(A \text{ and } B) = \tfrac{1}{36}$$ First roll 3 and the sum is 5 means only (3, 2)

$$P(B) = \tfrac{6}{36} = \tfrac{1}{6}$$ The six outcomes were listed in Example 1

Therefore:

$$P(A \text{ given } B) = \frac{P(A \text{ and } B)}{P(B)} = \frac{\tfrac{1}{36}}{\tfrac{1}{6}} = \tfrac{1}{6}$$ Same answer as before

The *unconditional* probability of rolling a sum of 5 is $\tfrac{1}{9}$ [happening on four outcomes, (1, 4), (2, 3), (3, 2), and (4, 1), out of the 36 possibilities], so *conditioning* on the first roll being 3 changed the probability from $\tfrac{1}{9}$ to $\tfrac{1}{6}$.

It is important to remember that when we speak of "event A given event B," we do not mean to imply that event B occurred first. We mean only that the random experiment has been performed, the outcome has been observed, and we are just discussing how the information that the outcome is part of event B influences the chance that it is also part of event A.

EXAMPLE 2 Finding Card Probabilities

You are playing cards with a friend, and have each been dealt five cards at random from a standard deck. If you have no face cards, what is the probability that your friend doesn't either?

Solution

Let A be the event that your friend has no face cards, and let B be the event that you have no face cards. The event A *and* B means that *neither* of you have face cards, and since the deck has 12 face cards, this means that the first 10 cards dealt were from the 40 nonface cards, so:

$$P(A \text{ and } B) = \frac{{}_{40}C_{10}}{{}_{52}C_{10}} \approx 0.0536$$

Ways of choosing 10 from ← the 40 nonface cards

← Ways of choosing 10 from 52

Using a calculator

$$P(B) = \frac{{}_{40}C_{5}}{{}_{52}C_{5}} \approx 0.253$$

Ways of choosing 5 from ← the 40 nonface cards

← Ways of choosing 5 from 52

Therefore:

$$P(A \text{ given } B) = \frac{P(A \text{ and } B)}{P(B)} \approx \frac{0.0536}{0.253} = 0.212$$

The probability that your friend has no face cards given that you don't either is about 21%.

We could also do this problem by looking directly at the restricted sample space. Given that you have five nonface cards, your friend's cards come from the remaining 47 cards, 12 of which are face cards and 35 of which are not. Therefore:

$$P(A \text{ given } B) = \frac{_{35}C_5}{_{47}C_5} \approx 0.212$$

5 from the remaining
← 35 nonface cards
← 5 from the remaining 47
⌐ Same answer as before

Solving the conditional probability formula for $P(A \text{ and } B)$ gives the following formula, which is very useful when a conditional probability is given.

Probability Conditioning Formula

$$P(A \text{ and } B) = P(A \text{ given } B) \cdot P(B)$$

In words, the probability of A and B is the probability of A *given* B times the probability of B. We may interpret this formula as finding the probability of A *and* B by *conditioning* on the event B.

EXAMPLE 3   Using Conditioning

You need to be somewhere within 30 minutes, and your parents are out with the car. If they come back soon (you give this a 50-50 chance), the probability that you will get there in time is 90%. Otherwise you will walk, with a 60% chance of arriving in time. What is the probability that you will arrive on time in their car?

Solution

Let H be the event that your parents come home in time, and A be the event that you arrive in time (by whatever means). You want to find $P(A \text{ and } H)$.

$$P(A \text{ and } H) = P(A \text{ given } H) \cdot P(H) \qquad \text{Conditioning on } H$$
$$= (0.90) \cdot (0.50) = 0.45$$

The probability that you will arrive in their car and on time is 0.45 or 45%.

PRACTICE PROBLEM 1

Using the information in the above example, what is the probability that you will go on foot and arrive on time?

Solution at the back of the book

Partitions and Total Probability

A set E such that $0 < P(E) < 1$ and its complement E^c are said to form a *partition* of a sample space S because they are disjoint and their union is the whole sample space: $S = E \cup E^c$. More generally, a *partition* of a sample space S is any collection of disjoint events $U_1, U_2, \ldots, U_m$ with positive probabilities such that their union is the whole sample space: $S = U_1 \cup U_2 \cup \cdots \cup U_m$. (Such events are said to be *mutually exclusive and collectively exhaustive*.) Since any outcome is contained in exactly *one* of the events of the partition, for any event A we have:

$$P(A) = \sum_{\substack{\text{All } U_i \text{ in} \\ \text{the partition}}} P(A \text{ and } U_i)$$

A partition can be illustrated in a Venn diagram by dividing the sample space into "strips." This formula simply says that any event A is the sum of the parts into which it is separated by the strips.

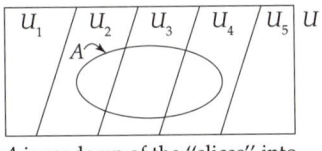

A is made up of the "slices" into
which the partition cuts it.

In the above formula, applying the probability conditioning formula (page 416) to each $P(A \text{ and } U_i)$ gives $P(A \text{ and } U_i) = P(A \text{ given } U_i) \cdot P(U_i)$, resulting in:

Total Probability Formula

$$P(A) = \sum_{\substack{\text{All } U_i \text{ in} \\ \text{the partition}}} P(A \text{ given } U_i) \cdot P(U_i)$$

This formula is so named because, like the previous formula, it gives the total probability of an event by separating the event into parts corresponding to the partition and then adding them up. It can be thought of as a *weighted average* of conditional probabilities formed by *conditioning* on the partition.

For instance, in the previous Example we found that your probability of arriving in time and by car was:

$$P(A \text{ given } H) \cdot P(H) = (0.90) \cdot (0.50) = 0.45$$

In Practice Problem 1 you found that the probability of arriving in time and on foot was:

$$P(A \text{ given } H^c) \cdot P(H^c) = (0.60) \cdot (0.50) = 0.30$$

These two together give $P(A)$, the *total probability* of arriving in time:

$$P(A) = P(A \text{ given } H) \cdot P(H) + P(A \text{ given } H^c) \cdot P(H^c) = 0.45 + 0.30 = 0.75$$

Therefore, your probability of arriving in time (by *any* means) is 75%, which we found by conditioning on whether your parents arrive home soon or not.

EXAMPLE 4   Sampling Without Replacement

A box contains three blue balls and two green balls. A ball is randomly chosen (and not replaced) and then a second ball is chosen. What is the probability that the second ball is green?

Solution

To find a probability for the second ball, we will *condition* on the color of the first ball. We write B_1 the event that the first ball is blue, G_1 for a green first ball and, similarly, B_2 and G_2 for the second ball. Since the first ball must be blue or green, B_1 and G_1 form a *partition* of the sample space. Their probabilities are $P(B_1) = \frac{3}{5}$ and $P(G_1) = \frac{2}{5}$ (since of the five balls, three are blue and two are green). We also have $P(G_2 \text{ given } B_1) = \frac{2}{4}$ (since given B_1 there are four balls left, of which two are green), and similarly $P(G_2 \text{ given } G_1) = \frac{1}{4}$. From the total probability formula with $A = G_2$, $U_1 = B_1$, and $U_2 = G_1$ we obtain:

$$P(G_2) = P(G_2 \text{ given } B_1) \cdot P(B_1) + P(G_2 \text{ given } G_1) \cdot P(G_1)$$

$$= \frac{2}{4} \cdot \frac{3}{5} + \frac{1}{4} \cdot \frac{2}{5} = \frac{3}{10} + \frac{1}{10} = \frac{2}{5}$$

The probability that the second ball is green is $\frac{2}{5}$, or 40%.

■

Bayes' Formula

From the definition of conditional probability (page 414) and the conditioning formula (page 416) we have

$$P(U_1 \text{ given } A) = \frac{P(A \text{ and } U_1)}{P(A)} = \frac{P(A \text{ given } U_1) \cdot P(U_1)}{P(A)}$$

Replacing the denominator by its equivalent according to the total probability formula (on the previous page) gives:

Bayes' Formula*

> Let $U_1, U_2, \ldots, U_m$ be a partition of the sample space S and let A be any event with positive probability. Then:
>
> $$P(U_1 \text{ given } A) = \frac{P(A \text{ given } U_1) \cdot P(U_1)}{\displaystyle\sum_{\substack{\text{All } U_i \text{ in} \\ \text{the partition}}} P(A \text{ given } U_i) \cdot P(U_i)}$$

Notice that Bayes' formula reverses the conditioning: U_1 given A on the left and A given U_i on the right. However, do not think of Bayes' formula as *reversing time*. $P(U_1 \text{ given } A)$ simply means the probability that the outcome is in U_1 given that it is in A.

EXAMPLE 5 Assessing Voting Patterns

Registered voters in Marlin County are 45% Democrats, 30% Republicans, and 25% Independents. In the last election for county supervisor, 70% of the Democrats voted, 80% of the Republicans voted, and 90% of the independents voted. What is the probability that a randomly selected voter in this last election was a Democrat?

Solution

Democrats, Republicans, and Independents form a *partition* of the voters, which we represent by D, R, and I. We have that

$$P(D) = 0.45, \quad P(R) = 0.30, \quad \text{and} \quad P(I) = 0.25$$

Let V be the event that a randomly selected voter voted in the last election. We are given:

$$P(V \text{ given } D) = 0.70, \quad P(V \text{ given } R) = 0.80, \quad P(V \text{ given } I) = 0.90$$

We are asked to find $P(D \text{ given } V)$. Applying Bayes' formula:

$P(D \text{ given } V)$

$$= \frac{P(V \text{ given } D) \cdot P(D)}{P(V \text{ given } D) \cdot P(D) + P(V \text{ given } R) \cdot P(R) + P(V \text{ given } I) \cdot P(I)}$$

$$= \frac{(0.70) \cdot (0.45)}{(0.70) \cdot (0.45) + (0.80) \cdot (0.30) + (0.90) \cdot (0.25)} \approx 0.404$$

The probability that a voter in the last election was a Democrat is about 40%.

---

* After Thomas Bayes (1702–1761), an English theologian and probabilist.

PRACTICE PROBLEM 2   Using the information in the previous Example, find the probability that the randomly selected voter in the last election was an Independent.

Solution at the back of the book

EXAMPLE 6 Management Training Programs

A corporation is reviewing the success rate of its management training program. It finds that 80% of those entering the program are college graduates, and that 90% of them successfully complete it, while just 60% of the noncollege graduates in the program successfully complete it. What is the probability that an employee who successfully completes the training program is not a college graduate?

Solution

Let C be the event that the employee entering the program is a college graduate. We are given that $P(C) = 0.80$. Let N be the event that the employee entering the program is *not* a college graduate. Since $N = C^c$, C and N form a partition of those entering the program, and $P(N) = 1 - P(C) = 0.20$. Let S be the event that the employee entering the program completes it successfully. We are told that $P(S$ given $C) = 0.90$ and that $P(S$ given $N) = 0.60$. The probability that an employee who successfully completes the program is *not* a college graduate is $P(N$ given $S)$, which we find using Bayes' formula:

$$P(N \text{ given } S) = \frac{P(S \text{ given } N) \cdot P(N)}{P(S \text{ given } N) \cdot P(N) + P(S \text{ given } C) \cdot P(C)}$$

$$= \frac{(0.60) \cdot (0.20)}{(0.60) \cdot (0.20) + (0.90) \cdot (0.80)} \approx 0.143$$

The probability that an employee who successfully completes the training program is not a college graduate is about 14%.

Independent Events

Roughly speaking, two events A and B are said to be *independent* if one has nothing to do with the other, so that the occurrence of one has no bearing on the probability of the other. In terms of conditional probability, events A and B with positive probabilities are *independent* if $P(A$ given $B)$ is the same as $P(A)$. From the definition of conditional probability, this equality can be written:

$$\frac{P(A \text{ and } B)}{P(B)} = P(A) \qquad\qquad P(A \text{ given } B) = P(A)$$

Multiplying each side by $P(B)$ gives the following equivalent condition, which we take as the definition of independence.

Independent Events

Events A and B are independent if:

$$P(A \text{ and } B) = P(A) \cdot P(B)$$

Events that are not independent are *dependent*.

Independent does not mean the same thing as *disjoint*, or *mutually exclusive*. If events A and B have positive probabilities and are disjoint, then $P(A \text{ and } B) = 0$ while $P(A) \cdot P(B)$ is positive, so A and B cannot be independent. Intuitively, disjointedness is a very strong kind of *dependence*, since the occurrence of one of two disjoint events guarantees the *non*-occurrence of the other, so they cannot be independent.

EXAMPLE 7   **Independent Coin Tosses**

A fair coin is tossed twice. Let A be the event that the first toss is heads and let B be the event that the second toss is heads. Are the events A and B independent?

Solution

The sample space is

$$S = \{(H, H), (H, T), (T, H), (T, T)\}$$

and

$$A = \{(H, H), (H, T)\}, \quad B = \{(H, H), (T, H)\}, \quad \text{and} \quad A \cap B = \{(H, H)\}$$

Since the outcomes are equally likely,

$$P(A) = \frac{2}{4} = \frac{1}{2}, \quad P(B) = \frac{2}{4} = \frac{1}{2}, \quad \text{and} \quad P(A \text{ and } B) = \frac{1}{4}$$

Since $P(A) \cdot P(B) = \frac{1}{2} \cdot \frac{1}{2} = \frac{1}{4}$ and $P(A \text{ and } B) = \frac{1}{4}$, we have $P(A \text{ and } B) = P(A) \cdot P(B)$, so the events A and B *are* independent.

■

If $P(A \text{ and } B)$ and $P(A) \cdot P(B)$ had *not* been equal, the events would have been *dependent*. The fact that successive coin tosses are independent is sometimes expressed by saying that the coin has "no memory of the last toss," and that each toss is a "new experience with no influence from the past."

EXAMPLE 8   **Assessing Independence**

A coin is tossed three times. Let A be the event that at most one head occurred and let B be the event that the tosses included both heads and tails. Are the events A and B independent?

Solution

The sample space is $S = \left\{ \begin{matrix} (H, H, H), & (H, H, T), & (H, T, H), & (H, T, T) \\ (T, H, H), & (T, H, T), & (T, T, H), & (T, T, T) \end{matrix} \right\}$

We then have:

$A = \{(H, T, T), \quad (T, H, T), \quad (T, T, H), \quad (T, T, T)\}$ At most 1 head

$B = \left\{ \begin{matrix} (H, H, T), & (H, T, H), & (H, T, T) \\ (T, H, H), & (T, H, T), & (T, T, H) \end{matrix} \right\}$ Both heads and tails

$A \cap B = \{(H, T, T), \quad (T, H, T), \quad (T, T, H)\}$ Intersection

Since the outcomes are equally likely,

$$P(A) = \frac{4}{8} = \frac{1}{2}, \quad P(B) = \frac{6}{8} = \frac{3}{4}, \quad \text{and} \quad P(A \text{ and } B) = \frac{3}{8}$$

Since $P(A \text{ and } B) = P(A) \cdot P(B)$ (each is $\frac{3}{8}$), the events A and B are independent.

The notion of independence can be extended to more than two events.

Many Independent Events

> A collection $E_1, E_2, \ldots, E_m$ of events is *independent* if every subcollection of them satisfies the multiplication formula:
>
> $P(E_i \text{ and } E_j \text{ and } \ldots \text{ and } E_k) = P(E_i) \cdot P(E_j) \cdot \ldots \cdot P(E_k)$

In practice, the most important applications of independence do not involve *proving* it but *assuming* it. That is, many applications involve events that clearly have nothing to do with each other, and so we will *assume* that they are independent, enabling us to find probabilities simply by multiplication. With the concept of independence, we can solve the Chevalier's second problem more easily than before.

EXAMPLE 9   **The Chevalier de Méré, Again**

What is the probability of at least one double six in 24 rolls of two fair dice?

Solution

The outcome of one roll of the dice does not affect the other rolls, so the rolls are independent. On any one of them, the probability of *not* getting a double six is $\frac{35}{36}$, so the probability of not getting a double six repeatedly on 24 rolls is

$$\underbrace{\frac{35}{36} \cdot \frac{35}{36} \cdot \ldots \cdot \frac{35}{36}}_{\text{24 terms}} = \left(\frac{35}{36}\right)^{24} \approx 0.509$$

Thus the probability of getting at least one double six is $1 - (\frac{35}{36})^{24} \approx 0.491$, agreeing with our calculation in Example 9 on page 409.

EXAMPLE 10   Used Car Sales

Each of five used car salesmen at an inner city car lot can sell a car within fifteen minutes of contact with a new customer 80% of the time. If all five salesmen have started talking with new customers, what is the probability that all will sell cars within the next fifteen minutes?

Solution

Since the salesmen work independently, the probability that all will make their sales is the product of the probabilities that each does:

$$\underbrace{\frac{80}{100} \cdot \frac{80}{100} \cdot \frac{80}{100} \cdot \frac{80}{100} \cdot \frac{80}{100}}_{\text{Five terms}} = \left(\frac{80}{100}\right)^5 \approx 0.328$$

The probability that all five will make sales is about 33%.

PRACTICE PROBLEM 3

For the situation described in the above example, find the probability that *none* of the salesmen sell cars to their customers within the next fifteen minutes.

Solution at the back of the book

SUMMARY

For events A and B with $P(B) > 0$, the *conditional probability* that A occurs given that B occurs is

$$P(A \text{ given } B) = \frac{P(A \text{ and } B)}{P(B)}$$

The definition of conditional probability may be rewritten as a probability *conditioned* on B:

$$P(A \text{ and } B) = P(A \text{ given } B) \cdot P(B)$$

A *partition* of a sample space S is any collection of events $U_1, U_2, \ldots , U_m$ with positive probabilities that are mutually exclusive ($U_i \cap U_j = \varnothing$ fn $i \neq j$) and collectively exhaustive ($S = U_1 \cup U_2 \cup \cdots \cup U_m$). For any event A we have the total probability formula:

$$P(A) = \sum_{\substack{\text{All } U_i \text{ in} \\ \text{the partition}}} P(A \text{ given } U_i) \cdot P(U_i)$$

and Bayes' formula:

$$P(U_1 \text{ given } A) = \frac{P(A \text{ given } U_1) \cdot P(U_1)}{\displaystyle\sum_{\substack{\text{All } U_i \text{ in} \\ \text{the partition}}} P(A \text{ given } U_i) \cdot P(U_i)} \qquad P(A) > 0$$

Two events A and B are *independent* if the probabilities multiply:

$$P(A \text{ and } B) = P(A) \cdot P(B)$$

and a collection $E_1, E_2, \ldots , E_m$ of events is *independent* if every subcollection of the events obeys a similar multiplication property.

EXERCISES 4.3

1. If $P(A) = 0.6$, $P(B) = 0.4$, and $P(A \text{ and } B) = 0.2$, find:

   a. $P(A \text{ given } B)$ b. $P(B \text{ given } A)$

2. If $P(A) = 0.5$, $P(B) = 0.3$, and $P(A \text{ and } B) = 0.1$, find:

   a. $P(A \text{ given } B)$ b. $P(B \text{ given } A)$

3. If $P(A) = 0.4$, $P(B) = 0.5$, and $P(A \cup B) = 0.6$, find:

   a. $P(A \text{ given } B)$ b. $P(B \text{ given } A)$

4. If $P(A) = 0.6$, $P(B) = 0.5$, and $P(A \cup B) = 0.8$, find:

   a. $P(A \text{ given } B)$ b. $P(B \text{ given } A)$

5. **Marbles**   A box contains 4 white, 2 red, and 4 black marbles. One marble is chosen at random and it is not black. Find the probability that it is white.

6. **Cards**   You select two cards at random from an ordinary deck. If the first card is a spade, what is the probability that the second card is a spade?

7. **Gender**   Your friend has two children, and you know that at least one is a girl. What is the probability that both are girls? (Assume that girls and boys are equally likely.)

8. **Eye Color**   Suppose that each of two children in a family have probability $\frac{1}{5}$ of having blue eyes. If at least one child has blue eyes, what is the probability that both have blue eyes?

9. **Cards**   A deck contains three cards: one is red on both sides, one is blue on both sides, and the third is red on one side and blue on the other. One card is chosen at random from the deck and the color on one side is observed. If this side

is blue, what is the probability that the other side is blue?*

10. **Credit Cards**   Looking back at Exercise 24 on page 412, if a customer has at least one credit card, what is the probability that the customer has a Visa card?

11. **Political Contributions**   Looking back at Exercise 23 on page 412, if a person is a contributor, what is the probability that the person contributes to the Republican party?

12. **Cards**   In the game of bridge, four players are each dealt 13 cards. If a player has no aces, what is the probability that that person's partner has no aces?

13. **Choosing Courses**   You will take either a basketweaving course or a philosophy course, depending on what your advisor decides. You estimate that the probability of your getting an A in basketweaving is 0.95, while in philosophy it is 0.70. However, the chance of your advisor choosing the basketweaving course is only 20%, whereas there is an 80% chance of his putting you in the philosophy course. What is the probability of your ending up with an A?

14. **Multiple-Choice Tests**   On a multiple-choice test you know the answers to 70% of the questions (and so get them right), and for the remaining 30% you choose randomly among the 5 answers. What percent of the answers will you get right?

15. **Driving**   Suppose that 70% of drivers are "careful" and 30% are "reckless." Suppose further that a careful driver has a 0.1 probability of being in an accident in a given year, whereas for a reckless driver the probability is 0.3. What is the probability that a randomly selected driver will have an accident within a year?

---

* Many people mistakenly believe that if one side is blue, then the probability that the other side is blue is $\frac{1}{2}$ since it can only be the blue-blue card or the blue-red card, and these are equally likely. The error comes from not realizing that the blue side is equally likely to be any of the *three* blue sides. In fact, more than one probability student has made money by knowing the correct conditional probabilities and offering bets on the outcome.

16. **Quality Control**   A computer manufacturer has assembly plants in three states. The Delaware plant produces 25% of the company's computers, the Michigan plant produces 35%, and the California plant produces the other 40%. The probabilities that a computer will pass inspection are 93% for the Delaware plant, 89% for the Michigan plant, and 94% for the California plant. What is the probability that a randomly selected computer from this company will pass inspection?

17. **Voting**   In a town, 60% of the citizens are Republicans and 40% are Democrats. In the last election, 55% of the Republicans voted and 65% of the Democrats voted. If a voter is randomly selected, what is the probability that the person is a Republican?

18. **Colorblindness**   An estimated 8% of men and 0.5% of women are colorblind. If a colorblind person is selected at random, what is the probability that the person is a male? Assume that men and women are occur in equal numbers.

19. **Medical Testing**   A new test is developed to test for a certain disease, giving "positive" or "negative" results to indicate that the person does or does not have the disease. For a person who actually has the disease, the test will give a positive result with probability 0.95 and a negative result with probability 0.05 (a so-called "false negative"). For a person who does *not* have the disease, the test will give a positive result with probability 0.05 (a "false positive") and a negative result with probability 0.95. Furthermore, only one person in 1000 actually has this disease. If a randomly selected person is given the test and tests positive, what is the probability that the person actually has the disease?

20. **Driving**   Looking back at Exercise 15 above, for a person who has had an accident within a year, what is the probability that the person is a safe driver?

21. **Manufacturing Defects**   A computer chip factory has three machines, A, B, and C, for producing memory chips. Machine A produces 50% of the factory's chips, machine B produces

30%, and machine C produces 20%. It is known that 3% of the chips produced by machine A are defective, as are 2% of the chips produced by machine B and 1% of the chips from machine C. If a randomly selected chip from the factory's output is found to be defective, what is the probability that it was produced by machine B?

22. **Manufacturing Defects** For the information in Exercise 21, if a randomly selected chip is found to be *not* defective, what is the probability that it came from machine B?

For the experiment of tossing a coin twice, find whether events A and B are independent or dependent.

23. *A*: heads on the first toss
    B: different results on the two tosses

24. *A*: heads on the second toss
    B: the same results on both tosses

For the experiment of rolling two dice, find whether events A and B are independent or dependent.

25. *A*: odd number on the first roll
    B: sum of the numbers is 4

26. *A*: even number on the first roll
    B: sum of the numbers is 10

27. **Dice** A pair of dice is rolled three times in succession. Find the probability that each of the rolls has a sum of 7.

28. **Brand Loyalty** Suppose that each time you buy a car, you choose between Ford and General Motors. Suppose that each time after the first, you stay with the same company with probability $\frac{2}{3}$ and switch with probability $\frac{1}{3}$. If

you are equally likely to choose either company for your first car, what is the probability that your first and second choices will be Ford cars and your third and fourth choices will be General Motors cars?

29. **Dice** Three dice are rolled. What is the probability of getting:
    a. All sixes
    b. All the same outcomes
    c. All different outcomes

Explorations and Excursions

The following problems extend and augment the material presented in the text.

30. If event *A* is such that $P(A) = 0$, and *B* is any other event, show that events *A* and *B* are independent.

31. If events *A* and *B* are independent, show that events *A* and B^c are independent.

32. If events *A* and *B* are independent, show that events A^c and B^c are independent. (*Hint:* Use the result of Exercise 31.)

33. In the experiment of tossing two coins, define events as follows.
    A: heads on first coin
    B: heads on second coin
    C: outcomes agree
    Show that *A*, *B*, and *C* are *pairwise* independent but do *not* satisfy $P(A \text{ and } B \text{ and } C) = P(A) \cdot P(B) \cdot P(C)$ and so are not independent.

34. To see why the events in Exercise 33 are not independent, calculate $P(C)$ and $P(C$ given $(A \text{ and } B))$.

4.4 Random Variables and Distributions

Introduction

Many times a random experiment results in a *number*, such as the sum of the numbers on two dice or your winnings in a lottery. Such numerical quantities are called *random variables* and are the central objects of probability and statistics. In this section we will use random variables to answer questions about chance events, and (also in the following chapter) discuss some of the random variables that have proven most useful in applications.

Random Variables

A *random variable* is an assignment of a numerical value to each outcome in the sample space in such a way that each outcome is associated with exactly one number. We will use capital letters like X and Y for random variables.

EXAMPLE 1 Defining A Random Variable

Let X be the number of heads in four tosses of a coin. Find the possible values for X and the outcomes corresponding to each possible value.

Solution

The number of heads in four tosses can be 0, 1, 2, 3, or 4, so these are the possible values for X. The sample space S consists of the following 16 sequences of H's and T's.

$$S = \begin{cases} (H, H, H, H), & (H, H, H, T), & (H, H, T, H), & (H, T, H, H) \\ (T, H, H, H), & (H, H, T, T), & (H, T, H, T), & (H, T, T, H) \\ (T, H, H, T), & (T, H, T, H), & (T, T, H, H), & (H, T, T, T) \\ (T, H, T, T), & (T, T, H, T), & (T, T, T, H), & (T, T, T, T) \end{cases}$$

Since X is the number of heads, for any particular outcome we can find the value of X by counting H's. For example, (H, T, H, H) gives $X = 3$. The following table lists the possible values of X and the outcomes for which it takes that value (as you should check by "counting heads").

Values of X	Outcomes
$X = 0$	(T, T, T, T)
$X = 1$	$(H, T, T, T), (T, H, T, T), (T, T, H, T), (T, T, T, H)$
$X = 2$	$(H, H, T, T), (H, T, H, T), (H, T, T, H), (T, H, H, T), (T, H, T, H), (T, T, H, H)$
$X = 3$	$(H, H, H, T), (H, H, T, H), (H, T, H, H), (T, H, H, H)$
$X = 4$	(H, H, H, H)

A random variable taking a particular value, such as $X = 2$ above, corresponds to a *subset* of the sample space, and so is an *event*. The probability of such an event is just the sum of the probabilities of the outcomes that it contains. The *probability distribution* of a random variable is the collection of the probabilities for its various values.

Random Variable

> A *random variable* X is an assignment of a number to each element in the sample space. The *probability distribution* of the random variable X is the collection of all probabilities $P(X = x)$ for each possible value x.

EXAMPLE 2   Finding a Probability Distribution

Find the probability distribution of the random variable in the Example 1.

Solution

From the table on the previous page, $X = 0$ occurs only for the outcome (T, T, T, T), which has probability $\frac{1}{16}$ (since the 16 outcomes are equally likely), so $P(X = 0) = \frac{1}{16}$. The event $X = 1$ corresponds to *four* outcomes in the table, each with probability $\frac{1}{16}$, so $P(X = 1) = \frac{4}{16} = \frac{1}{4}$. The probabilities of the other outcomes, $P(X = 2)$, $P(X = 3)$, and $P(X = 4)$, are similarly found by adding $\frac{1}{16}$ for each outcome, giving the probabilities in the following table.

x	0	1	2	3	4
$P(X = x)$	$\frac{1}{16}$	$\frac{1}{4}$	$\frac{3}{8}$	$\frac{1}{4}$	$\frac{1}{16}$

■

The probabilities sum to 1: $\frac{1}{16} + \frac{1}{4} + \frac{3}{8} + \frac{1}{4} + \frac{1}{16} = \frac{16}{16} = 1$, so the events $P(X = x)$ for the possible values x form a *partition* of the sample space. The relative sizes of these probabilities are most easily seen from a bar graph, with the height of each bar showing the probability that X takes that value.

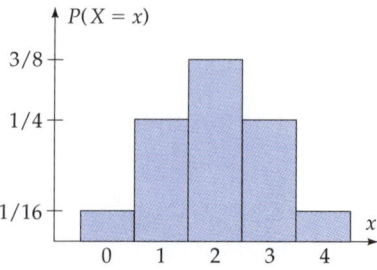

Probability distribution for the
number of heads in four tosses of a coin

The sum of the probabilities being one means that *the area under the graph is one.*

EXAMPLE 3 Sampling Without Replacement

A box contains four red balls and three green balls. Two balls are selected at random and removed from the box. Let X be the number of red balls removed. Find the probability distribution of X and graph it.

Solution

Let R_1 be the event that the first ball is red, R_2 be the event that the second ball is red, G_1 that the first ball is green, and G_2 that the second ball is green. Then $P(R_1) = \frac{4}{7}$ (since initially there are 4 red balls out of 7) and $P(R_2 \text{ given } R_1) = \frac{3}{6}$ (since after choosing a red, there are only 3 reds left out of 6). Finding the other probabilities similarly and using the conditioning formula (page 416), we obtain:

$$P(R_1 \text{ and } R_2) = P(R_2 \text{ given } R_1) \cdot P(R_1) = \frac{3}{6} \cdot \frac{4}{7} = \frac{2}{7} \qquad \text{Two red balls removed}$$

$$P(R_1 \text{ and } G_2) = P(G_2 \text{ given } R_1) \cdot P(R_1) = \frac{3}{6} \cdot \frac{4}{7} = \frac{2}{7}$$

$$P(G_1 \text{ and } R_2) = P(R_2 \text{ given } G_1) \cdot P(G_1) = \frac{4}{6} \cdot \frac{3}{7} = \frac{2}{7} \qquad \left.\begin{array}{l} \\ \\ \end{array}\right\} \text{One red ball removed}$$

$$P(G_1 \text{ and } G_2) = P(G_2 \text{ given } G_1) \cdot P(G_1) = \frac{2}{6} \cdot \frac{3}{7} = \frac{1}{7} \qquad \text{Zero red balls removed}$$

These probabilities, in view of the notations on the right, give the probabilities for X, the number of red balls removed. The probabilities are listed in the table and graphed below.

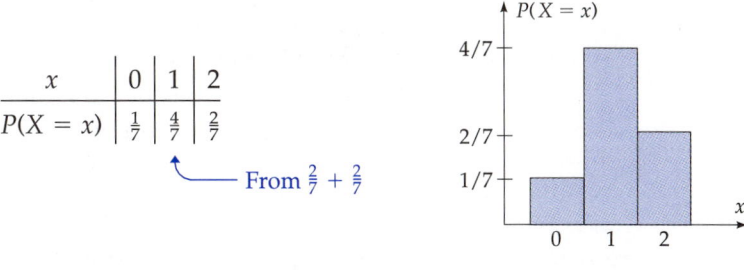

x	0	1	2
$P(X = x)$	$\frac{1}{7}$	$\frac{4}{7}$	$\frac{2}{7}$

From $\frac{2}{7} + \frac{2}{7}$

As is always the case for probability distributions, the area under the graph is one, and the sample space is *partitioned* by the values of the random variable.

EXAMPLE 4   Finding Gambling Probabilities

A die is rolled and your winnings are as follows: you win $5 if the die rolls evens, you win $1 if the die rolls a one or a three, and you lose $10 if the die comes up five. Find and graph the probability distribution of your winnings.

Solution

Let X represent your winnings, and let the sample space be $S = \{1, 2, 3, 4, 5, 6,\}$. The possible values for X are 5, 1, and -10 (indicating a loss). Since each face of the die has probability $\frac{1}{6}$, we obtain the probability distribution given in the table and graphed on the right.

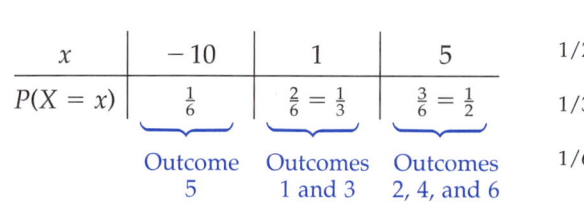

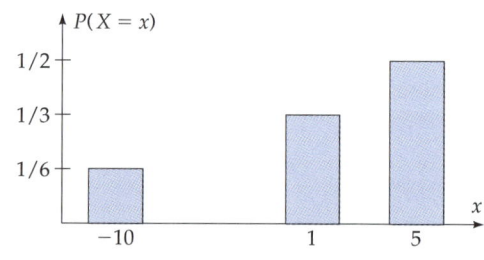

x	-10	1	5
$P(X = x)$	$\frac{1}{6}$	$\frac{2}{6} = \frac{1}{3}$	$\frac{3}{6} = \frac{1}{2}$

Outcome 5 | Outcomes 1 and 3 | Outcomes 2, 4, and 6

Expected Values

Is the game in Example 4 a *fair* game? That is, if you were to play it many times, in the long run would you expect to win money, lose money, or break even? Since you would win $5 with probability $\frac{1}{2}$, win $1 with probability $\frac{1}{3}$, and lose $10 with probability $\frac{1}{6}$ on each trial, you might reasonably expect that your *average* winnings might be:

$$(5) \cdot \tfrac{1}{2} + (1) \cdot \tfrac{1}{3} + (-10) \cdot \tfrac{1}{6} = \tfrac{15 + 2 - 10}{6} = \tfrac{7}{6}$$ Each value times its probability

Since this amount is positive, the game is *favorable* to you. If this amount were zero the game would be *fair*, and if it were negative, it would be *unfavorable* to you. For any random variable we can similarly calculate an *average* value.

Expected Value

The *expected value* $E(X)$ of a random variable X is defined as the sum of the possible values times their probabilities:

$$E(X) = \sum_{\substack{\text{All possible} \\ x\text{-values}}} x \cdot P(X = x)$$

The expected value is also called the *expectation* or the *mean* of the random variable (or of the probability distribution), and is also denoted μ.

(The Greek letter μ is pronounced "mu.") The expected value is a *weighted average* of the possible values, each value weighted by the probability that the random variable takes that value.

EXAMPLE 5   Average of the Roll of a Die

Find the expected value of one roll of a die.

Solution

The possible values are the integers 1 through 6, each with probability $\frac{1}{6}$, so:

$$E(X) = 1 \cdot \tfrac{1}{6} + 2 \cdot \tfrac{1}{6} + 3 \cdot \tfrac{1}{6} + 4 \cdot \tfrac{1}{6} + 5 \cdot \tfrac{1}{6} + 6 \cdot \tfrac{1}{6} = \tfrac{21}{6} = 3.5$$

If a random variable represents your winnings in a game, then the mean or expected value is often interpreted as the *fair price* for the game. That is, if you were to roll a die and win in dollars whatever the die showed, then a fair price for you to pay to play that game would be $3.50 since that would make your average winnings zero. Notice that the expected value need not be one of the possible values of the random variable.

EXAMPLE 6   Finding the Expected Value of a Raffle Ticket

Four hundred raffle tickets are sold for $100 each to raise money for a local children's hospital. First prize is a $3000 Florida vacation for two, second prize is a $1000 credit at the town's supermarket, and the five third prizes are $100 "dinner for two" gift certificates at a local restaurant. What is the expected value of a raffle ticket?

Solution

Let X represent the winnings of a ticket. Then $P(X = 3000) = \frac{1}{400}$, $P(X = 1000) = \frac{1}{400}$, $P(X = 100) = \frac{5}{400}$ (since there are five third prizes), and $P(X = 0) = \frac{395}{400}$ (since the rest of the tickets win nothing). The expected value is the sum of each prize times its probability:

$E(X) = 3000 \cdot \frac{1}{400} + 1000 \cdot \frac{1}{400} + 100 \cdot \frac{5}{400}$

$+ 0 \cdot \frac{395}{400} = 11.25$ Each winning times its probability

The expected value of the raffle ticket is $11.25.

This figure represents your average winnings, so for each $100 ticket that you buy, you are (on average), losing $88.75. On a more positive note, this latter figure represents your true generosity to the hospital.

PRACTICE PROBLEM 1   Suppose that the raffle prizes in Example 6 (above) are changed to one $3000 first prize, two $1000 second prizes, and ten $100 third prizes. What is the expected value of a raffle ticket now?

Solution at the back of the book

Standard Deviation

It is useful to summarize a random variable in just a few numbers to make it more comprehensible. To represent its "average" or "typical" value we use its expected value (although other choices are possible). However, its values may be tightly grouped near this mean, or widely spaced around it. For example, a random variable that is equally likely to be $+1$ or -1 is much more closely spaced around its mean of zero than a random variable that is equally likely to be $+100$ or -100, whose mean is also zero. To measure the *dispersion* or *spread* of a random variable, we calculate how far it is, on average, from its mean. However, to avoid negative deviations from canceling with positive ones, we square the deviations. This leads to the following definitions.

Variance and Standard Deviation

The *variance Var(X)* of a random variable X is defined as:

$$Var(X) = E[(X - \mu)^2] \qquad \mu = E(X)$$

The *standard deviation* σ or $\sigma(X)$ of X is the square root of the variance:

$$\sigma = \sqrt{Var(X)}$$

Thus:

$$\sigma = \sqrt{\sum_{\substack{\text{All} \\ \text{possible } x}} (x - \mu)^2 \cdot P(X = x)}$$

(The Greek letter σ is pronounced "sigma" and is the lowercase of the Greek letter Σ.) The variance is sometimes denoted σ^2. Finding the variance or standard deviation involves first finding the mean μ. The variance and standard deviation each measure the *spread* of the random variable away from its mean, but the standard deviation is easier to interpret because its square root means that its units (such as feet or seconds) are the same as the units of the original random variable.

EXAMPLE 7   Calculating a Standard Deviation

Find the standard deviation of the random variable X from Example 3.

Solution

This X is defined by the table of values and probabilities from page 429:

x	0	1	2
$P(X = x)$	$\frac{1}{7}$	$\frac{4}{7}$	$\frac{2}{7}$

First we find the mean.

$$\mu = E(X) = \sum_{\substack{\text{All} \\ \text{possible } x}} x \cdot P(X = x) = 0 \cdot \tfrac{1}{7} + 1 \cdot \tfrac{4}{7} + 2 \cdot \tfrac{2}{7} = \tfrac{8}{7}$$

Then:

$$\sigma = \sqrt{\sum_{\substack{\text{All} \\ \text{possible } x}} (x - \mu)^2 \cdot P(X = x)}$$

$$= \sqrt{(0 - \tfrac{8}{7})^2 \cdot \tfrac{1}{7} + (1 - \tfrac{8}{7})^2 \cdot \tfrac{4}{7} + (2 - \tfrac{8}{7})^2 \cdot \tfrac{2}{7}} \approx 0.639$$

■

The mean $\mu = \tfrac{8}{7}$ means that, on average, $1\tfrac{1}{7}$ red balls are removed each time. The standard deviation $\sigma = 0.639$ is harder to interpret by itself, except to say that a larger value would mean more spread, and a smaller value less spread. More on this point later in this section and in the following chapter.

Binomial Distributions

Suppose that an experiment can have only two outcomes, which we call "success" and "failure," and that success happens with probability p (for some $0 < p < 1$), so that failure occurs with probability $1 - p$. If we define a random variable to be $X = 1$ if the experiment is a success and $X = 0$ if it is a failure, then we have:

$$X = \begin{cases} 1 & \text{with probability } p \\ 0 & \text{with probability } 1 - p \end{cases}$$

$X = 1$ means "success"
$X = 0$ means "failure"

Such a random variable is called a *Bernoulli** random variable. Note that the term "success" can mean whatever we choose, such as *tossing heads,* or *catching a cold.* If such success-or-failure experiments are repeated, with successive repetitions being independent, the results are called *Bernoulli* trials. The number of successes in n Bernouilli trials is called a *binomial* random variable.

Binomial Distribution

Let X be the number of successes in n Bernoulli trials where p is the probability of success on each trial. Then X is called a *binomial random variable with parameters n and p,* and its distribution, called the *binomial distribution,* is:

$$P(X = x) = {}_nC_x\, p^x(1 - p)^{n-x} \qquad \text{for } x = 0, 1, \ldots, n$$

The mean is $\mu = np$ and the standard deviation is $\sigma = \sqrt{np(1 - p)}$.

To verify the formula for the probability distribution, recall that any sequence of n Bernoulli trials with exactly x successes (and so exactly $n - x$ failures) will, by independence, have probability $p^x(1 - p)^{n-x}$. How many such sequences are there? The successes may occur in any x of the n trials, with the other trials then being failures. But choosing the x positions (for the successes) out of a total of n can be done in exactly ${}_nC_x$ ways (see page 393). Therefore, adding ${}_nC_x$ probabilities, each of which is $p^x(1 - p)^{n-x}$, gives the binomial distribution formula in the box above. The formulas for the mean and standard deviation are proved in the Explorations and Excursions on pages 441–443.

EXAMPLE 8   A Binomial Distribution

A coin is tossed four times, and let X be the number of heads. Find the probability distribution for X.

Solution

Coin tossing is merely Bernoulli trials, so we want the probability distribution of a binomial random variable with $n = 4$ and $p = 0.5$.

---

* After the Swiss mathematician James Bernoulli (1654–1705), who first saw the importance of such random variables. His main work was *Ars Conjectandi* (*Art of Conjecturing*).

$$P(X = 0) = 1 \cdot \left(\tfrac{1}{2}\right)^0 \cdot \left(\tfrac{1}{2}\right)^4 = \tfrac{1}{16} \qquad\qquad {}_nC_xp^x(1-p)^{4-x} \text{ with } n = 4,$$
$$x = 0, \text{ and } p = \tfrac{1}{2}$$

$$P(X = 1) = 4 \cdot \left(\tfrac{1}{2}\right)^1 \cdot \left(\tfrac{1}{2}\right)^3 = \tfrac{4}{16} = \tfrac{1}{4} \qquad\qquad {}_nC_xp^x(1-p)^{4-x} \text{ with } n = 4,$$
$$x = 1, \text{ and } p = \tfrac{1}{2}$$

$$P(X = 2) = 6 \cdot \left(\tfrac{1}{2}\right)^2 \cdot \left(\tfrac{1}{2}\right)^2 = \tfrac{6}{16} = \tfrac{3}{8} \qquad\qquad {}_nC_xp^x(1-p)^{4-x} \text{ with } n = 4,$$
$$x = 2, \text{ and } p = \tfrac{1}{2}$$

$$P(X = 3) = 4 \cdot \left(\tfrac{1}{2}\right)^3 \cdot \left(\tfrac{1}{2}\right)^1 = \tfrac{4}{16} = \tfrac{1}{4} \qquad\qquad {}_nC_xp^x(1-p)^{4-x} \text{ with } n = 4,$$
$$x = 3, \text{ and } p = \tfrac{1}{2}$$

$$P(X = 4) = 1 \cdot \left(\tfrac{1}{2}\right)^4 \cdot \left(\tfrac{1}{2}\right)^0 = \tfrac{1}{16} \qquad\qquad {}_nC_xp^x(1-p)^{4-x} \text{ with } n = 4,$$
$$x = 4, \text{ and } p = \tfrac{1}{2}$$

These five probabilities make up the probability distribution X, the number of heads in four tosses. For the mean μ and standard deviation σ we use the formulas in the box above:

$$\mu = np = 4 \cdot \tfrac{1}{2} = 2 \qquad \sigma = \sqrt{npq} = \sqrt{4 \cdot \tfrac{1}{2} \cdot \tfrac{1}{2}} = 1$$

(The fact that the expected number of heads in four tosses of a fair coin is two should come as no surprise.) We found this same distribution in Example 2 (page 428). Here we found it using the binomial probability formula, and there we found it from basic principles, and the results agree. Binomial probabilities can also be found on many graphing calculators.

```
binompdf(4,.5)
(.0625 .25 .375...
Ans▶Frac
(1/16 1/4 3/8 1...
```

PRACTICE PROBLEM 2   Let X be the number of heads in six tosses of a coin. Find $P(X = 3)$ and the mean and standard deviation of X.   *Solution at the back of the book*

EXAMPLE 9 Employee Retention

A restaurant manager estimates that the probability that a newly hired waiter will still be working at the restaurant six months later is only 60%. For the five new waiters just hired, what is the probability that at least four of them will still be working at the restaurant in six months?

Solution

Assuming that the waiters decide independently of each other, they make up five Bernoulli trials. Counting a waiter who stays as a "success," the number X of waiters who stay is a binomial random variable with $n = 5$ and $p = 0.6$. We want $P(X \geq 4)$, which means $P(X = 4 \text{ or } 5) = P(X = 4) + P(X = 5)$.

$$P(X = 4) + P(X = 5) = \underbrace{{}_5C_4(0.6)^4(0.4)^1}_{5} + \underbrace{{}_5C_5(0.6)^5(0.4)^0}_{1} \approx 0.337$$

The probability that at least four of the new waiters will stay for six months is only about 0.337, or 34%.

```
binompdf(5,.6,4)
+binompdf(5,.6,5
)
              .33696
```

■

EXAMPLE 10   Money Back Guarantees

A manufacturer of computer diskettes sells them in packs of 10 with a "double your money back" guarantee if more than one diskette is defective. If each disk is defective with a 1% probability independently of the others, what proportion of packages will require refunds?

Solution

If X is the number of defective diskettes in a package, then X is a binomial random variable with $n = 10$ and $p = 0.01$. Rather than calculate the probability of a refund (2–10 defectives), we find the complementary probability:

$$P(X = 0) + P(X = 1) = \underbrace{{}_{10}C_0(0.01)^0(0.99)^{10}}_{1} + \underbrace{{}_{10}C_1(0.01)^1(0.99)^9}_{10} \approx 0.996$$

Subtracting this answer from 1 (since it is the complementary probability) gives 0.004. Therefore, the company will have to give refunds on only about 0.4% of the packages.

```
binompdf(10,.01,
0)+binompdf(10,.
01,1)
        .9957337998
1-Ans
        .0042662002
```

Graphing Calculator Exploration

The program BINOMIAL* graphs the binomial distribution for given values of n and p. For example, the following screens show the graph of the binomial probability distribution with $n = 10$ but with different values of p.

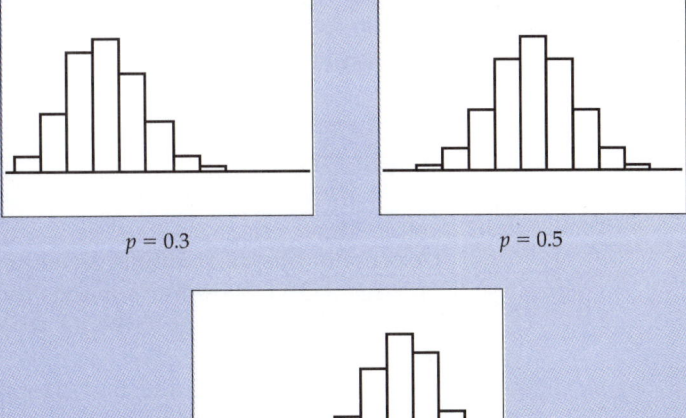

$p = 0.3$ $p = 0.5$

$p = 0.7$

Notice that the second graph is symmetric, and that values of p away from 0.5 skew the graph to one side or the other. This program also gives numerical values for the probabilities and the mean and standard deviation, and shows graphically where the mean and standard deviation fall, as in the following screen for $p = 0.5$.

---

* See the Preface for information on how to obtain this program.

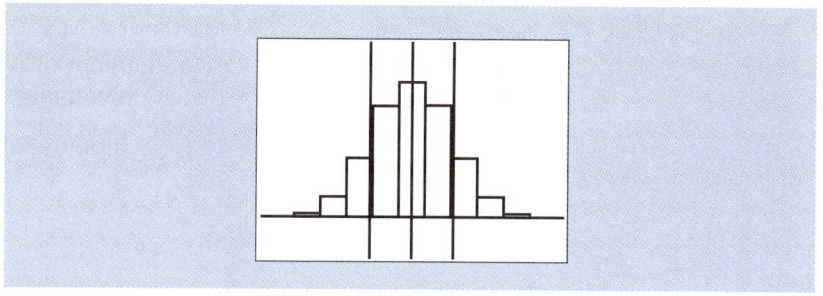

Chebyshev's Inequality

The standard deviation is designed to measure the *spread* or *dispersion* of a random variable away from its mean, but is there any general result that shows exactly how much spread a given standard deviation allows? The answer is yes, and the result is known as Chebyshev's inequality.*

Chebyshev's Inequality

The probability that the value of the random variable X is within k standard deviations on either side of the mean is at least $1 - \frac{1}{k^2}$.

(A proof of this inequality is given in the Explorations and Excursions on page 443.) Chebyshev's inequality says that for any random variable, the probability of its being within k standard deviations of the mean is at least $1 - \frac{1}{k^2}$. For example, for $k = 2, 3$, and 4 the values of $1 - \frac{1}{k^2}$ are 0.75, 0.89, and 0.94, respectively. Therefore, Chebyshev's inequality for these values of k says:

X will be within 2 standard deviations of its mean with probability at least 75%.

X will be within 3 standard deviations of its mean with probability at least 89%.

X will be within 4 standard deviations of its mean with probability at least 94%.

Chebyshev's inequality gives precision to the intuitive notion that a random variable is unlikely to be far (measured in standard deviation units) from its mean.

---

* After the Russian mathematician Pafnuti Lvovich Chebyshev (1821–1894), who first proved it.

SUMMARY

A *random variable* X is an assignment of numbers to the outcomes in the sample space so that each outcome is associated with exactly one number. The *expected value* (or *expectation* or *mean*), *variance*, and *standard deviation* of a random variable X are:

$$\mu = E(X) = \sum_{\substack{\text{All} \\ \text{possible } x}} x \cdot P(X = x)$$

$$Var(X) = \sum_{\substack{\text{All} \\ \text{possible } x}} (x - \mu)^2 \cdot P(X = x)$$

$$\sigma = \sqrt{\sum_{\substack{\text{All} \\ \text{possible } x}} (x - \mu)^2 \cdot P(X = x)}$$

A *Bernoulli* random variable takes only two values:

$$X = \begin{cases} 1 & \text{with probability } p \\ 0 & \text{with probability } 1 - p \end{cases} \qquad \begin{array}{l} \text{For "success"} \\ \text{For "failure"} \end{array}$$

Bernoulli trials are repeated independent experiments that result in "success" (with probability p) or "failure" (with probability $1 - p$). The number of successes in n Bernoulli trials is a *binomial* random variable with parameters n and p, and the *binomial probability distribution* is

$$P(X = x) = {}_nC_x \, p^x (1 - p)^{n-x} \qquad \text{for } x = 0, 1, \ldots, n$$

The mean is $\mu = np$ and the standard deviation is $\sigma = \sqrt{np(1 - p)}$.

Chebyshev's inequality states that the probability that the value of the random variable X is within k standard deviations of the mean is at least $1 - \frac{1}{k^2}$.

EXERCISES 4.4 (📈 helpful)

1. **Coins**   A coin is tossed three times, and X is the number of heads. Find and graph the probability distribution of X.

2. **Dice**   Two dice are rolled, and X is the sum of the faces. Find and graph the probability distribution of X.

3. **Coins**   You toss three coins and win $11 if they all agree (all heads or all tails), and otherwise you lose $1. Find and graph the probability distribution of your winnings.

4. **Marbles**   Two marbles are chosen at random (without replacement) from a box containing three red and five green marbles. Let X be the number of green marbles chosen. Find and graph the probability distribution of X.

5. **Dice**   A die is rolled, and you win $2 if it comes up odds, you lose $12 if it comes up 2, and you win $3 if it comes up 4 or 6. Find and graph the probability distribution of your winnings.

6. **Dice**   A die is rolled and you win $8 if it comes up 1, 3, or 5, you lose $15 if it comes up 2 or 4, and you win $6 if it comes up 6. Find and graph the probability distribution of your winnings.

7. **Dice**   Two dice are rolled, and X is the larger of the two numbers that come up. Find and graph the probability distribution of X.

8. Dice   Two dice are rolled, and X is the smaller of the two numbers that come up. Find and graph the probability distribution of X.

9. Mean   Find the mean of the random variable in Exercise 1.

10. Mean   Find the mean of the random variable in Exercise 2.

11. Mean and Standard Deviation   Find the mean and standard deviation of the random variable in Exercise 3.

12. Mean   Find the mean of the random variable in Exercise 4.

13. Mean and Standard Deviation   Find the mean and standard deviation of the random variable in Exercise 5.

14. Mean and Standard Deviation   Find the mean and standard deviation of the random variable in Exercise 6.

15. Mean   Find the mean of the random variable in Exercise 7.

16. Mean   Find the mean of the random variable in Exercise 8.

17. Raffle Tickets   One thousand raffle tickets are sold, and there is one first prize worth $2000, one second prize worth $250, and 20 third prizes worth $50. Find the expected value of a ticket.

18. Standard Deviation   Find the standard deviation of a random variable that is equally likely to be 1 or -1. Then find the standard deviation if the values are changed to 50 and -50 (still equally likely).

19. Standard Deviation   Find the standard deviation of a random variable that is equally likely to be 1 or -1. Then find the standard deviation if the values are changed to 100 and -100 (still equally likely).

20. Binomial Distribution   Find and graph the distribution of a binomial random variable with parameters $n = 4$ and $p = \frac{1}{5}$. Use the formulas on page 434 to find its mean and variance.

 Binomial Distribution   For a binomial random variable with the given parameters, find and graph

its probability distribution and use the formulas on page 434 to find its mean and variance.

21. $n = 20$ and $p = \frac{1}{2}$

22. $n = 25$ and $p = \frac{1}{2}$

23. $n = 20$ and $p = 0.8$

24. $n = 25$ and $p = 0.64$

25. Coins   For six tosses of a coin, find the probability that the number of heads is between two and four (inclusive).

26. Coins   What is the probability of getting exactly five heads in 10 tosses of a coin?

27. Coins   What is the probability of getting exactly four heads in eight tosses of a coin?

28. Coins   For an unfair coin whose probability of heads is 0.7, what is the most likely number of heads in 10 tosses, and what is its probability? What is the *least* likely number of heads, and what is its probability?

29. Coins   For an unfair coin whose probability of heads is $\frac{1}{3}$, what is the most likely number of heads in six tosses, and what is its probability? What is the *least* likely number of heads, and what is its probability?

30. Dice   In rolling a die 10 times, what is the most likely number of sixes, and what is its probability? What is the mean number of sixes?

31. Dice   In rolling a die 8 times, what is the most likely number of sixes, and what is its probability? What is the mean number of sixes?

32. Testing   You know that one of three batteries is dead. If you test them one at a time until you find the defective one, what is the expected number of tests?

33. Sales   An automobile salesperson predicts that a customer will buy a $30,000 car with probability $\frac{1}{10}$, a $25,000 car with probability $\frac{1}{5}$, a $20,000 car with probability $\frac{3}{10}$, and otherwise will buy nothing. What is the expected value of the sale?

34. Insurance   An insurance company estimates that on a typical policy it will have to pay out $10,000 with probability 0.05, $5000 with probability 0.1, and $1000 with probability 0.2. If the company wants to charge $200 more than the expected payout of the policy, what should it charge?

35. Product Quality   A company produces products of which 2% have hidden defects. If you buy 10 of them, then the number of defective ones in your purchase is a binomial random variable. Find its mean and standard deviation.

36. Multiple Choice Testing   A multiple choice test has five possible answers for each of ten questions. If a student guesses randomly:

   a. What are the mean and standard deviation of the number of right answers?
   b. What is the probability of getting four or more correct answers?

 37. ESP   A person claims to have ESP and calls 7 out of 10 tosses of a coin correctly. What is the probability of doing at least that well by guessing randomly?

 38. Juries   Suppose that on a civil court jury of 12 people it takes at least nine votes to convict. If each person decides correctly with probability 0.9, what is the probability that a guilty person is convicted? What is the probability that an innocent person is found innocent? What assumptions are you making about independence?

39. Communications   A one-digit message (0 or 1) is to be sent over a communications line that has a $\frac{1}{10}$ probability of changing the digit. Because of this, the message is to be transmitted in triplicate, 000 or 111, with decoding by *majority*, meaning that the digit that occurs two or three times will be taken to be the message. What is the probability of the message being received correctly? How does this answer compare to the probability if only *one* digit is transmitted?

40. Family Distribution   Assuming that boys and girls are equally probable, what is the probability that a family of six children consists of three boys and three girls?

 41. Cards   A bridge hand consists of 13 cards. What is the probability that you get no aces in three consecutive hands?

42. Target Practice   If the probability of hitting a target is $\frac{1}{4}$ and five shots are fired, what is the probability of hitting the target at least once?

43. Sports   Suppose that in a sports contest with no ties, the stronger team has probability $\frac{3}{5}$ of winning any particular game. What is the probability that the stronger team will win a two-out-of-three series? (*Hint:* For ease of calculation, assume that all three games are played. What assumptions about independence are you making?)

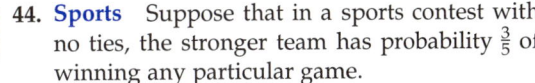 **44. Sports**   Suppose that in a sports contest with no ties, the stronger team has probability $\frac{3}{5}$ of winning any particular game.

   a. What is the probability that the stronger team will win a three-out-of-five series?
   b. A four-out-of-seven series?

   (*Hint:* For ease of calculation, assume that the entire series of games is played. What assumptions about independence are you making?)

Explorations and Excursions

The following exercises extend and augment the material presented in the text.

The mean and standard deviation of a binomial distribution   The following exercises prove the formulas $\mu = np$ and $\sigma = \sqrt{np(1 - p)}$ for binomial random variables with parameters n (number of trials) and p (probability of success on each trial). We establish these formulas by induction, first for $n = 1$, and then by showing that they hold for higher values of n by adding together independent binomial random variables to give a binomial random variable with higher values of n (since adding n independent Bernoulli trials to m Bernoulli trials simply results in $m + n$ Bernoulli trials). In what follows, X and Y are independent random variables taking values x and y.

Justify each of the following statements.

45. If X is binomial with $n = 1$, then $E(X) = 1 \cdot p + 0 \cdot (1 - p) = p$. Therefore, $\mu = np$ for $n = 1$.

46. Justify each numbered equals sign. For any independent X and Y:

$E(X + Y) \overset{1}{=} \sum_{\text{All } x, y} (x + y) \cdot P(X = x \text{ and } Y = y)$

$\overset{2}{=} \sum_{\text{All } x, y} (x + y) \cdot P(X = x) \cdot P(Y = y)$

$\overset{3}{=} \sum_{\text{All } x, y} x \cdot P(X = x) \cdot P(Y = y)$

$+ \sum_{\text{All } x, y} y \cdot P(X = x) \cdot P(Y = y)$

$\overset{4}{=} \left(\sum_{\text{All } x} x \cdot P(X = x) \right) \cdot \left(\sum_{\text{All } y} P(Y = y) \right)$

$+ \left(\sum_{\text{All } y} y \cdot P(Y = y) \right) \cdot \left(\sum_{\text{All } x} P(X = x) \right)$

$\overset{5}{=} \left(\sum_{\text{All } x} x \cdot P(X = x) \right) \cdot 1$

$+ \left(\sum_{\text{All } y} y \cdot P(Y = y) \right) \cdot 1 = E(X) + E(Y)$

That is, for independent random variables X and Y, $E(X + Y) = E(X) + E(Y)$ (in words, the expectation of the sum is the sum of the expectations).

47. If X and Y are independent binomial random variables, X with parameters n and p and Y with parameters 1 and p, then $Z = X + Y$ is binomial with parameters $n + 1$ and p, and

$$E(Z) = E(X + Y) = E(X) + E(Y)$$
$$= np + p = (n + 1)p$$

Therefore, if the formula for the mean holds for n, then it also holds for $n + 1$. This result, together with Exercise 45, shows that the formula $\mu = np$ holds for binomial random variables with *any* n and p.

48. We now prove the formula for the standard deviation. For simplicity, we show that the variance is $Var(X) = np(1 - p)$, so that then $\sigma = \sqrt{Var(X)} = \sqrt{np(1 - p)}$. We begin with X being binomial with $n = 1$ and any value of p. Then:

$$Var(X) = (1 - p)^2 \cdot p + (0 - p)^2 \cdot (1 - p)$$
$$= p(1 - p)(1 - p + p) = p(1 - p)$$

Therefore, $Var(X) = np(1 - p)$ for $n = 1$.

49. Justify each numbered equals sign. For any independent X and Y:

$Var(X + Y)$

$\overset{1}{=} \sum_{\text{All } x, y} ((x + y) - E(X + Y))^2 \cdot P(X = x \text{ and } Y = y)$

$\overset{2}{=} \sum_{\text{All } x, y} ((x - E(X)) + (y - E(Y)))^2$

$\cdot P(X = x) \cdot P(Y = y)$

$\overset{3}{=} \sum_{\text{All } x, y} ((x - E(X))^2 + 2(x - E(X)) \cdot (y - E(Y))$

$+ (y - E(Y))^2) \cdot P(X = x) \cdot P(Y = y)$

$\overset{4}{=} \sum_{\text{All } x, y} (x - E(X))^2 P(X = x) \cdot P(Y = y)$

$+ \sum_{\text{All } x, y} 2(x - E(X)) \cdot (y - E(Y))$

$\cdot P(X = x) \cdot P(Y = y)$

$+ \sum_{\text{All } x, y} (y - E(Y))^2 \cdot P(X = x) \cdot P(Y = y)$

$\overset{5}{=} \sum_{\text{All } x} (x - E(X))^2 P(X = x) \cdot \left(\sum_{\text{All } y} P(Y = y) \right)$

$+ 2 \cdot \left(\sum_{\text{All } x} (x - E(X)) \cdot P(X = x) \right)$

$\cdot \left(\sum_{\text{All } y} (y - E(Y)) \cdot P(Y = y) \right)$

$+ \sum_{\text{All } y} (y - E(Y))^2 P(Y = y) \cdot \left(\sum_{\text{All } x} P(X = x) \right)$

$\overset{6}{=} Var(X) \cdot 1 + 2 \cdot E(X - E(X)) \cdot E(Y - E(Y))$

$+ Var(Y) \cdot 1$

$\overset{7}{=} Var(X) + Var(Y)$

That is, for independent random variables, $Var(X + Y) = Var(X) + Var(Y)$ (in words, the variance of the sum is the sum of the variances).

50. If X and Y are independent binomial random variables, X with parameters n and p and Y with parameters 1 and p, then $Z = X + Y$ is binomial with parameters $n + 1$ and p, and

$$Var(Z) = Var(X + Y) = Var(X) + Var(Y)$$
$$= np(1 - p) + p(1 - p) = (n + 1)p(1 - p)$$

Therefore, if the formula for the variance holds for n, then it also holds for $n + 1$. This result, together with Exercise 48, shows that the formula for the variance, $np(1 - p)$, holds for binomial random variables with *any* values of n and p. By taking square roots, the formula for the standard deviation is established.

Proof of Chebyshev's inequality Exercises 51–55 prove Chebyshev's inequality. Justify each.

51. Justify each numbered equals sign or inequality:

$$\sigma^2 \overset{1}{=} \sum_{\text{All } x} (x - \mu)^2 P(X = x)$$

$$\overset{2}{=} \sum_{\substack{\text{Those } x \text{ with} \\ (x - \mu)^2 \,\geq\, (k\sigma)^2}} (x - \mu)^2 P(X = x)$$

$$+ \sum_{\substack{\text{Those } x \text{ with} \\ (x - \mu)^2 \,<\, (k\sigma)^2}} (x - \mu)^2 P(X = x)$$

$$\overset{3}{\geq} \sum_{\substack{\text{Those } x \text{ with} \\ (x - \mu)^2 \,\geq\, (k\sigma)^2}} (x - \mu)^2 P(X = x)$$

$$\overset{4}{\geq} (k\sigma)^2 \sum_{\substack{\text{Those } x \text{ with} \\ (x - \mu)^2 \,\geq\, (k\sigma)^2}} P(X = x)$$

52. $\sigma^2 \geq (k\sigma)^2 \, P((X - \mu)^2 \geq (k\sigma)^2)$

53. $P(|X - \mu| \geq k\sigma) \leq \dfrac{1}{k^2}$

54. $P(|X - \mu| < k\sigma) > 1 - \dfrac{1}{k^2}$

55. $P(\mu - k\sigma < X < \mu + k\sigma) > 1 - \dfrac{1}{k^2}$

56. Explain why the statement in Exercise 55 is Chebyshev's inequality.

57. Use Chebyshev's inequality to support the claim that "almost all of any probability distribution is within five standard deviations of the mean."

Chapter Summary with Hints and Suggestions

The reading and exercises of this chapter have helped you learn the following skills. For each skill, the section from which it came (in case you need to review it) and some exercises in this section that use it are indicated. Answers to all exercises are at the end of the book, and full solutions to all exercises are in the Student Solutions Manual.

4.1 Counting Techniques

- Read and interpret a Venn diagram. (*Review Exercise 1.*)

- Calculate a number of permutations or combinations. (*Review Exercise 2.*)

- Use the complementary or addition principles for counting or Venn diagrams to solve an applied problem. (*Review Exercise 3.*)

- Use the multiplication principle for counting to solve an applied problem. (*Review Exercises 4–5.*)

- Use the permutation and combination formulas (possibly using 🖩) to solve an applied problem. (*Review Exercise 6.*)

4.2 Probability Spaces

- Find an appropriate sample space for a random experiment. (*Review Exercises 7–8.*)

- Describe an event in terms of outcomes from a sample space. (*Review Exercise 9.*)

- Assign probabilities to outcomes in a sample space, and find probabilities of events. *(Review Exercises 10–14.)*

- Use the techniques of this section to find a probability in an applied problem. *(Review Exercises 15–23.)*

4.3   Conditional Probability and Independence

- Find a conditional probability using either the definition or a restricted sample space. *(Review Exercises 24–28.)*
- Use conditioning and the total probability formula to solve an applied problem. *(Review Exercises 29–31.)*
- Use Bayes' formula to solve an applied problem.   *(Review Exercises 32–34.)*
- Determine whether two events are independent or dependent.   *(Review Exercise 35.)*
- Use independence to find a probability in an applied problem.   *(Review Exercises 36–37.)*

4.4   Random Variables and Distributions

- Determine the values of a random variable for each outcome in a sample space. *(Review Exercise 38.)*
- Find and graph the probability distribution of a random variable.   *(Review Exercises 39, 41.)*
- Find the mean and standard deviation of a random variable. *(Review Exercises 40, 42, 43, 46, 47.)*
- For a binomial random variable, find and graph the probability distribution, find its mean and standard deviation, or find probabilities. *(Review Exercises 44–45.)*
- Use the binomial probability distribution to solve an applied problem. *(Review Exercises 48–50.)*

Hints and Suggestions

- Much of probability depends on counting, and there are several principles to simplify counting large numbers of objects. Roughly speak-

ing, the *complementary* principle says that you can count the *opposite* set and then subtract; the *addition* principle says that you can add numbers for two sets but you must then subtract what was double-counted; the *multiplication* principle says that the number of two-part choices is the number of first-part choices times the number of second-part choices.

- *Permutations* are arrangements where a different order means a different object (such as letters in a word or rankings of people). *Combinations* are arrangements where reordering does *not* make a new object (such as hands of cards or committees of people).

- Probabilities are assigned to possible outcomes of a random experiment so that each probability is between 0 and 1 (inclusive) and they add to 1. Equally likely outcomes should be assigned equal probabilities. Probabilities of more complicated events are found by adding up the probabilities of the outcomes in the event.

- Conditional probability is defined as the relative probability of both events compared to the probability of the given event. Conditional probabilities are sometimes more easily found from the restricted sample space than from the definition.

- A complicated probability can sometimes be found by *conditioning* on the occurrence of some underlying events (a *partition*).

- Bayes' formula is useful for finding events of the form B_1 given A in terms of events of the form A given B_i.

- Independent events have "nothing to do with each other," and the probability of both occurring is the product of their probabilities. That is not the same as *disjoint* events, where one *precludes* the other and which therefore cannot be independent.

- The mean of a random variable gives a *representative* or *typical* value, and the standard deviation measures the *spread* of its values about the mean.

- Bernoulli trials are repeated independent experiments with only two possible outcomes, success or failure. The number of successes in several Bernoulli trials is a *binomial* random variable.

- A graphing calculator that finds permutations, combinations, and binomial and other probabilities is very useful.

- Practice for test: Review Exercises 1, 2, 3, 4, 5, 7, 8, 15, 16, 17, 21, 22, 26, 29, 33, 37, 39, 40, 41, 43, 45, 46, 48, 50.

Review Exercises for Chapter 4   (*helpful*) *Practice test exercises are in blue.*

4.1 Counting Techniques

1. **Venn Diagrams** For the Venn diagram below find:

   a. $n(A)$ **b.** $n(A \cup B)$
   c. $n(B^c)$ **d.** $n(A^c \cap B)$

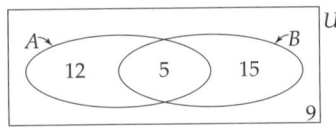

2. **Permutations and Combinations** Find:

   a. $_6P_3$ **b.** $_6C_3$

3. **Purchases** A survey of 1000 homeowners found that during the last year, 230 bought an automobile, 340 bought a major appliance, and 540 bought neither. How many homeowners bought both an automobile and a major appliance?

4. **Initials** How many three-letter initials are there if:

   a. Repeated letters are allowed?
   b. Repeated letters are not allowed?

5. **Committees** How many four-member committees can be formed from a club consisting of 20 students? What if the committee is to consist of a president, vice-president, secretary, and treasurer?

6. **Cards** How many 5-card hands are there that contain only spades?

4.2 Probability Spaces

7. **Wardrobes** Find the sample space for choosing an outfit consisting of one of two coats (C_1 or C_2), one of two scarves (S_1 or S_2), and one of two hats (H_1 or H_2).

8. **Marbles** A box contains three marbles, one blue, one yellow, and one red. A first marble is chosen, and then it is replaced and a second marble is chosen. Find the sample space. Then find the sample space if the first marble is *not* replaced before the second is chosen.

9. **Events** For the experiment of tossing a coin twice, describe each event as a subset of the sample space {(H, H), (H, T), (T, H), (T, T)}:

   a. At least one head   **b.** At most one head
   c. Different faces

10. **Probabilities of Events** For the sample space in Exercise 9:

    a. What probability should be assigned to each outcome?
    b. What is the probability of getting at least one head?
    c. What is the probability of getting different faces?

11. **Committees** A committee of 2 is to be chosen at random from a class of 15 students. What probability should be assigned to any particular committee? What if one of the two members is to be designated committee spokesperson?

Spinners Find the probability of the arrow landing in each numbered region. Assume that the probability of any region is proportional to its area, and that areas that look the same size *are* the same size.

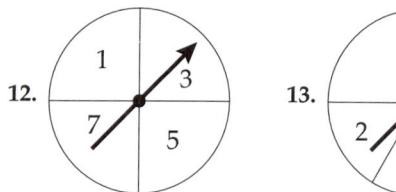

12.   13.

14. Marbles   A marble is drawn at random from a box containing 3 red, 6 green, and 9 black marbles, and the color noted. What is the probability of each outcome in the sample space {R, G, B}?

15. Smoking and Weight   For a randomly selected person, the probability of being a smoker is 0.35, the probability of being overweight is 0.40, and the probability of being both a smoker and overweight is 0.20. What is the probability of being *neither* a smoker nor overweight?

16. Dice   If you roll one die, find the probability of:

a. Rolling at most 2

b. Rolling an even number

17. Lottery   In a lottery you choose 4 numbers out of 40. Then 5 numbers are announced, and you win something if you have 4 of the 5 numbers. What is the probability that you win something?

18. Cards   You are dealt 5 cards at random from an ordinary deck. What is the probability of being dealt:

a. All face cards?   **b.** No face cards?

19. Cards   If you are dealt 13 cards at random from an ordinary deck, what is the probability of being dealt:

a. No spades?   **b.** No face cards?

20. Defective Products   You have thirty computer diskettes, and two of them contain computer viruses. If you lend five to a friend, what is the probability that they are virus-free?

21. Defective Products   A store shelf has 50 light bulbs, of which 2 are defective. If you buy 4, what is the probability that you get both defective bulbs?

22. Committees   A committee of 3 is to be formed from your class of 30. What is the probability that both you and your best friend are on it?

23. Memory   You remember the names of 6 of the 10 people you just met. If you run into 2 of the 10 on the street, what is the probability that you will remember their names?

4.3   Conditional Probability and Independence

24. If $P(A) = 0.5$, $P(B) = 0.4$, and $P(A \cap B) = 0.3$, find:

a. $P(A$ given $B)$ **b.** $P(B$ given $A)$

25. Dice   If you roll two dice, what is the probability of at least one six given that the sum of the numbers is seven?

26. Gender   If you know that a family of three children has at least one boy, what is the probability that all three are boys? Assume that boys and girls are equally likely.

27. Cards   You draw three cards at random from a standard deck. If the first is an ace, what is the probability that the other two are also aces?

28. Cards   If you are being dealt five cards, and the first four are hearts, what is the probability that the fifth is also a heart?

29. Market Share   An automobile salesperson estimates that with a male customer she can make a sale with probability 0.3, and with a female customer the probability is 0.4. If 80% of her customers are male and 20% are female, what is the probability that for a randomly selected customer she can make a sale?

30. Searches   An airplane is missing, and you estimate with probability 0.6 that it crashed in the mountains and with probability 0.4 that it crashed in the valley. If it is in the mountains then a search there will find it with probability 0.8, whereas if it is in the valley a search there will find it with probability 0.9. What is the probability that the plane will be found? (The complementary probabilities to the numbers 0.8

and 0.9 are called the *overlook* probabilities because they give the probabilities that it will *not* be found in a region if it is there.)

31. Births   Twelve percent of all births are by caesarean (surgical) section, and of these births 96% survive. Overall, 98% of all babies survive delivery. For a randomly chosen mother who did not have a caesarean, what is the probability that her baby survives?

32. Coins   You have two coins in your pocket, one with two heads, and one with a head and a tail. You choose one at random and toss it, and it comes up heads. What is the probability that it is the two-headed coin?

33. Manufacturing Defects   A soft drink bottler has three bottling machines, A, B, and C. Machine A bottles 25% of the company's output, machine B bottles 35%, and machine C bottles 40%. It is known that 4% of the bottles produced by machine A are defective, as are 3% of the bottles produced by machine B, and 2% of the bottles from machine C. If a randomly selected bottle from the factory's output is found to be defective, what is the probability that it was produced by machine A?

34. Manufacturing Defects   For the information in Exercise 33, if a randomly selected bottle is *not* defective, what is the probability that it came from machine A?

35. Coins   For the experiment of tossing a coin twice, with sample space $\{(H, H),\ (H, T),\ (T, H), (T, T)\}$, are the following events independent or dependent?

   a. *At least one head* and *at most one head*
   b. *Heads on first toss* and *same face on both tosses*

36. Computer Malfunctions   An airline reservations computer breaks down with probability 0.01, at which time a second computer takes over, but it also has probability 0.01 of failing. Find the probability that the airline will be able to take reservations.

37. Coins   Five coins are tossed. What is the probability of getting:

   a. All heads?
   b. All the same outcomes?
   c. Alternating outcomes?

4.4   Random Variables and Distributions

38. Coins   Five coins are tossed, and X is the number of heads. List the outcomes (from the usual sample space) corresponding to the events $X = 5$, $X = 1$, $X = 0$.

39. Coins   You toss two coins and win $34 if you get double heads, and otherwise you lose $2. Find and graph the probability distribution of your winnings.

40. Mean and Standard Deviation   Find the mean and standard deviation of the random variable in Exercise 39.

41. Marbles   Two marbles are chosen at random and without replacement from a box containing five red and four green marbles. If X is the number of green marbles chosen, find and graph the probability distribution of X.

42. Mean   Find the mean of the random variable in Exercise 41.

43. Raffle Tickets   Five hundred raffle tickets are sold, and there is one first prize worth $1000, two second prizes worth $150, and 10 third prizes worth $25. Find the expected value of a ticket.

44. Binomial Distribution   Find and graph the distribution of a binomial random variable with parameters $n = 4$ and $p = \frac{4}{5}$. Find its mean and standard deviation using the formulas developed for a binomial random variable.

45. Coins   For eight tosses of a coin, find the probability that the number of heads is between three and five, inclusive.

46. Sales   An television salesperson estimates that a customer will buy a $900 TV with probability $\frac{1}{10}$, a $500 TV with probability $\frac{1}{2}$, a $200 TV with probability $\frac{1}{10}$, and otherwise will buy nothing. If his commission is 10% of sales, what is his expected commission?

47. Testing   You know that one of three keys will work in a lock. If you try them one at a time until you find the right one, what is the expected number of tries?

48. Product Quality   A company produces products of which 1% have hidden defects. If you buy a dozen of them, what is the mean and standard deviation of the number of defective ones in your purchase?

 49. Product Quality   A company makes screws, of which 98% are usable. If you buy a box of 100 screws, what is the probability of at least 97 being usable?

50. Target Practice   You shoot at a target five times. If the probability of a hit on each shot is 75%, what is the probability that you hit it at least four times?

Projects and Essays

The following projects and essays are based on Chapter 4. Most have no right and wrong answers—the results depend only on your imagination and resourcefulness.

1. Research and write a page about the early history of probability, including the contributions of Pierre de Fermat, Blaise Pascal, and the Chevalier de Méré (and for more, include Gerolamo Cardano and Abraham de Moivre). Include a discussion of the Chevalier's dice questions (see pages 385, 406, 409, and 422–423), including the results of your own experiments with his dice games. (What is a chevalier anyway, and why did he have so much time to play dice?)

2. Reread the Application Preview about coincidences (pages 399–400) and then try to imagine some "one in a million" coincidences that have happened to you, like running into an old friend in a city, seeing something happen that you had dreamed of, or anything else. Then try to estimate the number of times that the coincidence did *not* occur before it actually *did* come true, and so on. Write about whether coincidences are as unlikely as they seem.

3. Imagine that you have an *unfair* coin, whose probability of heads is 0.6 and whose probability of tails is therefore 0.4. For the experiment of tossing it once, define X to be 1 for heads and 0 for tails. Find the mean and standard deviation of X. Is the standard deviation more or less than for a fair coin? Can you give an intuitive reason for this? What if the coin were even *more* unfair—how would the standard deviation change? What value of p makes the standard deviation the greatest (and the least), and why,

both mathematically and intuitively? Write a report discussing all of this.

4. Do Exercise 19 on page 425 about medical tests and false positives. Then try to find out the accuracy rate (including false positives and false negatives) for acquired immunodeficiency syndrome (AIDS) tests, especially for people in low-risk groups. Analyze the results using conditional probability. Discuss the consequences of false positives and false negatives on people's lives. Much information can be found in libraries, on the internet, or from calling AIDS information sources like the National AIDS Clearing House, many of which have 800 telephone numbers.

5. The "wallet paradox" can be described as follows: Al and Betty wonder whose wallet has more money. They agree that if they have different amounts of money in their wallets, they each have an equal chance of being the one with the larger amount. They decide to take out their wallets and look, and the person with more money gets to keep *both* amounts. Each of them agrees that if they exchange money, they have an equal chance of losing what they have or of getting *more* than they have. Therefore each feels that the game is favorable to them. However, such a game cannot be favorable to both. Write about ways to analyze, understand, and resolve this paradox.

6. There is a joke that goes as follows: A man is afraid that when he flies there will be a bomb on his plane. Therefore, whenever he flies he carries his own bomb in his suitcase because, he figures, the chances of there being *two* crazy people on the same plane with bombs is infinitesimal. Discuss in terms of probability and conditional probability.

7. If your state (or some other entity) has a lottery, find out about the probability of a ticket being a winner. Find the expected value of a ticket. Find the probability that, of 10 tickets, at least 1 is a winner. Find the probability that, of n tickets, at least one is a winner, for any value of n. Discuss two types of lotteries: one with a *fixed* number of winning tickets and the other with a *variable* number of winning tickets determined by how many chose the winning numbers. Discuss, for both types, whether different tickets have independent chances of winning. For each type, find a formula for the probability that, of n tickets, at least one is a winner. How do the two formulas compare for small and large values of n?

5

STATISTICS

5.1 Random Samples and Data Organization

5.2 Measures of Central Tendency

5.3 Measures of Variation

5.4 Normal Distributions

Surveys of just a few people can provide estimates of the result of asking everyone, and do so in less time and at lower cost. This chapter discusses the statistics of random samples.

5.1   Random Samples and Data Organization

APPLICATION PREVIEW

Roosevelt and Landon in 1936

1936 was an election year and Governor Alf Landon of Kansas was trying to unseat the incumbent President Franklin Delano Roosevelt. A highly respected magazine, *Literary Digest*, conducted a poll by sending 10 million questionnaires to names taken from telephone directories and automobile registration forms. Only a small percentage of the questionnaires were returned, but the results were overwhelming: Alf Landon was the favorite, and the magazine confidently predicted a Landon victory. The election results were equally overwhelming, but the other way: Roosevelt won, with 28 million votes to Landon's 17 million.

How could such a large poll have been so wrong? Besides the small return rate, basing the poll on the opinions of people who had telephones or owned automobiles during the depths of the Great Depression yielded a highly *unrepresentative* sample, one decidedly skewed toward the wealthy. Today we understand more about the need to use *random samples*, which is one of the subjects of this section.

Introduction

In life you have to make decisions based on incomplete information. For example, to find out how many Americans watch a television program, you can't ask everyone, and to find out how long a lightbulb lasts, a company can't test all of its bulbs, particularly if it wants to have any left to sell. These and many other "real world" problems depend upon taking samples and analyzing the data. Statistics is the branch of mathematics concerned with the collection, organization, analysis, and interpretation of numerical data.

Random Samples

A *statistical population* is the entire collection of whatever you are studying, such as the ages of citizens of the United States, the size of bank accounts in Illinois, or the brands of automobiles in California. While it is possible in principle to measure the entire population, the time, cost, or intrusiveness of such an undertaking usually dictates that we measure only part of the population. For that purpose we use a *random sample* to represent the population.

Random Sample

> A (simple) *random sample* of size n is a selection of n members of the population satisfying two requirements:
>
> 1. Every member of the population is equally likely to be included in the sample; and
> 2. Every possible sample of size n from the population is equally likely to be chosen.

A sample not meeting these criteria is not representative and cannot be used to infer characteristics of the entire population. For example, a telephone survey to homes at 10 A.M. on a weekday cannot produce valid information about work skills. Random samples can be chosen using random number tables or by other methods, and what we say in this chapter is based on the assumption that the data we are analyzing is from a random sample.

Levels of Measurement

We shall consider only numerical data, which we classify into four types: *nominal, ordinal, interval,* and *ratio* data.

Nominal data (*nominal* means *in name only*) means numbers used only to *identify* objects. For example, the numbers on the backs of football jerseys are nominal data, since they do not mean that "number 24" can run twice as fast or throw twice as far as "number 12."

Ordinal data (the word coming from *order*) means numbers that can be arranged in order, but differences between them do not have significance. For example, you might rate the teachers you have had in college as "best" (number 1), "second best" (number 2), and so on. However, the difference between "1" and "2" may not be the same as the difference between "7" and "8."

Interval data means numbers that can be put in order *and* differences between them can be compared, but ratios cannot. For example, it makes sense to talk about the *difference* between two temperatures, like 40° and 80°, but it makes no sense to say that 80° is *twice* as hot as 40°, since the zero point in the Fahrenheit (or centigrade) system is arbitrary and does not mean *zero heat*.

Ratio data means numbers that can be put in order, whose differences can be compared, *and* whose *ratios* have meaning. Examples of ratio data are amounts of money and weights, since 80 of either really *is* twice 40. In general, ratio data means that the zero level *does* mean "none" of something.

When we refer to the *type* of a data collection, we mean the highest of these levels that applies, from nominal (lowest) to ratio (highest).

EXAMPLE 1   Judging Levels of Data Measurement

Identify the level of measurement of each data collection.

a.   Ten vehicles passing through a turnpike toll booth were classified as 1 for a car, 2 for a bus, and 3 for a truck, giving data {1, 3, 2, 3, 1, 1, 1, 1, 3, 1}.

b.   A rock concert enthusiast rates the last eight performances she attended on a scale from 1 (awful) to 10 (awesome) as {8, 5, 7, 9, 7, 10, 6, 9}.

c.   The temperatures of five students at the college infirmary are {99.3, 102.1, 101.8, 101.5, 100.9} degrees Fahrenheit.

d.   The weights of twelve members of the cross-country team are {124, 151, 132, 153, 142, 147, 120, 127, 154, 119, 118, 116} pounds.

Solution

a.   *Nominal data:* The numbers are used only to identify the type of vehicle. The number 2 does not mean more or less than the number 3, so differences between these numbers have no meaning.

b.   *Ordinal data:* The rankings put the performances in relative order but no meaning can be given to differences between the rankings. The difference between ratings 5 and 6 may be more or less than the difference between ratings 9 and 10.

c.   *Interval data:* The temperatures can be put in order from "lowest" to "highest" and differences between them can be compared. However, this is not ratio data because the zero point in temperature is arbitrary—101° is not 1% hotter than 100°.

d.   *Ratio data:* The weights can be put in order, "10 pounds heavier" has meaning, *and* 200 pounds really is twice as heavy as 100 pounds.

■

PRACTICE PROBLEM 1   Before they left the final examination, a random sample of students were asked to complete a course evaluation form asking for the following information: (a) gender: 1—male, 2—female; (b) course rating: 1—poor, 2—acceptable, 3—good, 4—outstanding; and (c) current grade point average. Identify the level of measurement of the data gathered from each question.   *Solution at the back of the book*

Bar Chart

A *bar chart* provides a visual summary of data that contains just a few different values. The number of times each value appears in the data corresponds to the length of the bar for that value. The bars may be drawn vertically or horizontally. A bar chart is appropriate for *any* level of data measurement.

EXAMPLE 2   Constructing a Bar Chart

The colors of a random sample of thirty cars in the Country Corners Mall parking lot were recorded using: 1—white, 2—blue, 3—green, 4—yellow, 5—brown, and 6—black. Construct a bar chart for the data {3, 5, 2, 4, 1, 5, 1, 1, 1, 2, 4, 1, 6, 1, 6, 5, 6, 3, 1, 1, 1, 1, 1, 6, 3, 1, 2, 2, 4, 2}.

Solution

We first tally the frequency of each color and then sketch the bar chart:

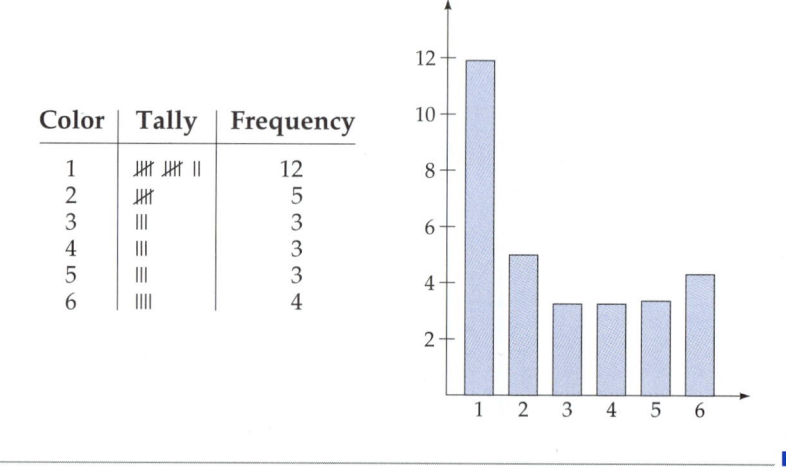

Color	Tally	Frequency
1	JHT JHT II	12
2	JHT	5
3	III	3
4	III	3
5	III	3
6	IIII	4

PRACTICE PROBLEM 2   Construct a bar chart for the data given in Example 1(a) (page 453).
          Solution at the back of the book

Stem-and-Leaf Display

A *stem-and-leaf display* is a quick way of visualizing data when the values consist of at least two digits. For each number, the "stem" is the leading digit and the "leaf" consists of the remaining digits. Thus 34 would have a stem of 3 and a leaf of 4, while in a different collection of data 1.5 would have a stem of 1 and a leaf of 5. The numbers are then listed by row, with a different stem beginning each row. No attempt is made to put the leaves in order—they are left in the order they came in. A stem-and-leaf display is appropriate for levels of measurement higher than the nominal level.

EXAMPLE 3   Constructing a Stem-and-Leaf Display

Construct a stem-and-leaf display of the data {34, 10, 53, 50, 80, 38, 39, 31, 52, 41, 46, 46, 41, 69, 73, 57, 40, 52, 47, 68, 22, 33, 51, 65, 23, 47, 64, 45, 26, 74}.

Solution

Since the values lie between 0 and 100, we choose the tens digit for the stem and the units digit for the leaf. Filling in the display in the order of the values, we obtain:

Stem	Leaf	
0		
1	0	
2	2, 3, 6	
3	4, 8, 9, 1, 3	← 34 is the first number entered
4	1, 6, 6, 1, 0, 7, 7, 5	
5	3, 0, 2, 7, 2, 1	← 53 is the third number entered
6	9, 8, 5, 4	
7	3, 4	← 74 is the last number entered
8	0	
9		

This stem-and-leaf display shows that the most frequent values are those in the forties, with the values ranging from the tens to the eighties.

PRACTICE PROBLEM 3   Construct a stem-and-leaf display for the data {7.8, 6.2, 5.4, 6.0, 6.4, 4.6, 6.2, 5.6, 3.7, 5.7, 2.5, 6.7, 4.4, 7.6, 8.1}.   *Solution at the back of the book*

Histogram

While a *histogram* is similar to a bar chart and to a stem-and-leaf display, now the *width* of each rectangle has meaning. The data is divided into *classes,* usually numbering from five to fifteen, one for each bar, and each class has the same *width*:

$$\begin{pmatrix} \text{Class} \\ \text{width} \end{pmatrix} \approx \frac{\begin{pmatrix} \text{Largest} \\ \text{data value} \end{pmatrix} - \begin{pmatrix} \text{Smallest} \\ \text{data value} \end{pmatrix}}{\begin{pmatrix} \text{Number} \\ \text{of classes} \end{pmatrix}}$$

Round the class width *up* so that all data will be covered

We then make a tally of the number of data values falling into each class. Some histograms use the convention that any value on the boundary between two classes belongs to the upper class, while others are designed so that no data value falls on a class boundary. The histogram is then drawn in the same way as a bar chart but with the sides of the rectangles touching. It is often helpful to make a quick stem-and-leaf display before constructing a histogram. A histogram is appropriate for both interval and ratio levels of data measurement.

EXAMPLE 4   Constructing a Histogram

Construct a histogram for the data values from Example 3 (page 455).

Solution

Since we know from the stem-and-leaf display of this data that the smallest value is 10 and the largest is 80, if we choose 6 classes, we obtain a class width of $\frac{80-10}{6} \approx 11.7$, which we round *up* to 12. Since we rounded up, we may choose our first class to start just below the lowest data point of 10, say at 9.5. We then add successively the class width of 12 to get the other class boundaries.

Class boundaries	Tally	Frequency
9.5–21.5	I	1
21.5–33.5	JHT	5
33.5–45.5	JHT II	7
45.5–57.5	JHT JHT	10
57.5–69.5	IIII	4
69.5–81.5	III	3

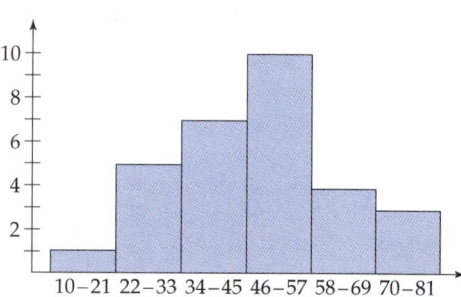

The base of each rectangle may be labeled with either the range of data values in that class or the class boundaries.

Graphing Calculator Exploration

Histograms can be drawn on graphing calculators using the STAT PLOT command. To explore several histograms of the data from Example 3 (page 455), proceed as follows.

a. Enter the data values as a list and store it in the list L_1.

```
{34,10,53,50,80,
38,39,31,52,41,4
6,46,41,69,73,57
,40,52,47,68,22,
33,51,65,23,47,6
4,45,26,74}→L₁
```

b. Turn off the axes in the window FORMAT menu, turn on STAT PLOT 1, select the histogram icon with Xlist L_1, and set the WINDOW parameters Xmin = 9.5, Xmax = 81.5, Xscl = 12 to match the start, finish, and class width we used in Example 4 (page 456).

c. GRAPH and then TRACE to explore your histogram.

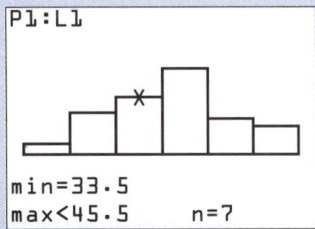

Of course, different choices of the class width, number of classes, and starting value will result in slightly different histograms. The following are several possibilities:

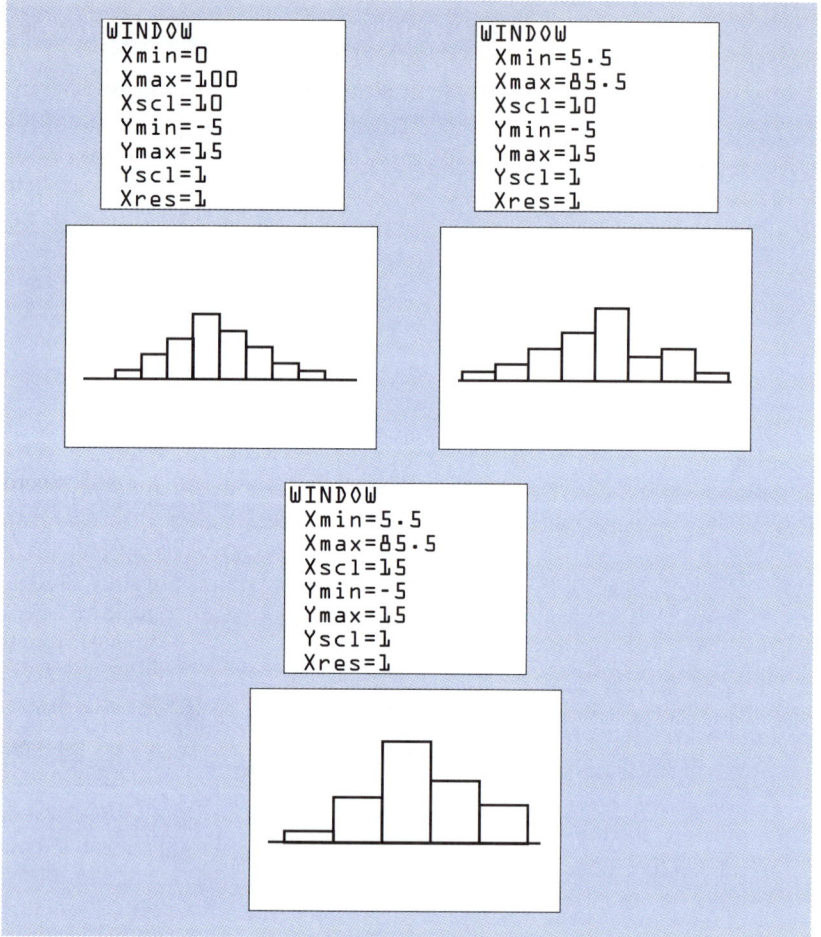

SUMMARY

A *random sample* of a statistical population requires that every member of the population is equally likely to be included, and every possible sample of the same size from the population is equally likely to be chosen.

There are four levels of data measurement:

Level of Measurement	Suitable Calculation
Nominal	Classify by name
Ordinal	Compare rank
Interval	Compare rank and differences
Ratio	Compare rank, differences, and ratios

Data of all levels of measurement may be displayed in a *bar chart* with the length of a bar indicating the frequency of the corresponding data value. A *stem-and-leaf display* shows the number of data values having the same stem, and is appropriate for data of at least the ordinal level. A *histogram* groups data values into classes and shows the frequency of values in each class, and is appropriate for data of at least the interval level.

EXERCISES 5.1

Identify the level of data measurement in each situation.

1. **Discount Brokers** A survey of ten randomly selected traders at a discount broker service records the numbers of transactions involving stocks, bonds, and mutual funds last Wednesday.

2. **Travel Agencies** An airline surveys twenty randomly selected travel agencies in and around San Francisco and records the numbers of tickets sold at each last month to Hawaii, Alaska, Mexico, and New York.

3. **Mortgage Banking** The home office of the First & National Bank sends questionnaires to two hundred randomly selected families who applied for mortgages last summer, asking for ratings (from "poor" to "outstanding") of the courtesy of the mortgage staff at the branch offices.

4. **Shoe Sales** The national personnel manager for the Feets-R-Us shoe store franchise reviews the supervisor reports on thirty randomly selected sales employees and counts the number of "below average," "average," and "above average" performance evaluations.

5. **SAT Scores** The test results on the verbal SAT administered by the Educational Testing Service are numerical scores from 200 to 800. A survey records the verbal SAT scores of twelve randomly selected seniors at J. T. Marshall High School.

6. **Midwest Aviation** A survey records the years of birth of fifty randomly selected commercial pilots residing in Missouri.

7. **IQ Scores** An educational research team working on a grant from the National Institute of Health records the IQ scores of eighty randomly selected fifth graders in the Baltimore County Public School system.

8. **Quality Control** A major cereal manufacturer randomly selects thirty boxes of Corn Chex Crunches from the shipping warehouse and records the actual net weights.

9. **Disposable Income** The Chamber of Commerce tabulates the disposable family incomes of ninety randomly selected households in Thomsonville.

10. **Financial Trends** The *Get Rich Slowly* investment newsletter reports the price-to-earnings ratios of sixty randomly selected technology stocks traded last month on the American Stock Exchange.

Construct a bar chart of the data in each situation.

11. **Marital Status** A random sample of shoppers in the Kingston Valley Mall counted 39 "never married," 97 "married," 16 "widowed," and 8 "divorced" individuals.

12. **Place of Birth** A random sample of skiers in Grand Junction last February counted 12 from Colorado, 27 from California, 19 from New York, 8 from Florida, and 6 from other states or nations.

13. **State Legislators** A random sample of candidates running for election to the Nevada legislature counted 4 business executives, 7 real estate developers, 18 lawyers, 1 doctor, 3 certified public accountants, and 2 retirees.

14. Voter Concerns A random sample of voters leaving the polls identified the most important issue in the Higginsville school board election as 8 for teacher tenure, 54 for budget, 11 for after-school day care, 19 for the bond proposal for a new football stand at the high school, and 3 making "no comment."

Construct a stem-and-leaf display of the data in each situation.

15. Carry-on Baggage At the Logan County Airport, a random sample of thirty passengers passing through the security checkpoint found the following total weights (in pounds) of their carry-on baggage:

```
22  29  23  34  27   0  21  38   9  35
48  29  18  32  32  42  29  50  25  19
43  12   0  27  25  21  21  25  24   0
```

16. Textbook Prices A random sample of eighteen college textbooks at the Student Book Exchange found the following prices (in dollars):

```
67  80  91  35  51  67  49  56  86
52  56  29  51  73  60  52  52  68
```

17. Starting Salaries A random sample of twenty-four recent graduates from the Gainsburg College School of Business found the following starting salaries (in thousands of dollars):

```
31  48  22  41  34  28  26  26
42  33  37  24  38  46  34  54
33  36  34  35  32  37  35  37
```

18. Life Expectancy At the Peaceful Valley Memorial Gardens cemetery, a random sample of forty headstones found the following ages at death (in years):

```
53  25  47  54  49  48  71  47  31  51
66  76  71  52  52  35  32  56  61  72
51   9  71  38  64  48  55  35  53  47
73  50  44  67  57  54  38  95  24  58
```

19. Car Speeds A random sample of thirty-two cars passing through a radar trap on I-95 in southern Georgia found the following speeds (in miles per hour):

```
52  59  68  76  74  46  66  65
40  70  49  52  42  76  53  55
56  74  60  56  78  66  59  77
61  54  52  71  58  63  75  59
```

20. Gasoline Consumption A random sample of twenty customers at Jack's Pump & Go Express Service found the following weekly gasoline usages (in gallons) for commuters to Evansville:

```
36  26  50  46  33  29  37  31  37  35
20  28  19  17  39  45  49  36  35  28
```

Construct a histogram of the data in each situation using an appropriate class width and starting class lower boundary.

You may use a graphing calculator, if your instructor permits it.

21. Shipping Weights A random sample of twenty packages shipped from the Holiday Values Corporation mail order center found the following weights (in ounces). (Use six classes of width 8 starting at 12.5.)

```
29  21  60  23  39  24  33  49  37  57
13  27  56  21  30  43  34  24  58  53
```

22. Blue Book Values A random sample of thirty-two cars parked at Emmett's Field last June found the following "blue book" values (in thousands of dollars). (Use six classes of width 4 starting at 2.5.)

```
15   5   8  17  18  17   8   9
 5  13   6   9   5  19   8  13
20   9   7  13  25  16   5  25
25  17  24  10  23  14  15   4
```

23. Web Sites A random sample of twenty-eight student accounts at the campus Academic Computing Center showed the following number of web site "hits" between 9 and 12 P.M. last Wednesday. (Use eight classes of width 10 starting at 10.5.)

```
38  29  16  35  69  63  13
58  11  58  34  73  45  57
70  16  45  37  39  88  69
47  30  40  12  47  20  61
```

24. Football Players A random sample of twenty-four defensive linemen in the East Coast Division II college conference found the following weights (in pounds). (Use eight classes of width 15 starting at 140.5.)

```
173  171  211  169  258  213  143  236
191  163  192  235  196  208  157  221
174  200  195  198  228  152  188  201
```

25. Airline Tickets   A random sample of forty passengers traveling economy class from New York to Los Angeles last Thanksgiving weekend found the following one-way ticket cost (in dollars):

425	236	481	230	258	473	423	480
249	451	244	440	240	565	445	569
255	515	412	377	453	425	490	301
379	439	229	422	465	252	219	237
565	477	440	227	518	232	230	437

26. Insurance Claims   A random sample of thirty accident repair claims received in December at the western division office of the Some States Insurance Company were for the following dollar amounts:

5197	1018	1859	6207	3497	1097
4785	1253	3652	6279	3639	4553
6249	1250	6601	1193	5961	4625
1308	4289	2948	1272	1329	5483
968	5297	1423	5159	5527	4543

27. Law Practice   A random sample of twenty-one attorneys hired within the past year at the U.S. Department of Justice found the following numbers of hours worked last week:

48	44	50	49	44	45	51
50	57	36	53	52	59	54
60	52	52	39	47	49	57

28. Lightbulbs   A random sample of thirty-six 75-watt lightbulbs found the following lifetimes (number of hours until burnout):

1070	1210	1280	1230	1340	1280
1200	1120	1310	1100	1170	1000
1290	1420	1190	1280	1160	1290
1250	1130	1180	1150	1110	1200
1230	1160	1240	1250	1230	1240
1000	1280	1180	1280	1270	1210

29. Medical Insurance   A random sample of twenty private practice physicians in the Lehigh Valley found the following numbers of insurance plans accepted by each:

16	21	9	23	19	23	20	23	18	19
19	29	20	21	20	23	20	14	11	19

30. Dining Out   A random sample of forty dinner checks at the Lotus East Chinese Restaurant last Saturday evening showed the following total costs (rounded to the nearest dollar):

43	104	45	60	25	103	33	96
108	41	67	53	43	48	99	45
35	40	66	103	27	110	60	33
46	126	112	84	42	90	74	46
102	41	48	41	54	44	82	33

5.2 Measures of Central Tendency

Introduction

We often hear of averages, from grade point averages to batting averages. Given a collection of data, we often use the *average* as a representative or typical value. In this section we will see that not all types of data are appropriately summarized by an average, and sometimes another kind of "typical" value is more representative. To avoid confusion with casual meanings of "average," these different kinds of typical values are called *measures of central tendency*.

Mode

The *mode* of a collection of values is the *most frequently occurring* value. If all of the values in a data set occur the same number of times, there is no mode, whereas if only a few of the values occur the same maximal number of times, there are several modes. The mode is an appropriate measure of central tendency for every level of data measurement.

EXAMPLE 1 **Finding Modes**

Find the mode of each collection of values.

a. {2, 2, 2, 3, 3, 3, 4, 4, 4}

b. {1, 2, 3, 3, 3, 3, 4, 5, 6, 7, 7, 7, 7, 8, 9, 9, 10}

c. {1, 2, 3, 3, 4, 4, 4, 5, 5, 6, 6, 6, 6, 7, 8, 8, 9, 50}

Solution

a. Since each value occurs the same number of times, there is no mode.

b. Since both 3 and 7 occur four times and every other value occurs fewer times, the modes are 3 and 7. A collection of values with two modes is said to be *bimodal.*

c. Since 6 occurs more often than any other value, the mode is 6. A collection of values with just one mode is said to be *unimodal.* Notice that the presence of one value that is very different from the others, in this case the 50, has no effect on the value of the mode. If the 50 were replaced by a 100, a 1000, or even a 1,000,000, the mode would still remain unchanged.

■

PRACTICE PROBLEM 1   Find the mode of the values {17, 20, 20, 13, 10, 17, 10, 20}.

Solution at the back of the book

Graphing Calculator Exploration

When finding the mode of a long list of values, it is often easier to first sort the list so that all the same values appear grouped within the list. For instance, to find the mode of {62, 56, 52, 53, 58, 64, 68, 67, 60, 61, 57, 58, 57, 63, 55}, proceed as follows.

a. Enter the data values as a list and STORE it in the list L_1 (or enter the numbers into L_1 directly using STAT and EDIT).

```
{62,56,52,53,58,
64,68,67,60,61,5
7,58,57,63,55}→L
1
```

b.   From the STAT or LIST menu, select SortA (to sort in *ascending* order). (You could also use SortD for descending order.)

```
EDIT CALC TESTS
1:Edit...
2:SortA(
3:SortD(
4:ClrList
5:SetUpEditor
```

c.   Complete the ascending sort command by entering the name of the list of values, L_1.

```
{62,56,52,53,58,
64,68,67,60,61,5
7,58,57,63,55}→L
1
{62 56 52 53 58...

SortA(L1)
```

d.   Examine the sorted list L_1 to identify the value(s) appearing most often. The modes of this collection of values are 57 and 58 because they appear twice and the others appear only once each.

```
SortA(L1)
            Done
L1
{52 53 55 56 57...
...57 58 58 60 61...
...62 63 64 67 68}
```

Median

The *median* is the *middle* value of a list of data values when sorted in ascending or descending order. If there is an even number of data values, the median is the number halfway between the two middle values. Since the median depends on arranging the values from "least" to "most," the median is not appropriate for nominal data. For instance, it would be meaningless to sort the data for car colors from Example 2 (page 454) from 1—white to 6—black, and find a "median color", since the assignment of numbers to colors was arbitrary. The median is an appropriate measure of central tendency for every level of data measurement except the nominal level.

EXAMPLE 2   Finding Medians

Find the median of each data set. Assume that each represents at least ordinal data.

a.   {1, 3, 4, 7, 11, 18, 39}

b.   {1, 3, 4, 7, 11, 18, 1000}

c.   {26, 16, 10, 6, 4, 2}

d.   {62, 56, 52, 53, 58, 64, 68, 67, 60, 61, 57, 58, 57, 63, 55}

Solution

a. Since these seven numbers are arranged in ascending order, the median is the fourth number, 7, because three numbers are smaller and three are larger.

b. Just as in (a), the median is 7. Notice that replacing the largest data value in the previous data set with a much larger value did not change the median.

c. Since these six values are arranged in descending order, the median is 8, the number halfway between the two middle values of 10 and 6. Thus the median need not be one of the data values.

d. Since these fifteen values are not in ascending or descending order, we first order them (as in the Graphing Calculator Exploration on pages 462–463):

$$\{52, 53, 55, 56, 57, 57, 58, 58, 60, 61, 62, 63, 64, 67, 68\}$$

The median of these is the middle value, 58, since there are seven data values no larger and seven values no smaller.

■

PRACTICE PROBLEM 2   Find the median of the following values representing nominal data: $\{4, 5, 6, 6, 7, 8\}$.   *Solution at the back of the book*

 Graphing Calculator Exploration

The median of an unsorted list of values can be found with a single calculator command. To find the median of $\{62, 56, 52, 53, 58, 64, 68, 67, 60, 61, 57, 58, 57, 63, 55\}$, the unordered values from Example 2(d) above, proceed as follows.

Enter the data values as a list and store it in the list L_1. From the LIST MATH menu, select the median command and apply it to the list L_1.

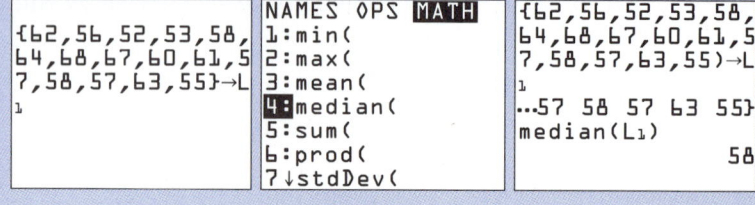

Mean

The mean is the usual arithmetic average found by summing the values and dividing by the number of them:

Mean

The *mean* $\bar{x}$ of the n values $x_1, x_2, \ldots, x_n$ is

$$\bar{x} = \frac{1}{n}(x_1 + x_2 + \cdots + x_n) = \frac{1}{n} \sum_{\text{All values}} x_k$$

The mean of a collection of numbers is equivalent to the mean of a random variable that is equally likely to be any of the numbers. The mean of a data set is sometimes called the *sample mean* to distinguish it from the concept of the mean μ of a probability distribution. Calculating the mean assumes that the distance between values is meaningful, so the mean is not an appropriate measure for ordinal or nominal data. For instance, it would be meaningless to find the mean of the concert ratings in Example 1(b) (page 453), from 1—awful to 10—awesome, because we do not know if a 6 and an 8 should average to a 7, since the intervals between them may not be the same. The mean is an appropriate measure of central tendency only for interval and ratio data.

EXAMPLE 3 Finding Means

Find the mean of each data set. Assume that each represents at least ordinal data.

a. {1, 3, 4, 7, 11, 18, 39}

b. {1, 3, 4, 7, 11, 18, 1000}

c. {62, 56, 52, 53, 58, 64, 68, 67, 60, 61, 57, 58, 57, 63, 55}

Solution

a. $\bar{x} = \frac{1}{7}(1 + 3 + 4 + 7 + 11 + 18 + 39) = \frac{1}{7}(83) = 11\frac{5}{7}.$

b. $\bar{x} = \frac{1}{7}(1 + 3 + 4 + 7 + 11 + 18 + 1000) = \frac{1}{7}(1044) = 149\frac{1}{7}.$
   Notice that the values are the same as in part (a) except that the last number changed from 39 to 1000, with the result that the mean increased substantially. Thus, unlike the median and the mode, the mean is sensitive to changes in just one of the values.

c. There is no need to arrange the values in any particular order when finding the mean:

$$\bar{x} = \tfrac{1}{15}(62 + 56 + 52 + 53 + 58 + 64 + 68 + 67 + 60 + 61 + 57$$
$$+ 58 + 57 + 63 + 55) = \tfrac{1}{15}(891) = 59\tfrac{2}{5}$$

Notice that the mean is usually not one of the original data values.

EXAMPLE 4   Baseball Salaries

In the 1995 baseball strike it was disclosed that the mean salary of major league baseball players was $1.2 million, while the median salary was only $500,000. What does this say about the distribution of baseball salaries?

Solution

If the *middle* salary is $500,000, then for the mean to be higher there must be a few very large salaries. That is, a few baseball players must be receiving enormously high salaries to skew the mean that much. In such cases, the median is usually considered a more representative value for most players than the mean.

PRACTICE PROBLEM 3

Find the mode, median, and mean of the values {1, 1, 2, 3, 8} representing ratio data.

Solution at the back of the book

 ## Graphing Calculator Exploration

Like the median, the mean of a list of values can be found with a single calculator command. To find the mean of {62, 56, 52, 53, 58, 64, 68, 67, 60, 61, 57, 58, 57, 63, 55}, the values from Examples 2(d) and 3(c) above, enter the data values as a list and store it in the list L_1. From the LIST MATH menu, select the mean command and apply it to the list L_1.

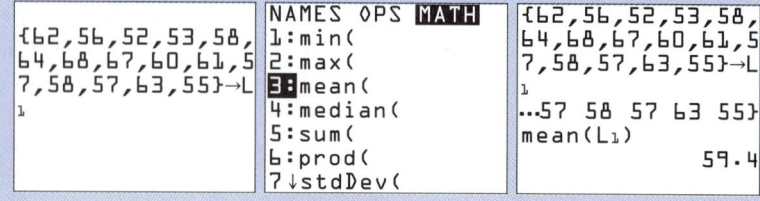

SUMMARY

Although there are three measures of central tendency, not all of them can be used with every type of data. For nominal data only the mode is appropriate, for ordinal data either the mode or the median is appropriate, and for interval or ratio data all three are appropriate.

Measure of Central Tendency	Definition	Appropriate Data Levels
Mode	Most frequent	All levels
Median	Middle value	Ordinal, interval, and ratio
Mean	$\bar{x} = \dfrac{1}{n} \displaystyle\sum_{\text{All values}} x_k$	Interval and ratio

The mean is sensitive to changes in the extremes, but the median and mode are not. The mean of a collection of numbers is equivalent to the mean of a random variable that is equally likely to be any of the numbers. The mean of a data set is sometimes called the *sample mean* to distinguish it from the concept of the mean of a probability distribution.

EXERCISES 5.2

(helpful throughout.)

Find the mode, median, and mean of each (ratio) data set.

1. {6, 17, 12, 10, 15, 16, 5, 8, 9, 14, 9}

2. {10, 7, 9, 13, 11, 14, 15, 14, 8, 9}

3. {15, 14, 5, 7, 5, 14, 9, 7, 7, 14, 7, 14, 12}

4. {11, 19, 20, 11, 12, 18, 11, 20, 16, 18, 17, 10, 13, 11, 15}

5. {19, 11, 10, 19, 18, 12, 19, 10, 14, 12, 13, 17, 20, 19, 15}

6. {12, 5, 10, 6, 7, 6, 15, 7, 6, 5, 6, 5, 13, 10, 7, 13, 8, 11, 12}

7. {8, 15, 10, 14, 13, 14, 5, 10, 7, 9, 13, 14, 15, 14, 15, 19, 13, 12, 8}

8. {11, 8, 23, 21, 14, 16, 15, 11, 9, 22, 15, 11, 12, 11, 10, 15, 10, 25, 18, 6, 11}

9. {19, 9, 15, 8, 18, 15, 12, 22, 16, 19, 15, 19, 20, 24, 7, 14, 21, 19, 20, 5, 19}

10. {10, 13, 8, 15, 8, 12, 12, 12, 14, 8, 12, 11, 14, 9, 8, 14, 14, 14, 10, 10}

APPLIED EXERCISES

Find the mode, median, and mean for the data in each situation.

11. **Emergency Services** A random sample from the records of the Farmington Volunteer Ambulance Corps found the following twenty response times (in minutes) to 911 calls received during February:

14	9	25	14	12	16	5	20	20	25
20	24	12	7	19	12	6	13	24	18

12. **Car Sales** A random sample of fifteen General Motors salesmen at Tri-State Dealers found the following earnings (in dollars) for the first week of September:

1200	1200	1100	1300	900
600	800	700	1200	500
1000	1200	600	900	900

13. Volunteer Recycling   A random sample of thirty volunteers helping sort newspapers, aluminum, and glass at the Parkerville Regional Recovery Center found the following hours contributed last week:

10	7	13	11	7	11	10	5	9	14
14	12	12	5	10	12	19	10	7	10
9	21	13	10	7	18	10	15	10	12

14. Commuting Times   A random sample of twenty-four employees at the Grover-Smith Forge found the following morning commute times (in minutes) for last Monday:

75	60	50	55	90	53	65	71
33	81	60	74	71	75	83	53
66	75	47	36	77	54	67	65

15. Unemployment   A random sample of nineteen newly unemployed workers in Wilmington last August reported the following times (in weeks) until they found new jobs:

12	6	9	9	11	14	8	15	12	13
9	8	9	7	9	5	6	13	15	

Explorations and Excursions

The following problems extend and augment the material presented in the text.

Mean of Grouped Data   Sometimes the original data are not available and only the class intervals and frequencies are known. The mean of these grouped data may be estimated by summing each midpoint value at the center of each class interval as many times as is indicated by the frequency and then dividing that total by the sum of the frequencies.

Verify the calculation of each mean of grouped data.

16. For the grouped data:

Class interval	14.5–19.5	19.5–24.5	24.5–29.5
Frequency	7	9	4

the mean is

$$\bar{x} = \frac{17 \cdot 7 + 22 \cdot 9 + 27 \cdot 4}{7 + 9 + 4}$$

$$= \frac{119 + 198 + 108}{20} = \frac{425}{20} = 21\frac{1}{4}$$

17. For the grouped data:

Class interval	6–12	13–19	20–26	27–33
Frequency	5	10	4	6

the mean is

$$\bar{x} = \frac{9 \cdot 5 + 16 \cdot 10 + 23 \cdot 4 + 30 \cdot 6}{5 + 10 + 4 + 6}$$

$$= \frac{477}{25} = 19.08$$

Use the midpoints of the class intervals to estimate the mean of each set of grouped data.

18. Truck Repairs   The Wink-Quick package service maintains its own statewide fleet of delivery trucks. A random sample of the repair records found the following number of days last year that fifty trucks were not available for use because their repairs could not be completed overnight:

Number of days	0–2	3–5	6–8	9–11
Frequency	30	7	12	1

19. Overtime Pay   A random sample of payroll records for one hundred craftspeople at the Old Fashion Vermont Country Furniture factory found the following amounts of overtime pay (in dollars) for the last week in November:

Overtime pay	0–19	20–39	40–59	60–79	80–99
Frequency	29	11	42	12	16

20. Charity Donations   Donors at the Elktonburg Hospital Charity Ball are designated as "friends" for donations from $51 to $100, "benefactors" for $101 to $150, and "founders" for $151 to $200. A random sample of thirty hospital supporters found the following designations:

Designation	Friends	Benefactors	Founders
Frequency	9	6	15

Measures of Variation

Pain, Suffering, and Statistics

In 1994, three secretaries sued the Digital Equipment Corporation, the manufacturer of the equipment on which they had been typing, claiming that the keyboards had caused numbness in their hands and pain in their arms and backs. In 1996, all three won large monetary judgments from the company, with separate amounts for "damages" (medical expenses and lost earnings) and for "pain and suffering" (noneconomic loss). In 1997, federal judge Jack B. Weinstein ruled that one of the pain and suffering awards was within reasonable limits, but another was unreasonably large, and that the case should be retried (for other reasons as well). (The third award was eliminated because the case had been filed too late.) On what reasoning did the judge base his conclusion about the size of a reasonable award? He employed a standard statistical measure called the *sample standard deviation,* which will be defined in this section.*

The judge asked the lawyers to find similar court cases from the past, and from these he selected a group of 22 cases as being applicable to one of the cases (that of Jeanette Rotolo), and another group of 27 as being comparable to the other case (that of Patricia Geressy), by calculating the means and standard deviations of the awards for each of these groups. He then decided that a reasonable award should be within two standard deviations of the mean. For the group similar to the Rotolo case the mean was $404,214 and the standard deviation was $465,489, which meant that her award should not exceed $404,214 + 2 \cdot 465,489 = \$1,335,192$. The award of $100,000 to Rotolo was far below this, and so was deemed not excessive. For the Geressy case the mean was $747,372 and the standard deviation was $606,873, so her award should not exceed $747,372 + 2 \cdot 606,873 = \$1,961,119$. The award of $3,490,000 to Geressy was far above this amount, so the judge ruled it to be excessive. Judge Weinstein's decision received much favorable attention as an innovative use of standard deviation to bring some rationality to the enormous disparity in pain and suffering awards.

Although this particular use of statistics was new, probability has a long history in the legal profession. In fact, Pierre de Fermat, one of the founders of probability (see page 385), was a lawyer and would be intrigued with this judicial application of the subject that he founded.

* U.S. District Court, Eastern District of New York; Geressy, Rotolo, *et al.* vs. Digital Equipment Corp.; Amended Judgment, Memorandum and Decision on post-trial motions; 94-CV-1427; Jack B. Weinstein, Senior District Judge.

Introduction

It is helpful to summarize a given data set by a few numbers, as we did with random variables in the previous chapter. For a *representative* or *typical* value, we use one of the measures of central tendency (mode, median, or mean, whichever is most appropriate for the level of data measurement). However, how do we measure whether these values are grouped tightly or spread widely about this central value? In this section we discuss three ways of describing the spread of the values about the center. Since all three use differences among data values, we must now restrict our discussion to data values at the interval or ratio level of measurement.

Range

The simplest measure of the spread of a data set is the *range,* which is the difference between the smallest and largest data values.

EXAMPLE 1   **Finding Ranges**

Find the range of each data set.

a.   {1, 5, 6, 7, 7, 7, 8, 15}
b.   {1, 2, 10, 10, 12, 13, 15}

Solution

a.   The largest value is 15 and the smallest is 1, so the range is $15 - 1 = 14$.

b.   Again, the range is $15 - 1 = 14$. Notice that the range is *not* sensitive to the distribution of values between the two extremes.

∎

PRACTICE PROBLEM 1   Find the range of the data values {15, 17, 27, 12, 30, 15, 10, 27, 25, 29}.
Solution at the back of the book

Box-and-Whisker Plot

A *box-and-whisker plot* provides a quick way to visualize how data values are distributed between the largest and smallest by graphically displaying a *five-point summary* of the data. The *minimum* is the smallest value, the *maximum* is the largest, and we have already discussed the *median*. The *first quartile* is the median of the data values below the median (but not including the median) and the *third quartile* is the me-

dian of the data values above the median (but not including the median). The median may then be called the *second quartile*. The box extends from the first quartile to the third quartile with a vertical bar at
the median. The whiskers extend from the box out to the minimum
and maximum data values.

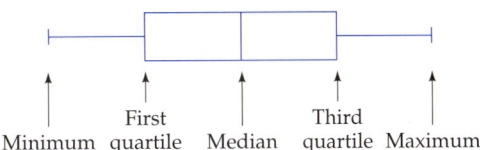

Notice that the distance between the extreme ends of the whiskers is
the *range* of the data, and that 50% of the data is contained in the box.
The length of the box, from the first quartile to the third quartile, is
sometimes called the *interquartile range*.

EXAMPLE 2   Constructing a Box-and-Whisker Plot

Make a five-point summary and draw the box-and-whisker plot for the
data values {20, 37, 65, 77, 78, 79, 81, 82, 83, 85, 87, 90}.

Solution

Since these values are already in order, the minimum is 20 and the
maximum is 90. Since there are twelve values, the median is halfway
between the sixth value (79) and the seventh (81), so the median is 80.
The values below the median are the six values {20, 37, 65, 77, 78, 79},
and the first quartile is the median of these values (that is, the value
halfway between 65 and 77), so the first quartile is 71. Similarly, the
values above the median are the six values {81, 82, 83, 85, 87, 90} and
the third quartile is the median 84 because it is the value halfway between 83 and 85. The five-point summary of this data set and the corresponding box-and-whisker plot are as follows:

Minimum = 20

First quartile = 71

Median = 80

Third quartile = 84

Maximum = 90

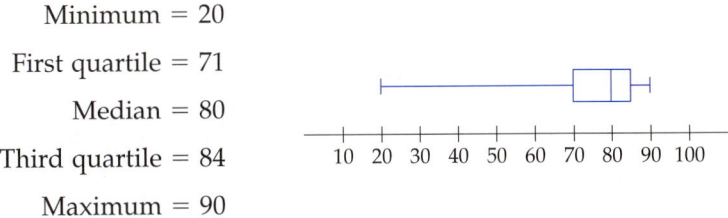

The box-and-whisker plot shows clearly that the bottom quarter of the
values are spread rather thinly from 20 to 71, whereas the entire top

half is concentrated between 80 and 90 (with a quarter of the values in the narrow range between 80 and 83). Such clustering is much easier to see from the box-and-whisker plot than from simply looking at the data values.

■

PRACTICE PROBLEM 2

Make a five-point summary and draw the box-and-whisker plot for the data values {8, 10, 13, 14, 15, 16, 17, 19, 23, 24, 29}. From your plot, which quarter is the most tightly clustered?

Solution at the back of the book

Graphing Calculator Exploration

Box-and-whisker plots can be drawn on graphing calculators using the STAT PLOT command. To explore several box-and-whisker plots, including one of the data from Example 2 (pages 471–472), proceed as follows.

a. Turn off the axes in the window FORMAT menu and set the WINDOW parameters to Xmin = 0 and Xmax = 100.

```
RectGC  PolarGC
CoordOn CoordOff
GridOff GridOn
AxesOn  AxesOff
LabelOff LabelOn
ExprOn  ExprOff
```

```
WINDOW
 Xmin=0
 Xmax=100
 Xscl=10
 Ymin=-10
 Ymax=10
 Yscl=1
 Xres=1
```

b. Enter the data values {20, 37, 65, 77, 78, 79, 81, 82, 83, 85, 87, 90} as the list L_1, the values {20, 31, 39, 41, 63, 80, 80, 80, 81, 87, 88, 90} as the list L_2, and {20, 55, 58, 60, 60, 61, 63, 64, 65, 79, 81, 90} as the list L_3.

```
{20,37,65,77,78,
79,81,82,83,85,8
7,90}→L₁
```

```
{20,31,39,41,63,
80,80,80,81,87,8
8,90}→L₂
```

```
{20,55,58,60,60,
61,63,64,65,79,8
1,90}→L₃
```

c. Turn on STAT PLOT 1 and select the box-and-whisker plot icon with Xlist L_1, and do the same for STAT PLOT 2 with Xlist L_2 and STAT PLOT 3 with Xlist L_3.

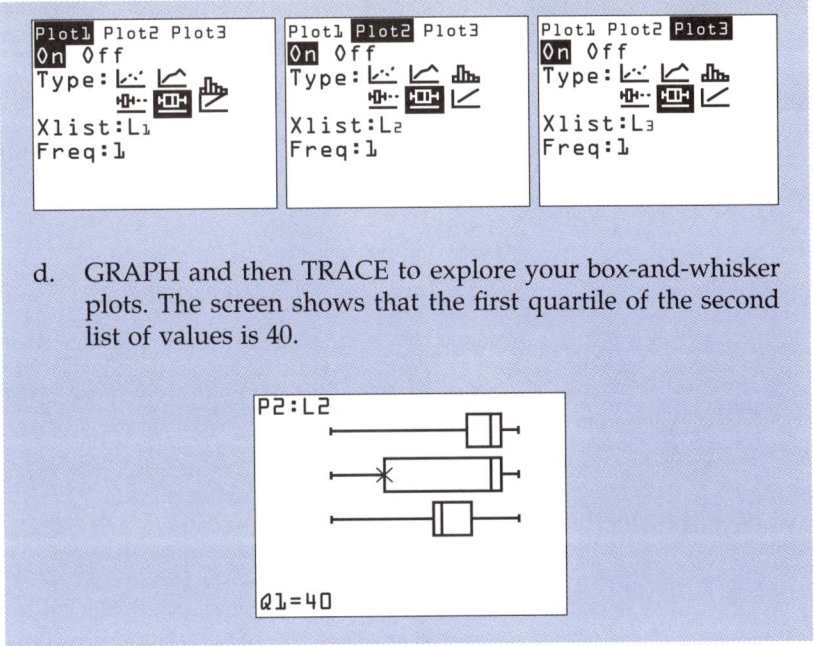

d.   GRAPH and then TRACE to explore your box-and-whisker plots. The screen shows that the first quartile of the second list of values is 40.

Sample Standard Deviation

The third way of measuring variation is the *sample standard deviation,* which estimates the typical variation of the data values from the (sample) mean of the data in much the same way that in probability theory the standard deviation σ measures the average variation from the expectation μ of a random variable.

Sample Standard Deviation

The *sample standard deviation* s of the data values $x_1, x_2 \ldots , x_n$ from a random sample of size n is

$$s = \sqrt{\frac{(x_1 - \bar{x})^2 + \cdots + (x_n - \bar{x})^2}{n - 1}}$$

$\bar{x} = \text{mean}$

$$= \sqrt{\frac{1}{n - 1} \sum_{\text{All } x_k \text{ values}} (x_k - \bar{x})^2}$$

Notice that for the sample standard deviation we divide by $n - 1$ instead of the n that we used for the standard deviation of a random variable in Chapter 4. This is because the n data values are related by the fact that their sum divided by n is the mean $\bar{x}$, so that any one value can be determined by the others together with the mean, leaving only $n - 1$ independent values in the sum.

EXAMPLE 3   Calculating a Sample Standard Deviation

Find the sample standard deviation of the data values {9, 13, 16, 18, 19}.

Solution

Since there are five values, $n = 5$. The mean is

$$\bar{x} = \tfrac{1}{5}(9 + 13 + 16 + 18 + 19) = \tfrac{1}{5}(75) = 15$$

Then:

$$s = \sqrt{\frac{(9 - 15)^2 + (13 - 15)^2 + (16 - 15)^2 + (18 - 15)^2 + (19 - 15)^2}{5 - 1}}$$

$$= \sqrt{\frac{36 + 4 + 1 + 9 + 16}{4}} = \sqrt{\frac{66}{4}} \approx 4.06$$

$$s = \sqrt{\frac{(x_1 - \bar{x})^2 + \cdots + (x_n - \bar{x})^2}{n - 1}}$$

The sample standard deviation s is (approximately) 4.06.

PRACTICE PROBLEM 3

Find the sample standard deviation of the data values {2, 6, 7, 8, 9, 15, 16, 17}.

Solution at the back of the book

Graphing Calculator Exploration

The sample standard deviation of a list of values can be found with a single calculator command. To find the sample standard deviation of {9, 13, 16, 18, 19}, the values from Example 3 above, enter the data values as a list, and store it in the list L_1. From the LIST MATH menu, select the standard deviation command and apply it to the list L_1.

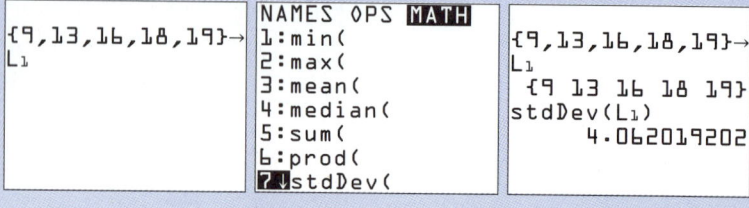

SUMMARY

There are three ways to measure the spread of a data set. The simplest is the *range*, which is the largest data value minus the smallest. The range gives no indication of the typical variation away from the mean, just the difference between the extremes.

A *box-and-whisker plot* graphically shows the range of each quarter of the data and can be drawn easily using a graphing calculator.

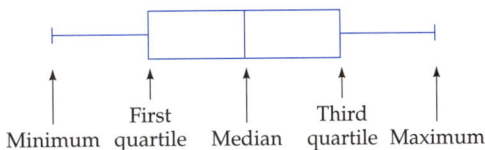

The *sample standard deviation* measures the typical spread of the data around the mean $\bar{x}$ as a single number and can be found easily using a calculator:

$$s = \sqrt{\frac{(x_1 - \bar{x})^2 + \cdots + (x_n - \bar{x})^2}{n - 1}} = \sqrt{\frac{1}{n - 1} \sum_{\text{All } x_k \text{ values}} (x_k - \bar{x})^2}$$

EXERCISES 5.3

(helpful throughout.)

Find the range of each (ratio) data set.

1. {21, 44, 48, 52, 83}

2. {26, 35, 47, 59, 86, 92, 116}

3. {5, 25, 6, 9, 7, 7, 21, 19, 23, 16}

4. {2.0, 3.8, 5.2, 1.6, 4.3, 2.3, 5.4, 3.5, 2.5, 3.1}

5. {121, 115, 147, 163, 171, 116, 167, 147, 169, 131}

Make a five-point summary and draw the box-and-whisker plot for each (ratio) data set.

6. {5, 8, 11, 13, 16, 20, 26}

7. {3, 10, 12, 14, 17, 21, 23}

8. {2, 3, 5, 9, 10, 12, 15, 17, 20}

9. {8, 9, 11, 12, 13, 14, 15, 19, 26}

10. {12, 8, 19, 17, 20, 8, 13, 14, 23, 5, 6, 9, 23, 12, 9}

11. {21, 13, 18, 14, 24, 8, 24, 24, 9, 12, 25, 20, 16, 17, 12}

12. {14, 17, 17, 13, 11, 6, 10, 21, 20, 18, 14, 11, 19, 14, 14}

13. {18, 7, 8, 18, 6, 15, 12, 15, 18, 24, 6, 18, 16, 17, 24}

14. {16, 9, 14, 12, 18, 15, 21, 19, 14, 18, 8, 4, 17, 15, 21, 14, 23, 13, 20, 17}

15. {19, 18, 7, 16, 5, 8, 9, 16, 18, 11, 20, 16, 17, 17, 8, 10, 15, 10, 8, 20}

Find the sample standard deviation of each (ratio) data set.

16. {8, 13, 14, 17, 13}

17. {9, 2, 17, 13, 14}

18. {18, 11, 18, 1, 5, 19}

19. {22, 19, 11, 2, 20, 4}

20. {14, 11, 11, 9, 2, 13}

APPLIED EXERCISES

Analyze the data in each situation by finding the range, the five-point summary, and the sample standard deviation, and then drawing the box-and-whisker plot.

21. Stock Prices   Last Thursday was a very active trading day for shares of the new DiNextron technology stock. A random sample of twenty purchase transactions found the following share selling prices (in dollars):

16   12   17   11   12   7   12   17   11   20

10   11   14   17   16   14   25   9   17   5

22. Kitchen Cabinets   A random sample of fifteen electric bills at Rick's Always Cooking cabinet factory found the following monthly electricity consumptions (in thousands of kilowatt hours):

10   12   6   15   13

15   12   9   14   6

5   8   9   12   9

23. Airline Cancellations   A random sample of thirty airplane departures from Dallas International found the following numbers of "no shows" at the boarding gates:

2   6   8   4   10   5   10   8   7   5

10   10   9   6   11   9   6   6   10   8

11   6   7   13   12   8   9   5   9   10

24. Malpractice Settlements   A random sample of twenty-four malpractice claims against a midwestern health maintenance organization (HMO) that were settled out of court found the following settlement amounts (in tens of thousands of dollars):

16   22   16   18   29   22   30   18

19   28   18   37   17   32   24   31

19   16   16   6   14   15   8   18

25. Trial Schedules   A random sample from the clerk's files at the Marksburg District Court found the following jail waiting times (in weeks) between arraignment and trial (or plea bargain) for twenty individuals charged with felonies:

11   35   15   33   20   21   2   21   23   24

18   16   9   25   12   20   20   25   21   33

Explorations and Excursions

The following problems extend and augment the material presented in the text.

Calculating Standard Deviations   While the formula on page 473 defines the sample standard deviation, there is a simpler way to calculate its value that is used in most computer programs and calculators.

26. Verify each step in the following transformation of the definition formula for the sample standard deviation.

$$s = \sqrt{\frac{1}{n-1} \sum (x_k - \bar{x})^2}$$

so:

$$
\begin{aligned}
(n-1) \cdot s^2 &= \sum (x_k - \bar{x})^2 \\
&= \sum (x_k^2 - 2x_k\bar{x} + \bar{x}^2) \\
&= \left(\sum x_k^2 \right) - 2\bar{x} \cdot \left(\sum x_k \right) \\
&\quad + \bar{x}^2 \cdot \left(\sum 1 \right) \\
&= \left(\sum x_k^2 \right) - 2 \cdot \left(\frac{1}{n} \sum x_k \right) \cdot \left(\sum x_k \right) \\
&\quad + \left(\frac{1}{n} \cdot \sum x_k \right)^2 \cdot n \\
&= \left(\sum x_k^2 \right) - \frac{1}{n} \cdot \left(\sum x_k \right)^2
\end{aligned}
$$

and then:

$$s = \sqrt{\frac{\left(\sum x_k^2 \right) - \frac{1}{n} \cdot \left(\sum x_k \right)^2}{n-1}}$$

That is, s can be found from the sum of the data values and the sum of their squares without calculating the mean $\bar{x}$.

27. Verify the following calculation of the sample standard deviation of the data {19, 18, 27, 5, 11}:

$$\sum x_k = 19 + 18 + 27 + 5 + 11 = 80$$

and

$$\sum x_k{}^2 = 19^2 + 18^2 + 27^2 + 5^2 + 11^2 = 1560$$

so

$$s = \sqrt{\frac{1560 - \frac{1}{5}(80)^2}{5 - 1}}$$

$$= \sqrt{\frac{1560 - 1280}{4}} = \sqrt{70} \approx 8.37$$

28. Verify the following calculation of the sample standard deviation of the data {11, 18, 29, 14, 24, 6}:

$$\sum x_k = 11 + 18 + 29 + 14 + 24 + 6 = 102$$

and

$$\sum x_k{}^2 = 11^2 + 18^2 + 29^2 + 14^2 + 24^2 + 6^2$$

$$= 2094$$

so

$$s = \sqrt{\frac{2094 - \frac{1}{6}(102)^2}{6 - 1}} = \sqrt{\frac{2094 - 1734}{5}}$$

$$= \sqrt{72} \approx 8.49$$

29. Use the formula from Exercise 26 to find the sample standard deviation of the data {9, 2, 17, 13, 14} and then compare your value with your answer to Exercise 17.

30. Use the formula from Exercise 26 to find the sample standard deviation of the data {22, 19, 11, 2, 20, 4} and then compare your value with your answer to Exercise 19.

5.4   Normal Distributions

APPLICATION   PREVIEW

The Bell-Shaped Curve

In the machine pictured on the following page, 30,000 steel balls fall, one at a time, from the center of the top, each ball making its way downward through the rows of pins, turning right or left at each pin, and finally falling into one of the slots at the bottom. A very few balls will turn right at each pin, ending up in the extreme right slot; a few will turn left at each pin, ending up in the extreme left slot; but most will make a combination of the right and left turns, ending up somewhere in the middle. As the picture shows, the balls fill out what is known as a *bell-shaped* or *normal curve*, highest in the middle and symmetrically lower further away on either side. This machine illustrates the principle that the cumulative effect of many small changes (right and left detours around each pin) often leads to the normal curve. In this section we will define the normal curve and use it to predict such quantities as people's heights and weights, lengths of hospital stays, and orders received by a company.

Introduction

The normal distribution, with its famous bell-shaped graph, is perhaps the most important of all probability distributions. It is used to predict everything from sizes of newborn babies to stock market fluctuations to retirement benefits. After we examine some of its most important features, we will see how it is used in applications and examine its relation to the binomial distribution.

Normal Distribution

The normal distribution depends upon two parameters, the mean μ and the standard deviation σ. The curve is given by the following formula, although we will not use it explicitly in what follows.

Normal Probability Distribution

The normal probability distribution with mean μ and standard deviation σ is given by the curve:

$$f(x) = \frac{1}{\sigma\sqrt{2\pi}} e^{-\frac{1}{2}\left(\frac{x-\mu}{\sigma}\right)^2} \qquad \text{for } -\infty < x < \infty$$

The curve has a central peak at μ, and then falls back symmetrically on either side to approach the x-axis. Different values of μ change the location of the peak, as is shown below by three normal distributions with different means.

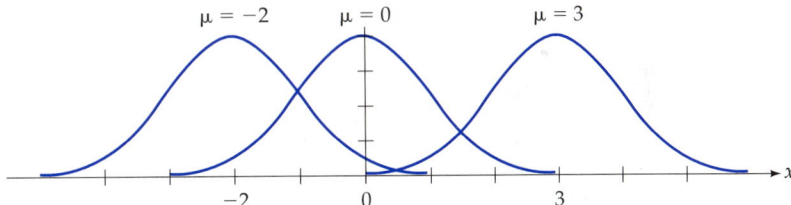

Three normal distributions with different means

The standard deviation σ measures how the values spread away from the mean. The distribution will be higher and more peaked if σ is small, and lower and more rounded if σ is large.

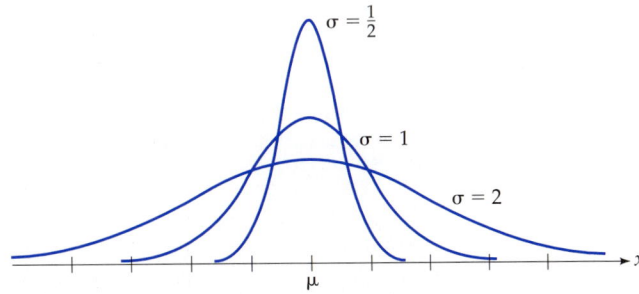

Three normal distributions with different standard deviations

As the distribution becomes "shorter" it also becomes "wider," and the area under the curve always stays at 1, although we will not prove this fact. The peaked shape of the normal distribution concentrates most of the probability (or area) near the center at μ. About 68% of the area under the normal curve is within one standard deviation of the mean, about 95% is within two standard deviations, and more than 99% is within three standard deviations. Notice also that the curve rises or falls most steeply at one standard deviation from the mean.

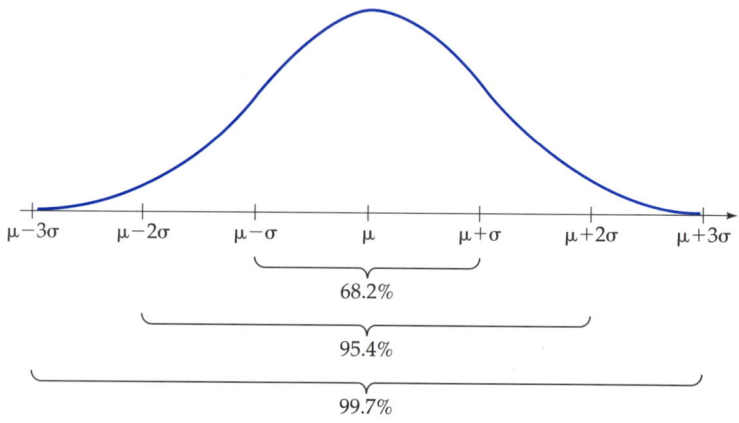

Area (probability) under the curve within 1, 2, and 3 standard deviations of the mean

The probability that a normal random variable has its value between two x-values corresponds to the *area* under the normal curve between those x-values. The probability is most easily found using a graphing calculator, but may also be found using tables of values of the normal distribution.

Readers who have graphing calculators that find normal probabilities should continue reading this section. Readers who do not have such calculators should now turn to page 1009 to read the appendix near the end of this book.

 ## Graphing Calculator Exploration

Areas under normal distribution curves can be found easily on a graphing calculator. To find the area under the normal distribution curve with mean $\mu = 20$ and standard deviation $\sigma = 5$ from 10 (which is $\mu - 2\sigma$) to 30 (which is $\mu + 2\sigma$), proceed as follows.

a. From the DISTRIBUTION menu, select the normalcdf(command to find the cumulative values of the normal distribution (that is, the area under the curve) between two given values.

```
DISTR DRAW
1:normalpdf(
2:normalcdf(
3:invNorm(
4:tpdf(
5:tcdf(
6:X²pdf(
7↓X²cdf(
```

b. Enter the (left) starting value and the (right) ending value as well as the mean and standard deviation. This area is the same as the probability that the normal random variable is between the two given values.

```
normalcdf(10,30,
20,5)
         .954499876
```

c. To see this area shaded under the normal curve, use the ShadeNorm(command from the DISTRIBUTION DRAW menu with an appropriate window.

```
WINDOW
 Xmin=0
 Xmax=40
 Xscl=5
 Ymin=-.05
 Ymax=.1
 Yscl=1
 Xres=1
```

```
DISTR DRAW
1:ShadeNorm(
2:Shade_t(
3:ShadeX²(
4:ShadeF(
```

```
ShadeNorm(10,30,
20,5)
```

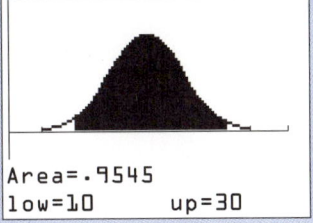

```
Area=.9545
low=10        up=30
```

EXAMPLE 1   Finding a Probability for a Normal Random Variable

The heights of American men are approximately normally distributed with mean $\mu = 68.1$ inches and standard deviation $\sigma = 2.7$ inches.* Find the proportion of men who are between 5 feet 9 inches and 6 feet tall.

* These and other data in this section are from *Handbook of Human Factors* by Gavriel Salvendy (John Wiley-Interscience, 1987).

Solution

Converting to inches, we want the probability of a man's height being between 69 inches and 72 inches. From the graphing calculator screen below, about 30% of American men are between 5 feet 9 inches and 6 feet tall.

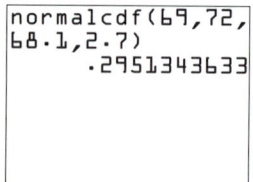

■

PRACTICE PROBLEM 1

The weights of American women are approximately normally distributed with mean 134.7 pounds and standard deviation 30.4 pounds. Find the proportion of women who weigh between 130 and 150 pounds.   *Solution at the back of the book*

z-Scores

The *standard normal distribution* is the normal distribution with mean $\mu = 0$ and standard deviation $\sigma = 1$ (the word "standard" indicates mean zero and standard deviation one). The letter z is traditionally used to denote the variable for this special distribution. Since $\mu = 0$, the highest point on the curve is at $z = 0$, and the curve is symmetric about this value.

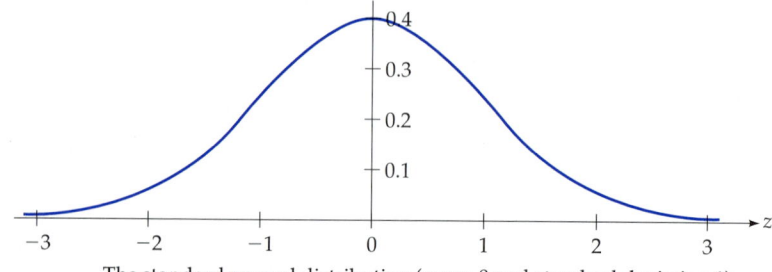

The standard normal distribution (mean 0 and standard deviation 1)

To change any x value for a normal distribution with mean μ and standard deviation σ into the corresponding z-score for the *standard normal distribution*, subtract the mean and then divide by the standard deviation.

z-Score

$$z = \frac{x - \mu}{\sigma}$$

The z-score shows how many standard deviations the value is away from its mean.

EXAMPLE 2 **Finding a z-Score**

Convert each x value into a z-score using the given values for μ and σ.

a. $x = 8$ with $\mu = 4$ and $\sigma = 2$ **b.** $x = 8$ with $\mu = 12$ and $\sigma = 4$

Solution

a. $z = \dfrac{8 - 4}{2} = 2$ **b.** $z = \dfrac{8 - 12}{4} = -1$

In part (a) the z-score $z = 2$ means that the original 8 was *two* standard deviations to the right of its mean. In (b) the z-score $z = -1$ means that the 8 in that distribution was one standard deviation to the *left* of *its* mean.

■

PRACTICE PROBLEM 2

Convert $x = 20$ with $\mu = 15$ and $\sigma = 1$ into a z-score. Is this value far from the mean?
Solution at the back of the book

The *central limit theorems* developed by Carl F. Gauss (1777–1855) and other mathematicians during the nineteenth century proved that the errors in observed values, the means of random samples, and many other statistical quantities were normally distributed. Extensive tables of the probabilities of collections of z-scores were laboriously calculated, and the results were worth the effort because they could be used for any normally distributed x values by first converting them to z-scores. Although graphing calculators can easily compute values for a normal distribution with any given mean and standard deviation, z-scores remain useful as a quick tool to place a particular x value in its proper position relative to μ and σ.

The Normal and Binomial Distributions

As we saw on page 437, the binomial distribution too has a kind of "bell" shape, with a peak at the expected value $\mu = np$ and falling away on both sides to very small probabilities several standard deviations $\sigma = \sqrt{np(1-p)}$ away.

$$n = 20, p = 0.4 \qquad n = 15, p = 0.5 \qquad n = 25, p = 0.7$$

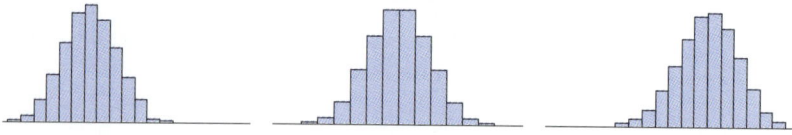

Several binomial distributions drawn using the program BINOMIAL

In the eighteenth century, Abraham de Moivre (1667–1754) and Pierre-Simon Laplace (1749–1827) discovered and proved that for any choice of p between 0 and 1 the binomial distribution *approaches* the normal distribution as n becomes large. This fundamental fact is known as the de Moivre–Laplace theorem.

 ### Graphing Calculator Exploration

The program SEENORML* draws binomial distributions for $p = \frac{1}{2}$ with n from 2 to 30. Run this program to watch the "boxlike" binomial distributions change into "bell-shaped" curves as n increases. The window for each graph is different so that each fills as much as possible of the viewing area. When n is a multiple of 5, the program also draws the corresponding normal distribution approximation.

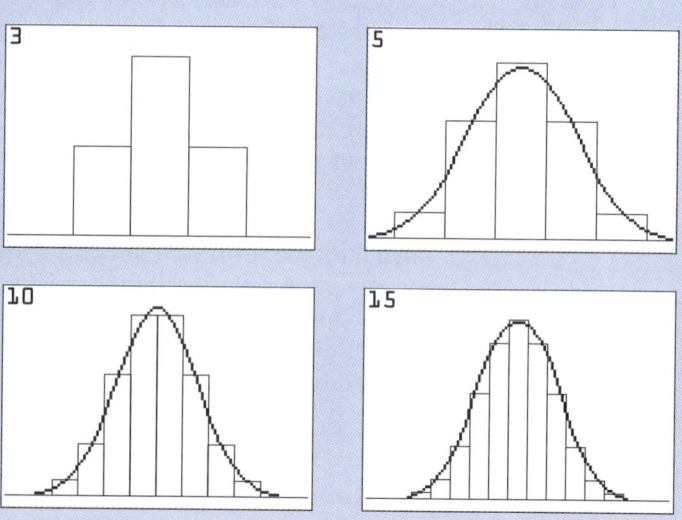

* See the Preface for information on how to obtain this program.

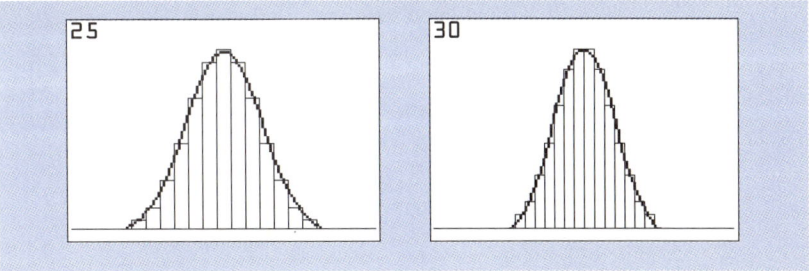

We may use the de Moivre–Laplace theorem to approximate binomial distributions by the normal distribution.

Normal Approximation to the Binomial

Let X be a binomial random variable with parameters n and p. If $np > 5$ and $n(1 - p) > 5$, then the distribution of X is approximately normal with mean $\mu = np$ and standard deviation $\sigma = \sqrt{np(1 - p)}$.

To have the same width for a "slice" of the normal distribution as for a slice of the binomial, we adopt the convention that the binomial probability $P(X = x)$ corresponds to the area under the normal distribution curve from $x - \frac{1}{2}$ to $x + \frac{1}{2}$:

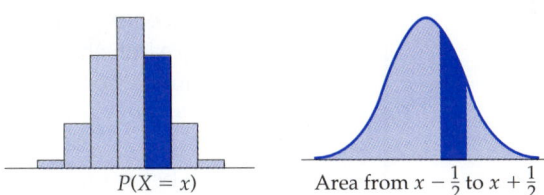

$P(X = x)$ Area from $x - \frac{1}{2}$ to $x + \frac{1}{2}$

EXAMPLE 3 Normal Approximation of a Binomial Probability

Estimate $P(X = 12)$ for the binomial random variable X with $n = 25$ and $p = 0.6$ using the corresponding normal distribution.

Solution

Since $np = (25)(0.6) = 15$ and $n(1 - p) = (25)(0.4) = 10$ are both greater than 5, we may use the normal distribution with $\mu = np = (25)(0.6) = 15$ and $\sigma = \sqrt{np(1 - p)} = \sqrt{(25)(0.6)(0.4)} = \sqrt{6}$. As stated above, we interpret the event $X = 12$ as $11\frac{1}{2} \le X \le 12\frac{1}{2}$ to include numbers that would round to 12. Using a graphing calculator, we have:

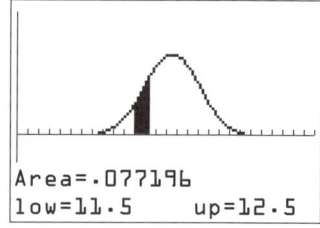

```
WINDOW
 Xmin=0
 Xmax=25
 Xscl=1
 Ymin=-.1
 Ymax=.25
 Yscl=1
 Xres=1
```

```
ShadeNorm(11.5,1
2.5,25*.6,√(25*.
6*.4))
```

```
Area=.077196
low=11.5      up=12.5
```

The required probability is (about) 0.077. This is a good approximation to the actual value of $_{25}C_{12}\, p^{12}(1 - p)^{13} \approx 0.0759667$.

∎

EXAMPLE 4   Management MBA's

At a major Los Angeles accounting firm, 73% of the managers have MBA degrees. In a random sample of 40 managers, what is the probability that between 27 and 32 will have MBA's?

Solution

Because each manager either has or does not have an MBA, presumably independently of each other, the question asks for the probability that a binomial random variable X with $n = 40$ and $p = 0.73$ satisfies $27 \le X \le 32$. Since $np = (40)(0.73) = 29.2$ and $n(1 - p) = (40)(0.27) = 10.8$ are both greater than 5, this probability can be approximated as the area under the normal distribution curve with mean $\mu = np = (40)(0.73) = 29.2$ and standard deviation $\sigma = \sqrt{np(1 - p)} = \sqrt{(40)(0.73)(0.27)} = \sqrt{7.884}$ from $26\frac{1}{2}$ to $32\frac{1}{2}$ (again to include values that would round to between 27 and 32):

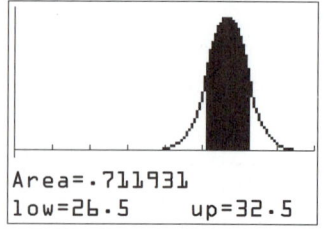

```
WINDOW
 Xmin=0
 Xmax=40
 Xscl=5
 Ymin=-.05
 Ymax=.15
 Yscl=1
 Xres=1
```

```
ShadeNorm(26.5,3
2.5,40*.73,√(40*
.73*.27))
```

```
Area=.711931
low=26.5      up=32.5
```

The probability is (about) 0.712, or about 71%. (The exact answer to this problem, from summing the binomial distribution from 27 to 32, is 71.5%, so the normal approximation is very accurate.)

∎

PRACTICE PROBLEM 3   A brand of imported VCR is known to have defective tape rewind mechanisms in 8% of the units imported last April. If Jerry's Discount Electronics received a shipment of 80 of these VCRs, what is the probability that 10 or more are defective?   *Solution at the back of the book*

SUMMARY

The graph of the normal distribution with mean μ and standard deviation σ is a bell-shaped curve that peaks at μ in the center and then falls back symmetrically on either side to approach the x-axis. The area under this curve between any two x values is the probability that the normal random variable lies between these values. These areas and probabilities can be found easily with a graphing calculator.

Any value x of a normal random variable with mean μ and standard deviation σ can be converted into a z-score using the formula:

$$z = \frac{x - \mu}{\sigma}$$

This z-score indicates the number of standard deviations the x value is away from the mean.

The normal distribution with mean $\mu = np$ and standard deviation $\sigma = \sqrt{np(1 - p)}$ is a good approximation to the binomial distribution provided both np and $n(1 - p)$ are greater than 5.

EXERCISES 5.4

( helpful throughout.)

Let X be the normal random variable with mean $\mu = 12$ and standard deviation $\sigma = 3$. Find each probability as an area under the normal curve.

1. $P(9 \leq X \leq 15)$

2. $P(6 \leq X \leq 18)$

3. $P(3 \leq X \leq 21)$

4. $P(0 \leq X \leq 24)$

5. $P(12 \leq X \leq 15)$

6. $P(9 \leq X \leq 12)$

7. $P(15 \leq X \leq 16)$

8. $P(7 \leq X \leq 10)$

9. $P(6\frac{1}{2} \leq X \leq 11)$

10. $P(10\frac{1}{2} \leq X \leq 14)$

Find the z-score of each x value of the normal random variable with mean μ and standard deviation σ.

11. $x = 6$ with $\mu = 4$ and $\sigma = 2$

12. $x = 20$ with $\mu = 15$ and $\sigma = 5$

13. $x = 10$ with $\mu = 10$ and $\sigma = 3$

14. $x = 6$ with $\mu = 6$ and $\sigma = 5$

15. $x = 15$ with $\mu = 20$ and $\sigma = 5$

16. $x = 6$ with $\mu = 8$ and $\sigma = 2$

17. $x = 15$ with $\mu = 20$ and $\sigma = 1$

18. $x = 6$ with $\mu = 8$ and $\sigma = 1$

19. $x = 130$ with $\mu = 100$ and $\sigma = 15$

20. $x = 90$ with $\mu = 100$ and $\sigma = 10$

Use the normal approximation to the binomial random variable X with $n = 20$ and $p = 0.7$ to find each probability as an area under the normal curve.

21. $P(X = 10)$

22. $P(X = 12)$

23. $P(11 \leq X \leq 12)$

24. $P(8 \leq X \leq 10)$

25. $P(3 \leq X \leq 8)$

26. $P(13 \leq X \leq 19)$

27. $P(15 \leq X \leq 20)$

28. $P(2 \leq X \leq 7)$

29. $P(0 \leq X \leq 20)$

30. $P(14 \leq X \leq 20)$

Answer each question using an appropriate normal distribution probability.

31. Weights   Weights of men are approximately normally distributed with mean 163 pounds and standard deviation 28 pounds. If the maximum and minimum weights in order to be a volunteer fireman in Martinsville are 128 and 254, what proportion of men meet this qualification?

32. Advertising   The percentage of early afternoon television viewers who watch episodes of "As the World Spins" is a normal random variable with mean 22% and standard deviation 3%. The producers guarantee advertisers that the viewer percentage will be better than 20% or else the advertisers will receive free air time. What is the probability that the producers will have to give free air time after any particular episode?

33. Airline Overbooking   In order to avoid empty seats, airlines generally sell more tickets than there are seats. Suppose that the number of ticketed passengers who show up for a flight with 100 seats is a normal random variable with mean 97 and standard deviation 6. What is the probability that at least one person will have to be "bumped" from the flight?

34. SAT Scores   The scores at Centerville High School on last year's mathematics SAT test were approximately normally distributed with mean 490 and standard deviation 140. What proportion of the scores were between 550 and 750?

35. Smoking   In a large-scale study carried out in the 1960s of male smokers 35–45 years old, the number of cigarettes smoked daily was approximately normally distributed with mean 28 and standard deviation 10. What proportion of the smokers smoked between 30 and 40 (two packs!) each day?

36. Coin Tosses   A fair coin is tossed fifty times. What is the probability that it comes up heads at least thirty times?

37. Actuarial Exams   The probability of passing the first actuarial examination is 60%. If 940 college students take the exam in March, what is the probability that the number that pass is between 575 and 725?

38. Flu Shots   While getting a flu shot in October is a good way of planning ahead for the winter flu season, approximately 1% of those getting the shots develop side effects. If 800 students at Micheles College get flu shots, what is the probability that between 6 and 12 of them will develop side effects?

39. Real Estate   Nationwide, 40% of the sales force at the Greener Pastures real estate franchises are college graduates. If two hundred sales personnel are chosen at random, what is the probability that between 85 and 100 of them are college graduates?

40. Old Age   If 3% of the population lives to age 90, what is the probability that between 10 and 15 of the 280 business majors at Henderson College will live to be that old?

Chapter Summary with Hints and Suggestions

The reading and exercises of this chapter have helped you learn the following skills. For each skill, the section from which it came (in case you need to review it) and some exercises in this section that use it are indicated. Answers to all exercises are at the end of the book, and full solutions to all exercises are in the Student Solutions Manual.

5.1 Random Samples and Data Organization

- Identify levels of data measurement (nominal, ordinal, interval, and ratio). (*Review Exercises 1–4.*)

- Construct a bar chart, a stem-and-leaf display or a histogram for a given data set. (*Review Exercises 5–10.*)

5.2 Measures of Central Tendency

- Find the mode, median, and mean of a data set respecting the level of data measurement. (*Review Exercises 10–15.*)

Mode: most frequent

Median: middle

$$\text{Mean: } \bar{x} = \frac{x_1 + \cdots + x_n}{n}$$

5.3 Measures of Variation

- Find the range of a data set. (*Review Exercise 16.*)

- Find the five-point summary of a data set and draw the box-and-whisker plot. (*Review Exercise 17.*)

- Find the sample standard deviation of a data set. (*Review Exercise 18.*)

$$s = \sqrt{\frac{(x_1 - \bar{x})^2 + \cdots + (x_n - \bar{x})^2}{n - 1}}$$

- Analyze the data in an application by finding the range, the five-point summary, and the sample standard deviation, and then drawing the box-and-whisker plot. (*Review Exercises 19–20.*)

5.4 Normal Distributions

- Find the probability that the normal random variable with mean μ and standard deviation σ is between two given values. (*Review Exercises 21–22.*)

- Find the z-score of an x value of the normal random variable with mean μ and standard deviation σ. (*Review Exercises 23–24.*)

$$z = \frac{x - \mu}{\sigma}$$

- Use the normal approximation to the binomial random variable X with n and p to find the probability that X takes given values. (*Review Exercises 25–26.*)

- Solve an applied problem involving probabilities using the normal distribution. (*Review Exercises 27–30.*)

Hints and Suggestions

- (*Overview*) Properties of "many" can be inferred from just "some" only if the "some" are a random sample from the population. There are four very different levels of data measurement: nominal, ordinal, interval, and ratio. Data may be organized into bar charts, stem-and-leaf displays, and histograms. Measures of central tendency (mode, median, and mean) summarize all the data as one typical value, while measures of variation (range, box-and-whisker plot, and sample standard deviation) explain how closely the values cluster about the center. The bell-shaped curve of the normal distribution applies to many situations, and z-scores give the number of standard deviations the values are from the mean. The binomial distribution is approximated by the normal distribution provided both n and p are sufficiently large.

- While bar charts are appropriate for any level of data measurement and the bars are separate, a histogram can be drawn only for interval or ratio data and the bars touch each other on both sides.

- While it is easiest to find the mode of nominal data by sorting it into order, this ordering of "by name only" data has no meaning, and you must *not* compute a median or mean for such data values.

- The mean is sensitive to changes in a few extreme values, while the mode and median are not.

- Box-and-whisker plots graphically show how the data are distributed among the four quartiles.

- Areas under the normal distribution curve are probabilities for the normal random variable.

- When using the normal distribution to approximate the binomial distribution, be sure to increase the largest value by $\frac{1}{2}$ and to decrease the smallest by $\frac{1}{2}$ to cover all the area represented by the boxes making up the binomial distribution.

- Practice for test: Review Exercises 1, 3, 5, 7, 9, 11, 19, 21, 23, 26, 28.

Review Exercises for Chapter 5   *Practice test exercises are in blue.*

(helpful throughout.)

5.1   Random Samples and Data Organization

Identify the level of data measurement in each situation.

1. **Chicago Commuters**   A random sample of seventy Chicago area commuters counts the number traveling to work last Thursday by car, taxi, bus, and train.

2. **Gourmet Coffee**   At the House of Java Coffee Emporium, a random sample of eighteen customers rates the new Turkish Sultan flavor on a scale from 1 ("I'll try something else") to 5 ("love at first sip").

3. **Fourth of July**   A random sample of five hundred suburban households records the backyard temperatures at 3 P.M. on Independence Day.

4. **Drinking Water**   A random sample of eighty small municipalities (less than 50,000 residents) lists the lead content (in parts per billion) of the tap water.

Construct a bar chart of the data in each situation.

5. **Veterans of Foreign Wars**   A random sample of forty members of the Hattiesburg V.F.W. identify their branches of military service as 14 in the Army, 12 in the Navy, 6 in the Marines, and 8 in the Air Force.

6. **Homicides**   A random sample of twenty homicides investigated last year by the Macon County Coroner's Office lists the causes as 12 by handgun, 3 by firearms other than handguns, 2 by knives, 2 by blunt instruments, and 1 by use of hands.

Construct a stem-and-leaf display of the data in each situation.

7. **Five-Mile Run**   At the Norrisville Cardiac Fitness cross country five-mile run, the finish times for twenty randomly selected entrants were (in minutes):

59	36	59	38	55	47	56	43	57	45
33	50	29	58	60	87	40	54	46	55

8. **Study Times**   A random sample of thirty business majors interviewed during spring break in Daytona Beach found that they studied the following numbers of hours the previous week:

36	60	39	31	36	30	14	4	57	9
48	45	18	49	52	41	38	13	32	19
58	3	25	21	17	37	27	26	51	53

Construct a histogram of the data in each situation using an appropriate class width and starting class lower boundary.

9. **Manufacturer's Rebates**   A random sample of twenty requests for the $5 rebate offered last summer on two pairs of Ultra-Cool Shades sunglasses found that the checks were received by the consumers after waits of the following numbers of days:

45	53	60	34	60	37	52	65	44	49
54	51	61	43	49	63	53	53	56	55

10. Checking Accounts   A random sample of thirty checking account balances on March 2 at the Lobsterman Trust Company branch in Bangor found the following amounts (rounded to the nearest dollar):

906	858	1168	1397	925	797
1036	1398	912	1059	931	815
698	787	711	1485	1048	937
1272	727	1339	1458	842	1264
1045	960	802	1204	1370	841

5.2   Measures of Central Tendency

Find the mode, median, and mean of each (ratio) data set.

11. {14, 11, 15, 8, 5, 4, 10, 3, 12, 11, 6}

12. {5, 6, 15, 13, 15, 6, 11, 13, 13, 6, 13, 6, 8}

13. {12, 13, 13, 5, 9, 14, 13, 8, 9, 14}

14. Stock Earnings   The third-quarter earnings per share for twelve randomly selected textile stocks on the New York Stock Exchange were (in dollars):

12	8	8	11	4	7
8	7	5	13	10	9

15. Personal Calls   A random sample of the telephone records for fifteen office clerks at the southern regional office of the Marston Glassware Products Company counted the following numbers of personal calls last week:

4	15	23	26	2
22	25	4	27	18
30	28	11	20	6

5.3   Measures of Variation

16. Find the range of the (ratio) data {38, 28, 32, 43, 41, 25, 35, 37, 43, 36}.

17. Find the five-point summary and draw the box-and-whisker plot for the (ratio) data {4, 5, 7, 9, 10, 13, 14, 15, 20}.

18. Find the sample standard deviation of the (ratio) data {5, 7, 9, 10, 13, 16}.

Analyze the data in each situation by finding the range, the five-point summary, and the sample standard deviation, and then drawing the box-and-whisker plot.

19. Bank Lines   A random sample of fifteen customers at the Middletown branch of the Peoples & First National Bank found the following waiting times (in minutes):

5	6	6	5	3
5	8	3	9	6
7	9	5	9	6

20. Mortgages   A random sample of twenty mortgages filed last month at the Seaford Town Clerk office had the following values (in thousands of dollars):

132	141	154	143	115
141	125	132	123	132
119	126	123	114	126
132	128	131	136	131

5.4   Normal Distributions

Let X be the normal random variable with mean $\mu = 50$ and standard deviation $\sigma = 10$. Find each probability as an area under the normal curve.

21. $P(30 \leq X \leq 60)$ **22.** $P(55 \leq X \leq 75)$

Find the z-score of each x value of the normal random variable with mean μ and standard deviation σ.

23. $x = 12$ with $\mu = 10$ and $\sigma = 1$

24. $x = 18$ with $\mu = 24$ and $\sigma = 3$

Use the normal approximation to the binomial random variable X with $n = 25$ and $p = 0.35$ to find each probability as an area under the normal curve.

25. $P(X = 14)$ **26.** $P(6 \leq X \leq 10)$

Answer each question using an appropriate normal distribution probability.

27. **Heights** Women's heights are approximately normally distributed with mean 63.2 inches and standard deviation 2.6 inches. Find the proportion of women with heights between 62 inches and 69 inches.

28. **Tire Wear** The life of Wear-Ever Super Tread automobile tires is normally distributed with mean 50,000 miles and standard deviation 5000 miles. What proportion of these tires last between 55,000 and 65,000 miles?

29. **Lost Luggage** During the Christmas rush, the chance that an airline will lose your suitcase increases to 1%. What is the probability that between 4 and 8 of the 690 suitcases checked in at Bellemeade Regional Airport this holiday season will be lost?

30. **Heart Disease** The leading cause of death in the United States is heart disease, which causes 32% of all deaths. In a random sample of 150 death certificates, what is the probability that between 50 and 60 list heart disease as the cause of death?

Projects and Essays

The following projects and essays are based on Chapter 5. Most have no right and wrong answers—the results depend only on your imagination and resourcefulness.

1. Write a one-page report on an interesting situation discussed in the book *How to Lie with Statistics* by Darrell Huff (W.W. Norton, 1954; paperback reissue 1993).

2. The bar chart was invented by William Playfair and first published by him in 1786. Write a one-page report on Playfair and his contributions to the graphical presentation of data after reading the articles "William Playfair: A Daring Worthless Fellow" and "Who Was Playfair?" by Ian Spence and Howard Wainer in the Winter 1997 issue of *Chance* magazine (Volume 10, Number 1), pages 31–34 and 35–37.

3. Stem-and-leaf displays and box-and-whisker plots are just two of many recent ideas for the presentation of numerical information. Write a one-page report on these and other methods after browsing through the book *Exploratory Data Analysis* by John Tukey (Addison-Wesley, 1977).

4. A student at your college has taken three examinations in another section of this course and received grades of 70, 80, and 90, and has asked her professor to drop the lowest grade. Should her professor agree to her request? The class means and standard deviations on these tests are listed in the table below. Write a one-page essay in support of your answer.

Test	Her Score	Mean	Standard Deviation
No. 1	70	60	5
No. 2	80	75	10
No. 3	90	95	10

5. Look over your notes, homework, and this text, and write a one-page report about how your graphing calculator helped you in this chapter. Include examples of how it helped you explore concepts and how it helped to simplify your work. What was the most helpful or interesting use? What was the least helpful or interesting use? Are there problems that can be done on a graphing calculator but that are easier to do by hand?

PART **II**

CALCULUS

6

Derivatives and Their Uses

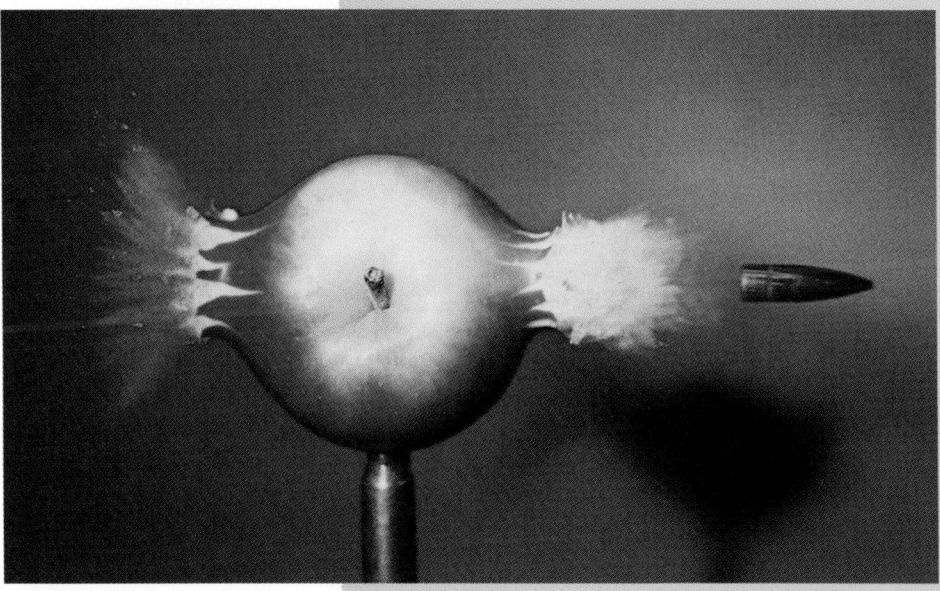

The instantaneous velocity of a bullet may be found by using calculus

6.1 Limits and Continuity

Temperature, Superconductivity, and Limits

It has long been known that there is a coldest possible temperature, called absolute zero, the temperature of an object if all heat could be removed from it. On the Fahrenheit temperature scale, absolute zero is 460 degrees below zero. On the "absolute" or "Kelvin" temperature scale (named after the nineteenth-century scientist Lord Kelvin), absolute zero temperature is assigned the value 0. Absolute zero is a temperature that can be *approached* but never actually *reached*. At temperatures approaching absolute zero, some metals become increasingly able to conduct electricity, with efficiencies approaching 100%, a state called *superconductivity*. The graph below gives the electrical conductivity of aluminum, showing the remarkable fact that as temperature approaches absolute zero, aluminum becomes *superconducting* (its conductivity approaches a limit of 100%).

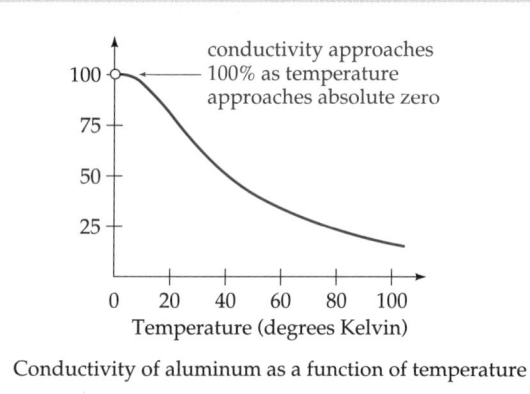

Conductivity of aluminum as a function of temperature

If $f(t)$ stands for the percent conductivity of aluminum at temperature t degrees Kelvin, then the fact that conductivity approaches 100 as temperature t approaches zero may be written

$$\lim_{t \to 0} f(t) = 100 \qquad \text{Limit as } t \to 0 \text{ of conductivity is 100}$$

This is an example of *limits*, which are discussed in this section. Superconductivity has many commercial applications, from high-speed "mag-lev" trains that "float" on magnetic fields above the tracks to supercomputers and medical imaging devices.

Introduction

In this chapter we begin the study of calculus and its applications. Calculus is, quite simply, the study of rates of change. We will use calculus to analyze rates of inflation, rates of learning, rates of growth of populations, and rates of consumption of natural resources. We begin by discussing two preliminary topics, limits and continuity, both of which will be treated intuitively rather than formally. These will be useful in the next section when we discuss the derivative, one of the central ideas of calculus.

Limits

The word "limit" is used in everyday conversation to describe the *ultimate* behavior of something, as in the "limit of one's endurance" or the "limit of one's patience." In mathematics, the word "limit" has a similar but more precise meaning.

The notation $x \rightarrow 3$ (read "x approaches 3") means that x takes values *arbitrarily close to 3 without ever reaching 3*. For example, the numbers 2.9, 2.99, 2.999, . . . approach 3 (from the left), and the numbers 3.1, 3.01, 3.001, . . . approach 3 (from the right). We can even let x approach 3 by alternating sides, taking values like 2.9, 3.01, 2.999, 3.0001, ... We emphasize that $x \rightarrow 3$ means that x takes values closer and closer to 3 *but never equaling 3*.

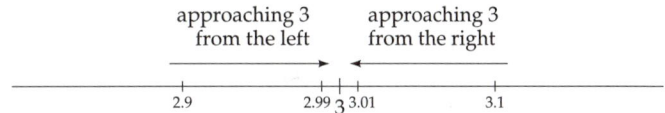

Given a function $f(x)$, if x approaching 3 causes the function to take values arbitrarily close to, say, 10, then we call 10 the *limit* of the function, and write

$$\lim_{x \to 3} f(x) = 10 \qquad \text{\textcolor{blue}{Limit of $f(x)$ as x approaches 3 is 10}}$$

The limit of a function may not exist (as we will see later), but if the limit *does* exist, it must be a *single* number. Limits can be defined formally,* but we will treat them intuitively, using the following definition.

---

* Formally, limits are defined as follows: $\lim_{x \to c} f(x) = L$ if for every number $\epsilon > 0$ there is a number $\delta > 0$ such that $|f(x) - L| < \epsilon$ whenever $0 < |x - c| < \delta$.

Limit of a Function

$$\lim_{x \to c} f(x) \qquad \text{Limit of } f(x) \text{ as } x \text{ approaches } c$$

is the number that $f(x)$ approaches as x approaches c ($x \neq c$).

Evaluating Limits

Limits may be found from tables of values of $f(x)$ for x near c.

EXAMPLE 1   Evaluating a Limit by Tables

Use tables to find $\lim\limits_{x \to 3} (2x + 4)$.   Limit of $2x + 4$ as x approaches 3

Solution

We choose x-values approaching 3, and calculate the resulting values of $f(x) = 2x + 4$. We make two tables, for $x < 3$ and for $x > 3$.

x	$2x + 4$
2.9	9.8
2.99	9.98
2.999	9.998

Numbers approaching 3 but less than 3 ↓

x	$2x + 4$
3.1	10.2
3.01	10.02
3.001	10.002

Numbers approaching 3 but greater than 3 ↓

Reading down these columns, the function values seem to approach 10

Choosing x-values even closer to 3 (like 2.9999 and 3.0001) would result in values of $2x + 4$ *even closer to 10*, so *the limit is 10:*

$$\lim_{x \to 3} (2x + 4) = 10$$

This limit may be seen graphically: As $x \to 3$ (on the x-axis), $f(x)$ approaches 10 (on the y-axis).

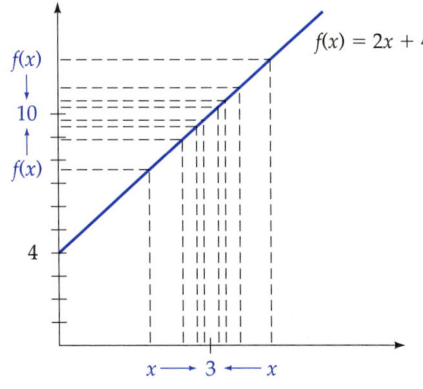

Graph showing $\lim\limits_{x \to 3} (2x + 4) = 10$

The correct limit in Example 1 could have been found simply by *evaluating* the function at $x = 3$:

$$f(3) = 2 \cdot 3 + 4 = 10 \qquad \text{\color{blue}$f(x) = 2x + 4$ evaluated at $x = 3$ gives the correct limit, 10}$$

However, finding limits by this technique of *direct substitution* is not always possible, as the next example shows.

EXAMPLE 2 Evaluating a Limit by Tables

Find $\lim\limits_{x \to 0} (1 + x)^{1/x}$ correct to three decimal places.

Solution

The function values in the following tables were found using a calculator.

x	$(1 + x)^{1/x}$	x	$(1 + x)^{1/x}$
0.1	2.594	-0.1	2.868
0.01	2.705	-0.01	2.732
0.001	2.717	-0.001	2.720
0.0001	2.718	-0.0001	2.718

To use a graphing calculator to find these numbers, enter the function as $(1 + x)\wedge(1/x)$ and use the TABLE feature of your calculator.

$$\lim\limits_{x \to 0} (1 + x)^{1/x} \approx 2.718 \qquad \text{\color{blue}From the agreement in the last columns of the tables}$$

PRACTICE PROBLEM 1   Evaluate $(1 + x)^{1/x}$ at $x = 0.000001$ and $x = -0.000001$. Based on the results, give a better approximation for $\lim_{x \to 0} (1 + x)^{1/x}$.

Solution at the back of the book

The actual value of this particular limit is the number $e \approx 2.71828$, which will be very important in our later work. Notice that $(1 + x)^{1/x}$ cannot be evaluated *at* $x = 0$ since the exponent would be $1/0$, which is undefined.

    Which limits can be evaluated by direct substitution (as in Example 1) and which cannot (as in Example 2)? The answer comes from the following "Rules of Limits."

Rules of Limits

For any constants a and c, and any positive integer n:

1. $\lim_{x \to c} a = a$

   The limit of a constant is just the constant

2. $\lim_{x \to c} x^n = c^n$

   The limit of a power is the power of the limit

3. $\lim_{x \to c} \sqrt[n]{x} = \sqrt[n]{c}$ $(c \geq 0$ if n is even$)$

   The limit of a root is the root of the limit

4. If $\lim_{x \to c} f(x)$ and $\lim_{x \to c} g(x)$ both exist, then

   a. $\lim_{x \to c} [f(x) + g(x)] = \lim_{x \to c} f(x) + \lim_{x \to c} g(x)$

   The limit of a sum is the sum of the limits

   b. $\lim_{x \to c} [f(x) - g(x)] = \lim_{x \to c} f(x) - \lim_{x \to c} g(x)$

   The limit of a difference is the difference of the limits

   c. $\lim_{x \to c} [f(x) \cdot g(x)] = [\lim_{x \to c} f(x)] \cdot [\lim_{x \to c} g(x)]$

   The limit of a product is the product of the limits

   d. $\lim_{x \to c} \dfrac{f(x)}{g(x)} = \dfrac{\lim_{x \to c} f(x)}{\lim_{x \to c} g(x)}$ $[\lim_{x \to c} g(x) \neq 0]$

   The limit of a quotient is the quotient of the limits

These rules, which may be proved from the definition of limit, may be summarized:

Summary of Rules of Limits

For functions composed of the operations of addition, subtraction, multiplication, division, and taking powers and roots, limits may be evaluated by direct substitution, provided that the resulting expression is defined.

$$\lim_{x \to c} f(x) = f(c) \qquad \text{Limit evaluated by direct substitution}$$

EXAMPLE 3 Evaluating Limits by Direct Substitution

a. $\displaystyle\lim_{x \to 4} \sqrt{x} = \sqrt{4} = 2$
Direct substitution of $x = 4$ (using Rule 3 or the Summary)

b. $\displaystyle\lim_{x \to 6} \frac{x^2}{x + 3} = \frac{6^2}{6 + 3} = \frac{36}{9} = 4$
Direct substitution of $x = 6$ (Rules 4, 2, and 1 or the Summary)

∎

PRACTICE PROBLEM 2

Find $\displaystyle\lim_{x \to 3} (2x^2 - 4x + 1)$. *Solution at the back of the book*

If direct substitution into a quotient gives the undefined expression $\dfrac{0}{0}$, factoring, simplifying, and *then* using direct substitution may help.

EXAMPLE 4 Simplifying a Limit

Evaluate $\displaystyle\lim_{x \to 1} \frac{x^2 - 1}{x - 1}$.

Solution

Direct substitution of $x = 1$ into $\dfrac{x^2 - 1}{x - 1}$ gives $\dfrac{0}{0}$, which is undefined. But simplifying the fraction gives

$$\lim_{x \to 1} \frac{x^2 - 1}{x - 1} = \lim_{x \to 1} \frac{(x + 1)(x - 1)}{x - 1} = \lim_{x \to 1} \frac{(x + 1)\cancel{(x - 1)}}{\cancel{x - 1}} = \lim_{x \to 1} (x + 1) = 2$$

Factoring the numerator | Canceling the $(x - 1)$s (since $x \neq 1$) | Now use direct substitution

Therefore, the limit *does* exist and equals 2.

∎

Graphing Calculator Exploration

The graph of $f(x) = \dfrac{x^2 - 1}{x - 1}$ (from Example 4) on a graphing calculator is shown below.

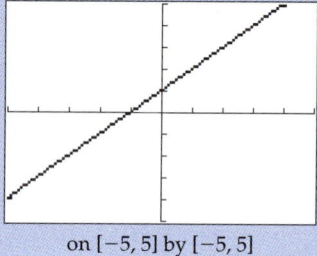

on $[-5, 5]$ by $[-5, 5]$

a. Can you explain why the graph appears to be a straight line?

b. Going beyond the calculator, using your knowledge of the function, is the correct graph really a (complete) line? (*Hint:* See page 53.) Try the window $[-3, 5]$ by $[-5, 5]$.

c. From the graph, can you verify that the limit as $x \to 1$ is 2? (*Hint:* Find the y-value near $x = 1$. Each tick mark is one unit.)

PRACTICE PROBLEM 3   Find $\lim\limits_{x \to 5} \dfrac{2x^2 - 10x}{x - 5}$.   *Solution at the back of the book*

If the values of the function do not approach a *single* number as x approaches a particular number c, then the limit $\lim\limits_{x \to c} f(x)$ *does not exist.*

 EXAMPLE 5   **Finding Whether a Limit Exists**

Find whether $\lim\limits_{x \to 2} \dfrac{1}{x - 2}$ exists by graphing $f(x) = \dfrac{1}{x - 2}$.

Solution

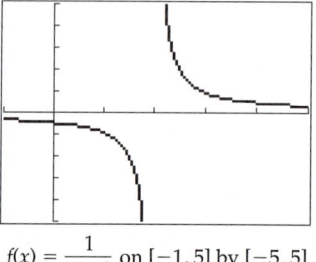

$f(x) = \dfrac{1}{x - 2}$ on $[-1, 5]$ by $[-5, 5]$

If your calculator shows a vertical line between the two curves, try graphing in "dot mode" or changing the viewing window so that $x = 2$ is in the middle.

The graph becomes extremely high just to the right of $x = 2$, and extremely low just to the left of $x = 2$, and so the function does not approach any *single* number as x approaches 2. Therefore:

$$\lim_{x \to 2} \frac{1}{x - 2} \quad \text{does not exist}$$

The fact that this limit does not exist can be seen from tables with x approaching 2.

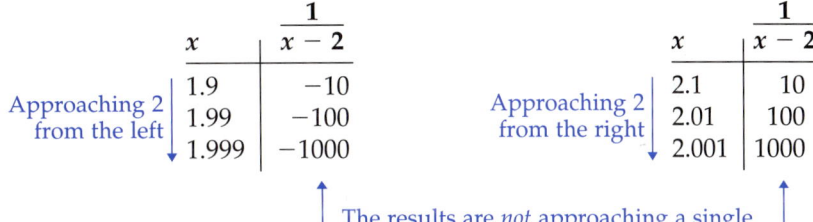

	x	$\dfrac{1}{x - 2}$		x	$\dfrac{1}{x - 2}$
Approaching 2 from the left	1.9	−10	Approaching 2 from the right	2.1	10
	1.99	−100		2.01	100
	1.999	−1000		2.001	1000

The results are *not* approaching a single number (*negative* 1000 and *positive* 1000 are very far apart), so the limit *does not exist*

Some limits involve two variables, with only one variable approaching a limit.

EXAMPLE 6   Finding a Limit for a Function of Two Variables

Find $\lim\limits_{h \to 0} (x^2 + xh + h^2)$.

Solution

Only h is approaching zero, so x remains unchanged. Since the function involves only powers of h, we may evaluate the limit by direct substitution of $h = 0$:

$$\lim_{h \to 0} (x^2 + xh + h^2) = x^2 + x \cdot 0 + 0^2 = x^2$$

$$\qquad\qquad\qquad\quad 0 \qquad 0$$

PRACTICE PROBLEM 4   Find $\lim\limits_{h \to 0} (3x^2 + 5xh + 1)$.   *Solution at the back of the book*

Continuity

Intuitively, a function is said to be *continuous at c* if its graph passes through the point at $x = c$ *without a "hole" or a "jump."* For example, the first function below is *continuous* at *c* (it has no hole or jump at *x* = *c*), while the second and third are *discontinuous* at *c*.

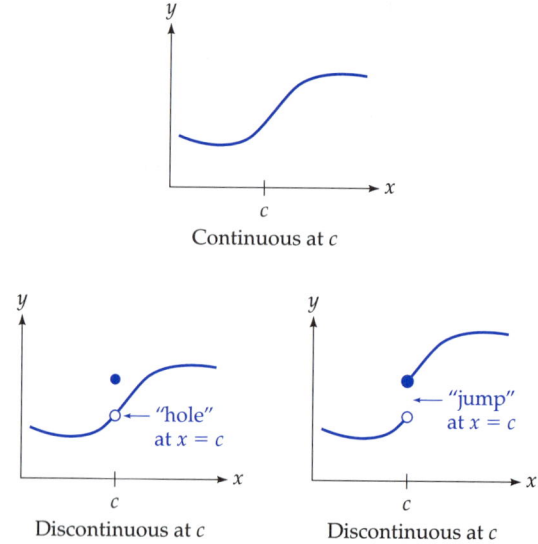

In other words, a function is *continuous at c if the curve approaches the point at x = c*, which may be stated in terms of limits:

$$\lim_{x \to c} f(x) = f(c)$$

This equation means that the quantities on both sides must exist and be *equal*, which we make explicit as follows:

Continuity

A function *f* is continuous at *c* if the following three conditions hold:

1. $f(c)$ is defined
2. $\lim_{x \to c} f(x)$ exists
3. $\lim_{x \to c} f(x) = f(c)$

f is *discontinuous* at *c* if one or more of these conditions *fails* to hold.

Condition 3, which is just the statement that the expressions in conditions 1 and 2 are equal to each other, may by itself be taken as the definition of continuity.

EXAMPLE 7 Finding Discontinuities from a Graph

Each function below is *discontinuous at c* for the indicated reason.

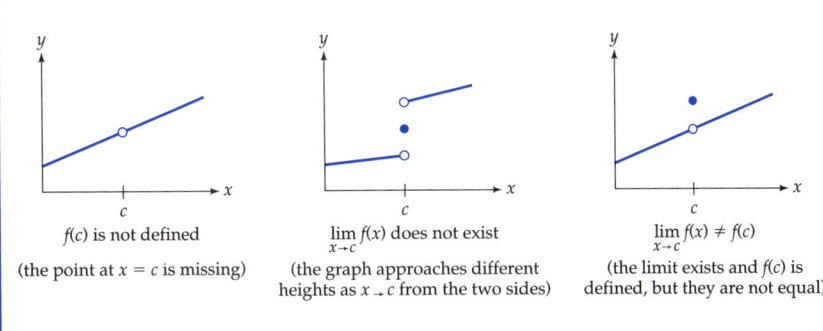

| $f(c)$ is not defined | $\lim\limits_{x \to c} f(x)$ does not exist | $\lim\limits_{x \to c} f(x) \neq f(c)$ |

(the point at $x = c$ is missing) · (the graph approaches different heights as $x \to c$ from the two sides) · (the limit exists and $f(c)$ is defined, but they are not equal)

PRACTICE PROBLEM 5

For each graph below, determine whether the function is continuous at *c*. If it is *not* continuous, indicate the *first* of the three conditions in the previous definition of continuity that is violated.

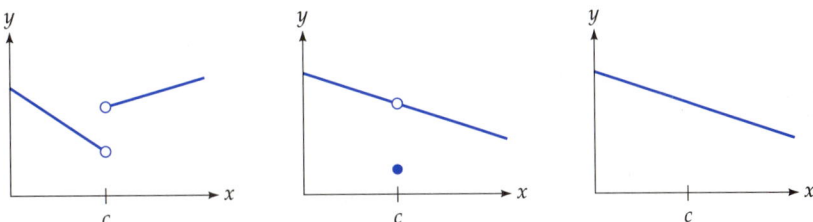

Solutions at the back of the book

A function is continuous on an *open interval* (a, b) if it is continuous at each point of the interval.* A function that is continuous on the entire real line $(-\infty, \infty)$ is said to be *continuous everywhere*, or simply *continuous*.

---

* For continuity on a *closed* interval (which will not be emphasized in this book) we first define "one-sided limits." The *limit from the right*, denoted $\lim\limits_{x \to c^+} f(x)$, means the limit as x approaches c through values *greater* than c. The limit *from the left*, denoted $\lim\limits_{x \to c^-} f(x)$, means the limit as x approaches c through values *less* than c. We then define a function f to be continuous on a *closed* interval $[a, b]$ if it is continuous on the open interval (a, b) and has "one-sided continuity" at the endpoints: $\lim\limits_{x \to a^+} f(x) = f(a)$ and $\lim\limits_{x \to b^-} f(x) = f(b)$.

Which Functions Are Continuous?

Which functions are continuous? Linear functions, the exponential function e^x, and the logarithmic function $\ln x$ (for $x > 0$) are continuous, since we saw in Chapter 0 that their graphs have no holes or jumps. These and other continuous functions can be combined as follows to give other continuous functions.

If functions f and g are continuous at c, then the following are also continuous at c:

1. $f \pm g$ — Sums and differences of continuous functions are continuous

2. $a \cdot f$ [for any constant a] — Constant multiples of continuous functions are continuous

3. $f \cdot g$ — Products of continuous functions are continuous

4. f/g [$g(c) \neq 0$] — Quotients of continuous functions are continuous

5. $f(g(x))$ [for f continuous at $g(c)$] — Compositions of continuous functions are continuous

---

These statements, which can be proved from the Rules of Limits, show that the following types of functions are continuous:

Polynomial functions, e^x, and $\ln x$ ($x > 0$) are continuous.

Rational functions are continuous except where their denominators are zero.

EXAMPLE 8   Determining Continuity

Determine whether each function is continuous.

a. $f(x) = x^3 - 3x^2 - x + 3$ **b.** $f(x) = \dfrac{1}{(x + 1)^2}$ **c.** $f(x) = e^{1+x^2}$

Solution

a. Continuous (it is a polynomial).
b. Discontinuous at $x = -1$ (where the denominator is zero).
c. Continuous (it is the composition of e^x and a polynomial).

■

Calculator-drawn graphs of the first two of these functions are shown below. The polynomial (on the left) is continuous, although you can't really tell from such a graph. The rational function (on the right) exhibits the discontinuity at $x = -1$.

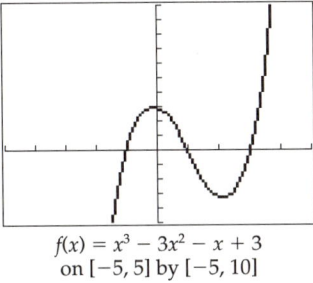

$f(x) = x^3 - 3x^2 - x + 3$
on $[-5, 5]$ by $[-5, 10]$

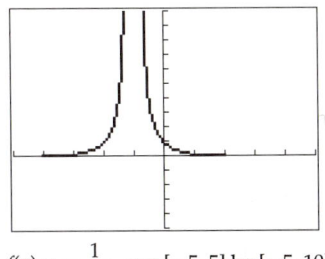

$f(x) = \dfrac{1}{(x + 1)^2}$ on $[-5, 5]$ by $[-5, 10]$

SUMMARY

The limit $\lim\limits_{x \to c} f(x)$ is the number (if it exists) that $f(x)$ approaches as x approaches c $(x \neq c)$. Limits can often be found from tables of function values for x near c, or by direct substitution of $x = c$ (provided that the function consists of additions, subtractions, multiplications, divisions, roots, and powers, and also provided that the resulting expression is defined). Sometimes it helps to simplify before direct substitution.

Continuity can be understood *geometrically* and *analytically*. Geometrically, a function is continuous at c if its graph has neither a hole nor a jump at $x = c$ (which is equivalent to the curve's being connected to the point *at* c). Analytically, a function f is continuous at c if both sides of the following equation are defined and equal:

$$\lim_{x \to c} f(x) = f(c)$$

This equation may be interpreted as saying that continuity is equivalent to being able to *move the limit operation through the function*:

$$\lim_{x \to c} f(x) = f(\overbrace{\lim_{x \to c} x}^{c}) = f(c)$$

Polynomials and e^x are continuous everywhere, $\ln x$ is continuous for $x > 0$, and rational functions are continuous everywhere except where the denominator is zero.

EXERCISES 6.1

Complete the tables and use them to find the given limit (or to find that the limit does not exist). Round calculations to 3 decimal places. A graphing calculator with a TABLE feature will be very helpful.

1. $\lim\limits_{x \to 2} (5x - 7)$

x	$5x - 7$	x	$5x - 7$
1.9		2.1	
1.99		2.01	
1.999		2.001	

2. $\lim\limits_{x \to 4} (2x + 1)$

x	$2x + 1$	x	$2x + 1$
3.9		4.1	
3.99		4.01	
3.999		4.001	

3. $\lim\limits_{x \to 5} \dfrac{1}{x - 5}$

x	$\dfrac{1}{x - 5}$	x	$\dfrac{1}{x - 5}$
4.9		5.1	
4.99		5.01	
4.999		5.001	

4. $\lim\limits_{x \to 3} \dfrac{x}{x - 3}$

x	$\dfrac{x}{x - 3}$	x	$\dfrac{x}{x - 3}$
2.9		3.1	
2.99		3.01	
2.999		3.001	

5. $\lim\limits_{x \to 1} \dfrac{x^3 - 1}{x - 1}$

x	$\dfrac{x^3 - 1}{x - 1}$	x	$\dfrac{x^3 - 1}{x - 1}$
0.9		1.1	
0.99		1.01	
0.999		1.001	

6. $\lim\limits_{x \to 1} \dfrac{x^4 - 1}{x - 1}$

x	$\dfrac{x^4 - 1}{x - 1}$	x	$\dfrac{x^4 - 1}{x - 1}$
0.9		1.1	
0.99		1.01	
0.999		1.001	

Find each limit (or state that it does not exist) by constructing tables similar to those in the previous exercises.

7. $\lim\limits_{x \to 4} \dfrac{x + 4}{x - 4}$

8. $\lim\limits_{x \to 0} (1 - x)^{1/x}$

9. $\lim\limits_{x \to 2} \dfrac{\dfrac{1}{x} - \dfrac{1}{2}}{x - 2}$

10. $\lim\limits_{x \to 1} \dfrac{\sqrt{x} - 1}{x - 1}$

Find each limit (or state that it does not exist) by using TRACE on a graphing calculator to examine the graph near the indicated x-value.

11. $\lim\limits_{x \to 1} \dfrac{\dfrac{1}{x} - 1}{x - 1}$

12. $\lim\limits_{x \to 1.5} \dfrac{2x^2 - 4.5}{x - 1.5}$

13. $\lim\limits_{x \to 4} \dfrac{x^{1.5} - 4x^{0.5}}{x^{1.5} - 2x}$

14. $\lim\limits_{x \to 5} \dfrac{x + 5}{x - 5}$

Evaluate the following limits *without* using a graphing calculator or making tables.

15. $\lim\limits_{x \to 3} (4x^2 - 10x + 2)$

16. $\lim\limits_{x \to 7} \dfrac{x^2 - x}{2x - 7}$

17. $\lim\limits_{x \to 5} \dfrac{3x^2 - 5x}{7x - 10}$

18. $\lim\limits_{t \to 3} \sqrt[3]{t^2 + t - 4}$

19. $\lim\limits_{x \to 3} \sqrt{2}$

20. $\lim\limits_{q \to 9} \dfrac{8 + 2\sqrt{q}}{8 - 2\sqrt{q}}$

21. $\lim\limits_{t \to 25} [(t + 5)t^{-1/2}]$

22. $\lim\limits_{s \to 4} (s^{3/2} - 3s^{1/2})$

23. $\lim\limits_{h \to 0} (5x^3 + 2x^2h - xh^2)$

24. $\lim\limits_{h \to 0} (2x^2 + 4xh + h^2)$

25. $\lim\limits_{x \to 2} \dfrac{x^2 - 4}{x - 2}$

26. $\lim\limits_{x \to 1} \dfrac{x - 1}{x^2 + x - 2}$

27. $\lim\limits_{x \to -1} \dfrac{3x^3 - 3x^2 - 6x}{x^2 + x}$

28. $\lim\limits_{x \to 0} \dfrac{x^2 - x}{x^2 + x}$

29. $\lim\limits_{h \to 0} \dfrac{2xh - 3h^2}{h}$

30. $\lim\limits_{h \to 0} \dfrac{5x^4h - 9xh}{h}$

31. $\lim\limits_{h \to 0} \dfrac{4x^2h + xh^2 - h^3}{h}$

32. $\lim\limits_{h \to 0} \dfrac{x^2h - xh^2 + h^3}{h}$

Determine whether each of the following functions is continuous or discontinuous at c. If it is discontinuous, indicate the *first* of the three conditions in the definition of continuity (page 504) that is violated.

33.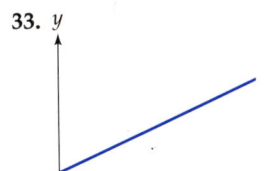

34.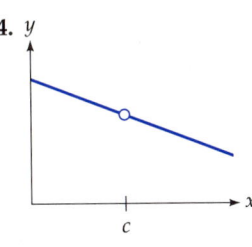

35.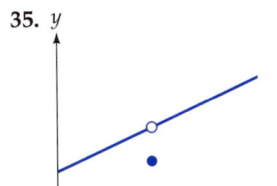

36.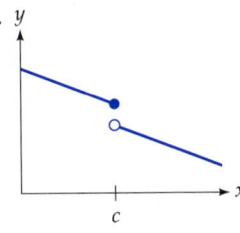

37.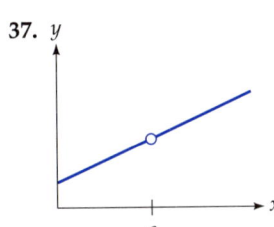

38.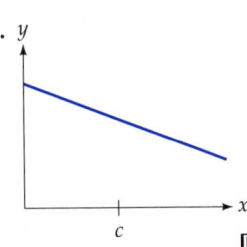

39.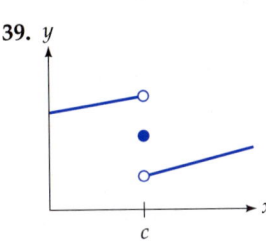

40.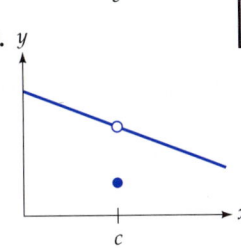

For each piecewise linear function (see pages 53–55) in Ex. 41–44:

a. Draw its graph (by hand or using a graphing calculator).
b. Is it continuous at $x = 3$? If not, indicate the first of the three conditions in the definition of continuity (page 504) that is violated.

41. $f(x) = \begin{cases} x & \text{if } x \le 3 \\ 6 - x & \text{if } x > 3 \end{cases}$

42. $f(x) = \begin{cases} 5 - x & \text{if } x \le 3 \\ x - 2 & \text{if } x > 3 \end{cases}$

43. $f(x) = \begin{cases} x & \text{if } x \le 3 \\ 7 - x & \text{if } x > 3 \end{cases}$

44. $f(x) = \begin{cases} 5 - x & \text{if } x \le 3 \\ x - 1 & \text{if } x > 3 \end{cases}$

Determine whether each function is continuous or discontinuous. If discontinuous, state where it is discontinuous.

45. $f(x) = 7x - 5$

46. $f(x) = 5x^3 - 6x^2 + 2x - 4$

47. $f(x) = \dfrac{x + 1}{x - 1}$

48. $f(x) = e^{-2x^2}$

49. $f(x) = \ln (x^2 + 1)$

50. $f(x) = \dfrac{12}{5x^3 - 5x}$

51. $f(x) = \dfrac{x + 2}{x^4 - 3x^3 - 4x^2}$

52. $f(x) = |x|$

Graph each function on a graphing calculator with viewing window $[-5, 5]$ by $[-5, 5]$ and determine from the graph the values of x at which it is discontinuous.

53. $f(x) = \text{INT}(x)$ On most graphing calculators, "INT(x)" is the greatest integer function, giving the greatest integer less than or equal to x. You may have to ignore some false lines.

54. $f(x) = \dfrac{|x|}{x}$ On most graphing calculators, enter: ABS(x) ÷ x

55. By canceling the common factor, $\dfrac{(x - 1)(x + 2)}{x - 1}$ simplifies to $x + 2$. At $x = 1$, however, the function $\dfrac{(x - 1)(x + 2)}{x - 1}$ is *discontinuous* (since it is undefined where its denominator is zero), whereas $x + 2$ is *continuous*. Are these two functions, one obtained from the other by simplification, equal to each other? Explain.

56. General: Relativity According to Einstein's special theory of relativity, under certain conditions a 1-ft-long object moving with velocity v will appear to a stationary observer to have length $\sqrt{1 - (v/c)^2}$, where c is a constant equal to the speed of light. Find the limiting value of the apparent length as the velocity of the object approaches the speed of light by finding

$$\lim_{v \to c} \sqrt{1 - \left(\frac{v}{c}\right)^2}$$

 57. Business: Interest Compounded Continuously If you deposit $1 into a bank paying 10% interest compounded continuously (see Section 1.2), a year later its value will be

$$\lim_{x \to 0} \left(1 + \frac{x}{10}\right)^{1/x}$$

a. Use a calculator to make tables to find this limit correct to two decimal places, thereby finding the value of the deposit in dollars and cents.

b. Compare your answer from part (a) to the answer found by using the formula Pe^{rt} (from page 126).

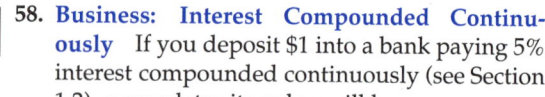 **58. Business: Interest Compounded Continuously** If you deposit $1 into a bank paying 5% interest compounded continuously (see Section 1.2), a year later its value will be

$$\lim_{x \to 0} \left(1 + \frac{x}{20}\right)^{1/x}$$

a. Use a calculator to make tables to find this limit correct to two decimal places, thereby finding the value of the deposit in dollars and cents.

b. Compare your answer from part (a) to the answer found by using the formula Pe^{rt} (from page 126).

6.2   Slopes, Rates of Change, and Derivatives

APPLICATION PREVIEW

The Confused Creation of Calculus

Calculus evolved from consideration of four major problems that were current in seventeenth-century Europe: how to define instantaneous velocity, how to define tangent lines to curves, how to find maximum and minimum values of functions, and how to calculate areas and volumes, all of which will be discussed in this and future chapters. While calculus is now presented as a logical mathematical system, it began instead as a collection of self-contradictory ideas and poorly understood techniques, accompanied by much controversy and criticism. It was developed by two people of diametrically opposed temperaments, Isaac Newton (1642–1727) and Gottfried Wilhelm Leibniz (1646–1716), each while in his twenties (Newton at age 22 and Leibniz at about age 29).

    Newton attended mediocre schools in England, entered Cambridge University (with a deficiency in Euclidean geometry), immersed himself in solitary studies, and graduated with no particular distinction. He then returned to his family's small farm where, during a 2-year period, he singlehandedly developed calculus, in addition to the science of optics and the theory of universal gravitation, all of which he kept to himself. He then returned to Cambridge for a master's degree and secured an appointment as a professor, after which time he became even more solitary and introverted. A contemporary described him as never taking "any recreation or pastime either in riding out to take the air, walking, bowling, or any other exercise whatever, thinking all hours lost that was not spent in his studies." He was often seen "with shoes down at heels, stockings untied, . . . , and his head scarcely combed." He was not a popular lecturer, sometimes lecturing to empty classrooms, and his mathematical writings were very difficult to understand (he told one friend that he wrote this way intentionally "to avoid being bated by little smatterers in mathematics").

Leibniz, on the other hand, was a diplomat, a lawyer, and a world traveler, corresponding with people as far away as China and Ceylon. He scorned universities as "monkish," believing that real knowledge came from more practical activities. While in Paris and London on political missions for his native Germany, he met scientists who stimulated his interest in mathematics, and he began to develop the ideas of calculus. Leibniz's writings, however, were as obscure as Newton's. His paper explaining the rules of calculus was deemed by another mathematician to be "an enigma rather than an explication." To define quantities approaching zero, Leibniz said they were to ordinary variables as the radius of the earth is to that of the heavens, clearly a statement with no mathematical meaning.

Criticism of calculus was not limited to mathematicians. One of its severest critics was Bishop George Berkeley, who, in *The Analyst, Or A Discourse Addressed to an Infidel Mathematician*, wrote that variables approaching zero were "neither finite quantities, nor quantities infinitely small, nor yet nothing," ridiculing them as "the ghosts of departed quantities." Toward the end of the eighteenth century, Voltaire summed up the study of calculus as "the art of numbering and measuring exactly a Thing whose Existence cannot be conceived."

In spite of such inauspicious beginnings, calculus was eventually placed on sound foundations, through the works of mathematicians such as the Bernoullis, Cauchy, Lagrange, and Weierstrass, and finally achieved logical status in the nineteenth century. The improvement in the explanation and understanding of calculus can be appreciated by its progress from a subject usable by only a few brilliant experts to one that is now routinely taught in colleges and understood by undergraduates.

Introduction

In this section we will define one of the most important concepts in all of calculus, the *derivative*, which measures the slope of a curve, or equivalently, the rate of change of a function. We begin by discussing *tangent lines*, first intuitively and later more precisely.

Tangent Lines and Slopes

The *steepness* or *slope* of a curve may vary from point to point. To measure the slope of a curve at a point like that shown below, we draw the *tangent line*, the line through the point whose steepness matches the steepness of the curve at that point.*

Tangent Lines and Slopes of Curves

The tangent line to a curve at a point P is the line through P whose steepness matches the steepness of the curve at P.

The slope of this tangent line gives the slope of the curve at P.

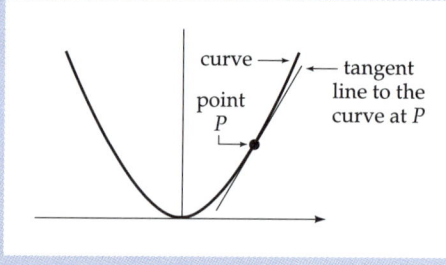

Because the tangent line fits the curve so closely, it is called *the best linear approximation to the curve* near the point of tangency. A graphing calculator can show a curve together with a tangent line. The first graph on the following page shows the curve $y = x^2$ along with its tangent line at $x = 1$. The next two graphs show the results of successively "zooming in" around $(1, 1)$, showing that the curve seems to straighten out and almost becomes its own tangent line (on a smaller and smaller viewing rectangle).

---

* The word "tangent" comes from the Latin *tangere*, "to touch," suggesting that the tangent line just "touches" the curve.

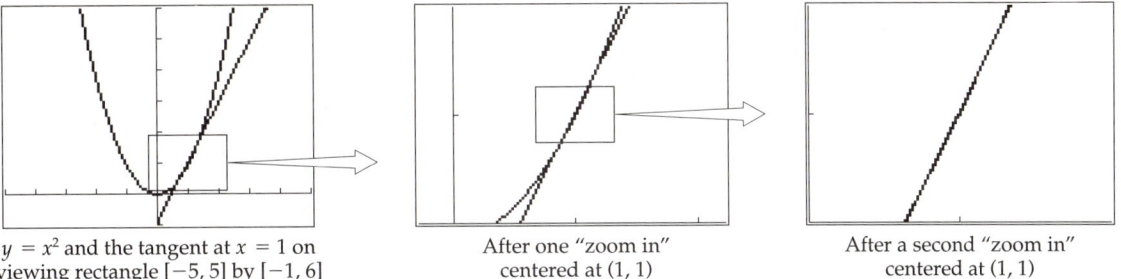

$y = x^2$ and the tangent at $x = 1$ on
viewing rectangle $[-5, 5]$ by $[-1, 6]$

After one "zoom in"
centered at $(1, 1)$

After a second "zoom in"
centered at $(1, 1)$

 Graphing Calculator Exploration

Use a graphing calculator to graph $y = x^3 - 2x^2 - 3x + 4$ (or any function of your choice). Then "zoom in" a few times around a point to see the curve staighten out and almost become its own tangent line.

How do we *define* the tangent line to a curve at a point P? We proceed as follows.

1.   Choose a second point Q on the curve and draw the line through the points P and Q. This line is called a *secant line.**

2.   Let Q slide along the curve toward P. The limiting position of the secant line through P and Q will be the *tangent* line at P.

The (blue) secant line
and (black) tangent line
are far apart

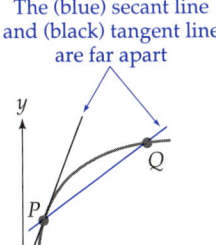

Lines are closer as
Q moves toward P

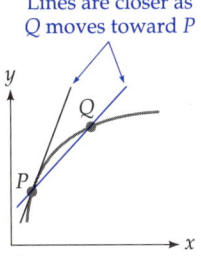

Lines are very close
when Q is near P

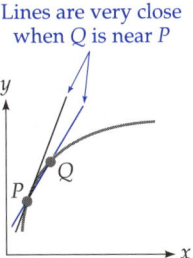

In the limit as $Q \to P$,
the secant line *becomes*
the tangent line at P

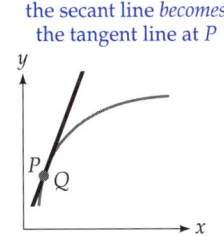

* The word "secant" comes from the Latin *secare*, "to cut," suggesting that the secant line "cuts" the curve at two points.

Suppose that the curve is defined by a function f, that the point P has x-coordinate x, and that Q has x-coordinate $x + h$ (h units further along the x-axis), as shown below.

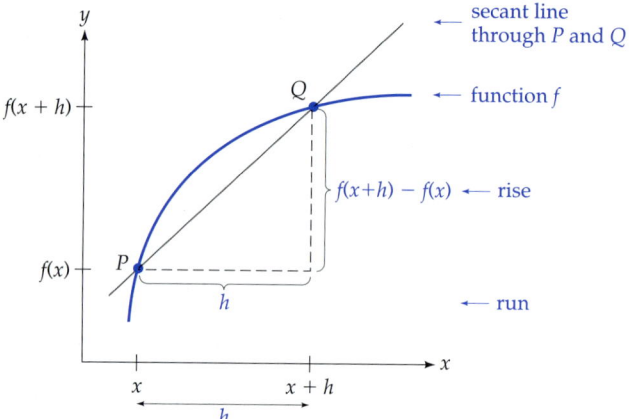

From this diagram, the *slope* (rise over run) of the secant line is

$$\begin{pmatrix} \text{Slope of secant} \\ \text{line through } P \text{ and } Q \end{pmatrix} = \frac{f(x + h) - f(x)}{h} \qquad \begin{array}{l} \leftarrow \text{Rise} \\[4pt] \leftarrow \text{Run} \end{array}$$

Since P and Q are h units apart (measured along the x-axis), letting h approach zero forces Q to approach P, making the slope of the *secant* line approach the slope of the *tangent* line:

$$\begin{pmatrix} \text{Slope of tangent} \\ \text{line at } P \end{pmatrix} = \lim_{h \to 0} \frac{f(x + h) - f(x)}{h} \qquad \text{Limit of } \frac{\text{rise}}{\text{run}}$$

The following example shows how to apply this formula to a particular function to find the slope of the tangent line.

EXAMPLE 1   Finding the Slope of a Tangent Line

Find the slope of the tangent line to $f(x) = x^2$ at the point where $x = 1$.

Solution

$$\text{Slope} = \lim_{h \to 0} \frac{f(1 + h) - f(1)}{h}$$

Formula for the slope of the tangent line, but with $x = 1$

$$= \lim_{h \to 0} \frac{(1 + h)^2 - (1)^2}{h}$$

$f(x) = x^2$ gives
$f(1 + h) = (1 + h)^2$
and $f(1) = 1^2$

$$= \lim_{h \to 0} \frac{1 + 2h + h^2 - 1}{h}$$

Expanding $(1 + h)^2$

$$= \lim_{h \to 0} \frac{\cancel{1} + 2h + h^2 - \cancel{1}}{h} = \lim_{h \to 0} \frac{2h + h^2}{h} \qquad \text{Simplifying}$$

$$= \lim_{h \to 0} \frac{h(2 + h)}{h} = \lim_{h \to 0} \frac{\cancel{h}(2 + h)}{\cancel{h}} \qquad \begin{array}{l}\text{Factoring out } h \text{ and} \\ \text{canceling (since } h \neq 0)\end{array}$$

$$= \lim_{h \to 0} (2 + h) = 2 \qquad \begin{array}{l}\text{Evaluating the limit} \\ \text{by direct substitution}\end{array}$$

The slope of the tangent line is 2.

The curve and the tangent line are shown here.

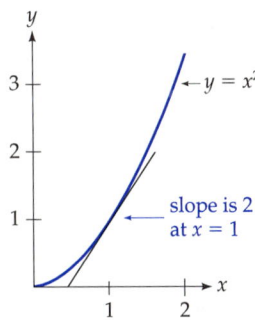

EXAMPLE 2   Finding a Tangent Line

Use the result of Example 1 to find the *equation* of the tangent line to $f(x) = x^2$ at $x = 1$.

Solution

From Example 1 we know that the *slope* of the tangent line is $m = 2$. The point on the curve at $x = 1$ is $(1, 1)$, the y-coordinate coming from $y = f(1) = 1^2 = 1$. The point-slope formula (page 10) then gives

$$y - 1 = 2(x - 1) \qquad \begin{array}{l}y - y_1 = m(x - x_1) \text{ with } m = 2, \\ x_1 = 1, \text{ and } y_1 = 1\end{array}$$

$$y - 1 = 2x - 2 \qquad \text{Multiplying out}$$

$$y = 2x - 1 \qquad \text{Simplifying}$$

↖ Equation of the
tangent line

Graphing Calculator Exploration

On a graphing calculator, graph the curve $y = x^2$ together with the line $y = 2x - 1$ to see that the line is indeed the tangent line to the curve at $x = 1$.

The Derivative

Instead of calculating the slope at just one number, like $x = 1$, if we use the same formula but keep x as a *variable*, we obtain a new function $f'(x)$ (read "f prime of x"), called the *derivative* of f at x, that gives the slope at *any* value of x. The derivative, being the slope, measures how steeply the *curve rises* at x, which means how rapidly the *function increases* at x; this is called the *instantaneous rate of change of the function at x*. To summarize:

Derivative

For a function f, the *derivative of f at x* is defined as

$$f'(x) = \lim_{h \to 0} \frac{f(x + h) - f(x)}{h}$$

(provided that the limit exists). The derivative $f'(x)$ gives the slope of the graph of f at x, and also the instantaneous rate of change of f at x.

The fraction $\dfrac{f(x + h) - f(x)}{h}$ (without the $\lim\limits_{h \to 0}$) is called the *difference quotient*, since it is a quotient (a fraction) and its numerator is a difference. It gives the *average* rate of change of the function over the interval of length h, from x to $x + h$. Taking the *limit* as $h \to 0$ makes the interval shrink to an "instant," giving the *instantaneous* rate of change at x.

| EXAMPLE 3   **Finding the Derivative from the Definition**

For the same function $f(x) = x^2$, find the derivative $f'(x)$ at an arbitrary x-value.

Solution

$$f'(x) = \lim_{h \to 0} \frac{f(x + h) - f(x)}{h}$$ Begin with the definition of the derivative

$$= \lim_{h \to 0} \frac{(x + h)^2 - (x)^2}{h}$$ $f(x + h) = (x + h)^2$ and $f(x) = x^2$

$$= \lim_{h \to 0} \frac{x^2 + 2xh + h^2 - x^2}{h}$$ Expanding $(x + h)^2$

$$= \lim_{h \to 0} \frac{\cancel{x^2} + 2xh + h^2 - \cancel{x^2}}{h} = \lim_{h \to 0} \frac{2xh + h^2}{h}$$ Simplifying

$$= \lim_{h \to 0} \frac{h(2x + h)}{h} = \lim_{h \to 0} \frac{\cancel{h}(2x + h)}{\cancel{h}}$$ Factoring and canceling the h (since $h \neq 0$)

$$= \lim_{h \to 0} (2x + h) = 2x$$ Evaluating the limit by direct substitution

Therefore, the derivative of $f(x) = x^2$ is $f'(x) = 2x$.

■

The derivative gives the slope or instantaneous rate of change at *any* value of x. For example:

At	the derivative is	Also found in Example 1
$x = 1$	$f'(1) = 2 \cdot 1 = 2$	$f'(x) = 2x$ with $x = 1$
$x = 2$	$f'(2) = 2 \cdot 2 = 4$	$f'(x) = 2x$ with $x = 2$
$x = 0$	$f'(0) = 2 \cdot 0 = 0$	$f'(x) = 2x$ with $x = 0$
$x = -1$	$f'(-1) = 2(-1) = -2$	$f'(x) = 2x$ with $x = -1$

These four slopes are shown on the following graph of $f(x) = x^2$.

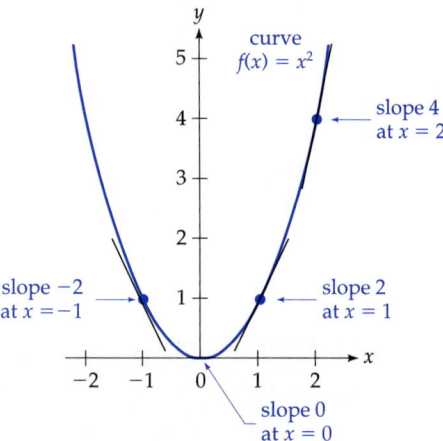

The curve $f(x) = x^2$ has slope
at x given by $f'(x) = 2x$.

The operation of calculating derivatives should be thought of as an operation on *functions,* taking one function [such as $f(x) = x^2$] and giving another function [$f'(x) = 2x$]. The resulting function is called the "derivative" because it is "derived" from the first, and the process of obtaining it is called "differentiation." If the derivative is defined at x, then the original function is said to be *differentiable at x.*

Leibniz's Notation for the Derivative

Since calculus was developed by two people in different countries, there naturally developed two different notations for the derivative. Newton denoted derivatives by a dot over the function, $\dot{f}$, a notation that has been largely replaced by our "prime" notation. Leibniz wrote the derivative of $f(x)$ by writing $\dfrac{d}{dx}$ in front of the function: $\dfrac{d}{dx} f(x)$. In Leibniz's notation, the fact that the derivative of x^2 is $2x$ would be written

$$\frac{d}{dx} x^2 = 2x \qquad \text{The derivative of } x^2 \text{ is } 2x$$

The following table shows equivalent expressions in the two notations.

Prime notation		Leibniz's notation	
$f'(x)$	$=$	$\dfrac{d}{dx} f(x)$	Prime and $\dfrac{d}{dx}$ both mean the derivative
y'	$=$	$\dfrac{dy}{dx}$	For y a function of x

Each notation has its own advantages, and we will use both.* Leibniz's notation comes from the definition of the derivative:

$$\frac{dy}{dx} = \lim_{\Delta x \to 0} \frac{f(x + \Delta x) - f(x)}{\Delta x} \qquad \begin{array}{l}\text{Definition of the derivative (page 516)} \\ \text{with the change in } x \text{ written as } \Delta x\end{array}$$

or

$$\frac{dy}{dx} = \lim_{\Delta x \to 0} \frac{\Delta y}{\Delta x} \qquad \begin{array}{l} f(x + \Delta x) - f(x) \text{ is the change in } y, \\ \text{and so can be written } \Delta y\end{array}$$

* Other notations for the derivative are $Df(x)$ and $D_x f(x)$, but we will not use them.

It is as if the limit turns Δ (the Greek letter D) into d, changing $\dfrac{\Delta y}{\Delta x}$ into $\dfrac{dy}{dx}$. That is, Leibniz's notation reminds us that the derivative $\dfrac{dy}{dx}$ is the limit of the slope $\dfrac{\Delta y}{\Delta x}$.

EXAMPLE 4   Finding a Rate of Change from the Definition of the Derivative

The refining of crude oil into various products like gasoline, heating oil, and plastics requires heating and cooling the oil. If the temperature of the oil at time x hours is $f(x) = x^2 - 7x + 15$ degrees (for $0 \le x \le 8$), find the instantaneous rate of change of temperature at times $x = 6$ hours and $x = 2$ hours.

Solution

The instantaneous rate of change means the derivative:

$$\frac{d}{dx} f(x) = \lim_{h \to 0} \frac{f(x+h) - f(x)}{h}$$

Definition of the derivative (using Leibniz's notation)

$$= \lim_{h \to 0} \frac{\overbrace{(x+h)^2 - 7(x+h) + 15}^{f(x+h)} - \overbrace{(x^2 - 7x + 15)}^{f(x)}}{h}$$

Using $f(x) = x^2 - 7x + 15$

$$= \lim_{h \to 0} \frac{x^2 + 2xh + h^2 - 7x - 7h + 15 - x^2 + 7x - 15}{h}$$

Expanding and simplifying

$$= \lim_{h \to 0} \frac{2xh + h^2 - 7h}{h} = \lim_{h \to 0} \frac{h(2x + h - 7)}{h}$$

Simplifying (since $h \ne 0$)

$$= \lim_{h \to 0} (2x + h - 7) = 2x - 7$$

Evaluating the limit by direct substitution

Derivative of $f(x) = x^2 - 7x + 15$

This derivative gives the instantaneous rate of change of the temperature at time x hours. Evaluating at $x = 6$ gives:

$$f'(6) = 2 \cdot 6 - 7 = 5 \qquad f(x) = 2x - 7 \text{ evaluated at } x = 6$$

Interpretation: At time $x = 6$ hours, the temperature is increasing at the rate of 5 degrees per hour.

Evaluating at $x = 2$ gives:

$$f'(2) = 2 \cdot 2 - 7 = -3 \qquad f(x) = 2x - 7 \text{ evaluated at } x = 2$$

Negative, so the temperature is *decreasing*

Interpretation: At time $x = 2$ hours, the temperature is *decreasing* at the rate of 3 degrees per hour.

—————————————————————————————— ∎

Since $f(x)$ gives the temperature (in degrees) at time x (hours), the derivative has the units *degrees per hour,* with a *positive* result meaning that the temperature is *rising,* and a *negative* result meaning that the temperature is *falling.* In general, the units of the derivative $f'(x)$ are *function units* per *x unit.*

PRACTICE PROBLEM 1   If $f(x)$ gives the population of a city in year x, what are the units of the derivative $f'(x)$?   *Solution at the back of the book*

The derivative gives the rate of change *at a particular instant,* not an actual change over a period of time. Instantaneous rates of change are like the speeds on an automobile speedometer—a reading of 50 mph at one moment does not mean that you will travel exactly 50 miles in the next hour, since the actual distance depends upon your speed during the entire hour. The derivative, however, may be interpreted as the *approximate* change resulting from a one-unit increase in the independent variable. For example, if your speedometer reads 50 mph, then you may say that you will travel *about* 50 miles during the next hour, meaning that this will be true provided that your speed remains steady throughout the hour. (In Chapter 9 we will see how to calculate actual changes from rates of change that do not stay constant.)

EXAMPLE 5   Finding the Derivative of a Rational Function from the Definition

Find the derivative of $f(x) = \dfrac{1}{x}$.

Solution

$$f'(x) = \lim_{h \to 0} \frac{f(x + h) - f(x)}{h}$$
    Definition of $f'(x)$

$$= \lim_{h \to 0} \frac{\dfrac{1}{x + h} - \dfrac{1}{x}}{h}$$
    $f(x + h) = \dfrac{1}{x + h}$ and $f(x) = \dfrac{1}{x}$

$$= \lim_{h \to 0} \frac{1}{h} \left(\frac{1}{x + h} - \frac{1}{x} \right)$$
    Since dividing by h is the same as multiplying by $\dfrac{1}{h}$

$$= \lim_{h \to 0} \frac{1}{h} \left[\frac{x - (x + h)}{(x + h)x} \right]$$
    Subtracting fractions, using the common denominator $(x + h)x$

$$= \lim_{h \to 0} \frac{1}{h} \frac{-h}{(x+h)x} = \lim_{h \to 0} \frac{1}{\cancel{h}} \frac{\cancel{-h}^{-1}}{(x+h)x}$$ Simplifying the numerator and canceling (since $h \neq 0$)

$$= \lim_{h \to 0} \frac{-1}{(x+h)x} = \frac{-1}{(x)x} = \frac{-1}{x^2}$$ Evaluating the limit by direct substitution

Therefore, the derivative of $f(x) = \dfrac{1}{x}$ is $f'(x) = -\dfrac{1}{x^2}$. ∎

Some functions are *not differentiable* (the derivative does not exist) at certain x-values. For example, the following diagram shows a function that has a "corner point" at $x = 1$. At this point the slope (and therefore the derivative) cannot be defined, so the function is not differentiable at $x = 1$. Other nondifferentiable functions will be discussed in Section 6.7.

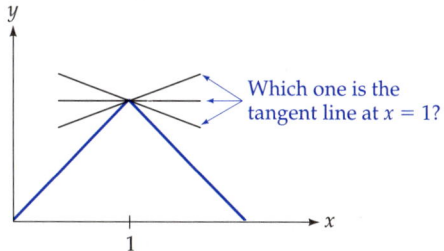

Since the tangent line cannot be uniquely defined at $x = 1$, the slope, and therefore the derivative, is undefined at $x = 1$.

The following diagram shows the geometric relationship between a function (upper graph) and its derivative (lower graph). Observe carefully how the *slope of f* is shown by the *sign of f'*.

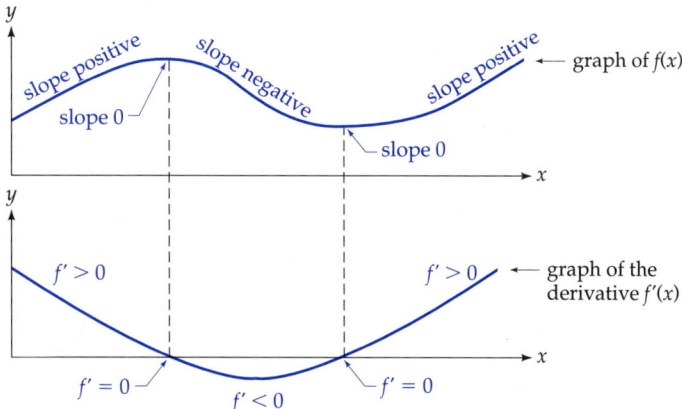

PRACTICE PROBLEM 2   The graph below shows a function and its derivative. Which is the original function (#1 or #2) and which is its derivative? (*Hint:* Which curve has *slope* zero where the other has *value* zero?)

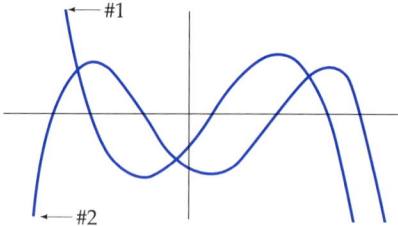

Solution at the back of the book

SUMMARY

A graphing calculator shows how a curve "staightens out" and almost *becomes* its own tangent line as we ZOOM IN around the point of tangency. We found the slope of a curve using the definition of the derivative of the function:

$$f'(x) = \lim_{h \to 0} \frac{f(x + h) - f(x)}{h} \qquad \text{Provided that the limit exists}$$

Some students remember the steps as **DESL** (pronounced "diesel"): write the **D**efinition, **E**xpress the numerator in terms of the function, **S**implify, and take the **L**imit.

The derivative $f'(x)$ gives both the *slope of the graph* of the function at x and the *instantaneous rate of change of the function* at x. In other words, "derivative," "slope," and "instantaneous rate of change" are merely the mathematical, the geometric, and the analytic versions of the same idea.

EXERCISES 6.2

By imagining tangent lines at points P_1, P_2, and P_3, state whether the slopes are positive, zero, or negative at these points.

1. *y*

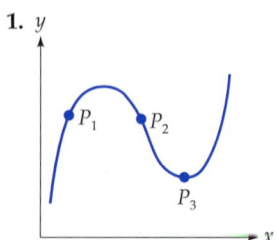

2. *y*

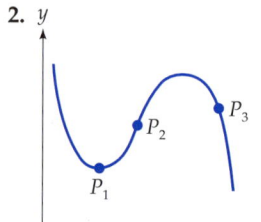

3.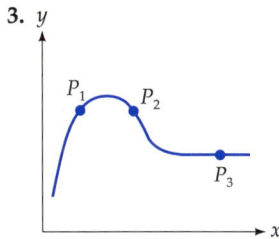

4.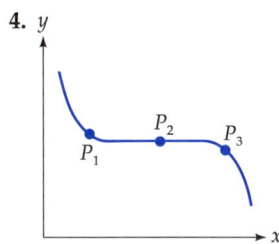

Use the tangent lines shown at points P_1 and P_2 to find the slopes of the curve at these points.

5.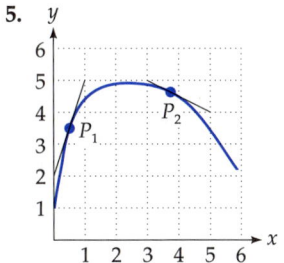

6.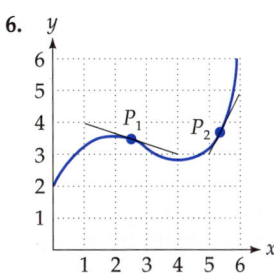

For each function $f(x)$ graphed below, make a rough sketch of the derivative $f'(x)$ showing where $f'(x)$

is positive, negative, and zero. (Omit scale on y-axis.)

7.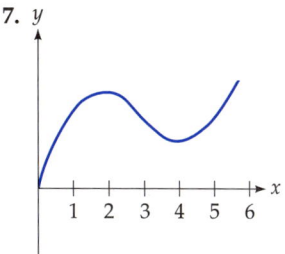

8.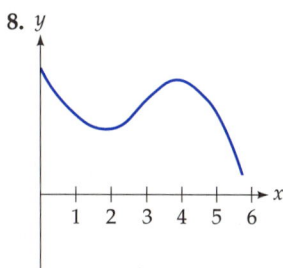

Find $f'(x)$ by using the definition of the derivative.

9. $f(x) = x^2 - 3x + 5$

10. $f(x) = 2x^2 - 5x + 1$

11. $f(x) = 1 - x^2$

12. $f(x) = \dfrac{1}{2}x^2 + 1$

13. $f(x) = 9x - 2$

14. $f(x) = -3x + 5$

15. $f(x) = \dfrac{x}{2}$

16. $f(x) = 0.01x + 0.05$

17. $f(x) = 4$

18. $f(x) = \pi$

19. $f(x) = ax^2 + bx + c$
(a, b, and c are constants)

20. $f(x) = (x + a)^2$
[a is a constant. *Hint:* First expand $(x + a)^2$.]

21. $f(x) = x^3$
[*Hint:* Use $(x + h)^3 = x^3 + 3x^2h + 3xh^2 + h^3$.]

22. $f(x) = x^4$
[*Hint:* Use $(x + h)^4 = x^4 + 4x^3h + 6x^2h^2 + 4xh^3 + h^4$.]

23. $f(x) = \dfrac{2}{x}$

24. $f(x) = \dfrac{1}{x^2}$

25. $f(x) = \sqrt{x}$

[*Hint:* Multiply the numerator and denominator of the difference quotient by $(\sqrt{x + h} + \sqrt{x})$ and then simplify.]

26. $f(x) = \dfrac{1}{\sqrt{x}}$

[*Hint:* Multiply the numerator and denominator of the difference quotient by $(\sqrt{x} + \sqrt{x + h})$ and then simplify.] ·

27. $f(x) = x^3 + x^2$

28. $f(x) = \dfrac{1}{2x}$

29. a. Using your answer to Exercise 9 above, find the equation for the tangent line to the curve $f(x) = x^2 - 3x + 5$ at $x = 2$, writing the equation in slope-intercept form.

b. Use a graphing calculator to graph the curve together with the tangent line to verify your answer.

30. a. Using your answer to Exercise 10 above, find the equation for the tangent line to the curve $f(x) = 2x^2 - 5x + 1$ at $x = 2$, writing the equation in slope-intercept form.

b. Use a graphing calculator to graph the curve together with the tangent line to verify your answer.

31. a. The tangent line to $y = x^2$ at $x = c$ passes through the point (c, c^2) and has slope $m = 2c$ (from Example 3 on pages 516–517). Substitute this point and slope into the point-slope equation for a line and simplify to show that the tangent line to the curve at $x = c$ can be written $y = 2cx - c^2$.

b. On a graphing calculator, graph the curve $y = x^2$ together with the tangent line $y = 2cx - c^2$ for several different values of c to verify that the equation for the tangent line is correct.

32. a. The tangent line to $y = 1/x$ at $x = c$ passes through the point $(c, 1/c)$ and has slope $m = -1/c^2$ (from Example 5 on pages 520–521, for $c \neq 0$). Substitute this point and slope into the point-slope equation for a line and simplify to show that the tangent line to $y = \dfrac{1}{x}$ at $x = c$ can be written

$$y = \frac{2}{c} - \frac{x}{c^2}.$$

b. On a graphing calculator, graph the function $y = 1/x$ together with the tangent line $y = \dfrac{2}{c} - \dfrac{x}{c^2}$ for several different values of c to verify that the equation for the tangent line is correct.

For each function in Exercises 33–38:

a. Find $f'(x)$ using the definition of the derivative.

b. Explain, by considering the original function, why the derivative is a constant.

33. $f(x) = 3x - 4$

34. $f(x) = 2x - 9$

35. $f(x) = 5$

36. $f(x) = 12$

37. $f(x) = mx + b$
(m and b are constants)

38. $f(x) = b$
(b is a constant)

39. a. Show that the definition of the derivative applied to the function $f(x) = \sqrt[4]{x} = x^{0.25}$ at $x = 1$ gives $f'(1) = \lim\limits_{h \to 0} \dfrac{(1 + h)^{0.25} - 1}{h}$.

b. Use a calculator to evaluate the difference quotient $\dfrac{(1 + h)^{0.25} - 1}{h}$ for the following values of h: 0.1, 0.01, and 0.001. (*Hint:* Enter the calculation into your calculator with h replaced by 0.1, and then change the value of h by inserting zeros.) From your answers, guess the value of the above limit, which will be the derivative of $f(x) = \sqrt[4]{x}$ at $x = 1$.

c. Evaluate the same difference quotient at the following *negative* values of h: -0.1, -0.01, and -0.001 (using the $(-)$ key). Do these answers seem to approach the same limit?

40. a. Show that the definition of the derivative applied to the function $f(x) = \sqrt[5]{x} = x^{0.2}$ at $x = 1$ gives $f'(1) = \lim\limits_{h \to 0} \dfrac{(1 + h)^{0.2} - 1}{h}$.

b. Use a calculator to evaluate the difference quotient $\dfrac{(1 + h)^{0.2} - 1}{h}$ for the following values of h: 0.1, 0.01, and 0.001. (*Hint:* Enter the calculation into your calculator with

h replaced by 0.1, and then change the value of h by inserting zeros.) From your answers, guess the value of the above limit, which will be the derivative of $f(x) = \sqrt[5]{x}$ at $x = 1$.

c. Evaluate the same difference quotient at the following *negative* values of h: -0.1, -0.01, and -0.001 (using the $\boxed{(-)}$ key). Do these answers seem to approach the same limit?

APPLIED EXERCISES

41. Business: Temperature The temperature in an industrial pasteurization tank is $f(x) = x^2 - 8x + 110$ degrees after x minutes (for $0 \le x \le 12$).

a. Find $f'(x)$ by using the definition of the derivative.

b. Use your answer to part (a) to find the instantaneous rate of change of the temperature after 2 minutes. Be sure to interpret the sign of your answer.

c. Use your answer to part (a) to find the instantaneous rate of change after 5 minutes.

42. General: Population The population of a town is $f(x) = 3x^2 - 12x + 200$ people after x weeks (for $0 \le x \le 20$).

a. Find $f'(x)$ by using the definition of the derivative.

b. Use your answer to part (a) to find the instantaneous rate of change of the population after 1 week. Be sure to interpret the sign of your answer.

c. Use your answer to part (a) to find the instantaneous rate of change of the population after 5 weeks.

43. Behavioral Science: Learning Theory In a psychology experiment a person could memorize x words in $f(x) = 2x^2 - x$ seconds (for $0 \le x \le 10$).

a. Find $f'(x)$ by using the definition of the derivative.

b. Evaluate your answer at $x = 5$ and interpret it as an instantaneous rate of change in the proper units.

44. Business: Advertising An automobile dealership finds that the number of cars that it sells on day x of an advertising campaign is $S(x) = -x^2 + 10x$ (for $0 \le x \le 7$).

a. Find $S'(x)$ by using the definition of the derivative.

b. Use your answer to part (a) to find the instantaneous rate of change on day $x = 3$.

c. Use your answer to part (a) to find the instantaneous rate of change on day $x = 6$.
Be sure to interpret the signs of your answers.

45. Biomedical: Temperature The temperature of a patient in a hospital on day x of an illness is given by $T(x) = -x^2 + 5x + 100$ (for $1 < x < 5$).

a. Find $T'(x)$ by using the definition of the derivative.

b. Use your answer to part (a) to find the instantaneous rate of change of temperature on day 2.

c. Use your answer to part (a) to find the instantaneous rate of change of temperature on day 3.

d. What do your answers to parts (b) and (c) tell you about the patient's health on those two days?

46. Biomedical: Bacteria The number of bacteria in a culture x hours after treatment with an antibiotic is given by $f(x) = -x^2 + 12x + 1000$ (for $1 < x < 30$).

a. Find $f'(x)$ by using the definition of the derivative.

b. Use your answer to part (a) to find the instantaneous rate of change after 2 hours.

c. Use your answer to part (a) to find the instantaneous rate of change after 20 hours.
Be sure to interpret the signs of your answers.

6.3 Some Differentiation Formulas

Calculus and Navigation

As calculus progressed from a collection of poorly understood ideas (see pages 510–511) to a coherent and understandable mathematical structure, it began to have a fundamental effect on many of the practical problems of the day. For example, one of the greatest problems was the absence of reliable means of navigation, hindering international communication, travel, and trade. A ship's position on the seas is determined by its latitude (its north-south position) and longitude (its east-west position). Latitude is easily found from sun and star charts, while determining longitude is far more difficult because of the earth's rotation. The costs of not knowing one's longitude could be deadly. In 1707, the British Admiral Sir Clowdisley Shovell was returning home after defeating the French in the Mediterranean. In fog for twelve days, and believing his navigators that he was safely west of the coast of France, he turned north to head home to England. His navigators, however, had misjudged the ship's longitude, and on the night of October 22, just a few miles from the coast of England, his fleet sailed into the rocks off the Isles of Scilly. The flagship *Association* hit first, sinking within minutes, drowning everyone on board. Before there was time to react, two more ships hit the rocks and quickly sank. In all, four of the five ships went down, taking 2000 sailors with them. Such misadventures were far from unique. Naval history of the time is filled with stories of ships that misjudged their longitude and ended up on the ocean bottom.

Calculus provided the basic tools for tackling navigation and other practical problems. By enabling scientists to calculate the effects of gravity, calculus led to more accurate predictions of the positions of the moon and planets for navigational tables. By providing precise definitions of concepts like velocity and acceleration, it led to improvements in pendulum clocks and chronometers, also vital to navigation. By defining tangents to curves, calculus allowed an understanding of how light rays bend as they travel through lenses, leading to better telescopes.

In these and many other ways, the simplified differentiation formulas (due mainly to Leibniz) of this and the following sections enabled calculus to be widely understood and used, laying the foundations for many of the technological advances from the seventeenth century to today.

Introduction

In Section 6.2 we defined the *derivative* of a function and used it to calculate slopes and instantaneous rates of change. Even for a function as simple as $f(x) = x^2$, however, calculating the derivative from the definition was rather involved. Calculus would be of limited usefulness if all derivatives had to be calculated in this way.

   In this section we will learn several *rules of differentiation* that will simplify our calculations. The rules are derived from the definition of the derivative, which is why we studied the definition first. We will also learn another important use for differentiation, one that will be very important in business applications: calculating "marginals" (marginal revenue, marginal cost, and marginal profit).

Derivative of a Constant

The first rule of differentiation shows how to differentiate a constant function.

For any constant c

$$\frac{d}{dx} c = 0 \qquad \text{Derivative of a constant is zero}$$

EXAMPLE 1 **Differentiating a Constant**

$$\frac{d}{dx} 7 = 0$$

∎

This rule is obvious geometrically. The graph of a constant function $f(x) = c$ is the horizontal line $y = c$. Since the slope of a horizontal line is zero, the derivative of $f(x) = c$ is zero.

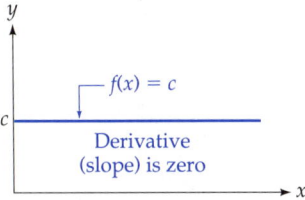

A constant function (a horizontal line) has derivative (slope) zero.

This rule follows immediately from the definition of the derivative. The constant function $f(x) = c$ has the same value c for *any* value of x, so in particular, $f(x + h) = c$ and $f(x) = c$. Substituting these into the definition of the derivative gives:

$$f'(x) = \lim_{h \to 0} \frac{\overbrace{f(x + h)}^{c} - \overbrace{f(x)}^{c}}{h} = \lim_{h \to 0} \frac{c - c}{h} = \lim_{h \to 0} \frac{\overbrace{0}^{0}}{h} = \lim_{h \to 0} 0 = 0$$

Therefore, the derivative of a constant function $f(x) = c$ is $f'(x) = 0$.

Power Rule

One of the most useful differentiation formulas in all of calculus is called the *power rule*. It tells how to differentiate powers like x^7 or x^{100}.

Power Rule

For any constant exponent n,

$$\frac{d}{dx} x^n = n \cdot x^{n-1}$$

To differentiate x^n, bring down the exponent as a multiplier and then decrease the exponent by 1

A derivation of the power rule for positive integer exponents is given at the end of this section. We will use the power rule for *all* real numbers n, since more general proofs will be given later (see pages 553, 669–670, and 707).

EXAMPLE 2   Using the Power Rule

a. $\quad \dfrac{d}{dx} x^7 = 7x^{7-1} = 7x^6$

Bring down the exponent   Decrease the exponent by 1

b. $\dfrac{d}{dx} x^{100} = 100x^{100-1} = 100x^{99}$

c. $\dfrac{d}{dx} x^{-2} = -2x^{-2-1} = -2x^{-3}$

The power rule holds for negative exponents

d. $\dfrac{d}{dx}\sqrt{x} = \dfrac{d}{dx}x^{\frac{1}{2}} = \dfrac{1}{2}x^{\frac{1}{2}-1} = \dfrac{1}{2}x^{-\frac{1}{2}}$ And for fractional exponents

e. $\dfrac{d}{dx}x^1 = 1x^{1-1} = x^0 = 1$

This last result is used so frequently that it should be remembered separately.

$$\frac{d}{dx}x = 1$$ The derivative of x is 1

From now on we will skip the middle step in these examples, differentiating powers in one step:

$$\frac{d}{dx}x^{50} = 50x^{49}$$

$$\frac{d}{dx}x^{2/3} = \frac{2}{3}x^{-1/3}$$ $\dfrac{2}{3}-1$

PRACTICE PROBLEM 1   Find

a. $\dfrac{d}{dx}x^2$ **b.** $\dfrac{d}{dx}x^{-5}$ **c.** $\dfrac{d}{dx}\sqrt[4]{x}$ *Solutions at the back of the book*

Constant Multiple Rule

The power rule shows how to differentiate a power like x^3. The *constant multiple rule* extends this result to functions like $5x^3$, a constant *times* a function. Briefly, to differentiate a constant times a function, we simply "carry along" the constant and differentiate the function.

Constant Multiple Rule

For any constant c,

$$\frac{d}{dx}[c \cdot f(x)] = c \cdot f'(x)$$ The derivative of a constant times a function is the constant times the derivative of the function

(provided, of course, that the derivative $f'(x)$ exists). A derivation of this rule is given at the end of this section.

EXAMPLE 3 Using the Constant Multiple Rule

a. $\dfrac{d}{dx}\,5x^3 = 5 \cdot 3x^2 = 15x^2$

Carry along the constant ⟶ ⟵ Derivative of x^3

b. $\dfrac{d}{dx}\,3x^{-4} = 3(-4)x^{-5} = -12x^{-5}$

Again we will skip the middle step, bringing down the exponent and immediately multiplying it by the number in front of the x.

EXAMPLE 4 Calculating Derivatives More Quickly

a. $\dfrac{d}{dx}\,8x^{-1/2} = -4x^{-3/2}$

$8\left(-\tfrac{1}{2}\right)$

b. $\dfrac{d}{dx}\,7x = 7 \cdot 1 = 7$

Derivative of x

This example, showing that the derivative of $7x$ is just 7, leads to a very useful general rule.

> For any constant c,
>
> $$\dfrac{d}{dx}\,(cx) = c$$
>
> The derivative of a constant times x is just the constant

EXAMPLE 5 Finding Derivatives Involving Constants

a. $\dfrac{d}{dx}\,(7x) = 7$ Using $\dfrac{d}{dx}\,(cx) = c$

b. $\dfrac{d}{dx}\,7 = 0$ For a constant alone, the derivative is zero

c. $\dfrac{d}{dx}\,(7x^2) = 7 \cdot 2x = 14x$ But for a constant times a function, the derivative is the constant times the derivative of the function

Sum Rule

The *sum rule* extends differentiation to sums of functions. Briefly, to differentiate a *sum* of two functions, just differentiate the functions separately and add the results.

Sum Rule

$$\frac{d}{dx}[f(x) + g(x)] = f'(x) + g'(x)$$ The derivative of a sum is the sum of the derivatives

(provided, of course, that both the derivatives $f'(x)$ and $g'(x)$ exist). A derivation of the sum rule is given at the end of this section. For example, the derivative of the sum

$$x^3 + x^5 \qquad \text{Sum of } x^3 \text{ and } x^5$$

is just the sum of the derivatives $\downarrow \quad \downarrow$

$$3x^2 + 5x^4 \qquad \text{Differentiating each separately and adding}$$

A similar rule holds for the *difference* of two functions,

$$\frac{d}{dx}[f(x) - g(x)] = f'(x) - g'(x)$$ The derivative of a difference is the difference of the derivatives

(provided that $f'(x)$ and $g'(x)$ exist). These two rules may be combined:

Sum–Difference Rule

$$\frac{d}{dx}[f(x) \pm g(x)] = f'(x) \pm g'(x)$$ Use both upper signs or both lower signs

Similar rules hold for sums and differences of any finite number of terms. Using these rules, we may differentiate any polynomial or, more generally, functions with variables raised to *any* constant powers.

EXAMPLE 6 Using the Sum–Difference Rule

a. $\dfrac{d}{dx}(x^3 - x^5) = 3x^2 - 5x^4$ Derivatives taken separately

b. $\dfrac{d}{dx}(5x^{-2} - 6x^{1/3} + 4) = -10x^{-3} - 2x^{-2/3}$ The constant 4 has derivative 0

∎

Leibniz's Notation and Evaluation of Derivatives

Leibniz's derivative notation, $\dfrac{d}{dx}$, is often read "the derivative with respect to x" to emphasize that the independent variable is x. To differentiate a function of some *other* variable, the x in $\dfrac{d}{dx}$ is replaced by the other variable. For example:

Function	Derivative	
$f(t)$	$\dfrac{d}{dt}f(t)$	Use $\dfrac{d}{dt}$ for the derivative with respect to t
w^3	$\dfrac{d}{dw}w^3$	Use $\dfrac{d}{dw}$ for the derivative with respect to w

The following two notations both mean the derivative *evaluated* at $x = 2$.

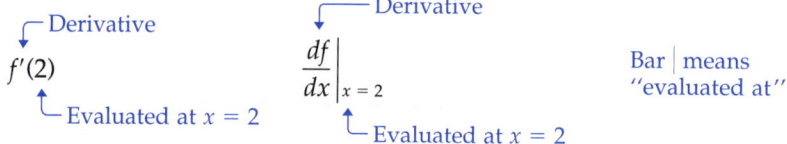

$f'(2)$ ⟵ Derivative, Evaluated at $x = 2$

$\left.\dfrac{df}{dx}\right|_{x=2}$ ⟵ Derivative, Evaluated at $x = 2$

Bar $|$ means "evaluated at"

Be careful! Both notations mean *first* differentiate and *then* evaluate.

EXAMPLE 7 Evaluating a Derivative

If $f(x) = x^4$, find $f'(2)$.

Solution

$f'(x) = 4x^3$ First differentiate

$f'(2) = 4 \cdot 2^3 = 4 \cdot 8 = 32$ Then evaluate

PRACTICE PROBLEM 2   If $f(x) = x^3$, find $\left.\dfrac{df}{dx}\right|_{x=-1}$.   *Solution at the back of the book*

Derivatives in Business and Economics: Marginals

There is another interpretation for the derivative, one that is particularly important in business and economics. Suppose that a company has calculated its revenue, cost, and profit functions, as defined below.

$$R(x) = \begin{bmatrix} \text{Total revenue (income)} \\ \text{from selling } x \text{ units} \end{bmatrix} \qquad \text{Revenue function}$$

$$C(x) = \begin{bmatrix} \text{Total cost of} \\ \text{producing } x \text{ units} \end{bmatrix} \qquad \text{Cost function}$$

$$P(x) = \begin{bmatrix} \text{Total profit from producing} \\ \text{and selling } x \text{ units} \end{bmatrix} \qquad \text{Profit function}$$

The term *marginal cost* means the additional cost of producing one more unit. The cost of one more unit is the *rate* at which costs are rising (measured in dollars per unit). If we calculate rates of change as *instantaneous* rates of change (that is, as derivatives), we see that the marginal cost function $MC(x)$ is the derivative of the cost function:

$$MC(x) = C'(x) \qquad \text{Marginal cost is the derivative of cost}$$

The marginal *revenue* function $MR(x)$ and the marginal *profit* function $MP(x)$ are similarly defined as the derivatives of the revenue and cost functions:

$$MR(x) = R'(x) \qquad \text{Marginal revenue is the derivative of revenue}$$
$$MP(x) = P'(x) \qquad \text{Marginal profit is the derivative of profit}$$

All of this can be summarized very briefly: "marginal" means "derivative of." We now have three interpretations for the derivative: *slopes*, *instantaneous rates of change*, and *marginals*.

EXAMPLE 8 Finding and Interpreting Marginal Cost

A company manufactures cordless telephones and finds that its cost function (the total cost of manufacturing x telephones) is

$$C(x) = 400\sqrt{x} + 500 \qquad \text{Cost function}$$

dollars, where x is the number of telephones produced.

a. Find the marginal cost function $MC(x)$.
b. Find the marginal cost when 100 telephones have been produced, and interpret your answer.

Solution

a. The marginal cost function is the derivative of the cost function $C(x) = 400x^{1/2} + 500$:

$$MC(x) = 200x^{-1/2} = \frac{200}{\sqrt{x}} \qquad \text{Derivative of } C(x)$$

b. To find the marginal cost when 100 telephones have been produced, we evaluate the marginal cost function at $x = 100$:

$$MC(100) = \frac{200}{\sqrt{100}} = \frac{200}{10} = \$20 \qquad MC(x) = \frac{200}{\sqrt{x}} \text{ evaluated at } x = 100$$

Interpretation: When 100 telephones have been produced, the marginal cost is \$20, meaning that to produce one more telephone costs about \$20.

■

EXAMPLE 9   Finding a Learning Rate

A psychology researcher finds that the number of names that a person can memorize in x minutes is approximately $f(x) = 6\sqrt[3]{x^2}$. Find the instantaneous rate of change of this function after 8 minutes and interpret your answer.

Solution

$$f(x) = 6x^{2/3} \qquad 6\sqrt[3]{x^2} \text{ in exponential form}$$

$$f'(x) = 6 \cdot \frac{2}{3} x^{-1/3} = 4x^{-1/3} \qquad \begin{array}{l}\text{The instantaneous rate} \\ \text{of change is } f'(x)\end{array}$$

$$f'(8) = 4(8)^{-1/3} = 4\left(\frac{1}{\sqrt[3]{8}}\right) = 4\left(\frac{1}{2}\right) = 2 \qquad \text{Evaluating at } x = 8$$

Interpretation: After 8 minutes the person can memorize about two additional names per minute.

■

Functions as Single Objects

You may have noticed that calculus requires a more abstract point of view than precalculus mathematics. In earlier courses you looked at functions and graphs as collections of individual points, to be plotted one at a time. Now, however, we are operating on *whole functions* all at once (for example, differentiating the function x^3 to obtain the function $3x^2$). In calculus, the basic objects of interest are *functions*, and a function should be thought of as a *single* object.

This is in keeping with a trend toward increasing abstraction as you learn mathematics. You first studied single numbers, then points (pairs of numbers), then functions (collections of points), and now collections of functions (polynomials, differentiable functions, and so on). Each stage has been a generalization of the previous stage as you reach higher levels of sophistication. This process of generalization or "chunking" of knowledge enables you to express ideas of wider applicability and power.

Derivatives on a Graphing Calculator

Graphing calculators have an operation called NDERIV (or something similar), standing for *numerical derivative*, which gives an *approximation* for the derivative of a function. Most do so by evaluating the *symmetric difference quotient,* $\dfrac{f(x + h) - f(x - h)}{2h}$ for a small value of h, like $h = 0.001$. The numerator represents the change in the function when x changes by $2h$ (from $x - h$ to $x + h$), and the denominator divides by this change in x. Geometrically, the symmetric difference quotient gives the slope of the secant line between two points on the curve equidistant on either side of the point at x. While NDERIV usually approximates the derivative quite closely, it sometimes gives erroneous results, as we will see in later sections. For this reason, using a graphing calculator effectively requires an understanding of both the calculus that underlies it and the technology that limits it.

Graphing Calculator Exploration

a. Use a graphing calculator to graph $y_1 = x^3 - x^2 - 6x + 3$ on $[-5, 5]$ by $[-10, 10]$.

b. Define y_2 as the derivative of y_1 (using NDERIV) and graph both functions.

c. Observe that where y_1 is horizontal, the *value* of y_2 is zero; where y_1 slopes *upward*, y_2 is *positive*; and where y_1 slopes *downward*, y_2 is *negative*. Would you be able to use these observations to identify which curve is the original function and which is the derivative?

d. Now check your answer to Example 9 as follows: Redefine y_1 as $y_1 = 6x^{2/3}$, reset the viewing window to $[0, 10]$ by $[-10, 30]$, GRAPH y_1 and y_2, and EVALUATE y_2 at $x = 8$. Your answer should agree with that of Example 9.

SUMMARY

Our development of calculus has followed two quite different lines—one technical (the *rules* of derivatives) and the other conceptual (the *meaning* of derivatives).

On the conceptual side, the derivative has three meanings:

- Instantaneous rates of change
- Slopes of curves
- Marginals

The fact that the derivative represents all three of these ideas simultaneously is one of the reasons that calculus is so useful.

On the technical side, although we have learned several differentiation rules, we really only know how to differentiate one kind of function, *x to a constant power*:

$$\frac{d}{dx} x^n = n x^{n-1}$$

The other rules:

$$\frac{d}{dx} [c \cdot f(x)] = c \cdot f'(x)$$

and

$$\frac{d}{dx} [f(x) \pm g(x)] = f'(x) \pm g'(x)$$

simply extend the power rule to sums, differences, and constant multiples of such powers. Therefore, any function to be differentiated must first be expressed in terms of powers. This is why we reviewed exponential notation so carefully in Chapter 0. (In Chapter 8 we will differentiate logarithmic and exponential functions.)

Verification of the Power Rule for Positive Integer Exponents

Multiplying $(x + h)$ times itself repeatedly gives:

$$(x + h)^2 = x^2 + 2xh + h^2$$

$$(x + h)^3 = x^3 + 3x^2h + 3xh^2 + h^3$$

and in general, for any positive integer n,

$$(x + h)^n = x^n + nx^{n-1}h + \tfrac{1}{2}n(n-1)x^{n-2}h^2 + \cdots + nxh^{n-1} + h^n$$

$$= x^n + nx^{n-1}h + h^2[\tfrac{1}{2}n(n-1)x^{n-2} + \cdots + nxh^{n-3} + h^{n-2}] \qquad \text{Factoring out } h^2$$

$$= x^n + nx^{n-1}h + h^2 \cdot P \qquad P \text{ stands for the polynomial in the square bracket above}$$

The resulting formula:

$$(x + h)^n = x^n + nx^{n-1}h + h^2 \cdot P$$

will be useful below. To prove the power rule for any positive integer n we use the definition of the derivative to differentiate $f(x) = x^n$.

$$f'(x) = \lim_{h \to 0} \frac{f(x + h) - f(x)}{h} \qquad \text{Definition of the derivative}$$

$$= \lim_{h \to 0} \frac{(x + h)^n - x^n}{h} \qquad \begin{array}{l}\text{Since } f(x + h) = (x + h)^n \text{ and} \\ f(x) = x^n\end{array}$$

$$= \lim_{h \to 0} \frac{x^n + nx^{n-1}h + h^2 \cdot P - x^n}{h} \qquad \begin{array}{l}\text{Expanding, using the formula} \\ \text{derived above}\end{array}$$

$$= \lim_{h \to 0} \frac{nx^{n-1}h + h^2 \cdot P}{h} \qquad \text{Canceling the } x^n \text{ and the } -x^n$$

$$= \lim_{h \to 0} \frac{h(nx^{n-1} + h \cdot P)}{h} \qquad \text{Factoring out an } h$$

$$= \lim_{h \to 0} (nx^{n-1} + h \cdot P) \qquad \text{Canceling the } h \text{ (since } h \neq 0)$$

$$\qquad\qquad\qquad\qquad \begin{array}{l}\text{Evaluating the limit by} \\ \text{direct substitution}\end{array}$$

$$= nx^{n-1}$$

This shows that for any positive integer n, the derivative of x^n is nx^{n-1}.

Verification of the Constant Multiple Rule

For a constant c and a function f, let $g(x) = c \cdot f(x)$. If $f'(x)$ exists, we may calculate the derivative $g'(x)$ as follows:

$$g'(x) = \lim_{h \to 0} \frac{g(x + h) - g(x)}{h} \qquad \text{Definition of the derivative}$$

$$= \lim_{h \to 0} \frac{c \cdot f(x + h) - c \cdot f(x)}{h} \qquad \begin{array}{l}\text{Since } g(x + h) = c \cdot f(x + h) \\ \text{and } g(x) = c \cdot f(x)\end{array}$$

$$= \lim_{h \to 0} \frac{c \cdot [f(x + h) - f(x)]}{h} \qquad \text{Factoring out the } c$$

$$= c \cdot \lim_{h \to 0} \frac{f(x + h) - f(x)}{h} \qquad \begin{array}{l}\text{Taking } c \text{ outside the limit} \\ \text{leaves just the definition of} \\ \text{the derivative } f'(x)\end{array}$$

$$= c \cdot f'(x) \qquad \text{Constant multiple rule}$$

This shows that the derivative of a constant times a function, $c \cdot f(x)$, is the constant times the derivative of the function, $c \cdot f'(x)$.

Verification of the Sum Rule

For two functions f and g, let their sum be $s = f + g$. If $f'(x)$ and $g'(x)$ exist, we may calculate $s'(x)$ as follows:

$$s'(x) = \lim_{h \to 0} \frac{s(x + h) - s(x)}{h}$$

Definition of the derivative

$$= \lim_{h \to 0} \frac{[f(x + h) + g(x + h)] - [f(x) + g(x)]}{h}$$

Since $s(x) = f(x) + g(x)$ and $s(x + h) = f(x + h) + g(x + h)$

$$= \lim_{h \to 0} \frac{f(x + h) + g(x + h) - f(x) - g(x)}{h}$$

Eliminating the brackets

$$= \lim_{h \to 0} \frac{f(x + h) - f(x) + g(x + h) - g(x)}{h}$$

Rearranging the numerator

$$= \lim_{h \to 0} \left[\frac{f(x + h) - f(x)}{h} + \frac{g(x + h) - g(x)}{h} \right]$$

Separating the fraction into two parts

$$= \underbrace{\lim_{h \to 0} \frac{f(x + h) - f(x)}{h}}_{f'(x)} + \underbrace{\lim_{h \to 0} \frac{g(x + h) - g(x)}{h}}_{g'(x)}$$

Using limit rule 4a on page 500

Recognizing the definition of the derivatives of f and g

$$= f'(x) + g'(x)$$

Sum rule

This shows that the derivative of a sum $f(x) + g(x)$ is the sum of the derivatives $f'(x) + g'(x)$.

EXERCISES 6.3

Find the derivative of each function.

1. $f(x) = x^4$ **2.** $f(x) = x^5$ **3.** $f(x) = x^{500}$

4. $f(x) = x^{1000}$ **5.** $f(x) = x^{1/2}$ **6.** $f(x) = x^{1/3}$

7. $g(x) = \dfrac{1}{2} x^4$ **8.** $f(x) = \dfrac{1}{3} x^9$

9. $g(w) = 6\sqrt[3]{w}$ **10.** $g(w) = 12 \sqrt{w}$

11. $h(x) = \dfrac{3}{x^2}$ **12.** $h(x) = \dfrac{4}{x^3}$

13. $f(x) = 4x^2 - 3x + 2$ **14.** $f(x) = 3x^2 - 5x + 4$

15. $g(x) = \sqrt{x} - \dfrac{1}{x}$

16. $g(x) = \sqrt[3]{x} - \dfrac{1}{x}$

17. $h(x) = 6\sqrt[3]{x_2} - \dfrac{12}{\sqrt[3]{x}}$ **18.** $h(x) = 8\sqrt{x^3} - \dfrac{8}{\sqrt[4]{x}}$

19. $f(x) = \dfrac{10}{\sqrt{x}} - 9\sqrt[3]{x^5} + 17$

20. $f(x) = \dfrac{9}{\sqrt[3]{x}} - 16\sqrt{x^5} - 14$

21. a. Find the derivative of $f(x) = 2$.
   b. Interpret your answer in terms of slope.
   c. Interpret your answer in terms of instantaneous rates of change.

22. a. Find the derivative of $f(x) = 3x$.
   b. Interpret your answer in terms of slope.
   c. Interpret your answer in terms of instantaneous rates of change.

Find the indicated derivatives.

23. If $f(x) = x^5$, find $f'(-2)$.

24. If $f(x) = x^4$, find $f'(-3)$.

25. If $f(x) = 6\sqrt[3]{x^2} - \dfrac{48}{\sqrt[3]{x}}$, find $f'(8)$.

26. If $f(x) = 12\sqrt[3]{x^2} + \dfrac{48}{\sqrt[3]{x}}$, find $f'(8)$.

27. If $f(x) = x^3$, find $\dfrac{df}{dx}\Big|_{x=-3}$.

28. If $f(x) = x^4$, find $\dfrac{df}{dx}\Big|_{x=-2}$.

29. If $f(x) = \dfrac{16}{\sqrt{x}} + 8\sqrt{x}$, find $\dfrac{df}{dx}\Big|_{x=4}$.

30. If $f(x) = \dfrac{54}{\sqrt{x}} + 12\sqrt{x}$, find $\dfrac{df}{dx}\Big|_{x=9}$.

 31. Use a graphing calculator to verify that the derivative of a constant is zero as follows. Define y_1 to be a constant (such as $y_1 = 5$) and then use NDERIV to define y_2 to be the derivative of y_1. Then graph the two functions together on an appropriate window and use TRACE to observe that the derivative y_2 is zero (graphed as a line along the x-axis), showing that the derivative of a constant is zero.

 32. Use a graphing calculator to verify that the derivative of a linear function is a constant as follows. Define y_1 to be a linear function (such as $y_1 = 3x - 4$) and then use NDERIV to define y_2 to be the derivative of y_1. Then graph the two functions together on an appropriate window and use TRACE to observe that the derivative y_2 is a constant (graphed as a horizontal line, such as $y_2 = 3$), verifying that the derivative of $y_1 = mx + b$ is $y_2 = m$.

APPLIED EXERCISES

33. Business: Marginal Profit   An electronics company finds that its total profit from selling x computer chips is $P(x) = 0.02x^{3/2} - 3000$ dollars.

   a. Find the company's marginal profit function.

   b. Find the marginal profit when 10,000 units have been sold and interpret your answer.

34. Business: Marginal Cost   A steel mill finds that its cost function is

$$C(x) = 8000\sqrt{x} - 6000\sqrt[3]{x}$$

dollars, where x is the (daily) production of steel (in tons).

   a. Find the marginal cost function.

   b. Find the marginal cost when 64 tons of steel are produced.

 35. Continuation of Exercise 33   Use a calculator to find the actual profit from the 10,001st computer chip, $P(10,001) - P(10,000)$, by evaluating the expression

$$\overbrace{[0.02(10{,}001)^{3/2} - 3000]}^{P(10{,}001)} - \overbrace{[0.02(10{,}000)^{3/2} - 3000]}^{P(10{,}000)}$$

Then return to the above calculation and replace 10,000 by 9999 and 10,001 by 10,000 to find $P(10,000) - P(9999)$, the actual profit from the 10,000th computer chip. Are your two answers close to the answer of \$3 found in Exercise 33b? Which way of finding the marginal profit was easier, using calculus (Exercise 33) or carrying out the calculation in this exercise?

 36. Continuation of Exercise 34   Use a calculator to find the actual cost of the 65th ton of steel, $C(65) - C(64)$, by evaluating the expression:

$$\overbrace{(8000\sqrt{65} - 6000\sqrt[3]{65})}^{C(65)} - \overbrace{(8000\sqrt{64} - 6000\sqrt[3]{64})}^{C(64)}$$

Then return to the above calculation and replace 64 by 63 and 65 by 64 to find $C(64) - C(63)$, the cost of the 64th ton of steel. Are your two answers close to the answer of $375 found in Exercise 34b? Which way of finding the marginal cost was easier, using calculus (Exercise 34) or carrying out the calculation in this exercise?

37. General: Population A company that makes games for teenage children forecasts that the teenage population in the United States x years from now will be

$$P(x) = 12,000,000 - 12,000x + 600x^2 + 100x^3$$

Find the rate of change of the teenage population

a. x years from now.
b. 1 year from now and interpret your answer.
c. 10 years from now and interpret your answer.

38. Biomedical: Flu Epidemic The number of people newly infected on day t of a flu epidemic is $f(t) = 13t^2 - t^3$ (for $0 \leq t \leq 13$). Find the instantaneous rate of change of this number on

a. day 5 and interpret your answer
b. day 10 and interpret your answer

39. Business: Advertising It has been estimated that the number of people who will see a newspaper advertisement that has run for x consecutive days is

$$N(x) = T - \frac{T/2}{x} \qquad \text{for } x \geq 1$$

where T is the total readership of the newspaper. If a newspaper has a circulation of 400,000, an ad that runs for x days will be seen by:

$$N(x) = 400,000 - \frac{200,000}{x}$$

people. Find how fast this pool of potential customers is growing when this ad has run for 5 days.

40. Environmental Science: Pollution An electrical generating plant burns high-sulfur oil, and the amount of sulfur dioxide pollution x miles downwind of the plant is $f(x) = 108x^{-2}$ parts

per million (ppm). Find the instantaneous rate of change of the pollution level 2 miles from the source. Interpret your answer in the proper units.

41. Biomedical: Blood Flow Nitroglycerin is often prescribed to enlarge blood vessels that have become too constricted. If the cross-sectional area of a blood vessel t hours after nitroglycerine is administered is $A(t) = 0.01t^2$ square centimeters (for $1 \leq t \leq 5$), find the instantaneous rate of change of the cross-sectional area 4 hours after the administration of nitroglycerine.

42. General: Hailstones Hailstones are frozen raindrops that increase in size as long as the updrafts keep them in the clouds. The weight of a typical hailstone that remains in a cloud for t minutes is $W(t) = 0.05t^3$ ounces. Find the instantaneous rate of change of the weight after 2 minutes.

43. Psychology: Learning Rates A language school has found that its students can memorize $p(t) = 24\sqrt{t}$ phrases in t hours of class (for $1 \leq t \leq 10$). Find the instantaneous rate of change of this quantity after 4 hours of class.

44. Environmental Science: Water Quality Downstream from a waste treatment plant the amount of dissolved oxygen in the water usually decreases for some distance (due to bacteria consuming the oxygen) and then increases (due to natural purification). A graph of the dissolved oxygen at various distances downstream looks like the curve below (known as the oxygen sag). The amount of dissolved oxygen is

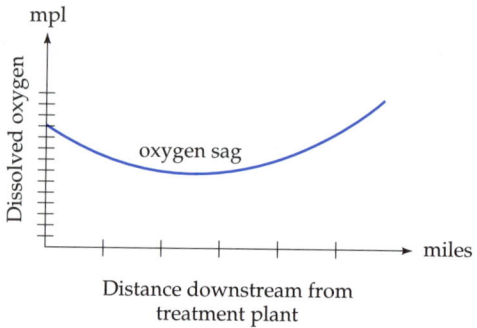

Distance downstream from treatment plant

usually taken as a measure of the health of the river. Suppose that the amount of dissolved oxygen x miles downstream is $D(x) = 0.2x^2 - 2x + 10$ mpl (milligrams per liter) for $0 \leq x \leq 20$. Use this formula to find the instantaneous rate of change of the dissolved oxygen

a. 1 mile downstream
b. 10 miles downstream

and interpret the signs of your answers.

45–46: Economics: Marginal Utility Generally, the more you have of something, the less valuable each additional unit becomes. For example, a dollar is less valuable to a millionaire than to a beggar. Economists define a person's "utility function" $U(x)$ for a product as the "perceived value" of having x units of that product. The *derivative* of $U(x)$ is called the *marginal utility function*, $MU(x) = U'(x)$. Suppose that a person's utility function for money is given by the function below. That is, $U(x)$ is the utility (perceived value) of x dollars.

a. Find the marginal utility function $MU(x)$.
b. Find $MU(1)$, the marginal utility of the first dollar.
c. Find $MU(1,000,000)$, the marginal utility of the millionth dollar.

45. $U(x) = 100\sqrt{x}$ **46.** $U(x) = 12\sqrt[3]{x}$

47. Business: Marginal Profit The profit function for Metropolitan Art Books Incorporated is $P(x) = 1.29x^{1.3} - 100{,}000$ dollars, where x is the number of books produced and sold.

a. On a graphing calculator, enter the profit function as y_1 and graph it on the viewing window $[0, 10{,}000]$ by $[-100{,}000, 100{,}000]$. Notice that the graph is not linear. Does it become steeper or less steep as x increases? What does this mean about how profit grows for larger values of x?
b. Trace along the curve to find how many books must be produced for the company to break even (to make zero profit). Give an ap-

proximate answer, to the nearest hundred books.

c. Define y_2 to be the marginal profit function by defining it as the derivative of the profit function y_1 (using NDERIV) and graph y_2 on the viewing window $[0, 10{,}000]$ by $[0, 30]$. Evaluate y_2 at $x = 6000$ and interpret your answer.
d. Does the marginal profit increase or decrease as x increases? How does this relate to your answer to part(a)?

48–49: General: College Tuition The following tables give the annual college tuition costs (in dollars) for a year at a public (Exercise 48) or private Exercise 49) college for the academic years ending in 1970–1998. To avoid large numbers, years are listed as years since 1970.

a. Enter these numbers into your graphing calculator and make a plot of the resulting points (Years Since 1970 on the x-axis and Tuition on the y-axis).
b. Have your calculator find the quadratic regression formula for these data. Then enter the result as function y_1. Plot the points together with the regression curve. Observe that the curve fits the points quite well.
c. Predict the tuition in the year 2005 by evaluating y_1 at $x = 35$ (years since 1970).
d. Define y_2 to be the derivative of y_1 (using NDERIV).
e. Predict the rate of change of tuition in the year 2005 by evaluating y_2 at $x = 35$. State your answer in the proper units.

48.

Years Since 1970	Tuition (Public College)
0	323
10	583
20	1356
28	2436

49.

Years Since 1970	Tuition (Private College)
0	1533
10	3130
20	8147
28	13,112

Sources: U.S. Department of Education; The College Board.

6.4   The Product and Quotient Rules

Introduction

In Section 6.3 we learned how to differentiate the sum and difference of two functions—we simply take the sum or difference of the derivatives. In this section we learn how to differentiate the *product* and *quotient* of two functions. Unfortunately, we do not simply take the product or quotient of the derivatives. Matters are a little more complicated.

Product Rule

To differentiate the product of two functions, $f(x) \cdot g(x)$, we use the *product rule*.

Product Rule

$$\frac{d}{dx}\,[f(x) \cdot g(x)] = f'(x) \cdot g(x) + f(x) \cdot g'(x)$$

The derivative of a product is the derivative of the first times the second plus the first times the derivative of the second

(provided, of course, that the derivatives $f'(x)$ and $g'(x)$ both exist). The formula is clearer if we write the functions simply as f and g.

$$\frac{d}{dx}\,(f \cdot g) = f' \cdot g + f \cdot g'$$

Derivative   Second   First   Derivative
of the first   of the second

A derivation of the product rule is given at the end of this section.

EXAMPLE 1   Using the Product Rule

Use the product rule to calculate $\dfrac{d}{dx}\,(x^3 \cdot x^5)$.

Solution

$$\frac{d}{dx}(x^3 \cdot x^5) = 3x^2 \cdot x^5 + x^3 \cdot 5x^4 = 3x^7 + 5x^7 = 8x^7$$

Derivative   Second   First   Derivative
of the first   of the second

We may check this answer by simplifying the original product, $x^3 \cdot x^5 = x^8$, and then differentiating:

$$\frac{d}{dx}(\underbrace{x^3 \cdot x^5}_{x^8}) = \frac{d}{dx}x^8 = 8x^7 \qquad \text{Agrees with above answer}$$

Notice that the derivative of a product is *not* the product of the derivatives: $(f \cdot g)' \neq f' \cdot g'$. For $x^3 \cdot x^5$ the product of the derivatives would be $3x^2 \cdot 5x^4 = 15x^6$, which is *not* the correct answer $8x^7$ that we found above. The product rule shows the correct way to differentiate a product.

EXAMPLE 2   Using the Product Rule

Use the product rule to find $\dfrac{d}{dx}[(x^2 - x + 2)(x^3 + 3)]$.

Solution

$$\frac{d}{dx}[(x^2 - x + 2)(x^3 + 3)]$$

$$= \underbrace{(2x - 1)}(x^3 + 3) + (x^2 - x + 2)\underbrace{(3x^2)}$$

Derivative   Derivative
of $x^2 - x + 2$ of $x^3 + 3$

$$= 2x^4 + 6x - x^3 - 3 + 3x^4 - 3x^3 + 6x^2 \qquad \text{Multiplying out}$$

$$= 5x^4 - 4x^3 + 6x^2 + 6x - 3 \qquad \text{Simplifying}$$

PRACTICE PROBLEM 1   Use the product rule to find $\dfrac{d}{dx}[x^3(x^2 - x)]$. *Solution at the back of the book*

Quotient Rule

The *quotient rule* shows how to differentiate a quotient of two functions.

Quotient Rule

$$\frac{d}{dx}\left(\frac{f(x)}{g(x)}\right) = \frac{g(x) \cdot f'(x) - g'(x) \cdot f(x)}{[g(x)]^2}$$

← The bottom times the derivative of the top, minus the derivative of the bottom times the top

└ The bottom squared

(provided that the derivatives $f'(x)$ and $g'(x)$ both exist and that $g(x) \neq 0$). A derivation of the quotient rule is given at the end of this section.

The quotient rule looks less formidable if we write the functions simply as f and g,

$$\frac{d}{dx}\left(\frac{f}{g}\right) = \frac{g \cdot f' - g' \cdot f}{g^2}$$

or even as

$$\frac{d}{dx}\left(\frac{\text{top}}{\text{bottom}}\right) = \frac{(\text{bottom}) \cdot \left(\frac{d}{dx}\text{top}\right) - \left(\frac{d}{dx}\text{bottom}\right) \cdot (\text{top})}{(\text{bottom})^2}$$

EXAMPLE 3 Using the Quotient Rule

Use the quotient rule to find $\dfrac{d}{dx}\left(\dfrac{x^9}{x^3}\right)$.

Solution

Bottom   Derivative of the top   Derivative of the bottom   Top

$$\frac{d}{dx}\left(\frac{x^9}{x^3}\right) = \frac{(x^3)(9x^8) - (3x^2)(x^9)}{(x^3)^2} = \frac{9x^{11} - 3x^{11}}{x^6} = \frac{6x^{11}}{x^6} = 6x^5$$

└ Bottom squared

We may check this answer by simplifying the original quotient and then differentiating:

$$\frac{d}{dx}\left(\frac{x^9}{x^3}\right) = \frac{d}{dx}x^6 = 6x^5 \qquad \text{Agrees with above answer}$$

└ x^6

Notice that the derivative of a quotient is *not* the quotient of the derivatives:

$$\left(\frac{f}{g}\right)' \neq \frac{f'}{g'}$$

For the quotient x^9/x^3, taking the quotient of the derivatives would give $(9x^8)/(3x^2) = 3x^6$, which is *not* the correct answer $6x^5$ that we found above. The quotient rule shows the correct way to differentiate a quotient.

EXAMPLE 4   Using the Quotient Rule

Find $\dfrac{d}{dx}\left(\dfrac{x^2}{x + 1}\right)$.

Solution   The quotient rule gives

Derivative of the top   Derivative of the bottom

Bottom   Top

$$\frac{d}{dx}\left(\frac{x^2}{x + 1}\right) = \frac{(x + 1)(2x) - (1)(x^2)}{(x + 1)^2} = \frac{2x^2 + 2x - x^2}{(x + 1)^2} = \frac{x^2 + 2x}{(x + 1)^2}$$

Bottom squared

$\blacksquare$

PRACTICE PROBLEM 2   Find $\dfrac{d}{dx}\left(\dfrac{2x^2}{x^2 + 1}\right)$.   *Solution at the back of the book*

Practically every city must purify its drinking water and treat its wastewater. The cost of the treatment rises steeply for higher degrees of purity.

EXAMPLE 5   Finding the Cost of Cleaner Water

For some methods, the cost of purifying a gallon of water to a purity of x percent is

$$C(x) = \frac{2}{100 - x} \qquad \text{for } 80 < x < 100$$

dollars. Find the rate of change of the purification costs when the purity is

a. 90%   and   **b.** 98%.

Solution

The rate of change of cost is the derivative of the cost function:

Derivative of 2

Derivative of $100 - x$

$$C'(x) = \frac{d}{dx}\left(\frac{2}{100 - x}\right) = \frac{(100 - x)(0) - (-1)(2)}{(100 - x)^2} \quad \text{Differentiating by the quotient rule}$$

$$= \frac{0 + 2}{(100 - x)^2} = \frac{2}{(100 - x)^2} \quad \text{Simplifying (the derivative is undefined at } x = 100)$$

a. For 90% purity we evaluate at $x = 90$:

$$C'(90) = \frac{2}{(100 - 90)^2} = \frac{2}{10^2} = \frac{2}{100} = 0.02 \qquad C'(x) = \frac{2}{(100 - x)^2} \text{ evaluated at } x = 90$$

Interpretation: At 90% purity, the rate of change of the cost is 0.02 dollar, meaning that the costs increase by about *2 cents for each additional percentage of purity.*

b. For 98% purity we evaluate $C'(x)$ at $x = 98$:

$$C'(98) = \frac{2}{(100 - 98)^2} = \frac{2}{2^2} = \frac{2}{4} = \frac{1}{2} = 0.50 \qquad C'(x) = \frac{2}{(100 - x)^2} \text{ evaluated at } x = 98$$

Interpretation: At 98% purity, the rate of change of the cost is 0.50 dollar, meaning that the costs increase by about *50 cents for each additional percentage of purity.*

Notice that an extra percentage of purity above the 98% level is 25 times as costly as an extra percentage above the 90% purity level. ∎

 Graphing Calculator Exploration

a. On a graphing calculator, enter the cost function from Example 5 as $y_1 = \dfrac{2}{(100 - x)}$. Then use NDERIV to define y_2 to be the derivative of y_1.

b. Graph both y_1 and y_2 on the window [80, 100] by [−1, 5]. Your graph should resemble the one below (but you may have an additional "false" vertical line on the right).

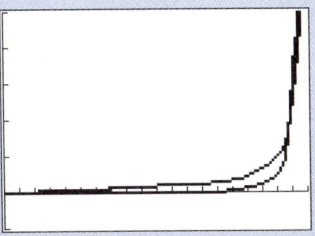

c. Verify the results of Example 5 by evaluating y_2 at $x = 90$ and at $x = 98$.

d. Evaluate y_2 at $x = 100$, giving (supposedly) the derivative of y_1 at $x = 100$. However, in Example 5 we saw that the derivative of y_1 is *undefined* at $x = 100$. Your calculator is giving you a "false value" for the derivative, resulting from NDERIV's using a symmetric difference quotient (see page 535) and a positive value for h. Therefore, to use your calculator effectively, you must also understand calculus.

Marginal Average Cost

It is often useful to calculate not just the *total* cost of producing x units of some product, but also the *average cost per unit*, denoted $AC(x)$, which is found by dividing the total cost $C(x)$ by the number of units x.

$$AC(x) = \frac{C(x)}{x} \qquad \text{Average cost per unit is total cost divided by the number of units}$$

The derivative of the average cost function is called the *marginal average cost*, MAC.*

$$MAC(x) = \frac{d}{dx}\left[\frac{C(x)}{x}\right] \qquad \text{Marginal average cost is the derivative of average cost}$$

Marginal average revenue MAR, and marginal average profit MAP, are defined similarly as the derivatives of average revenue per unit, $\dfrac{R(x)}{x}$, and average profit per unit, $\dfrac{P(x)}{x}$, respectively.

---

* The marginal average cost function is sometimes denoted $\overline{C}'(x)$, with similar notations used for marginal average revenue and marginal average profit.

$$MAR(x) = \frac{d}{dx}\left[\frac{R(x)}{x}\right]$$ Marginal average revenue is the derivative of average revenue

$$MAP(x) = \frac{d}{dx}\left[\frac{P(x)}{x}\right]$$ Marginal average profit is the derivative of average profit

EXAMPLE 6   Finding and Interpreting Marginal Average Cost

It costs a book publisher $12 to produce each book, and fixed costs are $1500. Therefore, the company's cost function is

$$C(x) = 12x + 1500$$ Total cost of producing x books

a. Find the average cost function.
b. Find the marginal average cost function.
c. Find the marginal average cost at $x = 100$ and interpret your answer.

Solution

a. The average cost function is

$$AC(x) = \underbrace{\frac{12x + 1500}{x}}_{\substack{\text{Total cost divided} \\ \text{by number of units}}} = \underbrace{12 + \frac{1500}{x}}_{\text{Simplifying}} = \underbrace{12 + 1500x^{-1}}_{\text{In power form}}$$

b. The *marginal* average cost is the derivative of average cost. We could use the quotient rule on the first expression above, but it is easier to use the power rule on the last expression:

$$MAC(x) = \frac{d}{dx}(12 + 1500x^{-1}) = -1500x^{-2} = -\frac{1500}{x^2}$$

c. Evaluating at $x = 100$:

$$MAC(100) = -\frac{1500}{100^2} = -\frac{1500}{10{,}000} = -0.15 \qquad -\frac{1500}{x^2} \text{ at } x = 100$$

Interpretation: When 100 books have been produced, the average cost per book is decreasing (because of the negative sign) by about *15 cents per additional book produced*. This reflects the fact that while *total* costs rise when you produce more, the *average cost per unit* decreases, because of the economies of mass production.

Graphing Calculator Exploration

Use a graphing calculator to investigate further the effects of mass production in Example 6.

a. Graph the average cost function [any of the expressions for $AC(x)$ from part (a) above] on the window [0, 400] by [0, 50]. Your graph should resemble that shown below. TRACE along the average cost curve to see how the average cost drops from the 20s down to the teens as the number of books increases from 100 to 400. Note that although average cost falls, it does so more slowly as the number of units increases.

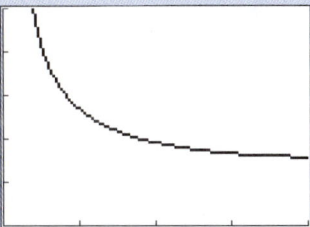

b. To see exactly how rapidly the average cost declines, graph the *marginal* average cost function on the window [0, 400] by [−1, 1]. TRACE along this curve to see how the marginal average cost (which is negative, since costs are decreasing) approaches to zero as the number of units increases (the law of diminishing returns).

EXAMPLE 7 **Finding Time Saved by Speeding**

A certain mathematics professor drives 25 miles to his office every day, mostly on highways. If he drives at constant speed v miles per hour, his travel time (distance divided by speed) is

$$T(v) = \frac{25}{v}$$

hours. Find $T'(55)$ and interpret this number.

Solution Since $T(v) = \frac{25}{v}$ is a quotient, we could differentiate it by the quotient rule. However, it is easier to write $\frac{25}{v}$ as a *power*,

$$T(v) = 25v^{-1}$$

and differentiate using the power rule:

$$T'(v) = -25v^{-2}$$

This gives the rate of change of the travel time with respect to driving speed. $T'(v)$ is negative, showing that as speed increases, travel time *decreases*. Evaluating this at speed $v = 55$ gives:

$$T'(55) = -25(55)^{-2} = \frac{-25}{(55)^2} \approx -0.00826 \qquad \text{Using a calculator}$$

This number, the rate of change of travel time with respect to driving speed, means that when driving at 55 miles per hour, you save only 0.00826 hour for each extra mile per hour of speed. Multiplying by 60 gives the savings in *minutes*:

$$(-0.00826)(60) \approx -0.50 = -\frac{1}{2}$$

That is, each extra mile per hour of speed saves only about half a minute, or 30 seconds. For example, speeding by 10 mph would save only about $\frac{1}{2} \cdot 10 = 5$ minutes. One must then decide whether this slight savings in time is worth the risk of an accident or a speeding ticket.

■

SUMMARY

The following is a list of the differentiation formulas that we have learned so far. The letters c and n stand for constants, and f and g stand for differentiable functions of x.

$$\frac{d}{dx} c = 0$$

$$\frac{d}{dx} x^n = nx^{n-1} \qquad \left[\text{special case: } \frac{d}{dx} x = 1 \right]$$

$$\frac{d}{dx} (c \cdot f) = c \cdot f' \qquad \left[\text{special case: } \frac{d}{dx} (cx) = c \right]$$

$$\frac{d}{dx} (f \pm g) = f' \pm g'$$

$$\frac{d}{dx} (f \cdot g) = f' \cdot g + f \cdot g'$$

$$\frac{d}{dx} \left(\frac{f}{g} \right) = \frac{g \cdot f' - g' \cdot f}{g^2} \qquad g \neq 0$$

These formulas are used extensively throughout calculus, and you should not proceed to Section 6.5 until you have mastered them.

Verification of the Differentiation Formulas

We conclude this section with derivations of the product and quotient rules, and the power rule in the case of *negative* integer exponents. First, however, we need to establish a preliminary result about an arbitrary function g:

$$\text{If } g'(x) \text{ exists, then } \lim_{h \to 0} g(x + h) = g(x).$$

We begin with $\lim_{h \to 0} g(x + h)$ and show that it is equal to $g(x)$:

$$\lim_{h \to 0} g(x + h) = \lim_{h \to 0} [g(x + h) - g(x) + g(x)] \qquad \text{Subtracting and adding } g(x)$$

$$= \lim_{h \to 0} \left[\frac{g(x + h) - g(x)}{h} \cdot h + g(x) \right] \qquad \text{Dividing and multiplying by } h$$

$$= \lim_{h \to 0} \left[\frac{g(x + h) - g(x)}{h} \cdot h \right] + \lim_{h \to 0} g(x) \qquad \begin{array}{l}\text{The limit of a sum is the sum} \\ \text{of the limits}\end{array}$$

$$= \underbrace{\lim_{h \to 0} \frac{g(x + h) - g(x)}{h}}_{g'(x)} \cdot \underbrace{\lim_{h \to 0} h}_{0} + g(x) \qquad \begin{array}{l}\text{The limit of a product is} \\ \text{the product of the limits}\end{array}$$

$$= g'(x) \cdot 0 + g(x) \qquad \begin{array}{l}\text{Since the first limit above is} \\ \text{the definition of } g'(x) \text{ and} \\ \text{the second limit is zero}\end{array}$$

$$= g(x) \qquad \text{Simplifying}$$

This proves the result that if $g'(x)$ exists, then $\lim_{h \to 0} g(x + h) = g(x)$.

Replacing $x + h$ by a new variable y, this equation becomes

$$\lim_{y \to x} g(y) = g(x) \qquad y = x + h, \text{ so } h \to 0 \text{ implies } y \to x$$

According to the definition of continuity on page 504 (but with different letters for variables), this equation means that the function g is *continuous* at x. Therefore, the result that we have shown can be stated simply:

If a function is *differentiable* at x, then it is *continuous* at x.

Or, even more briefly:

Differentiability implies continuity.

Verification of the Product Rule

For two functions f and g, let their product be $p(x) = f(x) \cdot g(x)$. If $f'(x)$ and $g'(x)$ exist, we may calculate $p'(x)$ as follows.

$$p'(x) = \lim_{h \to 0} \frac{p(x + h) - p(x)}{h} \qquad \text{Definition of the derivative}$$

$$= \lim_{h \to 0} \frac{f(x + h)g(x + h) - f(x)g(x)}{h} \qquad \begin{array}{l} p(x + h) = f(x + h) \cdot g(x + h) \\ \text{and } p(x) = f(x) \cdot g(x) \end{array}$$

$$= \lim_{h \to 0} \frac{f(x + h)g(x + h) - f(x)g(x + h) + f(x)g(x + h) - f(x)g(x)}{h} \qquad \begin{array}{l} \text{Subtracting and adding} \\ f(x)g(x + h) \end{array}$$

$$= \lim_{h \to 0} \left[\frac{f(x + h)g(x + h) - f(x)g(x + h)}{h} + \frac{f(x)g(x + h) - f(x)g(x)}{h} \right] \qquad \begin{array}{l} \text{Separating the fraction} \\ \text{into two parts} \end{array}$$

$$= \lim_{h \to 0} \frac{[f(x + h) - f(x)]g(x + h)}{h} + \lim_{h \to 0} \frac{f(x)[g(x + h) - g(x)]}{h} \qquad \begin{array}{l} \text{Using limit rule 4a on} \\ \text{page 500 and factoring} \end{array}$$

$$= \lim_{h \to 0} \underbrace{\frac{[f(x + h) - f(x)]}{h}}_{f'(x)} \underbrace{\lim_{h \to 0} g(x + h)}_{g(x)} + \underbrace{f(x)}_{f(x)} \underbrace{\lim_{h \to 0} \frac{[g(x + h) - g(x)]}{h}}_{g'(x)} \qquad \begin{array}{l} \text{Using limit rule 4c on} \\ \text{page 500} \\[4pt] \text{Recognizing the defini-} \\ \text{tions of } f'(x) \text{ and } g'(x) \end{array}$$

$$= f'(x)g(x) + f(x)g'(x) \qquad \text{Product rule}$$

Verification of the Quotient Rule

For two functions f and g with $g(x) \neq 0$, let the quotient be $q(x) = f(x)/g(x)$. If $f'(x)$ and $g'(x)$ exist, we may calculate $q'(x)$ as follows.

$$q'(x) = \lim_{h \to 0} \frac{q(x + h) - q(x)}{h} \qquad \text{Definition of the derivative}$$

$$= \lim_{h \to 0} \frac{\dfrac{f(x + h)}{g(x + h)} - \dfrac{f(x)}{g(x)}}{h} \qquad q(x + h) = \frac{f(x + h)}{g(x + h)} \text{ and } q(x) = \frac{f(x)}{g(x)}$$

$$= \lim_{h \to 0} \frac{1}{h} \left[\frac{f(x + h)}{g(x + h)} - \frac{f(x)}{g(x)} \right] \qquad \begin{array}{l} \text{Since dividing by } h \text{ is equivalent to} \\ \text{multiplying by } 1/h \end{array}$$

$$= \lim_{h \to 0} \left[\frac{1}{h} \cdot \frac{g(x)f(x + h) - g(x + h)f(x)}{g(x + h)g(x)} \right] \qquad \begin{array}{l} \text{Subtracting the fractions, using the} \\ \text{common denominator } g(x + h)g(x) \end{array}$$

$$= \lim_{h \to 0} \left[\frac{1}{h} \cdot \frac{g(x)f(x + h) - g(x)f(x) - [g(x + h)f(x) - g(x)f(x)]}{g(x + h)g(x)} \right] \qquad \begin{array}{l} \text{Subtracting and adding} \\ g(x)f(x) \end{array}$$

$$= \lim_{h \to 0} \left[\frac{1}{g(x + h)g(x)} \cdot \frac{g(x)[f(x + h) - f(x)] - [g(x + h) - g(x)]f(x)}{h} \right]$$

Factoring in the numerator; switching the denominators

$$= \lim_{h \to 0} \left[\frac{1}{g(x + h)g(x)} \left(g(x) \lim_{h \to 0} \underbrace{\frac{f(x + h) - f(x)}{h}}_{f'(x)} - \lim_{h \to 0} \underbrace{\frac{g(x + h) - g(x)}{h}}_{g'(x)} f(x) \right) \right]$$

Using limit rules 4b and 4c on page 500

$$\underbrace{\qquad}_{\substack{\text{Approaches} \\ g(x)}}$$

$$= \frac{1}{[g(x)]^2} [g(x)f'(x) - g'(x)f(x)]$$

Using limit rules 1 and 4d on page 500

$$= \frac{g(x)f'(x) - g'(x)f(x)}{[g(x)]^2}$$

Quotient rule

Verification of the Power Rule for Negative Integer Exponents

On page 537 we proved the power rule for *positive* integer exponents *n*. Using the quotient rule, we may now prove the power rule for *negative* integer exponents. Any negative integer *n* may be written as $n = -p$, where p is a *positive* integer. Then

$$\frac{d}{dx} x^n = \frac{d}{dx} \left(\frac{1}{x^p} \right)$$

Since $x^n = x^{-p} = \dfrac{1}{x^p}$

$$= \frac{x^p \cdot 0 - px^{p-1} \cdot 1}{x^{2p}}$$

Using the quotient rule, with $\dfrac{d}{dx} 1 = 0$ and $\dfrac{d}{dx} x^p = px^{p-1}$

$$= \frac{-px^{p-1}}{x^{2p}}$$

Simplifying

$$= -px^{p-1-2p} = \underbrace{-p}_{n} x^{\underbrace{-p-1}_{n-1}}$$

Subtracting exponents and simplifying

Since $-p = n$

$$= nx^{n-1}$$

Power rule

This proves the power rule, $\dfrac{d}{dx} x^n = nx^{n-1}$, for negative integer exponents *n*.

EXERCISES 6.4

Find the derivative of each function in two ways:

a. Using the *product* rule.
b. Multiplying out the function and using the *power* rule.

Your answers to parts (a) and (b) should agree.

1. $x^4 \cdot x^6$

2. $x^7 \cdot x^2$

3. $x^4(x^5 + 1)$

4. $x^5(x^4 + 1)$

Find the derivative of each function by using the product rule.

5. $f(x) = x^2(x^3 + 1)$ **6.** $f(x) = x^3(x^2 + 1)$

7. $f(x) = x(5x^2 - 1)$ **8.** $f(x) = 2x(x^4 + 1)$

9. $f(x) = (x^2 + 1)(x^2 - 1)$

10. $f(x) = (x^3 - 1)(x^3 + 1)$

11. $f(x) = (x^2 + x)(3x + 1)$

12. $f(x) = (x^2 + 2x)(2x + 1)$

13. $f(x) = (\sqrt{x} - 1)(\sqrt{x} + 1)$

14. $f(x) = (\sqrt{x} + 2)(\sqrt{x} - 2)$

15. $f(t) = 6t^{4/3}(3t^{2/3} + 1)$

16. $f(t) = 4t^{3/2}(2t^{1/2} - 1)$

17. $f(z) = (z^4 + z^2 + 1)(z^3 - z)$

18. $f(z) = (\sqrt[4]{z} + \sqrt{z})(\sqrt[4]{z} - \sqrt{z})$

Find the derivative of each function in two ways:

a. Using the *quotient* rule.
b. Simplifying the original function and using the *power* rule.

Your answers to parts (a) and (b) should agree.

19. $\dfrac{x^8}{x^2}$ **20.** $\dfrac{x^9}{x^3}$ **21.** $\dfrac{1}{x^3}$ **22.** $\dfrac{1}{x^4}$

Find the derivative of each function by using the quotient rule.

23. $f(x) = \dfrac{x^4 + 1}{x^3}$ **24.** $f(x) = \dfrac{x^5 - 1}{x^2}$

25. $f(x) = \dfrac{x + 1}{x - 1}$ **26.** $f(x) = \dfrac{x - 1}{x + 1}$

27. $f(t) = \dfrac{t^2 - 1}{t^2 + 1}$ **28.** $f(t) = \dfrac{t^2 + 1}{t^2 - 1}$

29. $f(s) = \dfrac{s^3 - 1}{s + 1}$ **30.** $f(s) = \dfrac{s^3 + 1}{s - 1}$

31. $f(x) = \dfrac{x^4 + x^2 + 1}{x^2 + 1}$ **32.** $f(x) = \dfrac{x^5 + x^3 + x}{x^3 + x}$

Find the instantaneous rate of change of the function at the given value of x.

33. $f(x) = \dfrac{x^2}{x - 1}$ at $x = 2$ **34.** $f(x) = \dfrac{x^2}{x - 2}$ at $x = 4$

35. $f(x) = \dfrac{x}{\sqrt{x}}$ at $x = 4$ **36.** $f(x) = \dfrac{x}{\sqrt{x}}$ at $x = 16$

37. Product Rule for Three Functions   Show that if f, g, and h are differentiable functions of x, then

$$\frac{d}{dx}(f \cdot g \cdot h) = f' \cdot g \cdot h + f \cdot g' \cdot h + f \cdot g \cdot h'$$

(*Hint:* Write the function as $f \cdot (g \cdot h)$ and apply the product rule twice.)

38. Derive the quotient rule from the product rule as follows.

a. Define the quotient to be a single function,
$$Q(x) = \frac{f(x)}{g(x)}.$$

b. Multiply both sides by $g(x)$ to obtain $Q(x) \cdot g(x) = f(x)$.

c. Differentiate each side, using the product rule on the left side.

d. Solve the resulting formula for the derivative $Q'(x)$.

e. Replace $Q(x)$ by $\dfrac{f(x)}{g(x)}$ and show that the resulting formula for $Q'(x)$ is the same as the quotient rule.

Note that in this derivation when we differentiate $Q(x)$ we are *assuming* that the derivative of the quotient exists, whereas the derivation on pages 552–553 we *proved* that the derivative exists.

39. Find a formula for $\dfrac{d}{dx}[f(x)]^2$ by writing it as $\dfrac{d}{dx}[f(x)f(x)]$ and using the product rule. Be sure to simplify your answer.

40. Find a formula for $\dfrac{d}{dx}[f(x)]^{-1}$ by writing it as $\dfrac{d}{dx}\left[\dfrac{1}{f(x)}\right]$ and using the quotient rule. Be sure to simplify your answer.

Find the derivative of each function.

41. $(x^3 + 2)\dfrac{x^2 + 1}{x + 1}$ **42.** $(x^5 + 1)\dfrac{x^3 + 2}{x + 1}$

43. $\dfrac{(x^2 + 3)(x^3 + 1)}{x^2 + 2}$ **44.** $\dfrac{(x^3 + 2)(x^2 + 2)}{x^3 + 1}$

45. $\dfrac{\sqrt{x} - 1}{\sqrt{x} + 1}$ **46.** $\dfrac{\sqrt{x} + 1}{\sqrt{x} - 1}$

APPLIED EXERCISES

47. Economics: Marginal Average Revenue Use the quotient rule to find a general expression for the marginal average revenue. That is, calculate $\dfrac{d}{dx}\left[\dfrac{R(x)}{x}\right]$ and simplify your answer.

48. Economics: Marginal Average Profit Use the quotient rule to find a general expression for the marginal average profit. That is, calculate $\dfrac{d}{dx}\left[\dfrac{P(x)}{x}\right]$ and simplify your answer.

49. Environmental Science: Water Purification If the cost of purifying a gallon of water to a purity of x percent is

$$C(x) = \frac{100}{100 - x} \text{ cents} \qquad \text{for } 50 \le x < 100$$

a. Find the instantaneous rate of change of the cost with respect to purity.
b. Evaluate this rate of change for a purity of 95% and interpret your answer.
c. Evaluate this rate of change for a purity of 98% and interpret your answer.

50. Business: Marginal Average Cost A toy company can produce plastic trucks at a cost of $8 each, while fixed costs are $1200 per day. Therefore, the company's cost function is $C(x) = 8x + 1200$.

a. Find the average cost function $AC(x) = C(x)/x$.
b. Find the marginal average cost function $MAC(x)$.
c. Evaluate $MAC(x)$ at $x = 200$ and interpret your answer.

 51. a. Continuation of Exercise 49 Use a graphing calculator to graph the cost function $C(x)$ from Exercise 49 on the graphing window [50, 100] by [0, 20]. TRACE along the curve to see how rapidly costs increase for purity (x-coordinate) increasing from 50 to near 100.
b. To check your answers to Exercise 49, use the "dy/dx" or SLOPE feature of your calculator to find the slope of the cost curve at $x = 95$ and at $x = 98$. The resulting rates of change of the cost should agree with your

answers to Exercises 49b and c. Note that further purification becomes increasingly expensive at higher purity levels.

52. a. Continuation of Exercise 50 Graph the average cost function $AC(x)$ that you found in Exercise 50a on the viewing rectangle [0, 400] by [0, 50]. TRACE along the average cost curve to see how the average cost falls from the 20s down to the teens as the number of trucks increases. Note that although average cost falls, it does so more slowly as the number of units increases.
b. To check your answer to Exercise 50, use the "dy/dx" or SLOPE feature of your calculator to find the slope of the average cost curve at $x = 200$. This slope gives the rate of change of the cost, which should agree with your answer to Exercise 50c. Find the slope (rate of change) for other x-values to see that the rate of change of average cost tends toward zero (the law of diminishing returns).

53. Business: Marginal Average Profit A company's profit function is $P(x) = 12x - 1800$ dollars.

a. Find the average profit function $AP(x) = P(x)/x$.
b. Find the marginal average profit function $MAP(x)$.
c. Evaluate $MAP(x)$ at $x = 300$ and interpret your answer.

54. Business: Sales The number of bottles of whiskey that a store will sell in a month at a price of p dollars per bottle is

$$N(p) = \frac{2250}{p + 7} \qquad (p \ge 5)$$

Find the rate of change of this quantity when the price is $8 and interpret your answer.

55. General: Body Temperature If a person's temperature after x hours of strenuous exercise is $T(x) = x^3(4 - x^2) + 98.6$ (for $0 \le x \le 2$), find the rate of change of the temperature after 1 hour.

56. Business: Record Sales After x months, monthly sales of a record are predicted to be

$S(x) = x^2(8 - x^3)$ thousand (for $0 \le x \le 2$). Find the rate of change of the sales after 1 month.

57. a. Continuation of Exercise 55   Graph the temperature function $T(x)$ given in Exercise 55 on the viewing rectangle [0, 2] by [90, 110]. TRACE along the temperature curve to see how the temperature rises and then falls as time increases.

   b. To check your answer to Exercise 55, use the "dy/dx" or SLOPE feature of your calculator to find the slope (rate of change) of the curve at $x = 1$. Your answer should agree with your answer to Exercise 55.

   c. TRACE along the temperature curve to estimate the maximum temperature.

58. a. Continuation of Exercise 56   Graph the sales function $S(x)$ given in Exercise 56 on the window [0, 2] by [0, 12]. TRACE along the sales curve to see how the sales rise and then fall as x, the number of months, increases.

   b. To check your answer to Exercise 56, use the "dy/dx" or SLOPE feature of your calculator to find the slope (rate of change) of the curve at $x = 1$. Your answer should agree with your answer to Exercise 56.

   c. TRACE along the curve to estimate the maximum sales.

59. Economics: National Debt   The following table gives the national debt (the amount of money that the federal government has borrowed from, and therefore owes to, its people), in millions of dollars, for the years 1970 to 1995. To avoid large numbers, years are listed in the table as years since 1970.

Years Since 1970	National Debt (millions of dollars)
0	370,100
5	533,200
10	907,700
15	1,823,100
20	3,233,300
25	4,974,000

(1970, 1975, 1980, 1985, 1990, 1995)

Source: U.S. Treasury Department

a. Enter these numbers into your graphing calculator and make a plot of the resulting points (Years Since 1970 on the x-axis and National Debt on the y-axis).

b. Have your calculator find the quadratic regression formula for these data. Then enter the result as function y_1. Plot the points together with the regression curve. Observe that the curve fits the points quite well.

c. The population of the United States (in millions) for these years is closely approximated by the linear function $y = 2.378x + 204.1$, where x is the number of years since 1970. Enter this function into your calculator as y_2.

d. Use your calculator to define the function y_3 to be $y_1 \div y_2$, the national debt divided by the number of people, giving the "per capita national debt." Graph the function on the viewing window [0, 40] by [0, 45,000] to show the per capita national debt for the years 1970 to 2010.

e. TRACE along the curve y_3. The y-values represent the amount of money that the federal government owes to each one of its citizens in year x after 1970 (if the debt were equally distributed). Observe how rapidly the amount grows.

f. Define function y_4 to be the derivative of y_3 (using the NDERIV feature of your calculator) and graph y_4 on the viewing window [0, 40] by [0, 2000]. Use TABLE to evaluate both y_3 and y_4 at $x = 40$ and interpret your answers.

60–61: Pitfalls of NDERIV on a Graphing Calculator

a. Find the derivative (by hand) of the function below, and observe that the derivative is undefined at $x = 0$.

b. Find the derivative of the function below by using NDERIV on a graphing calculator and evaluate the derivative at $x = 0$. If your calculator gives you an answer, this is a "false value" for the derivative, since in part (a) you showed that the derivative is undefined at $x = 0$. (For an explanation, see the Graphing Calculator Exploration part (d) on pages 546–547.)

60. $y = \dfrac{1}{x}$ **61.** $y = \dfrac{1}{x^2}$

6.5 Higher-Order Derivatives

APPLICATION   PREVIEW

AIDS

Acquired immunodeficiency syndrome (AIDS) is a disease related to infection with the human immunodeficiency virus (HIV). First reported in 1981, AIDS has become the leading cause of death among Americans aged 25–44. The cumulative number of AIDS cases in the United States for the years 1990 ($x = 0$) to 1996 ($x = 6$) is given by the function:

$$f(x) = 0.13x^4 - 2.6x^3 + 13.1x^2 + 51x + 200 \qquad \text{Thousand cases}$$

Differentiating $f(x)$ gives the *rate of growth* of AIDS cases:

$$f'(x) = 0.52x^3 - 7.8x^2 + 26.2x + 51 \qquad \begin{matrix}\text{Differentiating}\\ f(x) \text{ above}\end{matrix}$$

$$f'(5) = 0.52(5)^3 - 7.8(5)^2 + 26.2(5) + 51 = 52 \qquad \begin{matrix}\text{Evaluating at}\\ x = 5 \text{ (1995)}\end{matrix}$$

This means that in 1995, AIDS was growing by approximately 52 thousand new cases annually. Differentiating again gives what is called the *second* derivative of $f(x)$, which shows how the rate of growth is itself changing:

$$f''(x) = 1.56x^2 - 15.6x + 26.2 \qquad \begin{matrix}\text{Differentiating}\\ f'(x)\end{matrix}$$

$$f''(5) = 1.56(5)^2 - 15.6(5) + 26.2 = -12.8 \qquad \text{Evaluating at } x = 5$$

$$\uparrow$$
$$\text{Negative!}$$

The negative sign means that in 1995 the rate of growth of AIDS was *slowing*. That is, although the epidemic continued to grow (the first derivative was positive), it was growing at a slower rate (the second derivative is negative). This suggests that preventive measures are beginning to slow the spread of AIDS.

The annual *cost* of caring for AIDS and HIV patients is approximately

$$g(x) = 0.01x^3 + 0.19x^2 + 1.4x + 4.2 \qquad \text{Billion dollars}$$

Differentiating $g(x)$ gives the rate of growth of costs:

$$g'(x) = 0.03x^2 + 0.38x + 1.4 \qquad \begin{matrix}\text{Differentiating}\\ g(x) \text{ above}\end{matrix}$$

$$g'(5) = 0.03(5)^2 + 0.38(5) + 1.4 = 4.05 \qquad \text{Evaluating at } x = 5$$

Therefore, AIDS costs were increasing at the rate of $4.05 billion per year in 1995. The *second* derivative will show whether this growth was speeding up or slowing down:

$$g''(x) = 0.06x + 0.38$$

Differentiating
$g'(x) = 0.03x^2 + 0.38x + 1.4$

$$g''(5) = 0.06(5) + 0.38 = 0.68$$

Evaluating at $x = 5$

This second derivative being *positive* means that AIDS care costs were growing *increasingly rapidly* in 1995.

Second derivatives have shown that even though the growth in AIDS *cases* is slowing, the growth in AIDS *costs* is speeding up, due in part to the high cost of new medications. Note that the change in the rate of growth of AIDS is not immediately apparent from the graph on the left below but it is clear from the second derivative. In Exercise 48 you will find exactly when the growth in AIDS cases began to slow.

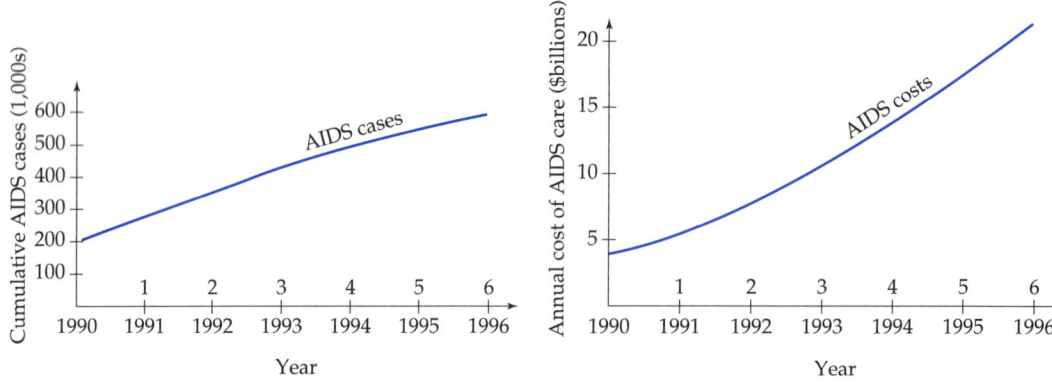

Source: Centers for Disease Control.

This section will give further examples of second (and higher) derivatives measuring how change is itself changing.

Introduction

We have seen that from one function we can calculate a new function, the *derivative* of the original function. This new function, however, can itself be differentiated, giving what is called the *second derivative* of the original function. Differentiating again gives the *third derivative* of the original function, and so on. In this section we will calculate and interpret such higher-order derivatives.

Calculating Higher-Order Derivatives

EXAMPLE 1   Finding Higher Derivatives of a Polynomial

From $f(x) = x^3 - 6x^2 + 2x - 7$ we may calculate:

$$f'(x) = 3x^2 - 12x + 2 \qquad \text{"First" derivative of } f$$

Differentiating again gives:

$$f''(x) = 6x - 12 \qquad \text{Second derivative of } f, \text{ read "} f \text{ double prime"}$$

and a third time:

$$f'''(x) = 6 \qquad \text{Third derivative of } f, \text{ read "} f \text{ triple prime"}$$

and a fourth time:

$$f''''(x) = 0 \qquad \text{Fourth derivative of } f, \text{ read "} f \text{ quadruple prime"}$$

All further derivatives of this function will, of course, be zero.

■

We also denote derivatives by replacing the primes by the number of derivatives *in parentheses*. For example, the fourth derivative would be denoted $f^{(4)}(x)$.

PRACTICE PROBLEM 1   If $f(x) = x^3 - x^2 + x - 1$, find:

a. $f'(x)$ **b.** $f''(x)$ **c.** $f'''(x)$ **d.** $f^{(4)}(x)$ *Solutions at the back of the book*

While Example 1 showed that a polynomial can be differentiated "down to zero," the same is not true for all functions.

EXAMPLE 2   Finding Higher Derivatives of a Rational Function

Find the first five derivatives of $f(x) = \dfrac{1}{x}$.

Solution

$$f(x) \quad = x^{-1} \qquad\qquad\qquad\text{\color{blue}{$f(x)$ in power form}}$$

$$f'(x) \quad = -x^{-2} \qquad\qquad\qquad\text{\color{blue}{First derivative}}$$

$$f''(x) \quad = 2x^{-3} \qquad\qquad\qquad\text{\color{blue}{Second derivative}}$$

$$f'''(x) \quad = -6x^{-4} \qquad\qquad\qquad\text{\color{blue}{Third derivative}}$$

$$f^{(4)}(x) = 24x^{-5} \qquad\qquad\qquad\text{\color{blue}{Fourth derivative}}$$

$$f^{(5)}(x) = -120x^{-6} \qquad\qquad\qquad\text{\color{blue}{Fifth derivative}}$$

Clearly, we will never get to zero no matter how many times we differentiate.

PRACTICE PROBLEM 2   If $f(x) = 16x^{-1/2}$, find:

a. $f'(x)$ **b.** $f''(x)$ *Solutions at the back of the book*

In Leibniz's notation, the second derivative $\dfrac{d}{dx}\dfrac{df}{dx}$ is written $\dfrac{d^2f}{dx^2}$. The superscript goes after the d in the numerator and after the dx in the denominator. The following table shows equivalent statements in the two notations.

Prime Notation		$\dfrac{d}{dx}$ Notation	
$f''(x)$	$=$	$\dfrac{d^2}{dx^2}f(x)$	Second derivative
y''	$=$	$\dfrac{d^2y}{dx^2}$	
$f'''(x)$	$=$	$\dfrac{d^3}{dx^3}f(x)$	Third derivative
y'''	$=$	$\dfrac{d^3y}{dx^3}$	
$f^{(n)}(x)$	$=$	$\dfrac{d^n}{dx^n}f(x)$	nth derivative
$y^{(n)}$	$=$	$\dfrac{d^ny}{dx^n}$	

Calculating higher derivatives merely requires repeated use of the same differentiation rules that we have been using.

EXAMPLE 3   **Finding the Second Derivative Using the Quotient Rule**

Find $\dfrac{d^2}{dx^2}\left(\dfrac{x^2+1}{x}\right)$.

Solution

$$\frac{d}{dx}\left(\frac{x^2+1}{x}\right) = \frac{x(2x)-(x^2+1)}{x^2} \qquad \text{The first derivative, using the quotient rule}$$

$$= \frac{2x^2-x^2-1}{x^2} = \frac{x^2-1}{x^2} \qquad \text{Simplifying}$$

Differentiating this answer gives the *second* derivative of the original:

$$\frac{d}{dx}\left(\frac{x^2-1}{x^2}\right) = \frac{x^2(2x)-2x(x^2-1)}{x^4} \qquad \text{Differentiating the first derivative}$$

$$= \frac{2x^3-2x^3+2x}{x^4} = \frac{2x}{x^4} = \frac{2}{x^3} \qquad \text{Simplifying}$$

■

The function in this example was a quotient, so it was perhaps natural to use the quotient rule. It is easier, however, to simplify the original function first,

$$\frac{x^2+1}{x} = \frac{x^2}{x} + \frac{1}{x} = x + x^{-1}$$

and then differentiate by the power rule. So, the first derivative of $x + x^{-1}$ is $1 - x^{-2}$, and differentiating again gives $2x^{-3}$, agreeing with the answer found by the quotient rule. *Moral:* **Always simplify before differentiating.**

PRACTICE PROBLEM 3   Find $f''(x)$ if $f(x) = \dfrac{x+1}{x}$.   *Solution at the back of the book*

EXAMPLE 4   **Evaluating a Second Derivative**

If $f(x) = \dfrac{1}{\sqrt{x}}$, find $f''\left(\dfrac{1}{4}\right)$.   First differentiate, then evaluate

Solution

$$f(x) = x^{-1/2} \qquad\qquad\qquad f(x) \text{ in power form}$$

$$f'(x) = -\frac{1}{2} x^{-3/2} \qquad\qquad \text{Differentiating once}$$

$$f''(x) = \frac{3}{4} x^{-5/2} \qquad\qquad \text{Differentiating again}$$

$$f''\left(\frac{1}{4}\right) = \frac{3}{4} \left(\frac{1}{4}\right)^{-5/2} = \frac{3}{4} (4)^{5/2} \qquad \text{Evaluating } f''(x) \text{ at } \frac{1}{4}$$

$$= \frac{3}{4} (\sqrt{4})^5 = \frac{3}{4} (2)^5 = \frac{3}{4} (32) = 24$$

■

PRACTICE PROBLEM 4   Find $\dfrac{d^2}{dx^2}(x^4 + x^3 + 1)\Big|_{x=-1}$.   *Solution at the back of the book*

Velocity and Acceleration

There is another important interpretation for the derivative, an interpretation that also gives a meaning to the *second* derivative. Imagine that you are driving along a straight road, and let $s(t)$ stand for your distance (in miles) from your starting point after t hours of driving. Then the derivative $s'(t)$ gives the instantaneous rate of change of distance with respect to time (miles per hour). However, "miles per hour" means speed or velocity, so the derivative of the *distance* function $s(t)$ is just the *velocity* function $v(t)$, giving your velocity at any time t.

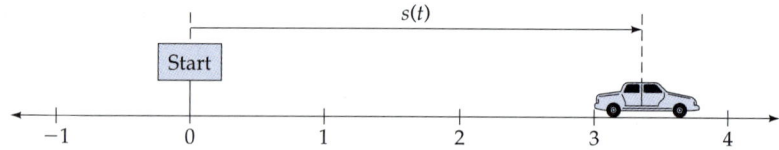

In general, for an object moving along a straight line, with distance measured from some fixed point, measured positively in one direction and negatively in the other (sometimes called "directed distance"):

$$\text{If} \qquad s(t) = \begin{pmatrix} \text{Distance} \\ \text{at time } t \end{pmatrix}$$

$$\text{then} \qquad s'(t) = \begin{pmatrix} \text{Velocity} \\ \text{at time } t \end{pmatrix}$$

More simply, letting $v(t)$ stand for the velocity at time t,

$$v(t) = s'(t)$$

The velocity function is the derivative of the distance function

The units of velocity come directly from the distance and time units of $s(t)$. For example, if distance is measured in feet and time in seconds, then the velocity is in *feet per second*, whereas if distance is in miles and time is in hours, then velocity is in *miles per hour*.

In everyday speech, the word "accelerating" means "speeding up." That is, acceleration means the rate of increase of speed, and since rates of increase are just derivatives, *acceleration is the derivative of velocity.* Since velocity is itself the derivative of distance, acceleration is the *second* derivative of distance. Letting $a(t)$ stand for the acceleration at time t, we have:

Distance, Velocity, and Acceleration

$$s(t) = \begin{pmatrix} \text{Distance} \\ \text{at time } t \end{pmatrix}$$

$v(t) = s'(t)$ Velocity is the derivative of distance

$a(t) = v'(t) = s''(t)$ Acceleration is the derivative of velocity, and the *second* derivative of distance

Therefore, we now have an interpretation for the *second* derivative: If $s(t)$ represents distance, then the *first* derivative represents velocity, and the *second* derivative represents *acceleration*. (In physics, there is even an interpretation for the third derivative, which gives the rate of change of acceleration: It is called the "jerk," since it is related to motion being "jerky."*)

REAL WORLD

EXAMPLE 5 Finding and Interpreting Velocity and Acceleration

A delivery truck is driving along a straight road, and after t hours its distance (in miles) east of its starting point is

$$s(t) = 24t^2 - 4t^3 \qquad \text{for } 0 \le t \le 6$$

Start

−1   0   1   2   3   4

* See T. R. Sandlin, "The Jerk," *Physics Teacher* **28**:36–40, January 1990.

a. Find the velocity of the truck after 2 hours.

b. Find the velocity of the truck after 5 hours.

c. Find the acceleration of the truck after 1 hour.

Solution

a. To find velocity, we differentiate distance:

$$v(t) = 48t - 12t^2$$ Differentiating
$s(t) = 24t^2 - 4t^3$

$$v(2) = 48 \cdot 2 - 12 \cdot (2)^2$$ Evaluating at $t = 2$

$$= 96 - 48 = 48 \text{ miles per hour}$$ Velocity after 2 hours

b. At $t = 5$ hours:

$$v(5) = 48 \cdot 5 - 12 \cdot (5)^2$$ Evaluating $v(t)$ at $t = 5$

$$= 240 - 300 = -60 \text{ miles per hour}$$ Velocity after 5 hours

What does the negative sign mean? Since distances are measured *eastward* (according to the original problem), the "positive" direction is east, so a negative velocity means a *westward* velocity. Therefore, at time $t = 5$ the truck is driving *westward at 60 miles per hour* (that is, back toward its starting point).

c. The acceleration is

$$a(t) = 48 - 24t$$ Differentiating $v(t) = 48t - 12t^2$

$$a(1) = 48 - 24 = 24$$ Acceleration after 1 hour

Therefore, after 1 hour the acceleration of the truck is 24 mi/hr² (it is speeding up).

 Graphing Calculator Exploration

Use a graphing calculator to graph the distance function $y_1 = 24x^2 - 4x^3$ (using x instead of t), the velocity function $y_2 = 48x - 12x^2$, and the acceleration function $y_3 = 48 - 24x$. (Alternatively, you could define y_2 and y_3 using NDERIV.) Your display should look like the one below. By looking at the graph, can you determine which curve represents distance, which represents velocity, and which represents acceleration? (*Hint:* Which curve gives the slope of which other curve?) Check your answer by using TRACE to identify functions 1, 2, and 3.

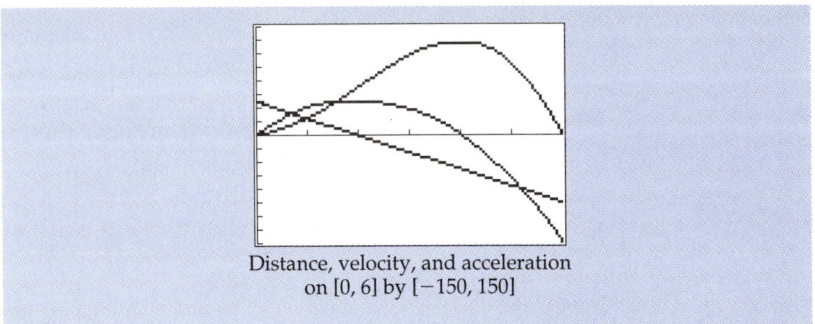

Distance, velocity, and acceleration
on $[0, 6]$ by $[-150, 150]$

In Example 5, velocity was in miles per hour (mi/hr), so acceleration, the rate of change of this with respect to time, is in miles per hour *per hour*, written mi/hr². In general, the units of acceleration are distance/time².

PRACTICE PROBLEM 5

A helicopter rises vertically, and after t seconds its height above the ground is $s(t) = 6t^2 - t^3$ feet (for $0 \le t \le 6$).

a. Find its velocity after 2 seconds.

b. Find its velocity after 5 seconds.

c. Find its acceleration after 1 second.

(*Hint:* Distances are measured *upward*, so a negative velocity means *downward*.)   *Solutions at the back of the book*

Other Interpretations of Second Derivatives

The second derivative has other meanings besides acceleration. In general, second derivatives measure how the rate of change is itself changing. That is, the second derivative tells whether growth is speeding up or slowing down.

EXAMPLE 6   Predicting Population Growth

Demographers predict that t years from now the population of a city will be:

$$P(t) = 2,000,000 + 28,800t^{1/3}$$

Find $P'(8)$ and $P''(8)$ and interpret these answers.

Solution   The derivative is

$$P'(t) = 9600t^{-2/3} \qquad \text{Derivative of } P(t)$$

so

$$P'(8) = 9600(8)^{-2/3} = 9600 \left(\tfrac{1}{4}\right) = 2400 \qquad \text{Evaluating at } t = 8$$

Interpretation: Eight years from now the population will be growing at the rate of 2400 people per year.

The second derivative is

$$P''(t) = -6400t^{-5/3} \qquad \text{Derivative of } P'(t) = 9600t^{-2/3}$$

$$P''(8) = -6400(8)^{-5/3} \qquad \text{Evaluating at } t = 8$$

$$= -6400 \left(\tfrac{1}{32}\right) = -200$$

That the second derivative (the rate of change of the rate of change) is *negative* means that the growth rate is slowing.

Interpretation: After 8 years, the growth rate is *decreasing* by about 200 people per year each year. In other words, in the following year the population will continue to grow, but at a slower rate, about $2400 - 200 = 2200$ people per year. ∎

Derivatives make the front page! If y represents the size of the economy, then this headline announces that

$$\frac{dy}{dx} > 0 \quad \text{and} \quad \frac{d^2y}{dx^2} < 0$$

U.S. ECONOMY GREW AT A SLOWER PACE IN THE 2D QUARTER

STILL BEATS EXPECTATIONS

Neither General Motors Strike Nor Asian Troubles Manage to Derail the Expansion

By Richard W. Stevenson
WASHINGTON, July 31— A powerful array of forces at home and abroad slowed the economy to a modest but still healthy pace during

Graphing Calculator Exploration

Use a graphing calculator to graph the population function $y_1 = 2{,}000{,}000 + 28{,}800x^{1/3}$ (using x instead of t). Your graph should resemble the one below. Can you see from the graph that the first derivative is positive (sloping upward), and that the second derivative is negative (slope decreasing)? You may check these facts numerically using NDERIV.

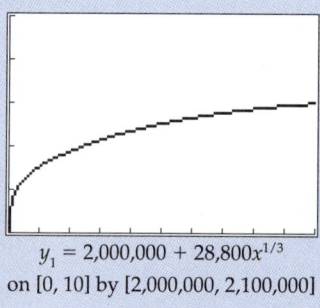

$y_1 = 2{,}000{,}000 + 28{,}800x^{1/3}$
on [0, 10] by [2,000,000, 2,100,000]

Statements about first and second derivatives occur frequently in everyday events. For example, the newspaper headline on the left (*New York Times*, August 1, 1998) says that the United States economy grew (the first derivative is positive) but more slowly (the second derivative is negative), following a curve like that shown below.

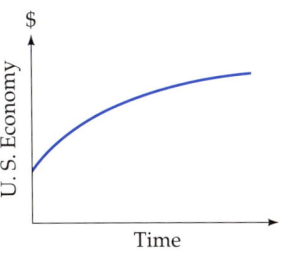

PRACTICE PROBLEM 6

The following headlines appeared recently in the *New York Times*. For each headline, sketch a curve representing the type of growth described and indicate the correct signs of the first and second derivatives.

a. Consumer Prices Rose in October at a Slower Rate

b. Households Still Shrinking, but Rate Is Slower

Solutions at the back of the book

SUMMARY

By simply repeating the process of differentiation we can calculate second, third, and higher derivatives. We also have another interpretation for the derivative, one that gives an interpretation for the second derivative as well. For distance measured along a straight line from some fixed point:

$$\text{If} \qquad s(t) \;=\; \textit{distance at time } t$$

$$\text{then} \qquad s'(t) = \textit{velocity at time } t$$

$$\text{and} \qquad s''(t) = \textit{acceleration at time } t$$

Therefore, whenever you are driving along a straight road, your speedometer reading is the derivative of your odometer reading.

Velocity is the derivative of distance

We now have four interpretations for derivatives: instantaneous rates of change, slopes, marginals, and velocity. It has been said that science is at its best when it unifies, and the derivative, unifying these four different concepts, is one of the most important ideas in all of science. We also saw that second derivatives, which measure the rate of change of the rate of change, can show whether growth is speeding up or slowing down.

Remember, however, that derivatives measure just what an automobile speedometer measures: the velocity at a particular *instant*. Although this statement may be obvious for velocities, it is easy to forget when dealing with marginals. For example, suppose that the marginal cost for a product is $15 when 100 units have been produced [which may be written $C'(100) = 15$]. Therefore, costs are increasing at the rate of $15 per additional unit, but only at the instant when $x = 100$. Although this may be used to *estimate* future costs (*about* $15 for each additional unit), it does not mean that one additional unit will increase costs by exactly $15, two more by exactly $30, and so on, since the marginal rate usually changes as production increases. A marginal cost is only an *approximate* predictor of future costs.

EXERCISES 6.5

For each function, find

a. $f'(x)$,   **b.** $f''(x)$,   **c.** $f'''(x)$,   **d.** $f^{(4)}(x)$.

1. $f(x) = x^4 - 2x^3 - 3x^2 + 5x - 7$

2. $f(x) = x^4 - 3x^3 + 2x^2 - 8x + 4$

3. $f(x) = 1 + x + \frac{1}{2}x^2 + \frac{1}{6}x^3 + \frac{1}{24}x^4 + \frac{1}{120}x^5$

4. $f(x) = 1 + x + \frac{1}{2}x^2 + \frac{1}{6}x^3 + \frac{1}{24}x^4$

5. $f(x) = \sqrt{x^5}$ 6. $f(x) = \sqrt{x^3}$

For each function, find **a.** $f''(x)$ and **b.** $f''(3)$.

7. $f(x) = \dfrac{x - 1}{x}$ 8. $f(x) = \dfrac{x + 2}{x}$

9. $f(x) = \dfrac{x + 1}{2x}$ 10. $f(x) = \dfrac{x - 2}{4x}$

11. $f(x) = \dfrac{1}{6x^2}$ 12. $f(x) = \dfrac{1}{12x^3}$

Find the *second* derivative of each function.

13. $f(x) = (x^2 - 2)(x^2 + 3)$

14. $f(x) = (x^2 - 1)(x^2 + 2)$

15. $f(x) = \dfrac{27}{\sqrt[3]{x}}$ 16. $f(x) = \dfrac{32}{\sqrt[4]{x}}$

17. $f(x) = \dfrac{x}{x - 1}$ 18. $f(x) = \dfrac{x}{x - 2}$

Evaluate each expression.

19. $\dfrac{d^2}{dr^2}(\pi r^2)$ 20. $\dfrac{d^2}{dr^2}\left(\dfrac{4}{3}\pi r^3\right)$

21. $\dfrac{d^2}{dx^2}x^{10}\bigg|_{x=-1}$ 22. $\dfrac{d^2}{dx^2}x^{11}\bigg|_{x=-1}$

23. $\dfrac{d^3}{dx^3}x^{10}\bigg|_{x=-1}$ 24. $\dfrac{d^3}{dx^3}x^{11}\bigg|_{x=-1}$

25. $\dfrac{d^2}{dx^2}\sqrt{x^3}\bigg|_{x=1/16}$ 26. $\dfrac{d^2}{dx^2}\sqrt[3]{x^4}\bigg|_{x=1/27}$

27. **General: Velocity**   Each of the following three "stories," labeled a, b, and c, matches one of the following velocity graphs labeled i, ii, and iii. For each story, choose the most appropriate graph.

a. I left my home and drove to meet a friend, but I got stopped for a speeding ticket. Afterward I drove on more slowly.
b. I started driving but then stopped to look at the map. Realizing that I was going the wrong way, I drove back the other way.
c. After driving for a while I got into some stop-and-go driving. Once past the tie-up I could speed up again.

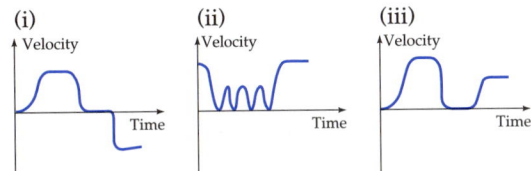

(i)   (ii)   (iii)

28. Business: Profit   Each of the following three descriptions of a company's profit over time matches one of the graphs below. For each description, choose the most appropriate graph.

a. Profits were growing increasingly rapidly.
b. Profits were declining but the rate of decline was slowing.
c. Profits were rising, but more and more slowly.

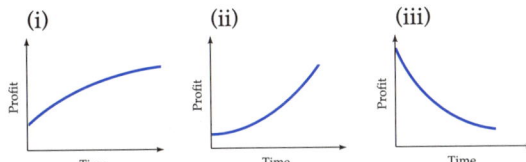

(i)   (ii)   (iii)

29. Find $\dfrac{d^{100}}{dx^{100}}(x^{99} - 4x^{98} + 3x^{50} + 6)$.

(*Hint:* No calculation is necessary. Think of what happens when an nth-degree polynomial is differentiated $n + 1$ times. For example, try differentiating x^3 four times.)

30. Find a general formula for $\dfrac{d^n}{dx^n}x^{-1}$.

[*Hint:* Calculate the first few derivatives and look for a pattern. You may use the factorial notation: $n! = n(n - 1) \cdots 1$. For example, $3! = 3 \cdot 2 \cdot 1 = 6$.]

31. Verify the following formula for the *second* derivative of a product, where f and g are differentiable functions of x:

$$\frac{d^2}{dx^2}(f \cdot g) = f'' \cdot g + 2f' \cdot g' + f \cdot g''$$

(*Hint:* Use the product rule repeatedly.)

32. Verify the following formula for the *third* derivative of a product, where f and g are differentiable functions of x:

$$\frac{d^3}{dx^3}(f \cdot g) = f''' \cdot g + 3f'' \cdot g' + 3f' \cdot g'' + f \cdot g'''$$

(*Hint:* Differentiate the formula in Exercise 31 by the product rule.)

APPLIED EXERCISES

33. General: Velocity   After t hours a freight train is $s(t) = 18t^2 - 2t^3$ miles due north of its starting point (for $t \le 9$).

a. Find its velocity at time $t = 3$ hours.
b. Find its velocity at time $t = 7$ hours.
c. Find its acceleration at time $t = 1$ hour.

34. General: Velocity   After t hours a passenger train is $s(t) = 24t^2 - 2t^3$ miles due west of its starting point (for $t \le 12$).

a. Find its velocity at time $t = 4$ hours.
b. Find its velocity at time $t = 10$ hours.
c. Find its acceleration at time $t = 1$ hour.

35. General: Velocity   A rocket can rise to a height of $h(t) = t^3 + 0.5t^2$ feet in t seconds. Find its velocity and acceleration 10 seconds after it is launched.

36. General: Velocity After t hours a car is a distance $s(t) = 60t + \dfrac{100}{t + 3}$ miles from its starting point. Find the velocity after 2 hours.

37. General: Impact Velocity A penny dropped from a building will fall a distance $s(t) = 16t^2$ feet in t seconds (neglecting air resistance).

   a. With what velocity will it hit the ground if it does so 5 seconds after it is dropped? (This is called the *impact velocity*.)

   b. Find the acceleration at any time t. (This number is the *acceleration due to gravity*.)

 38. General: Impact Velocity If a marble is dropped from the top of the Sears Tower in Chicago, its height above the ground t seconds after it is dropped $s(t) = 1454 - 16t^2$ feet (neglecting air resistance).

   a. How long will it take to reach the ground? (*Hint:* Find when the height equals zero.)

   b. Use your answer to part (a) to find the velocity with which it will strike the ground.

39. General: Maximum Height If a bullet from a 9-millimeter pistol is fired straight up from the ground, its height t seconds after it is fired will be $s(t) = -16t^2 + 1280t$ feet (neglecting air resistance), for $0 \le t \le 80$.

   a. Find the velocity function.

   b. Find the time t when the bullet will be at its maximum height. [*Hint:* At its maximum height the bullet is moving neither up nor down, and has velocity zero. Therefore, find the time when the velocity $v(t)$ equals zero.]

   c. Find the maximum height the bullet will reach. [*Hint:* Use the time found in part (b) together with the height function $s(t)$.]

 40. General: Maximum Height A frog can jump to a height of $s(t) = -4.9t^2 + 4.41t$ meters above the ground in t seconds. Find the maximum height that the frog will reach.

41. Economics: National Debt The national debt of a South American country t years from now is predicted to be $D(t) = 65 + 9t^{4/3}$ billion dollars. Find $D'(8)$ and $D''(8)$ and interpret your answers.

42. Environmental Science: Earth Temperature If the average temperature of the earth t years from now is predicted to be $T(t) = 65 - 4/t$ (for $t \ge 8$), find $T'(10)$ and $T''(10)$ and interpret your answers.

43–44: Environmental Science: Sea Level The burning of fossil fuels (such as coal and oil) generates carbon dioxide which traps heat in the atmosphere, thereby increasing the temperature of the earth (the "greenhouse effect").

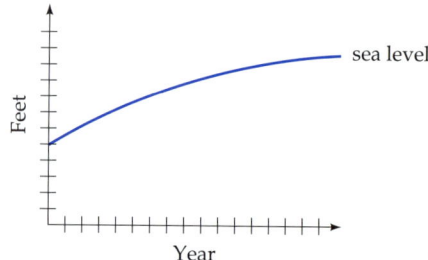

The higher temperature in turn melts the polar icecaps, raising the sea level. The precise results are very difficult to predict, but suppose the following:

43. In t years the average sea level on the east coast of the United States will be $L(t) = 30 - 4t^{-1/2}$ feet (for $t > 2$). Find $L'(4)$ and $L''(4)$ and interpret your answers.

44. In t years the average sea level on the west coast of the United States will be $L(t) = 35 - 8t^{-1/2}$ feet (for $t > 2$). Find $L'(4)$ and $L''(4)$ and interpret your answers.

 45. Business: Profit The annual profit of the Digitronics company x years from now is predicted to be $P(x) = 5.27x^{0.3} - 0.463x^{1.52}$ million dollars (for $0 \le x \le 8$). Evaluate the profit function and its first and second derivatives at $x = 3$ and interpret your answers.
[*Hint:* Enter the given function as y_1, define y_2 to be the derivative of y_1 (using NDERIV), and define y_3 to be the derivative of y_2. Then evaluate each at the stated x-value.]

46. General: Population The population of a country x years from now is predicted to be $P(x) = 3.17x^{1.3} + 0.192x^{0.74}$ million people (for $0 \le x \le 10$). Evaluate the population function and its first and second derivatives at $x = 3.75$ and interpret your answers. (See the hint in Exercise 45.)

 47. General: Windchill Index The windchill index for a temperature of 30° Fahrenheit and wind speed x miles per hour is

$$y_1 = 91.4 - 5.016(5.81 + 3.71\sqrt{x} - 0.25x)$$

for $4 \leq x \leq 45$.

a. Graph the windchill index on a graphing calculator on the window [4, 45] by [−10, 30]. Then find the windchill index for wind speeds of $x = 15$ and $x = 30$ mph.
b. Notice from your graph that the windchill index has first derivative negative and second derivative positive. What does this mean about how successive 1-mph increases in wind speed affect the windchill index?
c. Verify your answer to part (b) by defining y_2 to be the derivative of y_1 (using NDERIV),

evaluating it at $x = 15$ and $x = 30$, and interpreting your answers.

48. General: AIDS Reread the Application Preview on AIDS on pages 557–558 and find exactly when the growth of AIDS began to slow.

(*Hint:* The first derivative gives the rate of growth, and the second derivative tells how the growth rate is itself changing, so find when the second derivative first becomes negative. You may use a graphing calculator or do it "by hand." Then convert the x-value into a year.)

Find the second derivative of each function.

49. $(x^2 - x + 1)(x^3 - 1)$ **50.** $(x^3 + x - 1)(x^3 + 1)$

51. $\dfrac{x}{x^2 + 1}$ **52.** $\dfrac{x}{x^2 - 1}$

53. $\dfrac{2x - 1}{2x + 1}$ **54.** $\dfrac{3x + 1}{3x - 1}$

6.6 The Chain Rule and the Generalized Power Rule

Introduction

In this section we will learn the last of the general rules of differentiation, the *chain rule* for differentiating composite functions. We will then prove a very useful special case of it, the *generalized power rule* for differentiating powers of functions. We begin by reviewing composite functions.

Composite Functions

As we saw on page 56, composite functions are simply functions of functions: The composition of the functions $f(x)$ and $g(x)$ is $f(g(x))$.

EXAMPLE 1 Finding a Composite Function

For $f(x) = x^2$ and $g(x) = 4 - x$, find $f(g(x))$.

Solution

$$f(g(x)) = (4 - x)^2$$

$f(x) = x^2$ with x
replaced by $g(x) = 4 - x$

PRACTICE PROBLEM 1   For the same $f(x) = x^2$ and $g(x) = 4 - x$, find $g(f(x))$.

Solution at the back of the book

Graphing Calculator Exploration

Use a graphing calculator to verify that the above $f(g(x))$ and $g(f(x))$ are different.

a.   Enter $y_1 = x^2$, $y_2 = 4 - x$.

b.   Then define y_3 and y_4 to be the compositions in the two orders [on some calculators this is done by defining $y_3 = y_1(y_2)$ and $y_4 = y_2(y_1)$].

c.   Graph y_3 and y_4 (but turn "off" y_1 and y_2) on the standard window and notice that the graphs are very different.

Besides building compositions out of simpler functions, we can also *de*compose functions, expressing them as compositions of simpler functions.

EXAMPLE 2   Decomposing a Composite Function

Find functions $f(x)$ and $g(x)$ such that $(x^2 + 1)^5$ is the composition $f(g(x))$.

Solution   Think of $(x^2 + 1)^5$ as an inside function $x^2 + 1$ followed by an outside operation $(\quad)^5$. We match the "inside" and "outside" parts of $(x^2 + 1)^5$ and $f(g(x))$.

Outside function

$$(x^2 + 1)^5 \quad = \quad f(g(x))$$

Inside function

Therefore, $(x^2 + 1)^5$ can be written as $f(g(x))$ with $\begin{cases} f(x) = x^5 \\ g(x) = x^2 + 1 \end{cases}$.
(Other answers are possible.)

Notice that expressing a function as a composition involves thinking of the function in terms of "blocks," an inside block that starts the calculation and an outside block that completes it.

PRACTICE PROBLEM 2   Find $f(x)$ and $g(x)$ such that $\sqrt{x^5 - 7x + 1}$ is the composition $f(g(x))$.

Solution at the back of the book

The Chain Rule

The *chain rule* shows how to differentiate a composite function $f(g(x))$.

Chain Rule

$\dfrac{d}{dx} f(g(x)) = f'(g(x)) \cdot g'(x)$	To differentiate $f(g(x))$, differentiate $f(x)$, then replace each x by $g(x)$, and finally, multiply by the derivative of $g(x)$

(provided that the derivatives on the right-hand side of the equation exist). The name comes from thinking of compositions as "chains" of functions. A verification of the chain rule is given at the end of this section.

EXAMPLE 3 **Differentiating Using the Chain Rule**

Use the chain rule to find $\dfrac{d}{dx}(x^2 - 5x + 1)^{10}$.

Solution

$(x^2 - 5x + 1)^{10}$ is $f(g(x))$ with $\begin{cases} f(x) = x^{10} & \text{Outside function} \\ g(x) = x^2 - 5x + 1 & \text{Inside function} \end{cases}$

Since $f'(x) = 10x^9$, we have

$$f'(g(x)) = 10(g(x))^9 \qquad \begin{array}{l} f'(x) = 10x^9 \text{ with } x \\ \text{replaced by } g(x) \end{array}$$

$$= 10(x^2 - 5x + 1)^9 \qquad \text{Using } g(x) = x^2 - 5x + 1$$

Substituting this last expression into the chain rule gives:

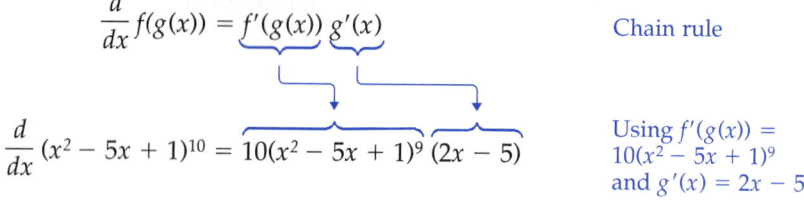

$$\frac{d}{dx} f(g(x)) = \underbrace{f'(g(x))}\ \underbrace{g'(x)}$$ Chain rule

$$\frac{d}{dx} (x^2 - 5x + 1)^{10} = \overbrace{10(x^2 - 5x + 1)^9}\ \overbrace{(2x - 5)}$$ Using $f'(g(x)) = 10(x^2 - 5x + 1)^9$ and $g'(x) = 2x - 5$

This result says that to differentiate $(x^2 - 5x + 1)^{10}$, we bring down the exponent 10, reduce the exponent to 9 (steps familiar from the power rule), and finally multiply by the derivative of the inside function.

$$\frac{d}{dx} \underbrace{(x^2 - 5x + 1)}_{\substack{\text{Inside} \\ \text{function}}}{}^{\underset{\substack{\text{Bring down} \\ \text{the power } n}}{10}} = 10(x^2 - 5x + 1)^{\underset{\substack{\text{Power} \\ n-1}}{9}}\ \underset{\substack{\text{Derivative of} \\ \text{the inside function}}}{(2x - 5)}$$

Generalized Power Rule

Example 3 suggests a general rule for differentiating a function to a power.

Generalized Power Rule

$$\frac{d}{dx} [g(x)]^n = n \cdot [g(x)]^{n-1} \cdot g'(x)$$

To differentiate a function to a power, bring down the power as a multiplier, reduce the exponent by 1, and then multiply by the derivative of the inside function

(provided, of course, that the derivative $g'(x)$ exists). The generalized power rule follows from the chain rule by reasoning similar to that of Example 3: The derivative of $f(x) = x^n$ is $f'(x) = nx^{n-1}$, so

$$\frac{d}{dx} f(g(x)) = \underbrace{f'(g(x))}\ \underbrace{g'(x)}$$ Chain rule

gives:

$$\frac{d}{dx} [g(x)]^n = \overbrace{n[g(x)]^{n-1}g'(x)}$$ Generalized power rule

EXAMPLE 4 Differentiating Using the Generalized Power Rule

Find $\dfrac{d}{dx} \sqrt{x^4 - 3x^3 - 4}$.

Solution

$$\frac{d}{dx} \underbrace{(x^4 - 3x^3 - 4)}_{\substack{\text{Inside} \\ \text{function}}} {}^{1/2} = \underbrace{\tfrac{1}{2}}_{\substack{\text{Bring down} \\ \text{the } n}} (x^4 - 3x^3 - 4)^{\overset{\uparrow}{-1/2}} \underbrace{(4x^3 - 9x^2)}_{\substack{\text{Derivative of} \\ \text{the inside function}}}$$

Power
$n - 1$

Think of the generalized power rule "from the outside in." That is, first bring down the outer exponent and then reduce it by 1, and only then multiply by the derivative of the inside function.

Be careful—it is the *original function* (not the differentiated function) that is raised to the power $n - 1$. Only at the end do you multiply by the derivative of the inside function.

EXAMPLE 5 Simplifying and Differentiating

Find $\dfrac{d}{dx} \left(\dfrac{1}{x^2 + 1} \right)^3$.

Solution   Writing the function as $(x^2 + 1)^{-3}$ gives:

$$\frac{d}{dx} \underbrace{(x^2 + 1)}_{\substack{\text{Inside} \\ \text{function}}} {}^{-3} = \underbrace{-3}_{\substack{\text{Bring down} \\ \text{the } n}} (x^2 + 1)^{\overset{\uparrow}{-4}} \underbrace{(2x)}_{\substack{\text{Derivative of} \\ \text{the inside function}}} = -6x(x^2 + 1)^{-4} = -\frac{6x}{(x^2 + 1)^4}$$

Power
$n - 1$

PRACTICE PROBLEM 3   Find $\dfrac{d}{dx} (x^3 - x)^{-1/2}$.   *Solution at the back of the book*

EXAMPLE 6 Finding the Growth Rate of an Oil Slick

An oil tanker hits a reef, and after t days the radius of the oil slick is $r(t) = \sqrt{4t + 1}$ miles. How fast is the radius of the oil slick expanding after 2 days?

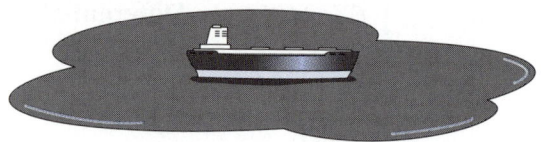

Solution   To find the rate of change of the radius, we differentiate:

$$\frac{d}{dt}(4t+1)^{1/2} = \frac{1}{2}(4t+1)^{-1/2}(4) = 2(4t+1)^{-1/2}$$

Derivative of $4t + 1$

and at $t = 2$ this is

$$2(4 \cdot 2 + 1)^{-1/2} = 2 \cdot 9^{-1/2} = 2 \cdot \frac{1}{3} = \frac{2}{3}$$

Interpretation: After 2 days the radius of oil slick is growing at the rate of $\frac{2}{3}$ of a mile per day.

Graphing Calculator Exploration

a. On a graphing calculator, enter the radius of the oil slick: $y_1 = \sqrt{4x + 1}$. Then use NDERIV to define y_2 to be the derivative of y_1. Graph both y_1 and y_2 on the window $[0, 6]$ by $[0, 5]$.

b. Verify your answer to Example 6 by evaluating y_2 at $x = 2$. Do you get $\frac{2}{3}$?

c. Notice from your graph that the (derivative) function y_2 is *decreasing*, meaning that the radius is growing more *slowly*.

Estimate when the radius will be growing by only $\frac{1}{2}$ mile per day. [*Hint:* One way is to use TRACE to follow along the curve y_2 to the x-value (number of days) when the y-coordinate is 0.5, zooming in if necessary. Your answer should be between 3 and 4 days. You can also use INTERSECT with $y_3 = .5$.]

Some problems require the generalized power rule in combination with another differentiation rule, such as the product or quotient rule.

EXAMPLE 7 **Differentiating Using Two Rules**

Find $\dfrac{d}{dx}[(5x-2)^4(9x+2)^7]$.

Solution   Since this is a product of powers, $(5x-2)^4$ times $(9x+2)^7$, we use the product rule together with the generalized power rule.

$$\frac{d}{dx}[(5x-2)^4(9x+2)^7] = \underbrace{4(5x-2)^3(5)(9x+2)^7}_{\substack{\text{Derivative of} \\ (5x-2)^4}} + (5x-2)^4\underbrace{[7(9x+2)^6(9)]}_{\substack{\text{Derivative of} \\ (9x+2)^7}}$$

$$= 20\underbrace{(5x-2)^3}_{4\,\cdot\,5}(9x+2)^7 + 63(5x-2)^4\underbrace{(9x+2)^6}_{7\,\cdot\,9} \qquad \text{Simplifying}$$

■

EXAMPLE 8 **Differentiating Using Two Rules**

Find $\dfrac{d}{dx}\left(\dfrac{x}{x+1}\right)^4$.

Solution   Since the function is a quotient raised to a power, we use the quotient rule together with the generalized power rule. Working from the outside in, we obtain

$$\frac{d}{dx}\left(\frac{x}{x+1}\right)^4 = 4\left(\frac{x}{x+1}\right)^3\underbrace{\frac{(x+1)(1)-(1)(x)}{(x+1)^2}}_{\substack{\text{Derivative of the} \\ \text{inside function } x/(x+1)}}$$

$$= 4\left(\frac{x}{x+1}\right)^3\frac{x+1-x}{(x+1)^2} = 4\left(\frac{x}{x+1}\right)^3\frac{1}{(x+1)^2} \qquad \text{Simplifying}$$

$$= 4\frac{x^3}{(x+1)^3}\frac{1}{(x+1)^2} = \frac{4x^3}{(x+1)^5} \qquad \text{Simplifying further}$$

■

EXAMPLE 9 **Differentiating Using a Rule Twice**

Find $\dfrac{d}{dz}[z^2+(z^2-1)^3]^5$.

Solution   Since this is a function to a power, where the inside function also contains a function to a power, we must use the generalized power rule *twice*.

$$\frac{d}{dz}[z^2 + (z^2 - 1)^3]^5 = 5[z^2 + (z^2 - 1)^3]^4\underbrace{[2z + 3(z^2 - 1)^2(2z)]}$$

Derivative of
$z^2 + (z^2 - 1)^3$

$$= 5[z^2 + (z^2 - 1)^3]^4[2z + 6z(z^2 - 1)^2]$$ Simplifying

$3 \cdot 2$

■

Chain Rule in Leibniz's Notation

A composite function $f(g(x))$ may be written in two parts:

$$y = f(g(x)) \quad \text{is equivalent to} \quad y = f(u) \text{ and } u = g(x)$$

The derivatives of these last two functions are:

$$\frac{dy}{du} = f'(u) \quad \text{and} \quad \frac{du}{dx} = g'(x)$$

The chain rule:

$$\frac{d}{dx}f(g(x)) = f'(g(x)) \cdot g'(x)$$ Chain rule

y

$\dfrac{dy}{dx}$ $\qquad$ $\dfrac{dy}{du}$ $\qquad$ $\dfrac{du}{dx}$

with the indicated substitutions then becomes:

Chain Rule in Leibniz's Notation

For $y = f(u)$ with $u = g(x)$,

$$\frac{dy}{dx} = \frac{dy}{du} \cdot \frac{du}{dx}$$

In this form the chain rule is easy to remember, since it looks as if the du in the numerator and the denominator cancel:

$$\frac{dy}{dx} = \frac{dy}{du} \cdot \frac{du}{dx}$$

However, since derivatives are not really fractions (they are *limits* of fractions), this is only a convenient device for remembering the chain rule.

In our discussion of the product rule on page 542, we saw that the derivative of a product is *not* the product of the derivatives. We now see where the product of the derivatives *does* appear: It appears in the chain rule, when differentiating composite functions. In other words, the product of the derivatives comes not from *products* but from *compositions* of functions.

A Simple Example of the Chain Rule

The derivation of the chain rule is rather technical, but we can show the basic idea in a simple example. Suppose that your company produces steel, and you want to calculate your company's total revenue in dollars per year. You would take the revenue from a ton of steel (dollars per ton) and multiply by your company's output (tons per year). In symbols:

$$\frac{\$}{\text{year}} = \frac{\$}{\text{ton}} \cdot \frac{\text{ton}}{\text{year}}$$

Note that "ton" cancels

If we were to express these rates as derivatives, the equation above would become the chain rule.

SUMMARY

To differentiate a composite function (a function of a function), we have the chain rule:

$$\frac{d}{dx} f(g(x)) = f'(g(x)) \cdot g'(x)$$

or in Leibniz's notation, writing $y = f(g(x))$ as $y = f(u)$ and $u = g(x)$,

$$\frac{dy}{dx} = \frac{dy}{du} \cdot \frac{du}{dx}$$

The derivative of a composite function is the product of the derivatives

To differentiate a function to a power, $[f(x)]^n$, we have the *generalized power rule* (a special case of the chain rule when the "outer" function is a power):

$$\frac{d}{dx} [f(x)]^n = n[f(x)]^{n-1} f'(x)$$

At the moment the generalized power rule is more useful than the chain rule, but in Chapter 8 we will make important use of the chain rule.

Verification of the Chain Rule

Let $f(x)$ and $g(x)$ be differentiable functions. We define k by:

$$k = g(x + h) - g(x)$$

or equivalently,

$$g(x + h) = g(x) + k$$

Then:

$$\lim_{h \to 0} \underbrace{[g(x + h) - g(x)]}_{k} = \lim_{h \to 0} k = 0$$

showing that $h \to 0$ implies $k \to 0$ (see page 551). With these relations we may calculate the derivative of the composition $f(g(x))$.

$$\frac{d}{dx} f(g(x)) = \lim_{h \to 0} \frac{f(g(x + h)) - f(g(x))}{h} \qquad \text{Definition of the derivative of } f(g(x))$$

$$= \lim_{h \to 0} \left[\frac{f(g(x + h)) - f(g(x))}{g(x + h) - g(x)} \cdot \frac{g(x + h) - g(x)}{h} \right] \qquad \text{Dividing and multiplying by } g(x + h) - g(x)$$

$$= \lim_{h \to 0} \frac{f(g(x + h)) - f(g(x))}{g(x + h) - g(x)} \lim_{h \to 0} \frac{g(x + h) - g(x)}{h} \qquad \text{The limit of a product is the product of the limits (limit rule 4c on page 500)}$$

$$= \underbrace{\lim_{k \to 0} \frac{f(g(x) + k) - f(g(x))}{k}}_{f'(g(x))} \underbrace{\lim_{h \to 0} \frac{g(x + h) - g(x)}{h}}_{g'(x)} \qquad \begin{array}{l} \text{Using the relations} \\ g(x + h) = g(x) + k \text{ and} \\ k = g(x + h) - g(x) \text{ and} \\ \text{that } h \to 0 \text{ implies } k \to 0 \end{array}$$

$$= f'(g(x)) \cdot g'(x) \qquad \text{Chain rule}$$

The last step comes from recognizing the first limit as the definition of the derivative f' at $g(x)$, and the second limit as the definition of the derivative $g'(x)$. This verifies the chain rule:

$$\frac{d}{dx} f(g(x)) = f'(g(x)) \cdot g'(x)$$

Strictly speaking, this verification requires an additional assumption, that the denominator $g(x + h) - g(x)$ is never zero. This assumption can be avoided by a more technical argument, which we omit.

EXERCISES 6.6

Find functions f and g such that the given function is the composition $f(g(x))$.

1. $\sqrt{x^2 - 3x + 1}$ **2.** $(5x^2 - x + 2)^4$

3. $(x^2 - x)^{-3}$ **4.** $\dfrac{1}{x^2 + x}$ **5.** $\dfrac{x^3 + 1}{x^3 - 1}$

6. $\dfrac{\sqrt{x} - 1}{\sqrt{x} + 1}$ **7.** $\left(\dfrac{x + 1}{x - 1}\right)^4$ **8.** $\sqrt{\dfrac{x - 1}{x + 1}}$

9. $\sqrt{x^2 - 9} + 5$ **10.** $\sqrt[3]{x^3 + 8} - 5$

Use the generalized power rule to find the derivative of each function.

11. $f(x) = (x^2 + 1)^3$ **12.** $f(x) = (x^3 + 1)^4$

13. $h(z) = (3z^2 - 5z + 2)^4$

14. $h(z) = (5z^2 + 3z - 1)^3$

15. $f(x) = \sqrt{x^4 - 5x + 1}$ **16.** $f(x) = \sqrt{x^6 + 3x - 1}$

17. $w(z) = \sqrt[3]{9z - 1}$ **18.** $w(z) = \sqrt[5]{10z - 4}$

19. $y = (4 - x^2)^4$ **20.** $y = (1 - x)^{50}$

21. $y = \left(\dfrac{1}{w^3 - 1}\right)^4$ **22.** $y = \left(\dfrac{1}{w^4 + 1}\right)^5$

23. $y = x^4 + (1 - x)^4$

24. $f(x) = (x^2 + 4)^3 - (x^2 + 4)^2$

25. $f(x) = \dfrac{1}{\sqrt[3]{(9x + 1)^2}}$ **26.** $f(x) = \dfrac{1}{\sqrt[3]{(3x - 1)^2}}$

27. $f(x) = [(x^2 + 1)^3 + x]^3$

28. $f(x) = [(x^3 + 1)^2 - x]^4$

29. $f(x) = 3x^2(2x + 1)^5$ **30.** $f(x) = 2x(x^3 - 1)^4$

31. $f(x) = (2x + 1)^3(2x - 1)^4$

32. $f(x) = (2x - 1)^3(2x + 1)^4$

33. $f(x) = \left(\dfrac{x + 1}{x - 1}\right)^3$ **34.** $f(x) = \left(\dfrac{x - 1}{x + 1}\right)^5$

35. $f(x) = x^2\sqrt{1 + x^2}$ **36.** $f(x) = x^2\sqrt{x^2 - 1}$

37. $f(x) = \sqrt{1 + \sqrt{x}}$ **38.** $f(x) = \sqrt[3]{1 + \sqrt[3]{x}}$

39. Find the derivative of $(x^2 + 1)^2$ in two ways:
   a. By the generalized power rule.
   b. By "squaring out" the original expression and then differentiating.

   Your answers should agree.

40. Find the derivative of $\dfrac{1}{x^2}$ in three ways:

   a. By the quotient rule.

   b. By writing $\dfrac{1}{x^2}$ as $(x^2)^{-1}$ and using the generalized power rule.

   c. By writing $\dfrac{1}{x^2}$ as x^{-2} and using the (ordinary) power rule.

   Your answers should agree.

41. Find the derivative of $\dfrac{1}{3x + 1}$ in two ways:

   a. By the quotient rule.
   b. By writing it as $(3x + 1)^{-1}$ and using the generalized power rule.

   Your answers should agree. Which way was easiest? Remember this for the future.

42. Find an expression for the derivative of the composition of *three* functions, $\dfrac{d}{dx} f(g(h(x)))$. (*Hint:* Use the chain rule twice.)

43. Suppose that $L(x)$ is a function such that $L'(x) = \dfrac{1}{x}$. Use the chain rule to show that the derivative of the composite function $L(g(x))$ is $\dfrac{d}{dx} L(g(x)) = \dfrac{g'(x)}{g(x)}$.

44. Suppose that $E(x)$ is a function such that $E'(x) = E(x)$. Use the chain rule to show that the derivative of the composite function $E(g(x))$ is $\dfrac{d}{dx} E(g(x)) = E(g(x)) \cdot g'(x)$.

Find the *second* derivative of each function.

45. $f(x) = (x^2 + 1)^{10}$ **46.** $f(x) = (x^3 - 1)^5$

APPLIED EXERCISES

47. Business: Cost   A company's cost function is $C(x) = \sqrt{4x^2 + 900}$ dollars, where x is the number of units. Find the marginal cost function and evaluate it at $x = 20$.

48. Continuation of Exercise 47   Graph the cost function $y_1 = \sqrt{4x^2 + 900}$ on the viewing window [0, 30] by [−10, 70]. Then use NDERIV to define y_2 as the derivative of y_1. Verify the answer to Exercise 47 by evaluating the marginal cost function y_2 at $x = 20$.

49. Continuation of Exercise 48   Find the number x of units at which the marginal cost is 1.75. (*Hint:* TRACE along the marginal cost function y_2 to find where the y-coordinate is 1.75, giving your answer as the x-coordinate rounded to the nearest whole number.)

50. Sociology: Educational Status   A study* estimated how a person's social status (rated on a scale where 100 indicates the status of a college graduate) depended on years of education. Based on this study, with e years of education, a person's status would be $S(e) = 0.22(e + 4)^{2.1}$. Find $S'(12)$ and interpret your answer.

51. Sociology: Income Status   A study* estimated how a person's social status (rated on a scale where 100 indicates the status of a college graduate) depended upon income. Based on this study, with an income of i thousand dollars, a person's status would be $S(i) = 17.5(i − 1)^{0.53}$. Find $S'(25)$ and interpret your answer.

52. Economics: Compound Interest   If $1000 is deposited in a bank paying $r\%$ interest compounded annually, 5 years later its value will be

$$V(r) = 1000(1 + 0.01r)^5 \quad \text{dollars}$$

Find $V'(6)$ and interpret your answer. (*Hint:* $r = 6$ corresponds to 6% interest.)

53. Biomedical: Drug Sensitivity   The strength of a patient's reaction to a dose of x milligrams

of a certain drug is $R(x) = 4x\sqrt{11 + 0.5x}$ for $0 \le x \le 140$. The derivative $R'(x)$ is called the *sensitivity* to the drug. Find $R'(50)$, the sensitivity to a dose of 50 mg.

54. General: Population   The population of a city x years from now is predicted to be $P(x) = \sqrt[4]{x^2 + 1}$ million people for $1 \le x \le 5$. Find when the population will be growing at the rate of a quarter of a million people per year. [*Hint:* On a graphing calculator, enter the given population function as y_1, use NDERIV to define y_2 to be the derivative of y_1, and graph both on the viewing window [1, 5] by [0, 3]. Then TRACE along y_2 to find the x-coordinate (rounded to the nearest tenth of a unit) at which the y-coordinate is 0.25. You may have to ZOOM IN to find the correct x-value.]

55. Continuation of Exercise 53   For the reaction function given in Exercise 53, find the dose at which the sensitivity is 25. [*Hint:* On a graphing calculator, enter the reaction function $y_1 = 4x\sqrt{11 + 0.5x}$, and use NDERIV to define y_2 to be the derivative of y_1. Then graph y_2 on the window [0, 140] by [0, 50] and TRACE along y_2 to find the x-coordinate (rounded to the nearest whole number) at which the y-coordinate is 25. You may have to ZOOM IN to find the correct x-value.]

56. Biomedical: Blood Flow   It follows from Poiseuille's Law that blood flowing through certain arteries will encounter a resistance of $R(x) = 0.25(1 + x)^4$, where x is the distance (in meters) from the heart. Find the instantaneous rate of change of the resistance at

a. 0 meters   **b.** 1 meter

57. Environmental Science: Pollution   The carbon monoxide level in a city is predicted to be $0.02x^{3/2} + 1$ ppm (parts per million), where x is the population in thousands. In t years the population of the city is predicted to be $x(t) = 12 + 2t$ thousand people. Therefore, in t years the carbon monoxide level will be

$$P(t) = 0.02(12 + 2t)^{3/2} + 1$$

* Robert L. Hamblin, "Mathematical Experimentation and Sociological Theory: A Critical Analysis," *Sociometry* 34:423–452, 1971.

Now h can approach zero through *positive* numbers (like 0.01, 0.001, 0.0001, . . .) or through *negative* numbers (like -0.01, -0.001, -0.0001, . . .). The limit as $h \to 0$ must be a *single* number, the same regardless of whether h is positive or negative. We will, however, see that $|h|/h$ approaches *two different* numbers, depending upon whether h is positive or negative. This will show that the limit, and hence the derivative, does not exist.

We will use the notation $\lim\limits_{h \to 0^+}$ to mean the limit as h approaches zero through *positive* numbers, and $\lim\limits_{h \to 0^-}$ to mean the limit as h approaches zero through *negative* numbers.

$$\text{For } \textit{positive } h: \lim_{h \to 0^+} \frac{|h|}{h} = \lim_{h \to 0^+} \frac{h}{h} = \lim_{h \to 0^+} 1 = 1$$

Since $|h| = h$ for $h > 0$ $\dfrac{h}{h} = 1$

$$\text{For } \textit{negative } h: \lim_{h \to 0^-} \frac{|h|}{h} = \lim_{h \to 0^-} \frac{-h}{h} = \lim_{h \to 0^-} (-1) = -1$$

For $h < 0, |h| = -h$ (the negative sign makes the negative h positive) $\dfrac{-h}{h} = -1$

These two different results,

$$\lim_{h \to 0^+} \frac{|h|}{h} = +1 \qquad \text{For positive } h, \text{ the limit is } +1$$

and

$$\lim_{h \to 0^-} \frac{|h|}{h} = -1 \qquad \text{For negative } h, \text{ the limit is } -1$$

show that as $h \to 0$, $\dfrac{|h|}{h}$ approaches two different numbers, $+1$ and -1, depending upon whether h is positive or negative. Therefore, the limit $\lim\limits_{h \to 0} \dfrac{|h|}{h}$ does not exist, so *the derivative does not exist*. This is what we wanted to show—that the absolute value function is not differentiable at $x = 0$.

■

Geometric Explanation of Nondifferentiability

We can give a geometric and intuitive reason why the absolute value function is not differentiable at $x = 0$. Its graph consists of two straight lines with slopes $+1$ and -1 that meet in a corner at the origin. To the right of the origin the slope is $+1$ and to the left of the origin the slope is -1, but *at* the origin the two conflicting slopes make it impossible to define a *single* slope. Therefore, the slope (and hence the derivative) is undefined at $x = 0$.

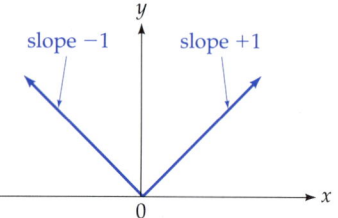

 ### Graphing Calculator Exploration

If your graphing calculator has an operation like ABS(or something similar for the absolute value function, graph $y_1 = \text{ABS}(x)$ on the graphing window $[-2, 2]$ by $[-1, 2]$. Use NDERIV to "find" the derivative of y_1 at $x = 0$. Your calculator may give a "false value" like 0, resulting from its *approximating* the derivative by the symmetric difference quotient (see page 535). The correct answer is that the derivative is *undefined* at $x = 0$, as we just showed. This is why it is important to understand nondifferentiability—your calculator may give a misleading answer.

Other Nondifferentiable Functions

For the same reason, at any "corner point" of a graph, where two different slopes conflict, the function will not be differentiable.

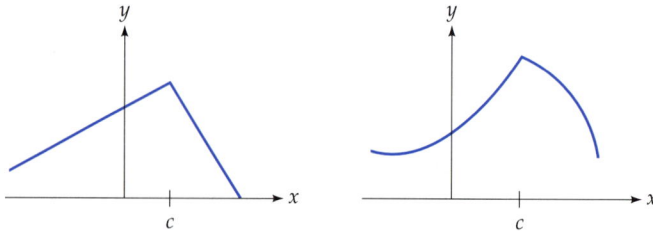

Each of the functions graphed here has a "corner point"
at $x = c$, and so is not differentiable at $x = c$.

There are other reasons, besides a corner point, why a function may not be differentiable. If a curve has a vertical tangent line at a point, the slope will not be defined at that x-value, since the slope of a vertical line is undefined.

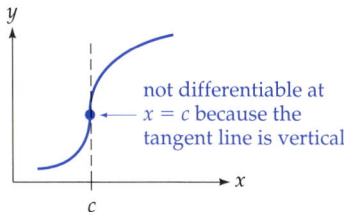

not differentiable at $x = c$ because the tangent line is vertical

We showed on page 551 that if a function is differentiable, then it is continuous. Therefore, if a function is discontinuous (has a "jump") at some point, then it will not be differentiable at that x-value.

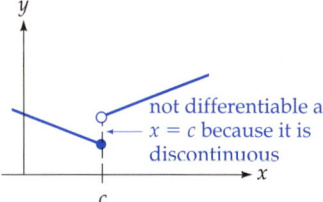

not differentiable at $x = c$ because it is discontinuous

Therefore, if a function f satisfies any of the following conditions:

1. f has a corner point at $x = c$
2. f has a vertical tangent at $x = c$
3. f is discontinuous at $x = c$

then f will not be differentiable at c. In Chapter 7, when we use calculus for graphing, it will be important to remember these three conditions that make the derivative fail to exist.

PRACTICE PROBLEM 1

For the function graphed below, find the x-values at which the derivative is undefined.   *Solution at the back of the book*

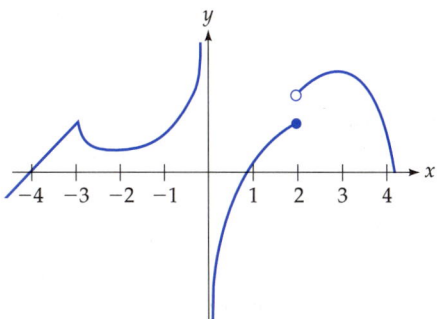

All differentiable functions are continuous (see page 551), but *not* all continuous functions are differentiable (for example, $f(x) = |x|$). These facts are shown in the following diagram.

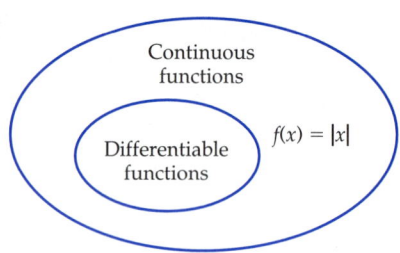

EXERCISES 6.7

For each function graphed below, state the x-values at which the derivative does not exist.

1.

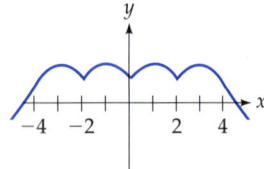

2.

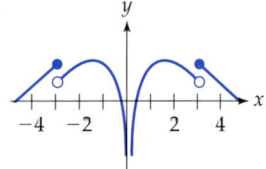

3.

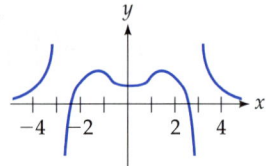

4.

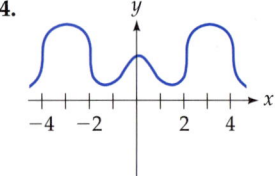

Use the definition of the derivative to show that the following functions are not differentiable at $x = 0$.

5. $f(x) = |2x|$ **6.** $f(x) = |3x|$

(Hint for Exercises 5 and 6: Modify the calculations on pages 584–585 to apply to these functions.)

7. $f(x) = x^{2/5}$ **8.** $f(x) = x^{4/5}$

 Use NDERIV to "find" the derivative of each function at $x = 0$. Is the result correct?

9. $f(x) = \dfrac{1}{x}$ **10.** $f(x) = \dfrac{1}{x^2}$

 11. a. Show that the definition of the derivative applied to the function $f(x) = \sqrt{x}$ at $x = 0$ gives
$$f'(0) = \lim_{h \to 0} \frac{\sqrt{h}}{h}.$$

b. Use a calculator to evaluate the difference quotient $\dfrac{\sqrt{h}}{h}$ for the following values of h: 0.1, 0.001, and 0.00001. (*Hint:* Enter the calculation into your calculator with h replaced by 0.1, and then change the value of h by inserting zeros.)

c. From your answers to part (b), does the limit exist? Does the derivative of $f(x) = \sqrt{x}$ at $x = 0$ exist?

d. Graph $f(x) = \sqrt{x}$ on the viewing window [0, 1] by [0, 1]. Do you see why the slope at $x = 0$ does not exist?

12. a. Show that the definition of the derivative applied to the function $f(x) = \sqrt[3]{x}$ at $x = 0$ gives

$$f'(0) = \lim_{h \to 0} \frac{\sqrt[3]{h}}{h}.$$

b. Use a calculator to evaluate the difference quotient $\dfrac{\sqrt[3]{h}}{h}$ for the following values of h: 0.1, 0.0001, and 0.0000001. (*Hint:* Enter the calculation into your calculator with h replaced by 0.1, and then change the value of h by inserting zeros.)

c. From your answers to part (b), does the limit exist? Does the derivative of $f(x) = \sqrt[3]{x}$ at $x = 0$ exist?

d. Graph $f(x) = \sqrt[3]{x}$ on the viewing window $[-1, 1]$ by $[-1, 1]$. Do you see why the slope at $x = 0$ does not exist?

Chapter Summary with Hints and Suggestions

The reading and exercises of this chapter have helped you to learn the following skills. For each skill, the section from which it came (in case you need to review it) and some exercises in this section that use it are indicated. Answers to all exercises are at the end of the book, and full solutions to all exercises are in the Student Solutions Manual.

6.1   *Limits and Continuity*

- Find the limit of a function from tables, some using .   (*Review Exercises 1–4.*)

- Find the limit of a function by direct substitution.   (*Review Exercises 5–12.*)

- Determine whether a function is continuous or discontinuous by looking at its algebraic form. (*Review Exercises 13–20.*)

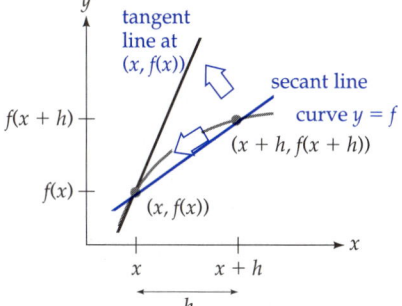

6.2   *Slopes, Rates of Change, and Derivatives*

- Find the derivative of a function from the *definition* of the derivative.
(*Review Exercises 21–24.*)

$$f'(x) = \lim_{h \to 0} \frac{f(x + h) - f(x)}{h}$$

6.3   *Some Differentiation Formulas*

- Find the derivative of a function using the rules of differentiation.   (*Review Exercises 25–30.*)

$$\frac{d}{dx} c = 0 \qquad \frac{d}{dx} x^n = nx^{n-1}$$

$$\frac{d}{dx} (c \cdot f) = c \cdot f' \qquad \frac{d}{dx} (f \pm g) = f' \pm g'$$

- Calculate and interpret a company's marginal cost.   *(Review Exercise 31.)*

$$MC(x) = C'(x)   \qquad MR(x) = R'(x)$$

$$MP(x) = P'(x)$$

- Find and interpret the derivative of a learning curve.   *(Review Exercise 32.)*

- Find and interpret the derivative of an area or volume formula.   *(Review Exercises 33–34.)*

6.4   *The Product and Quotient Rules*

- Find the derivative of a function using the product or quotient rule.
(Review Exercises 35–45.)

$$\frac{d}{dx}(f \cdot g) = f' \cdot g + f \cdot g'$$

$$\frac{d}{dx}\left(\frac{f}{g}\right) = \frac{g \cdot f' - g' \cdot f}{g^2}$$

- Use differentiation to solve an applied problem and interpret the answer.
(Review Exercises 46–48.)

6.5   *Higher-Order Derivatives*

- Calculate the second derivative of a function.
(Review Exercises 49–58.)

- Find and interpret the first and second derivatives in an application.   *(Review Exercise 59.)*

- Find the velocity and acceleration of a rocket.
(Review Exercise 60.)

$$v(t) = s'(t)   \qquad a(t) = v'(t) = s''(t)$$

- Find the maximum height of a projectile, checking by 📊.   *(Review Exercise 61.)*

- Use 📊 to find a company's profit function, then calculate and interpret the first and second derivatives.   *(Review Exercise 62.)*

6.6   *The Chain Rule and the Generalized Power Rule*

- Find the derivative of a function using the generalized power rule.   *(Review Exercises 63–72.)*

$$\frac{d}{dx}f^n = n \cdot f^{n-1} \cdot f'$$

$$\frac{d}{dx}f(g(x)) = f'(g(x)) \cdot g'(x)$$

$$\frac{dy}{dx} = \frac{dy}{du} \cdot \frac{du}{dx}$$

- Find the derivative of a function using *two* differentiation rules.   *(Review Exercises 73–84.)*

- Find the *second* derivative of a function using the generalized power rule.
(Review Exercises 85–88.)

- Find the derivative of a function in several different ways.   *(Review Exercises 89–90.)*

- Use the generalized power rule to find the derivative in an applied problem and interpret the answer.   *(Review Exercises 91–92.)*

- Use 📊 to compare the profit from one unit to the marginal profit found by differentiation.
(Review Exercise 93.)

- Use 📊 to find where the marginal profit equals a given number.   *(Review Exercise 94.)*

- Use the generalized power rule to solve an applied problem and interpret the answer.
(Review Exercises 95–96.)

6.7   *Nondifferentiable Functions*

- See from a graph where the derivative is undefined.   *(Review Exercises 97–100.)*

$$f' \text{ is undefined at } \begin{cases} \text{corner points} \\ \text{vertical tangents} \\ \text{discontinuities} \end{cases}$$

- Prove that a function is not differentiable at a given value.   *(Review Exercises 101–102.)*

Hints and Suggestions

- (Overview) This chapter introduced one of the most important concepts in all of mathematics, the *derivative*. First we defined it (using limits), then we developed several "rules of differentiation" to simplify its calculation.

- Remember the four interpretations of the derivative—*slopes, instantaneous rates of change, marginals,* and *velocities*.

- The *second* derivative gives the rate of change of the rate of change, and acceleration.

- Graphing calculators help to find limits, graph curves and their tangent lines, and calculate derivatives (using NDERIV) and second deriva-

tives (using NDERIV twice). NDERIV, however, provides only an *approximation* to the derivative, and therefore sometimes gives a misleading result.

- The units of the derivative are important in applications. For example, if $f(x)$ gives the temperature in degrees at time x hours, then the derivative $f'(x)$ is in *degrees per hour*. In general, the units of the derivative $f'(x)$ are "f-units" per "x-unit."

- Practice for test: Review Exercises 1, 4, 7, 9, 11, 17, 21, 25, 31, 37, 43, 47, 59, 62, 67, 77, 83, 91, 94, 97.

Review Exercises for Chapter 6   *Practice test exercises are in blue.*

6.1   Limits and Continuity

For each limit, complete the limit table and evaluate the limit (or state that the limit does not exist). Round calculations to three decimal places.

1. $\lim\limits_{x \to 2} (4x + 2)$

x	$4x + 2$
1.9	
1.99	
1.999	

x	$4x + 2$
2.1	
2.01	
2.001	

2. $\lim\limits_{x \to 8} \dfrac{1}{x - 8}$

x	$\dfrac{1}{x - 8}$
7.9	
7.99	
7.999	

x	$\dfrac{1}{x - 8}$
8.1	
8.01	
8.001	

3. $\lim\limits_{x \to 0} \dfrac{x}{|x|}$

x	$\dfrac{x}{|x|}$
-0.1	
-0.01	
-0.001	

x	$\dfrac{x}{|x|}$
0.1	
0.01	
0.001	

4. $\lim\limits_{x \to 0} \dfrac{\sqrt{x + 1} - 1}{x}$

x	$\dfrac{\sqrt{x + 1} - 1}{x}$
-0.1	
-0.01	
-0.001	

x	$\dfrac{\sqrt{x + 1} - 1}{x}$
0.1	
0.01	
0.001	

Evaluate the following limits (*without* using limit tables).

5. $\lim\limits_{x \to 4} \sqrt{x^2 + x + 5}$

6. $\lim\limits_{x \to 0} \pi$

7. $\lim\limits_{s \to 16} (\tfrac{1}{2}s - s^{1/2})$

8. $\lim\limits_{r \to 8} \dfrac{r}{r^2 - 30\sqrt[3]{r}}$

9. $\lim\limits_{x \to 1} \dfrac{x^2 - x}{x^2 - 1}$

10. $\lim\limits_{x \to -1} \dfrac{3x^3 - 3x}{2x^2 + 2x}$

11. $\lim\limits_{h \to 0} \dfrac{2x^2h - xh^2}{h}$

12. $\lim\limits_{h \to 0} \dfrac{6xh^2 - x^2h}{h}$

For each function, state whether it is continuous or discontinuous. If it is discontinuous, state the values of x at which it is discontinuous.

13. $f(x) = 2x + 5$

14. $f(x) = e^{-x}$

15. $\ln(e^x + 1)$

16. $f(x) = \dfrac{1}{x^2 + 1}$

17. $f(x) = \dfrac{x - 1}{x^2 + x}$

18. $f(x) = \dfrac{1}{|x| - 3}$

19. $f(x) = \begin{cases} 3 - x & \text{if } x > 0 \\ 3 + x & \text{if } x \le 0 \end{cases}$ (*Hint*: Sketch the graph.)

20. $f(x) = \begin{cases} 4 - x & \text{if } x > 0 \\ x + 2 & \text{if } x \le 0 \end{cases}$ (*Hint*: Sketch the graph.)

6.2 Slopes, Rates of Change, and Derivatives

Find the derivative of each function using the definition of the derivative (that is, as you did in Section 6.2).

21. $f(x) = 2x^2 + 3x - 1$ **22.** $f(x) = 3x^2 + 2x - 3$

23. $f(x) = \dfrac{3}{x}$ **24.** $f(x) = 4\sqrt{x}$

6.3 Some Differentiation Formulas

Find the derivative of each function.

25. $f(x) = 6\sqrt[3]{x^5} - \dfrac{4}{\sqrt{x}} + 1$

26. $f(x) = 4\sqrt{x^5} - \dfrac{6}{\sqrt[3]{x}} + 1$

Evaluate each expression.

27. If $f(x) = \dfrac{1}{x^2}$, find $f'(\tfrac{1}{2})$.

28. If $f(x) = \dfrac{1}{x}$, find $f'(\tfrac{1}{3})$.

29. If $f(x) = 12\sqrt[3]{x}$, find $f'(8)$.

30. If $f(x) = 6\sqrt[3]{x}$, find $f'(-8)$.

31. Business: Marginal Cost A company's cost function is $C(x) = 20 + 3x + \dfrac{54}{\sqrt{x}}$ dollars (for $5 \le x \le 20$).

   a. Find the marginal cost function.

   b. Find the marginal cost at $x = 9$ and interpret your answer.

32. Learning Curves in Industry From page 27, the learning curve for building Boeing 707 airplanes is $f(x) = 150x^{-0.322}$, where $f(x)$ is the time (in thousands of hours) that it took to build the xth Boeing 707. Find the instantaneous rate of change of this production time for the tenth plane, and interpret your answer.

33. General: Geometry The formula for the area of a circle is $A = \pi r^2$, where r is the radius of the circle and π is a constant.

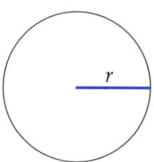

   a. Show that the derivative of the area formula is $2\pi r$, the formula for the circumference of a circle.

   b. Given an explanation for this in terms of rates of change.

34. General: Geometry The formula for the volume of a sphere is $V = \tfrac{4}{3}\pi r^3$, where r is the radius of the sphere and π is a constant.

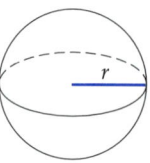

   a. Show that the derivative of the volume formula is $4\pi r^2$, the formula for the surface area of a sphere.

   b. Give an explanation for this in terms of rates of change.

6.4   The Product and Quotient Rules

Find the derivative of each function.

35. $f(x) = 2x(5x^3 + 3)$

36. $f(x) = x^2(3x^3 - 1)$

37. $f(x) = (x^2 + 5)(x^2 - 5)$

38. $f(x) = (x^2 + 3)(x^2 - 3)$

39. $y = (x^4 + x^2 + 1)(x^5 - x^3 + x)$

40. $y = (x^5 + x^3 + x)(x^4 - x^2 + 1)$

41. $y = \dfrac{x - 1}{x + 1}$ **42.** $y = \dfrac{x + 1}{x - 1}$

43. $y = \dfrac{x^5 + 1}{x^5 - 1}$ **44.** $y = \dfrac{x^6 - 1}{x^6 + 1}$

45. Find the derivative of $f(x) = \dfrac{2x + 1}{x}$ in three different ways, and check that the answers agree:

   a. By the quotient rule
   b. By writing the function in the form $f(x) = (2x + 1)\,(x^{-1})$ and using the product rule.
   c. By thinking of another way, which is the easiest of all.

46. Business: Sales The manager of an electronics store estimates that the number of cassette tapes that a store will sell at a price of x dollars is

$$S(x) = \frac{2250}{x + 9}$$

Find the rate of change of this quantity when the price is \$6 per tape, and interpret your answer.

47. Business: Marginal Average Profit A company's profit function is $P(x) = 6x - 200$ dollars, where x is the number of units.

   a. Find the average profit function.
   b. Find the marginal average profit function.
   c. Evaluate the marginal average profit function at $x = 10$ and interpret your answer.

48. Business: Marginal Average Cost A company's cost function is $C(x) = 5x + 100$ dollars, where x is the number of units.

   a. Find the average cost function.
   b. Find the marginal average cost function.
   c. Evaluate the marginal average cost function at $x = 20$ and interpret your answer.

6.5   Higher-Order Derivatives

Find the *second* derivative of each function.

49. $f(x) = 12\sqrt{x^3} - 9\sqrt[3]{x}$

50. $f(x) = 18\sqrt[3]{x^2} - 4\sqrt{x^3}$

51. $f(x) = \dfrac{1}{3x^2}$ **52.** $f(x) = \dfrac{1}{2x^3}$

Evaluate each expression.

53. If $f(x) = \dfrac{2}{x^3}$, find $f''(-1)$.

54. If $f(x) = \dfrac{3}{x^4}$, find $f''(-1)$.

55. $\dfrac{d^2}{dx^2} x^6 \Big|_{x=-2}$ **56.** $\dfrac{d^2}{dx^2} x^{-2} \Big|_{x=-2}$

57. $\dfrac{d^2}{dx^2} \sqrt{x^5} \Big|_{x=16}$ **58.** $\dfrac{d^2}{dx^2} \sqrt{x^7} \Big|_{x=4}$

59. General: Population The population of a city t years from now is predicted to be $P(t) = 0.25t^3 - 3t^2 + 5t + 200$ thousand people. Find $P(10)$, $P'(10)$, and $P''(10)$ and interpret your answers.

60. General: Velocity A rocket rises $s(t) = 8t^{5/2}$ feet in t seconds. Find its velocity and acceleration after 25 seconds.

61. General: Velocity The fastest baseball pitch on record (thrown by Lynn Nolan Ryan of the California Angels on August 20, 1974) was clocked at 100.9 miles per hour (148 feet per second).

   a. If this pitch had been thrown straight up, its height after t seconds would have been $s(t) = -16t^2 + 148t + 5$ feet. Find the maximum height the ball would have reached.

   b. Verify your answer to part (a) by graphing the height function $y_1 = -16x^2 + 148x + 5$ on the graphing window $[0, 10]$ by $[0, 400]$. Then TRACE along the curve to find its highest point (or use the MAXIMUM feature of your calculator).

 62. Business: Profit The table below gives a company's annual profit at the end of the year for years 1 through 6.

Year	Annual Profit (millions of dollars)
1	3.4
2	3.1
3	3.0
4	3.1
5	4.1
6	5.2

a. Enter the data in the table into your graphing calculator and make a plot of the resulting points (Years on the x-axis and Annual Profit on the y-axis).

b. Have your calculator find the quadratic regression formula for these data. Then enter the result as y_1, giving the annual profit for year x. Plot the points together with the regression curve (on an appropriate viewing window). Observe that the curve fits the points quite well.

c. Predict the annual profit at the end of year 7 by evaluating $y_1(7)$.

d. Define y_2 to be the derivative of y_1 (using NDERIV). Find $y_2(7)$ and interpret this number.

e. Define y_3 to be the derivative of y_2 (so y_3 is the *second* derivative of y_1). Find and interpret $y_3(7)$.

f. Graph all three functions and the points on the viewing window $[0, 7]$ by $[-1,7]$. Can you explain why y_3 appears to be a horizontal straight line?

6.6   The Chain Rule and the Generalized Power Rule

Find the derivative of each function.

63. $h(z) = (4z^2 - 3z + 1)^3$

64. $h(z) = (3z^2 - 5z - 1)^4$

65. $g(x) = (100 - x)^5$

66. $g(x) = (1000 - x)^4$

67. $f(x) = \sqrt{x^2 - x + 2}$

68. $f(x) = \sqrt{x^2 - 5x - 1}$

69. $w(z) = \sqrt[3]{6z - 1}$

70. $w(z) = \sqrt[3]{3z + 1}$

71. $h(x) = \dfrac{1}{\sqrt[5]{(5x + 1)^2}}$

72. $h(x) = \dfrac{1}{\sqrt[5]{(10x + 1)^3}}$

73. $g(x) = x^2(2x - 1)^4$

74. $g(x) = 5x(x^3 - 2)^4$

75. $y = x^3\sqrt[3]{x^3 + 1}$

76. $y = x^4\sqrt{x^2 + 1}$

77. $f(x) = [(2x^2 + 1)^4 + x^4]^3$

78. $f(x) = [(3x^2 - 1)^3 + x^3]^2$

79. $f(x) = \sqrt{(x^2 + 1)^4 - x^4}$

80. $f(x) = \sqrt{(x^3 + 1)^2 + x^2}$

81. $f(x) = (3x + 1)^4(4x + 1)^3$

82. $f(x) = (x^2 + 1)^3(x^2 - 1)^4$

83. $f(x) = \left(\dfrac{x + 5}{x}\right)^4$

84. $f(x) = \left(\dfrac{x + 4}{x}\right)^5$

Find the *second* derivative of each function.

85. $h(w) = (2w^2 - 4)^5$

86. $h(w) = (3w^2 + 1)^4$

87. $g(z) = z^3(z + 1)^3$

88. $g(z) = z^4(z + 1)^4$

89. Find the derivative of $(x^3 - 1)^2$ in two ways:

a. By the generalized power rule.

b. By "squaring out" the original expression and then differentiating.

Your answers should agree.

90. Find the derivative of $g(x) = \dfrac{1}{x^3 + 1}$ in two ways:

a. By the quotient rule.

b. By the generalized power rule.

Your answers should agree.

91. Business: Marginal Profit A company's profit from producing x tons of polyurethane is $P(x) = \sqrt{x^3 - 3x + 34}$ thousand dollars (for $0 \le x \le 10$). Find $P'(5)$ and interpret your answer.

92. General: Compound Interest If $500 is deposited in an account earning interest at r percent annually, after 3 years its value will be $V(r) = 500(1 + 0.01r)^3$ dollars. Find $V'(8)$ and interpret your answer.

93. Continuation of Exercise 91 Using a graphing calculator, enter the profit function from Exercise 91 on the graphing window [0, 10] by [0, 30].

  a. Evaluate the profit function at 4, 5, and 6 and calculate the following actual costs:

$P(5) - P(4)$ (actual cost of the fifth ton)
$P(6) - P(5)$ (actual cost of the sixth ton)

  Compare these results to the "instantaneous" marginal profit at $x = 5$ found in Exercise 91.

  b. Define another function to be the derivative of the previously entered profit function (using NDERIV) and graph both together. Find (to the nearest tenth of a unit) the x-value where the marginal profit reaches 4.

 94. On a graphing calculator, graph the cost function $C(x) = \sqrt[3]{x^2 + 25x} + 8$ as y_1 on the viewing window [0, 20] by [−2, 10]. Use NDERIV to define y_2 to be its derivative, graphing both. TRACE along the derivative to find the x-value (to the nearest unit) where the marginal cost is 0.25.

95. Biomedical: Blood Flow Blood flowing through an artery encounters a resistance of $R(x) = 0.25 (0.01x + 1)^4$ where x is the distance (in centimeters) from the heart. Find the instantaneous rate of change of the resistance 100 centimeters from the heart.

96. General: Survival Rate Suppose that for a group of 10,000 people, the number who survive to age x is $N(x) = 1000\sqrt{100 - x}$. Find $N'(96)$ and interpret your answer.

6.7 *Nondifferentiable Functions*

For each function graphed below, state the values of x at which the derivative does not exist.

97.

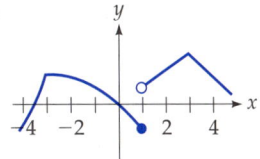

98.

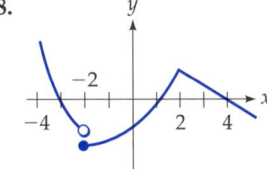

99.

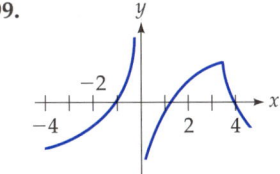

100.

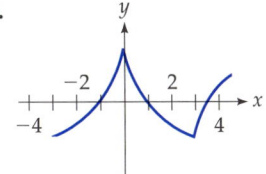

Use the definition of the derivative to show that the following functions are not differentiable at $x = 0$.

101. $f(x) = |5x|$ **102.** $f(x) = x^{3/5}$

Projects and Essays

The following projects and essays are based on Chapter 6. There are no right and wrong answers—the results depend only on your imagination and resourcefulness.

1. A student once said: "The limit is where a function is going, even if it never gets there." Write a page explaining what the student meant. Include examples and graphs.

2. The same student also said: "A continuous function gets where it's going; a discontinuous function doesn't." Write a page explaining what the student meant. Include examples and graphs.

3. Write a page about the relationships between the four interpretations of the derivative: instantaneous rates of change, slopes, marginals, and velocity. (*Hint:* Show how the other three come from the most general one, instantaneous rates of change.) Include graphs and diagrams.

4. Write about the differences between the *analytic* interpretation of the derivative (instantaneous rate of change) and the *geometric* interpretation (slope). Which is clearer to you? Why? Which do you think came first, and why?

5. For a function $s(t)$ giving the distance (in miles) from a starting point at time t (hours), write a paragraph explaining how interpreting the derivative as an instantaneous rate of change leads to $s'(t)$ being the velocity and $s''(t)$ being the acceleration.

6. Draw an interesting graph of a function, interpret the horizontal axis as time and the vertical axis as velocity of a car, and make up a story to explain the graph. (*Hint:* See Exercise 27 on page 568.) Be imaginative.

7. Draw an interesting graph of a function, interpret the horizontal axis as time and the vertical axis as a company's annual profit, and make up a story to explain the graph. (*Hint:* See Exercise 28 on page 569, but make your graph much more interesting.)

8. Look up Newton and Leibniz in a book on the history of mathematics and write about their discovery (or invention) of calculus. Also write about the feud that developed between their supporters as to who "really" discovered it.

9. Look up Weierstrass in a book on the history of mathematics and write about his discovery of a function that is everywhere continuous but nowhere differentiable. Also try to find out about "fractals" and how they relate to this topic.

10. Look over your notes, your homework, and the text, and write a page about how a graphing calculator has helped you in this chapter. Include examples of how it has helped you to *explore concepts* and how it has helped to *simplify your work*. What was its *most* helpful or interesting use? What was its *least* helpful or interesting use? Are there problems that can be done on a graphing calculator but that are easier to do "by hand"?

7

FURTHER APPLICATIONS OF DERIVATIVES

The St. Louis Arch. Note that at the highest point, the slope (steepness) of the curve is zero.

7.1 Graphing Using the First Derivative

Graphing Global Warming

Human activities affect the environment in many ways. The burning of coal, oil, and other "fossil fuels" increases the carbon dioxide in the atmosphere, which then absorbs infrared radiation from the sun (the "greenhouse effect"). There is growing evidence that this will significantly raise global temperatures over the coming decades, with possible consequences of more violent weather patterns, loss of usable farmland, melting of polar ice, higher sea levels, and flooding of lowland areas. Global warming is very difficult to predict, since atmospheric chemistry is not well understood, but scientists have developed three models of global warming, based on different levels of fossil fuel consumption. The graph below for shows the maximum temperature increase for each of these three models.

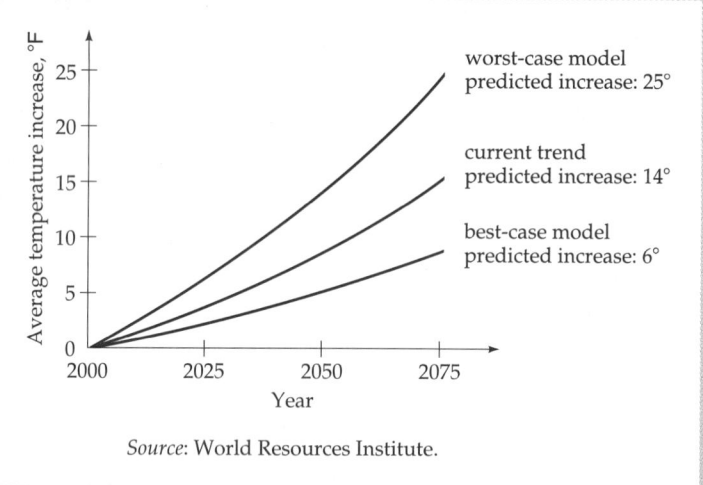

Source: World Resources Institute.

The predictions of the three models are shown very dramatically by their graphs. For example, the temperatures in the worst-case model not only are higher but are increasing more rapidly than in the other models. Graphs convey such information much more readily than would the functions describing these curves. Graphs enable one to "visualize" functions, and in this chapter we will see that calculus provides some very powerful tools for drawing graphs.

Incidentally, the United States, with only 5% of the world's population, generates 26% of the world's industrial carbon dioxide.

Introduction

In this chapter we will put derivatives to two major uses: graphing and optimization. Graphing involves using calculus to find the most important points on a curve, then sketching the curve either by hand or using a graphing calculator. Optimization means finding the largest or smallest values of a function (for example, maximizing profit or minimizing risk). We begin with graphing, since it will form the basis for optimization later in the chapter.

We saw in Chapter 6 that the derivative of a function gives the slope of its graphs.

If $f' > 0$ on an interval, then f is *increasing* on that interval.

If $f' < 0$ on an interval, then f is *decreasing* on that interval.

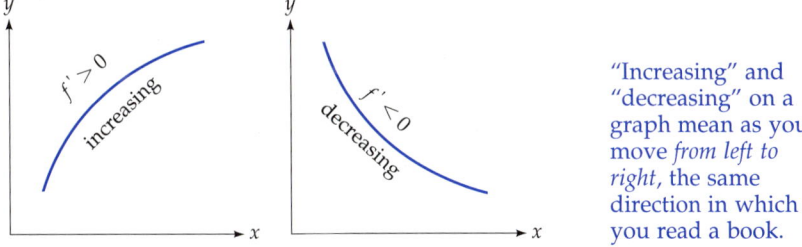

"Increasing" and "decreasing" on a graph mean as you move *from left to right*, the same direction in which you read a book.

Relative Extreme Points and Critical Values

On a graph, a *relative maximum point* is a point that is at least as *high* as the neighboring points of the curve on either side, and a *relative minimum point* is a point that is at least as *low* as the neighboring points on either side.

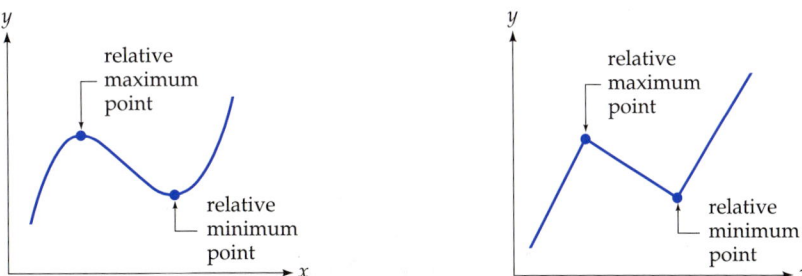

The word "relative" means that although these points may not be the highest and lowest on the *entire* curve, they are the highest and lowest *relative to points nearby*. (Later we will use the terms "absolute maximum" and "absolute minimum" to mean the highest and lowest points on the entire curve.) A curve may have any number of relative maximum and minimum points (collectively, "relative extreme points"), even none. For a function *f*, the relative extreme points may be defined more formally in terms of the values of *f*.

> *f* has a *relative maximum value* at *c* if $f(c) \geq f(x)$ for all values of *x* near* *c*.

> *f* has a *relative minimum value* at *c* if $f(c) \leq f(x)$ for all values of *x* near* *c*.

In the first of the graphs below, the relative extreme points occur where the slope *zero* (where the tangent line is horizontal), and in the second graph they occur where the slope is *undefined* (at corner points). Such points are called *critical points* of the curve, determining *critical values* of the function.

Critical Value

> A *critical value* of a function *f* is an *x*-value in the domain of *f* at which either
>
> $$f'(x) = 0 \qquad \text{Derivative is zero}$$
> or
> $$f'(x) \text{ is undefined} \qquad \text{or undefined}$$

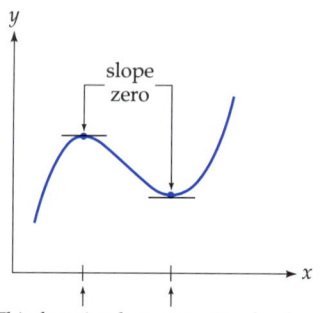

This function has two critical values (where the derivative is zero).

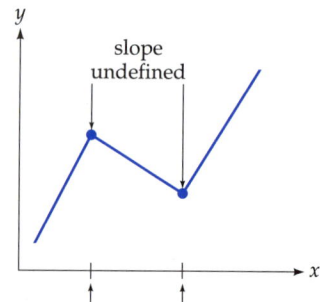

This function also has two critical values (but where the derivative is undefined).

Graphing Functions

We graph a function by finding its critical values, making a "sign diagram" for the derivative to show the relative extreme points, and then

---

* "Near" *c* means in some open interval containing *c*.

drawing the curve on a graphing calculator or "by hand." Obtaining a reasonable graph even with a graphing calculator requires more than just pushing buttons, as the following graph of $f(x) = x^3 - 12x^2 - 60x + 36$ shows. We will improve on this graph in the following example.

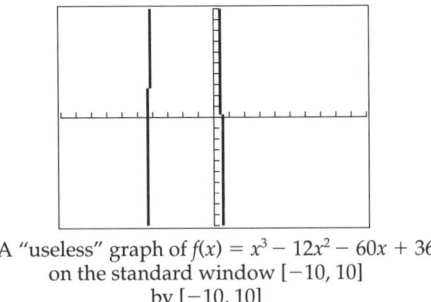

A "useless" graph of $f(x) = x^3 - 12x^2 - 60x + 36$
on the standard window $[-10, 10]$
by $[-10, 10]$

EXAMPLE 1   Graphing a Function

Graph the function $f(x) = x^3 - 12x^2 - 60x + 36$.

Solution

Step 1: Find critical values.

$$f'(x) = 3x^2 - 24x - 60 \qquad \text{Derivative of } f(x) = x^3 - 12x^2 - 60x + 36$$

$$= 3(x^2 - 8x - 20) = 3(x - 10)(x + 2) = 0 \qquad \text{Factoring and setting equal to zero}$$

Zero at   Zero at
$x = 10$ $x = -2$

The derivative is zero at $x = 10$ and at $x = -2$, and there are no points where the derivative is undefined (it is a polynomial), so the critical values (CVs) are

$$\text{CV} \begin{cases} x = 10 \\ x = -2 \end{cases} \qquad \text{Both are in the domain of the original function}$$

Step 2: Make a sign diagram for the derivative f'. A sign diagram for f' begins with a copy of the x-axis with the critical values written below it and the behavior of f' indicated above it.

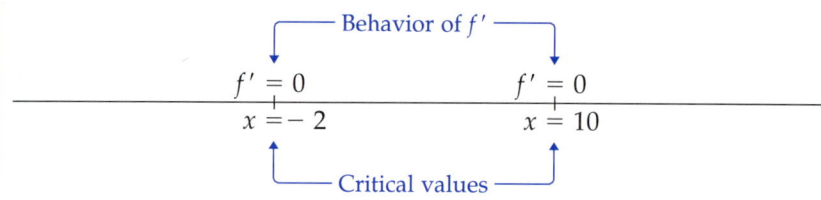

Since f' is continuous, it can change sign only at critical values, so f' must keep the same sign between two consecutive critical values. We determine the sign of f' in each interval by choosing a "test point" in each interval and substituting it into f'. We indicate the sign of f' above each interval, and we use arrows to indicate the slope of f: ↗ for positive slope, → for zero slope, and ↘ for negative slope.

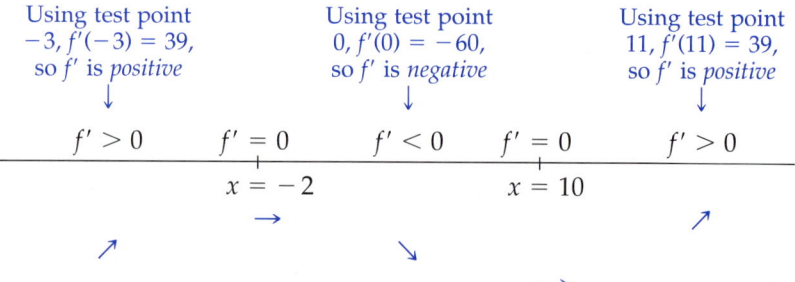

[*Hint:* To find the sign of f', use its factored form: $f'(x) = 3(x - 10)(x + 2)$.   For   example, $f'(-3) = 3(-3 - 10)(-3 + 2) = 3(\text{negative})(\text{negative}) = (\text{positive})$.]

Arrows ↗ → ↘ indicate a relative maximum point, and ↘ → ↗ indicate a relative minimum point. We then list these points under the critical values.

$f' > 0$		$f' = 0$		$f' < 0$		$f' = 0$		$f' > 0$
		$x = -2$				$x = 10$		
		→				rel min		
↗		rel max		↘		$(10, -764)$		↗
		$(-2, 100)$				→		

The y-coordinates of the points were found by evaluating the *original* function at the x-coordinate: $f(-2) = 100$ and $f(10) = -764$ (*Hint:* Use ⊞ with TABLE or EVALUATE.)

Step 3: **Sketch the graph.** The arrows ↗ → ↘ → ↗ show the general shape of the curve: going up, over, down, over, and up again. The x-coordinates of the critical points show that we want to include x-values from before $x = -2$ to after $x = 10$, say the interval $[-10, 20]$. The y-coordinates show that we want to go from above 100 to below -764, suggesting an interval of y-values like $[-800, 200]$.

| If you are using a graphing calculator, you would graph on the window $[-10, 20]$ by $[-800, 200]$ (or some other reasonable window), as shown below. | If you are graphing by hand, you would plot the relative maximum point (with a "cap" ⌒) and the relative minimum point (with a "cup" ⌣), and then draw an "up-down-up" curve through them. |

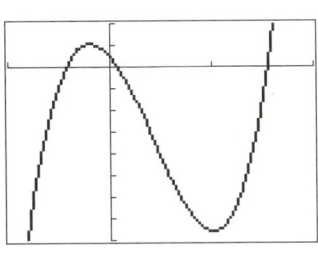 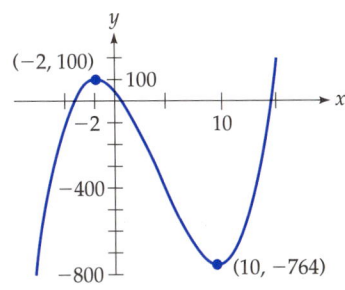

Two important observations:

1. Even with a graphing calculator, you still need calculus to find the relative extreme points for determining an appropriate viewing window.

2. Given a calculator-drawn graph like that on the left above, you should be able to make a hand-drawn graph like that on the right, including numbers on the axes and coordinates of important points (using TRACE or TABLE if necessary).

 Graphing Calculator Exploration

Do you see where the "useless" graph on page 601 fits into the "useful" graph on the left above? Check this by graphing $y_1 = x^3 - 12x^2 - 60x + 36$ first on the window $[-10, 20]$ by $[-800, 200]$ (obtaining the graph shown above) and then on the standard window $[-10, 10]$ by $[-10, 10]$.

PRACTICE PROBLEM 1   Find the critical values of $f(x) = x^3 - 12x + 8$.

Solution at the back of the book

In Example 1 there were no critical values where the derivative was *undefined* (it was a polynomial). For a function with a critical value where the derivative is *undefined*, think of the absolute value function

$f(x) = |x|$. The graph has a "corner point" at the origin, and so the derivative is undefined at the critical value $x = 0$.

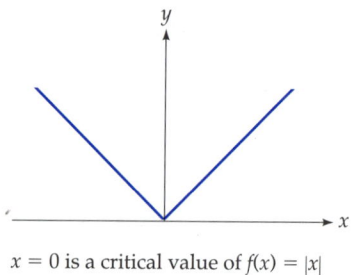

$x = 0$ is a critical value of $f(x) = |x|$

First Derivative Test for Relative Extreme Values

The graphical idea from the sign diagram, that ↗ → ↘ (up, over, and down) indicates a relative maximum and ↘ → ↗ (down, over, and up) indicates a relative minimum, can be stated more formally in terms of the derivative.

First Derivative Test

If a function f has a critical value at $x = c$, then at $x = c$ the function has

a *relative maximum* if $f' > 0$ just before c and $f' < 0$ just after c

a *relative minimum* if $f' < 0$ just before c and $f' > 0$ just after c

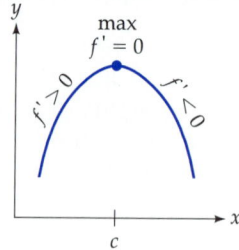

If f' is positive then negative,
then f has a relative *maximum* at c.

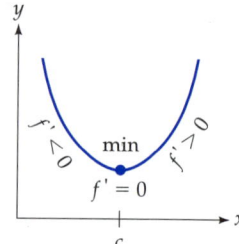

If f' is negative then positive,
then f has a relative *minimum* at c.

If the derivative has the *same* sign on both sides of c, then the function has *neither* a relative maximum nor a relative minimum at $x = c$.

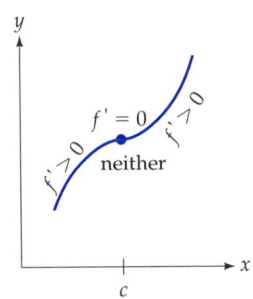

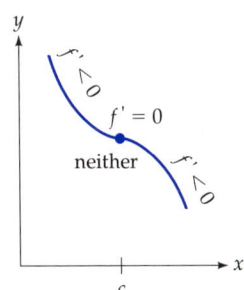

f' is positive on both sides, so
f has *neither* at $x = c$.

f' is negative on both sides, so
f has *neither* at $x = c$.

The following diagrams show that the first derivative test applies even at critical values where the derivative is *undefined* (abbreviated: f' und).

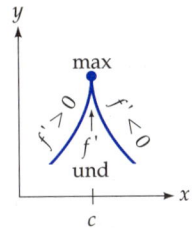

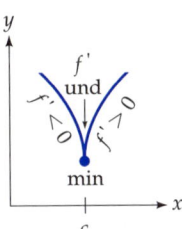

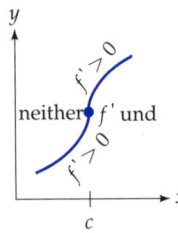

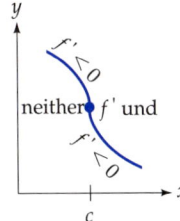

f' is positive then
negative, so f has a
relative *maximum* at
$x = c$.

f' is negative then
positive, so f has a
relative *minimum* at
$x = c$.

f' is positive on both
sides, so f has
neither at
$x = c$.

f' is negative on both
sides, so f has
neither at
$x = c$.

EXAMPLE 2   **Graphing a Function**

Graph $f(x) = -x^4 + 4x^3 - 20$.

Solution

$$f'(x) = -4x^3 + 12x^2 \qquad \text{The derivative}$$
$$= -4x^2(x - 3) = 0 \qquad \text{Factoring and setting equal to zero}$$

Critical values:

$$\text{CV} \begin{cases} x = 0 & \text{From } 4x^2 = 0 \\ x = 3 & \text{From } (x - 3) = 0 \end{cases}$$

We make a sign diagram for the derivative:

$f' = 0$	$f' = 0$	Behavior of f'
$x = 0$	$x = 3$	Critical values

We determine the sign of $f'(x) = -4x^2(x - 3)$ using test points in each interval, and then add arrows.

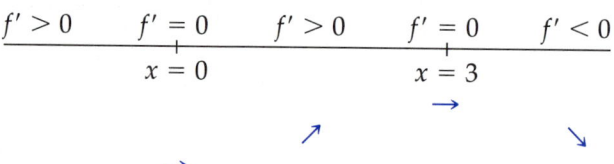

Finally, we interpret the arrows to describe the critical points, which we list at the bottom.

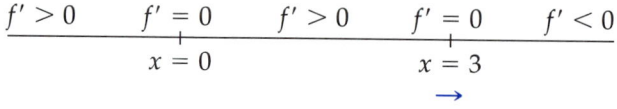

Using a graphing calculator, we would choose a viewing window like $[-2, 5]$ by $[-30, 10]$ to include the critical points, and graph the function.

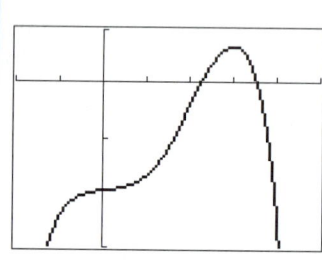

Graphing by hand, we would plot the two critical points and join them by a curve that goes, according to the arrows, "up-over-up-over-down."

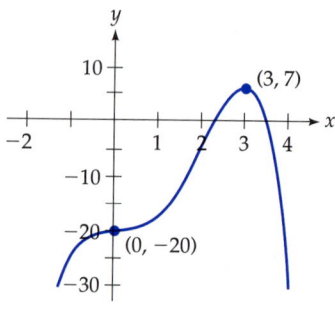

With a graphing calculator, an incomplete sign diagram may be enough to find an appropriate viewing window.

EXAMPLE 3   Graphing a Rational Function

Graph the rational function $f(x) = \dfrac{1}{x^2 - 4x}$.

Solution

$$f(x) = \frac{1}{x^2 - 4x} = \frac{1}{x(x-4)}$$

Original function with factored denominator

The denominator is zero at $x = 0$ and at $x = 4$, so the function is *undefined* at these numbers. The derivative is

$$f'(x) = \frac{(x^2 - 4x) \cdot 0 - (2x - 4) \cdot 1}{(x^2 - 4x)^2}$$

Using the quotient rule on $\dfrac{1}{x^2 - 4x}$

$$= \frac{-(2x - 4)}{(x^2 - 4x)^2} = \frac{-2(x-2)}{(x^2 - 4x)^2}$$

Simplifying, and factoring the numerator

Zero at $x = 2$

The derivative is zero at $x = 2$ and undefined (abbreviated: f' und) at $x = 0$ and at $x = 4$ (as was the original function). We list all of these on a sign diagram.

f' und	$f' = 0$	f' und
$x = 0$	$x = 2$	$x = 4$
	$(2, -\frac{1}{4})$	

y-coordinate from

$$f(x) = \frac{1}{x^2 - 4x} \text{ at } x = 2$$

For the viewing window, we might choose x-values $[-2, 6]$ (to include the x-values in the sign diagram) and y-values $[-2, 2]$ (narrow, so that the $-\frac{1}{4}$ will be distinct from the x-axis). The graph is shown below (you may need to ignore some false vertical lines). If the right and left parts of the curve did not appear, we would have used TABLE or TRACE to find the y-values and adjusted the y-window accordingly. A hand-drawn graph, which you should be able to make from the calculator graph, is next to it.

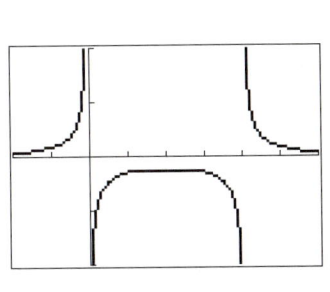

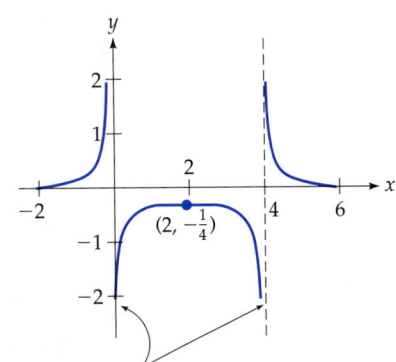

Lines like these that the curve approaches but does not touch are called *asymptotes*.

Graphing Calculator Exploration

Graph the function $f(x) = \dfrac{1}{x^2 - 4x}$ on the standard graphing window $[-10, 10]$ by $[-10, 10]$. Is the graph as clear as when graphed on the window we chose in Example 3?

PRACTICE PROBLEM 2   Explain why, for a rational function, the function and its derivative will be undefined at the same x-values.   *Solution at the back of the book*

SUMMARY

To sketch a graph:

- Find the domain (excluding any x-values where the function is undefined).

- Find the critical values (where f' is zero or undefined, but where f *is* defined).

- List all of these on a sign diagram for the derivative, indicating the behavior of f' on each interval. Add arrows and relative extreme points.

- Finally, sketch the curve. If you use a graphing calculator, the sign diagram will suggest an appropriate viewing window. If you graph by hand, your sign diagram will show you the shape of the curve.

If there are no relative extreme points, choose a few x-values, including any of special interest. On a graphing calculator, use these x-values to determine an x-interval, and use TABLE or TRACE to find a y-interval; then graph the function on the resulting window. By hand, plot the points corresponding to the chosen x-values and draw an appropriate curve through the points.

EXERCISES 7.1

For the functions graphed in Exercises 1 and 2:

a. Find the intervals on which the derivative is positive.

b. Find the intervals on which the derivative is negative.

3. Which of the numbers 1, 2, 3, 4, 5, and 6 are critical values of the function graphed below?

1.
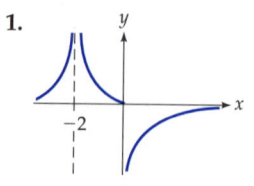

2.
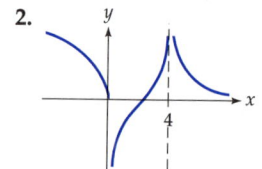

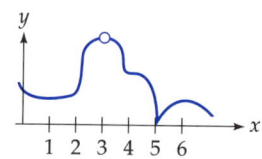

4. The first column in parts (a) through (d) shows the graphs of four functions, and the second column shows the graphs of their derivatives, but not necessarily in the same order. Write below each derivative the correct function from which it came.

a.

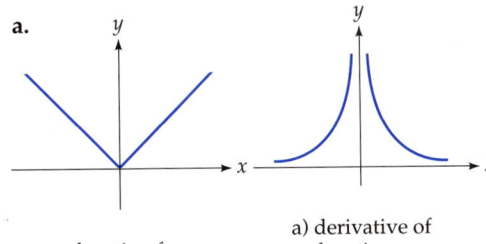

function f_1

a) derivative of function: _____

b.

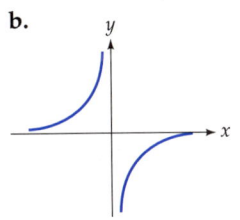

function f_2

b) derivative of function: _____

c.

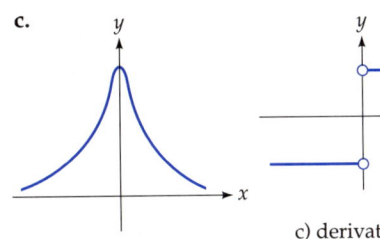

function f_3

c) derivative of function: _____

d.

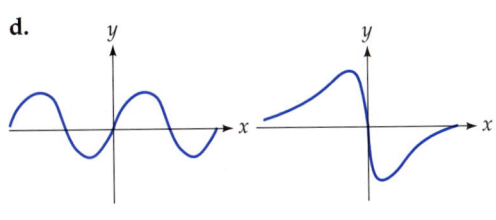

function f_4

d) derivative of function: _____

Find the critical values of each function.

5. $f(x) = x^3 - 48x$

6. $f(x) = x^3 - 27x$

7. $f(x) = x^4 + 4x^3 - 8x^2 + 1$

8. $f(x) = x^4 + 4x^3 - 20x^2 - 12$

9. $f(x) = (2x - 6)^4$

10. $f(x) = (x^2 + 6x - 7)^2$

11. $f(x) = 3x + 5$

12. $f(x) = 4x - 12$

13. $f(x) = x^3 - 2x^2 + x + 11$

14. $f(x) = x^3 - x^2 + 15$

Sketch the graph of each function "by hand" after making a sign diagram for the derivative and plotting all critical points.

15. $f(x) = x^4 + 4x^3 - 8x^2 + 64$

16. $f(x) = x^4 - 4x^3 - 8x^2 + 64$

17. $f(x) = -x^4 + 4x^3 - 4x^2 + 1$

18. $f(x) = -x^4 - 4x^3 - 4x^2 + 1$

19. $f(x) = 3x^4 - 8x^3 + 6x^2$

20. $f(x) = 3x^4 + 8x^3 + 6x^2$

21. $f(x) = (x - 1)^6$

22. $f(x) = (x - 1)^5$

23. $f(x) = (x^2 - 4)^2$

24. $f(x) = (x^2 - 2x - 8)^2$

25. $f(x) = x^2(x - 4)^2$

26. $f(x) = x(x - 4)^3$

27. $f(x) = x^2(x - 5)^3$

28. $f(x) = x^3(x - 5)^2$

Graph each function using a graphing calculator by first making a sign diagram for the derivative. (The relative extreme points will help you to find an appropriate viewing window. Answers may vary depending on the window chosen.)

29. $f(x) = x^3 - 300x$ **30.** $f(x) = x^3 - 243x$
For Exercises 29 and 30, also make a sketch from the screen, showing numbers on the axes and coordinates of relative extreme points.

31. $f(x) = x^4 - 50x^2 - 25$ **32.** $f(x) = x^4 - 72x^2 - 4$
For Exercises 31 and 32, also make a sketch from the screen, showing numbers on the axes and coordinates of relative extreme points.

33. $f(x) = x^5 - 5x^4 + 5x^3 - 23$

34. $f(x) = x^5 + 5x^4 + 5x^3 - 2$

35. $f(x) = 0.01x^5 - 0.05x$

36. $f(x) = 0.02x^5 - 0.1x$

37. $f(x) = x^3 - 2x^2 + x + 11$

38. $f(x) = x^3 - x^2 + 12$

39. $f(x) = \sqrt{400 - x^2}$ **40.** $f(x) = \sqrt{2x - x^2}$

 For each function, find all critical points and graph the function on an appropriate window using a graphing calculator. (Answers may vary depending on the window chosen.)

41. $f(x) = \dfrac{1}{x^2 - 2x - 8}$ **42.** $f(x) = \dfrac{1}{x^2 - 2x - 3}$
For Exercises 41 and 42, also make a sketch from the screen, showing numbers on the axes and coordinates of relative extreme points. You may need to ignore some false lines.

43. $f(x) = \dfrac{8}{x^2 + 4}$
This curve is called the Witch of Agnesi.

44. $f(x) = \dfrac{4x}{x^2 + 4}$
This curve is called Newton's serpentine.

45. $f(x) = \dfrac{x^2}{x^2 + 1}$ **46.** $f(x) = \dfrac{1}{x^2 + 1}$

47. $f(x) = \dfrac{x^2}{x - 3}$ **48.** $f(x) = \dfrac{x^2 - 3x - 1}{x - 5}$

49. $f(x) = \dfrac{10x^2}{x^2 - 5}$ **50.** $f(x) = \dfrac{x^2 + 4}{x}$

51. $f(x) = \dfrac{2x^2}{x^4 + 1}$ **52.** $f(x) = \dfrac{x^4 - 2x^2 + 1}{x^4 + 2}$

 Graph each function on an appropriate window using a graphing calculator. (Answers may vary depending on the window chosen.)

53. $f(x) = \dfrac{1}{x - 5}$ **54.** $f(x) = \dfrac{1}{2 + x}$

55. $f(x) = \dfrac{1}{0.1 + 2^{-0.5x}}$ **56.** $f(x) = \dfrac{3x}{\sqrt{x^2 + 1}}$

57. Derive the formula $x = \dfrac{-b}{2a}$ for the x-coordinate of the vertex of the parabola $f(x) = ax^2 + bx + c$. (*Hint:* The slope is zero at the vertex, so finding the vertex means finding the critical value.)

58. Derive the formula $x = -b$ for the x-coordinate of the vertex of the parabola $f(x) = a(x + b)^2 + c$. (*Hint:* The slope is zero at the vertex, so finding the vertex means finding the critical value.)

APPLIED EXERCISES

59. Biomedical: Bacterial Growth A population of bacteria grows to size $p(x) = x^3 - 9x^2 + 24x + 10$ after x hours (for $x \geq 0$). Graph this population curve (based on, if you wish, a calculator graph), showing the coordinates of the relative extreme points.

60. Behavioral Science: Learning Curves A learning curve is a function $L(x)$ that gives the amount of time that a person requires to learn x pieces of information. Many learning curves take the form $L(x) = (x - a)^n + b$ (for $x \geq 0$), where a, b, and n are constants. Graph the learning curve $L(x) = (x - 2)^3 + 8$ (based on, if you wish, a calculator graph), showing the coordinates of all critical points.

61. Business: Marginal and Average Cost A company's cost function is $C(x) = x^2 + 2x + 4$ dollars, where x is the number of units.

a. Enter the above cost function as y_1 on a graphing calculator.

b. Define y_2 to be the *marginal* cost function by defining y_2 to be the derivative of y_1 (using NDERIV).

c. Define y_3 to be the company's *average* cost function, $AC(x) = \dfrac{C(x)}{x}$, by defining y_3 to be $y_3 = y_1/x$.

d. Turn off the function y_1 so that it will not be graphed, but graph the marginal cost function y_2 and the average cost function y_3 on the window [0 ,10] by [0, 10]. Observe that the marginal cost function pierces the average cost function at its minimum point (use TRACE to see which curve is which function).

e. To see that the previous sentence is true in general, change the coefficients in the cost function $C(x)$, or change it to a cubic, or some other function [so that $C(x)/x$ has a minimum]. Again turn off the cost function and graph the other two to see that the marginal cost function pierces the average cost function at its minimum. We will return to this observation later.

62. **General: Airplane Flight Path** A plane is to take off and reach a level cruising altitude of 5 miles after a horizontal distance of 100 miles, as shown in the diagram below. Find a polynomial flight path of the form $f(x) = ax^3 + bx^2 + cx + d$ by following steps i to iv to determine the constants a, b, c, and d.

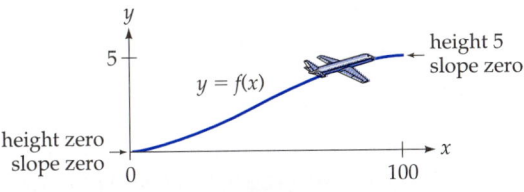

i. Use the fact that the plane is on the ground at $x = 0$ [that is, $f(0) = 0$] to determine the value of d.

ii. Use the fact that the path is horizontal at $x = 0$ [that is, $f'(0) = 0$] to determine the value of c.

iii. Use the fact that at $x = 100$ the height is 5 and the path is horizontal to determine the values of a and b. State the function $f(x)$ that you have determined.

iv. Use a graphing calculator to graph your function on the window [0, 100] by [0, 10] to verify its shape.

63. **General: Drug Interception** Suppose that the cost of a border patrol that intercepts x percent of the illegal drugs crossing a state border is $C(x) = \dfrac{600}{100 - x}$ million dollars (for $x < 100$).

a. Graph this function on [0, 100] by [0, 100].

b. Observe that the curve is at first rather flat, but then rises increasingly steeply as x nears 100. Predict what the graph of the derivative would look like.

c. Check your prediction by defining y_2 to be the derivative of y_1 (using NDERIV) and graphing both y_1 and y_2.

64. **General: Aspirin** Clinical studies have shown that the analgesic (pain-relieving) effect of aspirin is approximately $f(x) = \dfrac{100x^2}{x^2 + 0.02}$ where $f(x)$ is the percentage of pain relief from x grams of aspirin.

a. Graph this "dose-response" curve for doses up to 1 gram, that is, on the window [0, 1] by [0, 100].

b. TRACE along the curve to see that the curve is very close to its maximum height of 100%, or 1, by the time x reaches 0.65. This means that there is very little added effect in going above 650 milligrams, the amount of aspirin in two regular tablets, notwithstanding the aspirin companies' promotion of "extra strength" tablets.
(*Note:* Aspirin's dose-response curve is extremely unusual in that it levels off quite early, and, even more unusual, aspirin's effect in protecting against heart attacks even *decreases* as the dosage x increases above about 80 milligrams.)

7.2 Graphing Using the First and Second Derivatives

APPLICATION PREVIEW

Stevens' Law of Psychophysics

Suppose that you are given two weights and asked to judge how much heavier one is than the other. If one weight is actually *twice* as heavy as the other, most people will judge the heavier weight as being *less* than twice as heavy. This is one of the oldest problems in experimental psychology—how sensation (perceived weight) varies with stimulus (actual weight). Similar experiments can be performed for perceived brightness of a light compared to actual brightness, perceived effort compared to actual work, and so on. The results will vary somewhat from person to person, but the following diagram shows some typical stimulus-response curves (in arbitrary units).

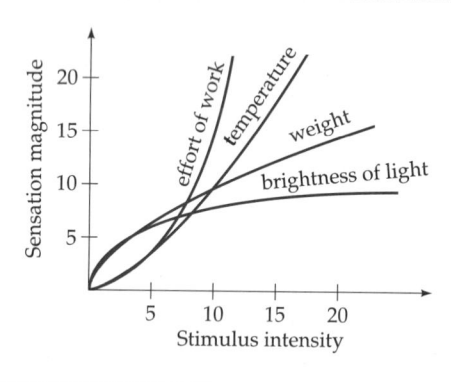

Notice, for example, that perceived effort increases more rapidly than actual work, which suggests that a 10% increase in an employee's work should be rewarded with a *greater* than 10% increase in pay.

Such stimulus-response curves were studied by the psychologist S. S. Stevens* at Harvard, who expressed them as "power functions"

$$\text{Response} = a(\text{stimulus})^b \qquad \text{For constants } a \text{ and } b$$

or
$$f(x) = ax^b$$

Functions of this form either "curl upward" (like the work and temperature curves above) or "curl downward" (like the weight and brightness curves). In this section we will graph such functions, using the terms *concave up* and *concave down* to describe their curl.

* See S. S. Stevens, "On the Psychophysical Law," *Psychol. Rev.*, **64**:153–181, 1957.

Introduction

In the previous section we used the first derivative to find a function's slope and relative extreme points and to draw its graph. In this section we will use the *second* derivative to find the "curl" or *concavity* of the curve, and to define the important concept of *inflection point*. The second derivative also gives a very useful way to distinguish between maximum and minimum points of a curve.

Concavity and Inflection Points

A curve that curls upward is said to be *concave up,* and a curve that curls downward is said to be *concave down.* A point where the concavity *changes* (from up to down or down to up) is called an *inflection point.*

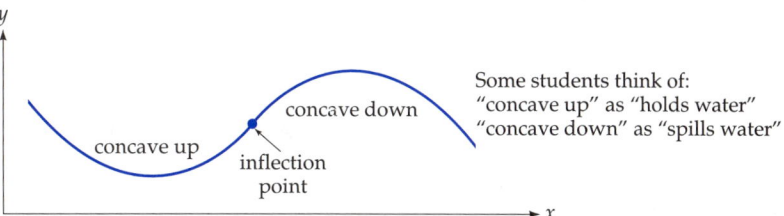

Concavity shows how a curve *curls* or *bends* away from straightness.

A straight line (with any slope) has *no concavity.*

However, bending the two ends *upward* makes it *concave up,*

and bending the two ends *downward* makes it *concave down.*

As these pictures show, a curve that is concave *up* lies *above* its tangent, whereas a curve that is concave *down* lies *below* its tangent (except, of course, at the point of tangency).

PRACTICE PROBLEM 1

For each of the following curves, label the parts that are concave up and the parts that are concave down. Then find all inflection points.

a. y

b. y

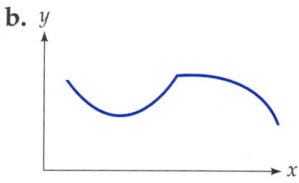

Solutions at the back of the book

How can we use calculus to determine concavity? The key is the *second* derivative. The second derivative, being the derivative of the derivative, gives the rate of change of slope, showing whether the slope is increasing or decreasing. That is, $f'' > 0$ means that the slope is increasing, and so the curve must be *concave up* (as in the diagram on the left below). Similarly, $f'' < 0$ means that the slope is decreasing, and so the curve must be *concave down* (as on the right below).

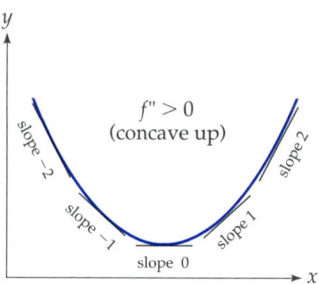

$f'' > 0$ means that the slope is
increasing, so f is *concave up*.

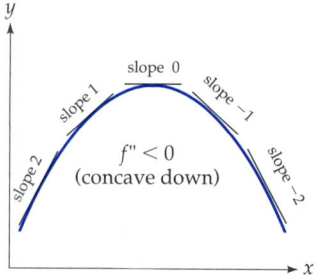

$f'' < 0$ means that the slope is
decreasing, so f is *concave down*.

   Since an inflection point is where the concavity changes, the second derivative must be negative on one side and positive on the other. Therefore, *at* an inflection point, f'' must be either zero or undefined. All of this may be summarized as follows.

Concavity and Inflection Points

$f'' > 0$ on an interval means that f is *concave up* (curls upward) on
   that interval.
$f'' < 0$ on an interval means that f is *concave down* (curls downward) on that interval.
An *inflection point* is where the concavity *changes* (f'' must be zero or undefined).

Graphing Calculator Exploration

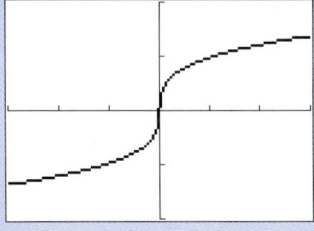

a. Use a graphing calculator to graph $y_1 = \sqrt[3]{x}$ on the window $[-3, 3]$ by $[-2, 2]$. Observe where the curve is concave up and where it is concave down.

b. Use NDERIV to define y_2 to be the derivative of y_1, and y_3 to be the derivative of y_2. Graph y_1 and y_3 (but turn off y_2 so that it will not be graphed).

c. Verify that y_3 (the second derivative) is positive where y_1 is concave up, and negative where y_1 is concave down.

d. Now change y_1 to $y_1 = \dfrac{x^2 + 2}{x^2 + 1}$ and observe that where this curve is concave up or down agrees with where y_3 is positive or negative. According to y_3, how many inflection points does y_1 have? Can you see them on y_1?

To find inflection points, we make a sign diagram for the *second* derivative to show where the concavity changes (where f'' changes sign). An example will make the method clear.

EXAMPLE 1 **Graphing and Interpreting a Company's Annual Profit Function**

A company's annual profit after x years is $f(x) = x^3 - 9x^2 + 24x$ million dollars (for $x \geq 0$). Graph this function, showing all relative extreme points and inflection points. Interpret the inflection points.

Solution

$$f'(x) = 3x^2 - 18x + 24 \qquad \text{The derivative}$$
$$= 3(x^2 - 6x + 8) = 3(x - 2)(x - 4) \qquad \text{Factoring}$$

The critical values are $x = 2$ and $x = 4$, and the sign diagram for f' (found in the usual way) is

$$
\begin{array}{ccccc}
f' > 0 & f' = 0 & f' < 0 & f' = 0 & f' > 0 \\
\hline
 & x = 2 & & x = 4 & \\
 & \rightarrow & & & \\
\nearrow & \text{rel max} & \searrow & & \nearrow \\
 & (2, 20) & & \rightarrow & \\
 & & & \text{rel min} & \\
 & & & (4, 16) &
\end{array}
$$

To find the inflection points, we calculate the second derivative,

$$f''(x) = 6x - 18 = 6(x - 3)$$ Differentiating $f'(x) = 3x^2 - 18x + 24$

This is zero at $x = 3$, which we enter on a sign diagram for the *second* derivative.

$$f'' = 0$$ ← Behavior of f''

$$x = 3$$ ← Where f'' is zero or undefined

We use test points to determine the sign of $f''(x) = 6(x - 3)$ on either side of $x = 3$, just as we did for the first derivative.

$$f''(2) = 6(2 - 3) < 0 \qquad f''(4) = 6(4 - 3) > 0$$

$$f'' < 0 \qquad f'' = 0 \qquad f'' > 0$$
$$x = 3$$

con dn   con up   Concave down, concave up (so concavity *does* change)

IP (3, 18)   IP means inflection point. The 18 comes from substituting $x = 3$ into $f(x) = x^3 - 9x^2 + 24x$

Using a graphing calculator, we would choose an x-interval like $[0, 6]$ (to include the x-values on the sign diagrams) and a y-interval like $[0, 30]$ (to include the origin and the y-values on the sign diagrams), and graph the function.

By hand, we would plot the relative maximum ($\frown$), minimum ($\smile$), and inflection point and sketch the curve according to the sign diagrams, being sure to show the concavity changing at the inflection point.

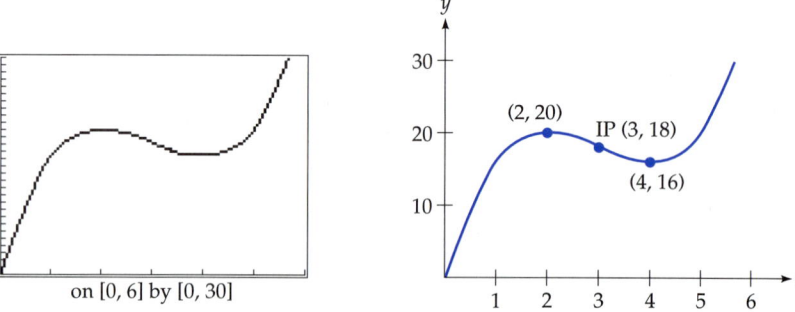

on [0, 6] by [0, 30]

Interpretation of the inflection point: Observe what the graph shows: The company's profit increased (up to year 2), then decreased (up to year 4), and then increased again. The inflection point at $x = 3$ is where the profit *first began to show signs of improvement.* It marks the end of the period of increasingly steep decline and the first sign of an "upturn," where a clever investor might begin to "buy in."

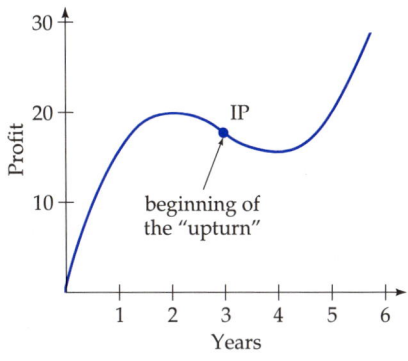

At an inflection point, the concavity (that is, the sign of f'') must *actually change*. For example, a second derivative sign diagram like this:

$$\underset{\text{con dn}}{f'' < 0} \quad \underset{\substack{| \\ x = 3}}{f'' = 0} \quad \underset{\text{con dn}}{f'' < 0}$$

Sign of f'' (concavity) does *not* change

would mean that there is *not* an inflection point, since the concavity is the same on both sides. For there to be an inflection point, the sign of f'' must actually change.

PRACTICE PROBLEM 2

For each curve, is there an inflection point? (*Hint:* Does the concavity change?)   *Solutions at the back of the book*

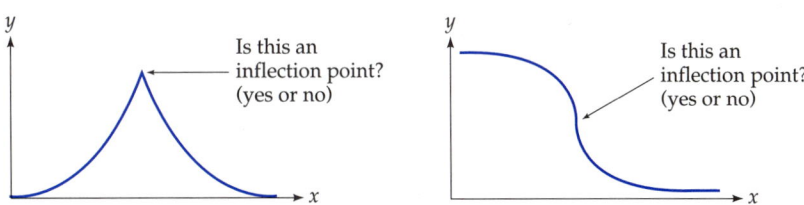

Inflection Points in Everyday Life

Inflection points have important interpretations besides the "first sign of the upturn" in Example 1.*

---

* For an interesting application of concavity to economics, see Harry M. Markowitz, "The Utility of Wealth," *Journal of Applied Political Economy* **60**(2):151–158, 1952. In this article, "concave" means "concave up" and "convex" means "concave down." Markowitz won the Nobel Memorial Prize in Economics in 1990.

The graph below shows a person's temperature during an illness, and the inflection point is where the temperature ends its steepest rise and begins to moderate, that is, the point at which the illness is *first brought under control.*

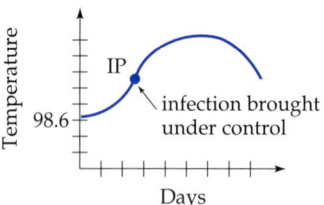

The graph below gives a company's total sales after x days of an advertising campaign. The inflection point gives the *point of diminishing returns:* Advertising beyond this point will still bring additional sales, but at a slower rate.

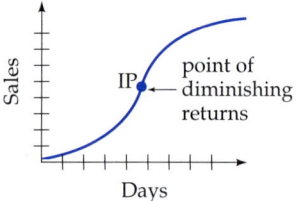

Distinguish carefully between slope and concavity: *Slope* measures *steepness,* whereas *concavity* measures *curl.* All combinations of slope and concavity are possible. A graph may be

Increasing and concave *up* ($f' > 0,\ f'' > 0$), such as

Increasing and concave *down* ($f' > 0, f'' < 0$), such as

Decreasing and concave *up* ($f' < 0, f'' > 0$), such as

Decreasing and concave *down* ($f' < 0, f'' < 0$), such as

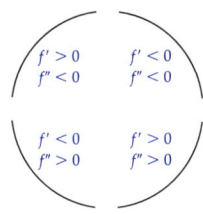

$f' > 0$ $f' < 0$
$f'' < 0$ $f'' < 0$

$f' < 0$ $f' > 0$
$f'' > 0$ $f'' > 0$

The four quarters of a circle illustrate all four possibilities, as shown on the left.

EXAMPLE 2   Graphing a Fractional Power Function

Graph $f(x) = 18x^{1/3}$.

Solution   The derivative is

$$f'(x) = 6x^{-2/3} = \frac{6}{\sqrt[3]{x^2}}$$
   Undefined at $x = 0$
   (zero denominator)

The sign diagram for f' is

$f' > 0$ f' und $f' > 0$
_____+_____
          $x = 0$ ↗
          neither
    ↗ $(0, 0)$

f' is undefined at $x = 0$ and positive on either side (using test points)

The *second* derivative is

$$f''(x) = -4x^{-5/3} = \frac{-4}{\sqrt[3]{x^5}}$$
   Also undefined at $x = 0$

The sign diagram for f'' is

$f'' > 0$ f'' und $f'' < 0$
_____+_____
con up $x = 0$ con dn
         IP $(0, 0)$

Concavity is different on either side of $x = 0$ (using test points), so $x = 0$ *is* an inflection point

Based on this information, we may graph the function with a graphing calculator or by hand:

Using a graphing calculator, we would experiment with viewing windows centered at the inflection point $(0, 0)$ until we found one that shows the curve effectively. The graph on $[-2, 2]$ by $[-20, 20]$ is shown below.

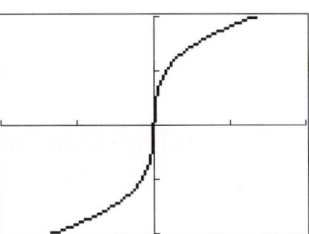

By hand, we would use the sign diagrams to draw the curve to the *left* of $x = 0$ as *increasing* and concave *up*, and to the right of $x = 0$ as increasing but concave down, with the two parts meeting at the origin. The scale comes from calculating the points $(1, 18)$ and $(-1, -18)$.

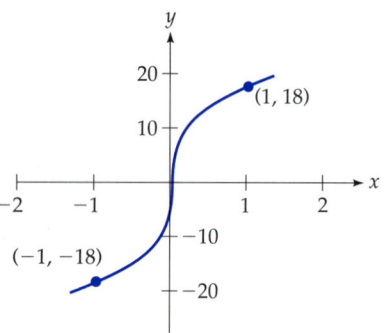

The fact that the derivative is *undefined* at $x = 0$ is shown in the graph by the *vertical tangent* at the origin (since the slope of a vertical line is undefined). This function is the stimulus-response curve for brightness of light (for $x \geq 0$; see page 612), and the vertical tangent indicates the disproportionately large effect of small increases in dim light.

 ### EXAMPLE 3   Graphing Using a Graphing Calculator

Use a graphing calculator to graph $f(x) = 36\sqrt[3]{(x-1)^2}$.

Solution   Using a standard window $[-10, 10]$ by $[-10, 10]$ gives the graph below, which is useless.

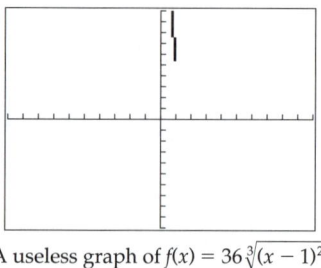

A useless graph of $f(x) = 36\sqrt[3]{(x-1)^2}$
on window $[-10, 10]$ by $[-10, 10]$

Instead, we begin by differentiating $f(x) = 36(x - 1)^{2/3}$ and making its sign diagram.

$$f'(x) = 24(x - 1)^{-1/3} = \frac{24}{\sqrt[3]{x - 1}} \qquad \text{Undefined at } x = 1$$

$$f' < 0 \qquad f' \text{ und} \qquad f' > 0$$

$$x = 1$$

↘   rel min   ↗
(1, 0)

Using test points on either side of $x = 1$

The y-coordinate in $(1, 0)$ comes from evaluating the original function at $x = 1$

From the sign diagram we choose an x-interval centered at the critical value $x = 1$, such as $[-4, 6]$. After some experimentation, we choose the y-interval $[0, 100]$ (beginning at $x = 0$ because the minimum point has y-coordinate 0) so that the curve fills the screen. A sharp point on a graph like the one at $x = 1$ is called a *cusp*, where the function is not differentiable.

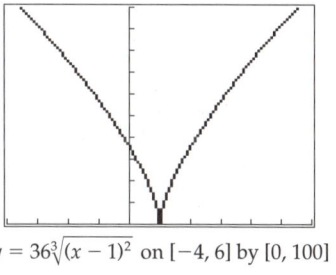

$y = 36\sqrt[3]{(x-1)^2}$ on $[-4, 6]$ by $[0, 100]$

Graphing Calculator Exploration

a.   In the graph at the very beginning of Example 3, why didn't the curve reach the x-axis, as the later graph shows that it does? (*Hint:* It has to do with pixels.)

b.   Try graphing the same function $36\sqrt[3]{(x-1)^2}$ but entered in exponential form as $y_1 = 36(x-1)\text{^}(2/3)$. If your calculator displays only half of the curve, try separating the exponent $2/3$ into two parts, like $36((x-1)\text{^}2)\text{^}(1/3)$ or $36((x-1)\text{^}(1/3))\text{^}2$. You may need to make similar changes in the future.

Second-Derivative Test

Determining whether a twice-differentiable function has a relative maximum or minimum at a critical value is merely a question of concavity: concave *up* means a relative *minimum*, and concave *down* means a relative *maximum*.

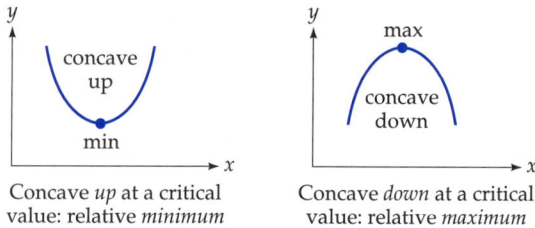

Concave *up* at a critical   Concave *down* at a critical
value: relative *minimum*   value: relative *maximum*

Since the second derivative determines concavity, we have the following *second-derivative test*, which will be very useful in the next two sections.

Second-Derivative Test for Relative Extreme Points

If $x = c$ is a critical value of f at which f'' is defined, then

$f''(c) > 0$ means that f has a relative *minimum* at $x = c$.

$f''(c) < 0$ means that f has a relative *maximum* at $x = c$.

To use the second-derivative test, we first find all critical values, substituting each into the second derivative and determining the sign of the result: A *positive* result means a *minimum* at the critical value, and a *negative* result means a *maximum*. [If the second derivative is zero, then the test is inconclusive, and you should use the first derivative test (page 604) or make a sign diagram for f'.]

EXAMPLE 4   Using the Second Derivative Test

Use the second derivative test to find all relative extreme points of $f(x) = x^3 - 9x^2 + 24x$.

Solution

$$f'(x) = 3x^2 - 18x + 24 \qquad \text{The derivative}$$

$$= 3(x^2 - 6x + 8) = 3(x - 2)(x - 4) \qquad \text{Factoring}$$

$$\text{CV} \begin{cases} x = 2 \\ x = 4 \end{cases} \qquad \text{Critical values}$$

We substitute each critical value into $f''(x) = 6x - 18$.

$$f''(2) = 6 \cdot 2 - 18 = -6 \quad \text{(negative)} \qquad \begin{array}{l} f''(x) = 6x - 18 \\ \text{at } x = 2 \end{array}$$

Therefore, f has a relative *maximum* at $x = 2$.

$$f''(4) = 6 \cdot 4 - 18 = 6 \quad \text{(positive)} \qquad \begin{array}{l} f''(x) = 6x - 18 \\ \text{at } x = 4 \end{array}$$

Therefore, f has a relative *minimum* at $x = 4$.

The relative maximum at $x = 2$ and the relative minimum at $x = 4$ are exactly what we found when we graphed this function on page 616.

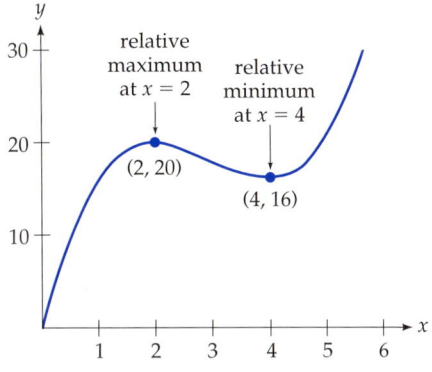

Be careful: The second derivative test tells us *nothing* if the second derivative is *zero* at a critical value. This is shown by the three functions below: Each has a critical value $x = 0$ at which f'' is zero (as you may check), but one has a maximum, one has a minimum, and one has neither.

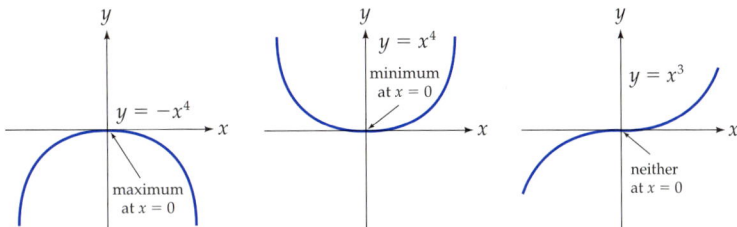

Three functions showing that a critical point where $f'' = 0$
may be a maximum, a minimum, or neither.

SUMMARY

The main developments of this section were the use of the second derivative to determine the *concavity* or *curl* of a function, and finding and interpreting *inflection points.*

$f'' > 0$ on an interval means that f is concave *up* on that interval.

$f'' < 0$ on an interval means that f is concave *down* on that interval.

To locate inflection points (points where the concavity changes), we find where the second derivative is zero or undefined, and then make a sign diagram for f'' to see whether the concavity *actually changes.*

To graph a function, we use the *first*-derivative sign diagram to find *slope* and *relative extreme points*, and the *second*-derivative sign diagram to find *concavity* and *inflection points*. Then we graph the function, either by hand (using the sign diagrams to show the shape) or on a graphing calculator (using the critical points and inflection points to determine the viewing window).

The second-derivative test shows whether a function has a relative maximum or minimum at a critical value (provided that f'' is defined and not zero):

$$f'' > 0 \text{ means a relative } minimum.$$

$$f'' < 0 \text{ means a relative } maximum.$$

The second-derivative test should be thought of as a simple application of concavity.

EXERCISES 7.2

For each graph, which of the numbered points are inflection points?

1.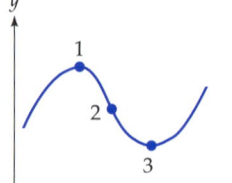

2.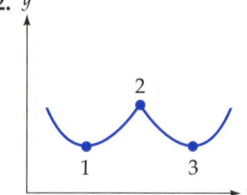

3.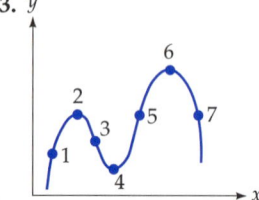

4.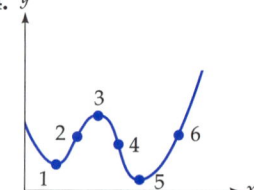

5.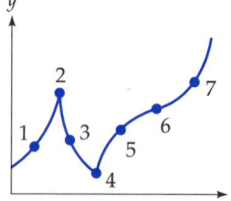

6.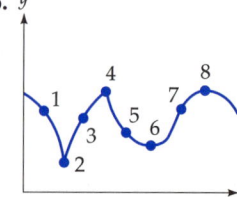

For each function:

    a. Make a sign diagram for the first derivative.
    b. Make a sign diagram for the second derivative.
    c. Sketch the graph by hand, showing all relative extreme points and inflection points.

7. $f(x) = x^3 + 3x^2 - 9x + 5$

8. $f(x) = x^3 - 3x^2 - 9x + 7$

9. $f(x) = x^3 - 3x^2 + 3x + 4$

10. $f(x) = x^3 + 3x^2 + 3x + 6$

11. $f(x) = x^4 - 8x^3 + 18x^2 + 2$

12. $f(x) = x^4 + 8x^3 + 18x^2 + 8$

13. $f(x) = 5x^4 - x^5$

14. $f(x) = (x - 2)^3 + 2$

15. $f(x) = (2x + 4)^5$

16. $f(x) = (3x - 6)^6 + 1$

17. $f(x) = x(x - 3)^2$

18. $f(x) = x^3(x - 4)$

19. $f(x) = x^{3/5}$

20. $f(x) = x^{1/5}$

21. $f(x) = \sqrt[5]{x^4} + 2$

22. $f(x) = \sqrt[5]{x^2} - 1$

23. $f(x) = \sqrt[4]{x^3}$

24. $f(x) = \sqrt{x^5}$

25. $f(x) = \sqrt[3]{(x - 1)^2}$

26. $f(x) = \sqrt[5]{x + 2} + 3$

 Graph each function using a graphing calculator by first making sign diagrams for the first and second derivatives. Graphs may vary depending on the window chosen.

27. $f(x) = x^3 - 18x^2 + 60x + 20$

For Exercises 27 and 28, also make a sketch from the screen, showing the coordinates of all relative extreme points and inflection points.

28. $f(x) = x^3 - 300x$ *(See instructions for Ex. 27.)*

29. $f(x) = x^4 - 16x^3$

30. $f(x) = x^4 - 20x^3$

31. $f(x) = x^3 - 9x^2 - 48x + 48$

32. $f(x) = x^3 - 6x^2 - 63x + 42$

33. $f(x) = x^3 - 2x^2 + x + 5$

34. $f(x) = x^3 + 2x^2 + x - 4$

35. $f(x) = 36\sqrt[3]{x - 1}$

36. $f(x) = 45\sqrt[3]{(x - 1)^2}$

 Graph each function using a graphing calculator by first making a sign diagram for just the first derivative. Graphs may vary depending on the window chosen.

37. $f(x) = x^{1/2}$

38. $f(x) = x^{3/2}$

39. $f(x) = x^{-1/2}$

40. $f(x) = x^{-3/2}$

41. $f(x) = 9x^{2/3} - 6x$

42. $f(x) = 30x^{1/3} - 10x$

For Exercises 41 and 42, also make a sketch from the screen, showing the coordinates of all relative extreme points and inflection points.

43. $f(x) = 8x - 10x^{4/5}$

44. $f(x) = 6x - 10x^{3/5}$

45. $f(x) = 3x^{2/3} - x^2$

46. $f(x) = 3x^{4/3} - 2x^2$

47. Concavity of a Parabola Show that the quadratic function $f(x) = ax^2 + bx + c$ is concave up if $a > 0$ and is concave down if $a < 0$. (Therefore, the rule that a parabola opens up if $a > 0$ and down if $a < 0$ is merely an application of concavity. *Hint:* Find the second derivative.)

48. Inflection Point of a Cubic Show that the general "cubic" function $f(x) = ax^3 + bx^2 + cx + d$ (with $a \neq 0$) has an inflection point at $x = -b/(3a)$.

49. Inflection Points Explain why, at an inflection point, a curve must cross its tangent line (assuming that the tangent line exists).

50. Inflection Points For a twice-differentiable function, explain why the slope must have a relative maximum or minimum value at an inflection point. (*Hint:* Use the fact that the concavity changes at an inflection point, and then interpret concavity in terms of increasing and decreasing slope.)

 51–52: Finding Inflection Points Use a graphing calculator to estimate the x-coordinates of the inflection points of each function, rounding your answer to two decimal places.
(*Hint:* Graph the second derivative, either calculating it directly or using NDERIV twice, and see where it crosses the x-axis).

51. $f(x) = x^5 - 2x^3 + 3x + 4$

52. $f(x) = x^5 - 3x^3 + 6x + 2$

APPLIED EXERCISES

53. Business: Revenue A company's annual revenue after x years is $f(x) = x^3 - 9x^2 + 15x + 25$ thousand dollars (for $x \geq 0$).

  a. Make sign diagrams for the first and second derivatives.

  b. Sketch the graph of the revenue function, showing all relative extreme points and inflection points.

54. Business: Sales A company's weekly sales (in thousands) after x weeks are given by $f(x) = -x^4 + 4x + 70$ (for $0 \leq x \leq 3$).

  a. Make sign diagrams for the first and second derivatives.

  b. Sketch the graph of the sales function, showing all relative extreme points and inflection points.

55. General: Temperature The temperature in a refining tower after x hours is $f(x) = x^4 - 4x + 112$ degrees Fahrenheit (for $x \geq 0$).

  a. Make sign diagrams for the first and second derivatives.

  b. Sketch the graph of the temperature function, showing all relative extreme points and inflection points.

56. Biomedical: Dosage Curve The dose–response curve for x grams of a drug is $f(x) = 8(x - 1)^3 + 8$ (for $x \geq 0$).

  a. Make sign diagrams for the first and second derivatives.

  b. Sketch the graph of the response function, showing all relative extreme points and inflection points.

57. Psychology: Stimulus and Response Sketch the graph of the brightness response curve $f(x) = x^{2/5}$ for $x \geq 0$, showing all relative extreme points and inflection points.

58. Psychology: Stimulus and Response Sketch the graph of the loudness response curve $f(x) = x^{4/5}$ for $x \geq 0$, showing all relative extreme points and inflection points.

59–60: Sociology: Status Sociologists have estimated how a person's "status" in society (as per-ceived by others) depends upon the person's income and education level. One estimate is that status S depends upon income i according to the formula $S(i) = 16\sqrt{i}$ (for $i \geq 0$), and that status depends upon education level e according to the formula $S(e) = \frac{1}{4}e^2$ (for $e \geq 0$).

59. a. Sketch the graph of the function $S(i)$ above.

  b. Is the curve concave up or down? What does this signify about the rate at which status increases at higher income levels?

60. a. Sketch the graph of the function $S(e)$ above.

  b. Is the curve concave up or down? What does this signify about the rate at which status increases at higher education levels?

61–62: Psychology: Response Curves The data in the tables below were collected from two stimulus-response experiments. The table in Exercise 61 gives weight (in pounds, actual and perceived), and the table in Exercise 62 gives temperature (degrees Centigrade above room temperature, actual and perceived).

61.

Actual Weight	Perceived Weight
1	2
2	3
5	6
10	8
20	15

62.

Actual Temperature	Perceived Temperature
2	1
5	3
10	9
12	12
15	16

  a. Enter the data from the table you are using into a graphing calculator and make a plot of the resulting points (Actual on the x-axis and Perceived on the y-axis).

b. Have your calculator find the power regression formula (of the form ax^b) for these data. Then enter the results as y_1, which gives a stimulus-response curve. Plot the points together with the power regression curve. Observe that the curve fits the points quite well.

c. Is the stimulus-response curve concave up or down?

d. Use your stimulus-response curve to predict the *perceived* value if the *actual* value were 25.

e. Use your stimulus-response curve to predict the *actual* value if the *perceived* value were 8.

63. General: Airplane Flight Path In Exercise 62 on page 611 it was found that the flight path that satisfies the conditions in the diagram shown below was $y = -0.00001x^3 + 0.0015x^2$. Find the inflection point of this path, and explain why it represents the point of steepest ascent of the airplane.

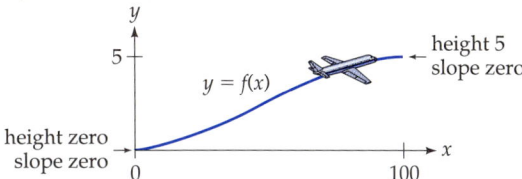

7.3 Optimization

Introduction

Many problems consist of "optimizing" a function, that is, finding its maximum or minimum value. For example, you might want to maximize your profit, or to minimize the time required to do a task. If you could express your happiness as a function, you would want to maximize it.* One of the principal uses of calculus is that it provides a very general technique for optimizing functions.

We will concentrate on *applications* of optimization. Accordingly, we will optimize continuous functions that are defined on closed intervals, or functions that have only one critical value in their domains. Most applications fall into these two categories, and the wide range of examples and exercises in these sections will demonstrate the power of these techniques.

Absolute Extreme Values

The *absolute maximum value* of a function is the *largest* value of the function on its domain. Similarly, the *absolute minimum value* of a function is the *smallest* value of the function on its domain. An absolute *extreme* value is a value that is either the absolute maximum or the absolute minimum value of the function. (This use of the word "absolute" has nothing to do with its use in the *absolute value function* discussed on pages 54–55.) The maximum and minimum values of the function correspond to the highest and lowest points on its graph.

---

* Expressing happiness in numbers has an honorable past: Plato (*Republic* IX, 587) calculated that a king is exactly 729 (= 3^6) times as happy as a tyrant.

For a given function, both absolute extreme values may exist, or one or both may fail to exist, as the following graphs show.

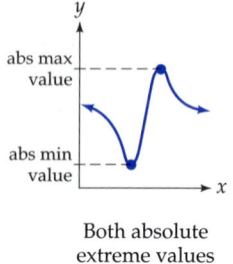

Both absolute
extreme values
exist.

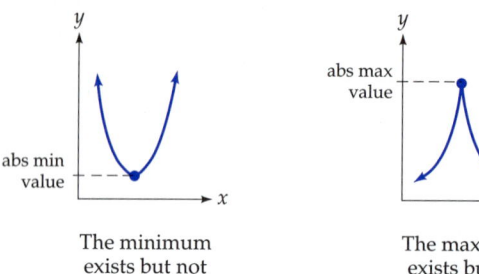

The minimum
exists but not
the maximum.

The maximum
exists but not
the minimum.

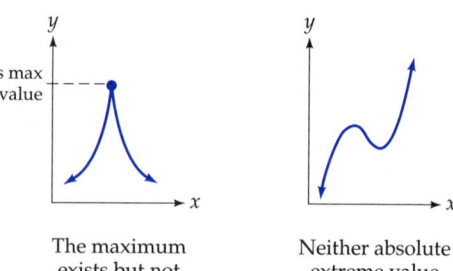

Neither absolute
extreme value
exists.

When will both extreme values exist? For a *continuous* function (one whose graph is a single unbroken curve) defined on a *closed* interval, the absolute maximum and minimum values are *guaranteed* to exist. We know from graphing functions that maximum and minimum values can occur only at critical values, unless they occur at endpoints, where the curve is "cut off" from rising or falling further. This observation leads to the following procedure for optimizing functions.

Optimizing Continuous Functions on Closed Intervals

To optimize a continuous function f on $[a, b]$:

1. Find all critical values of f in $[a, b]$.

2. Evaluate f at these critical values and at the endpoints.

The largest and smallest values found in step 2 will be the absolute maximum and minimum values of f on $[a, b]$.

Simply stated, to find absolute extreme values, we need to consider only *critical values and endpoints*.

EXAMPLE 1   Optimizing a Continuous Function on a Closed Interval

Find the absolute extreme values of $f(x) = x^3 - 9x^2 + 15x$ on $[0, 3]$.

Solution

The function is continuous (it is a polynomial) and the interval is closed, so both extreme values exist. First we find the critical values.

$$f'(x) = 3x^2 - 18x + 15 \qquad \text{The derivative}$$
$$= 3(x^2 - 6x + 5) = 3(x - 1)(x - 5) \qquad \text{Factoring}$$

$$\text{CV} \begin{cases} x = 1 \\ x = 5 \quad \leftarrow \text{Not in the given domain} \\ \qquad\quad [0, 3], \text{ so we eliminate it} \end{cases}$$

We evaluate f at the remaining critical value and at the endpoints (EP).

CV: $x = 1$ $\quad f(1) = 1 - 9 + 15 \qquad\qquad = \quad 7 \quad \leftarrow$ Largest function value

$$\text{EP} \begin{cases} x = 0 \quad f(0) = 0 - 0 + 0 \qquad\qquad\quad = \quad 0 \\ x = 3 \quad f(3) = 27 - 9 \cdot 9 + 15 \cdot 3 = -9 \quad \leftarrow \text{Smallest function value} \end{cases}$$

The largest (7) and the smallest (-9) of the resulting values of the function are the absolute extreme values of f on [0, 3]:

Maximum value of f is $\quad$ 7 (occurring at $x = 1$).

Minimum value of f is -9 (occurring at $x = 3$).

∎

The graph of the function shows these absolute extreme values, one occurring at a critical value ($x = 1$), and the other occurring at an endpoint ($x = 3$).

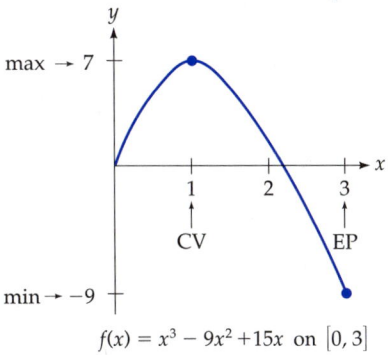

$$f(x) = x^3 - 9x^2 + 15x \text{ on } [0, 3]$$

In other problems, both extreme values might occur at critical values or both might occur at endpoints.

Notice how calculus helped in this example. The absolute extreme values could have occurred at *any* x-value in [0, 3]. Calculus reduced this infinite list of possibilities (*all* numbers between 0 and 3, not just integers) to a mere three numbers (0, 1, and 3). We then had only to "test" these numbers by substituting them into the function to find which made f largest and smallest.

Second-Derivative Test for Absolute Extreme Values

Earlier we used the second-derivative test to find *relative* maximum and minimum values. However, if a continuous function has only one critical value in its domain, then the second derivative test can be used to find *absolute* extreme values (since without a second critical value the function must continue to increase or decrease away from the relative extreme point). Some students refer to this as "the only critical point in town" test.

Applications of Optimization

If a timber forest is allowed to grow for t years, the value of the timber increases in proportion to the square root of t, while maintenance costs are proportional to t. Therefore, the value of the forest after t years is

$$V(t) = a\sqrt{t} - bt \qquad \text{\textcolor{blue}{a and b are constants}}$$

EXAMPLE 2   Optimizing the Value of a Timber Forest

The value of a timber forest after t years is $V(t) = 96\sqrt{t} - 6t$ thousand dollars (for $t > 0$). Find when its value is maximized.

Solution

$$V(t) = 96t^{1/2} - 6t \qquad \text{\textcolor{blue}{$V(t)$ in exponential form}}$$

$$V'(t) = 48t^{-1/2} - 6 = 0 \qquad \text{\textcolor{blue}{The derivative, set equal to zero}}$$

$$48t^{-1/2} = 6 \qquad \text{\textcolor{blue}{Adding 6 to each side of $48t^{-1/2} - 6 = 0$}}$$

$$t^{-1/2} = \frac{6}{48} = \frac{1}{8} \qquad \text{\textcolor{blue}{Dividing by 48}}$$

$$\frac{1}{\sqrt{t}} = \frac{1}{8} \qquad \text{\textcolor{blue}{Expressing $t^{-1/2}$ in radical form}}$$

$$\sqrt{t} = 8 \qquad \text{\textcolor{blue}{Inverting both sides}}$$

$$t = 64 \qquad \text{\textcolor{blue}{Squaring both sides}}$$

Since there is a *single* critical value ($t = 0$, which makes the derivative undefined, is not in the domain), we may use the second derivative test. The second derivative is

$$V''(t) = -24t^{-3/2} = -\frac{24}{\sqrt{t^3}} \qquad \text{\textcolor{blue}{Differentiating $V'(t) = 48t^{-1/2} - 6$}}$$

At $t = 64$ this is clearly negative, so by the second derivative test $V(t)$ is *maximized* at $t = 64$. Finally, we state the answer clearly:

The value of the forest is maximized after 64 years. The maximum value is $V(64) = 96\sqrt{64} - 6 \cdot 64 = 384$ thousand dollars, or $384,000.

∎

Maximizing Profit

The famous economist John Maynard Keynes said, "The engine that drives Enterprise is Profit." Many management problems consist of maximizing profit, and require *constructing* the profit function before maximizing it. Such problems have three economic ingredients. The first is that profit is defined as *revenue minus cost*:

$$\text{Profit} = \text{Revenue} - \text{Cost}$$

The second ingredient is that revenue is *price times quantity*. For example, if a company sells 100 toasters for $25 each, the revenue will obviously be $25 \cdot 100 = 2500$.

$$\text{Revenue} = \begin{pmatrix} \text{Unit} \\ \text{price} \end{pmatrix} \cdot (\text{Quantity})$$

The third economic ingredient reflects the fact that, in general, price and quantity are inversely related: Increasing the price decreases sales, whereas decreasing the price increases sales. To put this another way, "flooding the market" with a product drives the price down, whereas creating a shortage drives the price up. If the relationship between the price p and the quantity x that consumers will buy at that price is expressed as a function $p(x)$, it is called the *price function*.*

Price Function

> $p(x)$ gives the price p at which consumers will buy exactly x units of the product

The price function, relating price and quantity, may be linear or curved (as shown on the following page), but it will always be a *decreasing* function.

---

* We will use *lowercase p* for price and *capital P* for profit.

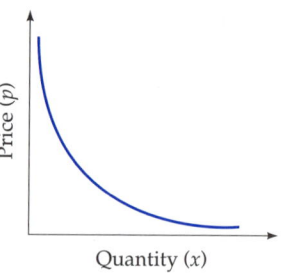

The price function $p(x)$ shows
the inverse relation between
price p and quantity x.

In actual practice, price functions are very difficult to determine, requiring extensive (and expensive) market research. In this section we will be given the price function. In the next section we will see how to do without price functions, at least in simple cases.

EXAMPLE 3   Maximizing a Company's Profit

It costs the American Automobile Company $8000 to produce each automobile, and fixed costs (rent and other costs that do not depend on the amount of production) are $20,000 per week. The company's price function is $p(x) = 22{,}000 - 70x$, where p is the price at which exactly x cars will be sold.

a. How many cars should be produced each week to maximize profit?

b. For what price should they be sold?

c. What is the company's maximum profit?

Solution   **Revenue** is price times quantity, $R = p \cdot x$:

$$R = p \cdot x = \underbrace{(22{,}000 - 70x)}_{p(x)}x = 22{,}000x - 70x^2$$

Replacing p by the
price function
$p = 22{,}000 - 70x$

Revenue
function $R(x)$

Cost is the cost per car ($8000) times the number of cars (x) plus the fixed cost ($20,000):

$$C(x) = 8000x + 20{,}000 \qquad \text{(Unit cost)} \cdot \text{(quantity)} + \text{(fixed costs)}$$

Profit is revenue minus cost:

$$P(x) = \underbrace{(22{,}000x - 70x^2)}_{R(x)} - \underbrace{(8000x + 20{,}000)}_{C(x)}$$

$$= -70x^2 + 14{,}000x - 20{,}000 \qquad \text{Profit function (after simplification)}$$

a. We maximize the profit by setting its derivative equal to zero:

Differentiating

$$P'(x) = -140x + 14{,}000 = 0 \qquad P = -70x^2 + 14{,}000x - 20{,}000$$

$$-140x = -14{,}000 \qquad \text{Solving}$$

$$x = \frac{-14{,}000}{-140} = 100 \qquad \text{Only one critical value}$$

$$P''(x) = -140 \qquad \text{From } P'(x) = -140x + 14{,}000$$

The second derivative is negative, so profit is maximized at the critical value. (If the second derivative had involved x, we would have substituted the critical value $x = 100$.) Since x is the number of cars, the company should produce 100 cars per week (the time period stated in the problem).

b. The selling price p is found from the price function:

$$p = 22{,}000 - 70 \cdot 100 = \$15{,}000 \qquad \begin{array}{l} p(x) = 22{,}000 - 70x \\ \text{evaluated at } x = 100 \end{array}$$

c. The maximum profit is found from the profit function:

$$P(100) = -70(100)^2 + 14{,}000(100) - 20{,}000 \qquad \begin{array}{l} P(x) = -70x^2 + \\ 14{,}000x - 20{,}000 \\ \text{evaluated at } x = 100 \end{array}$$

$$= \$680{,}000$$

Finally, state the answer clearly in words.

The company should make 100 cars per week and sell them for $15,000 each. The maximum profit will be $680,000.

Actually, automobile dealers seem to prefer prices like $14,999, as if $1 makes a difference.

■

Graphs of the Revenue, Cost, and Profit Functions

The graphs of the revenue and cost functions are shown on the following page. At x-values where revenue is above cost, there is a profit, and where the cost is above the revenue, there is a loss.

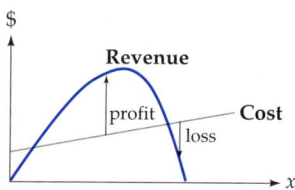

The height of the profit function at any x is the amount by which the revenue is above the cost in the graph above.

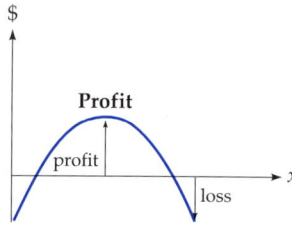

Since profit equals revenue minus cost, we may differentiate each side of $P(x) = R(x) - C(x)$, obtaining

$$P'(x) = R'(x) - C'(x)$$

This shows that setting $P'(x) = 0$ (which we do to maximize profit) is equivalent to setting $R'(x) - C'(x) = 0$, which is equivalent to $R'(x) = C'(x)$. This last equation may be expressed in marginals, MR = MC, which is a classic economic criterion for maximum profit.

Classic Economic Criterion for Maximum Profit

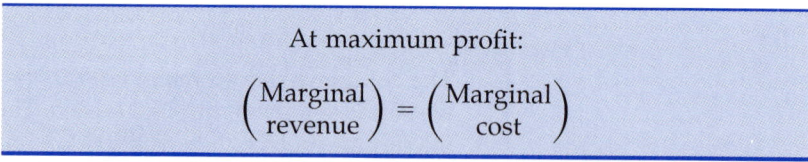

At maximum profit:

$$\begin{pmatrix} \text{Marginal} \\ \text{revenue} \end{pmatrix} = \begin{pmatrix} \text{Marginal} \\ \text{cost} \end{pmatrix}$$

EXAMPLE 4   **Maximizing the Area of an Enclosure**

A farmer has 1000 feet of fence and wants to build a rectangular enclosure along a straight wall. If the side along the wall needs no fence, find the dimensions that make the enclosure as large as possible. Also find the maximum area.

Solution   The largest enclosure means, of course, the largest area.

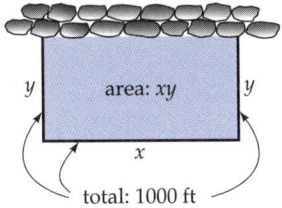

To maximize the area of the rectangle, we let variables stand for the length and width:

$$x = \text{length (parallel to wall)}$$

$$y = \text{width (perpendicular to wall)}$$

The problem becomes

Maximize $A = xy$ Area is length times width

subject to $x + 2y = 1000$ One x side and two y sides from 1000 feet of fence

We must express the area $A = xy$ in terms of one variable. We use

$x = 1000 - 2y$ Solving $x + 2y = 1000$ for x

$$A = xy = \underbrace{(1000 - 2y)}_{x}y = 1000y - 2y^2$$ Substituting $x = 1000 - 2y$ into $A = xy$

$A' = 1000 - 4y = 0$ Maximizing $A = 1000y - 2y^2$ by setting the derivative equal to zero

$y = 250$ Solving $1000 - 4y = 0$ for y

Since $A'' = -4$, the second derivative test shows that the area is indeed *maximized* when $y = 250$. The length x is

$x = 1000 - 2 \cdot 250 = 500$ Evaluating $x = 1000 - 2y$ at $y = 250$

Length (parallel to the wall) is 500 feet, width (perpendicular to the wall) is 250 feet, and area (length times width) is 125,000 square feet.

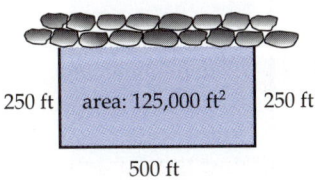

Graphing Calculator Exploration

In the previous example on maximizing the area, you might think that it does not matter how the fence is laid out as long as all of the 1000 feet is used. That the area really does change may be seen as follows:

a. Enter the area formula found in the previous example (it is easier to enter it with y replaced by x) as $y_1 = 1000x - 2x^2$.

b. Set the table to begin at $x = 240$ and change by $\Delta = 1$.

c. Press TABLE and see how the area y_1 increases as x gets to 250, and then decreases when x passes 250. You are seeing *numerically* that the area y_1 peaks at $x = 250$.

d. Notice that the amount by which y_1 changes each time becomes quite small for x near 250. This shows again numerically that the derivative (rate of change) becomes zero as x approaches 250.

e. Based on your table, how much area will the farmer lose if he mistakenly makes the width 249 or 251 feet instead of 250 (and therefore the length 502 or 498 feet)? Is this loss significant based on an area of 125,000 square feet? This is characteristic of maximization problems where the slope is zero — being *near* the maximizing value is essentially as good as being *at* it.

EXAMPLE 5   Maximizing the Volume of a Box

An open-top box is to be made from a square sheet of metal 12 inches on each side by cutting a square from each corner and folding up the sides, as in the diagram below. Find the volume of the largest box that can be made in this way.

Square sheet

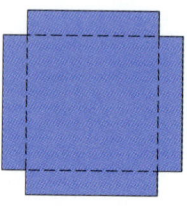

Corners removed

Side flaps folded up to make open-top box

Solution Let x = the length of the side of the square cut from each corner.

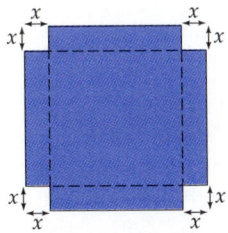

The 12" by 12" square with four x by x corners removed

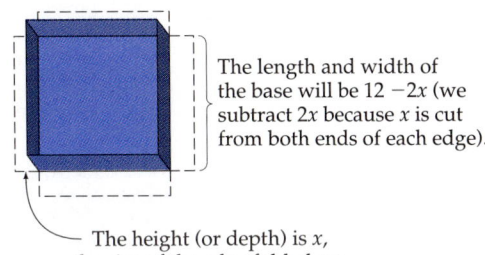

The length and width of the base will be $12 - 2x$ (we subtract $2x$ because x is cut from both ends of each edge).

The height (or depth) is x, the size of the edge folded up.

Therefore, the volume is

$$V(x) = (12 - 2x)(12 - 2x)x \qquad \text{(length)} \cdot \text{(width)} \cdot \text{(height)}$$

Since x is a length, $x > 0$, and since x inches are cut from *both* sides of each 12-inch edge, we must have $2x < 12$, so $x < 6$. The problem becomes

$$\text{Maximize} \quad V(x) = (12 - 2x)(12 - 2x)x \qquad \text{On } 0 < x < 6$$

$$V(x) = (144 - 48x + 4x^2)x \qquad \text{Multiplying out}$$

$$= 4x^3 - 48x^2 + 144x \qquad \text{Multiplying out}$$

$$V'(x) = 12x^2 - 96x + 144 \qquad \text{Differentiating}$$

$$= 12(x^2 - 8x + 12) \qquad \text{Factoring}$$

$$= 12(x - 2)(x - 6) \qquad \text{Factoring}$$

$$\text{CV} \begin{cases} x = 2 \\ x = \cancel{6} \end{cases} \qquad \text{(Not in the domain, so we eliminate it)}$$

The second derivative is $V''(x) = 24x - 96$, which at $x = 2$ is

$$V''(2) = 48 - 96 < 0$$

Therefore, the volume is *maximized* at $x = 2$.

Maximum volume is 128 cubic inches.

From $V(x)$ evaluated at $x = 2$

■

Graphing Calculator Exploration

Do the previous example (and then modify it) on a graphing calculator as follows:

a. Enter the volume as $y_1 = (12 - 2x)(12 - 2x)x$ and graph it on [0, 6] by [0, 150].

b. Use MAXIMUM to maximize y_1. Your answer should agree with that found in Example 5. (While this may seem faster than doing the problem "by hand," you would still need to have found the volume function and the domain, which was most of the work.)

c. What if the beginning size were not a 12-inch by 12-inch square, but a standard 8.5-inch by 11-inch sheet of paper? Go back to y_1 and replace the two 12s by 8.5 and 11 and find the x and the maximum volume. (*Answer:* Volume about 66 cubic inches)

d. What about a 3-by-5 card?

Parts (c) and (d), which would be more difficult to solve by hand, show how useful a graphing calculator is for solving related problems after one has been analyzed using calculus.

SUMMARY

There is no single, all-purpose procedure for solving word problems. You must think about the problem, draw a picture if possible, and express the quantity to be maximized or minimized in terms of some appropriate variable. With practice you can become good at it!

We have two procedures for optimizing continuous functions on intervals:

1. If the function has only one critical value in the interval, we find the critical value and use the second derivative test to show whether the function is maximized or minimized there.

2. If the interval is closed, we evaluate the function at all critical values and endpoints in the interval; the largest and smallest resulting values will be the maximum and minimum values of the function, respectively.

For functions satisfying neither of these two conditions, graph the function. The maximum (or minimum) value will be the y-coordinate of the highest (or lowest) point on the graph.

EXERCISES 7.3

Find (*without* using a calculator) the absolute extreme values of each function on the given interval.

1. $f(x) = x^3 - 6x^2 + 9x + 8$ on $[-1, 2]$

2. $f(x) = x^3 - 6x^2 + 22$ on $[-2, 2]$

3. $f(x) = x^3 - 12x$ on $[-3, 3]$

4. $f(x) = x^3 - 27x$ on $[-2, 2]$

5. $f(x) = x^4 + 4x^3 + 4x^2$ on $[-2, 1]$

6. $f(x) = x^4 - 4x^3 + 4x^2$ on $[0, 3]$

7. $f(x) = 5 - x$ on $[0, 5]$

8. $f(x) = x(100 - x)$ on $[0, 100]$

9. $f(x) = (x^2 - 1)^2$ on $[-1, 1]$

10. $f(x) = \sqrt[3]{x^2}$ on $[-1, 8]$

11. $f(x) = \dfrac{x}{x^2 + 1}$ on $[-3, 3]$

12. $f(x) = \dfrac{1}{x^2 + 1}$ on $[-3, 3]$

13. Find the number in the interval $[0, 3]$ such that the number minus its square is

   a. As large as possible
   b. As small as possible

14. Find the number in the interval $[\frac{1}{3}, 2]$ such that the sum of the number and its reciprocal is

   a. As large as possible
   b. As small as possible

 15. **One Function, Different Domains**

   a. Graph the function $y_1 = x^3 - 15x^2 + 63x$ on the window $[0, 10]$ by $[0, 130]$. By visual inspection of this function on this domain, where do the absolute maximum and minimum values occur: both at critical values, both at endpoints, or one at a critical value and one at an endpoint?

   b. Now change the domain to $[0, 8]$ and answer the same question.

   c. Now change the domain to $[2, 8]$ and answer the same question.

   d. Can you find a domain such that the minimum occurs at a critical value and the maximum at an endpoint?

16. **Existence of Extreme Values**

   a. Graph the function $y_1 = 5/x$ on the window $[1, 10]$, by $[0, 10]$. By visual inspection, does the function have an absolute maximum value on this domain? An absolute minimum value?

   b. Now change the x window to $[0, 10]$ and answer the same questions. (We cannot now call $[0, 10]$ the domain, since the function is not defined at 0.)

   c. Based on the screen display, answer the same questions for the domain $(0, \infty)$.

   d. Is there a domain on which this function has an absolute maximum but no absolute minimum?

   e. It was stated on page 628 that a continuous function on a closed interval will always have an absolute maximum and minimum value. Is this claim violated by the function $f(x) = 5/x$ on $[0, 10]$ [part (b)]? Is it violated by $f(x) = 5/x$ on $(0, \infty)$ [part (c)]?

APPLIED EXERCISES

17. **Biomedical: Pollen Count** The average pollen count in New York City on day x of the pollen season is $P(x) = 8x - 0.2x^2$ (for $0 \le x \le 40$). On which day is the pollen count highest?

18. **General: Fuel Economy** The fuel economy (in miles per gallon) of an average American compact car is $E(x) = -0.015x^2 + 1.14x + 8.3$, where x is the driving speed (in miles per hour,

$20 \le x \le 60$). At what speed is fuel economy greatest?

19. **General: Fuel Economy** The fuel economy (in miles per gallon) of an average American midsized car is $E(x) = -0.01x^2 + 0.62x + 10.4$, where x is the driving speed (in miles per hour, $20 \le x \le 60$). At what speed is fuel economy greatest?

20. General: Water Power The proportion of a river's energy that can be obtained from an undershot waterwheel is $E(x) = 2x^3 - 4x^2 + 2x$, where x is the speed of the waterwheel relative to the speed of the river. Find the maximum value of this function on the interval $[0, 1]$, thereby showing that only about 30% of a river's energy can be captured. Your answer should agree with the old millwright's rule that the speed of the wheel should be about one-third of the speed of the river.

21. General: Timber Value The value of a timber forest after t years is $V(t) = 480\sqrt{t} - 40t$ (for $0 \leq t \leq 50$). Find when its value is maximized.

22. General: Longevity and Exercise A recent study of the exercise habits of 17,000 Harvard alumni found that the death rate (deaths per 10,000 person-years) was approximately $R(x) = 5x^2 - 35x + 104$, where x is the weekly amount of exercise in thousands of calories $(0 \leq x \leq 4)$. Find the exercise level that minimizes the death rate.

23. Environmental Science: Pollution Two chemical factories are polluting a large lake, and the pollution level at a point x miles from factory A toward factory B is $P(x) = 3x^2 - 72x + 576$ parts per million (for $0 \leq x \leq 50$). Find where the pollution is the least.

24. Business: Maximum Profit City Cycles Incorporated finds that it costs $70 to manufacture each bicycle, and fixed costs are $100 per day. The price function is $p(x) = 270 - 10x$, where p is the price (in dollars) at which exactly x bicycles will be sold. Find the quantity City Cycles should produce and the price it should charge to maximize profit. Also find the maximum profit.

25. Business: Maximum Profit Country Motorbikes Incorporated finds that it costs $200 to produce each motorbike, and that fixed costs are $1500 per day. The price function is $p(x) = 600 - 5x$, where p is the price (in dollars) at which exactly x motorbikes will be sold. Find the quantity Country Motorbikes should produce and the price it should charge to maximize profit. Also find the maximum profit.

26. Business: Maximum Profit A retired potter can produce china pitchers at a cost of $5 each. She estimates her price function to be $p = 17 - 0.5x$, where p is the price at which exactly x pitchers will be sold per week. Find the number of pitchers that she should produce and the price that she should charge in order to maximize profit. Also find the maximum profit.

27. General: Parking Lot Design A company wants to build a parking lot along the side of one of its buildings using 800 feet of fence. If the side along the building needs no fence, what are the dimensions of the largest possible parking lot?

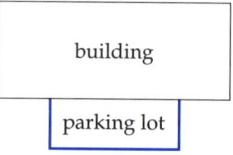

28. General: Area A farmer wants to make two identical rectangular enclosures along a straight river, as in the diagram shown below. If he has 600 yards of fence, and if the sides along the river need no fence, what should be the dimensions of each enclosure if the total area is to be maximized?

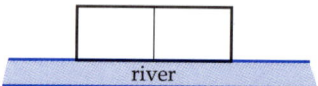

29. General: Area A farmer wants to make three identical rectangular enclosures along a straight river, as in the diagram shown below. If he has 1200 yards of fence, and if the sides along the river need no fence, what should be the dimensions of each enclosure if the total area is to be maximized?

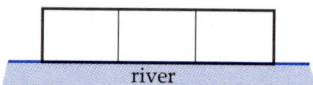

30. General: Area What is the area of the largest rectangle whose perimeter is 100 feet?

31. General: Package Design An open-top box is to be made from a square piece of cardboard that measures 18 inches by 18 inches by removing a square from each corner and folding up the sides. What are the dimensions and volume of the largest box that can be made in this way?

32. General: Gutter Design A long gutter is to be made from a 12-inch-wide strip of metal by folding up the two edges. How much of each edge should be folded up in order to maximize the capacity of the gutter? (*Hint:* Maximizing the capacity means maximizing the cross-sectional area, shown below.)

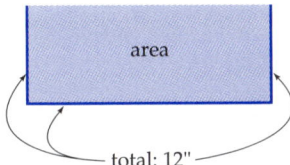

total: 12"

33. General: Maximizing a Product Find the two numbers whose sum is 50 and whose product is a maximum.

34. General: Maximizing Area Show that the largest rectangle with a given perimeter is a square.

35. Biomedical: Coughing When you cough, you are using a high-speed stream of air to clear your trachea (windpipe). During a cough your trachea contracts, forcing the air to move faster, but also increasing the friction. If a trachea contracts from a normal radius of 3 centimeters to a radius of r centimeters, the velocity of the airstream is $V(r) = c(3 - r)r^2$, where c is a constant depending on the length and the elasticity of the trachea. Find the radius r that maximizes this velocity. (X-ray pictures verify that the trachea does indeed contract to this radius.)

36. General: "Efishency" At what speed should a fish swim upstream so as to reach its destination with the least expenditure of energy? The energy depends on the friction of the fish through the water and on the duration of the trip. If the fish swims with velocity v, the energy has been found experimentally to be proportional to v^k (for constant $k > 2$) times the dura-

tion of the trip. A distance of s miles against a current of speed c requires time $s/(v - c)$ (distance divided by speed). The energy required is then proportional to $v^k s/(v - c)$. For $k = 3$, minimizing energy is equivalent to minimizing

$$E(v) = \frac{v^3}{v - c}$$

Find the speed v with which the fish should swim in order to minimize its energy expenditure $E(v)$. (Your answer will depend on c, the speed of the current.)

37. General: Athletic Fields A running track consists of a rectangle with a semicircle at each end, as shown below. If the perimeter is to be exactly 440 yards, find the dimensions (x and r) that maximize the area of the rectangle. (*Hint:* The perimeter is $2x + 2\pi r$.)

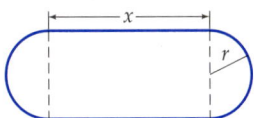

38. General: Window Design A Norman window consists of a rectangle topped by a semicircle, as shown below. If the perimeter is to be 18 feet, find the dimensions (x and r) that maximize the area of the window. (*Hint:* The perimeter is $2x + 2r + \pi r$.)

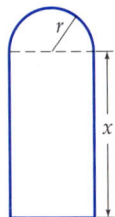

39–40: General: Maximizing Capacity of a Computer Disk Personal computers store information on disks by magnetically writing data on concentric circular "tracks" on the disk. The capacity of the disk depends on the radius x of the innermost track according to the formulas given on the next page (in which x is in inches and y is in megabytes). Enter the formula into a graphing calculator and use

MAXIMUM to find the inner radius x that maximizes the disk's capacity y. Also find the maximum capacity. (Computers actually use an inner radius slightly larger then the optimum value, which reduces the capacity slightly.)

39. $y = 0.988x(2.25 - x)$ (For 5.25-inch high-
        for $0 \le x \le 2.25$ density double-sided
                                    disks)

40. $y = 1.06x(1.68 - x)$ (For 3.5-inch double-
        for $0 \le x \le 1.68$ density double-sided
                                    disks)

 41–42: General: Value of Pulpwood Forest Suppose that the value (in thousands of dollars) of a pulpwood forest after t years is given by the follow-

ing formula. Enter the formula into a graphing calculator and use MAXIMUM to find when its value is maximized. Also find the maximum value of the forest.

41. $y = 150x^{0.7} - 50x$ for $0 \le x \le 50$

42. $y = 400x^{0.4} - 40x$ for $0 \le x \le 50$

 43–44: General: Package Design Use a graphing calculator (as explained on page 638) to find the side of the square removed and the volume of the box described in Example 5 (pages 636–637) if the square piece of metal is replaced by a:

43. 5-by-7 card (5 inches by 7 inches)

44. 6-by-8 card (6 inches by 8 inches)
        You might try constructing such a box.

 # 7.4 Further Applications of Optimization

Introduction

In this section we continue to solve optimization problems. In particular, we will see how to maximize a company's profit if we are not given the price function, provided that we are given information describing how price changes will affect sales. We will also see that sometimes x should be chosen as something *other* than quantity sold.

EXAMPLE 1   Finding Price and Quantity Functions

A store can sell 20 bicycles per week at a price of $400 each. The manager estimates that for each $10 price reduction she can sell two more bicycles per week. The bicycles cost the store $200 each. If x stands for *the number of $10 price reductions*, express the price p and the quantity q as functions of x.

Solution   Let

$$x = \text{the number of \$10 price reductions}$$

For example, $x = 4$ means that the price is reduced by $40 (four $10 price reductions). Therefore, in general, if there are x $10 price reductions from the original $400 price, then the price $p(x)$ is

$$p(x) = 400 - 10x \qquad\qquad\qquad \text{Price}$$

Original
price   └─ Less x $10 price reductions

The quantity sold $q(x)$ will be

$$q(x) = \underbrace{20}_{\substack{\text{Original} \\ \text{quantity}}} + 2x \qquad\qquad\qquad\qquad \text{Quantity}$$

Plus two for each price reduction

We will return to this example and maximize the store's profit after a practice problem.

PRACTICE PROBLEM

A computer manufacturer can sell 1500 personal computers per month at a price of $3000 each. The manager estimates that for each $200 price reduction he will sell 300 more each month. If x stands for *the number of $200 price reductions*, express the price p and the quantity q as functions of x.   *Solution at the back of the book*

EXAMPLE 2   Maximizing Profit (Continuation of Example 1)

Using the information in Example 1, find the price of the bicycles and the quantity that maximizes profit. Also find the maximum profit.

Solution   In Example 1 we found

$$p(x) = 400 - 10x \qquad\qquad \text{Price}$$

$$q(x) = 20 + 2x \qquad\qquad \text{Quantity sold at that price}$$

Revenue is price times quantity, $p(x)q(x)$:

$$R(x) = (400 - 10x)(20 + 2x) \qquad p(x)q(x)$$
$$= 8000 + 600x - 20x^2 \qquad \text{Multiplying out and simplifying}$$

The cost function is unit cost times quantity:

$$C(x) = \underbrace{200}_{\substack{\text{Unit} \\ \text{cost}}} \underbrace{(20 + 2x)}_{\substack{\text{Quantity} \\ q(x)}} = 4000 + 400x \qquad \begin{array}{l}\text{If there were a fixed cost,} \\ \text{we would add it}\end{array}$$

Profit is revenue minus cost:

$$P(x) = \underbrace{(8000 + 600x - 20x^2)}_{R(x)} - \underbrace{(4000 + 400x)}_{C(x)}$$
$$= 4000 + 200x - 20x^2 \qquad\qquad\qquad \text{Simplifying}$$

We maximize profit by setting the derivative equal to zero:

$$P'(x) = 200 - 40x = 0 \qquad \text{Differentiating } P = 4000 + 200x - 20x^2$$

The critical value is $x = 5$. The second derivative, $P''(x) = -40$, shows that the profit is *maximized* at $x = 5$. Since $x = 5$ is the number of $10 price reductions, the original price of $400 should be lowered by $50 ($10 five times), from $400 to $350. The quantity sold is found from the quantity function:

$$q(5) = 20 + 2 \cdot 5 = 30 \qquad \text{\color{blue}} q(x) = 20 + 2x \text{ at } x = 5$$

Finally, we state the answer clearly.
   Sell the bicycles for $350 each.
   Quantity sold: 30 per week
   Maximum profit: $4500   From $P(x) = 4000 + 200x - 20x^2$ at $x = 5$

Exercise 25 will show how a graphing calculator enables you to modify the problem (such as changing the cost per bicycle) and then immediately recalculate the new answer.

Choosing Variables

Notice that in Example 2 we did not choose x to be the quantity sold, but instead, to be *the number of $10 price reductions*. (Therefore, a negative x would have meant a price *increase*). We chose this x because from it we could easily calculate both the new price and the new quantity. Other choices for x are also possible, but in situations where a price change will make one quantity rise and another fall, it is often easiest to choose x to be the *number of such changes*.

EXAMPLE 3   Maximizing Harvest Size

An orange grower finds that if he plants 80 orange trees per acre, each tree will yield 60 bushels of oranges. He estimates that for each additional tree that he plants per acre, the yield of each tree will decrease by 2 bushels. How many trees should he plant per acre to maximize his harvest?

Solution   We take x equal to the number of "changes," that is, let

$$x = \text{the number of added trees per acre}$$

With x extra trees per acre,

   Trees per acre: $80 + x$ Original 80 plus x more
   Yield per tree: $60 - 2x$ Original yield less 2 per extra tree

Therefore, the total yield per acre will be

$$Y(x) = \underbrace{(60 - 2x)}_{\substack{\text{Yield} \\ \text{per tree}}}\underbrace{(80 + x)}_{\substack{\text{Trees} \\ \text{per acre}}} = 4800 - 100x - 2x^2$$

We maximize this by setting the derivative equal to zero:

$$-100 - 4x = 0 \qquad \text{Differentiating } Y = 4800 - 100x - 2x^2$$

$$x = -25 \qquad \text{Negative!}$$

The number of *added* trees is negative, meaning that the grower should plant 25 *fewer* trees per acre. The second derivative, $Y''(x) = -4$, shows that the yield is indeed maximized at $x = -25$. Therefore:

   Plant 55 trees per acre. $\qquad\qquad\qquad\qquad\qquad 80 - 25 = 55$

Earlier problems involved maximizing areas and volumes using only a fixed amount of material (such as a fixed length of fence). Instead, we could minimize the amount of materials for a fixed area or volume.

EXAMPLE 4   **Minimizing Package Materials**

A moving company wishes to design an open-top box with a square base whose volume is exactly 32 cubic feet. Find the dimensions of the box requiring the least amount of materials.

Solution   The base is square, so we define

x = length of side of base

y = height

The volume (length · width · height) is $x \cdot x \cdot y$ or x^2y, which (according to the problem) must equal 32 cubic feet:

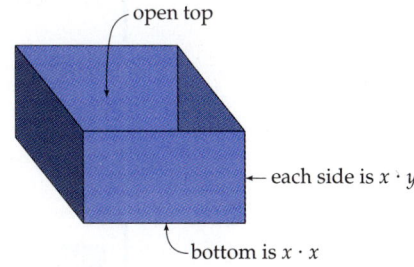

open top

each side is $x \cdot y$

bottom is $x \cdot x$

$$x^2y = 32$$

The box consists of a bottom (area x^2) and four sides (each of area xy). Minimizing the amount of materials means minimizing the *surface area* of the bottom and four sides:

$$A = x^2 + 4xy$$

$$\left(\begin{array}{c}\text{Area of}\\\text{bottom}\end{array}\right) + \left(\begin{array}{c}\text{Area of}\\\text{four sides}\end{array}\right)$$

As usual, we must express this area in terms of just *one* variable:

$$y = \frac{32}{x^2}$$

Solving $x^2y = 32$ for y

The area function becomes

$$A = x^2 + 4x\,\frac{32}{x^2}$$

$A = x^2 + 4xy$ with y replaced by $\dfrac{32}{x^2}$

$$= x^2 + \frac{128}{x}$$

Simplifying

$$= x^2 + 128x^{-1}$$

Writing $\dfrac{1}{x}$ as x^{-1}

We minimize this by differentiating:

$$A'(x) = 2x - 128x^{-2}$$

Derivative of $A = x^2 + 128x^{-1}$

$$2x - \frac{128}{x^2} = 0$$

Setting the derivative equal to zero

$$2x^3 - 128 = 0$$

Multiplying by x^2 (since $x > 0$)

$$x^3 = 64$$

Adding 128 and then dividing by 2

$$x = 4$$

Taking cube roots

The second derivative

$$A''(x) = 2 + 256x^{-3} = 2 + \frac{256}{x^3}$$

From $A'(x) = 2x - 128x^{-2}$

is positive at $x = 4$, so the area is minimized. Therefore:

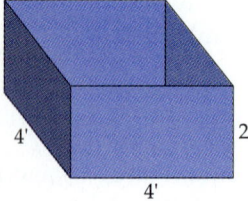

Base: 4 feet on each side
Height: 2 feet

Height from $y = 32/x^2$ at $x = 4$

Graphing Calculator Exploration

Use a graphing calculator to solve the previous example as follows:

a. Enter the area function to be minimized: $y_1 = x^2 + 128/x$.

b. Graph y_1 for x-values [0, 10], using TABLE or TRACE to find where y_1 seems to "bottom out" to determine an appropriate y-interval.

c. Graph the function on the viewing window determined in (b) and then use MINIMUM to find the minimum value. Your answer should agree with that found in Example 4. (Your calculator may give an inexact answer that needs to be rounded.)

Notice that either way required first finding the function to be minimized.

Minimizing the Cost of Materials

How would this problem have changed if the material for the bottom of the box had been more costly than the material for the sides? If, for example, the material for the sides cost $2 per square foot and the material for the base, needing greater strength, cost $4 per square foot, then instead of simply minimizing the surface area, we would minimize *total cost:*

$$\text{Cost} = \binom{\text{Area of}}{\text{bottom}}\binom{\text{Cost of bottom}}{\text{per square foot}} + \binom{\text{Area of}}{\text{sides}}\binom{\text{Cost of sides}}{\text{per square foot}}$$

Since the areas would be just as before, this cost would be

$$\text{Cost} = (x^2)(4) + (4xy)(2) = 4x^2 + 8xy$$

From here on we would proceed just as before, eliminating the y (using the volume relationship $x^2y = 32$) and then setting the derivative equal to zero.

Maximizing Tax Revenue

Governments raise money by collecting taxes. If a sales tax or an import tax is too high, trade will be discouraged and tax revenues will fall. If, on the other hand, the tax rate is too low, trade may flourish but tax revenues will again fall. Economists often want to determine the tax rate that maximizes revenue for the government. To do this, they must first predict the relationship between the tax on an item and the total sales of the item.

Suppose, for example, that the relationship between the tax rate t on an item and its total sales, S, is

$$S(t) = 9 - 20\sqrt{t}$$

t = tax rate $(0 \leq t < 0.20)$
S = total sales (millions of dollars)

If the tax rate is $t = 0$ (0%), then the total sales will be

$$S(0) = 9 - 20\sqrt{0} = 9 \qquad\qquad \text{\$9 million}$$

If the tax rate is raised to $t = 0.16$ (16%), then sales will be

$$S(0.16) = 9 - 20\sqrt{0.16}$$
$$= 9 - (20)(0.4) = 9 - 8 = 1 \qquad \text{\$1 million}$$

That is, raising the tax rate from 0% to 16% has discouraged \$8 million worth of sales. The graph of $S(t)$ below shows how total sales decrease as the tax rate increases.

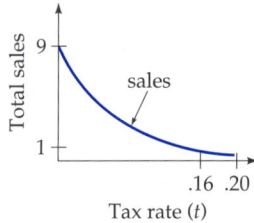

With such information (which may be found from historical data), one can find the tax rate that maximizes revenue.

EXAMPLE 5   Maximizing Tax Revenue

If economists predict that the relationship between the tax rate t on an item and the total sales S of that item (in millions of dollars) is

$$S(t) = 9 - 20\sqrt{t} \qquad\qquad \text{For } 0 \leq t \leq 0.20$$

find the tax rate that maximizes revenue to the government.

Solution

The government's revenue R is the tax rate t times the total sales $S(t) = 9 - 20\sqrt{t}$:

$$R(t) = t\underbrace{(9 - 20t^{1/2})}_{S(t)} = 9t - 20t^{3/2}$$

The graph of this function is shown on the right. To maximize it, we set its derivative equal to zero:

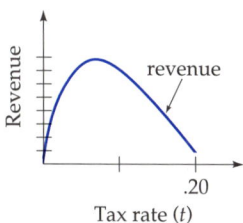

Revenue

revenue

.20

Tax rate (*t*)

$$9 - 30t^{1/2} = 0$$ Derivative of $9t - 20t^{3/2}$

$$9 = 30t^{1/2}$$ Adding $30t^{1/2}$ to each side

$$t^{1/2} = 9/30 = 0.3$$ Switching sides and dividing by 30

$$t = 0.09$$ Squaring both sides

This gives a tax rate of $t = 9\%$. The second derivative:

$$R''(t) = -30 \cdot \tfrac{1}{2}t^{-1/2} = -\frac{15}{\sqrt{t}}$$ From $R' = 9 - 30t^{1/2}$

is negative at $t = 0.09$, showing that the revenue is maximized. Therefore:

A tax rate of 9% maximizes revenue for the government.   ■

Graphing Calculator Exploration

The graph of the function from Example 5, $y_1 = 9x - 20x^{3/2}$ (written in x instead of t for ease of entry), is shown on the right, using the standard window $[-10, 10]$ by $[-10, 10]$. This might lead you to believe, erroneously, that the function is maximized at the endpoint $(0, 0)$.

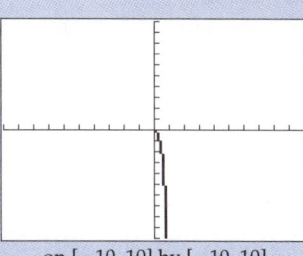

on $[-10, 10]$ by $[-10, 10]$

a.   Why does this graph not look like the graph at the top of this page? (*Hint:* Look at the scale.)

b.   Can you find a window on which your graphing calculator will show a graph like the one at the top of this page?

This example illustrates one of the pitfalls of graphing calculators—the part of the curve where the "action" takes place may entirely be hidden in one pixel. Calculus, on the other hand, will *always* find the critical value, no matter where it is, and then a graphing calculator can be used to confirm your answer by showing the graph on an appropriate window.

EXERCISES 7.4

1. **Business: Maximum Profit** An automobile dealer can sell 12 cars per day at a price of $15,000 each. He estimates that for each $300 price reduction he can sell two more cars per day. If each car costs him $12,000, and fixed costs are $1000, what price should he charge to maximize his profit? How many cars will he sell at this price? (*Hint:* Let x = the number of $300 price reductions.)

2. **Business: Maximum Profit** An automobile dealer can sell four cars per day at a price of $12,000 each. She estimates that for each $200 price reduction she can sell two more cars per day. If each car costs her $10,000, and her fixed costs are $1000, what price should she charge to maximize her profit? How many cars will she sell at this price? (*Hint:* Let x = the number of $200 price reductions.)

3. **Business: Maximum Revenue** An airline finds that if it prices a cross-country ticket at $200, it will sell 300 tickets per day. It estimates that each $10 price reduction will result in 30 more tickets sold per day. Find the ticket price (and the number of tickets sold) that will maximize the airline's revenue.

4. **Economics: Oil Prices** An oil-producing country can sell 1 million barrels of oil a day at a price of $25 per barrel. If each $1 price increase will result in a sales decrease of 50,000 barrels per day, what price will maximize the country's revenue? How many barrels will it sell at that price?

5. **Business: Maximum Revenue** Rent-A-Reck Incorporated finds that it can rent 60 cars if it charges $80 for a weekend. It estimates that for each $5 price increase it will rent three fewer cars. What price should it charge to maximize its revenue? How many cars will it rent at this price?

6. **General: Maximum Yield** A peach grower finds that if he plants 40 trees per acre, each tree will yield 60 bushels of peaches. He also estimates that for each additional tree that he plants per acre, the yield of each tree will decrease by 2 bushels. How many trees should he plant per acre to maximize his harvest?

7. **General: Maximum Yield** An apple grower finds that if she plants 20 trees per acre, each tree will yield 90 bushels of apples. She also estimates that for each additional tree that she plants per acre, the yield of each tree will decrease by 3 bushels. How many trees should she plant per acre to maximize her harvest?

8. **General: Fencing** A farmer has 1200 feet of fence and wishes to build two identical rectangular enclosures, as in the diagram. What should be the dimensions of each enclosure if the total area is to be a maximum?

9. **General: Minimum Materials** An open-top box with a square base is to have a volume of 4 cubic feet. Find the dimensions of the box that can be made with the smallest amount of materials.

10. **General: Minimum Materials** An open-top box with a square base is to have a volume of 108 cubic inches. Find the dimensions of the box that can be made with the smallest amount of materials.

11. **General: Largest Postal Package** The U.S. Postal Service will accept a package if its length plus its girth (the distance all the way around) does not exceed 84 inches. Find the dimensions and volume of the largest package with a square base that can be mailed.

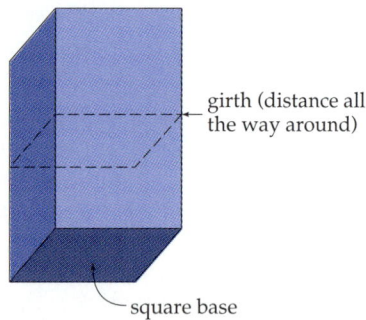

girth (distance all the way around)

square base

12. General: Fencing A homeowner wants to build, along his driveway, a garden surrounded by a fence. If the garden is to be 800 square feet, and the fence along the driveway costs $6 per foot whereas on the other three sides it costs only $2 per foot, find the dimensions that will minimize the cost. Also find the minimum cost.

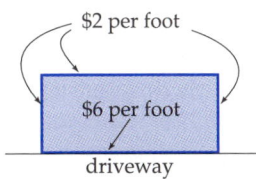

$2 per foot

$6 per foot

driveway

13. General: Fencing A homeowner wants to build, along her driveway, a garden surrounded by a fence. If the garden is to be 5000 square feet, and the fence along the driveway costs $6 per foot whereas on the other three sides it costs only $2 per foot, find the dimensions that will minimize the cost. Also find the minimum cost. (See the diagram for Exercise 12.)

14–15: Economics: Tax Revenue Suppose that the relationship between the tax rate t on imported shoes and the total sales S (in millions of dollars) is given by the function below. Find the tax rate t that maximizes revenue for the government.

14. $S = 4 - 6\sqrt[3]{t}$ **15.** $S = 8 - 15\sqrt[3]{t}$

16. Biomedical: Drug Concentration If the amount of a drug in a person's blood after t hours is $f(t) = t/(t^2 + 9)$, when will the drug concentration be the greatest?

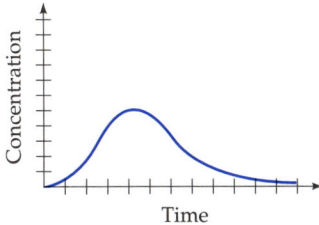

17. General: Wine Storage A case of vintage wine appreciates in value each year, but there is also an annual storage charge. The net value

of the wine after t years is $V(t) = 2000 + 96\sqrt{t} - 12t$ dollars (for $0 \le t \le 25$). Find the storage time that will maximize the value of the wine.

18. General: Bus Shelter Design A bus stop shelter, consisting of two square sides, a back, and a roof, as shown below, is to have volume 1024 cubic feet. What are the dimensions that require the least amount of materials?

square side

19. General: Area Show that the rectangle of fixed area whose perimeter is a minimum is a square.

20. Political Science: Campaign Expenses A politician estimates that by campaigning in a county for x days, she will gain $2x$ (thousand) votes, but her campaign expenses will be $5x^2 + 500$ dollars. She wants to campaign for the number of days that maximizes the number of votes per dollar,

$$f(x) = \frac{2x}{5x^2 + 500}$$

For how many days should she campaign?

21. General: Page Design A page of 96 square inches is to have margins of 1 inch on either side and $1\frac{1}{2}$ inches at the top and bottom, as in the diagram. Find the dimensions of the page that maximize the (inner) print area.

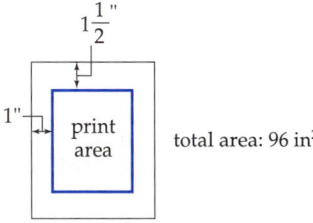

$1\frac{1}{2}$"

1"

print area

total area: 96 in²

22. Biomedical: Contagion If an epidemic spreads through a town at a rate that is proportional to the number of infected people and to the number of uninfected people, then the rate is $R(x) = cx(p - x)$ where x is the number of infected people and c and p (the population) are positive constants. Show that the rate $R(x)$ is greatest when half of the population is infected.

23. **Biomedical: Contagion**   If an epidemic spreads through a town at a rate that is proportional to the number of uninfected people and to the square of the number of infected people, then the rate is $R(x) = cx^2(p - x)$ where x is the number of infected people and c and p (the population) are positive constants. Show that the rate $R(x)$ is greatest when two thirds of the population is infected.

24. **Business: Maximizing Profit**   An electronics store can sell 35 cellular telephones per week at a price of $200 each. The manager estimates that for each $20 price reduction she can sell 9 more per week. The telephones cost the store $100 each, and fixed costs are $700 per week.

a. If x is the number of $20 price reductions, find the price $p(x)$ and enter it as y_1. Then enter the quantity function $q(x)$ as y_2.

b. Make y_3 the revenue function by defining $y_3 = y_1 y_2$ (price times quantity).

c. Make y_4 the cost function by defining y_4 as unit cost times y_2 plus fixed costs.

d. Make y_5 the profit function by defining $y_5 = y_3 - y_4$ (revenue minus cost).

e. Turn off y_1, y_2, y_3, and y_4 and graph the profit function y_5 for x-values $[-10, 10]$, using TABLE or TRACE to find an appropriate y-interval. Then use MAXIMUM to maximize it.

f. Use EVALUATE to find the best price and the quantity for this maximum profit.

25. **Business: Exploring a Profit Maximization Problem**   Use a graphing calculator to further explore Example 2 (page 643) as follows:

a. Enter the price function $y_1 = 400 - 10x$ and the quantity function $y_2 = 20 + 2x$ into your graphing calculator.

b. Make y_3 the revenue function by defining $y_3 = y_1 y_2$ (price times quantity).

c. Make y_4 the cost function by defining $y_4 = 200 y_2$ (unit cost times quantity).

d. Make y_5 the profit function by defining $y_5 = y_3 - y_4$ (revenue minus cost).

e. Turn off y_1, y_2, y_3, and y_4 and graph the profit function y_5 on the window $[0, 10]$ by $[0, 10,000]$, and then use MAXIMUM to maximize it. Your answer should agree with that found in Example 2.

Now change the problem!

f. What if the store finds that it can buy the bicycles from another wholesaler for $150 instead of $200? In y_4, change the 200 to 150. Then graph the profit y_5 (you may have to turn off y_4 again) and maximize it. Find the new price and quantity by evaluating y_1 and y_2 (using EVALUATE) at the new x-value.

g. What if cycling becomes more popular and the manager estimates that she can sell 30 instead of 20 bicycles per week at the original $400 price? Go back to y_2 and change 20 to 30 (keeping the change made earlier) and graph and find the price and quantity that maximize profit now.

Notice how flexible this setup is for changing any of the numbers.

7.5   Optimizing Lot Size and Harvest Size

Introduction

In this section we discuss two important applications of optimization, one economic and one ecological. The first concerns the most efficient way for a business to order merchandise (or for a manufacturer to produce merchandise), and the second concerns the preservation of animal populations that are harvested by people. Either of these applications can be read independently of the other.

Minimizing Inventory Costs

A business encounters two kinds of costs in maintaining inventory: storage costs (warehouse and insurance costs for merchandise not yet sold) and reorder costs (delivery and bookkeeping costs for each order). For example, if a furniture store expects to sell 250 sofas in a year, it could order all 250 at once (incurring high storage costs), or it could order them in many small lots, say 50 orders of five each, spaced throughout the year (incurring high reorder costs). Obviously, the best order size (or "lot" size) is the one that minimizes the total of storage plus reorder costs.

EXAMPLE 1   Minimizing Inventory Costs

A furniture showroom expects to sell 250 sofas a year. Each sofa costs the store $300, and there is a fixed charge of $500 per order. If it costs $100 to store a sofa for a year, how large should each order be and how often should orders be placed to minimize inventory costs?

Solution   Let

$$x = \text{lot size} \qquad \textcolor{blue}{x \text{ is the number of sofas in each order}}$$

Storage Costs: If the sofas sell steadily throughout the year, and if the store reorders x more whenever the stock runs out, then its inventory during the year looks like the following graph.

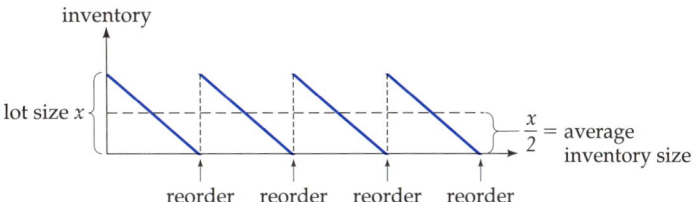

Notice that the inventory level varies from the lot size x down to zero, with an average inventory of $x/2$ sofas throughout the year. Because it costs $100 to store a sofa for a year, the total (annual) storage costs are

$$\begin{pmatrix} \text{Storage} \\ \text{costs} \end{pmatrix} = \begin{pmatrix} \text{Storage} \\ \text{per item} \end{pmatrix} \cdot \begin{pmatrix} \text{Average num-} \\ \text{ber of items} \end{pmatrix}$$

$$= \quad 100 \cdot \frac{x}{2} \quad = 50x$$

Reorder Costs: Each sofa costs $300, so an order of lot size x costs $300x$, plus the fixed order charge of $500:

$$\begin{pmatrix} \text{Cost} \\ \text{per order} \end{pmatrix} = 300x + 500$$

The yearly supply of 250 sofas, with x sofas in each order, requires $\frac{250}{x}$ orders. (For example, 250 sofas at 5 per order requires $\frac{250}{5} = 50$ orders.) Therefore, the yearly reorder costs are

$$\begin{pmatrix} \text{Reorder} \\ \text{costs} \end{pmatrix} = \begin{pmatrix} \text{Cost} \\ \text{per order} \end{pmatrix} \cdot \begin{pmatrix} \text{Number} \\ \text{of orders} \end{pmatrix}$$

$$= (300x + 500) \cdot \left(\frac{250}{x} \right)$$

Total Cost: $C(x)$ is storage costs plus reorder costs:

$$C(x) = \begin{pmatrix} \text{Storage} \\ \text{costs} \end{pmatrix} + \begin{pmatrix} \text{Reorder} \\ \text{costs} \end{pmatrix}$$

$$= 100\frac{x}{2} + (300x + 500)\left(\frac{250}{x} \right) \qquad \text{Using the storage and reorder costs found earlier}$$

$$= 50x + 75{,}000 + 125{,}000x^{-1} \qquad \text{Simplifying}$$

To minimize $C(x)$, we differentiate:

$$C'(x) = 50 - 125{,}000x^{-2} = 50 - \frac{125{,}000}{x^2} \qquad \begin{array}{l}\text{Differentiating} \\ C = 50x + 75{,}000 + \\ 125{,}000x^{-1}\end{array}$$

$$50 - \frac{125{,}000}{x^2} = 0 \qquad \begin{array}{l}\text{Setting the deriva-} \\ \text{tive equal to zero}\end{array}$$

$$50x^2 = 125{,}000 \qquad \begin{array}{l}\text{Multiplying by } x^2 \\ \text{and adding 125,000} \\ \text{to each side}\end{array}$$

$$x^2 = \frac{125{,}000}{50} = 2500 \qquad \begin{array}{l}\text{Dividing each} \\ \text{side by 50}\end{array}$$

$$x = 50$$

Taking square roots $(x > 0)$ gives lot size 50

$$C''(x) = 250{,}000x^{-3} = 250{,}000\,\frac{1}{x^3}$$

C'' is positive, so C is minimized at $x = 50$

At 50 sofas per order, the yearly 250 will require $\dfrac{250}{50} = 5$ orders. Therefore:

Lot size is 50 sofas, with orders placed five times a year.

Graphing Calculator Exploration

a. Verify the answer to the previous example by graphing the total cost function $y_1 = 50x + 75{,}000 + 125{,}000/x$ on the window $[0, 200]$ by $[0, 150{,}000]$ and using MINIMUM.

b. Notice that the curve is rather flat to the right of $x = 50$. From this observation, if you cannot order exactly 50 at a time, would it be better to order somewhat more than 50 or somewhat less than 50?

Modifications and Assumptions

If the number of orders per year is not a whole number, say 7.5 orders per year, we just interpret it as 15 orders in two years, and handle it accordingly.

We made two major assumptions in Example 1. We assumed that there was a steady demand, and that orders were equally spaced throughout the year. These are reasonable assumptions for many products, while for seasonal products like bathing suits or fur coats, separate calculation can be done for the "on" and "off" seasons.

Production Runs

Similar analysis applies to manufacturing. For example, if a book publisher can estimate the yearly demand for a book, it may print the yearly total all at once, incurring high storage costs, or it may print the books in several smaller runs throughout the year, incurring setup costs for each run. Here the setup costs for each printing run play the role of the reorder costs for a store.

EXAMPLE 2   Minimizing Inventory Costs For a Publisher

A publisher estimates the annual demand for a book to be 4000 copies. Each book costs $8 to print, and setup costs are $1000 for each printing. If storage costs are $2 per book per year, find how many books should be printed per run and how many printings will be needed if costs are to be minimized.

Solution   Let

$$x = \text{the number of books in each run}$$

Storage costs: As before, an average of $\dfrac{x}{2}$ books are stored throughout the year, at a cost of $2 each, so annual storage costs are

$$\begin{pmatrix} \text{Storage} \\ \text{costs} \end{pmatrix} = \left(\frac{x}{2}\right) \cdot 2 = x$$

Production costs: The cost per run is

$$\begin{pmatrix} \text{Costs} \\ \text{per run} \end{pmatrix} = 8x + 1000 \qquad \textcolor{red}{x \text{ books at } \$8 \text{ each, plus } \$1000 \text{ setup costs}}$$

The 4000 books at x books per run will require $4000/x$ runs. Therefore, production costs are

$$\begin{pmatrix} \text{Production} \\ \text{costs} \end{pmatrix} = (8x + 1000)\left(\frac{4000}{x}\right) \qquad \textcolor{red}{\text{Cost per run times number of runs}}$$

Total cost: The total cost is storage costs plus production costs:

$$C(x) = x + (8x + 1000)\left(\frac{4000}{x}\right) \qquad \textcolor{red}{\text{Storage + production}}$$

$$= x + 32{,}000 + 4{,}000{,}000x^{-1} \qquad \textcolor{red}{\text{Multiplying out}}$$

We differentiate, set the derivative equal to zero, and solve, just as before (omitting the details), obtaining $x = 2000$. The second derivative test will show that costs are minimized at 2000 books per run. The 4000 books require $\dfrac{4000}{2000} = 2$ printings. Therefore:

Print 2000 books per run, with two printings.   ∎

Maximum Sustainable Yield

The next application involves industries like fishing, in which a naturally occurring animal population is "harvested." Taking too large a harvest will kill off the animal population (like the bowhead whale, hunted almost to extinction in the nineteenth century). We want the "maximum sustainable yield" that may be taken, year after year, while preserving a stable animal population.

For some animals one can determine a *reproduction function*, $f(p)$, which gives the expected population a year from now if the present population is p.

Reproduction Function

A reproduction function $f(p)$ gives the population a year from now if the current population is p.

For example, the reproduction function $f(p) = -\frac{1}{4} p^2 + 3p$ (where p and $f(p)$ are measured in thousands) means that if the population is now $p = 6$ (thousand), then a year from now the population will be

$$f(6) = -\frac{1}{4} 6^2 + 3 \cdot 6 = -9 + 18 = 9 \qquad \text{(thousand)}$$

Therefore, during the year the population will increase from 6000 to 9000.

If, on the other hand, the present population is $p = 10$ (thousand), a year later the population will be

$$f(10) = -\frac{1}{4} \cdot 10^2 + 3 \cdot 10 = -25 + 30 = 5 \qquad \text{(thousand)}$$

That is, during the year the population will decline from 10,000 to 5000 (perhaps because of inadequate food to support such a large population). In actual practice, reproduction functions are very difficult to calculate, but can sometimes be estimated by analyzing previous population and harvest data.*

Suppose that we have a reproduction function f and a current population of size p, which will therefore grow to size $f(p)$ next year. The *amount of growth* in the population during that year is

$$\begin{pmatrix} \text{Amount} \\ \text{of growth} \end{pmatrix} = f(p) - p$$

Current population

Next year's population

---

*For more information, see J. Blower, L. Cook, and J. Bishop, *Estimating the Size of Animal Populations* (George Allen and Unwin Ltd., London, 1981).

Harvesting this amount removes only the *growth,* returning the population to its former size p. The population will then repeat this growth, and taking the same harvest $f(p) - p$ will cause this situation to repeat itself year after year. The quantity $f(p) - p$ is called the *sustainable yield.*

Sustainable Yield

For reproduction function $f(p)$, the sustainable yield is

$$Y(p) = f(p) - p$$

We want the population size p that maximizes the sustainable yield $Y(p)$. To maximize $Y(p)$, we set its derivative equal to zero:

$$Y'(p) = f'(p) - 1 = 0 \qquad \text{Derivative of } Y = f(p) - p$$

$$f'(p) = 1 \qquad \text{Solving for } f'(p)$$

For a given reproduction function $f(p)$, we find the maximum sustainable yield by solving this equation [provided that the second derivative test gives $Y''(p) = f''(p) < 0$].

Maximum Sustainable Yield

For reproduction function $f(p)$, the population p that results in the maximum sustainable yield is the solution to

$$f'(p) = 1$$

[provided that $f''(p) < 0$]. The maximum sustainable yield is then

$$Y(p) = f(p) - p$$

Once we calculate the population p that gives the maximum sustainable yield, we wait until the population reaches this size and then harvest, year after year, an amount $Y(p)$.

Note that to find the maximum sustainable yield, we set the derivative $f'(p)$ equal to 1, not 0. This is because we are maximizing not the reproductive function $f(p)$ but the yield function $Y(p) = f(p) - p$.

EXAMPLE 3 **Finding Maximum Sustainable Yield**

The reproduction function for the American lobster in an east coast fishing area is $f(p) = -0.02p^2 + 2p$ (where p and $f(p)$ are in thousands). Find the population p that gives the maximum sustainable yield and the size of the yield.

Solution

We set the derivative of the reproduction function equal to 1:

$$f'(p) = -0.04p + 2 = 1 \qquad \text{Differentiating } f(p) = -0.02p^2 + 2p$$

$$-0.04p = -1 \qquad \text{Subtracting 2 from each side}$$

$$p = \frac{-1}{-0.04} = 25 \qquad \text{Dividing by } -0.04$$

The second derivative $f''(p) = -0.04$ is negative, showing that $p = 25$ (thousand) is the population that gives the maximum sustainable yield. The actual yield is found from the yield function $Y(p) = f(p) - p$:

$$Y(p) = \underbrace{-0.02p^2 + 2p}_{f(p)} - p = -0.02p^2 + p$$

$$Y(25) = -0.02(25)^2 + 25 \qquad \text{Evaluating at } p = 25$$

$$= -12.5 + 25 = 12.5 \qquad \text{(thousand)}$$

The population size for the maximum sustainable yield is 25,000, and the yield is 12,500 lobsters. $\qquad$ 25,000 from $p = 25$

Graphing Calculator Exploration

Solve the previous example on a graphing calculator as follows:

a. Enter the reproduction function as $y_1 = -0.02x^2 + 2x$.

b. Define y_2 to be the derivative of y_1 (using NDERIV).

c. Define $y_3 = 1$.

d. Turn off y_1 and graph y_2 and y_3 on [0, 40] by [0, 2].

e. Use INTERSECT to find where y_2 and y_3 meet, thereby solving $f' = 1$. (You should find $x = 25$, as above.)

f. Find the yield by evaluating $y_1 - x$ at the x-value found in part (e).

In this problem, solving "by hand" was probably easier, but this graphing calculator method may be preferable if the reproduction function is more complicated.

EXERCISES 7.5

Lot Size

1. A supermarket expects to sell 4000 boxes of sugar in a year. Each box costs the store $2, and there is a fixed delivery charge of $20 per order. If it costs $1 to store a box for a year, what is the order size and how many times a year should the orders be placed to minimize inventory costs?

2. A supermarket expects to sell 5000 boxes of rice in a year. Each box costs the store $2, and there is a fixed delivery charge of $50 per order. If it costs $2 to store a box for a year, what is the order size and how many times a year should the orders be placed to minimize inventory costs?

3. A liquor warehouse expects to sell 10,000 bottles of scotch whiskey in a year. Each bottle costs the store $12, plus a fixed charge of $125 per order. If it costs $10 to store a bottle for a year, how many bottles should be ordered at a time and how many orders should the warehouse place in a year to minimize inventory costs?

4. A wine warehouse expects to sell 30,000 bottles of wine in a year. Each bottle costs the store $9, plus a fixed charge of $200 per order. If it costs $3 to store a bottle for a year, how many bottles should be ordered at a time and how many orders should the warehouse place in a year to minimize inventory costs?

5. An automobile dealer expects to sell 800 cars a year. The cars cost $9000 each plus a fixed charge of $1000 per delivery. If it costs $1000 to store a car for a year, find the order size and the number of orders that minimize inventory costs.

6. An automobile dealer expects to sell 400 cars a year. The cars cost $11,000 each plus a fixed charge of $500 per delivery. If it costs $1000 to store a car for a year, find the order size and the number of orders that minimize inventory costs.

Production Runs

7. A toy manufacturer estimates the demand for a game to be 2000 per year. Each game costs $3 to manufacture, plus setup costs of $500 for each production run. If a game can be stored for a year for a cost of $2, how many should be manufactured at a time and how many production runs should there be to minimize costs?

8. A toy manufacturer estimates the demand for a doll to be 10,000 per year. Each doll costs $5 to manufacture, plus setup costs of $800 for each production run. If it costs $4 to store a doll for a year, how many should be manufactured at a time and how many production runs should there be to minimize costs?

9. A producer of audio tapes estimates the yearly demand for a popular tape to be 1,000,000. It costs $800 to set the machinery for the tape, plus $10 for each tape produced. If it costs the company $1 to store a tape for a year, how many should be produced at a time and how many production runs will be needed to minimize costs?

10. A compact disc manufacturer estimates the yearly demand for a CD to be 10,000. It costs $400 to set the machinery for the CD, plus $3 for each CD produced. If it costs the company $2 to store a CD for a year, how many should be pressed at a time and how many production runs will be needed to minimize costs?

Maximum Sustainable Yield

11. Marine ecologists estimate the reproduction curve for swordfish in the Georges Bank fishing grounds to be $f(p) = -0.01p^2 + 5p$, where p and $f(p)$ are in hundreds. Find the population that gives the maximum sustainable yield, and the size of the yield.

12. The reproduction function for the Hudson Bay lynx is estimated to be $f(p) = -0.02p^2 + 5p$, where p and $f(p)$ are in thousands. Find the pop-

ulation that gives the maximum sustainable yield, and the size of the yield.

13. The reproduction function for the Antarctic blue whale is estimated to be $f(p) = -0.0004p^2 + 1.06p$, where p and $f(p)$ are in thousands. Find the population that gives the maximum sustainable yield, and the size of the yield.

14. The reproduction function for the Canadian snowshoe hare is estimated to be $f(p) = -0.025p^2 + 4p$, where p and $f(p)$ are in thousands. Find the population that gives the maximum sustainable yield, and the size of the yield.

15. A fisherman estimates the reproduction function for rainbow trout in a large lake to be $f(p) = 50\sqrt{p}$, where p and $f(p)$ are in thousands. Find the population that gives the maximum sustainable yield, and the size of the yield.

16. The reproduction function for oysters in a large bay is $f(p) = 30\sqrt[3]{p^2}$, where p and $f(p)$ are in pounds. Find the size of the population that gives the maximum sustainable yield, and the size of the yield.

17. Suppose that the reproduction function for Pacific salmon in a northwest fishing area is

$f(p) = 24\sqrt[3]{p} - 9\sqrt{p}$ (for $0 \le p < 50$), where p and $f(p)$ are in thousands. Use a graphing calculator to find the population that gives the maximum sustainable yield, and the size of the yield. (*Hint:* Follow the steps in the Graphing Calculator Exploration on page 659 using this reproduction function and x-interval $[0, 50]$.)

18. Business: Exploring a Lot Size Problem Use a graphing calculator to explore Example 1 (page 653) as follows:

a. Enter the total cost function, unsimplified, as $y_1 = 100(x/2) + (300x + 500)(250/x)$.

b. Graph y_1 on the window $[0, 200]$ by $[0, 150{,}000]$ and use MINIMUM to minimize it. Your answer should agree with the $x = 50$ found in Example 1.

c. Suppose that business improves, and the showroom expects to sell 350 per year instead of 250. In y_1, change the 250 to 350 and minimize it.

d. Suppose that a modest recession decreases sales (so change the 350 back to 250), and that inflation has driven the cost of storage up to \$125 (so change the 100 to 125), and minimize y_1.

7.6   Implicit Differentiation and Related Rates

Introduction

A function written in the form $y = f(x)$ is said to be defined *explicitly*, meaning that y is defined by a rule or *formula* $f(x)$ *in* x alone. A function may instead be defined *implicitly*, meaning that y is defined by an *equation in* x *and* y, such as $x^2 + y^2 = 25$. In this section we will see how to differentiate such *implicit functions* when ordinary "explicit" differentiation is difficult or impossible. We will then use implicit differentiation to find rates of change.

Implicit Differentiation

The equation $x^2 + y^2 = 25$ defines a circle. While a circle is not the graph of a function (it violates the vertical line test), the top half by itself defines a function, as does the bottom half by itself. To find these two functions, we solve $x^2 + y^2 = 25$ for y:

$$y^2 = 25 - x^2$$

Subtracting x^2 from each side of $x^2 + y^2 = 25$

$$y = \pm\sqrt{25 - x^2}$$

Plus or minus since either one, when squared, gives $25 - x^2$

The *positive* square root defines the top half of the circle (where y is positive), and the *negative* square root defines the bottom half (where y is negative). The equation $x^2 + y^2 = 25$ defines *both* functions at the same time.

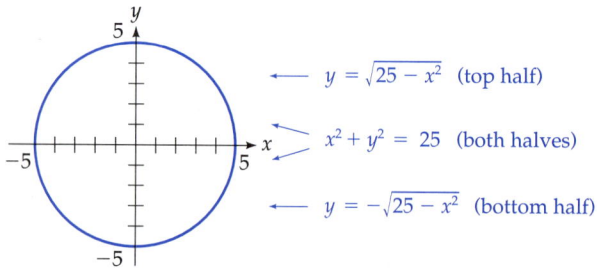

$\longleftarrow$ $y = \sqrt{25 - x^2}$ (top half)

$x^2 + y^2 = 25$ (both halves)

$\longleftarrow$ $y = -\sqrt{25 - x^2}$ (bottom half)

To find the slope anywhere on the circle, we could differentiate the "top" and "bottom" functions separately. However, it is easier to differentiate *implicitly* by differentiating both sides of the equation $x^2 + y^2 = 25$ with respect to x. Remember, however, that y is a *function* of x, so differentiating y^2 means differentiating a *function* squared, which requires the generalized power rule:

$$\frac{d}{dx} y^n = n \cdot y^{n-1} \frac{dy}{dx}$$

EXAMPLE 1   Differentiating Implicitly

Use implicit differentiation to find $\dfrac{dy}{dx}$ when $x^2 + y^2 = 25$.

Solution

We differentiate both sides of the equation with respect to x:

$$\underbrace{\frac{d}{dx} x^2} + \underbrace{\frac{d}{dx} y^2} = \underbrace{\frac{d}{dx} 25}$$

Differentiating $x^2 + y^2 = 25$

$$2x + 2y \frac{dy}{dx} = 0$$

Using the generalized power rule on y^2

Solving for $\dfrac{dy}{dx}$:

$$2y\,\frac{dy}{dx} = -2x \qquad\qquad \text{Subtracting } 2x$$

$$\frac{dy}{dx} = -\frac{x}{y} \qquad\qquad \begin{array}{l}\text{Canceling the 2s}\\\text{and dividing by } y\end{array}$$

Therefore, $\dfrac{dy}{dx} = -\dfrac{x}{y}$ when x and y are related by $x^2 + y^2 = 25$.

Notice that the formula for $\dfrac{dy}{dx}$ involves both x and y. Implicit differentiation enables us to find derivatives that would otherwise be difficult or impossible to calculate, but at a "cost"—the result may depend on both x and y.

Remember that x and y play different roles: x is the *independent* variable, and y is a *function*. Therefore, we must include a $\dfrac{dy}{dx}$ (from the generalized power rule) when differentiating y^n, but not when differentiating x^n (since $\dfrac{dx}{dx} = 1$).

EXAMPLE 2   Evaluating an Implicit Derivative *(Continuation of Example 1)*

Find the slope of the circle $x^2 + y^2 = 25$ at the points $(3, 4)$ and $(3, -4)$.

Solution

We simply evaluate the derivative $\dfrac{dy}{dx} = -\dfrac{x}{y}$ (found in Example 1) at the given points.

At $(3, 4)$: $\dfrac{dy}{dx} = -\dfrac{3}{4}$ ← x
                                 ← y

At $(3, -4)$: $\dfrac{dy}{dx} = -\dfrac{3}{-4} = \dfrac{3}{4}$

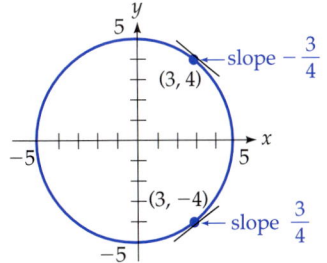

Note that the negative sign in $\dfrac{dy}{dx} = -\dfrac{x}{y}$ gives the slope the correct sign: negative at $(3, 4)$ and positive at $(3, -4)$.

Graphing Calculator Exploration

a. Graph the entire circle by graphing $y_1 = \sqrt{25 - x^2}$ and $y_2 = -\sqrt{25 - x^2}$. You may have to adjust the window to make the circle look "circular."

b. Verify the answer to Example 2 by finding the derivatives of y_1 and y_2 at $x = 3$.

c. Can you find the derivative at $x = 5$? Why not?

dy/dx=-.75

Derivatives should be evaluated only at points on the curve, so we evaluate $\dfrac{dy}{dx}$ only at x- and y-values *satisfying the original equation*. (It is easy to check that $x = 3$ and $y = \pm 4$ *do* satisfy $x^2 + y^2 = 25$.) Evaluating at a point not on the curve, like $(2, 3)$, would give a meaningless result.

The following are typical "pieces" that might appear in implicit differentiation problems.

EXAMPLE 3   Finding Derivatives—Implicit and Explicit

a. $\dfrac{d}{dx} y^3 = 3y^2 \dfrac{dy}{dx}$

Differentiating y^3, so include $\dfrac{dy}{dx}$

b. $\dfrac{d}{dx} x^3 = 3x^2$

Differentiating x^3, so no $\dfrac{dx}{dx}$

c. $\dfrac{d}{dx} (x^3 y^5) = 3x^2 \cdot y^5 + x^3 \cdot 5y^4 \dfrac{dy}{dx}$

$\underbrace{\dfrac{d}{dx} x^3} \qquad \underbrace{\dfrac{d}{dx} y^5}$

Using the product rule

$= 3x^2 y^5 + 5x^3 y^4 \dfrac{dy}{dx}$

Try to do problems like this in one step, putting the constants in front from the start

PRACTICE PROBLEM   Find:

a. $\dfrac{d}{dx} x^4$ **b.** $\dfrac{d}{dx} y^2$ **c.** $\dfrac{d}{dx} (x^2 y^3)$ *Solutions at the back of the book*

In general, finding $\dfrac{dy}{dx}$ from an equation that defines y implicitly involves three steps:

1. Differentiate both sides of the equation *with respect to x*.

2. Collect all terms involving $\dfrac{dy}{dx}$ on one side, and all others on the other side.

3. Factor out the $\dfrac{dy}{dx}$ and solve for it.

EXAMPLE 4   Finding and Evaluating an Implicit Derivative

For $y^4 + x^4 - 2x^2y^2 = 9$, find $\dfrac{dy}{dx}$ and evaluate it at $x = 2$, $y = 1$.

Solution

$$4y^3 \frac{dy}{dx} + 4x^3 - 4xy^2 - 4x^2y \frac{dy}{dx} = 0$$

Differentiating with respect to x, putting constants first

$$4y^3 \frac{dy}{dx} - 4x^2y \frac{dy}{dx} = -4x^3 + 4xy^2$$

Collecting $\dfrac{dy}{dx}$ terms on the left, others on the right

$$(4y^3 - 4x^2y) \frac{dy}{dx} = -4x^3 + 4xy^2$$

Factoring out $\dfrac{dy}{dx}$

$$\frac{dy}{dx} = \frac{-4x^3 + 4xy^2}{4y^3 - 4x^2y}$$

Dividing by $4y^3 - 4x^2y$ to solve for $\dfrac{dy}{dx}$

$$= \frac{-x^3 + xy^2}{y^3 - x^2y}$$

Dividing by 4

$$\frac{dy}{dx} = \frac{-(2)^3 + (2)(1)^2}{(1)^3 - (2)^2(1)} = \frac{-6}{-3} = 2$$

Evaluating at $x = 2$, $y = 1$

Note that in the above example the given point *is* on the curve, since $x = 2$ and $y = 1$ satisfy the original equation:

$$1^4 + 2^4 - 2 \cdot 2^2 \cdot 1^2 = 1 + 16 - 8 = 9$$

In economics, a *demand equation* is the relationship between the price p of an item and the quantity x that consumers will demand at that price. (All prices are in dollars, unless otherwise stated.)

EXAMPLE 5   Finding and Interpreting an Implicit Derivative

For the demand equation $x = \sqrt{1900 - p^3}$, use implicit differentiation to find dp/dx. Then evaluate it at $x = 30$, $p = 10$ and interpret your answer.

Solution

$$x = (1900 - p^3)^{1/2}$$ In power form

$$1 = \frac{1}{2}(1900 - p^3)^{-1/2}(-3p^2 p')$$ Differentiating with respect to x, using the generalized power rule (writing p' for dp/dx)

$$1 = \frac{-3p^2}{2\sqrt{1900 - p^3}} p'$$ Simplifying and factoring out p'

$$p' = -\frac{2\sqrt{1900 - p^3}}{3p^2}$$ Solving for p'

$$p' = -\frac{2\sqrt{1900 - (10)^3}}{3(10)^2}$$ Evaluating at $p = 10$, $x = 30$ (except that there is no x)

$$= -\frac{2\sqrt{900}}{300} = -\frac{60}{300} = -0.2$$ Simplifying

Interpretation: $dp/dx = -0.2$ says that the rate of change of price with respect to quantity is -0.2, so that increasing the quantity by 1 means decreasing the price by 0.20 (or 20¢). Therefore, each 20¢ price decrease brings approximately one more sale (at the given values of x and p). ∎

Notice that this particular demand function $x = \sqrt{1900 - p^3}$ can be solved *explicitly* for p:

$$x^2 = 1900 - p^3$$ Squaring

$$p^3 = 1900 - x^2$$ Adding p^3 and subtracting x^2

$$p = (1900 - x^2)^{1/3}$$ Taking cube roots

We can differentiate this *explicitly* with respect to x:

$$p' = \frac{1}{3}(1900 - x^2)^{-2/3}(-2x)$$ Using the generalized power rule

$$= -\frac{2}{3}x(1900 - x^2)^{-2/3}$$ Simplifying

Evaluating at the given values of $x = 30$ and $p = 10$ gives

$$p' = -\frac{2}{3} \cdot 30(1900 - 30^2)^{-2/3}$$ Substituting $x = 30$

$$= -20(1000)^{-2/3} = -\frac{20}{100} = -0.2$$

This agrees with the answer by implicit differentiation.

Related Rates

Sometimes in an equation, *both* variables will be functions of a *third* variable, usually t for time. For example, for a seasonal product like fur coats, the price p and weekly sales x will be related by an equation (the demand equation), and both price p and quantity x will depend on the time of year. Differentiating both sides of the demand equation with respect to time t will give an equation relating the derivatives dp/dt and dx/dt. Such "related rate" equations show how fast one quantity is changing relative to another. First, an "everyday" example.

EXAMPLE 6   Finding Related Rates

A pebble thrown into a pond causes circular ripples to radiate outward. If the radius of the outer ripple is growing by 2 feet per second, how fast is the area of its circle growing at the moment when the radius is 10 feet?

Solution   The formula for the area of a circle is $A = \pi r^2$. Both the area A and the radius r of the circle increase with time, so both are functions of t. We are told that the radius is increasing by 2 feet per second ($dr/dt = 2$), and we want to know how fast the area is changing (dA/dt). To find the 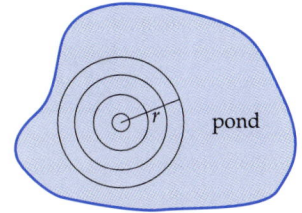 relationship between dA/dt and dr/dt, we differentiate both sides of $A = \pi r^2$ with respect to t.

$$\frac{dA}{dt} = 2\pi r \cdot \frac{dr}{dt}$$ Writing the 2 before the π

$$\frac{dA}{dt} = 2\pi \cdot 10 \cdot 2 = 40\pi \approx 125.6$$ Substituting $r = 10$ and $dr/dt = 2$

$$\underbrace{}_{r}\;\underbrace{}_{\frac{dr}{dt}}\quad\underbrace{}_{\substack{\text{Using}\\ \pi \approx 3.14}}$$

Therefore at the moment when the radius is 10 feet, the area of the circle is growing at the rate of about 126 square feet per second.

We should be ready to interpret any *rate* as a derivative, just as we interpreted the radius growing by 2 feet per second as $dr/dt = 2$.

EXAMPLE 7   Using Related Rates to Find Profit Growth

A boat yard's total profit from selling x outboard motors is $P = -x^2 + 1000x - 2000$. If the outboards are selling at the rate of 20 per week, how fast is the profit changing when 400 motors have been sold?

Solution   Profit P and quantity x both change with time, so both are functions of t. We differentiate both sides of $P = -x^2 + 1000x - 2000$ with respect to t and then substitute the given data.

$$\frac{dP}{dt} = -2x\frac{dx}{dt} + 1000\frac{dx}{dt} \qquad \text{Differentiating with respect to } t$$

$$= -2\underbrace{(400)}_{x}\underbrace{20}_{dx/dt} + 1000 \cdot \underbrace{20}_{dx/dt} \qquad \begin{array}{l}\text{Substituting } x = 400 \text{ (number sold)}\\\text{and } dx/dt = 20 \text{ (sales per week)}\end{array}$$

$$= -16{,}000 + 20{,}000 = 4000$$

Therefore, the company's profits are growing at the rate of $4000 per week.   ∎

EXAMPLE 8   Using Related Rates to Predict Pollution

A study of urban pollution predicts that sulfur oxide emissions in a city will be $S = 2 + 20x + 0.1x^2$ tons, where x is the population (in thousands). The population of the city t years from now is expected to be $x = 800 + 20\sqrt{t}$ thousand people. Find how rapidly sulfur oxide pollution will be increasing 4 years from now.

Solution   Finding the rate of increase of pollution means finding $\dfrac{dS}{dt}$.

$$\frac{dS}{dt} = 20\frac{dx}{dt} + 0.2x\frac{dx}{dt} \qquad \begin{array}{l}S = 2 + 20x + 0.1\,x^2\\\text{differentiated with respect to } t\\(x \text{ is also a function of } t)\end{array}$$

We then find dx/dt:

$$\frac{dx}{dt} = 10t^{-1/2} \qquad \begin{array}{l}x = 800 + 20t^{1/2} \text{ differentiated}\\\text{with respect to } t\end{array}$$

$$= 10 \cdot 4^{-1/2} = 10 \cdot \frac{1}{2} = 5 \qquad \begin{array}{l}\text{Substituting the given } t = 4\\\text{gives } dx/dt = 5\end{array}$$

$\dfrac{dS}{dt}$ then becomes

$$\dfrac{dS}{dt} = 20 \cdot 5 + 0.2(800 + 20\sqrt{4})\,5$$

$$\underbrace{}_{\dfrac{dx}{dt}} \quad \underbrace{\phantom{800 + 20\sqrt{4}}}_{x} \quad \underbrace{}_{\dfrac{dx}{dt}}$$

$$\dfrac{dS}{dt} = 20\,\dfrac{dx}{dt} + 0.2x\,\dfrac{dx}{dt}$$
with $dx/dt = 5$ and
$x = 800 + 20t^{1/2}$ at $t = 4$

$$= 100 + 0.2(840)5 = 100 + 840 = 940$$

Therefore, in 4 years, sulfur oxide emissions will be increasing at the rate of 940 tons per year.

Graphing Calculator Exploration

Verify the answer to Example 8 on a graphing calculator as follows:

a. Define $y_1 = 2 + 20x + 0.1x^2$ (the sulfur oxide function).

b. Define $y_2 = 800 + 20\sqrt{x}$ (the population function, using x for ease of entry).

c. Define $y_3 = y_1(y_2)$ (the composition of y_1 and y_2, giving pollution in year x).

d. Define y_4 to be the derivative of y_3 (using NDERIV, giving rate of change of pollution).

e. Graph these on the window $[0, 10]$ by $[0, 1500]$. (Which function does *not* appear on the screen?)

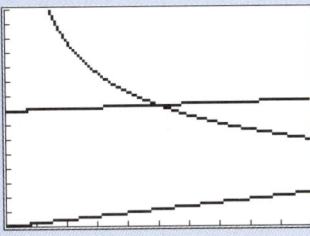

f. Evaluate y_4 at $x = 4$ to verify the answer to Example 8.

Verification of the Power Rule for Rational Powers

On page 528 we stated the power rule for differentiation:

$$\dfrac{d}{dx}\,x^n = nx^{n-1}$$

Although we proved it only for *integer* powers, we have been using the power rule for *all* constant powers n. Using implicit differentiation, we may now prove the power rule for *rational* powers. (Recall that a rational number is of the form p/q, where p and q are integers with $q \neq 0$.) Let $y = x^n$ for a rational exponent $n = p/q$, and let $x^{p/q}$ be differentiable at x. Then

$y = x^n = x^{\frac{p}{q}}$	Since $n = p/q$
$y^q = x^p$	Raising each side to the power q
$qy^{q-1} \dfrac{dy}{dx} = px^{p-1}$	Differentiating each side implicitly with respect to x
$\dfrac{dy}{dx} = \dfrac{px^{p-1}}{qy^{q-1}}$	Dividing each side by qy^{q-1}
$= \dfrac{px^{p-1}}{q\left(x^{\frac{p}{q}}\right)^{q-1}} = \dfrac{p}{q} \dfrac{x^{p-1}}{x^{p-\frac{p}{q}}}$	Using $y = x^{p/q}$ and multiplying out the exponents in the denominator
$= \dfrac{p}{q} x^{\underbrace{p-1-\left(p-\frac{p}{q}\right)}_{-1+\frac{p}{q}}} = \dfrac{p}{q} x^{p-\frac{p}{q}-1} = nx^{n-1}$	Subtracting powers, simplifying, and replacing p/q by n (twice)

This is what we wanted to show, that the derivative of $y = x^n$ is $dy/dx = nx^{n-1}$ for any rational exponent $n = p/q$. This proves the power rule for rational exponents.

SUMMARY

An equation in x and y may define one or more functions $y = f(x)$. Instead of solving the equation for y, which may be difficult or impossible, we may differentiate *implicitly*, differentiating both sides of the equation with respect to x (writing a dy/dx or y' whenever we differentiate y) and solving for the derivative dy/dx. The derivative at any point on the curve may then be found by substituting the coordinates of that point.

Implicit differentiation is especially useful when several variables in an equation depend on an underlying variable, usually t for time. Differentiating the equation implicitly with respect to this underlying variable gives an equation involving the rates of change of the original variables. Numbers may then be substituted into this "related rate equation" to find a particular rate of change.

EXERCISES 7.6

For each equation, use implicit differentiation to find dy/dx.

1. $y^3 - x^2 = 4$

2. $y^2 = x^4$

3. $x^3 = y^2 - 2$

4. $x^2 + y^2 = 1$

5. $y^4 - x^3 = 2x$

6. $y^2 = 4x + 1$

7. $(x + 1)^2 + (y + 1)^2 = 18$

8. $xy = 12$

9. $x^2 y = 8$

10. $x^2 y + xy^2 = 4$

11. $xy - x = 9$

12. $x^3 + 2xy^2 + y^3 = 1$

13. $x(y - 1)^2 = 6$

14. $(x - 1)(y - 1) = 25$

15. $y^3 - y^2 + y - 1 = x$

16. $x^2 + y^2 = xy + 4$

17. $\dfrac{1}{x} + \dfrac{1}{y} = 2$

18. $\sqrt[3]{x} + \sqrt[3]{y} = 2$

19. $x^3 = (y - 2)^2 + 1$

20. $\sqrt{xy} = x + 1$

For each equation, find dy/dx evaluated at the given values.

21. $y^2 - x^3 = 1$ at $x = 2$, $y = 3$

22. $x^2 + y^2 = 25$ at $x = 3$, $y = 4$

23. $y^2 = 6x - 5$ at $x = 1$, $y = -1$

24. $xy = 12$ at $x = 6$, $y = 2$

25. $x^2 y + y^2 x = 0$ at $x = -2$, $y = 2$

26. $y^2 + y + 1 = x$ at $x = 1$, $y = -1$

27. $x^2 + y^2 = xy + 7$ at $x = 3$, $y = 2$

28. $\sqrt[3]{x} + \sqrt[3]{y} = 3$ at $x = 1$, $y = 8$

For each demand equation, use implicit differentiation to find dp/dx.

29. $p^2 + p + 2x = 100$

30. $p^3 + p + 6x = 50$

31. $12p^2 + 4p + 1 = x$

32. $8p^2 + 2p + 100 = x$

33. $xp^3 = 36$

34. $xp^2 = 96$

35. $(p + 5)(x + 2) = 120$

36. $(p - 1)(x + 5) = 24$

APPLIED EXERCISES ON IMPLICIT DIFFERENTIATION

37. Business: Demand Equation A company's demand equation is $x = \sqrt{68 - p^2}$, where p is the price in dollars. Find dp/dx when $p = 2$ and interpret your answer.

38. Business: Demand Equation A company's demand equation is $x = \sqrt{650 - p^2}$, where p is the price in dollars. Find dp/dx when $p = 5$ and interpret your answer.

39. Business: Sales If a company spends r million dollars on research, its sales will be s million dollars, where r and s are related by $s^2 = r^3 - 55$.

    a. Find ds/dr by implicit differentiation and evaluate it at $r = 4$, $s = 3$.
(*Hint:* Differentiate the equation with respect to r.)

    b. Find dr/ds by implicit differentiation and evaluate it at $r = 4$, $s = 3$.
(*Hint:* Differentiate the original equation with respect to s.)

    c. Interpret your answers to parts (a) and (b) as rates of change.

40. Biomedical: Muscle Contraction When a muscle lifts a load, it does so according to the "fundamental equation of muscle contraction" $(L + m)(V + n) = k$, where L is the load that the muscle is lifting, V is the velocity of contraction of the muscle, and m, n, and k are constants. Use implicit differentiation to find dV/dL.

Exercises on Related Rates: In each equation, x and y are functions of t. Differentiate with respect to t to find a relation between dx/dt and dy/dt.

41. $x^3 + y^2 = 1$

42. $x^5 - y^3 = 1$

43. $x^2 y = 80$

44. $xy^2 = 96$

45. $3x^2 - 7xy = 12$

46. $2x^3 - 5xy = 14$

47. $x^2 + xy = y^2$

48. $x^3 - xy = y^3$

APPLIED EXERCISES ON RELATED RATES

49. General: Snowballs   A large snowball is melting so that its radius is decreasing at the rate of 2 inches per hour. How fast is the volume decreasing at the moment when the radius is 3 inches?
(*Hint:* The volume of a sphere of radius r is $V = \frac{4}{3}\pi r^3$.)

50. General: Hailstones   A hailstone (a small sphere of ice) is forming in the clouds so that its radius is growing at the rate of 1 mm (millimeter) per minute. How fast is its volume growing at the moment when the radius is 2 mm?
(*Hint:* The volume of a sphere of radius r is $V = \frac{4}{3}\pi r^3$.)

51. Biomedical: Tumors   The radius of a spherical tumor is growing by $\frac{1}{2}$ cm (centimeter) per week. Find how rapidly the volume is increasing at the moment when the radius is 4 cm.
(*Hint:* The volume of a sphere of radius r is $V = \frac{4}{3}\pi r^3$.)

52. Business: Profit   A company's profit from selling x units of an item is $P = 1000x - \frac{1}{2}x^2$ dollars. If sales are growing at the rate of 20 per day, find how rapidly profit is growing (in dollars per day) when 600 units have been sold.

53. Business: Revenue   A company's revenue from selling x units of an item is $R = 1000x - x^2$ dollars. If sales are increasing at the rate of 80 per day, find how rapidly revenue is growing (in dollars per day) when 400 units have been sold.

54. Social Science: Accidents   The number of traffic accidents per year in a city of population p is predicted to be $T = 0.002p^{3/2}$. If the population is growing by 500 people a year, find the rate at which traffic accidents will be rising when the population is $p = 40,000$.

55. Social Science: Welfare   The number of welfare cases in a city of population p is expected to be $W = 0.003p^{4/3}$. If the population is growing by 1000 people per year, find the rate at which welfare cases will be rising when the population is $p = 1,000,000$.

56. General: Rockets   A rocket fired straight up is being tracked by a radar station 3 miles from the launching pad. If the rocket is traveling at 2 miles per second, how fast is the distance between the rocket and the tracking station changing at the moment when the rocket is 4 miles up?
(*Hint:* The distance D in the illustration satisfies $D^2 = 9 + y^2$. To find the value of D, solve $D^2 = 9 + 4^2$.)

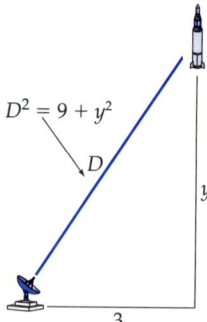

57. Biomedical: Poiseuille's Law   Blood flowing through an artery flows faster in the center of the artery and more slowly near the sides (because of friction). The speed of the blood is $V = c(R^2 - r^2)$, where R is the radius of the artery, r is the distance of the blood from the center of the artery, and c is a constant. Suppose that arteriosclerosis is narrowing the artery at the rate of $dR/dt = -0.01$ mm (millimeter) per year.

Artery

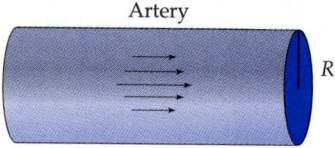

Find the rate at which blood flow is being reduced in an artery whose radius is $R = 0.05$ mm.

(*Hint:* Using $c = 500$, find dV/dt, considering r to be a constant. The units of dV/dt will be mm/sec per year.)

58. Implicit and Explicit Differentiation   The equation $x^2 + 4y^2 = 100$ describes an ellipse.

   a. Use implicit differentiation to find its slope at the points $(8, 3)$ and $(8, -3)$.

   b. Solve the equation for y, obtaining *two* functions, and differentiate both to find the slopes at $x = 8$. [Answers should agree with part (a).]

   c. Use a graphing calculator to graph the two functions found in part (b) on an appropriate window. Then use NDERIV to find the derivatives at (or near) $x = 8$. [Answers should agree with parts (a) and (b).]

   Notice that differentiating implicitly was easier than solving for y and then differentiating.

59. Related Rates: Speeding A traffic patrol helicopter is stationary a quarter of a mile directly

above a highway, as shown in the diagram below. Its radar detects a car whose line-of-sight distance from the helicopter is half a mile and is increasing at the rate of 57 mph. Is the car exceeding the highway's speed limit of 60 mph?

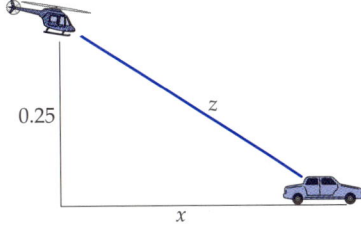

60. Related Rates: Continuation of Exercise 59 In Exercise 59 you found that the car's speed was $\dfrac{(0.5)(57)}{\sqrt{(0.5)^2 - (0.25)^2}}$ mph. Enter this expression into a graphing calculator and then replace both occurrences of the line-of-sight distance 0.5 by 0.4 and calculate the new speed of the car. What if the line-of-sight distance were 0.3 mile?

Chapter Summary with Hints and Suggestions

The reading and exercises of this chapter have helped you to learn the following skills. For each skill, the section from which it came (in case you need to review it) and some exercises in this section that use it are indicated. Answers to all exercises are at the end of the book, and full solutions to all exercises are in the Student Solutions Manual.

7.1   Graphing Using the First Derivative
7.2   Graphing Using the First and Second Derivatives

• Graph a polynomial, showing all relative extreme points and inflection points. (*Review Exercises 1–8.*)

   f' gives slope, f'' gives concavity.

• Graph a fractional power function, showing all

relative extreme points and inflection points. (*Review Exercises 9–12.*)

• Graph a rational function, showing all relative extreme points.   (*Review Exercises 13–18.*)

7.3   Optimization

• Find the absolute extreme values of a given function on a given interval. (*Review Exercises 19–28.*)

- Show that the marginal cost function pierces the average cost function where the average cost is minimized.   *(Review Exercises 29–30.)*

- Maximize the efficiency of a tugboat or a flying bird.   *(Review Exercises 31, 38.)*

7.4   Further Applications of Optimization

- Solve a geometric optimization problem. *(Review Exercises 32–37.)*

- Maximize profit for a company. *(Review Exercise 39.)*

- Maximize revenue from an orchard. *(Review Exercise 40.)*

- Minimize the cost of a power cable. *(Review Exercise 41.)*

- Use 📊 to maximize tax revenue to the government.   *(Review Exercise 42.)*

- Use 📊 to minimize the materials used in a container.   *(Review Exercises 43–45.)*

7.5   Optimizing Lot Size and Harvest Size

- Find the lot size to minimize production costs or inventory costs.   *(Review Exercises 46–47.)*

- Find the population size that allows the maximum sustainable yield. *(Review Exercises 48–49.)*

7.6   Implicit Differentiation and Related Rates

- Find a derivative by implicit differentiation. *(Review Exercises 50–53.)*

- Find a derivative by implicit differentiation and evaluate it.   *(Review Exercises 54–57.)*

- Solve a geometric related rate problem. *(Review Exercise 58.)*

- Use related rates to find the growth in profit or revenue.   *(Review Exercises 59–60.)*

Hints and Suggestions

- Do not confuse *relative* and *absolute* extremes: A function may have several *relative* extreme points (high points compared to their neighbors), but it can have at most one *absolute* maximum value (the largest value of the function on its entire domain). *Relative* extremes are used in graphing, and *absolute* extremes are used in optimization.

- Graphing calculators can be very helpful for graphing functions. However, you must first find a window that shows the interesting parts of the curve (relative extreme points and inflection points), and that is where calculus comes in.

- We have two procedures for optimizing continuous functions. Both begin by finding all critical values of the function in the domain. Then:

  1. If the function has only one critical value, find the sign of the second derivative there: a *positive* sign means an absolute *minimum*, and a *negative* sign means an absolute *maximum*.

  2. If the interval is closed, the maximum and minimum values may be found by evaluating the function at all critical values in the interval and at the endpoints—the largest and smallest resulting values are the maximum and minimum values of the function.

  A good strategy is this: Find all critical values in the domain. If there is only one, use procedure (1) above. Otherwise, try to define the function on a closed interval and use procedure (2). If all else fails, make a sketch of the graph.

- Don't forget to use the second derivative test in applied problems. Your employer will not be happy if you accidentally *minimize* your company's profits.

- In implicit differentiation problems, remember which is the *function* (usually y) and which is the independent variable (usually x).

- In related rate problems, begin by looking for an equation that relates the variables, and then differentiate it with respect to the underlying variable (usually t).

- Practice for test: Review Exercises 3, 11, 21, 25, 33, 35, 39, 43, 45, 47, 53, 55.

Review Exercises for Chapter 7

Practice test exercises are in blue.

7.1 and 7.2   Graphing

Graph each function "by hand," showing all relative extreme points and inflection points.

1. $f(x) = x^3 - 3x^2 - 9x + 12$

2. $f(x) = x^3 + 3x^2 - 9x - 7$

3. $f(x) = x^4 - 4x^3 + 15$

4. $f(x) = x^4 + 4x^3 + 17$

5. $f(x) = x(x + 3)^2$ **6.** $f(x) = x(x - 6)^2$

7. $f(x) = x(x - 4)^3$ **8.** $f(x) = x(x + 4)^3$

9. $f(x) = \sqrt[7]{x^5} + 1$ **10.** $f(x) = \sqrt[7]{x^6} + 1$

11. $f(x) = \sqrt[7]{x^4} + 1$ **12.** $f(x) = \sqrt[7]{x^3} + 1$

Graph each function, showing all relative extreme points. If you use a graphing calculator, make a hand-drawn sketch showing all relative extreme points.

13. $f(x) = \dfrac{1}{x^2 - 6x}$ **14.** $f(x) = \dfrac{1}{x^2 + 4x}$

15. $f(x) = \dfrac{x^2}{x^4 + 1}$ **16.** $f(x) = \dfrac{2x^2}{1 + x^2}$

17. $f(x) = \dfrac{1 - 2x}{x^2}$ **18.** $f(x) = \dfrac{1 - x}{x^2}$

7.3 and 7.4   Optimization

Find the absolute extreme values of each function on the given interval.

19. $f(x) = 2x^3 - 6x$ on $[0, 5]$

20. $f(x) = 2x^3 - 24x$ on $[0, 5]$

21. $f(x) = x^4 - 4x^3 - 8x^2 + 64$ on $[-1, 5]$

22. $f(x) = x^4 - 4x^3 + 4x^2 + 1$ on $[0, 10]$

23. $h(x) = (x - 1)^{2/3}$ on $[0, 9]$

24. $f(x) = \sqrt{100 - x^2}$ on $[-10, 10]$

25. $g(w) = (w^2 - 4)^2$ on $[-3, 3]$

26. $g(x) = x(8 - x)$ on $[0, 8]$

27. $f(x) = \dfrac{x}{x^2 + 1}$ on $[-3, 3]$

28. $f(x) = \dfrac{x}{x^2 + 4}$ on $[-4, 4]$

29. Business: Average and Marginal Cost   For the cost function $C(x) = 10{,}000 + x^2$, graph the marginal cost function $MC(x)$ and also the average cost function $AC(x) = C(x)/x$ on the same graph, showing that they intersect at the point where the average cost is minimized.

30. Prove that the marginal cost function intersects the average cost function at the point where the average cost is minimized (as shown in the diagram on the following page) by justifying each numbered step in the following series of equations.

At the x-value where the average cost function $AC(x) = \dfrac{C(x)}{x}$ is minimized, we must have

$$0 \overset{①}{=} \frac{d}{dx}\,AC(x) \overset{②}{=} \frac{xC'(x) - C(x)}{x^2} \overset{③}{=} \frac{1}{x}\left[\frac{xC'(x) - C(x)}{x}\right]$$

$$\overset{④}{=} \frac{1}{x}\left[C'(x) - \frac{C(x)}{x}\right] \overset{⑤}{=} \frac{1}{x}\,[MC(x) - AC(x)].$$

⑥ Finally, explain why the equality of the first and last expressions in the string of equations proves the result.

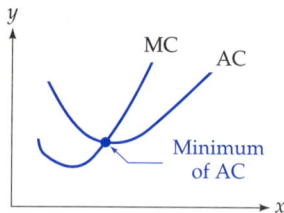

The marginal cost function pierces the average cost function at its minimum.

31. General: Fuel Efficiency At what speed should a tugboat travel upstream so as to use the least amount of fuel to reach its destination? If the tugboat's speed through the water is v, and if the speed of the current (relative to the land) is c, then the energy used is proportional to $E(v) = \dfrac{v^2}{v - c}$. Find the velocity v that minimizes the energy $E(v)$. Your answer will depend upon c, the speed of the current.

32. General: Fencing A homeowner wants to enclose three adjacent rectangular pens of equal size along a straight wall, as in the diagram below. If the side along the wall needs no fence, what is the largest total area that can be enclosed using only 240 feet of fence?

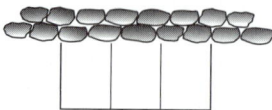

33. General: Maximum Area A homeowner wants to enclose three adjacent rectangular pens of equal size, as in the diagram below. What is the largest total area that can be enclosed using only 240 feet of fence?

34. General: Unicorns To celebrate the acquisition of Styria in 1261, Ottokar II sent hunters into the Bohemian woods to capture a unicorn.

To display the unicorn at court, the king built a rectangular cage. The material for three sides of the cage cost three ducats per running cubit, while the fourth was to be gilded and cost 51 ducats per running cubit. In 1261 it was well known that a happy unicorn requires an area of 2025 square cubits. Find the dimensions that would keep the unicorn happy at the lowest cost.

35. General: Box Design An open-top box with a square base is to have a volume of exactly 500 cubic inches. Find the dimensions of the box that can be made with the smallest amount of materials.

36. General: Packaging Find the dimensions of the cylindrical tin can with volume 16π cubic inches that can be made from the least amount of tin. (*Note:* $16\pi \approx 50$ cubic inches.)

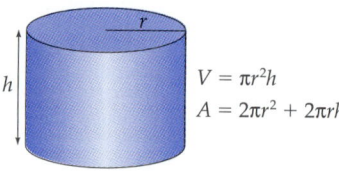

37. General: Packaging Find the dimensions of the open-top cylindrical tin can with volume 8π cubic inches that can be made from the least amount of tin. (*Note:* $8\pi \approx 25$ cubic inches.)

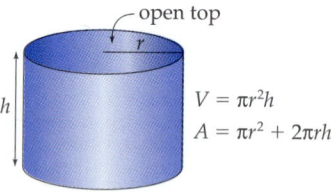

38. General: Bird Flight Let v be the flying speed of a bird, and let w be its weight. The power P that the bird must maintain during flight is $P = \dfrac{aw^2}{v} + bv^3$, where a and b are positive constants depending on the shape of the bird and the density of the air. Find the speed v that minimizes the power P.

39. Business: Maximum Profit A computer dealer can sell 12 personal computers per week at a price of $2000 each. He estimates that each $400 price decrease will result in three more sales per week. If the computers cost him $1200 each, what price should he charge to maximize his profit? How many will he sell at that price?

40. General: Farming A peach tree will yield 100 pounds of peaches now, which will sell for 40¢ a pound. Each week that the farmer waits will increase the yield by 10 pounds, but the selling price will decrease by 2¢ per pound. How long should the farmer wait to pick the fruit in order to maximize her revenue?

41. General: Minimum Cost A cable is to connect a power plant to an island, which is 1 mile offshore and 3 miles downshore from the power plant. It costs $5000 per mile to lay a cable underwater and $3000 per mile to lay it underground. If the cost of laying the cable is to be minimized, find the distance x downshore from the island where the cable should meet the land.

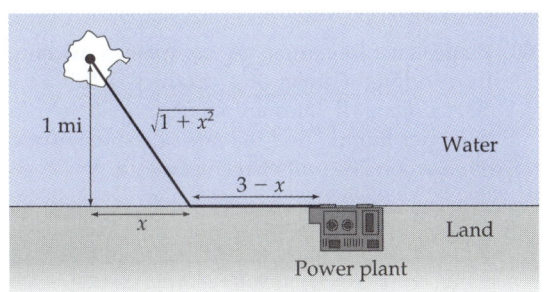

42. Economics: Tax Revenue Economists[*] have found that if cigarettes are taxed at a rate t, then cigarette sales will be $S(t) = 64 - 51.26t$ (billion dollars annually).

a. Use a graphing calculator to find the tax rate t that maximizes revenue to the government.

[*] Michael Grossman, Gary Becker (winner of 1992 Nobel Memorial Prize in Economics), Frank Chaloupka, and Kevin Murphy.

b. Multiply this tax rate times $2.02 (the average price of a premium pack of cigarettes) to find the actual tax per pack.

43. General: Packaging A 12-ounce soft drink can has volume 21.66 cubic inches. If the top and bottom are twice as thick as the sides, find the dimensions (radius and height) that minimize the amount of metal used in the can.

44. General: Box Design A standard 8.5- by 11-inch piece of paper can be made into a box with a lid by cutting x by x squares from two corners and x by 5.5-inch rectangles from the other corners and then folding, as shown below. What value of x maximizes the volume of the box, and what is the maximum volume?

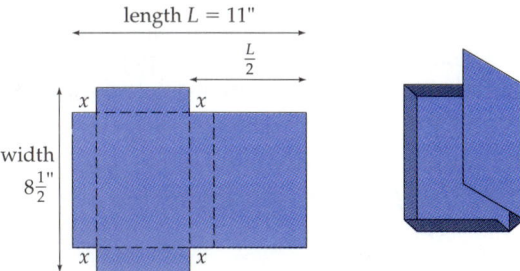

45. General: Box Design A standard 6 by 8 card (6 inches by 8 inches) can be made into a box with a lid by cutting x by x squares from two corners and x by 4-inch rectangles from the other corners and then folding. (See the diagram above, but use length 8 and width 6.) What value of x maximizes the volume of the box, and what is the maximum volume?

7.5 Optimizing Lot Size and Harvest Size

46. Business: Production Runs A wallpaper company estimates the demand for a certain pattern to be 900 rolls per year. It costs $800 to set up the presses to print the pattern, plus $200 to print each roll. If the company can store a roll of wallpaper for a year at a cost of $4, how many rolls should it print at a time and how many printing runs will it need in a year to minimize production costs?

47. Business: Lot Size A motorcycle shop estimates that it will sell 500 motorbikes in a year. Each bike costs the store $300, plus a fixed charge of $500 per order. If it costs the store $200 to store a motorbike for a year, what is the order size and how many orders will be needed in a year to minimize inventory costs?

48. Maximum Sustainable Yield The reproduction function for the North American duck is estimated to be $f(p) = -0.02p^2 + 7p$, where p and $f(p)$ are measured in thousands. Find the size of the population that allows the maximum sustainable yield, and also find the size of the yield.

49. Maximum Sustainable Yield Ecologists estimate the reproduction function for striped bass in an East Coast fishing ground to be $f(p) = 60\sqrt{p}$, where p and $f(p)$ are measured in thousands. Find the size of the population that allows the maximum sustainable yield, and also the size of the yield.

7.6 Implicit Differentiation and Related Rates

For each equation, use implicit differentiation to find dy/dx.

50. $6x^2 + 8xy + y^2 = 100$

51. $8xy^2 - 8y = 1$

52. $2xy^2 - 3x^2y = 0$ **53.** $\sqrt{x} - \sqrt{y} = 10$

For each equation, find dy/dx evaluated at the given values.

54. $x + y = xy$ at $x = 2$, $y = 2$

55. $y^3 - y^2 - y = x$ at $x = 2$, $y = 2$

56. $xy^2 = 81$ at $x = 9$, $y = 3$

57. $x^2y^2 - xy = 2$ at $x = -1$, $y = 1$

58. General: Melting Ice A cube of ice is melting so that each edge is decreasing at the rate of 2 inches per hour. Find how fast the volume of the ice is decreasing at the moment when each edge is 10 inches long.

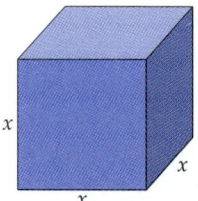

59. Business: Profit A company's profit from selling x units of a product is $P = 2x^2 - 20x$ dollars. If sales are growing at the rate of 30 per day, find the rate of change of profit when 40 units have been sold.

60. Business: Revenue A company's revenue from selling x units of a product is $R = x^2 + 500x$ dollars. If sales are increasing at the rate of 50 per month, find the rate of change of revenue when 200 units have been sold.

Projects and Essays

The following projects and essays are based on Chapter 7. Most have no right and wrong answers—the results depend only on your imagination and resourcefulness.

1. The graphs on the following page show the weight and height of an average child from age 2 to 15. Which curve is concave up and which is concave down, and what does concavity mean about growth? That is, does weight gain speed up or slow down as children become older? What about height gain? Can you give a physical reason for why height and weight increase so differently? If extended further, which curve will have an inflection point? Can you give any physical meaning to the inflection point?

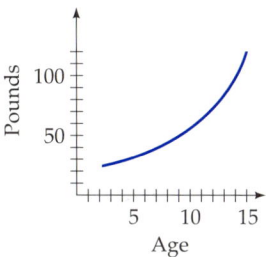

Weight from 2 to 15

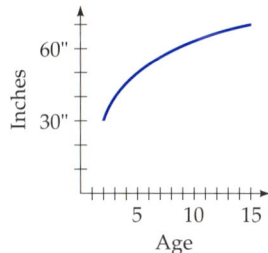

Height from 2 to 15

2. On pages 617–618 are three examples of curves representing various quantities, with interpretations for their inflection points. Think of three more quantities, draw graphs (with inflection points) that might represent these quantities, and interpret the meaning of each inflection point in terms of the quantity. Can you draw a graph with *two* inflection points? Your examples may be realistic or fanciful.

3. Read Review Exercise 30 on page 675 and try to think of an intuitive reason why the marginal cost function intersects the average cost at the point where average cost is minimized (*Hints:* If additional "marginal" units are cheaper to make than "average" units, should a manufacturer increase or decrease production? If it does this, what will happen to average cost? Can you give an example to illustrate your explanation?)

4. The physicist and Nobel laureate Richard Feynman told the following story.*

I often liked to play tricks on people when I was at MIT. One time, in a mechanical drawing class, some joker picked up a French curve (a piece of plastic for drawing smooth curves—a curly funny-looking thing) and said, "I wonder if the curves on this thing have some special formula?" I thought for a moment and said, "Sure they do. The curves are very special curves. Lemme show ya," and I picked up my French curve and began to turn it slowly. "The French curve is made so that at the lowest point on each curve, no matter how you turn it, the tangent is horizontal." All the guys in the class were holding up their French curve at different angles, holding their pencil up to it at the lowest point and laying it along, and discovering that, sure enough, the tangent is horizontal. They were all excited by this "discovery"—even though they had already gone through a certain amount of calculus. . . . They didn't put two and two together. They didn't even know what they "knew."

Explain Feynman's joke, and what he meant by "They didn't even know what they 'knew.' "

5. Explain why, in general, price and quantity are inversely related (when price rises, quantity falls, and vice versa). Is this always so—might there sometimes be a "snob" effect causing lower sales at lower prices and higher sales at higher prices? What kind of sales would you expect for the lowest-priced wine in an expensive restaurant? What might happen to sales if the restaurant raised the wine's price so that it wasn't the lowest? What sort of goods, both high-priced and low-priced, might be subject to a "snob effect"? Write a paragraph about these questions and include relevant information from anyone you know who has worked in a restaurant or in sales.

6. Explain why the second derivative test can be used for *absolute* extreme values only if there is just *one* critical value in the domain. Can you draw a curve that has two critical values and that has a relative maximum point that is *not* the absolute maximum point? Can you do this if there is only *one* critical value? Why not?

* Richard P. Feynman, *Surely You're Joking, Mr. Feynman* (W.W. Norton & Co., New York, 1985), p. 36.

7. Look over your notes, your homework, and the text, and write a page about how a graphing calculator has helped you in this chapter, both in graphing and optimization. Include examples of how it has helped you to *explore concepts* and how it has helped to *simplify your work*. What was its *most* helpful or interesting use? What was its *least* helpful or interesting use? Are there problems that can be done on a graphing calculator but that are easier to do "by hand"?

Cumulative Review—Chapters 6 and 7

The following exercises review some of the basic techniques that you learned in Chapters 6 and 7. Answers to all of these cumulative review exercises are given in the answer section at the end of the book.

1. Find, correct to 3 decimal places:
$$\lim_{x \to 0} (1 + 3x)^{1/x}.$$

2. Use the definition of the derivative, $f'(x) = \lim_{h \to 0} \dfrac{f(x + h) - f(x)}{h}$, to find the derivative of $f(x) = 2x^2 - 5x + 7$.

3. Find the derivative of $f(x) = 8\sqrt{x^3} - \dfrac{3}{x^2} + 5$.

4. Find the derivative of $f(x) = (x^5 - 2)(x^4 + 2)$

5. Find the derivative of $f(x) = \dfrac{2x - 5}{3x - 2}$.

6. The population of a city x years from now is predicted to be $P(x) = 3600x^{2/3} + 250{,}000$ million people. Find $P'(8)$ and $P''(8)$ and interpret your answers.

7. Find $\dfrac{d}{dx}\sqrt{2x^2 - 5}$ and write your answer in radical form.

8. Find $\dfrac{d}{dx}((3x + 1)^4(4x + 1)^3)$

9. Find $\dfrac{d}{dx}\left(\dfrac{x - 2}{x + 2}\right)^3$ and simplify your answer.

10. Make sign diagrams for the first and second derivatives and graph the function $f(x) = x^3 - 12x^2 - 60x + 400$. Show on your graph all relative extreme points and inflection points. (Do this "by hand" or using [calculator icon].)

11. Make sign diagrams for the first and second derivatives and graph the function $f(x) = \sqrt[3]{x^2} - 1$. Show on your graph all relative extreme points and inflection points.

12. A homeowner wishes to use 600 feet of fence to enclose two identical adjacent pens, as in the diagram below. Find the largest total area that can be enclosed.

13. A store can sell 12 telephone answering machines per day at a price of $200 each. The manager estimates that for each $10 price reduction she can sell 2 more per day. The answering machines cost the store $80 each. Find the price and the quantity sold per day to maximize the store's profit.

14. For y defined implicitly by $x^3 + 9xy^2 + 3y = 43$, find $\dfrac{dy}{dx}$ and evaluate it at the point $(1, 2)$.

15. A large spherical balloon is being inflated at the rate of 32 cubic feet per minute. Find how fast the radius is increasing at the moment when the radius is 2 feet.

8

EXPONENTIAL AND LOGARITHMIC FUNCTIONS

Populations are often modeled by exponential functions.

8.1 Review of Exponential and Logarithmic Functions

8.2 Differentiation of Exponential and Logarithmic Functions

8.3 Two Applications to Economics: Relative Rates and Elasticity of Demand

8.1 Review of Exponential and Logarithmic Functions

A P P L I C A T I O N   P R E V I E W

The Power of Poetry and Music

Exponential notation has applications to the arts as well as the sciences. The French novelist and poet Raymond Queneau published a 10-page book of sonnets entitled *A Hundred Thousand Billion Poems* (*Cent Mille Milliards de Poèmes*) in which the pages were cut between the lines of the poems so that any first line could be combined with any second line, and then with any third line, and so on, down to the fourteenth line. Each of the resulting 10^{14} possible combinations, according to Queneau, made a meaningful sonnet. It is unlikely, however, that his claim could ever be checked by reading them all. If the 10^{14} sonnets were read in succession, one each minute, 24 hours a day, 365 days a year, it would take 10^{14} minutes, which is more than 190 million years, until the last one would be read.

As another example of exponential notation, Mozart, toward the end of his brief life, wrote a piece entitled *A Musical Dice Game* (*Musikalisches Würfelspiel*, K516F). The music consists of 16 bars, and for 14 of them Mozart gave 11 possible ways to play the bar, and for 1 he gave 2 ways. Therefore, the total number of ways in which the piece can be played is $2 \cdot 11^{14}$. Playing all the variations would be a formidable task. If one variation were played each minute, 24 hours a day, 365 days a year, it would take $2 \cdot 11^{14}$ minutes, which is more than a billion years, before the last one would be played.

Introduction

This section reviews certain aspects of exponential and logarithmic functions that will be useful in the next section when we differentiate these functions. This section should serve as a review and continuation of, but not as a replacement for, Sections 0.5 and 0.6 (pages 66–90).

Exponential Functions

A function with a variable in an exponent, such as $f(x) = 2^x$, is called an *exponential function*. The number being raised to the power is called the *base*, so $f(x) = 2^x$ has base 2. We may use any positive number as a base for an exponential function, and one of the most useful bases is the constant e defined on pages 71–72:

$$e = \lim_{n \to \infty} \left(1 + \frac{1}{n} \right)^n = 2.71828 \cdots$$

The following graphs of exponential functions with bases 2, $\frac{1}{2}$, and e can be obtained either from a graphing calculator or by plotting points.

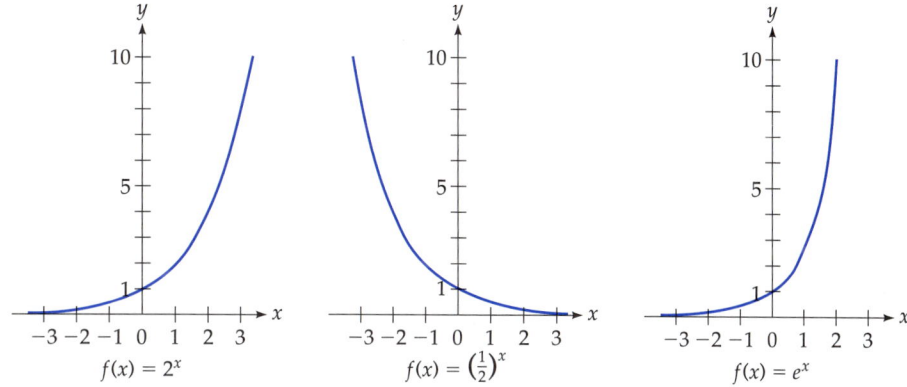

$f(x) = 2^x$ $f(x) = \left(\frac{1}{2}\right)^x$ $f(x) = e^x$

Compound Interest

On pages 69 and 117 we saw that a principal of P dollars that increases in value (appreciates) at rate r each year will grow in t years to $P(1 + r)^t$ dollars. In terms of compound interest, adding interest to the value at the end of each year is called *annual compounding*. If the compounding is done more frequently, say m times a year, then the interest rate per period is r/m, and the number of compoundings in t years (the exponent) is $m \cdot t$. As we saw on page 118, this gives:

Compound Interest

The value of P dollars at interest rate r compounded m times a year for t years is

$$P \left(1 + \frac{r}{m} \right)^{mt}$$

(Note that regardless of the compounding, the given rate r is always an *annual* interest rate unless otherwise stated.) If the compounding is done more and more frequently, with the number m of compounding periods approaching infinity, the result is called *continuous compounding*. The formula for continuously compounded interest is found by taking the limit of the above formula as $m \to \infty$ and using the limit definition of e, resulting in the following formula (see page 125).

Interest Compounded Continuously

The value of P dollars at interest rate r compounded continuously for t years is

$$Pe^{rt}$$

Powers of e are easily found on a calculator by using the 2nd and LN keys.

```
e^(1)
              2.718281828
e^(2)
              7.389056099
e^(.5)
              1.648721271
```

EXAMPLE 1   **Finding Value Under Compound Interest**

Find the value of $2000 invested at 6.4% interest for 9 years if the compounding is done:

a.   annually   **b.**   quarterly   **c.**   continuously

Solution

a.   The compound interest formula with $m = 1$ (annual compounding) simplifies to $P(1 + r)^t$:

$$2000(1 + 0.064)^9 = 2000 \cdot 1.064^9 \approx 3495.46$$

Using a calculator

$P(1 + r)^t$ with $P = 2000$, $r = 0.064$, and $t = 9$

The value is $3495.46.

b.   For quarterly compounding:

$$2000\left(1 + \frac{0.064}{4}\right)^{4 \cdot 9} = 2000 \cdot 1.016^{36} \approx 3541.63$$

$P\left(1 + \dfrac{r}{m}\right)^{mt}$ with $P = 2000$, $r = 0.064$, $m = 4$, and $t = 9$

The value is $3541.63 (better than annual compounding by $46.17).

c. For continuous compounding:

$$2000e^{0.064 \cdot 9} = 2000 \cdot e^{0.576} \approx 3557.82$$

Pe^{rt} with $P = 2000$, $r = 0.064$, and $t = 9$

The value is $3557.82 (better than quarterly compounding by $16.19).

Note that continuous compounding is best because small amounts of interest are added to the principal continuously, with no waiting until the end of a period, so the interest starts earning more interest immediately.

Depreciation by a Fixed Percentage

Most equipment, such as an industrial printing press or a personal automobile, wears out over time, and its value for accounting or tax purposes can be decreased or "depreciated." One of the most widely used methods of depreciation decreases the value by a fixed percentage each year, resulting in a large dollar drop in value in the first years (when the value is high) and a smaller dollar decrease in later years. For depreciation by a fixed percentage we use the compound interest formula with $m = 1$, that is, $P(1 + r)^t$, but with interest rate r being *negative* (for a decrease in value).

EXAMPLE 2 **Depreciating an Asset**

A rental car company buys a fleet of new cars for $240,000, and depreciates the fleet by 18% per year. Find the value of the fleet after:
a. 4 years   **b.** 6 months

a.   After 4 years the value will be

$$240{,}000(1 - 0.18)^4 = 240{,}000 \cdot 0.82^4 \approx 108{,}509$$

$P(1 + r)^t$ with $P = 240{,}000$, $r = -0.18$, and $t = 4$

After 4 years the value will be $108,509.

b.   After 6 months (half a year) the value will be

$$240{,}000(1 - 0.18)^{0.5} = 240{,}000 \cdot 0.82^{0.5} \approx 217{,}329$$

$P(1 + r)^t$ with $P = 240{,}000$, $r = -0.18$, and $t = 0.5$

After 6 months the value will be $217,329.

Behavioral Science: Learning Theory

One's ability to do a task generally improves with practice. Frequently, one's skill after t units of practice is given by a function of the form

$$S(t) = c(1 - e^{-kt})$$

where c and k are positive constants.

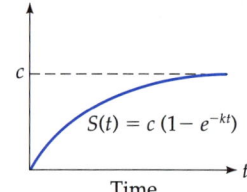

EXAMPLE 3 **Predicting Skill Levels**

A secretarial school guarantees that after t weeks of training a person will type

$$S(t) = 100(1 - e^{-0.25t})$$

words per minute. If you hire a secretary who has been in the program for 3 weeks, how fast will he type?

Solution

We evaluate $S(t)$ at $t = 3$:

$$100(1 - e^{-0.25 \cdot 3}) = 100(1 - e^{-0.75}) \approx 53$$

$100(1 - e^{-0.25t})$
with $t = 3$

He will type approximately 53 words per minute.   ■

Logarithmic Functions

On pages 81 and 85 we defined logarithms as exponents. The *common logarithm* of a number x, written $\log x$, is the exponent of 10 that gives that number.

Common Logarithms

> $\log x = y$ is equivalent to $10^y = x$ $\log x$ means $\log_{10} x$

EXAMPLE 4 **Finding Common Logarithms**

a. $\log 1000 = 3$ Since $10^3 = 1000$

b. $\log \dfrac{1}{10} = -1$ Since $10^{-1} = \dfrac{1}{10}$

c. $\log \sqrt{10} = \dfrac{1}{2}$ Since $10^{1/2} = \sqrt{10}$

∎

Graphing Calculator Exploration

On a graphing calculator, common loga-
rithms are easily found using the $\boxed{\text{LOG}}$
key. Verify the common logarithms from
Example 4 on your calculator.

```
log(1000)
                3
log(1/10)
               -1
log(√(10))
               .5
```

Natural Logarithms

In calculus we work almost exclusively with *natural* (base e) logarithms.
The natural logarithm of a number x, written ln x, is the power of e
that gives that number.

Natural Logarithms

$\ln x = y$ is equivalent to $e^y = x$ $\ln x$ means $\log_e x$

EXAMPLE 5   **Finding Natural Logarithms**

a. $\ln e = 1$ Since $e^1 = e$
b. $\ln 1 = 0$ Since $e^0 = 1$
c. $\ln \sqrt{e} = \dfrac{1}{2}$ Since $e^{1/2} = \sqrt{e}$

∎

Graphing Calculator Exploration

On a graphing calculator, natural loga-
rithms are found using the $\boxed{\text{LN}}$ key. Verify the natural logarithms from Example
5 on your calculator.

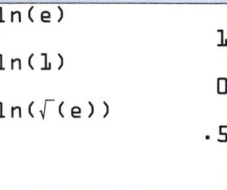

```
ln(e)
               1
ln(1)
               0
ln(√(e))
              .5
```

The following properties of natural logarithms were justified on pages 85 and 89.

Properties of Natural Logarithms

1. $\ln 1 = 0$	The natural log of 1 is 0 (since $e^0 = 1$)
2. $\ln e = 1$	The natural log of e is 1 (since $e^1 = e$)
3. $\ln e^x = x$	The natural log of e to a power is just the power (since $e^x = e^x$)
4. $e^{\ln x} = x$	e raised to the natural log of a number is just the number
5. $\ln (M \cdot N) = \ln M + \ln N$	The natural log of a product is the sum of the logs
6. $\ln \left(\dfrac{1}{N}\right) = -\ln N$	The natural log of 1 over a number is minus the log of the number
7. $\ln \left(\dfrac{M}{N}\right) = \ln M - \ln N$	The natural log of a quotient is the difference of the logs
8. $\ln (M^N) = N \cdot \ln M$	The natural log of a number to a power is the power times the log

Properties 3 and 4 show that $y = \ln x$ and $y = e^x$ are *inverse functions*, so their graphs are reflections of each other in the diagonal line $y = x$.

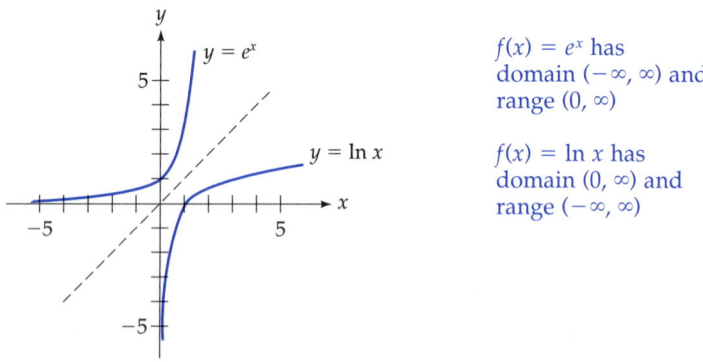

$f(x) = e^x$ has domain $(-\infty, \infty)$ and range $(0, \infty)$

$f(x) = \ln x$ has domain $(0, \infty)$ and range $(-\infty, \infty)$

$\ln x$ and e^x are inverse functions

The properties of natural logarithms are useful for simplifying functions.

EXAMPLE 6   Simplifying a Function

$$f(x) = \ln(x^2) - \ln\sqrt{x}$$

$$= \ln(x^2) - \ln(x^{1/2}) \qquad \text{Since } \sqrt{x} = x^{1/2}$$

$$= 2\ln x - \tfrac{1}{2}\ln x \qquad \text{Using property 8 (twice)}$$

$$= \tfrac{3}{2}\ln x \qquad \text{Combining}$$

■

EXAMPLE 7   Simplifying a Function

$$f(x) = x + \ln(e^{-0.05x})$$

$$= x - 0.05x \qquad \text{Using property 3}$$

$$= 0.95x \qquad \text{Combining}$$

■

Drug Dosage

The amount of a drug that remains in a person's bloodstream decreases exponentially with time. If the initial concentration is c (milligrams per milliliter of blood), the concentration t hours later will be

$$C(t) = ce^{-kt}$$

where the "absorption constant" k measures how rapidly the drug is absorbed.

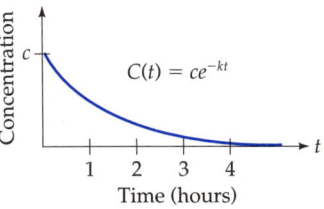

Drug concentration in the blood

Every medicine has a minimum concentration below which it is not effective. When the concentration falls to this level, another dose should be administered. If doses are administered regularly every T hours over a period of time, the concentration will look like the following graph:

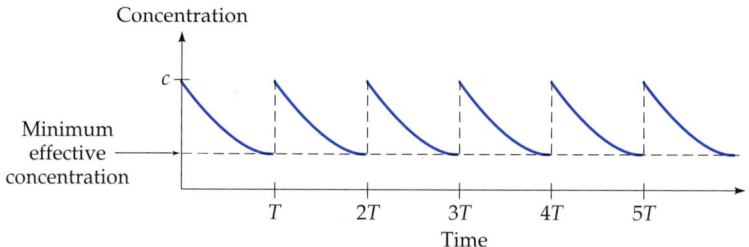

Drug concentration with repeated doses

The problem is to determine the time T at which the dose should be repeated so as to maintain an effective concentration.

EXAMPLE 8   Calculating Drug Dosage

The absorption constant for penicillin is $k = 0.11$, and the minimum effective concentration is 2 mg. If the original concentration is $c = 5$ mg, find when another dose should be administered in order to maintain an effective concentration.

Solution   The concentration formula $C(t) = ce^{-kt}$ with $c = 5$ and $k = 0.11$ is

$$C(t) = 5e^{-0.11t}$$

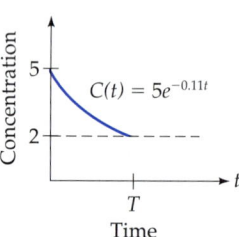

To find the time when this concentration reaches the minimum effective level of 2, we solve

$$5e^{-0.11t} = 2$$

$$e^{-0.11t} = 0.4 \qquad \text{Dividing by 5}$$

$$\underbrace{\ln e^{-0.11t}}_{-0.11t} = \ln 0.4 \qquad \text{Taking natural logs to bring down the exponent}$$

$$-0.11t = \ln 0.4$$

$$t = \frac{\ln 0.4}{-0.11} \approx \frac{-0.9163}{-0.11} \approx 8.3$$ Solving for t using a calculator

The concentration will reach the minimum effective level in 8.3 hours, so the dose should be repeated approximately every 8 hours.

■

EXAMPLE 9   Estimating Learning Time (Continuation of Example 3)

After t weeks of training, your secretary can type

$$S(t) = 100(1 - e^{-0.25t})$$

words per minute. How many weeks will he take to reach a speed of 80 words per minute?

Solution   We solve for t in the following equation:

$$100(1 - e^{-0.25t}) = 80$$ Setting $S(t)$ equal to 80

$$1 - e^{-0.25t} = 0.80$$ Dividing by 100

$$-e^{-0.25t} = -0.20$$ Subtracting 1

$$e^{-0.25t} = 0.20$$ Multiplying by -1

$$-0.25t = \ln 0.20$$ Taking natural logs

$$t = \frac{\ln 0.20}{-0.25}$$ Solving for t

$$\approx \frac{-1.6094}{-0.25} \approx 6.4$$ Using a calculator

He will reach a speed of 80 words per minute in about $6\frac{1}{2}$ weeks.

■

The last two examples have involved solving for t in an equation of the form $e^{at} = b$. Such equations occur frequently in applications, and are solved by taking logarithms of both sides to bring down the exponent. On a graphing calculator such equations can be solved by graphing the function (using x for ease of entry) together with a constant function and using INTERSECT.

 Graphing Calculator Exploration

In Example 9, find when the typing speed $S(t)$ reaches 80 by entering $y_1 = 100(1 - e^{-0.25x})$ and $y_2 = 80$, graphing both (you should be able to find an appropriate window), and finding the value where they INTERSECT.

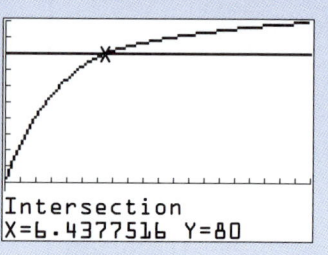

Intersection
X=6.4377516 Y=80

Social Science: Diffusion of Information by Mass Media

When a news bulletin is repeatedly broadcast over radio and television, the proportion of people who hear the bulletin within t hours is

$$p(t) = 1 - e^{-kt}$$

for some constant k.

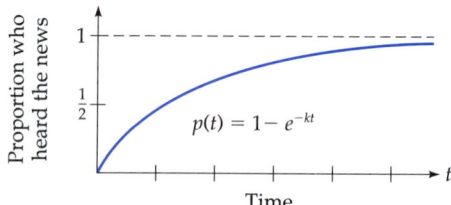

   *EXAMPLE 10*   **Predicting the Spread of Information**

A storm warning is broadcast, and the proportion of people who hear the bulletin within t hours of its first broadcast is $p(t) = 1 - e^{-0.30t}$. When will 75% of the people have heard the bulletin?

Solution

Equating the proportions gives $1 - e^{-0.30t} = 0.75$. Solving this equation as in the other examples (omitting the details) gives $t \approx 4.6$. Therefore, it takes about $4\frac{1}{2}$ hours for 75% of the people to hear the news.

SUMMARY

The value of P dollars invested for t years at interest rate r is given by one of the following exponential functions, depending on how frequently the interest is compounded.

Compounding	Value	
Annual	$P(1 + r)^t$	
m times a year	$P\left(1 + \dfrac{r}{m}\right)^{mt}$	P = principal r = interest rate m = compoundings per year t = number of years
Continuous	Pe^{rt}	

In calculus, the most useful base for exponential and logarithmic functions is the constant e, since it makes the differentiation formulas as simple as possible (as we will see in the next section).

$$e = \lim_{n \to \infty}\left(1 + \frac{1}{n}\right)^n = 2.71828 \cdots$$

The natural logarithm of a number x, written $\ln x$, is the power of e that gives the number.

$$\ln x = y \quad \text{is equivalent to} \quad e^y = x \qquad \ln x \text{ means } \log_e x$$

Natural logarithms and exponentials are easily found on a calculator by using the $\boxed{\text{LN}}$ and $\boxed{\text{2nd}}$ keys. The properties of natural logarithms (see page 688) come directly from familiar properties of exponents, and are useful for solving for variables in exponents and for simplifying functions.

Logarithmic Property	Exponential Property
$\ln 1 = 0$	$e^0 = 1$
$\ln (M \cdot N) = \ln M + \ln N$	$e^x \cdot e^y = e^{x+y}$
$\ln\left(\dfrac{1}{N}\right) = -\ln N$	$\dfrac{1}{e^y} = e^{-y}$
$\ln\left(\dfrac{M}{N}\right) = \ln M - \ln N$	$\dfrac{e^x}{e^y} = e^{x-y}$

EXERCISES 8.1

Graph each function. If you are using , make a hand-drawn sketch from the screen.

1. $f(x) = 2^{x/2}$

2. $f(x) = (\frac{1}{2})^{x/2}$

3. $f(x) = \ln(x + 1)$

(*Hint for Exercises 3 and 4:* You may need to sketch parts not shown on your graphing calculator.)

4. $f(x) = 1 - \ln x$

Evaluate each expression *without* using a calculator.

5. a. $e^{\ln 99}$ **b.** $\ln\left(\dfrac{1}{\sqrt{e}}\right)$ **c.** $\ln(e^{\ln 1})$

6. a. $\ln(e^{-500})$ **b.** $e^{\ln(e^2)}$ **c.** $\ln(e^{\ln e})$

 Find each natural logarithm using a calculator and then raise e to the resulting power (include all decimal places shown on your screen) to check that you get back (approximately) the original number.

7. $\ln 2.75$ **8.** $\ln 5.85$

Use the properties of natural logarithms to simplify each function.

9. $f(x) = \ln(10x) - \ln 10$

10. $f(x) = \ln(\frac{x}{2}) + \ln 2 - \ln 1$

11. $f(x) = \ln(\frac{1}{x}) + \ln x + \ln e$

12. $f(x) = \ln(x^3) - \ln(x^2) - \ln x$

13. $f(x) = \dfrac{e^{\ln(3x^2)}}{x} - x^{\ln e}$

14. $f(x) = \dfrac{\ln(e^{4x^2})}{x} - e^{\ln x}$

APPLIED EXERCISES (or required) ——————————————

15. Business: Interest   Find the value of $1000 invested at 10% interest for 8 years if the compounding is done:

a. annually   **b.** quarterly   **c.** monthly
d. continuously

16. Business: Interest   Find the value of $1000 invested at 8% interest for 10 years if the compounding is done:

a. annually   **b.** quarterly   **c.** monthly
d. continuously

17. General: Interest   You have $8000 to deposit in a bank. The First City Trust Company offers 6.5% interest compounded quarterly, and the Nation's Central Bank offers 6.4% interest compounded continuously. Which bank will give you more (and by how much) if you leave your money in for 5 years?

18. General: Interest   You have $10,000 to deposit in a bank. The People's Fiduciary Bank offers 6.2% interest compounded monthly, and the Citizen's Loan and Trust Company offers 6.1% interest compounded continu-ously. Which bank will give you more (and by how much) if you leave your money in for 8 years?

19. General: Interest   A loan shark lends you $100 at 2% compound interest per week (this is a *weekly* rate, not an annual rate).

a. How much will you owe after 3 years?
b. In "street" language, the profit on such a loan is called the "vigorish" or the "vig." Find the shark's vig.

20. General: Interest   A loan shark lends you $200 at 10% compound interest per month (this is a *monthly* rate, not an annual rate).

a. How much will you owe after 3 years?
b. In "street" language, the profit on such a loan is called the "vigorish" or the "vig." Find the shark's vig.

21. Business: Depreciation   A hospital buys a $2.4 million magnetic resonance imaging (MRI) scanner, and it depreciates by 14% per year. Find its value after:

a. 3 years   **b.** 6 months

22. Business: Depreciation A farming coopera- tive buys a $350,000 combine harvester and it depreciates by 16% per year. Find its value af- ter:

a. 5 years   **b.** 6 months

23. (Continuation of Exercise 21) For the MRI scanner in Exercise 21, find when it will be worth only $1 million.

24. (Continuation of Exercise 22) For the combine harvester in Exercise 22, find when it will be worth only $100,000.

25. General: Depreciation Your family buys a $20,000 automobile and it depreciates by 18% per year. Find its value after:

a. 5 years   **b.** 6 months

26. General: Depreciation Your $5000 com- puter depreciates by 24% per year. Find its value after:

a. 4 years   **b.** 6 months

27. (Continuation of Exercise 25) For the auto- mobile in Exercise 25, find when it will be worth only $2000.

28. (Continuation of Exercise 26) For the com- puter in Exercise 26, find when it will be worth only $500.

29. General: Population In 1994 the population of Washington, D.C. was 567,094 and declining by 1.7% annually. At this rate, when will the population fall to 450,000? (*Hint:* Losing a per- centage each year is like depreciation by a fixed percentage.)

30. General: Population In 1994 the population of St. Louis, Missouri, was 368,215 and declin- ing by 1.8% annually. At this rate, when will the population fall to 300,000? (*Hint:* Losing a percentage each year is like depreciation by a fixed percentage.)

31. Business: Advertising After t days of adver- tisements for a new laundry detergent, the pro- portion of shoppers in a town who have seen the ad is $1 - e^{-0.04t}$. How long must the ad run to reach:

a. 50% of the shoppers?
b. 90% of the shoppers?

 c. Graph $p(x) = 1 - e^{-0.04x}$ on a graphing cal- culator (using an appropriate window) to- gether with horizontal lines representing 50% and 90% and verify your answers to parts (a) and (b) using INTERSECT.

32. Business: Advertising After t days of adver- tisements for a new television program, the pro- portion of viewers in a city who have seen the ad is $1 - e^{-0.07t}$. How long must the ad run to reach:

a. 60% of the viewers?
b. 90% of the viewers?

 c. Graph $p(x) = 1 - e^{-0.07x}$ on a graphing cal- culator (using an appropriate window) to- gether with horizontal lines representing 60% and 90% and verify your answers to parts (a) and (b) using INTERSECT.

33. Behavioral Science: Memory In a psychol- ogy experiment the proportion of students who could remember an eight-digit number cor- rectly for t minutes was $e^{-0.08t}$.

a. What proportion remembered the number correctly after 5 minutes?
b. How long did it take for the proportion to fall to 20%?

 c. Graph $r(x) = e^{-0.08x}$ on a graphing calculator (using an appropriate window) together with a horizontal line representing 20% and verify your answer to part (b) using INTER- SECT.

34. Behavioral Science: Learning In a psychol- ogy experiment, the proportion of information from a photograph that a student could remem- ber after t minutes was $e^{-0.18t}$.

a. What proportion was remembered after 2 minutes?
b. How long did it take for the proportion to fall to 50%?

 c. Graph $r(x) = e^{-0.18x}$ on a graphing calculator (using an appropriate window) together with a horizontal line representing 50% and verify your answer to part (b) using INTER- SECT.

35. Social Science: Diffusion of Information by Mass Media Election returns are broadcast in a town of 50,000 people, and the number of people who have heard the news within t hours is $50,000(1 - e^{-0.3t})$.

a. How many people will have heard the news within 3 hours?

b. How long will it take for 40,000 people to hear the news?

 c. Graph $p(x) = 50,000(1 - e^{-0.3x})$ on a graphing calculator (using an appropriate window) together with a horizontal line representing 40,000 and verify your answer to part (b) using INTERSECT.

36. Social Science: Diffusion of Information by Mass Media   A bulletin about school closings is broadcast in a city of 200,000 people, and the number of people who have heard the bulletin within t hours is $200,000(1 - e^{-0.25t})$.

a. How many people will have heard the bulletin within 2 hours?

b. How long will it take for three quarters of the city to hear the bulletin?

 c. Graph $p(x) = 200,000(1 - e^{-0.25x})$ on a graphing calculator (using an appropriate window) together with a horizontal line representing three quarters of the population and verify your answer to part (b) using INTERSECT.

37–38: Biomedical: Drug Dosage   If the original concentration of a drug in a patient's bloodstream is c (milligrams per milliliter), then after t hours the concentration will be ce^{-kt}, where k is the absorption constant.

37. a. If the original concentration is 4 and the absorption constant is 0.25, when should the drug be readministered so that the concentration does not fall below the minimum effective concentration of 1.4?

 b. Graph the concentration function on a graphing calculator (using an appropriate window) together with a horizontal line representing the minimum effective concentration and verify your answer to part (a) using INTERSECT.

38. a. If the original concentration is 5 and the absorption constant is 0.15, when should the drug be readministered so that the concentration does not fall below the minimum effective concentration of 2.7?

 b. Graph the concentration function on a graphing calculator (using an appropriate window) together with a horizontal line representing the minimum effective concentration and verify your answer to part (a) using INTERSECT.

39. Environmental Science: Radioactive Waste   Hospitals use radioactive tracers in many medical tests. After the tracer is used, it must be stored as radioactive waste until its radioactivity has decreased enough to be disposed of as ordinary chemical waste. For the radioactive isotope of potassium, the proportion of radioactivity remaining after t days is $e^{-0.05t}$. How soon will the proportion of radioactivity decrease to 0.001 so that it can be disposed of as ordinary chemical waste?

40. Environmental Science: Rain Forests   It has been estimated* that the world's tropical rain forests are disappearing at the rate of 1.8% per year. If this rate continues, how soon will the rain forests be reduced to 50% of their present size? (Rain forests not only generate much of the oxygen that we breathe, but also contain plants with unique medical properties, such as the rosy periwinkle, which has revolutionized the treatment of leukemia.)

41. Biomedical: Half-Life of a Drug   The time required for the amount of a drug in one's body to decrease by half is called the "half-life" of the drug.

a. For a drug with absorption constant k, derive the following formula for its half-life:

$$\left(\begin{array}{c}\text{Half-} \\ \text{life}\end{array}\right) = \frac{\ln 2}{k}$$

(*Hint:* Solve the equation $e^{-kt} = \frac{1}{2}$ for t and use the properties of logarithms.)

b. Find the half-life of the cardioregulator digoxin if its absorption constant is $k = 0.018$ and time is measured in hours.

*Paul R. Ehrlich and Edward O. Wilson, "Biodiversity Studies: Science and Policy," *Science* **253**:758–762, August 16, 1991.

42. Environmental Science: Nuclear Waste More than half a century after the beginning of the nuclear age, not a single country has found a safe or permanent way of disposing of long-lived radioactive waste. Among the most haz- ardous radioactive waste is irradiated fuel from nuclear power plants, totaling 143,000 tons in 1995 and growing by 11.3% annually. At this rate, how long will it take for this amount to double? (*Source:* Worldwatch Institute.)

8.2 Differentiation of Exponential and Logarithmic Functions

Introduction

In the previous section we reviewed exponential and logarithmic functions, and in this section we differentiate these functions and use their derivatives for graphing, optimization, and finding rates of change. Verifications of the differentiation rules are given at the end of the section.

Derivatives of Logarithmic Functions

The rule for differentiating the natural logarithm function is as follows:

Derivative of ln x

$$\frac{d}{dx} \ln x = \frac{1}{x}$$ The derivative of ln x is 1 over x

EXAMPLE 1 **Differentiating a Logarithmic Function**

Differentiate $f(x) = x^3 \ln x$.

Solution The function is a *product*, x^3 times ln x, so we use the product rule.

$$\frac{d}{dx}(x^3 \ln x) = 3x^2 \ln x + x^3 \frac{1}{x} = 3x^2 \ln x + x^2$$

From $x^3 \frac{1}{x} = x^2$

Derivative of the first   Second left alone   First left alone   Derivative of ln x

PRACTICE PROBLEM 1   Differentiate $f(x) = \dfrac{\ln x}{x}$.   *Solution at the back of the book*

The preceding rule, together with the chain rule, shows how to differentiate the natural logarithm of a *function*. For any differentiable function $f(x)$,

Derivative of ln $f(x)$

$\dfrac{d}{dx} \ln f(x) = \dfrac{f'(x)}{f(x)}$	The derivative of the natural log of a function is the derivative of the function over the function

Notice that the right-hand side does not involve logarithms at all.

EXAMPLE 2   Differentiating a Logarithmic Function

$$\frac{d}{dx} \ln (x^2 + 1) = \frac{2x}{x^2 + 1}$$

← Derivative of $x^2 + 1$
← Original function (without the ln)

As we observed, the answer does not involve logarithms.

PRACTICE PROBLEM 2   Find $\dfrac{d}{dx} \ln (x^3 - 5x + 1)$.   *Solution at the back of the book*

EXAMPLE 3   Differentiating a Logarithmic Function

Find the derivative of $f(x) = \ln (x^4 - 1)^3$.

Solution For this problem we need the rule for differentiating the natural logarithm of a function, together with the generalized power rule [for differentiating $(x^4 - 1)^3$].

$$\frac{d}{dx} \ln (x^4 - 1)^3 = \frac{\dfrac{d}{dx}(x^4 - 1)^3}{(x^4 - 1)^3} \qquad \text{Using } \frac{d}{dx} \ln f = \frac{f'}{f}$$

$$= \frac{3(x^4 - 1)^2 4x^3}{(x^4 - 1)^3} \qquad \text{Using the generalized power rule}$$

$$= \frac{12x^3}{x^4 - 1} \qquad \text{Dividing top and bottom by } (x^4 - 1)^2$$

Alternative Solution   It is easier if we simplify first, using property 8 of logarithms (see the inside back cover) to bring down the exponent 3:

$$\ln (x^4 - 1)^3 = 3 \ln (x^4 - 1) \qquad \text{Using } \ln (M^N) = N \ln M$$

Now we differentiate the simplified expression:

$$\frac{d}{dx} 3 \ln (x^4 - 1) = 3 \frac{4x^3}{x^4 - 1} = \frac{12x^3}{x^4 - 1} \qquad \text{Same answer as before}$$

∎

Moral: Changing $\ln (\cdots)^n$ to $n \ln (\cdots)$ simplifies differentiation.

Derivatives of Exponential Functions

The rule for differentiating the exponential function e^x is as follows:

Derivative of e^x

$$\frac{d}{dx} e^x = e^x \qquad \text{The derivative of } e^x \text{ is simply } e^x$$

This shows the rather surprising fact that e^x is its own derivative. Stated another way, the function e^x is unchanged by the operation of differentiation.

This rule can be interpreted graphically. If $y = e^x$, then $y' = e^x$, so that $y = y'$. This means that on the graph of $y = e^x$, the slope y' always equals the y-coordinate, as shown below.

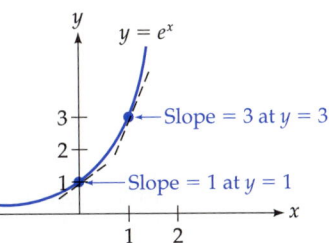

For the function $y = e^x$, $y' = y$.

Graphing Calculator Exploration

a. Define $y_1 = e^x$ and y_2 as the derivative of y_1 (using NDERIV) and graph them together on the window $[-3, 3]$ by $[-1, 10]$.

b. Why does the screen show only one curve?

c. Use TRACE to compare the values of y_1 and y_2 at some chosen x-value. Do the y-values agree *exactly*? If not, explain the slight discrepancy. (*Hint:* Is NDERIV really the derivative?)

EXAMPLE 4   Differentiating an Exponential Function

Find $\dfrac{d}{dx}\left(\dfrac{e^x}{x}\right)$.

Solution   Since the function is a quotient, we use the quotient rule:

Derivative of e^x
Derivative of x

$$\frac{d}{dx}\left(\frac{e^x}{x}\right) = \frac{x \cdot e^x - 1 \cdot e^x}{x^2} = \frac{xe^x - e^x}{x^2}$$

■

EXAMPLE 5   Finding a Derivative Involving e^x

If $f(x) = x^2 e^x$, find $f'(1)$.

Solution

$$f'(x) = 2xe^x + x^2 e^x \qquad \text{Using the product rule on } x^2 \cdot e^x$$

$$f'(1) = 2(1)e^1 + (1)^2 e^1 \qquad \text{Substituting } x = 1$$

$$= 2e + e = 3e \qquad \text{Simplifying}$$

■

In these problems we leave our answers in "exact" form, leaving e as e. Later, in applied problems, we will approximate our answers using $e \approx 2.718$ or a calculator.

PRACTICE PROBLEM 3   If $f(x) = xe^x$, find $f'(1)$.   *Solution at the back of the book*

The derivative of $e^{f(x)}$ is given by the following rule. For any differentiable function $f(x)$,

Derivative of $e^{f(x)}$

$$\frac{d}{dx}\, e^{f(x)} = e^{f(x)} \cdot f'(x)$$ The derivative of e to a function is e to the function times the derivative of the function

That is, to differentiate $e^{f(x)}$ we simply "copy" the original $e^{f(x)}$ and then multiply by the derivative of the exponent.

EXAMPLE 6   Differentiating an Exponential Function

$$\frac{d}{dx}\, e^{x^4+1} = \underbrace{e^{x^4+1}}_{\text{Copied}}(4x^3) = 4x^3 e^{x^4+1}$$ Reversing the order

Derivative of the exponent

■

EXAMPLE 7   Differentiating an Exponential Function

Find $\dfrac{d}{dx}\, e^{x^2/2}$.

Solution   It is better to write the exponent $x^2/2$ as $\frac{1}{2}x^2$, a constant times x to a power, since then its derivative is easily seen to be x.

$$\frac{d}{dx}\, e^{x^2/2} = \frac{d}{dx}\, e^{\frac{1}{2}x^2} = e^{\frac{1}{2}x^2}(x) = xe^{\frac{1}{2}x^2} = xe^{x^2/2}$$

Derivative of the exponent

■

PRACTICE PROBLEM 4   Find $\dfrac{d}{dx}\, e^{1-4x^3}$.   *Solution at the back of the book*

The formulas for differentiating natural logarithmic and exponential functions are summarized as follows, with $f(x)$ written simply as f.

Logarithmic Formulas	Exponential Formulas	
$\dfrac{d}{dx}\ln x = \dfrac{1}{x}$	$\dfrac{d}{dx}e^x = e^x$	Top formulas apply only to $\ln x$ and e^x
$\dfrac{d}{dx}\ln f = \dfrac{f'}{f}$	$\dfrac{d}{dx}e^f = e^f \cdot f'$	Bottom formulas apply to $\ln$ and e of a *function*

e^x versus x^n

Notice that we do *not* take the derivative of e^x by the power rule,

$$\frac{d}{dx}x^n = nx^{n-1}$$

This is because the power rule applies to x^n, *a variable to a constant power*, whereas e^x is *a constant to a variable power*. The two types of functions are quite different, as their graphs show.

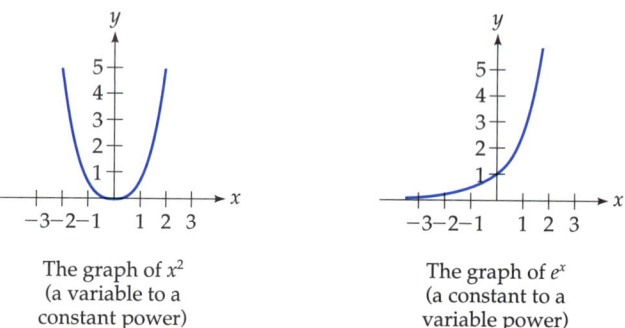

The graph of x^2
(a variable to a
constant power)

The graph of e^x
(a constant to a
variable power)

Each type of function has its own differentiation formula.

$$\frac{d}{dx}x^n = nx^{n-1} \qquad\qquad \frac{d}{dx}e^x = e^x$$

For a variable
x to a constant
power n

For the constant
e to a variable
power x

EXAMPLE 8 Differentiating a Logarithmic and Exponential Function

Find the derivative of $\ln(1 + e^x)$.

Solution

$$\frac{d}{dx} \ln (1 + e^x) = \underbrace{\frac{\frac{d}{dx}(1 + e^x)}{1 + e^x}}_{\text{Using } \frac{d}{dx} \ln f = \frac{f'}{f}} = \underbrace{\frac{e^x}{1 + e^x}}_{\substack{\text{Working out} \\ \text{the numerator}}}$$

■

Functions of the form e^{kx} (for constant k) arise in many applications. The derivative of e^{kx} is as follows.

$$\frac{d}{dx} e^{kx} = e^{kx}k = ke^{kx} \qquad\qquad \text{Using } \frac{d}{dx} e^f = e^f \cdot f'$$

Derivative of the exponent

This result is so useful that we record it as a separate formula.

Derivative of e^{kx}

$$\frac{d}{dx} e^{kx} = ke^{kx} \qquad\qquad \text{For any constant } k$$

This formula says that the rate of change (the derivative) of e^{kx} is proportional to itself. That is, the function $y = e^{kx}$ satisfies the *differential equation*

$$y' = ky$$

We noted this earlier when we observed that in exponential growth a quantity *grows in proportion to itself* (as in populations and savings accounts).

These differentiation formulas enable us to find instantaneous rates of change of logarithmic and exponential functions. In many applications the variable stands for time, so we use t instead of x.

EXAMPLE 9   Finding a Rate of Improvement of a Skill

After t weeks of practice a pole vaulter can vault

$$H(t) = 14(1 - e^{-0.10t})$$

feet. Find the rate of change of the athlete's jumps after

a. 0 weeks (at the beginning of training)   **b.** 12 weeks

Solution

$$H(t) = 14(1 - e^{-0.10t}) = 14 - 14e^{-0.10t} \qquad \text{\textcolor{blue}{$H(t)$ multiplied out}}$$

We differentiate to find the rate of change:

$$H'(t) = -14(-0.10)e^{-0.10t} = 1.4e^{-0.10t} \qquad \text{\textcolor{blue}{Differentiating}}$$
$$\text{\textcolor{blue}{$14 - 14e^{-0.10t}$}}$$

$$\text{\textcolor{blue}{Using } \frac{d}{dt}\, e^{kt} = ke^{kt} \qquad \text{Simplifying}}$$

a. For the rate of change after 0 weeks:

$$H'(0) = 1.4e^{-0.10(0)} = 1.4e^0 = 1.4 \qquad \text{\textcolor{blue}{$H'(t) = 1.4e^{-0.10t}$}}$$
$$\text{\textcolor{blue}{with } t = 0}$$

b. After 12 weeks:

$$H'(12) = 1.4e^{-0.10(12)}$$
$$= 1.4e^{-1.2} \approx 1.4(0.30) = 0.42 \qquad \text{\textcolor{blue}{$H'(t) = 1.4e^{-0.10t}$}}$$
$$\text{\textcolor{blue}{with } t = 12}$$

$$\text{\textcolor{blue}{Using a calculator}}$$

At first the vaults increased by 1.4 feet per week. After 12 weeks, the gain was only 0.42 foot (about 5 inches) per week.

This result is typical of learning a new skill: Early improvement is rapid, later improvement is slower.

Maximizing Consumer Expenditure

The amount of a commodity that consumers will buy depends on the price of the commodity. For a commodity whose price is p, let the consumer demand be given by a function $D(p)$. Multiplying the number of units $D(p)$ by the price p gives the total *consumer expenditure* for the commodity.

Consumer Demand and Expenditure

Let $D(p)$ be the consumer demand at price p. Then the consumer expenditure is

$$E(p) = p \cdot D(p)$$

EXAMPLE 10   Maximizing Consumer Expenditure

If consumer demand for a commodity is $D(p) = 10{,}000e^{-0.02p}$ units per week, where p is the selling price, find the price that maximizes consumer expenditure.

Solution Using the previous formula for consumer expenditure,

$$E(p) = p \cdot 10,000e^{-0.02p} = 10,000pe^{-0.02p} \qquad\qquad E(p) = p \cdot D(p)$$

To maximize $E(p)$ we differentiate:

$$E'(p) = \underbrace{10,000e^{-0.02p}}_{\substack{\text{Derivative} \\ \text{of } 10,000p}} + \underbrace{10,000p(-0.02)e^{-0.02p}}_{\substack{\text{Derivative} \\ \text{of } e^{-0.02p}}}$$

Using the product rule to differentiate $E(p) = 10,000p \cdot e^{-0.02p}$

$$= 10,000e^{-0.02p} - 200pe^{-0.02p} \qquad\qquad \text{Simplifying}$$
$$= 200e^{-0.02p}(50 - p) \qquad\qquad \text{Factoring}$$

$$\text{CV:}\quad p = 50$$

Critical value from $(50 - p)$ (since e to a power is never zero)

We calculate E'' for the second derivative test:

$$E''(p) = 200(-0.02)e^{-0.02p}(50 - p) + 200e^{-0.02p}(-1)$$

From $E'(p) = 200e^{-0.02p}(50 - p)$ using the product rule

$$= -4e^{-0.02p}(50 - p) - 200e^{-0.02p} \qquad\qquad \text{Simplifying}$$

At the critical value $p = 50$,

$$E''(50) = -4e^{-0.02(50)}(50 - 50) - 200e^{-0.02(50)} \qquad \text{Substituting } p = 50$$

$$= -200e^{-1} = \frac{-200}{e} \qquad\qquad \text{Simplifying}$$

E'' is negative, so the expenditure $E(p)$ is maximized at $p = 50$.
Consumer expenditure is maximized at price $p = \$50$. ∎

Graphing Calculator Exploration

Use a graphing calculator to verify the answer to the above example by graphing $E(x) = 10,000xe^{-0.02x}$ (using x instead of p) on the window $[0, 200]$ by $[0, 200,000]$ and using MAXIMUM.

How did we choose the graphing window? We first find the critical value "by hand" (setting the derivative equal to zero, finding, in this example, 50) and choose x-values around it. For the

> y-values, we evaluate the function at the critical value (using EVALUATE, finding in this example $y \approx 184{,}000$) and choose y-values including it. Notice how graphing calculators and calculus are most effective when used together.

Graphing Logarithmic and Exponential Functions

To graph logarithmic and exponential functions using a graphing calculator, we first find critical points and possible inflection points and then graph the function on a window including these points. (If graphing "by hand," we make sign diagrams for the first and second derivatives and then sketch the graph, as on pages 615–620.)

EXAMPLE 11   Graphing an Exponential Function

Graph $f(x) = e^{-x^2/2}$.

Solution   As before, we write the function as $f(x) = e^{-\frac{1}{2}x^2}$. The derivative is

$$f(x) = e^{-\frac{1}{2}x^2}(-x) = -xe^{-\frac{1}{2}x^2} \qquad \text{Using } \frac{d}{dx}\,e^f = e^f \cdot f'$$

Derivative of the exponent

$$\text{CV:} \quad x = 0 \qquad\qquad\qquad\qquad \text{Critical value is 0}$$

$$y = 1 \qquad\qquad\qquad\qquad \text{From } y = e^{-\frac{1}{2}x^2} \text{ evaluated at } x = 0$$

The second derivative is

$$f''(x) = (-1)e^{-\frac{1}{2}x^2} - xe^{-\frac{1}{2}x^2}(-x) \qquad \begin{array}{l}\text{From } f'(x) = -x \cdot e^{-\frac{1}{2}x^2}\\ \text{using the product rule}\end{array}$$

$$= -e^{-\frac{1}{2}x^2} + x^2 e^{-\frac{1}{2}x^2} \qquad \text{Simplifying}$$

$$= e^{-\frac{1}{2}x^2}(-1 + x^2) \qquad \text{Factoring}$$

$$= (x^2 - 1)e^{-\frac{1}{2}x^2} \qquad \text{Rearranging}$$

$$= (x + 1)(x - 1)e^{-\frac{1}{2}x^2} \qquad \text{Factoring}$$

$$x = \pm 1 \qquad\qquad \text{Where } f'' = 0$$

$$y = e^{-\frac{1}{2}} \approx 0.6 \qquad \text{From } y = e^{-\frac{1}{2}x^2} \text{ evaluated at } x = \pm 1$$

Based on these values, we choose the graphing window as follows: For the x-values we choose $[-3, 3]$ (to include 0 and ± 1 and beyond), and for the y-values we choose $[-1, 2]$ (to include 1 and 0.6 and above and below). This window gives the graph shown on the following page.

(Many other windows would be just as good, and after seeing the graph you might want to adjust the window.)

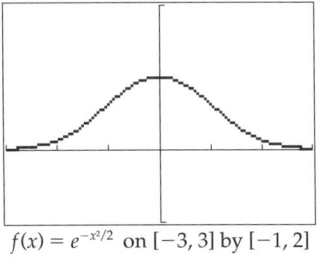

$f(x) = e^{-x^2/2}$ on $[-3, 3]$ by $[-1, 2]$

This function, multiplied by the constant $1/\sqrt{2\pi}$, is the famous "bell-shaped curve" of statistics (see Section 5.4). It is used for statistical predictions of many things from the IQs of newborn babies to stock prices.

Verification of the Power Rule for Arbitrary Powers

On pages 536–537 we proved the power rule $\dfrac{d}{dx} x^n = nx^{n-1}$ for positive integer exponents, and on pages 669–670 we extended it to rational exponents. We can now show that the power rule holds for *all* exponents. We begin by using one of the "inverse" properties of logarithms, $x = e^{\ln x}$, but with x replaced by x^n:

$$f(x) = x^n = e^{\ln x^n} = e^{n \ln x}$$

Using the property that logs bring down exponents

Differentiating:

$$f'(x) = \underbrace{e^{n \ln x}}_{x^n} \underbrace{\left(n \frac{1}{x} \right)}_{\text{Derivative of the exponent}}$$

Differentiating $f(x) = e^{n \ln x}$ by the $e^{f(x)}$ formula (page 701)

$$= \underbrace{x^n \cdot \frac{1}{x}}_{x^{n-1}} \cdot n = nx^{n-1}$$

Replacing $e^{n \ln x}$ by x^n, reordering, and combining x's

This is what we wanted to show—that the derivative of $f(x) = x^n$ is $f'(x) = nx^{n-1}$ for *all* exponents n. (The equation $x^n = e^{n \ln x}$ can be taken as a *definition* of x to a power. A definition of logarithms without recourse to exponents will be given in Exercise 69 on page 775.)

Verification of the Differentiation Formulas

The formula for the derivative of the natural logarithm function comes from applying the definition of the derivative to $f(x) = \ln x$.

$$f'(x) = \lim_{h \to 0} \frac{f(x + h) - f(x)}{h} = \lim_{h \to 0} \frac{\ln (x + h) - \ln x}{h} \qquad \text{Definition of } f'(x)$$

$$= \lim_{h \to 0} \frac{1}{h} [\ln (x + h) - \ln x] \qquad \begin{array}{l}\text{Dividing by } h \text{ is equivalent to} \\ \text{multiplying by } 1/h\end{array}$$

$$= \lim_{h \to 0} \frac{1}{h} \ln \left(\frac{x + h}{x} \right) = \lim_{h \to 0} \frac{1}{h} \ln \left(1 + \frac{h}{x} \right) \qquad \begin{array}{l}\text{Using property 7 of} \\ \text{logarithms (inside back} \\ \text{cover), and simplifying}\end{array}$$

$$= \lim_{h \to 0} \frac{1}{x} \frac{x}{h} \ln \left(1 + \frac{h}{x} \right) = \lim_{h \to 0} \frac{1}{x} \ln \left(1 + \frac{h}{x} \right)^{x/h} \qquad \begin{array}{l}\text{Dividing and multiplying by} \\ x, \text{ and then using property 8} \\ \text{of logarithms}\end{array}$$

$$= \lim_{n \to \infty} \frac{1}{x} \ln \underbrace{\left(1 + \frac{1}{n} \right)^{n}}_{\substack{\text{Approaches } e \\ \text{as } n \to \infty}} \qquad \begin{array}{l}\text{Defining } n = x/h, \text{ so that} \\ h \to 0 \text{ implies } n \to \infty \text{ (for} \\ h > 0)\end{array}$$

$$= \frac{1}{x} \ln e = \frac{1}{x} \qquad \begin{array}{l}\text{Since } \ln e = 1 \text{ (the same} \\ \text{conclusion follows if } h < 0)\end{array}$$

This is the result that we wanted to show—that the derivative of $f(x) = \ln x$ is $f'(x) = 1/x$.

For the rule to differentiate the natural logarithm of a *function*, we begin with

$$\frac{d}{dx} f(g(x)) = f'(g(x))g'(x) \qquad \begin{array}{l}\text{Chain rule (from page 573) for} \\ \text{differentiable functions } f \text{ and } g\end{array}$$

$$\frac{d}{dx} \ln (g(x)) = \frac{1}{g(x)} g'(x) = \frac{g'(x)}{g(x)} \qquad \text{Taking } f(x) = \ln x, \text{ so } f'(x) = 1/x$$

Replacing g by f, this is exactly the formula we wanted to show:

$$\frac{d}{dx} \ln f(x) = \frac{f'(x)}{f(x)}$$

To derive the rule for differentiating e^x we begin with

$$\ln e^x = x \qquad \text{Property 3 of natural logarithms}$$

and differentiate both sides:

$$\frac{\frac{d}{dx} e^x}{e^x} = 1 \qquad \text{Using } \frac{d}{dx} \ln f = \frac{f'}{f} \text{ on the left side}$$

Multiplying each side by e^x gives

$$\frac{d}{dx} e^x = e^x$$

This is the rule for differentiating e^x. This rule together with the chain rule gives the rule for differentiating $e^{f(x)}$, to a function, just as before.

EXERCISES 8.2

Find the derivative of each function.

1. $x^2 \ln x$ **2.** $\dfrac{\ln x}{x^3}$ **3.** $\ln x^2$

4. $\ln (x^3 + 1)$ **5.** $\ln \sqrt{x}$ **6.** $\sqrt{\ln x}$

7. $\ln (x^2 + 1)^3$ **8.** $\ln (x^4 + 1)^2$

9. $\ln (-x)$ **10.** $\ln (5x)$ **11.** $\dfrac{e^x}{x^2}$

12. $x^3 e^x$ **13.** $e^{x^3 + 2x}$ **14.** $2e^{7x}$

15. $e^{x^3/3}$ **16.** $\ln (e^x - 2x)$

17. $x - e^{-x}$ **18.** $x \ln x - x$

19. $\ln e^{2x}$ **20.** $\ln e^x$ **21.** $e^{1 + e^x}$

22. $\ln (e^x + e^{-x})$ **23.** x^e

24. ex **25.** e^3 **26.** $\sqrt{e}$

27. $\ln (x^4 + 1) - 4e^{x/2} - x$

28. $x^2 e^x - 2 \ln x + (x^2 + 1)^3$

29. $x^2 \ln x - \frac{1}{2}x^2 + e^{x^2} + 5$

30. $e^{-2x} - x \ln x + x - 7$

For each function, find the indicated expressions.

31. $f(x) = \dfrac{\ln x}{x^5}$,   find   **a.** $f'(x)$ **b.** $f'(1)$

32. $f(x) = x^4 \ln x$,   find   **a.** $f'(x)$ **b.** $f'(1)$

33. $f(x) = \ln (x^4 + 48)$,   find   **a.** $f'(x)$ **b.** $f'(2)$

34. $f(x) = x^2 \ln x - x^2$,   find   **a.** $f'(x)$ **b.** $f'(e)$

35. $f(x) = \ln (e^x - 3x)$,   find   **a.** $f'(x)$ **b.** $f'(0)$

36. $f(x) = \ln (e^x + e^{-x})$,   find   **a.** $f'(x)$ **b.** $f'(0)$

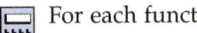 For each function:

a. Find $f'(x)$.

b. Evaluate the given expression and approximate it to three decimal places.

37. $f(x) = 5x \ln x$, find and approximate $f'(2)$.

38. $f(x) = e^{x^2/2}$, find and approximate $f'(2)$.

39. $f(x) = \dfrac{e^x}{x}$, find and approximate $f'(3)$.

40. $f(x) = \ln (e^x - 1)$, find and approximate $f'(3)$.

Find the *second* derivative of each function.

41. $f(x) = e^{-x^5/5}$ **42.** $f(x) = e^{-x^6/6}$

By calculating the first few derivatives, find a formula for the nth derivative of each function (k is a constant).

43. $f(x) = e^{kx}$ **44.** $f(x) = e^{-kx}$

 Use your graphing calculator to graph each function on a window that includes all relative extreme points and inflection points, and give the coordinates of these points (rounded to two decimal places). *Hint:* Use NDERIV once or twice with ZERO.

45. $f(x) = e^{-2x^2}$ **46.** $f(x) = 1 - e^{-x^2/2}$

47. $f(x) = \ln (1 + x^2)$ **48.** $f(x) = e^x + e^{-x}$

 Use your graphing calculator to graph each function on the indicated interval, and give the coordinates of all relative extreme points and inflection points (rounded to two decimal places). *Hint:* Use NDERIV once or twice with ZERO.

49. $f(x) = \dfrac{x^2}{e^x}$ for $-1 \le x \le 8$

50. $f(x) = \dfrac{x}{e^x}$ for $-1 \le x \le 5$

51. $f(x) = x \ln |x|$ for $-2 \le x \le 2$
[*Hint for Exercises 51–52:* $|x|$ is sometimes entered as ABS (x).]

52. $f(x) = x^2 \ln |x|$ for $-2 \le x \le 2$

APPLIED EXERCISES (Most require or)

53. **General: Compound Interest**   A sum of $1000 at 5% interest compounded continuously will grow to $V(t) = 1000e^{0.05t}$ dollars in t years. Find the rate of growth after

   a. 0 years (the time of the original deposit)
   b. 10 years

   (*Hint:* The rate of growth means the derivative.)

54. **General: Depreciation**   A $10,000 automobile depreciates so that its value after t years is $V(t) = 10{,}000e^{-0.35t}$ dollars. Find the rate of change of its value

   a. when it is new ($t = 0$)
   b. after 2 years

   (*Hint:* The rate of change means the derivative.)

55. **General: Population**   The world population (in billions) is predicted to be $P(t) = 5.7e^{0.0175t}$, where t is the number of years after 1995. Find the rate of change of the population in the year 2005. (*Hint:* The rate of change means the derivative.)

56. **Behavioral Science: Ebbinghaus Memory Model**   According to the Ebbinghaus model of memory, if one is shown a list of items, the percentage of items that one will remember t time units later is $P(t) = (100 - a)e^{-bt} + a$, where a and b are constants. For $a = 25$ and $b = 0.2$ this function becomes $P(t) = 75e^{-0.2t} + 25$. Find the rate of change of this percentage

   a. at the beginning of the test ($t = 0$)
   b. after 3 time units

   (*Hint:* The rate of change means the derivative.)

57. **Biomedical: Drug Dosage**   A patient receives an injection of 1.2 cc of a drug, and the amount remaining in the bloodstream t hours later is $A(t) = 1.2e^{-0.05t}$. Find the rate of change of this amount

   a. just after the injection (at time $t = 0$)
   b. after 2 hours

   (*Hint:* The rate of change means the derivative.)

58. **General: Temperature**   A covered cup of coffee at 200 degrees, if left in a 70-degree room, will cool to $T(t) = 70 + 130e^{-2.5t}$ degrees in t hours. Find the rate of change of the temperature

   a. at time $t = 0$ **b.** after 1 hour

59. **Business: Sales**   The weekly sales (in thousands) of a new product after x weeks of advertising are predicted to be $S(x) = 1000 - 900e^{-0.1x}$. Find the rate of change of sales after

   a. 1 week   **b.** 10 weeks

60. **Social Science: Diffusion of Information by Mass Media**   The number of people in a town of 50,000 who have heard an important news bulletin within t hours of its first broadcast is $N(t) = 50{,}000(1 - e^{-0.4t})$. Find the rate of change of the number of informed people

   a. at time $t = 0$ **b.** after 8 hours

61–62: **Economics: Consumer Expenditure**   If consumer demand for a commodity is given by the function below (where p is the selling price in dollars), find the price that maximizes consumer expenditure.

61. $D(p) = 5000e^{-0.01p}$ 62. $D(p) = 8000e^{-0.05p}$

63–64: **Business: Maximizing Revenue**   The function below is a company's price function, where p is the price (in dollars) at which quantity x (in thousands) will be sold.

   a. Find the revenue function $R(x)$. (*Hint:* Revenue is price times quantity, $p \cdot x$.)
   b. Find the quantity and price that will maximize revenue.

63. $p = 400 e^{-0.20x}$ 64. $p = 4 - \ln x$

65. **Biomedical: Reynolds Number**   An important characteristic of blood flow is the "Reynolds number." As the Reynolds number increases, blood flows less smoothly. For blood flowing through certain arteries, the Reynolds number is

$$R(r) = a \ln r - br$$

where a and b are positive constants and r is the radius of the artery. Find the radius r that maximizes the Reynolds number R. (Your answer will involve the constants a and b.)

66. Biomedical: Drug Concentration   If a drug is injected intramuscularly, the concentration of the drug in the bloodstream after t hours will be

$$A(t) = \frac{c}{b - a} (e^{-at} - e^{-bt})$$

If the constants are $a = 0.4$, $b = 0.6$, and $c = 0.1$, find the time of maximum concentration.

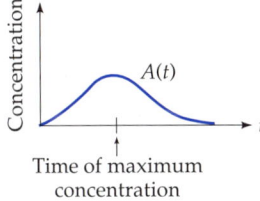

Time of maximum
concentration

67. General: Temperature   A mug of beer chilled to 40 degrees, if left in a 70-degree room, will warm to a temperature of $T(t) = 70 - 30e^{-3.5t}$ degrees in t hours. Enter this temperature function as y_1 (using x for ease of entry), define y_2 as its derivative (using NDERIV), and graph them on the window $[0, 2]$ by $[0, 80]$.
  a. Evaluate y_1 and y_2 at $x = 0.25$ (using EVALUATE) and interpret your answers.
  b. Evaluate y_1 and y_2 at $x = 1$ and interpret your answers.

68. Social Science: Diffusion of Information by Mass Media   The number of people in a city of 200,000 who have heard a weather bulletin within t hours of its first broadcast is $N(t) = 200,000(1 - e^{-0.5t})$. Enter this function as y_1 (using x for ease of entry), define y_2 as its derivative (using NDERIV), and graph them on the window $[0, 4]$ by $[0, 200,000]$.
  a. Evaluate y_1 and y_2 at $x = 0.5$ (using VALUE) and interpret your answers.
  b. Evaluate y_1 and y_2 at $x = 3$ and interpret your answers.

69–70: General: World's Record 100-Meter Run   In 1987 Carl Lewis set a new world's record of 10.14 seconds for the 100-meter run. The distance that he ran in the first x seconds was

$$10.64[x - 0.739(1 - e^{-x/0.739})] \text{ meters}$$

for $0 \le x \le 10.14$.* Enter this function as y_1, define y_2 as its derivative (using NDERIV), and graph them on the window $[0, 10.14]$ by $[0, 100]$.

69. Trace along the velocity curve to verify that Lewis's maximum speed was about 10.64 meters per second. Find how quickly he reached a speed of 10 meters per second, which is 94% of his maximum speed.

70. Define y_3 as the derivative of y_2 (using NDERIV) so that y_3 gives the acceleration after x seconds, and graph y_2 and y_3 on the window $[0, 10.14]$ by $[0, 20]$. Evaluate y_2 and y_3 at $x = 0.1$ and also at $x = 10$ (using EVALUATE). Interpret your answers.

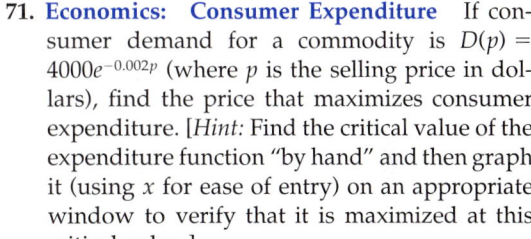
71. Economics: Consumer Expenditure   If consumer demand for a commodity is $D(p) = 4000e^{-0.002p}$ (where p is the selling price in dollars), find the price that maximizes consumer expenditure. [*Hint:* Find the critical value of the expenditure function "by hand" and then graph it (using x for ease of entry) on an appropriate window to verify that it is maximized at this critical value.]

72. Business: Maximizing Revenue   An electronics company finds that the price function for mobile telephones is $p(x) = 400e^{-0.005x}$, where p is the price in dollars at which quantity x (in thousands) will be sold. Find the quantity x that maximizes revenue. (*Hint:* Revenue is price times quantity, $p \cdot x$. Find the critical value "by hand" and then graph the revenue function on an appropriate window to verify that it is maximized at this critical value.)

73. Biomedical: Drug Concentration   If a certain drug is injected intramuscularly, the concentration of the drug in the bloodstream after t hours will be $C(t) = 0.75(e^{-0.2t} - e^{-0.6t})$. Find the time of maximum concentration.

* See W. G. Pritchard, "Mathematical Models of Running," *SIAM Review* **35**(3):359–379, September 1993.

74. Business: Maximizing Profit   A company estimates that if it spends x million dollars on advertising, its profit will be $P(x) = 5x - e^{0.5x} + 2$ million dollars (for $0 \leq x \leq 8$). Find the advertising expenditure x that maximizes profit. Also find the maximum profit.

EXPONENTIAL AND LOGARITHMIC FUNCTIONS TO OTHER BASES

The rules for differentiating exponential functions with (positive) base a are as follows:

Derivative of a^x and $a^{f(x)}$

$$\frac{d}{dx}\, a^x = (\ln a)a^x$$

$$\frac{d}{dx}\, a^{f(x)} = (\ln a)a^{f(x)}f'(x) \qquad \text{For a differentiable function } f$$

For example,

$$\frac{d}{dx}\, 2^x = (\ln 2)2^x$$

$$\frac{d}{dx}\, 5^{3x^2+1} = (\ln 5)5^{3x^2+1}(6x) = 6(\ln 5)x\; 5^{3x^2+1}$$

These formulas are more complicated than the corresponding base e formulas (pages 699 and 701), which is why e is called the "natural" base: It makes the derivative formulas simplest. These formulas reduce to the natural (base e) formulas if $a = e$.

Use the above formulas to find the derivative of each function.

75. a. $f(x) = 10^x$ **b.** $f(x) = 3^{x^2+1}$ **c.** $f(x) = 2^{3x}$ **76. a.** $f(x) = 5^x$ **b.** $f(x) = 2^{x^2-1}$ **c.** $f(x) = 3^{4x}$
    d. $f(x) = 5^{3x^2}$ **e.** $f(x) = 2^{4-x}$ **d.** $f(x) = 9^{5x^2}$ **e.** $f(x) = 10^{1-x}$

The rules for differentiating logarithmic functions with (positive) base a are as follows:

Derivative of $\log_a x$ and $\log_a f(x)$

$$\frac{d}{dx}\, \log_a x = \frac{1}{(\ln a)x}$$

$$\frac{d}{dx}\, \log_a f(x) = \frac{f'(x)}{(\ln a)f(x)} \qquad \text{For a differentiable function } f$$

For example,

$$\frac{d}{dx} \log_5 x = \frac{1}{(\ln 5)x}$$

$$\frac{d}{dx} \log_2 (x^3 + 1) = \frac{3x^2}{(\ln 2)(x^3 + 1)}$$

These formulas are more complicated than the corresponding base e formulas (pages 697 and 698), and again the simplicity of the base e formulas is why e is called the "natural" base. As before, these formulas reduce to the natural (base e) formula if $a = e$.

Use the formulas on the previous page to find the derivative of each function.

77. a. $\log_2 x$ **b.** $\log_{10} (x^2 - 1)$ **c.** $\log_3 (x^4 - 2x)$ **78. a.** $\log_3 x$ **b.** $\log_2 (x^2 + 1)$ **c.** $\log_{10} (x^3 - 4x)$

8.3   Two Applications to Economics: Relative Rates and Elasticity of Demand

APPLICATION PREVIEW

Elasticity, Heroin, and Marijuana

It is common knowledge that if the price of an item rises, demand for it will usually fall. But for a given price increase, will the drop in demand be large or small? To answer this question, economists have developed the concept of *elasticity of demand*. For a particular product at a particular price, the elasticity is a number $E \geq 0$ (calculated from a formula given later in this section) that measures the *responsiveness* of demand to price changes. For elasticities *greater* than 1, demand is said to be *elastic*, meaning that demand responds relatively *strongly* as price changes. For elasticities *less* than 1, demand is said to be *inelastic*, meaning that demand changes relatively *weakly* as price changes. To be specific, an elasticity $E = 2$ (elastic) means that a 1% increase in price would cause a drop of about 2% in demand, whereas an elasticity of $E = \frac{1}{2}$ (inelastic) means that a 1% increase in price would cause only about a $\frac{1}{2}$% decrease in demand.

Consider two particular cases. For heroin sold in Detroit, elasticity has been estimated to be 0.267 (inelastic), meaning that demand for heroin will respond only slightly to price changes.* Therefore, a price increase will cause only a slight decrease in demand, and so the drug dealers' revenue, which is price times quantity, will actually *increase*. The increased revenues to drug dealers might also bring about an increase in crime to support the increased amount that addicts will have to pay. On the other hand, for marijuana sold on a college campus, elasticity has been estimated to be 1.013 (elastic), meaning that a small price increase will result in a larger decrease in demand, so that revenue to the drug dealers would fall.†

These different elasticities lead to different conclusions: In Detroit, police might decide to concentrate their efforts on heroin *users* rather than on dealers, since concentrating on dealers would cause a shortage, and the resulting price increases would paradoxically *increase* revenue to drug dealers and possibly increase crime. On college campuses, authorities might decide to concentrate on drug *suppliers*, since the resulting shortage would cause prices to rise, which, as we saw above, would *decrease* revenue to the dealers.

Elasticity of demand clearly has economic and social implications, and will be discussed further in this section.

Introduction

In this section we define *relative* rates of change and see how they are used in economics. (*Relative* rates are not the same as the *related* rates discussed in Section 7.6.) We then define the very important economic concept of *elasticity of demand*.

Relative Versus Absolute Rates

The derivative of a function gives its rate of change. For example, if $f(t)$ is the cost of a pair of shoes at time t years, then $f'(t)$ is the rate of change of cost (in dollars per year). That is, $f' = 3$ would mean that the price of shoes is increasing at the rate of $3 per year. Similarly, if $g(t)$ is the price of a new automobile at time t years, then $g' = 300$

---

* Lester P. Silverman and Nancy L. Spruill, "Urban Crime and the Price of Heroin," *Journal of Urban Economics* 4:80–103.
† Charles T. Nisbit and Firouz Vakil, "Some Estimates of Price and Expenditure Elasticities of Demand for Marijuana Among U.C.L.A. Students," *Review of Economics and Statistics* **54**:473–475.

would mean that automobile prices are increasing at the rate of $300 per year.

Does this mean that car prices are rising 100 times as fast as shoe prices? In absolute terms, yes. However, this does not take into account the enormous price difference between automobiles and shoes.

Relative Rates of Change

If shoe prices are increasing at the rate of $3 per year, and if the current price of a pair of shoes is $60, the *relative* rate of increase is $\frac{3}{60} = \frac{1}{20} = 0.05$, which means that shoe prices are increasing at the *relative* rate of 5% per year. Similarly, if the price of an average automobile is $15,000, then an increase of $300 relative to this price is $\frac{300}{15,000} = \frac{1}{50} = 0.02$, for a relative rate of 2% per year. Therefore, in a *relative* sense (that is, as a fraction of the current price), car prices are increasing *less* rapidly than shoe prices.

In general, if $f(t)$ is the price of an item at time t, then the rate of change is $f'(t)$, and the *relative* rate of change is $f'(t)/f(t)$, the derivative divided by the function. We will sometimes call the derivative $f'(x)$ the "absolute" rate of change to distinguish it from the relative rate of change $f'(x)/f(x)$.

Relative rates are often more meaningful than absolute rates. For example, it is easier to grasp the fact that the gross domestic product is growing at the relative rate of 4% per year than that it is growing at the absolute rate of $300,000,000,000 per year.

The expression $f'(x)/f(x)$ is the derivative of the natural logarithm of $f(x)$:

$$\frac{d}{dx} \ln f(x) = \frac{f'(x)}{f(x)}$$

This provides an alternative expression for the relative rate of change, in terms of logarithms.

Relative Rate of Change

$$\left(\begin{array}{c} \text{Relative rate of} \\ \text{change of } f(t) \end{array} \right) = \frac{d}{dt} \ln f(t) = \frac{f'(t)}{f(t)} \qquad \text{For a differentiable function } f > 0$$

We use the variable t since it often stands for time. The middle expression is sometimes called the *logarithmic derivative*, since it is found by first taking the logarithm and then differentiating.

The relative rate of change, being a ratio or a percent, does not depend on the units of the function (whether dollars or rubles, pounds or kilos). Therefore, relative rates can be compared between different products, and even between different nations. This is in contrast to absolute rates of change (that is, derivatives), which *do* depend upon the units (for example, dollars per year).

EXAMPLE 1   Finding a Relative Rate of Change

If the gross domestic product t years from now is predicted to be $G(t) = 8.2e^{\sqrt{t}}$ trillion dollars, find the relative rate of change 25 years from now.

We give two solutions, showing the use of both formulas.

Solution   [Using the $\dfrac{d}{dt} \ln f(t)$ formula]

$$\ln G(t) = \ln 8.2e^{\sqrt{t}} \qquad \text{Taking natural logs}$$

$$= \ln 8.2 + \ln e^{\sqrt{t}} \qquad \begin{array}{l}\text{Log of a product is}\\\text{the sum of the logs}\end{array}$$

$$= \ln 8.2 + \sqrt{t} \qquad \begin{array}{l}\ln e^{\sqrt{t}} = \sqrt{t} \text{ by property 3}\\\text{of logs (see the inside back cover)}\end{array}$$

$$= \ln 8.2 + t^{1/2} \qquad \text{In exponent form}$$

Then we differentiate:

$$\frac{d}{dt}(\ln 8.2 + t^{1/2}) = 0 + \frac{1}{2}t^{-1/2} = \frac{1}{2}t^{-1/2} \qquad \begin{array}{l}\ln 8.2 \text{ is a constant, so its}\\\text{derivative is zero}\end{array}$$

Finally, we evaluate at the given time $t = 25$:

$$\frac{1}{2}(25)^{-1/2} = \frac{1}{2}\frac{1}{\sqrt{25}} = \frac{1}{2}\frac{1}{5} = \frac{1}{10} = 0.10 \qquad \tfrac{1}{2}t^{-1/2} \text{ evaluated at } t = 25$$

Therefore, in 25 years the gross national product will be increasing at the relative rate of 0.10, or 10% per year. ∎

Alternative Solution   [using the $\dfrac{f'(t)}{f(t)}$ formula]

$$G(t) = 8.2e^{\sqrt{t}} = 8.2e^{t^{1/2}} \qquad \text{Writing } G(t) \text{ with fractional exponents}$$

$$G'(t) = 8.2e^{t^{1/2}}\left(\frac{1}{2}t^{-1/2}\right) \qquad \text{Differentiating}$$

Derivative of the exponent

Therefore, the relative rate of change $\dfrac{G'(t)}{G(t)}$ is

$$\frac{G'(t)}{G(t)} = \frac{8.2e^{t^{1/2}}\left(\dfrac{1}{2}t^{-1/2}\right)}{8.2e^{t^{1/2}}} = \frac{1}{2}t^{-1/2} \qquad \text{Same result as with the first formula}$$

At $t = 25$,

$$\frac{1}{2}(25)^{-1/2} = \frac{1}{2}\frac{1}{\sqrt{25}} = \frac{1}{2}\frac{1}{5} = \frac{1}{10} = 0.10 \qquad \text{Again the same}$$

Therefore, the relative growth rate is 10%, just as we found before.

Both formulas give the same answer, so you should use the one that is easier to apply in any particular problem. The $\dfrac{d}{dx}\ln f(t)$ formula sometimes allows simplification *before* the differentiation, whereas the $\dfrac{f'(t)}{f(t)}$ formula often involves simplification afterward.

PRACTICE PROBLEM 1

An investor estimates that if a piece of land is held for t years, it will be worth $f(t) = 300 + t^2$ thousand dollars. Find the relative rate of change at time $t = 10$ years. [*Hint:* Use the $\dfrac{f'(t)}{f(t)}$ formula.]

Solution at the back of the book

Graphing Calculator Exploration

Enter the function from Practice Problem 1 into a graphing calculator as $y_1 = 300 + x^2$, and then turn off the function so that it will not graph. Then graph $y_2 = \dfrac{d}{dx}\ln y_1$

and $y_3 = \dfrac{y_1{}'}{y_1}$ (using NDERIV) to-

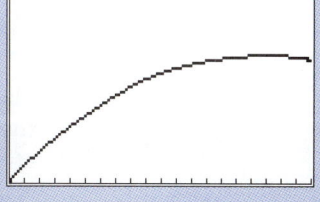

gether on the graphing window [0, 20] by [0, 0.1]. Why do you get only one curve for the two functions?

Elasticity of Demand

Farmers are aware of the paradox that an abundant harvest usually brings *lower* total revenue than a poor harvest. The reason is simply that the larger quantities in an abundant harvest result in lower prices, which in turn cause increased demand, but the demand does *not* increase enough to compensate for the lower prices.

Revenue is price times quantity, and when one of these quantities increases, the other generally decreases. The question is whether the increase in one is enough to compensate for the decrease in the other. The concept of *elasticity of demand* was invented to answer this question.

Intuitively, elasticity of demand is a measure of how *responsive* demand is to price changes. Think of "elastic" as meaning "very responsive." If demand is elastic, a small price cut will bring a large increase in demand, so total revenue will rise. On the other hand, if demand is *in*elastic, a price cut will bring only a slight increase in demand, so total revenue will fall.

In general, *elastic* demand means that consumers will purchase *significantly* more or less in response to price changes. *Inelastic* demand means that consumers will buy only *slightly* more or less in response to price changes. (This is the cause of the farmers' difficulties: Demand for farm products is inelastic.)

Demand Functions

In general, if the price of an item rises, the demand will fall, and vice versa. If the relationship between the price p of an item and the quantity x that will be sold at that price can be expressed with x as a function of p, $x = D(p)$, the function is called the *demand function*.

Demand Function

The demand function

$$x = D(p)$$

gives the quantity x of an item that will be demanded by consumers at price p.

Since, in general, demand falls as prices rise, the slope of the demand function will be negative, as shown here. This is known as the *law of downward-sloping demand*.

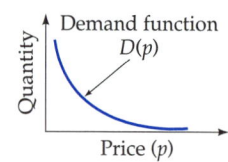

Law of downward-sloping demand

Calculating Elasticity of Demand

For a demand function $D(p)$, let us calculate the relative rate of change of demand divided by the relative rate of change of price. Using the derivative-of-the-logarithm formula,

$$\frac{\left(\begin{array}{c}\text{Relative rate of}\\ \text{change of demand}\end{array}\right)}{\left(\begin{array}{c}\text{Relative rate of}\\ \text{change of price}\end{array}\right)} = \frac{\dfrac{d}{dp}\ln D(p)}{\dfrac{d}{dp}\ln p} = \frac{\dfrac{D'(p)}{D(p)}}{\dfrac{1}{p}} = \underbrace{\frac{pD'(p)}{D(p)}}_{\text{Simplified}}$$

Because most demand functions are downward-sloping, the derivative $D'(p)$ is generally negative. Economists prefer to work with positive numbers, so the *elasticity of demand* is taken to be the negative of this quantity (in order to make it positive).*

Elasticity of Demand

> For a demand function $D(p)$ the elasticity of demand is
>
> $$E(p) = \frac{-p \cdot D'(p)}{D(p)}$$
>
> Demand is *elastic* if $E(p) > 1$ and *inelastic* if $E(p) < 1$.

Elasticity, being composed of *relative* rates of change, does not depend on the units of the demand function. Therefore, elasticities can be compared between different products, and even between different countries.

EXAMPLE 2   Finding Elasticity of Demand for Commuter Bus Service

A bus line estimates the demand function for its daily commuter tickets to be $D(p) =$ $81 - p^2$ (in thousands of tickets), where p is the price in dollars $(0 < p < 9)$. Find the elasticity of demand when the price is

a. \$3   **b.** \$6

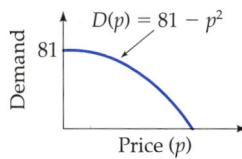

*Some economists omit the negative sign.

Solution

$$E(p) = \frac{-pD'(p)}{D(p)}$$ Definition of elasticity

$$= \frac{-p(-2p)}{81 - p^2}$$ Substituting $D(p) = 81 - p^2$
so $D'(p) = -2p$

$$= \frac{2p^2}{81 - p^2}$$ Simplifying

a. Evaluating at $p = 3$ gives

$$E(3) = \frac{2(3)^2}{81 - (3)^2} = \frac{18}{81 - 9} = \frac{18}{72} = \frac{1}{4}$$ $E(p) = \frac{2p^2}{81 - p^2}$ with $p = 3$

Interpretation: The elasticity is less than 1, so demand for tickets is *inelastic* at a price of $3. This means that small price changes (up or down from this level) will cause only *slight* changes in demand. More precisely, elasticity of $\frac{1}{4}$ means that a 1% price change will cause only about a $\frac{1}{4}$% change in demand.

b. At the price of $6, the elasticity of demand is

$$E(6) = \frac{2(6)^2}{81 - (6)^2} = \frac{72}{81 - 36} = \frac{8}{5} = 1.6$$ $E(p) = \frac{2p^2}{81 - p^2}$ with $p = 6$

Interpretation: The elasticity is greater than 1, so demand is *elastic* at a price of $6. This means that small changes in price (up or down from this level) will cause a relatively large change in demand. In particular, an elasticity of 1.6 means that a price change of 1% will cause about a 1.6% change in demand.

■

The changes in demand are, of course, in the opposite direction from the changes in price. That is, if prices are *raised* by 1% (from the $6 level), demand will *fall* by 1.6%, whereas if prices are *lowered* by 1%, demand will *rise* by 1.6%. In the future we will assume that the *direction* of the change is clear, and say simply that a 1% change in price will cause about a 1.6% change in demand.

PRACTICE PROBLEM 2   For the demand function $D(p) = 90 - p$, find the elasticity of demand $E(p)$ and evaluate it at $p = 30$ and $p = 75$. (Be sure to complete this Practice Problem, as the results will be used shortly.)

Solution at the back of the book

Graphing Calculator Exploration

For any demand function y_1 (written in terms of x), define y_2 as the elasticity function for y_1 by defining $y_2 = -x \cdot y_1'/y_1$ (using NDERIV). Try entering the demand function from Example 2 or Practice Problem 2 as y_1 and then evaluating y_2 at appropriate numbers to check the answers found there.

Using Elasticity to Increase Revenue

In Example 2 we found that at a price of \$3, demand is inelastic ($E = \frac{1}{4} < 1$), and so demand responds only *weakly* to price changes. Therefore, to increase revenue the company should *raise* prices, since the higher prices will drive away only a relatively small number of customers. On the other hand, at a price of \$6, demand is elastic ($E = 1.6 > 1$), and so demand is very responsive to price changes. In this case, to increase revenue the company should *lower* prices, since this will attract more than enough new customers to compensate for the price decrease. In general,

Using Elasticity to Increase Revenue

To increase revenue:

Raise prices if demand is *inelastic* ($E < 1$).

Lower prices if demand is *elastic* ($E > 1$).

This statement shows why elasticity of demand is important to any company that cuts prices in an attempt to boost revenues, or to any utility that raises prices in order to increase revenues. Elasticity shows whether the strategy will succeed or fail.

The borderline case, $E = 1$ (called *unitary elastic*), is where revenue cannot be raised, which will be the case if revenue is at its maximum. Therefore, elasticity must be unitary when revenue is maximized.

At maximum revenue, elasticity of demand must equal 1.

We could use this fact as a basis for a new method for maximizing revenue, but instead we will stick with our earlier (and easier) method of maximizing functions by finding critical values.

Verification of the Relationship Between Elasticity and Revenue

We may verify the relationship between elasticity of demand and revenue as follows: Revenue is price p times quantity x:

$$R = px = p \cdot D(p) \qquad \text{Using } x = D(p)$$

Differentiating will show how revenue responds to price changes.

$$R'(p) = D(p) + pD'(p) \qquad \begin{array}{l} \text{Using the product rule on} \\ p \cdot D(p) \end{array}$$

$$= D(p)\left[1 + \frac{pD'(p)}{D(p)}\right] \qquad \text{Factoring out } D(p)$$

$$= D(p)\left[1 - \frac{-pD'(p)}{D(p)}\right] \qquad \begin{array}{l} \text{Replacing the plus sign by two} \\ \text{minus signs} \end{array}$$

This is the definition of elasticity $E(p)$

$$= D(p)[1 - E(p)] \qquad \text{Replacing } \frac{-pD'(p)}{D(p)} \text{ by } E(p)$$

If demand is *elastic*, $E > 1$, then the quantity in brackets is negative, and so the derivative $R'(p)$ is negative, showing that revenue *decreases* as price increases. Therefore, to increase revenue, one should *lower* prices. On the other hand, if the demand is *inelastic*, $E < 1$, then the quantity in brackets is positive, and so the derivative $R'(p)$ is positive, showing that revenue *increases* as price increases. In this case, to increase revenue, one should *raise* prices. This proves the statements on the previous page under "Using Elasticity to Increase Revenue."

Elasticity Is Not the Same as Slope

Do not confuse elasticity of demand with the slope of the demand curve. The two ideas are quite different. For example, in Practice Problem 2 on page 720 you found that the linear demand function $D(p) = 90 - p$ (which has slope -1 all along it) has elasticity $\frac{1}{2}$ at one point and elasticity 5 at another.

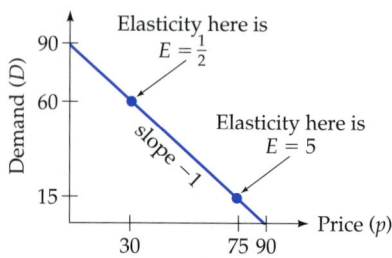

Demand function $D(p) = 90-p$,
showing the elasticities found in Practice Problem 2

To understand why the elasticity changes, notice that the upper point represents a high demand (vertical axis) and a low price (horizontal axis), compared to the lower point, which represents a low demand and a high price. The demand function having slope -1 means that each \$1 price increase lowers demand by 1 unit, but losing one low-priced sale out of many is less important than losing one high-priced sale out of a few. That is, to see how revenue changes, we must look not at the absolute rates of change of quantity and price, but at their *relative* rates of change, which is what elasticity is all about.

EXERCISES 8.3

For each function:

a. Find the relative rate of change.
b. Evaluate the relative rate of change at the given values of t.

1. $f(t) = t^2$, $t = 1$ and $t = 10$

2. $f(t) = t^3$, $t = 1$ and $t = 10$

3. $f(t) = 100e^{0.2t}$, $t = 5$ **4.** $f(t) = 100e^{-0.5t}$, $t = 4$

5. $f(t) = e^{t^2}$, $t = 10$ **6.** $f(t) = e^{t^3}$, $t = 5$

7. $f(t) = e^{-t^2}$, $t = 10$ **8.** $f(t) = e^{-t^3}$, $t = 5$

9. $f(t) = 25\sqrt{t-1}$, $t = 6$

10. $f(t) = 100\sqrt[3]{t+2}$, $t = 8$

APPLIED EXERCISES ON RELATIVE RATES

11–12: Economics: National Debt   If the national debt of a country (in trillions of dollars) t years from now is given by the formula below, find the relative rate of change of the debt 10 years from now.

11. $N(t) = 0.5 + 1.1e^{0.01t}$

12. $N(t) = 0.4 + 1.2e^{0.01t}$

13–14: General: Population   The population (in millions) of a city t years from now is given by the formula below.

a. Find the relative rate of change of the population 8 years from now.
b. Will the relative rate of change ever reach 1.5%?

13. $P(t) = 4 + 1.3e^{0.04t}$ **14.** $P(t) = 6 + 1.7e^{0.05t}$

EXERCISES ON ELASTICITY OF DEMAND

For each demand function $D(p)$:

a. Find the elasticity of demand $E(p)$.
b. Determine whether the demand is elastic, inelastic, or unitary elastic at the given price p.

15. $D(p) = 200 - 5p$, $p = 10$

16. $D(p) = 60 - 8p$, $p = 5$

17. $D(p) = 300 - p^2$, $p = 10$

18. $D(p) = 100 - p^2$, $p = 5$

19. $D(p) = \dfrac{300}{p}$, $p = 4$

20. $D(p) = \dfrac{500}{p}$, $p = 2$

21. $D(p) = \sqrt{175 - 3p}$, $p = 50$

22. $D(p) = \sqrt{100 - 2p}$, $p = 20$

23. $D(p) = \dfrac{100}{p^2}$, $p = 40$

24. $D(p) = \dfrac{600}{p^3}$, $p = 25$

25. $D(p) = 4000e^{-0.01p}$, $p = 200$

26. $D(p) = 6000e^{-0.05p}$, $p = 100$

APPLIED EXERCISES ON ELASTICITY OF DEMAND

27. **Automobile Sales** An automobile dealer is selling cars at a price of $12,000 each. The demand function is $D(p) = 2(15 - 0.001p)^2$ where p is the price of a car. Should the dealer raise or lower the price to increase revenue?

28. **Liquor Sales** A liquor distributor wants to increase its revenues by discounting its best-selling liquor. If the demand function for this liquor is $D(p) = 60 - 3p$, where p is the price per bottle, and if the current price is $15, will the discounts succeed?

29. **City Bus Revenues** A city bus line estimates its demand function to be $D(p) = 150,000\sqrt{1.75 - p}$, where p is the fare in dollars. The bus line currently charges a fare of $1.25, and it plans to raise the fare to increase its revenues. Will it succeed?

30. **Newspaper Sales** The demand function for a newspaper is $D(p) = 80,000\sqrt{75 - p}$, where p is the price in cents. The publisher currently charges 50 cents, and it plans to raise the price to increase revenues. Will the publisher succeed?

31. **Electricity Rates** An electrical utility asks the Federal Regulatory Commission for permission to raise rates to increase revenues. The utility's demand function is

$$D(p) = \frac{120}{10 + p}$$

where p is the price (in cents) of a kilowatt-hour of electricity. If the utility currently charges 6 cents per kilowatt-hour, should the commission grant the request?

32. **Oil Prices** A Middle Eastern oil-producing country estimates that the demand for oil (in millions of barrels) is $D(p) = 28e^{-0.04p}$, where p is the price of a barrel of oil. To raise its revenues, should it raise or lower its price from its current level of $20 per barrel?

33. **Oil Prices** A European oil-producing country estimates that the demand for its oil (in millions of barrels) is $D(p) = 41e^{-0.06p}$ where p is the price of a barrel of oil. To raise its revenues, should it raise or lower its price from its current level of $20 per barrel?

34–35: **Liquor and Beer** The demand functions for distilled spirits and for beer are given below, where p is the retail price and $D(p)$ is the demand in gallons per capita.* For each demand function, find the elasticity of demand for *any* price p. (*Note:* You will find, in each case, that demand is inelastic. This means that taxation, which acts like a price increase, is an ineffective way of discouraging liquor consumption, but is an effective way of raising revenue.)

34. $D(p) = 3.509p^{-0.859}$ (for distilled spirits)

35. $D(p) = 7.881p^{-0.112}$ (for beer)

36. **Constant Elasticity**

    a. Show that for a demand function of the form $D(p) = \dfrac{c}{p^n}$, where c and n are positive constants, the elasticity is constant.

    b. What type of demand function has elasticity equal to 1 for every value of p?

37. **Linear Elasticity** Show that for a demand function of the form $D(p) = ae^{-cp}$ where a and c are positive constants, the elasticity of demand is $E(p) = cp$.

38–39: **Elasticity of Supply** A supply function $S(p)$ gives the total amount of a product that producers are willing to supply at a given price p. The elasticity of supply is defined as

$$E_s(p) = \frac{p \cdot S'(p)}{S(p)}$$

Elasticity of supply measures the relative increase in supply resulting from a small relative increase in price. It is less useful than elasticity of demand, however, since it is not related to total revenue.

38. Use the formula above to find the elasticity of supply for a supply function of the form $S(p) = ae^{cp}$, where a and c are positive constants.

*Stanley Ornstein and Dominique Hanssens, "Alcohol Control Laws and the Consumption of Distilled Spirits and Beer," *Journal of Consumer Research* **12**: 200–213, September 1985. Variables in this study other than price have been ignored.

39. Use the formula above Exercise 38 to find the elasticity of supply for a supply function of the form $S(p) = ap^n$, where a and n are positive constants.

 40–41: Automobile Sales The demand function for automobiles in a dealership is given below, where p is the selling price.

a. Use the method described in the Graphing Calculator Exploration on page 721 to

find the elasticity of demand at a price of $12,000.

b. Should the dealer raise or lower the price from this level to increase revenue?

c. Find the price at which elasticity equals 1. (*Hint:* Use INTERSECT.)

40. $D(p) = 3\sqrt{20 - 0.001p}$

41. $D(p) = \dfrac{200}{8 + e^{0.0001p}}$

Chapter Summary with Hints and Suggestions

The reading and exercises of this chapter have helped you to learn the following skills. For each skill, the section from which it came (in case you need to review it) and some exercises in this section that use it are indicated. Answers to all exercises are at the end of the book, and full solutions to all exercises are in the Student Solutions Manual.

8.1 Review of Exponential and Logarithmic Functions

- Find the value of money invested at compound interest. (*Review Exercise 1.*)

$$P(1 + r)^t \qquad P\left(1 + \frac{r}{m}\right)^{mt} \qquad Pe^{rt}$$

- Determine which of two banks gives a better return. (*Review Exercise 2.*)

- Depreciate an asset. (*Review Exercise 3.*)

$$P(1 + r)^t \qquad \text{(for depreciation, } r \text{ is negative)}$$

- Determine which of two drugs provides more medication. (*Review Exercise 4.*)

- Use to predict the world's largest city. (*Review Exercise 5.*)

- Predict how productivity improves with practice. (*Review Exercise 6.*)

- Find how soon an investment will reach a given level. (*Review Exercises 7–8.*)

- Use the properties of natural logarithms to simplify a function. (*Review Exercise 9.*)

  (See the logarithm properties on page 688.)

- Find and check a natural logarithm. (*Review Exercise 10.*)

- Predict the spread of information, or oil demand. (*Review Exercises 11–12.*)

- Use and analyze the accuracy of the "rule of 72" for compound interest. (*Review Exercises 13–14.*)

- Find the k-multiple time for an investment. (*Review Exercises 15–16.*)

- Use to find a doubling time, or the effectiveness of advertising. (*Review Exercises 17–18.*)

8.2 Differentiation of Exponential and Logarithmic Functions

- Find the derivative of a logarithmic or exponential function. (*Review Exercises 19–34.*)

$$\frac{d}{dx}\ln x = \frac{1}{x} \qquad \frac{d}{dx}e^x = e^x \qquad \frac{d}{dx}\ln f = \frac{f'}{f}$$

$$\frac{d}{dx}e^f = e^f \cdot f' \qquad \frac{d}{dx}e^{kx} = ke^{kx}$$

- Graph an exponential or a logarithmic function. *(Review Exercises 35–36.)*

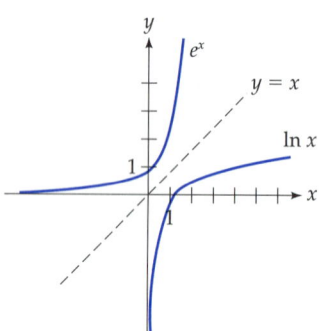

- Find the rate of change of sales, or amount of medication, learning, temperature, or information. *(Review Exercises 37–41.)*

- Find when a company maximizes its present value. *(Review Exercise 42.)*

- Maximize a company's revenue. *(Review Exercises 43–44.)*

- Maximize consumer expenditure for a product. *(Review Exercise 45.)*

- Use 📈 to graph an exponential or logarithmic function. *(Review Exercises 46–47.)*

- Use 📈 to maximize consumer expenditure or to maximize revenue. *(Review Exercises 48–49.)*

8.3   Two Applications to Economics: Relative Rates and Elasticity of Demand

- Find the relative rate of change of a country's GDP. *(Review Exercises 50–51.)*

$$\left(\begin{array}{c}\text{Relative}\\\text{rate}\end{array}\right) = \frac{d}{dt}\ln f = \frac{f'}{f}$$

- Find the elasticity of demand for a product, and its consequences. *(Review Exercises 52–54.)*

$$\left(\begin{array}{c}\text{Elasticity}\\\text{of demand}\end{array}\right) = \frac{-p \cdot D'(p)}{D(p)}$$

- Use 📈 to find the relative rate of change of population.   *(Review Exercise 55.)*

- Use 📈 to find the elasticity of demand for a product, how it affects revenue, and what price gives unitary elasticity. *(Review Exercise 56.)*

Hints and Suggestions

- (Overview) Exponential and logarithmic functions should be thought of as just other types of functions, like polynomials, but having their own differentiation rules. In fact, in a sense they are more "natural" than polynomials because they give natural growth rates.

- (Graphing Calculators) A graphing calculator helps by drawing graphs of exponential and logarithmic functions, and finding intersection points (for example, where one population exceeded another). It is also useful for checking derivatives and graphically verifying maximum and minimum values.

- Interest rates are always *annual* rates unless stated otherwise.

- When do you use Pe^{rt} and when $P(1 + r/m)^{mt}$? Use Pe^{rt} if the word "continuous" occurs, and $P(1 + r/m)^{mt}$ if it does not. $P(1 + r)^t$ with *negative r* gives depreciation.

- The power rule $\dfrac{d}{dx}x^n = nx^{n-1}$ is for differentiating a *variable to a constant power*, and $\dfrac{d}{dx}e^x = e^x$ is for differentiating the *constant e to a variable power*.

- Practice for test: Review Exercises 1, 3, 5, 7, 8, 11, 17, 19, 23, 25, 27, 31, 35, 37, 41, 45, 49, 51, 53, 55.

Review Exercises for Chapter 8   *Practice test exercises are in blue.*

8.1   Review of Exponential and Logarithmic Functions (Most require .)

1. General: Interest   Find the value of $10,000 invested for 8 years at 8% interest if the compounding is done

a. quarterly   b. continuously

2. General: Interest   You have $4000 to deposit, and are considering two banks. The East Side Credit and Trust offers 6% compounded quarterly, and the West Side Savings and Loan offers 5.9% compounded continuously. Which bank does better for you if you leave your money in for 6 years, and by how much?

3. Business: Depreciation   An $800,000 computer loses 20% of its value each year.

a. Give a formula for its value after t years.
b. Find its value after 4 years.

4. Biomedical: Drug Concentration   If the concentration of a drug in a patient's bloodstream is c (milligrams per milliliter), then t hours later the concentration will be $C(t) = ce^{-kt}$, where k is a constant (the "elimination constant"). Two drugs are being compared: drug A (with initial concentration $c = 2$ and elimination constant $k = 0.2$) and drug B (with initial concentration $c = 3$ and elimination constant $k = 0.25$). Which drug will have the greater concentration 4 hours later?

 5. General: Population   The largest city in the world is Tokyo, with São Paulo (Brazil) smaller but growing faster. According to the Census Bureau, x years after 1995 the population of Tokyo will be $27e^{0.0034x}$ and the population of São Paulo will be $16.5e^{0.011x}$ (both in millions). Graph both functions on a calculator on the window $[0, 100]$ by $[0, 50]$. When will São Paulo overtake Tokyo as the world's largest city (assuming that these growth rates continue to hold)?

6. Business: Productivity   A factory worker assembles portable tape recorders, and after t days of practice his productivity is $30(1 - e^{-0.12t})$ tape recorders per hour. Find his productivity after:

a. 5 days of practice   b. 15 days of practice
 c. Check your answers to parts (a) and (b) by graphing the productivity function on a graphing calculator on an appropriate window and then using TRACE or EVALUATE.

7. General: Interest   You invest $4000 in real estate that you expect to increase in value by 12% annually, and you want to sell when the value reaches $12,000. How long will you have to wait?

8. General: Interest   You make a $5000 investment that you expect to increase in value by 12% compounded continuously. How long will you have to wait for your investment to grow to $15,000?

9. General: Natural Logarithms   Simplify:

$$f(x) = \ln\left(\frac{1}{x}\right) + \ln(x^2) + \ln e$$

10. General: Natural Logarithms

a. Use a calculator to find $\ln 9.3$.
b. Raise e to the resulting power (include all decimal places shown) to verify that you return (approximately) to 9.3.

11. Social Science: Diffusion of Information by Mass Media   In a city of a million people, news of election results broadcast over radio and television will reach $N(t) = 1,000,000(1 - e^{-0.3t})$ people within t hours. Find when the news will have reached 500,000 people.

12. Economics: Oil Demand   The demand for oil in the United States is increasing by 3% per year. Assuming that this rate continues, how soon will demand increase by 50%?

13–14: Business: Rule of 72   If a sum is invested at interest rate r compounded continuously, the doubling time (the time in which it will double in value) is found by solving the equation $Pe^{rt} = 2P$. The solution (by the usual method of canceling the P and taking logs) is $t = \frac{\ln 2}{r} \approx \frac{0.69}{r}$. For *annual* compounding, the doubling time should be somewhat longer, and may be estimated by replacing 69 by 72.

Rule of 72

> For $r\%$ interest compounded annually, the doubling time is approximately $\frac{72}{r}$ years.

For example, to estimate the doubling time for an investment at 8% compounded annually we would divide 72 by 8, giving $\frac{72}{8} = 9$ years. The 72, however, is only a rough "upward adjustment" of 69, and the rule is most accurate for interest rates around 9%. For each interest rate:

a. Use the rule of 72 to estimate the doubling time for annual compounding.

b. Use the compound interest formula $P(1 + r)^t$ to find the actual doubling time for annual compounding.

13. 6%

14. 1% (This shows that for interest rates very different from 9% the rule of 72 is less accurate.)

15. General: Interest   Find a formula for the time required for an investment to grow to k times its original size if it grows at interest rate r compounded annually.

16. General: Interest   Find a formula for the time required for an investment to grow to k times its original size if it grows at interest rate r compounded continuously.

 17. General: Interest   If a bank offers 6.5% interest, in how many years will a deposit of $1000 increase by 50% if the compounding is done

a. quarterly?   **b.** continuously?

 18. Business: Advertising   After the opening of a new store has been advertised for t days, the proportion of people in a city who have seen

the ad is $p(t) = 1 - e^{-0.032t}$. How long must the ad run to reach

a. 30% of the people?   **b.** 40% of the people?

8.2   Differentiation of Exponential and Logarithmic Functions

Find the derivative of each function.

19. $\ln 2x$

20. $\ln (x^2 - 1)^2$

21. $\ln (1 - x)$

22. $\ln \sqrt{x^2 + 1}$

23. $\ln \sqrt[3]{x}$

24. $\ln e^x$

25. $\ln x^2$

26. $x \ln x - x$

27. e^{-x^2}

28. e^{1-x}

29. $\ln e^{x^2}$

30. $e^{x^2 \ln x - x^2/2}$

31. $5x^2 + 2x \ln x + 1$

32. $2x^3 + 3x \ln x - 1$

33. $2x^3 - 3xe^{2x}$

34. $4x - 2x^2e^{2x}$

Graph each function, showing all relative extreme points and inflection points.

35. $f(x) = \ln (x^2 + 4)$ **36.** $f(x) = 16e^{-x^2/8}$

37. Business: Sales   The weekly sales (in thousands) of a new product after x weeks of advertising is $S(x) = 2000 - 1500e^{-0.1x}$. Find the rate of change of sales after

a. 1 week   **b.** 10 weeks

38. Biomedical: Drug Dosage   A patient receives an injection of 1.5 cc of a drug, and the amount remaining in the bloodstream t hours later is $A(t) = 1.5e^{-0.08t}$. Find the instantaneous rate of change of this amount:

a. immediately after the injection (time $t = 0$)

b. after 5 hours

39. Behavioral Science: Learning   In a test of short-term memory, the percent of subjects who remember an eight-digit number for at least t seconds is $P(t) = 100 - 200 \ln (t + 1)$. Find the rate of change of this percent after 5 seconds.

40. General: Temperature   A thermos bottle that is chilled to 35 degrees and then left in a 70-degree room will warm to a temperature of $T(t) = 70 - 35e^{-0.1t}$ degrees after t hours. Find the rate of change of the temperature

a. at time $t = 0$ **b.** after 5 hours

41. Social Science: Diffusion of Information by Mass Media The number of people in a town of 30,000 who have heard an important news bulletin within t hours of its first broadcast is $N(t) = 30,000(1 - e^{-0.3t})$. Find the instantaneous rate of change of the number of informed people after

a. 1 hour   **b.** 8 hours

42. Business: Maximizing Present Value A new company is growing so that its value t years from now will be $50t^2$ dollars. Therefore, its present value (at the rate of 8% compounded continuously) is

$$V(t) = 50t^2e^{-0.08t} \text{ dollars (for } t > 0)$$

Find the number of years that maximizes the present value.

43–44: Business: Maximizing Revenue The function given below is a company's price function, where x is the quantity (in thousands) that will be sold at price p dollars.

a. Find the revenue function $R(x)$. (*Hint:* Revenue is price times quantity, $p \cdot x$.)
b. Find the quantity and the price that will maximize revenue.

43. $p = 200e^{-0.25x}$ **44.** $p = 5 - \ln x$

45. Economics: Maximizing Consumer Expenditure Consumer demand for a commodity is estimated to be $D(p) = 25,000e^{-0.02p}$ units per month, where p is the selling price in dollars. Find the selling price that maximizes consumer expenditure.

 46–47: Use your graphing calculator to graph each function on a window that includes all relative extreme points and inflection points, and give the coordinates of these points (rounded to two decimal places). *Hint:* Use NDERIV once or twice with ZERO. (Answers may vary depending on the window chosen.)

46. $f(x) = \dfrac{x^4}{e^x}$ **47.** $f(x) = x^3 \ln |x|$

 48. Economics: Consumer Expenditure If consumer demand for a commodity is $D(p) = 200e^{-0.0013p}$ (where p is the selling price in dollars), find the price that maximizes consumer expenditure.

49. Business: Maximizing Revenue A manufacturer finds that the price function for auto-focus cameras is $p(x) = 20e^{-0.0077x}$, where p is the price in dollars at which quantity x (in thousands) will be sold. Find the quantity x that maximizes revenue.

8.3 Two Applications to Economics: Relative Rates and Elasticity of Demand

50–51: Economics: Relative Rate of Change The gross national product of a developing country is forecast to be $G(t) = 5 + 2e^{0.01t}$ million dollars t years from now. Find the relative rate of change

50. 20 years from now   **51.** 10 years from now

52. Economics: Elasticity of Demand A South American country exports coffee and estimates the demand function to be $D(p) = 63 - 2p^2$. If the country wants to raise revenues to improve its balance of payments, should it raise or lower the price from the present level of $3 per pound?

53. Economics: Elasticity of Demand A South African country exports gold and estimates the demand function to be $D(p) = 400\sqrt{600 - p}$. If the country wants to raise revenues to improve its balance of payments, should it raise or lower the price from the present level of $350 per ounce?

54. Economics: Elasticity of Demand The demand function for cigarettes is of the form $D(p) = 1.2p^{-0.44}$, where p is the price of a pack of cigarettes and $D(p)$ is the demand measured in packs per day per capita.* Find the elasticity of demand. (*Note:* You will find that demand is inelastic. This means that taxation, which acts like a price increase, is an ineffective way of discouraging smoking, but is an effective way of raising revenue.)

 55. General: Relative Rate of Change The population of a city x years from now is projected to be $P(x) = 3.25 + 0.04x + 0.002x^3$ million people (for $0 \le x \le 10$). Find the relative rate of change 9 years from now.

*Jeffrey E. Harris, "Taxing Tar and Nicotine," *American Economic Review* **70**(3):300–311.

56. General: Elasticity of Demand   A boat dealer finds that the demand function for outboard motor boats near a large lake is $D(p) = 200 - 20p + p^2 - 0.03p^3$ (for $0 \le p \le 15$), where p is the selling price in thousands of dollars.

a. Use a graphing calculator to find the elasticity of demand at a price of $10,000. (*Hint:* What value of p corresponds to $10,000?)
b. Should the dealer raise or lower the price from this level to increase revenue?
c. Find the price at which elasticity equals 1.

Projects and Essays

The following projects and essays are based on Chapter 8. Most have no right and wrong answers—the results depend only on your imagination and resourcefulness.

1. Look back at the applications that involve the constant e (such as compound interest, population growth, and others) and write about how each involves the idea of something growing (or shrinking) in proportion to its present size. Which application is most interesting to you, and why?

2. The constant e is defined as $e = \lim\limits_{n \to \infty} \left(1 + \frac{1}{n}\right)^n \approx 2.71828$. Use a calculator to evaluate $\left(1 + \frac{1}{n}\right)^n$ for large values of n. Compare e to $\lim\limits_{n \to \infty} \left(1 + \frac{1}{100}\right)^n$ and $\lim\limits_{n \to \infty} \left(1 + \frac{1}{n}\right)^{100}$ where, in each case, one of the n's is replaced by a constant. Estimate $\lim\limits_{n \to \infty} \left(1 + \frac{1}{n}\right)^{2n}$ and $\lim\limits_{n \to \infty} \left(1 + \frac{1}{2n}\right)^n$ and try to determine how their values are related to e. What if 2 is replaced by another number?

3. Compare *daily* compounding to *continuous* compounding. For example, find the value of $1000 after 1 year at 8% interest compounded **a.** daily and **b.** continuously. Is there much of a difference? Will there be much of a difference if the period is longer? How long (work out some examples)?

4. Does a 10% increase followed by another 10% increase result in a 20% increase? Suppose that your originally weighed 100 pounds and then gained 10% (10% of 100 is 10 pounds), thereby weighing $100 + 10 = 110$ pounds. If your weight then increased again by 10% (10% of 110 is 11 pounds), you would weigh $110 + 11 = 121$. However, a 20% increase in your original weight of 100 pounds would bring you to only 120, not 121. Where did the extra pound come from? Explain why successive percentage increases don't "add up." Will they always add up to *more* than just adding the percentages? What about two successive 10% *decreases?* What about a 10% increase followed by a 10% decrease, or vice versa? Explain.

5. The "Money Angles" column in *Time* Magazine (May 17, 1993) recommended buying wine by the case to save an estimated 10% every 12 weeks. The column then suggested that repeating this purchase every 12 weeks would lead to a saving of "more than 40% per year." Questions: If you did this for two years, would you save 80%? For three years, would you save 120% (so the wine would be free, or even better)? Explain what is wrong with *Time*'s reasoning.

6. Find out how logarithms (base 10) were used for calculating products and quotients before the invention of pocket calculators. Which properties of logs enable multiplication problems to be changed into addition problems, and division problems into subtraction problems? Look up "slide rule" and explain how slide rules were used for calculations, and how they are based on the addition and subtraction of logarithms.

7. Explain carefully how the chain rule (page 573) leads from the differentiation rule $\frac{d}{dx} \ln x = \frac{1}{x}$ to $\frac{d}{dx} \ln f(x) = \frac{f'(x)}{f(x)}$, and from $\frac{d}{dx} e^x = e^x$ to $\frac{d}{dx} e^{f(x)} = e^{f(x)} \cdot f'(x)$. [It may be easier to replace $f(x)$ by $g(x)$ in the "function" rules.]

8. Write about the difference between straight-line depreciation (page 16) and depreciation by a fixed percentage, describing the advantages of each. For example, if the line (from straight-line depreciation) and the curve (from fixed percentage depreciation) both begin at the same point and end at the same point, which method of depreciation provides the greater dollar drop in value in the first year? In the last year? If you had to pay income tax on the value each year, which method would you prefer? If you had to sell the asset midway, which method would you prefer? If you had to buy the asset midway?

9. Read the Application Preview on pages 713–714 and write about what the elasticity of demand for a drug suggests about the most productive law enforcement strategy for combating that drug. What does the elasticity mean to the *dealers?* Can you think of any reason why demand for heroin would be inelastic, while demand for marijuana would be elastic? Similarly, discuss the elasticities of beer and liquor (see Exercises 34–35 in Section 8.3) and cigarettes (Exercise 54 in the Review Exercises).

10. Look over your notes, your homework, and the text, and write a page about how a graphing calculator has helped you in this chapter. Include examples of how it has helped you to *explore concepts* and how it has helped to *simplify your work.* What was its *most* helpful or interesting use? What was its *least* helpful or interesting use? Are there problems that can be done on a graphing calculator but that are easier to do "by hand"?

9 INTEGRATION AND ITS APPLICATIONS

A roller coaster. The area under the curve can be estimated from the rectangular grid.

9.1 Antiderivatives

APPLICATION PREVIEW

Trade Deficits, Automobile Safety, and Integrals

This chapter introduces a new concept, the *integral*, which enables us to calculate areas between curves. Areas between curves have important uses.

For example, the upper curve in the following graph represents *imports* (that is, foreign goods sold to America, paid for by dollars sent overseas) and the lower curve represents *exports* (that is, American goods sold abroad, bought by foreign funds paid to America) during the last few years. The area *between* the import and export curves represents the cumulative U.S. *trade deficit* during these years. This area is a measure of how much of America has become owned by foreign countries.

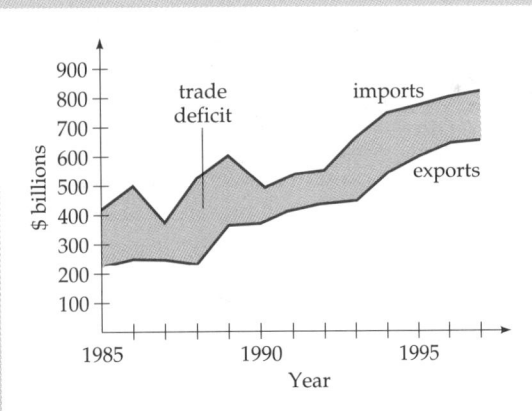

Source: U.S. Department of Commerce.

As another example, the upper curve in the following graph shows the annual traffic fatalities during the last few years, and the lower curve the *predicted* fatalities if everyone used seat belts. The area *between* the curves then represents the *total number of lives that could have been saved by seat belts* during this period.

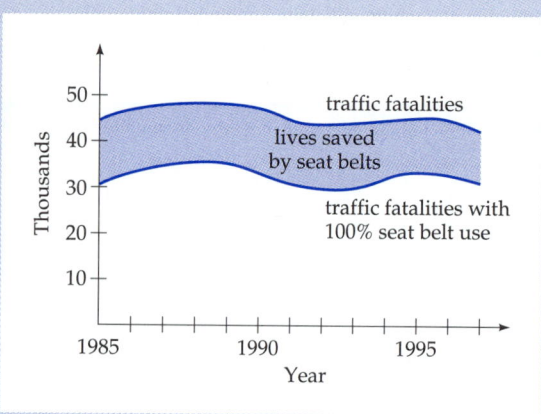

Source: American Automobile Association.

Integrals can be used to calculate these and many other important quantities, as we will see in this chapter.

Introduction

We have been studying differentiation and its uses. We now consider the reverse process, *anti*differentiation, which, for a given derivative, essentially recovers the original function. Antidifferentiation has many uses. For example, differentiation turns a cost function into a marginal cost function, and so antidifferentiation turns marginal cost back into cost. Later we will use antidifferentiation for other purposes, such as finding areas.

Antiderivatives and Indefinite Integrals

We begin with a simple example of antidifferentiation. Since the derivative of x^2 is $2x$, an *anti*derivative of $2x$ is x^2:

an antiderivative of $2x$ is x^2 Since the derivative of x^2 is $2x$

There are, however, other antiderivatives of $2x$. Each of the following is an antiderivative of $2x$:

$x^2 + 1$ $x^2 - 17$ $x^2 + e$ Since the derivative of each is $2x$

Clearly, we may add *any* constant to x^2 and the derivative will still be $2x$. Therefore, $x^2 + C$ is an antiderivative of $2x$ for *any* constant C. Furthermore, it can be shown that there are no other antiderivatives of $2x$, and so the *most general* antiderivative of $2x$ is $x^2 + C$. The most general antiderivative of $2x$ is called the *indefinite integral* of $2x$, and is written with the $2x$ between an *integral sign* $\int$ and a dx:

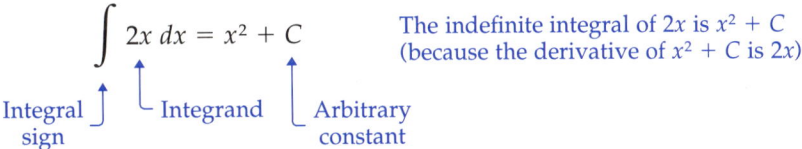

$$\int 2x \, dx = x^2 + C$$

The indefinite integral of $2x$ is $x^2 + C$ (because the derivative of $x^2 + C$ is $2x$)

Integral sign

Integrand

Arbitrary constant

The function to be integrated (here, $2x$) is called the *integrand*. The dx reminds us that the variable of integration is x. The constant C is called an *arbitrary constant* because it may take any value, positive, negative, or zero.

An indefinite integral (sometimes called simply an *integral*) can always be checked by differentiation: The derivative of the answer must equal the integrand (as is the case with $x^2 + C$ and $2x$).

Indefinite Integral

$$\int f(x) \, dx = g(x) + C$$

The integral of $f(x)$ is $g(x) + C$

if and only if

if and only if

$$g'(x) = f(x)$$

the derivative of $g(x)$ is $f(x)$

Integration Rules

There are several "rules" that simplify integration. The first, which is one of the most useful rules in all of calculus, shows how to integrate x to a constant power.

Power Rule for Integration

To integrate x to a power, add 1 to the power and multiply by 1 over the new power:

$$\int x^n \, dx = \frac{1}{n+1} x^{n+1} + C \qquad (n \neq -1)$$

EXAMPLE 1   Finding an Indefinite Integral

$$\int x^3 \, dx = \frac{1}{4} x^4 + C$$

Using the power rule with $n = 3$

$n = 3$ $n + 1 = 4$

$$\frac{1}{n+1} = \frac{1}{4}$$

Differentiating the answer $\frac{1}{4}x^4 + C$ immediately gives the integrand x^3, so the answer is correct.

PRACTICE PROBLEM 1   Find $\displaystyle\int x^2\,dx$ and check your answer by differentiation.

Solution at the back of the book

The proof of the power rule for integration consists simply of differentiating the right-hand side.

$$\frac{d}{dx}\left(\frac{1}{n+1}x^{n+1} + C\right) = \frac{1}{n+1}(n+1)x^n = x^n$$

The power $n + 1$ brought down   The power decreased by 1   Simplified

Since the derivative is the integrand x^n, the power rule for integration is correct.

To integrate functions like $\sqrt{x}$ and $\frac{1}{x^2}$ we first express them as powers.

EXAMPLE 2   **Expressing as a Power before Integrating**

$$\int \sqrt{x}\,dx = \int x^{1/2}\,dx = \frac{2}{3}x^{3/2} + C$$

Using the power rule for integration with $n = \frac{1}{2}$

$n = \frac{1}{2}$

$n + 1 = \frac{1}{2} + 1 = \frac{3}{2}$

$\dfrac{1}{n+1} = \dfrac{1}{\frac{3}{2}} = \dfrac{2}{3}$

Note that in the answer the multiple $\frac{2}{3}$ and the exponent $\frac{3}{2}$ are reciprocals of each other. Can you see, from the Power Rule, why this will always be so?

EXAMPLE 3   **Expressing as a Power before Integrating**

$$\int \frac{1}{x^2}\,dx = \int x^{-2}\,dx = \frac{1}{-1}x^{-1} + C = -x^{-1} + C$$

$n = -2$

Simplified answer

$n + 1 = -1$

PRACTICE PROBLEM 2   Find $\int \dfrac{dx}{x^3}$. $\left(Hint:\ \text{Equivalent to } \int \dfrac{1}{x^3}\,dx.\right)$

Solution at the back of the book

EXAMPLE 4   Integrating 1

$$\int 1\,dx = \int x^0\,dx = \frac{1}{1}x^1 + C = x + C \qquad \text{Using } x^0 = 1$$

$$\underset{n=0}{\uparrow} \qquad \underset{n+1=1}{\uparrow}$$

This result is so useful that it should be memorized.

$$\int 1\,dx = x + C \qquad \text{The integral of 1 is } x + C$$

EXAMPLE 5   Integrating with Other Variables

a. $\displaystyle\int t^3\,dt = \frac{1}{4}t^4 + C$

Using the power rule ($n = 3$) with dt since the variable is t ("integrating with respect to t")

b. $\displaystyle\int u^{-4}\,du = \frac{1}{-3}u^{-3} + C = -\frac{1}{3}u^{-3} + C$

Using the power rule ($n = -4$) with du since the variable is u ("integrating with respect to u")

PRACTICE PROBLEM 3   Find $\displaystyle\int z^{-1/2}\,dz$.

Solution at the back of the book

Notice what happens if we try to integrate x^{-1}.

$$\int x^{-1}\,dx = \frac{1}{0}x^0 + C \qquad \text{Undefined because of the } \frac{1}{0}$$

$$\underset{n=-1}{\uparrow} \qquad\qquad \underset{n+1=0}{\uparrow}$$

The power rule for integration fails for the exponent -1 because it leads to the undefined expression $\frac{1}{0}$. For this reason, the power rule for integration includes the restriction "$n \neq -1$." It can integrate any power of x *except* x^{-1}.

The sum and constant multiple rules for differentiation (pages 531 and 529, respectively) lead immediately to analogous rules for simplifying integrals. The first rule says that the sum of two functions may be integrated one at a time.

Sum Rule for Integration

$$\int [\, f(x) + g(x)]\, dx = \int f(x)\, dx + \int g(x)\, dx$$

The integral of a sum is the sum of the integrals

EXAMPLE 6   Using the Sum Rule

$$\int (x^2 + x^3)\, dx = \int x^2\, dx + \int x^3\, dx$$

Using the sum rule to break the integral into two integrals

$$= \frac{1}{3} x^3 + \frac{1}{4} x^4 + C$$

Using the power rule on each (one + C is enough)

The second rule says that a constant may be moved across the integral sign.

Constant Multiple Rule for Integration

For any constant k,

$$\int k \cdot f(x)\, dx = k \int f(x)\, dx$$

The integral of a constant times a function is the constant times the integral of the function

EXAMPLE 7   Using the Constant Multiple Rule

$$\int 6x^2\, dx = 6 \int x^2\, dx$$

Using the constant multiple rule to move the 6 across the integral sign

$$= 6 \cdot \frac{1}{3} x^3 + C$$

Using the power rule with $n = 2$

$$= 2x^3 + C$$

Simplifying

EXAMPLE 8 **Integrating a Constant**

$$\int 7 \, dx = 7 \int 1 \, dx = 7x + C$$

Moving the constant outside

The integral of 1 is x (plus C)

This leads to a very useful general rule. For any constant k,

Integral of a Constant

$$\int k \, dx = kx + C$$

The integral of a constant is the constant times x (plus C)

The sum rule can be extended to integrate the sum or difference of *any* number of terms, writing only one C at the end, since any number of arbitrary constants can be added together to give just one.

EXAMPLE 9 **Integrating a Sum of Powers**

$$\int (6x^2 - 3x^{-2} + 5) \, dx$$

$$= 6 \int x^2 \, dx - 3 \int x^{-2} \, dx + 5 \int 1 \, dx$$

Breaking up the integral and moving constants outside

$$= 6 \cdot \frac{1}{3} x^3 - 3 \frac{1}{-1} x^{-1} + 5x + C$$

Integrating each separately

From integrating the 1

$$= 2x^3 + 3x^{-1} + 5x + C$$

Simplifying

EXAMPLE 10 **Simplifying before Integrating**

$$\int \left(\frac{3\sqrt{x}}{2} - \frac{2}{\sqrt{x}} \right) dx = \int \left(\frac{3}{2} x^{1/2} - 2x^{-1/2} \right) dx$$

Written as powers of x

$$= \frac{3}{2} \frac{2}{3} x^{3/2} - 2 \frac{2}{1} x^{1/2} + C$$

Integrating each term separately

$$= x^{3/2} - 4x^{1/2} + C$$

Simplifying

PRACTICE PROBLEM 4   Find $\int \left(\sqrt[3]{w} - \dfrac{4}{w^3} \right) dw.$ *Solution at the back of the book*

Some integrals are so simple that they can be integrated "at sight."

EXAMPLE 11   **Integrating at Sight**

a. $\int 4x^3 \, dx = x^4 + C$ By remembering that $4x^3$ is the derivative of x^4

b. $\int 7x^6 \, dx = x^7 + C$ By remembering that $7x^6$ is the derivative of x^7

■

PRACTICE PROBLEM 5   Integrate "at sight" by noticing that each integrand is of the form nx^{n-1} and integrating to x^n without working through the power rule.

a. $\int 5x^4 \, dx$ **b.** $\int 3x^2 \, dx$ *Solutions at the back of the book*

Algebraic Simplification of Integrals

Sometimes an integrand needs to be multiplied out or otherwise simplified before it can be integrated.

EXAMPLE 12   **Simplifying before Integrating**

Find $\int x^2(x + 6)^2 \, dx.$

Solution

$$\int x^2(x + 6)^2 \, dx = \int x^2\underbrace{(x^2 + 12x + 36)}_{(x + 6)^2} \, dx \qquad \text{"Squaring out" the } (x + 6)^2$$

$$= \int (x^4 + 12x^3 + 36x^2) \, dx \qquad \text{Multiplying out}$$

$$= \frac{1}{5}x^5 + 12 \cdot \frac{1}{4}x^4 + 36 \cdot \frac{1}{3}x^3 + C \qquad \text{Integrating each term separately}$$

$$= \frac{1}{5}x^5 + 3x^4 + 12x^3 + C \qquad \text{Simplifying}$$

■

PRACTICE PROBLEM 6   Find $\int \dfrac{6t^2 - t}{t} \, dt.$ (*Hint:* First simplify the integrand.)

Solution at the back of the book

Since differentiation turns a cost function into a marginal cost function, integration turns a marginal cost function back into a cost function. To evaluate the constant, however, we need the fixed costs.

EXAMPLE 13   Recovering Cost from Marginal Cost

A company's marginal cost function is $MC(x) = 6\sqrt{x}$ and the fixed cost is \$1000. Find the cost function.

Solution

We integrate the marginal cost to find the cost function.

$$C(x) = \int MC(x)\, dx = \int 6\sqrt{x}\, dx = 6\int x^{1/2}\, dx \qquad \text{Integrating}$$

$$= 6 \cdot \frac{2}{3} x^{3/2} + K = 4x^{3/2} + K \qquad \begin{array}{l}\text{From the power rule}\\ \text{(using } K \text{ to avoid}\\ \text{confusion with } C \text{ for}\\ \text{cost)}\end{array}$$

It remains to find the value of the constant K. The cost function evaluated at $x = 0$ always gives the fixed cost (because when nothing is produced, only the fixed cost remains):

$$\underbrace{C(0) = K}_{\text{Fixed cost}} \qquad \begin{array}{l}\text{Evaluating } C(x) = 4x^{3/2} + K\\ \text{at } x = 0 \text{ gives } K\end{array}$$

Therefore, K equals the fixed cost, which is given as 1000, so $K = 1000$. The completed cost function is obtained by replacing K by 1000.

$$C(x) = 4x^{3/2} + 1000 \qquad \begin{array}{l}C(x) = 4x^{3/2} + K\\ \text{with } K = 1000\end{array}$$

Notice that finding the cost function involved two steps:

a.   Integrating the marginal cost to find the cost function

b.   Using the fixed cost to evaluate the arbitrary constant

Integration, being the reverse of differentiation, can recover *any* quantity from its rate of change (together with one fixed value). For example given the size of an economy and its rate of growth, we can integrate to find the size of the economy at any time in the future.

EXAMPLE 14   Recovering a Quantity from Its Rate of Change

The gross domestic product (GDP) of a country is \$78 billion and growing at the rate of $4.4t^{-1/3}$ billion dollars per year after t years. Find a formula for GDP after t years. Then use this formula to find the GDP after 8 years.

Solution

$$G(t) = \int 4.4t^{-1/3}\, dt = 4.4 \int t^{-1/3}\, dt \qquad \text{Integrating the rate of change}$$

$$= 4.4 \left(\frac{3}{2}\right) t^{2/3} + C = 6.6t^{2/3} + C \qquad \text{Using the power rule and then simplifying}$$

As before, evaluating $G(t) = 6.6t^{2/3} + C$ at $t = 0$ gives $G(0) = C$, which shows that the constant C is the GDP at time $t = 0$, which is given as \$78 billion. Therefore

$$G(t) = 6.6t^{2/3} + 78 \qquad \text{\textcolor{blue}{$G(t) = 6.6t^{2/3} + C$ with $C = 78$}}$$

For the GDP after 8 years, we evaluate $G(t)$ at $t = 8$:

$$G(8) = 6.6 \cdot \underbrace{8^{2/3}}_{4} + 78 = 6.6(4) + 78 = 104.4 \qquad \textcolor{blue}{G(t) = 6.6t^{2/3} + 78 \text{ at } t = 8}$$

Therefore, the GDP after 8 years is \$104.4 billion.

■

Graphing Calculator Exploration

A graphing calculator can show the geometric meaning of the arbitrary constant. In $\int 2x\, dx = x^2 + C$, the solution $x^2 + C$ is actually a whole collection of functions, a different curve for each value of C. Examine these curves for various values of C as follows:

a. Graph the five functions $x^2 - 2$, $x^2 - 1$, x^2, $x^2 + 1$, and $x^2 + 2$ on the graphing window $[-4, 4]$ by $[-2, 5]$. Use TRACE to see how the constant shifts the curve vertically.

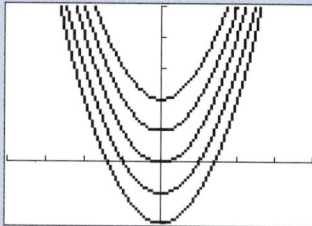

b. Predict what the curves $x^2 + 4$ and $x^2 - 4$ (the solution with $C = \pm 4$) would look like. Then check your predictions by graphing them.

c. Find the slope (using NDERIV or dy/dx) of several of the curves at a particular x-value and check that the slope is twice the x-value. This verifies that the derivative of each curve is $2x$, so each is an integral of $2x$.

Integrals as Continuous Sums

We have seen that differentiation is a kind of "breaking down," changing total cost into marginal (per unit) cost. Integration, the reverse process, is therefore a kind of "adding up," recovering the total from its parts. For example, integration recovers the total cost from marginal (per unit) costs, and integration adds up the growth of an economy over several years to give its total size.

To put this another way, ordinary addition is for summing "chunks," like $2 plus $3 equals $5, whereas integration is for summing *continuous* change, like the slow but steady growth of an economy over time.

Integration is continuous summation.

SUMMARY

In this section we have discussed both the *techniques* of integration and the *meaning* or *uses* of integration.

On the *technical* side, we defined indefinite integration as the reverse process of differentiation. The power rule enables us to integrate powers.

$$\int x^n \, dx = \frac{1}{n+1} x^{n+1} + C \qquad n \neq -1$$

(Increasing the exponent by 1 should seem quite reasonable: Differentiation lowers the exponent by 1, and so integration, the reverse process, should raise it by 1.) The sum and constant multiple rules extended integration to more complicated functions such as polynomials. Sometimes algebra is necessary to express a function in power form. *Reminder:* Don't forget the C.

As for the *uses* of integration, it recovers cost, revenue, and profit from their marginals, and in general it recovers any quantity from its rate of change (together with a particular value). The inverse relationship between integration and differentiation can be shown in a diagram.

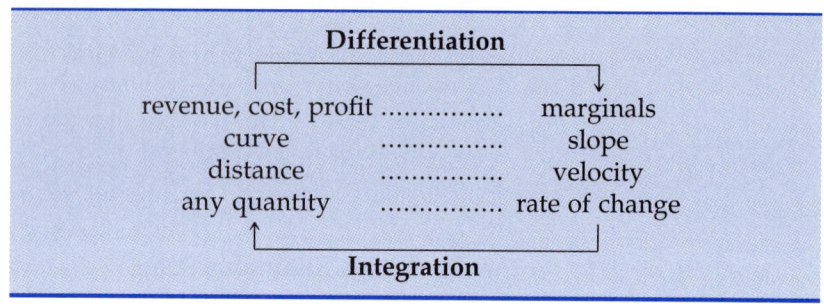

EXERCISES 9.1

Find each integral.

1. $\displaystyle\int x^4 \, dx$

2. $\displaystyle\int x^7 \, dx$

3. $\displaystyle\int x^{2/3} \, dx$

4. $\displaystyle\int x^{3/2} \, dx$

5. $\displaystyle\int \sqrt{u} \, du$

6. $\displaystyle\int \sqrt[3]{u} \, du$

7. $\displaystyle\int \frac{dw}{w^4}$

8. $\displaystyle\int \frac{dw}{w^2}$

9. $\displaystyle\int \frac{dz}{\sqrt{z}}$

10. $\displaystyle\int \frac{dz}{\sqrt[3]{z}}$

11. $\displaystyle\int 6x^5 \, dx$

12. $\displaystyle\int 9x^8 \, dx$

13. $\displaystyle\int (8x^3 - 3x^2 + 2) \, dx$

14. $\displaystyle\int (12x^3 + 3x^2 - 5) \, dx$

15. $\displaystyle\int \left(6\sqrt{x} + \frac{1}{\sqrt[3]{x}} \right) dx$

16. $\displaystyle\int \left(3\sqrt{x} + \frac{1}{\sqrt{x}} \right) dx$

17. $\displaystyle\int \left(16 \sqrt[3]{x^5} - \frac{16}{\sqrt[3]{x^5}} \right) dx$

18. $\displaystyle\int \left(14 \sqrt[4]{x^3} - \frac{3}{\sqrt[4]{x^3}} \right) dx$

19. $\displaystyle\int \left(10 \sqrt[3]{t^2} + \frac{1}{\sqrt[3]{t^2}} \right) dt$

20. $\displaystyle\int \left(21 \sqrt{t^5} + \frac{6}{\sqrt{t^5}} \right) dt$

21. $\displaystyle\int (x - 1)^2 \, dx$

22. $\displaystyle\int (x + 2)^2 \, dx$

23. $\displaystyle\int (1 + 10w) \sqrt{w} \, dw$

24. $\displaystyle\int (1 - 7w) \sqrt[3]{w} \, dw$

25. $\displaystyle\int \frac{6x^3 - 6x^2 + x}{x} \, dx$

26. $\displaystyle\int \frac{4x^4 + 4x^2 - x}{x} \, dx$

27. $\displaystyle\int (x - 2)(x + 4) \, dx$

28. $\displaystyle\int (x + 5)(x - 3) \, dx$

29. $\displaystyle\int (r - 1)(r + 1) \, dr$

30. $\displaystyle\int (3s + 1)(3s - 1) \, ds$

31. $\displaystyle\int \frac{x^2 - 1}{x + 1} \, dx$

32. $\displaystyle\int \frac{x^2 - 1}{x - 1} \, dx$

33. $\displaystyle\int (t + 1)^3 \, dt$

34. $\displaystyle\int (t - 1)^3 \, dt$

35. Evaluate

a. $\displaystyle\int \frac{1}{x^3} \, dx$

b. $\displaystyle\frac{\int 1 \, dx}{\int x^3 \, dx}$

Notice that the answers are not the same, showing that you do not integrate a fraction by integrating the numerator and denominator separately.

36. Evaluate

a. $\displaystyle\int x\,dx$ **b.** $x\displaystyle\int 1\,dx$

Notice that the two answers are not the same, showing that a variable cannot be moved across the integral sign.

37. a. Verify that $\int x^2\,dx = \frac{1}{3}x^3 + C.$

b. Graph the five functions $\frac{1}{3}x^3 - 2$, $\frac{1}{3}x^3 - 1$, $\frac{1}{3}x^3$, $\frac{1}{3}x^3 + 1$, and $\frac{1}{3}x^3 + 2$ (the solution for five different values of C) on the graphing window $[-3, 3]$ by $[-5, 5]$. Use TRACE to see how the constant shifts the curve vertically.

c. Find the slope (using NDERIV or dy/dx) of several of the curves at a particular x-value and check that in each case the slope is the square of the x-value. This verifies that the derivative of each curve is x^2, and so each is an integral of x^2.

38. a. Graph the five functions $\ln x - 2$, $\ln x - 1$, $\ln x$, $\ln x + 1$, and $\ln x + 2$ on the graphing window $[0, 4]$ by $[-3, 3]$.

b. Find the slope (using NDERIV or dy/dx) of several of the curves at a particular x-value and check that in each case the slope is the reciprocal of the x-value. This suggests that the derivative of each function is $\frac{1}{x}$.

c. Based on part (b), conjecture the indefinite integral of the function $\frac{1}{x}$ (for $x > 0$).

APPLIED EXERCISES

39. Business: Cost A company's marginal cost function is $MC = 20x^{3/2} - 15x^{2/3} + 1$, where x is the number of units, and fixed costs are $4000. Find the cost function.

40. Business: Cost A company's marginal cost function is $MC = 21x^{4/3} - 6x^{1/2} + 50$, where x is the number of units, and fixed costs are $3000. Find the cost function.

41. Business: Revenue A company's marginal revenue function is $MR = 12\sqrt[3]{x} + 3\sqrt{x}$, where x is the number of units. Find the revenue function. (Evaluate C so that revenue is zero when nothing is produced.)

42. Business: Revenue A company's marginal revenue function is $MR = 15\sqrt{x} + 4\sqrt[3]{x}$, where x is the number of units. Find the revenue function. (Evaluate C so that revenue is zero when nothing is produced.)

43. General: Velocity A Porsche 928 can accelerate from a standing start to a speed of $v(t) = -0.09t^2 + 8t$ feet per second after t seconds (for $t < 35$).

a. Find a formula for the distance that it will travel from its starting point in the first t seconds. (*Hint:* Integrate velocity to find distance, and then use the fact that distance is 0 at time $t = 0$.)

b. Use the formula that you found in part (a) to find the distance that the car will travel in the first 10 seconds.

44. General: Velocity A BMW 733i can accelerate from a standing start to a speed of $v(t) = -0.09t^2 + 6t$ feet per second after t seconds (for $t < 40$).

a. Find a formula for the distance that it will travel from its starting point in the first t seconds. (*Hint:* Integrate velocity to find distance, and then use the fact that distance is 0 at time $t = 0$.)

b. Use the formula that you found in part (a) to find the distance that the car will travel in the first 10 seconds.

45. General: Learning A person can memorize words at the rate of $\frac{3}{\sqrt{t}}$ words per minute.

a. Find a formula for the total number of words that can be memorized in t minutes. (*Hint:* Evaluate C so that 0 words have been memorized at time $t = 0$.)

b. Use the formula that you found in part (a) to find the total number of words that can be memorized in 25 minutes.

46. **Biomedical: Temperature**   A patient's temperature is 108 degrees and is changing at the rate of $t^2 - 4t$ degrees per hour, where t is the number of hours since taking fever-reducing medication ($t \leq 3$).

    a. Find a formula for the patient's temperature after t hours. (*Hint:* Evaluate the constant C so that the temperature is 108 at time $t = 0$.)

    b. Use the formula that you found in part (a) to find the patient's temperature after 3 hours.

47. **Environmental Science: Pollution**   A chemical plant is adding pollution to a lake at the rate of $40\sqrt{t^3}$ tons per year, where t is the number of years that the plant has been in operation.

    a. Find a formula for the total amount of pollution that will enter the lake in the first t years of the plant's operation. (*Hint:* Evaluate C so that no pollution has been added at time $t = 0$.)

    b. Use the formula that you found in part (a) to find how much pollution will enter the lake in the first 4 years of the plant's operation.

    c. If all life in the lake will cease when 400 tons of pollution have entered the lake, will the lake "live" beyond 4 years?

48. **Business: Appreciation**   A \$20,000 art collection is increasing in value at the rate of $300\sqrt{t}$ (dollars per year) after t years.

    a. Find a formula for its value after t years. (*Hint:* Evaluate C so that its value at time $t = 0$ is \$20,000.)

    b. Use the formula that you found in part (a) to find its value after 25 years.

REVIEW EXERCISE

(This exercise will be important in the next section.)

49. Find $\dfrac{d}{dx} \ln (-x)$.

9.2 Integration Using Logarithmic and Exponential Functions

Introduction

In Section 9.1 we defined integration as the reverse of differentiation, and we introduced several integration formulas. In this section we develop integration formulas involving logarithmic and exponential functions. One of these formulas will answer a question that we could not answer earlier, namely, how to integrate x^{-1}, the only power not covered by the power rule.

The Integral $\int e^{ax}\, dx$

For any constant $a \neq 0$, the integral of the exponential function e^{ax} is

$$\int e^{ax}\, dx = \frac{1}{a} e^{ax} + C$$

The integral of e to a constant times x is one over the constant times the original function (plus C)

EXAMPLE 1   Integrating an Exponential Function

$$\int e^{2x}\, dx = \frac{1}{2} e^{2x} + C \qquad \text{Using the formula with } a = 2$$

$$\underset{a\,=\,2}{\uparrow} \quad \frac{1}{a} \quad \underset{\text{Original function}}{\uparrow}$$

As always, we may check the answer by differentiation.

$$\frac{d}{dx}\left(\frac{1}{2} e^{2x} + C\right) = \frac{1}{2} \cdot (2)e^{2x} = e^{2x} \qquad \text{Using } \frac{d}{dx} e^{ax} = ae^{ax}$$

The result is the integrand e^{2x}, so the integration is correct.

The proof of this rule consists simply of differentiating the right-hand side.

$$\frac{d}{dx}\left(\frac{1}{a} e^{ax} + C\right) = \frac{1}{a} ae^{ax} = e^{ax} \qquad \text{Using } \frac{d}{dx} e^{ax} = ae^{ax}$$

The result is the integrand, so the integration formula is correct.

EXAMPLE 2   Integrating Exponential Functions

a. $\displaystyle\int e^{\frac{1}{2}x}\, dx = 2e^{\frac{1}{2}x} + C \qquad\qquad a = \frac{1}{2} \quad \text{so} \quad \frac{1}{a} = \frac{1}{\frac{1}{2}} = 2$

b. $\displaystyle\int 6e^{-3x}\, dx = 6\int e^{-3x}\, dx = 6\left(-\frac{1}{3}\right)\cdot e^{-3x} + C = -2e^{-3x} + C$

$$\underset{a\,=\,-3}{\underset{\llcorner}{}} \qquad \underset{\frac{1}{a}\,=\,-\frac{1}{3}}{\underset{\llcorner}{}}$$

c. $\displaystyle\int e^{x}\, dx = 1e^{x} + C = e^{x} + C \qquad\qquad a = 1, \quad \text{so} \quad \frac{1}{a} = 1$

Each of these answers may be checked by differentiation. Example 2c says that e^x is the integral of itself, just as e^x is the derivative of itself.

$$\int e^x \, dx = e^x + C \qquad \text{The integral of } e^x \text{ is } e^x \text{ (plus } C\text{)}$$

PRACTICE PROBLEM 1   Find: **a.** $\int 12e^{4x} \, dx$ **b.** $\int e^{\frac{1}{3}x} \, dx$ *Solutions at the back of the book*

When using these new integration formulas to solve applied problems, be careful to evaluate the constant C correctly. It will not always be equal to the initial value of the function.

EXAMPLE 3   Finding Total Flu Cases from the Rate of Change

An influenza epidemic hits a large city and spreads at the rate of $12e^{0.2t}$ new cases per day, where t is the number of days since the epidemic began. The epidemic began with 4 cases.

a. Find a formula for the total number of flu cases in the first t days of the epidemic.

b. Use your formula to find the number of cases during the first 30 days.

Solution

a. To find the total number of cases, we integrate the growth rate $12e^{0.2t}$.

$$\underbrace{f(t)}_{\substack{\text{Total cases} \\ \text{in first } t \text{ days}}} = \int 12e^{0.2t} \, dt = 12 \int e^{0.2t} \, dt \qquad \text{Taking out the constant}$$

$$= 12 \, \frac{1}{0.2} \, e^{0.2t} + C = 60e^{0.2t} + C \qquad \text{Using the } \int e^{ax} \, dx \text{ formula}$$
$$\underset{5}{\big\lfloor}$$

Evaluating $f(t)$ at $t = 0$ must give the initial number of cases:

$$f(0) \quad = 60e^{0.2(0)} + C = 60 + C$$

$\underbrace{}$ $\underbrace{}$
Initial $e^0 = 1$
number
of cases

$f(t) = 60e^{0.2t} + C$ evaluated at $t = 0$ (the beginning of the epidemic)

This initial number must equal the given initial number, 4:

$$60 + C = 4$$

Initial number from the formula set equal to the given initial number

$$C = -56$$

Solving for C

Replacing C by -56 gives the total number of flu cases within t days:

$$f(t) \qquad = 60e^{0.2t} - 56$$

$\underbrace{}$ $\underbrace{\phantom{60e^{0.2t} - 56}}$
Number of cases   Answer to
in first t days   part (a)

$f(t) = 60e^{0.2t} + C$ with $C = -56$

b. To find the number within 30 days, we evaluate at $t = 30$:

$$f(30) = 60e^{0.2(30)} - 56$$

$f(t) = 60e^{0.2t} - 56$ at $t = 30$

$$= 60e^6 - 56 \approx 24{,}150$$

Using a calculator

Therefore, within 30 days the epidemic will have grown to more than 24,000 cases. ∎

Evaluating the Constant C

Notice that we did *not* simply replace the constant C by the initial number of cases, 4. (This would have given the wrong initial number of cases.) Instead, we evaluated the function at the initial time $t = 0$ and set it equal to the initial number of cases:

$$60e^0 + C \quad = \quad 4$$

$\underbrace{}$ $\underbrace{}$
$f(t)$ evaluated   Given
at $t = 0$ initial value

We then solved to find the correct value of C, $C = -56$, which we then substituted into the formula.

In general, to evaluate the constant C:

1. Evaluate the integral at the given number (usually $t = 0$) and set the result equal to the stated initial value.

2. Solve for C.

3. Write the answer with C replaced by its correct value.

The Integral $\int \dfrac{1}{x}\,dx$

The differentiation formula $\dfrac{d}{dx}\ln x = \dfrac{1}{x}$ can be read "backwards" as an integration formula

$$\int \frac{1}{x}\,dx = \ln x + C$$ The integral of 1 over x is the natural log of x (plus C)

This formula, however, is restricted to $x > 0$, for only then is $\ln x$ defined. For $x < 0$ we can differentiate $\ln(-x)$, giving

$$\frac{d}{dx}\ln(-x) = \frac{-1}{-x} = \frac{1}{x}$$ Using $\dfrac{d}{dx}\ln f = \dfrac{f'}{f}$ and simplifying

This result says that for $x < 0$, the integral of $\dfrac{1}{x}$ is $\ln(-x)$. The negative sign in $\ln(-x)$ serves only to make the already negative x positive, and this could be accomplished just as well with absolute value bars.

$$\int \frac{1}{x}\,dx = \ln|x| + C$$ The integral of 1 over x is the natural logarithm of the absolute value of x

This formula holds for negative *and* positive values of x, since in both cases $\ln|x|$ is defined. The integral can be written in three different ways, all of which have the same answer.

$$\int \frac{1}{x}\,dx \;=\; \int \frac{dx}{x} \;=\; \int x^{-1}\,dx \;=\; \ln|x| + C$$

EXAMPLE 4   Integrating Using the ln Rule

$$\int \frac{5}{2x}\,dx = \int \frac{5}{2}\frac{1}{x}\,dx = \frac{5}{2}\int \frac{1}{x}\,dx = \frac{5}{2}\ln|x| + C$$

Taking out   Using the
the constant   ln formula

■

EXAMPLE 5   Integrating Negative Powers

$$\int (x^{-1} + x^{-2})\,dx = \int x^{-1}\,dx + \int x^{-2}\,dx = \ln|x| - x^{-1} + C$$

From the   From the
natural log   power rule
formula   with $n = -2$

■

PRACTICE PROBLEM 2   Find $\int \dfrac{3}{4x}\,dx$.   *Solution at the back of the book*

EXAMPLE 6   Finding Total Sales from the Sales Rate

An electronics dealer estimates that during month t of a sale, a discontinued computer will sell at a rate of approximately $\frac{25}{t}$ per month, where $t = 1$ corresponds to the beginning of the sale, at which time none have been sold. Find a formula for the total number of computers that will be sold up to month t. Will the store's inventory of 64 computers be sold by month $t = 12$?

Solution

To find the total sales, we integrate the sales rate $\frac{25}{t}$:

$$\underbrace{S(t)}_{\substack{\text{Total sales in}\\ \text{first } t \text{ months}}} = \int \frac{25}{t}\,dt = 25\int \frac{1}{t}\,dt = 25\ln t + C \qquad \begin{array}{l}\text{Omitting absolute}\\ \text{values, since } t > 0\end{array}$$

To evaluate C, we evaluate at the given starting time $t = 1$:

$$S(1) \quad = 25 \ln 1 + C = C$$

$\underbrace{}$ Initial number of sales $\underbrace{}$ 0

The initial number of sales must be zero, giving $C = 0$, and substituting this into the sales function gives

$$S(t) = 25 \ln t \qquad\qquad\qquad S(t) = 25 \ln t + C \text{ with } C = 0$$

Total number sold up to month t

To find the number sold up to month 12, we evaluate at $t = 12$:

$$S(12) = 25 \ln 12 \approx 62 \qquad\qquad \text{Using a calculator}$$

Therefore, all but two of the 64 computers will be sold by month 12.

■

In this example the initial time (the beginning of the sale) was given as $t = 1$ rather than the more usual $t = 0$. The initial time will be clear from the problem.

Consumption of Natural Resources

Just as the world population grows exponentially, so does the world's annual consumption of natural resources. We can estimate the total consumption at any time in the future by integrating the rate of consumption, and from this predict when the known reserves will be exhausted.

EXAMPLE 7   **Finding a Formula for Total Consumption from the Rate**

According to estimates from the United Nations Development Programme, the annual world consumption of tin will be $0.23e^{0.01t}$ million metric tons per year, where t is the number of years since 1995. Find a formula for the total tin consumption within t years of 1995 and estimate when the known world reserves of 7 million metric tons will be exhausted.*

* *World Resources 1996–97* (Oxford University Press, 1996). See also H. E. Goeller and A. Zucker, "Infinte Resources: The Ultimate Strategy," *Science* **223:** 456–462, 1984.

Solution

To find the total consumption, we integrate the rate $0.23e^{0.01t}$.

$$\underbrace{C(t)}_{} = \int 0.23e^{0.01t}\, dt = 0.23 \int e^{0.01t}\, dt \qquad \text{Taking out constant}$$

Total tin consumed
in first t years
after 1995

$$= 0.23\, \frac{1}{\underbrace{0.01}_{100}}\, e^{0.01t} + C = 23e^{0.01t} + C \qquad \text{Using } \int e^{ax}dx \text{ formula}$$

The total consumed in the first *zero* years must be zero, so $C(0) = 0$.

$$\underbrace{C(0)}_{0} = 23e^0 + C = \underbrace{23 + C}_{} \qquad\qquad C(t) = 23e^{0.01t} + C \text{ with } t = 0$$

Must equal zero

From $23 + C = 0$ we find $C = -23$. We substitute this into the formula for $C(t)$:

$$C(t) = \underbrace{23e^{0.01t} - 23}_{} \qquad\qquad C(t) = 23e^{0.01t} + C \text{ with } C = -23$$

Formula for total consumption
within first t years since 1995

To predict when the total world reserves of 7 million metric tons will be exhausted, we set this function equal to 7 and solve for t:

$$23e^{0.01t} - 23 = 7 \qquad\qquad\qquad C(t) \text{ set equal to 7}$$

$$23e^{0.01t} = 30 \qquad\qquad\qquad \text{Adding 23}$$

$$e^{0.01t} = \frac{30}{23} \approx 1.304 \qquad\qquad \text{Dividing by 23}$$

$$\underbrace{\ln e^{0.01t}}_{0.01t} = \underbrace{\ln 1.304}_{0.266} \qquad\qquad \text{Taking natural logs and using } \ln e^x = x$$

$$t = \frac{0.266}{0.01} = 26.6 \qquad\qquad \text{Solving } 0.01t = 0.266 \text{ for } t$$

Therefore, the known world supply of tin will be exhausted in about 27 years after 1995, which means in about the year 2022 (assuming that consumption continues at the predicted rate). ∎

Graphing Calculator Exploration

a. Graph the tin consumption function found in Example 7, $y_1 = 23e^{0.01x} - 23$, on the graphing window [0, 30] by [0, 10].

b. Verify that the curve is not straight by drawing the TANGENT at some point of the curve.

c. Show the world's known resources by graphing $y_2 = 7$.

d. Use TRACE to find the world's current position on the consumption curve. (*Hint:* Find the point with *x*-value equal to the current year minus 1995.)

e. TRACE along the curve to find (roughly) how much tin will remain 10 years from now.

Incidentally, tin is used mainly for coating steel. For example, a "tin" can is actually a steel can with a thin protective coating of tin to prevent rust. The predicted unavailability of tin is already causing major changes in the food-packaging industry.

Power Rule for Integration, Revisited

Our new integration formula shows how to integrate x^{-1}, the only power not covered by the power rule. Therefore, we may now write one "combined" formula for the integral $\int x^n \, dx$ for *any* power *n*.

Integral of Powers of x

$$\int x^n \, dx = \begin{cases} \dfrac{1}{n+1} x^{n+1} + C & \text{if } n \neq -1 \\[2ex] \ln |x| + C & \text{if } n = -1 \end{cases}$$

Use the power rule if *n* is *other* than -1

Use the ln formula if *n* equals -1

It is a curious fact that every power of *x* integrates to another power of *x*, with the single exception of x^{-1}, which integrates to an entirely different kind of function, the natural logarithm.

SUMMARY

We have three integration formulas.

$$\int x^n\, dx = \frac{1}{n+1} x^{n+1} + C \qquad\qquad n \neq -1$$

$$\int \frac{1}{x}\, dx = \int \frac{dx}{x} = \int x^{-1}\, dx = \ln |x| + C$$

$$\int e^{ax}\, dx = \frac{1}{a} e^{ax} + C \qquad\qquad a \neq 0$$

The absolute value bars in the middle formula should be omitted if x is positive. In applications we now evaluate C by setting the function (evaluated at the given number) equal to the stated initial value, and solving for C.

EXERCISES 9.2

Find each integral.

1. $\displaystyle\int e^{3x}\, dx$

2. $\displaystyle\int e^{4x}\, dx$

3. $\displaystyle\int e^{\frac{1}{4}x}\, dx$

4. $\displaystyle\int e^{\frac{1}{3}x}\, dx$

5. $\displaystyle\int e^{0.05x}\, dx$

6. $\displaystyle\int e^{0.02x}\, dx$

7. $\displaystyle\int e^{-2y}\, dy$

8. $\displaystyle\int e^{-3y}\, dy$

9. $\displaystyle\int e^{-0.5x}\, dx$

10. $\displaystyle\int e^{-0.4x}\, dx$

11. $\displaystyle\int 6e^{\frac{2}{3}x}\, dx$

12. $\displaystyle\int 24e^{-\frac{2}{3}u}\, du$

13. $\displaystyle\int -5x^{-1}\, dx$

14. $\displaystyle\int -\frac{1}{2}x^{-1}\, dx$

15. $\displaystyle\int \frac{3\, dx}{x}$

16. $\displaystyle\int \frac{dx}{2x}$

17. $\displaystyle\int \frac{3}{2v}\, dv$

18. $\displaystyle\int \frac{2}{3v}\, dv$

19. $\displaystyle\int \left(e^{3x} - \frac{3}{x}\right) dx$

20. $\displaystyle\int \left(e^{2x} - \frac{2}{x}\right) dx$

21. $\displaystyle\int (3e^{0.5t} - 2t^{-1})\, dt$

22. $\displaystyle\int (5e^{0.5t} - 4t^{-1})\, dt$

23. $\displaystyle\int (x^2 + x + 1 + x^{-1} + x^{-2})\, dx$

24. $\displaystyle\int (x^{-2} - x^{-1} + 1 - x + x^2)\, dx$

25. $\displaystyle\int (5e^{0.02t} - 2e^{0.01t})\, dt$

26. $\displaystyle\int (3e^{0.05t} - 2e^{0.04t})\, dt$

APPLIED EXERCISES (Most require **.)** _____

27–28: Biomedical: Epidemics A flu epidemic hits a college community, beginning with five cases on day $t = 0$. The rate of growth of the epidemic (new cases per day) is given by the function $r(t)$

on the next page, where t is the number of days since the epidemic began.
a. Find a formula for the total number of cases of flu in the first t days.

b. Use your answer to part (a) to find the total number of cases in the first 20 days.

27. $r(t) = 18e^{0.05t}$　　　　**28.** $r(t) = 20e^{0.04t}$

29. Business: Sales In an effort to reduce its inventory, a music store runs a sale on its least popular compact disk (CD). The sales rate (CDs sold per day) on day t of the sale is predicted to be $\dfrac{50}{t}$ (for $t \geq 1$), where $t = 1$ corresponds to the beginning of the sale, at which time none of the inventory of 200 CDs has been sold.

a. Find a formula for the total number of CDs sold up to day t.

b. Will the store have sold its inventory of 200 CDs by day $t = 30$?

30. Business: Sales In an effort to reduce its inventory, a book store runs a sale on its least popular mathematics books. The sales rate (books sold per day) on day t of the sale is predicted to be $\dfrac{60}{t}$ (for $t \geq 1$), where $t = 1$ corresponds to the beginning of the sale, at which time none of the inventory of 350 books has been sold.

a. Find a formula for the number of books sold up to day t.

b. Will the store have sold its inventory of 350 books by day $t = 30$?

31. General: Consumption of Natural Resources World consumption of silver is running at the rate of $16e^{0.02t}$ thousand metric tons per year, where t is measured in years and $t = 0$ corresponds to 1995.

a. Find a formula for the total amount of silver that will be consumed within t years of 1995.

b. When will the known world resources of 420 thousand metric tons of silver be exhausted? (Silver is used extensively in photography.)

32. General: Consumption of Natural Resources World consumption of copper is running at the rate of $13e^{0.04t}$ million metric tons per year, where t is measured in years and $t = 0$ corresponds to 1995.

a. Find a formula for the total amount of copper that will be used within t years of 1995.

b. When will the known world resources of 610 million metric tons of copper be exhausted?

33. General: Cost of Maintaining a Home The cost of maintaining a home generally increases as the home becomes older. Suppose that the rate of cost (dollars per year) for a home that is x years old is $200e^{0.4x}$.

a. Find a formula for the total maintenance cost during the first x years. (Total maintenance should be zero at $x = 0$.)

b. Use your answer to part (a) to find the total maintenance cost during the first 5 years.

34. Biomedical: Cell Growth A culture of bacteria is growing at the rate of $20e^{0.8t}$ cells per day, where t is the number of days since the culture was started. Suppose that the culture began with 50 cells.

a. Find a formula for the total number of cells in the culture after t days.

b. If the culture is to be stopped when the population reaches 500, when will this occur?

35. General: Freezing of Ice An ice cube tray filled with tap water is placed in the freezer, and the temperature is changing at the rate of $-12e^{-0.2t}$ degrees per hour after t hours. The temperature of the tap water is 70 degrees.

a. Find a formula for the temperature of water that has been in the freezer for t hours.

b. When will the ice be ready? (Water freezes at 32 degrees.)

36. Social Science: Divorces Since 1995 there have been approximately $1.2e^{0.01t}$ million divorces per year, where t is measured in years and $t = 0$ corresponds to 1995. Find a formula for the total number of divorces expected within t years of 1995.

37. Business: Total Savings A factory installs new equipment that is expected to generate savings at the rate of $800e^{-0.2t}$ dollars per year, where t is the number of years that the equipment has been in operation.

a. Find a formula for the total savings that the equipment will generate during its first t years.

b. If the equipment originally cost $2000, when will it "pay for itself"?

38. Business: Total Savings A company installs a new computer that is expected to generate savings at the rate of $20{,}000e^{-0.02t}$ dollars per year, where t is the number of years that the computer has been in operation.

a. Find a formula for the total savings that the computer will generate during its first t years.

b. If the computer originally cost $250,000, when will it "pay for itself"?

Find each integral. (*Hint:* Use some algebra first.)

39. $\displaystyle\int \frac{(x+1)^2}{x}\, dx$

40. $\displaystyle\int \frac{(x-1)^2}{x}\, dx$

41. $\displaystyle\int \frac{(t-1)(t+3)}{t^2}\, dt$

42. $\displaystyle\int \frac{(t+2)(t-4)}{t^2}\, dt$

43. $\displaystyle\int \frac{(x-2)^3}{x}\, dx$

44. $\displaystyle\int \frac{(x+2)^3}{x}\, dx$

45. General: Consumption of Natural Resources
World consumption of lead is running at the rate of $5.3e^{0.01t}$ million metric tons per year,

where t is measured in years and $t = 0$ corresponds to 1995.

a. Find a formula for the total amount of lead that will be consumed within t years of 1995.

b. Use a graphing calculator to find when the world's known resources of 130 million metric tons of lead will be exhausted. (*Hint:* Use INTERSECT. Lead has many uses, from batteries to shields against radioactivity.)

46. General: Total Savings   A homeowner installs a solar water heater that is expected to generate savings at the rate of $70e^{0.03t}$ dollars per year, where t is the number of years since it was installed.

a. Find a formula for the total savings within the first t years of operation.

b. Use a graphing calculator to find when the heater will "pay for itself" if it costs $800. (*Hint:* Use INTERSECT.)

9.3 Definite Integrals and Areas

Cigarette Smoking

Most cigarettes today have filters to absorb some of the toxic material or "tar" in the smoke before it is inhaled. The tobacco near the filter acts like an additional filter, absorbing tar, until it is itself smoked, at which time it releases all of its accumulated toxins. A typical cigarette consists of 8 centimeters of tobacco followed by a filter.

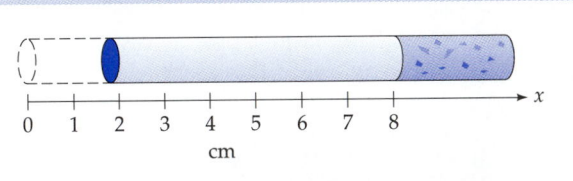

As the cigarette is smoked, tar is typically inhaled at the rate of $r(x) = 300e^{0.025x} - 240e^{0.02x}$ milligrams (mg) of tar per centimeter (cm) of tobacco, where x is the distance along the cigarette.* The amount of tar inhaled from any particular segment of the cigarette is the area under the graph of this function over that interval, which can be calculated by a process called *definite integration*, as explained in this section.

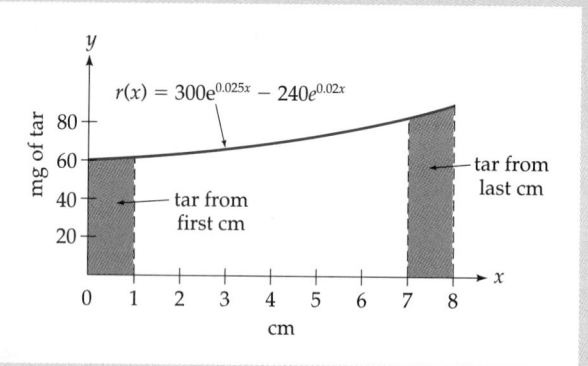

The results of integrating over the first and last centimeter of the cigarette are as follows (omitting the details—see Exercise 92). The numbers represent milligrams of tar from the beginning and end of the cigarette.

$$\begin{pmatrix} \text{Tar from} \\ \text{first cm} \end{pmatrix} = \int_0^1 (300e^{0.025x} - 240e^{0.02x})\, dx = 61 \qquad \begin{array}{l} \text{Integrate from 0} \\ \text{to 1 for the first} \\ \text{centimeter} \end{array}$$

$$\begin{pmatrix} \text{Tar from} \\ \text{last cm} \end{pmatrix} = \int_7^8 (300e^{0.025x} - 240e^{0.02x})\, dx = 83 \qquad \begin{array}{l} \text{Integrate from 7} \\ \text{to 8 for the last} \\ \text{centimeter} \end{array}$$

Notice that the last centimeter releases significantly more tar (about 36% more) than the first centimeter.

Moral: The Surgeon General has determined that smoking is hazardous to your health, and the last puffs are 36% more hazardous than the first.

This is just one of several applications of definite integration discussed in this section.

---

* The actual rate depends upon the type of cigarette and the proportion of time that it is smoked rather than left to burn in the air. For further information, see Helen Marcus-Roberts and Maynard Thompson (eds.), *Life Science Models* (Modules in Applied Mathematics, vol. 4) (Springer-Verlag, 1976), pp. 238–249.

Introduction

We begin this section by calculating areas under curves, leading to a definition of the *definite integral* of a function. The *Fundamental Theorem of Integral Calculus* then provides an easier way to calculate definite integrals using *indefinite* integrals. Finally, we will illustrate the wide variety of applications of definite integrals.

Area Under a Curve

The diagram below shows a continuous nonnegative function. We want to calculate the area under the curve and above the *x*-axis from $x = a$ to $x = b$.

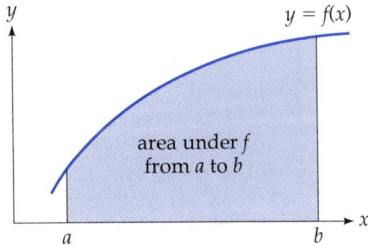

We begin by *approximating* the area by rectangles. On the left below, the area under the curve is approximated by five rectangles with equal bases and with heights equal to the height of the curve at the left-hand edge of the rectangle. These five rectangles, however, do not give a very accurate approximation for the area under the curve: They underestimate the actual area by the small white spaces just above the rectangles.

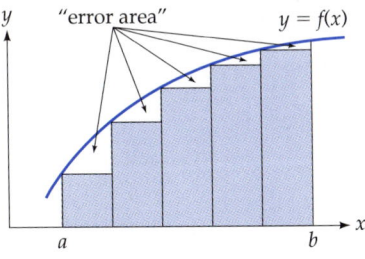

Area under *f* from *a* to *b*
approximated by 5 rectangles

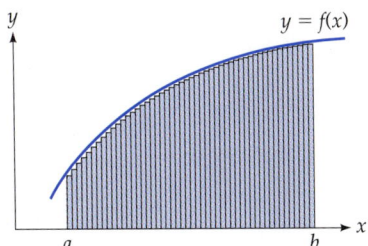

Area under *f* from *a* to *b*
approximated by 50 rectangles

On the right above, this same area is approximated by *fifty* rectangles, giving a much better approximation: The white "error area" between the curve and the rectangles is so small as to be almost invisible.

These diagrams suggest that more rectangles give a better approximation. In fact, the *exact* area under the curve is defined as the *limit* of the approximations as the number of rectangles *approaches infinity.* The following example shows how to carry out such an approximation for the area under a given curve, and afterward we will find the *exact* area by letting the number of rectangles approach infinity.

EXAMPLE 1   Approximating Area by Rectangles

Approximate the area under the curve $f(x) = x^2$ from 1 to 2 by five rectangles. Use rectangles with equal bases and with heights equal to the height of the curve at the left-hand edge of the rectangle.

Solution

For five rectangles, we divide the distance from $a = 1$ to $b = 2$ into five equal parts, so that each rectangle has width

$$\Delta x = \frac{2 - 1}{5} = \frac{1}{5} = 0.2 \qquad \text{For } n \text{ rectangles, } \Delta x = \frac{b - a}{n}$$

Along the x-axis beginning at $x = 1$ we mark successive points with spacing $\Delta x = 0.2$, giving points 1, 1.2, 1.4, 1.6, 1.8, and 2, as shown below. Above each of the resulting subintervals we draw a rectangle whose height is the height of the curve at the left-hand edge of that rectangle. The curve is $y = x^2$, so these heights are the squares of the corresponding x-values.

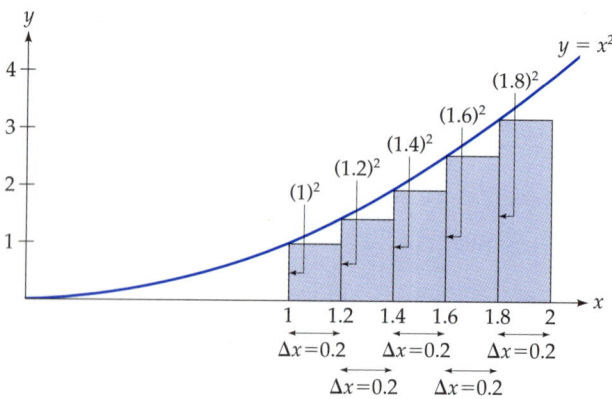

Area from 1 to 2
approximated by 5 rectangles

An enlarged view of the middle rectangle is shown below.

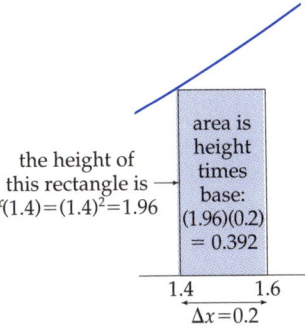

the height of
this rectangle is →
$f(1.4)=(1.4)^2=1.96$

area is
height
times
base:
(1.96)(0.2)
= 0.392

1.4   1.6
$\Delta x = 0.2$

Middle rectangle

The sum of the areas of the five rectangles, height times base $\Delta x = 0.2$ for each rectangle, is

| First rectangle | Second rectangle | Third rectangle | Fourth rectangle | Fifth rectangle | Adding height times base for each rectangle |

$$(1)^2 \cdot (0.2) + (1.2)^2 \cdot (0.2) + (1.4)^2 \cdot (0.2) + (1.6)^2 \cdot (0.2) + (1.8)^2 \cdot (0.2)$$

$$= \ 0.2 + 0.288 + 0.392 + 0.512 + 0.648 = 2.04$$

Multiplying out and summing

Therefore, the area under the curve is approximately 2.04 square units.

■

As we saw earlier, using only five rectangles does not give a very accurate approximation for the true area under the curve. For greater accuracy we use more rectangles, calculating the area in the same way. The following table gives the "rectangular approximation" for the area under the curve in Example 1 for larger numbers of rectangles. The calculations were carried out on a graphing calculator using the program* on page 774, rounding answers to three decimal places.

Number of Rectangles	Sum of Areas of Rectangles	
5	2.04	← Found in Example 1
10	2.185	The sum of the
100	2.318	areas is
1,000	2.332	approaching
10,000	2.333	2.333...

* See the Preface for information on how to obtain this and other programs.

The areas in the right-hand column are approaching 2.333... $= 2\frac{1}{3}$, which is the *exact* area under the curve (as we will verify later). Therefore:

The area under the curve $f(x) = x^2$ from 1 to 2 is $2\frac{1}{3}$ square units.

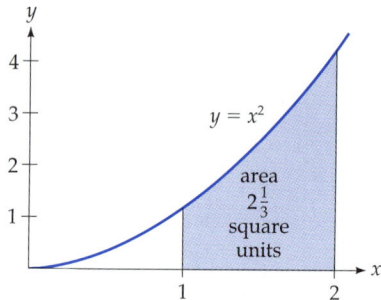

Areas are given in "square units," meaning that if the units on the graph are inches, feet, or some other unit, then the area is in *square* inches, *square* feet, or, in general, some other *square* units.

Definite Integral

Approximating the area under a nonnegative function f by n rectangles means multiplying heights $f(x)$ by widths Δx and adding, obtaining

$$f(x_1) \cdot \Delta x + f(x_2) \cdot \Delta x + f(x_3) \cdot \Delta x + \cdots + f(x_n) \cdot \Delta x$$

The general procedure is as follows:

Area Under f from a to b Approximated by n Rectangles

1. Calculate the rectangle width $\Delta x = \dfrac{b - a}{n}$.

2. Find x-values $x_1, x_2, \cdots, x_n$ by successive additions of Δx beginning with $x_1 = a$.

3. Calculate the sum:

$$f(x_1) \cdot \Delta x + f(x_2) \cdot \Delta x + f(x_3) \cdot \Delta x + \cdots + f(x_n) \cdot \Delta x$$

The sum in step 3 is called a *Riemann sum*, after the great German mathematician Georg Bernhard Riemann (1826–1866).* The *limit* of the

---

* Actually, Riemann sums are slightly more general, allowing the subintervals to have different widths, with each width multiplied by the function evaluated at *any* x value within that subinterval. For a continuous function, any Riemann sum will approach the same limiting value as the rectangles become arbitrarily narrow, so we may restrict our attention to the particular Riemann sums defined above (sometimes called "left Riemann sums").

Riemann sum as the number n of rectangles approaches infinity gives the *area under the curve*, and is called *the definite integral of the function f from a to b*, written $\int_a^b f(x)\, dx$. Formally:

Definite Integral

Let f be a continuous function on an interval $[a, b]$. The definite integral of f from a to b is defined as

$$\int_a^b f(x)\, dx = \lim_{n\to\infty}\left[f(x_1)\cdot\Delta x + f(x_2)\cdot\Delta x + \cdots + f(x_n)\cdot\Delta x \right]$$

where $\Delta x = \dfrac{b-a}{n}$, and $x_1, x_2, \ldots, x_n$ are x-values beginning with $x_1 = a$ and obtained by successive additions of Δx. If f is nonnegative on $[a, b]$, then the definite integral gives the *area under the curve from a to b*.

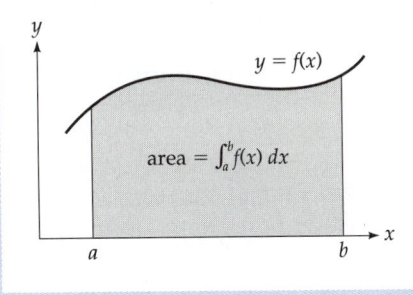

The numbers a and b are called the *lower* and *upper limits of integration*.

Fundamental Theorem of Integral Calculus

A function followed by a vertical bar $\Big|_a^b$ with numbers a and b means evaluate the function at the *upper* number b and then subtract the evaluation at the *lower* number a.

$$F(x)\,\Big|_a^b \;=\; F(b) \;-\; F(a)$$

Evaluation at   Evaluation at
upper number   lower number

EXAMPLE 2 **Using the Evaluation Notation**

$$x^2 \Big|_3^5 = (5)^2 - (3)^2 = 25 - 9 = 16$$

Evaluation notation

x^2 at $x = 5$ x^2 at $x = 3$ Simplifying

PRACTICE PROBLEM 1

Evaluate $\sqrt{x} \Big|_4^{25}$.

Solution at the back of the book

The following *Fundamental Theorem of Integral Calculus* shows how to evaluate definite integrals by using *indefinite* integrals. A geometric and intuitive justification of the theorem is given at the end of this section.

Fundamental Theorem of Integral Calculus

For a continuous function f on an interval $[a, b]$,

$$\int_a^b f(x)\, dx = F(b) - F(a)$$

The right-hand side may be written $F(x) \Big|_a^b$

where F is any antiderivative of f.

The theorem is "fundamental" in that it establishes a deep and unexpected connection between definite integrals (limits of Riemann sums) and antiderivatives. It says that definite integrals may be evaluated in two simple steps:

1. Find an *indefinite* integral of the function (omitting the C).
2. *Evaluate* the result at b and *subtract* the evaluation at a.

EXAMPLE 3 **Finding a Definite Integral by the Fundamental Theorem**

Find $\displaystyle\int_1^2 x^2\, dx$.

Solution

Because x^2 is continuous on [1, 2], we may use the Fundamental Theorem.

$$\int_1^2 x^2\,dx = \underbrace{\frac{1}{3}x^3}\bigg|_1^2 = \underbrace{\frac{1}{3}2^3} - \underbrace{\frac{1}{3}1^3} = \underbrace{\frac{8}{3} - \frac{1}{3}} = \frac{7}{3}$$

Integrating x^2 Evaluating $\frac{1}{3}x^3$ at $x = 2$ And at $x = 1$ Subtracting

Since definite integrals give areas, this result means that the area under the (nonnegative) function $f(x) = x^2$ is $\frac{7}{3}$ square units.

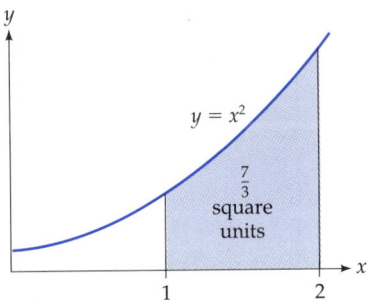

Earlier we found this same area by the much more laborious process of calculating Riemann sums and taking the limit as the number of rectangles approached infinity, and our answer here, $\frac{7}{3}$, agrees with the answer there, $2\frac{1}{3}$ square units. Whenever possible, we will calculate definite integrals and find areas in this much simpler way, using the Fundamental Theorem of Integral Calculus.

Graphing Calculator Exploration

a. Graph $y_1 = x^2$ on the graphing window [0, 2] by [-1, 4].

b. Verify the result of Example 3 by having your graphing calculator find the definite integral of x^2 from 1 to 2. [*Hint:* Use a command like FnInt or $\int f(x)\,dx$.]

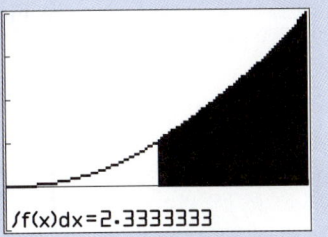

PRACTICE PROBLEM 2   Find the area under $y = x^3$ from 0 to 2 by evaluating $\int_0^2 x^3 \, dx$.

Solution at the back of the book

EXAMPLE 4   Finding the Area Under a Curve

Find the area under $y = e^{2x}$ from $x = 0$ to $x = 1$.

Solution   Because e^{2x} is nonnegative and continuous on $[0, 1]$, the area is given by the definite integral:

$$\int_0^1 e^{2x} \, dx = \underbrace{\left. \frac{1}{2} e^{2x} \right|_0^1}_{\substack{\text{Indefinite integral} \\ \text{(using the formula} \\ \text{for } \int e^{ax} \, dx)}} = \underbrace{\frac{1}{2} e^2 - \frac{1}{2} e^0}_{\substack{\text{Evaluating} \\ \text{at } x = 1 \text{ and} \\ \text{at } x = 0}} = \underbrace{\frac{1}{2} e^2 - \frac{1}{2}}_{\substack{\text{Simplifying} \\ \text{(using } e^0 = 1)}}$$

Therefore, the area is $\frac{1}{2} e^2 - \frac{1}{2}$ square units.

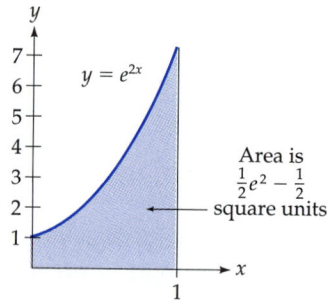

Area is
$\frac{1}{2} e^2 - \frac{1}{2}$
square units

We leave the answer in this "exact" form (in terms of the number e). In an application we would use a calculator to approximate this answer as 3.19 square units.

EXAMPLE 5   Finding the Area Under a Curve

Find the area under $f(x) = \frac{1}{x}$ from $x = 1$ to $x = e$.

Solution   Because $\frac{1}{x}$ is nonnegative and continuous on $[1, e]$, the area is:

$$\int_1^e \frac{1}{x} \, dx = \left. \ln x \right|_1^e = \underbrace{\ln e}_{1} - \underbrace{\ln 1}_{0} = 1$$

Therefore, the area is 1 square unit.

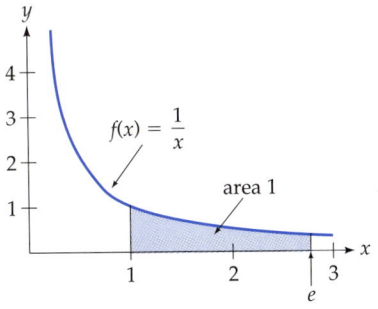

![calculator icon] **Graphing Calculator Exploration**

Use a graphing calculator to evaluate the definite integral $\int_0^1 e^{-x^2}\,dx$. (This definite integral, which is important in probability and statistics, cannot be evaluated by the Fundamental Theorem because the function e^{-x^2} has no simple antiderivative. Your calculator may find the answer by approximating the area under the curve by modified Riemann sums.)

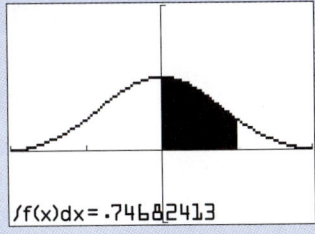

Definite integrals have many of the properties of indefinite integrals. These properties follow from interpreting definite integrals as limits of Riemann sums.

Properties of Definite Integrals

$$\int_a^b c \cdot f(x)\,dx = c \int_a^b f(x)\,dx$$

A constant may be moved across the integral sign

$$\int_a^b \left[f(x) \pm g(x) \right] dx = \int_a^b f(x)\,dx \pm \int_a^b g(x)\,dx$$

⎣ Read both upper signs or both lower signs ⎦

The integral of a sum is the sum of the integrals (and similarly for differences)

EXAMPLE 6   Using the Properties of Definite Integrals

Find the area under $y = 24 - 6x^2$ from -1 to 1.

Solution

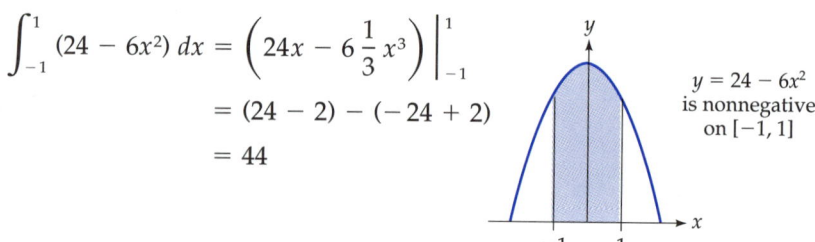

$$\int_{-1}^{1} (24 - 6x^2)\, dx = \left(24x - 6\frac{1}{3} x^3 \right) \Big|_{-1}^{1}$$

$$= (24 - 2) - (-24 + 2)$$

$$= 44$$

$y = 24 - 6x^2$ is nonnegative on $[-1, 1]$

Therefore, the area is 44 square units.

Total Cost of a Succession of Units

Given a marginal cost function, to find the *total* cost of producing, say, units 100 to 400, we could proceed as follows: *integrate* the marginal cost to find the total cost, *evaluate* at 400 to find the total cost up to unit 400, and *subtract* the evaluation at 100 to leave just the cost of units 100 to 400. However, these steps of integrating, evaluating, and subtracting are just the steps in evaluating a definite integral by the Fundamental Theorem. Therefore, the cost of a succession of units is equal to the definite integral of the marginal cost function.

Cost of a Succession of Units

For a marginal cost function $MC(x)$

$$\binom{\text{Total cost of}}{\text{units } a \text{ to } b} = \int_{a}^{b} MC(x)\, dx$$

EXAMPLE 7   Finding the Cost of a Succession of Units

A company's marginal cost function is $MC(x) = \frac{75}{\sqrt{x}}$, where x is the number of units. Find the total cost of producing units 100 to 400.

Solution

$$\begin{pmatrix} \text{Total cost of} \\ \text{units 100 to 400} \end{pmatrix} = \int_{100}^{400} \frac{75}{\sqrt{x}}\, dx = 75 \int_{100}^{400} x^{-1/2}\, dx \qquad \text{Integrating marginal cost}$$

$$= (75 \cdot 2 \cdot x^{1/2}) \Big|_{100}^{400}$$

$$= 150 \cdot (400)^{1/2} - 150 \cdot (100)^{1/2}$$

$$= 150 \cdot 20 - 150 \cdot 10 = 1500$$

The cost of producing units 100 to 400 is $1500.

■

Similarly, integrating *any* rate from a to b gives the *total accumulation* at that rate between a and b.

Total Accumulation at a Given Rate

$$\begin{pmatrix} \text{Total accumulation at} \\ \text{rate } f \text{ from } a \text{ to } b \end{pmatrix} = \int_a^b f(x)\, dx$$

The diagrams below illustrate this idea. In each case, the *curve* represents a *rate*, and the *area under the curve*, given by the definite integral, gives the *total accumulation* at that rate.

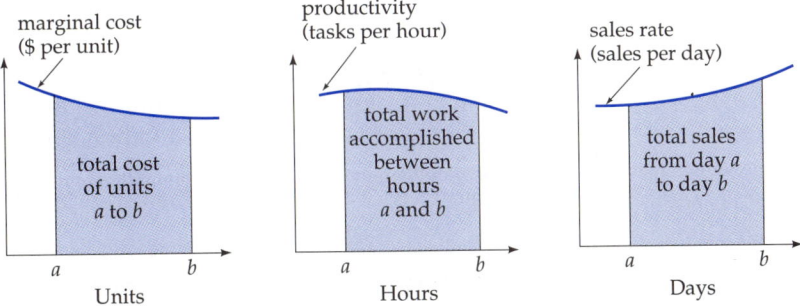

In repetitive tasks, a person's productivity usually increases with practice until it is slowed by monotony.

EXAMPLE 8   **Finding Total Productivity from a Rate**

A technician can test computer chips at the rate of $-3t^2 + 18t + 15$ chips per hour (for $t \le 6$), where t is the number of hours after 9:00 A.M. How many chips can be tested between 10:00 A.M. and 1:00 P.M.?

Solution

The total work accomplished is the integral of this rate from $t = 1$ (10 A.M.) to $t = 4$ (1 P.M.):

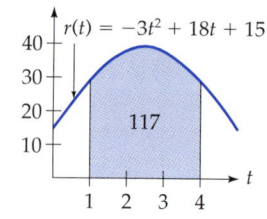

$$\int_1^4 (-3t^2 + 18t + 15)\, dt$$

$$= \left(-t^3 + 18 \cdot \frac{1}{2} t^2 + 15t \right)\Big|_1^4$$

$$= -64 + 144 + 60 - (-1 + 9 + 15) = 117$$

That is, between 10 A.M. and 1 P.M., 117 chips can be tested.

■

Integration Notation

The symbol Σ (the Greek letter S) is used in mathematics to indicate a *sum*, and so the Riemann sum can be written $\sum_1^n f(x_k)\, \Delta x$. The fact that the Riemann sum approaches the definite integral can be expressed:

$$\sum_1^n f(x_k)\, \Delta x \longrightarrow \int_a^b f(x)\, dx$$

as $n \to \infty$
Σ becomes $\int$
Δ becomes d

That is, the n approaching infinity changes the Σ (a Greek S) into an integral sign $\int$ (a "stretched out" S), and the Δ (a Greek D) into a d. In other words, the integral notation reminds us that definite integrals represent *sums of rectangles*, with $f(x)$ and dx representing the height and base of the rectangles being added.

Verification of the Fundamental Theorem of Integral Calculus

The following is a geometric and intuitive justification of the Fundamental Theorem of Integral Calculus. For a continuous and nonnegative function f on an interval $[a, b]$, we define a new function $A(x)$ as the *area under f from a to x*.

$$A(x) = \left(\begin{array}{c}\text{Area under } f \\ \text{from } a \text{ to } x\end{array}\right)$$

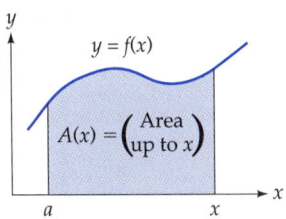

Therefore

$$A(b) = \left(\begin{array}{c}\text{Area under } f \\ \text{from } a \text{ to } b\end{array}\right)$$

Replacing x by b gives the entire area from a to b

and

$$A(a) = 0$$

The area "from a to a" must be zero

Subtracting these last two expressions, $A(b) - A(a)$, gives the total area from a to b [since $A(b)$ *is* the total area, and $A(a)$ is zero]. This same area can be expressed as the definite integral of f from a to b, leading to the equation

$$\int_a^b f(x)\,dx = A(b) - A(a)$$

Area under f from a to b expressed in two ways

If we can show that $A(x)$ is an antiderivative of $f(x)$, then the above equation will show that the definite integral can be evaluated by an antiderivative evaluated at the upper and lower limits and subtracted, which will verify the Fundamental Theorem. To show that $A(x)$ is an antiderivative of $f(x)$, we show that $A'(x) = f(x)$, differentiating $A(x)$ by the definition of the derivative:

$$A'(x) = \lim_{h \to 0} \frac{A(x+h) - A(x)}{h}$$

In the numerator, $A(x + h)$ is the area under the curve up to $x + h$ and $A(x)$ is the area up to x, and subtracting them, $A(x + h) - A(x)$, leaves just the area from x to $x + h$, as shown below.

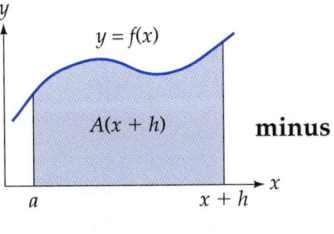

Area up to $x + h$

minus

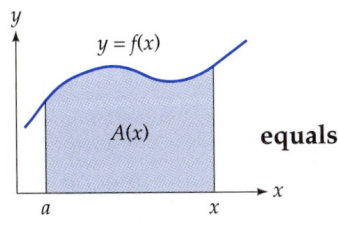

Area up to x

equals

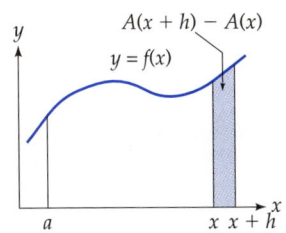

Area from x to $x + h$

When h is small, this last area can be approximated by a rectangle of base h and height $f(x)$, where the approximation becomes exact as h approaches zero. Therefore, in the limit we may replace $A(x + h) - A(x)$ by the area of the rectangle, $h \cdot f(x)$:

$$A'(x) = \lim_{h \to 0} \overbrace{\frac{A(x + h) - A(x)}{h}}^{\approx\, h\,\cdot\, f(x)} = \lim_{h \to 0} \frac{h \cdot f(x)}{h} = f(x)$$

This equation says that $A'(x) = f(x)$, showing that $A(x)$ *is* an antiderivative of $f(x)$. This completes the verification of the Fundamental Theorem of Integral Calculus.

SUMMARY

We began by approximating the area under a curve from a to b by Riemann sums (sums of rectangles). We then defined the *definite integral* $\int_a^b f(x)\, dx$ as the *limit* of the Riemann sum as the number of rectangles approaches infinity. The Fundamental Theorem of Integral Calculus showed how to evaluate definite integrals much more simply by evaluating *indefinite* integrals:

$$\int_a^b f(x)\, dx = F(x)\Big|_a^b \qquad \textcolor{blue}{F \text{ is any antiderivative of } f}$$

Distinguish carefully between definite and indefinite integrals: A *definite* integral is a *number*, whereas an *indefinite* integral is a function plus an arbitrary constant.

As for the *uses* of definite integrals, the definite integral of a nonnegative function gives the *area under the curve:*

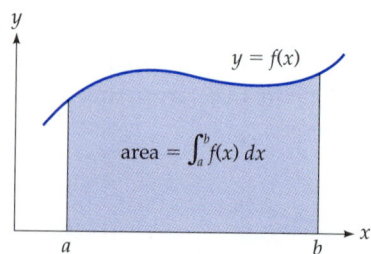

The definite integral of a *rate* gives the *total accumulation* at that rate:

$$\int_a^b f(x)\, dx = \left(\begin{array}{c}\text{Total accumulation at} \\ \text{rate } f \text{ from } a \text{ to } b\end{array}\right)$$

Riemann sums can be calculated for *any* continuous function, even one that takes negative values. The following diagram illustrates a Riemann sum for a continuous function that takes positive and negative values.

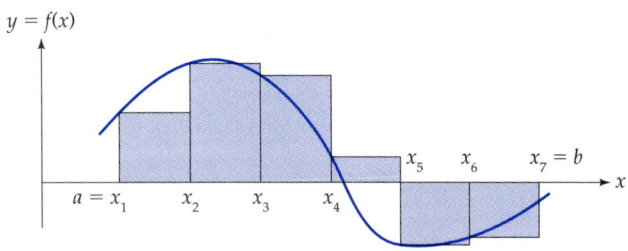

EXERCISES 9.3

For each function:

i. *Approximate* the area under the curve from a to b by calculating a Riemann sum with the given number of rectangles. Use the method described in Example 1 on pages 760–761, rounding to three decimal places.

ii. Find the *exact* area under the curve from a to b by evaluating an appropriate definite integral by the Fundamental Theorem.

1. $f(x) = 2x$ from $a = 1$ to $b = 2$. For part (i), use 5 rectangles.

2. $f(x) = x^2 + 1$ from $a = 0$ to $b = 1$. For part (i), use 5 rectangles.

 3. $f(x) = \sqrt{x}$ from $a = 1$ to $b = 4$. For part (i), use 6 rectangles.

 4. $f(x) = e^x$ from $a = -1$ to $b = 1$. For part (i), use 8 rectangles.

 5. $f(x) = \dfrac{1}{x}$ from $a = 1$ to $b = 2$. For part (i), use 10 rectangles.

 6. $f(x) = \dfrac{1}{\sqrt{x}}$ from $a = 1$ to $b = 4$. For part (i), use 6 rectangles.

On the next page are programs for calculating Riemann sums (one program for the TI-83 graphing calculator and one in BASIC). Use one of these programs (modified as necessary for your calculator or computer) or some other program to calculate the following Riemann sums.

i. Calculate the Riemann sum for each function for the following values of n: 10, 100, and 1000 ($n = 10{,}000$ is optional). Make a table of the answers, rounded to three decimal places.

ii. Find the *exact* value of the area under the curve by evaluating an appropriate definite integral using the Fundamental Theorem. The values of the Riemann sums from part (i) should approach this number.

7. $f(x) = 2x$ from $a = 1$ to $b = 2$

8. $f(x) = x^2 + 1$ from $a = 0$ to $b = 1$

9. $f(x) = \sqrt{x}$ from $a = 1$ to $b = 4$

10. $f(x) = e^x$ from $a = -1$ to $b = 1$

11. $f(x) = \dfrac{1}{x}$ from $a = 1$ to $b = 2$

12. $f(x) = \dfrac{1}{\sqrt{x}}$ from $a = 1$ to $b = 4$

BASIC Program for Riemann Sums	TI-83 Program† for Riemann Sums
(Lines 2, 3, 4, and 8 to be completed as indicated in small type)	*(Enter the function in y_1 before executing)*

BASIC Program for Riemann Sums

(Lines 2, 3, 4, and 8 to be completed as indicated in small type)

Riemsum = 0
a =   fill in beginning x-value
b =   fill in ending x-value
n =   fill in number of rectangles
delta = (b − a)/n
x = a
FOR i = 1 to n

Riemsum = Riemsum + $\left(\begin{smallmatrix}\text{fill in}\\\text{function}\end{smallmatrix}\right)$*delta

x = x + delta
NEXT i
PRINT "RIEMANN SUM IS ", Riemsum
END

TI-83 Program† for Riemann Sums

(Enter the function in y_1 before executing)

$0 \rightarrow S$
Disp "USES FUNCTION Y_1"
Prompt A, B, N
$(B-A)/N \rightarrow D$
$A \rightarrow X$
For(I, 1, N)
$S+Y_1{*}D \rightarrow S$
$X+D \rightarrow X$
End
Disp "RIEMANN SUM IS ", S

Find the area under each curve between the given *x*-values. For Exercises 13–18 also make a sketch of the curve, showing the region.

13. $f(x) = x^2$ from $x = 0$ to $x = 3$

14. $f(x) = x$ from $x = 0$ to $x = 4$

15. $f(x) = 4 - x$ from $x = 0$ to $x = 4$

16. $f(x) = 1 - x^2$ from $x = -1$ to $x = 1$

17. $f(x) = \dfrac{1}{x}$ from $x = 1$ to $x = 2$

18. $f(x) = e^x$ from $x = 0$ to $x = 1$

19. $f(x) = 8x^3$ from $x = 1$ to $x = 3$

20. $f(x) = 6x^2$ from $x = 2$ to $x = 3$

21. $f(x) = 6x^2 + 4x - 1$ from $x = 1$ to $x = 2$

22. $f(x) = 27 - 3x^2$ from $x = 1$ to $x = 3$

23. $f(x) = \dfrac{1}{\sqrt{x}}$ from $x = 4$ to $x = 9$

24. $f(x) = \dfrac{1}{x^2}$ from $x = 1$ to $x = 3$

25. $f(x) = 8 - 4\sqrt[3]{x}$ from $x = 0$ to $x = 8$

26. $f(x) = 9 - 3\sqrt{x}$ from $x = 0$ to $x = 9$

27. $f(x) = \dfrac{1}{x}$ from $x = 1$ to $x = 5$

28. $f(x) = \dfrac{1}{x}$ from $x = e$ to $x = e^3$

29. $f(x) = x^{-1} - x^2$ from $x = 1$ to $x = 2$

30. $f(x) = 6e^{2x}$ from $x = 0$ to $x = 2$

31. $f(x) = 2e^x$ from $x = 0$ to $x = \ln 3$

32. $f(x) = e^{-x}$ from $x = 0$ to $x = 1$

33. $f(x) = e^{\frac{1}{2}x}$ from $x = 0$ to $x = 2$

34. $f(x) = e^{x/3}$ from $x = 0$ to $x = 3$

 For each function:

a. Integrate ("by hand") to find the area under the curve between the given *x*-values.

b. Verify your answer to part (a) by having your calculator graph the function and find the area [using a command like FnInt or $\int f(x)\,dx$].

35. $f(x) = 12 - 3x^2$ from $x = 1$ to $x = 2$

36. $f(x) = 9x^2 - 6x + 1$ from $x = 1$ to $x = 2$

37. $f(x) = \dfrac{1}{x^3}$ from $x = 1$ to $x = 4$

38. $f(x) = \dfrac{1}{\sqrt[3]{x}}$ from $x = 8$ to $x = 27$

39. $f(x) = 2x + 1 + x^{-1}$ from $x = 1$ to $x = 2$

40. $f(x) = e^x$ from $x = 0$ to $x = 3$

---

† See the Preface for information on how to obtain this and other programs.

Evaluate each definite integral.

41. $\int_0^1 (x^{99} + x^9 + 1)\, dx$

42. $\int_2^4 (1 + x^{-2})\, dx$

43. $\int_1^2 (6t^2 - 2t^{-2})\, dt$

44. $\int_{-2}^2 (3w^2 - 2w)\, dw$

45. $\int_1^4 \frac{1}{y^2}\, dy$

46. $\int_1^4 \frac{1}{\sqrt{z}}\, dz$

47. $\int_1^e \frac{dx}{x}$

48. $\int_1^{e^2} \frac{3}{x}\, dx$

49. $\int_1^3 (9x^2 + x^{-1})\, dx$

50. $\int_1^2 (x^{-1} - 4x^2)\, dx$

51. $\int_{-2}^{-1} 3x^{-1}\, dx$

52. $\int_{-3}^{-1} (1 + x^{-1})\, dx$

53. $\int_0^1 12e^{3x}\, dx$

54. $\int_0^2 3e^{\frac{1}{2}x}\, dx$

55. $\int_{-1}^1 5e^{-x}\, dx$

56. $\int_0^1 (6x^2 - 4e^{2x})\, dx$

57. $\int_{\ln 2}^{\ln 3} e^x\, dx$

58. $\int_0^{\ln 5} e^x\, dx$

59. $\int_1^2 \frac{(x+1)^2}{x}\, dx$

60. $\int_1^2 \frac{(x+1)^2}{x^2}\, dx$

 Use a graphing calculator to evaluate each definite integral, rounding answers to three decimal places. [*Hint:* Use a command like FnInt or $\int f(x)\, dx$.]

61. $\int_0^2 \frac{1}{x^2 + 1}\, dx$

62. $\int_{-1}^1 \sqrt{x^4 + 1}\, dx$

63. $\int_{-1}^1 e^{x^2}\, dx$

64. $\int_{-2}^2 e^{(-1/2)x^2}\, dx$

65. $\int_0^4 \sqrt{x}\, e^x\, dx$

66. $\int_1^4 x^x\, dx$

67. Omitting the C in Definite Integrals

   a. Evaluate the definite integral $\int_0^3 x^2\, dx$.

   b. Evaluate the same definite integral by completing the following calculation, in which the antiderivative includes a constant C.

$$\int_0^3 x^2\, dx = \left(\frac{1}{3}x^3 + C\right)\Big|_0^3 = \cdots$$

[The constant C should cancel out, giving the same answer as in part (a)].

   c. Explain why the constant will cancel out of *any* definite integral. (We therefore omit the constant in definite integrals. However, be sure to keep the +C in *indefinite* integrals.)

68. Evaluate $\int_1^1 \dfrac{x^{43}e^{\sqrt[3]{x^2}} + x^{-39}}{\ln \sqrt[17]{x^3 + 11} + x^{199}}\, dx$

(*Hint:* No work necessary.)

69. Show that for any number $a > 0$,

$$\int_1^a \frac{1}{x}\, dx = \ln a$$

This equation is often used as a *definition* of natural logarithms, defining $\ln a$ as the area under the curve $y = \dfrac{1}{x}$ between 1 and a.

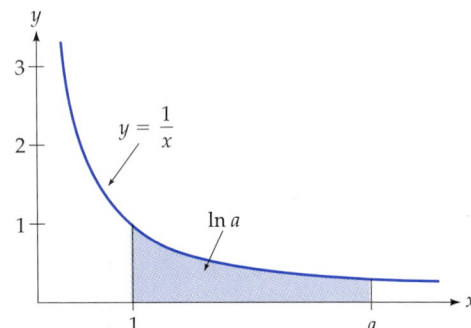

70. a. Try to evaluate the integral $\int_{-1}^1 \dfrac{1}{x^2}\, dx$.

   b. Explain why the answer is negative in spite of the fact that the integrand is nonnegative, and so the integral should be positive. (*Hint:* In order to use the Fundamental Theorem, or even to *define* the definite integral, the integrand must be continuous on the interval. Is it?)

71–72: Geometric Integration   Find $\int_0^4 f(x)\, dx$ for the function $f(x)$ graphed on the next page. (*Hint:* No calculus is necessary—just as we used integrals to find areas, we may use areas to find integrals. The curves shown are quarter circles.)

71.

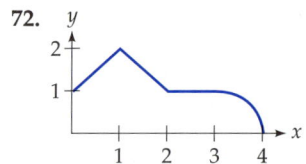

72.

APPLIED EXERCISES (Most require **)**

73. General: Electrical Consumption On a hot summer afternoon, a city's electricity consumption is $-3t^2 + 18t + 10$ units per hour, where t is the number of hours after noon ($t \leq 6$). Find the total consumption of electricity between the hours of 1 and 5 P.M.

74. General: Weight An average child of age x years gains weight at the rate of $3.9x^{1/2}$ pounds per year (for $x \leq 16$). Find the total weight gain from age 1 to age 9.

75–76: General: Repetitive Tasks After t hours of work a bank clerk can process checks at the rate of $r(t)$ checks per hour for the function $r(t)$ given below. How many checks will the clerk process during the first three hours (time 0 to time 3)?

75. $r(t) = -t^2 + 90t + 5$

76. $r(t) = -t^2 + 60t + 9$

77–78: Business: Cost A company's marginal cost function is MC(x) (given below), where x is the number of units. Find the total cost of the first hundred units ($x = 0$ to $x = 100$).

77. MC(x) $= 6e^{-0.02x}$ **78.** MC(x) $= 8e^{-0.01x}$

79. General: Price Increase The price of a double-dip ice cream cone is increasing at the rate of $13e^{0.08t}$ cents per year, where t is measured in years and $t = 0$ corresponds to 1995. Find the total change in price between the years 1995 and 2005.

80. Business: Sales An automobile dealer estimates that the newest model car will sell at the rate of $\dfrac{30}{t}$ cars per month, where t is measured in months and $t = 1$ corresponds to the beginning of January. Find the number of cars that will be sold from the beginning of January to the beginning of May.

81. Business: Tin Consumption World consumption of tin is running at the rate of $0.23e^{0.01t}$ million tons per year, where t is measured in years and $t = 0$ corresponds to the beginning of 1995. Find the total consumption of tin from the beginning of 1995 to the beginning of the year 2005.

82. Sociology: Marriages There are approximately $2.3e^{0.01t}$ million marriages per year in America, where t is the number of years since 1995. Assuming that this rate continues, find the number of marriages from the year 1995 to the year 2005.

83. Behavioral Science: Learning A student can memorize words at the rate of $6e^{-t/5}$ words per minute after t minutes. Find the total number of words that the student can memorize in the first 10 minutes.

84. Biomedical: Epidemics An epidemic is spreading at the rate of $12e^{0.2t}$ new cases per day, where t is the number of days since the epidemic began. Find the total number of new cases in the first 10 days of the epidemic.

85. Economics: Pareto's Law The economist Vilfredo Pareto estimated that the number of people who have an income between A and B dollars ($A < B$) is given by a definite integral of the form

$$N = \int_A^B ax^{-b}\,dx \qquad (b \neq -1)$$

where a and b are constants. Solve this integral.

86. Biomedical: Poiseuille's Law According to Poiseuille's law, the speed of blood in a blood vessel is given by $V = \frac{p}{4Lv}(R^2 - r^2)$, where R is the radius of the blood vessel, r is the distance of the blood from the center of the blood vessel, and p, L, and v are constants determined by the pressure and viscosity of the blood and the length of the vessel. The total blood flow is then given by

$$\left(\begin{array}{c}\text{Total} \\ \text{blood flow}\end{array}\right) = \int_0^R 2\pi \frac{p}{4Lv}(R^2 - r^2)r\, dr$$

Find the total blood flow by finding this integral (p, L, v, and R are constants).

87–88: Business: Capital Value of an Asset The *capital value* of an asset (such as an oil well) that produces a continuous stream of income is the sum of the present value of all future earnings from the asset. Therefore, the capital value of an asset that produces income at the rate of $r(t)$ dollars per year (at a continuous interest rate i) is

$$\left(\begin{array}{c}\text{Capital} \\ \text{value}\end{array}\right) = \int_0^T r(t)e^{-it}\, dt$$

where T is the expected life (in years) of the asset.

87. Use the above formula to find the capital value (at interest rate $i = 0.06$) of an oil well that produces income at the constant rate of $r(t) = 240{,}000$ dollars per year for 10 years.

 88. Use the above formula to find the capital value (at interest rate $i = 0.05$) of a uranium mine that produces income at the rate of $r(t) = 560{,}000t^{1/2}$ dollars per year for 20 years.

 89. **General: Area**
a. Use your graphing calculator to find the area between 0 and 1 under the following curves: $y = x$, $y = x^2$, $y = x^3$, and $y = x^4$.
b. Based on your answers to part (a), conjecture a formula for the area under $y = x^n$ between 0 and 1 for any value of $n > 0$.
c. Prove your conjecture by evaluating an appropriate definite integral "by hand."

 90. **General: Dam Construction** Ever since the Johnstown Dam burst in 1889, killing 2200 people, dam construction has become increasingly scientific.
a. Estimate the amount of concrete needed to build the Snake River Dam by finding the area of the cross section shown in the next column and multiplying it by the 574-foot length of the dam. All dimensions are in feet.

(*Hint:* Find the cross-sectional area by integrating and using area formulas.)
b. If a mixing truck carries 300 cubic feet of concrete, about how many truckloads would be needed to build the dam?

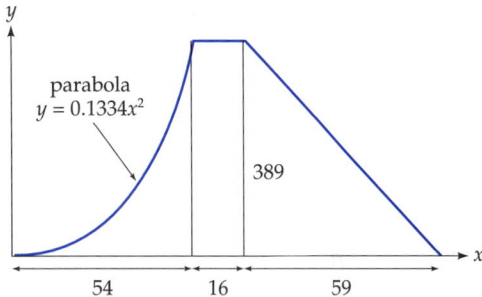

91. **Business: Sales** A dealer predicts that new cars will sell at the rate of $8xe^{-0.1x}$ sales per week in week x. Find the total sales in the first half year (week 0 to week 26).

92. **General: Cigarette Smoking** Reread, if necessary, the Application Preview on pages 757–758.
a. Evaluate the definite integrals

$$\int_0^1 \left(300e^{0.025x} - 240e^{0.02x}\right)dx$$

and $$\int_7^8 \left(300e^{0.025x} - 240e^{0.02x}\right)dx$$

to verify the answers given there for the amount of tar inhaled from the first and last centimeters of the cigarette.
b. Evaluate the definite integral

$$\int_0^8 (300e^{0.025x} - 240e^{0.02x})\, dx$$

to find the amount of tar inhaled from smoking the entire cigarette.

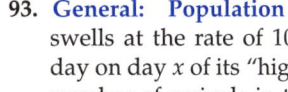 93. **General: Population** A resort community swells at the rate of $100e^{0.4\sqrt{x}}$ new arrivals per day on day x of its "high season." Find the total number of arrivals in the first two weeks (day 0 to day 14).

94. **General: Repetitive Tasks** After t hours of work a medical technician can carry out T-cell counts at the rate of $2x^2e^{-x/4}$ tests per hour. How many tests will the clerk process during the first 8 hours (time 0 to time 8)?

9.4 Further Applications of Definite Integrals: Average Value and Area Between Curves

Introduction

In this section we will use definite integrals for two important purposes: finding average values of functions and finding areas between curves. Average values are used everywhere. Birth weights of babies are compared to average weights, and retirement benefits are determined by average income. Averages eliminate fluctuations, reducing a collection of numbers to a single "representative" number. Areas between curves are used to find quantities from trade deficits to lives saved by seat belts, as we saw on pages 733–734.

Average Value of a Function

The average value of n numbers is found by adding the numbers and dividing by n. For example,

$$\begin{pmatrix} \text{Average of} \\ a, b, \text{ and } c \end{pmatrix} = \frac{1}{3}(a + b + c)$$

How can we find the average value of a *function* on an interval? For example, if a function gives the temperature over a 24-hour period, how can we calculate the *average temperature?* We could, of course, just take the temperature at every hour and then average these 24 values, but this would ignore the temperature at all of the intermediate times. Intuitively, the average should represent a "leveling off" of the curve to a uniform height, the dashed line shown below.

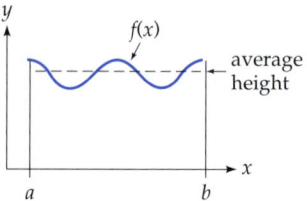

This leveling should use the "hills" to fill in the "valleys," maintaining the same total area under the curve. Therefore, the area under the line (a rectangle with base $(b - a)$ and height up to the line) must equal the area under the curve (the definite integral of the function).

$$\underbrace{(b - a) \left(\begin{array}{c} \text{Average} \\ \text{height} \end{array}\right)}_{\text{Area under line}} = \underbrace{\int_a^b f(x)\,dx}_{\text{Area under curve}} \qquad \text{Equating the two areas}$$

Therefore:

$$\left(\begin{array}{c} \text{Average} \\ \text{height} \end{array}\right) = \frac{1}{b - a} \int_a^b f(x)\,dx \qquad \text{Dividing by } (b - a)$$

This formula gives the average (or "mean") value of a continuous function on an interval.

Average Value of a Function

$$\left(\begin{array}{c} \text{Average value} \\ \text{of } f \text{ on } [a, b] \end{array}\right) = \frac{1}{b - a} \int_a^b f(x)\,dx$$

Average value is the definite integral of the function divided by the length of the interval

Finding the average value of a function by integrating and dividing by $b - a$ is analogous to averaging n numbers by adding and dividing by n (since integrals are continuous sums). A derivation of this formula by Riemann sums is given at the end of this section.

EXAMPLE 1 **Finding the Average Value of a Function**

Find the average value of $f(x) = \sqrt{x}$ from $x = 0$ to $x = 9$.

Solution

$$\left(\begin{array}{c} \text{Average} \\ \text{value} \end{array}\right) = \frac{1}{9 - 0} \int_0^9 \sqrt{x}\,dx = \frac{1}{9} \int_0^9 x^{1/2}\,dx \qquad \begin{array}{l} \text{Integral divided} \\ \text{by the length of the} \\ \text{interval} \end{array}$$

$$= \frac{1}{9} \frac{2}{3} x^{3/2} \Big|_0^9 = \frac{2}{27} 9^{3/2} - \frac{2}{27} 0^{3/2} \qquad \begin{array}{l} \text{Integrating and} \\ \text{evaluating} \end{array}$$

$$= \frac{2}{27} (\sqrt{9})^3 - 0 = \frac{2}{27} 27 = 2 \qquad \text{Simplifying}$$

The average value of $f(x) = \sqrt{x}$ over the interval $[0, 9]$ is 2.

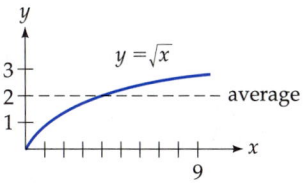

EXAMPLE 2 Finding Average Population

The population of the United States is predicted to be $P(t) = 263e^{0.01t}$ million people, where t is the number of years since 1995. Find the average population between the year 2000 and 2010.

Solution   We integrate from $t = 5$ (year 2000) to $t = 15$ (year 2010).

$$\left(\begin{array}{c}\text{Average} \\ \text{value}\end{array}\right) = \frac{1}{15 - 5} \int_5^{15} 263e^{0.01t} \, dt \qquad \begin{array}{l}\text{Integral divided} \\ \text{by the length of the} \\ \text{interval}\end{array}$$

$$= \frac{263}{10} \int_5^{15} e^{0.01t} \, dt$$

$$= 26.3 \, \frac{1}{0.01} e^{0.01t} \, \Big|_5^{15} \qquad \begin{array}{l}\text{Integrating by the} \\ \int e^{ax} \, dx \text{ formula}\end{array}$$

$$= 2630 \, e^{0.15} - 2630 \, e^{0.05} \approx 291 \qquad \begin{array}{l}\text{Evaluating, using a} \\ \text{calculator}\end{array}$$

The average population of the United States during the first decade of the twenty-first century will be about 291 million people.   ∎

PRACTICE PROBLEM 1   Find the average value of $f(x) = 3x^2$ from $x = 0$ to $x = 2$.

Solution at the back of the book

Areas Between Curves: Integrating "Upper Minus Lower"

We know that definite integrals give areas under curves. To calculate the area *between* two curves, we take the area under the *upper* curve and subtract the area under the *lower* curve.

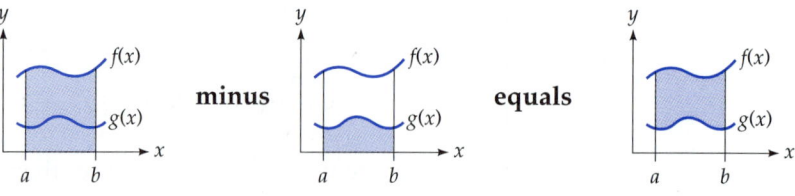

In terms of integrals:

$$\int_a^b f(x) \, dx \qquad - \qquad \int_a^b g(x) \, dx \qquad = \qquad \int_a^b [\, f(x) - g(x)] \, dx$$

$$\underset{\substack{\uparrow \\ \text{Upper} \\ \text{function}}}{} \quad \underset{\substack{\uparrow \\ \text{Lower} \\ \text{function}}}{}$$

Therefore, the area between the curves can be written as a single integral:

Area Between Curves

The area between two continuous curves $f(x) \geq g(x)$ on $[a, b]$ is

$$\left(\begin{array}{c} \text{Area between} \\ f \text{ and } g \text{ on } [a, b] \end{array} \right) = \int_a^b [f(x) - g(x)] \, dx \qquad \begin{array}{l} \text{Integrate "upper} \\ \text{minus lower"} \end{array}$$

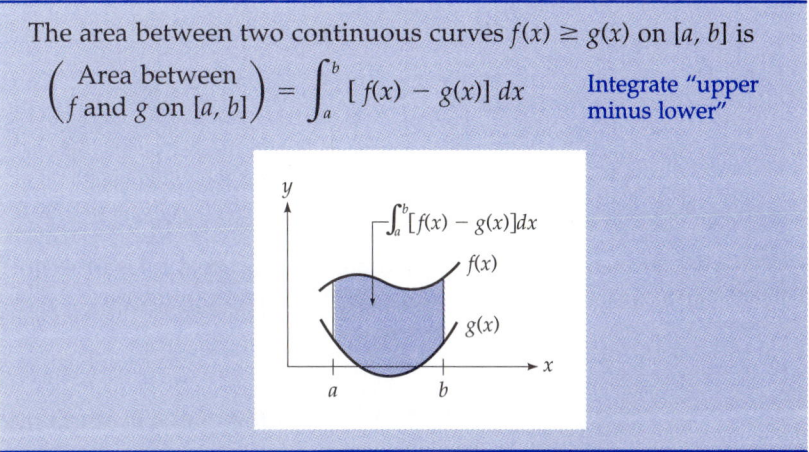

Integrating "upper minus lower" works regardless of whether one or both curves dip below the x-axis.

EXAMPLE 3   Finding Area Between Curves

Find the area between $y = 3x^2 + 4$ and $y = 2x - 1$ from $x = -1$ to $x = 2$.

Solution   The area is shown in the diagram on the right. (You may need to make a similar rough sketch for each problem to see which curve is "upper" and which is "lower.") We integrate "upper minus lower" between the given x-values.

$$\int_{-1}^{2} [\underbrace{(3x^2 + 4)}_{\text{Upper}} - \underbrace{(2x - 1)}_{\text{Lower}}] \, dx = \int_{-1}^{2} (3x^2 + 4 - 2x + 1) \, dx$$

$$= \int_{-1}^{2} (3x^2 - 2x + 5) \, dx \qquad \text{Simplifying}$$

$$= (x^3 - x^2 + 5x) \Big|_{-1}^{2} \qquad \text{Integrating}$$

$$= (8 - 4 + 10) - (-1 - 1 - 5) \qquad \text{Evaluating}$$

$$= 21 \text{ square units}$$

PRACTICE PROBLEM 2   Find the area between $y = 2x^2 + 1$ and $y = -x^2 - 1$ from $x = -1$ to $x = 1$.

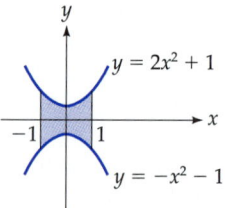

Solution at the back of the book

If the two curves represent *rates* (one unit per another unit), then the area between the curves gives the *total accumulation* at the upper rate minus the lower rate.

REAL WORLD

EXAMPLE 4   Finding Sales From Extra Advertising

A company marketing high-definition televisions expects to sell them at the rate of $2e^{0.05t}$ thousand sets per month, where t is the number of months since they became available. However, with additional advertising using a sports celebrity, they should sell at the rate of $3e^{0.1t}$ thousand sets per month. How many additional sales would result from the celebrity endorsement during the first year?

Solution

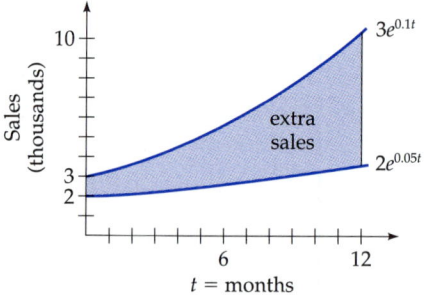

We integrate the difference of the rates from month $t = 0$ (the beginning of the first year) to month $t = 12$ (the end of the year):

$$\int_0^{12} (3e^{0.1t} - 2e^{0.05t})\, dt = \left(3\,\underbrace{\frac{1}{0.1}}_{30}\, e^{0.1t} - 2\,\underbrace{\frac{1}{0.05}}_{40}\, e^{0.05t} \right)\Big|_0^{12}$$

$$= \quad \underbrace{(30e^{1.2} - 40e^{0.6})}_{\text{Evaluating at } t = 12} \quad - \quad \underbrace{(30e^0 - 40e^0)}_{\text{Evaluating at } t = 0}$$

$$\approx \quad \underbrace{26.7 - (-10)}_{\text{Using a calculator}} \quad = 36.7 \qquad\qquad \text{In thousands}$$

The celebrity endorsement should result in 36.7 thousand, or 36,700 additional sales during the first year. (The profits from these additional sales must then be compared to the cost of the celebrity endorsement to decide whether it is worthwhile.)

■

Area Between Curves That Cross

At a point where two curves cross, the upper curve becomes the lower curve and the lower becomes the upper. The area between them must then be calculated by two (or more) integrals, upper minus lower on each interval.

EXAMPLE 5   Finding Area Between Curves That Cross

Find the area between the curves $y = 12 - 3x^2$ and $y = 4x + 5$ from $x = 0$ to $x = 3$.

Solution

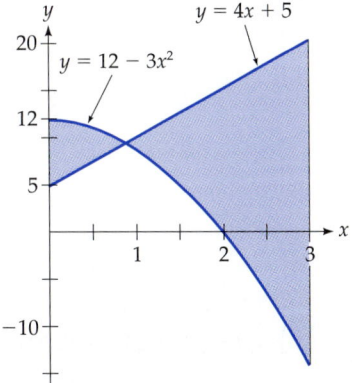

A sketch like this shows that the curves *do* cross. To find where they cross, we set the functions equal to each other and solve.

$$4x + 5 = 12 - 3x^2 \qquad \text{Equating the two functions}$$

$$3x^2 + 4x - 7 = 0 \qquad \text{Combining all terms on the left}$$

$$(3x + 7)(x - 1) = 0 \qquad \text{Factoring}$$

$$x = 1 \qquad \text{The other solution, } x = -7/3,$$
$$\text{is not in the interval } [0, 3]$$

The curves cross at $x = 1$, so we must integrate separately over the intervals $[0, 1]$ and $[1, 3]$. The diagram shows which curve is upper and which is lower on each interval.

$$\int_0^1 [\underbrace{(12 - 3x^2)}_{\text{Upper}} - \underbrace{(4x + 5)}_{\text{Lower}}]\, dx + \int_1^3 [\underbrace{(4x + 5)}_{\text{Upper}} - \underbrace{(12 - 3x^2)}_{\text{Lower}}]\, dx \qquad \begin{array}{l}\text{Integrating upper}\\ \text{minus lower on}\\ \text{each interval}\end{array}$$

$$= \int_0^1 (-3x^2 - 4x + 7)\, dx + \int_1^3 (3x^2 + 4x - 7)\, dx \qquad \begin{array}{l}\text{Simplifying the}\\ \text{integrands}\end{array}$$

$$= (-x^3 - 2x^2 + 7x)\Big|_0^1 + (x^3 + 2x^2 - 7x)\Big|_1^3 \qquad \begin{array}{l}\text{Integrating and}\\ \text{simplifying}\end{array}$$

$$= -1 - 2 + 7 + 27 + 18 - 21 - (1 + 2 - 7) = 32 \qquad \text{Evaluating}$$

Therefore, the area between the curves is 32 square units. ∎

Graphing Calculator Exploration

In Example 5 we used *two* integrals, since "upper" and "lower" switched at $x = 1$.

a. To see what happens if you integrate *without* regard to upper and lower, enter $y_1 = 12 - 3x^2$ and $y_2 = 4x + 5$ and graph them on the window $[0, 3]$ by $[-20, 20]$. Have your calculator find the definite integral of $y_1 - y_2$ on the interval $[0, 3]$. [Use a command like FnInt or $\int f(x)\, dx$. You should get a negative answer, which cannot be correct for an area.]

b. Explain why the answer was negative. (*Hint:* Look at the graph.)

c. Finally, obtain the correct answer for the area by returning to the calculation in part (a) and integrating the *absolute value* of the difference, $|y_1 - y_2|$ [on some calculators, entered as ABS $(y_1 - y_2)$].

Areas Bounded by Curves

Some problems ask for the *area bounded by two curves,* without giving the starting and ending x-values. In such problems the curves completely enclose an area, and the x-values for the upper and lower limits of integration are found by setting the functions equal to each other and solving.

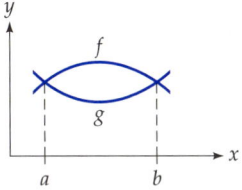

The x-values a and b are
where the curves meet.

EXAMPLE 6   Finding Area Bounded by Curves

Find the area bounded by the curves

$$y = 3x^2 - 12 \text{ and } y = 12 - 3x^2$$

Solution   The x-values for the upper and lower limits of integration are not given, so we find them by setting the functions equal to each other and solving.

$3x^2 - 12 = 12 - 3x^2$	Setting the functions equal to each other
$6x^2 - 24 = 0$	Combining everything on one side
$6(x^2 - 4) = 0$	Factoring
$6(x + 2)(x - 2) = 0$	Factoring further
$x = -2 \quad \text{and} \quad x = 2$	Solving

The smaller of these, $x = -2$, is the lower limit of integration and the larger, $x = 2$, is the upper limit. To determine which function is "upper" and which is "lower," we choose a "test value" between $x = -2$ and $x = 2$ ($x = 0$ will do), which we substitute into each function to see which is larger. Evaluating each of the original functions at the test point $x = 0$ yields

$$3x^2 - 12 = 3(0)^2 - 12 = -12 \quad \text{(Smaller)} \qquad \begin{array}{l} 3x^2 - 12 \\ \text{at } x = 0 \end{array}$$

$$12 - 3x^2 = 12 - 3(0)^2 = 12 \quad \text{(Larger)} \qquad \begin{array}{l} 12 - 3x^2 \\ \text{at } x = 0 \end{array}$$

Therefore, $y = 12 - 3x^2$ is the "upper" function (since it gives a higher y-value) and $y = 3x^2 - 12$ is the "lower" function. We then integrate upper minus lower between the x-values found earlier.

$$\int_{-2}^{2} [\underbrace{12 - 3x^2}_{\text{Upper}} - \underbrace{(3x^2 - 12)}_{\text{Lower}}] \, dx = \int_{-2}^{2} (24 - 6x^2) \, dx \qquad \text{Simplifying}$$

$$= (24x - 2x^3) \Big|_{-2}^{2} \qquad \text{Integrating}$$

$$= (48 - 16) - (-48 + 16) = 64 \qquad \text{Evaluating}$$

The area bounded by the two curves is 64 square units.

■

The two curves $y = 12 - 3x^2$ and $y = 3x^2 - 12$ are shown below. Notice that we were able to calculate the area between them without having to graph them.

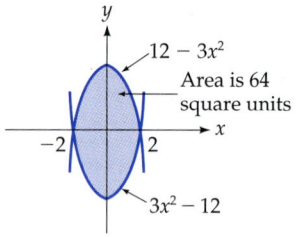

PRACTICE PROBLEM 3

Find the area bounded by $y = 2x^2 - 1$ and $y = 2 - x^2$.

Solution at the back of the book

For curves that intersect at *more* than two points, several integrals may be needed, integrating "upper minus lower" on each interval, as in Example 5. Test points in each interval will determine the upper and lower curves on that interval.

SUMMARY

The average value of a continuous function over an interval is defined as the definite integral of the function divided by the length of the interval:

$$\left(\begin{array}{c} \text{Average value} \\ \text{of } f \text{ on } [a, b] \end{array} \right) = \frac{1}{b - a} \int_{a}^{b} f(x) \, dx$$

To find the area between two curves:

1. If the x-values are not given, set the functions equal to each other and solve for the points of intersection.

2. Use a test point within each interval to determine which curve is "upper" and which is "lower."

3. Integrate "upper minus lower" on each interval.

If a curve lies *below* the x-axis, as in the diagram below, then the "upper" curve is the x-axis ($y = 0$) and the "lower" is the curve $y = f(x)$, and so integrating "upper minus lower" results in integrating the *negative* of the function:

$$\int_a^b \underbrace{[0}_{\text{Upper}} - \underbrace{f(x)]}_{\text{Lower}} \, dx = \int_a^b [-f(x)] \, dx = -\int_a^b f(x) \, dx$$

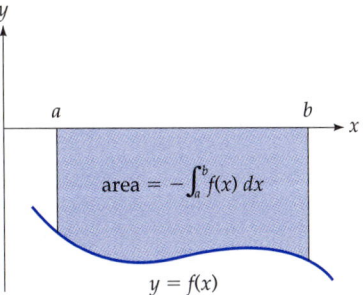

This case need not be remembered separately if you simply remember always to integrate "upper minus lower" over each interval.

Definite integration provides a very powerful method for finding areas, taking us far beyond the few formulas (for rectangles, triangles, and circles) that we knew before calculus.

Average Value Formula Derived from Riemann Sums

The formula for the average value of a function (page 779) can be derived using Riemann sums. For a continuous function f we could define a "sample average" by "sampling" the function at n points and averaging the results. From the interval $[a, b]$, we choose n numbers $x_1, x_2, \cdots, x_n$ from successive subintervals of length $\Delta x = \dfrac{b - a}{n}$, sum the resulting values of the function, and divide by n. This gives a "sample average" of the following form:

$$\frac{1}{n}[f(x_1) + f(x_2) + \cdots + f(x_n)] \qquad \begin{array}{l}\text{Sum of } n \text{ values} \\ \text{divided by } n\end{array}$$

$$= \frac{1}{b-a} \cdot \underbrace{\frac{b-a}{n}}_{\Delta x}[f(x_1) + f(x_2) + \cdots + f(x_n)] \qquad \begin{array}{l}\text{Dividing and} \\ \text{multiplying by } b-a\end{array}$$

$$= \frac{1}{b-a}\underbrace{[f(x_1) + f(x_2) + \cdots + f(x_n)] \cdot \Delta x}_{} \qquad \begin{array}{l}\text{Moving } \Delta x = \frac{b-a}{n} \\ \text{to the right}\end{array}$$

This is a Riemann sum for $\displaystyle\int_a^b f(x)\,dx$

To get a more "representative" average, we increase the number n of sample points. Letting n approach infinity makes the above Riemann sum approach the definite integral $\displaystyle\int_a^b f(x)\,dx$, leading to the definition of the average value of a function:

$$\left(\begin{array}{l}\text{Average value} \\ \text{of } f \text{ over } [a,\, b]\end{array}\right) = \frac{1}{b-a}\int_a^b f(x)\,dx$$

EXERCISES 9.4

Average Value

Find the average value of each function over the given interval.

1. $f(x) = x^2$ on $[0, 3]$ 2. $f(x) = x^3$ on $[0, 2]$

3. $f(x) = 3\sqrt{x}$ on $[0, 4]$ 4. $f(x) = \sqrt[3]{x}$ on $[0, 8]$

5. $f(x) = \frac{1}{x^2}$ on $[1, 5]$ 6. $f(x) = \frac{1}{x^2}$ on $[1, 3]$

7. $f(x) = 2x + 1$ on $[0, 4]$

8. $f(x) = 4x - 1$ on $[0, 10]$

9. $f(x) = 36 - x^2$ on $[-2, 2]$

10. $f(x) = 9 - x^2$ on $[-3, 3]$

11. $f(x) = 3$ on $[10, 50]$ 12. $f(x) = 2$ on $[5, 100]$

13. $f(x) = e^{\frac{1}{2}x}$ on $[0, 2]$ 14. $f(x) = e^{-2x}$ on $[0, 1]$

15. $f(x) = \frac{1}{x}$ on $[1, 2]$ 16. $f(x) = \frac{1}{x}$ on $[1, 10]$

17. $f(x) = x^n$ on $[0, 1]$, where n is a constant $(n > -1)$

18. $f(x) = e^{kx}$ on $[0, 1]$, where k is a constant $(k \neq 0)$

19. $f(x) = ax + b$ on $[0, 2]$, where a and b are constants

20. $f(x) = \frac{1}{x}$ on $[1, c]$, where c is a constant $(c > 1)$

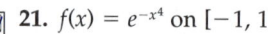 21. $f(x) = e^{-x^4}$ on $[-1, 1]$

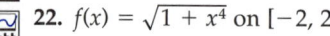 22. $f(x) = \sqrt{1 + x^4}$ on $[-2, 2]$

APPLIED EXERCISES ON AVERAGE VALUE

23–24: Business: Sales A store's sales on day x are given by the function $S(x)$ below. Find the average sales during the first 3 days (day 0 to day 3).

23. $S(x) = 200x + 6x^2$ **24.** $S(x) = 400x + 3x^2$

25. General: Temperature The temperature at time t hours is $T(t) = -0.3t^2 + 4t + 60$ (for $t \leq 12$). Find the average temperature between time 0 and time 10.

26. Behavioral Science: Practice After x practice sessions, a person can accomplish a task in $f(x) = 12x^{-1/2}$ minutes. Find the average time required from the end of session 1 to the end of session 9.

27. Environmental Science: Pollution The amount of pollution in a lake x years after the closing of a chemical plant is $P(x) = \dfrac{100}{x}$ tons (for $x \geq 1$). Find the average amount of pollution between 1 and 10 years after the closing.

28. General: Population The population of the United States is predicted to be $P(t) = 276e^{0.01t}$ million, where t is the number of years after the year 2000. Find the average population between the years 2000 and 2050.

29. Business: Compound Interest A deposit of $1000 at 5% interest compounded continuously will grow to $V(t) = 1000e^{0.05t}$ dollars after t years. Find the average value during the first 40 years (that is, from time 0 to time 40).

30. Biomedical: Bacteria A colony of bacteria is of size $S(t) = 300e^{0.1t}$ after t hours. Find the average size during the first 12 hours (that is, from time 0 to time 12).

31. Business: Profit The MediGenics Corporation expects its profits to be $1.4e^{0.05x^2}$ million dollars per year x years from now. Find the company's average profit over the next decade (year 0 to year 10).

32. General: Population The population of a town is predicted to be $5.2e^{\sqrt{x}}$ thousand people x years from now. Find the average population over the next decade (year 0 to year 10).

AREA

33. Find the area between the curve $y = x^2 + 1$ and the straight line $y = 2x - 1$ (shown below) from $x = 0$ to $x = 3$.

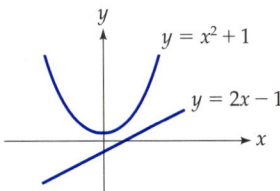

34. Find the area between the curve $y = x^2 + 3$ and the straight line $y = 2x$ (shown below) from $x = 0$ to $x = 3$.

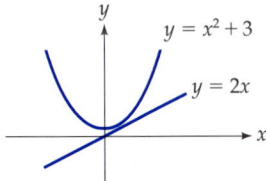

35. Find the area between the curves $y = e^x$ and $y = e^{2x}$ (shown below) from $x = 0$ to $x = 2$. (Leave the answer in its exact form.)

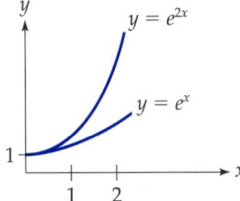

36. Find the area between the curves $y = e^x$ and $y = e^{-x}$ from $x = 0$ to $x = 1$. (Leave the answer in its exact form.)

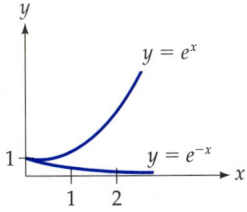

37. a. Sketch the parabola $y = x^2 + 4$ and the straight line $y = 2x + 1$ on the same graph.
b. Find the area between them from $x = 0$ to $x = 3$.

38. a. Sketch the parabola $y = x^2 + 5$ and the straight line $y = 2x + 3$ on the same graph.
b. Find the area between them from $x = 0$ to $x = 3$.

39. a. Sketch the parabola $y = 3x^2 - 3$ and the line $y = 2x + 5$ on the same graph.
b. Find the area between them from $x = 0$ to $x = 3$.

40. a. Sketch the parabola $y = 3x^2 - 12$ and the line $y = 2x - 11$ on the same graph.
b. Find the area between them from $x = 0$ to $x = 3$.

Find the area bounded by the given curves.

41. $y = x^2 - 1$ and $y = 2 - 2x^2$

42. $y = x^2 - 4$ and $y = 8 - 2x^2$

43. $y = 6x^2 - 10x - 8$ and $y = 3x^2 + 8x - 23$

44. $y = 3x^2 - x - 1$ and $y = 5x + 8$

45. $y = x^2$ and $y = x^3$ **46.** $y = x^3$ and $y = x^4$

47. $y = 7x^3 - 36x$ and $y = 3x^3 + 64x$

48. $y = x^n$ and $y = x^{n-1}$ (for $n > 1$)

49. $y = e^x$ and $y = x + 3$
(*Hint for Exercises 49–50:* Use INTERSECT to find the intersection points for the upper and lower limits of integration.)

50. $y = \ln x$ and $y = x - 2$

APPLIED EXERCISES ON AREA

51. General: Population The birth rate in Africa has increased from $15e^{0.02t}$ million births per year to $20e^{0.02t}$ million births per year, where t is the number of years since 1995. Find the total increase in population that will result from this higher birth rate between 1995 ($t = 0$) and 2015 ($t = 20$).

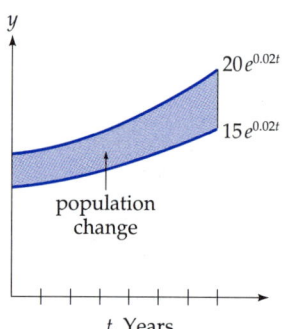

52. Business: Profit from Expansion A company expects profits of $60e^{0.02t}$ thousand dollars per month, but predicts that if it builds a new and larger factory, its profits will be $80e^{0.04t}$ thousand dollars per month, where t is the number of months from now. Find the extra profits resulting from the new factory during

the first two years ($t = 0$ to $t = 24$). If the new factory will cost $1,000,000, will this cost be paid off during the first two years?

53. Business: Net Savings A factory installs new machinery that saves $S(x) = 1200 - 20x$ dollars per year, where x is the number of years since installation. However, the cost of maintaining the new machinery is $C(x) = 100x$ dollars per year.

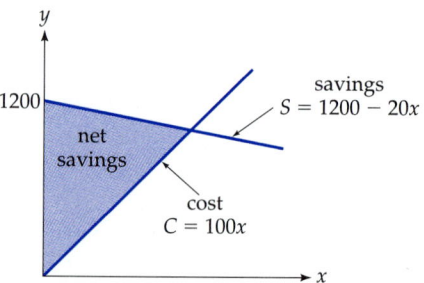

a. Find the year x at which the maintenance costs $C(x)$ will equal the savings $S(x)$. (At this time the new machinery should be replaced.)
b. Find the accumulated net savings [savings $S(x)$ minus cost $C(x)$] during the period $t = 0$ to the replacement time found in part (a).

54. General: Design A graphic design consists of a white square with a blue shape inside, as shown below. Find the area of the blue interior.

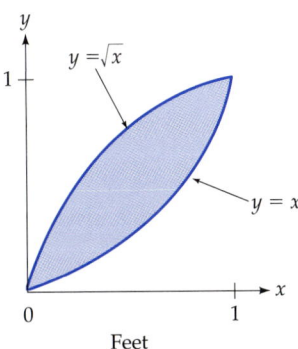

Feet

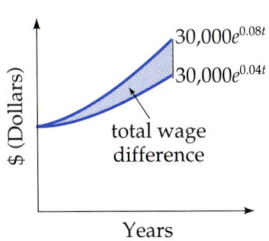

57–58: Business: Cumulative Profit A company's marginal revenue function is $MR(x) = 700x^{-1}$ and its marginal cost function is $MC(x) = 500x^{-1}$ (both in thousands of dollars), where x is the number of units ($x > 1$). Find the total profit from

57. $x = 100$ to $x = 200$ **58.** $x = 200$ to $x = 300$

 55. Economics: Balance of Trade A country's annual imports are $I(t) = 30e^{0.2t}$ and its exports are $E(t) = 25e^{0.1t}$, both in billions of dollars, where t is measured in years and $t = 0$ corresponds to the beginning of 1995. Find the country's accumulated trade deficit (imports minus exports) for the 10 years beginning with 1995.

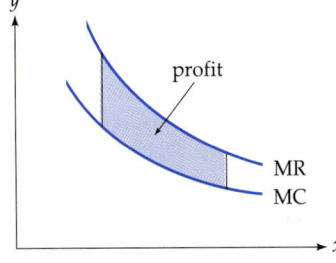

59. General: Lives Saved by Seat Belts Seat-belt use in the United States has now risen to 66%, but nonusers still risk needless expense, injury, and death. The table below gives the number of automobile fatalities per year, and the predicted number of fatalities if everyone wore seat belts. To avoid large numbers, years are listed as years since 1992.

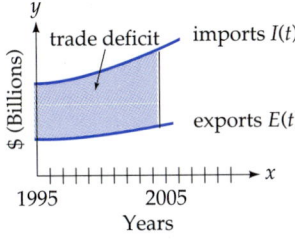

Years

56. Economics: Cost of Labor Contracts An employer offers to pay workers at the rate of $30,000e^{0.04t}$ dollars per year, while the union demands payment at the rate of $30,000e^{0.08t}$ dollars per year, where $t = 0$ corresponds to the beginning of the contract. Find the accumulated difference in pay between these two rates over the 10-year life of the contract. (See the diagram in the next column.)

Years since 1992	Automobile Fatalities	Predicted Fatalities with 100% Seat-belt Use
1992	0	40,980	36,200
1993	1	41,893	37,000
1994	2	42,700	37,700
1995	3	43,900	38,800

Source: National Safety Council.

a. Enter the first two columns of numbers into your graphing calculator and make a plot of the resulting points (Years since 1992 on the x-axis and Fatalities on the y-axis).

b. Have your calculator find the linear regression formula for these data. Then enter the result as y_1, which gives a formula for fatalities each year. Plot the points together with the regression line.

c. Enter the last column of numbers into your calculator and make a plot of the resulting points (Years since 1992 on the x-axis and Predicted Fatalities on the y-axis), keeping the earlier points, too.

d. Have your calculator find the linear regression formula for the new points found in (c). Then enter the result as y_2, which gives a formula for predicted fatalities each year. Plot both sets of points together with both regression lines.

e. Extend your viewing window to [0, 10] by [0, 55,000] to see what the lines predict for fatalities from 1992 to 2002. The area between these lines represents the lives that could be saved by seat belts.

f. Have your calculator find the area between these lines from 0 to 10, predicting the lives that would be saved by 100% seat-belt use between 1992 and 2002.

Review Exercises

These exercises review material that will be helpful in Section 9.6.

Find the derivative of each function.

60. $e^{x^3 + 6x}$

61. $e^{x^2 + 5x}$

62. $\ln (x^3 + 6x)$

63. $\ln (x^2 + 5x)$

9.5 Two Applications to Economics: Consumers' Surplus and Income Distribution

Introduction

In this section we discuss several important economic concepts—consumers' and producers' surplus, and the Gini Index of income distribution—each of which is defined as the area between two curves.

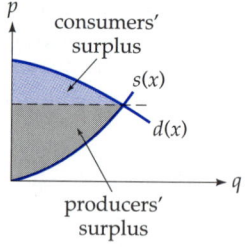

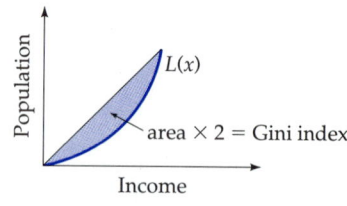

Consumers' Surplus

Imagine that you really liked pizza and were willing to pay \$12 for a pizza pie. If, in fact, a pizza costs only \$8, then you have, in some sense, "saved" \$4, the \$12 that you were willing to pay minus the \$8 market price. If one were to add up this difference for all pizzas sold in a given period of time (the price that each consumer was willing to pay minus the price actually paid), the total savings would be called the *consumers' surplus* for that product. The consumers' surplus measures the benefit that consumers derive from an economy in which competition keeps prices low.

Demand Functions

Price and quantity are inversely related: If the price of an item rises, the quantity sold generally falls, and vice versa. Through market research, economists can determine the relationship between price and quantity for an item. This relationship can often be expressed as a *demand function* (or demand curve) $d(x)$, so called because it gives the price at which exactly x units will be demanded.

Demand Function

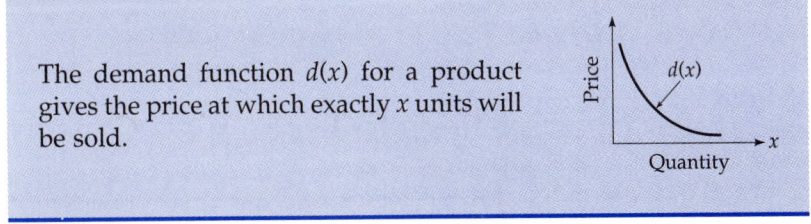

The demand function $d(x)$ for a product gives the price at which exactly x units will be sold.

In Section 7.3 we called $d(x)$ the *price function*.

Mathematical Definition of Consumers' Surplus

The demand curve gives the price that consumers are *willing* to pay, and the *market price* is what they *do* pay, so the amount by which the demand curve is above the market price measures the benefit or "surplus" to consumers. We add up these benefits by integrating, so the area between the demand curve and the market price line gives the *total benefit* that consumers derive from being able to buy at the market price. This total benefit (the shaded area in the diagram) is called the *consumers' surplus*.

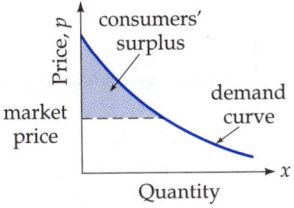

Consumers' Surplus

For a demand function $d(x)$ and demand level A, the market price B is the demand function evaluated at $x = A$, so that $B = d(A)$:

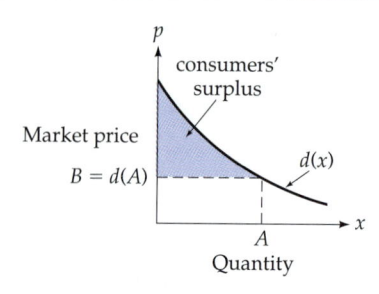

The consumers' surplus is the area between the demand curve and the market price.

$$\begin{pmatrix} \text{Consumers'} \\ \text{surplus} \end{pmatrix} = \int_0^A [d(x) - B]\, dx$$

Demand   Market
function   price

EXAMPLE 1   **Finding Consumers' Surplus for Electricity**

If the demand function for electricity is $d(x) = 1100 - 10x$ dollars (where x is in million kilowatt-hours), find the consumers' surplus at the demand level $x = 80$.

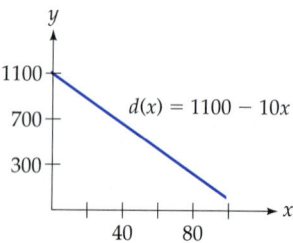

Solution   The market price is the demand function $d(x)$ evaluated at $x = 80$.

$$\begin{pmatrix} \text{Market} \\ \text{price } B \end{pmatrix} = d(80) = 1100 - 10 \cdot 80 = 300 \qquad \begin{array}{l} d(x) = 1100 - 10x \\ \text{at } x = 80 \end{array}$$

The consumers' surplus is the area between the demand curve and the market price line.

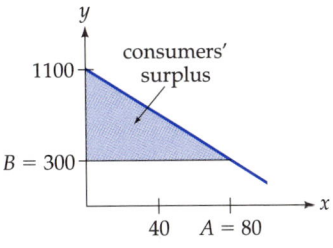

$$\begin{pmatrix} \text{Consumers'} \\ \text{surplus} \end{pmatrix} = \int_0^{80} (\underbrace{1100 - 10x}_{\substack{\text{Demand} \\ \text{function}}} - \underbrace{300}_{\substack{\text{Market} \\ \text{price}}}) \, dx$$

$$= \int_0^{80} (800 - 10x) \, dx = (800x - 5x^2) \Big|_0^{80}$$

$$= (64{,}000 - 32{,}000) - (0) = 32{,}000$$

Therefore, the consumers' surplus for electricity is $32,000.

∎

How Consumers' Surplus Is Used

In Example 1, at demand level $x = 80$ the consumers' surplus was $32,000. If electricity usage were to increase to $x = 90$, the market price would then drop to $d(90) = 1100 - 10 \cdot 90 = 200$. We could then calculate the consumers' surplus at this higher demand level (the answer is $40,500). Therefore, a price decrease from $300 to $200 would mean that consumers would benefit by an additional $40,500 − $32,000 = $7500. This benefit would then be compared to the cost of a new generator to decide whether the expenditure would be worthwhile.

Graphing Calculator Exploration

a. Verify the solution to Example 1 on your graphing calculator by entering $y_1 = 1100 - 10x$ and then using FnInt or $\int f(x)dx$ to integrate $y_1(x) - y_1(80)$ from 0 to 80.

b. Find the consumers' surplus at demand level 90 by integrating $y_1(x) - y_1(90)$ from 0 to 90. [*Hint:* Simply return to the calculation of part (a) and replace 80 by 90. Your answer should agree with that in the above paragraph.]

Producers' Surplus

Just as the consumers' surplus measures the total benefit to consumers, the *producers' surplus* measures the total benefit that producers derive from being able to sell at the market price. Returning to our pizza example, if a pizza producer might just be willing to remain in business if the price of pizzas dropped to $5, the fact that he can sell them for $8 gives him a "benefit" of $3. The sum of all such benefits is the *producers' surplus* for a product.

Supply Functions

Clearly, as the price of an item rises, so will the quantity that producers are willing to supply at that price. The relationship between the price of an item and the quantity that producers are willing to supply at that price can be expressed as a supply function (or supply curve) $s(x)$.

Supply Function

The supply function $s(x)$ for a product gives the price at which exactly x units will be supplied.

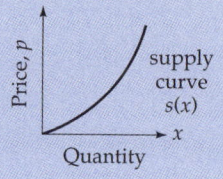

Mathematical Definition of Producers' Surplus

As before, we integrate to find the total benefit, but now "upper" is the market price and "lower" is the supply curve $s(x)$.

Producers' Surplus

For supply function $s(x)$ and demand level A, the market price is $B = s(A)$. The producers' surplus is the area between the market price and the supply curve.

$$\begin{pmatrix} \text{Producers'} \\ \text{surplus} \end{pmatrix} = \int_0^A [B - s(x)]\, dx$$

EXAMPLE 2 Finding Producers' Surplus

For the supply function $s(x) = 0.09x^2$ dollars and the demand level $x = 200$, find the producers' surplus.

Solution   The market price is the supply function $s(x)$ evaluated at $x = 200$.

$$\binom{\text{Market}}{\text{price } B} = s(200) = 0.09(200)^2 = 3600 \qquad \begin{array}{l} s(x) = 0.09x^2 \\ \text{at } x = 200 \end{array}$$

The producers' surplus is the area between the market price line and the supply curve.

$$\binom{\text{Producers'}}{\text{surplus}} = \int_0^{200} (\underbrace{3600}_{\substack{\text{Market} \\ \text{price}}} - \underbrace{0.09x^2}_{\substack{\text{Supply} \\ \text{function}}}) \, dx$$

$$= (3600x - 0.03x^3) \Big|_0^{200}$$

$$= (720{,}000 - 240{,}000) - (0)$$

$$= 480{,}000$$

Therefore, the producers' surplus is $480,000.

∎

Consumers' and Producers' Surplus

The demand x at which the supply and demand curves intersect is called the *market demand*. The consumers' and the producers' surplus can be shown together on the same graph. These two areas together give a numerical measure to the total benefit that consumers and producers derive from competition, showing that both consumers and producers benefit from an open market.

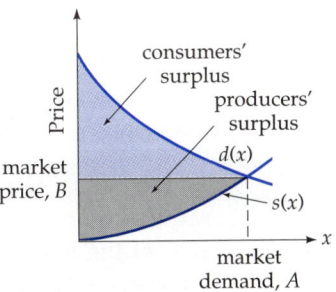

Gini Index of Income Distribution

In any society, some people make more money than others. To measure the "gap" between the rich and the poor, economists calculate the proportion of the total income that is earned by the lowest 20% of the population, and then the proportion that is earned by the lowest 40% of the population, and so on. This information (for the year 1996) is given in the table below (with percentages written as decimals), and graphed on the right.

Proportion of Population	Proportion of Income
0.20	0.03
0.40	0.12
0.60	0.27
0.80	0.50
1.00	1.00

Source: U.S. Bureau of the Census.

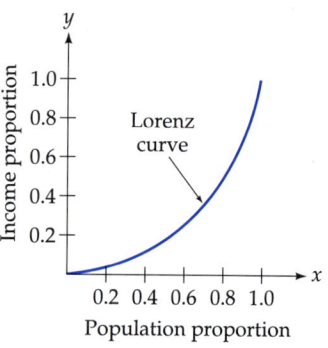

For example, the lowest 20% of the population earns only 3% of the total income, the lowest 40% earns only 12% of the total income, and so on. The curve is known as the *Lorenz curve* (after the American statistician Max Otto Lorenz).

Lorenz Curve

> The Lorenz curve $L(x)$ gives the proportion of total income earned by the lowest proportion x of the population.

Gini Index

The Lorenz curve may be compared to two extreme cases of income distribution.

1. *Absolute equality of income* means that everyone earns exactly the same income, and so the lowest 10% of the population earns exactly 10% of the total income, the lowest 20% earns exactly 20% of the income, and so on. This gives the Lorenz curve $y = x$ shown on the left on the following page.

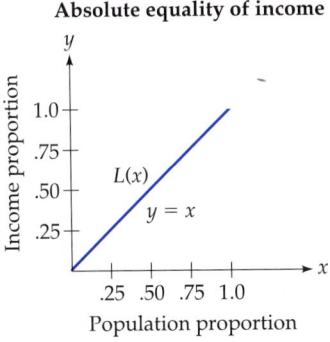

Absolute equality of income

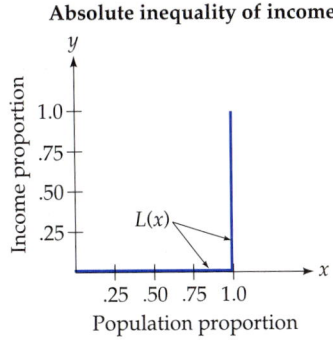

Absolute inequality of income

2.   *Absolute inequality of income* means that nobody earns any income except one person, who earns all the income. This gives the Lorenz curve shown above on the right.

To measure how the actual distribution differs from absolute equality, we calculate the area between the actual distribution and the line of absolute equality $y = x$. Since this area can be at most $\frac{1}{2}$ (the area of the entire lower triangle), economists multiply the area by 2 to get a number between 0 (absolute equality) and 1 (absolute inequality). This measure is called the *Gini index*. Note that a higher Gini index means greater *in*equality (greater deviation from the line of absolute equality).

Gini Index

For a Lorenz curve $L(x)$, the Gini index is

$$\left(\begin{matrix} \text{Gini} \\ \text{index} \end{matrix}\right) = 2\int_0^1 [x - L(x)]\, dx$$

$$\text{area} \times 2 = \left[\begin{matrix} \text{Gini} \\ \text{index} \end{matrix}\right]$$

The Gini index varies from 0 (absolute equality) to 1 (absolute inequality).

EXAMPLE 3   Finding the Gini Index

The Lorenz curve for income distribution in the United States in 1996 is approximately $L(x) = x^{2.7}$. Find the Gini index.

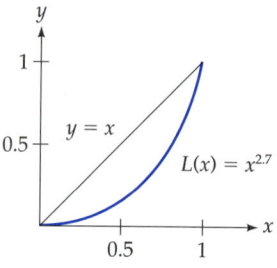

Solution   First we calculate the area between the curve of absolute equality, $y = x$, and the Lorenz curve $y = x^{2.7}$.

$$\int_0^1 (x - x^{2.7})\, dx = \left(\frac{1}{2} x^2 - \frac{1}{3.7} x^{3.7} \right) \Big|_0^1$$

$$= \frac{1}{2} \cdot 1^2 - \frac{1}{3.7} \cdot 1^{3.7} - 0$$

$$\approx 0.5 - 0.270 = 0.23$$

Multiplying by 2 gives the Gini index.

$$\left(\begin{array}{c} \text{Gini} \\ \text{index} \end{array} \right) = 0.46 \qquad \text{Rounding to two decimal places}$$

PRACTICE PROBLEM 1

In 1993 the Gini index for income was 0.39. Since 1993, has income distribution become more equal or less equal?

Solution at the back of the book

Graphing Calculator Exploration

In Example 3, how was the Lorenz function of the form x^n found? It was found by a method called "least squares" (discussed more fully in Section 11.4), which minimizes the squared differences between the income proportions (from the table on page 798) and the curve x^n at the x-values 0.2, 0.4, 0.6, and 0.8. This amounts to minimizing the function $(0.2^x - 0.03)^2 + (0.4^x - 0.12)^2 + (0.6^x - 0.27)^2 + (0.8^x - 0.50)^2$. (Notice that here we are using x for the *exponent*.)

a. Graph this function on your calculator on the window $[0, 5]$ by $[-0.2, 1]$.

b. Find the x that minimizes it. The resulting value is the exponent. Your answer should agree with the exponent in Example 3.

The following graph, on the window $[0, 1]$ by $[0, 1]$, shows that the function $x^{2.7}$ fits the points from the table on page 798 reasonably well. (Other types of functions besides x^n could also be fit to these data.)

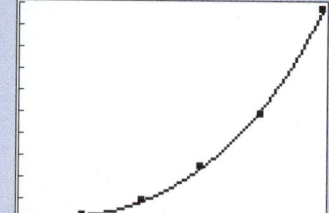

The Gini index for total wealth can be calculated similarly.

PRACTICE PROBLEM 2   The Lorenz curve for total wealth in the United States during 1996 was $L(x) = x^{6.9}$. Calculate the Gini index.   *Solution at the back of the book*

Practice Problem 2 shows that the Gini index for wealth is greater than the Gini index for income. That is, total wealth in the United States is distributed more unequally than total income. One reason for this is that we have an income tax but no "wealth" tax.

EXERCISES 9.5 (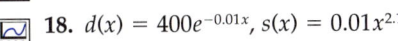 helpful.)

For each demand function $d(x)$ and demand level x, find the consumers' surplus.

1. $d(x) = 4000 - 12x, x = 100$

2. $d(x) = 500 - x, x = 400$

3. $d(x) = 300 - \frac{1}{2}x, x = 200$

4. $d(x) = 200 - \frac{1}{2}x, x = 300$

5. $d(x) = 350 - 0.09x^2, x = 50$

6. $d(x) = 840 - 0.06x^2, x = 100$

 7. $d(x) = 200e^{-0.01x}, x = 100$

 8. $d(x) = 400e^{-0.02x}, x = 75$

For each supply function $s(x)$ and demand level x, find the producers' surplus.

9. $s(x) = 0.2x, x = 100$

10. $s(x) = 0.4x, x = 200$

11. $s(x) = 0.03x^2, x = 200$

12. $s(x) = 0.06x^2, x = 50$

For each demand function $d(x)$ and supply function $s(x)$:

a. Find the market demand (the positive value of x at which the demand function intersects the supply function).
b. Find the consumers' surplus at the market demand found in part (a).
c. Find the producers' surplus at the market demand found in part (a).

13. $d(x) = 300 - 0.4x, s(x) = 0.2x$

14. $d(x) = 120 - 0.16x, s(x) = 0.08x$

15. $d(x) = 300 - 0.03x^2, s(x) = 0.09x^2$

16. $d(x) = 360 - 0.03x^2, s(x) = 0.006x^2$

17. $d(x) = 300e^{-0.01x}, s(x) = 100 - 100e^{-0.02x}$

18. $d(x) = 400e^{-0.01x}, s(x) = 0.01x^{2.1}$

Find the Gini index for the given Lorenz curve.

19. $L(x) = x^{3.2}$ (the Lorenz curve for U.S. income in 1929)

20. $L(x) = x^3$ (the Lorenz curve for U.S. income in 1935)

21. $L(x) = x^{2.1}$ (the Lorenz curve for income in Sweden)

22. $L(x) = x^{15.3}$ (the Lorenz curve for wealth in Great Britain)

23. $L(x) = 0.4x + 0.6x^2$

24. $L(x) = 0.2x + 0.8x^3$

25. $L(x) = x^n$ (for $n > 1$)

26. $L(x) = \frac{1}{2}x + \frac{1}{2}x^n$ (for $n > 1$)

27. $L(x) = \dfrac{e^{x^2} - 1}{e - 1}$

28. $L(x) = 1 - \sqrt{1 - x}$

29. $L(x) = \dfrac{x + x^2 + x^3}{3}$

30. $L(x) = 0.62x^{7.15} + 0.38x^{9.47}$

 31–32: The following tables give the distribution of family income in the United States: Exercise 31 is for the year 1977 and Exercise 32 is for 1989. Use the procedure described in the Graphing Calculator Exploration on page 800 to find the Lorenz function of the form x^n for the data. Then find the Gini index. If you do both problems, did family income become more concentrated or less concentrated from 1977 to 1989?

31.

Proportion (Lowest) of Families	Proportion of Income (1977)
0.20	0.06
0.40	0.18
0.60	0.34
0.80	0.57

Source: Congressional Budget Office; *New York Times*, May 11, 1992.

32.

Proportion (Lowest) of Families	Proportion of Income (1989)
0.20	0.04
0.40	0.14
0.60	0.29
0.80	0.51

Source: Congressional Budget Office; *New York Times*, May 11, 1992.

Review Exercises

(These exercises are useful for Section 9.6.)

Find the derivative of each function.

33. $(x^5 - 3x^3 + x - 1)^4$ **34.** $(x^4 - 2x^2 - x + 1)^5$

35. $\ln(x^4 + 1)$ **36.** $\ln(x^3 - 1)$

37. e^{x^3} **38.** e^{x^4}

9.6 Integration by Substitution

Introduction

The chain rule on page 573 greatly expanded the range of functions that we could differentiate. In this section we will learn a similar technique for integration, called the substitution method, which will greatly expand the range of functions that we can integrate. First, however, we must define differentials.

Differentials

One of the notations for the derivative of a function $f(x)$ is $\frac{df}{dx}$. Although written as a fraction, $\frac{df}{dx}$ was not defined as the quotient of two quantities df and dx, but was defined as a single object, the *derivative*. We will now define df and dx separately (they are called *differentials*) so that their quotient $df \div dx$ is equal to the derivative $\frac{df}{dx}$. We begin with

$$\frac{df}{dx} = f' \qquad\qquad \text{Since } df/dx \text{ and } f' \text{ are both notations for the derivative}$$

$$df = f' \, dx \qquad\qquad \text{Multiplying each side by } dx$$

This leads to a definition for the differential df.

Differential

For a differentiable function $f(x)$, the differential df is

$$df = f'(x) \, dx \qquad\qquad df \text{ is the derivative times } dx$$

Note that df does *not* mean d times f. The dx is just the notation that appears at the end of integrals, arising from the Δx in the Riemann sum. The reason for finding differentials will be clear shortly.

EXAMPLE 1 **Finding Differentials**

Function $f(x)$	Differential df	
$f(x) = x^2$	$df = 2x \, dx$	
$f(x) = \ln x$	$df = \dfrac{1}{x} \, dx$	Each differential is the derivative of the function times dx
$f(x) = e^{x^2}$	$df = e^{x^2}(2x) \, dx$	
$f(x) = x^4 - 5x + 2$	$df = \underbrace{(4x^3 - 5)}_{f'(x)} \, dx$	

■

PRACTICE PROBLEM 1

For $f(x) = x^3 - 4x - 2$, find the differential df.

Solution at the back of the book

The differential formula $df = f' \cdot dx$ is easy to remember since dividing both sides by dx gives

$$\frac{df}{dx} = f'$$

which simply says "the derivative equals the derivative." We may use other letters besides f and x.

EXAMPLE 2   **Calculating Differentials in Other Variables**

Function	Differential
$u = x^3 + 1$	$du = 3x^2\, dx$
$u = e^{2t} + 1$	$du = 2e^{2t}\, dt$

Differentials end
with *d* followed
by the variable

■

PRACTICE PROBLEM 2   For $u = e^{-5t}$, find the differential du.   *Solution at the back of the book*

Substitution Method

Using differential notation, we can state three very useful integration formulas.

(A) $\displaystyle\int u^n\, du = \frac{1}{n+1} u^{n+1} + C$ $n \neq -1$

(B) $\displaystyle\int e^u\, du = e^u + C$

(C) $\displaystyle\int \frac{du}{u} = \int \frac{1}{u}\, du = \int u^{-1}\, du = \ln |u| + C$

These formulas are easy to remember since they are exactly the formulas that we learned earlier (see pages 748 and 755) except that here we use the letter u to stand for a function. The du is the differential of the function. Each of these formulas may be justified by differentiating the right-hand side (see Exercises 61–63). A few examples illustrate their use.

EXAMPLE 3   **Integrating by Substitution**

Find $\displaystyle\int (x^2 + 1)^3\, 2x\, dx$.

Solution

The integral involves a function to a power:

$$\int (x^2 + 1)^3\, 2x\, dx$$

$(x^2 + 1)^3$ is a function to a power

$\updownarrow$

as does formula (A): $\displaystyle\int u^n\, du$

u^n is a function to a power in

$\displaystyle\int u^n\, du = \frac{1}{n+1} u^{n+1} + C$

To make the integral "fit" the formula we take $u = x^2 + 1$ and $n = 3$.

$$\int \underbrace{(x^2 + 1)^3}_{u^3} 2x \, dx \qquad\qquad u = x^2 + 2 \text{ and } n = 3$$

For $u = x^2 + 1$ the differential is $du = 2x \, dx$, which is exactly the remaining part of the integral. We then write the integral with each x-expression replaced by its equivalent u-expression.

$$\int \underbrace{(x^2 + 1)^3}_{u^3} \underbrace{2x \, dx}_{du} = \int \underbrace{u^3 \, du}_{\text{Written in terms of } u} \qquad \begin{array}{l}\text{Using } u = x^2 + 1 \text{ and } du = \\ 2x \, dx\end{array}$$

This last integral we solve by formula (A):

$$\int u^3 \, du = \frac{1}{4} u^4 + C \qquad\qquad \int u^n \, du = \frac{1}{n + 1} u^{n+1} + C$$
$$\text{(formula A) with } n = 3$$

Finally, we substitute back to the original variable x, using our relationship $u = x^2 + 1$, to get the answer:

$$\frac{1}{4}(x^2 + 1)^4 + C \qquad\qquad \frac{1}{4} u^4 + C \text{ with } u = x^2 + 1$$

The procedure is not as complicated as it might seem. All of these steps may be written together as follows.

$$\int \underbrace{(x^2 + 1)^3}_{u^3} \underbrace{2x \, dx}_{du} = \int u^3 \, du = \frac{1}{4} u^4 + C = \frac{1}{4}(x^2 + 1)^4 + C$$

u^3	du	Substituting	Integrating	Substituting
Choosing $u = x^2 + 1$		$u^3 = (x^2 + 1)^3$	by formula (A)	back to x using
therefore $du = 2x \, dx$		$du = 2x \, dx$	with $n = 3$	$u = x^2 + 1$

We may check this answer by differentiation (using the generalized power rule).

$$\frac{d}{dx}\left[\underbrace{\frac{1}{4}(x^2 + 1)^4 + C}_{\text{Answer}}\right] = \frac{1}{4} \cdot 4(x^2 + 1)^3 \underbrace{2x}_{\substack{\text{Derivative} \\ \text{of the inside}}} = \underbrace{(x^2 + 1)^3 \, 2x}_{\text{Integrand}}$$

Since the result of the differentiation agrees with the original integrand, the integration is correct.

Multiplying Inside and Outside by Constants

If the integral does not exactly match the form $\int u^n \, du$, we may sometimes still solve the integral by multiplying by constants.

EXAMPLE 4   Inserting Constants Before Substituting

Find $\displaystyle\int (x^2 + 1)^3 x \, dx$.   Same as Example 3 but without the 2

Solution   As before, we use formula (A) with $u = x^2 + 1$, which gives $du = 2x \, dx$. But the integral has only an $x \, dx$, not the $\underline{2x} \, dx$, which would allow us to substitute du.

$$\int \underbrace{(x^2 + 1)^3}_{u^3} \underbrace{x \, dx}_{} \qquad\qquad u = x^2 + 1, \text{ so } du = 2x \, dx$$

not $du = 2x \, dx$
because there is no 2

Therefore, the integral is *not* in the form $\int u^3 \, du$. (The integral must fit the formula exactly: *everything* in the integral must be accounted for either by the u^n or by the du.) However, we may multiply inside the integral by 2 as long as we compensate by also multiplying by $\frac{1}{2}$, and the $\frac{1}{2}$ may be written outside the integral (since constants may be moved across the integral sign), leading to the solution:

Multiplying by $\frac{1}{2}$ and 2

$$\frac{1}{2} \int \underbrace{(x^2 + 1)^3}_{u^3} \underbrace{2x \, dx}_{du} = \frac{1}{2} \int u^3 \, du = \frac{1}{2}\frac{1}{4} u^4 + C = \frac{1}{8}(x^2 + 1)^4 + C$$

$u = x^2 + 1$ Substituting   Integrating   Substituting back
$du = 2x \, dx$ by formula (A)   to x using
                                                   $u = x^2 + 1$

∎

This method of multiplying inside and outside by a constant is very useful, and may be used with the other substitution formulas as well.

EXAMPLE 5   Using Other Substitution Formulas

Find $\displaystyle\int e^{x^3} x^2 \, dx$.

Solution

The integral involves e to a function:

$$\int e^{x^3} x^2 \, dx \qquad\qquad e^{x^3} \text{ is } e \text{ to a function}$$

$$\updownarrow$$

as does formula (B) on page 804:

$$\int e^u \, du = e^u + C$$

Matching exponents of e gives $u = x^3$, and the differential of u is $du = 3x^2 \, dx$. The differential requires a 3, which is not in the integral, so we multiply inside by 3 and outside by $\frac{1}{3}$.

$$\underbrace{\int e^{x^3} x^2 \, dx}_{\substack{u = x^3 \\ du = 3x^2 \, dx}} = \underbrace{\frac{1}{3} \int e^{x^3} \, 3x^2 \, dx}_{\substack{e^u \quad du \\ \text{Multiplying} \\ \text{by } \frac{1}{3} \text{ and } 3}} = \underbrace{\frac{1}{3} \int e^u \, du}_{\text{Substituting}} = \underbrace{\frac{1}{3} e^u + C}_{\substack{\text{Integrating} \\ \text{using} \\ \text{formula (B)}}} = \underbrace{\frac{1}{3} e^{x^3} + C}_{\substack{\text{Substituting} \\ \text{back to } x \\ \text{using } u = x^3}}$$

■

Why did the $\frac{1}{3}$ become part of the answer, but the 3 disappeared? The 3 became part of the du ($du = 3x^2 \, dx$), which was then "used up" in the integration along with the integral sign.

EXAMPLE 6   Recovering Cost from Marginal Cost

A company's marginal cost function is $MC(x) = \dfrac{x^3}{x^4 + 1}$ and fixed costs are \$1000. Find the cost function.

Solution   Cost is the integral of marginal cost.

$$C(x) = \int \frac{x^3 \, dx}{x^4 + 1}$$

The differential of the denominator is $4x^3 \, dx$, which except for the 4 is just the numerator. This suggests formula (C), $\displaystyle\int \frac{du}{u} = \ln |u| + C$ with

$u = x^4 + 1$. We multiply inside by 4 (to complete the $du = 4x^3\,dx$ in the numerator) and outside by $\frac{1}{4}$.

$$\int \frac{x^3\,dx}{x^4+1} = \frac{1}{4}\int \frac{4x^3\,dx}{x^4+1} = \frac{1}{4}\int \frac{du}{u} = \frac{1}{4}\ln|u| + C = \frac{1}{4}\ln(x^4+1) + C$$

$u = x^4 + 1$ $du = 4x^3\,dx$	Multiplying by 4 and $\frac{1}{4}$	Substituting	Integrating by formula (C)	Substituting back to x

We dropped the absolute value bars since $x^4 + 1$ is positive. To evaluate the constant C, we set the cost function (evaluated at $x = 0$) equal to the given fixed cost.

$$\frac{1}{4}\underbrace{\ln(1)}_{0} + C = 1000 \qquad \begin{array}{l}\ln(x^4+1) + C \text{ at } x = 0 \\ \text{set equal to } 1000\end{array}$$

This gives $C = 1000$. Therefore, the company's cost function is

$$C(x) = \frac{1}{4}\ln(x^4+1) + 1000 \qquad \begin{array}{l} C(x) = \ln(x^4+1) + C \\ \text{with } C = 1000\end{array}$$

■

Which Formula to Use

The three formulas apply to three different types of integrals.

(A) $\displaystyle\int u^n\,du = \frac{1}{n+1}u^{n+1} + C \qquad (n \neq -1)$

Integrates a *function to a constant power* (except −1) times the differential of the function

(B) $\displaystyle\int e^u\,du = e^u + C$

Integrates *e to a power* times the differential of the exponent

(C) $\displaystyle\int \frac{du}{u} = \int u^{-1}\,du = \ln|u| + C$

Integrates a *fraction whose top is the differential of the bottom*, or equivalently, a *function to the power −1* times the differential of the function

To solve an integral by substitution, choose the formula whose left-hand side has the same form as the given integral.

PRACTICE PROBLEM 3

For each of the following integrals, choose the most appropriate formula: (A), (B), or (C). (Do not solve the integral.)

a. $\displaystyle\int e^{5x^2-1}x\,dx$ **b.** $\displaystyle\int \frac{x\,dx}{x^2+1}$ **c.** $\displaystyle\int (x^4-12)^4x^3\,dx$

d. $\displaystyle\int (x^4-12)^{-1}x^3\,dx$ *Solutions at the back of the book*

Only Constants Can Be Adjusted

We may multiply inside and outside only by constants, not variables, since only constants can be moved across the integral sign. Therefore, the *du* in a problem must already be "complete" except for adjusting the constant. Otherwise, the problem cannot be found by a substitution. For example, the following integral cannot be found by a substitution.

$$\int \underbrace{e^{x^3}}_{\substack{e^u \\ u=x^3 \\ du=3x^2\,dx}}\ \underbrace{x\,dx}_{\substack{\uparrow \\ \text{not } du\ (\text{does not have an } x^2)}}$$

PRACTICE PROBLEM 4

Which of these integrals can be found by a substitution? (*Hint:* See whether only a constant is needed to complete the *du*.)

a. $\displaystyle\int (x^3+1)^3x^3\,dx$ **b.** $\displaystyle\int e^{x^2}\,dx$ *Solution at the back of the book*

EXAMPLE 7 **Integrating by Substitution**

Find $\displaystyle\int \sqrt{x^3-3x}\,(x^2-1)\,dx$.

Solution Since $\sqrt{x^3-3x}=(x^3-3x)^{1/2}$ is a *function to a power*, we use the formula for $\int u^n\,du$ with $u=x^3-3x$. Comparing the differential $du=(3x^2-3)dx=3(x^2-1)dx$ with the problem shows that we need to multiply by 3.

$$\int (x^3-3x)^{1/2}(x^2-1)\,dx=\frac{1}{3}\int \underbrace{(x^3-3x)^{1/2}}_{u^{1/2}}\,\underbrace{3(x^2-1)\,dx}_{du}\qquad \begin{array}{l}\text{Multiplying}\\ \text{by 3 and }\frac{1}{3}\end{array}$$

$$u=x^3-3x$$
$$du=(3x^2-3)\,dx$$
$$\quad=3(x^2-1)\,dx$$

$$=\underbrace{\frac{1}{3}\int u^{1/2}\,du}_{\substack{\text{Substituting}}}=\underbrace{\frac{1}{3}\frac{2}{3}u^{3/2}+C}_{\substack{\text{Integrating}\\ \text{by formula (A)}}}\quad=\underbrace{\frac{2}{9}(x^3-3x)^{3/2}+C}_{\substack{\text{Substituting}\\ \text{back to } x}}$$

■

EXAMPLE 8   **Integrating by Substitution**

Find $\displaystyle\int e^{\sqrt{x}}\, x^{-1/2}\, dx$.

Solution   The integral involves $e^{\sqrt{x}}$, which is e to a function, so we use the formula $\int e^u\, du = e^u + C$ with $u = x^{1/2}$.

$$\int e^{x^{1/2}} x^{-1/2}\, dx = 2 \int \underbrace{e^{x^{1/2}}}_{e^u} \underbrace{\tfrac{1}{2} x^{-1/2}\, dx}_{du} \qquad\qquad \text{Multiplying by } \tfrac{1}{2} \text{ and } 2$$

$u = x^{1/2}$

$du = \dfrac{1}{2} x^{-1/2}\, dx$

$$= 2 \int e^u\, du \;=\; 2e^u + C \;=\; 2e^{x^{1/2}} + C$$

Substituting   Integrating   Substituting
back to x

Evaluating Definite Integrals by Substitution

Sometimes a definite integral requires a substitution. In such cases changing from x to u also requires changing the limits of integration from x-values to u-values, using the substitution formula for u.

EXAMPLE 9   **Evaluating a Definite Integral by Substitution**

Evaluate $\displaystyle\int_4^5 \frac{dx}{3 - x}$.

Solution   The differential of the denominator $3 - x$ is $-1 \cdot dx$, which except for the -1 is just the numerator. This suggests formula (C), $\int \dfrac{du}{u} = \ln |u| + C$ with $u = 3 - x$ (from equating the denominators). We multiply inside and outside by -1.

$$\int_4^5 \frac{dx}{3 - x} = -\int_4^5 \frac{-dx}{3 - x} = -\int_{-1}^{-2} \frac{du}{u}$$

$u = 3 - 5 = -2$

$u = 3 - 4 = -1$

New upper and lower limits of integration for u are found by evaluating $u = 3 - x$ at the old x limits

$u = 3 - x$

$du = -dx$

$$= -\ln|u|\,\Big|_{-1}^{-2} = -\ln|-2| - (-\ln|-1|) = -\ln 2 + \ln 1 = -\ln 2$$

Graphing Calculator Exploration

Why is the answer to Example 9 *negative?* Graph $f(x) = \dfrac{1}{3-x}$ on [3, 6] by [−2, 2] to see why. Have your calculator find the definite integral from 4 to 5. How would you change the integral so that it gives the *area* between the curve and the *x*-axis?

Definite integrals are used to find areas, total accumulations, and average values, and any of these uses may require a substitution.

EXAMPLE 10 Finding Total Pollution from the Rate

Pollution is being discharged into a lake at the rate of $r(t) = 400te^{t^2}$ tons per year, where t is the number of years since measurements began. Find the total amount of pollutant discharged into the lake during the first 2 years.

Solution The total accumulation is the definite integral from $t = 0$ (the beginning) to $t = 2$ (2 years later). Since the integral involves e to a function, we use the formula for $\int e^u \, du$ with $u = t^2$ (by equating exponents).

$$\int_0^2 400\, te^{t^2} \, dt \;=\; 400 \cdot \frac{1}{2} \int_0^2 2te^{t^2} \, dt = 200 \int_0^4 e^u \, du$$

$u = 2^2 = 4$

e^u Taking out the constant

$u = t^2$
$du = 2t\, dt$

e^u

du

$u = 0^2 = 0$

Changing the limits to *u*-values using $u = t^2$

$$= \; 200e^u \Big|_0^4 = 200e^4 - 200e^0 \approx 10{,}720$$

Therefore, during the first 2 years 10,720 tons of pollutant were discharged into the lake.

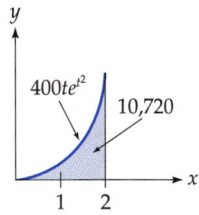

Notice that the *du* does not need to be all together, but can be in several separate pieces, as long as it is all *there*.

EXAMPLE 11   Finding Average Water Depth

After x months the water level in a newly built reservoir is $L(x) = 40x(x^2 + 9)^{-1/2}$ feet. Find the average depth during the first 4 months.

Solution   The average value is the definite integral from $x = 0$ to $x = 4$ (the end of month 4) divided by the length of the interval.

$$u = 4^2 + 9 = 25$$

$$\frac{1}{4}\int_0^4 40x\underbrace{(x^2 + 9)^{-1/2}}_{u^{-1/2}}\,dx = \frac{1}{4}\cdot 40 \cdot \frac{1}{2}\int_0^4 \underbrace{2x(x^2 + 9)^{-1/2}\,dx}_{u^{-1/2} \qquad du} = 5\int_9^{25} u^{-1/2}\,du$$

Changing the limits to *u*-values using $u = x^2 + 9$

$$u = x^2 + 9$$
$$du = 2x\,dx$$

$$u = 0^2 + 9 = 9$$

$$= 5 \cdot 2u^{1/2}\Big|_9^{25} = 10(25)^{1/2} - 10(9)^{1/2} = 10 \cdot 5 - 10 \cdot 3 = 20$$

That is, the average depth of the reservoir over the last 4 months was 20 feet.

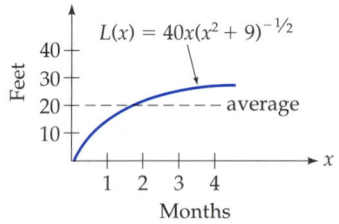

SUMMARY

The three substitution formulas are listed on the inside back cover. Most of the work in using these formulas is making a problem "fit" the left-hand side of one of the formulas (choosing the *u* and adjusting constants to complete the *du*). Once a problem fits a left-hand side, the right-hand side immediately gives the answer (except for substituting back to the original variable).

Note that the *du* now plays a very important role: The *du* must be correct if the answer is to be correct. For example, the formula $\int e^u\,du = e^u + C$ should not be thought of as the formula for integrating e^u, but as the formula for integrating $e^u\,du$. The *du* is just as important as the e^u.

EXERCISES 9.6

Find each integral. (Integration formulas (A), (B), and (C) are on the inside back cover, numbered 5, 6, and 7, respectively.)

1. $\int (x^2 + 1)^9\, 2x\, dx$

(*Hint:* Use $u = x^2 + 1$ and formula 5.)

2. $\int (x^3 + 1)^4\, 3x^2\, dx$

(*Hint:* Use $u = x^3 + 1$ and formula 5.)

3. $\int (x^2 + 1)^9 x\, dx$

(*Hint:* Use $u = x^2 + 1$ and formula 5.)

4. $\int (x^3 + 1)^4 x^2\, dx$

(*Hint:* Use $u = x^3 + 1$ and formula 5.)

5. $\int e^{x^5} x^4\, dx$

(*Hint:* Use $u = x^5$ and formula 7.)

6. $\int e^{x^4} x^3\, dx$

(*Hint:* Use $u = x^4$ and formula 7.)

7. $\int \dfrac{x^5\, dx}{x^6 + 1}$

(*Hint:* Use $u = x^6 + 1$ and formula 6.)

8. $\int \dfrac{x^4\, dx}{x^5 + 1}$

(*Hint:* Use $u = x^5 + 1$ and formula 6.)

Show that each integral *cannot* be found by our substitution formulas.

9. $\int \sqrt{x^3 + 1}\, x\, dx$

10. $\int \sqrt{x^5 + 9}\, x^2\, dx$

11. $\int e^{x^4} x^5\, dx$

12. $\int e^{x^3} x^4\, dx$

Find each integral by the substitution method or state that it cannot be found by our substitution formulas.

13. $\int (x^4 - 16)^5 x^3\, dx$

14. $\int (x^5 - 25)^6 x^4\, dx$

15. $\int e^{-x^2} x\, dx$

16. $\int e^{-x^4} x^3\, dx$

17. $\int e^{3x}\, dx$

18. $\int e^{5x}\, dx$

19. $\int e^{x^2} x^2\, dx$

20. $\int e^{x^3} x\, dx$

21. $\int \dfrac{dx}{1 + 5x}$

22. $\int \dfrac{dx}{1 + 3x}$

23. $\int (x^2 + 1)^9 5x\, dx$

24. $\int (x^2 - 4)^6 3x\, dx$

25. $\int \sqrt[4]{z^4 + 16}\, z^3\, dz$

26. $\int \sqrt[3]{z^3 - 8}\, z^2\, dz$

27. $\int \sqrt[4]{x^4 + 16}\, x^2\, dx$

28. $\int \sqrt[3]{x^3 - 8}\, x\, dx$

29. $\int (2y^2 + 4y)^5 (y + 1)\, dy$

30. $\int (3y^2 - 6y)^3 (y - 1)\, dy$

31. $\int e^{x^2 + 2x + 5}(x + 1)\, dx$

32. $\int e^{x^3 - 3x + 7}(x^2 - 1)\, dx$

33. $\int \dfrac{x^3 + x^2}{3x^4 + 4x^3}\, dx$

34. $\int \dfrac{x^2 - x}{2x^3 - 3x^2}\, dx$

35. $\int \dfrac{x^3 + x^2}{(3x^4 + 4x^3)^2}\, dx$

36. $\int \dfrac{x^2 - x}{(2x^3 - 3x^2)^3}\, dx$

37. $\int \dfrac{x}{1 - x^2}\, dx$

38. $\int \dfrac{1}{1 - x}\, dx$

39. $\int (2x - 3)^7\, dx$

40. $\int (5x + 9)^9\, dx$

41. $\int \dfrac{e^{2x}}{e^{2x} + 1}\, dx$

42. $\int \dfrac{e^{3x}}{e^{3x} - 1}\, dx$

43. $\int \dfrac{\ln x}{x}\, dx$ (*Hint:* Let $u = \ln x$.)

44. $\int \dfrac{(\ln x)^2}{x}\, dx$ (*Hint:* Let $u = \ln x$.)

45. $\int \dfrac{e^{\sqrt{x}}}{\sqrt{x}}\, dx$ (*Hint:* Let $u = \sqrt{x}$.)

46. $\int \dfrac{e^{1/x}}{x^2}\, dx$ $\left(\textit{Hint:} \text{ Let } u = \dfrac{1}{x}.\right)$

Find each integral. (*Hint:* Try some algebra.)

47. $\int (x + 1)x^2\, dx$

48. $\int (x + 4)(x - 2)\, dx$

49. $\int (x + 1)^2 x^3\, dx$

50. $\int (x - 1)^2 \sqrt{x}\, dx$

For each definite integral:

a. Evaluate it "by hand."

b. Check your answer by using a graphing calculator.

51. $\int_0^3 e^{x^2} x\, dx$

52. $\int_0^2 e^{x^3} x^2\, dx$

53. $\int_0^1 \dfrac{x}{x^2 + 1}\, dx$

54. $\int_2^3 \dfrac{x^2}{x^3 - 7}\, dx$

55. $\int_0^4 \sqrt{x^2 + 9}\, x\, dx$

56. $\int_0^3 \sqrt{x^2 + 16}\, x\, dx$

57. $\int_2^3 \dfrac{dx}{1 - x}$

58. $\int_3^4 \dfrac{dx}{2 - x}$

59. $\int_1^8 \dfrac{e^{\sqrt[3]{x}}}{\sqrt[3]{x^2}}\, dx$

60. $\int_1^4 \dfrac{e^{\sqrt{x}}}{\sqrt{x}}\, dx$

61. Prove the integration formula
$$\int u^n\, du = \frac{1}{n + 1} u^{n+1} + C \ (n \neq -1) \text{ as follows.}$$

a. Differentiate the right-hand side of the formula with respect to x (remembering that u is a function of x).

b. Verify that this agrees with the integrand in the formula (after replacing du in the formula by $u'\, dx$).

62. Prove the integration formula $\int e^u\, du = e^u + C$ by following the steps in Exercise 61.

63. Prove the integration formula $\displaystyle\int \frac{du}{u} = \ln u + C$ ($u > 0$) by following the steps in Exercise 61. (Absolute value bars come from applying the same argument to $-u$ for $u < 0$.)

64. Find $\int (x + 1)\, dx$ by

a. using the formula for $\int u^n\, du$ with $n = 1$,

b. dropping the parentheses and integrating directly.

c. Can you reconcile the two seemingly different answers? (*Hint:* Think of the arbitrary constant.)

APPLIED EXERCISES

65. Business: Cost   A company's marginal cost function is $MC(x) = \dfrac{1}{2x + 1}$ and its fixed costs are 50. Find the cost function.

66. Business: Cost   A company's marginal cost function is $MC(x) = \dfrac{1}{\sqrt{2x + 25}}$ and its fixed costs are 100. Find the cost function.

67. General: Average Value   The population of a city is expected to be $P(x) = x(x^2 + 36)^{-1/2}$ million people after x years. Find the average population between year $x = 0$ and year $x = 8$.

68. General: Area   Find the area between the curve $y = xe^{x^2}$ and the x-axis from $x = 1$ to $x = 3$. (Leave the answer in its exact form.)

69. Business: Average Sales   A company's sales (in millions) during week x are given by $S(x) = \dfrac{1}{x + 1}$. Find the average sales from week $x = 1$ to week $x = 4$.

70. Behavioral Science: Repeated Tasks   A subject can perform a task at the rate of $\sqrt{2t + 1}$ tasks per minute at time t minutes. Find the total number of tasks performed from time $t = 0$ to time $t = 12$.

71. Biomedical: Cholesterol   An experimental drug lowers a patient's blood serum cholesterol at the rate of $t\sqrt{25 - t^2}$ units per day, where t is the number of days since the drug was administered ($t \le 5$). Find the total change during the first 3 days.

72. Business: Total Sales   During an automobile sale, cars are selling at the rate of $\dfrac{12}{x + 1}$ cars per day, where x is the number of days since the sale began. How many cars will be sold during the first 7 days of the sale?

73. Business: Total Sales A real estate office is selling condominiums at the rate of $100e^{-(1/4)x}$ per week after x weeks. How many condominiums will be sold during the first 8 weeks?

74. Business: Revenue An aircraft company estimates its marginal revenue function for helicopters to be $MR(x) = \sqrt{x^2 + 80x}(x + 40)$ thousand dollars, where x is the number of helicopters sold. Find the total revenue from the sale of the first 10 helicopters.

75–76: Environmental Science: Pollution A factory is discharging pollution into a lake at the rate $r(t)$ tons per year given below, where t is the number of years that the factory has been in operation. Find the total amount of pollution discharged during the first 3 years of operation.

75. $r(t) = \dfrac{t}{t^2 + 1}$ **76.** $r(t) = t\sqrt{t^2 + 16}$

Chapter Summary with Hints and Suggestions

The reading and exercises of this chapter have helped you to learn the following skills. For each skill, the section from which it came (in case you need to review it) and some exercises in this section that use it are indicated. Answers to all exercises are at the end of the book, and full solutions to all exercises are in the Student Solutions Manual.

9.1   Antiderivatives

- Find an indefinite integral using the power rule. *(Review Exercises 1–8.)*

$$\int x^n \, dx = \frac{1}{n + 1} x^{n+1} + C \qquad (n \neq -1)$$

- Solve an applied problem involving integration. *(Review Exercises 9–10.)*

9.2   Integration Using Logarithmic and Exponential Functions

- Find an indefinite integral involving e^x or $\frac{1}{x}$.

(Review Exercises 11–18.)

$$\int e^{ax} \, dx = \frac{1}{a} e^{ax} + C$$

$$\int \frac{1}{x} \, dx = \int x^{-1} \, dx = \ln |x| + C$$

- Solve an applied problem involving integration (some using 📈).

(Review Exercises 19–22.)

9.3   Definite Integrals and Areas

- Evaluate a definite integral. *(Review Exercises 23–30.)*

$$\int_a^b f(x) \, dx = F(b) - F(a)$$

- Find the area under a curve (and check using 📈).   *(Review Exercises 31–36.)*

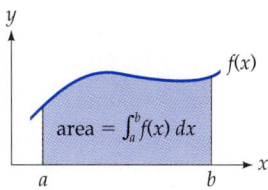

- Use ⊡ to graph a function and find the area under it.   *(Review Exercises 37–38.)*

- Solve a word problem using definite integration (some using ⊡).   *(Review Exercises 39–44.)*

- Approximate the area under a curve by rectangles (Riemann sum).
  (Review Exercises 45–46.)

$$\int_a^b f(x)\,dx = \lim_{n\to\infty}\,[f(x_1)\cdot\Delta x + \cdots + f(x_n)\cdot\Delta x]$$

- Use ⊡ and a Riemann sum program to find Riemann sums.   *(Review Exercises 47–48.)*

9.4   *Further Applications of Definite Integrals: Average Value and Area Between Curves*

- Find the area bounded by two curves (some using ⊡).   *(Review Exercises 49–56.)*

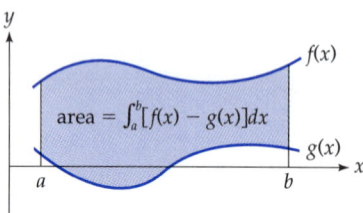

area = $\int_a^b[f(x) - g(x)]dx$

- Find the average value of a function on an interval (some using ⊡).
  (Review Exercises 57–60.)

$$\left(\begin{array}{c}\text{Average of } f\\ \text{from } a \text{ to } b\end{array}\right) = \frac{1}{b-a}\int_a^b f(x)\,dx$$

- Solve an applied problem involving average value (some using ⊡).
  (Review Exercises 61–64.)

- Solve an applied problem involving area between curves (some using ⊡).
  (Review Exercises 65–68.)

9.5   *Two Applications to Economics: Consumers' Surplus and Income Distribution*

- Find consumers' surplus for a product (some using ⊡).   *(Review Exercises 69–72.)*

$$\left(\begin{array}{c}\text{Consumers'}\\ \text{surplus}\end{array}\right) = \int_0^A [d(x) - B]\,dx$$

- Find the Gini index of income distribution (some using ⊡).   *(Review Exercises 73–76.)*

$$\left(\begin{array}{c}\text{Gini}\\ \text{index}\end{array}\right) = 2\cdot\int_0^1 [x - L(x)]\,dx$$

9.6   *Integration by Substitution*

- Use a substitution to find an integral.
  (Review Exercises 77–92.)

$$\int u^n\,du = \frac{1}{n+1}u^{n+1} + C \qquad (n \ne -1)$$

$$\int e^u\,du = e^u + C \qquad \int \frac{1}{u}\,du = \ln|u| + C$$

- Use a substitution to find a definite integral. Verify using ⊡.   *(Review Exercises 93–100.)*

- Use a substitution to find the area under a curve.   *(Review Exercises 101–102.)*

- Use a substitution to find the average value of a function.   *(Review Exercises 103–104.)*

- Use a substitution to solve an applied problem.
  (Review Exercises 105–106.)

Hints and Suggestions

- An *indefinite* integral is a function plus a constant, while a *definite* integral is a *number* (the area between the curve and the *x*-axis if the function is nonnegative). The Fundamental Theorem shows how to evaluate one in terms of the other.

- To integrate any power of *x except* x^{-1}, use the power rule; for x^{-1}, use the ln rule.

- *Indefinite* integrals have a $+C$. *Definite* integrals do not.

- Differentiation *breaks things down into parts,* for example, turning cost into marginal cost (cost per unit). Integration *combines back into a whole,* for example, combining all of the per-unit costs back into a total cost. In fact, the word "integrate" means "make whole."

- To find the area between two curves, integrate *upper* minus *lower.*

- For average values, don't forget the $\frac{1}{b-a}$.

- The substitution method can only "fix up" the *constant;* the variable part must already be correct (or else the function cannot be integrated by this technique).

- A graphing calculator helps by graphing functions, evaluating definite integrals, and finding areas under curves and total accumulations. With an appropriate program it can calculate Riemann sums with many rectangles.

- Practice for test: Review Exercises 1, 9, 15, 21, 27, 31, 37, 39, 43, 49, 55, 61, 69, 73, 81, 83, 93.

Review Exercises for Chapter 9   *Practice test exercises are in blue.*

9.1 Antiderivatives

Find each integral.

1. $\int (24x^2 - 8x + 1)\, dx$

2. $\int (12x^3 + 6x - 3)\, dx$

3. $\int (6\sqrt{x} - 5)\, dx$

4. $\int (8\sqrt[3]{x} - 2)\, dx$

5. $\int (10\sqrt[3]{x^2} - 4x)\, dx$

6. $\int (5\sqrt{x^3} - 6x)\, dx$

7. $\int (x + 4)(x - 4)\, dx$

8. $\int \frac{3x^3 + 2x^2 + 4x}{x}\, dx$

9. **Business: Cost** A company's marginal cost function is $MC(x) = x^{-1/2} + 4$, where x is the number of units. If fixed costs are 20,000, find the company's cost function.

10. **General: Population** The population of a town is now 40,000 and t years from now will be growing at the rate of $300\sqrt{t}$ people per year.

   a. Find a formula for the population of the town t years from now.

   b. Use your formula to find the population of the town 16 years from now.

9.2 Integration Using Logarithmic and Exponential Functions

Find each integral.

11. $\int e^{\frac{1}{2}x}\, dx$

12. $\int e^{-2x}\, dx$

13. $\int 4x^{-1}\, dx$

14. $\int \frac{2}{x}\, dx$

15. $\int \left(6e^{3x} - \frac{6}{x}\right) dx$

16. $\int (x - x^{-1})\, dx$

17. $\int (9x^2 + 2x^{-1} + 6e^{3x})\, dx$

18. $\int \left(\frac{1}{x^2} + \frac{1}{x} + e^{-x}\right) dx$

19. **General: Consumption of Natural Resources** World consumption of aluminum is running at the rate of $40e^{0.05t}$ thousand tons per year, where t is the number of years since 1995.

   a. Find a formula for the total amount of aluminum consumed within t years of 1995.

   b. If consumption continues at this rate, when will the known resources of 7800 thousand tons of aluminum be exhausted?

20. **General: Total Savings** A homeowner installs a solar heating system, which is expected

to generate savings at the rate of $200e^{0.1t}$ dollars per year, where t is the number of years since the system was installed.

a. Find a formula for the total savings in the first t years.

b. If the system originally cost $1500, when will it "pay for itself?"

21. General: Consumption of Natural Resources World consumption of zinc is running at the rate of $8e^{0.02t}$ million metric tons per year, where t is the number of years since 1995.

a. Find a formula for the total amount of zinc consumed within t years of 1995.

b. If consumption continues at this rate, when will the known resources of 140 million metric tons of zinc be exhausted? (Zinc is used to make protective coatings for iron and steel.)

22. Business: Profit A company's profit is growing at the rate of $200x^{-1}$ thousand dollars per month after x months, for $x \geq 1$.

a. Find a formula for the total growth in the profit from month 1 up to month x.

b. When will the total growth in profit reach 600 thousand dollars?

9.3   Definite Integrals and Areas

Evaluate each definite integral.

23. $\displaystyle\int_1^9 \left(x - \frac{1}{\sqrt{x}} \right) dx$

24. $\displaystyle\int_2^5 (3x^2 - 4x + 5)\, dx$

25. $\displaystyle\int_1^{e^4} \frac{dx}{x}$ **26.** $\displaystyle\int_1^5 \frac{dx}{x}$

27. $\displaystyle\int_0^2 e^{-x}\, dx$ **28.** $\displaystyle\int_0^2 e^{\frac{1}{2}x}\, dx$

29. $\displaystyle\int_0^{100} (e^{0.05x} - e^{0.01x})\, dx$

30. $\displaystyle\int_0^{10} (e^{0.04x} - e^{0.02x})\, dx$

For each function:

a. Find the area under the curve between the given x-values.

b. Verify your answer to part (a) by using a graphing calculator to find the area.

31. $f(x) = 6x^2 - 1$, $x = 1$ to $x = 2$

32. $f(x) = 9 - x^2$, $x = -3$ to $x = 3$

33. $f(x) = 12e^{2x}$, $x = 0$ to $x = 3$

34. $f(x) = e^{x/2}$, $x = 0$ to $x = 4$

35. $f(x) = \dfrac{1}{x}$, $x = 1$ to $x = 100$

36. $f(x) = x^{-1}$, $x = 1$ to $x = 1000$

 Use a calculator to graph each function and find the area under it between the given x-values.

37. $f(x) = \dfrac{10}{x^4 + 1}$ from $x = -2$ to $x = 2$

38. $f(x) = e^{x^4}$ from $x = -1$ to $x = 1$

39. General: Weight Gain An average child of age t years gains weight at the rate of $1.7t^{1/2}$ kilograms per year. Find the total weight gain from age 1 to age 9.

40. Behavioral Science: Learning A student can memorize foreign vocabulary words at the rate of $\dfrac{2}{\sqrt[3]{t}}$ words per minute, where t is the number of minutes since the studying began. Find the number of words that can be memorized in the first 8 minutes.

41. Environmental Science: Global Warming The temperature of the earth is rising, as a result of the "greenhouse effect," in which carbon dioxide prevents the escape of heat from the atmosphere. If the temperature is rising at the rate of $0.15e^{0.1t}$ degrees per year, find the total rise in temperature over the next 10 years.

42. Business: Cost A company's marginal cost function is $MC(x) = x^{-1/2} + 4$ dollars, where x is the number of units. Find the total cost of the first 400 units (units $x = 0$ to $x = 400$).

43. Business: Cost A company's marginal cost function is $MC(x) = 22e^{-\sqrt{x}/5}$ dollars, where x is the number of units. Find the total cost of the first hundred units (units $x = 0$ to $x = 100$).

44. Behavioral Science: Repetitive Tasks A proofreader can read $15xe^{-0.25x}$ pages per hour, where x is the number of hours worked. Find the total number of pages that can be proofread in 8 hours.

Exercises on Riemann Sums

45. a. Approximate the area under the curve $f(x) = x^2$ from 0 to 2 using 10 inscribed rectangles with equal bases.

b. Find the *exact* area under the curve between the given x-values by evaluating an appropriate definite integral using the Fundamental Theorem.

 46. a. Approximate the area under the curve $f(x) = \sqrt{x}$ from 0 to 4 using 10 inscribed rectangles with equal bases. (Round calculations to three decimal places.)

b. Find the *exact* area under the curve between the given x-values by evaluating an appropriate definite integral using the Fundamental Theorem.

47–48: Use one of the programs on page 774 (modified as necessary for your calculator or computer), or some similar program, to calculate the following Riemann sums.

a. Calculate the Riemann sum for each function below for the following values of n: 10, 100, 1000 ($n = 10{,}000$ is optional). Make a table of the answers, keeping four decimal places of accuracy.

b. Find the *exact* value of the area under the curve in each exercise below by evaluating an appropriate definite integral using the Fundamental Theorem. Your answers from part (a) should approach the number found from this integral.

47. $f(x) = e^x$ from $a = -2$ to $b = 2$

48. $f(x) = \dfrac{1}{x}$ from $a = 1$ to $b = 4$

9.4 Further Applications of Definite Integrals: Average Value and Area Between Curves

Find the area bounded by each pair of curves.

49. $y = x^2 + 3x$ and $y = 3x + 1$

50. $y = 12x - 3x^2$ and $y = 6x - 24$

51. $y = x^2$ and $y = x$

52. $y = x^4$ and $y = x$

53. $y = 4x^3$ and $y = 12x^2 - 8x$

54. $y = x^3 + x^2$ and $y = x^2 + x$

55. $y = e^x$ and $y = x + 5$

56. $y = \ln x$ and $y = \dfrac{x^2}{10}$

Find the average value of the function on the given interval.

57. $f(x) = \dfrac{1}{x^2}$ on $[1, 4]$

58. $f(x) = 6\sqrt{x}$ on $[1, 4]$

59. $f(x) = \sqrt{x^3 + 1}$ on $[0, 5]$

60. $f(x) = \ln(e^{x^2} + 10)$ on $[-2, 2]$

61. General: Average Population The population of the world is predicted to be $P(t) = 6e^{0.01t}$ billion people, where t is the number of years after the year 2000. Find the average population between the years 2000 and 2100.

62. General: Compound Interest A deposit of $3000 in a bank paying 6% interest compounded continuously will grow to $V(t) = 3000e^{0.06t}$ dollars after t years. Find the average value during the first 20 years ($t = 0$ to $t = 20$).

63. General: Real Estate The value of a suburban plot of land being considered for rezoning is assessed at $4.3e^{0.01x^2}$ hundred thousand dollars x years from now. Find the average value over the next 10 years (year 0 to year 10).

64. General: Stock Price The price of a share of stock is expected to be $28e^{0.01x^{1.2}}$ dollars, where x is the number of weeks from now. Find the average price over the next year (week 0 to week 52).

65. General: Art An artist wants to paint the interior shape shown on the next page on the side of a building. If the shape is to be painted blue, how much area (in square meters) will the artist need to paint?

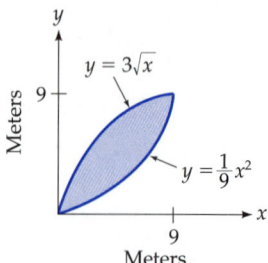

a. Enter these numbers into your graphing calculator and make a plot of the resulting points (Years after 1987 on the x-axis and Births on the y-axis).

b. Have your calculator find the linear regression formula for these data. Then enter the result as y_1, which gives a formula for births for each year. Plot the points together with the regression line, and observe that the fit is reasonably good.

c. Enter $y_2 = 1.09e^{0.012x}$, the population (in billions) for China x years after 1987.

d. Enter $y_3 = 23.3*y_2$, the number of births (in millions) at the 1987 rate, x years after 1987. (*Note:* * means multiplication.)

e. Enter $y_4 = y_1*y_2$, the number of births (in millions) at the actual rate (based on the linear regression from the preceding table), x years after 1987.

f. Turn off y_1 and y_2 so that they will not graph, and graph y_3 and y_4 on [0, 13] by [0, 40]. The area between the graphs represents the actual reduction in births (in millions) resulting from the lowered birth rates projected from 1987 to 2000.

g. Find the area between the curves from 0 to 13, thereby predicting the decrease in births (in millions) during the years 1987 to 2000.

 66. Economics: Balance of Trade   A country's annual exports will be $E(t) = 40e^{0.2t}$ and its imports will be $I(t) = 20e^{0.1t}$ (both in billions of dollars per year), where t is the number of years from now. Find the accumulated trade surplus (exports minus imports) over the next 10 years.

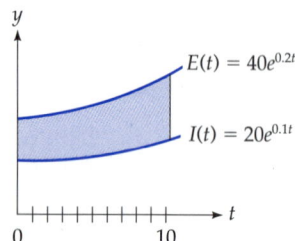

 67. Business: Profit   A company's annual revenue and annual costs are expected to be $R(t) = 50e^{0.08t}$ and $C(t) = 20e^{0.04t}$ million dollars per year, where t is the number of years from now. Find the cumulative profit over the next 8 years (year 0 to year 8).

 68. General: Population Reduction   China, with 22% of the world's population on only 7% of the world's arable land, has taken drastic measures to reduce its population. The following table shows how the birth rate (births per 1000 population) has decreased recently. To avoid large numbers, years are listed as years after 1987.

Years after 1987	Births per 1000 Population
1987	0	23.3
1988	1	22.4
1989	2	21.6
1990	3	21.1
1991	4	19.7
1992	5	18.2

Source: New York Times, April 25, 1993.

9.5   Two Applications to Economics: Consumers' Surplus and Income Distribution

69–72: Economics: Consumers' Surplus   For each demand function $d(x)$ and the demand level x, find the consumers' surplus.

69. $d(x) = 8000 - 24x$, $x = 200$

70. $d(x) = 1800 - 0.03x^2$, $x = 200$

 71. $d(x) = 300e^{-0.2\sqrt{x}}$, $x = 120$

 72. $d(x) = \dfrac{100}{1 + \sqrt{x}}$, $x = 100$

73–76: Economics: Gini Index   For the Lorenz curve given below, find the Gini index.

 73. $L(x) = x^{3.5}$ **74.** $L(x) = x^{2.5}$

75. $L(x) = \dfrac{x}{2 - x}$ **76.** $L(x) = \dfrac{e^{x^4} - 1}{e - 1}$

9.6 Integration by Substitution

Find each integral or state that it cannot be evaluated by our substitution formulas.

77. $\displaystyle\int x^2 \sqrt[3]{x^3 - 1}\, dx$ **78.** $\displaystyle\int x^3 \sqrt{x^4 - 1}\, dx$

79. $\displaystyle\int x \sqrt[3]{x^3 - 1}\, dx$ **80.** $\displaystyle\int x^2 \sqrt{x^4 - 1}\, dx$

81. $\displaystyle\int \frac{dx}{9 - 3x}$ **82.** $\displaystyle\int \frac{dx}{1 - 2x}$

83. $\displaystyle\int \frac{dx}{(9 - 3x)^2}$ **84.** $\displaystyle\int \frac{dx}{(1 - 2x)^2}$

85. $\displaystyle\int \frac{x^2}{\sqrt[3]{8 + x^3}}\, dx$ **86.** $\displaystyle\int \frac{x}{\sqrt{9 + x^2}}\, dx$

87. $\displaystyle\int \frac{w + 3}{(w^2 + 6w - 1)^2}\, dw$

88. $\displaystyle\int \frac{t - 2}{(t^2 - 4t + 1)^2}\, dt$

89. $\displaystyle\int \frac{(1 + \sqrt{x})^2}{\sqrt{x}}\, dx$ **90.** $\displaystyle\int \frac{(1 + \sqrt[3]{x})^2}{\sqrt[3]{x^2}}\, dx$

91. $\displaystyle\int \frac{e^x}{e^x - 1}\, dx$ **92.** $\displaystyle\int \frac{1}{x \ln x}\, dx$

For each definite integral:

a. Evaluate it ("by hand") or state that it cannot be evaluated by our substitution formulas.

b. Verify your answer to part (a) by using a graphing calculator.

93. $\displaystyle\int_0^3 x \sqrt{x^2 + 16}\, dx$ **94.** $\displaystyle\int_0^4 \frac{dz}{\sqrt{2z + 1}}$

95. $\displaystyle\int_0^4 \frac{w}{\sqrt{25 - w^2}}\, dw$ **96.** $\displaystyle\int_1^2 \frac{x + 1}{(x^2 + 2x - 2)^2}\, dx$

97. $\displaystyle\int_9^3 \frac{dx}{x - 2}$ **98.** $\displaystyle\int_4^5 \frac{dx}{x - 6}$

99. $\displaystyle\int_0^1 x^3 e^{x^4}\, dx$ **100.** $\displaystyle\int_0^1 x^4 e^{x^5}\, dx$

Find the area under the given curve between the given x-values.

101. $y = \dfrac{x^2 + 6x}{\sqrt[3]{x^3 + 9x^2 + 17}}$ from $x = 1$ to $x = 3$

102. $y = \dfrac{x + 6}{\sqrt{x^2 + 12x + 4}}$ from $x = 0$ to $x = 3$

Find the average value of the function on the given interval.

103. $f(x) = xe^{-x^2}$ on $[0, 2]$

104. $f(x) = \dfrac{x}{x^2 - 3}$ on $[2, 4]$

105. Business: Cost A company's marginal cost function is $MC(x) = \dfrac{1}{\sqrt{2x + 9}}$ and fixed costs are 100. Find the cost function.

106. Biomedical: Temperature An experimental drug changes a patient's temperature at the rate $\dfrac{3x^2}{x^3 + 1}$ degrees per milligram of the drug, where x is the amount of the drug administered. Find the total change in temperature resulting from the first 3 milligrams of the drug. (*Note:* The rate of change of temperature with respect to dosage is called the "drug sensitivity.")

Projects and Essays

The following projects and essays are based on Chapter 9. Most have no right and wrong answers—the results depend only on your imagination and resourcefulness.

1. Look back over the examples and applied exercises in the chapter and find at least ten applications where integration can be interpreted as a "continuous sum." Explain what is being summed in each case. Explain what this has to do with Riemann sums.

2. Look back over the chapter and find the two most interesting (in your opinion) uses of integration (definite or indefinite). Explain why they are your favorites.

3. In Example 7 on pages 752–753 it was assumed that consumption of tin would continue at the same rate right up to its point of exhaustion. Discuss whether this is a reasonable assumption. How will scarcity affect the price of tin, and how will the price affect consumption? Draw a graph of what you would expect for the consumption curve as the supply dwindles. Can shortages induce new innovations, efficiencies, or substitutions? Can you think of specific examples? What is the effect of recycling? How might these ideas affect your graph of consumption? Finally, discuss the value of the original assumption of unchanging consumption in providing a worst-case prediction.

4. Look up Georg Bernard Riemann in a book on the history of mathematics and write about his contributions to mathematics, including the development of Riemann sums and the definite integral.

5. Can the curve below be a Lorenz curve? (*Hint:* What proportion of income would be held by the lowest 20% of the population?) Characterize Lorenz curves in terms of their values at 0 and at 1, their slope, and their concavity.

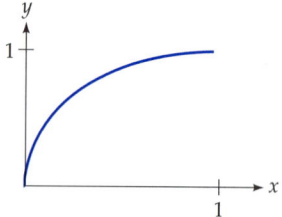

6. The Lorenz curves for the proportion of the world's wealth held by various proportions of the world's people are as shown below for various years. Calculate the Gini indices for these years and discuss what trend they indicate about the concentration of wealth in the world. Would this trend tend to increase or decrease world stability? Can you think of any recent world events related to this trend?

Year	Lorenz Curve
1960	$x^{5.42}$
1970	$x^{6.87}$
1980	$x^{6.45}$
1990	$x^{7.94}$

Source: United Nations Development Programme.

7. Write about differentiation and (indefinite) integration as inverse operations. That is, does integrating a function and then differentiating the result give back the original function? What about the other way around? What about the $+ C$? Include a discussion (with several examples) of how differentiation "breaks things down into parts" and integration "combines them back into a whole" (see the Hints and Suggestions on page 817).

8. Look over your notes, your homework, and the text, and write a page about how a graphing calculator has helped you in this chapter. Include examples of how it has helped you to *explore concepts* and how it has helped to *simplify your work*. What was its *most* helpful or interesting use? What was its *least* helpful or interesting use? Are there problems that can be done on a graphing calculator but that are easier to do "by hand"?

10

INTEGRATION TECHNIQUES AND DIFFERENTIAL EQUATIONS

The length of a cable in a suspension bridge can be calculated by integration.

10.1 Integration by Parts

Introduction

In this chapter we introduce further techniques for finding integrals: integration by parts, integration by tables, and numerical integration. We also discuss improper integrals (integrals over infinite intervals) and differential equations.

First we discuss the method of *integration by parts*, which comes from interpreting the product rule as an integration formula.

Integration by Parts

For two differentiable functions $u(x)$ and $v(x)$, hereafter denoted simply u and v, the product rule is

$$(uv)' = u'v + uv'$$

> The derivative of a product is the derivative of the first times the second, plus the first times the derivative of the second

If we integrate both sides of this equation, integrating the left side "undoes" the differentiation.

$$uv = \int u'v \, dx + \int uv' \, dx$$

$$\underbrace{}_{du} \qquad \underbrace{}_{dv}$$

> Using differential notation $du = u' \, dx$, $dv = v' \, dx$

$$uv = \int v \, du + \int u \, dv$$

> Above formula in differential notation

Solving this equation for the second integral $\int u \, dv$ gives

$$\int u \, dv = uv - \int v \, du$$

This formula is the basis for a technique called "integration by parts."

Integration by Parts

For differentiable functions u and v,

$$\int u \, dv = uv - \int v \, du$$

We use this formula to solve integrals by a "double substitution," substituting u for part of the given integral and dv for the rest, and then expressing the integral in the form $uv - \int v\,du$. The point is to choose the u and the dv so that the resulting integral $\int v\,du$ is *simpler* than the original integral $\int u\,dv$. A few examples will make the method clear.

EXAMPLE 1 **Integrating by Parts**

Use integration by parts to find $\displaystyle\int xe^x\,dx$.

Solution

$$\int \underset{u\quad dv}{\underbrace{xe^x\,dx}}$$

Original integral

We choose $u = x$ and $dv = e^x\,dx$

$$\begin{bmatrix} u = x & dv = e^x\,dx \\ \downarrow & \downarrow \\ du = 1\,dx = dx & v = \int e^x\,dx = e^x \end{bmatrix}$$

Differentiating u to find du and integrating dv to find v (omitting the C) (the arrows show which part leads to which other part)

$$= \underset{u\ v}{xe^x} - \int \underset{v\quad du}{e^x\,dx}$$

Replacing the original integral $\int u\,dv$ by the right-hand side of the formula, $u \cdot v - \int v\,du$, with $u = x$, $v = e^x$, and $du = dx$

$$= xe^x - e^x + C$$
$$\underset{\text{From } \int e^x\,dx}{}$$

Finding the new integral $\int e^x\,dx = e^x$ to give the final answer (with $+C$)

The procedure is not as complicated as it might seem. All the steps may be written together as follows:

$$\int \underset{u\quad dv}{xe^x\,dx} \qquad = \underset{uv}{xe^x} - \int \underset{v\quad du}{e^x\,dx} = xe^x - e^x + C$$
$$\underset{\text{From } \int e^x\,dx}{}$$

$$\begin{bmatrix} u = x & dv = e^x\,dx \\ du = dx & v = \int e^x\,dx = e^x \end{bmatrix}$$

We may check this answer by differentiation.

$$\frac{d}{dx}(xe^x - e^x + C) = e^x + xe^x - e^x = xe^x$$

Cancel

Differentiating xe^x by the product rule

Agrees with the original integrand, so the integration is correct

Remarks on the Integration by Parts Procedure

i. The differentials du and dv include the dx.

ii. We omit the constant C when we integrate dv to get v because one C at the end is enough.

iii. The integration by parts formula does not give a "final answer," but rather expresses the given integral as $uv - \int v \, du$, a product $u \cdot v$ (already integrated) and a new integral $\int v \, du$. That is, integration by parts "exchanges" the original integral $\int u \, dv$ for another integral $\int v \, du$. The hope is that the second integral will be simpler than the first. In our example we "exchanged" $\int xe^x \, dx$ for the simpler $\int e^x \, dx$, which could be integrated immediately by formula 3 (inside back cover).

iv. Integration by parts is rather complicated, so it should be used only if formulas 1 through 7 (inside back cover) fail to solve the integral.

How to Choose the *u* and the *dv*

In Example 1, the choice of $\begin{cases} u = x \\ dv = e^x \, dx \end{cases}$ "exchanged" the original integral $\int xe^x \, dx$ for the simpler integral $\int e^x \, dx$. If we had instead chosen $\begin{cases} u = e^x \\ dv = x \, dx \end{cases}$ we would have "exchanged" the original integral for $\int x^2 e^x \, dx$ (as you may check), which is *more* difficult than the original (because of the x^2). Therefore, the first choice was the "right" choice in that it led to a solution. Generally, one choice for u and dv will be "best," and finding it may involve some trial and error. While there is no foolproof rule for finding the best u and dv, the following guidelines often help.

Guidelines for Choosing *u* and *dv*

1. Choose dv to be the most complicated part of the integral that can be integrated easily.

2. Choose u so that u' is simpler than u.

EXAMPLE 2 **Integrating by Parts**

Find $\displaystyle\int x^2 \ln x \, dx$.

Solution   None of the easier formulas (1 through 7 inside the back cover) solve the integral, as you may easily check. Therefore, we try integration by parts. The integrand is a product, x^2 times $\ln x$. The guidelines say to choose dv to be the most complicated part that can be easily integrated. We can integrate x^2 but not $\ln x$ (we know how to *differentiate* $\ln x$, but not how to *integrate* it), so we choose $dv = x^2\, dx$, and therefore, $u = \ln x$.

$$\int x^2 \ln x\, dx = \ln x\, \frac{1}{3}x^3 - \int \frac{1}{3}x^3 \frac{1}{x}\, dx = \frac{1}{3}x^3 \ln x - \frac{1}{3}\int x^2\, dx$$

$$\underbrace{}_{\substack{u \\ dv}} \qquad \underbrace{}_{u}\,\underbrace{}_{v} \qquad \underbrace{}_{v}\,\underbrace{}_{du} \qquad \underbrace{}_{\text{Moving the } \frac{1}{3}}$$

outside and simplifying

$\frac{1}{x}x^3$ to x^2

$$\left[\begin{array}{ll} u = \ln x & dv = x^2\, dx \\[2mm] du = \dfrac{1}{x}\, dx & v = \int x^2\, dx = \dfrac{1}{3}x^3 \end{array} \right]$$

$$= \frac{1}{3}x^3 \ln x - \frac{1}{3}\frac{1}{3}x^3 + C = \frac{1}{3}x^3 \ln x - \frac{1}{9}x^3 + C$$

We may check this answer by differentiation.

$$\frac{d}{dx}\left(\frac{1}{3}x^3 \ln x - \frac{1}{9}x^3 + C \right) = x^2 \ln x + \frac{1}{3}x^3 \frac{1}{x} - \frac{1}{3}x^2$$

$$\underbrace{}$$

Differentiating $\frac{1}{3}x^3 \ln x$

by the product rule

$$= x^2 \ln x + \frac{1}{3}x^2 - \frac{1}{3}x^2 = x^2 \ln x$$

Agrees with the original integrand, so the integration is correct

From $x^3\, \dfrac{1}{x}$ Cancel

PRACTICE PROBLEM 1   Use integration by parts to find: $\displaystyle\int x^3 \ln x\, dx$.

Solution at the back of the book

Integration by parts is also useful for integrating products of powers of linear functions.

EXAMPLE 3 **Integrating by Parts**

Use integration by parts to find: $\displaystyle\int (x - 2)(x + 4)^8\, dx$.

Solution   The guidelines recommend that dv be the most complicated part that can be integrated. Both $x - 2$ and $(x + 4)^8$ can be integrated.

For example,

$$\int (x + 4)^8 \, dx = \frac{1}{9} (x + 4)^9 + C \qquad \text{By the substitution method with } u = x + 4 \text{ (omitting the details)}$$

Since $(x + 4)^8$ is more complicated than $(x - 2)$, we take $dv = (x + 4)^8 \, dx$.

$$\int \underbrace{(x - 2)}_{u} \underbrace{(x + 4)^8 \, dx}_{dv} = \underbrace{(x - 2)}_{u} \underbrace{\frac{1}{9} (x + 4)^9}_{v} - \int \underbrace{\frac{1}{9} (x + 4)^9}_{v} \underbrace{dx}_{du}$$

$$\begin{bmatrix} u = x - 2 & dv = (x + 4)^8 \, dx \\ du = dx & v = \int (x + 4)^8 \, dx \\ & = \frac{1}{9} (x + 4)^9 \end{bmatrix}$$

$$= \frac{1}{9} (x - 2)(x + 4)^9 - \frac{1}{9} \int (x + 4)^9 \, dx$$

$$\uparrow$$
$$\text{Taking out the } \tfrac{1}{9}$$

$$= \frac{1}{9} (x - 2)(x + 4)^9 - \frac{1}{9} \underbrace{\frac{1}{10} (x + 4)^{10} + C}_{\text{Integrating by the substitution method}}$$

$$= \frac{1}{9} (x - 2)(x + 4)^9 - \frac{1}{90} (x + 4)^{10} + C$$

Again we could check this answer by differentiation. Do you see how the product rule, applied to the first part of the answer, will give a piece that will cancel with the derivative of the second part?

PRACTICE PROBLEM 2   Use integration by parts to find: $\displaystyle\int (x + 1)(x - 1)^3 \, dx$.

Solution at the back of the book

Present Value of a Continuous Stream of Income

If a business or some other asset generates income continuously at the rate $C(t)$ dollars per year, where t is the number of years from now, then $C(t)$ is called a *continuous stream of income*.* On pages 126–127 we saw that to find the *present value* of a sum (the amount now that will later yield the stated sum) under continuous compounding we multi-

---

* $C(t)$ must be continuous, meaning that the income is being paid continuously rather than in "lump-sum" payments. However, even lump-sum payments can be approximated by a continuous stream if the payments are frequent enough.

ply by e^{-rt}, where r is the interest rate and t is the number of years. (We will refer to a rate with continuous compounding as a "continuous interest rate.") Therefore, the present value of the continuous stream $C(t)$ is found by multiplying by e^{-rt} and summing (integrating) over the time period.

Present Value of a Continuous Stream of Income

> The present value of the continuous stream of income $C(t)$ dollars per year, where t is the number of years from now, for T years at continuous interest rate r is
>
> $$\begin{pmatrix} \text{Present} \\ \text{value} \end{pmatrix} = \int_0^T C(t)e^{-rt}\, dt$$

EXAMPLE 4 Finding the Present Value of a Continuous Stream of Income

A business generates income at the rate of $2t$ million dollars per year, where t is the number of years from now. Find the present value of this continuous stream for the next five years at the continuous interest rate of 10%.

Solution

$$\begin{pmatrix} \text{Present} \\ \text{value} \end{pmatrix} = \int_0^5 2te^{-0.1t}\, dt \qquad \begin{array}{l}\text{Multiplying } C(t) = 2t \text{ by } e^{-0.1t} \\ \text{(since } 10\% = 0.1\text{) and integrating} \\ \text{from 0 to 5 years}\end{array}$$

This is a *definite* integral, but we will ignore the limits of integration until after we have found the *indefinite* integral. None of the formulas 1 through 7 on the inside back cover will find this integral, so we try integration by parts with $u = 2t$ and $dv = e^{-0.1t}\, dt$. (Do you see why the guidelines on page 826 suggest this choice?)

$$\int \underbrace{2t}_{u}\underbrace{e^{-0.1t}\, dt}_{dv} = \underbrace{(2t)}_{u}\underbrace{(-10e^{-0.1t})}_{v} - \int \underbrace{(-10e^{-0.1t})}_{v}\underbrace{2dt}_{du}$$

$$\begin{bmatrix} u = 2t & dv = e^{-0.1t}\, dt \\ du = 2\, dt & v = \int e^{-0.1t}\, dt \\ & \quad = -10e^{-0.1t} \end{bmatrix}$$

$$= -20te^{-0.1t} + 20 \int e^{-0.1t}\, dt$$

$$= -20te^{-0.1t} + 20(-10)e^{-0.1t} + C$$

$$= -20te^{-0.1t} - 200e^{-0.1t} + C$$

For the *definite* integral, we evaluate this from 0 to 5:

$$[-20te^{-0.1t} - 200e^{-0.1t}]\Big|_0^5 = \underbrace{-20 \cdot 5e^{-0.5} - 200e^{-0.5}}_{\text{Evaluation at } t = 5} - [\overbrace{-20 \cdot 0e^0}^{0} - \overbrace{200e^0}^{200}]$$

$$= -300e^{-0.5} + 200 \approx 18 \qquad \text{In millions of dollars (using a calculator)}$$

The present value of the stream of income over 5 years is approximately $18 million.

This answer means that $18 million at 10% interest compounded continuously would generate the continuous stream $C(t) = 2t$ million dollars for 5 years. This method is often used to determine the fair value of a business or some other asset, since it gives the present value of its future income.

Graphing Calculator Exploration

a. Verify the answer to Example 4 by graphing $2xe^{-0.1x}$ and finding the area under the curve from 0 to 5.

b. Can you explain why the curve increases less steeply further to the right? (*Hint:* Think of the present value of money to be paid in the more distant future.)

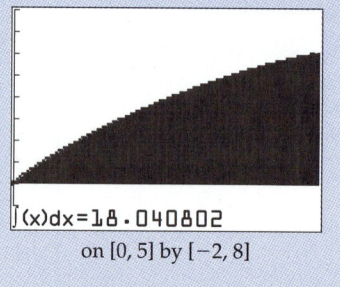

$\int(x)dx=18.040802$

on [0, 5] by [−2, 8]

Why is it necessary to learn integration by parts if integrals like the one in Example 4 can be evaluated on graphing calculators? One answer is that you should have a *variety* of ways to approach a problem—sometimes geometrically, sometimes analytically, and sometimes numerically (using a calculator). These various approaches mutually support one another; for example, you can solve a problem in one way and check it another way. Furthermore, not all applications involve *definite* integrals. Example 4 might have asked for a *formula* for the present value up to any time t, and such formulas may be found by integrating "by hand" but not from graphing calculators.

Remember that you should use integration by parts *only* if the "easier" formulas (1 through 7 on the inside back cover) fail to solve the integral.

PRACTICE PROBLEM 3

Which of the following integrals requires integration by parts, and which can be found by the substitution formula $\int e^u \, du = e^u + C$? (Do not solve the integrals.)

a. $\displaystyle\int xe^x \, dx$ **b.** $\displaystyle\int xe^{x^2} \, dx$ *Solutions at the back of the book*

SUMMARY

The integration by parts formula

$$\int u \, dv = uv - \int v \, du$$

is simply the integration version of the product rule. The guidelines on page 826 and the examples lead to the following particular suggestions for choosing u and dv.

For Integrals of the Form:	Choose:	
$\displaystyle\int x^n e^{ax} \, dx$	$u = x^n$	$dv = e^{ax} \, dx$
$\displaystyle\int x^n \ln x \, dx$	$u = \ln x$	$dv = x^n \, dx$
$\displaystyle\int (x + a)(x + b)^n \, dx$	$u = x + a$	$dv = (x + b)^n \, dx$

Integration by parts can be useful in any situation involving integrals, such as recovering total cost from marginal cost, calculating areas, average values (the definite integral divided by the length of the interval), continuous accumulations, or present values of continuous income streams.

EXERCISES 10.1

Integration by parts often involves finding an integral like one of the following when integrating dv to find v. Find the following integrals *without* using integration by parts (using formulas 1 through 7 on the inside back cover). Be ready to find similar integrals during the integration by parts procedure.

1. $\displaystyle\int e^{2x} \, dx$

2. $\displaystyle\int x^5 \, dx$

3. $\displaystyle\int (x + 2) \, dx$

4. $\displaystyle\int (x - 1) \, dx$

5. $\displaystyle\int \sqrt{x} \, dx$

6. $\displaystyle\int e^{-0.5t} \, dt$

7. $\int (x + 3)^4 \, dx$

8. $\int (x - 5)^6 \, dx$

Use integration by parts to find each integral.

9. $\int xe^{2x} \, dx$

10. $\int xe^{3x} \, dx$

11. $\int x^5 \ln x \, dx$

12. $\int x^4 \ln x \, dx$

13. $\int (x + 2)e^x \, dx$ (*Hint:* Take $u = x + 2$.)

14. $\int (x - 1)e^x \, dx$ (*Hint:* Take $u = x - 1$.)

15. $\int \sqrt{x} \ln x \, dx$

16. $\int \sqrt[3]{x} \ln x \, dx$

17. $\int (x - 3)(x + 4)^5 \, dx$ **18.** $\int (x + 2)(x - 5)^5 \, dx$

19. $\int te^{-0.5t} \, dt$

20. $\int te^{-0.2t} \, dt$

21. $\int \frac{\ln t}{t^2} \, dt$

22. $\int \frac{\ln t}{\sqrt{t}} \, dt$

23. $\int s(2s + 1)^4 \, ds$

24. $\int \frac{x + 1}{e^{3x}} \, dx$

25. $\int \frac{x}{e^{2x}} \, dx$

26. $\int \frac{\ln (x + 1)}{\sqrt{x + 1}} \, dx$

27. $\int \frac{x}{\sqrt{x + 1}} \, dx$

28. $\int x\sqrt{x + 1} \, dx$

29. $\int xe^{ax} \, dx$ $(a \neq 0)$

30. $\int (x + b)e^{ax} \, dx$ $(a \neq 0)$

31. $\int x^n \ln ax \, dx$ $(a \neq 0, n \neq -1)$

32. $\int (x + a)^n \ln (x + a) \, dx$ $(n \neq -1)$

33. $\int \ln x \, dx$ (*Hint:* Take $u = \ln x$, $dv = dx$.)

34. $\int \ln x^2 \, dx$ (*Hint:* Take $u = \ln x^2$, $dv = dx$.)

35. $\int x^3 e^{x^2} \, dx$ (*Hint:* Take $u = x^2$, $dv = xe^{x^2}$, using a substitution to find v from dv.)

36. $\int x^3 (x^2 - 1)^6 \, dx$ (*Hint:* Use $u = x^2$, $dv = x(x^2 - 1)^6 \, dx$, using a substitution to find v from dv.)

Find each integral by whatever means are necessary (integration by parts or a substitution).

37. a. $\int xe^{x^2} \, dx$

b. $\int \frac{(\ln x)^3}{x} \, dx$

c. $\int x^2 \ln 2x \, dx$

d. $\int \frac{e^x}{e^x + 4} \, dx$

38. a. $\int \sqrt{\ln x} \, \frac{1}{x} \, dx$

b. $\int x^2 e^{x^3} \, dx$

c. $\int x^7 \ln 3x \, dx$

d. $\int xe^{4x} \, dx$

Evaluate each definite integral using integration by parts. (Leave answers in exact form.)

39. $\int_0^2 xe^x \, dx$

40. $\int_0^3 xe^x \, dx$

41. $\int_1^3 x^2 \ln x \, dx$

42. $\int_1^2 x \ln x \, dx$

43. $\int_0^2 z(z - 2)^4 \, dz$

44. $\int_0^4 z(z - 4)^6 \, dz$

45. $\int_0^{\ln 4} te^t \, dt$

46. $\int_1^e \ln x \, dx$

Find in two different ways.

47. $\int x(x - 2)^5 \, dx$

a. Use integration by parts.
b. Use the substitution $u = x - 2$ (so x is replaced by $u + 2$) and then multiply out the integrand.

48. $\int x(x + 4)^6 \, dx$

a. Use integration by parts.
b. Use the substitution $u = x + 4$ (so x is replaced by $u - 4$) and then multiply out the integrand.

Derive each formula by using integration by parts on the left-hand side. (Assume $n > 0$.)

49. $\displaystyle\int x^n e^x \, dx = x^n e^x - n \int x^{n-1} e^x \, dx$

50. $\displaystyle\int (\ln x)^n \, dx = x(\ln x)^n - n \int (\ln x)^{n-1} \, dx$

51. Use the formula in Exercise 49 to find the integral $\int x^2 e^x \, dx$. (*Hint:* Apply the formula twice.)

52. Use the formula in Exercise 50 to find the integral $\int (\ln x)^2 \, dx$. (*Hint:* Apply the formula twice.)

53. a. Find the integral $\int x^{-1} \, dx$ by integration by parts (using $u = x^{-1}$ and $dv = dx$), obtaining

$$\int x^{-1} \, dx = x^{-1} x - \int (-x^{-2}) x \, dx$$

which gives

$$\int x^{-1} \, dx = 1 + \int x^{-1} \, dx$$

b. Subtract the integral from both sides of this last equation, obtaining $0 = 1$. Explain this apparent contradiction.

54. We omit the constant of integration when we integrate dv to get v. Including the constant C in this step simply replaces v by $v + C$, giving the formula

$$\int u \, dv = u(v + C) - \int (v + C) \, du$$

Multiplying out the parentheses and expanding the last integral into two gives

$$\int u \, dv = uv + Cu - \int v \, du - C \int du$$

Show that the second and fourth terms on the right cancel, giving the "old" integration by parts formula $\int u \, dv = u \cdot v - \int v \, du$. This shows that including the constant in the dv to v step gives the same formula. One constant of integration at the end is enough.

APPLIED EXERCISES (Most require . Use [calculator icon] only when indicated.) ————————

55. Business: Revenue If a company's marginal revenue function is $MR(x) = xe^{\frac{1}{4}x}$ find the revenue function. (*Hint:* Evaluate the constant C so that revenue is 0 at $x = 0$.)

56. Business: Cost A company's marginal cost function is $MC(x) = xe^{-\frac{1}{2}x}$ and fixed costs are 200. Find the cost function. (*Hint:* Evaluate the constant C so that the cost is 200 at $x = 0$.)

57. Business: Present Value of a Continuous Stream of Income An electronics company generates a continuous stream of income of $4t$ million dollars per year, where t is the number of years that the company has been in operation. Find the present value of this stream of income over the first 10 years at a continuous interest rate of 10%.

For Exercises 58–59:

    a. Solve *without* using a graphing calculator.
    b. Verify your answer to part (a) using a graphing calculator.

58. Business: Present Value of a Continuous Stream of Income An oil well generates a continuous stream of income of $60t$ thousand dollars per year, where t is the number of years that the rig has been in operation. Find the present value of this stream of income over the first 20 years at a continuous interest rate of 5%. (*Instructions in left column*)

59. Biomedical: Drug Dosage A drug taken orally is absorbed into the bloodstream at the rate of $te^{-0.5t}$ milligrams per hour, where t is the number of hours since the drug was taken. Find the total amount of the drug absorbed during the first 5 hours. (*Instructions in left column*)

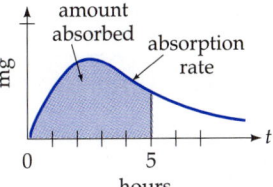

60. Environmental Science: Pollution Contamination is leaking from an underground waste-disposal tank at the rate of $t \cdot \ln t$ thousand gallons per month, where t is the number of months since the leak began. Find the total leakage between months $t = 1$ and $t = 4$.

61. General: Area Find the area under the curve $y = x \cdot \ln x$ and above the x-axis from $x = 1$ to $x = 2$.

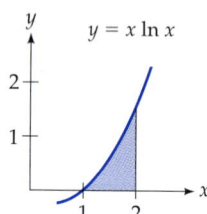

62. Political Science: Fund Raising A politician can raise campaign funds at the rate of $50te^{-0.1t}$ thousand dollars per week during the first t weeks of a campaign. Find the average amount raised during the first 5 weeks.

63. Business: Product Recognition A company begins advertising a new product and finds that after t months the product is gaining customer recognition at the rate of $t^2 \ln t$ thousand customers per week (for $t \geq 1$). Find the total gain in recognition between weeks $t = 1$ and $t = 6$.

64. General: Population The population of a town is increasing at the rate of $400te^{0.02t}$ people per year, where t is the number of years from now. Find the total gain in population during the next 5 years.

Repeated Integration by Parts

Sometimes an integral requires two or more integrations by parts. As an example, we apply integration by parts to the integral $\int x^2 e^x \, dx$.

$$\int \underbrace{x^2}_{u} \underbrace{e^x \, dx}_{dv} = \underbrace{x^2}_{u} \underbrace{e^x}_{v} - \int \underbrace{e^x}_{v} \underbrace{2x \, dx}_{du} = x^2 e^x - 2 \int x e^x \, dx$$

$$\begin{bmatrix} u = x^2 & dv = e^x \, dx \\ du = 2x \, dx & v = \int e^x \, dx = e^x \end{bmatrix}$$

The new integral $\int xe^x \, dx$ is solved by a second integration by parts. Continuing with the previous solution:

$$= x^2 e^x - 2 \left(\underbrace{\int xe^x \, dx}_{u \ dv} \right) \qquad \begin{bmatrix} u = x & dv = e^x \, dx \\ du = dx & v = e^x \end{bmatrix}$$

$$= x^2 e^x - 2 \left(\underbrace{xe^x}_{uv} - \int \underbrace{e^x dx}_{v \ du} \right)$$

$$= x^2 e^x - 2(xe^2 - e^x) + C$$

$$= x^2 e^x - 2xe^x + 2e^x + C$$

After reading the explanation above, find each integral by repeated integration by parts.

65. $\int x^2 e^{-x} \, dx$ **66.** $\int x^2 e^{2x} \, dx$

67. $\int (x + 1)^2 e^x \, dx$ **68.** $\int (\ln x)^2 \, dx$

69. $\int x^2 (\ln x)^2 \, dx$ **70.** $\int x^3 e^x \, dx$

71–72: For each definite integral:

a. Evaluate it by integration by parts. (Give answer in its *exact* form.)

b. Verify your answer to part (a) using a graphing calculator.

71. $\int_0^2 x^2 e^x \, dx$ **72.** $\int_1^5 (\ln x)^2 \, dx$

Repeated Integration by Parts Using a Table

The solution to an integration by parts problem can be organized in a table. As an example, we solve $\int x^2 e^{3x} \, dx$. We begin by choosing

$$u = x^2 \qquad dv = v' \, dx = e^{3x} \, dx$$

We then make a table consisting of the following three columns:

Alternating Signs	$u = x^2$ and Its Derivatives	$v' = e^{3x}$ and Its Antiderivatives
+	x^2	e^{3x}
−	$2x$	$\frac{1}{3} e^{3x}$
+	2	$\frac{1}{9} e^{3x}$
−	0	$\frac{1}{27} e^{3x}$

Using the formula for $\int e^{ax} dx$

Stop when you get to 0

Finally, the solution is found from the *signed* products of the above diagonals:

$$\int x^2 e^{3x}\, dx = \frac{1}{3} x^2 e^{3x} - \frac{2}{9} xe^{3x} + \frac{2}{27} e^{3x} + C$$

After reading the preceding example, find each integral by integration by parts using a table.

73. $\displaystyle\int x^2 e^{-x}\, dx$ **74.** $\displaystyle\int x^2 e^{2x}\, dx$

75. $\displaystyle\int x^3 e^{2x}\, dx$ **76.** $\displaystyle\int x^3 e^{-x}\, dx$

77. $\displaystyle\int (x-1)^3 e^{3x}\, dx$

78. $\displaystyle\int (x+1)^2 (x+2)^5\, dx$

10.2 Integration Using Tables

Introduction

There are many techniques of integration, and only a few of the most useful ones will be discussed in this book. Many of the advanced techniques lead to integration formulas, which can then be collected into a "table of integrals." In this section we will see how to find integrals by choosing an appropriate formula from such a table.

Inside the back cover of this book is a short table of integrals which we shall use. The formulas are grouped according to the type of integrand (for example, "Forms Involving $x^2 - a^2$"). Look at the table now (formulas 9 through 23 on the inside back cover) to see how it is organized.

Using Integral Tables

Given a particular integral, we first look for a formula that fits it exactly.

EXAMPLE 1   Integrating Using an Integral Table

Find $\displaystyle\int \frac{1}{x^2 - 4}\, dx$.

Solution   The denominator $x^2 - 4$ is of the form $x^2 - a^2$ (with $a = 2$), so we look under "Forms Involving $x^2 - a^2$" (inside back cover). Formula 15:

$$\int \frac{1}{x^2 - a^2}\, dx = \frac{1}{2a} \ln \left| \frac{x - a}{x + a} \right| + C \qquad \text{Formula 15}$$

with $a = 2$ becomes

$$\int \frac{1}{x^2 - 4}\, dx = \frac{1}{4} \ln \left| \frac{x - 2}{x + 2} \right| + C \qquad \begin{array}{l}\text{Formula 15 with } a = 2 \\ \text{substituted on both sides}\end{array}$$

∎

Note that the expression $x^2 - a^2$ does not require that the last number be a "perfect square." For example, $x^2 - 3$ can be written $x^2 - a^2$ with $a = \sqrt{3}$.

Integral tables are useful in many applications, such as integrating a rate to find the total accumulation.

EXAMPLE 2   Finding Total Sales From the Sales Rate

A company's sales rate is $\dfrac{x}{\sqrt{x + 9}}$ sales per week after x weeks. Find a formula for the total sales after x weeks.

Solution   To find the *total* sales $S(x)$ we integrate the *rate* of sales.

$$S(x) = \int \frac{x}{\sqrt{x + 9}}\, dx$$

In the table under "Forms Involving $\sqrt{ax + b}$" we find

$$\int \frac{x}{\sqrt{ax + b}}\, dx = \frac{2ax - 4b}{3a^2} \sqrt{ax + b} + C \qquad \text{Formula 13}$$

This formula with $a = 1$ and $b = 9$ gives the integral

$$S(x) = \int \frac{x}{\sqrt{x + 9}}\, dx = \frac{2x - 36}{3} \sqrt{x + 9} + C \qquad \begin{array}{l}\text{Formula 13 with } a = 1 \\ \text{and } b = 9\end{array}$$

$$= \left(\frac{2}{3}x - 12 \right) \sqrt{x + 9} + C \qquad \text{Simplifying}$$

To evaluate the constant C we use the fact that total sales at time $x = 0$ must be zero: $S(0) = 0$.

$$(-12)\sqrt{9} + C = 0 \qquad \begin{array}{l}\left(\frac{2}{3}x - 12 \right) \sqrt{x + 9} + C \\ \text{at } x = 0 \text{ set equal to zero}\end{array}$$

$$-36 + C = 0 \qquad \text{Simplifying}$$

Therefore, $C = 36$. Substituting this into $S(x)$ gives the formula for the total sales in the first x months.

$$S(x) = \left(\frac{2}{3}x - 12\right)\sqrt{x + 9} + 36$$

$S(x) = \left(\frac{2}{3}x - 12\right)\sqrt{x + 9} + C$

with $C = 36$

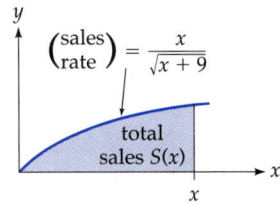

Modern methods of biotechnology are being used to develop many new products, including powerful antibiotics, disease-resistant crops, and bacteria that literally "eat" oil spills. These "gene splicing" techniques require solving definite integrals such as the following.

EXAMPLE 3   Genetic Engineering

Under certain circumstances, the number of generations of bacteria needed to increase the frequency of a gene from 0.2 to 0.5 is

$$n = 2.5 \int_{0.2}^{0.5} \frac{1}{q^2(1 - q)}\, dq$$

Find n (rounded to the nearest integer).

Solution   Formula 12 (inside back cover) integrates a similar-looking fraction.

$$\int \frac{1}{x^2(ax + b)}\, dx = -\frac{1}{b}\left(\frac{1}{x} + \frac{a}{b}\ln\left|\frac{x}{ax + b}\right|\right) + C \qquad \text{Formula 12}$$

To make $(ax + b)$ into $(1 - x)$, we take $a = -1$ and $b = 1$, so the left-hand side of the formula becomes

$$\int \frac{1}{x^2(-x + 1)}\, dx \qquad \text{or} \qquad \int \frac{1}{x^2(1 - x)}\, dx \qquad \begin{array}{l}\text{From formula 12 with}\\ a = -1, b = 1\end{array}$$

Except for replacing x by q, this is the same as our integral. Therefore, the indefinite integral is found by formula 12 with $a = -1$ and $b = 1$ (which we express in the variable q).

$$\int \frac{1}{q^2(1-q)}\,dq = -\left(\frac{1}{q} - \ln\left|\frac{q}{1-q}\right|\right) + C \qquad \text{Formula 12 with } a = -1 \text{ and } b = 1$$

For the *definite* integral from 0.2 to 0.5, we evaluate and subtract.

$$\underbrace{-\left(\frac{1}{0.5} - \ln\left|\frac{0.5}{1-0.5}\right|\right)}_{\text{Evaluation at } q = 0.5} - \underbrace{\left[-\left(\frac{1}{0.2} - \ln\left|\frac{0.2}{1-0.2}\right|\right)\right]}_{\text{Evaluation at } q = 0.2}$$

$$= -\underbrace{(2 - \ln 1)}_{0} + \left(5 - \underbrace{\ln\frac{0.2}{0.8}}_{\ln 0.25}\right) = -2 + 5 - \ln 0.25 \approx 4.38$$

(In the last step we used a calculator.) We multiply this by the 2.5 in front of the original integral.

$$(2.5)(4.38) = 10.95$$

Therefore, 11 generations are needed to raise the gene frequency from 0.2 to 0.5.   ■

 ## Graphing Calculator Exploration

a. Verify the answer to Example 3 by finding the definite integral of $y = \frac{2.5}{x^2(1-x)}$ from 0.2 to 0.5.

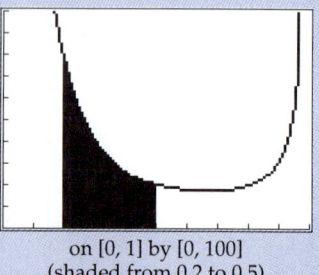
on [0, 1] by [0, 100]
(shaded from 0.2 to 0.5)

b. From the graph of the function shown on the right, which would require more generations: increasing the gene frequency from 0.1 to 0.2, or from 0.5 to 0.6?

c. Can you think of a reason for this? (*Hint:* Genes reproduce from other similar genes.)

Sometimes a substitution is needed to transform a formula to fit a given integral. In such cases both the x and the dx must be transformed. A few examples will make the method clear.

EXAMPLE 4 Using a Table with a Substitution

Find $\displaystyle\int \frac{x}{\sqrt{x^4 + 1}}\, dx$.

Solution The table has no formula involving x^4. However, $x^4 = (x^2)^2$, so a formula involving x^2, along with a substitution, might work. Formula 18 looks promising:

$$\int \frac{1}{\sqrt{x^2 \pm a^2}}\, dx = \ln\left| x + \sqrt{x^2 \pm a^2}\right| + C \qquad \text{The } \pm \text{ means use either the upper sign or the lower sign on } both \text{ sides}$$

$$\int \frac{1}{\sqrt{x^2 + 1}}\, dx = \ln\left| x + \sqrt{x^2 + 1}\right| + C \qquad \text{Formula 18 with } a = 1 \text{ and the upper sign}$$

With the substitution:

$$x = z^2$$

$$dx = 2z\, dz \qquad \text{Differential of } x = z^2$$

this becomes

$$\int \frac{1}{\sqrt{z^4 + 1}}\, 2z\, dz = \ln\left| z^2 + \sqrt{z^4 + 1}\right| + C \qquad \begin{array}{l}\text{Above formula with} \\ x = z^2 \text{ and} \\ dx = 2z\, dz\end{array}$$

Dividing by 2 and replacing z by x gives the integral that we wanted:

$$\int \frac{x}{\sqrt{x^4 + 1}}\, dx = \frac{1}{2}\ln\left(x^2 + \sqrt{x^4 + 1}\right) + C \qquad \begin{array}{l}\text{Dropping the absolute} \\ \text{value bars since} \\ x^2 + \sqrt{x^4 + 1} \text{ is positive}\end{array}$$

∎

Given a particular integral, how do we choose a formula?

How to Choose a Formula

Find a formula that matches the *most complicated part* of the integral, making appropriate substitutions to change the formula into the given integral.

For example, in Example 4 we matched the $\sqrt{x^4 + 1}$ in the given integral to the $\sqrt{x^2 \pm a^2}$ in the formula, and the rest of the integral followed from the differential.

PRACTICE PROBLEM 1   Find $\int \dfrac{t}{9t^4 - 1}\, dt$. (*Hint:* Use formula 15 with the substitution $x = 3t^2$.)

Solution at the back of the book

EXAMPLE 5   Using a Table with a Substitution

Find $\int \dfrac{e^{-2t}}{e^{-t} + 1}\, dt$.

Solution   Looking in the table under "Forms Involving e^{ax} and ln x", (inside back cover) none of the formulas looks anything like this integral. However, replacing e^{-t} by x would make the denominator of our integral into $x + 1$, so formula 9 might help. This formula with $a = 1$ and $b = 1$ is

$$\int \frac{x}{x + 1}\, dx = x - \ln\ |x + 1|\ + C \qquad \text{Formula 9 with } a = 1 \text{ and } b = 1$$

With the substitution

$$x = e^{-t}$$

$$dx = -e^{-t}\, dt \qquad \text{Differential of } x = e^{-t}$$

formula 9 becomes

$$\int \frac{e^{-t}}{e^{-t} + 1}\, (-e^{-t})\, dt = e^{-t} - \ln\ |e^{-t} + 1|\ + C$$

or

$$-\int \frac{e^{-2t}}{e^{-t} + 1}\, dt = e^{-t} - \ln\ (e^{-t} + 1) + C$$

Except for the negative sign this is the given integral. Multiplying through by -1 gives the final answer.

$$\int \frac{e^{-2t}}{e^{-t} + 1}\, dt = -e^{t} + \ln\ (e^{-t} + 1) + C$$

∎

Reduction Formulas

Sometimes we must apply a formula several times to simplify an integral in stages.

EXAMPLE 6 Using a Reduction Formula

Find $\int x^3 e^{-x}\, dx$.

Solution In the integral table, we find formula 21.

$$\int x^n e^{ax}\, dx = \frac{1}{a} x^n e^{ax} - \frac{n}{a} \int x^{n-1} e^{ax}\, dx$$

Formula 21. With $n = 3$ and $a = -1$, the left side fits our integral.

The right-hand side of this formula involves a new integral, but with a *lower* power of x. We will apply formula 21 several times, each time reducing the power of x until we eliminate it completely. Applying formula 21 with $n = 3$ and $a = -1$:

$$\int x^3 e^{-x}\, dx = -x^3 e^{-x} + 3 \int x^2 e^{-x}\, dx$$

After 1 application

The power has been reduced

$$= -x^3 e^{-x} + 3\left(-x^2 e^{-x} + 2 \int x^1 e^{-x}\, dx\right)$$

Applying formula 21 again (now with $n = 2$) to the last integral above

$$= -x^3 e^{-x} - 3x^2 e^{-x} + 6 \int x^1 e^{-x}\, dx$$

Multiplying out

$$= -x^3 e^{-x} - 3x^2 e^{-x} + 6\left(-xe^{-x} + \int x^0 e^{-x}\, dx\right)$$

Using formula 21 a third time (now with $n = 1$)

1

Now solve this last integral by the formula $\int e^{ax}\, dx = \frac{1}{a} e^{ax} + C$

$$= -x^3 e^{-x} - 3x^2 e^{-x} - 6xe^{-x} + 6 \int e^{-x}\, dx$$

The solution, after three applications of formula 21

$$= -x^3 e^{-x} - 3x^2 e^{-x} - 6xe^{-x} - 6e^{-x} + C$$

$$= -e^{-x}(x^3 + 3x^2 + 6x + 6) + C$$

Factoring

■

We used formula 21 three times, reducing the x^3 in steps, first down to x^2, then to x^1, and finally to $x^0 = 1$, at which point we could solve the integral easily. If the power of x in the integral had been higher, more applications of formula 21 would have been necessary. Formulas like 21 and 22 are called *reduction formulas*, since they express an integral in terms of a similar integral but with a reduced power of x.

SUMMARY

To find a formula that "fits" a given integral, we look for the formula whose left-hand side matches most closely the most complicated part of the integral. Then we choose constants (and possibly a substitution) to make the formula fit exactly. Although we have been using a very brief table, the technique is the same with a more extensive table. Many integral tables have been published; some of them are book-length, containing several thousand formulas.*

EXERCISES 10.2

For each integral, state the number of the integration formula (from the inside back cover) and the values of the constants a and b so that the formula fits the integral. (Do not evaluate the integral.)

1. $\displaystyle\int \frac{1}{x^2(5x-1)}\,dx$ **2.** $\displaystyle\int \frac{x}{2x-3}\,dx$

3. $\displaystyle\int \frac{1}{x\sqrt{-x+7}}\,dx$ **4.** $\displaystyle\int \frac{x}{\sqrt{-2x+1}}\,dx$

5. $\displaystyle\int \frac{x}{1-x}\,dx$ **6.** $\displaystyle\int \frac{1}{x\sqrt{1-4x}}\,dx$

Find each integral by using the integral table on the inside back cover.

7. $\displaystyle\int \frac{1}{9-x^2}\,dx$ (*Hint:* Use formula 16 with $a = 3$.)

8. $\displaystyle\int \frac{1}{x^2-25}\,dx$ (*Hint:* Use formula 15 with $a = 5$.)

9. $\displaystyle\int \frac{1}{x^2(2x+1)}\,dx$
   (*Hint:* Use formula 12 with $a = 2$, $b = 1$.)

10. $\displaystyle\int \frac{x}{x+2}\,dx$
   (*Hint:* Use formula 9 with $a = 1$, $b = 2$.)

11. $\displaystyle\int \frac{x}{1-x}\,dx$ (*Hint:* Use formula 9.)

12. $\displaystyle\int \frac{x}{\sqrt{1-x}}\,dx$ (*Hint:* Use formula 13.)

13. $\displaystyle\int \frac{1}{(2x+1)(x+1)}\,dx$

14. $\displaystyle\int \frac{x}{(x+1)(x+2)}\,dx$

15. $\displaystyle\int \sqrt{x^2-4}\,dx$

16. $\displaystyle\int \frac{1}{\sqrt{x^2-1}}\,dx$

17. $\displaystyle\int \frac{1}{z\sqrt{1-z^2}}\,dz$

18. $\displaystyle\int \frac{\sqrt{4+z^2}}{z}\,dz$

19. $\displaystyle\int x^3 e^{2x}\,dx$

20. $\displaystyle\int x^{99}\ln x\,dx$

21. $\displaystyle\int x^{-101}\ln x\,dx$

22. $\displaystyle\int (\ln x)^2\,dx$

23. $\displaystyle\int \frac{1}{x(x+3)}\,dx$

24. $\displaystyle\int \frac{1}{x(x-3)}\,dx$

25. $\displaystyle\int \frac{z}{z^4-4}\,dz$

26. $\displaystyle\int \frac{z}{9-z^4}\,dz$

27. $\displaystyle\int \sqrt{9x^2+16}\,dx$

28. $\displaystyle\int \frac{1}{\sqrt{16x^2-9}}\,dx$

29. $\displaystyle\int \frac{1}{\sqrt{4-e^{2t}}}\,dt$

30. $\displaystyle\int \frac{e^t}{9-e^{2t}}\,dt$

31. $\displaystyle\int \frac{e^t}{e^{2t}-1}\,dt$

32. $\displaystyle\int \frac{e^{2t}}{1-e^t}\,dt$

33. $\displaystyle\int \frac{x^3}{\sqrt{x^8-1}}\,dx$

34. $\displaystyle\int x^2\sqrt{x^6+1}\,dx$

35. $\displaystyle\int \frac{1}{x\sqrt{x^3+1}}\,dx$

36. $\displaystyle\int \frac{\sqrt{1-x^6}}{x}\,dx$

37. $\displaystyle\int \frac{e^t}{(e^t-1)(e^t+1)}\,dt$ **38.** $\displaystyle\int \frac{e^{2t}}{(e^t-1)(e^t+1)}\,dt$

39. $\displaystyle\int xe^{x/2}\,dx$

40. $\displaystyle\int \frac{x}{e^x}\,dx$

41. $\displaystyle\int \frac{1}{e^{-x}+4}\,dx$

42. $\displaystyle\int \frac{1}{\sqrt{e^{-x}+4}}\,dx$

---

*A useful table of integrals containing more than 400 formulas is found in the *CRC Standard Mathematical Tables,* published by CRC Press, Boca Raton, Florida.

For each definite integral:

a. Evaluate it using the table of integrals on the inside back cover. (Leave answers in *exact* form.)

 b. Use a graphing calculator to verify your answer to part (a).

43. $\int_4^5 \sqrt{x^2 - 16} \, dx$ **44.** $\int_0^4 \frac{1}{\sqrt{x^2 + 9}} \, dx$

45. $\int_2^3 \frac{1}{x^2 - 1} \, dx$ **46.** $\int_2^4 \frac{1}{1 - x^2} \, dx$

47. $\int_3^5 \frac{\sqrt{25 - x^2}}{x} \, dx$ **48.** $\int_3^4 \frac{1}{x\sqrt{25 - x^2}} \, dx$

Find each integral by whatever means are necessary (either substitution or tables).

49. $\int \frac{1}{2x + 6} \, dx$ **50.** $\int \frac{x}{x^2 - 4} \, dx$

51. $\int \frac{x}{2x + 6} \, dx$ **52.** $\int \frac{1}{4 - x^2} \, dx$

53. $\int x\sqrt{1 - x^2} \, dx$ **54.** $\int \frac{x}{\sqrt{1 - x^2}} \, dx$

55. $\int \frac{\sqrt{1 - x^2}}{x} \, dx$ **56.** $\int \frac{1}{\sqrt{x^2 - 1}} \, dx$

Find each integral. (*Hint:* Separate each integral into two integrals, using the fact that the numerator is a sum or difference, and find the two integrals by two different formulas.)

57. $\int \frac{x - 1}{(3x + 1)(x + 1)} \, dx$ **58.** $\int \frac{x - 1}{x^2(x + 1)} \, dx$

59. $\int \frac{x + 1}{x\sqrt{1 + x^2}} \, dx$ **60.** $\int \frac{x - 1}{x\sqrt{x^2 + 4}} \, dx$

61. $\int \frac{x + 1}{x - 1} \, dx$ (*Hint:* After separating into two integrals, find one by a formula and the other by substitution.)

62. $\int \frac{x + 1}{\sqrt{x^2 + 1}} \, dx$ (*Hint:* After separating into two integrals, find one by a formula and the other by substitution.)

APPLIED EXERCISES (Use only when indicated.) _____

63. Business: Total Sales A company's sales rate is $x^2 e^{-x}$ million sales per month after x months. Find a formula for the total sales in the first x months. (*Hint:* Integrate the sales rate to find the total sales and determine the constant C so that total sales are zero at time $x = 0$.)

64. General: Population The population of a city is expected to grow at the rate of $x/\sqrt{x + 9}$ thousand people per year after x years. Find the total change in population from year 0 to year 27.

 65. Biomedical: Gene Frequency Under certain circumstances, the number of generations necessary to increase the frequency of a gene from 0.1 to 0.3 is

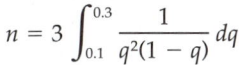
$$n = 3 \int_{0.1}^{0.3} \frac{1}{q^2(1 - q)} \, dq$$

Find n (rounded to the nearest integer).

66. Behavioral Science A subject in a psychology experiment gives responses at the rate of $t/\sqrt{t + 1}$ correct answers per minute after t minutes.

a. Find the total number of correct responses between times $t = 0$ and $t = 15$.

 b. Verify your answer to part (a) using a graphing calculator.

67. Business: Cost The marginal cost function for a computer chip manufacturer is $MC(x) = 1/\sqrt{x^2 + 1}$, and fixed costs are $2000. Find the cost function.

68. Social Science: Employment An urban job placement center estimates that the number of residents seeking employment t years from now will be $t/(2t + 4)$ million people.

a. Find the average number of job seekers during the period $t = 0$ to $t = 10$.

 b. Verify your answer to part (a) using a graphing calculator.

10.3 Improper Integrals

APPLICATION PREVIEW

Improper Integrals and Eternal Recognition

Suppose that after you become rich and famous, you decide to commission a statue of yourself for your hometown. Your town, however, will accept this selfless gesture only if you pay for the perpetual upkeep of the statue by establishing a fund that will generate $2000 annually for every year in the future. Before deciding whether or not to accept this condition, you of course want to know how much it will cost. (Interestingly, we will see that a fund that will generate income forever does *not* require an infinite amount of money.) On pages 126–127 we found that to realize a yield of $2000 t years from now requires only its *present value* deposited now in a bank. At an interest rate of, say, 5% compounded continuously, $2000 in t years requires a deposit now of

$$\begin{pmatrix}\text{Present value of \$2000 at 5\%} \\ \text{compounded continuously}\end{pmatrix} = 2000e^{-0.05t} \qquad \begin{array}{l}\text{Amount times} \\ e^{-0.05t}\end{array}$$

Therefore, the size of the fund needed to generate an annual $2000 *forever* is found by summing (integrating) this present value over the infinite time interval from zero to infinity (∞).

$$\begin{pmatrix}\text{Size of fund to yield an} \\ \text{annual \$2000 forever}\end{pmatrix} = \int_0^\infty 2000e^{-0.05t}\,dt \qquad \begin{array}{l}\text{Integrating out} \\ \text{to infinity}\end{array}$$

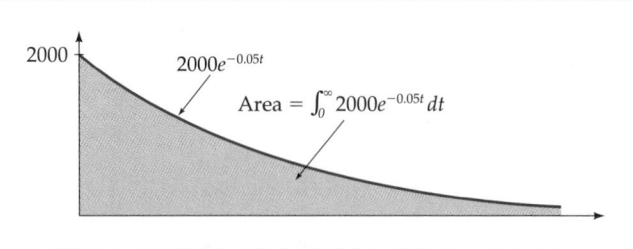

In this section we will learn how to evaluate such "improper integrals." We will find that the value of this integral is $40,000, meaning that $40,000 deposited in a bank at 5% compounded continuously will generate $2000 every year *forever*. That is, $40,000 would pay for the perpetual upkeep of your statue, buying you (or at least your likeness) a kind of immortality.

Incidentally, to generate $2000 annually for only the first *hundred* years would require $\displaystyle\int_{0}^{100} 2000e^{-0.05t}\, dt \approx \$39{,}730$ (integrating from 0 to 100). This amount is only $270 less than the amount needed to generate the same sum *forever*. This small additional cost shows that the short term is expensive, but eternity is cheap.

Introduction

In this section we define integrals over intervals that are infinite in length. Such integrals are called *improper* integrals, and have many applications, such as in the preceding Application Preview.

Limits as *x* Approaches ±∞

The notation $x \to \infty$ ("*x* approaches infinity") means that *x* takes on arbitrary large values.

$x \to \infty$ means:	*x* takes values arbitrarily far to the *right* on the number line.
$x \to -\infty$ means:	*x* takes values arbitrarily far to the *left* on the number line.

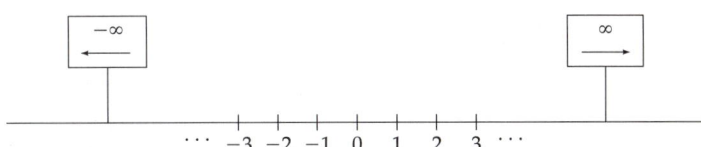

Evaluating limits as *x* approaches positive or negative infinity is simply a matter of thinking about large and small numbers. The reciprocal of a large number is a small number. For example:

$$\frac{1}{1{,}000{,}000} = 0.000001 \qquad \text{One over a million is one one-millionth}$$

Similarly:

$$\lim_{x \to \infty} \frac{1}{x^2} = 0$$

$$\lim_{x \to \infty} \frac{1}{e^x} = \lim_{x \to \infty} e^{-x} = 0$$

As the denominator approaches infinity (with the numerator constant), the value approaches zero

These examples illustrate the following general rules.

$$\lim_{x \to \infty} \frac{1}{x^n} = 0 \qquad (n > 0)$$

As x approaches infinity, 1 over x to a *positive* power approaches zero

$$\lim_{x \to \infty} e^{-ax} = 0 \qquad (a > 0)$$

As x approaches infinity, e to a *negative* number times x approaches zero

EXAMPLE 1   Evaluating Limits

a. $\lim\limits_{b \to \infty} \dfrac{1}{b^2} = 0$ Using the first rule in the box above

b. $\lim\limits_{b \to \infty} \left(3 - \dfrac{1}{b} \right) = 3$ Because the $\dfrac{1}{b}$ approaches zero

c. $\lim\limits_{b \to \infty} (e^{-2b} - 5) = -5$ Because the e^{-2b} approaches zero

Similar rules hold for x approaching *negative* infinity.

$$\lim_{x \to -\infty} \frac{1}{x^n} = 0 \qquad \text{(for integer } n > 0)$$

As x approaches *negative* infinity, 1 over x to a positive integer approaches zero

$$\lim_{x \to -\infty} e^{ax} = 0 \qquad (a > 0)$$

As x approaches *negative* infinity, e to a positive number times x approaches zero (because the exponent is approaching $-\infty$)

Graphing Calculator Exploration

a. Define $y_1 = \frac{1}{x^2}$ and $y_2 = e^{-x}$ and use the TABLE feature of your calculator to evaluate these functions at x-values like 1, 2, 3, 5, 10, 100, and 1000. Which of the previous limit rules do the results verify? (*Note:* Set your calculator so that *you* choose the x-values. An answer like 3E − 5 means $3 \cdot 10^{-5} = 0.00003$.)

b. Change y_2 to be $y_2 = e^x$ and use the TABLE to evaluate y_1 and y_2 at *negative* x-values like − 1, − 2, − 3, − 5, − 10, and − 100. Which limit rules do these results verify?

Some quantities become arbitrarily large, and so do not have limits. (For a limit to exist, it must be *finite*.)

The following limits do not exist:

$\lim\limits_{x \to \infty} x^n$ $(n > 0)$ As x approaches infinity, x to a positive power has no limit

$\lim\limits_{x \to \infty} e^{ax}$ $(a > 0)$ As x approaches infinity, e to a positive number times x has no limit

$\lim\limits_{x \to \infty} \ln x$ As x approaches infinity, the natural logarithm of x has no limit

EXAMPLE 2   Finding Whether a Limit Exists

a. $\lim\limits_{b \to \infty} b^3$ does not exist.   Because b^3 becomes arbitrarily large as b approaches infinity

b. $\lim\limits_{b \to \infty} (\sqrt{b} - 1)$ does not exist.   Because $\sqrt{b}$ becomes arbitrarily large as b approaches infinity

PRACTICE PROBLEM 1

Evaluate the following limits (if they exist).

a. $\lim\limits_{b \to \infty} \left(1 - \frac{1}{b}\right)$ b. $\lim\limits_{b \to \infty} (\sqrt[3]{b} + 3)$ *Solutions at the back of the book*

Improper Integrals

As a first example, we evaluate the improper integral $\int_1^\infty \dfrac{1}{x^2}\,dx$, which gives the area under the curve $y = \dfrac{1}{x^2}$ from $x = 1$ arbitrarily far to the right.

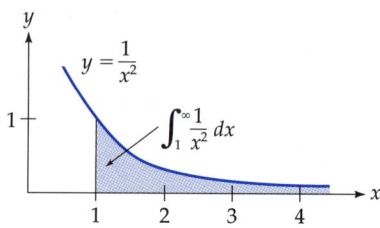

EXAMPLE 3   Evaluating an Improper Integral

Evaluate $\displaystyle\int_1^\infty \dfrac{1}{x^2}\,dx$.

Solution   To integrate to infinity, we first integrate over a *finite* interval, from 1 to some number b (think of b as some very large number), and then take the limit as b approaches ∞. First integrate from 1 to b.

$$\int_1^b \frac{1}{x^2}\,dx = \int_1^b x^{-2}\,dx = (-x^{-1})\Big|_1^b = \left(-\frac{1}{x}\right)\Big|_1^b \qquad \text{Using the power rule}$$

$$= \underbrace{-\frac{1}{b}}_{\text{at } x = b} - \underbrace{\left(-\frac{1}{1}\right)}_{\text{at } x = 1} = -\frac{1}{b} + 1 \qquad \begin{array}{l}\text{Evaluating and}\\ \text{simplifying}\end{array}$$

Then take the limit of this answer as $b \to \infty$.

$$\lim_{b\to\infty}\left(-\frac{1}{b} + 1\right) = 1 \qquad \begin{array}{l}\text{Limit as } b \to \infty \\ (\text{the } \frac{1}{b} \text{ approaches zero})\end{array}$$

This gives the answer:

$$\int_1^\infty \frac{1}{x^2}\,dx = 1 \qquad \begin{array}{l}\text{Integral from}\\ 1 \text{ to } \infty \text{ equals } 1\end{array}$$

Since the limit exists, we say that the improper integral is *convergent*. Geometrically, this procedure amounts to finding the area under the curve from 1 to some number b, shown on the left below, and then letting $b \to \infty$ to find the area.

Integrating to b

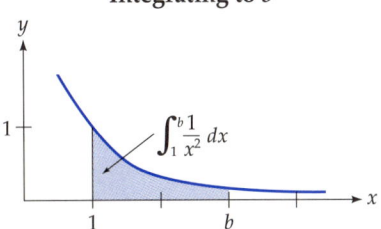

Integrating to ∞

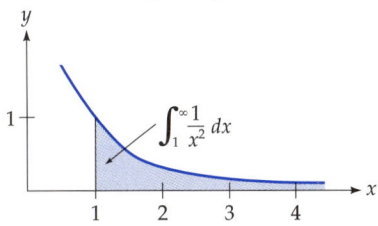

Improper Integrals (Integrating to ∞)

For f continuous and nonnegative for $x \geq a$, we define

$$\int_a^\infty f(x)\, dx = \lim_{b \to \infty} \int_a^b f(x)\, dx$$

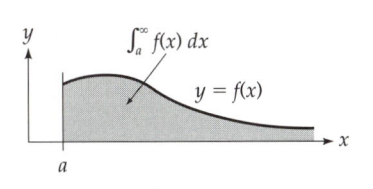

provided that the limit exists. The improper integral is said to be *convergent* if the limit exists, and *divergent* if the limit does not exist.

It is possible to define improper integrals for functions that take negative values, and even for discontinuous functions, but we shall not do so in this book, since most applications involve functions that are positive and continuous.

PRACTICE PROBLEM 2   Evaluate $\displaystyle\int_2^\infty \frac{1}{x^2}\, dx$.
Solution at the back of the book

Divergent Integrals

If we are asked to evaluate an integral that turns out to be divergent (that is, whose limit does not exist), we simply say that *the integral is divergent*.

EXAMPLE 4   Finding Whether an Integral Diverges

Evaluate $\int_1^\infty \dfrac{1}{\sqrt{x}}\,dx$.

Solution   Integrating up to b:

$$\int_1^b \dfrac{1}{\sqrt{x}}\,dx = \int_1^b x^{-1/2}\,dx = \underbrace{2 \cdot x^{1/2}\Big|_1^b}_{\substack{\text{Integrating by}\\ \text{the power rule}}} = \underbrace{2\sqrt{b} - 2\sqrt{1}}_{\text{Evaluating}} = 2\sqrt{b} - 2$$

Letting b approach infinity:

$$\lim_{b\to\infty} (2\sqrt{b} - 2) \text{ does not exist} \qquad \text{Because } \sqrt{b} \text{ becomes infinite as } b \to \infty$$

Therefore,

$$\int_1^\infty \dfrac{1}{\sqrt{x}}\,dx \text{ is } divergent \qquad \text{The integral cannot be evaluated}$$

Notice from Examples 3 and 4 that $\int_1^\infty \dfrac{1}{x^2}\,dx$ was convergent, whereas $\int_1^\infty \dfrac{1}{\sqrt{x}}\,dx$ was divergent.

Area under $\dfrac{1}{x^2}$ is convergent

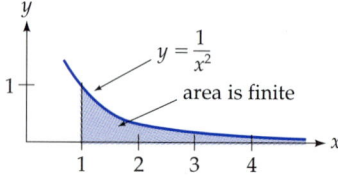

Area under $\dfrac{1}{\sqrt{x}}$ is divergent

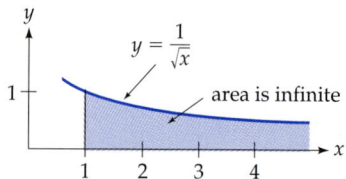

Intuitively, this difference is because the curve $\dfrac{1}{x^2}$ lies much closer to the x-axis than does the curve $\dfrac{1}{\sqrt{x}}$ for large values of x, as shown above, and so has a smaller area under it.

Graphing Calculator Exploration

With a graphing calculator you can "see," numerically, one of these integrals converging while the other diverges.

a.   Define y_1 to be the definite integral (using FnInt) of $\frac{1}{x^2}$ from 1 to x.

b.   Define y_2 to be the definite integral of $\frac{1}{\sqrt{x}}$ from 1 to x.

X	Y_1	Y_2
1	0	0
2	.5	.82843
10	.9	4.3246
100	.99	18
1000	.999	61.246
10000	.9999	198
100000	.99999	630.46

X=100000

c. y_1 and y_2 then give the *areas* under these curves out to any number x. Make a TABLE of values of y_1 and y_2 for x-values like 1, 2, 10, 100, 1000, 10,000, and 100,000. Is it clear which of these integrals (areas) is converging and which is diverging?

We may combine the two steps of integrating up to b and letting $b \to \infty$ into a single line. We show how to do this by evaluating the integral from Example 3 again, but more briefly.

$$\int_1^\infty x^{-2}\, dx = \lim_{b \to \infty} \int_1^b x^{-2}\, dx = \lim_{b \to \infty} (-x^{-1}) \Big|_1^b = \lim_{b \to \infty} \left[-\frac{1}{b} - \left(-\frac{1}{1} \right) \right] = 1$$

Integrating;   Approaches
now use   0
$x^{-1} = \dfrac{1}{x}$

Permanent Endowments

Funds that generate steady income forever are called *permanent endowments*.

EXAMPLE 5   Finding the Size of a Permanent Endowment

In the Application Preview on pages 844–845 we found that the size of the fund necessary to generate $2000 annually forever (at 5% interest compounded continuously) is $\int_0^\infty 2000e^{-0.05t}\, dt$. Find the size of this permanent endowment by evaluating the integral.

Solution

$$\int_0^\infty \underbrace{2000}e^{-0.05t}\,dt \;=\; \lim_{b\to\infty}\left(\underbrace{2000}\int_0^b e^{-0.05t}\,dt\right)$$

Annual
income   ⌣ Continuous
interest rate   Moved
outside

$$= \lim_{b\to\infty}\left[\,2000(-20)e^{-0.05t}\Big|_0^b\,\right]$$

Integrating by

$$\int e^{ax}\,dx = \frac{1}{a}e^{ax}$$

$$= \lim_{b\to\infty}\,(-40{,}000\underbrace{e^{-0.05b}} + 40{,}000\underbrace{e^0}) = 40{,}000$$

Approaches
0   1

Therefore, the size of the permanent endowment that will pay the $2000 annual maintenance forever is $40,000.

Permanent endowments are used to estimate the ultimate cost of anything that requires continuous long-term funding, from buildings to government agencies to toxic waste storage sites.

Finding Improper Integrals Using Substitutions

Solving an improper integral may require a substitution. In such cases we apply the substitution not only to the integrand but also to the differential and the upper and lower limits of integration.

EXAMPLE 6   Finding an Improper Integral Using a Substitution

Evaluate $\displaystyle\int_2^\infty \frac{x}{(x^2+1)^2}\,dx$.

Solution   We use the substitution $u = x^2 + 1$, so $du = 2x\,dx$, requiring multiplication by 2 and by $\frac{1}{2}$. Notice how the substitution changes the limits.

as $x \to \infty$, $u = x^2 + 1 \to \infty$

$$\int_2^\infty \frac{x}{(x^2+1)^2}\,dx = \frac{1}{2}\int_2^\infty \frac{2x}{(x^2+1)^2}\,dx = \frac{1}{2}\int_5^\infty \frac{du}{u^2} = \frac{1}{2}\lim_{b\to\infty}\int_5^b u^{-2}\,du$$

$$\begin{bmatrix} u = x^2 + 1 \\ du = 2x\,dx \end{bmatrix}$$

$$u = 2^2 + 1 = 5$$

$$= \frac{1}{2} \lim_{b \to \infty} \underbrace{\left[-u^{-1} \right]\Big|_5^b}_{\text{Integrating}} = \frac{1}{2} \lim_{b \to \infty} \left[\underbrace{-\frac{1}{b}}_{\text{Approaches } 0} - \left(-\frac{1}{5} \right) \right] = \frac{1}{2} \cdot \frac{1}{5} = \frac{1}{10}$$

∎

To integrate over an interval that extends arbitrarily far to the *left,* we again integrate over a finite interval and then take the limit.

Improper Integrals (Integrating to $-\infty$)

For f continuous and nonnegative for $x \leq b$, we define

$$\int_{-\infty}^{b} f(x)\, dx = \lim_{a \to -\infty} \int_{a}^{b} f(x)\, dx$$

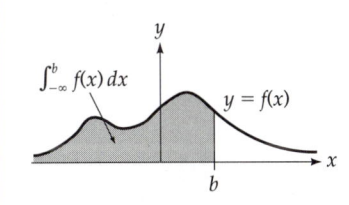

provided that the limit exists. The improper integral is *convergent* if the limit exists, and *divergent* if the limit does not exist.

To integrate over the *entire* x-axis, from $-\infty$ to ∞, we use two integrals, one from $-\infty$ to 0, and the other from 0 to ∞, and then add the results.

Improper Integrals (Integrating from $-\infty$ to ∞)

For f continuous and nonnegative for *all* values of x, we define

$$\int_{-\infty}^{\infty} f(x)\, dx = \lim_{a \to -\infty} \int_{a}^{0} f(x)\, dx + \lim_{b \to \infty} \int_{0}^{b} f(x)\, dx$$

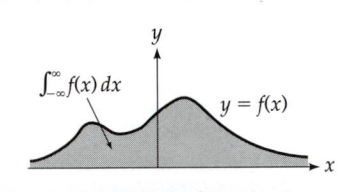

The improper integral is *convergent* if both limits exist, and *divergent* if either limit does not exist.

EXAMPLE 7 **Integrating to** $-\infty$

Evaluate $\displaystyle\int_{-\infty}^{3} 4e^{2x}\,dx$.

Solution

$$\int_{-\infty}^{3} 4e^{2x}\,dx = \lim_{a\to-\infty} \int_{a}^{3} 4e^{2x}\,dx$$

$$= \lim_{a\to-\infty} \left[4\cdot\frac{1}{2}\cdot e^{2x}\Big|_{a}^{3} \right] \qquad \text{Integrating}$$

$$= \lim_{a\to-\infty} (2e^{6} - \underbrace{2e^{2a}}) = 2e^{6}$$

$$\text{Approaches } 0$$
$$\text{as } a\to-\infty$$

■

PRACTICE PROBLEM 3   Evaluate the improper integral $\displaystyle\int_{-\infty}^{1} 12e^{3x}\,dx$.

Solution at the back of the book

SUMMARY

A definite integral in which one or both limits of integration are infinite is called an "improper" integral. The improper integral of a continuous nonnegative function is defined as the *limit* of the integral over a finite interval. The integral is *convergent* if the limit exists, and *divergent* otherwise. This idea of dealing with the infinite by "dropping back to the finite and then taking the limit" is a standard technique in mathematics.

    Several particular limits are helpful in evaluating improper integrals.

Approaching Infinity	*Approaching Negative Infinity*
$\displaystyle\lim_{x\to\infty} \frac{1}{x^{n}} = 0 \quad (n>0)$	$\displaystyle\lim_{x\to-\infty} \frac{1}{x^{n}} = 0 \quad \text{(for integer } n>0)$
$\displaystyle\lim_{x\to\infty} e^{-ax} = 0 \quad (a>0)$	$\displaystyle\lim_{x\to-\infty} e^{ax} = 0 \quad (a>0)$

Improper integrals give continuous sums over infinite intervals. For example, the total future output of an oil well can be found by integrating the production rate out to infinity, and the value of an asset that lasts indefinitely (like land) can be found by integrating the present value of future income out to infinity. Even when infinite duration is unrealistic, ∞ is used to represent "long-term behavior."

EXERCISES 10.3

Evaluate each limit (or state that it does not exist).

1. $\lim\limits_{x \to \infty} \dfrac{1}{x^2}$

2. $\lim\limits_{b \to \infty} \left(\dfrac{1}{\sqrt{b}} - 8 \right)$

3. $\lim\limits_{b \to \infty} (1 - 2e^{-5b})$

4. $\lim\limits_{b \to \infty} (3e^{3b} - 4)$

5. $\lim\limits_{x \to \infty} (2 - e^{x/2})$

6. $\lim\limits_{x \to \infty} (1 - e^{-x/3})$

7. $\lim\limits_{b \to \infty} (3 + \ln b)$

8. $\lim\limits_{b \to \infty} (2 - \ln b^2)$

Evaluate each improper integral or state that it is divergent.

9. $\displaystyle\int_1^\infty \dfrac{1}{x^3}\, dx$

10. $\displaystyle\int_1^\infty \dfrac{1}{\sqrt[3]{x^4}}\, dx$

11. $\displaystyle\int_2^\infty 3x^{-4}\, dx$

12. $\displaystyle\int_0^\infty e^{-t}\, dt$

13. $\displaystyle\int_2^\infty \dfrac{1}{x}\, dx$

14. $\displaystyle\int_1^\infty \dfrac{1}{x^{0.99}}\, dx$

15. $\displaystyle\int_1^\infty \dfrac{1}{x^{1.01}}\, dx$

16. $\displaystyle\int_{10}^\infty e^{-x/5}\, dx$

17. $\displaystyle\int_0^\infty e^{-0.05t}\, dt$

18. $\displaystyle\int_0^\infty e^{0.01t}\, dt$

19. $\displaystyle\int_5^\infty \dfrac{1}{(x-4)^3}\, dx$

20. $\displaystyle\int_0^\infty \dfrac{x}{(x^2+1)^2}\, dx$

21. $\displaystyle\int_0^\infty \dfrac{x}{x^2+1}\, dx$

22. $\displaystyle\int_0^\infty \dfrac{x^2}{x^3+1}\, dx$

23. $\displaystyle\int_0^\infty x^2 e^{-x^3}\, dx$

24. $\displaystyle\int_e^\infty (\ln x)^{-2} \dfrac{1}{x}\, dx$

25. $\displaystyle\int_{-\infty}^0 e^{3x}\, dx$

26. $\displaystyle\int_{-\infty}^0 \dfrac{x^4}{(x^5-1)^2}\, dx$

27. $\displaystyle\int_{-\infty}^1 \dfrac{1}{2-x}\, dx$

28. $\displaystyle\int_{-\infty}^0 \dfrac{1}{1-x}\, dx$

29. $\displaystyle\int_{-\infty}^\infty \dfrac{e^x}{(1+e^x)^2}\, dx$

30. $\displaystyle\int_{-\infty}^\infty \dfrac{e^{-x}}{(1+e^{-x})^3}\, dx$

31. $\displaystyle\int_{-\infty}^\infty \dfrac{e^x}{1+e^x}\, dx$

32. $\displaystyle\int_{-\infty}^\infty \dfrac{e^{-x}}{1+e^{-x}}\, dx$

33. Use a graphing calculator to find the improper integrals (if they exist) $\displaystyle\int_0^\infty e^{\sqrt{x}}\, dx$ and $\displaystyle\int_0^\infty e^{-x^2}\, dx$ as follows:

  a. Define y_1 to be the definite integral (using FnInt) of $e^{\sqrt{x}}$ from 0 to x.

  b. Define y_2 to be the definite integral of e^{-x^2} from 0 to x.

  c. y_1 and y_2 then give the *areas* under these curves out to any number x. Make a TABLE of values of y_1 and y_2 for x-values like 1, 10, 100, and 1000. Which integral converges (and to what number, approximated to five decimal places) and which diverges?

34. Use a graphing calculator to find the improper integrals (if they exist) $\displaystyle\int_0^\infty \dfrac{1}{x^2+1}\, dx$ and $\displaystyle\int_0^\infty \dfrac{1}{\sqrt{x}+1}\, dx$ as follows:

  a. Define y_1 to be the definite integral (using FnInt) of $\dfrac{1}{x^2+1}$ from 0 to x.

  b. Define y_2 to be the definite integral of $\dfrac{1}{\sqrt{x}+1}$ from 0 to x.

  c. y_1 and y_2 then give the *areas* under these curves out to any number x. Make a TABLE of values of y_1 and y_2 for x-values like 1, 10, 100, 1000, and 10,000. Which integral converges (and to what number, approximated to five decimal places) and which diverges?

APPLIED EXERCISES

35. General: Permanent Endowments Find the size of the permanent endowment needed to generate an annual $12,000 forever at continuous interest rate 6%.

36. General: Permanent Endowments Show that the size of the permanent endowment needed to generate an annual C dollars forever at interest rate r compounded continuously is C/r dollars.

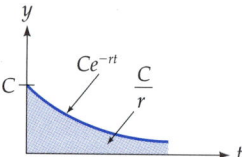

37. General: Permanent Endowments
a. Find the size of the permanent endowment needed to generate an annual $1000 forever at a continuous interest rate of 10%.
b. At this same interest rate, the size of the fund needed to generate an annual $1000 for precisely 100 years is $\int_0^{100} 1000e^{-0.1t}\, dt$. Evaluate this integral (it is not an improper integral), approximating your answer using a calculator.
c. Notice that the cost for the first hundred years is almost the same as the cost forever. This illustrates again the principle that in endowments, the short term is expensive, but eternity is cheap.

38–40: Business: Capital Value of an Asset The capital value of an asset is defined as the present value of all future earnings. For an asset that may last indefinitely (like real estate or a corporation), the capital value is

$$\begin{pmatrix} \text{Capital} \\ \text{value} \end{pmatrix} = \int_0^\infty C(t)e^{-rt}\, dt$$

where $C(t)$ is the income per year and r is the continuous interest rate. Find the capital value of a piece of property that will generate an annual in-

come of $C(t)$, for the function $C(t)$ given below, at a continuous interest rate of 5%.

38. $C(t) = 8000$ dollars

39. $C(t) = 50\sqrt{t}$ thousand dollars

40. $C(t) = 59\, t^{0.1}$ thousand dollars

41. Business: Oil Well Output An oil well is expected to produce oil at the rate $50e^{-0.05t}$ thousand barrels per month indefinitely, where t is the number of months that the well has been in operation. Find the total output over the lifetime of the well by integrating this rate from 0 to ∞. (*Note:* The owner will shut down the well when production falls too low, but it is convenient to estimate the total output as if production continued forever.)

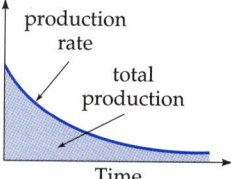

42. General: Duration of Telephone Calls Studies have shown that the proportion of telephone calls that last longer than t minutes is approximately $\int_t^\infty 0.3e^{-0.3s}\, ds$. Use this formula to find the proportion of telephone calls that last longer than 4 minutes.

43. Area Find the area between the curve $y = \frac{1}{x^{3/2}}$ and the x-axis from $x = 1$ to ∞.

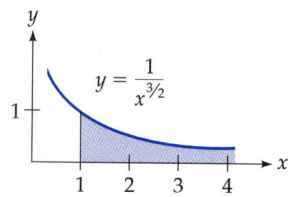

44. Area   Find the area between the curve $y = e^{-4x}$ and the x-axis from $x = 0$ to ∞.

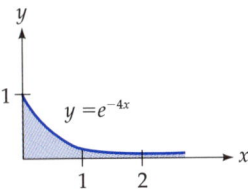

45. Area   Find the area between the curve $y = e^{-ax}$ (for $a > 0$) and the x-axis from $x = 0$ to ∞.

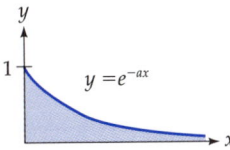

46. Area   Find the area between the curve $y = \dfrac{1}{x^n}$ (for $n > 1$) and the x-axis from $x = 1$ to ∞.

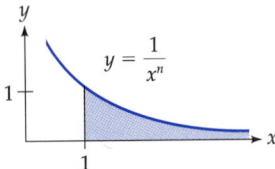

47. Behavioral Science: Mazes   In a psychology experiment, rats were placed in a T-maze, and the proportion of rats who required more than t seconds to reach the end was $\displaystyle\int_{t}^{\infty} 0.05e^{-0.05s}\, ds$. Use this formula to find the proportion of the rats that required more than 10 seconds.

48. Sociology: Prison Terms   If the proportion of prison terms that are longer than t years is given by the improper integral $\displaystyle\int_{t}^{\infty} 0.2e^{-0.2s}\, ds$, find the proportion of prison terms that are longer than 5 years.

49. Business: Product Reliability   The proportion of light bulbs that last longer than t hours is predicted to be $\displaystyle\int_{t}^{\infty} 0.001e^{-0.001s}\, ds$. Use this formula to find the proportion of light bulbs that will last longer than 1200 hours.

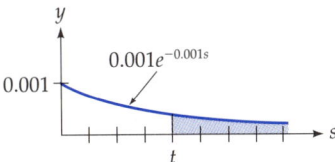

50. Business: Warranties   When a company sells a product with a lifetime guarantee, the number of items returned for repair under the guarantee usually decreases with time. A company estimates that the annual rate of returns after t years will be $800e^{-0.2t}$. Find the total number of returns by summing (integrating) this rate from 0 to ∞.

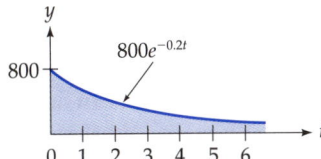

51. Business: Sales   A publisher estimates that a book will sell at the rate of $16{,}000e^{-0.8t}$ books per year t years from now. Find the total number of books that will be sold by summing (integrating) this rate from 0 to ∞.

52. Biomedical: Drug Absorption   To determine how much of a drug is absorbed into the body, researchers measure the difference between the dosage D and the amount of the drug excreted from the body. The total amount excreted is found by integrating the excretion rate $r(t)$ from 0 to ∞. Therefore, the amount of the drug absorbed by the body is

$$D - \int_{0}^{\infty} r(t)\, dt.$$

If the initial dose is $D = 200$ mg (milligrams), and the excretion rate is $r(t) = 40e^{-0.5t}$ mg per hour, find the amount of the drug absorbed by the body.

53–54: General: Permanent Endowments   The formula for integrating the exponential function a^{bx} for constants $a > 0$ and b is:

$$\int a^{bx}\,dx = \frac{1}{b\ln a}\,a^{bx} + C$$

as may be verified by using the differentiation formulas on page 712.

53. Use the above formula to find the size of the permanent endowment needed to generate an annual $2000 forever at 5% interest compounded annually. *Hint:* Find $\int_0^{\infty} 2000 \cdot 1.05^{-x}\,dx$. (Compare your answer to

that found in Example 5 for the same interest rate but compounded continuously.)

54. Use the previous formula to find the size of the permanent endowment needed to generate an annual $12,000 forever at 6% interest compounded annually. *Hint:* Find $\int_0^{\infty} 12000 \cdot 1.06^{-x}\,dx$. (Compare your answer to that found in Exercise 35 for the same interest rate but compounded continuously.)

10.4   Numerical Integration

Introduction

In spite of the many techniques of integration, there are still integrals that cannot be found by *any* method (as finite combinations of elementary functions). One example is the integral $\int e^{-x^2}\,dx$, which is closely related to the famous "bell-shaped curve" of probability and statistics that was discussed on pages 477–487. For *definite* integrals, however, it is always possible to *approximate* the actual value by interpreting it as the area under a curve, a process called *numerical integration.* We will discuss two of the most useful methods of numerical integration, based on approximating areas by *trapezoids* and by *parabolas* (known as "Simpson's rule"). Both methods may be programmed on a calculator or computer. In fact, when you evaluate a definite integral on a graphing calculator using a command like FnInt, the calculator is using a numerical integration procedure similar to those described here. This section explains the mathematics behind such operations.

Rectangular Approximation and Riemann Sums

In Section 9.3 we approximated the area under a curve by rectangles, and we called the sum of the areas of the rectangles a *Riemann sum.* However, these rectangles underestimate the true area under the curve by the white "error area" just above the rectangles, as shown on the following page. For greater accuracy we could increase the number of rectangles (as we did in Section 9.3), but this would involve more calculation and consequently more roundoff errors. Instead, we will replace the rectangles by shapes that fit the curve more closely.

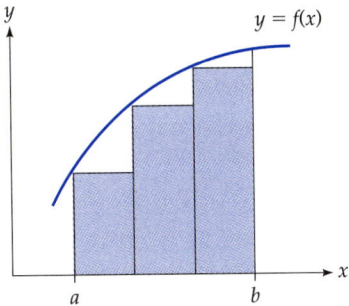

Area under $y = f(x)$ from a to b
approximated by three rectangles

Trapezoidal Approximation

We modify the approximating rectangles by allowing their tops to slant with the curve, as shown below left, giving a much better "fit" to the curve. Such shapes, in which two sides are parallel, are called *trapezoids* (below right). Clearly, approximating the area under the curve by trapezoids is more accurate than approximating it by rectangles: The white "error area" is much smaller.

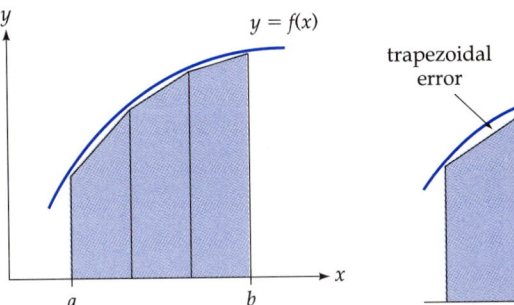

Area under $y = f(x)$ from a to b
approximated by three trapezoids

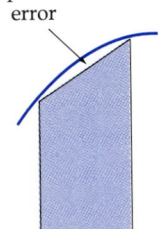

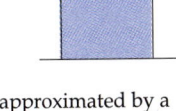

Area under a curve approximated by a
trapezoid (left) and by a rectangle (right).

The area of a trapezoid is the average of the two heights times the width Δx.

$$\binom{\text{Area of a}}{\text{trapezoid}} = \underbrace{\frac{h_1 + h_2}{2}}_{\text{Average height}} \cdot \underbrace{\Delta x}_{\text{Width}}$$

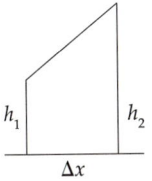

If we use trapezoids that all have the same width Δx, we may add up all of the average heights first and multiply by Δx at the end. Furthermore, averaging the heights means dividing each height by 2 $\left[\text{since } \frac{h_1 + h_2}{2} = \frac{h_1}{2} + \frac{h_2}{2}\right]$, but a side *between* two rectangles will be counted twice, once for the trapezoid on either side, thereby canceling the division by 2. Therefore, in adding up the heights, only the two outside heights, at a and b, should be divided by 2. This leads to the following procedure for trapezoidal approximation.

Trapezoidal Approximation

To approximate $\displaystyle\int_a^b f(x)\, dx$ by using n trapezoids:

1. Calculate $\Delta x = \dfrac{b - a}{n}$.   Trapezoid width

2. Find numbers $x_1, x_2, \ldots, x_{n+1}$ starting with $x_1 = a$ and successively adding Δx, ending with $x_{n+1} = b$.

3. The approximation for the integral is

$$\int_a^b f(x)\, dx \approx \left[\frac{1}{2}f(x_1) + f(x_2) + \cdots + f(x_n) + \frac{1}{2}f(x_{n+1})\right]\Delta x$$

This formula calculates the total area of the n trapezoids shown below.

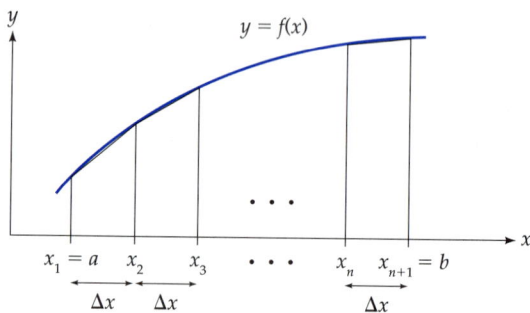

It is easiest to carry out the calculation in a table.

EXAMPLE 1   Approximating an Integral Using Trapezoids

Approximate $\displaystyle\int_1^2 x^2\, dx$ using four trapezoids.

Solution   The limits of integration are $a = 1$ and $b = 2$, and we are using $n = 4$ trapezoids. The method consists of the following six steps.

1. Calculate the trapezoid width $\Delta x = \dfrac{b - a}{n} = \dfrac{2 - 1}{4} = 0.25$.

2. List the x-values
   a through b with
   spacing Δx.

3. Apply $f(x)$ to
   each x-value.

x	$f(x) = x^2$	
Initial point a 1	$(1)^2$ $= 1$ 0.5	
add Δx 1.25	$(1.25)^2 = 1.56$	4. Take half of first and last entries.
add Δx 1.5	$(1.5)^2 = 2.25$	
add Δx 1.75	$(1.75)^2 = 3.06$	5. Sum last column.
add Δx Final point b 2.0	$(2)^2 = 4$ 2	6. Multiply by Δx.

$9.37 \cdot (0.25) \approx 2.34$

Final answer

Therefore, the estimate using four trapezoids is $\displaystyle\int_{1}^{2} x^2 \, dx \approx 2.34$.

■

The integral $\displaystyle\int_{1}^{2} x^2 \, dx$ can be found exactly, and the answer is $\frac{7}{3} \approx 2.33$ (as you may easily check). Therefore, the trapezoidal approximation of 2.34 is very accurate in spite of the fact that we used only four trapezoids. The *relative* error (the actual error, 0.01, divided by the actual value, $\frac{7}{3}$) is $\dfrac{0.01}{\frac{7}{3}} \approx 0.004$. Expressed as a percentage, this is 0.4% (four tenths of one percent), which is also remarkably small.

Error in Trapezoidal Approximation

The maximum error in trapezoidal approximation obeys the following formula.

Trapezoidal Error

For the trapezoidal approximation of $\int_a^b f(x)\,dx$ with n trapezoids,

$$(\text{Error}) \leq \frac{(b-a)^3}{12n^2}\left[\max_{a \leq x \leq b} f''(x)\right]$$

This formula is very difficult to use, as it involves maximizing the second derivative. We will not make further use of it except to observe that the n^2 in the denominator means that doubling the number of trapezoids reduces the maximum error by a factor of *four*.

The trapezoidal approximation procedure is easily programmed for a calculator or a computer. Below are two programs for trapezoidal approximation, one for the TI-83 calculator and one in BASIC. They may be adapted for other calculators or computers. The small print is explanation, not part of the program.

TI-83 Program for Trapezoidal Approximation
(Enter the function in y_1 before executing.)

Disp "USES FUNCTION Y₁"
Prompt A, B, N
(B−A)/N→D
(Y₁(A)+Y₁(B))/2→S
A+D→X
For(I,1,N−1)
S+Y₁→S
X+D→X
End
S*D→S
Disp "TRAP APPROX IS ",S

BASIC Program for Trapezoidal Approximation
(Lines 2, 3, 4, and 8 to be completed as indicated in small type.)

```
approx=0
a=                    fill in beginning x-value
b=                    fill in ending x-value
n=                    fill in number of trapezoids
delta=(b−a)/n
x=a
FOR i=1 TO n+1
f=( fill in the
    function )
IF i=1 OR i=n+1 THEN f=f/2
approx=approx+f
x=x+delta
NEXT i
approx=approx*delta
PRINT "TRAP APPROX IS", approx
END
```

On a computer or calculator, the error formula given above is of limited usefulness, since maximizing the second derivative is itself subject to error. What is often done instead is to choose a given accuracy (for example, 0.001) and repeat the trapezoidal approximation for larger and larger values of n until two successive approximations differ by less than the given accuracy. While this does not guarantee the desired accuracy, it is often accepted in practice.

Graphing Calculator Exploration

a. Approximate $\int_1^2 x^2\,dx$ using one of the programs on the previous page (or another trapezoidal program) with $n = 4$ trapezoids. Your answer should agree with Example 1.

b. Run the program again but with $n = 10$ trapezoids, and then with $n = 100$ trapezoids. Do the results approach the exact value of $2\frac{1}{3}$ calculated "by hand"?

c. Use FnInt (or a similar command) on your graphing calculator to evaluate the definite integral. How do the answers compare? (FnInt is probably faster because it is designed into the calculator, but the methods are similar.)

Simpson's Rule (Parabolic Approximation)

For even greater accuracy, we could increase n (resulting in more calculation and more roundoff errors) or we could change the trapezoids, replacing the tops by *curves* chosen to fit the given curve even more closely. Replacing the tops of the trapezoids by *parabolas* that pass through three points of the given curve leads to an even more accurate method of approximation, called Simpson's rule.* The following diagram shows such a curve and its approximation by two parabolas.

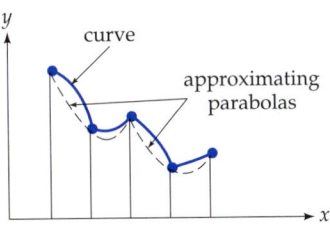

A curve approximated by two parabolas

Since each parabola spans two intervals, the number of intervals must be even. The area under the approximating parabolas is easily found by integration. The procedure is described on the following page and illustrated in Example 2. A justification of Simpson's rule is given in Exercise 35.

* Named after Thomas Simpson (1701–1761), an early user, but not the discoverer, of the formula.

Simpson's Rule (Parabolic Approximation)

To approximate $\int_a^b f(x)\,dx$ by Simpson's rule using n intervals:

1. Calculate $\Delta x = \dfrac{b-a}{n}$. $\qquad\qquad$ *n must be even*

2. Find numbers $x_1, x_2, x_3, \ldots, x_{n+1}$ starting with $x_1 = a$ and successively adding Δx, ending with $x_{n+1} = b$.

3. The approximation for the integral is

$$\int_a^b f(x)\,dx \approx [\,f(x_1) + \underbrace{4f(x_2) + 2f(x_3) + \cdots + 4f(x_n)} + f(x_{n+1})\,]\frac{\Delta x}{3}$$

$$\text{Alternating 4s and 2s}$$

The function values are multiplied by "weights," which, written out by themselves, are

$$1 \quad 4 \quad 2 \quad 4 \quad 2 \quad 4 \quad \ldots \quad 4 \quad 2 \quad 4 \quad 1$$

$\qquad\qquad\uparrow$ $\qquad\qquad$ Alternating 4s and 2s, $\qquad\qquad \uparrow$
$\qquad$ Initial 1 $\qquad$ beginning and ending with 4 $\qquad$ Final 1

EXAMPLE 2 **Approximating an Integral Using Simpson's Rule**

Approximate $\displaystyle\int_3^5 \frac{1}{x}\,dx$ by Simpson's rule with $n = 4$. $\quad$ *n = 4 means 2 parabolas*

Solution The method consists of the following six steps.

1. Calculate $\Delta x = \dfrac{b-a}{n} = \dfrac{5-3}{4} = 0.5$ $\quad$ (*n must be even*)

2. List x-values $\quad$ 3. Apply $f(x)$ $\quad$ 4. Multiply by the
   a through b $\qquad$ to each $\qquad\qquad$ weights to get
   with spacing Δx. $\quad$ x-value.

x	$f(x) = \dfrac{1}{x}$	weights	$f(x) \cdot$ **weight**
a $\quad$ 3	0.3333	1 ← Initial 1	0.3333
Δx $\quad$ 3.5	0.2857	4	1.1428
Δx $\quad$ 4	0.25	2 $\rbrace$ Alternating 4s and 2s	0.5000
Δx $\quad$ 4.5	0.2222	4	0.8888
b $\quad$ 5	0.2000	1 ← Final 1	0.2000

$$3.0649 \cdot \left(\frac{0.5}{3}\right) = 0.5109$$

5. Sum last column.

6. Multiply by $\dfrac{\Delta x}{3}$. $\qquad\qquad$ Final answer

Therefore $\displaystyle\int_3^5 \frac{1}{x}\,dx \approx 0.5109.$

This integral too can be found exactly. The answer is $\ln 5 - \ln 3 \approx$ 0.5108, for an error of only 0.0001. The *relative* error (actual error divided by actual value) is

$$\frac{0.0001}{0.5108} \approx 0.0002 = 0.02\%$$

which is also small.

Below are programs for Simpson's rule, one for the TI-83 calculator and one in BASIC.

TI-83 Program for Simpson's Rule	BASIC Program for Simpson's Rule
(Enter the function in y_1 before executing.)	(Lines 1, 2, 3, and 9 to be completed as indicated in small type.)

TI-83 Program for Simpson's Rule
(Enter the function in y_1 before executing.)

```
Disp "USES FUNCTION Y1"
Disp "N MUST BE EVEN"
Prompt A, B, N
(B−A)/N→D
Y1(A)−Y1(B)→S
A→X
For(I,1,N/2)
S+4*Y1(X+D)+2*Y1(X+2D)→S
X+2D→X
End
S*D/3→S
Disp "SIMP APPROX IS ", S
```

BASIC Program for Simpson's Rule
(Lines 1, 2, 3, and 9 to be completed as indicated in small type.)

```
a=                    fill in beginning x-value
b=                    fill in ending x-value
n=                    fill in number of intervals (even)
delta=(b−a)/n
approx=0
x=a
FOR i=1 TO n+1
IF i=1 OR i=n+1 THEN weight=1
approx=approx+(fill in the function) * weight
x=x+delta
IF weight=4 THEN weight=2 ELSE weight=4
NEXT i
approx=approx*delta/3
PRINT "SIMP APPROX IS", approx
END
```

IQ Distribution

Although it is increasingly clear that human intelligence cannot be measured by a single number, IQ tests are still widely used. (IQ stands for Intelligence Quotient, and is defined as mental age divided by chronological age, multiplied by 100.) The average American IQ is 100, and the distribution of IQs follows the famous "bell-shaped curve" so often used in statistics.*

---

* For those familiar with Chapter 5, IQ scores are normally distributed with mean 100 and standard deviation 15.

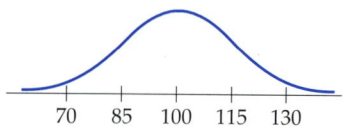

The proportion of Americans with IQs between two numbers A and B (with $A < B$) is given by the integral:

$$\left(\begin{array}{l}\text{Proportion of Americans} \\ \text{with IQs between } A \text{ and } B\end{array}\right) \approx \frac{1}{\sqrt{2\pi}} \int_{(A-100)/15}^{(B-100)/15} e^{-\frac{1}{2}x^2}\, dx$$

For example, the proportion of Americans who have IQs between 115 and 145 is found by substituting $A = 115$ and $B = 145$ into the lower and upper limits in the above formula (which, in the language of Section 5.4, give the *z-scores*).

$$\left(\begin{array}{l}\text{Proportion of IQs} \\ \text{between 115 and 145}\end{array}\right) = \frac{1}{\sqrt{2\pi}} \int_{1}^{3} e^{-\frac{1}{2}x^2}\, dx$$

3 from $\dfrac{B - 100}{15} = \dfrac{145 - 100}{15} = \dfrac{45}{15} = 3$

1 from $\dfrac{A - 100}{15} = \dfrac{115 - 100}{15} = \dfrac{15}{15} = 1$

This integral can be approximated by Simpson's rule, just as in Example 2.

Graphing Calculator Exploration

Evaluate the integral $\dfrac{1}{\sqrt{2\pi}} \int_{1}^{3} e^{-\frac{1}{2}x^2}\, dx$ by Simpson's rule using one of the programs on the previous page (or another program) as follows:

a. Enter $y_1 = \dfrac{e^{-x^2/2}}{\sqrt{2\pi}}$. (Use parentheses in the exponent.)

b. Run the program with $n = 4$ intervals.

c. Run the program again but with $n = 10$ intervals, and then with $n = 100$ intervals. To how many decimal places do the results agree?

d. Use FnInt (or a similar command) on your graphing calculator to evaluate the definite integral. How do the answers compare? (Answer: about 0.1573.)

Converting the answer to a percentage means that about 16% of all Americans have IQs in the range 115 to 145.

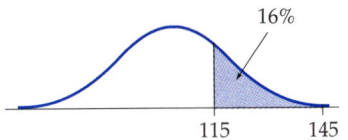

The Error in Simpson's Rule

The maximum error in Simpson's rule obeys the following formula.

Error in Simpson's Rule

In approximating $\int_a^b f(x)\, dx$ by Simpson's rule with n intervals,

$$(\text{Error}) \le \frac{(b - a)^5}{180n^4}\left[\max_{a \le x \le b} f^{(4)}(x)\right]$$

This formula is difficult to use because it involves maximizing the fourth derivative. It does, however, show that Simpson's rule is *exact* for cubics (third-degree polynomials), since these have zero fourth derivatives, and that doubling the value of n reduces the maximum error by a factor of *sixteen* (because of the n^4).

SUMMARY

Some definite integrals cannot be evaluated exactly because it is impossible (or very difficult) to find an antiderivative. However, any definite integral may be *approximated* by numerical integration, and two of the most useful methods are trapezoidal approximation and Simpson's rule (parabolic approximation). Each method involves choosing a number n of intervals and calculating a "weighted average" of function values at the endpoints of these intervals. Higher values of n generally give greater accuracy, but also involve more calculation (and therefore more roundoff errors). In practice, Simpson's rule is generally the method of choice, since it gives greater accuracy for only slightly more effort. Trapezoidal approximation, however, has the advantage of having a simpler formula for its error. It is particularly appropriate to seek an approximation if the "exact" answer involves logarithms or exponentials, since these functions will probably be approximated anyway in the evaluation step.

It is curious that in Chapter 9 we used integrals to evaluate areas, and here we are using areas to evaluate integrals. This is typical of mathematics, in which any equivalence (such as definite integrals and areas) is exploited in both directions.

EXERCISES 10.4

Trapezoidal Approximation

For each definite integral:

a. Approximate it "by hand," using trapezoidal approximation with $n = 4$ trapezoids. Round calculations to three decimal places.

b. Evaluate the integral exactly using antiderivatives.

c. Find the actual error (the difference between the actual value and the approximation).

d. Find the relative error (the actual error divided by the actual value, expressed as a percent).

1. $\displaystyle\int_1^3 x^2\,dx$ 2. $\displaystyle\int_1^2 x^3\,dx$

3. $\displaystyle\int_2^4 \frac{1}{x}\,dx$ 4. $\displaystyle\int_1^3 \frac{1}{x}\,dx$

Approximate each integral using trapezoidal approximation "by hand" with the given value of n. Round all calculations to three decimal places.

5. $\displaystyle\int_0^1 \sqrt{1+x^2}\,dx$, $n = 3$

6. $\displaystyle\int_0^1 \sqrt{1+x^3}\,dx$, $n = 3$

7. $\displaystyle\int_0^1 e^{-x^2}\,dx$, $n = 4$ 8. $\displaystyle\int_0^1 e^{x^2}\,dx$, $n = 4$

 Use a trapezoidal approximation program (see page 862) to approximate each integral for $n = 10$, $n = 100$, and $n = 500$, rounding calculations to four decimal places. Give an estimate for the value of the integral, stating your answer to as many decimal places as the last two approximations agree.

9. $\displaystyle\int_1^2 \sqrt{\ln x}\,dx$ 10. $\displaystyle\int_0^1 \ln(x^2+1)\,dx$

11. $\displaystyle\int_{-1}^1 \sqrt{16+9x^2}\,dx$ 12. $\displaystyle\int_{-1}^1 \sqrt{25-9x^2}\,dx$

13. $\displaystyle\int_{-1}^1 e^{x^2}\,dx$ 14. $\displaystyle\int_0^4 \sqrt{1+x^4}\,dx$

 15–16: General: IQs   Use the formula on page 866 and a trapezoidal approximation program (see page 862) to find the proportion of Americans with IQs between the following two numbers. Use suc-

cessively higher values of n until answers agree to four decimal places.

15. 100 and 130   16. 130 and 145

 17–18: Business: Investment Growth   An investment grows at a rate of $3.2\,e^{\sqrt{t}}$ thousand dollars per year, where t is the number of years since the beginning of the investment. Use a trapezoidal approximation program to estimate the total growth of the investment during the period stated below. Use successively higher values of n until answers agree to two decimal places.

17. In the first 2 years (year 0 to year 2)

18. In the first 3 years (year 0 to year 3)

Simpson's Rule (Parabolic Approximation)

Estimate each definite integral "by hand," using Simpson's rule with $n = 4$. Round all calculations to three decimal places. Exercises 19–22 correspond to Exercises 1–4, in which the same integrals were estimated using trapezoids. If you did the corresponding exercise, compare your Simpson's rule answer with your trapezoidal answer.

19. $\displaystyle\int_1^3 x^2\,dx$ 20. $\displaystyle\int_1^2 x^3\,dx$

21. $\displaystyle\int_2^4 \frac{1}{x}\,dx$ 22. $\displaystyle\int_1^3 \frac{1}{x}\,dx$

Approximate each integral using Simpson's rule "by hand" with $n = 4$. Round all calculations to three decimal places.

23. $\displaystyle\int_0^1 \sqrt{1+x^2}\,dx$ 24. $\displaystyle\int_0^1 \sqrt{1+x^3}\,dx$

25. $\displaystyle\int_0^1 e^{-x^2}\,dx$ 26. $\displaystyle\int_0^1 e^{x^2}\,dx$

 Use a Simpson's rule program (see page 865) to approximate each integral for $n = 10$, $n = 100$, and $n = 500$, rounding calculations to four decimal places. Give an estimate for the value of the integral, stating your answer to as many decimal places as the last two approximations agree.

27. $\displaystyle\int_1^2 \sqrt{\ln x}\,dx$ 28. $\displaystyle\int_0^1 \ln(x^2+1)\,dx$

29. $\displaystyle\int_{-1}^{1} \sqrt{16 + 9x^2}\, dx$ **30.** $\displaystyle\int_{-1}^{1} \sqrt{25 - 9x^2}\, dx$

31. $\displaystyle\int_{-1}^{1} e^{x^2}\, dx$ **32.** $\displaystyle\int_{0}^{4} \sqrt{1 + x^4}\, dx$

33. General: Suspension Bridges The cable of a suspension bridge hangs in a parabolic curve. The equation of the cable shown below is $y = \dfrac{x^2}{2000}$. Its length is given by the integral

$$\int_{-400}^{400} \sqrt{1 + \left(\frac{x}{1000}\right)^2}\, dx$$

Approximate this integral using Simpson's rule using successively higher values of n until answers agree to the nearest whole number.

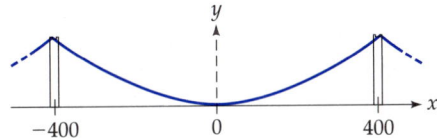

34. Approximation of π The number π is the ratio of the circumference of a circle to its diameter (since $C = \pi D$). It can be shown that the following definite integral is equal to π.

$$\int_{0}^{1} \frac{4}{x^2 + 1}\, dx = \pi$$

Find π by approximating this integral using a Simpson's rule program, using successively higher values of n until answers agree to four decimal places.

35. Justification of Simpson's Rule Justify Simpson's rule by carrying out the following steps, which lead to the formula for Simpson's rule (page 864) in a simple case.

   i. Observe that if the three points shown in the following diagram lie on the parabola $f(x) = ax^2 + bx + c$, then the following three equations hold:

$$a(-d)^2 + b(-d) + c = y_1 \quad \text{Since } f(-d) = y_1$$
$$a(0)^2 + b(0) + c = y_2 \quad \text{Since } f(0) = y_2$$
$$a(d)^2 + b(d) + c = y_3 \quad \text{Since } f(d) = y_3$$

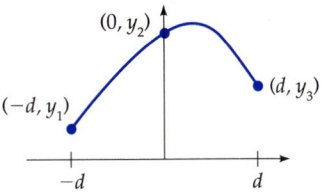

   ii. Simplify these three equations to obtain

$$ad^2 - bd + c = y_1$$
$$c = y_2$$
$$ad^2 + bd + c = y_3$$

   iii. Add the first and last equation plus four times the middle equation to obtain

$$2ad^2 + 6c = y_1 + 4y_2 + y_3$$

   You will use this equation in step v.

   iv. Evaluate the integral $\displaystyle\int_{-d}^{d} (ax^2 + bx + c)\, dx$ and simplify to show that the area under the parabola from $-d$ to d is

$$\text{Area} = \frac{2}{3} ad^3 + 2cd = \left(2ad^2 + 6c\right) \frac{d}{3}$$

   v. Use the equation found in step iii to write this area as

$$\text{Area} = (y_1 + 4y_2 + y_3) \frac{d}{3}$$

   vi. Use $y_1 = f(-d)$, $y_2 = f(0)$, $y_3 = f(d)$ and the fact that the spacing Δx is equal to d to rewrite this last equation as

$$\text{Area} = [f(-d) + 4f(0) + f(d)] \frac{\Delta x}{3}$$

   This is exactly the formula for Simpson's rule using one parabola ($n = 2$). For several parabolas placed next to each other, the function values *between* two neighboring parabolas are added twice (once for each side), and so should have weight 2. Therefore, successive function values are multiplied by the weights given on page 864:

$$1 \quad 4 \quad 2 \quad 4 \quad 2 \quad 4 \quad \ldots \quad 4 \quad 1$$

10.5 Differential Equations

APPLICATION PREVIEW

Personal Wealth and Differential Equations

Most people's income comes from two sources: salary and personal investments (such as interest on savings). From this income, they pay "necessary" expenses (housing and food), spend some of the remainder on "luxuries" (vacations and entertainment), and save the rest (increasing their investments). Given some simplifying assumptions, we can construct a mathematical model for your wealth at any time t. Suppose that after paying for necessities, you spend a fixed percentage (typically 75%) of the remainder on luxuries, and that your investments generate income at a fixed rate. Let

s = your salary
$W(t)$ = your wealth (savings), which is a function of time t
r = rate of interest on your wealth (savings)
n = amount spent on necessities
p = proportion of your income after necessities that you spend on luxuries

A model of your wealth is then given by the following "differential equation," which states that income must equal outflow:

$$\underbrace{s + r \cdot W}_{\text{Income}} = \underbrace{n + p(s + r \cdot W - n) + \frac{dW}{dt}}_{\text{Outflow}}$$

| Salary | Interest on wealth (savings) | Necessities | Luxuries (a fixed proportion of income after necessities) | Changes in wealth |

With a little algebra, this "differential equation" can be written

$$\frac{dW}{dt} = (1 - p)(s - n + r \cdot W)$$

The *solution* to this differential equation (assuming zero initial wealth) is the "wealth" function

$$W(t) = \frac{s - n}{r} (e^{(1-p)rt} - 1)$$

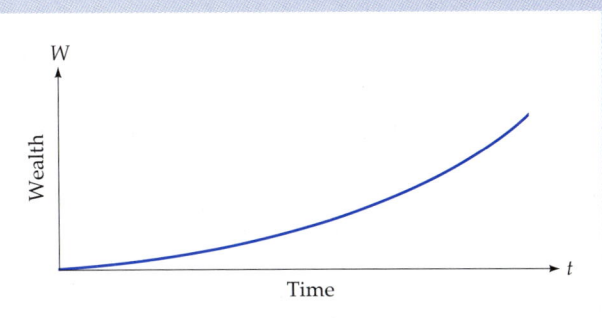

Given numbers for salary, necessities, interest rate, and proportion spent on luxuries, this solution could be used to predict when your savings will reach a particular level (for example, enough to buy a house). It could also be used to show how your wealth would change if you lowered your "luxury proportion" from, say, 75% to 70%.

This is but one illustration of the wide variety of applications of differential equations, which are discussed in the following sections.

Introduction

A differential equation is simply an equation involving derivatives. Practically any relationship involving rates of change can be expressed in a differential equation, and many differential equations can be solved by a technique called "separation of variables."

We consider a function $y = f(x)$, which we will sometimes write as $y(x)$ to indicate that y depends on x. We will write the derivative of y as either y' or $\frac{dy}{dx}$.

Differential Equation $y' = f(x)$

We have actually been solving differential equations since the beginning of Chapter 9. For example, the differential equation

$$y' = 2x$$

which states that the derivative of a function is $2x$, is solved simply by integrating:

$$y = \int 2x \, dx = x^2 + C \qquad\qquad y = x^2 + C \text{ satisfies } y' = 2x$$

In general, a differential equation of the form

$$y' = f(x)$$

is solved by integrating:

$$y = \int f(x)\, dx$$

Therefore, whenever we integrated a marginal cost function to find cost, or integrated a rate to find the total accumulation, we were solving a differential equation of the form $y' = f(x)$.

General and Particular Solutions

The solution to the differential equation $y' = 2x$ is $y = x^2 + C$, with an arbitrary constant C. We call $y = x^2 + C$ the *general solution* because taking all possible values of the constant C gives *all* solutions of the differential equation. If we take C to be a particular number, we get a *particular solution*. Some particular solutions to the differential equation $y' = 2x$ are

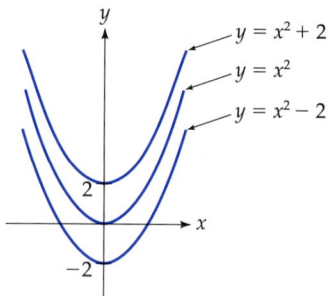

$y = x^2 + 2$ (taking $C = 2$)

$y = x^2$ (taking $C = 0$)

$y = x^2 - 2$ (taking $C = -2$)

Three particular solutions from the general solution $y = x^2 + C$

The different values of the arbitrary constant C give a "family" of curves, and the general solution $y = x^2 + C$ may be thought of as the entire family.

The solution to a differential equation is a *function*. The *general* solution contains an arbitrary constant, and a *particular* solution has this constant replaced by a particular number.

Verifying Solutions

Verifying that a function is a solution of a differential equation is simply a matter of calculating the necessary derivatives, substituting them into the differential equation, and checking that the resulting expressions are equal.

EXAMPLE 1   **Verifying a Solution to a Differential Equation**

Verify that $y = e^{2x} + e^{-x} - 1$ is a solution to the differential equation

$$y'' - y' - 2y = 2$$

Solution   The differential equation involves y, y', and y'', so we calculate

$$y = e^{2x} + e^{-x} - 1 \qquad \text{Given function}$$

$$y' = 2e^{2x} - e^{-x} \qquad \text{Derivative}$$

$$y'' = 4e^{2x} + e^{-x} \qquad \text{Second derivative}$$

Then we substitute these into the differential equation.

$$y'' - y' - 2y = 2$$

$$(4e^{2x} + e^{-x}) - (2e^{2x} - e^{-x}) - 2(e^{2x} + e^{-x} - 1) \stackrel{?}{=} 2 \qquad \stackrel{?}{=} \text{ means the equation may not be true}$$

$$4e^{2x} + e^{-x} - 2e^{2x} + e^{-x} - 2e^{2x} - 2e^{-x} + 2 \stackrel{?}{=} 2 \qquad \text{Expanding}$$

$$4e^{2x} + e^{-x} - 2e^{2x} + e^{-x} - 2e^{2x} - 2e^{-x} + 2 \stackrel{?}{=} 2 \qquad \text{Canceling}$$

$$2 = 2 \qquad \text{It checks!}$$

Since the equation is satisfied, the given function is indeed a solution of the differential equation.   ∎

If the two sides had not turned out to be exactly the same, the given function y would *not* have been a solution to the differential equation.

PRACTICE PROBLEM 1

Verify that $y = e^{-x} + e^{3x}$ is a solution to the differential equation

$$y'' - 2y' - 3y = 0 \qquad \textit{Solution at the back of the book}$$

The differential equations in Example 1 and Practice Problem 1 are called *second-order* differential equations because they involve second derivatives and no higher. From here on we will restrict our attention to *first-order* differential equations, that is, differential equations involving only the *first* derivative. Many first-order differential equations can be solved by a method called *separation of variables*.

Separation of Variables

A differential equation is said to be "separable" if the variables can be "separated" by moving every x to one side of the equation and every y to the other side. We may then solve the differential equation by integrating both sides. Several examples will make the method clear.

EXAMPLE 2   **Finding a General Solution**

Find the general solution to the differential equation $\dfrac{dy}{dx} = 2xy^2$.

Solution

$$dy = 2xy^2\, dx$$ Multiplying both sides of the differential equation by dx

$$\frac{dy}{y^2} = 2x\, dx$$ Dividing each side by y^2 ($y \neq 0$); the variables are now separated: y's on one side, x's on the other

$$y^{-2}\, dy = 2x\, dx$$ In power form

$$\int y^{-2}\, dy = \int 2x\, dx$$ Integrating both sides

$$-y^{-1} = x^2 + C$$ Integrating by the power rule (writing one C for both integrations)

$$\frac{1}{y} = -x^2 - C$$ Writing y^{-1} as $\dfrac{1}{y}$ and multiplying by -1

$$y = \frac{1}{-x^2 - C}$$ Taking reciprocals of both sides

This is the general solution to the differential equation. The solution may be left in this form, but if we replace the arbitrary constant C by $-c$, another constant but with the opposite sign, this solution may be written

$$y = \frac{1}{-x^2 + c}$$

or, reversing the order,

$$y = \frac{1}{c - x^2}$$ The general solution

(The differential equation $y' = 2xy^2$ has another solution: the function that is identically zero, $y(x) \equiv 0$. This is known as a *singular solution* and cannot be obtained from the general solution. We will ignore singular solutions in this book.)

Separable Differential Equations

A first-order differential equation is *separable* if it can be written in the following form for some functions $f(x)$ and $g(y)$:

$$\frac{dy}{dx} = \frac{f(x)}{g(y)} \qquad\qquad [g(y) \neq 0]$$

It is solved by separating variables and integrating:

$$\int g(y)\, dy = \int f(x)\, dx \qquad\qquad \text{Multiplying by } dx \text{ and } g(y) \\ \text{and integrating}$$

The previous example asked for the *general* solution to a differential equation. Sometimes we will be given a differential equation together with some additional information that selects a *particular* solution from the general solution, information that determines the value of the arbitrary constant in the general solution. This additional information is called an *initial condition*.

EXAMPLE 3   Finding a Particular Solution

Solve the differential equation $y' = \dfrac{6x}{y^2}$ with the initial condition $y(1) = 2$.

Solution   First we find the general solution by separating variables.

$$\frac{dy}{dx} = \frac{6x}{y^2} \qquad\qquad \text{Replacing } y' \text{ by } \frac{dy}{dx}$$

$$y^2\, dy = 6x\, dx \qquad\qquad \begin{array}{l}\text{Multiplying both sides by}\\ dx \text{ and } y^2 \text{ (the variables}\\ \text{are separated)}\end{array}$$

$$\int y^2\, dy = \int 6x\, dx \qquad\qquad \text{Integrating both sides}$$

$$\frac{1}{3} y^3 = 3x^2 + C \qquad\qquad \text{Using the power rule}$$

$$y^3 = 9x^2 + \underbrace{3C}_{c} \qquad\qquad \text{Multiplying by 3}$$

$$ \qquad\qquad \begin{array}{l} c \quad (3 \text{ times a constant is}\\ \text{just another constant)}\end{array}$$

$$y^3 = 9x^2 + c \qquad\qquad \text{Replacing } 3C \text{ by } c$$

$$y = \sqrt[3]{9x^2 + c} \qquad\qquad \text{Taking cube roots}$$

This is the general solution, with arbitrary constant c. The initial condition $y(1) = 2$ says that $y = 2$ when $x = 1$. We substitute these into the general solution $y = \sqrt[3]{9x^2 + c}$ and solve for c.

$$2 = \sqrt[3]{9 + c} \qquad\qquad y = \sqrt[3]{9x^2 + c} \text{ with } x = 1 \text{ and } y = 2$$

$$8 = 9 + c \qquad\qquad \text{Cubing each side}$$

$$-1 = c \qquad\qquad \text{Solving for } c \text{ gives } c = -1$$

Therefore, we replace c by -1 in the general solution to obtain the particular solution:

$$y = \sqrt[3]{9x^2 - 1} \qquad\qquad y = \sqrt[3]{9x^2 + c} \text{ with } c = -1$$

■

This solution $y = \sqrt[3]{9x^2 - 1}$ with $x = 1$ gives $y = \sqrt[3]{9 - 1} = \sqrt[3]{8} = 2$, so the initial condition $y(1) = 2$ is indeed satisfied. We could also verify that the differential equation is satisfied by substituting the solution into it.

In general, solving a differential equation with an initial condition just means finding the general solution and then using the initial condition to evaluate the constant. Several solutions to the differential equation $y' = \frac{6x}{y^2}$ are shown below, with the particular solution $y = \sqrt[3]{9x^2 - 1}$ shown in color.

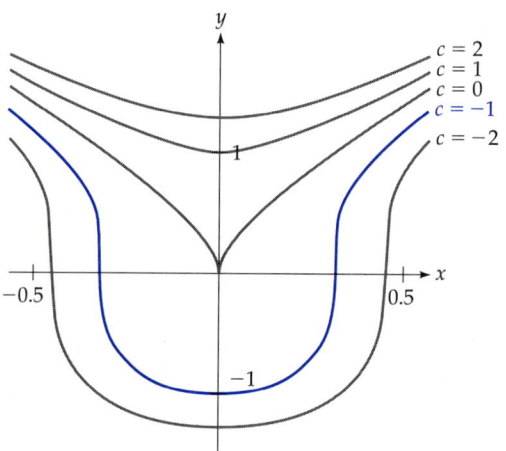

The solution $y = \sqrt[3]{9x^2 + c}$
for various values of c.

Slope Fields

A graphing calculator can help you "see" what a differential equation is saying. The differential equation of Example 3,

$$\frac{dy}{dx} = \frac{6x}{y^2} \qquad\qquad \text{Slope equals } \frac{6x}{y^2}$$

says that the slope at any point (x, y) in the plane is given by the formula $\frac{6x}{y^2}$. There are programs* (often with names like "slopefield") that draw on the screen little line segments with the specified slope. The slope field from a graphing calculator for the above differential equation is shown below.

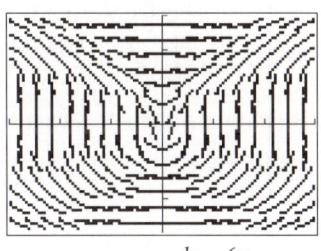

Slopefield for $\frac{dy}{dx} = \frac{6x}{y^2}$ on
$[-0.6, 0.6]$ by $[-1.5, 1.5]$

Compare this slope field with the graphs on the previous page of the solutions of the differential equation. Observe that the curves *do* follow the prescribed slope lines. In fact starting at any point in the plane, you could draw a curve following the indicated slopes, thereby geometrically constructing a "solution curve."

PRACTICE PROBLEM 2

Solve the differential equation and initial condition
$$\begin{cases} y' = \dfrac{6x^2}{y^4} \\ y(0) = 2 \qquad \textit{Solution at the back of the book} \end{cases}$$

Recall that to solve for y in the logarithmic equation

$$\ln y = f(x)$$

we exponentiate both sides and simplify

$$y = e^{f(x)} \qquad\qquad \text{Using } e^{\ln y} = y \\ \text{on the left side}$$

This idea will be useful in the next example.

---

* See the Preface for information how to obtain this and other programs.

EXAMPLE 4   Finding a Particular Solution

Solve the differential equation and initial condition:

$$\begin{cases} \dfrac{dy}{dx} = xy \\[2mm] y(0) = 2 \end{cases}$$

Solution

$dy = xy\ dx$	Multiplying by dx
$\dfrac{dy}{y} = x\ dx$	Dividing by y ($y \neq 0$). The variables are separated.
$\displaystyle\int \dfrac{dy}{y} = \int x\ dx$	Integrating
$\ln y = \dfrac{1}{2}x^2 + C$	Evaluating the integrals (assuming $y > 0$)
$y = e^{\frac{1}{2}x^2 + C}$	Solving for y by exponentiating
$y = e^{\frac{1}{2}x^2}\ \underbrace{e^C}$	e to a sum can be expressed as a product
$\qquad\qquad c$	(A constant to a constant is just another constant)
$y = ce^{\frac{1}{2}x^2}$	Replacing e^C by c (moved to the front)

This is the general solution. To satisfy the initial condition $y(0) = 2$, we substitute $y = 2$ and $x = 0$ and solve for c.

$2 = ce^0$	$y = ce^{\frac{1}{2}x^2}$ with $y = 2$ and $x = 0$
$2 = c$	Since $e^0 = 1$

Substituting $c = 2$ into the general solution gives the particular solution

$$y = 2e^{\frac{1}{2}x^2} \qquad\qquad\qquad y = ce^{\frac{1}{2}x^2} \text{ with } c = 2$$

∎

We wrote the solution to the integral $\displaystyle\int \dfrac{dy}{y}$ as $\ln y$, without absolute value bars. Although this amounts to assuming that $y > 0$, we have not lost any generality because the constant c in $y = ce^{\frac{1}{2}x^2}$ allows for positive and negative solutions. We will make similar simplifying assumptions in similar situations.

Note that the constant that was *added* in the integration step became a constant *multiplier* in the solution $y = ce^{\frac{1}{2}x^2}$. In general, the constant may appear *anywhere* in the solution. The following diagram shows solutions $y = ce^{\frac{1}{2}x^2}$ for various values of c, with the particular solution $y = 2e^{\frac{1}{2}x^2}$ shown in color.

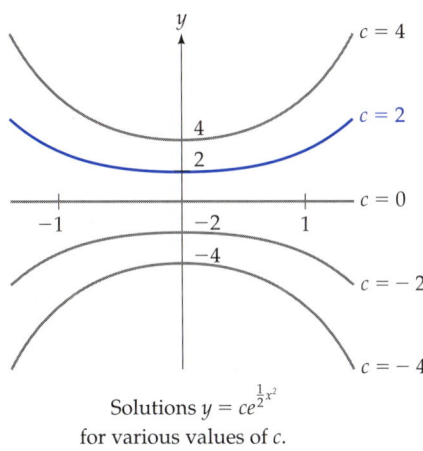

Solutions $y = ce^{\frac{1}{2}x^2}$
for various values of c.

 Graphing Calculator Exploration

Use a slope field program to have your calculator draw the slope field for the above differential equation. (Enter $y_1 = xy$, use the window $[-2, 2]$ by $[-6, 6]$, and run the program.) Observe how the slopes are determined by the function $x \cdot y$: In the first quadrant they increase as x or y increases, are positive when x 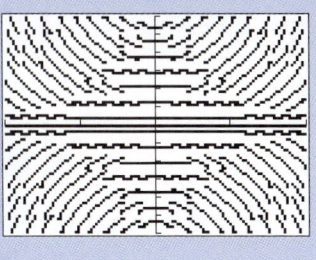 and y have the same sign, and are zero when either x or y is zero. Then try to see how the curves in the above graph follow the slope segments.

PRACTICE PROBLEM 3   Solve the differential equation and initial condition
$$\begin{cases} \dfrac{dy}{dx} = x^2 y \\[2mm] y(0) = 2 \end{cases}$$
Solution at the back of the book

EXAMPLE 5   **Finding a General Solution**

Find the general solution to the differential equation

$$yy' - x = 0$$

Solution   We are asked for the general solution rather than a particular solution.

$$y\frac{dy}{dx} = x$$ Replacing y' by $\frac{dy}{dx}$ and moving the x to the other side

$$y\,dy = x\,dx$$ Multiplying by dx

$$\int y\,dy = \int x\,dx$$ Integrating

$$\frac{1}{2}y^2 = \frac{1}{2}x^2 + C$$ Using the power rule

$$y^2 = x^2 + \underbrace{2C}_{c}$$ Multiplying by 2

To solve for y we take the square root of each side. However, there are *two* square roots, one positive and one negative.

$$y = \sqrt{x^2 + c} \qquad y = -\sqrt{x^2 + c}$$

These two solutions together are the general solution to the differential equation:

$$y = \pm\sqrt{x^2 + c}$$

■

 Graphing Calculator Exploration

Use a slope field program to have your calculator draw the slope field for the above differential equation, $\frac{dy}{dx} = \frac{x}{y}$, using the window $[-4, 4]$ by $[-4, 4]$. Try to draw a curve through the point $(0, 1)$ following the slopes. This would be the particular solution satisfying $y(0) = 1$.

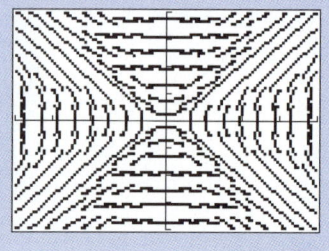

For some differential equations the integration step requires a substitution.

EXAMPLE 6 Finding a Particular Solution Using a Substitution

Solve the differential equation and initial condition $\begin{cases} y' = xy - x \\ y(0) = 4 \end{cases}$

Solution

$$\frac{dy}{dx} = xy - x \qquad\qquad \text{Replacing } y' \text{ by } \frac{dy}{dx}$$

$$\frac{dy}{dx} = x(y - 1) \qquad\qquad \text{Factoring (to separate variables)}$$

$$\frac{dy}{y - 1} = x\,dx \qquad\qquad \text{Dividing by } y - 1 \\ \text{and multiplying by } dx$$

$$\int \frac{dy}{y - 1} = \int x\,dx \qquad\qquad \text{Integrating}$$

$$\begin{bmatrix} u = y - 1 \\ du = dy \end{bmatrix} \qquad\qquad \text{Using a substitution}$$

$$\int \frac{du}{u} = \int x\,dx \qquad\qquad \text{Substituting}$$

$$\ln u = \frac{1}{2}x^2 + C \qquad\qquad \text{Integrating (assuming } u > 0)$$

$$\ln(y - 1) = \frac{1}{2}x^2 + C \qquad\qquad \text{Substituting back to } y \\ \text{using } u = y - 1$$

$$y - 1 = e^{\frac{1}{2}x^2 + C} \qquad\qquad \text{Solving for } y - 1 \text{ by ex-} \\ \text{ponentiating}$$

$$y - 1 = e^{\frac{1}{2}x^2} \cdot \underbrace{e^C = c \cdot e^{\frac{1}{2}x^2}}_{c} \qquad \text{Replacing } e^C \text{ by } c$$

$$y = c \cdot e^{\frac{1}{2}x^2} + 1 \qquad\qquad \text{Adding 1 to each side}$$

This is the general solution. To satisfy the initial condition $y(0) = 4$, we substitute $y = 4$ and $x = 0$.

$$4 = ce^0 + 1 \qquad\qquad y = ce^{\frac{1}{2}x^2} + 1 \text{ with} \\ y = 4 \text{ and } x = 0$$

$$4 = c + 1 \qquad\qquad \text{Since } e^0 = 1$$

This gives $c = 3$. Therefore, the particular solution is

$$y = 3e^{\frac{1}{2}x^2} + 1 \qquad\qquad y = ce^{\frac{1}{2}x^2} + 1 \\ \text{with } c = 3$$

■

Accumulation of Wealth

The examples so far have *given* us a differential equation to solve. In this application we will first *derive* a differential equation and then solve it.

EXAMPLE 7   Predicting Wealth

Suppose that you have saved $5000, and that you expect to save an additional $3000 during each year. If you deposit these savings in a bank paying 5% interest compounded continuously, find a formula for your bank balance after t years.

Solution   Let $y(t)$ stand for your bank balance (in thousands of dollars) after t years. Each year $y(t)$ grows by 3 (thousand dollars) plus 5% interest. This growth can be modeled by a differential equation:

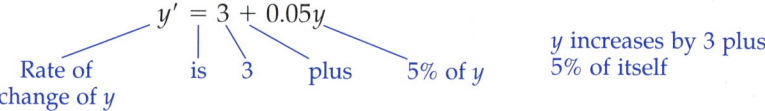

					y increases by 3 plus

$$y' = 3 + 0.05y$$

Rate of change of y is   3   plus   5% of y y increases by 3 plus 5% of itself

Before continuing, be sure that you understand how this differential equation models the changes due to savings and interest.
    We solve it by separating variables.

$$\frac{dy}{dt} = 3 + 0.05y$$
     Replacing y' by $\frac{dy}{dt}$

$$\int \frac{dy}{3 + 0.05y} = \int dt$$
     Dividing by $3 + 0.05y$, multiplying by dt, and then integrating

$$\left[\begin{array}{l} u = 3 + 0.05y \\ du = 0.05dy \end{array} \right]$$
     Using a substitution

$$20 \int \frac{du}{u} = \int dt$$
     Substituting (the 20 comes from $\frac{1}{0.05} = 20$)

$$20 \ln u = t + C$$
     Integrating (assuming $u > 0$)

$$\ln (3 + 0.05y) = 0.05t + \underbrace{0.05C}_{c}$$
     Substituting $u = 3 + 0.05y$ and dividing by 20

$$\ln (3 + 0.05y) = 0.05t + c$$
     Replacing $0.05C$ by c

$$3 + 0.05y = e^{0.05t+c} = e^{0.05t}\,\underbrace{e^c}_{k} = ke^{0.05t}$$

Exponentiating and then simplifying constants

$$0.05y = ke^{0.05t} - 3$$

Subtracting 3

$$y = \underbrace{20k}_{b}e^{0.05t} - 60$$

Dividing by 0.05

b (Always simplify arbitrary constants)

$$y = be^{0.05t} - 60$$

With arbitrary constant b

You began at time $t = 0$ with 5 thousand dollars, which gives the initial condition $y(0) = 5$. We substitute $y = 5$ and $t = 0$.

$$5 = be^0 - 60$$

$y = be^{0.05t} - 60$ with $y = 5$ and $t = 0$

$$5 = b - 60$$

Since $e^0 = 1$

Therefore, $b = 65$, which we substitute into the general solution, obtaining a formula for your accumulated wealth after t years.

$$y = 65e^{0.05t} - 60$$

$y = be^{0.05t} - 60$ with $b = 65$

For example, to find your wealth after 10 years, we evaluate the solution at $t = 10$.

$$y = 65e^{0.5} - 60 \approx 47.167$$

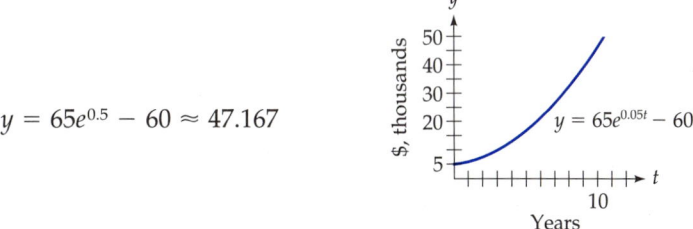

Therefore, after 10 years you will have $47,167 in the bank.

SUMMARY

We solved separable differential equations by separating the variables (moving every y to one side and every x to the other, with dx and dy in the numerators) and integrating both sides. The *general* solution to a differential equation includes an arbitrary constant, and a *particular* solution results from evaluating the constant, usually by an initial condition. The *slope field* of a differential equation allows us to geometrically construct a solution from any point by following the slopes away from the point.

We also saw how to *derive* a differential equation from information about how a quantity changes. For example, we can "read" the differential equation as follows.

$$y' \quad = \quad ay + b$$

Rate of change   is   a constant times the amount present   plus a constant amount added

The applied exercises will show how differential equations lead to important formulas in a wide variety of fields.

EXERCISES 10.5

Verify that the function y satisfies the given differential equation.

1. $y = e^{2x} - 3e^x + 2$
$\quad y'' - 3y' + 2y = 4$

2. $y = e^{5x} - 4e^x + 1$
$\quad y'' - 6y' + 5y = 5$

3. $y = ke^{ax} - \dfrac{b}{a}$ (for constants a, b, and k)
$\quad y' = ay + b$

4. $y = ax^2 + bx$ (for constants a and b)
$\quad y' = \dfrac{y}{x} + ax$

Find the general solution of each differential equation or state that the differential equation is not separable. If the exercise says "check," verify that your answer is a solution.

5. $y^2 y' = 4x$

6. $y^4 y' = 8x$

7. $y' = x + y$

8. $y' = xy - 1$

9. $y' = 6x^2 y$ and check

10. $y' = 12x^3 y$ and check

11. $y' = \dfrac{y}{x}$ and check

12. $y' = \dfrac{y^2}{x^2}$ and check

13. $yy' = 4x$

14. $yy' = 6x^2$

15. $y' = e^{xy}$

16. $y' = e^x + y$

17. $y' = 9x^2$

18. $y' = 6e^{-2x}$

19. $y' = \dfrac{x}{x^2 + 1}$

20. $y' = xy^2$

21. $y' = x^2 y$

22. $y' = \dfrac{x}{y}$

23. $y' = x^m y^n$ (for $m > 0$, $n \neq 1$)

24. $y' = x^m y$ (for $m > 0$)

25. $y' = 2\sqrt{y}$

26. $y' = 5 + y$

27. $xy' = x^2 + y^2$

28. $y' = \sqrt{x + y}$

29. $y' = xy + x$

30. $y' = x - 2xy$

31. $y' = ye^x - e^x$

32. $y' = ye^x - y$

33. $y' = ay^2$ (for constant $a > 0$)

34. $y' = axy$ (for constant a)

35. $y' = ay + b$ (for constants a and b)

36. $y' = (ay + b)^2$ (for constants $a \neq 0$ and b)

Solve each differential equation and initial condition. If the exercise says "and check," verify that your answer satisfies both the differential equation and the initial condition.

37. $\begin{cases} y^2 y' = 2x \\ y(0) = 2 \end{cases}$

38. $\begin{cases} y^4 y' = 3x^2 \\ y(0) = 1 \end{cases}$

39. $\begin{cases} y' = xy \\ y(0) = -1 \end{cases}$ and check

40. $\begin{cases} y' = y^2 \\ y(2) = -1 \end{cases}$ and check

41. $\begin{cases} y' = 2xy^2 \\ y(0) = 1 \end{cases}$ and check

42. $\begin{cases} y' = 2xy^4 \\ y(0) = 1 \end{cases}$ and check

43. $\begin{cases} y' = \dfrac{y}{x} \\ y(1) = 3 \end{cases}$ and check

44. $\begin{cases} y' = \dfrac{2y}{x} \\ y(1) = 2 \end{cases}$ and check

45. $\begin{cases} y' = 2\sqrt{y} \\ y(1) = 4 \end{cases}$

46. $\begin{cases} y' = \sqrt{y}e^x - \sqrt{y} \\ y(0) = 1 \end{cases}$

49. $\begin{cases} y' = ax^2y \quad \text{(for constant } a > 0\text{)} \\ y(0) = 2 \end{cases}$

47. $\begin{cases} y' = y^2e^x + y^2 \\ y(0) = 1 \end{cases}$

48. $\begin{cases} y' = xy - 5x \\ y(0) = 4 \end{cases}$

50. $\begin{cases} y' = axy \quad \text{(for constant } a > 0\text{)} \\ y(0) = 4 \end{cases}$

APPLIED EXERCISES

51. **Business: Elasticity**   For a demand function $D(p)$, the elasticity of demand (see page 719) is defined as $E = \dfrac{-pD'}{D}$. Find demand functions $D(p)$ that have constant elasticity by solving the differential equation $\dfrac{-pD'}{D} = k$, where k is a constant.

52. **Biomedical: Cell Growth**   A cell receives nutrients through its surface, and its surface area is proportional to the two-thirds power of its weight. Therefore, if $w(t)$ is the cell's weight at time t, $w(t)$ satisfies $w' = aw^{2/3}$, where a is a positive constant. Solve this differential equation with the initial condition $w(0) = 1$ (initial weight 1 unit).

53–54: Business: Annuities   An annuity is a fund into which one makes equal payments at regular intervals. If the fund earns interest at rate r compounded continuously, and deposits are made continuously at the rate of d dollars per year (a "continuous annuity"), then the value $y(t)$ of the fund after t years satisfies the differential equation $y' = d + ry$. (Do you see why?)

53. Solve this differential equation for the continuous annuity $y(t)$ with deposit rate $d = \$1000$ and continuous interest rate $r = 0.05$, subject to the initial condition $y(0) = 0$ (zero initial value).

54. Solve the above differential equation for the continuous annuity $y(t)$, where d and r are unknown constants, subject to the initial condition $y(0) = 0$ (zero initial value).

55. **General: Crime**   A medical examiner called to the scene of a murder will usually take the temperature of the body. A body cools at a rate proportional to the difference between its temperature and the temperature of the room. If $y(t)$ is the temperature of the body t hours after the murder, and if the room temperature is 70, then y satisfies

$$y' = -0.32(y - 70) \text{ with } y(0) = 98.6$$

(body temperature initially 98.6)

a. Solve this differential equation and initial condition.

b. Use your answer to part (a) to estimate how long ago the murder took place if the temperature of the body when it was discovered was 80. (*Hint:* Find the value of t that makes your solution equal 80.)

56. **Biomedical: Glucose Levels**   Hospital patients are often given glucose (blood sugar) through a tube connected to a bottle suspended over their beds. Suppose that this "drip" supplies glucose at the rate of 25 mg per minute, and each minute 10% of the accumulated glucose is consumed by the body. Then the amount $y(t)$ of glucose (in excess of the normal level) in the body after t minutes satisfies

$$y' = 25 - 0.1y \quad \text{(Do you see why?)}$$

$$y(0) = 0 \quad \text{(zero excess glucose at } t = 0\text{)}$$

Solve this differential equation and initial condition.

57. **General**   Suppose that you meet 30 new people each year, but each year you forget 20% of all of the people that you know. If $y(t)$ is the total number of people who you remember after t years, then y satisfies the differential equation $y' = 30 - 0.2y$. (Do you see why?) Solve this differential equation subject to the condition $y(0) = 0$ (you knew no one at birth).

58. **Environmental Science: Pollution**   For more than 75 years the Flexfast Rubber Company in Massachusetts discharged toxic toluene solvents into the ground at a rate of 5 tons per year.

Each year approximately 10% of the accumulated pollutants evaporated into the air. If $y(t)$ is the total accumulation of pollution in the ground after t years, then y satisfies

$$y' = 5 - 0.1y$$

$$y(0) = 0 \quad \text{(initial accumulation zero)}$$

Solve this differential equation and initial condition to find a formula for the accumulated pollutant after t years.

59. General: Accumulation of Wealth   Suppose that you now have $6000, you expect to save an additional $3000 during each year, and all of this is deposited in a bank paying 10% interest compounded continuously. Let $y(t)$ be your bank balance (in thousands of dollars) t years from now.

a. Write a differential equation that expresses the fact that your balance will grow by 3 (thousand dollars) and also by 10% of itself. (*Hint:* See Example 7.)

b. Write an initial condition to say that at time zero the balance is 6 (thousand dollars).

c. Solve your differential equation and initial condition.

d. Use your solution to find your bank balance $t = 25$ years from now.

60. General: Accumulation of Wealth   Suppose that you now have $2000, you expect to save an additional $6000 during each year, and all of this is deposited in a bank paying 4% interest compounded continuously. Let $y(t)$ be your bank balance (in thousands of dollars) t years from now.

a. Write a differential equation that expresses the fact that your balance will grow by 6 (thousand dollars) and also by 4% of itself. (*Hint:* See Example 7).

b. Write an initial condition to say that at time zero the balance is 2 (thousand dollars).

c. Solve your differential equation and initial condition.

d. Use your solution to find your bank balance $t = 20$ years from now.

61. Business: Sales   Your company has developed a new product, and your marketing department has predicted how it will sell. Let $y(t)$ be the (monthly) sales of the product after t months.

a. Write a differential equation that says that the rate of growth of the sales will be four times the one-half power of the sales.

b. Write an initial condition that says that at time $t = 0$ sales were 10,000.

c. Solve this differential equation and initial value.

d. Use your solution to predict the sales at time $t = 12$ months.

62. Business: Sales   Your company has developed a new product, and your marketing department has predicted how it will sell. Let $y(t)$ be the (monthly) sales of that product after t months.

a. Write a differential equation that says that the rate of growth of the sales will be six times the two-thirds power of the sales.

b. Write an initial condition that says that at time $t = 0$ sales were 1000.

c. Solve this differential equation and initial value.

d. Use your solution to predict the sales at time $t = 12$ months.

63. Biomedical: Bacterial Colony   Let $y(t)$ be the size of a colony of bacteria after t hours.

a. Write a differential equation that says that the rate of growth of the colony is equal to eight times the three-fourths power of its present size.

b. Write an initial condition that says that at time zero the colony is of size 10,000.

c. Solve the differential equation and initial condition.

d. Use your solution to find the size of the colony at time $t = 6$ hours.

64. General: Value of a Building   Let $y(t)$ be the value of a commercial building (in millions of dollars) after t years.

a. Write a differential equation that says that the rate of growth of the value is equal to two times the one-half power of its present value.

b. Write an initial condition that says that at time zero the value of the building is 9 million dollars.

c. Solve the differential equation and initial condition.

d. Use your solution to find the value of the building at time $t = 5$ years.

 The following exercises require the use of a slope field program.

For each differential equation, use a slope field program to graph the slope field on the window $[-5, 5]$ by $[-5, 5]$ (for most programs, enter the right-hand side of the differential equation in y_1). From the resulting slope field, make a rough sketch of the solution curve passing through the indicated point. (You may want to alter the window to see more detail.)

65. $\dfrac{dy}{dx} = 2x - y$

point: $(0, -1)$

66. $\dfrac{dy}{dx} = x^2 y$

point: $(0, 2)$

67. $\dfrac{dy}{dx} = y - x^2$

point: $(0, 1)$

68. $\dfrac{dy}{dx} = x - y^2$

point: $(0, 2)$

69. $\dfrac{dy}{dx} = \dfrac{x}{y^2 + 1}$

point: $(0, -1)$

70. $\dfrac{dy}{dx} = x \ln (y^2 + 1)$

point: $(0, -2)$

10.6 Further Applications of Differential Equations: Three Models of Growth

Introduction

This section continues our study of differential equations, but with a different approach. Instead of solving individual differential equations, we will solve three important classes of differential equations (for *unlimited*, *limited*, and *logistic* growth) and remember their solutions. This will enable us to solve many problems by determining the appropriate differential equation, and then immediately writing the solution. In this section we begin to *think in terms of differential equations*.

Proportion

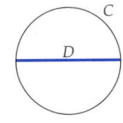

The circumference of a circle is proportional to its diameter.

We say that one quantity is *proportional to* another quantity if the first quantity is a *constant multiple* of the second. That is, y is proportional to x if $y = ax$ for some "proportionality constant" a. For example, the formula $C = \pi D$ for the circumference of a circle shows that the circumference C is proportional to the diameter D.

Unlimited Growth

In many situations the growth of a quantity is proportional to its present size. For example, a population of cells will grow in proportion to its present size, and a bank account earns interest in proportion to its size. If a quantity y grows so that its rate of growth y' is proportional to its present size y, then y satisfies the differential equation

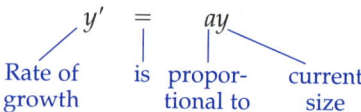

$$y' = ay$$

Rate of growth · is proportional to · current size

We solve this differential equation by separating variables.

$$\frac{dy}{dt} = ay \qquad\qquad \text{Replacing } y' \text{ by } \frac{dy}{dt}$$

$$\int \frac{dy}{y} = \int a\, dt \qquad\qquad \text{Dividing by } y \ (y \neq 0), \text{ multiplying by } dt, \text{ and integrating}$$

$$\ln y = at + C \qquad\qquad \text{Integrating}$$

$$y = e^{at+C} = e^{at}e^{C} = ce^{at} \qquad \begin{array}{l}\text{Solving for } y \text{ by exponentiating and} \\ \text{then simplifying } e^{C} = c\end{array}$$

$$\underset{c}{\underbrace{}}$$

At time $t = 0$ this becomes

$$y(0) = ce^{0} = c \qquad\qquad y(t) = ce^{at} \text{ with } t = 0$$

Summarizing:

Unlimited Growth

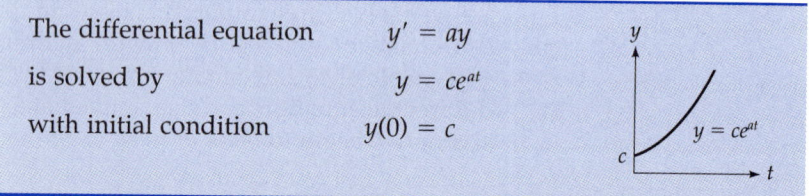

The differential equation $\qquad y' = ay$

is solved by $\qquad\qquad y = ce^{at}$

with initial condition $\qquad y(0) = c$

$y = ce^{at}$

Such growth, where the rate of growth is proportional to the present size, is called *unlimited growth* because the solution y grows arbitrarily large. Given this result, whenever we encounter a differential equation of the form $y' = ay$, we can immediately write the solution $y = ce^{at}$, where c is the initial value.

EXAMPLE 1　Predicting Art Appreciation (Unlimited Growth)

An art collection, initially worth $25,000, continuously grows in value at the rate of 5% per year. Express this growth rate as a differential equation and find a formula for the value of the collection after t years. Then estimate its value after 10 years.

Solution   Growing continuously at the rate of 5% means that the value $y(t)$ grows by 5% *of itself*.

$$y' = 0.05y$$

Rate of growth   is   5% of the current value

This differential equation is of the form $y' = ay$ (unlimited growth) with $a = 0.05$ and initial value 25,000, so we may immediately write its solution.

$$y(t) = 25,000e^{0.05t} \qquad \qquad y = ce^{at} \text{ with } a = 0.05 \text{ and } c = 25,000$$

This formula gives the value of the art collection after t years. To find the value after 10 years, we evaluate at $t = 10$:

$$y(10) = 25,000e^{0.05 \cdot 10} \qquad \qquad y(t) = 25,000e^{0.05t} \text{ with } t = 10$$
$$= 25,000e^{0.5} \approx 41,218 \qquad \qquad \text{Using a calculator}$$

In 10 years the art collection will be worth \$41,218.

The solution ce^{at} is the same as the continuous compounding formula Pe^{rt} (except for different letters). On page 125 we derived the formula Pe^{rt} rather laboriously, using the discrete interest formula $P(1 + r/m)^{mt}$ and taking the limit as $m \to \infty$. The present derivation, using differential equations, is much simpler and shows that ce^{at} applies to *any* situation governed by the differential equation $y' = ay$.

Limited Growth

No real population can undergo unlimited growth for very long. Restrictions of food and space would soon slow its growth. If a quantity $y(t)$ cannot grow larger than a certain fixed maximum size M, and if its growth rate y' is proportional to how far it is from its upper limit, then y satisfies

$$y' = a(M - y) \qquad \qquad a > 0$$

Rate of growth   is   proportional to   distance below upper bound M

We solve this by separating variables.

$$\frac{dy}{dt} = a(M - y) \qquad \qquad \text{Replacing } y' \text{ by } \frac{dy}{dt}$$

$$\int \frac{dy}{M - y} = \int a \, dt$$

Dividing by $M - y$, multiplying by dt, and integrating

$$\left[\begin{array}{l} u = M - y \\ du = -dy \end{array} \right]$$

Using a substitution

$$-\int \frac{du}{u} = \int a \, dt$$

Substituting

$$-\ln u = at + C$$

Integrating ($u > 0$)

$$\ln (M - y) = -at - C$$

Multiplying by -1 and replacing u by $(M - y)$

$$M - y = e^{-at-C} = e^{-at}e^{-C} = ce^{-at}$$

Solving for $M - y$ by exponentiating, and simplifying exponents

$$y = M - ce^{-at}$$

Subtracting M and multiplying by -1

We impose the initial condition $y(0) = 0$ (size zero at time $t = 0$).

$$0 = M - ce^0 = M - c \qquad\qquad y = M - ce^{-at} \text{ with } y = 0 \text{ and } t = 0$$

Therefore, $c = M$, which gives the solution

$$y = M - Me^{-at} = M(1 - e^{-at}) \qquad\qquad y = M - ce^{-at} \text{ with } c = M$$

Summarizing:

Limited Growth

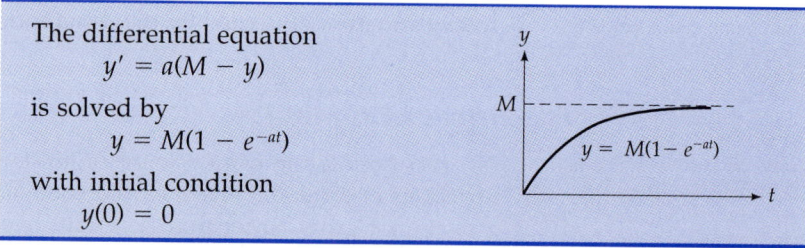

The differential equation
$$y' = a(M - y)$$
is solved by
$$y = M(1 - e^{-at})$$
with initial condition
$$y(0) = 0$$

This type of growth, in which the rate of growth is proportional to the distance below an upper limit M, is called *limited growth* because the solution $y = M(1 - e^{-at})$ approaches the limit M as $t \to \infty$. Given this result, whenever we encounter a differential equation of the form

$$y' = a(M - y)$$

with initial value zero, we may immediately write the solution

$$y = M(1 - e^{-at})$$

Diffusion of Information by Mass Media

If a news bulletin is repeatedly broadcast over radio and television, the news spreads quickly at first, but later more slowly when most people have already heard it. Sociologists often assume that the rate at which news spreads is proportional to the number who have not yet heard the news. Let M be the population of a city, and $y(t)$ be the number of people who have heard the news within t time units. Then y satisfies the differential equation

$$y' = a(M - y)$$

Rate of growth   is   propor-   number who have
                        tional to   not heard the news

We recognize this as the differential equation for limited growth, whose solution is $y = M(1 - e^{-at})$. It remains only to determine the values of the constants M and a.

EXAMPLE 2   Predicting Spread of Information (Limited Growth)

An important news bulletin is broadcast to a town of 50,000 people, and after 2 hours 30,000 people have heard the news. Find a formula for the number of people who have heard the bulletin within t hours. Then find how many people will have heard the news within 6 hours.

Solution   If $y(t)$ is the number of people who have heard the news within t hours, then y satisfies

$$y' = a(50,000 - y) \qquad \text{$y' = a(M - y)$ with $M = 50,000$}$$

Number who have
not heard the news

This is the differential equation for limited growth with $M = 50,000$, so the solution (from the preceding box) is

$$y = 50,000(1 - e^{-at}) \qquad \text{$y' = a(M - y)$ with $M = 50,000$}$$

To find the value of the constant a, we use the given information that 30,000 people have heard the news within 2 hours.

$30,000 = 50,000(1 - e^{-a \cdot 2})$ $y = 50,000(1 - e^{-at})$ with $y = 30,000$ and $t = 2$

$0.6 = 1 - e^{-2a}$ Dividing each side by 50,000

$0.4 = e^{-2a}$ Subtracting 1 and then multiplying by -1

$-0.916 \approx -2a$ Taking natural logs (using $\ln e^x = x$ on the right)

$a \approx 0.46$ Dividing by -2

Therefore, the number of people who have heard the news within t hours is

$$y(t) = 50,000(1 - e^{-0.46t})$$ $y = 50,000(1 - e^{-at})$ with $a = 0.46$

To find the number who have heard the news within 6 hours, we evaluate this solution at $t = 6$.

$$y(6) = 50,000(1 - e^{-(0.46)(6)}) = 50,000(1 - e^{-2.76}) \approx 46,835$$ Using a calculator

Within 6 hours about 46,800 people have heard the news.

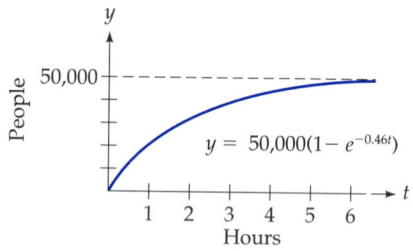

News spread by mass media

We found the value of the constant a by substituting the given data into the solution, simplifying, and taking logs. Similar steps will be required in many other problems, and in the future we shall omit the details.

Learning Theory

Psychologists have found that there seems to be an upper limit to the number of meaningless words that a person can memorize, and that memorizing becomes increasingly difficult approaching this upper limit. If M is this upper limit, and if $y(t)$ is the number of words that can be memorized in t minutes, then the situation is modeled by the differential equation for limited growth.

$$y' = a(M - y)$$

Rate of   is   propor-   upper limit M minus
increase   tional to   number already memorized

This may be interpreted as saying that the rate at which new words can be memorized is proportional to the "unused memory capacity."

EXAMPLE 3   Predicting Memorization (Limited Growth)

Suppose that a person can memorize at most 100 meaningless words, and that after 15 minutes 10 words have been memorized. How long will it take to memorize 50 words?

Solution   The number $y(t)$ of words that can be memorized in t minutes satisfies

$$y' = a(100 - y) \qquad\qquad y' = a(M - y) \text{ with } M = 100$$

This is the differential equation for limited growth, so the solution is

$$y = 100(1 - e^{-at}) \qquad\qquad y = M(1 - e^{-at}) \text{ with } M = 100$$

To evaluate the constant a, we use the given information that 10 words have been memorized in 15 minutes.

$$10 = 100(1 - e^{-a \cdot 15}) \qquad y = 100(1 - e^{-at}) \text{ with } y = 10 \text{ and } t = 15$$

Solving this for the constant a (omitting the details, which are the same as in Example 2) gives $a = 0.007$.

$$y(t) = 100(1 - e^{-0.007t}) \qquad y = 100(1 - e^{-at}) \text{ with } a = 0.007$$

This gives the number of words that can be memorized in t minutes. To find how long it takes to memorize 50 words, we set this solution equal to 50 and solve for t.

$$50 = 100(1 - e^{-0.007t}) \qquad y = 100(1 - e^{-0.007t}) \text{ with } y = 50$$

Solving for t (again the details are similar to those in Example 2) gives $t = 99$ minutes. Therefore, 50 words can be memorized in about 1 hour and 39 minutes.

The solution is graphed below, showing how the number of words approaches 100.

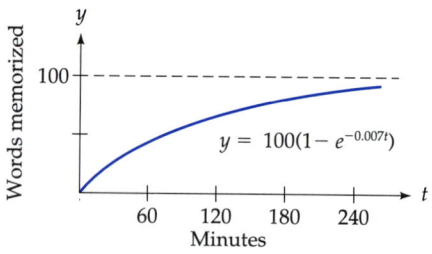

The slope (the rate at which additional
words can be memorized) decreases
near the upper limit.

Logistic Growth

Some quantities grow in proportion to both their present size *and* their distance from an upper limit M.

$$y' \;=\; ay(M - y)$$

Rate of ——— is   propor-   present   upper limit
growth   tional to   size   minus present

This differential equation can be solved by separation of variables (see Exercise 55) to give the solution

$$y = \frac{M}{1 + ce^{-aMt}}$$

where c and a are positive constants. This function is called the *logistic* function, governing *logistic growth.*

Logistic Growth

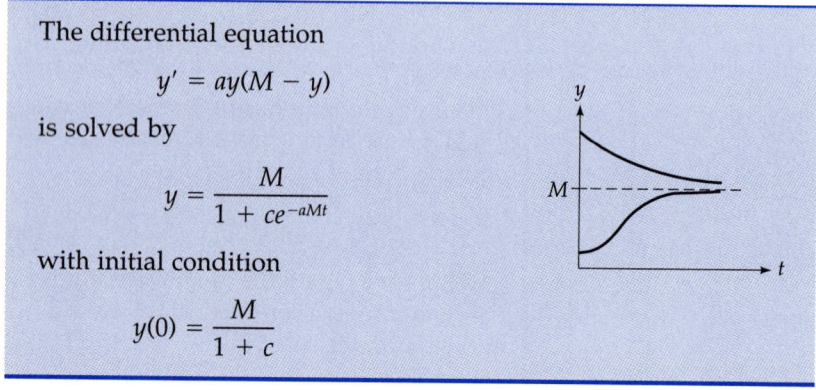

The differential equation

$$y' = ay(M - y)$$

is solved by

$$y = \frac{M}{1 + ce^{-aMt}}$$

with initial condition

$$y(0) = \frac{M}{1 + c}$$

(The upper and lower curves represent solutions whose initial values are, respectively, greater than or less than *M*.) As before, this result enables us to solve a differential equation "at sight," leaving only the evaluation of constants.

This curve that rises to *M* is called a *sigmoidal* or *S-shaped curve*, and is used to model growth that begins slowly, then becomes more rapid, and finally slows again near the upper limit. Many different quantities grow according to sigmoidal curves.

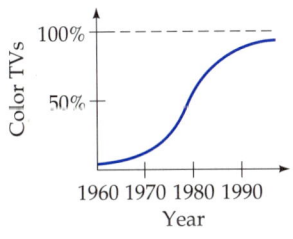

Percent of households owning a color television (*Source:* Census Bureau.)

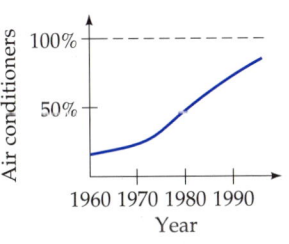

Percent of households with air conditioning (*Source:* Worldwatch.)

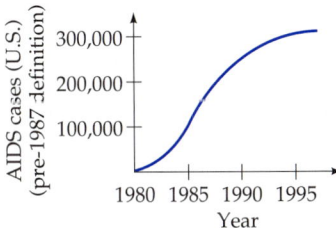

Total AIDS cases by year of diagnosis (*Source:* Centers for Disease Control.)

Environmental Science

For an animal environment (like a lake or a forest), the population that it can support will have an upper limit, called the *carrying capacity of the environment*. Ecologists often assume that an animal population grows in proportion to both its present size and its distance below its carrying capacity.

$$y' = ay(M - y)$$

Rate of growth   is   propor- tional to   current size   carrying minus present

Since this is the logistic differential equation, we know that the solution is the logistic function

$$y = \frac{M}{1 + ce^{-aMt}}$$

EXAMPLE 4   Predicting an Animal Population (Logistic Growth)

Ecologists estimate that an artificial lake can support a maximum of 2500 fish. The lake is initially stocked with 500 fish, and after 6 months the fish population is estimated to be 1500. Find a formula for the number of fish in the lake after t months, and estimate the fish population at the end of the first year.

Solution   Letting $y(t)$ stand for the number of fish in the lake after t months, the situation is modeled by the logistic differential equation

$$y' = ay(2500 - y)$$

Rate of growth   is   propor-tional to   current size   carrying capacity

$y' = ay(M - y)$ with $M = 2500$

The solution is the logistic function with $M = 2500$.

$$y = \frac{2500}{1 + ce^{-a2500t}} \qquad y = \frac{M}{1 + ce^{-aMt}} \quad \text{with} \quad M = 2500$$

$$= \frac{2500}{1 + ce^{-bt}} \qquad \text{Replacing } a \cdot 2500 \text{ by another constant } b$$

To evaluate the constants c and b, we use the fact that the lake was originally stocked with 500 fish.

$$500 = \frac{2500}{1 + ce^0} \qquad y = \frac{2500}{1 + ce^{-bt}} \quad \text{with} \quad y = 500, t = 0$$

$$500 = \frac{2500}{1 + c} \qquad \text{Simplifying}$$

Solving this for c (omitting the details—the first step is to multiply both sides by $1 + c$) gives $c = 4$, so the logistic function becomes

$$y = \frac{2500}{1 + 4e^{-bt}} \qquad y = \frac{2500}{1 + ce^{-bt}} \quad \text{with} \quad c = 4$$

To evaluate b, we substitute the information that the population is $y = 1500$ at $t = 6$.

$$1500 = \frac{2500}{1 + 4e^{-b \cdot 6}} \qquad y = \frac{2500}{1 + 4e^{-bt}} \quad \text{with} \quad \begin{array}{l} y = 1500 \\ t = 6 \end{array}$$

Solving for b (again omitting the details—the first step is to multiply both sides by $1 + 4e^{-b \cdot 6}$) gives $b = 0.30$, so the logistic function becomes

$$y(t) = \frac{2500}{1 + 4e^{-0.3t}} \qquad\qquad y = \frac{2500}{1 + 4e^{-bt}} \quad \text{with} \quad b = 0.3$$

This is the formula for the population after t months. To find the population after a year, we evaluate at $t = 12$.

$$y(12) = \frac{2500}{1 + 4e^{-0.3(12)}} = \frac{2500}{1 + 4e^{-3.6}} \approx 2254 \qquad \textcolor{blue}{\text{Using a calculator}}$$

Therefore, the population at the end of the first year is 2254, which is about 90% of the carrying capacity of the lake.

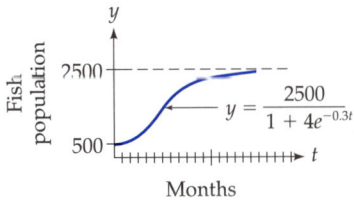

Epidemics

Many epidemics spread at a rate proportional to both the number of people already infected (the "carriers") and also the number who have yet to catch the disease (the "susceptibles"). If $y(t)$ is the number of infected people at time t from a population of size M, then $y(t)$ satisfies the logistic differential equation:

$$y' = ay(M - y)$$

Rate of is propor- number number
growth   tional to infected susceptible

Therefore, the size of the infected population is given by the logistic function

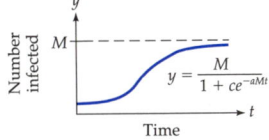

$$y = \frac{M}{1 + ce^{-aMt}}$$

The constants are evaluated just as in Example 4, using the (initial) number of cases reported at time $t = 0$, and also the number of cases at some later time.

Spread of Rumors

Sociologists have found that rumors spread through a population of size M at a rate proportional to the number who have heard the rumor (the "informed") and the number who have not heard the rumor (the "uninformed"). Therefore, the number $y(t)$ who have heard the rumor within t time units satisfies the logistic differential equation:

$$y' \;=\; ay(M - y)$$

Rate of growth | is | propor- tional to | number informed | number uninformed

The solution $y(t)$ is then the logistic function

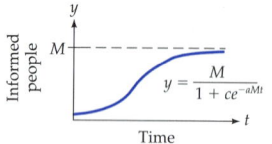

$$y = \frac{M}{1 + ce^{-aMt}}$$

It remains only to evaluate the constants. The spread of a rumor is analogous to the spread of a disease, with an "informed" person being one who has been "infected."

Limited and Logistic Growth of Sales

The sales of a product whose total sales will approach an upper limit (market saturation) can be modeled by either the limited or the logistic equation. Which do you use when? For a product advertised over mass media, sales will at first grow rapidly, indicating a *limited* model. For a product becoming known only through "word of mouth," sales will at first grow slowly, indicating a *logistic* model.

PRACTICE PROBLEM

The graphs below show the total sales through day t for two different products, A and B. Which of these products was advertised and which became known only by "word of mouth"? State an appropriate differential equation for each curve.

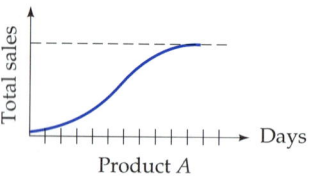

Product A

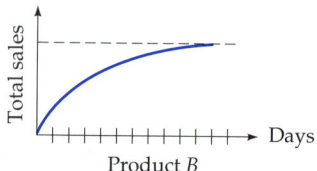

Product B

Solution at the back of the book

SUMMARY

The unlimited, limited, and logistic growth models are summarized in the following table. If one of these differential equations governs a particular situation, we can write the solution immediately, evaluating the constants from the given data.

THREE MODELS OF GROWTH

Type	Differential Equation	Solution	Graph	Examples
Unlimited Growth is proportional to present size.	$y' = ay$	$y = ce^{at}$		Investments Bank accounts Unlimited populations
Limited Growth (starting at 0) is proportional to maximum size M minus present size.	$y' = a(M - y)$	$y = M(1 - e^{-at})$		Information spread by mass media Memorizing random information Total sales (advertised)
Logistic Growth is proportional to present size and to maximum size M minus present size.	$y' = ay(M - y)$	$y = \dfrac{M}{1 + ce^{-aMt}}$		Confined populations Epidemics Rumors Total sales (unadvertised)

To decide which (if any) of the three models applies in a given situation, think of whether the growth is proportional to size, to unused capacity, or to both (as shown in the chart). Notice that the differential equation gives much more insight into how the growth occurs than does the solution. This is what we meant at the beginning of the section by "thinking in terms of differential equations."

Graphing Calculator Exploration

If you have a slope field program, graph the slope field for the following differential equations (one at a time) on the window [0, 5] by [0, 5]:

a. $y' = y$ (unlimited)   c. $y' = y(3 - y)$ (logistic)

b. $y' = 3 - y$ (limited)

Do you see how the slope field gives a "picture" of the differential equation? Each graph is on window [0, 5] by [0, 5].

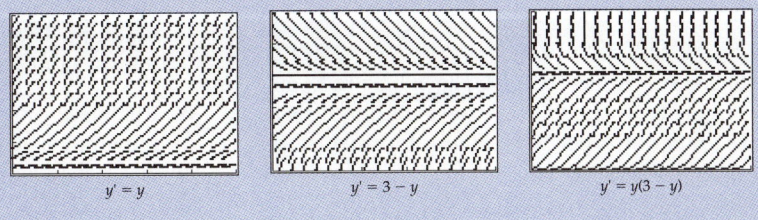

EXERCISES 10.6

1. Verify that $y(t) = ce^{at}$ solves the differential equation for unlimited growth, $y' = ay$, with initial condition $y(0) = c$.

2. Verify that $y(t) = M(1 - e^{-at})$ solves the differential equation for limited growth, $y' = a(M - y)$, with initial condition $y(0) = 0$.

Determine the type of each differential equation: *unlimited* growth, *limited* growth, *logistic* growth, or *none* of these. (Do not solve, just identify the type.)

3. $y' = 0.02y$

4. $y' = 5(100 - y)$

5. $y' = 30(0.5 - y)$

6. $y' = 0.4y(0.01 - y)$

7. $y' = 2y^2(0.5 - y)$

8. $y' = 6y$

9. $y' = y(6 - y)$

10. $y' = 0.01(100 - y^2)$

11. $y' = 4y(0.04 - y)$

12. $y' = 4500(1 - y)$

Find the solution $y(t)$ by recognizing each differential equation as determining unlimited, limited, or logistic growth, and then determining the constants.

13. $y' = 6y$
$y(0) = 1.5$

14. $y' = 0.25y$
$y(0) = 4$

15. $y' = -y$
$y(0) = 100$

16. $y' = \frac{y}{2}$
$y(0) = 8$

17. $y' = -0.45y$
$y(0) = -1$

18. $y' = 0$
$y(0) = 5$

19. $y' = 2(100 - y)$
$y(0) = 0$

20. $y' = 48(2 - y)$
$y(0) = 0$

21. $y' = 0.05(0.25 - y)$
$y(0) = 0$

22. $y' = \frac{2}{3}(1 - y)$
$y(0) = 0$

23. $y' = 80 - 2y$
$y(0) = 0$

24. $y' = 27 - 3y$
$y(0) = 0$

Hint for Exercises 23–26: Use factoring to write the differential equation in the form $y' = a(M - y)$.

25. $y' = 2 - 0.01y$
$y(0) = 0$

26. $y' = 6 - 8y$
$y(0) = 0$

27. $y' = 5y(100 - y)$
$y(0) = 10$

28. $y' = y(1 - y)$
$y(0) = \frac{1}{2}$

29. $y' = 0.25y(0.5 - y)$
$y(0) = 0.1$

30. $y' = \frac{1}{3}y(\frac{1}{2} - y)$
$y(0) = \frac{1}{6}$

31. $y' = 3y(10 - y)$
$y(0) = 20$

32. $y' = y(2 - y)$
$y(0) = 4$

33. $y' = 6y - 2y^2$
$y(0) = 1$

34. $y' = 3y - 6y^2$
$y(0) = \frac{1}{6}$

Hint for Exercises 33–34: Use factoring to write the differential equation in the form $y' = ay(M - y)$.

APPLIED EXERCISES

Write the differential equation (unlimited, limited, or logistic) that applies to the situation described. Then use its solution to solve the problem.

35. **General: Appreciation** The value of a stamp collection, initially worth $1500, grows continuously at the rate of 8% per year. Find a formula for its value after t years.

36. **General: Appreciation** The value of a home, originally worth $25,000, grows continuously at the rate of 6% per year. Find a formula for its value after t years.

37. **Business: Total Sales** A manufacturer estimates that he can sell a maximum of 100,000 digital tape recorders in a city. His total sales grow at a rate proportional to the distance below this upper limit. If after 5 months total sales are 10,000, find a formula for the total sales after t months. Then use your answer to estimate the total sales at the end of the first year.

38. **Business: Product Recognition** Let $p(t)$ be the number of people in a city who have heard of a new product after t weeks of advertising.

The city is of size 1,000,000, and $p(t)$ grows at a rate proportional to the number of people in the city who have *not* heard of the product. If after 8 weeks 250,000 people have heard of the product, find a formula for $p(t)$. Use your formula to estimate the number of people who will have heard of the product after 20 weeks of advertising.

39. General: Fund Raising   In a drive to raise $5000, fundraisers estimate that the rate of contributions is proportional to the distance from the goal. If $1000 was raised in 1 week, find a formula for the amount raised in t weeks. How many weeks will it take to raise $4000?

40. General: Learning   A person can memorize at most 40 two-digit numbers. If the person can memorize 15 numbers in the first 20 minutes, find a formula for the number that can be memorized in t minutes. Use your answer to estimate how long it will take to memorize 30 numbers.

41. Business: Sales   A telephone company estimates the maximum market for car phones in a city to be 10,000. Total sales are proportional to both the number already sold and the size of the remaining market. If 100 phones have been sold at time $t = 0$ and after 6 months 2000 have been sold, find a formula for the total sales after t months. Use your answer to estimate the total sales at the end of the first year.

42. Biomedical: Epidemics   During a flu epidemic in a city of 1,000,000, a flu vaccine sells in proportion to both the number of people already inoculated and the number not yet inoculated. If 100 doses have been sold at time $t = 0$ and after 4 weeks 2000 doses have been sold, find a formula for the total number of doses sold within t weeks. Use your formula to predict the sales after 10 weeks.

43. Sociology: Rumors   One person at an airport starts a rumor that a plane has been hijacked, and within 10 minutes 200 people have heard the rumor. If there are 800 people in the airport, find a formula for the number who have heard the rumor within t minutes. Use your answer to estimate how many will have heard the rumor within 15 minutes.

44. General: Epidemics   A flu epidemic on a college campus of 4000 students begins with 12 cases, and after 1 week has grown to 100 cases. Find a formula for the size of the epidemic after t weeks. Use your answer to estimate the size of the epidemic after 2 weeks.

45. Environmental Science: Deer Population   A wildlife refuge is initially stocked with 100 deer, and can hold at most 800 deer. If 2 years later the deer population is 160, find a formula for the deer population after t years. Use your answer to estimate when the deer population will reach 400.

46. Political Science: Voting   Suppose that a bill in the U.S. Senate gains votes in proportion to the number of votes that it already has and to the number of votes that it does not have. If it begins with one vote (from its sponsor) and after 3 days it has 30 votes, find a formula for the number of votes that it will have after t days. (*Note:* The number of votes in the Senate is 100.) When will the bill have "majority support" of 51 votes?

Solve each exercise by recognizing that the differential equation is one of the three types whose solution we know.

47. Biomedical: Drug Absorption   A drug injected into a vein is absorbed by the body at a rate proportional to the amount remaining in the blood. For a certain drug, the amount $y(t)$ remaining in the blood after t hours satisfies $y' = -0.15y$ with $y(0) = 5$. Find $y(t)$ and use your answer to estimate the amount present after 2 hours.

48. Business: Stock Value   One model for the growth of the value of stock in a corporation assumes that the stock has a limiting "market value" L, and that the value $v(t)$ of the stock on day t satisfies the differential equation $v' = a(L - v)$ for some constant a. Find a formula for the value $v(t)$ of a stock whose market value is $L = 40$ if on day $t = 10$ it was selling for $v = 30$.

49. General: Dam Sediment   A hydroelectric dam generates electricity by forcing water through turbines. Sediment accumulating behind the dam, however, will reduce the flow and eventually require dredging. Let $y(t)$ be the amount of sediment (in thousands of tons) accumulated in t years. If sediment flows in from the river at the constant rate of 20 thousand tons annually, but each year 10% of the accumulated sediment passes through the turbines, then the amount of sediment remaining satisfies the differential equation $y' = 20 - 0.1y$.

a. By factoring the right-hand side, write this differential equation in the form $y' = a(M - y)$. Note the value of M, the maximum amount of sediment that will accumulate.

b. Solve this (factored) differential equation together with the initial condition $y(0) = 0$ (no sediment until the dam was built).

c. Use your solution to find when the accumulated sediment will reach 95% of the value of M found in step (a). This is when dredging is required.

50. Biomedical: Glucose Levels Solve Exercise 56 on page 885 by factoring the right-hand side of the differential equation to write it in the form $y' = a(M - y)$.

OTHER GROWTH MODELS

Solve each differential equation by separation of variables.

51. General: Population If $y(t)$ is the size of a population at time t, then $\dfrac{y'}{y}$ is the population growth rate divided by the size of the population, and is called the *individual birthrate*. Suppose that the individual birthrate is proportional to the size of the population, $\dfrac{y'}{y} = ay$, for some constant a. Find a formula for the size of the population after t years.

52. General: Gompertz Curve Another differential equation that is used to model the growth

of a population $y(t)$ is $y' = bye^{-at}$, where a and b are constants. Solve this differential equation.

53. General: Allometry Solve the differential equation of allometric growth: $y' = \dfrac{ay}{x}$ (where a is a constant). This differential equation governs the relative growth rate of different parts of the same animal.

54. General: Population Suppose that a population $y(t)$ in a certain environment grows in proportion to the *square* of the difference between the carrying capacity M and the present population, $y' = a(M - y)^2$, where a is a constant. Solve this differential equation.

LOGISTIC GROWTH FUNCTION

55. Solve the logistic differential $y' = ay(M - y)$ as follows:

a. Separate variables to obtain

$$\frac{dy}{y(M - y)} = a\, dt$$

b. Integrate, using on the left-hand side the integration formula

$$\int \frac{dy}{y(M - y)} = \frac{1}{M} \ln\left(\frac{y}{M - y}\right)$$

(which may be checked by differentiation).

c. Exponentiate to solve for $\dfrac{y}{(M - y)}$ and then solve for y.

d. Show that the solution can be expressed as

$$y = \frac{M}{1 + ce^{-aMt}}$$

56. Find the inflection point of the logistic curve

$$f(x) = \frac{M}{1 + ce^{-aMx}}$$

and show that it occurs at midheight between $y = 0$ and the upper limit $y = M$.

57. General: Raindrops (*Requires Slope Field Program*) Why do larger-sized raindrops fall faster than smaller ones? It depends on the resistance they encounter as they fall through the air. For large raindrops, the resistance to gravity's acceleration is proportional to the *square* of the velocity, whereas for small droplets, the resistance is proportional to the *first power* of the velocity. More precisely, their velocities obey the following differential equations, with each differential equation leading to a different *terminal velocity* for the raindrop:

i. $\dfrac{dv}{dt} = 32.2 - 0.1115v^2$ Downpour droplets, about 0.05 inch in diameter

ii. $\dfrac{dv}{dt} = 32.2 - 52.6v$ Drizzle droplets, about 0.003 inch in diameter

iii. $\dfrac{dv}{dt} = 32.2 - 5260v$ Fog droplets, about 0.0003 inch in diameter

(The 32.2 represents the force of gravity, and the other constant is determined experimentally.*)

a. Use a slope field program to graph the slope field of differential equation (i) on the window [0, 3] by [0, 20] (using x and y instead of t and v). From the slope field, must the solution curves rising from the bottom level off at a particular y-value? Estimate the value. This number is the terminal velocity (in feet per second) for a downpour droplet.

b. Do the same for differential equation (ii), but on the window [0, 0.1] by [0, 1]. What is the terminal velocity for a drizzle droplet?

c. Do the same for differential equation (iii), but on the window [0, 0.001] by [0, 0.01]. What is the terminal velocity for a fog droplet?

d. At this speed [from part (c)], how long would it take a fog droplet to fall 1 foot? This shows why fog clears so slowly.

Chapter Summary with Hints and Suggestions

The reading and exercises of this chapter have helped you to learn the following skills. For each skill, the section from which it came (in case you need to review it) and some exercises in this section that use it are indicated. Answers to all exercises are at the end of the book, and full solutions to all exercises are in the Student Solutions Manual.

10.1   Integration By Parts

- Find an integral using integration by parts. (*Review Exercises 1–14.*)

$$\int u \, dv = uv - \int v \, du$$

- Find an integral by whatever technique is necessary.   (*Review Exercises 15–22.*)

- Solve an applied problem using integration by parts (and verify using).

(*Review Exercises 23–24.*)

$$\left(\begin{array}{c}\text{Present} \\ \text{value}\end{array}\right) = \int_0^T C(t)e^{-rt} \, dt$$

$$\left(\begin{array}{c}\text{Total} \\ \text{accumulation}\end{array}\right) = \int_0^T r(t) \, dt$$

* R. Gunn and Gilbert D. Kinzer, "The Terminal Velocity of Fall for Water Droplets in Stagnant Air," *Journal of Meteorology* **6:**243, 1949.

10.2   Integration Using Tables

- Find an integral using a table of integrals. (*Review Exercises 25–36.*)

- Solve an applied problem using an integral table (and verify using 📊).

  (*Review Exercises 37–38.*)

10.3   Improper Integrals

- Evaluate an improper integral (if it is convergent).   (*Review Exercises 39–56.*)

$$\int_a^\infty f(x)\, dx = \lim_{b \to \infty} \int_a^b f(x)\, dx$$

- Solve an applied problem involving an improper integral.   (*Review Exercises 57–60.*)

- Use 📊 to predict whether an improper integral converges, and then check by evaluating it.   (*Review Exercises 61–62.*)

10.4   Numerical Integration

- Approximate an integral using trapezoidal approximation "by hand." (*Review Exercises 63–68.*)

$$\int_a^b f(x)\, dx \approx \left[\frac{1}{2} f(x_1) + f(x_2) + \right.$$
$$\left. \cdots + \frac{1}{2} f(x_{n+1}) \right] \cdot \Delta x$$

- Approximate an integral by trapezoidal approximation using 📊 .

  (*Review Exercises 69–74.*)

- Approximate an integral using Simpson's rule "by hand."   (*Review Exercises 75–80.*)

$$\int_a^b f(x)\, dx \approx [f(x_1) + 4f(x_2) + 2f(x_3) +$$
$$\cdots + 4f(x_n) + f(x_{n+1})] \cdot \frac{\Delta x}{3}$$

- Approximate an integral by Simpson's rule using 📊 .

  (*Review Exercises 81–86.*)

10.3 and 10.4   Numerical Integration of Improper Integrals

- Evaluate an improper integral using trapezoidal approximation.   (*Review Exercises 87–88.*)

10.5   Differential Equations

- Find the general solution of a differential equation.   (*Review Exercises 89–98.*)

- Find the particular solution of a differential equation with an initial condition. (*Review Exercises 99–102.*)

- Solve an applied problem involving a differential equation.   (*Review Exercises 103–106.*)

- Use 📊 to graph the slope field of a differential equation, and sketch the solution through a given point.   (*Review Exercises 107–108.*)

10.6   Further Applications of Differential Equations: Three Models of Growth

- Choose an appropriate differential equation for an applied problem, and use it to solve the problem.   (*Review Exercises 109–115.*)

Hints and Suggestions

- The unifying idea of this chapter is extensions of the concept of integration.

- Integration by parts takes one integral and gives another integral which, it is hoped, is simpler than the original integral. The formula is simply an integration version of the product rule. When using it, try choosing *dv* (including the *dx*) to be the most complicated part of the integrand that you can integrate, and, if possible, the *u* to be something that simplifies when differentiated.

- There are tables of integrals that are much longer than the one inside the back cover of this book. Longer tables, however, require much more time spent searching for the "right" formula. Other techniques (such as a substitution, integration by parts, or using a formula more than once) may be used with an integral table.

- To find an integral, try the following methods: First try the "basic" formulas 1 through 4 on the inside back cover of this book. Then try a substitution (formulas 5 through 7). If these fail, try integration by parts or an integral table. Remember that some integrals *cannot* be integrated (in terms of elementary functions). A *definite* integral can always be approximated by numerical methods.

- Before "evaluating" an improper integral, be sure that the integrand is defined over the interval, and that the integral is convergent. If the integral diverges, then it has no value and we simply state that it is divergent.

- Numerical integration involves approximating the area under a curve by geometric figures such as trapezoids or parabolas (Simpson's rule). In practice, the calculations are usually carried out on a calculator or computer, but doing some "by hand" helps to make the method clear.

- A graphing calculator is very helpful for approximating definite integrals by trapezoidal approximation or Simpson's rule for large values of n. Graphing calculators also have their own built-in numerical procedures for approximating integrals when you use FnInt.

- A differential equation is an equation involving derivatives (rates of change). Separation of variables involves separating the x's and y's to opposite sides of the equation and integrating both sides. Many useful differential equations can be solved by separation of variables, but many cannot. In fact, many differential equations cannot be solved by *any* method.

- For a differential equation, a solution involving an arbitrary constant is called a *general solution,* and a solution with the arbitrary constant replaced by a number is called a *particular solution.* The constant is usually evaluated by an *initial condition,* specifying the value of the solution at a particular point.

- A graphing calculator with a slope field program can show a "picture" of a differential equation of the form $dy/dx = f(x, y)$, drawing little slanted dashes with the correct slope at many points of the plane. A solution can then be drawn through a given point by following the indicated slopes.

- Practice for test: Review Exercises 3, 5, 13, 25, 27, 33, 39, 41, 57, 63, 71, 79, 85, 89, 101, 103, 107, 113, 115.

Review Exercises for Chapter 10   *Practice test exercises are in blue.*

10.1   Integration by Parts

Find each integral using integration by parts.

1. $\displaystyle\int xe^{2x}\, dx$

2. $\displaystyle\int xe^{-x}\, dx$

3. $\displaystyle\int x^8 \ln x\, dx$

4. $\displaystyle\int \sqrt[4]{x} \ln x\, dx$

5. $\displaystyle\int (x - 2)(x + 1)^5\, dx$

6. $\displaystyle\int (x + 3)(x - 1)^4\, dx$

7. $\displaystyle\int \frac{\ln t}{\sqrt{t}}\, dt$

8. $\displaystyle\int x^7 e^{x^4}\, dx$

9. $\displaystyle\int x^2 e^x\, dx$

10. $\displaystyle\int (\ln x)^2\, dx$

11. $\displaystyle\int x(x + a)^n\, dx$ (for constants a and $n > 0$)

12. $\displaystyle\int x(1 - x)^n\, dx$ (for constant $n > 0$)

13. $\displaystyle\int_0^5 xe^x\, dx$

14. $\displaystyle\int_1^e x \ln x\, dx$

Find each integral by whatever means are necessary.

15. $\displaystyle\int \frac{dx}{1-x}\,dx$

16. $\displaystyle\int xe^{-x^2}\,dx$

17. $\displaystyle\int x^3 \ln 2x\,dx$

18. $\displaystyle\int \frac{dx}{(1-x)^2}$

19. $\displaystyle\int \frac{\ln x}{x}\,dx$

20. $\displaystyle\int \frac{e^{2x}}{e^{2x}+1}\,dx$

21. $\displaystyle\int \frac{e^{\sqrt{x}}}{\sqrt{x}}\,dx$

22. $\displaystyle\int (e^{2x}+1)^3 e^{2x}\,dx$

23. Business: Present Value of a Continuous Stream of Income   A company generates a continuous stream of income of $25t$ million dollars per year, where t is the number of years that the company has been in operation.

a. Find the present value of this stream for the first 10 years at 5% interest compounded continuously. (Do not use a graphing calculator.)

b. Verify your answer to part (a) using FnInt on a graphing calculator.

24. Environmental Science: Pollution   Radioactive waste is leaking out of cement storage vessels at the rate of $te^{0.2t}$ hundred gallons per month, where t is the number of months since the leak began.

a. Find the total leakage during the first 3 months. (Do not use a graphing calculator.)

b. Verify your answer to part (a) using FnInt on a graphing calculator.

10.2   Integration Using Tables

Use the integral table inside the back cover to find each integral.

25. $\displaystyle\int \frac{1}{25-x^2}\,dx$

26. $\displaystyle\int \frac{1}{x^2-4}\,dx$

27. $\displaystyle\int \frac{x}{(x-1)(x-2)}\,dx$

28. $\displaystyle\int \frac{1}{(x-1)(x-2)}\,dx$

29. $\displaystyle\int \frac{1}{x\sqrt{x+1}}\,dx$

30. $\displaystyle\int \frac{x}{\sqrt{x+1}}\,dx$

31. $\displaystyle\int \frac{1}{\sqrt{x^2+9}}\,dx$

32. $\displaystyle\int \frac{1}{\sqrt{x^2+16}}\,dx$

33. $\displaystyle\int \frac{z^3}{\sqrt{z^2+1}}\,dz$

34. $\displaystyle\int \frac{e^{2t}}{e^t+2}\,dt$

35. $\displaystyle\int x^2 e^{2x}\,dx$

36. $\displaystyle\int (\ln x)^4\,dx$

37. Business: Cost   A company's marginal cost function is $MC(x) = \dfrac{1}{(2x+1)(x+1)}$ and fixed costs are 1000. Find the company's cost function.

38. General: Population   The population of a town is growing at the rate of $\sqrt{t^2+1600}$ people per year, where t is the number of years from now.

a. Find the total increase in population during the first 30 years. (Do not use a graphing calculator.)

b. Verify your answer to part (a) using FnInt on a graphing calculator.

10.3   Improper Integrals

Find the value of each improper integral or state that it is divergent.

39. $\displaystyle\int_1^\infty \frac{1}{x^5}\,dx$

40. $\displaystyle\int_1^\infty \frac{1}{x^6}\,dx$

41. $\displaystyle\int_1^\infty \frac{1}{\sqrt[5]{x}}\,dx$

42. $\displaystyle\int_1^\infty \frac{1}{\sqrt[6]{x}}\,dx$

43. $\displaystyle\int_0^\infty e^{-2x}\,dx$

44. $\displaystyle\int_4^\infty e^{-0.5x}\,dx$

45. $\displaystyle\int_0^\infty e^{2x}\,dx$

46. $\displaystyle\int_4^\infty e^{0.5x}\,dx$

47. $\displaystyle\int_0^\infty e^{-t/5}\,dt$

48. $\displaystyle\int_{100}^\infty e^{-t/10}\,dt$

49. $\displaystyle\int_0^\infty \frac{x^3}{(x^4+1)^2}\,dx$

50. $\displaystyle\int_0^\infty \frac{x^4}{(x^5+1)^2}\,dx$

51. $\displaystyle\int_{-\infty}^0 e^{2t}\,dt$

52. $\displaystyle\int_{-\infty}^0 e^{4t}\,dt$

53. $\displaystyle\int_{-\infty}^4 \frac{1}{(5-x)^2}\,dx$

54. $\displaystyle\int_{-\infty}^8 \frac{1}{(9-x)^2}\,dx$

55. $\displaystyle\int_{-\infty}^\infty \frac{e^{-x}}{(1+e^{-x})^4}\,dx$

56. $\displaystyle\int_{-\infty}^\infty \frac{e^{-x}}{(1+e^{-x})^3}\,dx$

57. General: Permanent Endowments Find the size of the permanent endowment needed to generate an annual $6000 forever at an interest rate of 10% compounded continuously.

58. General: Automobile Age Insurance records indicate that the proportion of cars on the road that are more than x years old is approximated by the integral $\int_x^\infty 0.21e^{-0.21t}\, dt$. Find the proportion of cars that are more than 5 years old.

59. Business: Book Sales A publisher estimates that the demand for a certain book will be $12e^{-0.05t}$ thousand copies per year, where t is the number of years since its publication. Find the total number of books that will be sold from its publication onward.

60. General: Resource Consumption If the rate of consumption of a certain mineral is $300e^{-0.04t}$ million tons per year (where t is the number of years from now), find the total amount of the mineral that will be consumed from now on.

 For each improper integral below, use a graphing calculator to evaluate it (or to show that it diverges) as follows:

a. Define y_1 to be the definite integral (using FnInt) of the integrand from 1 to x.
b. Make a TABLE of values of y_1 for x-values like 1, 10, 100, and 1000. Does the integral converge (and if so, to what number) or does it diverge?
c. Verify your answers to (b) by evaluating the improper integral "by hand."

61. $\int_1^\infty \dfrac{1}{x^3}\, dx$

62. $\int_1^\infty \dfrac{1}{\sqrt[3]{x}}\, dx$

10.4 Numerical Integration

 Estimate each integral using trapezoidal approximation with the given value of n. (Round all calculations to three decimal places.)

63. $\int_0^1 \sqrt{1 + x^4}\, dx,$ $n = 3$

64. $\int_0^1 \sqrt{1 + x^5}\, dx,$ $n = 3$

65. $\int_0^1 e^{\frac{1}{2}x^2}\, dx,$ $n = 4$

66. $\int_0^1 e^{-\frac{1}{2}x^2}\, dx,$ $n = 4$

67. $\int_{-1}^1 \ln (1 + x^2)\, dx,$ $n = 4$

68. $\int_{-1}^1 \ln (x^3 + 2)\, dx,$ $n = 4$

Use a trapezoidal approximation program to approximate each integral. Use successively higher values of n until the results agree to three decimal places (rounded).

69. $\int_0^1 \sqrt{1 + x^4}\, dx$

70. $\int_0^1 \sqrt{1 + x^5}\, dx$

71. $\int_0^1 e^{\frac{1}{2}x^2}\, dx$

72. $\int_0^1 e^{-\frac{1}{2}x^2}\, dx$

73. $\int_{-1}^1 \ln (1 + x^2)\, dx$

74. $\int_{-1}^1 \ln (x^3 + 2)\, dx$

Estimate each integral using Simpson's rule (parabolic approximation) with the given value of n. (Round all calculations to four decimal places.)

75. $\int_0^1 \sqrt{1 + x^4}\, dx,$ $n = 4$

76. $\int_0^1 \sqrt{1 + x^5}\, dx,$ $n = 4$

77. $\int_0^1 e^{\frac{1}{2}x^2}\, dx,$ $n = 4$

78. $\int_0^1 e^{-\frac{1}{2}x^2}\, dx,$ $n = 4$

79. $\int_{-1}^1 \ln (1 + x^2)\, dx,$ $n = 4$

80. $\int_{-1}^1 \ln (x^3 + 2)\, dx,$ $n = 4$

Use a Simpson's rule approximation program to approximate each integral. Use successively higher values of n until the results agree to six decimal places (rounded).

81. $\displaystyle\int_0^1 \sqrt{1 + x^4}\, dx$ **82.** $\displaystyle\int_0^1 \sqrt{1 + x^5}\, dx$

83. $\displaystyle\int_0^1 e^{\frac{1}{2}x^2}\, dx$ **84.** $\displaystyle\int_0^1 e^{-\frac{1}{2}x^2}\, dx$

85. $\displaystyle\int_{-1}^1 \ln(1 + x^2)\, dx$ **86.** $\displaystyle\int_{-1}^1 \ln(x^3 + 2)\, dx$

10.3 and 10.4   *Numerical Integration of Improper Integrals*

 For each improper integral:

a. Make it a "proper" integral by using the substitution $x = \dfrac{1}{t}$ and simplifying.

b. Approximate the proper integral using trapezoidal approximation with $n = 4$. Keep three decimal places.

87. $\displaystyle\int_1^\infty \frac{1}{x^2 + 1}\, dx$ **88.** $\displaystyle\int_1^\infty \frac{x^2}{x^4 + 1}\, dx$

10.5   *Differential Equations*

Find the general solution for each differential equation.

89. $y^2 y' = x^2$ **90.** $y' = x^2 y$ **91.** $y' = \dfrac{x^3}{x^4 + 1}$

92. $y' = x e^{-x^2}$ **93.** $y' = y^2$ **94.** $y' = y^3$

95. $y' = 1 - y$ **96.** $y' = \dfrac{1}{y}$

97. $y' = xy - y$ **98.** $y' = x^2 + x^2 y$

Solve each differential equation and initial condition.

99. $y^2 y' = 3x^2$ **100.** $y' = \dfrac{y}{x^2}$
     $y(0) = 1$ $\quad\quad y(1) = 1$

101. $y' = \dfrac{y}{x^3}$ **102.** $y' = \sqrt[3]{y}$
      $y(1) = 1$ $\quad\quad y(1) = 0$

103. General: Accumulation of Wealth   Suppose that you now have \$10,000 and that you expect to save an additional \$4000 during each year, and all of this is deposited in a bank paying 5% interest compounded continuously. Let $y(t)$ be your bank balance (in thousands of dollars) after t years.

a. Write a differential equation and initial condition to model your bank balance.

b. Solve your differential equation and initial condition.

c. Use your solution to find your bank balance after 10 years.

104. Environmental Science: Pollution   A town discharges 4 tons of pollutant into a lake annually, and each year bacterial action removes 25% of the accumulated pollution.

a. Write a differential equation and initial condition for the amount of pollution in the lake.

b. Solve your differential equation to find a formula for the amount of pollution in the lake after t years.

105. General: Fever Thermometers   How long should you keep a thermometer in your mouth to take your temperature? *Newton's law of cooling* says that the thermometer reading rises at a rate proportional to the difference between your actual temperature and the present reading. For a fever of 106°, the thermometer reading $y(t)$ after t minutes in your mouth satisfies $y' = 2.3(106 - y)$ with $y(0) = 70$ (initially at room temperature). (The constant 2.3 is typical for household thermometers.) Solve this differential equation and initial condition.

 106. Continuation of Exercise 105   Use your solution $y(t)$ to Exercise 105 to calculate $y(1), y(2),$ and $y(3)$, the thermometer readings after 1, 2, and 3 minutes. Do you see why 3 minutes is usually the recommended time for keeping the thermometer in your mouth?

 107–108: (*Requires Slope Field Program*) For each differential equation below, display the slope field on the window $[-5, 5]$ by $[-5, 5]$ (for most programs, enter the right-hand side of the differential equation in y_1). From the resulting slope field, make a rough sketch of the solution curve passing through the indicated point. (You may want to alter the window to see more detail.)

107. $\dfrac{dy}{dx} = \dfrac{x^2}{y^2}$

point: $(0, 1)$

108. $\dfrac{dy}{dx} = xy^2$

point: $(0, -2)$

10.6 Further Applications of Differential Equations: Three Models of Growth

For each situation, write an appropriate differential equation (unlimited, limited, or logistic). Then find its solution and solve the problem. (All require [calculator] .)

109. General: Postage Stamps The price of a first-class postage stamp grows continuously at the rate of 7% each year (on the average). If in 1999 the price was 33 cents, estimate the price in the year 2005.

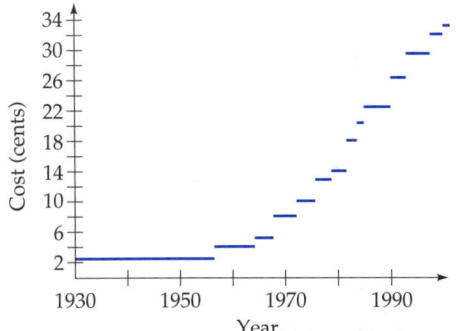

110. Economics: Computer Expenditures The amount spent by Americans on computers and related hardware is increasing continuously at the rate of 22% per year. If in 1995 the amount was $10.9 billion, estimate the amount in the year 2005. (*Source: World Almanac.*)

111. Biomedical: Epidemics A virus spreads through a university community of 8000 people at a rate proportional to both the number already infected and the number not yet infected. If it begins with 10 cases and grows in a week to 150 cases, estimate the size of the epidemic after 2 weeks.

112. Social Science: Rumors A rumor spreads through a school of 500 students at a rate proportional to both those who know and those who do not know the rumor. If the rumor began with 2 students and within a day had spread to 75, how many students will have heard the rumor within 2 days?

113. Business: Total Sales A manufacturer estimates that he can sell a maximum of 10,000 videocassette recorders in a city. His total sales grow at a rate proportional to how far they are below this upper limit. If after 7 months the total sales are 3000, find a formula for the total sales after t months. Then use your answer to estimate the total sales at the end of the first year.

114. General: Learning Suppose that the maximum rate at which a mail carrier can sort letters is 60 letters per minute, and that she learns at a rate proportional to her distance from this upper limit. If after 2 weeks on the route she can sort 25 letters per minute, how many weeks will it take her to sort 50 letters per minute?

115. Business: Advertising A new product is advertised extensively on television to a city of 500,000 people, and the number of people who have seen the ads increases at a rate proportional to the number who have not yet seen the ads. If within 2 weeks 200,000 have seen the ads, how long must the product be advertised to reach 400,000 people?

Projects and Essays

The following projects and essays are based on Chapter 10. Most have no right and wrong answers—the results depend only on your imagination and resourcefulness.

1. The integration by parts formula is based on the product rule. Try to invent an integration formula by integrating the *quotient* rule. (*Hint:* Study the derivation of the integration by parts formula on page 824 and try to use the same ideas with the quotient rule.) Then try to make up integrals that can be integrated by your formula but that are otherwise difficult.

2. Look in a college or scientific library for a longer integral table (some are book length). How many integration formulas are there in the longest table that you can find? (You may have to estimate.) Use an integral table longer than the one in this book to solve as many of the following integrals as you can:

$$\int \frac{dx}{1 + e^x} \qquad \int \frac{dx}{\sqrt{e^x + 1}}$$

$$\int \frac{\sqrt{x}}{1 - \sqrt{x}} \, dx \qquad \int (1 + x)^2 \ln x \, dx$$

3. Look up "improper integrals" in a more theoretical calculus book. If a function f takes both positive and negative values, what extra restriction must be put on f so that $\int_{-\infty}^{\infty} f(x) \, dx$ is defined? To see why this restriction is necessary, find (if possible)

$$\lim_{b \to \infty} \int_{-b}^{b} \frac{2x \, dx}{x^2 + 1} \qquad \text{and also}$$

$$\lim_{a \to \infty} \int_{a}^{0} \frac{2x \, dx}{x^2 + 1} + \lim_{b \to \infty} \int_{0}^{b} \frac{2x \, dx}{x^2 + 1}$$

4. The following *modified trapezoidal rule* is often more accurate than Simpson's rule:

$$\int_{a}^{b} f(x) \, dx \approx T - \frac{[f'(b) - f'(a)](\Delta x)^2}{12}$$

where T is the usual trapezoidal approximation. Use this formula with $n = 8$ to approximate $\int_{1}^{3} x^4 \, dx$ and compare the approximation to the exact value of the integral (which you should find).

5. Read the Application Preview on pages 870–871. Make up some reasonable numbers for s and n that might apply to you, and substitute them together with $r = 0.06$ and $p = 0.75$ into the formula for $W(t)$. Then use $W(t)$ with $t = 20$ to predict your wealth in 20 years. Change p to 0.70 and see how your wealth would change. Explore the model by making other reasonable changes (such as "giving yourself a raise" by increasing s or "buying a house" by increasing n).

6. Imagine an animal population (deer in a forest, rats in a garbage dump, etc.) and make up some appropriate numbers to model your population by the logistic model (see pages 894–897, especially Example 4). Find your population function and graph it. Explore the model by changing the numbers (for example, suppose that the initial population is halved, or the carrying capacity is doubled, or both). How does the population (at some fixed later time) change?

7. Look over your notes, your homework, and the text, and write a page about how a graphing calculator has helped you in this chapter. Include examples of how it has helped you to *explore concepts* and how it has helped to *simplify your work*. What was its *most* helpful or interesting use? What was its *least* helpful or interesting use? Are there problems that can be done on a graphing calculator but that are easier to do "by hand"?

CALCULUS OF SEVERAL VARIABLES

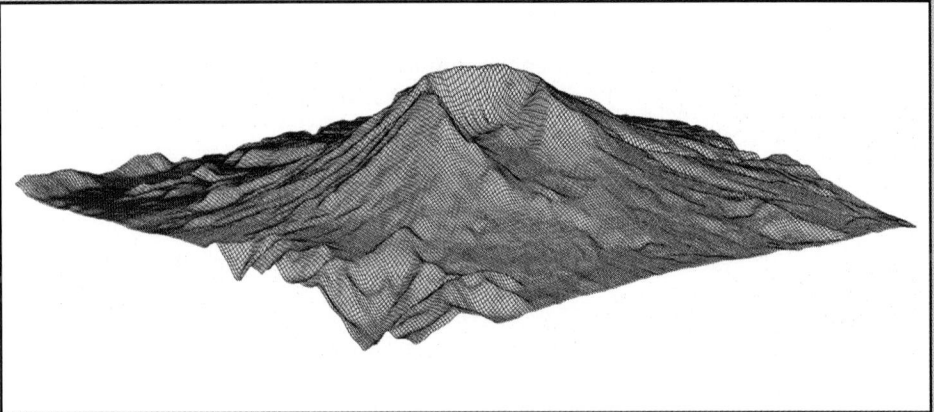

Mount Saint Helens in Washington, shortly after its May 1980 eruption, which blew away a third of the mountain, as may be seen in the topographical map above. The elevation h at any point depends on two variables, latitude and longitude, shown as grid lines, and may be written as a function of two variables, h(x, y). Such elevation functions make it possible to discern changes and predict future eruptions.

911

11.1 Functions of Several Variables

Introduction

Many quantities depend on *several* variables. For example, the "wind-chill" factor announced by the weather bureau during the winter depends on two variables: temperature and wind speed. The cost of a telephone call depends on *three* variables: distance, duration, and time of day.

In this chapter we define functions of two or more variables and learn how to differentiate and integrate them. We use derivatives for calculating rates of change and optimizing functions, and integrals for finding volumes, continuous sums, and average values.

Graphing calculators will be less useful in this chapter because of their limited screens and computing power, but computer-drawn pictures of three-dimensional surfaces will be very useful.

Functions of Two Variables

A function f that depends on *two* variables, x and y, is written $f(x, y)$ (read: f of x and y). The *domain* of $f(x, y)$ is the set of all ordered pairs (x, y) for which the function is defined. The *range* is the set of all resulting values of the function. Formally:

Function of Two Variables

> A function f of two variables is a rule such that to each ordered pair (x, y) in the domain of f there corresponds one and only one number $f(x, y)$.

If the domain is not stated, it will always be taken to be the largest set of ordered pairs for which the function is defined (the "natural domain").

EXAMPLE 1   Finding the Domain of a Function $f(x, y)$

For $f(x, y) = \dfrac{\sqrt{x}}{y^2}$, find   **a.** the domain   **b.** $f(9, -1)$

Solution

a. $\{(x, y) \mid x \geq 0,\ y \neq 0\}$ In $\sqrt{x}/y^2$, x cannot be negative (because of the $\sqrt{}$), and y cannot be zero

b. $f(9, -1) = \dfrac{\sqrt{9}}{(-1)^2} = \dfrac{3}{1} = 3$ $f(x, y) = \sqrt{x}/y^2$ with $x = 9$ and $y = -1$

■

EXAMPLE 2 Finding the Domain of a Function Involving Logarithms and Exponentials

For $g(u, v) = e^{uv} - \ln u$, find   **a.** the domain   **b.** $g(1, 2)$

Solution

a. $\{(u, v) \mid u > 0\}$ u must be positive so that its logarithm is defined

b. $g(1, 2) = e^{1 \cdot 2} - \underbrace{\ln 1}_{0} = e^2 - 0 = e^2$ $g(u, v) = e^{uv} - \ln u$ with $u = 1$ and $v = 2$

■

PRACTICE PROBLEM 1   For $f(x, y) = \dfrac{\ln x}{e^{\sqrt{y}}}$, find   **a.** the domain   **b.** $f(e, 4)$

Solutions at the back of the book

Functions of two variables are used in many applications.

EXAMPLE 3 Finding a Company's Cost Function

A company manufactures three-speed and ten-speed bicycles. It costs $100 to make each three-speed bicycle, it costs $150 to make each ten-speed bicycle, and fixed costs are $2500. Find the cost function and use it to find the cost of producing 15 three-speed bicycles and 20 ten-speed bicycles.

Solution

Let: x = the number of three-speed bicycles

y = the number of ten-speed bicycles

The cost function is

$$C(x, y) = 100x \quad + \quad 150y \quad + \quad 2500$$

Unit Quan- Unit Quan- Fixed
cost tity   cost tity   costs

The cost of producing 15 three-speed bicycles and 20 ten-speed bicycles is found by evaluating $C(x, y)$ at $x = 15$ and $y = 20$:

$$C(15, 20) = 100 \cdot 15 + 150 \cdot 20 + 2500$$
$$= 1500 + 3000 + 2500 = 7000$$

Producing 15 three-speed and 20 ten-speed bicycles costs $7000.

■

The variables x and y in the preceding example stand for numbers of bicycles, and so should take only integer values. Instead, however, we will allow x and y to be "continuous" variables, and round to integers at the end if necessary.

Some other "everyday" examples of functions of two variables are as follows:

$A(l, w) = lw$ Area of a rectangle of length l and width w

$f(w, v) = kwv^2$ Length of the skid marks for a car of weight w and velocity v skidding to a stop (k is a constant depending on the road surface)

Cobb–Douglas Production Functions

A function used to model the output of a company or a nation is called a *production function*, and the most famous is the Cobb–Douglas production function*

$$P(L, K) = aL^bK^{1-b}$$ For constants $a > 0$ and $0 < b < 1$

This function expresses the total production P as a function of L, the number of units of labor, and K, the number of units of capital. (Labor is measured in work-hours, and capital means *invested* capital, including the cost of buildings, equipment, and raw materials.)

EXAMPLE 4   Evaluating a Cobb–Douglas Production Function

Cobb and Douglas modeled the output of the American economy by the function $P(L, K) = L^{0.75}K^{0.25}$. Find $P(150, 220)$.

Solution

$$P(150, 220) = (150)^{0.75}(220)^{0.25}$$ $P(L, K) = L^{0.75}K^{0.25}$ with $L = 150$ and $K = 220$

$$\approx (42.9)(3.85) \approx 165$$ Using a calculator

That is, 150 units of labor and 220 units of capital should result in approximately 165 units of production.

---

* First used by Charles Cobb and Paul Douglas in a landmark study of the American economy published in 1928.

Graphing Calculator Exploration

The windchill index, announced by the weather bureau during the winter to measure the combined effect of wind and cold, is calculated from the formula below, where x is wind speed (mph) and y is temperature (degrees Fahrenheit).

$$W(x, y) = 0.0817(5.81 + 3.71\sqrt{x} - 0.25x)(y - 91.4) + 91.4$$
$$\text{for } 4 \le x \le 45$$

a. Enter this function into your graphing calculator, but with y (temperature) replaced by 30 so that it becomes a function of one variable, x (wind speed).

b. Graph the function on the window [4, 45] by [−10, 40]. Your graph shows how the perceived temperature decreases as wind speed increases.

c. Find the wind speed that makes it feel like 0 degrees (that is, find x where $y = 0$).

d. Notice that the graph drops more steeply for low wind speeds than for high wind speeds. What does this mean about the effect of an extra 5 mph of wind on a calm day as opposed to a windy day? (Exercises 36 and 37 continue this analysis.)

Functions of Three or More Variables

Functions of three (or more) variables are defined analogously. Some examples are:

$$V(l, w, h) = lwh$$

Volume of a rectangular solid of length l, width w, and height h

$$V(P, r, t) = Pe^{rt}$$

Value of P dollars invested at a continuous interest rate r for t years

$$f(w, x, y, z) = \frac{w + x + y + z}{4}$$

Average of four numbers

EXAMPLE 5 **Finding the Domain of a Function $f(x, y, z)$**

For $f(x, y, z) = \dfrac{\sqrt{x}}{y} + \ln\dfrac{1}{z}$, find   **a.** the domain   **b.** $f(4, -1, 1)$

Solution

a. In $f(x, y, z) = \frac{\sqrt{x}}{y} + \ln \frac{1}{z}$ we must have $x \geq 0$ (because of the square root), $y \neq 0$ (since it is the denominator), and $z > 0$ (so that $\frac{1}{z}$ has a logarithm). Therefore, the domain is

$$\{(x, y, z) \mid x \geq 0, y \neq 0, z > 0\}$$

b. $f(4, -1, 1) = \frac{\sqrt{4}}{-1} + \ln \frac{1}{1} = \underbrace{\frac{2}{-1} + \ln 1}_{0} = -2$

EXAMPLE 6   Finding the Volume and Area of a Divided Box

An open-top box is to have a center divider, as shown in the diagram. Find formulas for the volume V of the box and for the total amount of materials M needed to construct the box.

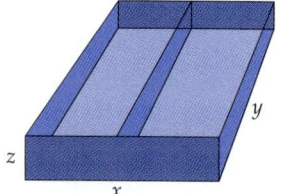

Solution   The volume is length times width times height.

$$V = xyz$$

bottom area xy

side area yz (3 of these)

front area xz (2 of these)

The box consists of a bottom, a front and back, two sides, and a divider, whose areas are shown in the diagram. Therefore, the total amount of materials (the area) is

$$M = xy + 2xz + 3yz$$

Bottom   Back and front   Sides and divider

PRACTICE PROBLEM 2

Find a formula for the total amount of materials M needed to construct an open-top box with three parallel dividers. Use the variables shown in the diagram.

Solution at the back of the book

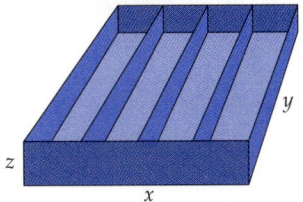

Graph of a Function of Two Variables

Graphing a function $f(x, y)$ requires a three-dimensional coordinate system. We draw three perpendicular axes as shown on the left.* We will usually draw only the positive half of each axis, although each axis extends infinitely far in the negative direction as well. The plane at the base is called the *x-y plane*.

A point in a three-dimensional coordinate system is specified by three coordinates, giving its distances from the origin in the x, y, and z directions. For example, the point

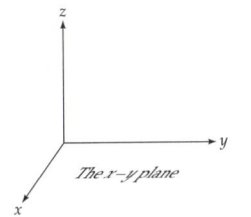

The three-dimensional ("right-handed") coordinate system

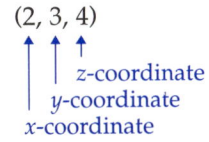

is plotted by starting at the origin, moving 2 units in the x direction, 3 units in the y direction, and then 4 units in the (vertical) z direction.

To graph a function $f(x, y)$, we choose values for x and y, calculate z-values from $z = f(x, y)$, and plot the points (x, y, z).

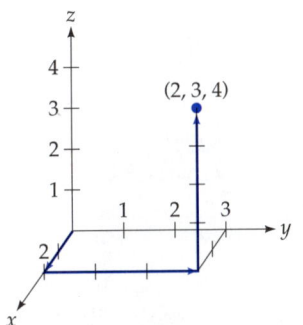

The point $(2, 3, 4)$

EXAMPLE 7 Graphing a Function of Two Variables

To graph $f(x, y) = 18 - x^2 - y^2$, we set z equal to the function.

$$z = 18 - x^2 - y^2 \qquad \text{z replaces } f(x, y)$$

Then we choose values for x and y. Choosing $x = 1$ and $y = 2$ gives

$$z = 18 - 1^2 - 2^2 = 13 \qquad \begin{array}{l} z = 18 - x^2 - y^2 \\ \text{with } x = 1 \text{ and } y = 2 \end{array}$$

*This is called a "right-handed" coordinate system because the x, y, and z axes correspond to the first two fingers and thumb of the right hand.

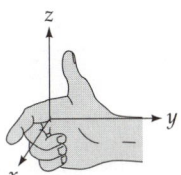

for the point

$$(1, 2, 13)$$

The chosen $x = 1$, $y = 2$, and the calculated z

Choosing $x = 2$ and $y = 3$ gives

$$z = 18 - 2^2 - 3^2 = 5$$

$z = 18 - x^2 - y^2$ with $x = 2$ and $y = 3$

for the point

$$(2, 3, 5)$$

The chosen $x = 2$, $y = 3$, and the calculated z

These points $(1, 2, 13)$ and $(2, 3, 5)$ are plotted on the graph on the left below. The completed graph of the function is shown on the right below.

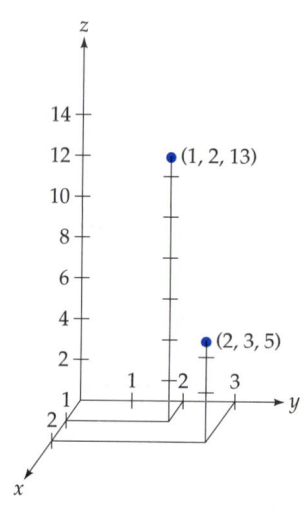

The points $(1, 2, 13)$ and $(2, 3, 5)$ of the function $f(x, y) = 18 - x^2 - y^2$

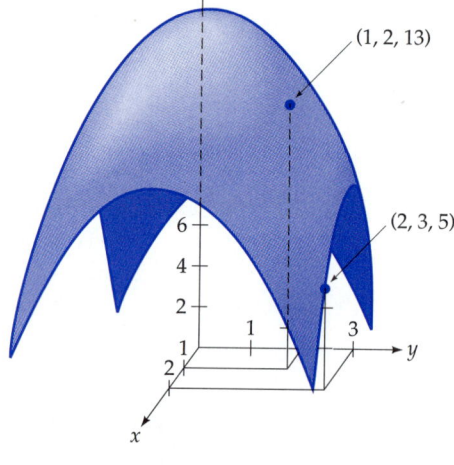

The graph of $f(x, y) = 18 - x^2 - y^2$

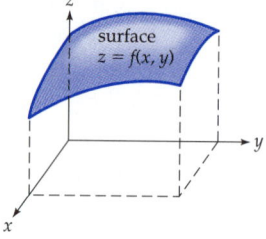

The graph of $f(x, y)$ is a surface whose height above the point (x, y) in the x-y plane is given by $z = f(x, y)$.

In general, the graph of a function $f(x, y)$ of *two* variables is a *surface* above or below the x-y plane, just as the graph of a function $f(x)$ of *one* variable is a *curve* above or below the x-axis.

Graphing functions of two variables involves drawing three-dimensional graphs, which is very difficult. Graphing functions of *more* than two variables requires *more* than three dimensions, and is impossible. For this reason we will not graph functions of several variables.

We will, however, often speak of a function $f(x, y)$ as a *surface* in three-dimensional space.

Relative Extreme Points and Saddle Points

Certain points on a surface $z = f(x, y)$ are of special importance.

Relative Maximum Point

A point (a, b, c) on a surface $z = f(x, y)$ is a *relative maximum point* if $f(a, b) \geq f(x, y)$ for all (x, y) in some region surrounding (a, b).

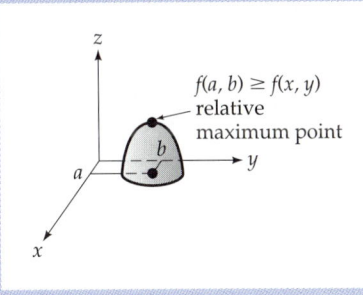

Relative Minimum Point

A point (a, b, c) on a surface $z = f(x, y)$ is a *relative minimum point* if $f(a, b) \leq f(x, y)$ for all (x, y) in some region surrounding (a, b).

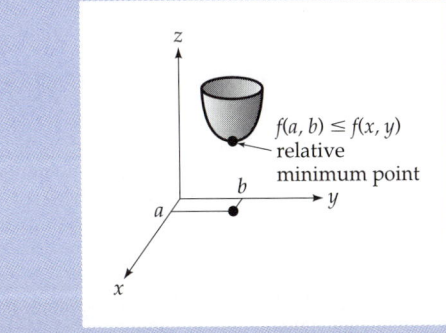

As before, the term *relative extreme point* means a point that is either a relative maximum or a relative minimum point. A surface may have any number of relative extreme points, even none.

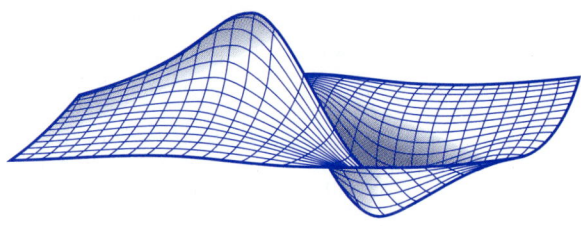

A surface with two relative extreme points: one relative
maximum and one relative minimum.

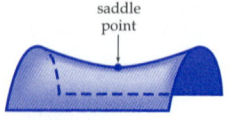

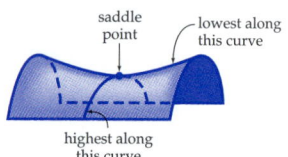

The point shown on the left is called a *saddle point* (so called because
the diagram resembles a saddle).

A saddle point is a point that is the highest point along one curve
of the surface and the lowest point along another curve. A saddle point
is *not* a relative extreme point.

If we think of a surface $z = f(x, y)$ as a landscape, then relative
maximum and minimum points correspond to "hilltops" and "valley
bottoms," and a saddle point corresponds to a "mountain pass" between two peaks.

Gallery of Surfaces

The following are the graphs of a few functions.

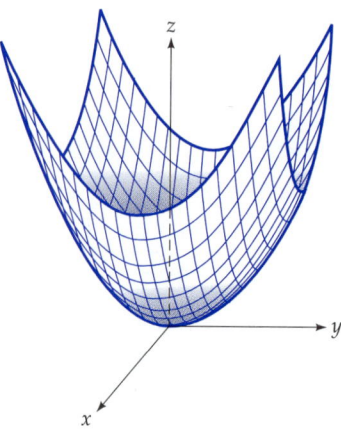

The surface $f(x, y) = x^2 + y^2$ has a
relative minimum point at the origin.

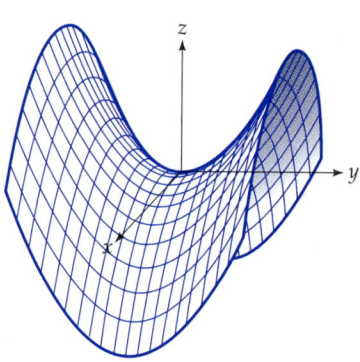

The surface $f(x, y) = y^2 - x^2$ has a
saddle point at the origin.

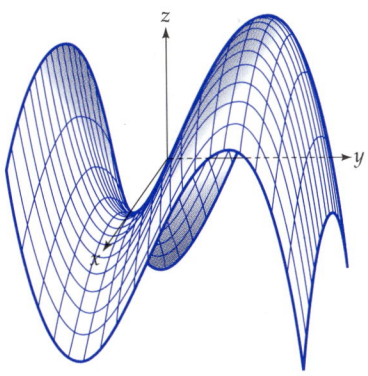

The surface $f(x, y) = 12y + 6x - x^2 - y^3$ has a saddle point and a relative maximum point.

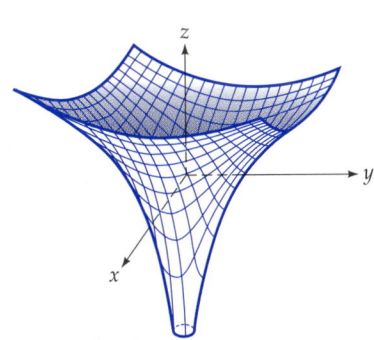

The surface $f(x, y) = \ln (x^2 + y^2)$ has no relative extreme points. It is undefined at $(0, 0)$.

EXERCISES 11.1

For each function, find the domain.

1. $f(x, y) = \dfrac{1}{xy}$

2. $f(x, y) = \dfrac{\sqrt{x}}{\sqrt{y}}$

3. $f(x, y) = \dfrac{1}{x - y}$

4. $f(x, y) = \dfrac{\sqrt[3]{x}}{\sqrt[3]{y}}$

5. $f(x, y) = \dfrac{\ln x}{y}$

6. $f(x, y) = \dfrac{x}{\ln y}$

7. $f(x, y, z) = \dfrac{e^{1/y} \ln z}{x}$

8. $f(x, y, z) = \dfrac{\sqrt{x} \ln y}{z}$

For each function, evaluate the given expression.

9. $f(x, y) = \sqrt{99 - x^2 - y^2}$, find $f(3, -9)$.

10. $f(x, y) = \sqrt{75 - x^2 - y^2}$, find $f(5, -1)$.

11. $g(x, y) = \ln (x^2 + y^4)$, find $g(0, e)$.

12. $g(x, y) = \ln (x^3 - y^2)$, find $g(e, 0)$.

13. $w(u, v) = \dfrac{1 + 2u + 3v}{uv}$, find $w(-1, 1)$.

14. $w(u, v) = \dfrac{2u + 4u}{v - u}$, find $w(1, -1)$.

15. $h(x, y) = e^{xy + y^2 - 2}$, find $h(1, -2)$.

16. $h(x, y) = e^{x^2 - xy - 4}$, find $h(1, -2)$.

17. $f(x, y) = xe^y - ye^x$, find $f(1, -1)$.

18. $f(x, y) = xe^y + ye^x$, find $f(-1, 1)$.

19. $f(x, y, z) = xe^y + ye^z + ze^x$, find $f(1, -1, 1)$.

20. $f(x, y, z) = xe^y + ye^z + ze^x$, find $f(-1, 1, -1)$.

21. $f(x, y, z) = z \ln \sqrt{xy}$, find $f(-1, -1, 5)$.

22. $f(x, y, z) = z\sqrt{x} \ln y$, find $f(4, e, -1)$.

APPLIED EXERCISES

23. Business: Stock Yield The *yield* of a stock is defined as $Y(d, p) = \dfrac{d}{p}$, where d is the dividend per share and p is the price of a share of stock. Find the yield of a stock that sells for $140 and offers a dividend of $2.20.

24. Business: Price-Earnings Ratio The price-earnings ratio of a stock is defined as $R(P, E) = \dfrac{P}{E}$, where P is the price of a share of stock and E is its earnings. Find the price-earnings ratio of a stock that is selling for $140 with earnings of $1.70.

25. General: Scuba Diving   The maximum duration of a scuba dive (in minutes) can be estimated from the formula

$$T(v, d) = \frac{33v}{d + 33}$$

where v is the volume of air (at sea-level pressure) in the tank and d is the depth of the dive. Find $T(90, 33)$.

26. Social Science: Cephalic Index   Anthropologists define the *cephalic index* to distinguish the head shapes of different races. For a head of width W and length L (measured from above), the cephalic index is

$$C(W, L) = 100\,\frac{W}{L}$$

Calculate the cephalic index for a head of width 8 inches and length 10 inches.

 27. Economics: Cobb–Douglas Functions   A company's production is estimated to be $P(L, K) = 2L^{0.6}K^{0.4}$. Find $P(320, 150)$.

 28. Biomedical: Body Area   The surface area (in square feet) of a person of weight w pounds and height h feet is approximated by a function of two variables: $A(w, h) = 0.55\,w^{0.425}h^{0.725}$. Use this function to estimate the surface area of a person who weighs 160 pounds and who is 6 feet tall. (Such estimates are important in certain medical procedures.)

29. Economics: Cobb–Douglas Functions   Show that the Cobb–Douglas function $P(L, K) = aL^bK^{1-b}$ satisfies the equation $P(2L, 2K) = 2 \cdot P(L, K)$. This shows that doubling the amounts of labor and capital doubles production, a property called *returns to scale*.

30. Economics: Cobb–Douglas Functions   Show that the Cobb–Douglas function $P(L, K) = aL^bK^{1-b}$ with $0 < b < 1$ satisfies

$$P(2L, K) < 2P(L, K) \quad \text{and} \quad P(L, 2K) < 2P(L, K)$$

This shows that doubling the amount of either labor or capital alone results in *less* than double production, a property called *diminishing returns*.

31. General: Telephone Calls   For two cities with populations x and y that are d miles apart, the number of telephone calls per hour between them can be estimated by the function of three variables

$$f(x, y, d) = \frac{3xy}{d^{2.4}}$$

(This is called the *gravity model*.) Use the gravity model to estimate the number of calls between two cities of populations 40,000 and 60,000 that are 600 miles apart.

32. Environmental Science: Tag and Recapture Estimates   Ecologists estimate the size of animal populations by capturing and tagging a few animals, and then releasing them. After the first group has mixed with the population, a second group of animals is captured, and the number of tagged animals in this group is counted. If originally T animals were tagged, and the second group is of size S and contains t tagged animals, then the population is estimated by the function of three variables

$$P(T, S, t) = \frac{TS}{t}$$

Estimate the size of a deer population if 100 deer were tagged, and then a second group of 250 contained 20 tagged deer.

33. Business: Cost Function   It costs an appliance company $210 to manufacture each washer and $180 to manufacture each dryer, and fixed costs are $4000. Find the company's cost function $C(x, y)$, using x and y for the numbers of washers and dryers, respectively.

34–35: General: Box Design   For each open-top box shown below and on the next page, find formulas for

a. the volume
b. the total amount of materials (the area)

34.

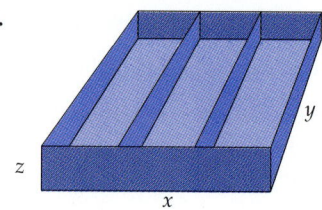

35.

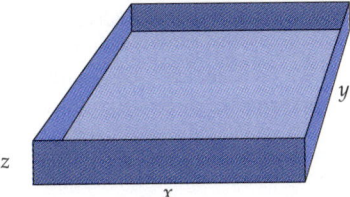

36. General: Windchill Enter the formula for windchill from the Graphing Calculator Exploration on page 915, but with y (temperature) replaced by 30 so that it becomes a function of one variable, x (wind speed).

  a. Graph the function on the window $[4, 45]$ by $[-10, 40]$.

  b. Use NDERIV or dy/dx to find the slope of this curve at $x = 10$. Interpret the answer.

  c. Repeat part (b) but for $x = 20$, interpreting the answer.

  d. What do the answers indicate about the windchill effect of additional wind on a

calm day as opposed to on an already windy day?

37. General: Windchill Enter the formula for windchill from the Graphing Calculator Exploration on page 915, but with y (temperature) replaced by *several* temperatures, like 20, 30, 40, and 50. (On some calculators, this is done simply by replacing y by the *set* $\{10, 20, 30, 40, 50\}$. On others, you may need to enter the formula several times with different values for y.)

  a. Graph the functions on the window $[4, 45]$ by $[-30, 50]$.

  b. Notice that the lower temperature curves slope downward more steeply than the others. What does this mean about the effect of wind on a colder day?

  c. Use NDERIV or dy/dx to find the slope of the lowest and the highest curves at $x = 10$. Interpret the answers.

  d. Do your answers to part (c) support your conclusion in part (b)?

11.2 Partial Derivatives

Introduction

Functions of several variables have several derivatives, one for each variable. In this section we will learn how to calculate and interpret these derivatives.

Partial Derivatives

Before calculating partial derivatives, we review the rules governing derivatives and constants.

$$\frac{d}{dx} c = 0$$

For a constant standing alone, the derivative is zero.

$$\frac{d}{dx}(cx^3) = c \cdot 3x^2$$

For a constant multiplying a function, the constant is carried along.

Carry along the constant — Derivative of x^3

These ideas will be very useful in this section.

A function $f(x, y)$ has two derivatives, called *partial derivatives*, one with respect to x and the other with respect to y.

Partial Derivatives

$$\frac{\partial}{\partial x} f(x, y) = \lim_{h \to 0} \frac{f(x + h, y) - f(x, y)}{h}$$

Partial derivative of f with respect to x
(x is increased by h, y is held constant)

$$\frac{\partial}{\partial y} f(x, y) = \lim_{h \to 0} \frac{f(x, y + h) - f(x, y)}{h}$$

Partial derivative of f with respect to y
(y is increased by h, x is held constant)

(provided that the limits exist). Partial derivatives are written with a "curly" ∂, $\partial / \partial x$ instead of d/dx, and are often called "partials."

Rewriting the first formula, but without the y, gives

$$\lim_{h \to 0} \frac{f(x + h) - f(x)}{h}$$

$$\lim_{h \to 0} \frac{f(x + h, y) - f(x, y)}{h}$$
but omitting the y's

which is just the "ordinary derivative" from Chapter 6. Therefore, the partial derivative with respect to x is just the ordinary derivative with respect to x with y held constant. Similarly, the partial with respect to y is just the ordinary derivative with respect to y, but now with x held constant.

$$\frac{\partial}{\partial x} f(x, y) = \left(\begin{matrix} \text{Derivative of } f \text{ with respect} \\ \text{to } x, \text{ with } y \text{ held constant} \end{matrix} \right)$$

$$\frac{\partial}{\partial y} f(x, y) = \left(\begin{matrix} \text{Derivative of } f \text{ with respect} \\ \text{to } y, \text{ with } x \text{ held constant} \end{matrix} \right)$$

EXAMPLE 1   Finding a Partial Derivative with Respect to x

Find $\dfrac{\partial}{\partial x} x^3 y^4$.

Solution   The $\partial / \partial x$ means differentiate with respect to x, holding y (and therefore y^4) constant. We therefore differentiate the x^3 and carry along the "constant" y^4:

Derivative of x^3

$$\frac{\partial}{\partial x}\, x^3 y^4 = 3x^2 y^4$$

Carry along the "constant" y^4

$\partial/\partial x$ means differentiate with respect to x, treating y like a constant

---

EXAMPLE 2 **Finding a Partial with Respect to *y***

Find $\dfrac{\partial}{\partial y}\, x^3 y^4$.

Solution

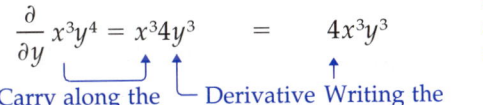

$$\frac{\partial}{\partial y}\, x^3 y^4 = x^3 4y^3 \qquad = \qquad 4x^3 y^3$$

Carry along the "constant" x^3 Derivative of y^4 Writing the constant first

$\partial/\partial y$ means differentiate with respect to y, treating x (and therefore x^3) like a constant

---

PRACTICE PROBLEM 1

Find

a. $\dfrac{\partial}{\partial x}\, x^4 y^2$ **b.** $\dfrac{\partial}{\partial y}\, x^4 y^2$

Solutions at the back of the book

EXAMPLE 3 **Finding a Partial Derivative**

Find $\dfrac{\partial}{\partial x}\, y^4$.

Solution

$$\frac{\partial}{\partial x}\, y^4 = 0$$

Partial with respect to x Function of y alone

The derivative of a constant is zero (since $\partial/\partial x$ means hold y constant)

PRACTICE PROBLEM 2   Find $\dfrac{\partial}{\partial y}\, x^2$.

Solution at the back of the book

EXAMPLE 4   **Finding a Partial of a Polynomial in Two Variables**

Find $\dfrac{\partial}{\partial x}(2x^4 - 3x^3y^3 - y^2 + 4x + 1)$.

Solution

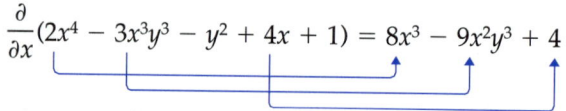

$$\dfrac{\partial}{\partial x}(2x^4 - 3x^3y^3 - y^2 + 4x + 1) = 8x^3 - 9x^2y^3 + 4$$

Differentiating with respect to x, so each y is held constant

■

PRACTICE PROBLEM 3   Find $\dfrac{\partial}{\partial y}(2x^4 - 3x^3y^3 - y^2 + 4x + 1)$.   *Solution at the back of the book*

Subscript Notation for Partial Derivatives

Partial derivatives are often denoted by subscripts: a subscript x means the partial with respect to x, and a subscript y means the partial with respect to y.*

$$f_x(x, y) = \dfrac{\partial}{\partial x} f(x, y)$$ f_x means the partial of f with respect to x

$$f_y(x, y) = \dfrac{\partial}{\partial y} f(x, y)$$ f_y means the partial of f with respect to y

EXAMPLE 5   **Using Subscript Notation**

Find $f_x(x, y)$ if $f(x, y) = 5x^4 - 2x^2y^3 - 4y^2$.

Solution

$$f_x(x, y) = 20x^3 - 4xy^3$$ Differentiating with respect to x, holding y constant

■

*Sometimes subscripts 1 and 2 are used to indicate partial derivatives with respect to the first and second variables: $f_1(x, y) = f_x(x, y)$ and $f_2(x, y) = f_y(x, y)$. We will not use this notation in this book.

EXAMPLE 6 Finding a Partial Involving Logs and Exponentials

Find both partials of $f = e^x \ln y$.

Solution

$$f_x = e^x \ln y$$

The derivative of e^x is e^x
(times the "constant" $\ln y$)

$$f_y = e^x \frac{1}{y}$$

The derivative of $\ln y$ is $\frac{1}{y}$
(times the "constant" e^x)

EXAMPLE 7 Finding a Partial of a Function to a Power

Find f_y if $f = (xy^2 + 1)^4$.

Solution

$$f_y = 4(xy^2 + 1)^3(x2y)$$

Partial of the inside
with respect to y

Using the generalized power rule (the derivative of f^n is $nf^{n-1}f'$, but with f' meaning a *partial*)

$$= 8xy(xy^2 + 1)^3$$

Simplifying

EXAMPLE 8 Finding a Partial of a Quotient

Find $\dfrac{\partial g}{\partial x}$ if $g = \dfrac{xy}{x^2 + y^2}$.

Solution

Partial of the top with respect to x
Partial of the bottom with respect to x

$$\frac{\partial g}{\partial x} = \frac{(x^2 + y^2)\,y - 2x \cdot xy}{(x^2 + y^2)^2}$$

Using the quotient rule

Bottom squared

$$= \frac{x^2y + y^3 - 2x^2y}{(x^2 + y^2)^2}$$

Simplifying

EXAMPLE 9   Finding a Partial of the Logarithm of a Function

Find $f_x(x, y)$ if $f(x, y) = \ln(x^2 + y^2)$.

Solution

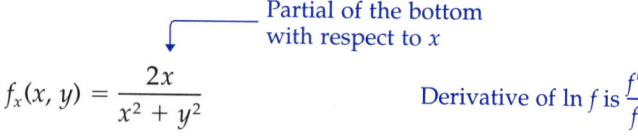

$$f_x(x, y) = \frac{2x}{x^2 + y^2}$$

Partial of the bottom with respect to x

Derivative of $\ln f$ is $\dfrac{f'}{f}$

◼

An expression like $f_x(2, 5)$, which involves both differentiation and evaluation, means *first differentiate and then evaluate.**

EXAMPLE 10   Evaluating a Partial Derivative

Find $f_y(1, 3)$ if $f(x, y) = e^{x^2+y^2}$.

Solution

$$f_y(x, y) = e^{x^2+y^2}(2y)$$

Derivative of e^f is $e^f \cdot f'$

Partial of the exponent with respect to y

$$f_y(1, 3) = e^{1^2+3^2}(2 \cdot 3)$$

$$= 6e^{10}$$

Evaluating at $x = 1$ and $y = 3$

Simplifying

◼

PRACTICE PROBLEM 4   Find $f_y(1, 2)$ if $f(x, y) = e^{x^3+y^3}$.   *Solution at the back of the book*

Partial Derivatives in Three or More Variables

Partial derivatives in three or more variables are defined similarly. That is, the partial derivative of $f(x, y, z)$ with respect to any one variable is the "ordinary" derivative with respect to that variable, holding all other variables constant.

---

*$f_x(2, 5)$ may also be written $\dfrac{\partial f}{\partial x}(2, 5)$ or $\dfrac{\partial f}{\partial x}\Big|_{(2, 5)}$, again meaning first differentiate, then evaluate.

EXAMPLE 11 Finding a Partial of a Function of Three Variables

$$\frac{\partial}{\partial x}(x^3y^4z^5) = 3x^2y^4z^5$$

$\partial/\partial x$ means differentiate with respect to x, holding y and z constant

Hold constant — Derivative of x^3

PRACTICE PROBLEM 5 Find $\dfrac{\partial}{\partial y}(x^3y^4z^5)$. *Solution at the back of the book*

EXAMPLE 12 Evaluating a Partial in Three Variables

Find $f_z(1, 1, 1)$ if $f(x, y, z) = e^{x^2+y^2+z^2}$.

Solution

$$f_z(x, y, z) = e^{x^2+y^2+z^2}(2z)$$ Partial with respect to z

$$= 2ze^{x^2+y^2+z^2}$$ Writing the $2z$ first

$$f_z(1, 1, 1) = 2e^{1^2+1^2+1^2} = 2e^3$$ Evaluating

Interpreting Partial Derivatives as Rates of Change

Since partials are just "ordinary" derivatives with the other variable held constant, they give *instantaneous rates of change* with respect to one variable at a time.

Partials as Rates of Change

$$f_x(x, y) = \begin{pmatrix} \text{Instantaneous rate of change of } f \\ \text{with respect to } x \text{ when } y \text{ is held constant} \end{pmatrix}$$

$$f_y(x, y) = \begin{pmatrix} \text{Instantaneous rate of change of } f \\ \text{with respect to } y \text{ when } x \text{ is held constant} \end{pmatrix}$$

This is why they are called *partial* derivatives: not all the variables are changed at once, only a "partial" change is made.

Cobb–Douglas Production Functions

Recall that a Cobb–Douglas production function $P(L, K) = aL^bK^{1-b}$ expresses production P as a function of L (units of labor) and K (units of capital). The partials therefore give the rate of increase of production with respect to one of these variables while the other is held constant.

EXAMPLE 13   Interpreting Partials of a Cobb–Douglas Production Function

Find and interpret $P_L(120, 200)$ and $P_K(120, 200)$ for the Cobb–Douglas function $P(L, K) = 20L^{0.6}K^{0.4}$.

Solution

$$P_L = 12L^{-0.4}K^{0.4}$$

Partial with respect to L (the 12 is 20 times 0.6)

$$P_L(120, 200) = 12(120)^{-0.4}(200)^{0.4} \approx 14.7$$

Substituting $L = 120$, $K = 200$, and evaluating using a calculator

Interpretation: $P_L = 14.7$ means that production increases by about 14.7 units for each additional unit of labor (when $L = 120$ and $K = 200$). This is called the *marginal productivity of labor*.

$$P_K = 8L^{0.6}K^{-0.6}$$

Partial with respect to K (the 8 is 20 times 0.4)

$$P_K(120, 200) = 8(120)^{0.6}(200)^{-0.6} \approx 5.9$$

Substituting $L = 120$, $K = 200$, and evaluating using a calculator

Interpretation: $P_K = 5.9$ means that production increases by about 5.9 units for each additional unit of capital. This is called the *marginal productivity of capital*.

■

These numbers show that to increase production, additional units of labor are more than twice as effective as additional units of capital (at the levels $L = 120$ and $K = 200$).

Partial derivatives give the marginals for one product at a time.

Interpreting Partials as Marginals

Let $C(x, y)$ be the (total) cost function for x units of product 1 and y units of product 2. Then

$$C_x(x, y) = \left(\begin{array}{c}\text{Marginal cost function for product 1}\\ \text{when production of product 2 is held constant}\end{array}\right)$$

$$C_y(x, y) = \left(\begin{array}{c}\text{Marginal cost function for product 2}\\ \text{when production of product 1 is held constant}\end{array}\right)$$

Similar statements hold, of course, for revenue and profit functions: The partials give the marginals for one variable at a time when the other variables are held constant.

EXAMPLE 14   Interpreting Partials of a Profit Function

A company's profit from producing x radios and y televisions per day is $P(x, y) = 4x^{3/2} + 6y^{3/2} + xy$. Find the marginal profit functions. Then find and interpret $P_y(25, 36)$.

Solution

$$P_x(x, y) = 6x^{1/2} + y$$

Marginal profit for radios when television production is held constant

$$P_y(x, y) = 9y^{1/2} + x$$

Marginal profit for televisions when radio production is held constant

$$P_y(25, 36) = 9\underbrace{(36)^{1/2}}_{6} + 25 = 79$$

Evaluating at $x = 25$ and $y = 36$

Interpretation: Profit increases by about \$79 per additional television (when producing 25 radios and 36 televisions per day).

Interpreting Partials Geometrically

A function $f(x, y)$ represents a surface in three-dimensional space, and the partial derivatives are the slopes along the surface in different directions: $\partial f / \partial x$ gives the slope of the surface "in the x direction," and $\partial f / \partial y$ gives the slope of the surface "in the y direction" at the point (x, y).

To put this colloquially, if you were on the surface $z = f(x, y)$, then $\partial f / \partial x$ would be the steepness of the surface *in the x direction*, and $\partial f / \partial y$ would be the steepness of the surface *in the y direction* from the point (x, y).

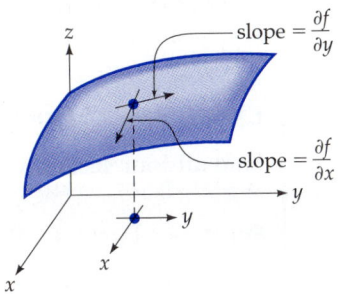

Partial derivatives are slopes.

PRACTICE PROBLEM 6   From looking at the previous graph at the indicated point:

a.   Is $\dfrac{\partial f}{\partial x}$ positive or negative?   (*Hint:* Would walking along the surface "in the x-direction" mean climbing or descending?)

b.   Is $\dfrac{\partial f}{\partial y}$ positive or negative?   (*Hint:* Same question, but now "in the y-direction.")

Solutions at the back of the book

Higher-Order Partial Derivatives

We can differentiate a function more than once to obtain *higher-order* partials.

Second-Order Partials

Subscript Notation	∂ Notation	In Words
f_{xx}	$\dfrac{\partial^2}{\partial x^2} f$	Differentiate twice with respect to x
f_{yy}	$\dfrac{\partial^2}{\partial y^2} f$	Differentiate twice with respect to y
f_{xy}	$\dfrac{\partial^2}{\partial y\, \partial x} f$	Differentiate first with respect to x, then with respect to y
f_{yx}	$\dfrac{\partial^2}{\partial x\, \partial y} f$	Differentiate first with respect to y, then with respect to x

Each notation means differentiate first with respect to the letter *closest* to f.

Calculating a "second partial" such as f_{xy} is a two-step process: First calculate f_x, and then differentiate the *result* with respect to y.

EXAMPLE 15   Finding Second-Order Partials

Find all four second-order partials of $f = x^4 + 2x^2y^2 + x^3y + y^4$.

Solution   First we calculate

$$f_x = 4x^3 + 4xy^2 + 3x^2y \qquad \text{Partial of } f \text{ with respect to } x$$

Then from this we find f_{xx} and f_{xy}:

$$f_{xx} = 12x^2 + 4y^2 + 6xy$$

Differentiating $f_x = 4x^3 + 4xy^2 + 3x^2y$ with respect to x

$$f_{xy} = 8xy + 3x^2$$

Differentiating $f_x = 4x^3 + 4xy^2 + 3x^2y$ with respect to y

Now, returning to the original function $f = x^4 + 2x^2y^2 + x^3y + y^4$, we calculate

$$f_y = 4x^2y + x^3 + 4y^3$$ Partial of f with respect to y

Then, from this,

$$f_{yy} = 4x^2 + 12y^2$$

Differentiating $f_y = 4x^2y + x^3 + 4y^3$ with respect to y

$$f_{yx} = 8xy + 3x^2$$

Differentiating $f_y = 4x^2y + x^3 + 4y^3$ with respect to x

■

Notice that in Example 15 the "mixed partials" are equal:

$$f_{xy} = 8xy + 3x^2$$
$$f_{yx} = 8xy + 3x^2$$

Equal

That is, $f_{xy} = f_{yx}$, so reversing the order of differentiation (first x, then y, or first y, then x) made no difference. This is *not* true for all functions, but it is true for all the functions that we will encounter in this book, and it is also true for all functions that you are likely to encounter in applications.*

PRACTICE PROBLEM 7   For the function $f(x, y) = x^3 - 3x^2y^4 + y^3$, find

a. f_x **b.** f_{xy} **c.** f_y **d.** f_{yx} *Solutions at the back of the book*

EXERCISES 11.2

For each function, find the partials **a.** $f_x(x,y)$ and **b.** $f_y(x,y)$.

1. $f(x, y) = x^3 + 3x^2y^2 - 2y^3 - x + y$

2. $f(x, y) = 2x^4 - 7x^3y^2 - xy + 1$

3. $f(x, y) = 12x^{1/2}y^{1/3} + 8$

4. $f(x, y) = x^{-1}y + xy^{-2}$

5. $f(x, y) = 100x^{0.05}y^{0.02}$ **6.** $f(x, y) = x/y$

7. $f(x, y) = (x + y)^{-1}$

8. $f(x, y) = (x^2 + xy + 1)^4$

9. $f(x, y) = \ln(x^3 + y^3)$ **10.** $f(x, y) = x^2e^y$

11. $f(x, y) = 2x^3e^{-5y}$ **12.** $f(x, y) = e^{x+y}$

13. $f(x, y) = e^{xy}$ **14.** $f(x, y) = \ln(xy^3)$

15. $f(x, y) = \ln\sqrt{x^2 + y^2}$ **16.** $f(x, y) = \dfrac{xy}{x + y}$

*$f_{xy} = f_{yx}$ if these partials are continuous. A more detailed statement can be found in an advanced calculus book.

For each function, find **a.** $\dfrac{\partial w}{\partial u}$ and **b.** $\dfrac{\partial w}{\partial v}$.

17. $w = (uv - 1)^3$ **18.** $w = (u - v)^3$

19. $w = e^{\frac{1}{2}(u^2 - v^2)}$ **20.** $w = \ln (u^2 + v^2)$

For each function, evaluate the stated partials.

21. $f(x, y) = 4x^3 - 3x^2y^2 - 2y^2$, find $f_x(-1, 1)$ and $f_y(-1, 1)$.

22. $f(x, y) = 2x^4 - 5x^2y^3 - 4y$, find $f_x(1, -1)$ and $f_y(1, -1)$.

23. $f(x, y) = e^{x^2 + y^2}$, find $f_x(0, 1)$ and $f_y(0, 1)$.

24. $g(x, y) = (xy - 1)^5$, find $g_x(1, 0)$ and $g_y(1, 0)$.

25. $h(x, y) = x^2y - \ln (x + y)$, find $h_x(1, 1)$.

26. $f(x, y) = \sqrt{x^2 + y^2}$, find $f_y(8, -6)$.

For each function, find the second-order partials **a.** f_{xx}, **b.** f_{xy}, **c.** f_{yx}, and **d.** f_{yy}.

27. $f(x, y) = 5x^3 - 2x^2y^3 + 3y^4$

28. $f(x, y) = 4x^2 - 3x^3y^2 + 5y^5$

29. $f(x, y) = 9x^{1/3}y^{2/3} - 4xy^3$

30. $f(x, y) = 32x^{1/4}y^{3/4} - 5x^3y$

31. $f(x, y) = ye^x - x \ln y$

32. $f(x, y) = y \ln x + xe^y$

For each function, calculate the third-order partials **a.** f_{xxy}, **b.** f_{xyx}, and **c.** f_{yxx}.

33. $f(x, y) = x^4y^3 - e^{2x}$ **34.** $f(x, y) = x^3y^4 - e^{2y}$

For each function of three variables, find the partials **a.** f_x, **b.** f_y, and **c.** f_z.

35. $f = xy^2z^3$ **36.** $f = x^2y^3z^4$

37. $f = (x^2 + y^2 + z^2)^4$ **38.** $f = (xyz + 1)^3$

39. $f = e^{x^2 + y^2 + z^2}$ **40.** $f = \ln (x^2 - y^3 + z^4)$

For each function, evaluate the stated partial.

41. $f = 3x^2y - 2xz^2$, find $f_x(2, -1, 1)$.

42. $f = 2yz^3 - 3x^2z$, find $f_z(2, -1, 1)$.

43. $f = e^{x^2 + 2y^2 + 3z^2}$, find $f_y(-1, 1, -1)$.

44. $f = e^{2x^3 + 3y^3 + 4z^3}$, find $f_y(1, -1, 1)$.

APPLIED EXERCISES

45–46: Business: Marginal Profit   An electronics company's profit from making x tape decks and y CD players per day is given by the profit function $P(x, y)$ (given below).

a. Find the marginal profit function for tape decks.

b. Evaluate your answer to part (a) at $x = 200$ and $y = 300$ and interpret the result.

c. Find the marginal profit function for CD players.

d. Evaluate your answer to part (c) at $x = 200$ and $y = 100$ and interpret the result.

45. $P(x, y) = 2x^2 - 3xy + 3y^2 + 150x + 75y + 200$

46. $P(x, y) = 3x^2 - 4xy + 4y^2 + 80x + 100y + 200$

47–48: Business: Cobb–Douglas   Production Functions   A company's production is given by the Cobb–Douglas function $P(L, K)$ (given in Exercises 47–48), where L is the number of units of labor and K is the number of units of capital.

a. Find $P_L(27, 125)$ and interpret this number.

b. Find $P_K(27, 125)$ and interpret this number.

c. From your answers to parts (a) and (b), which will increase production more: an additional unit of labor or an additional unit of capital?

47. $P(L, K) = 270L^{1/3}K^{2/3}$

48. $P(L, K) = 225L^{2/3}K^{1/3}$

49. Business: Sales   A store's television sales depend on x, the price of the televisions, and y, the amount spent on advertising, according to the function $S(x, y) = 200 - 0.1x + 0.2y^2$. Find and interpret the marginals S_x and S_y.

50. Economics: Value of an MBA   A 1973 study found that a businessperson with an MBA (master's degree in business administration) earned an average salary of $S(x, y) = 10,990 + 1120x + 873y$ dollars, where x is the number of years of work experience before the MBA, and y is the number of years of work experience after the MBA. Find and interpret the marginals S_x and S_y.

51. Sociology: Status   A study found that a person's status in a community depends on the person's income and education according to the function $S(x, y) = 7x^{1/3}y^{1/2}$, where x is income (in thousands of dollars) and y is years of education beyond high school.

a. Find $S_x(27, 4)$ and interpret this number.

b. Find $S_y(27, 4)$ and interpret this number.

52. Biomedical: Blood Flow The resistance of blood flowing through an artery of radius r and length L (both in centimeters) is $R(r, L) = 0.08Lr^{-4}$.

a. Find $R_r(0.5, 4)$ and interpret this number.
b. Find $R_L(0.5, 4)$ and interpret this number.

53. General: Automobile Safety The length of the skid marks from a truck of weight w (tons) traveling at velocity v (miles per hour) skidding to a stop on a dry road is $S(w, v) = 0.027wv^2$.

a. Find $S_w(4, 60)$ and interpret this number.
b. Find $S_v(4, 60)$ and interpret this number.

54. General: Windchill Temperature The windchill temperature announced by the weather bureau during the cold weather measures how cold it "feels" for a given temperature and wind speed. The formula is $C(t, w) = (0.475 + 0.304\sqrt{w} - 0.0203w)(t - 91.4) + 91.4$ where t is the temperature (Fahrenheit) and w is the wind speed (mph). Find and interpret $C_w(30, 20)$.

COMPETITIVE AND COMPLEMENTARY COMMODITIES

55. Economics: Competitive Commodities Certain commodities (like butter and margarine) are called "competitive" or "substitute" commodities because one may substitute for the other. If $B(b, m)$ gives the daily sales of butter as a function of b, the price of butter and m, the price of margarine:

a. Give an interpretation of $B_b(b, m)$.
b. Would you expect $B_b(b, m)$ to be positive or negative? Explain.
c. Give an interpretation of $B_m(b, m)$.
d. Would you expect $B_m(b, m)$ to be positive or negative? Explain.

56. Economics: Complementary Commodities Certain commodities, such as washing machines and clothes dryers, are called "complementary" commodities because they are often used together. If $D(d, w)$ gives the monthly sales of dryers as a function of d, the price of dryers and w, the price of washers:

a. Give an interpretation of $D_d(d, w)$.
b. Would you expect $D_d(d, w)$ to be positive or negative? Explain.
c. Give an interpretation of $D_w(d, w)$.
d. Would you expect $D_w(d, w)$ to be positive or negative? Explain.

11.3 Optimizing Functions of Several Variables

Introduction

A function $f(x, y)$ represents a *surface*, with relative maximum and minimum points ("hilltops" and "valley bottoms") and saddle points.

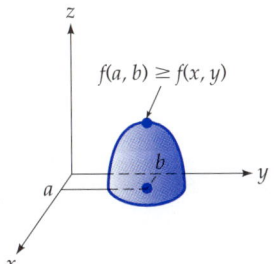

f has a relative *maximum* value at (a, b).

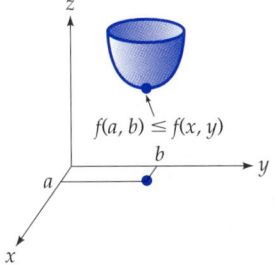

f has a relative *minimum* value at (a, b) (view from below).

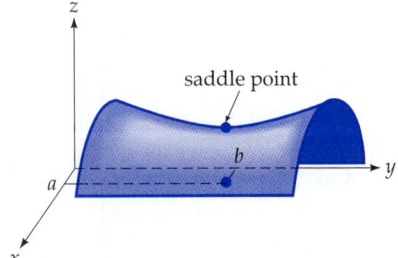

f has a saddle point at (a, b) (neither a maximum nor a minimum).

In this section we will see how to maximize and minimize such functions by finding critical points and using a two-variable version of the second derivative test. For simplicity, we will consider only functions whose first and second partials are defined everywhere. Such optimization techniques have many applications, such as maximizing profit for a company that makes several products.

Critical Points

At the very top of a smooth hill, the slope or steepness in any direction must be zero. That is, a straight stick would balance horizontally at the top. The partials f_x and f_y are the slopes in the x and y directions, so these partials must both be zero at a relative maximum or minimum point. A point (a, b) at which both partials are zero is called a *critical point* of the function.

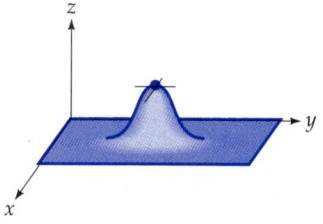

A critical point (both partials are zero)

Critical Point

(a, b) is a critical point of $f(x, y)$ if

$$f_x(a, b) = 0 \quad \text{and} \quad f_y(a, b) = 0$$

Both first partials are zero

Relative maximum and minimum values can occur only at critical points.

EXAMPLE 1   Finding Critical Points

Find all critical points of

$$f(x, y) = 3x^2 + y^2 + 3xy + 3x + y + 6$$

Solution   We want all points (a, b) at which both partials are zero.

$$f_x = 6x + 3y + 3 \quad \text{and} \quad f_y = 2y + 3x + 1 \qquad \text{Partials}$$

$$6x + 3y + 3 = 0$$
$$3x + 2y + 1 = 0$$

} Partials
set equal to zero

Reordered so the x
and y terms line up

To solve these equations simultaneously, we multiply the second by -2 so that the x-terms drop out when we add.

$6x + 3y + 3 = 0$	First equation
$-6x - 4y - 2 = 0$	Second equation times -2
$-y + 1 = 0$	Adding (x drops out)
$y = 1$	From solving $-y + 1 = 0$

Substituting $y = 1$ into either equation gives x:

$6x + 3 + 3 = 0$	Substituting $y = 1$ into $6x + 3y + 3 = 0$
$6x = -6$	Simplifying
$x = -1$	Solving

These x- and y-values give one critical point.

CP: $(-1, 1)$	From $x = -1, y = 1$

■

Second Derivative Test for Functions $f(x, y)$
The D-Test

To determine whether $f(x, y)$ has a relative maximum, a relative minimum, or a saddle point at a critical point, we use the following D-test.

D-test

If (a, b) is a critical point of the function $f(x, y)$, then for D defined by

$$D = f_{xx}(a, b) \cdot f_{yy}(a, b) - [f_{xy}(a, b)]^2$$

More briefly
$D = f_{xx}f_{yy} - (f_{xy})^2$

i. f has a relative *maximum* at (a, b) if $D > 0$ and $f_{xx}(a, b) < 0$.

ii. f has a relative *minimum* at (a, b) if $D > 0$ and $f_{xx}(a, b) > 0$.

Different signs
for $f_{xx}(a, b)$

iii. f has a *saddle point* at (a, b) if $D < 0$.

The following observations may help in understanding the D-test.

1. The D-test is used only *after* finding the critical points. The test is then applied to each critical point, one at a time.

2. $D > 0$ appears only in parts i and ii, and so guarantees that the function has either a relative maximum or a relative minimum. Then all that remains to be done is to use the "old" second derivative test (checking the sign of f_{xx}) to determine which one (maximum or minimum).

3. $D < 0$ means a saddle point, regardless of the sign of f_{xx}. (A saddle point is neither a maximum nor a minimum.)

4. $D = 0$ means that the D-test is inconclusive—the function may have a maximum, a minimum, or a saddle point at the critical point.

EXAMPLE 2   Finding Relative Extreme Values of a Polynomial

Find the relative extreme values of

$$f(x, y) = 3x^2 + y^2 + 3xy + 3x + y + 6$$

Solution   We find critical points by setting the two partials equal to zero and solving. But we did this for the same function in Example 1, finding one critical point, $(-1, 1)$. For the D-test, we calculate the partials.

$$f_{xx} = 6 \qquad \text{From } f_x = 6x + 3y + 3$$

$$f_{yy} = 2 \qquad \text{From } f_y = 2y + 3x + 1$$

$$f_{xy} = 3 \qquad \text{From } f_x = 6x + 3y + 3$$

Calculating D:

$$D = 6 \cdot 2 - (3)^2 = 12 - 9 = 3 \qquad\qquad D = f_{xx}f_{yy} - (f_{xy})^2$$

$$\underbrace{}_{f_{xx}} \quad \underbrace{}_{f_{yy}} \; \underbrace{}_{f_{xy}} \qquad\qquad \text{Positive!}$$

D is positive and f_{xx} is positive (since $f_{xx} = 6$), so f has a *relative minimum* (part ii of the D-test) at the critical point $(-1, 1)$. (If f_{xx} had been negative, there would have been a relative maximum.) The relative minimum *value* is found by evaluating $f(x, y)$ at $(-1, 1)$.

$$f(-1, 1) = 3 + 1 - 3 - 3 + 1 + 6 \qquad f = 3x^2 + y^2 + 3xy + 3x + y + 6$$
$$= 5 \qquad\qquad\qquad \text{evaluated at } x = -1, y = 1$$

Relative minimum value: $f = 5$ at $x = -1, y = 1$.

EXAMPLE 3 **Finding Relative Extreme Values of an Exponential Function**

Find the relative extreme values of $f(x, y) = e^{x^2 - y^2}$.

Solution

$$f_x = e^{x^2 - y^2}(2x)$$
$$f_y = e^{x^2 - y^2}(-2y)$$
$\left.\right\}$ Partials

$$e^{x^2 - y^2}(2x) = 0$$
$$e^{x^2 - y^2}(-2y) = 0$$
$\left.\right\}$ Partials set equal to zero

CP: (0, 0)   Since $\begin{cases} e^{x^2-y^2}(2x) = 0 \\ e^{x^2-y^2}(2y) = 0 \end{cases}$ only at $x = 0, y = 0$

For the *D*-test we calculate the second partials:

$$f_{xx} = e^{x^2 - y^2}(2x)(2x) + e^{x^2 - y^2}(2)$$

From $f_x = e^{x^2-y^2}(2x)$ using the product rule

$$= 4x^2 e^{x^2 - y^2} + 2e^{x^2 - y^2}$$

Simplifying

$$f_{yy} = e^{x^2 - y^2}(-2y)(-2y) + e^{x^2 - y^2}(-2)$$

From $f_y = e^{x^2-y^2}(-2y)$ using the product rule

$$= 4y^2 e^{x^2 - y^2} - 2e^{x^2 - y^2}$$

Simplifying

$$f_{xy} = e^{x^2 - y^2}(2x)(-2y) = -4xy e^{x^2 - y^2}$$

From $f_x = e^{x^2-y^2}(2x)$ using the product rule

Evaluating at the critical point (0, 0):

$f_{xx}(0, 0) = 0e^0 + 2e^0 = 0 + 2 = 2$ $f_{xx} = 4x^2 e^{x^2-y^2} + 2e^{x^2-y^2}$ at (0, 0)

$f_{yy}(0, 0) = 0e^0 - 2e^0 = 0 - 2 = -2$ $f_{yy} = 4y^2 e^{x^2-y^2} - 2e^{x^2-y^2}$ at (0, 0)

$f_{xy}(0, 0) = 0e^0 = 0$ $f_{xy} = -4xy e^{x^2-y^2}$ at (0, 0)

Therefore *D* is

$$D = (2)(-2) - (0)^2 = -4 - 0 = -4 \qquad\qquad D = f_{xx}f_{yy} - (f_{xy})^2$$

$\nearrow$ $\searrow$ $\searrow$ Negative!

f_{xx} f_{yy} f_{xy}

Since *D* is negative, the function has a *saddle point* (part iii of the *D*-test) at the critical point (0, 0).

     f has no relative extreme values
     (it has a saddle point at $x = 0, y = 0$). ∎

Maximizing Profit

If a company makes too few of its products, the resulting lost sales will lower the company's profits. On the other hand, making too much will "flood the market" and depress prices, again resulting in lower profits. Therefore, any realistic profit function must have a maximum at some "intermediate" point (x, y), which therefore must be a *relative maximum* point. Many applied problems are solved in this way: by knowing that the *absolute* extreme values (the highest and lowest values on the entire domain) must exist, and finding them as *relative* extreme points.

Suppose that a company produces two products, A and B, and that the two price functions are

$$p(x) = \left(\begin{array}{c} \text{The price at which exactly } x \\ \text{units of product A will be sold} \end{array} \right)$$

and

$$q(y) = \left(\begin{array}{c} \text{The price at which exactly } y \\ \text{units of product B will be sold} \end{array} \right)$$

If $C(x, y)$ is the (total) cost function, then the company's profit will be

$$\underbrace{P(x, y)}_{\text{Profit}} = \underbrace{p(x) \cdot x}_{\substack{\text{Price times} \\ \text{quantity for} \\ \text{product A}}} + \underbrace{q(y) \cdot y}_{\substack{\text{Price times} \\ \text{quantity for} \\ \text{product B}}} - \underbrace{C(x, y)}_{\text{Cost}} \quad \begin{array}{l} \text{Revenue for each} \\ \text{product (price times} \\ \text{quantity) minus the} \\ \text{cost function} \end{array}$$

EXAMPLE 4   Maximizing Profit for a Company

Universal Motors makes compact and midsized cars. The price function for compacts is $p = 17 - 2x$ (for $x \leq 8$), and the price function for midsized cars is $q = 20 - y$ (for $y \leq 20$), both in thousands of dollars, where x and y are, respectively, the number of compact and midsized cars produced per hour. If the company's cost function is

$$C(x, y) = 15x + 16y - 2xy + 5$$

thousand dollars, find how many of each car should be produced and the prices that should be charged in order to maximize profit. Also find the maximum profit.

Solution   The profit function is

$$P(x, y) = \underbrace{(17 - 2x)x}_{} + \underbrace{(20 - y)y}_{} - \underbrace{(15x + 16y - 2xy + 5)}_{\text{Cost}}$$

Price Quantity Price Quantity

For compacts   For midsized

$$= 17x - 2x^2 + 20y - y^2 - 15x - 16y + 2xy - 5 \qquad \text{Multiplying out}$$
$$= -2x^2 - y^2 + 2xy + 2x + 4y - 5 \qquad \text{Simplifying}$$

We maximize $P(x, y)$ in the usual way:

$$\left. \begin{array}{l} P_x = -4x + 2y + 2 \\[6pt] P_y = -2y + 2x + 4 \end{array} \right\} \text{Partials}$$

$$-4x + 2y + 2 = 0 \qquad \text{Partials set equal to zero}$$
$$\underline{2x - 2y + 4 = 0} \qquad \text{Rearranged to line up } x\text{'s and } y\text{'s}$$
$$-2x \qquad + 6 = 0 \qquad \text{Adding (the } y\text{'s cancel)}$$
$$x = 3 \qquad \text{From solving } -2x + 6 = 0$$
$$y = 5 \qquad \begin{array}{l}\text{From substituting } x = 3 \text{ into either} \\ \text{equation (omitting the details)}\end{array}$$

These give one critical point.

$$\text{CP: } (3, 5)$$

For the D-test we calculate the partials:

$$P_{xx} = -4 \qquad P_{xy} = 2 \qquad P_{yy} = -2 \qquad \begin{array}{l}\text{From } P_x = -4x + 2y + 2 \\ \text{and } P_y = -2y + 2x + 4\end{array}$$

$$D = (-4)(-2) - (2)^2 = 4 \qquad D = P_{xx}P_{yy} - (P_{xy})^2$$

D is positive and $P_{xx} = -4$ is negative, so profit is indeed *maximized* at $x = 3$ and $y = 5$. To find the prices, we evaluate the price functions:

$$p = 17 - 2 \cdot 3 = 11 \qquad \text{(thousand dollars)} \qquad \begin{array}{l}p = 15 - 2x \\ \text{evaluated at } x = 3\end{array}$$

$$q = 20 - 5 = 15 \qquad \text{(thousand dollars)} \qquad \begin{array}{l}q = 20 - y \\ \text{evaluated at } y = 5\end{array}$$

The profit comes from the profit function:

$$P(3, 5) = \underbrace{-2 \cdot 3^2 - 5^2 + 2 \cdot 3 \cdot 5 + 2 \cdot 3 + 4 \cdot 5 - 5}$$

$$\begin{array}{c}P = -2x^2 - y^2 + 2xy + 2x + 4y - 5 \\ \text{evaluated at } x = 3, y = 5\end{array}$$

$$= 8 \text{ (thousand dollars)}$$

Profit is maximized when the company produces 3 compacts per hour, selling them for \$11,000 each, and 5 midsized cars per hour, selling them for \$15,000 each. The maximum profit will be \$8000 per hour.

Some functions have *more* than one critical point.

EXAMPLE 5   Finding Relative Extreme Values

Find the relative extreme values of

$$f(x, y) = x^2 + y^3 - 6x - 12y$$

Solution

$$2x - 6 = 0 \qquad\qquad f_x = 0$$

$$3y^2 - 12 = 0 \qquad\qquad f_y = 0$$

The first gives

$$x = 3 \qquad\qquad \text{Solving } 2x - 6 = 0$$

and the second gives

$$3y^2 = 12 \qquad\qquad \text{Adding 12 to each side} \\ \text{of } 3y^2 - 12 = 0$$

$$y^2 = 4 \qquad\qquad \text{Dividing by 3}$$

$$y = \pm 2 \qquad\qquad \text{Taking square roots}$$

From $x = 3$ and $y = \pm 2$ we get *two* critical points:

$$\text{CP: } (3, 2) \quad \text{and} \quad (3, -2)$$

Calculating the second partials and substituting them into D:

$$D = (2)(6y) - (0)^2 = 12y \qquad \begin{array}{l} D = (f_{xx})(f_{yy}) - (f_{xy})^2 \\ \text{with } f_{xx} = 2, f_{yy} = 6y, f_{xy} = 0 \end{array}$$

We apply the D-test to the critical points one at a time.

at (3, 2): $D = 12 \cdot 2 > 0$ $D = 12y$ evaluated at (3, 2)

$$f_{xx} = 2 > 0$$

relative minimum at $x = 3$, $y = 2$ Since D and f_{xx} are both positive

at (3, -2): $D = 12 \cdot (-2) < 0$ $D = 12y$ evaluated at (3, -2)

saddle point at $x = 3$, $y = -2$ Since D is negative

Answer:

Relative minimum value: $f = -25$ at $\begin{cases} x = 3 \\ y = 2 \end{cases}$ $\begin{array}{l} f = -25 \text{ from} \\ f = x^2 + y^3 - 6x - 12y \\ \text{at } (3, 2) \end{array}$

(saddle point at $x = 3$, $y = -2$) ∎

Competition and Collusion

In 1938, the French economist Antoine Cournot published the following comparison of a monopoly (a market with only one supplier) and a duopoly (a market with two suppliers).

Monopoly. Imagine that you are selling spring water from your own spring (or any product whose cost of production is negligible). Since you are the only supplier in town, you have a "monopoly." Suppose that your price function is $p = 6 - 0.01x$, where p is the price at which you will sell precisely x gallons ($x \leq 600$). Your revenue is then

$$R(x) = (6 - 0.01x)x = 6x - 0.01x^2 \qquad \text{Price } (6 - 0.01x) \text{ times quantity } x$$

You maximize revenue by setting its derivative equal to zero:

$$R'(x) = 6 - 0.02x = 0$$

$$x = \frac{6}{0.02} = 300 \qquad \text{Solving } 6 - 0.02x = 0$$

Therefore, you should sell 300 gallons per day. (The second derivative test will verify that revenue is maximized.) The price will be

$$p = 6 - 0.01 \cdot 300 = 6 - 3 = 3 \qquad \begin{array}{l} p = 6 - 0.01x \text{ evaluated} \\ \text{at } x = 300 \end{array}$$

or $3 per gallon. Your maximum revenue will be

$$R(300) = 6 \cdot 300 - 0.01(300)^2 = \$900 \qquad \begin{array}{l} R(x) = 6x - 0.01x^2 \text{ evaluated} \\ \text{at } x = 300 \end{array}$$

Duopoly. Suppose now that your neighbor opens a competing spring water business. (A market like this with two suppliers is called a "duopoly.") Now both of you must share the same market. If he sells y gallons per day (and you sell x), you must both sell at price

$$p = 6 - 0.01(x + y) = 6 - 0.01x - 0.01y \qquad \begin{array}{l} \text{Price function } p = 6 - 0.01x \\ \text{with } x \text{ replaced by the} \\ \text{combined quantity } x + y \end{array}$$

Each of you calculates revenue as price times quantity:

$$\left(\begin{array}{c} \text{Your} \\ \text{revenue} \end{array} \right) = p \cdot x = (6 - 0.01x - 0.01y)x = 6x - 0.01x^2 - 0.01xy$$

$$\left(\begin{array}{c} \text{His} \\ \text{revenue} \end{array} \right) = p \cdot y = (6 - 0.01x - 0.01y)y = 6y - 0.01xy - 0.01y^2$$

You each want to maximize revenue, so you set the partials equal to zero:

$$6 - 0.02x - 0.01y = 0$$

Partial of $6x - 0.01x^2 - 0.01xy$ with respect to x, set equal to zero

$$6 - 0.01x - 0.02y = 0$$

Partial of $6y - 0.01xy - 0.01y^2$ with respect to y, set equal to zero

These are easily solved by multiplying one of them by 2 and subtracting the other. The solution (omitting the details) is

$$x = 200 \qquad y = 200$$

so each of you sells 200 gallons per day. The selling price for both will be

$$p = 6 - 0.01(200 + 200) = 6 - 4 = 2$$

$p = 6 - 0.01(x + y)$ evaluated at $x = 200$, $y = 200$

or $2 per gallon. Revenue is price ($2) times quantity (200), resulting in $400 for each of you.

Comparison of the Monopoly and the Duopoly

The two systems may be compared as follows:

	Monopoly	**Duopoly**
Quantity:	300 gallons	200 gallons each, 400 total
Price:	$3 per gallon	$2 per gallon
Revenue:	$900	$400 each

Notice that the duopoly produces *more* than the monopoly (400 gallons versus only 300 in the monopoly), and does so at a *lower price* ($2 versus $3 in the monopoly). Cournot therefore concluded that consumers benefit more from a duopoly than from a monopoly.

However, the smart duopolist will notice that his revenue of $400 is less than half of the monopoly revenue of $900. Therefore, it benefits each duopolist to cooperate and share the market as a single monopoly. This is called *collusion*. It is for this reason that markets with only a few suppliers tend toward collusion rather than competition.

Optimizing Functions of Three or More Variables

Functions of *more* than two variables are optimized in the same way: by finding critical points (where all of the first partials are zero). For example, to optimize a function $f(x, y, z)$ we would set the three partials equal to zero and solve:

$$f_x = 0$$

$$f_y = 0$$

$$f_z = 0$$

The second derivative test for functions of three or more variables is very complicated, and we shall not discuss it.

SUMMARY

To optimize a function $f(x, y)$ of two variables, first find all critical points, the points (a, b) where both first partials are zero.

$$f_x(a, b) = 0$$

$$f_y(a, b) = 0$$

Then apply the D-test (page 937) to each critical point to determine whether the function has a relative maximum, a relative minimum, or a saddle point at that critical point.

EXERCISES 11.3

Find the relative extreme values of each function.

1. $f(x, y) = x^2 + 2y^2 + 2xy + 2x + 4y + 7$

2. $f(x, y) = 2x^2 + y^2 + 2xy + 4x + 2y + 5$

3. $f(x, y) = 2x^2 + 3y^2 + 2xy + 4x - 8y$

4. $f(x, y) = 3x^2 + 2y^2 + 2xy + 8x - 4y$

5. $f(x, y) = 3xy - 2x^2 - 2y^2 + 14x - 7y - 5$

6. $f(x, y) = 2xy - 2x^2 - 3y^2 + 4x - 12y + 5$

7. $f(x, y) = xy + 4x - 2y + 1$

8. $f(x, y) = 5xy - 2x^2 - 3y^2 + 5x - 7y + 10$

9. $f(x, y) = 3x - 2y - 6$

10. $f(x, y) = 5x - 4y + 5$

11. $f(x, y) = e^{\frac{1}{2}(x^2 + y^2)}$ **12.** $f(x, y) = e^{5(x^2 + y^2)}$

13. $f(x, y) = \ln(x^2 + y^2 + 1)$

14. $f(x, y) = \ln(2x^2 + 3y^2 + 1)$

15. $f(x, y) = -x^3 - y^2 + 3x - 2y$

16. $f(x, y) = x^3 - y^2 - 3x + 6y$

17. $f(x, y) = y^3 - x^2 - 2x - 12y$

18. $f(x, y) = -x^2 - y^3 - 6x + 3y + 4$

19. $f(x, y) = x^3 - 2xy + 4y$

20. $f(x, y) = y^3 - 2xy - 4x$

APPLIED EXERCISES

21. Business: Maximum Profit   A company manufactures two products. The price function for product A is $p = 12 - \dfrac{1}{2}x$ (for $x \le 24$), and that for product B is $q = 20 - y$ (for $y \le 20$), both in thousands of dollars, where x and y are the amounts of products A and B, respectively. If the cost function is

$$C(x, y) = 9x + 16y - xy + 7$$

thousand dollars, find the quantities and the prices of the two products that maximize profit. Also find the maximum profit.

22. Business: Maximum Profit   A company manufactures two products. The price function for product A is $p = 16 - x$ (for $x \le 16$), and that for product B is $q = 19 - \frac{1}{2}y$ (for $y \le 38$), both in thousands of dollars, where x and y are the amounts of product A and B, respectively. If the cost function is

$$C(x, y) = 10x + 12y - xy + 6$$

thousand dollars, find the quantities and the prices of the two products that maximize profit. Also find the maximum profit.

23. **Business: Price Discrimination** An automobile manufacturer sells cars in America and Europe, charging different prices in the two markets. The price function for cars sold in America is $p = 20 - 0.2x$ thousand dollars (for $x \le 100$), and the price function for cars sold in Europe is $q = 16 - 0.1y$ thousand dollars (for $y \le 160$), where x and y are the numbers of cars sold per day in America and Europe, respectively. The company's cost function is

$$C = 20 + 4(x + y) \qquad \text{thousand dollars}$$

a. Find the company's profit function. (*Hint:* Profit is revenue from America plus revenue from Europe, minus costs, where each revenue is price times quantity.)
b. Find how many cars should be sold in each market to maximize profit. Also find the price for each market.

24. **Biomedical: Drug Dosage** In a laboratory test the combined antibiotic effect of x mg (milligrams) of medicine A and y mg of medicine B is given by the function

$$f(x, y) = xy - 2x^2 - y^2 + 110x + 60y$$

(for $x \le 55$, $y \le 60$). Find the amounts of the two medicines that maximize the antibiotic effect.

25. **Psychology: Practice and Rest** A subject in a psychology experiment who practices a skill for x hours and then rests for y hours is predicted to achieve a test score of $f(x, y) = xy - x^2 - y^2 + 11x - 4y + 120$ (for $x \le 10$, $y \le 4$). Find the number of hours of practice and rest that maximize the subject's score.

26. **Sociology: Absenteeism** The number of office workers near a beach resort who call in "sick" on a warm summer day is

$$f(x, y) = xy - x^2 - y^2 + 110x + 50y - 5200$$

where x is the air temperature ($70 \le x \le 100$) and y is the water temperature ($60 \le y \le 80$). Find the air and water temperatures that maximize the number of absentees.

27–28: Economics: Competition and Collusion Compare the output of a monopoly and a duopoly for the price function given below (repeating the analysis on pages 943–944). That is:

a. For a monopoly, calculate the quantity x that maximizes revenue. Also calculate the price p and the revenue R.
b. For a duopoly, calculate the quantities x and y that maximize revenue for each duopolist. Calculate the price p and the two revenues.
c. Are more goods produced under a monopoly or a duopoly?
d. Is the price lower under a monopoly or a duopoly?

27. $p = 12 - 0.005x$ $\qquad (x \le 2400)$

28. $p = a - bx$ $\qquad$ (for positive numbers a and b with $x \le a/b$)

OPTIMIZING FUNCTIONS OF THREE VARIABLES

29. **Business: Price Discrimination** An automobile manufacturer sells cars in America, Europe, and Asia, charging different prices in each of the three markets. The price function for cars sold in America is $p = 20 - 0.2x$ (for $x \le 100$), the price function for cars sold in Europe is $q = 16 - 0.1y$ (for $y \le 160$), and the price function for cars sold in Asia is $r = 12 - 0.1z$ (for $z \le 120$), all in thousands of dollars, where $x, y,$ and z are the numbers of cars sold in America, Europe, and Asia, respectively. The company's cost function is $C = 22 + 4(x + y + z)$ thousand dollars.

a. Find the company's profit function $P(x, y, z)$. (*Hint:* The profit will be revenue from America plus revenue from Europe plus revenue from Asia, minus costs, where each revenue is price times quantity.)

b. Find how many cars should be sold in each market to maximize profit. (*Hint:* Set the three partials P_x, P_y, and P_z equal to zero and solve. Assuming that the maximum exists, it must occur at this point.)

30. Economic: Competition and Collusion   Suppose that in the discussion of competition and collusion (pages 943–944), *two* of your neighbors began selling spring water. Use the price function $p = 36 - 0.01x$ (for $x \le 3600$) and repeat the analysis, but now comparing a monopoly to competition among *three* suppliers (a "triopoly"). That is:

a. For a monopoly, calculate the quantity x that maximizes your profit. Also calculate the price p and the revenue R.
b. For a triopoly, find the quantities x, y, and z for the three suppliers that maximize revenue for each. Also calculate the price p and the three revenues. (*Hint:* Find the three revenue functions, one for each supplier, and maximize each with respect to that supplier's variable.)
c. Are more goods produced under a monopoly or under a triopoly?
d. Is the price lower under a monopoly or under a triopoly?

EXERCISES WITH MORE THAN ONE CRITICAL POINT _____

Find the relative extreme values of each function.

31. $f(x, y) = x^3 + y^3 - 3xy$

32. $f(x, y) = x^5 + y^5 - 5xy$

33. $f(x, y) = 12xy - x^3 - 6y^2$

34. $f(x, y) = 6xy - x^3 - 3y^2$

35. $f(x, y) = 2x^4 + y^2 - 12xy$

36. $f(x, y) = 16xy - x^4 - 2y^2$

Least Squares

A P P L I C A T I O N   P R E V I E W

Safe Cars, Unsafe Streets

Since 1970, improvements in automobile design and increased use of seat belts have lowered the number of automobile-related fatalities in the United States. On the other hand, during this same period, the greater availability of handguns, together with other socioeconomic causes, has increased the number of gunshot fatalities. The data are shown in the table below and graphed on the next page.

	Years since 1970	Automobile Fatalities	Gunshot Fatalities
1970	0	55,800	26,800
1975	5	47,100	33,400
1980	10	53,200	32,900
1985	15	45,800	31,800
1990	20	47,600	36,000
1995	25	43,900	37,900

(*Source:* National Center for Health Statistics.)

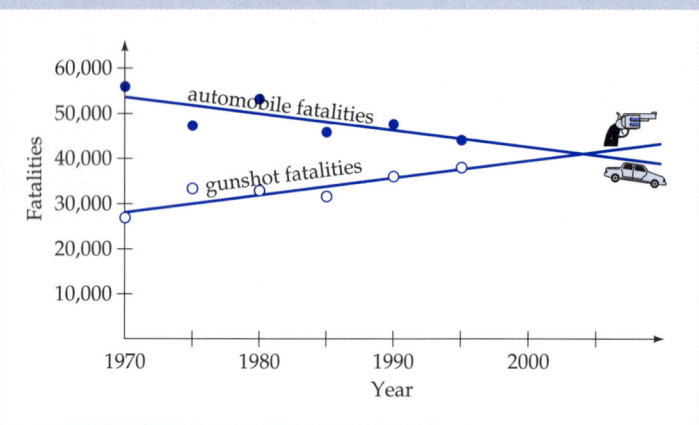

Two lines have been fit to these data points by a technique called "least squares." The lines show that at some time guns will overtake automobiles as the leading cause of accidental death in America. (The year is approximately 2004.)

In this section we will discuss the least squares technique for fitting lines or curves to data points.

Introduction

You may have wondered how the mathematical models in this book were developed. For example, how are the constants a and b in the Cobb–Douglas production function $P = aL^bK^{1-b}$ determined for a particular company or nation? The problem of finding the function that fits a collection of data can be viewed geometrically as the problem of fitting a curve to a collection of points. The simplest case of this problem is fitting a straight line to a collection of points, and the most widely used method for doing this is called *least squares*. Least squares is the method that graphing calculators use to find regression lines and curves. Even if you continue to use your calculator to carry out the calculations, this section is important in that it explains what you are actually doing.

Least squares lines are used extensively in forecasting and for detecting underlying trends in data. On pages 258–263 we found least squares lines by matrix methods, but now with calculus we can *prove* that such lines provide the "best fit" to a collection of data points (in the least squares sense, to be explained shortly).

A First Example

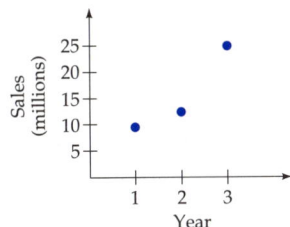

The graph on the left shows a company's annual sales (in millions) over a 3-year period. How can we fit a straight line to these three points? Clearly, these points do not lie exactly on a line, and so rather than an "exact" fit, we want the line $y = ax + b$ that fits these three points most closely.

Let d_1, d_2, and d_3 stand for the vertical distances between the three points and the line $y = ax + b$. The line that minimizes the sum of the squares of these vertical deviations is called the *least squares line* or the *regression line.** Squaring the deviations ensures that none are negative, so that a deviation below the line does not "cancel" one above the line.

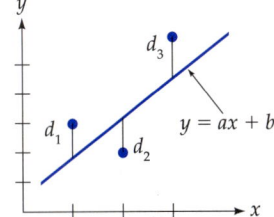

EXAMPLE 1   **Minimizing the Squared Deviations**

The table below gives a company's sales (in millions) in year x. Find the least squares line for the data. Then use the line to predict sales in year 4.

x	y
1	10
2	12
3	25

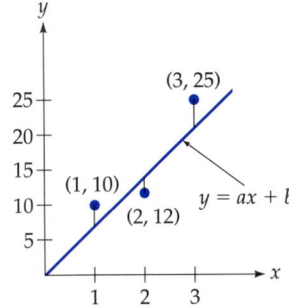

Solution   The vertical deviations are the heights of the line $y = ax + b$ at each x-value, minus the y-values from the table. These differences are then squared and summed.

---

*The word "regression" comes from an early use of this technique to determine whether unusually tall parents have unusually tall children. It seems that tall parents do have tall offspring, but not quite as tall, with later generations exhibiting a "regression" toward the average height of the population.

$$S = (a \cdot 1 + b - 10)^2 + (a \cdot 2 + b - 12)^2 + (a \cdot 3 + b - 25)^2$$

| Height of the line $y = ax + b$ at $x = 1$ | y-value of the point at $x = 1$ | Height of the line $y = ax + b$ at $x = 2$ | y-value of the point at $x = 2$ | Height of the line $y = ax + b$ at $x = 3$ | y-value of the point at $x = 3$ |

This sum S depends on a and b, the numbers that determine the line $y = ax + b$. To minimize S, we set its partials with respect to a and b equal to zero:

$$\frac{\partial S}{\partial a} = 2(a + b - 10) + 2(2a + b - 12) \cdot 2 + 2(3a + b - 25) \cdot 3 \qquad \text{Differentiating each part of } S \text{ by the generalized power rule}$$

$$= 2a + 2b - 20 + 8a + 4b - 48 + 18a + 6b - 150 \qquad \text{Multiplying out}$$

$$= 28a + 12b - 218 \qquad \text{Combining terms}$$

$$\frac{\partial S}{\partial b} = 2(a + b - 10) + 2(2a + b - 12) + 2(3a + b - 25) \qquad \text{Differentiating each part of } S \text{ by the generalized power rule}$$

$$= 2a + 2b - 20 + 4a + 2b - 24 + 6a + 2b - 50 \qquad \text{Multiplying out}$$

$$= 12a + 6b - 94 \qquad \text{Combining terms}$$

We set the two partials equal to zero and solve:

$$\left. \begin{array}{l} 28a + 12b - 218 = 0 \\ 12a + 6b - 94 = 0 \end{array} \right\} \quad \text{Partials set equal to zero}$$

$$28a + 12b - 218 = 0 \qquad \text{First equation}$$

$$\underline{-24a - 12b + 188 = 0} \qquad \text{Second multiplied by } -2$$

$$4a \qquad\quad - 30 = 0 \qquad \text{Adding (the } b\text{'s drop out)}$$

$$a = \frac{30}{4} = 7.5 \qquad \text{Solving } 4a - 30 = 0$$

$$b = \frac{4}{6} \approx 0.67 \qquad \begin{array}{l} \text{From substituting } a = 7.5 \text{ into} \\ 12a + 6b - 94 \text{ and solving for } b \end{array}$$

These values for a and b give the least squares line. (The D-test would show that S has indeed been minimized.)

$$y = 7.5x + 0.67$$

$y = ax + b$ with
$a = 7.5$ and $b = 0.67$

least squares line
$y = 7.5x + 0.67$

To predict the sales in year 4, we evaluate the least squares line at $x = 4$:

$$y = 7.5(4) + 0.67 = 30.67 \qquad y = 7.5x + 0.67 \text{ evaluated at } x = 4$$

Prediction: 30.67 million sales in year 4.

■

The slope of the line is 7.5, meaning that the linear trend in the company's sales is a growth of 7.5 million sales per year.

 Graphing Calculator Exploration

Verify that your graphing calculator gives the same least squares line $y = 7.5x + 0.67$ that we found in Example 1 by entering the data from the table on page 949 into your calculator and finding the linear regression line. The r that your calculator might find is a measure of "goodness of fit," with values closer to 1 indicating a better fit.

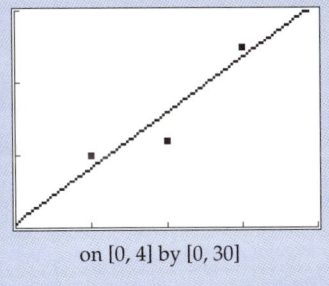

on [0, 4] by [0, 30]

Least Squares Line for *n* Points

Example 1 used only three points, which is too small for most realistic applications. If we carry out the same steps for *n* points, we would obtain the following formulas (in which Σ stands for sum).

Least Squares Line

For data:

x	y
x_1	y_1
x_2	y_2
.	.
.	.
.	.
x_n	y_n

the least squares line is

$$y = ax + b$$

with a and b given by

$$a = \frac{n \sum xy - \left(\sum x\right)\left(\sum y\right)}{n \sum x^2 - \left(\sum x\right)^2} \qquad b = \frac{1}{n}\left(\sum y - a \sum x\right)$$

$\sum x =$ sum of x's
$\sum y =$ sum of y's
$\sum xy =$ sum of products $x \cdot y$
$\sum x^2 =$ sum of squares of x's
$n =$ number of data points

From now on we will find least squares lines by using these formulas, a derivation of which is given at the end of this section.

EXAMPLE 2   Finding The Least Squares Line

A 1955 study compared cigarette smoking to the mortality rate for lung cancer in several countries. Find the least squares line that fits these data. Then use the line to predict lung cancer deaths if per capita cigarette consumption is 600 cigarettes per month (a pack a day).

	Cigarette Consumption (per capita)	Lung Cancer Deaths (per million)
Norway	250	90
Sweden	300	120
Denmark	350	170
Canada	500	150

Solution   The procedure for calculating a and b consists of six steps, beginning with the following table.

1. List the x- and y-values   2. Multiply $x \cdot y$ 3. Square each x

x	y	xy	x^2
250	90	22,500	62,500
300	120	36,000	90,000
350	170	59,500	122,500
500	150	75,000	250,000
1400	530	193,000	525,000
‖	‖	‖	‖
$\Sigma\, x$	$\Sigma\, y$	$\Sigma\, xy$	$\Sigma\, x^2$

4. Sum each column

5. Calculate a and b using the formulas on the previous page.

$$a = \frac{(4)(193{,}000) - (1400)(530)}{(4)(525{,}000) - 1400^2}$$

$$a = \frac{n\,\Sigma\, xy - (\Sigma\, x)(\Sigma\, y)}{n\,\Sigma\, x^2 - (\Sigma\, x)^2}$$

$$= \frac{30{,}000}{140{,}000} = 0.21$$

$$b = \frac{1}{4}[530 - 0.21(1400)] = 59$$

$$b = \frac{1}{n}[\Sigma\, y - a\,\Sigma\, x]$$

n = number of points

6. The least squares line is $y = ax + b$ with the above a and b

$$y = 0.21x + 59$$

$y = ax + b$ with $a = 0.21$ and $y = 59$

This line is graphed below, showing how lung cancer mortality increases with cigarette smoking.

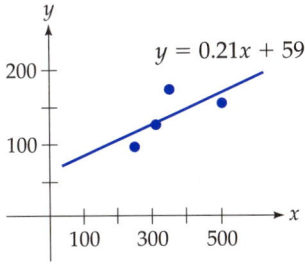

To predict the lung cancer deaths if per capita cigarette consumption reaches 600, we evaluate the least squares line at $x = 600$.

$$y = 0.21 \cdot 600 + 59 = 185 \qquad \text{\textcolor{blue}{$y = 0.21x + 59$ with $x = 600$}}$$

Predicted annual mortality: 185 deaths per million.

■

Criticism of Least Squares

Least squares is the most widely used method for fitting lines to points, but it does have one weakness: The vertical deviations from the line are squared, so one large deviation, when squared, can have an unexpectedly large influence on the line. For example, the graph on the right shows four points and their least squares line. The line fits the points quite closely.

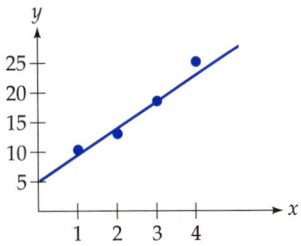

The second graph adds a fifth point, one quite out of line with the others, and shows the least squares line for the five points. The added point has an enormous effect on the line, causing it to slope downward even though all of the other points suggest an upward slope. In actual applications, one should inspect the points for such "outliers" before calculating the least squares line.

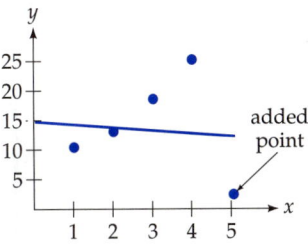

Fitting Exponential Curves by Least Squares

It is not always appropriate to fit a straight line to a set of points. Sometimes a collection of points will suggest a curve rather than a line. Example 3 shows how to fit an exponential curve of the form $y = Be^{Ax}$ to a collection of points.

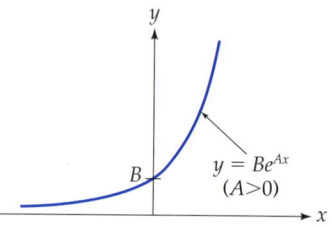

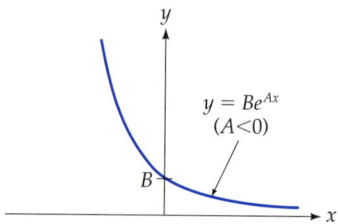

Least squares can be used to fit an exponential curve of the form $y = Be^{Ax}$ to a collection of points as follows. Taking natural logs of both sides of $y = Be^{Ax}$ gives

$$\ln y = \ln (Be^{Ax}) = \ln B + \ln e^{Ax} = \ln B + Ax \qquad \text{Using the properties of natural logarithms}$$

If we introduce a new variable $Y = \ln y$, and let $b = \ln B$, then $\ln y = \ln B + Ax$ becomes

$$Y = b + Ax \qquad \text{Or equivalently } Y = Ax + b$$

Therefore, we fit a straight line to the *logarithms* of the y-values. (Now we are minimizing not the squared deviations, but the squared deviations of the logarithms). The procedure consists of the eight steps shown in the following example.

EXAMPLE 3   Fitting an Exponential Curve by Least Squares

The world population since 1960 is shown in the table on the left below. Fit an exponential curve to these data and predict the world population in the year 2010.

Year	Population (billions)
1960	3.03
1970	3.68
1980	4.48
1990	5.32

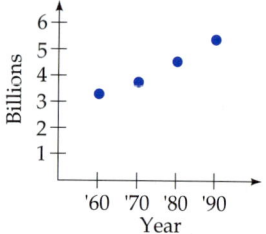

Solution We number the years 1 through 4 (since they are evenly spaced).

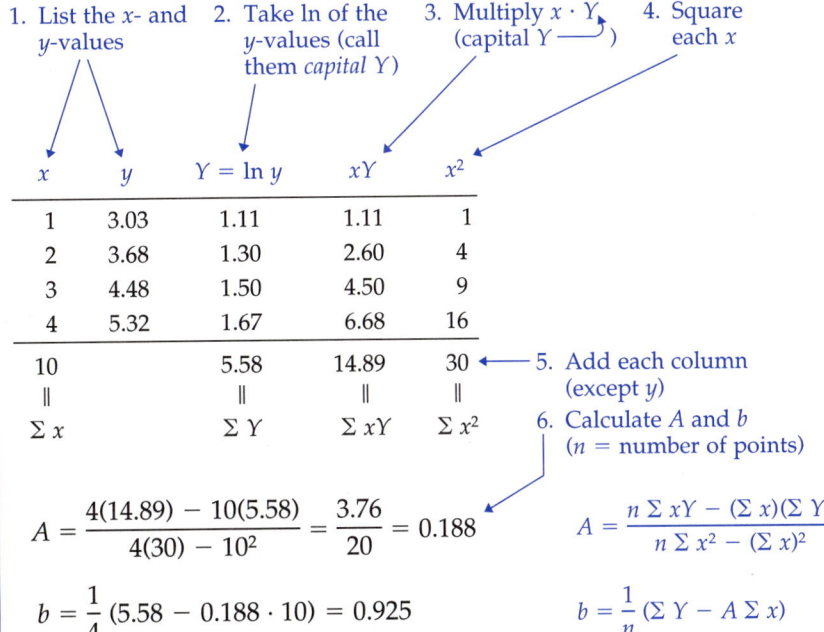

1. List the x- and y-values
2. Take ln of the y-values (call them *capital Y*)
3. Multiply $x \cdot Y$ (capital Y)
4. Square each x

x	y	$Y = \ln y$	xY	x^2
1	3.03	1.11	1.11	1
2	3.68	1.30	2.60	4
3	4.48	1.50	4.50	9
4	5.32	1.67	6.68	16
10		5.58	14.89	30
$\parallel$	$\parallel$	$\parallel$	$\parallel$	
Σx		ΣY	ΣxY	Σx^2

5. Add each column (except y)

6. Calculate A and b (n = number of points)

$$A = \frac{4(14.89) - 10(5.58)}{4(30) - 10^2} = \frac{3.76}{20} = 0.188$$

$$A = \frac{n \, \Sigma xY - (\Sigma x)(\Sigma Y)}{n \, \Sigma x^2 - (\Sigma x)^2}$$

$$b = \frac{1}{4}(5.58 - 0.188 \cdot 10) = 0.925$$

$$b = \frac{1}{n}(\Sigma Y - A \, \Sigma x)$$

$$B = e^b = e^{0.925} \approx 2.52$$

7. Calculate $B = e^b$ (since $b = \ln B$)

The exponential curve is

$$y = 2.52 e^{0.188x}$$

8. The curve is $y = Be^{Ax}$ with the A and B found above

The population in 2010 is this evaluated at $x = 6$:

$$y = 2.52 e^{0.188(6)} \approx 7.8$$

Therefore, world population in the year 2010 is predicted to be 7.8 billion.

Graphing Calculator Exploration

Verify that your graphing calculator gives the same curve $y = 2.52 e^{0.188x}$ that we found in Example 3 by entering the data from the table on page 954 into your calculator and finding the exponential regression curve. [You may find $y = 2.52(1.207)^x$, which is the same curve, since $e^{0.188} \approx 1.207$.]

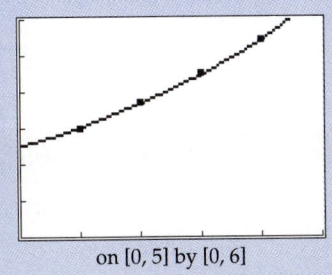

on [0, 5] by [0, 6]

SUMMARY

The technique of least squares can be used to determine the equation of a line that "best fits" a collection of points ("best" in the sense that it minimizes the sum of the squared vertical deviations). The method is described on pages 951–953. Least squares can also be used to fit a *curve* to a collection of points. The procedure for fitting an exponential curve $y = Be^{Ax}$ is described on pages 954–955.

To determine whether to fit a line or a curve to a collection of points, you should first graph the points. For example, the points on the left below suggest a line, and the graph on the right suggests an exponential curve.

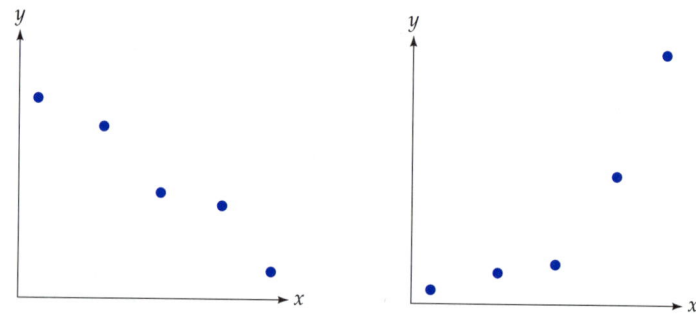

Derivation of the Formula for the Least Squares Line

The formulas for a and b in the least squares line come from minimizing the squared vertical deviations, just as in Example 1, but now for the n points $(x_1, y_1), (x_2, y_2), \ldots, (x_n, y_n)$.

$$S = (ax_1 + b - y_1)^2 + (ax_2 + b - y_2)^2 + \cdots + (ax_n + b - y_n)^2$$

The partials are

$$\frac{\partial S}{\partial a} = 2(ax_1 + b - y_1)x_1 + 2(ax_2 + b - y_2)x_2 + \cdots + 2(ax_n + b - y_n)x_n$$

$$= 2ax_1^2 + 2bx_1 - 2x_1y_1 + 2ax_2^2 + 2bx_2 - 2x_2y_2 + \cdots \qquad \text{Multiplying out}$$
$$+ 2ax_n^2 + 2bx_n - 2x_ny_n$$

$$= 2a(x_1^2 + x_2^2 + \cdots + x_n^2) + 2b(x_1 + x_2 + \cdots + x_n) \qquad \text{Regrouping}$$
$$- 2(x_1y_1 + x_2y_2 + \cdots + x_ny_n)$$

$$= 2a \sum x^2 + 2b \sum x - 2 \sum xy \qquad \text{Using } \Sigma \text{ for sum}$$

$$\frac{\partial S}{\partial b} = 2(ax_1 + b - y_1) + 2(ax_2 + b - y_2) + \cdots + 2(ax_n + b - y_n)$$

$$= 2ax_1 + 2b - 2y_1 + 2ax_2 + 2b - 2y_2 + \cdots + 2ax_n + 2b - 2y_n \qquad \text{Multiplying out}$$

$$= 2a(x_1 + x_2 + \cdots + x_n) + 2b(1 + 1 + \cdots + 1) \qquad \text{Regrouping}$$
$$- 2(y_1 + y_2 + \cdots + y_n)$$

$$= 2a \sum x + 2bn - 2 \sum y \qquad \text{Using } \Sigma \text{ for sum}$$

We set the partials equal to zero:

$$a \sum x^2 + b \sum x - \sum xy = 0 \qquad \text{Dividing each by 2}$$

$$a \sum x + bn - \sum y = 0$$

To solve for the number a, we multiply the first equation by n, multiply the second by $\sum x$, subtract, and solve the resulting equation. The result is

$$a = \frac{n \sum xy - \left(\sum x\right)\left(\sum y\right)}{n \sum x^2 - \left(\sum x\right)^2}$$

and

$$b = \frac{1}{n}\left(\sum y - a \sum x\right) \qquad a \sum x + nb - \sum y = 0 \text{ solved for } b$$

These are the formulas in the box on page 951.

EXERCISES 11.4

You may use if your instructor permits.

Least Squares Line

Find the least squares line for the following points.

1.

x	y
1	2
2	5
3	9

2.

x	y
1	2
2	4
3	7

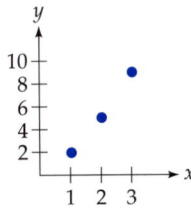

3.

x	y
1	6
3	4
6	2

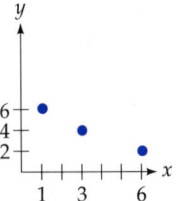

4.

x	y
1	9
4	6
5	1

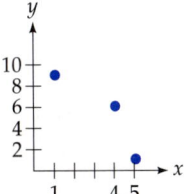

5.

x	y
0	7
1	10
2	10
3	15

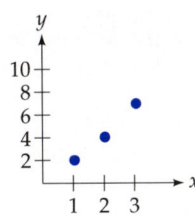

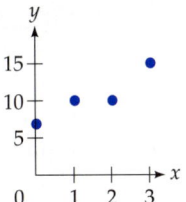

6.

x	y
0	5
1	8
2	8
3	12

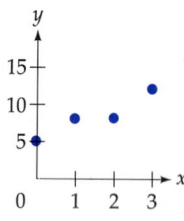

8.

x	y
−2	12
0	10
2	6
4	0
5	−3

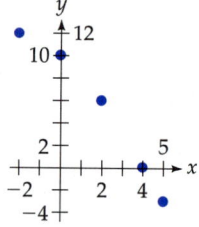

7.

x	y
−1	10
0	8
1	5
3	0
5	−2

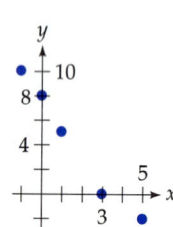

APPLIED EXERCISES

9. Business: Sales A company's annual sales are shown in the following table. Find the least squares line. Use your line to predict the sales in the next year ($x = 5$).

Year	1	2	3	4
Sales (millions)	7	10	11	14

10. General: Automobile Costs The following table gives the cost per mile of operating a compact car, depending upon the number of miles driven per year. Find the least squares line for these data. Use your answer to predict the cost per mile for a car driven 25,000 miles annually ($x = 5$).

Annual Mileage	x	Cost per Mile (cents)
5,000	1	50
10,000	2	35
15,000	3	27
20,000	4	25

11. Sociology: Crime A sociologist finds the following data for the number of felony arrests per year in a town. Find the least squares line. Then use it to predict the number of felony arrests in the next year.

Year	1	2	3	4
Arrests	120	110	90	100

12. General: Farming A farmer's wheat yield (bushels per acre) depends upon the amount of fertilizer (hundreds of pounds per acre) according to the following table. Find the least squares line. Then use the line to predict the yield using 3 hundred pounds of fertilizer per acre.

Fertilizer	1.0	1.5	2	2.5
Yield	30	35	38	40

13. Baseball: The Disappearance of the .400 Hitter Between 1901 and 1930, baseball boasted several .400 hitters (Lajoie, Cobb, Jackson, Sisler, Heilmann, Hornsby, and Terry), but only one since then (Ted Williams in 1941). The decline of the "heavy hitter" is evidenced by the following data, showing the differences between the average for the top five hitters and the major league average over several decades. Find the least squares line for these data and use it to predict the difference for the period 1981–2000 ($x = 5$).

x	High Average Minus League Average
1901–1920	1	82
1921–1940	2	76
1941–1960	3	68
1961–1980	4	59

14. Economics: Consumer Price Index   The consumer price index (CPI) is shown in the following table. Fit a least squares line to the data. Then use the line to predict the CPI in the year 2005 ($x = 6$).

	x	CPI
1980	1	82.4
1985	2	107.6
1990	3	130.7
1995	4	152.4

15–16: General: Percentage of Smokers   The following tables show the percentage of the adult population in the United States who smoke (the first table is males, the second females). Find the least squares line for these data, and use your answer to predict the percentage of that sex who will smoke in the year 2005 ($x = 6$).

15.

	x	Percent Males
1980	1	38
1985	2	32
1990	3	28
1995	4	27

16.

	x	Percent Females
1980	1	30
1985	2	28
1990	3	23
1995	4	22

(*Source:* National Center for Health Statistics, Office on Smoking and Health.)

17. General: Smoking and Longevity   The following data show the life expectancies of a 25-year-old male based on the number of cigarettes smoked daily. Find the least squares line for these data. The slope of the line estimates the years lost per extra cigarette per day.

Cigarettes Smoked Daily	Life Expectancy
0	73.6
5	69.0
15	68.1
30	67.4
40	65.3

18. General: Pollution and Absenteeism   The following table shows the relationship between the sulfur dust content of the air (in $\mu g/m^3$) and the number of female absentees in industry. (Only absences of at least 7 days were counted.) Find the least squares line for these data. Use your answer to predict absences in a city with a sulfur dust content of 25.

	Sulfur	Absences per 1000 Employees
Cincinnati	7	19
Indianapolis	13	44
Woodbridge	14	53
Camden	17	61
Harrison	20	88

Fitting Exponential Curves

Use least squares to find the exponential curve $y = Be^{Ax}$ for the following points.

19.

x	y
1	2
2	4
3	7

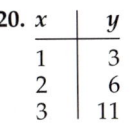

20.

x	y
1	3
2	6
3	11

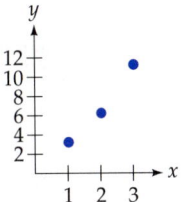

21.

x	y
1	10
3	5
6	1

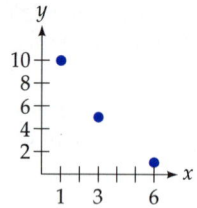

24.

x	y
0	2
1	4
2	7
3	15

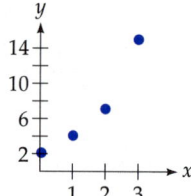

22.

x	y
1	12
4	3
5	2

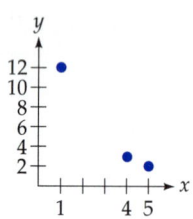

25.

x	y
−1	20
0	18
1	15
3	4
5	1

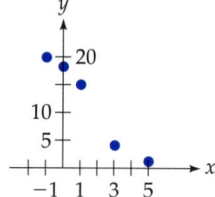

23.

x	y
0	1
1	2
2	5
3	10

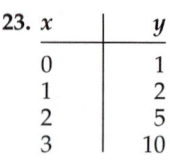
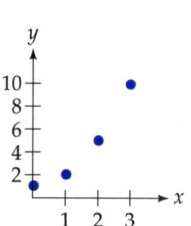

26.

x	y
−2	20
0	12
2	9
4	6
5	5

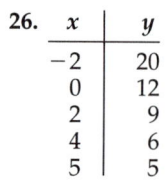
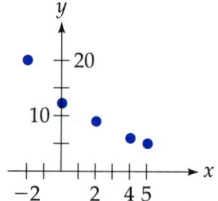

APPLIED EXERCISES

27. Political Science: Cost of a Congressional Victory   The following table shows the average amount spent by the winner of a seat in the House of Representatives. Fit an exponential curve to these data and use it to predict the cost of a House seat in the year 2002.

	x	Cost (in $1000s)
1990	1	408
1992	2	544
1994	3	516
1996	4	674

(*Source:* Center for Responsive Politics.)

28. General: Drunk Driving   The following table shows how a driver's blood-alcohol level (% by weight) affects the probability of being in a collision. A collision factor of 3 means that the

probability of a collision is 3 times greater than normal. Fit an exponential curve to the data. Then use your curve to estimate the collision factor for a blood-alcohol level of 15.

Blood-Alcohol Level (%)	Collision Factor
0	1
6	1.1
8	3
10	6

29. Business: Super Bowl Advertising   The following table shows the cost of advertising during the Super Bowl in *thousands of dollars per second.* Fit an exponential curve to the data and then use it to predict the cost in the year 2002.

	x	Cost ($1000 per Second)
1982	1	11
1987	2	20
1992	3	28.3
1997	4	40

(*Source*: Nielsen Media Research.)

30. General: Stamp Prices   The following table shows the rise in the cost of a first-class postage stamp. Fit an exponential curve to these data and use your answer to predict the cost of a stamp in the year 2008.

	x	Postage (cents)
1968	1	6
1978	2	15
1988	3	25
1998	4	32

31. General: Cost of College Education   The following table shows the cost of one year of tuition with room and board at a private four-year college. Fit an exponential curve to the data and then use it to predict the cost for the academic year 2005–2006.

	x	Cost ($1000)
91–92	1	13.9
92–93	2	14.6
93–94	3	15.5
94–95	4	16.2

(*Source*: College Board.)

Explorations and Excursions

The following exercises extend and augment the material presented in the text.

On pages 258–263 we found least squares lines by a method that used matrices. How does that matrix method relate to the calculus method of this section? Exercise 32 shows that, for a particular set of

points, the two methods give the same result. Exercise 33 shows that for *any* points the two methods lead to the same result, thereby justifying the matrix method.

32. Use the matrix method of Section 2.6 to find a least squares line $y = ax + b$ for a particular set of points by carrying out the following steps.

a. Check that the matrix method described on page 263, applied to the points

x	1	2	3
y	10	12	25

leads to the following matrix equation for a and b:

$$\begin{pmatrix} 1 & 2 & 3 \\ 1 & 1 & 1 \end{pmatrix} \begin{pmatrix} 1 & 1 \\ 2 & 1 \\ 3 & 1 \end{pmatrix} \begin{pmatrix} a \\ b \end{pmatrix} = \begin{pmatrix} 1 & 2 & 3 \\ 1 & 1 & 1 \end{pmatrix} \begin{pmatrix} 10 \\ 12 \\ 25 \end{pmatrix}$$

b. Use matrix multiplication to simplify each side of this equation and obtain:

$$\begin{pmatrix} 14 & 6 \\ 6 & 3 \end{pmatrix} \begin{pmatrix} a \\ b \end{pmatrix} = \begin{pmatrix} 109 \\ 47 \end{pmatrix}$$

c. Solve this matrix equation for $\begin{pmatrix} a \\ b \end{pmatrix}$ and show that the resulting values for a and b give the least squares line $y = 7.5x + 0.67$. Verify that this is the same line that was found by calculus methods in Example 1 on pages 949–951.

33. Why does the matrix method of Section 2.6 find the least squares line? Because it leads to the same equations that are found by setting the partial derivatives of the sum of squared deviations equal to zero. To see this, carry out the following steps.

a. Check that the matrix method described on page 263 for the least squares line $y = ax + b$, when applied to the points

x	x_1	x_2	$\cdots$	x_n
y	y_1	y_2	$\cdots$	y_n

n is the number of points

leads to the following matrix equation for a and b.

$$\begin{pmatrix} x_1 & x_2 & \cdots & x_n \\ 1 & 1 & \cdots & 1 \end{pmatrix} \begin{pmatrix} x_1 & 1 \\ x_2 & 1 \\ \vdots & \vdots \\ x_n & 1 \end{pmatrix} \begin{pmatrix} a \\ b \end{pmatrix}$$

$$= \begin{pmatrix} x_1 & x_2 & \cdots & x_n \\ 1 & 1 & \cdots & 1 \end{pmatrix} \begin{pmatrix} y_1 \\ y_2 \\ \vdots \\ y_n \end{pmatrix}$$

b. Use matrix multiplication to simplify each side of this equation and obtain:

$$\begin{pmatrix} (\sum x^2)a + (\sum x)b \\ (\sum x)a + nb \end{pmatrix} = \begin{pmatrix} \sum xy \\ \sum y \end{pmatrix}$$

c. Verify that this matrix equation is equivalent to the following two equations, from which the formulas for a and b for the least squares line were found (see page 957):

$$a \sum x^2 + b \sum x - \sum xy = 0$$

$$a \sum x + bn - \sum y = 0$$

Therefore, the matrix method leads to the same equations as setting the partial derivatives of the sum of the squares equal to zero. This completes the justification of the matrix method for finding the least squares line.

11.5 Lagrange Multipliers and Constrained Optimization

Introduction

In Section 11.3 we optimized functions of several variables. Some problems, however, involve maximizing or minimizing a function subject to a *constraint*. For example, a company might want to maximize production subject to the constraint of staying within its budget, or a beer distributor might want to design the least expensive aluminum can subject to the constraint that it hold exactly twelve ounces of beer. In this section we solve such "constrained optimization" problems by the method of Lagrange multipliers. The method was invented by the French mathematician Joseph Louis Lagrange (1736–1813).

Constraints

If a constraint can be written as an equation, such as

$$x^2 + y^2 = 100$$

then by moving all the terms to the left-hand side, it can be written with zero on the right-hand side:

$$\underbrace{x^2 + y^2 - 100}_{g(x,\, y)} = 0$$

In general, any equation can be written with all terms moved to the left-hand side, and we will write all constraints in the form

$$g(x,\, y) = 0$$

PRACTICE PROBLEM 1   Write the constraint $2y^3 = 3x - 1$ in the form $g(x, y) = 0$.

Solution at the back of the book

First Example of Lagrange Multipliers

We illustrate the method of Lagrange multipliers by an example. The method requires a new variable, and it is customary to use λ ("lambda," the Greek letter L, as in Lagrange).

EXAMPLE 1   **Maximizing the Area of an Enclosure**

A cattle rancher wants to build a rectangular holding pen along an existing stone wall. If the side along the wall needs no fence, find the dimensions of the largest pen that can be made using only 400 feet of fence.

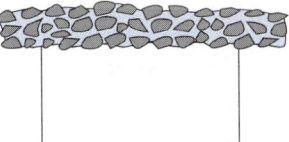

Solution

We want the pen of largest *area*. Let

    x = width (perpendicular to the wall)

    y = length (parallel to the wall)

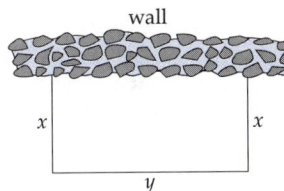

Two widths and one length must be made from the 400 feet of fence, so the constraint is

$$2x + y = 400$$

Therefore, the problem becomes:

    Maximize $A = xy$ Area is length times width

    Subject to $\underbrace{2x + y - 400}_{g(x,y)} = 0$ Constraint $2x + y = 400$ written with zero on the right

We write a new function $F(x, y, \lambda)$, called the *Lagrange function*, which consists of the function to be maximized plus λ times the constraint function:

$$F(x, y, \lambda) = xy + \lambda(2x + y - 400) \qquad F(x, y, \lambda) = \begin{pmatrix} \text{Function to} \\ \text{be optimized} \end{pmatrix}$$

$$+ \lambda \begin{pmatrix} \text{Constraint} \\ \text{function} \end{pmatrix}$$

$$= xy + \lambda 2x + \lambda y - \lambda 400 \qquad \text{Multiplied out}$$

The Lagrange function $F(x, y, \lambda)$ is a function of three variables, x, y, and λ, and we begin as usual by setting its partials with respect to each variable equal to zero.

$$F_x = y + 2\lambda \qquad = 0 \qquad \begin{array}{l} \text{Partial of } xy + \lambda 2x + \lambda y - \lambda 400 \\ \text{with respect to } x \end{array}$$

$$F_y = x + \lambda \qquad = 0 \qquad \begin{array}{l} \text{Partial of } xy + \lambda 2x + \lambda y - \lambda 400 \\ \text{with respect to } y \end{array}$$

$$F_\lambda = \underbrace{2x + y - 400}_{} = 0 \qquad \begin{array}{l} \text{Partial of } xy + \lambda 2x + \lambda y - \lambda 400 \\ \text{with respect to } \lambda \end{array}$$

Constraint $g = 0$

We solve the first two of these equations for λ:

$$\lambda = -\frac{1}{2}y \qquad\qquad\qquad \text{Solving } y + 2\lambda = 0 \text{ for } \lambda$$

$$\lambda = -x \qquad\qquad\qquad \text{Solving } x + \lambda = 0 \text{ for } \lambda$$

Then we set these two expressions for λ equal to each other:

$$-\frac{1}{2}y = -x \qquad\qquad \text{Equating } \lambda = -\frac{1}{2}y \text{ and } \lambda = -x$$

$$y = 2x \qquad\qquad\qquad \text{Multiplying by } -2$$

We use this relationship $y = 2x$ to eliminate the y in the equation $2x + y - 400 = 0$ (which came from the third partial $F_\lambda = 0$):

$$2x + 2x - 400 = 0 \qquad 2x + y - 400 = 0 \text{ with } y \text{ replaced by } 2x$$

$$4x = 400 \qquad\qquad \text{Simplifying}$$

$$x = 100 \qquad\qquad \text{Dividing by 4}$$

$$y = 200 \qquad\qquad \text{From } y = 2x \text{ with } x = 100$$

Width is 100 feet (perpendicular to the wall); length is 200 feet (parallel to the wall), as shown in the following diagram.

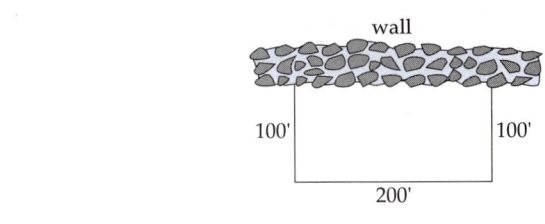

Method of Lagrange Multipliers

In general, the function to be maximized or minimized is called the *objective function,* because the "objective" of the whole procedure is to optimize it. (In Example 1 the objective function was the area function $A = xy$ to be maximized.) The variable λ is called the *Lagrange multiplier.* The entire method may be summarized as follows. (A justification of Lagrange's method is given at the end of this section.)

Lagrange Multipliers

To optimize $f(x, y)$ subject to $g(x, y) = 0$:

Objective function and constraint

1. Write $F(x, y, \lambda) = f(x, y) + \lambda g(x, y)$.

Objective function plus λ times the constraint function

2. Set the partials of F equal to zero:

$$F_x = 0 \qquad F_y = 0 \qquad F_\lambda = 0$$

and solve for the critical points.

3. The solution (if it exists) to the original problem will occur at one of these critical points.

It is important to realize that Lagrange's method only finds *critical points*—it does not tell whether the function is maximized, minimized, or neither at the critical point. [The *D*-test, involving $D = f_{xx}f_{yy} - (f_{xy})^2$, is for *unconstrained* optimization, and cannot be used with constraints.] In each problem we must *know* that the maximum or minimum (whichever is requested) *does* exist, and it then follows that it must occur at a critical point (found by Lagrange multipliers).

How do we know that the maximum or minimum does exist? Most reasonable applied problems *do* have solutions. In Example 1, for example, making the width or length too small would make the area (length times width) small, and so the area must be largest for some "intermediate" length and width. Therefore, the problem *does* have a solution, which is what Lagrange multipliers find.

We could have solved Example 1 *without* Lagrange multipliers by solving the constraint equation $2x + y = 400$ for y and using it to eliminate the y in the objective function, as we did in Sections 7.3 and 7.4. However, Lagrange's method has two advantages: (1) It eliminates the need to solve the constraint equation (which can sometimes be difficult), and (2) it is symmetric in the variables, thereby allowing you to solve for whichever variable is easier to find.

Hints for Solving the Partial Equations

Solving the constraint equations $F_x = 0$, $F_y = 0$, and $F_\lambda = 0$ can sometimes be difficult. The following strategy (used in each example in this section) may be helpful.

1. Solve each of $F_x = 0$ and $F_y = 0$ for λ.
2. Set the two expressions for λ equal to each other.
3. Solve the equation resulting from (2) together with $F_\lambda = 0$ to find x and y.

EXAMPLE 2   Designing the Most Efficient Container

A container company wants to design an aluminum can requiring the least amount of aluminum, but that contains exactly 12 fluid ounces (21.3 cubic inches). Find the radius and height of the can.

Solution   Minimizing the amount of aluminum means minimizing the surface area (top + bottom + sides) of the cylindrical can. If we let r and h stand for the radius and height (in inches) of the can, then the following diagram shows that the area is:

$$A = \underbrace{2\pi r^2}_{\substack{\text{Top and}\\\text{bottom}}} + \underbrace{2\pi rh}_{\substack{\text{Side}\\\text{area}}}$$

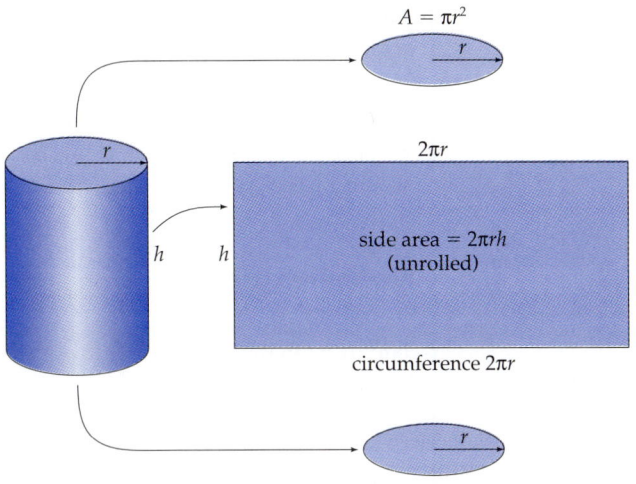

The volume is

$$V = \pi r^2 h$$

Therefore, the problem becomes

minimize $A = 2\pi r^2 + 2\pi rh$ Area = (top and bottom) + side

subject to $\pi r^2 h = 21.3$ Volume must equal 21.3

The Lagrange function is

$$F = 2\pi r^2 + 2\pi rh + \lambda(\pi r^2 h - 21.3)$$ Objective function A plus λ times the constraint

$$F_r = 4\pi r + 2\pi h + \lambda 2\pi rh = 0$$

$$F_h = 2\pi r + \lambda\pi r^2 \qquad\quad = 0$$ Partials set equal to zero

$$F_\lambda = \pi r^2 h - 21.3 \qquad\quad = 0$$

Solving, we have

$$\lambda = -\frac{4\pi r + 2\pi h}{2\pi rh} = -\frac{2r + h}{rh}$$ $4\pi r + 2\pi h + \lambda 2\pi rh = 0$ solved for λ

$$\lambda = -\frac{2\pi r}{\pi r^2} = -\frac{2}{r}$$ $2\pi r + \lambda\pi r^2 = 0$ solved for λ

Equating λ's yields

$$\frac{2r + h}{rh} = \frac{2}{r}$$ Equating the two expressions for λ and multiplying each side by -1

$$2r + h = 2h$$ Multiplying each side by rh

$$2r = h$$ Subtracting h from each side

Therefore,

$$\pi r^2 (2r) = 21.3$$

$\pi r^2 h - 21.3 = 0$ (the third partial equation) with $h = 2r$

$$2\pi r^3 = 21.3$$

Simplifying

$$r^3 = \frac{21.3}{2\pi} \approx 3.39$$

Dividing by 2π (using a calculator)

$$r \approx \sqrt[3]{3.39} \approx 1.5$$

Taking cube roots

$$h = 3$$

From $h = 2r$ with $r = 1.5$

The most economical 12-fluid-ounce can has radius $r = 1.5$ inches and height $h = 3$ inches.

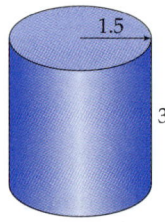

Notice that the height (3 inches) is twice the radius (1.5 inches), so *the height equals the diameter.* This shows that the most efficient can (least area for given volume) has a "squarish" shape.

It is interesting to consider why so few cans are shaped like this. The most common 12-ounce can for beer or soft drinks is about twice as tall as it is across, requiring about 67% more aluminum than the most efficient can. This results in an enormous waste for the millions of cans manufactured each year. It seems that beverage companies prefer taller cans because they have more area for advertising, and they are easier to hold. Some products, however, are sold in "efficient" cans, with the height equal to the diameter.

PRACTICE PROBLEM 2   Which of the cans pictured below is most "efficiently" proportioned?

cola   paint   motor oil   tuna

Solution at the back of the book

With Lagrange multipliers we can maximize a company's output subject to a budget constraint.

EXAMPLE 3   Maximizing Production

A company's output is given by the Cobb–Douglas production function $P = 600L^{2/3}K^{1/3}$, where L and K are the numbers of units of labor and capital, respectively. Each unit of labor costs the company \$40, and each unit of capital costs \$100. If the company has a total of \$3000 for labor and capital, how much of each should it use to maximize production?

Solution   The budget constraint is

$$40L + 100K = 3000$$ Unit cost times the number of units for labor and capital

Cost of   Cost of   Total
labor   capital   budget

Therefore, the problem becomes:

maximize $P = 600L^{2/3}K^{1/3}$ Constraint with zero
Subject to $40L + 100K - 3000 = 0$ on the right

$$F(L, K, \lambda) = 600L^{2/3}K^{1/3} + \lambda(40L + 100K - 3000)$$ Lagrange function

$$F_L = 400L^{-1/3}K^{1/3} + 40\lambda = 0$$
$$F_K = 200L^{2/3}K^{-2/3} + 100\lambda = 0$$ Partials set equal to zero
$$F_\lambda = 40L + 100K - 3000 = 0$$

$$\lambda = -10L^{-1/3}K^{1/3}$$ Solving the first for λ

$$\lambda = -2L^{2/3}K^{-2/3}$$ Solving the second for λ

$$-10L^{-1/3}K^{1/3} = -2L^{2/3}K^{-2/3}$$ Equating the two expressions for λ

$$5L^{-1/3}K^{1/3} = L^{2/3}K^{-2/3}$$ Dividing each side by -2

$$5K = L$$ Multiplying by $L^{1/3}K^{2/3}$ on each side

$$40(5K) + 100K - 3000 = 0$$ Substituting $L = 5K$ into $40L + 100K - 3000 = 0$

$$300K = 3000$$ Simplifying

$$K = 10$$ Solving for K

$$L = 50$$ From $L = 5K$ with $K = 10$

The company should use 50 units of labor and 10 units of capital.

Meaning of the Lagrange Multiplier

The Lagrange multiplier λ has a useful interpretation. If we call the units of the objective function "objective units" and the units of the constraint function "constraint units," then λ (or, more precisely, its absolute value) has the following interpretation, which will be justified in the next section.

Interpretation of λ

$$|\lambda| = \left(\begin{array}{c}\text{Number of additional objective units}\\ \text{for each additional constraint unit}\end{array}\right)$$

As an illustration, in Example 3 we maximized production subject to a budget constraint (dollars), so λ gives *the number of additional units produced per additional dollar*, or, in the jargon of economics, *the marginal productivity of money*. We can calculate the value of λ from either of the expressions for it.

$$\lambda = -10L^{-1/3}K^{1/3} = -10(50)^{-1/3}(10)^{1/3} \approx -5.8$$

From the previous page

From Example 3   Substituting $L = 50, K = 10$ Using a calculator

This number (ignoring the negative sign) means that production increases by about 5.8 units for each additional dollar. Therefore, an additional \$100 would result in about 580 extra units of production.

PRACTICE PROBLEM 3 In Example 1 we maximized the area of a pen subject to the constraint of having only 400 feet of fence.

a. Interpret the meaning of λ in Example 1. (*Hint:* The objective function is area, and the constraint units are feet of fence.)

b. Calculate the absolute value of λ in Example 1. (*Hint:* Use either expression for λ on page 964.)

c. Use this number to approximate the additional area that could be enclosed by an additional 5 feet of fence.

Solutions at the back of the book

Geometry of Constrained Optimization

A constrained optimization problem may be visualized as a surface (the objective function), with a curve along it determined by the constraint. The *constrained* maximum is the highest point on the *curve*, whereas the *unconstrained* maximum is the highest point on the *entire surface*, as shown in the following diagram.

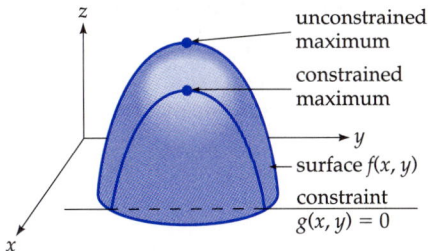

If there are several critical points, we evaluate the objective function at each. The maximum and minimum values (which we assume to exist) are the largest and smallest of the resulting values of the function. In this way we can find both the maximum and minimum values in a constrained optimization problem.

EXAMPLE 4 Finding Both Extreme Values

Maximize *and* minimize $f(x, y) = 4xy$ subject to the constraint $x^2 + y^2 = 50$.

Solution

$$F(x, y, \lambda) = 4xy + \lambda(x^2 + y^2 - 50) \qquad \text{Lagrange function}$$

$$F_x = 4y + \lambda 2x \quad = 0$$

$$F_y = 4x + \lambda 2y \quad = 0 \qquad \left.\begin{array}{r}\end{array}\right\} \text{Partials set equal to zero}$$

$$F_\lambda = x^2 + y^2 - 50 = 0$$

$$\lambda = -\frac{4y}{2x} = -\frac{2y}{x} \qquad \begin{array}{l}\text{Solving } 4y + \lambda 2x = 0 \text{ for } \lambda \\ \text{(and simplifying)}\end{array}$$

$$\lambda = -\frac{4x}{2y} = -\frac{2x}{y} \qquad \begin{array}{l}\text{Solving } 4x + \lambda 2y = 0 \text{ for } \lambda \\ \text{(and simplifying)}\end{array}$$

$$\frac{2y}{x} = \frac{2x}{y} \qquad \begin{array}{l}\text{Equating the two } \lambda\text{'s} \\ \text{(multiplying by } -1)\end{array}$$

$$2y^2 = 2x^2 \qquad \begin{array}{l}\text{Multiplying both sides by } xy \\ \text{(or "cross-multiplying")}\end{array}$$

$$y^2 = x^2 \qquad \text{Dividing by 2}$$

$$y = \pm x \qquad \begin{array}{l}\text{Taking square roots} \\ (+ \text{ and } -)\end{array}$$

$$x^2 + x^2 - 50 = 0 \qquad \begin{array}{l}x^2 + y^2 - 50 = 0 \text{ (the third} \\ \text{partial equation) with } y = \pm x\end{array}$$

$$2x^2 - 50 = 0 \qquad \text{Simplifying}$$

$$2(x^2 - 25) = 0 \qquad \text{Factoring}$$

$$\underbrace{}$$

$$(x + 5)(x - 5)$$

$$x = \pm 5 \qquad \text{Solving}$$

$$y = \pm 5 \qquad \text{From } y = \pm x$$

These give *four* critical points (all possible combinations of $x = \pm 5$ and $y = \pm 5$), and we evaluate the objective function $f(x, y) = 4xy$ at each.

Critical Point	$f(x, y) = 4xy$	
(5, 5)	100	$\left.\begin{array}{l}\\\end{array}\right\}$ 100 is the highest value of $f = 4xy$, occurring at (5, 5) and at (−5, −5)
(−5, −5)	100	
(5, −5)	−100	$\left.\begin{array}{l}\\\end{array}\right\}$ −100 is the lowest value of $f = 4xy$, occurring at (−5, 5) and at (5, −5)
(−5, 5)	−100	

Maximum value of f is 100, occurring at $\begin{cases} x = 5 \\ y = 5 \end{cases}$ and $\begin{cases} x = -5 \\ y = -5 \end{cases}$.

Minimum value of f is -100, occurring at $\begin{cases} x = -5 \\ y = 5 \end{cases}$ and $\begin{cases} x = 5 \\ y = -5 \end{cases}$.

■

Constrained Optimization of Functions of Three or More Variables

To optimize a function f of any number of variables subject to a constraint $g = 0$, we proceed just as before, writing the Lagrange function $F = f + \lambda g$ and setting the partial with respect to each variable equal to zero. For example, to maximize $f(x, y, z)$ subject to $g(x, y, z) = 0$, we would write

$$F(x, y, z, \lambda) = f(x, y, z) + \lambda g(x, y, z) \qquad \text{Lagrange function}$$

and solve the partial equations

$$F_x = 0$$
$$F_y = 0$$
$$F_z = 0$$
$$F_\lambda = 0$$

Justification of Lagrange's Method

The following is a geometric justification of Lagrange's method for maximizing an objective function $f(x, y)$ subject to a constraint $g(x, y) = 0$. In the graph below, each of the parallel curves represents a curve along which the objective function f takes a constant value, with successively higher curves corresponding to higher values of the constant. Maximizing f subject to $g = 0$ means finding the highest of these "objective" curves that still meets the constraint curve.

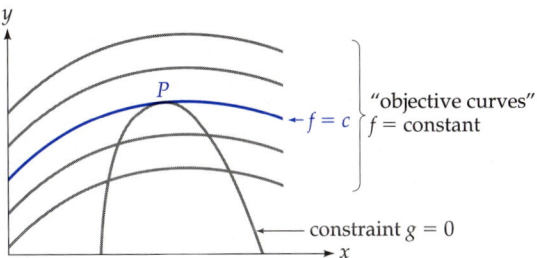

Observe that at the point where the highest objective curve meets the constraint curve (the point P), *the two curves have the same slope*. In Exercise 37 on page 986 we will show that the slope of the curve $f = c$ is $-f_x/f_y$, and the slope of the curve $g = 0$ is $-g_x/g_y$. Equating these slopes gives

$$-\frac{f_x}{f_y} = -\frac{g_x}{g_y} \qquad \text{or equivalently} \qquad -\frac{f_x}{g_x} = -\frac{f_y}{g_y} \qquad \begin{array}{l}\text{Multiplying by } f_y \\ \text{and dividing by } g_x\end{array}$$

The two sides of this last equation being equal is equivalent to equating each to a number λ:

$$\lambda = -\frac{f_x}{g_x} \qquad \text{and} \qquad \lambda = -\frac{f_y}{g_y}$$

With a little algebra (clearing fractions, and moving everything to one side), these become

$$f_x + \lambda g_x = 0 \qquad \text{and} \qquad f_y + \lambda g_y = 0$$

However, these two equations, together with the constraint $g = 0$, are precisely the equations obtained from Lagrange's method by setting the partials of $F = f + \lambda g$ equal to zero. This completes the justification of Lagrange's method, and also exhibits its simplicity and power by showing all that is accomplished by the familiar steps of setting the partials equal to zero.

EXERCISES 11.5

Use Lagrange multipliers to maximize each function $f(x, y)$ subject to the constraint. (The maximum values *do* exist.)

1. $f(x, y) = 3xy, \quad x + 3y = 12$

2. $f(x, y) = 2xy, \quad 2x + y = 20$

3. $f(x, y) = 6xy, \quad 2x + 3y = 24$

4. $f(x, y) = 3xy, \quad 3x + 2y = 60$

5. $f(x, y) = xy - 2x^2 - y^2, \quad x + y = 8$

6. $f(x, y) = 12xy - 3y^2 - x^2, \quad x + y = 16$

7. $f(x, y) = x^2 - y^2 + 3, \quad 2x + y = 3$

8. $f(x, y) = y^2 - x^2 - 5, \quad x + 2y = 9$

9. $f(x, y) = \ln(xy), \quad x + y = 2e$

10. $f(x, y) = e^{xy}, \quad x + 2y = 8$

Use Lagrange multipliers to minimize each function $f(x, y)$ subject to the constraint. (The minimum values *do* exist.)

11. $f(x, y) = x^2 + y^2, \quad 2x + y = 15$

12. $f(x, y) = x^2 + y^2, \quad x + 2y = 30$

13. $f(x, y) = xy, \quad y = x + 8$

14. $f(x, y) = xy, \quad y = x + 6$

15. $f(x, y) = x^2 + y^2, \quad 2x + 3y = 26$

16. $f(x, y) = 5x^2 + 6y^2 - xy, \quad x + 2y = 24$

17. $f(x, y) = \ln(x^2 + y^2), \quad 2x + y = 25$

18. $f(x, y) = \sqrt{x^2 + y^2 + 5}, \quad 2x + y = 10$

19. $f(x, y) = e^{x^2 + y^2}, \quad x + 2y = 10$

20. $f(x, y) = 2x + y, \quad 2 \ln x + \ln y = 12$

Use Lagrange multipliers to maximize *and* minimize each function subject to the constraint. (The maximum and minimum values *do* exist.)

21. $f(x, y) = 2xy, \quad x^2 + y^2 = 8$

22. $f(x, y) = 2xy, \quad x^2 + y^2 = 18$

23. $f(x, y) = x + 2y, \quad 2x^2 + y^2 = 72$

24. $f(x, y) = 12x + 30y, \quad x^2 + 5y^2 = 81$

APPLIED EXERCISES

Solve each using Lagrange multipliers. (The stated extreme values *do* exist.)

25. General: Fences A parking lot, divided into two equal parts, is to be constructed against a building, as shown in the diagram. Only 6000 feet of fence are to be used, and the side along the building needs no fence.

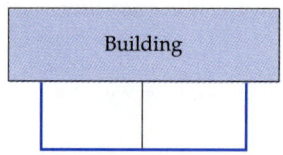

a. What are the dimensions of the largest area that can be so enclosed?

b. Evaluate and give an interpretation for λ.

26. General: Fences Three adjacent rectangular lots are to be fenced in, as shown in the dia-

gram, using 12,000 feet of fence. What is the largest total area that can be so enclosed?

27–28: General: Container Design A cylindrical tank without a top is to be constructed with the least amount of materials (bottom plus side area). Find the dimensions if the volume is to be:

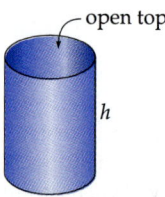

27. 160 cubic feet

28. 120 cubic feet

29. General: Postal Regulations   The U.S. Postal Service will accept a package if its length plus its girth is not more than 84 inches. Find the dimensions and volume of the largest package with a square end that can be mailed.

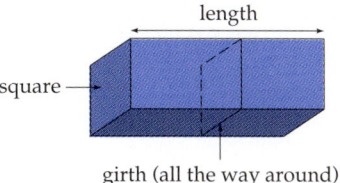

length

square

girth (all the way around)

30. General: Postal Regulations   Solve Exercise 29, but now for a package with a round end, so that the package is a cylinder rather than a rectangular solid. Compare the volume with that of Exercise 29.

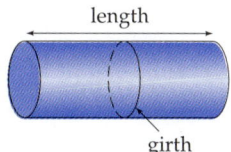

length

girth

 31. Business: Maximum Production   A company's output is given by the Cobb–Douglas production function $P = 200L^{3/4}K^{1/4}$, where L and K are the numbers of units of labor and capital, respectively. Each unit of labor costs $50 and each unit of capital costs $100, and $8000 is available to pay for labor and capital.

a. How many units of each should be used to maximize production?

b. Evaluate and give an interpretation for λ.

32. Business: Production Possibilities   A company manufactures two products, in quantities x and y. Because of limited materials and capital, the quantities produced must satisfy the equation $2x^2 + 5y^2 = 32{,}500$. (This curve is called a *production possibilities curve*.) If the company's profit function is $P = 4x + 5y$ dollars, how many of each product should be made

to maximize profit? Also find the maximum profit.

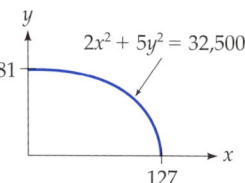

$2x^2 + 5y^2 = 32{,}500$

81

127

33. General: Package Design   A metal box with a square base is to have a volume of 45 cubic inches. If the top and bottom cost 50 cents per square inch and the sides cost 30 cents per square inch, find the dimensions that minimize the cost. [*Hint:* The cost of the box is the area of each part (top, bottom, and sides) times the cost per square inch for that part. Minimize this subject to the volume constraint.]

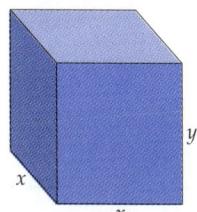

y

x

x

34. General: Building Design   A one-story building is to have 8000 square feet of floor space. The front of the building is to be made of brick, which costs $120 per linear foot, and the back and sides are to be made of cinderblock, which costs only $80 per linear foot.

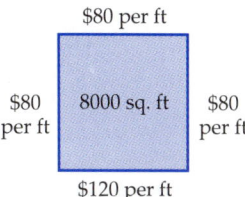

$80 per ft

$80 per ft   8000 sq. ft $80 per ft

$120 per ft

a. Find the length and width that minimize the cost of the building. (*Hint:* The cost of the building is the length of the front, back, and sides, each times the cost per foot for that part. Minimize this subject to the area constraint.)

b. Evaluate and give an interpretation for λ.

Functions of Three Variables

(The stated extreme values *do* exist.)

35. Minimize $f(x, y, z) = x^2 + y^2 + z^2$
subject to $2x + y - z = 12$.

36. Minimize $f(x, y, z) = x^2 + y^2 + z^2$
subject to $x - y + 2z = 6$.

37. Maximize $f(x, y, z) = x + y + z$
subject to $x^2 + y^2 + z^2 = 12$.

38. Maximize $f(x, y, z) = xyz$
subject to $x^2 + y^2 + z^2 = 12$.

39. General: Building Design   The one-story storage building shown in the following diagram is to have a volume of 250,000 cubic feet. The roof costs \$32 per square foot, the walls \$10 per square foot, and the floor \$8 per square foot. Find the dimensions that minimize the cost of the building.

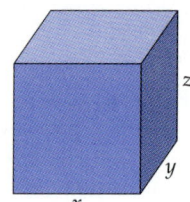

40. General: Container Design   An open-top box with two parallel partitions, as in the diagram, is to have volume 64 cubic inches. Find the dimensions that require the least amount of materials.

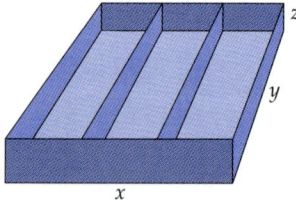

Total Differentials and Approximate Changes

Introduction

In this section we define the *total differential* of a function of several variables, and use it to approximate the change in the function resulting from changes in the independent variables. In addition to giving several applications, we will justify the interpretation of the Lagrange variable λ introduced in the previous section.

Total Differential of a Function of Two Variables

For a function $f(x)$ of one variable, we defined the differential to be the *derivative multiplied by dx*, $dy = f'(x) \cdot dx$. For a function $f(x, y)$ of *two* variables, we define the total differential analogously as the partial derivatives multiplied by dx and dy and added.*

---

*Technically, the partials f_x and f_y must be continuous. However, since we have not defined continuity for functions of two variables, we will not discuss this requirement further, except to say that it is satisfied by all functions in this section and by most functions encountered in applications.

Total Differential of $f(x, y)$

For a function $f(x, y)$, the total differential df is

$$df = f_x(x, y) \cdot dx + f_y(x, y) \cdot dy$$

More briefly,
$df = f_x \cdot dx + f_y \cdot dy$

The total differential may be written with the partials in the ∂ notation:

$$df = \frac{\partial f}{\partial x} \cdot dx + \frac{\partial f}{\partial y} \cdot dy$$
Total differential in ∂ notation

EXAMPLE 1 Finding a Total Differential

Find the total differential df of $f(x, y) = 5x^3 - 4xy^{-1} + 3y^4$.

Solution The partials are

$$f_x = 15x^2 - 4y^{-1}$$
Partial of $5x^3 - 4xy^{-1} + 3y^4$ with respect to x

$$f_y = 4xy^{-2} + 12y^3$$
Partial of $5x^3 - 4xy^{-1} + 3y^4$ with respect to y

Then df is

$$df = \underbrace{(15x^2 - 4y^{-1})}_{f_x(x, y)} \cdot dx + \underbrace{(4xy^{-2} + 12y^3)}_{f_y(x, y)} \cdot dy$$
$df = f_x \, dx + f_y \, dy$ with the above partials

If dx and dy are considered as new variables, then the total differential df is a function of the *four* variables x, y, dx, and dy. The total differential of $z = f(x, y)$ is denoted dz.

EXAMPLE 2 Finding the Total Differential of a Logarithmic Function

Find the total differential of $z = \ln (x^2 + y^3)$.

Solution The partials are

Partials of denominator

$$z_x = \frac{2x}{x^2 + y^3} \qquad z_y = \frac{3y^2}{x^2 + y^3}$$
Derivative of $\ln f$ is $\dfrac{f'}{f}$

Therefore, dz is

$$dz = \frac{2x}{x^2 + y^3} \cdot dx + \frac{3y^2}{x^2 + y^3} \cdot dy$$

Partials times dx and dy

■

PRACTICE PROBLEM

Find the total differential of $g(x, y) = x^2 e^{5y}$. *Solution at the back of the book*

Approximating Changes by Total Differentials

For a function $f(x, y)$, changing the values of x and y generally changes the value of the function. The change Δf in the function is found by evaluating f at the "changed" values and subtracting f evaluated at the original values.

Change in $f(x, y)$

$$\Delta f = \underbrace{f(x + \Delta x, y + \Delta y)}_{\substack{f \text{ at the} \\ \text{"new" values}}} - \underbrace{f(x, y)}_{\substack{f \text{ at the} \\ \text{"old" values}}}$$

For *independent* variables (like x and y) we will use "Δ" and "d" interchangeably to denote changes. That is,

$$\Delta x = dx \quad \text{and} \quad \Delta y = dy$$

For independent variables,
"Δ" = "d"

However, for *dependent* variables, "Δ" and "d" have different meanings: Δ indicates the *actual* change, and d indicates the total differential.

For some functions, calculating the actual change Δf can be complicated. The total differential provides a simple *approximation* for the actual change. The partials f_x and f_y give the rate of change of f per unit change in x and y, respectively. Therefore, changing x by Δx units changes f by approximately $f_x \cdot \Delta x$ units, and changing y by Δy units changes f by approximately $f_y \cdot \Delta y$ units. Therefore, changing *both* x and y should change f by the *sum* of these approximate changes. In symbols:

Differential Approximation Formula

$$\underbrace{f(x + \Delta x, y + \Delta y) - f(x, y)}_{\text{Change in } f} \approx \underbrace{f_x \cdot \Delta x + f_y \cdot \Delta y}_{\substack{\text{Total differential of } f \\ \text{(since } \Delta x = dx \text{ and } \Delta y = dy)}}$$

Written more compactly:

$$\Delta f \approx df \qquad \begin{aligned} &\text{Since } \Delta f = f(x + \Delta x, y + \Delta y) - f(x, y) \\ &\text{and } df = f_x(x, y) \cdot dx + f_y(x, y) \cdot dy \end{aligned}$$

The partials in the first formula are evaluated at (x, y), and the approximation improves as Δx and Δy approach zero.

EXAMPLE 3 Approximating an Actual Change by a Differential

For $f(x, y) = x^2 + 4xy + y^3$ and values $x = 3$, $y = 2$, $\Delta x = 0.2$, and $\Delta y = -0.1$, find: **a.** Δf **b.** df

Solution

a. From the given values,

$$x + \Delta x = 3 + 0.2 = 3.2 \qquad y + \Delta y = 2 - 0.1 = 1.9$$

The change Δf is

$$\Delta f = f(3.2, 1.9) - f(3, 2) \qquad \qquad \Delta f = f(x + \Delta x, y + \Delta y) - f(x, y)$$

$$= \overbrace{3.2^2 + 4 \cdot (3.2) \cdot (1.9) + 1.9^3}^{f(3.2, 1.9)} - \overbrace{(3^2 + 4 \cdot 3 \cdot 2 + 2^3)}^{f(3, 2)} \quad \begin{aligned} &\text{Using } f(x, y) = \\ &x^2 + 4xy + y^3 \end{aligned}$$

$$= 10.24 + 24.32 + 6.859 - (9 + 24 + 8) \qquad \text{Evaluating}$$

$$= 41.419 - 41 = 0.419 \qquad \begin{aligned} &\text{Change in } f \text{ is} \\ &\Delta f = 0.419 \end{aligned}$$

b. The total differential df is

$$df = \overbrace{(2x + 4y)}^{f_x(x, y)} \cdot dx + \overbrace{(4x + 3y^2)}^{f_y(x, y)} \cdot dy \qquad \begin{aligned} &\text{Partials of} \\ &x^2 + 4xy + y^3 \\ &\text{times } dx \text{ and } dy \end{aligned}$$

$$= (2 \cdot 3 + 4 \cdot 2) \cdot (0.2) + (4 \cdot 3 + 3 \cdot 2^2) \cdot (-0.1) \qquad \begin{aligned} &\text{Evaluating at } x = 3, \\ &y = 2, dx = 0.2, \text{ and} \\ &dy = 0.1 \end{aligned}$$

$$= 2.8 - 2.4 = 0.4 \qquad \begin{aligned} &\text{Total differential} \\ &\text{is } df = 0.4 \end{aligned}$$

We found $\begin{cases} \Delta f = 0.419 \\ df = 0.4 \end{cases}$. The total differential $df = 0.4$ provides an easier and reasonably accurate approximation for the actual change $\Delta f = 0.419$. The approximation would be even more accurate for smaller values of Δx and Δy.

EXAMPLE 4   Estimating Additional Profit

The American Farm Machinery Company finds that if it manufactures x economy tractors and y heavy-duty tractors per month, then its profit (in thousands of dollars) will be $P(x, y) = 3x^{4/3} + 0.05xy + 4y$. If the company now manufactures 125 economy tractors and 100 heavy-duty tractors per month, find an approximation for the additional profit from manufacturing 3 more economy tractors and 2 more heavy-duty tractors per month.

Solution   We want an estimate for the change in the profit $P(x, y)$ when production increases above the levels $x = 125$ and $y = 100$ by amounts $\Delta x = 3$ and $\Delta y = 2$. The partials are

$$P_x = 4x^{1/3} + 0.05y \qquad P_y = 0.05x + 4$$

Partials of
$P = 3x^{4/3} + 0.05xy + 4y$
with respect to x and y

The total differential is

$$dP = (4x^{1/3} + 0.05y) \cdot dx + (0.05x + 4) \cdot dy \qquad\qquad dP = P_x \cdot dx + P_y \cdot dy$$

$$= (4 \cdot 125^{1/3} + 0.05 \cdot 100) \cdot 3 + (0.05 \cdot 125 + 4) \cdot 2$$
Substituting
$x = 125, y = 100,$
$\Delta x = 3, \Delta y = 2$

$$\underbrace{\phantom{(4 \cdot 125^{1/3}}}$$
$$\sqrt[3]{125} = 5$$

$$= (20 + 5) \cdot 3 + (6.25 + 4) \cdot 2 = 75 + 20.5 = 95.5 \qquad\text{In thousands of dollars}$$

Therefore, producing 3 more economy tractors and 2 more heavy-duty tractors will generate about \$95,500 in additional profit.

■

The actual change in the profit function, found by applying the Δf formula on page 978, gives $\Delta P = P(125 + 3, 100 + 2) - P(125, 100) = 96$ so the approximation of 95.5 is indeed quite accurate.

Estimating Errors

No physical measurement can ever be made with perfect accuracy. If you can estimate the maximum error in a measurement, then you can use differentials to estimate the resulting error in a calculation. The measurement errors may be expressed as *actual* numbers (as in Example 6, to be discussed shortly), or as *percentage* errors. The following example estimates the percentage error in calculating the volume of a cylinder. Such calculations have applications from predicting variations in soft-drink cans to assuring safety margins in volumes of artificial arteries.

EXAMPLE 5   **Estimating the Error in Calculating Volume**

A cylinder is measured to have radius r and height h, but these measurements may be in error by up to 1%. Estimate the resulting percentage error in calculating the volume of the cylinder.

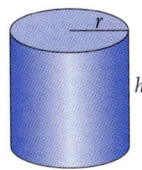

Solution   The height and radius being in error by 1% means that

$$\Delta r = 0.01r$$

$$\Delta h = 0.01h$$
For each, the change is 1% of the value

The volume of a cylinder is $V = \pi r^2 h$, and the total differential is

$$dV = 2\pi rh \cdot dr + \pi r^2 \cdot dh$$
Partials of $V = \pi r^2 h$ times dr and dh

$$= 2\pi rh \cdot 0.01r + \pi r^2 \cdot 0.01h$$
Substituting $dr = \Delta r = 0.01r$ and $dh = \Delta h = 0.01h$

$$= \pi r^2 h(\underbrace{2 \cdot 0.01 + 0.01}_{0.03})$$
Factoring out $\pi r^2 h$

$$= 0.03\underbrace{\pi r^2 h}_{} = 0.03 \cdot V$$
Change is 3% of volume

— Volume V of the cylinder

This result, $dV = 0.03 \cdot V$, means that the volume may be in error by as much as 3% if the radius and height are "off" by 1%.

∎

Geometric Visualization of *df* and Δ*f*

A function $f(x, y)$ represents a *surface* in three-dimensional space, and the change Δf represents the change in height when moving from one point to another along the surface, as shown below.

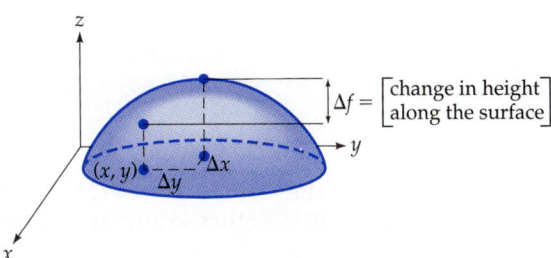

The total differential df represents the change in height when moving from one point to another along the *plane* that best fits the surface at the first point, as shown below. This plane is called the *tangent plane* to the surface at the point.

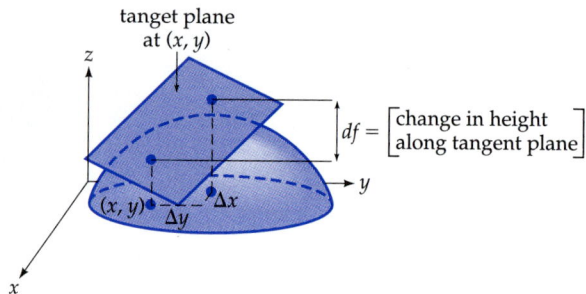

Since the tangent plane fits the surface closely near the point (x, y), changes df in height along the *tangent plane* should be very close to changes Δf in height along the *surface*, which is why the total differential df closely approximates the actual change Δf.

Total Differential of a Function of Three Variables

The total differential may be generalized to apply to functions of three (or more) variables, multiplying each partial by "d" of that variable and adding.

Total Differential of $f(x, y, z)$

For a function $f(x, y, z)$, the total differential df is

$$df = f_x(x, y, z) \cdot dx + f_y(x, y, z) \cdot dy + f_z(x, y, z) \cdot dz$$

Partials times dx, dy, and dz

The total differential of a function $f(x, y, z)$ gives an estimate for the change Δf when the variables are changed by amounts Δx, Δy, and Δz:

$$\underbrace{f(x + \Delta x, y + \Delta y, z + \Delta z) - f(x, y, z)}_{\text{Change in } f} \approx \underbrace{\frac{\partial f}{\partial x} \cdot \Delta x + \frac{\partial f}{\partial y} \cdot \Delta y + \frac{\partial f}{\partial z} \cdot \Delta z}_{\substack{\text{Total differential of } f \\ (\text{since } \Delta x = dx, \\ \Delta y = dy, \Delta z = dz)}}$$

or more briefly:
$$\Delta f \approx df$$

The partials in the above formula are evaluated at (x, y, z), and the approximation is increasingly accurate for smaller values of Δx, Δy, and Δz.

EXAMPLE 6   Estimating the Error in Calculating Volume

A rectangular box is measured to be 30 inches long, 24 inches wide, and 10 inches high. If the maximum error in measuring its length is 0.3 inch, in its width is 0.2 inch, and in its height is 0.1 inch, estimate the maximum error in calculating its volume.

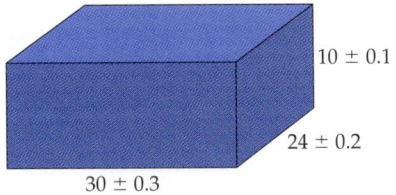

10 ± 0.1

24 ± 0.2

30 ± 0.3

Solution   The volume of the box is length times width times height, $V = x \cdot y \cdot z$. We want to estimate the change in volume resulting from changing the dimensions $x = 30$, $y = 24$, and $z = 10$ by amounts $\Delta x = 0.3$, $\Delta y = 0.2$, and $\Delta z = 0.1$. The total differential is

$$df = \underbrace{y \cdot z}_{V_x} \cdot dx + \underbrace{x \cdot z}_{V_y} \cdot dy + \underbrace{x \cdot y}_{V_z} \cdot dz$$
Partials of $V = x \cdot y \cdot z$ times dx, dy, and dz

$$= 24 \cdot 10 \cdot 0.3 + 30 \cdot 10 \cdot 0.2 + 30 \cdot 24 \cdot 0.1$$
Substituting $x = 30$, $y = 24$, $z = 10$, $dx = 0.3$, $dy = 0.2$, $dz = 0.1$

$$= 72 + 60 + 72 = 204$$
Multiplying out and adding

That is, the maximum error in calculating the volume is approximately 204 cubic inches. ∎

While an error of 204 cubic inches may seem large, the *relative* error (the error divided by the volume, expressed as a percent) is only

$$\frac{204}{30 \cdot 24 \cdot 10} = \frac{204}{7200} \approx 0.028 = 2.8\%$$
Relative error is 2.8%

Justification of the Lagrange Multiplier λ Interpretation

In the previous section we used Lagrange multipliers to optimize an objective function $f(x, y)$ subject to a constraint $g(x, y) = 0$. We did so by setting the partials of $F(x, y, \lambda) = f(x, y) + \lambda g(x, y)$ equal to zero, and we interpreted the value of the "Lagrange multiplier" λ as *the number of additional objective units per additional constraint unit*. This interpretation may be justified as follows. Setting the partials of F with respect to x and y equal to zero gives

$$\begin{cases} f_x + \lambda g_x = 0 \\ f_y + \lambda g_y = 0 \end{cases} \quad \text{or equivalently} \quad \begin{cases} f_x = -\lambda g_x \\ f_y = -\lambda g_y \end{cases}$$

If we increase x and y by amounts Δx and Δy, the resulting change in the objective function $f(x, y)$ can be approximated by the total differential:

$$\underbrace{f(x + \Delta x, y + \Delta y) - f(x, y)}_{\Delta f} \approx f_x \, \Delta x + f_y \, \Delta y \qquad \Delta f \approx df$$

$$= -\lambda g_x \, \Delta x - \lambda g_y \, \Delta y \qquad \begin{array}{l}\text{Substituting} \\ f_x = -\lambda g_x \text{ and} \\ f_y = -\lambda g_y\end{array}$$

$$= -\lambda \underbrace{(g_x \, \Delta x + g_y \, \Delta y)}_{dg} \qquad \begin{array}{l}\text{Factoring out } -\lambda \\ \text{leaves the total} \\ \text{differential}\end{array}$$

$$\approx -\lambda \underbrace{[g(x + \Delta x, y + \Delta y) - g(x, y)]}_{\Delta g} \qquad \begin{array}{l}\text{Replacing } dg \text{ by } \Delta g \\ (\text{since } dg \approx \Delta g)\end{array}$$

These equations, read from beginning to end, say that $\Delta f \approx -\lambda \cdot \Delta g$, or

$$\frac{\Delta f}{\Delta g} \approx -\lambda \qquad \text{Dividing by } \Delta g$$

The left-hand side of this is the ratio of the change in the objective function f to the change in the constraint function g. Taking absolute values (to eliminate the negative sign) and letting Δx and Δy approach zero (to make the approximation exact) shows that $|\lambda|$ is *the number of objective units per additional constraint unit*, as stated on page 970.

SUMMARY

For a function $f(x, y)$, the total differential df is defined as the partials multiplied by dx and dy and added:

$$df = f_x(x, y) \cdot dx + f_y(x, y) \cdot dy \qquad df = \frac{\partial f}{\partial x} dx + \frac{\partial f}{\partial y} dy$$

The actual change in the function when x and y change by amounts Δx and Δy is

$$\Delta f = f(x + \Delta x, y + \Delta y) - f(x, y)$$

For small values of $\Delta x = dx$ and $\Delta y = dy$, the actual change Δf can be approximated by the total differential df:

$$f(x + \Delta x, y + \Delta y) - f(x, y) \approx f_x(x, y) \cdot dx + f_y(x, y) \cdot dy \qquad \Delta f \approx df$$

The actual change Δf may depend on Δx and Δy in very complicated ways, but the total differential df is *linear* in $\Delta x = dx$ and $\Delta y = dy$, and is therefore easier to calculate.

EXERCISES 11.6

Find the total differential of each function.

1. $f(x, y) = x^2 y^3$

2. $f(x, y) = x^4 y^{-1}$

3. $f(x, y) = 6x^{1/2} y^{1/3} + 8$

4. $f(x, y) = 100x^{0.05} y^{0.02} - 7$

5. $g(x, y) = \dfrac{x}{y}$

6. $g(x, y) = \dfrac{x}{x + y}$

7. $g(x, y) = (x - y)^{-1}$

8. $g(x, y) = \sqrt{x^2 + y^2}$

9. $z = \ln(x^3 - y^2)$

10. $z = x^2 \ln y$

11. $z = xe^{2y}$

12. $z = e^{3x - 2y}$

13. $w = 2x^3 + xy + y^2$

14. $w = 3x^2 - xy^{-1} + y^3$

15. $f(x, y, z) = 2x^2 y^3 z^4$

16. $f(x, y, z) = xy + yz + xz$

17. $f(x, y, z) = \ln(xyz)$

18. $f(x, y, z) = \ln(x^2 + y^2 + z^2)$

19. $f(x, y, z) = e^{xyz}$

20. $f(x, y, z) = e^{x^2 + y^2 + z^2}$

For the given function and values, find:
a. Δf **b.** df

21. $f(x, y) = x^2 + xy + y^3$, $x = 4$, $\Delta x = dx = 0.2$, $y = 2$, $\Delta y = dy = -0.1$

22. $f(x, y) = x^3 + xy + y^3$, $x = 5$, $\Delta x = dx = 0.01$, $y = 3$, $\Delta y = dy = -0.01$

23. $f(x, y) = e^x + xy + \ln y$, $x = 0$, $\Delta x = dx = 0.05$, $y = 1$, $\Delta y = dy = 0.01$

24. $f(x, y) = \ln(x^2 + y^2)$, $x = 6$, $\Delta x = dx = 0.1$, $y = 8$, $\Delta y = dy = 0.2$

25. $f(x, y, z) = xy + z^2$, $x = 3$, $\Delta x = dx = 0.03$, $y = 2$, $\Delta y = dy = 0.02$, $z = 1$, $\Delta z = dz = 0.01$

26. $f(x, y, z) = x^2 + y^2 + z^2$, $x = 3$, $\Delta x = dx = 0.1$, $y = 4$, $\Delta y = dy = 0.1$, $z = 5$, $\Delta z = dz = 0.1$

APPLIED EXERCISES

Use total differentials to solve the following exercises.

27. General: Measurement Errors   A rectangle is measured to be 150 feet by 100 feet, but each measurement may be "off" by half a foot. Estimate the error in calculating the area.

28. General: Telephone Calls   For two cities with populations x and y (in thousands) that are 500 miles apart, the number of telephone calls per day between them can be modeled by the function $12xy$. For two cities with populations 40 thousand and 60 thousand, estimate the number of additional telephone calls if each city grows by 1 thousand people.

29–30: Business: Profit   An electronics company's profit from making x tape decks and y CD players per day is given by the following profit function $P(x, y)$. If the company currently produces 200 tape decks and 300 CD players, estimate the extra profit that would result from producing 5 more tape decks and 4 more CD players.

29. $P(x, y) = 2x^2 - 3xy + 3y^2$

30. $P(x, y) = 3x^2 - 4xy + 4y^2$

31. General: Automobile Safety   The emergency stopping distance for a truck of weight w tons traveling at v miles per hour on a dry road is $S = 0.027wv^2$. For a truck that weighs 4 tons and is usually driven at 60 mph, estimate the extra stopping distance if it has an extra half ton of load and is traveling 5 mph faster than usual.

32. General: Scuba Diving   The maximum duration (in minutes) of a scuba dive can be estimated by the formula $T = \dfrac{33v}{d + 33}$, where v is the volume of air in the tank (in cubic feet at sea-level pressure) and d is the depth (in feet) of the dive. For values $v = 100$ and depth 67 feet, estimate the change in duration if an extra 20 cubic feet of air are added and the dive is 10 feet deeper.

33. General: Relative Error in Calculating Area

A rectangle is measured to have length x and width y, but each measurement may be in error by 1%. Estimate the percentage error in calculating the area.

34. General: Relative Error in Calculating Volume   A rectangular solid is measured to have length x, width y, and height z, but each measurement may be in error by 1%. Estimate the percentage error in calculating the volume.

35. Biomedical: Cardiac Output   Medical researchers calculate the quantity of blood pumped through the lungs (in liters per minute) by the formula $C = \dfrac{x}{y - z}$, where x is the amount of oxygen absorbed by the lungs (in milliliters per minute), and y and z are, respectively, the concentrations of oxygen in the blood just after and just before passing through the lungs (in milliliters of oxygen per liter of blood).

Typical measurements are $x = 250$, $y = 160$, and $z = 150$. Estimate the error in calculating the cardiac output C if each measurement may be "off" by 5 units.

36. General: Windchill Temperature   The windchill temperature announced during the winter by the weather bureau measures how cold it "feels" for a given temperature t (in degrees Fahrenheit) and wind speed w (in mph). It is calculated by the formula $C(t, w) = (0.475 + 0.304\sqrt{w})(t - 91.4)$. If the temperature is 30° and the wind speed is 16 mph, estimate the change in the windchill temperature if the wind speed increases by 4 mph and the temperature drops by 5°.

37. The Slope of $f(x, y) = c$ On page 973 we used the fact that the slope in the x-y plane of the curve defined by $f(x, y) = c$ (for constant c) is given by the formula $-\dfrac{f_x}{f_y}$. Verify this formula by justifying the following five steps (a) through (e).

a. If $f(x, y) = c$ can be solved explicitly for a function $y = F(x)$, then we may write $f(x, F(x)) = c$.
Justify: $f(x + \Delta x, F(x + \Delta x)) - f(x, F(x)) = 0$.

b. Justify: $f(x + \Delta x, F(x + \Delta x) - F(x) + F(x)) - f(x, F(x)) = 0$.

c. Defining ΔF by $\Delta F = F(x + \Delta x) - F(x)$, we may write the above equation as

$$f(x + \Delta x, \Delta F + F(x)) - f(x, F(x)) = 0$$

Then, writing F for $F(x)$, this becomes

$$f(x + \Delta x, F + \Delta F) - f(x, F) = 0$$

Justify: $f_x \, \Delta x + f_y \, \Delta F \approx 0$.

d. Justify: $\dfrac{\Delta F}{\Delta x} \approx -\dfrac{f_x}{f_y}$.

e. Justify: $\dfrac{dF}{dx} = -\dfrac{f_x}{f_y}$.

This shows that the slope of $F(x)$, and, therefore, the slope of $f(x, y) = c$, is $-\dfrac{f_x}{f_y}$.

11.7 Multiple Integrals

Introduction

This section discusses *integration* of functions of several variables. We define the *double integral* of a function by considering the volume under a surface $z = f(x, y)$. We then evaluate double integrals by "iterated" (repeated) single integrations. Finally, we use double integrals to calculate volumes, average values, and total accumulations. We restrict our attention to continuous functions (surfaces that have no holes or breaks), since most functions encountered in applications satisfy this restriction.

Rectangular Regions, Volumes, and Double Integrals

The points (x, y) in the plane with x taking values between numbers a and b and y between numbers c and d determine a *rectangular region R.*

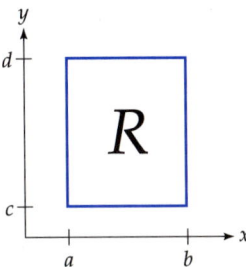

Rectangular region $R = \{(x,y) \mid a \le x \le b, c \le y \le d\}$

The graph below shows a nonnegative function $f(x, y)$ defined on a rectangular region R. We want to find the *volume* under the surface f above the region R.

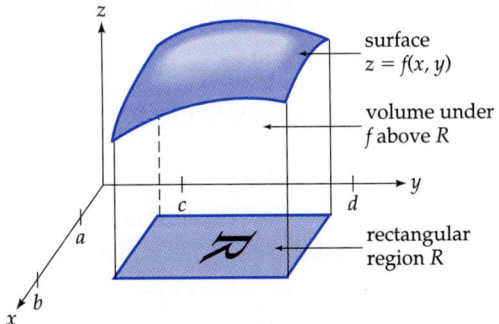

Volume under the surface $z = f(x,y)$ lying above
a rectangular region R

We begin by *approximating* the volume by rectangular solids ("boxes") extending from R up to the surface. We divide R into small rectangles by drawing lines parallel to the x- and y-axes with spacing Δx and Δy, as shown below.* On each of these small rectangles we erect a rectangular solid with height $f(x_i, y_j)$, the height of the surface at some point (x_i, y_j) in the base rectangle.

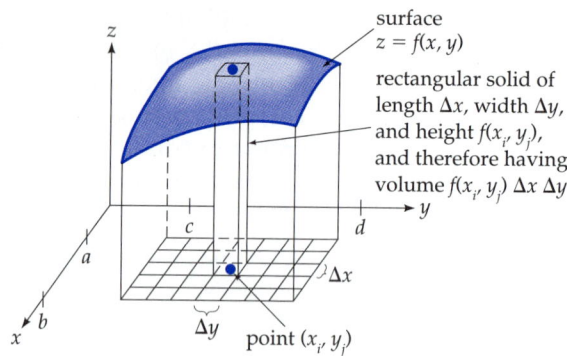

Volume under f over R showing one of the rectangular solids

The volume of the rectangular solid is $f(x_i, y_j) \cdot \Delta x \cdot \Delta y$ (height times length times width), and the sum of all such rectangular solids approximates the volume under f above R:

$$\begin{pmatrix} \text{Volume under} \\ f \text{ above } R \end{pmatrix} \approx \sum f(x_i, y_j) \cdot \Delta x \cdot \Delta y \qquad \begin{array}{l} \Sigma \text{ means sum over} \\ \text{all base rectangles} \end{array}$$

The *limit* of this sum as both Δx and Δy approach zero (so that the base rectangles become smaller and more numerous) gives the *exact* volume, and is called the *double integral of f over R*, denoted $\iint\limits_{R} f(x, y)\, dx\, dy$.

Double Integrals

> The double integral of a continuous function $f(x, y)$ on a rectangular region R is
>
> $$\iint\limits_{R} f(x, y)\, dx\, dy = \lim_{\Delta x, \Delta y \to 0} \sum f(x_i, y_j) \cdot \Delta x \cdot \Delta y \qquad \begin{array}{l} \text{The sum is over} \\ \text{all rectangles in} \\ R, \text{ each contain-} \\ \text{ing one } (x_i, y_j) \end{array}$$
>
> If $f(x, y)$ is nonnegative on R, then the double integral gives the *volume under f over R*.

---

* Technically, the parallel lines need not have equal spacing. However, we will be letting the spacings approach zero, and the final results are the same for equal or unequal spacing.

Iterated Integrals

Evaluating double integrals from the definition is difficult. Fortunately, double integrals can be evaluated by two separate "single" integrations, integrating with respect to one variable at a time while holding the other variable constant. (This is analogous to partial differentiation with respect to one variable, holding the other variable constant.) Such repeated integrals are called *iterated* integrals ("iterated" means "repeated"). A proof that double integrals can be evaluated as iterated integrals can be found in a more theoretical calculus book.

EXAMPLE 1   Evaluating an Iterated Integral

Evaluate the iterated integral $\displaystyle\int_0^1 \int_0^2 (3x^2 + 6xy^2)\, dx\, dy$.

Solution   The two separate integrations will be clearer if we use parentheses:

$$\int_0^1 \left(\int_0^2 (3x^2 + 6xy^2)\, dx \right) dy \qquad \text{An inner } x\text{-integral and an outer } y\text{-integral}$$

The inner integral gives

$$\int_0^2 (3x^2 + 6xy^2)\, dx = \left(x^3 + 6 \cdot \frac{1}{2} x^2y^2 \right) \Bigg|_{x=0}^{x=2}$$

dx means integrate with respect to x holding y constant   Integral of $3x^2$ Integral of x Held constant

$$= \underbrace{2^3 + 3 \cdot 2^2 y^2}_{\substack{\text{Evaluated} \\ \text{at } x = 2}} - \underbrace{(0)}_{\substack{\text{And at} \\ x = 0}} = \underbrace{8 + 12y^2}_{\text{Simplified}}$$

We now apply the outer y-integral to this result:

$$\int_0^1 (8 + 12y^2)\, dy = (8y + 4y^3) \Bigg|_{y=0}^{y=1} = \underbrace{8 + 4}_{\substack{\text{Evaluated} \\ \text{at } y = 1}} - \underbrace{(0)}_{\substack{\text{And} \\ \text{at } y = 0}} = \underbrace{12}_{\substack{\text{Final} \\ \text{answer}}}$$

Result of the inner integral dy means integrate with respect to y $12 \cdot \frac{1}{3}$

Therefore: $\displaystyle\int_0^1 \int_0^2 (3x^2 + 6xy^2)\, dx\, dy = 12$ The iterated integral equals 12

■

Always solve an iterated integral "from the inside out."

$$\int_0^1 \left(\int_0^2 (3x^2 + 6xy^2)\, dx \right) dy$$

Limits for y Limits for x First integrate with respect to x Then with respect to y

In Example 1 we integrated first with respect to x and then with respect to y. The next example shows that switching the order of integration gives the same answer, provided that we also switch the x and y limits of integration. That is,

$$\int_0^1 \int_0^2 (3x^2 + 6xy^2)\, dx\, dy$$

is equal to

$$\int_0^2 \int_0^1 (3x^2 + 6xy^2)\, dy\, dx$$

EXAMPLE 2   Reversing the Order of Integration

Evaluate $\displaystyle\int_0^2 \int_0^1 (3x^2 + 6xy^2)\, dy\, dx.$ Same as Example 1, but with the order of integration reversed

Solution   First we evaluate the inner y-integral:

$$\int_0^1 (3x^2 + 6xy^2)\, dy = \left. (3x^2y + 2xy^3) \right|_{y=0}^{y=1} = 3x^2 + 2x - (0)$$

Integrate with respect to y x held constant $\frac{1}{3} \cdot 6$ Evaluated at $y = 1$ And at $y = 0$

$$= 3x^2 + 2x$$

Then we apply the outer x-integral to this expression:

$$\int_0^2 (3x^2 + 2x)\, dx = \left. (x^3 + x^2) \right|_{x=0}^{x=2} = 8 + 4 - (0) = 12$$

From inner integration   Evaluated at $x = 2$ And at $x = 0$ Final answer

Therefore:

$$\int_0^2 \int_0^1 (3x^2 + 6xy^2)\, dy\, dx = 12$$

Notice that Examples 1 and 2 (in which the order of integration was reversed) gave the same answer, 12. Reversing the order of integration *always* gives the same answer, provided that the function is continuous.

For a continuous $f(x, y)$

$$\int_c^d \int_a^b f(x, y) \, dx \, dy$$

is equal to

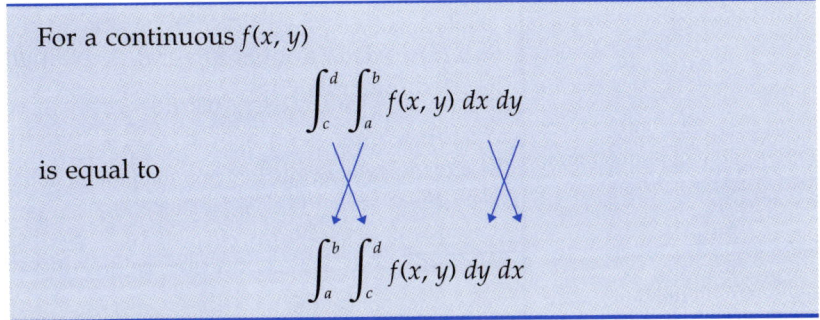

$$\int_a^b \int_c^d f(x, y) \, dy \, dx$$

Double Integrals and Volumes

Earlier, we defined double integrals over rectangular regions. Double integrals can be evaluated by *either* of two iterated integrals (integrating in either order). The limits of integration are taken directly from the region R.

Evaluating Double Integrals

The double integral $\displaystyle\iint\limits_R f(x, y) \, dx \, dy$

with $R = \{(x, y) \mid a \le x \le b, c \le y \le d\}$
is evaluated by finding *either* of the iterated integrals

$$\int_c^d \int_a^b f(x, y) \, dx \, dy \qquad \text{or} \qquad \int_a^b \int_c^d f(x, y) \, dy \, dx$$

The limits of integration
come from the definition of
the rectangle R

The rectangle $R = \{(x, y) \mid a \le x \le b, c \le y \le d\}$ is shown below.

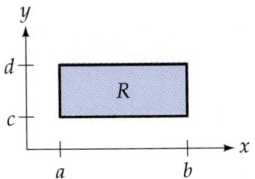

EXAMPLE 3   Evaluating a Double Integral

Evaluate $\iint\limits_R y^2\,e^{-x}\,dx\,dy$ with $R = \{(x, y) \mid 0 \le x \le 2, -1 \le y \le 1\}$.

Solution   This double integral is evaluated by solving either of the iterated integrals

$$\int_{-1}^{1}\int_{0}^{2} y^2\,e^{-x}\,dx\,dy \qquad \text{or} \qquad \int_{0}^{2}\int_{-1}^{1} y^2\,e^{-x}\,dy\,dx \qquad \text{Limits of integration come from } R$$

We will solve the second one, beginning with the inner integral.

$$\int_{-1}^{1} y^2\,e^{-x}\,dy = \left(\frac{1}{3}y^3\,e^{-x}\right)\bigg|_{y=-1}^{y=1} = \underbrace{\frac{1}{3}e^{-x}}_{\substack{\text{Evaluated} \\ \text{at } y=1}} - \underbrace{\left(\frac{1}{3}(-1)e^{-x}\right)}_{\substack{\text{And at} \\ y = -1}}$$

Integrated ⎯⎯ Held constant

$$= \frac{1}{3}e^{-x} + \frac{1}{3}e^{-x} = \frac{2}{3}e^{-x}$$

Then we integrate this with respect to x:

$$\int_{0}^{2}\frac{2}{3}e^{-x}\,dx = -\frac{2}{3}e^{-x}\bigg|_{x=0}^{x=2} = -\frac{2}{3}e^{-2} - \left(-\frac{2}{3}e^{0}\right) = -\frac{2}{3}e^{-2} + \frac{2}{3}$$

■

PRACTICE PROBLEM

Show that evaluating this same double integral by the iterated integral in the *other* order gives the same answer. That is, evaluate the iterated integral

$$\int_{-1}^{1}\int_{0}^{2} y^2\,e^{-x}\,dx\,dy$$

Solution at the back of the book

The volume under a surface can be found by evaluating a double integral (since that is how double integrals were defined).

Volume by Double Integrals

For a nonnegative continuous function $f(x, y)$, the volume under the surface $z = f(x, y)$ and above a rectangular region R in the x-y plane is

$$\begin{pmatrix} \text{Volume under} \\ f \text{ above } R \end{pmatrix} = \iint\limits_R f(x, y)\, dx\, dy$$

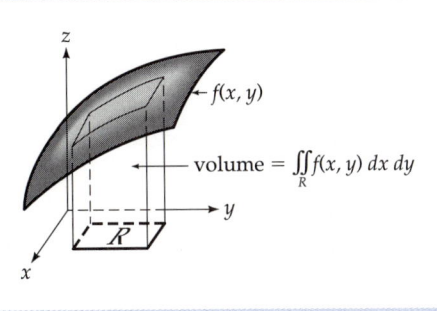

$$\text{volume} = \iint\limits_R f(x, y)\, dx\, dy$$

If the surface lies *below* the x-y plane, this integral gives the *negative* of the volume.

EXAMPLE 4   Finding the Volume Under a Surface

A modernistic tent with closed sides is constructed according to the function shown below. To design ventilation and heating systems, it is necessary to know the tent's volume. Find the volume under the tent on the indicated rectangle.

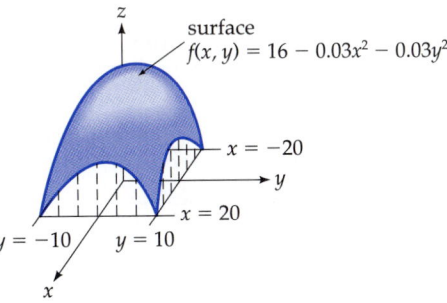

surface
$f(x, y) = 16 - 0.03x^2 - 0.03y^2$

$x = -20$

$x = 20$

$y = -10$

$y = 10$

Solution   The volume is the integral of the function over the rectangle R:

$$\iint\limits_R f(x, y)\, dx\, dy = \int_{-10}^{10} \int_{-20}^{20} (16 - 0.03x^2 - 0.03y^2)\, dx\, dy \qquad \begin{array}{l}\text{Limits of} \\ \text{integration} \\ \text{come from } R\end{array}$$

The inner integral is

$$\int_{-20}^{20} (16 - 0.03x^2 - 0.03y^2)\, dx = (16x - 0.01x^3 - 0.03y^2x)\,\Big|_{x=-20}^{x=20}$$

$$= \underbrace{320 - 80 - 0.6y^2}_{\text{Evaluated at } x = 20} - \underbrace{(-320 + 80 + 0.6y^2)}_{\text{Evaluated at } x = -20} = 480 - 1.2y^2$$

Integrating this with respect to y:

From integrating $1.2y^2$

$$\int_{-10}^{10} (480 - 1.2y^2)\, dy = (480y - 0.4y^3)\,\Big|_{-10}^{10}$$

$$= 4800 - 400 - (-4800 + 400) = 8800$$

Therefore, the volume under the tent is 8800 cubic feet.   ∎

Average Value

Recall that the average value of a function of *one* variable over an interval was defined as the definite integral of the function divided by the length of the interval. For similar reasons, the average value of a function $f(x, y)$ of *two* variables over a region is defined as the *double* integral divided by the *area* of the region.

Average Value

$$\begin{pmatrix}\text{Average value} \\ \text{of } f \text{ over } R\end{pmatrix} = \frac{1}{\text{area of } R} \iint_R f(x, y)\, dx\, dy \qquad \begin{array}{l}\text{Double integral} \\ \text{over the region} \\ \text{divided by the} \\ \text{area of the region}\end{array}$$

For a rectangular region R, the area of R is just length times width.

EXAMPLE 5   Finding the Average Temperature Over a Region

The temperature x miles east and y miles north of a weather station is $T(x, y) = 60 + 2x - 4y$ degrees. Find the average temperature over the rectangular region R shown in the following diagram.

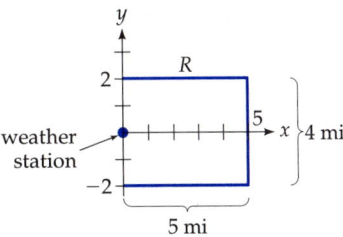

Solution   The area of the region R is $4 \cdot 5 = 20$ square miles (length times width). The average temperature is the double integral divided by 20:

$$\text{Average} = \frac{1}{20} \int_{-2}^{2} \int_{0}^{5} (60 + 2x - 4y) \, dx \, dy$$

The inner integral is

$$\int_{0}^{5} (60 + 2x - 4y) \, dx = (60x + x^2 - 4yx) \Big|_{x=0}^{x=5}$$

$$= \underbrace{300 + 25 - 20y} - (0) = 325 - 20y$$

Evaluating at $x = 5$

The outer integral of this is

$$\int_{-2}^{2} (325 - 20y) \, dy = (325y - 10y^2) \Big|_{y=-2}^{y=2} = 650 - 40 - (-650 - 40)$$

$$= 610 + 690 = 1300$$

Finally, for the average we divide by 20 (the area of the region):

$$\frac{1300}{20} = 65$$

The average temperature over the region is 65°.

■

Integrating Over More General Regions

We may also integrate over regions R that are bounded by curves, provided that the curves, as well as the function being integrated, are continuous.

Double Integrals Over Regions Between Curves

Let R be the region bounded by a lower curve $y = g(x)$ and an upper curve $y = h(x)$ from $x = a$ to $x = b$, as shown below. Then the double integral of $f(x, y)$ over R is defined by

$$\iint\limits_R f(x, y)\ dx\ dy = \int_a^b \int_{g(x)}^{h(x)} f(x, y)\ dy\ dx$$

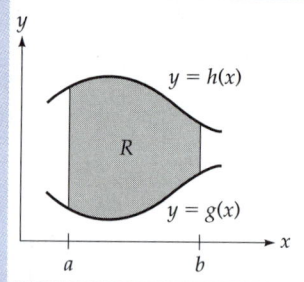

If f is nonnegative, this double integral gives the volume under the surface $f(x, y)$ above R.

EXAMPLE 6   Finding the Volume Under a Surface

Find the volume under the surface $f(x, y) = 12xy$ and above the region R shown below.

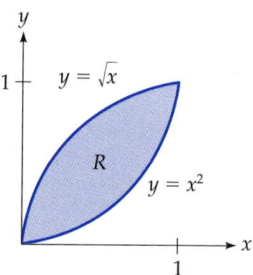

Solution   The region R is bounded by the upper curve $h(x) = \sqrt{x}$ and the lower curve $g(x) = x^2$ from $x = 0$ to $x = 1$. From the boxed definition above, the volume is given by the following integral:

$$\text{Volume} = \int_0^1 \int_{x^2}^{\sqrt{x}} 12xy\ dy\ dx$$

We find the inner integral first:

$$\int_{x^2}^{\sqrt{x}} 12xy\, dy = 12x\,\frac{1}{2}\,y^2\,\Big|_{y=x^2}^{y=\sqrt{x}} = 6xy^2\,\Big|_{y=x^2}^{y=\sqrt{x}}$$

$$= \underbrace{6x(\sqrt{x})^2}_{\substack{\text{Evaluating}\\ \text{at } y = \sqrt{x}}} - \underbrace{6x(x^2)^2}_{\substack{\text{Evaluating}\\ \text{at } y = x^2}} = \underbrace{6x^2 - 6x^5}_{\text{Simplified}}$$

Now we integrate the result with respect to x:

$$\int_0^1 (6x^2 - 6x^5)\, dx = \left(6 \cdot \frac{1}{3}x^3 - x^6\right)\Big|_0^1$$

$$= (2x^3 - x^6)\,\Big|_0^1 = 2 - 1 - (0) = 1$$

Therefore, the volume under the surface and above the region R is 1 cubic unit. ∎

SUMMARY

We defined the double integral of a nonnegative function $f(x, y)$ over a region R as the *volume* under the surface above R. [This was analogous to defining the definite integral of a function $f(x)$ of *one* variable as the *area* under the curve.]

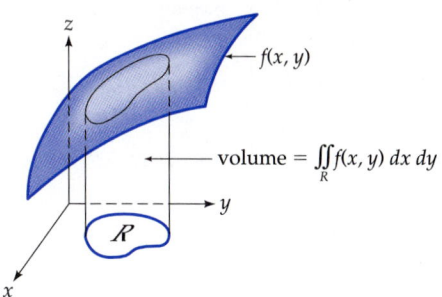

We evaluated double integrals by evaluating iterated (repeated) integrals:

$$\iint_R f(x, y)\, dx\, dy \qquad \text{over} \qquad R = \{(x, y) \mid a \le x \le b, c \le y \le d\}$$

is found by evaluating either of the two iterated integrals:

$$\int_c^d \int_a^b f(x, y)\, dx\, dy \qquad \text{or} \qquad \int_a^b \int_c^d f(x, y)\, dy\, dx \qquad \text{\textcolor{blue}{Integrating in either order}}$$

Note the distinction: *Double* integrals are written with R (which must be specified) under the double integral, and *iterated* integrals are written with upper and lower *limits* on each integral sign.

Double integrals do more than just find volumes; they give *continuous sums* (as illustrated in the exercises), and when divided by the area of the region they give the *average value of the function over the region*.

The exercises also discuss "triple" integrals of functions $f(x, y, z)$ of three variables. Triple integrals are evaluated by iterated integrals (but *three* of them), integrating successively with respect to one variable at a time, holding all others constant.

EXERCISES 11.7

Evaluate each (single) integral.

1. $\int_1^{x^2} 8xy^3 \, dy$

2. $\int_1^{y^2} 10x^4 \, dx$

3. $\int_{-y}^{y} 9x^2y \, dx$

4. $\int_{-x}^{x} 6xy^2 \, dy$

5. $\int_0^x (6y - x) \, dy$

6. $\int_0^y (4x - y) \, dx$

Evaluate each iterated integral.

7. $\int_0^2 \int_0^1 4xy \, dx \, dy$

8. $\int_0^2 \int_0^1 8xy \, dy \, dx$

9. $\int_0^2 \int_0^1 x \, dy \, dx$

10. $\int_0^4 \int_0^3 y \, dx \, dy$

11. $\int_0^1 \int_0^2 x^3y^7 \, dx \, dy$

12. $\int_0^1 \int_0^3 x^8y^2 \, dy \, dx$

13. $\int_1^3 \int_0^2 (x + y) \, dy \, dx$

14. $\int_1^2 \int_0^4 (x - y) \, dx \, dy$

15. $\int_{-1}^1 \int_0^3 (x^2 - 2y^2) \, dx \, dy$

16. $\int_{-1}^1 \int_0^3 (2x^2 + y^2) \, dy \, dx$

17. $\int_{-3}^3 \int_0^3 y^2e^{-x} \, dy \, dx$

18. $\int_{-2}^2 \int_0^2 xe^{-y} \, dx \, dy$

19. $\int_{-2}^2 \int_{-1}^1 ye^{xy} \, dx \, dy$

20. $\int_{-1}^1 \int_{-1}^1 xe^{xy} \, dy \, dx$

21. $\int_0^2 \int_x^1 12xy \, dy \, dx$

22. $\int_0^1 \int_y^1 4xy \, dx \, dy$

23. $\int_3^5 \int_0^y (2x - y) \, dx \, dy$

24. $\int_2^4 \int_0^x (x - 2y) \, dy \, dx$

25. $\int_{-3}^3 \int_0^{4x} (y - x) \, dy \, dx$

26. $\int_{-1}^1 \int_0^{2y} (x + y) \, dx \, dy$

27. $\int_0^1 \int_{-y}^y (x + y^2) \, dx \, dy$

28. $\int_0^2 \int_{-x}^x (x^2 - y) \, dy \, dx$

For each double integral:

a. Write the *two* iterated integrals that are equal to it.
b. Evaluate *both* iterated integrals (the answers should agree).

29. $\iint\limits_R 3xy^2 \, dx \, dy$

with $R = \{(x, y) \mid 0 \le x \le 2, 1 \le y \le 3\}$

30. $\iint\limits_R 6x^2y \, dx \, dy$

with $R = \{(x, y) \mid 0 \le x \le 1, 1 \le y \le 2\}$

31. $\iint\limits_R ye^x \, dx \, dy$

with $R = \{(x, y) \mid -1 \le x \le 1, 0 \le y \le 2\}$

32. $\displaystyle\iint\limits_{R} xe^y\,dx\,dy$

with $R = \{(x, y) \mid 0 \le x \le 1, -2 \le y \le 2\}$

Use integration to find the volume under each surface $f(x, y)$ above the region R.

33. $f(x, y) = x + y$
$R = \{(x, y) \mid 0 \le x \le 2, 0 \le y \le 2\}$

34. $f(x, y) = 8 - x - y$
$R = \{(x, y) \mid 0 \le x \le 4, 0 \le y \le 4\}$

35. $f(x, y) = 2 - x^2 - y^2$
$R = \{(x, y) \mid 0 \le x \le 1, 0 \le y \le 1\}$

36. $f(x, y) = x^2 + y^2$
$R = \{(x, y) \mid 0 \le x \le 2, 0 \le y \le 2\}$

37. $f(x, y) = 2xy$
for the region R
shown on the right:

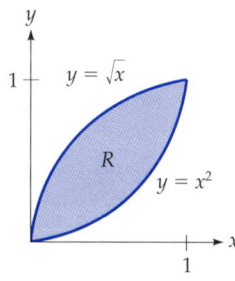

38. $f(x, y) = 3xy^2$
for the region R
shown on the right:

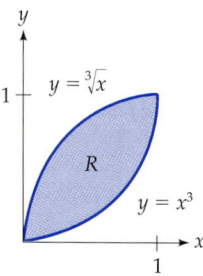

39. $f(x, y) = e^y$
for the region R
shown on the right:

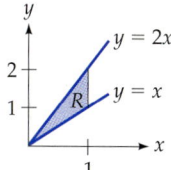

40. $f(x, y) = e^y$
for the region R
shown on the right:

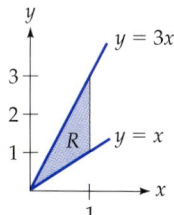

APPLIED EXERCISES

41. General: Average Temperature   The temperature x miles east and y miles north of a weather station is given by the function $f(x, y) = 48 + 4x - 2y$. Find the average temperature over the region R shown below.

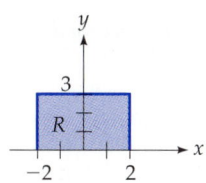

42. Environmental Science: Average Air Pollution   The air pollution near a chemical refinery is $f(x, y) = 20 + 6x^2y$ ppm (parts per million), where x and y are the numbers of miles east and north of the refinery. Find the average pollution levels for the region R shown below.

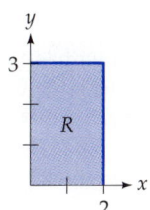

43. General: Total Population of a Region   The population density (people per square mile) x miles east and y miles north of the center of a city is $P(x, y) = 12{,}000e^{x-y}$. Find the total population of the region R shown below. (*Hint:* Integrate the population density over the region R. This is an example of a double integral as a *sum*, giving a *total* population over a region.)

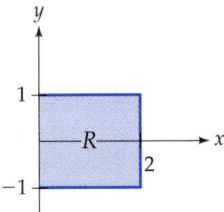

44. Business: Value of a Mineral Deposit   The value of an offshore mineral deposit x miles east and y miles north of a certain point is $f(x, y) = 4x + 6y^2$ million dollars per square mile. Find the total value of the tract shown below. (*Hint:* Integrate the function over the region R. This is an example of a double integral as a *sum*, giving a *total* value over a region.)

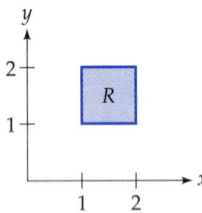

45–46: General: Volume of a Building   To estimate heating and air-conditioning costs, it is necessary to know the volume of a building.

45. A conference center has a curved roof of height $f(x, y) = 40 - 0.006x^2 + 0.003y^2$. The building sits on a rectangle extending from $x = -50$ to $x = 50$ and $y = -100$ to $y = 100$. Use integration to find the volume of the building.

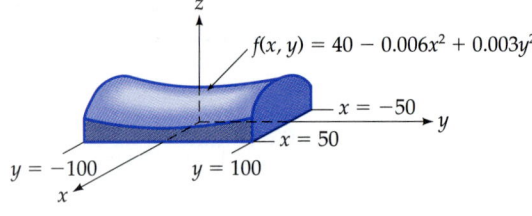

46. An airplane hangar has a curved roof of height $f(x, y) = 40 - 0.03x^2$. The building sits on a rectangle extending from $x = -20$ to $x = 20$ and $y = -100$ to $y = 100$. Use integration to find the volume of the building.

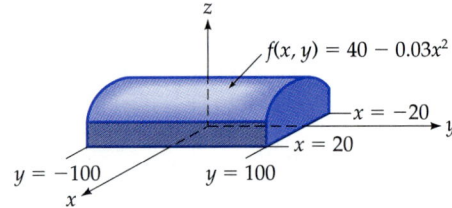

Triple Integrals

Evaluate each triple iterated integral. (*Hint:* Integrate with respect to one variable at a time, treating the other variables like constants, working from the inside out.)

47. $\displaystyle \int_{1}^{2} \int_{0}^{3} \int_{0}^{1} (2x + 4y - z^2) \, dx \, dy \, dz$

48. $\displaystyle \int_{1}^{2} \int_{0}^{3} \int_{0}^{2} (6x - 2y + z^2) \, dx \, dy \, dz$

49. $\displaystyle \int_{1}^{2} \int_{0}^{2} \int_{0}^{1} 2xy^2z^3 \, dx \, dy \, dz$

50. $\displaystyle \int_{1}^{3} \int_{0}^{1} \int_{0}^{2} 12x^3y^2z \, dx \, dy \, dz$

Chapter Summary with Hints and Suggestions

The reading and exercises of this chapter have helped you to learn the following skills. For each skill, the section from which it came (in case you need to review it) and some exercises in this section that use it are indicated. Answers to all exercises are at the end of the book, and full solutions to all exercises are in the Student Solutions Manual.

11.1   Functions of Several Variables

● Find the domain of a function of two variables. (*Review Exercises 1–4.*)

11.2   Partial Derivatives

● Find the first and second partials of a function of two variables.   (*Review Exercises 5–12.*)

$$\frac{\partial}{\partial x} f(x, y) = f_x(x, y) = \lim_{h \to 0} \frac{f(x + h, y) - f(x, y)}{h}$$

$$\frac{\partial}{\partial y} f(x, y) = f_y(x, y) = \lim_{h \to 0} \frac{f(x, y + h) - f(x, y)}{h}$$

● Evaluate the first partials of a function of two variables.   (*Review Exercises 13–16.*)

● Solve an applied problem involving partials, and interpret the answer. (*Review Exercises 17–18.*)

$$\frac{\partial}{\partial x} f(x, y) = \begin{pmatrix} \text{Rate of change of } f \text{ with respect} \\ \text{to } x \text{ when } y \text{ is held constant} \end{pmatrix}$$

$$\frac{\partial}{\partial y} f(x, y) = \begin{pmatrix} \text{Rate of change of } f \text{ with respect} \\ \text{to } y \text{ when } x \text{ is held constant} \end{pmatrix}$$

11.3   Optimizing Functions of Several Variables

● Find the relative extreme values of a function. (*Review Exercises 19–30.*)

$$\begin{cases} f_x = 0 \\ f_y = 0 \end{cases} \quad D = f_{xx} f_{yy} - (f_{xy})^2$$

● Solve an applied problem by optimizing a function of two variables. (*Review Exercises 31–32.*)

11.4   Least Squares

● Find a least squares line "by hand" (and verify using ⌨).   (*Review Exercises 33–34.*)

● Find the least squares line for actual data, and use it to make a prediction. (*Review Exercises 35–36.*)

11.5   Lagrange Multipliers and Constrained Optimization

● Solve a constrained maximum or minimum problem using Lagrange multipliers. (*Review Exercises 37–42.*)

$$F(x, y, \lambda) = f(x, y) + \lambda g(x, y) \quad \begin{cases} F_x = 0 \\ F_y = 0 \\ F_\lambda = 0 \end{cases}$$

● Find *both* extreme values in a constrained optimization problem using Lagrange multipliers.   (*Review Exercises 43–44.*)

● Solve an applied problem using Lagrange multipliers.   (*Review Exercises 45–48.*)

11.6   Total Differentials and Approximate Changes

- Find the total differential of a function. *(Review Exercises 49–54.)*

$$df = \frac{\partial f}{\partial x}\,dx + \frac{\partial f}{\partial y}\,dy$$

- Solve an applied problem by using the total differential to estimate an actual change. *(Review Exercise 55.)*

$$\Delta f = f(x + \Delta x, y + \Delta y) - f(x, y) \qquad \Delta f \approx df$$

- Estimate the relative error in an area calculation.   *(Review Exercise 56.)*

11.7   Multiple Integrals

- Evaluate an iterated integral. *(Review Exercises 57–60.)*

$$\int_c^d \int_a^b f(x, y)\,dx\,dy = \int_a^b \int_c^d f(x, y)\,dy\,dx$$

- Find the volume under a surface above a region using a double integral. *(Review Exercises 61–64.)*

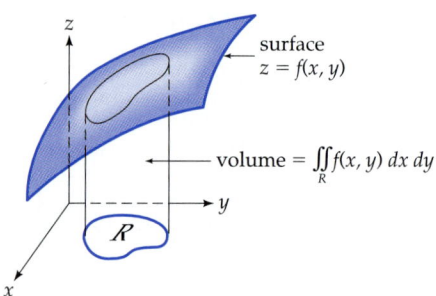

surface
$z = f(x, y)$

volume $= \iint_R f(x, y)\,dx\,dy$

- Find the average population over a region using a double integral.   *(Review Exercise 65.)*

- Find the total value of a region using a double integral.   *(Review Exercise 66.)*

Hints and Suggestions

- The graph of a function of two variables is represented by a *surface* above (or below) the *x-y* plane. Such three-dimensional graphs are difficult to draw "by hand" but can be shown on some graphing calculator and computer screens. Such surfaces have maximum and minimum points (just as with functions of one variable), but also a new phenomenon, *saddle points* (see page 920).

- The graph of a function of *three* or more variables would require *four* or more dimensions, and so cannot be drawn.

- When finding partial derivatives, remember which is the variable of differentiation and then treat the other variable like a constant. Other than this, the differentiation formulas are the same.

- Partials give instantaneous rates of change with respect to one variable (while the other is held constant). They also give *marginals* with respect to one product while production of the other is held constant.

- The *D*-test (page 937) applies only to critical points, where the first partials are zero. These critical points must be found first, and then the *D*-test is applied to each, one at a time.

- Least squares is carried out automatically on a graphing calculator or computer when it finds the linear regression line.

- Solving a constrained optimization problem by Lagrange multipliers is often easier than eliminating one of the variables (as was done in Sections 7.3 and 7.4).

- Multiple integrals give volume under a function only if the function is nonnegative.

- When finding the average value of a function of two variables, don't forget to divide by the area of the region that you are integrating over.

- Practice for test: Review Exercises 3, 7, 15, 17, 19, 33, 41, 45, 49, 51, 57, 65.

Review Exercises for Chapter 11   *Practice test exercises are in blue.*

11.1   Functions of Several Variables

For each function, state the domain.

1. $f(x, y) = \dfrac{\sqrt{x}}{\sqrt[3]{y}}$

2. $f(x, y) = \dfrac{\sqrt[3]{x}}{\sqrt{y}}$

3. $f(x, y) = e^{1/x} \ln y$

4. $f(x, y) = \dfrac{\ln y}{x}$

11.2   Partial Derivatives

For each function f, calculate **a.** f_x, **b.** f_y, **c.** f_{xy}, and **d.** f_{yx}.

5. $f(x, y) = 2x^5 - 3x^2y^3 + y^4 - 3x + 2y + 7$

6. $f(x, y) = 3x^4 + 5x^3y^2 - y^6 - 6x + y - 9$

7. $f(x, y) = 18x^{2/3}y^{1/3}$ **8.** $f(x, y) = \ln(x^2 + y^3)$

9. $f(x, y) = e^{x^3 - 2y^3}$ **10.** $f(x, y) = 3x^2e^{-5y}$

11. $f(x, y) = ye^{-x} - x \ln y$

12. $f(x, y) = x^2e^y + y \ln x$

For each function, calculate **a.** $f_x(1, -1)$ and **b.** $f_y(1, -1)$.

13. $f(x, y) = \dfrac{x + y}{x - y}$ **14.** $f(x, y) = \dfrac{x}{x^2 + y^2}$

15. $f(x, y) = (x^3 + y^2)^3$ **16.** $f(x, y) = (2xy - 1)^4$

17. Business: Marginal Productivity   A company's production is given by the Cobb–Douglas function $P(L, K) = 160L^{3/4}K^{1/4}$, where L is the number of units of labor and K is the number of units of capital.

a. Find $P_L(81, 16)$ and interpret this number.
b. Find $P_K(81, 16)$ and interpret this number.
c. From your answers to parts (a) and (b), which will increase production more, an additional unit of labor or an additional unit of capital?

18. Business: Advertising   A clothing designer's sales S depend on x, the amount spent on television advertising, and y, the amount spent on print advertising (both in thousands of dollars), according to the function

$S(x, y) = 60x^2 + 90y^2 - 6xy + 200$

Find $S_x(2, 3)$, $S_y(2, 3)$, and interpret these numbers.

11.3   Optimizing Functions of Several Variables

For each function, find all relative extreme values.

19. $f(x, y) = 2x^2 - 2xy + y^2 - 4x + 6y - 3$

20. $f(x, y) = x^2 - 2xy + 2y^2 - 6x + 4y + 2$

21. $f(x, y) = 2xy - x^2 - 5y^2 + 2x - 10y + 3$

22. $f(x, y) = 2xy - 5x^2 - y^2 + 10x - 2y + 1$

23. $f(x, y) = 2xy + 6x - y + 1$

24. $f(x, y) = 4xy - 4x + 2y - 4$

25. $f(x, y) = e^{-(x^2 + y^2)}$ **26.** $f(x, y) = e^{2(x^2 + y^2)}$

27. $f(x, y) = \ln(5x^2 + 2y^2 + 1)$

28. $f(x, y) = \ln(4x^2 + 3y^2 + 10)$

29. $f(x, y) = x^3 - y^2 - 12x - 6y$

30. $f(x, y) = y^2 - x^3 + 12x - 4y$

31. Business: Maximum Profit   A boatyard builds 18-foot and 22-foot sailboats. Each 18-foot boat costs \$3000 to build, each 22-foot boat costs \$5000 to build, and the company's fixed costs are \$6000. The price function for the 18-foot boats is $p = 7000 - 20x$, and that for the 22-foot boats is $q = 8000 - 30y$ (both in dollars, where x and y are the numbers of 18-foot and 22-foot boats, respectively).

a. Find the company's cost function $C(x, y)$.
b. Find the company's revenue function $R(x, y)$.
c. Find the company's profit function $P(x, y)$.
d. Find the quantities and prices that maximize profit. Also find the maximum profit.

32. Business: Price Discrimination   A company sells farm equipment in America and Europe, charging different prices in the two markets. The price function for harvesters sold in America is $p = 80 - 0.2x$, and the price function for harvesters sold in Europe is $q = 64 - 0.1y$ (both in thousands of dollars), where x and y are the numbers sold per day in America and Europe, respectively. The company's cost function is $C = 100 + 12(x + y)$ thousand dollars.

a. Find the company's profit function.
b. Find how many harvesters should be sold in each market to maximize profit. Also find the price for each market.

11.4 Least Squares

For each exercise:

a. Find the least squares line "by hand" (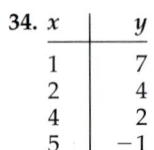 will be helpful).
b. Check your answer using a graphing calculator.

33.

x	y
1	-1
3	6
4	6
5	10

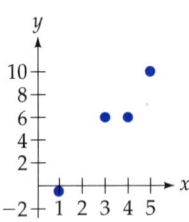

34.

x	y
1	7
2	4
4	2
5	-1

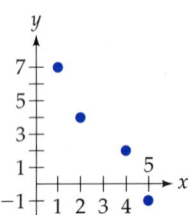

35. General: The Aging of America The population of Americans who are over 65 years old is growing faster than the population at large. Find the least squares line for the following data for the over-65 population, and use it to predict the size of the over-65 population in the year 2010 ($x = 7$).

	x	y = Population (millions)
1950	1	12.4
1960	2	16.7
1970	3	20.1
1980	4	25.5
1990	5	31.2

36. Economics: Unemployment The unemployment rate in the United States is given by the following table. Find the least squares line for these data and use it to predict the average unemployment rate for the year 2005 ($x = 7$).

	x	y = Percent Unemployed
1975	1	7.2
1980	2	7.1
1985	3	7.2
1990	4	5.6
1995	5	5.6

11.5 Lagrange Multipliers and Constrained Optimization

Use Lagrange multipliers to optimize each function subject to the constraint. (The stated extreme values *do* exist.)

37. Maximize $f(x, y) = 6x^2 - y^2 + 4$ subject to $3x + y = 12$.

38. Maximize $f(x, y) = 4xy - x^2 - y^2$ subject to $x + 2y = 26$.

39. Minimize $f(x, y) = 2x^2 + 3y^2 - 2xy$ subject to $2x + y = 18$.

40. Minimize $f(x, y) = 12xy - 1$ subject to $y - x = 6$.

41. Minimize $f(x, y) = e^{x^2 + y^2}$ subject to $x + 2y = 15$.

42. Maximize $f(x, y) = e^{-x^2 - y^2}$ subject to $2x + y = 5$.

Use Lagrange multipliers to find the maximum *and* minimum values of each function subject to the constraint. (Both extreme values *do* exist.)

43. $f(x, y) = 6x - 18y$ subject to $x^2 + y^2 = 40$

44. $f(x, y) = 4xy$ subject to $x^2 + y^2 = 32$

45. Business: Maximum Profit A company's profit is $P = 300x^{2/3}y^{1/3}$, where x and y are, respectively, the amounts spent on production and advertising. The company has a total of $60,000 to spend.

a. Use Lagrange multipliers to find the amounts for production and advertising that maximize profit.

b. Evaluate and give an interpretation for λ.

 46. Biomedical: Nutrition A nursing home uses two vitamin supplements, and the nutritional value of x ounces of the first together with y ounces of the second is $4x + 2xy + 8y$. The first costs $2 per ounce, and the second costs $1 per ounce, and the nursing home can spend only $8 per patient per day.

a. Use Lagrange multipliers to find how much of each supplement should be used to maximize the nutritional value subject to the budget constraint.

b. Evaluate and give an interpretation for λ.

 47. Economics: Least Cost Rule A company's production is given by the Cobb–Douglas function $P = 60L^{2/3}K^{1/3}$, where L and K are the numbers of units of labor and capital, respectively. Each unit of labor costs $25 and each unit of capital costs $100. The company wants to produce exactly 1920 units.

a. Find the numbers of units of labor and capital that meet the production requirements at the lowest cost.

b. Find the marginal productivity of labor and the marginal productivity of capital. (*Hint:* This means the partials of P with respect to L and K.)

c. Show that at the values found in part (a), the following relationship holds:

$$\frac{\text{Marginal productivity of labor}}{\text{Marginal productivity of capital}}$$

$$= \frac{\text{Price of labor}}{\text{Price of capital}}$$

This is called the "least cost rule."

48. General: Container Design An open-top box with a square base and two perpendicular dividers, as in the diagram, is to have volume 576 cubic inches. Use Lagrange multipliers to find the dimensions that require the least amount of materials.

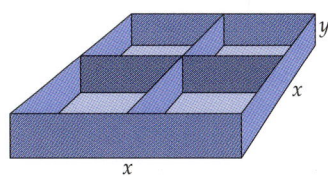

11.6 Total Differentials and Approximate Changes

Find the total differential of each function.

49. $f(x, y) = 3x^2 + 2xy + y^2$

50. $f(x, y) = x^2 + xy - 3y^2$

51. $g(x, y) = \ln(xy)$ **52.** $g(x, y) = \ln(x^3 + y^3)$

53. $z = e^{x-y}$ **54.** $z = e^{xy}$

55. Business: Sales A clothing designer's sales S depend on x, the amount spent on television advertising, and y, the amount spent on print advertising (both in thousands of dollars) according to the formula $S(x, y) = 60x^2 - 6xy + 90y^2 + 200$. If the company now spends 2 thousand dollars on television advertising and 3 thousand dollars on print advertising, use the total differential to estimate the change in sales if television advertising is increased by $500 and print advertising is decreased by $500. (*Hint:* Δx and Δy must be in thousands of dollars, just as x and y are.)

56. General: Relative Error in Calculating Area A triangular piece of real estate is measured to have length x feet and altitude y feet, but each measurement may be in error by 1%. Estimate the percentage error in calculating the area by using a total differential.

11.7 Multiple Integrals

Evaluate each iterated integral.

57. $\displaystyle\int_0^4 \int_{-1}^1 2xe^{2y}\, dy\, dx$

58. $\displaystyle\int_{-1}^{1}\int_{0}^{3}(x^2 - 4y^2)\,dx\,dy$

59. $\displaystyle\int_{-1}^{1}\int_{-y}^{y}(x + y)\,dx\,dy$

60. $\displaystyle\int_{-2}^{2}\int_{-x}^{x}(x + y)\,dy\,dx$

Find the volume under the surface $f(x, y)$ above the region R.

61. $f(x, y) = 8 - x - y$
  $R = \{(x, y) \mid 0 \le x \le 2, 0 \le y \le 4\}$

62. $f(x, y) = 6 - x - y$
  $R = \{(x, y) \mid 0 \le x \le 4, 0 \le y \le 2\}$

63. $f(x, y) = 12xy^3$
  over the region R
  shown on the right:

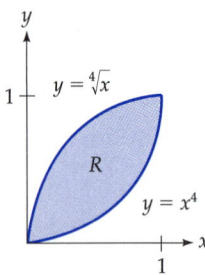

64. $f(x, y) = 15xy^4$
  over the region R
  shown on the right:

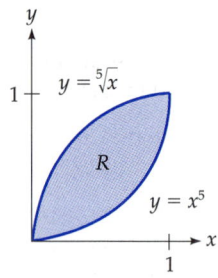

65. General: Average Population of a Region
The population x miles east and y miles north of the center of a city is $P(x, y) = 12{,}000 + 100x - 200y$. Find the *average* population over the region shown below.

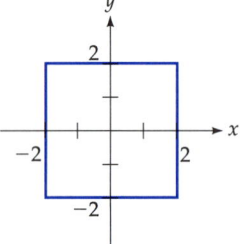

66. General: Total Value of a Region The value of land in a city x blocks east and y blocks north of the center of a town is $V(x, y) = 40 - 4x - 2y$ hundred thousand dollars per block. Find the *total* value of the parcel of land shown above.

Projects and Essays

The following projects and essays are based on Chapter 11. Most have no right and wrong answers—the results depend only on your imagination and resourcefulness.

1. Find a computer with a program for graphing functions of two variables and have it graph some interesting functions (like, but not limited to, those on pages 920–921). You will have to choose an appropriate domain and "point of view" for each graph. Print out several of the most interesting graphs and label them with their functions.

2. Find out about Cobb–Douglas production functions and write about their history and their use in economics.

3. Look up "tesseract" in an encyclopedia and write about four-dimensional cubes and how they can be represented.

4. A function $f(x, y)$ of two variables has two first partials and four second partials, but really only three "different" second partials, since $f_{xy} = f_{yx}$ (assuming continuity). How many *third* partials are there, and how many are "different"? What about for a function of *three* variables?

5. Go to a library and find out about the mathematician Carl Friedrich Gauss. Write a report about his life and, in particular, his invention of the least squares method.

6. Go to a library and find out about the mathe-matician Joseph Louis Lagrange. Write a report about his life and, in particular, his invention of the method of Lagrange multipliers.

7. Go to a supermarket and find examples of the most efficient and least efficient cans (see page 968). Estimate what percentage of metal is wasted in the inefficient cans.

8. Look in more theoretical calculus books for a function $f(x, y)$ whose iterated integrals (in the two orders) are *not* equal. Find both iterated integrals and show that they differ. Try to draw the graph of the function, using a computer with a program for graphing functions of two variables.

Cumulative Review—Chapters 0 and 6–11

The following exercises review some of the basic techniques that you learned in Chapters 0 and 6–11. Answers to all of these cumulative review exercises are given in the answer section near the end of the book. A graphing calculator is suggested but not required.

1. Draw the graph of the function

$$f(x) = \begin{cases} 2x - 1 & \text{if } x \le 3 \\ 7 - x & \text{if } x > 3 \end{cases}.$$

2. Simplify: $\left(\dfrac{1}{8}\right)^{-2/3}$

3. Use the definition of the derivative

$$f'(x) = \lim_{h \to 0} \frac{f(x + h) - f(x)}{h}$$

to find the derivative of $f(x) = \dfrac{1}{x}$.

4. If $f(x) = 12\sqrt[3]{x^2} - 4$, find $f'(8)$.

5. A camera store finds that if it sells disposable cameras for p dollars each, it will sell $S(p) = \dfrac{800}{p + 8}$ of them per week. Find $S'(12)$ and inter-pret your answer.

6. Find $\dfrac{d}{dx} [x^2 + (2x + 1)^4]^3$.

7. Make sign diagrams for the first and second de-rivatives and graph the function $f(x) = x^3 + 9x^2 - 48x - 148$, showing all relative extreme points and inflection points.

8. Make a sign diagram for the first derivative and graph the function $f(x) = \dfrac{1}{x^2 - 4x}$, showing all asymptotes and relative extreme points.

9. A homeowner wants to use 80 feet of fence to make a rectangular enclosure along an existing long stone wall. If the side along the existing wall needs no fence, what are the dimensions of the enclosure that has the largest possible area?

10. An open-top box with a square base is to have a volume of 108 cubic feet. Find the dimensions of the box of this type that can be made with the least amount of material.

11. A spherical balloon is being inflated at the con-stant rate of 128 cubic feet per minute. Find how fast the radius is increasing at the moment when the radius is 4 feet.

12. A sum of $1000 is deposited into a bank paying 8% interest. Find the value of the account after 3 years if the interest is compounded:

    a. quarterly   b. continuously

13. In t years the population of a county is predicted to be $P(t) = 12{,}000e^{0.02t}$, and the population of a neighboring county is predicted to be $Q(t) = 9000e^{0.04t}$. In how many years will the second county overtake the first in population?

14. A $12,000 automobile depreciates by 15% per year. When will it be worth only $4500?

15. Make sign diagrams for the first and second derivatives and graph the function $f(x) = e^{-\frac{1}{2}x^2}$, showing all relative extreme points and inflection points.

16. Find $\int (12x^2 - 4x + 1)\, dx$.

17. A city plans to discharge pollution into a lake at the rate of $18e^{0.02t}$ million gallons per year, where t is the number of years from now. Find a formula for the total amount of pollution that will be discharged into the lake during the next t years.

18. Find the area bounded by $y = 20 - x^2$ and $y = 8 - 4x$.

19. Find the average value of $f(x) = 12\sqrt{x}$ over the interval $[0, 4]$.

20. Find: a. $\displaystyle\int \frac{x^2\, dx}{x^3 + 1}$ b. $\displaystyle\int \frac{e^{\sqrt{x}}\, dx}{\sqrt{x}}$

21. Find by integration by parts: $\int xe^{4x}\, dx$.

22. Use the integral table in the back of the book to find $\displaystyle\int \frac{\sqrt{4 - x^2}}{x}\, dx$.

23. Evaluate $\displaystyle\int_1^\infty \frac{1}{x^3}\, dx$.

24. Use trapezoidal approximation with $n = 4$ trapezoids to approximate $\int_0^1 \sqrt{x^2 + 1}\, dx$. (If you use a graphing calculator, compare your answer to the answer using FnInt.)

25. Use Simpson's rule with $n = 4$ to approximate $\int_0^1 \sqrt{x^2 + 1}\, dx$. (If you use a graphing calculator, compare your answer to the answer using FnInt.)

26. a. Find the general solution to the differential equation $y' = x^3y$.
    b. Then find the particular solution that satisfies $y(0) = 2$.

27. Find the first partial derivatives of $f(x, y) = x \ln y + ye^{2x}$.

28. Find the relative extreme values of $f(x, y) = 2x^2 - 2xy + y^2 + 4x - 6y + 12$.

29. Find the least squares line for the following points:

x	y
1	-3
3	1
5	3
7	8

30. Use Lagrange multipliers to find the minimum value of $f(x, y) = 3x^2 + 2y^2 - 2xy$ subject to the constraint $x + 2y = 18$.

31. Find the total differential of $f(x, y) = 2x^2 + xy - 3y^2 + 4$.

32. Find the volume under the surface $f(x, y) = 12 - x - 2y$ over the rectangle $R = \{(x, y) \mid 0 \le x \le 2, 0 \le y \le 3\}$.

Appendix

Normal Probabilities Using Tables

This appendix, showing how to find normal probabilities from tables, replaces pages 481–486 of Chapter 5 (Statistics) for readers who do not have graphing calculators that calculate normal probabilities.

z-Scores

The *standard normal distribution* is the normal distribution with mean $\mu = 0$ and standard deviation $\sigma = 1$ (the word "standard" indicates mean zero and standard deviation one). The letter z is traditionally used to denote the variable for this special distribution. Since $\mu = 0$, the highest point on the curve is at $z = 0$, and the curve is symmetric about this value.

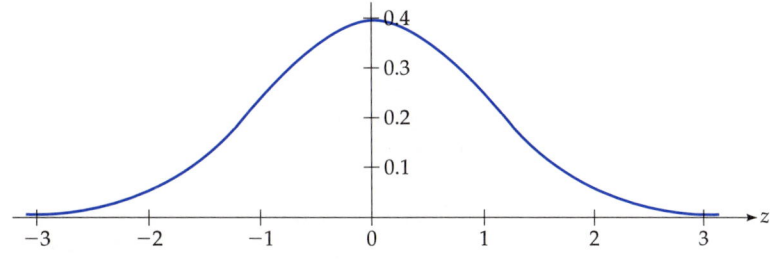

The standard normal distribution (mean 0 and standard deviation 1)

To change any x-value for a normal distribution with mean μ and standard deviation σ into the corresponding z-score for the *standard* normal distribution, subtract the mean and then divide by the standard deviation.

z-Score

$$z = \frac{x - \mu}{\sigma}$$

The z-score shows how many standard deviations the value is away from its mean.

EXAMPLE 1 **Finding a *z*-Score**

Convert each x-value into a z-score using the given values for μ and σ.

a. $x = 8$ with $\mu = 4$ and $\sigma = 2$ **b.** $x = 8$ with $\mu = 12$ and $\sigma = 4$

Solution

a. $z = \dfrac{8 - 4}{2} = 2$ **b.** $z = \dfrac{8 - 12}{4} = -1$

In part (a) the z-score $z = 2$ means that the original 8 is *two* standard deviations to the right of its mean. In (b) the z-score $z = -1$ means that the 8 in that distribution is one standard deviation to the *left* of *its* mean.

■

PRACTICE PROBLEM 1

Convert $x = 20$ with $\mu = 15$ and $\sigma = 1$ into a z-score. Is this value far from the mean?   *Solution at the back of the book*

The *central limit theorems* developed by Carl F. Gauss (1777–1855) and other mathematicians during the nineteenth century proved that the errors in observed values, the means of random samples, and many other statistical quantities were normally distributed. Extensive tables of the probabilities of collections of z-scores were laboriously calculated, and the results were worth the effort because they could be used for *any* normally distributed x-values by first converting them to z-scores.

On page 1016 of this appendix is a brief table for the normal distribution, giving the probability that a standard normal random variable takes a value between 0 and any given positive number. For example, to find $P(0 \le Z \le 1.24)$ (in words, the probability that a standard normal random variable has a value between 0 and 1.24), we locate in the table the row headed **1.2** and the column headed **0.04** (the second decimal place), and $P(0 \le Z \le 1.24)$ is the number where this row and column intersect, **0.3925**.

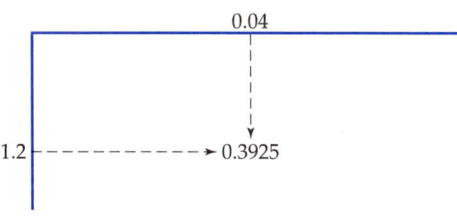

In the table on page 1016, the row headed **1.2** and the column headed **0.04** intersect at the table value **0.3925**

Therefore:

$$P(0 \le Z \le 1.24) \approx 0.3925$$

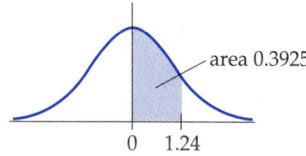

area 0.3925

Probabilities for other intervals can be found by adding and subtracting such areas and using the symmetry of the normal curve, as the following examples illustrate.

REAL WORLD

EXAMPLE 2   Finding a Probability for a Normal Random Variable

The heights of American men are approximately normally distributed with mean $\mu = 68.1$ inches and standard deviation $\sigma = 2.7$ inches.* Find the proportion of men who are between 5 feet 9 inches and 6 feet tall.

Solution

Converting to inches, we want the probability of a man's height being between 69 inches and 72 inches. We must convert these numbers to the corresponding z-scores for a *standard* normal random variable by subtracting the mean and dividing by the standard deviation:

$x = 69$ corresponds to: $z = \dfrac{69 - 68.1}{2.7} \approx 0.33$ Using $z = \dfrac{x - \mu}{\sigma}$ with

$x = 72$ corresponds to: $z = \dfrac{72 - 68.1}{2.7} \approx 1.44$ $\mu = 68.1$ and $\sigma = 2.7$

---

* These and other data in this appendix are from *Handbook of Human Factors* by Gavriel Salvendy (John Wiley-Interscience, 1987).

Using these values, we then want $P(0.33 \leq Z \leq 1.44)$, which is equivalent to the shaded area in the first graph below, which is equal to the difference between next two areas on the right. These two areas (or probabilities) are found from the table on page 1016, with the calculations shown below.

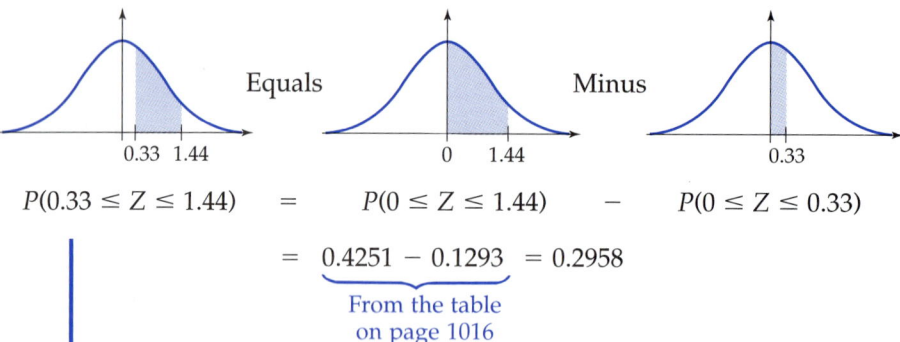

$$P(0.33 \leq Z \leq 1.44) \quad = \quad P(0 \leq Z \leq 1.44) \quad - \quad P(0 \leq Z \leq 0.33)$$

$$= \quad \underbrace{0.4251 - 0.1293}_{\text{From the table on page 1016}} = 0.2958$$

Therefore, about 30% of American men are between 5 feet 9 inches and 6 feet tall.

■

PRACTICE PROBLEM 2   The weights of American women are approximately normally distributed with mean 134.7 pounds and standard deviation 30.4 pounds. Find the proportion of women who weigh between 130 and 150 pounds.   *Solution at the back of the book*

The Normal and Binominal Distributions

As we saw on page 437, the binomial distribution too has a kind of "bell" shape, with a peak at the expected value $\mu = np$ and falling away on both sides to very small probabilities several standard deviations $\sigma = \sqrt{np(1 - p)}$ away.

$n = 20, p = 0.4$ $\qquad$ $n = 15, p = 0.5$ $\qquad$ $n = 25, p = 0.7$

Several binomial distributions with different values for n and p

In the eighteenth century, Abraham de Moivre (1667–1754) and Pierre-Simon Laplace (1749–1827) discovered and proved that for any choice of p between 0 and 1 the binomial distribution *approaches* the normal distribution as n becomes large. This fundamental fact is known as the *de Moivre–Laplace theorem*.

We may use the de Moivre–Laplace theorem to approximate binomial distributions by the normal distribution.

Normal Approximation to the Binomial

> Let X be a binomial random variable with parameters n and p. If $np > 5$ and $n(1 - p) > 5$, then the distribution of X is approximately normal with mean $\mu = np$ and standard deviation $\sigma = \sqrt{np(1 - p)}$.

To have the same width for a "slice" of the normal distribution as for the binomial, we adopt the convention that the binomial probability $P(X = x)$ corresponds to the area under the normal distribution curve from $x - \frac{1}{2}$ to $x + \frac{1}{2}$. This is called the *continuous correction*.

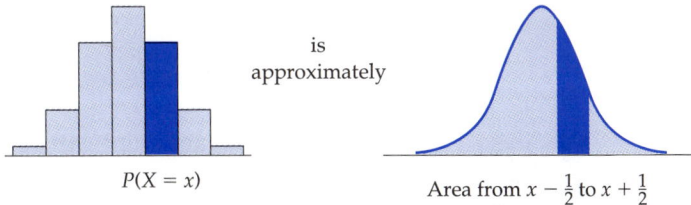

$P(X = x)$ is approximately   Area from $x - \frac{1}{2}$ to $x + \frac{1}{2}$

EXAMPLE 3   Normal Approximation of a Binomial Probability

Estimate $P(X = 12)$ for the binomial random variable X with $n = 25$ and $p = 0.6$ using the corresponding normal distribution.

Solution

Since $np = (25)(0.6) = 15$ and $n(1 - p) = (25)(0.4) = 10$ are both greater than 5, we may use the normal distribution with $\mu = np = (25)(0.6) = 15$ and $\sigma = \sqrt{np(1 - p)} = \sqrt{(25)(0.6)(0.4)} = \sqrt{6} \approx 2.45$. Using the continuous correction we interpret the event $X = 12$ as $11.5 \leq X \leq 12.5$ to include numbers that would round to 12. Converting these x-values into z-scores:

$$x = 11.5 \text{ corresponds to:} \quad z = \frac{11.5 - 15}{2.45} \approx -1.43$$

Using $z = \dfrac{x - \mu}{\sigma}$

$$x = 12.5 \text{ corresponds to:} \quad z = \frac{12.5 - 15}{2.45} \approx -1.02$$

with $\mu = 15$ and $\sigma = 2.45$

The probability $P(-1.43 \leq Z \leq -1.02)$ is represented on the following page by the shaded area in the first graph which, by symmetry, is equivalent to the second graph, which in turn is equivalent to the difference between the third and fourth graphs. The calculation with the probabilities found from the normal table is shown below.

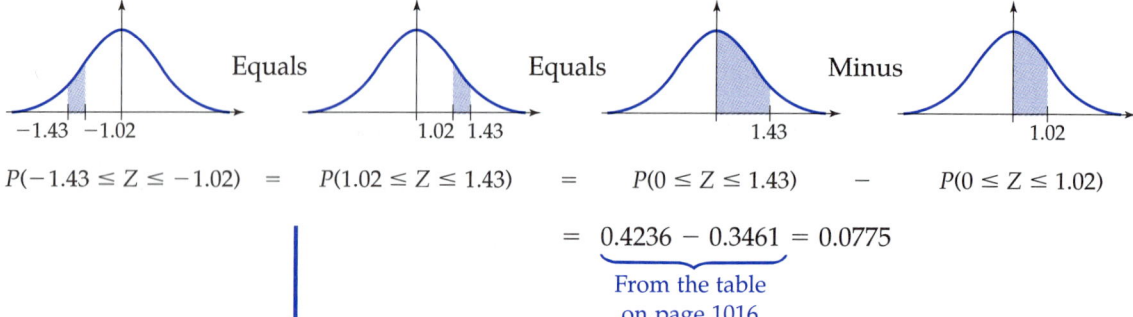

$P(-1.43 \leq Z \leq -1.02)$ = $P(1.02 \leq Z \leq 1.43)$ = $P(0 \leq Z \leq 1.43)$ − $P(0 \leq Z \leq 1.02)$

$$= \underbrace{0.4236 - 0.3461}_{\text{From the table on page 1016}} = 0.0775$$

The required probability is (about) 0.078. This is a good approximation to the actual value of $_{25}C_{12}p^{12}(1 - p)^{13} \approx 0.0759667$.

EXAMPLE 4   Management MBAs

At a major Los Angeles accounting firm, 73% of the managers have MBA degrees. In a random sample of 40 managers, what is the probability that between 27 and 32 will have MBAs?

Solution

Because each manager either has or does not have an MBA, presumably independently of each other, the question asks for the probability that a binomial random variable X with $n = 40$ and $p = 0.73$ satisfies $27 \leq X \leq 32$. Since $np = (40)(0.73) = 29.2$ and $n(1 - p) = (40)(0.27) = 10.8$ are both greater than 5, this probability can be approximated as the area under the normal distribution curve with mean $\mu = np = (40)(0.73) = 29.2$ and standard deviation $\sigma = \sqrt{np(1 - p)} = \sqrt{(40)(0.73)(0.27)} = \sqrt{7.884} \approx 2.81$ from $26\frac{1}{2}$ to $32\frac{1}{2}$ (again to include values that would round to between 27 and 32). Converting the x-values to z-scores:

$x = 26.5$ corresponds to: $z = \dfrac{26.5 - 29.2}{2.81} \approx -0.96$ Using $z = \dfrac{x - \mu}{\sigma}$

$x = 32.5$ corresponds to: $z = \dfrac{32.5 - 29.2}{2.81} \approx 1.17$ with $\mu = 29.2$ and $\sigma = 2.81$

The probability $P(-0.96 \leq Z \leq 1.17)$ is represented on the following page by the shaded area in the first graph which is equivalent to the sum of the next two areas, with the calculation shown below.

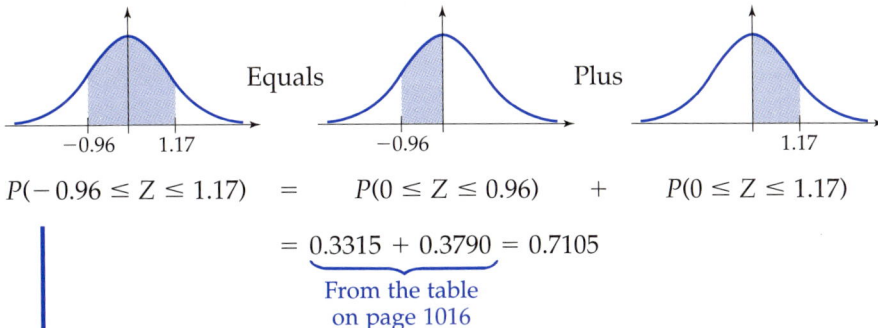

$$P(-0.96 \leq Z \leq 1.17) \quad = \quad P(0 \leq Z \leq 0.96) \quad + \quad P(0 \leq Z \leq 1.17)$$

$$= \underbrace{0.3315 + 0.3790}_{\substack{\text{From the table} \\ \text{on page 1016}}} = 0.7105$$

The probability that the number of MBAs will be between 27 and 32 is 0.711, or about 71%. (The exact answer to this problem, from summing the binomial distribution from 27 to 32, is 71.5%, so the normal approximation is very accurate.) ∎

PRACTICE PROBLEM 3

A brand of imported VCR is known to have defective tape rewind mechanisms in 8% of the units imported last April. If Jerry's Discount Electronics received a shipment of 80 of these VCRs, what is the probability that 10 or more are defective?   *Solution at the back of the book*

After completing Practice Problem 3, you should return to page 487 to read the Summary and do the Exercises in Section 5.4.

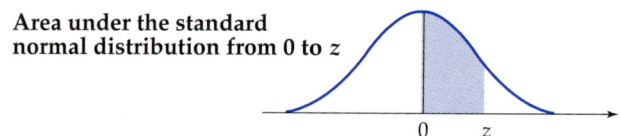

Area under the standard normal distribution from 0 to z

x	0.00	0.01	0.02	0.03	0.04	0.05	0.06	0.07	0.08	0.09
0.0	0.0000	0.0040	0.0080	0.0120	0.0160	0.0199	0.0239	0.0279	0.0319	0.0359
0.1	0.0398	0.0438	0.0478	0.0517	0.0557	0.0596	0.0636	0.0675	0.0714	0.0754
0.2	0.0793	0.0832	0.0871	0.0910	0.0948	0.0987	0.1026	0.1064	0.1103	0.1141
0.3	0.1179	0.1217	0.1255	0.1293	0.1331	0.1368	0.1406	0.1443	0.1480	0.1517
0.4	0.1554	0.1591	0.1628	0.1664	0.1700	0.1736	0.1772	0.1808	0.1844	0.1879
0.5	0.1915	0.1950	0.1985	0.2019	0.2054	0.2088	0.2123	0.2157	0.2190	0.2224
0.6	0.2258	0.2291	0.2324	0.2357	0.2389	0.2422	0.2454	0.2486	0.2518	0.2549
0.7	0.2580	0.2612	0.2642	0.2673	0.2704	0.2734	0.2764	0.2794	0.2823	0.2852
0.8	0.2881	0.2910	0.2939	0.2967	0.2996	0.3023	0.3051	0.3078	0.3106	0.3133
0.9	0.3159	0.3186	0.3212	0.3238	0.3264	0.3289	0.3315	0.3340	0.3365	0.3389
1.0	0.3413	0.3438	0.3461	0.3485	0.3508	0.3531	0.3554	0.3577	0.3599	0.3621
1.1	0.3643	0.3665	0.3686	0.3708	0.3729	0.3749	0.3770	0.3790	0.3810	0.3820
1.2	0.3849	0.3869	0.3888	0.3907	0.3925	0.3944	0.3962	0.3980	0.3997	0.4015
1.3	0.4032	0.4049	0.4066	0.4082	0.4099	0.4115	0.4131	0.4147	0.4162	0.4177
1.4	0.4192	0.4207	0.4222	0.4236	0.4251	0.4265	0.4279	0.4292	0.4306	0.4319
1.5	0.4332	0.4345	0.4357	0.4370	0.4382	0.4394	0.4406	0.4418	0.4429	0.4441
1.6	0.4452	0.4463	0.4474	0.4484	0.4495	0.4505	0.4515	0.4525	0.4535	0.4545
1.7	0.4554	0.4564	0.4573	0.4582	0.4591	0.4599	0.4608	0.4616	0.4625	0.4633
1.8	0.4641	0.4649	0.4656	0.4664	0.4671	0.4678	0.4686	0.4693	0.4699	0.4706
1.9	0.4713	0.4719	0.4726	0.4732	0.4738	0.4744	0.4750	0.4756	0.4761	0.4767
2.0	0.4772	0.4778	0.4783	0.4788	0.4793	0.4798	0.4803	0.4808	0.4812	0.4817
2.1	0.4821	0.4826	0.4830	0.4834	0.4838	0.4842	0.4846	0.4850	0.4854	0.4857
2.2	0.4861	0.4864	0.4868	0.4871	0.4875	0.4878	0.4881	0.4884	0.4887	0.4890
2.3	0.4893	0.4896	0.4898	0.4901	0.4904	0.4906	0.4909	0.4911	0.4913	0.4916
2.4	0.4918	0.4920	0.4922	0.4925	0.4927	0.4929	0.4931	0.4932	0.4934	0.4936
2.5	0.4938	0.4940	0.4941	0.4943	0.4945	0.4946	0.4948	0.4949	0.4951	0.4952
2.6	0.4953	0.4955	0.4956	0.4957	0.4959	0.4960	0.4961	0.4962	0.4963	0.4964
2.7	0.4965	0.4966	0.4967	0.4968	0.4969	0.4970	0.4971	0.4972	0.4973	0.4974
2.8	0.4974	0.4975	0.4976	0.4977	0.4977	0.4978	0.4979	0.4979	0.4980	0.4981
2.9	0.4981	0.4982	0.4982	0.4983	0.4984	0.4984	0.4985	0.4985	0.4986	0.4986
3.0	0.4987	0.4987	0.4987	0.4988	0.4988	0.4989	0.4989	0.4989	0.4990	0.4990
3.1	0.4990	0.4991	0.4991	0.4991	0.4992	0.4992	0.4992	0.4992	0.4993	0.4993
3.2	0.4993	0.4993	0.4994	0.4994	0.4994	0.4994	0.4994	0.4995	0.4995	0.4995
3.3	0.4995	0.4995	0.4995	0.4996	0.4996	0.4996	0.4996	0.4996	0.4996	0.4997
3.4	0.4997	0.4997	0.4997	0.4997	0.4997	0.4997	0.4997	0.4997	0.4997	0.4998
3.5	0.4998	0.4998	0.4998	0.4998	0.4998	0.4998	0.4998	0.4998	0.4998	0.4998

Solutions to Practice Problems

SECTION 0.1

1. $-1,000,000$ [the negative sign makes it less than (to the left of) the positive number $\frac{1}{100}$]

2. a. $\{x \mid x \geq -7\}$

b. the set of all x such that x is less than -1

3. a. S_1

b. S_4

4. $m = \dfrac{7 - 1}{4 - 2} = \dfrac{6}{2} = 3$ From points (2, 1) and (4, 7)

$y - 1 = 3(x - 2)$ Using the point-slope form with

$y - 1 = 3x - 6$ $(x_1, y_1) = (2, 1)$

$y = 3x - 5$

5. $x = -2$

6. $x - \dfrac{y}{3} = 2$

$-\dfrac{y}{3} = -x + 2$ Subtracting x from each side

$y = 3x - 6$ Multiplying each side by -3

Slope is $m = 3$ and y-intercept is $(0, -6)$.

SECTION 0.2

1. a. $\dfrac{x^5 \cdot x}{x^2} = \dfrac{x^6}{x^2} = x^4$

b. $[(x^3)^2]^2 = x^{3 \cdot 2 \cdot 2} = x^{12}$

2. a. $2^0 = 1$

b. $2^{-4} = \dfrac{1}{2^4} = \dfrac{1}{16}$

3. $\left(\dfrac{2}{3}\right)^{-2} = \left(\dfrac{3}{2}\right)^2 = \dfrac{9}{4}$

4. a. $(-27)^{1/3} = \sqrt[3]{-27} = -3$

b. $\left(\dfrac{16}{81}\right)^{1/4} = \sqrt[4]{\dfrac{16}{81}} = \dfrac{2}{3}$

5. a. $16^{3/2} = (\sqrt{16})^3 = 4^3 = 64$

b. $(-8)^{2/3} = (\sqrt[3]{-8})^2 = (-2)^2 = 4$

6. a. $25^{-3/2} = \dfrac{1}{25^{3/2}} = \dfrac{1}{(\sqrt{25})^3} = \dfrac{1}{5^3} = \dfrac{1}{125}$

b. $\left(\dfrac{1}{4}\right)^{-1/2} = \left(\dfrac{4}{1}\right)^{1/2} = \sqrt{4} = 2$

c. $5^{1.3} \approx 8.103$

SECTION 0.3

1. Domain: $\{x \mid x \leq 0 \text{ or } x \geq 3\}$, Range: $\{y \mid y \geq 0\}$

2. a. $g(27) = \sqrt{27 - 2} = \sqrt{25} = 5$

b. Domain: $\{z \mid z \geq 2\}$

c. Range: $\{y \mid y \geq 0\}$

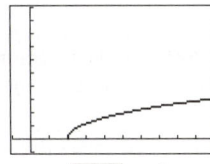

$y_1 = \sqrt{x - 2}$ on $[-1, 10]$ by $[-1, 10]$

A1

3. $D(x) = 25 + 0.05x$

4.
$$9x - 3x^2 = -30$$
$$-3x^2 + 9x + 30 = 0$$
$$-3(x^2 - 3x - 10) = 0$$
$$-3(x - 5)(x + 2) = 0$$
$$x = 5, x = -2 \quad \text{or from}$$

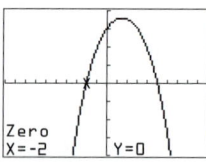

Zero
X=-2   Y=0

SECTION 0.4

1.
$$2x^3 - 4x^2 - 48x = 0$$
$$2x(x^2 - 2x - 24) = 0$$
$$2x(x + 4)(x - 6) = 0$$
$$x = 0, x = -4, x = 6$$

2. Domain: $\{x \mid x \neq 0, x \neq -10\}$
Range: $\{y \mid y > 0 \text{ or } y \leq -7200\}$

3. a. $f(g(x)) = [g(x)]^2 + 1 = (\sqrt[3]{x})^2 + 1$ or $x^{2/3} + 1$
b. $g(f(x)) = \sqrt[3]{f(x)} = \sqrt[3]{x^2 + 1}$ or $(x^2 + 1)^{1/3}$

4. $\dfrac{f(x + h) - f(x)}{h}$

$$= \frac{3(x + h)^2 - 2(x + h) + 1 - (3x^2 - 2x + 1)}{h}$$

$$= \frac{3x^2 + 6xh + 3h^2 - 2x - 2h + 1 - 3x^2 + 2x - 1}{h}$$

$$= \frac{h(6x + 3h - 2)}{h} = 6x + 3h - 2$$

SECTION 0.5

1. $10,000(1 + 0.07)^{30} = 10,000 \cdot 1.07^{30} \approx 76,122.55$
The value will be $76,122.55.

2. $50,000(1 - 0.20)^4 = 50,000(0.8)^4 = \$20,480$

SECTION 0.6

1. $\log 10,000 = 4$ (Since $10^4 = 10,000$)

2. $\ln 8.34 \approx 2.121$ (Using a calculator)

SECTION 1.1

1. $I = (50,000)(0.198)(92/360) = 2530$, so the Banker's rule interest is $2530. A 3-month loan would have interest $I = (50,000)(0.198)(3/12) = 2475$, which means the Banker's rule gives the lender a $2530 - 2475 = \$55$ advantage.

2. $PV = \dfrac{5000}{1 + (4)(0.12)} = \dfrac{5000}{1.48} \approx 3378.38$

3. $r_s = \dfrac{0.06}{1 - (0.06)(3)} \approx 0.0732$. The effective simple interest rate is 7.32%. $r_s = \dfrac{0.06}{1 - (0.06)(5)} \approx 0.0857$. The effective simple interest rate is 8.57%. Can you think of an intuitive reason for the effective rate to be higher if the term is longer? (*Hint:* Think of how much earlier the lender gets the money, or of how much less the borrower really gets.)

SECTION 1.2

1. $A = \$3500\left(1 + \dfrac{0.051}{6}\right)^{(6)(8)} \approx \$5254.26$.
The amount due is $5254.26.

2. $2P = P\left(1 + \dfrac{0.06}{4}\right)^{4t}$ which simplifies to $2 = (1 + 0.06/4)^{4t}$. Taking logarithms: $\log (2) = \log (1 + 0.06/4)^{4t} = 4t \log (1 + 0.06/4)$ so
$$4t = \frac{\log (2)}{\log (1 + 0.06/4)} \approx 46.6 \text{ (quarters!)}.$$
Rounding up gives 47 quarters, or $11\frac{3}{4}$ years.

3. $r_e = \left(1 + \dfrac{0.062}{12}\right)^{12} - 1 \approx 0.0638$. The effective rate is 6.38%.

SECTION 1.3

1. $A = 40 \dfrac{\left(1 + \dfrac{0.05}{52}\right)^{(52)(35)} - 1}{\dfrac{0.05}{52}} \approx 197,590.27$. The final balance is $197,590.27. Since the deposits total $40 \times 52 \times 35 = \$72,800$, the final balance contains $124,790.27 interest.

2. $P = 18,000 \dfrac{\dfrac{0.045}{26}}{\left(1 + \dfrac{0.045}{26}\right)^{(26)(3)} - 1} \approx 215.742$.
$215.75 should be deposited every other week. (Compare double this amount, $431.50, to the

$467.95 monthly payment found in Example 2 on page 139.)

3. $12t = \dfrac{\log\left(\dfrac{18{,}000}{250}\dfrac{0.045}{12} + 1\right)}{\log\left(1 + \dfrac{0.045}{12}\right)} \approx 63.9$ months,

which rounds up to 64 months. It will take 5 years 4 months.

SECTION 1.4

1. $PV = 850\,\dfrac{1 - \left(1 + \dfrac{0.0753}{12}\right)^{-(12)(20)}}{\dfrac{0.0753}{12}} \approx$

105,272.468. The present value is $105,272.47.

2. $P = 150{,}000\,\dfrac{\dfrac{0.086}{12}}{1 - \left(1 + \dfrac{0.086}{12}\right)^{-(12)(25)}} \approx 1217.97.$

The required payment is $1217.97 each month. The borrower will pay a total of $1217.97 × 12 × 25 = $365,391.

3. The monthly payment to amortize this loan

is $P = 12{,}000\,\dfrac{\dfrac{0.047}{12}}{1 - \left(1 + \dfrac{0.047}{12}\right)^{-(12)(4)}} \approx 274.724,$

which rounds up to $274.73. The remaining payments form a 3-year annuity at 4.7% and the present value of this annuity is $PV =$

$274.73\,\dfrac{1 - \left(1 + \dfrac{0.047}{12}\right)^{-(12)(3)}}{\dfrac{0.047}{12}} \approx 9207.884.$ The

amount still owed is $9207.88.

SECTION 2.1

1. Let x be the number of pennies in the jar and y be the number of nickels. Then the first statement may be expressed as "$x + y = 80$" and the second as "$0.01x + 0.05y = 1.60$" since each penny is worth $0.01 and each nickel $0.05 (if

you want to write the second statement in cents rather than dollars, you would have "$x + 5y = 160$"). The situation may be represented as

$$\begin{cases} x + y = 80 \\ 0.01x + 0.05y = 1.60 \end{cases}.$$

2. Since the first equation can be solved for y as $y = 10 - 2x$, we can substitute $10 - 2x$ for y in the second equation: $x + 2(10 - 2x) = 8$. Multiplying out and collecting like terms, $x + 20 - 4x = 8$ so $-3x = -12$ so that $x = 4$. Substituting $x = 4$ into $y = 10 - 2x$ gives $y = 10 - 2(4) = 10 - 8 = 2$. The solution is $x = 4$, $y = 2$. There are several other ways of solving this problem by the substitution method and all reach the same conclusion.

3. $\begin{cases} x + y = 100 \\ x - y = -20 \end{cases} \xrightarrow[\text{to first}]{\substack{\text{Add} \\ \text{second}}} \begin{cases} 2x + 0y = 80 \\ x - y = -20 \end{cases} \xrightarrow[\text{by 2}]{\substack{\text{Divide} \\ \text{first}}}$

$\begin{cases} 1x + 0y = 40 \\ x - y = -20 \end{cases} \xrightarrow[\substack{\text{from} \\ \text{second}}]{\substack{\text{Subtract} \\ \text{first}}} \begin{cases} 1x + 0y = 40 \\ 0x - y = -60 \end{cases} \xrightarrow[\text{by } -1]{\substack{\text{Multiply} \\ \text{second}}}$

$\begin{cases} 1x + 0y = 40 \\ 0x + 1y = 60 \end{cases}$

The solution is $x = 40$, $y = 60$. There are many other possible sequences of equivalent systems that solve this problem and all reach the same conclusion.

SECTION 2.2

1. a. $\begin{pmatrix} 2 & -1 & 14 \\ 1 & 3 & 21 \end{pmatrix}$

  b. $\begin{cases} 3x + 2y = 35 \\ x + 3y = 21 \end{cases}$

2. a. $\begin{pmatrix} 6 & 3 & 42 \\ 1 & -3 & 21 \end{pmatrix} R'1 = 3R1$

  b. $\begin{pmatrix} 7 & 0 & 63 \\ 1 & -3 & 21 \end{pmatrix} R'1 = R1 + R2$

  c. $\begin{pmatrix} 1 & 0 & 9 \\ 1 & -3 & 21 \end{pmatrix} R'1 = \frac{1}{7}R1$

  d. No: you can multiply only by a *nonzero* number.

  e. No: Multiplying row 1 by 5 must still give row 1 (not row 2).

3. Continuing from the solutions to Practice Problem 2, $\begin{pmatrix} 1 & 0 & 9 \\ 0 & -3 & 12 \end{pmatrix}$ $R'2 = R2 - R1$ and then $\begin{pmatrix} 1 & 0 & 9 \\ 0 & 1 & -4 \end{pmatrix}$ $R'2 = -\frac{1}{3}R2$. The solution is $x = 9$, $y = -4$ and the system of equations is independent and consistent. The are many other sequences of row operations to reduce this augmented matrix and all reach the same conclusion.

4. For $x = -5$, $y = -20$, the first equation becomes $6(-5) - 3(-20) = -30 + 60 = 30$ and the second becomes $-8(-5) + 4(-20) = 40 - 80 = -40$ as needed. For $x = 15$, $y = 20$, the first equation becomes $6(15) - 3(20) = 90 - 60 = 30$ and the second becomes $-8(15) + 4(20) = -120 + 80 = -40$ as needed.

SECTION 2.3

1. The system is independent (there are no zero rows) and inconsistent (there is a row of zeros ending in a 1). The system of equations has no solution.

2. We found in Example 2 that the corresponding augmented matrix $\begin{pmatrix} 5 & 5 & 0 & 5 & 50 \\ 2 & 3 & 1 & 0 & 17 \\ 2 & 2 & 1 & -1 & 9 \\ 2 & 3 & 1 & 1 & 22 \end{pmatrix}$ row reduces to $\begin{pmatrix} 1 & 0 & 0 & 0 & 2 \\ 0 & 1 & 0 & 0 & 3 \\ 0 & 0 & 1 & 0 & 4 \\ 0 & 0 & 0 & 1 & 5 \end{pmatrix}$. The system is independent (there are no zero rows) and consistent (there is no row of zeros ending in a 1). The solution, read from the last column, is $x_1 = 2$, $x_2 = 3$, $x_3 = 4$, $x_4 = 5$. We check this solution by substitution into the original equations:

$$5(2) + 5(3) + 0(4) + 5(5) = 50$$

$$2(2) + 3(3) + 1(4) + 0(5) = 17$$

$$2(2) + 2(3) + 1(4) - 1(5) = 9$$

$$2(2) + 3(3) + 1(4) + 1(5) = 22$$

It checks!

SECTION 2.4

1. $(1 \quad 2)\begin{pmatrix} 3 \\ 4 \end{pmatrix} = (1\cdot3 + 2\cdot4) = (11)$

2. $\begin{pmatrix} 1.00 & 1.50 & 0.75 \\ 1.25 & 1.75 & 0.50 \end{pmatrix}\begin{pmatrix} 4 \\ 2 \\ 5 \end{pmatrix}$

$$= \begin{pmatrix} 1.00\cdot4 + 1.50\cdot2 + 0.75\cdot5 \\ 1.25\cdot4 + 1.75\cdot2 + 0.50\cdot5 \end{pmatrix} = \begin{pmatrix} 10.75 \\ 11.00 \end{pmatrix}$$

SECTION 2.5

1. $A^{-1}A = \begin{pmatrix} 1 & 2 & -2 \\ -1 & 0 & 1 \\ 0 & -1 & 1 \end{pmatrix}\begin{pmatrix} 1 & 0 & 2 \\ 1 & 1 & 1 \\ 1 & 1 & 2 \end{pmatrix}$

$$= \begin{pmatrix} 1+2-2 & 0+2-2 & 2+2-4 \\ -1+0+1 & 0+0+1 & -2+0+2 \\ 0-1+1 & 0-1+1 & 0-1+2 \end{pmatrix}$$

$$= \begin{pmatrix} 1 & 0 & 0 \\ 0 & 1 & 0 \\ 0 & 0 & 1 \end{pmatrix} = I$$

and

$AA^{-1} = \begin{pmatrix} 1 & 0 & 2 \\ 1 & 1 & 1 \\ 1 & 1 & 2 \end{pmatrix}\begin{pmatrix} 1 & 2 & -2 \\ -1 & 0 & 1 \\ 0 & -1 & 1 \end{pmatrix}$

$$= \begin{pmatrix} 1+0+0 & 2+0-2 & -2+0+2 \\ 1-1+0 & 2+0-1 & -2+1+1 \\ 1-1+0 & 2+0-2 & -2+1+2 \end{pmatrix}$$

$$= \begin{pmatrix} 1 & 0 & 0 \\ 0 & 1 & 0 \\ 0 & 0 & 1 \end{pmatrix} = I.$$

2. Using $X = A^{-1}B$:

$$\begin{pmatrix} x_1 \\ x_2 \\ x_3 \end{pmatrix} = \begin{pmatrix} 1 & 2 & -2 \\ -1 & 0 & 1 \\ 0 & -1 & 1 \end{pmatrix}\cdot\begin{pmatrix} -5 \\ 10 \\ 0 \end{pmatrix} = \begin{pmatrix} 15 \\ 5 \\ -10 \end{pmatrix},$$

so $\begin{cases} x_1 = 15 \\ x_2 = 5 \\ x_3 = -10 \end{cases}$.

SECTION 3.1

1. The boundary is the line $3x - 5y = 60$. The x-intercept ratio is $x = 60/3 = 20$ and the x-

intercept is (20, 0). The y-intercept ratio is $y = 60/(-5) = -12$ and the y-intercept is $(0, -12)$. Since $3 \cdot 0 - 5 \cdot 0$ is ≤ 60, the origin $(0, 0)$ is on the correct side of the boundary line.

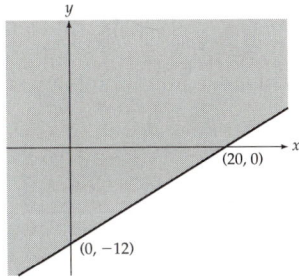

2. The boundaries are the lines $x + 2y = 20$ (with intercepts $(20, 0)$ and $(0, 10)$), $x + y = 10$ (with intercepts $(10, 0)$ and $(0, 10)$), and $x = 10$ (a vertical line with x-intercept $(10, 0)$ and no y-intercept). The origin $(0, 0)$ is on the correct side of $x + 2y \le 20$ and $x \le 10$ but not on the correct side of $x + y \ge 10$. This *is* a feasible system of linear inequalities.

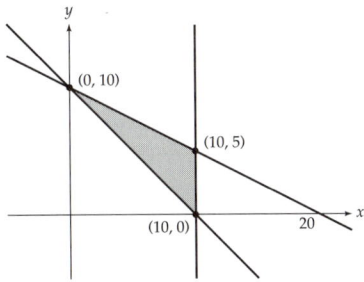

3. Two of the three vertices of the region are known from the x- and y-intercepts of the boundary lines in the previous sketch for Practice Problem 2. The third is the intersection of the lines $x + 2y = 20$ and $x = 10$. Substituting $x = 10$ into the first equation, $(10) + 2y = 20$ means $2y = 20 - 10 = 10$ so $y = 10/2 = 5$ and this vertex is $(10, 5)$. The vertices of the region are $(10, 0)$, $(10, 5)$, and $(0, 10)$. The region is bounded.

SECTION 3.2

1.

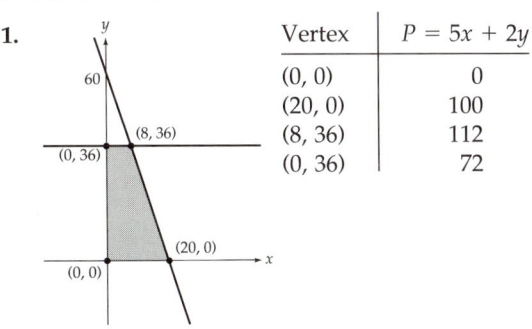

Vertex	$P = 5x + 2y$
(0, 0)	0
(20, 0)	100
(8, 36)	112
(0, 36)	72

The maximum value is 112 when $x = 8$ and $y = 36$.

2.

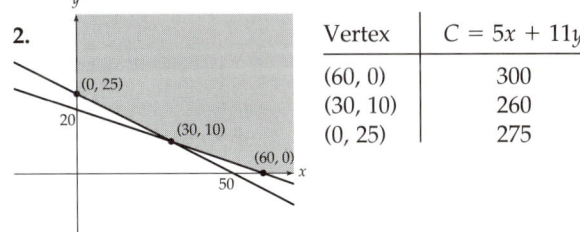

Vertex	$C = 5x + 11y$
(60, 0)	300
(30, 10)	260
(0, 25)	275

The minimum value is 260 when $x = 30$ and $y = 10$.

SECTION 3.3

1. The matrix form of the problem is

$$\text{Maximize } P = \begin{pmatrix} 9 & -1 & 10 & 12 \end{pmatrix} \begin{pmatrix} x_1 \\ x_2 \\ x_3 \\ x_4 \end{pmatrix}$$

$$\text{Subject to} \begin{cases} \begin{pmatrix} 1 & 1 & 2 & 2 \\ 2 & 2 & 1 & 1 \\ 1 & 2 & 2 & 1 \end{pmatrix} \begin{pmatrix} x_1 \\ x_2 \\ x_3 \\ x_4 \end{pmatrix} \le \begin{pmatrix} 16 \\ 20 \\ 18 \end{pmatrix} \\[2pt] \qquad\qquad\qquad\quad \uparrow \\ \qquad\qquad\quad \text{Standard since} \\ \text{and } \begin{pmatrix} x_1 \\ x_2 \\ x_3 \\ x_4 \end{pmatrix} \ge 0 \quad b = \begin{pmatrix} 16 \\ 20 \\ 18 \end{pmatrix} \ge 0 \end{cases}$$

The simplex tableau is

	x_1	x_2	x_3	x_4	s_1	s_2	s_3	
s_1	1	1	2	2	1	0	0	16
s_2	2	2	1	1	0	1	0	20
s_3	1	2	2	1	0	0	1	18
P	-9	1	-10	-12	0	0	0	0

2. a. The smallest negative entry in the bottom row is -6, so the pivot column is column 3. The ratios are $\frac{4}{1} = 4$ and $\frac{9}{3} = 3$, so the smallest non-negative ratio is 3 and the pivot row is row 2. The pivot element is the 3 in column 3 and row 2 of the simplex tableau.

	x_1	x_2	x_3	x_4	s_1	s_2	
s_1	-2	0	0	$\frac{1}{3}$	1	$-\frac{1}{3}$	1
x_3	1	2	1	$\frac{2}{3}$	0	$\frac{1}{3}$	3
P	1	9	0	2	0	2	18

$R_1^{new} = R_1 - (1)R_{pivot}^{new}$
$R_{pivot}^{new} = R_{pivot}/3$
$R_3^{new} = R_3 - (-6)R_{pivot}^{new}$

b. The smallest negative entry in the bottom row is -4, so the pivot column is column 3. The first row may not be considered for the pivot row because the pivot column entry is negative. The second row may not be considered for the pivot row because the pivot column entry is zero. There is no pivot row for this simplex tableau and therefore there is no pivot element.

3. The basic variables take the values at the right ends of their rows, so $x_2 = 6$, $s_2 = 6$, and $x_1 = 4$. The nonbasic variables are zero, so $s_1 = 0$ and $s_3 = 0$. The value of the objective function appears in the bottom right corner, so $P = 62$. *Solution:* The maximum is 62 when $x_1 = 4$ and $x_2 = 6$.

SECTION 3.4

1. In matrix form, this problem is:

$$\text{Minimize } C = (11 \quad 9 \quad 7)\begin{pmatrix} y_1 \\ y_2 \\ y_3 \end{pmatrix}$$

$$\text{Subject to}\begin{cases}\begin{pmatrix} 2 & 1 & 1 \\ 1 & 1 & 0 \\ 0 & 1 & 1 \\ 1 & 1 & 3 \end{pmatrix}\begin{pmatrix} y_1 \\ y_2 \\ y_3 \end{pmatrix} \geq \begin{pmatrix} 4 \\ 7 \\ 5 \\ 6 \end{pmatrix} \\ \text{and } \begin{pmatrix} y_1 \\ y_2 \\ y_3 \end{pmatrix} \geq 0\end{cases}$$

and the dual maximum problem is:

$$\text{Maximize } P = (4 \quad 7 \quad 5 \quad 6)\begin{pmatrix} x_1 \\ x_2 \\ x_3 \\ x_4 \end{pmatrix}$$

$$\text{Subject to}\begin{cases}\begin{pmatrix} 2 & 1 & 0 & 1 \\ 1 & 1 & 1 & 1 \\ 1 & 0 & 1 & 3 \end{pmatrix}\begin{pmatrix} x_1 \\ x_2 \\ x_3 \\ x_4 \end{pmatrix} \leq \begin{pmatrix} 11 \\ 9 \\ 7 \end{pmatrix} \\ \text{and } \begin{pmatrix} x_1 \\ x_2 \\ x_3 \\ x_4 \end{pmatrix} \geq 0\end{cases}$$

The initial simplex tableau is

	x_1	x_2	x_3	x_4	s_1	s_2	s_3	
s_1	2	1	0	1	1	0	0	11
s_2	1	1	1	1	0	1	0	9
s_3	1	0	1	3	0	0	1	7
P	-4	-7	-5	-6	0	0	0	0

2. The bottom row of the final tableau for the dual maximum problem displays the values for the slack variables ($t_1 = 0$, $t_2 = 0$, $t_3 = 0$), the variables ($y_1 = 7$, $y_2 = 3$, $y_3 = 1$, $y_4 = 0$), and the objective function ($C = 3390$). The minimum value is 3390 when $y_1 = 7$, $y_2 = 3$, $y_3 = 1$, and $y_4 = 0$.

SECTION 3.5

1. a. The dual pivot row is row 1 because -10 is the smallest negative entry in the rightmost column (omitting the bottom row). The dual pivot column is column 2 because the ratio $\frac{-6}{-1} = 6$ is greater than the ratio $\frac{-5}{-1} = 5$ for the first column and the ratio $\frac{-8}{-2} = -4$ for the fourth column; the other columns may not be considered since their dual pivot row entries are zero or positive, and the rightmost column is never considered. The dual pivot element is the -1 in row 1 and column 2.

b. The tableau does not have a dual pivot element. The dual pivot row is row 4 because the smallest negative entry in the rightmost column (omitting the bottom row) is -20. Since the other entries in row 4 are either zero or positive, there is no dual pivot column. (This means that the constraints are infeasible. The fourth row represents the inequality $2x_1 + x_2 + x_3 + 4x_4 \leq -20$ and this is impossible because the variables are nonnegative. The problem has no solution.

2. Rewriting the second inequality as $-2x_1 - x_2 - 2x_3 \leq -10$, this problem may be written in matrix form as:

$$\text{Maximize } P = (4 \quad 1 \quad 3) \begin{pmatrix} x_1 \\ x_2 \\ x_3 \end{pmatrix}$$

Subject to
$$\begin{cases} \begin{pmatrix} 1 & 2 & 1 \\ -2 & -1 & -2 \end{pmatrix} \begin{pmatrix} x_1 \\ x_2 \\ x_3 \end{pmatrix} \leq \begin{pmatrix} 50 \\ -10 \end{pmatrix} \\ \\ \text{and } \begin{pmatrix} x_1 \\ x_2 \\ x_3 \end{pmatrix} \geq 0 \end{cases}$$

The initial simplex tableau is

	x_1	x_2	x_3	s_1	s_2	
s_1	1	2	1	1	0	50
s_2	-2	-1	-2	0	1	-10
P	-4	-1	-3	0	0	0

This is not feasible because $s_2 = -10$. Pivoting on the dual pivot element in row 2 and column 1, the tableau becomes feasible:

	x_1	x_2	x_3	s_1	s_2	
s_1	0	3/2	1	1	1/2	45
x_1	1	1/2	1	0	$-1/2$	5
P	0	1	1	0	-2	20

Pivoting on the (regular) pivot element in column 5 and row 1, the tableau becomes optimal:

	x_1	x_2	x_3	s_1	s_2	
s_2	0	3	0	2	1	90
x_1	1	2	1	1	0	50
P	0	7	1	4	0	200

This is the final tableau since it is both feasible and optimal. The maximum is $P = 200$ when $x_1 = 50$ and $x_2 = 0$.

SECTION 4.1

1. Let D be the set of businesses offering dental insurance and let V be those offering vision insurance. Then

$$\begin{aligned} n(D \cup V) &= n(D) + n(V) - n(D \cap V) \\ &= 150 + 150 - 100 = 200 \end{aligned}$$

Two hundred of the three hundred businesses surveyed offer dental or vision insurance.

2. Since
$$\underbrace{26 \cdot 26 \cdot 26 \cdot 26}_{\text{Four letters}} \cdot \underbrace{10 \cdot 10 \cdot 10 \cdot 10}_{\text{Four digits}} = 4{,}569{,}760{,}000,$$
there are 4,569,760,000 different passwords.

3. Since the winners must be selected in order (first, second, and third place), there are $_{35}P_3 = 35 \cdot 34 \cdot 33 = 39{,}270$ different ways of choosing the winners.

4. Since the courses can be taken in any order, there are $_{10}C_6 = \frac{10 \cdot 9 \cdot 8 \cdot 7 \cdot 6 \cdot 5}{6 \cdot 5 \cdot 4 \cdot 3 \cdot 2 \cdot 1} = 210$ different ways to minor in computer science.

SECTION 4.2

1. "No heads" becomes $0 = \{(T, T)\}$, "one head" becomes $1 = \{(H, T), (T, H)\}$ and "two heads" becomes $2 = \{(H, H)\}$.

2. $P(A) = \frac{60°}{360°} = \frac{1}{6}$, $P(B) = \frac{1}{2}$, $P(C) = \frac{120°}{360°} = \frac{1}{3}$.

3. In three rolls there are $6 \cdot 6 \cdot 6 = 6^3 = 216$ possible outcomes, each with probability $\frac{1}{216}$. If the event D represents at least one six (in three rolls), then D^c (no sixes in three rolls) contains

$5 \cdot 5 \cdot 5 = 5^3 = 125$ possible outcomes, each having probability $\frac{1}{216}$. Using the summation formula, $P(D^c) = 125 \cdot \frac{1}{216} = \frac{125}{216} \approx 0.579$. Then, by the complementary probability principle, $P(D) = 1 - P(D^c) \approx 1 - 0.579 = 0.421$. Therefore, betting on at least one six in *three* rolls would win only about 42% of the time.

SECTION 4.3

1. If H is the event that your parents arrive home soon, then H^c is the event that they don't, and you go on foot. So $P(H^c) = 0.50$ and $P(A$ given $H^c) = 0.60$. Then $P(A$ and $H^c) = P(A$ given $H^c) \cdot P(H^c) = (0.60) \cdot (0.50) = 0.30$. The probability of arriving on foot and on time is 30%.

2. Using the notation from the solution to Example 5 (page 419):

$P(I$ given $V)$

$$= \frac{P(V \text{ given } I) \cdot P(I)}{P(V \text{ given } D) \cdot P(D) + P(V \text{ given } R) \cdot P(R) + P(V \text{ given } I) \cdot P(I)}$$

$$= \frac{(0.90) \cdot (0.25)}{(0.70) \cdot (0.45) + (0.80) \cdot (0.30) + (0.90) \cdot (0.25)} \approx 0.288$$

The probability that a voter in the last election was an independent is about 29%.

3. Since the probability that any one does not make the sale is 0.20, the probability that all five do not is

$$\underbrace{\frac{20}{100} \cdot \frac{20}{100} \cdot \frac{20}{100} \cdot \frac{20}{100} \cdot \frac{20}{100}}_{\text{Five terms}} = \left(\frac{20}{100}\right)^5 = 0.00032$$

SECTION 4.4

1. $E(X) = 3000 \cdot \frac{1}{400} + 1000 \cdot \frac{2}{400} + 100 \cdot \frac{10}{400} + 0 \cdot \frac{387}{400} = 15$
The expected value is $15.00.

2. $_6C_3(\frac{1}{2})^3(\frac{1}{2})^3 = \frac{6 \cdot 5 \cdot 4}{3 \cdot 2 \cdot 1}(\frac{1}{2})^6 = 20 \cdot \frac{1}{64} = \frac{5}{16}$.

$\mu = n \cdot p = 6 \cdot \frac{1}{2} = 3$

$\sigma = \sqrt{np(1-p)}$

$= \sqrt{6 \cdot \frac{1}{2} \cdot \frac{1}{2}} \approx 1.22$

```
binompdf(6,.5,3)
          .3125
Ans▶Frac
          5/16
√(6*.5*.5)
       1.224744871
```

SECTION 5.1

1. **a.** Nominal   **b.** Ordinal   **c.** Ratio

2.

Vehicle	Tally	Frequency
1	JHT I	6
2	I	1
3	III	3

3. Choosing the stem to be the units digit and the leaf to be the first decimal place, we obtain:

Stem	Leaf
0	
1	
2	5
3	7
4	6, 4
5	4, 6, 7
6	2, 0, 4, 2, 7
7	8, 6
8	1
9	

SECTION 5.2

1. Since 20 occurs three times and no other value occurs this often, the mode is 20.

2. The median of nominal data values is not defined.

3. The mode is 1, the median is 2, and the mean is

$$\bar{x} = \frac{1}{5}(1 + 1 + 2 + 3 + 8) = \frac{1}{5}(15) = 3$$

```
{1,1,2,3,8}→L₁
      {1 1 2 3 8}
median(L₁)
             2
mean(L₁)
             3
```

SECTION 5.3

1. The largest value is 30 and the smallest is 10, so the range is $30 - 10 = 20$.

2. The minimum is 8 and the maximum is 29. Since there are eleven values, the median is 16, the

sixth value. From the first five values, {8, 10, 13, 14, 15}, the first quartile is 13, while from the last five values, {17, 19, 23, 24, 29}, the third quartile is 23. From this five-point summary, the box-and-whisker plot is

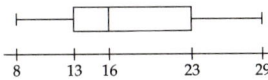

The tightest clustering is the quarter of the values between 13 and 16.

3. First, find the mean: $\bar{x} = \frac{1}{8}(2 + 6 + 7 + 8 + 9 + 15 + 16 + 17) = \frac{1}{8}(80) = 10$. Then

$$s = \sqrt{\frac{(2-10)^2 + \cdots + (17-10)^2}{8-1}}$$

$$= \sqrt{\frac{64 + 16 + 9 + 4 + 1 + 25 + 36 + 49}{7}}$$

$$= \sqrt{\frac{204}{7}} \approx 5.4$$

SECTION 5.4

1.
```
normalcdf(130,15
0,134.7,30.4)
        .2540533807
```

About 25% of American women weigh between 130 and 150 pounds. (For a solution using tables, see page A14.)

2. $z = \dfrac{20 - 15}{1} = 5$. Since more than 99% of the z-scores are within 3 of the mean, this value is *very* far from the mean!

3. Since $np = (80)(0.08) = 6.4 > 5$ and $n(1 - p) = (80)(0.92) = 73.6 > 5$, we may use the normal distribution with $\mu = np = (80)(0.08)$ and $\sigma = \sqrt{np(1-p)} = \sqrt{(80)(0.08)(0.92)}$ to approximate $P(X \geq 10)$ as the area under the normal curve from 9.5 to 80.5 (corresponding to all 80 being defective).

```
normalcdf(9.5,80
.5,80*.08,√(80*.
08*.92))
        .1007041942
```

The probability is (about) 0.10. (For a solution using tables, see pages A14–A15.)

SECTION 6.1

1. $\displaystyle\lim_{x \to 0} (1 + x)^{1/x} \approx 2.71828$

2. $\displaystyle\lim_{x \to 3} (2x^2 - 4x + 1) = 2 \cdot 3^2 - 4 \cdot 3 + 1$
$$= 18 - 12 + 1 = 7$$

3. $\displaystyle\lim_{x \to 5} \frac{2x^2 - 10x}{x - 5} = \lim_{x \to 5} \frac{2x(x - 5)}{x - 5}$

$$= \lim_{x \to 5} \frac{2x(x - 5)}{x - 5} = \lim_{x \to 5} 2x = 10$$

4. $\displaystyle\lim_{h \to 0} (3x^2 + 5xh + 1) = 3x^2 + 5x \cdot 0 + 1$
$$= 3x^2 + 1 \quad \text{(Using direct substitution)}$$

5. a. Discontinuous, $f(c)$ is not defined
   b. Discontinuous, $\displaystyle\lim_{x \to c} f(x) \neq f(c)$
   c. Continuous

SECTION 6.2

1. The units of $f'(x)$ are people per year, measuring the rate of growth of the population.

2. #2 is the original function and #1 is its derivative.

SECTION 6.3

As is shown in a longer way on pages 516–517

1. a. $\dfrac{d}{dx} x^2 = 2x^{2-1} = 2x$

b. $\dfrac{d}{dx} x^{-5} = -5x^{-5-1} = -5x^{-6}$

c. $\dfrac{d}{dx} \sqrt[4]{x} = \dfrac{d}{dx} x^{1/4} = \dfrac{1}{4} x^{(1/4)-1} = \dfrac{1}{4} x^{-3/4}$

2. $\dfrac{df}{dx} = 3x^2$

$$\left.\dfrac{df}{dx}\right|_{x=-1} = 3(-1)^2 = 3$$

SECTION 6.4

1. $\dfrac{d}{dx}\,[x^3(x^2 - x)] = 3x^2(x^2 - x) + x^3(2x - 1)$

$$= 3x^4 - 3x^3 + 2x^4 - x^3$$
$$= 5x^4 - 4x^3$$

2. $\dfrac{d}{dx}\left(\dfrac{2x^2}{x^2 + 1}\right) = \dfrac{(x^2 + 1)4x - 2x \cdot 2x^2}{(x^2 + 1)^2}$

$$= \dfrac{4x^3 + 4x - 4x^3}{(x^2 + 1)^2} = \dfrac{4x}{(x^2 + 1)^2}$$

SECTION 6.5

1. a. $f'(x) = 3x^2 - 2x + 1$
  b. $f''(x) = 6x - 2$
  c. $f'''(x) = 6$
  d. $f^{(4)}(x) = 0$

2. a. $f'(x) = -8x^{-3/2}$
  b. $f''(x) = 12x^{-5/2}$

3. $f(x) = \dfrac{x}{x} + \dfrac{1}{x} = 1 + x^{-1}$ Simplifying first

  $f'(x) = -x^{-2}$

  $f''(x) = 2x^{-3}$

4. $\dfrac{d}{dx}\,(x^4 + x^3 + 1) = 4x^3 + 3x^2$

  $\dfrac{d^2}{dx^2}\,(x^4 + x^3 + 1) = \dfrac{d}{dx}\,(4x^3 + 3x^2) = 12x^2 + 6x$

  $(12x^2 + 6x)\big|_{x=-1} = 12 - 6 = 6$

5. a. $v(t) = 12t - 3t^2$
    $v(2) = 24 - 12 = 12$ ft/sec
  b. $v(5) = 60 - 75 = -15$ ft/sec or 15 ft/sec *down-ward*
  c. $a(t) = 12 - 6t$
    $a(1) = 12 - 6 = 6$ ft/sec^2

6. a.   **b.**

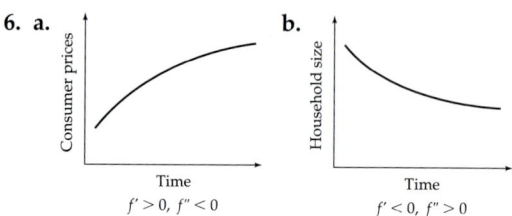

$f' > 0,\ f'' < 0$ $f' < 0,\ f'' > 0$

SECTION 6.6

1. $g(f(x)) = 4 - f(x) = 4 - x^2$

2. $f(x) = \sqrt{x},\ g(x) = x^5 - 7x + 1$

3. $\dfrac{d}{dx}\,(x^3 - x)^{-1/2} = -\dfrac{1}{2}\,(x^3 - x)^{-3/2}(3x^2 - 1)$

SECTION 6.7

1. $x = -3,\ x = 0,$ and $x = 2$

SECTION 7.1

1. $f'(x) = 3x^2 - 12 = 3(x^2 - 4) = 3(x + 2)(x - 2)$

  $\text{CV}\begin{cases} x = -2 \\ x = 2 \end{cases}$

2. Because the denominator of f' will be the square of the denominator of f (from the quotient rule).

SECTION 7.2

1. a.

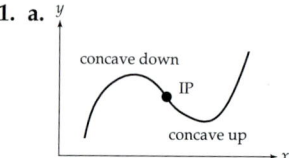

b.

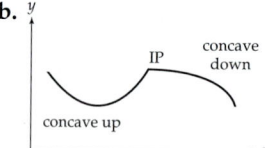

2. No (concave up on both sides);
  Yes (concave down then concave up)

SECTION 7.4

Quantity $q(x) = 1500 + 300x$;
price $p(x) = 3000 - 200x$

SECTION 7.6

a. $\dfrac{d}{dx} x^4 = 4x^3$ **b.** $\dfrac{d}{dx} y^2 = 2y \dfrac{dy}{dx}$

c. $\dfrac{d}{dx} (x^2 y^3) = 2xy^3 + 3x^2 y^2 \dfrac{dy}{dx}$

SECTION 8.2

1. $f'(x) = \dfrac{d}{dx}\left(\dfrac{\ln x}{x}\right) = \dfrac{x(\frac{1}{x}) - 1 \ln x}{x^2} = \dfrac{1 - \ln x}{x^2}$

Derivative of $\ln x$ — Derivative of x

2. $\dfrac{d}{dx} \ln (x^3 - 5x + 1) = \dfrac{3x^2 - 5}{x^3 - 5x + 1}$

3. $f'(x) = e^x + xe^x$
$f'(1) = e^1 + 1e^1 = 2e$ (which is approximately $2 \cdot 2.718 = 5.436$)

4. $\dfrac{d}{dx} e^{1 - 4x^3} = e^{1 - 4x^3}(-12x^2) = -12x^2 e^{1 - 4x^3}$

SECTION 8.3

1. $\dfrac{f'(t)}{f(t)} = \dfrac{2t}{300 + t^2}$; At $t = 10$:

$\dfrac{2 \cdot 10}{300 + (10)^2} = \dfrac{20}{400} = \dfrac{1}{20} = 0.05$ or 5%

2. $E(p) = \dfrac{-pD'(p)}{D(p)} = \dfrac{-p(-1)}{90 - p} = \dfrac{p}{90 - p}$

At $p = 30$, $E(30) = \dfrac{30}{90 - 30} = \dfrac{30}{60} = \dfrac{1}{2} = 0.5$

(demand is inelastic)

At $p = 75$, $E(75) = \dfrac{75}{90 - 75} = \dfrac{75}{15} = 5$

(demand is elastic)

SECTION 9.1

1. $\displaystyle\int x^2 \, dx = \dfrac{1}{3} x^3 + C$ Check: $\dfrac{d}{dx}\left(\dfrac{1}{3} x^3 + C\right) = x^2$

2. $\displaystyle\int x^{-3} \, dx = -\dfrac{1}{2} x^{-2} + C$

3. $\displaystyle\int z^{-1/2} \, dz = \dfrac{1}{1/2} z^{1/2} + C = 2z^{1/2} + C$

4. $\displaystyle\int (w^{1/3} - 4w^{-3}) \, dw$

$= \displaystyle\int w^{1/3} \, dw - 4 \int w^{-3} \, dw$

$= \dfrac{3}{4} w^{4/3} - 4 \left(-\dfrac{1}{2}\right) w^{-2} + C$

$= \dfrac{3}{4} w^{4/3} + 2w^{-2} + C$

5. a. $\displaystyle\int 5x^4 \, dx = x^5 + C$ **b.** $\displaystyle\int 3x^2 \, dx = x^3 + C$

6. $\displaystyle\int \dfrac{6t^2 - t}{t} \, dt = \int \dfrac{t(6t - 1)}{t} \, dt$

$= \displaystyle\int (6t - 1) \, dt$

$= 6 \displaystyle\int t \, dt - \int 1 \, dt = 6 \cdot \dfrac{1}{2} t^2 - t + C$

$= 3t^2 - t + C$

SECTION 9.2

1. a. $\displaystyle\int 12e^{4x} \, dx = 12 \int e^{4x} \, dx = 12 \cdot \dfrac{1}{4} e^{4x} + C$

$= 3e^{4x} + C$

b. $\displaystyle\int e^{\frac{1}{3}x} \, dx = 3e^{\frac{1}{3}x} + C$

2. $\displaystyle\int \dfrac{3}{4x} \, dx = \dfrac{3}{4} \int \dfrac{1}{x} \, dx = \dfrac{3}{4} \ln |x| + C$

SECTION 9.3

1. $\sqrt{x} \,\Big|_{4}^{25} = \sqrt{25} - \sqrt{4} = 5 - 2 = 3$

2. $\displaystyle\int_{0}^{2} x^3 \, dx = \dfrac{1}{4} x^4 \,\Big|_{0}^{2} = \dfrac{1}{4} 16 - \dfrac{1}{4} 0 = 4$ square units

SECTION 9.4

1. $\dfrac{1}{2} \displaystyle\int_{0}^{2} 3x^2 \, dx = \dfrac{1}{2} \cdot x^3 \,\Big|_{0}^{2} = \dfrac{1}{2} \cdot 2^3 - \dfrac{1}{2} \cdot 0^3 = \dfrac{1}{2} \cdot 8 = 4$

2. $\displaystyle\int_{-1}^{1} [(2x^2 + 1) - (-x^2 - 1)] \, dx$

$$= \int_{-1}^{1} (2x^2 + 1 + x^2 + 1)\, dx$$

$$= \int_{-1}^{1} (3x^2 + 2)\, dx = (x^3 + 2x)\; \Big|_{-1}^{1}$$

$$= (1 + 2) - (-1 - 2) = 6 \text{ square units}$$

3.
$$2x^2 - 1 = 2 - x^2$$
$$3x^2 - 3 = 0$$
$$3(x^2 - 1) = 0$$
$$3(x + 1)(x - 1) = 0$$
$$x = 1 \quad \text{and} \quad x = -1$$

Test value $x = 0$ shows that $2 - x^2$ is "upper" and $2x^2 - 1$ is "lower."

$$\int_{-1}^{1} [(2 - x^2) - (2x^2 - 1)]\, dx$$

$$= \int_{-1}^{1} (3 - 3x^2)\, dx = (3x - x^3)\; \Big|_{-1}^{1}$$

$$= (3 - 1) - (-3 + 1) = 4 \text{ square units}$$

SECTION 9.5

1. Less equal

2. $\int_{0}^{1} (x - x^{6.9})\, dx$

$$= \left(\frac{1}{2} x^2 - \frac{1}{7.9} x^{7.9} \right) \Big|_{0}^{1}$$

$$= \frac{1}{2} - \frac{1}{7.9} - 0 \approx 0.5 - 0.127 = 0.373$$

Gini index for wealth is 0.75 (multiplying by 2 and rounding).

SECTION 9.6

1. $df = (3x^2 - 4)\, dx$

2. $du = -5e^{-5t}\, dt$

3. a. B   **b.** C   **c.** A   **d.** C

4. Neither.

a. Try formula A with $u = x^3 + 1$, so $du = 3x^2\, dx$. The problem has an x^3 for the differential instead of the needed x^2.

b. Try formula B with $u = x^2$, so $du = 2x\, dx$. The problem does not have the x that is needed for the differential.

SECTION 10.1

1. $\int x^3 \ln x\, dx = (\ln x) \left(\frac{1}{4} x^4 \right) - \int \frac{1}{x} \frac{1}{4} x^4\, dx$

$$\left[\begin{array}{ll} u = \ln x & dv = x^3\, dx \\[2mm] du = \frac{1}{x}\, dx & v = \int x^3\, dx = \frac{1}{4} x^4 \end{array} \right]$$

$$= (\ln x) \left(\frac{1}{4} x^4 \right) - \frac{1}{4} \int x^3\, dx$$

$$= \frac{1}{4} x^4 \ln x - \frac{1}{16} x^4 + C$$

2. $\int (x + 1)(x - 1)^3\, dx$

$$\left[\begin{array}{ll} u = x + 1 & dv = (x - 1)^3\, dx \\[2mm] du = dx & v = \int (x - 1)^3\, dx = \frac{1}{4} (x - 1)^4 \end{array} \right]$$

$$= (x + 1) \frac{1}{4} (x - 1)^4 - \int \frac{1}{4} (x - 1)^4\, dx$$

$$= \frac{1}{4} (x + 1)(x - 1)^4 - \frac{1}{4} \cdot \frac{1}{5} (x - 1)^5 + C$$

$$= \frac{1}{4} (x + 1)(x - 1)^4 - \frac{1}{20} (x - 1)^5 + C$$

3. a. Requires integration by parts
b. Can be solved by the substitution $u = x^2$

SECTION 10.2

1. Formula 15 with the substitution $x = 3t^2$, $dx = 6t\, dt$, and $a = 1$ becomes

$$\int \frac{1}{9t^4 - 1} 6t\, dt = \frac{1}{2} \ln \left| \frac{3t^2 - 1}{3t + 1} \right| + C$$

Dividing each side by 6 gives the answer.

$$\int \frac{t}{9t^4 - 1}\, dt = \frac{1}{12} \ln \left| \frac{3t^2 - 1}{3t^2 + 1} \right| + C$$

SECTION 10.3

1. a. $\lim\limits_{b \to \infty} \left(1 - \frac{1}{b} \right) = 1$

b. $\lim\limits_{b \to \infty} (\sqrt[3]{b} + 3)$ does not exist.

2. $\displaystyle\int_2^b x^{-2}\, dx = (-x^{-1}) \Big|_2^b = -\dfrac{1}{b} - \left(-\dfrac{1}{2}\right)$

$$= \dfrac{1}{2} - \dfrac{1}{b}$$

$\lim\limits_{b \to \infty} \left(\dfrac{1}{2} - \dfrac{1}{b}\right) = \dfrac{1}{2}$

Therefore, $\displaystyle\int_2^\infty \dfrac{1}{x^2}\, dx = \dfrac{1}{2}.$

3. $\displaystyle\int_{-\infty}^1 12e^{3x}\, dx = \lim\limits_{a \to -\infty} \int_a^1 12e^{3x}\, dx$

$$= \lim\limits_{a \to -\infty} \left[12 \cdot \dfrac{1}{3} e^{3x} \Big|_a^1\right]$$

$$= \lim\limits_{a \to -\infty} (4e^3 - 4e^{3a}) = 4e^3$$

SECTION 10.5

1. $y = e^{-x} + e^{3x}$

$y' = -e^{-x} + 3e^{3x}$

$y'' = e^{-x} + 9e^{3x}$

Substituting these into the differential equation:

$(e^{-x} + 9e^{3x}) - 2(-e^{-x} + 3e^{3x}) - 3(e^{-x} + e^{3x}) \stackrel{?}{=} 0$

$e^{-x} + 9e^{3x} + 2e^{-x} - 6e^{3x} - 3e^{-x} - 3e^{3x} \stackrel{?}{=} 0$

$0 = 0$ (it checks)

2. $\dfrac{dy}{dx} = \dfrac{6x^2}{y^4}$

$y^4\, dy = 6x^2\, dx$

$\displaystyle\int y^4\, dy = \int 6x^2\, dx$

$\dfrac{1}{5} y^5 = 2x^3 + C$

$y^5 = 10x^3 + 5C = 10x^3 + c$

$y = \sqrt[5]{10x^3 + c}$

The initial condition gives

$2 = \sqrt[5]{0 + c}$

$2 = \sqrt[5]{c}$

$32 = c$ (raising each side to the fifth power)

Solution: $y = \sqrt[5]{10x^3 + 32}$

3. $\dfrac{dy}{y} = x^2\, dx$

$\displaystyle\int \dfrac{dy}{y} = \int x^2\, dx$

$\ln y = \dfrac{1}{3} x^3 + C$

$y = e^{\frac{1}{3}x^3 + C} = e^{\frac{1}{3}x^3} e^C = c e^{\frac{1}{3}x^3}$

The initial condition gives

$2 = ce^0 = c$

Solution: $y = 2e^{\frac{1}{3}x^3}$

SECTION 10.6

Product A, whose growth begins slowly, was not advertised, and the differential equation is logistic: $y' = ay(M - y)$. Product B, whose growth begins rapidly, was advertised, and the differential equation is limited: $y' = a(M - y)$.

SECTION 11.1

1. a. $\{(x, y) \mid x > 0,\, y \geq 0\}$

b. $f(e, 4) = \dfrac{\ln e}{e^{\sqrt{4}}} = \dfrac{1}{e^2} = e^{-2}$

2. $M = xy + 2xz + 5yz$

SECTION 11.2

1. a. $\dfrac{\partial}{\partial x} x^4 y^2 = 4x^3 y^2$

b. $\dfrac{\partial}{\partial y} x^4 y^2 = x^4 2y = 2x^4 y$

2. $\dfrac{\partial}{\partial y}\, x^2 = 0$

3. $\dfrac{\partial}{\partial y}\,(2x^4 - 3x^3y^3 - y^2 + 4x + 1)$
$= -3x^3 3y^2 - 2y = -9x^3y^2 - 2y$

4. $f_y(x, y) = e^{x^3 + y^3}(3y^2) = 3y^2 e^{x^3 + y^3}$
$f_y(1, 2) = 3 \cdot 2^2 \cdot e^{1^3 + 2^3} = 12e^9$

5. $\dfrac{\partial}{\partial y}\,(x^3y^4z^5) = x^3 4y^3 z^5 = 4x^3y^3z^5$

6. a. Negative
   b. Positive

7. a. $f_x = 3x^2 - 6xy^4$
   b. $f_{xy} = -6x4y^3 = -24xy^3$
   c. $f_y = -3x^2 4y^3 + 3y^2 = -12x^2y^3 + 3y^2$
   d. $f_{yx} = -24xy^3$

SECTION 11.5

1. $2y^3 - 3x + 1 = 0$ Moving all terms to the left-hand side

2. The paint can

3. a. The approximate additional area for each additional foot of fence
   b. $|\lambda| = 100$ (using $\lambda = -x$ with $x = 100$)
   c. Approximately 500 more square feet of area (from $5 \cdot 100$)

SECTION 11.6

$g_x = 2xe^{5y}$ $g_y = x^2 5e^{5y}$ Partials

$dg = 2xe^{5y} \cdot dx + 5x^2 e^{5y} \cdot dy$ Total differential

SECTION 11.7

$\displaystyle\int_0^2 y^2 e^{-x}\, dx = -y^2 e^{-x}\Big|_{x=0}^{x=2} = -y^2 e^{-2} + y^2 e^0$

$= -y^2 e^{-2} + y^2$

$\displaystyle\int_{-1}^1 (-y^2 e^{-2} + y^2)\, dy = \left(-\dfrac{1}{3}y^3 e^{-2} + \dfrac{1}{3}y^3\right)\Big|_{-1}^1$

$= -\dfrac{1}{3}e^{-2} + \dfrac{1}{3} - \left(\dfrac{1}{3}e^{-2} - \dfrac{1}{3}\right) = -\dfrac{2}{3}e^{-2} + \dfrac{2}{3}$

APPENDIX

1. $z = \dfrac{20 - 15}{1} = 5$. Since more than 99% of the z-scores are within 3 of the mean, this value is *very* far from the mean!

2. We first change the weights into z-scores:

$x = 130$ corresponds to: $z = \dfrac{130 - 134.7}{30.4} \approx -0.15$

Using $z = \dfrac{x - \mu}{\sigma}$
with $\mu = 134.7$ and $\sigma = 30.4$

$x = 150$ corresponds to: $z = \dfrac{150 - 134.7}{30.4} \approx 0.50$

Using these values, we then want $P(-0.15 \le Z \le 0.50)$, which is equivalent to the shaded area shown in the graph on the left below, which is equal to the *sum* of the following two shaded areas. The two areas are found from the table on page 1016, with the calculation shown below.

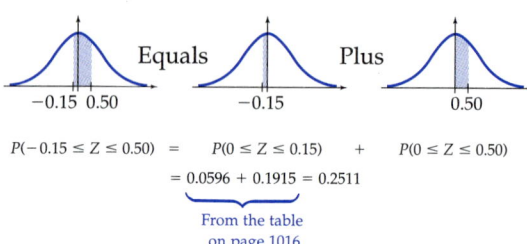

$P(-0.15 \le Z \le 0.50) =$ $P(0 \le Z \le 0.15)$ $+$ $P(0 \le Z \le 0.50)$
$= 0.0596 + 0.1915 = 0.2511$
From the table on page 1016

Therefore, about 25% of American women weigh between 130 and 150 pounds.

3. Since $np = (80)(0.08) = 6.4$ and $n(1 - p) = (80)(0.92) = 73.6$ are each greater than 5, we may use the normal distribution with $\mu = np = (80)(0.08) \approx 6.4$ and $\sigma = \sqrt{np(1 - p)} = \sqrt{(80)(0.08)(0.92)} \approx 2.43$ to approximate $P(X \ge 10)$ as the area under the normal curve from 9.5 to 80.5 (corresponding to all 80 being defective, and again including rounding). Converting the *x*-values into *z*-scores:

$x = 9.5$ corresponds to: $\quad z = \dfrac{9.5 - 6.4}{2.43} \approx 1.28$

$$\text{Using } z = \frac{x - \mu}{\sigma}$$

with $\mu = 6.4$ and
$\sigma = 2.43$

$x = 80.5$ corresponds to: $\quad z = \dfrac{80.5 - 6.4}{2.43} \approx 30.49$

The probability $P(1.28 \le Z \le 30.49)$ is represented by the shaded area on the left, which is equivalent to the difference between the two areas on the right with the calculation shown below.

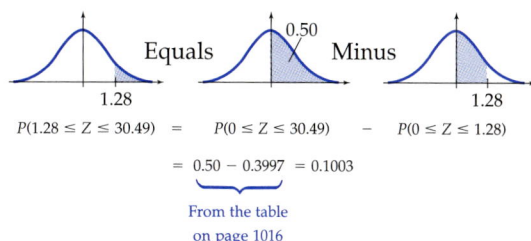

$$P(1.28 \le Z \le 30.49) = P(0 \le Z \le 30.49) - P(0 \le Z \le 1.28)$$

$$= 0.50 - 0.3997 = 0.1003$$

From the table
on page 1016

The probability of at least ten defective VCRs is (about) 0.10, or 10%.

Answers to Selected Exercises

EXERCISES 0.1 page 13

1. $\{x \mid 0 \le x < 6\}$ 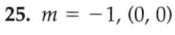 **3.** $\{x \mid x \le 2\}$ **5. a.** Increase by 15 units

b. Decrease by 10 units **7.** $m = -2$ **9.** $m = \dfrac{1}{3}$ **11.** $m = 0$ **13.** Slope is undefined.

15. $m = 3$, $(0, -4)$ **17.** $m = -\dfrac{1}{2}$, $(0, 0)$ **19.** $m = 0$, $(0, 4)$

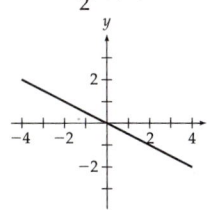
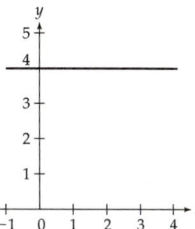

21. Slope and y-intercept do not exist **23.** $m = \dfrac{2}{3}$, $(0, -4)$ **25.** $m = -1$, $(0, 0)$

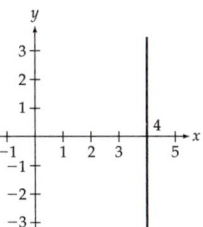
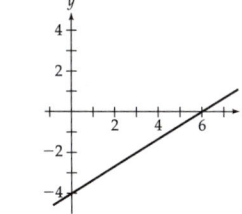
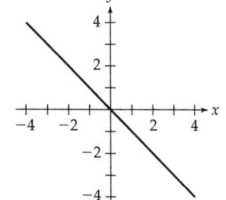

27. $m = 1$, $(0, 0)$ **29.** $m = \dfrac{1}{3}$, $\left(0, \dfrac{2}{3}\right)$ **31.** $m = \dfrac{2}{3}$, $(0, -1)$

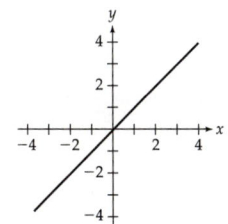
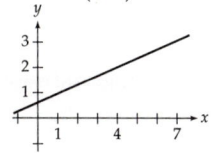
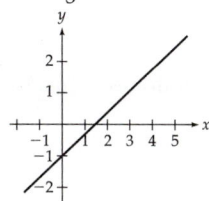

33.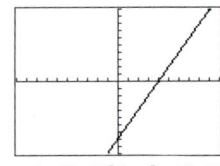
on $[-10, 10]$ by $[-10, 10]$

35.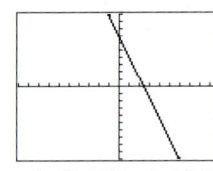
on $[-10, 10]$ by $[-10, 10]$

37.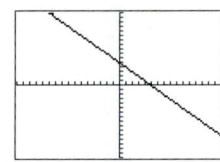
on $[-160, 160]$ by $[-160, 160]$

39. $y = -2.25x + 3$ **41.** $y = 5x + 3$ **43.** $y = -4$ **45.** $x = 1.5$ **47.** $y = -2x + 13$ **49.** $y = -1$

51. $y = -2x + 1$ **53.** $y = \dfrac{3}{2}x - 2$ **55.** $y = -x + 5, y = -x - 5, y = x + 5, y = x - 5$

57. Substituting $(0, b)$ into $y - y_1 = m(x - x_1)$ gives $y - b = m(x - 0)$, or $y = mx + b$.   **59.** $(-b/m, 0), m \neq 0$

61. a. 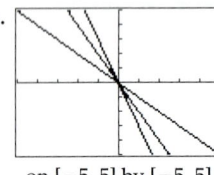   **b.**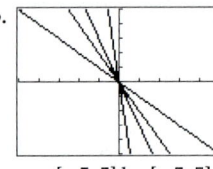
on $[-5, 5]$ by $[-5, 5]$ on $[-5, 5]$ by $[-5, 5]$

63. Low: $[0, 8)$; average: $[8, 20)$; high: $[20, 40)$, critical $[40, \infty)$ **65. a.** 3 minutes 38.19 seconds   **b.** the year 2033

67. a. $y = 4x + 2$ **b.** \$10 million   **c.** \$22 million   **69. a.** $y = \dfrac{9}{5}x + 32$ **b.** 68°   **71. a.** $V = 50,000 - 2200t$

b. \$39,000   **c.**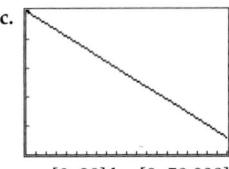
on $[0, 20]$ by $[0, 50,000]$

73. b.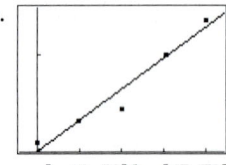
on $[-10, 50]$ by $[65, 72]$

$y_1 = 0.158x + 65.06$ **c.** 76.9 years

EXERCISES 0.2 page 28

1. 64   **3.** $\dfrac{1}{16}$ **5.** 8   **7.** $\dfrac{8}{5}$ **9.** $\dfrac{1}{32}$ **11.** $\dfrac{8}{27}$ **13.** 1   **15.** $\dfrac{4}{9}$ **17.** 5   **19.** 125   **21.** 8   **23.** 4   **25.** -32

27. $\dfrac{125}{216}$ **29.** $\dfrac{9}{25}$ **31.** $\dfrac{1}{4}$ **33.** $\dfrac{1}{2}$ **35.** $\dfrac{1}{8}$ **37.** $\dfrac{1}{4}$ **39.** $-\dfrac{1}{2}$ **41.** $\dfrac{1}{4}$ **43.** $\dfrac{4}{5}$ **45.** $\dfrac{64}{125}$ **47.** -243 **49.** 2.14

51. 274.37   **53.** -128 **55.** 6.25   **57.** 0.5   **59.** 0.4   **61.** 0.977 (rounded)   **63.** 2.720 (rounded)   **65.** x^{10}

67. z^{27} **69.** x^8 **71.** w^5 **73.** y^5/x **75.** $27y^4$ **77.** $u^2v^2w^2$ **79.** 25.6 ft   **81.** Costs will be multiplied by 2.3

83.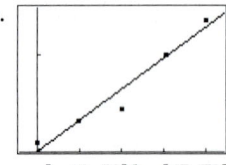
on $[0, 5]$ by $[0, 3]$

Capacity can be multiplied by about 3.2   **85.** 125 beats per minute

87.
on $[0, 200]$ by $[0, 150]$

Heart rate decreases more slowly as body weight increases.

Answers to Selected Exercises

EXERCISES 0.1 page 13

1. $\{x \mid 0 \le x < 6\}$ **3.** $\{x \mid x \le 2\}$ **5. a.** Increase by 15 units

b. Decrease by 10 units **7.** $m = -2$ **9.** $m = \dfrac{1}{3}$ **11.** $m = 0$ **13.** Slope is undefined.

15. $m = 3$, $(0, -4)$ **17.** $m = -\dfrac{1}{2}$, $(0, 0)$ **19.** $m = 0$, $(0, 4)$

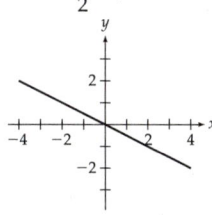 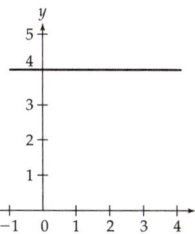

21. Slope and y-intercept do not exist **23.** $m = \dfrac{2}{3}$, $(0, -4)$ **25.** $m = -1$, $(0, 0)$

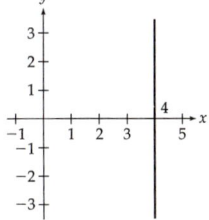 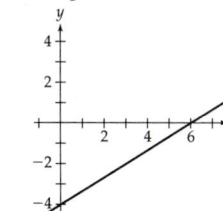 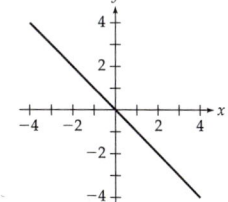

27. $m = 1$, $(0, 0)$ **29.** $m = \dfrac{1}{3}$, $\left(0, \dfrac{2}{3}\right)$ **31.** $m = \dfrac{2}{3}$, $(0, -1)$

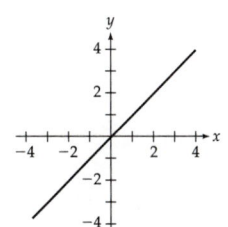 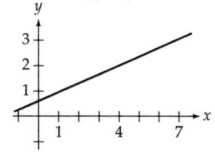 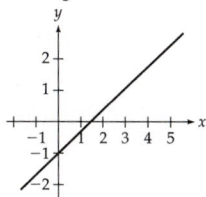

33.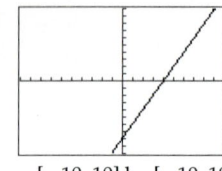

35.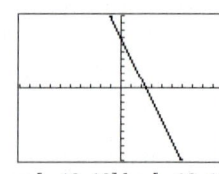

37.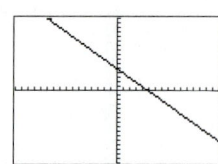

on $[-10, 10]$ by $[-10, 10]$　　on $[-10, 10]$ by $[-10, 10]$　　on $[-160, 160]$ by $[-160, 160]$

39. $y = -2.25x + 3$　　**41.** $y = 5x + 3$　　**43.** $y = -4$　　**45.** $x = 1.5$　　**47.** $y = -2x + 13$　　**49.** $y = -1$

51. $y = -2x + 1$　　**53.** $y = \frac{3}{2}x - 2$　　**55.** $y = -x + 5, y = -x - 5, y = x + 5, y = x - 5$

57. Substituting $(0, b)$ into $y - y_1 = m(x - x_1)$ gives $y - b = m(x - 0)$, or $y = mx + b$.　　**59.** $(-b/m, 0), m \neq 0$

61. a. 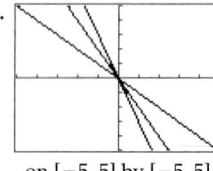　　**b.**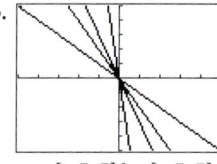

on $[-5, 5]$ by $[-5, 5]$　　on $[-5, 5]$ by $[-5, 5]$

63. Low: $[0, 8)$; average: $[8, 20)$; high: $[20, 40)$, critical $[40, \infty)$　　**65. a.** 3 minutes 38.19 seconds　　**b.** the year 2033

67. a. $y = 4x + 2$　　**b.** \$10 million　　**c.** \$22 million　　**69. a.** $y = \frac{9}{5}x + 32$　　**b.** $68°$　　**71. a.** $V = 50,000 - 2200t$

b. \$39,000　　**c.** 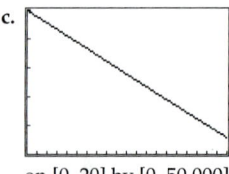　　**73. b.** 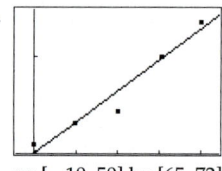　　$y_1 = 0.158x + 65.06$　　**c.** 76.9 years

on $[0, 20]$ by $[0, 50,000]$　　on $[-10, 50]$ by $[65, 72]$

EXERCISES 0.2 page 28

1. 64　　**3.** $\frac{1}{16}$　　**5.** 8　　**7.** $\frac{8}{5}$　　**9.** $\frac{1}{32}$　　**11.** $\frac{8}{27}$　　**13.** 1　　**15.** $\frac{4}{9}$　　**17.** 5　　**19.** 125　　**21.** 8　　**23.** 4　　**25.** -32

27. $\frac{125}{216}$　　**29.** $\frac{9}{25}$　　**31.** $\frac{1}{4}$　　**33.** $\frac{1}{2}$　　**35.** $\frac{1}{8}$　　**37.** $\frac{1}{4}$　　**39.** $-\frac{1}{2}$　　**41.** $\frac{1}{4}$　　**43.** $\frac{4}{5}$　　**45.** $\frac{64}{125}$　　**47.** -243　　**49.** 2.14

51. 274.37　　**53.** -128　　**55.** 6.25　　**57.** 0.5　　**59.** 0.4　　**61.** 0.977 (rounded)　　**63.** 2.720 (rounded)　　**65.** x^{10}

67. z^{27}　　**69.** x^8　　**71.** w^5　　**73.** y^5/x　　**75.** $27y^4$　　**77.** $u^2v^2w^2$　　**79.** 25.6 ft　　**81.** Costs will be multiplied by 2.3

83. 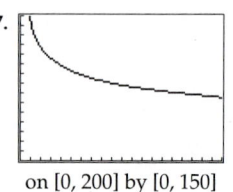　　Capacity can be multiplied by about 3.2　　**85.** 125 beats per minute

on $[0, 5]$ by $[0, 3]$

87. 　Heart rate decreases more slowly as body weight increases.

on $[0, 200]$ by $[0, 150]$

89. About 42.6 thousand work-hours, or 42,600 work-hours, rounded to the nearest hundred hours
91. a. About 32 times more ground motion   **b.** About 8 times more ground motion   **93.** About 312 mph

95.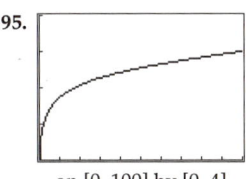

$x \approx 18.2$. Therefore, the land area must be increased by a factor of more than 18 to double the number of species.

on [0, 100] by [0, 4]

97. b.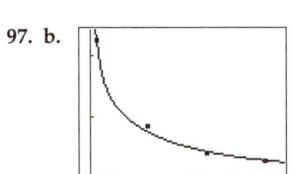

$y = 3261x^{-0.267}$ **c.** 1147 work-hours

on [−2, 32] by [1000, 3500]

EXERCISES 0.3 page 44

1. Yes   **3.** No   **5.** No   **7.** No   **9.** Domain = $\{x \mid x \le 0 \text{ or } x \ge 1\}$; Range = $\{y \mid y \ge -1\}$
11. a. $f(10) = 3$ **b.** $\{x \mid x \ge 1\}$ **c.** $\{y \mid y \ge 0\}$ **13. a.** $h(-5) = -1$ **b.** $\{z \mid z \ne -4\}$ **c.** $\{y \mid y \ne 0\}$
15. a. $h(81) = 3$ **b.** $\{x \mid x \ge 0\}$ **c.** $\{y \mid y \ge 0\}$ **17. a.** $f(-8) = 4$ **b.** $\mathbb{R}$ **c.** $\{y \mid y \ge 0\}$
19. a. $f(0) = 2$ **b.** $\{x \mid -2 \le x \le 2\}$ **c.** $\{y \mid 0 \le y \le 2\}$ **21. a.** $f(-25) = 5$ **b.** $\{x \mid x \le 0\}$ **c.** $\{y \mid y \ge 0\}$

23. 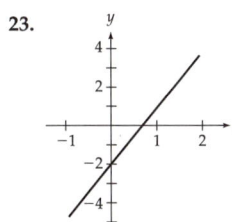   **25.** 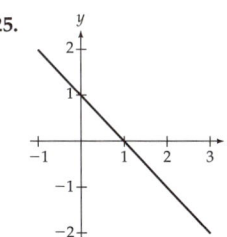   **27.** 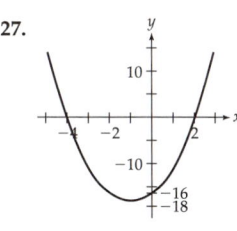   **29.**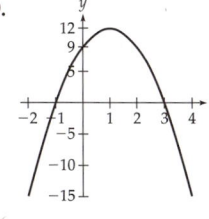

31. a. (20, 100)   **b.** 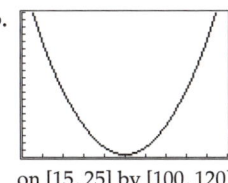   **33. a.** (−40, −200)   **b.**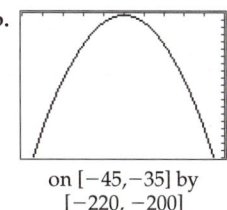

on [15, 25] by [100, 120]   on [−45, −35] by [−220, −200]

35. $x = 7, x = -1$ **37.** $x = 3, x = -5$ **39.** $x = 4, x = 5$ **41.** $x = 0, x = 10$ **43.** $x = 5, x = -5$ **45.** $x = -3$
47. $x = 1, x = 2$ **49.** No solutions   **51.** No solutions   **53.** $x = -4, x = 5$ **55.** $x = 4, x = 5$ **57.** $x = -3$
59. No (real) solutions   **61.** $x = 1.14, x = -2.64$ **63. a.** The slopes are all 2, but the y-intercepts differ
     b. $y = 2x - 8$
65. a. Shifted 4 units to the right; vertex: (4, 0)   **b.** Shifted 3 units to the left; vertex: (−3, 0)
     c. Shifted a units to the right, with vertex $(a, 0)$. A plus sign means a shift to the left.
67. a. Shifted 3 units to the right and 2 units up; vertex: (3, 2)
     b. Shifted 2 units to the left and 5 units down; vertex: (−2, −5)
     c. Shifted a units to the right and b units up, with vertex (a, b)
69. $C(x) = 4x + 20$ **71.** $P(x) = 15x + 500$ **73. a.** 17.7 lb/in.²   **b.** 15,765 lb/in.²   **75.** 132 ft
77. a. 400   **b.** 5200   **79.** About 208 mph   **81.** 2.92 seconds

83. a. Break even at 40 and 200 units.   **b.** Profit maximized at 120 units. Maximum profit is $12,800.
85. a. Break even at 20 and 80 units.   **b.** Profit maximized at 50 units. Maximum profit is $1800.
87. b. 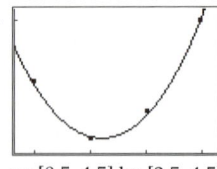   **c.** $y_1(5) = 4.575$, so about 4.6 million units

on [0.5, 4.5] by [3.5, 4.5]

EXERCISES 0.4 page 62

1. Domain: $\{x \mid x > 0 \text{ or } x < -4\}$; Range: $\{y \mid y > 0 \text{ or } y < -2\}$ **3. a.** $f(-3) = 1$ **b.** $\{x \mid x \neq -4\}$ **c.** $\{y \mid y \neq 0\}$
5. a. $f(-1) = -\dfrac{1}{2}$ **b.** $\{x \mid x \neq 1\}$ **c.** $\{y \mid y \leq 0 \text{ or } y \geq 4\}$

7. a. $f(2) = 1$ **b.** $\{x \mid x \neq 0, x \neq -4\}$ **c.** $\{y \mid y > 0 \text{ or } y \leq -3\}$ **9. a.** $g(-5) = 3$ **b.** $\mathbb{R}$ **c.** $\{y \mid y \geq 0\}$
11. $x = 0, x = -3, x = 1$ **13.** $x = 0, x = 2, x = -2$ **15.** $x = 0, x = 3$ **17.** $x = 0, x = 5$ **19.** $x = 0, x = 3$
21. $x = -2, x = 0, x = 4$ **23.** $x = -1, x = 0, x = 3$ **25.** $x = 0, x = 3$ **27.** $x = -5, x = 0$ **29.** $x = 0, x = 1$

31. $x \approx -1.79, x = 0, x \approx 2.79$ **33.** 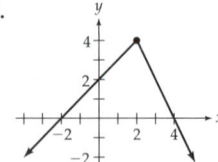   **35.**

37. 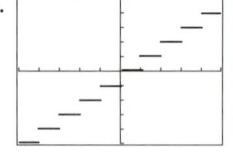   **39.** Polynomial   **41.** Piecewise linear   **43.** Polynomial   **45.** Rational

47. Piecewise linear   **49.** Polynomial   **51.** None (not a polynomial because of the fractional exponent)

53. a. y_3 **b.** y_1 **c.** [graph]   **d.** (10, 1000).

55. a. [graph] $y = \text{INT}(x)$ on $[-5, 5]$ by $[-5, 5]$ Note that each line segment in this graph includes its left endpoint but excludes its right endpoint, and so should be drawn like ●———○.

b. Domain: $\mathbb{R}$ Range: $\{\ldots, -3, -2, -1, 0, 1, 2, 3, \ldots\}$; that is, the set of all integers
57. a. $(7x - 1)^5$ **b.** $7x^5 - 1$ **59. a.** $\dfrac{1}{x^2 + 1}$ **b.** $\left(\dfrac{1}{x}\right)^2 + 1$ **61. a.** $(\sqrt{x} - 1)^3 - (\sqrt{x} - 1)^2$ **b.** $\sqrt{x^3 - x^2 - 1}$

63. a. $\dfrac{(x^2 - x)^3 - 1}{(x^2 - x)^3 + 1}$ **b.** $\left(\dfrac{x^3 - 1}{x^3 + 1}\right)^2 - \dfrac{x^3 - 1}{x^3 + 1}$ **65. a.** $f(g(x)) = acx + ad + b$ **b.** Yes

67. $5x^2 + 10xh + 5h^2$ or $5(x^2 + 2xh + h^2)$ **69.** $2x^2 + 4xh + 2h^2 - 5x - 5h + 1$ **71.** $10x + 5h$ or $5(2x + h)$

73. $4x + 2h - 5$ **75.** $14x + 7h - 3$ **77.** $3x^2 + 3xh + h^2$ **79.** $\dfrac{-2}{(x + h)x}$

81. $\dfrac{-2x - h}{x^2(x + h)^2}$ or $\dfrac{-2x - h}{x^2(x^2 + 2xh + h^2)}$ or $\dfrac{-2x - h}{x^4 + 2x^3h + x^2h^2}$ **83. a.** 2.70481   **b.** 2.71815   **c.** 2.71828   **d.** Yes, 2.71828

85. a. \$300   **b.** \$500   **c.** \$2000   **d.**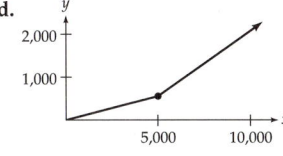

87. $R(v(t)) = 2(60 + 3t)^{0.3}$, $R(v(10)) \approx 7.714$ million dollars   **89. a.** 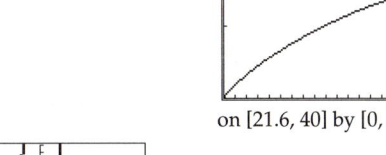   **b.** About $x = 27.9$ mpg

on [21.6, 40] by [0, 2000]

91. Shifted left 3 units and up 6 units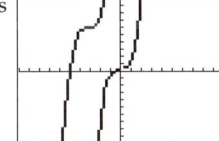

EXERCISES 0.5 page 75

1. 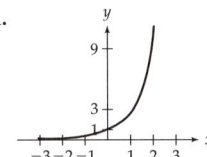   **3.** 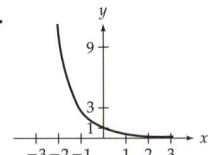   **5.** 5.697 (rounded to 3 decimal places)

7. a. e^x **b.** e^x **c.** e^x **d.** e^x **e.** e^x will exceed any power of x for large enough x.
9. \$1,096,000 (approx.)   **11.** \$1,989,300 (approx.)   **13.** \$101,257   **15. a.** \$4463   **b.** \$20,156
17. 7.1 billion   **19. a.** 0.53 (the chances are better than 50–50)   **b.** 0.70 (quite likely)
21. a. 0.267 or 26.7%   **b.** 0.012 or 1.2%   **23. a.** 1.3 mg   **b.** 0.84 mg   **25.** 208
27. a. About 153°   **b.** About 123°   **29.** 38   **31.** 6.5%
33. b. Texas, Florida, New York   **c.** Florida, Texas, New York   **d.** about 2016 (from $x \approx 26$ years after 1990)

EXERCISES 0.6 page 90

1. a. 5   **b.** -2 **c.** $\frac{1}{2}$ **3. a.** 5   **b.** -1 **c.** $\frac{1}{3}$ **5. a.** 0   **b.** 1   **c.** $\frac{2}{3}$ **7. a.** 2   **b.** -1 **c.** $\frac{1}{2}$
9. a. 1.348   **b.** 3.105   **11.** $\ln x$ **13.** $2 \ln x$ or $\ln x^2$ **15.** $\ln x$ **17.** $3x$ **19.** $7x$

21.

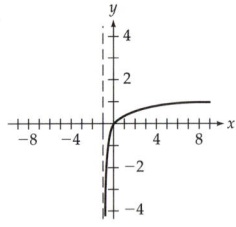

23. Domain: $\{x \mid x > 1 \text{ or } x < -1\}$
Range: $\mathbb{R}$

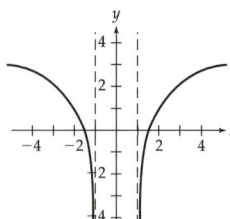

25. 7.27 years. It will have doubled in 8 years.   **27.** 33.35 years. It will have doubled in 34 years.
29. 29.23 years. It will have doubled in 30 years.   **31.** About 31,400 years   **33.** About 1.7 million years
35. 5 years (from 4.67)   **37.** 13 years   **39.** 228 million years   **41. a.** $\log_b b^x = x$, property 8 of logarithms
b. Follows directly from (a)   **c.** Using the change of base formula, cancellation

CHAPTER 0 REVIEW EXERCISES page 94

1. $\{x \mid 2 < x \le 5\}$ ⟵○————●⟶
    2   5

2. $\{x \mid -2 \le x < 0\}$ ⟵●————○⟶
    −2   0

3. $\{x \mid x \ge 100\}$ ⟵●————⟶
    100

4. $\{x \mid x \le 6\}$ ⟵————●————⟶
    6

5. Hurricane: $[74, \infty)$; storm: $[55, 74)$; gale: $[38, 55)$; small craft warning: $[21, 38)$

6. a. $(0, \infty)$ **b.** $(-\infty, 0)$ **c.** $[0, \infty)$ **d.** $(-\infty, 0]$ **7.** $y = 2x - 5$ **8.** $y = -3x + 3$ **9.** $x = 2$ **10.** $y = 3$

11. $y = -2x + 1$ **12.** $y = 3x - 5$ **13.** $y = 2x - 1$ **14.** $y = -\frac{1}{2}x + 1$ **15. a.** $V = 25,000 - 3000t$ **b.** \$13,000
16. a. $V = 78,000 - 5000t$ **b.** \$38,000

17. b.

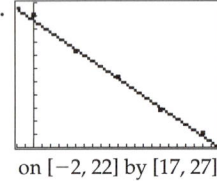

on $[-2, 22]$ by $[17, 27]$

The regression line $y_1 = -0.396x + 25.68$ fits the data well.

c. 13.8 million tons in the year 2005 [from $y_1(30)$]; 11.8 million tons in the year 2010 [from $y_1(35)$]

18. b.

on $[-10, 40]$ by $[25, 65]$

The regression line $y_1 = 1.03x + 26.3$ fits the data reasonably well.

c. 73 to 1 in the year 2005 [from $y_1(45) = 72.65$]; 78 to 1 in the year 2010 [from $y_1(50) = 77.8$]

19. 36   **20.** $\frac{3}{4}$ **21.** 8   **22.** 10   **23.** $\frac{1}{27}$ **24.** $\frac{1}{1000}$ **25.** $\frac{9}{4}$ **26.** $\frac{64}{27}$ **27.** 13.97   **28.** 112.32

29. a. $f(11) = 2$ **b.** $\{x \mid x \ge 7\}$ **c.** $\{y \mid y > 0\}$ **30. a.** $g(-1) = \frac{1}{2}$ **b.** $\{t \mid t \ne -3\}$ **c.** $\{y \mid y \ne 0\}$

31. a. $h(16) = \frac{1}{8}$ **b.** $\{w \mid w > 0\}$ **c.** $\{y \mid y > 0\}$ **32. a.** $w(8) = \frac{1}{16}$ **b.** $\{z \mid z \ne 0\}$ **c.** $\{y \mid y > 0\}$ **33.** Yes

34. No   **35.** 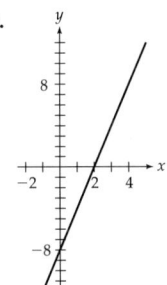   **36.** 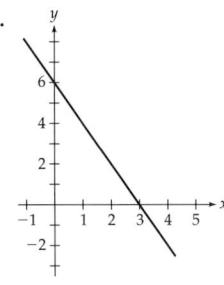   **37.** 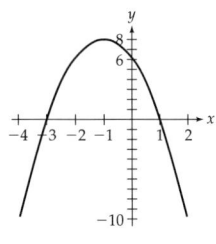   **38.**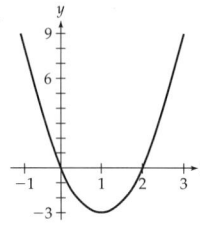

39. $x = 0, x = -3$ **40.** $x = 5, x = -1$ **41.** $x = -2, x = 1$ **42.** $x = 1, x = -1$

43. a. Vertex: $(5, -50)$ **b.**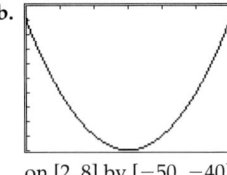
on $[2, 8]$ by $[-50, -40]$

44. a. Vertex: $(-7, -64)$ **b.**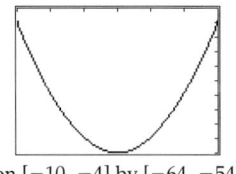
on $[-10, -4]$ by $[-64, -54]$

45. $C(x) = 45 + 0.12x$ **46.** $I(t) = 800t$ **47.** $T(x) = 70 - \dfrac{x}{300}$

48. $C(t) = 24 + 0.46t$; in about 13 years from 1995, so in about 2008.

49. a. Break even at 15 and 65 units.   **b.** Profit maximized at 40 units. Maximum profit: $1250.

50. a. Break even at 150 and 450 units.   **b.** Profit maximized at 300 units. Maximum profit: $67,500.

51. a. $f(-1) = 1$ **b.** $\{x \mid x \neq 0 \text{ and } x \neq 2\}$ **c.** $\{y \mid y > 0 \text{ or } y \leq -3\}$

52. a. $f(-8) = \dfrac{1}{2}$ **b.** $\{x \mid x \neq 0 \text{ and } x \neq -4\}$ **c.** $\{y \mid y > 0 \text{ or } y \leq -4\}$ **53. a.** $g(-4) = 0$ **b.** $\mathbb{R}$ **c.** $\{y \mid y \geq -2\}$

54. a. $g(-5) = -10$ **b.** $\mathbb{R}$ **c.** $\{y \mid y \leq 0\}$ **55.** $x = 0, x = 1, x = -3$ **56.** $x = 0, x = 2, x = -4$ **57.** $x = 0, x = 5$

58. $x = 0, x = 2$ **59.** 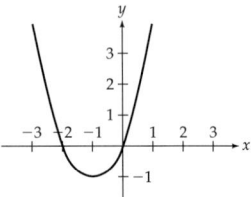   **60.** 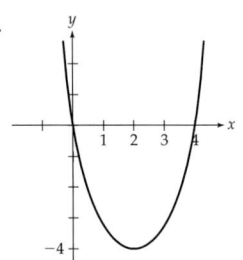   **61.**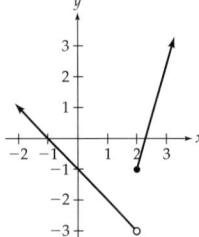

62. 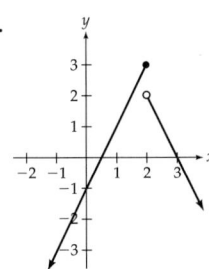   **63. a.** $f(g(x)) = \left(\dfrac{1}{x}\right)^2 + 1 = \dfrac{1}{x^2} + 1$ **b.** $g(f(x)) = \dfrac{1}{x^2 + 1}$

64. a. $f(g(x)) = \sqrt{5x - 4}$ **b.** $g(f(x)) = 5\sqrt{x} - 4$ **65. a.** $f(g(x)) = \dfrac{x^3 + 1}{x^3 - 1}$ **b.** $g(f(x)) = \left(\dfrac{x + 1}{x - 1}\right)^3$

66. a. $f(g(x)) = |x + 2|$ **b.** $g(f(x) = |x| + 2$ **67.** $4x + 2h - 3$ **68.** $\dfrac{-5}{(x + h)x}$

69. $A(p(t)) = 2(18 + 2t)^{0.15}$, $A(p(4)) \approx \$3.26$ million   **70. a.** $x = -1, x = 0, x = 3$ **b.**

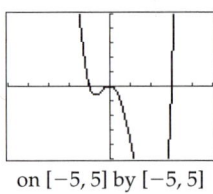

on $[-5, 5]$ by $[-5, 5]$

71. a. $x = -3, x = 0, x = 1$ **b.**

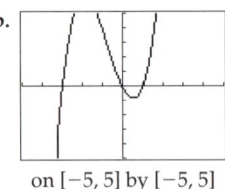

on $[-5, 5]$ by $[-5, 5]$

72. a. The points suggest a parabolic (quadratic) curve.   **b.**

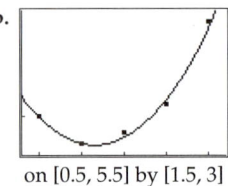

on $[0.5, 5.5]$ by $[1.5, 3]$ **c.** $\$3.6$ million, $\$4.8$ million

73.

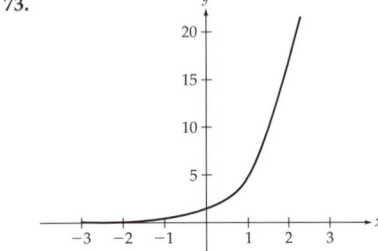

74.

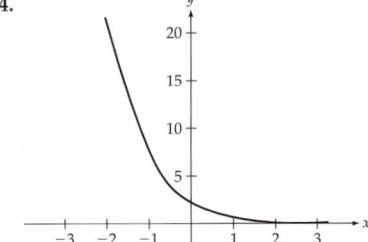

75. $\$15.3$ million

76. $\$3.49$ million   **77. a.** $800{,}000 \, (0.8)^t$ **b.** $\$327{,}680$ **78. a.** $5.4 \, (0.88)^t$ **b.** $\$1.5$ million
79. a. In about 2004 (from $x \approx 8.8$)   **b.** In about 2050 (from $x \approx 55$)
80. 1024 megabits (which is enough to hold four 16-volume encyclopedias on one chip)
81. a. 3   **b.** -3 **c.** 3   **d.** $\frac{1}{4}$ **82. a.** $\frac{1}{2}$ **b.** 8   **c.** -1 **d.** $\frac{3}{2}$ **83.** $f(x) = \ln x$ **84.** $f(x) = 2x - 1$
85. 4.2 years. It will have doubled in 5 years.   **86.** In about 13 years
87. 22.7 years. It will have doubled in 23 years.   **88.** 29.2 years. It will have doubled in 30 years.
89. About 50.7 million years   **90.** About 1.85 million years

EXERCISES 1.1 page 113

1. $\$1050$ **3.** $\$3120$ **5.** $\$298.57$ **7.** $\$15.36$ **9.** $\$15.70$ **11.** $\$2550$ **13.** $\$7751.58$
15. $\$3250.26$ **17.** 6.4% **19.** $\$2500$ **21.** 4 years   **23.** $\$7500$ **25.** $\$9450.90$
27. $\$4669.41$ **29.** 6 years   **31.** 8 years 9 months   **33.** 20 years   **35.** 5.48%
37. No; since the $\$1000$ is the same as the interest, you would be agreeing to pay $\$1000$ of interest on a loan of $\$0$.
39. $\$950.40$ **41.** 50% **43.** 58.25% **45.** 17.89%

EXERCISES 1.2 page 129

1. $2950.73   **3.** $10,009.04   **5.** $1181.26   **7.** $1477.83   **9.** $1706.98   **11.** $7625.24   **13.** $1869.50
15. $9919.10   **17.** $11,027.27   **19.** $13,152.79   **21.** 17 years and 9 months   **23.** 15 years and 11 months
25. 8 years and 9 months   **27.** 5 years and 11 months   **29.** 5 years   **31.** 8 years; 8.04 years rounds up to 9 years
33. 12 years; 11 years and 9 months   **35.** 9 years; 9 years   **37.** 35 years; 33 years and 11 months
39. 12 years; 11 years and 40 weeks   **41.** 4.39%   **43.** 8.91%   **45.** 10.17%   **47.** 9.90%
49. 3.79%   **51.** $3020.63; $69.90 more interest than annual compounding
53. $10,092.17; $83.13 more interest than semiannual compounding
55. $1184.75; $3.49 more interest than quarterly compounding
57. $1478.25; 42¢ more interest than monthly compounding
59. $1707.21; 23¢ more interest than weekly compounding
61. $7448.78; $176.46 less value than annual compounding
63. $1856.35; $13.15 less value than semiannual compounding
65. $9859.65; $59.45 less value than quarterly compounding
67. $10,885; $142.27 less value than monthly compounding
69. $13,145.15; $7.64 less value than weekly compounding
71. 17.50 years; about 3 months shorter   **73.** 15.85 years; about 1 month shorter
75. 8.64 years; about 1 month shorter   **77.** 5.89 years; about half a month shorter
79. 4.71 years; about 3 months shorter   **81.** 4.39%; about the same
83. 8.95%; about 0.04% greater   **85.** 10.30%; about 0.13% greater   **87.** 33.38%   **89.** 100%
107. $6827.69   **109.** $14,467.34   **111.** 30 years and 6 months   **113.** 17.45%
115. The bond gives the greater return (6.80% while the CD returns only 6.63%).   **117.** 32.89 years
119. People's State Bank offers the higher effective rate (4.27% while Statewide Federal's effective rate is just 4.19%).

EXERCISES 1.3 page 143

1. $20,724.67   **3.** $68,852.94   **5.** $31,405.51   **7.** $514,647.37   **9.** $69,796.84   **11.** $32.20   **13.** $172.81
15. $312.46   **17.** $345.85   **19.** $35.37   **21.** 24 years   **23.** 22 years and 6 months   **25.** 7 years and 3 months
27. 7 years and 2 months   **29.** 1 year and 45 weeks   **31.** 8.40%   **33.** 12.20%   **35.** 7.40%
37. 6.70%   **39.** 4.60% with 365 days per year; 4.64% with 360 days per year
49. a. 242 **b.** 242 [part (b) is just part (a) "backward"] **c.** 333,333. The sum on page 136 has the correct value of
$10,819.57.   **51.** Joe will have $316,781.40 while Jill will have $364,548.28.   **53.** $4001.27   **55.** $6.39
57. 1 year and 9 months   **59.** 5.65%

EXERCISES 1.4 page 156

1. $105,353.72   **3.** $95,896.47   **5.** $70,405.00   **7.** $212,572.07   **9.** $6242.74   **11.** $584.59   **13.** $116
15. $7518.84   **17.** $872.41   **19.** $20.77   **21.** $85,934.02   **23.** $2719.70   **25.** $38,879.53   **27.** $10,284.16
29. $986.86   **35.** This unpaid balance formula gives slightly different answers because the payments are not rounded
to the upper penny before continuing the calculation.
39. The amortization table uses an annual payment of $7518.83 and the unpaid balance after 3 years is $38,879.49, a
difference of 4¢ compared to Exercise 25. The final payment is $7518.87, a correction of 4¢ from the others in the
table.
41. $4,306,638   **43.** $14,204,375   **45.** $70,777.95   **47.** $468,407.35; John should buy $500,000 of life insurance.
49. a. $3334.58 **b.** It saves $334.25 each month. **c.** The longer mortgage costs an extra $2,640,702.24.   **51.** $31.53
53. $115.66   **55.** $41,640.59   **57.** 11.9%   **59.** $2579.31

CHAPTER 1 REVIEW EXERCISES page 162

1. $217.50   **2.** $52.90   **3.** $2250   **4.** $336.88   **5.** $33.86   **6.** $56.63   **7.** $16.53   **8.** $188.08   **9.** $10,212.75
10. $1491.53   **11.** $2138.56   **12.** $4288.61   **13.** 6.8%   **14.** $1800   **15.** 1 year 6 months   **16.** 4 years
17. 5 years   **18.** $4500   **19.** $9394.08   **20.** $47,267.91   **21.** 42.86%   **22.** 21.43%   **23.** 6.6%
24. $500,000   **25.** 11.1%; $1110   **26.** $31,535.24   **27.** $264,247.79   **28.** $370,893.09   **29.** $10,749.24
30. No; it would only be worth $112 billion.   **31.** $10,198.43   **32.** $18,563.50   **33.** $71,056.01   **34.** $12,702.50
35. $8466.50   **36.** 18 years and 6 months   **37.** 4 years and 2 months   **38.** 7.35 years   **39.** 3 years
40. 18 years   **41.** 9 years (9.01 years)   **42.** 12 years; 11 years and 7 months   **43.** 6 years; 5 years and 44 weeks
44. 6 years (6.12 years)   **45.** 7 years (7.27 years)   **46.** 13.92% if quarterly and 14.17% if continuously   **47.** 28.19%

48. 106.64%   **49.** 13.51%   **50.** 2.94%   **51.** $96,304.25   **52.** $19,541.30   **53.** $33,322.04   **54.** $935,211.12
55. $715,609.58   **56.** $1559.49   **57.** $205.98   **58.** $106.28   **59.** $168.76   **60.** $19.51   **61.** 26 years
62. 24 years and 6 months   **63.** 5 years and 1 month   **64.** 15 years and 7 months   **65.** 13 years and 6 months
66. 9.80%   **67.** 2.97%   **68.** 9.42%   **69.** 6.58%   **70.** 3.00% with 365 days per year; 3.12% with 360 days per year
71. $319,583.90   **72.** $20,923,408   **73.** $30,147,921   **74.** $7,082,196   **75.** $158,891
76. $106.12   **77.** $46.14   **78.** $423.17   **79.** $1180.06
80. $548.25; do not confuse paying off a current debt with accumulating money in the future.
81. $75,148.12   **82.** $142,934.54   **83.** $5449.20   **84.** $523,692.95   **85.** $143,929.12

EXERCISES 2.1 page 180

1. $\begin{cases} x + y = 18 \\ x - y = 2 \end{cases}$ **3.** $\begin{cases} x - y = 6 \\ x + y = 40 \end{cases}$ **5.** $\begin{cases} x + y = 30 \\ x + 5y = 70 \end{cases}$ **7.** $\begin{cases} x + y = 100 \\ 10x + 5y = 650 \end{cases}$ **9.** $\begin{cases} x + y = 225 \\ x - 2y = 0 \end{cases}$

11. The solution is $x = 4$, $y = 2$. The equations are independent and consistent.

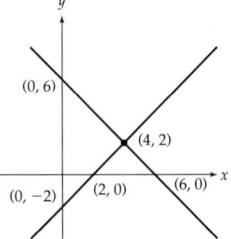

13. The solution is $x = 3$, $y = 2$. The equations are independent and consistent.

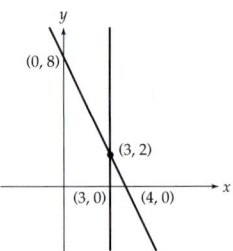

15. The solution is $x = 2$, $y = -2$. The equations are independent and consistent.

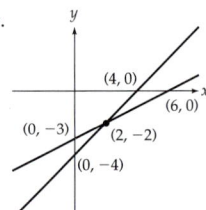

17. There is no solution. The equations are independent and inconsistent.

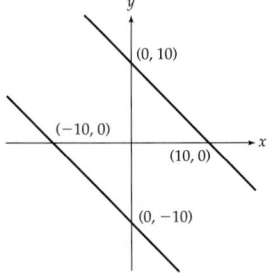

19. There are infinitely many solutions that may be parameterized as $x = 10 - t$, $y = t$. The equations are dependent.

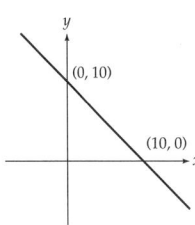

21. $x = 4$, $y = 3$. The equations are independent and consistent.
23. $x = 5$, $y = 10$. The equations are independent and consistent.
25. $x = 10$, $y = -10$. The equations are independent and consistent.
27. $x = 4$, $y = 9$. The equations are independent and consistent.
29. There are infinitely many solutions that may be parameterized as $x = 10 + t$, $y = t$. The equations are dependent.
31. $x = 3$, $y = 8$. The equations are independent and consistent.
33. $x = 4$, $y = 3$. The equations are independent and consistent.
35. $x = 8$, $y = 3$. The equations are independent and consistent.
37. $x = 15$, $y = 6$. The equations are independent and consistent.
39. There is no solution. The equations are independent and inconsistent.
41. There are 34 nickels and 26 dimes in the jar.
43. The retired couple should invest $2000 in the money market account and $8000 in the stock mutual fund.
45. The concession stand sold 1800 sodas and 1200 hot dogs.
47. The federal tax is $4900 and the state tax is $900.
49. The required calcium and phosphorus can be provided by just 5 tablets of supplement B (with none of supplement A) each day.
51. The equations are inconsistent and dependent.

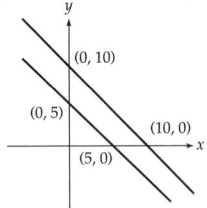

53. The equations are consistent and independent. The solution is $x = 15$, $y = 4$.

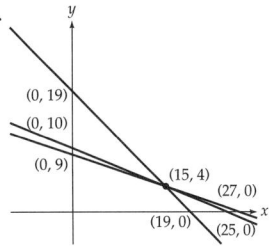

55. The equations are inconsistent and independent.

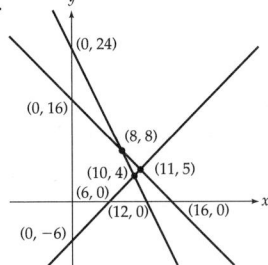

EXERCISES 2.2 page 194

1. $3 \times 2; 1; 6$ **3.** $3 \times 4; 3; -6$ **5.** $4 \times 4; 1; 1; 1; 1; 0$ **7.** $1 \times 4; 6$ **9.** $5 \times 1; 8; 6$ **11.** $\begin{pmatrix} 1 & 2 & 2 \\ 3 & 4 & 12 \end{pmatrix}$

13. $\begin{pmatrix} -4 & 3 & 84 \\ 5 & -2 & 70 \end{pmatrix}$ **15.** $\begin{pmatrix} 3 & -2 & 24 \\ 1 & 0 & 6 \end{pmatrix}$ **17.** $\begin{pmatrix} 5 & -15 & 30 \\ -4 & 12 & 24 \end{pmatrix}$ **19.** $\begin{pmatrix} 1 & 0 & 20 \\ 0 & 1 & 30 \end{pmatrix}$ **21.** $\begin{cases} x + y = 9 \\ y = 4 \end{cases}$

23. $\begin{cases} -4x + 3y = -60 \\ x - 2y = 20 \end{cases}$ **25.** $\begin{cases} x - 3y = -70 \\ x + y = 10 \end{cases}$ **27.** $\begin{cases} 2x + y = 6 \\ x + 2y = -6 \end{cases}$ **29.** $\begin{cases} 20x - 15y = 60 \\ -16x + 12y = -48 \end{cases}$

31. $\begin{pmatrix} 5 & 6 & 30 \\ 3 & 4 & 24 \end{pmatrix} \begin{matrix} R'1 = R2 \\ R'2 = R1 \end{matrix}$ **33.** $\begin{pmatrix} 2 & 2 & -4 \\ 6 & 5 & 60 \end{pmatrix} R'1 = R1 - R2$ **35.** $\begin{pmatrix} 5 & 6 & 30 \\ 5 & 10 & 90 \end{pmatrix} R'2 = 5R2$

37. $\begin{pmatrix} 6 & 6 & -12 \\ 0 & 1 & -72 \end{pmatrix} R'2 = R1 - R2$ **39.** $\begin{pmatrix} 1 & -2 & -42 \\ 0 & 1 & 15 \end{pmatrix} R'2 = \frac{1}{8}R2$

41. $x = 7, y = -3$. The equations are independent and consistent.

43. No solution. The equations are independent and inconsistent.

45. No solution. The equations are independent and inconsistent.

47. $x = 3 - 2t, y = t$ (infinitely many solutions). The equations are dependent.

49. $x = t, y = -3$ (infinitely many solutions). The equations are dependent.

51. $x = 3, y = 2$. The equations are independent and consistent.

53. $x = 3, y = 1$. The equations are independent and consistent.

55. $x = 1, y = 2$. The equations are independent and consistent.

57. $x = 3, y = 2$. The equations are independent and consistent.

59. $x = 9, y = 2$. The equations are independent and consistent.

61. No solution. The equations are independent and inconsistent.

63. $x = 8, y = -15$. The equations are independent and consistent.

65. $x = 4, y = 15$. The equations are independent and consistent.

67. $x = 9 + 3t, y = t$ (infinitely many solutions). The equations are dependent.

69. $x = 21, y = -4$. The equations are independent and consistent.

71. The commodities speculator invested $10,000 in soybean futures and $5000 in corn futures.

73. The older brother receives $4.8 million and the younger brother receives $2.4 million, leaving $4.8 million for their sister.

75. There are 175 nickels and 112 quarters in the jar.

77. Since the dietician probably wants positive whole numbers for the solution, the possibilities are:

Cans of NutraDrink:	12	8	4	0
Tablets of VitaPills:	0	5	10	15

79. The campaign manager should use 7 TV ads and 30 radio ads.

81. $x = 18, y = 16$. The equations are independent and consistent.

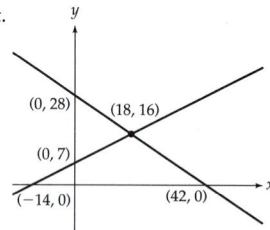

83. No solution. The equations are independent and inconsistent.

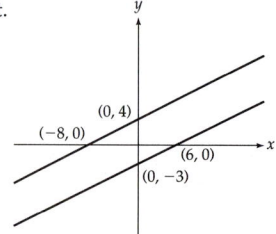

85. No solution. The equations are independent and inconsistent.

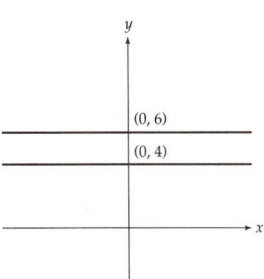

87. $x = 14 - \frac{7}{2}t$, $y = t$ (infinitely many solutions). The equations are dependent.

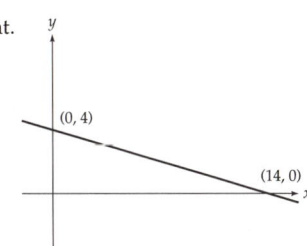

89. $x = t$, $y = 6$ (infinitely many solutions). The equations are dependent.

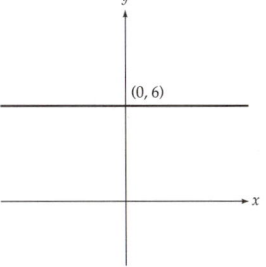

91. $\begin{pmatrix} 3 & -10 & -65 \\ -4 & 13 & 84 \end{pmatrix} \rightarrow \begin{pmatrix} -4 & 13 & 84 \\ 3 & -10 & -65 \end{pmatrix} \begin{matrix} R'1 = R2 \\ R'2 = R1 \end{matrix} \rightarrow \begin{pmatrix} 3 & -10 & -65 \\ -4 & 13 & 84 \end{pmatrix} \begin{matrix} R'1 = R2 \\ R'2 = R1 \end{matrix}$

93. $\begin{pmatrix} 3 & -10 & -65 \\ -4 & 13 & 84 \end{pmatrix} \rightarrow \begin{pmatrix} 15 & -50 & -325 \\ -4 & 13 & 84 \end{pmatrix} R'1 = 5R1 \rightarrow \begin{pmatrix} 3 & -10 & -65 \\ -4 & 13 & 84 \end{pmatrix} R'1 = \frac{1}{5}R1$

95. $\begin{pmatrix} 3 & -10 & -65 \\ -4 & 13 & 84 \end{pmatrix} \rightarrow \begin{pmatrix} 3 & -10 & -65 \\ -1 & 13/4 & 21 \end{pmatrix} R'2 = \frac{1}{4}R2 \rightarrow \begin{pmatrix} 3 & -10 & -65 \\ -4 & 13 & 84 \end{pmatrix} R'2 = 4R2$

97. $\begin{pmatrix} 3 & -10 & -65 \\ -4 & 13 & 84 \end{pmatrix} \rightarrow \begin{pmatrix} -1 & 3 & 19 \\ -4 & 13 & 84 \end{pmatrix} R'1 = R1 + R2 \rightarrow \begin{pmatrix} 3 & -10 & -65 \\ -4 & 13 & 84 \end{pmatrix} R'1 = R1 - R2$

99. $\begin{pmatrix} 3 & -10 & -65 \\ -4 & 13 & 84 \end{pmatrix} \rightarrow \begin{pmatrix} 7 & -23 & -149 \\ -4 & 13 & 84 \end{pmatrix} R'1 = R1 - R2 \rightarrow \begin{pmatrix} 3 & -10 & -65 \\ -4 & 13 & 84 \end{pmatrix} R'1 = R1 + R2$

EXERCISES 2.3 page 210

1. $\begin{pmatrix} 1 & 1 & 1 & 4 \\ 1 & 2 & 1 & 3 \\ 1 & 2 & 2 & 5 \end{pmatrix}$ **3.** $\begin{pmatrix} 2 & -1 & 2 & 11 \\ -1 & 1 & -3 & -12 \\ 2 & -2 & 7 & 27 \end{pmatrix}$ **5.** $\begin{pmatrix} 2 & 1 & 5 & 4 & 5 & 2 \\ 1 & 1 & 3 & 3 & 3 & -1 \end{pmatrix}$ **7.** $\begin{pmatrix} 6 & 3 & 5 & 8 \\ 1 & 2 & 2 & 1 \\ 4 & 3 & 4 & 5 \\ 5 & 1 & 3 & 7 \end{pmatrix}$

9. $\begin{pmatrix} 3 & 4 & 2 & 4 & 12 \\ 1 & 2 & 1 & 1 & 4 \\ 4 & 5 & 2 & 5 & 14 \\ 6 & 6 & 1 & 6 & 15 \end{pmatrix}$

11. $\begin{cases} 4x_1 + 3x_2 + 2x_3 = 11 \\ 3x_1 + 3x_2 + x_3 = 6 \\ x_1 - 2x_2 + 3x_3 = 13 \end{cases}$

13. $\begin{cases} 2x_1 + x_2 + x_3 = 7 \\ 2x_1 + 2x_2 + x_3 = 6 \\ 3x_1 + 3x_2 + 2x_3 = 10 \end{cases}$

15. $\begin{cases} 8x_1 + 3x_2 - 2x_3 + 19x_4 = 15 \\ 3x_1 + x_2 - x_3 + 7x_4 = 6 \end{cases}$

17. $\begin{cases} 2x_1 + 3x_2 + 2x_3 = 5 \\ 3x_1 + 5x_2 + 3x_3 = 8 \\ x_1 + 2x_2 + 2x_3 = 2 \\ 4x_1 + 7x_2 + 5x_3 = 9 \end{cases}$

19. $\begin{cases} 3x_1 + 3x_2 + 5x_3 + 4x_4 = 11 \\ 2x_1 + 2x_2 + 3x_3 + 3x_4 = 9 \\ 2x_1 + x_2 + 2x_3 + 2x_4 = 7 \\ 3x_1 + 2x_2 + 3x_3 + 3x_4 = 11 \end{cases}$

21. The system of equations is independent and consistent with solution $x_1 = 4$, $x_2 = 5$, $x_3 = -4$.

23. The system of equations is independent and consistent with solution $x_1 = 2$, $x_2 = -1$, $x_3 = 3$, $x_4 = 1$.

25. The system of equations is independent and inconsistent with no solution.

27. The system of equations is dependent and consistent with solution $x_1 = -5 + t$, $x_2 = 5 - t$, $x_3 = t$.

29. The system of equations is dependent and consistent with solution $x_1 = 8 + t_1 - t_2$, $x_2 = t_1$, $x_3 = 4 + t_2$, $x_4 = t_2$.

31. $\begin{pmatrix} 1 & 0 & 0 & 1 \\ 0 & 1 & 0 & 2 \\ 0 & 0 & 1 & 3 \end{pmatrix}$

33. $\begin{pmatrix} 1 & 0 & 0 & 1 \\ 0 & 1 & 0 & 2 \\ 0 & 0 & 1 & 3 \end{pmatrix}$

35. $\begin{pmatrix} 1 & 2 & 0 & 3 \\ 0 & 0 & 1 & -3 \\ 0 & 0 & 0 & 0 \end{pmatrix}$

37. $\begin{pmatrix} 1 & 0 & 0 & -2 & 0 \\ 0 & 1 & 0 & -1 & 0 \\ 0 & 0 & 1 & 1 & 0 \\ 0 & 0 & 0 & 0 & 1 \end{pmatrix}$

39. $\begin{pmatrix} 1 & 0 & 0 & 0 & 2 \\ 0 & 1 & 0 & 0 & 2 \\ 0 & 0 & 1 & 0 & 3 \\ 0 & 0 & 0 & 1 & 1 \end{pmatrix}$

41. The system of equations is independent and consistent with solution $x_1 = 1$, $x_2 = -2$, $x_3 = 3$.

43. The system of equations is independent and consistent with solution $x_1 = 1$, $x_2 = -2$, $x_3 = 3$.

45. The system of equations is independent and consistent with solution $x_1 = 1$, $x_2 = 1$, $x_3 = 1$, $x_4 = 1$.

47. The system of equations is independent and inconsistent with no solution.

49. The system of equations is dependent and consistent with solution $x_1 = 3 + 7t$, $x_2 = 4 - 10t$, $x_3 = t$.

51. The system of equations is independent and consistent with solution $x_1 = 2$, $x_2 = 1$, $x_3 = -2$, $x_4 = 3$.

53. The system of equations is independent and consistent with solution $x_1 = -1$, $x_2 = 1$, $x_3 = 2$, $x_4 = -2$.

55. The system of equations is independent and consistent with solution $x_1 = 1$, $x_2 = 2$, $x_3 = -2$, $x_4 = -1$, $x_5 = 1$.

57. The system of equations is dependent and inconsistent with no solution.

59. The system of equations is dependent and consistent with solution $x_1 = 2 + t_1 - t_2$, $x_2 = 1 - 2t_1 - t_2$, $x_3 = t_1$, $x_4 = t_2$.

61. The gardener should use 2 bags of GrowRite, 1 bag of MiracleMix, and 3 bags of GreatGreen.

63. Letting Q be the number of quarters, there are $200 + 3Q$ nickels and $500 - 4Q$ dimes. Since the number of each can never be negative, there can be no more than 125 quarters in the jar.

65. a. The federal tax is $80,510, the state tax is $16,490, and the city tax is $2490.
  b. The effective combined tax rate is (about) 55.3%.

67. a. Letting B be the number of blouses and S be the number of skirts, the shop can make $60 + \frac{6}{5}B - \frac{4}{5}S$ scarves and $60 - \frac{9}{10}B - \frac{2}{5}S$ dresses (where none of these quantities are negative).   **b.** 76 scarves and 38 dresses

69. Letting R be the number of radio ads, the number of TV ads is $20 - \frac{1}{5}R$, and the number of newspaper ads is 60. The promotional director may choose R to be 0, 5, 10, 15, 20, . . . , 85, 90, 95, or 100 because these are the only values for R that will result in a whole number of TV ads while ensuring that neither is negative.

71. $\begin{cases} x_1 + 2x_2 = 3 \\ x_2 = 1 \end{cases}$

73. $\begin{cases} x_1 + x_2 + x_3 = 4 \\ x_2 + x_3 = 3 \\ x_3 = 2 \end{cases}$

75. All are equivalent to $\begin{pmatrix} 1 & 0 & 1 \\ 0 & 1 & 1 \end{pmatrix}$ and so to each other.

77. All are equivalent to $\begin{pmatrix} 1 & 0 & 0 & 1 \\ 0 & 1 & 0 & 1 \\ 0 & 0 & 1 & 2 \end{pmatrix}$ and so to each other.   **79.** $\begin{pmatrix} 1 & 0 & 1 \\ 0 & 1 & 1 \end{pmatrix}$ **81.** $x_1 = 1$, $x_2 = 1$

83. $x_1 = 1$, $x_2 = 1$ **85.** $x_1 = 1$, $x_2 = 1$, $x_3 = 2$ **87.** $x_1 = 3$, $x_2 = 2$, $x_3 = 4$, $x_4 = 1$ **89.** $x_1 = 3$, $x_2 = -1$, $x_3 = 2$

EXERCISES 2.4 page 226

1. $\begin{pmatrix} 1 & 4 & 7 \\ 2 & 5 & 8 \\ 3 & 6 & 9 \end{pmatrix}$

3. $\begin{pmatrix} 3 & 18 & 24 \\ 12 & 6 & 21 \\ 27 & 15 & 9 \end{pmatrix}$

5. $\begin{pmatrix} -9 & -8 & -7 \\ -6 & -5 & -4 \\ -3 & -2 & -1 \end{pmatrix}$

7. $\begin{pmatrix} 2 & 8 & 11 \\ 8 & 7 & 13 \\ 16 & 13 & 12 \end{pmatrix}$

9. $\begin{pmatrix} -1 & 4 & 5 \\ 0 & -4 & 1 \\ 2 & -3 & -7 \end{pmatrix}$

11. (6)   **13.** (0)   **15.** $\begin{pmatrix} -2 \\ 5 \end{pmatrix}$ **17.** $\begin{pmatrix} 6 & 6 \\ 7 & 7 \end{pmatrix}$ **19.** $\begin{pmatrix} 11 & 7 & 11 \\ 6 & 7 & 6 \\ -1 & 3 & -1 \end{pmatrix}$ **21.** $\begin{pmatrix} 3 & -1 & 4 \\ 2 & 1 & 3 \end{pmatrix}$ **23.** $\begin{pmatrix} 2 & 8 \\ 4 & 1 \\ 5 & -4 \end{pmatrix}$

25. $\begin{pmatrix} -1 & -4 & -1 \\ -2 & 6 & -1 \end{pmatrix}$

27. $\begin{pmatrix} 7 & 1 & 4 \\ 0 & -1 & 0 \\ 4 & 2 & 2 \end{pmatrix}$

29. $\begin{pmatrix} 5 & 5 & 6 \\ 7 & -7 & 4 \end{pmatrix}$

31. $\begin{pmatrix} 1 & 5 & 4 \\ 1 & 1 & 1 \\ 2 & 3 & 3 \end{pmatrix} \begin{pmatrix} x_1 \\ x_2 \\ x_3 \end{pmatrix} = \begin{pmatrix} 6 \\ 4 \\ 9 \end{pmatrix}$

33. $\begin{pmatrix} 4 & 3 & -1 \\ 3 & 3 & 2 \\ 2 & 1 & -3 \end{pmatrix} \begin{pmatrix} x_1 \\ x_2 \\ x_3 \end{pmatrix} = \begin{pmatrix} 2 \\ 9 \\ -6 \end{pmatrix}$

35. $\begin{pmatrix} 5 & 2 & -4 & 1 & 5 \\ 3 & 1 & -3 & 1 & 3 \end{pmatrix} \begin{pmatrix} x_1 \\ x_2 \\ x_3 \\ x_4 \\ x_5 \end{pmatrix} = \begin{pmatrix} 7 \\ 5 \end{pmatrix}$

37. $\begin{cases} 5x_1 + 9x_2 + 9x_3 = 11 \\ 4x_1 + 7x_2 + 6x_3 = 9 \\ 3x_1 + 5x_2 + 3x_3 = 8 \\ 4x_1 + 7x_2 + 5x_3 = 10 \end{cases}$

39. $\begin{cases} 5x_1 + 4x_2 + 7x_3 + 6x_4 = 18 \\ 2x_1 + 2x_2 + 3x_3 + 3x_4 = 9 \\ 4x_1 + 3x_2 + 5x_3 + 5x_4 = 16 \\ 3x_1 + 2x_2 + 3x_3 + 3x_4 = 11 \end{cases}$

41. $\begin{pmatrix} 1 & 0 & 0 & 0 \\ 0 & 0 & 0 & 1 \\ 0 & 0 & 1 & 0 \\ 0 & 1 & 0 & 0 \end{pmatrix}$

43. $\begin{pmatrix} 1 & 0 & 0 & 0 \\ 0 & 1 & 0 & 0 \\ 0 & 0 & 1 & 0 \\ 0 & 0 & 0 & 3 \end{pmatrix}$

45. $\begin{pmatrix} 1 & 0 & -1 & 0 \\ 0 & 1 & 0 & 0 \\ 0 & 0 & 1 & 0 \\ 0 & 0 & 0 & 1 \end{pmatrix}$

47. $\begin{pmatrix} 1 & 0 & 0 & 0 \\ 0 & 1 & 0 & 0 \\ 0 & 0 & 1 & -2 \\ 0 & 0 & 0 & 1 \end{pmatrix}$

49. $\begin{pmatrix} 1 & 0 & -3 & 0 \\ 0 & 1 & -2 & 0 \\ 0 & 0 & 1 & 0 \\ 0 & 0 & -1 & 1 \end{pmatrix}$

51. Let P be the "price" matrix of selling prices and C be the "commission" matrix so that $C = 0.15P$. Choosing P to be a 3×3 matrix with the rows representing the manufacturers SlumberKing, DreamOn, and RestEasy in that order and the columns representing the models "economy," "best," and "deluxe" in that order, $P = \begin{pmatrix} 300 & 350 & 500 \\ 350 & 400 & 550 \\ 400 & 500 & 700 \end{pmatrix}$ and $C = \begin{pmatrix} 45.00 & 52.50 & 75.00 \\ 52.50 & 60.00 & 82.50 \\ 60.00 & 75.00 & 105.00 \end{pmatrix}$.

53. Let D be the "dealer invoice" matrix and S be the "sticker price" matrix so that the "markup" matrix M is $M = S - D$. Choosing D to be a 2×4 matrix with the rows representing the sales lots in Oakdale and Roanoke in that order and the columns representing the vehicle models "sedan," "station wagon," "van," and "pickup truck" in that order, $D = \begin{pmatrix} 15,000 & 19,000 & 23,000 & 25,000 \\ 15,000 & 19,000 & 23,000 & 25,000 \end{pmatrix}$ and $S = \begin{pmatrix} 18,900 & 22,900 & 26,900 & 29,900 \\ 19,900 & 21,900 & 27,900 & 28,900 \end{pmatrix}$ so then $M = \begin{pmatrix} 3900 & 3900 & 3900 & 4900 \\ 4900 & 2900 & 4900 & 3900 \end{pmatrix}$.

55. Let L be the "labor costs" matrix and M be the "materials cost" matrix so that the "total cost" matrix T is $T = L + M$. Choosing L to be a 2×3 matrix of values in pennies with the rows representing the countries Costa Rica and Honduras in that order and the columns representing the apparel items "shorts," "tee-shirts," and "caps" in that order, $L = \begin{pmatrix} 75 & 25 & 45 \\ 80 & 20 & 55 \end{pmatrix}$ and $M = \begin{pmatrix} 160 & 95 & 115 \\ 150 & 80 & 110 \end{pmatrix}$ so then $T = \begin{pmatrix} 235 & 120 & 160 \\ 230 & 100 & 165 \end{pmatrix}$.

57. Let P be the "sale price" row matrix with values in dollars and N be the "number of items" column matrix so that the "total cost" matrix C is $C = P \cdot N$. Choosing the columns of P and the rows of N to represent the items "bottles of soda," "bottles of pickles," "packages of hot dogs," and "bags of chips" in that order, $P = (0.89 \quad 1.29 \quad 2.39 \quad 1.69)$ and $N = \begin{pmatrix} 12 \\ 2 \\ 3 \\ 4 \end{pmatrix}$ so then $C = (27.19)$. The total cost of these items at these prices is $27.19.

59. Let T be the "time" matrix and L be the "labor hourly cost" matrix so that the "production cost" matrix C is $C = T \cdot L$. Choosing T to be a 3×3 matrix with the rows representing the furniture items "table," "chair," and "desk" in that order and the columns representing the manufacturing steps "cutting and milling," "assembly," and "finishing" in that order, $T = \begin{pmatrix} 2 & 1 & 2 \\ 1.5 & 1 & 0.5 \\ 3 & 2 & 3 \end{pmatrix}$, and choosing L to be a 3×2 matrix with the rows representing the manufacturing steps "cutting and milling," "assembly," and "finishing" in that order and the columns representing the factory

locations "Wytheville" and "Andersen" in that order, $L = \begin{pmatrix} 9 & 10 \\ 14 & 13 \\ 13 & 12 \end{pmatrix}$, so then $C = \begin{pmatrix} 58 & 57 \\ 34 & 34 \\ 94 & 92 \end{pmatrix}$. For these choices of

T and L, the rows of C represent the furniture items "table," "chair," and "desk" in that order and the columns represent the factory locations "Wytheville" and "Andersen" in that order.

61. $A^t \cdot A = \begin{pmatrix} 17 & 22 & 27 \\ 22 & 29 & 36 \\ 27 & 36 & 45 \end{pmatrix}$ and $A \cdot A^t = \begin{pmatrix} 14 & 32 \\ 32 & 77 \end{pmatrix}$; both are symmetric.

63. $A^t \cdot A = \begin{pmatrix} 1 & -1 & 2 \\ -1 & 5 & -4 \\ 2 & -4 & 6 \end{pmatrix}$ and $A \cdot A^t = \begin{pmatrix} 6 & -4 & 2 \\ -4 & 5 & -1 \\ 2 & -1 & 1 \end{pmatrix}$; both are symmetric.

67. $f(g(x)) = \dfrac{8x - 1}{5x}$ and $F \cdot G = \begin{pmatrix} 8 & -1 \\ 5 & 0 \end{pmatrix}$ **69.** $f(g(x)) = \dfrac{4x - 1}{2x + 1}$ and $F \cdot G = \begin{pmatrix} 4 & -1 \\ 2 & 1 \end{pmatrix}$ **81.** $\begin{pmatrix} 1 & \frac{1}{2} & \frac{1}{2} \\ 0 & \frac{1}{2} & \frac{3}{2} \\ 0 & \frac{1}{2} & \frac{1}{2} \end{pmatrix}$

83. $\begin{pmatrix} 1 & 0 & -2 \\ 0 & 1 & 4 \\ 0 & 0 & -1 \end{pmatrix}$ **85.** $(2)(\frac{1}{2})(-1) = -1$ **87.** $(3)(\frac{-19}{3})(\frac{-5}{19}) = 5$ **89.** 1 (switched rows 1 and 2)

91. -5 (multiplied second row by 5)   **93.** -1 (added row 1 to row 2 five times)   **95.** -1 (transposed the matrix)

EXERCISES 2.5 page 242

1. $\begin{pmatrix} 1 & 2 \\ -1 & -1 \end{pmatrix} \begin{pmatrix} -1 & -2 \\ 1 & 1 \end{pmatrix} = \begin{pmatrix} 1 & 0 \\ 0 & 1 \end{pmatrix}$ so this pair of matrices is a matrix and its inverse.

3. $\begin{pmatrix} 1 & 1 & 0 \\ 2 & 1 & 1 \\ 1 & 0 & 0 \end{pmatrix} \begin{pmatrix} 0 & 0 & 1 \\ 1 & 0 & -1 \\ -1 & 1 & -1 \end{pmatrix} = \begin{pmatrix} 1 & 0 & 0 \\ 0 & 1 & 0 \\ 0 & 0 & 1 \end{pmatrix}$ so this pair of matrices is a matrix and its inverse.

5. $\begin{pmatrix} 4 & 6 & 3 \\ 3 & 4 & 1 \\ 5 & 7 & 3 \end{pmatrix} \begin{pmatrix} -5 & -3 & 6 \\ 4 & 3 & -5 \\ -1 & -2 & 2 \end{pmatrix} = \begin{pmatrix} 1 & 0 & 0 \\ 0 & 1 & 0 \\ 0 & 0 & 1 \end{pmatrix}$ so this pair of matrices is a matrix and its inverse.

7. $\begin{pmatrix} 10 & -4 & -7 \\ -7 & 3 & 5 \\ 4 & -1 & -3 \end{pmatrix} \begin{pmatrix} 4 & 5 & -1 \\ 1 & 2 & 1 \\ 5 & 6 & 2 \end{pmatrix} = \begin{pmatrix} 1 & 0 & -28 \\ 0 & 1 & 20 \\ 0 & 0 & -11 \end{pmatrix}$ so this pair of matrices is not a matrix and its inverse.

9. $\begin{pmatrix} 2 & 0 & 1 & 0 \\ 1 & 1 & 1 & 0 \\ -2 & 0 & -1 & 1 \\ 1 & 0 & 0 & 1 \end{pmatrix} \begin{pmatrix} -1 & 0 & -1 & 1 \\ -2 & 1 & -1 & 1 \\ 3 & 0 & 2 & -2 \\ 1 & 0 & 1 & 0 \end{pmatrix} = \begin{pmatrix} 1 & 0 & 0 & 0 \\ 0 & 1 & 0 & 0 \\ 0 & 0 & 1 & 0 \\ 0 & 0 & 0 & 1 \end{pmatrix}$ so this pair of matrices is a matrix and its inverse.

11. $\begin{pmatrix} 1 & -3 \\ 0 & 1 \end{pmatrix}$ **13.** $\begin{pmatrix} -1 & 2 \\ 6 & -11 \end{pmatrix}$ **15.** $\begin{pmatrix} 0 & 1 & -2 \\ 1 & -1 & 2 \\ 0 & -1 & 3 \end{pmatrix}$ **17.** Singular matrix   **19.** $\begin{pmatrix} 1 & 0 & 0 & -1 \\ 0 & 1 & 0 & 0 \\ -1 & 0 & 1 & 1 \\ 0 & -1 & 0 & 1 \end{pmatrix}$

21. $\begin{pmatrix} 1 & -3 & 1 \\ 0 & 2 & -1 \\ -1 & 0 & 1 \end{pmatrix}$ **23.** $\begin{pmatrix} 1 & 2 & 0 \\ -2 & -2 & 1 \\ -5 & -7 & 2 \end{pmatrix}$ **25.** $\begin{pmatrix} 3 & -5 & 1 & 1 \\ -1 & 1 & 0 & 0 \\ -4 & 5 & -1 & 1 \\ 0 & 1 & 0 & -1 \end{pmatrix}$ **27.** Singular matrix

29. $\begin{pmatrix} 0 & -2 & 0 & 1 & 0 \\ 3 & 3 & -1 & -1 & 1 \\ 1 & 6 & -1 & -3 & 2 \\ -5 & -5 & 2 & 2 & -2 \\ 0 & -4 & 1 & 2 & -2 \end{pmatrix}$ **31.** $\begin{pmatrix} x_1 \\ x_2 \end{pmatrix} = \begin{pmatrix} -1 & 2 \\ 6 & -11 \end{pmatrix} \begin{pmatrix} 9 \\ 5 \end{pmatrix} = \begin{pmatrix} 1 \\ -1 \end{pmatrix}$ so $\begin{cases} x_1 = 1 \\ x_2 = -1 \end{cases}$

33. $\begin{pmatrix} x_1 \\ x_2 \\ x_3 \end{pmatrix} = \begin{pmatrix} 0 & 1 & -2 \\ 1 & -1 & 2 \\ 0 & -1 & 3 \end{pmatrix} \begin{pmatrix} 2 \\ 5 \\ 2 \end{pmatrix} = \begin{pmatrix} 1 \\ 1 \\ 1 \end{pmatrix}$ so $\begin{cases} x_1 = 1 \\ x_2 = 1 \\ x_3 = 1 \end{cases}$ **35.** $x_1 = 1, x_2 = 1, x_3 = 1$ **37.** $x_1 = 4, x_2 = -1, x_3 = -2$

39. $x_1 = 2, x_2 = -1, x_3 = 2, x_4 = 1$ **41.** $x_1 = 1, x_2 = -2, x_3 = 1, x_4 = 2$ **43.** $x_1 = 1, x_2 = -2, x_3 = 2, x_4 = -1, x_5 = 1$

45. $x_1 = 1, x_2 = -2, x_3 = 1, x_4 = -2, x_5 = 1$ **47.** $x_1 = 2, x_2 = 4, x_3 = 1$ **49.** $x_1 = -5, x_2 = 10, x_3 = -20, x_4 = 15$

51. The multiplex sold 150 adult tickets and 350 child tickets for Film No. 1; 200 adult tickets and 200 child tickets for Film No. 2; 250 adult tickets and 200 child tickets for Film No. 3; 400 adult tickets and 100 child tickets for Film No. 4; and 600 adult tickets and no child tickets for Film No. 5.

53. The red jar contains 250 pennies, 150 nickels, and 100 dimes; the green jar contains 150 pennies, 350 nickels, and 200 dimes; and the blue jar contains 325 pennies, 115 nickels, and 160 dimes.

55. Billy needs 4 drops of Supplement No. 1, no drops of Supplement No. 2, 5 drops of Supplement No. 3, and 2 drops of Supplement No. 4; Susie needs 2 drops of Supplement No. 1, 2 drops of Supplement No. 2, 3 drops of Supplement No. 3, and 6 drops of Supplement No. 4; and Jimmy needs 3 drops of Supplement No. 1, 1 drop of Supplement No. 2, no drops of Supplement No. 3, and 8 drops of Supplement No. 4.

57. Mr. and Mrs. Jordan should invest $100,000 in the stock fund, $150,000 in the money market fund, and $50,000 in the bond fund; Mr. and Mrs. French should invest $78,300 in the stock fund, $51,100 in the money market fund, and $105,500 in the bond fund; and Mrs. Daimen should invest $90,000 in the stock fund, $105,000 in the money market fund, and $75,000 in the bond fund.

59. The mass transit manager should assign 120 subway cars, 50 buses, and 10 jitneys to Brighton; 100 subway cars, 30 buses, and 20 jitneys to Conway; 100 subway cars, 40 buses, and 10 jitneys to Longwood; and 110 subway cars, 40 buses, and 15 jitneys to Oakley.

61. $X = (A + I)^{-1} \cdot C; X = \begin{pmatrix} 60 \\ 64 \\ -12 \end{pmatrix}$ **63.** $X = (A + B)^{-1} \cdot (C + D); X = \begin{pmatrix} -60 \\ 100 \\ 180 \end{pmatrix}$

EXERCISES 2.6 page 264

1. $\begin{pmatrix} 0.30 & 0.50 \\ 0.40 & 0.10 \end{pmatrix}$ for sectors A and L **3.** $\begin{pmatrix} 0.20 & 0.10 & 0.15 \\ 0.10 & 0.30 & 0.20 \\ 0.40 & 0.15 & 0.10 \end{pmatrix}$ for sectors C, E, and L

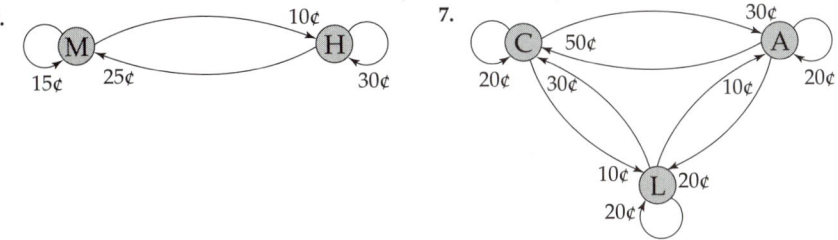

5. **7.**

9. $Y = \begin{pmatrix} 71 \\ 37 \end{pmatrix}$ **11.** $Y = \begin{pmatrix} 89 \\ 118 \\ 101 \end{pmatrix}$ **13.** $X = \begin{pmatrix} 150 \\ 120 \end{pmatrix}$ **15.** $X = \begin{pmatrix} 80 \\ 60 \\ 100 \end{pmatrix}$

17. Heavy industry production must be $300 million and light industry production must be $280 million.

19. The current GNP from these sectors is $115 million (made up of $20 million from heavy industry, $50 million from light industry, and $45 million from the railroads). Each $10 million increase in heavy industry production raises the GNP by $1 million (but since the heavy industry production consumes both light industry and railroad production, this increase is composed of an $8 million increase in heavy industry production together with decreases of $5 million from light industry and $2 million from the railroads). When the heavy industry production level reaches $200 million, all light industry excess production will be consumed by the heavy industry sector and no further expansion will be possible without expanding the light industry production.

21. $\begin{pmatrix} 0.60 & 0.25 & 0.15 \\ 0.50 & 0.40 & 0.10 \\ 0.30 & 0.20 & 0.50 \end{pmatrix}$ **23.** $\begin{pmatrix} 0.10 & 0.20 & 0 & 0.70 \\ 0.40 & 0.20 & 0.40 & 0 \\ 0 & 0.90 & 0.10 & 0 \\ 0 & 0.30 & 0.40 & 0.30 \end{pmatrix}$ **25.** Ergodic **27.** Ergodic **29.** Regular

31. $X = (\frac{1}{2}\ \frac{1}{2})$ **33.** $X = (\frac{1}{3}\ \frac{1}{3}\ \frac{1}{3})$ **35.** $X = (\frac{1}{4}\ \frac{1}{4}\ \frac{1}{4}\ \frac{1}{4})$
37. There are 4400 residents of Lucas County, 4200 residents of Marion County, and 2600 residents of Warren County.
39. There were 185 people in the Clyberg display room, 315 people in the Stevensen display room, 195 people in the Georgan display room, and 360 people in the central refreshments room.   **41.** $x = 5, y = -4$
43. $x_1 = 2, x_2 = 3, x_3 = 4$ **45.** Inconsistent, yet "almost" satisfied by $x = 5, y = 5$
47. Inconsistent, yet "almost" satisfied by $x_1 = 1, x_2 = 3, x_3 = 3$
49. Inconsistent, yet "almost" satisfied by $x_1 = 5, x_2 = 3, x_3 = 2, x_4 = 1$
51. $y = 9x + 34$ **53.** $y = -29x + 92$ **55.** $y = 9x + 165$ **57.** His sales in the fifth month may be expected to be $910,000.   **59.** At 79¢ per eight-ounce bag, the manufacturer can expect 1711 sales per 20,000 customers.

CHAPTER 2 REVIEW EXERCISES page 272

1. Let x be the number of 30-day advance sale tickets and let y be the number of full-fare tickets: $\begin{cases} x + y = 30 \\ 79x + 159y = 3970 \end{cases}$

2. Let x be the number of cows and let y be the number of horses: $\begin{cases} x + y = 420 \\ x - 2y = 0 \end{cases}$

3. The solution is $x = 13, y = 5$. The equations are independent and consistent.

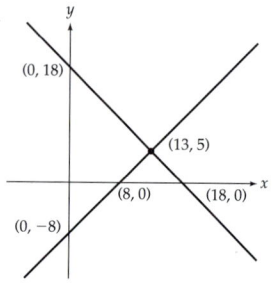

4. There are infinitely many solutions that may be parameterized as $x = 18 + 2t, y = t$. The equations are dependent.

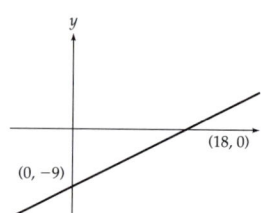

5. $x = 14, y = 30$. The equations are independent and consistent.
6. No solution. The equations are independent and inconsistent.
7. $x = 9, y = 3$. The equations are independent and consistent.
8. $x = -15, y = 24$. The equations are independent and consistent.
9. He has 10 rosebushes and 15 tomato plants in his garden.
10. The fraternity should use 2 cars and 2 vans for the trip to Orlando.   **11.** 3×3; 5; 6; 4   **12.** 4×4; 7; 10; 13

13. $\begin{pmatrix} 3 & 4 & 12 \\ 3 & 6 & 6 \end{pmatrix} R'2 = 3R2$ **14.** $\begin{pmatrix} 1 & -1 & 8 \\ 2 & -3 & 6 \end{pmatrix} R'1 = R1 + R2$

15. $x = 15, y = 16$. The equations are independent and consistent.
16. $x = 20, y = 30$. The equations are independent and consistent.
17. No solution. The equations are independent and inconsistent.
18. $x = 3 + \frac{3}{4}t, y = t$ (infinitely many solutions). The equations are dependent.
19. The pharmacist filled 63 prescriptions for antibiotics and 29 prescriptions for cough suppressants.
20. The price of each old *Life* magazine is $6 while each old copy of *The New Yorker* costs $5.
21. The system of equations is independent and consistent with solution $x_1 = 3, x_2 = -3, x_3 = 6$.
22. The system of equations is dependent and consistent with solution $x_1 = 4 - t, x_2 = t, x_3 = 2$.

23. $\begin{pmatrix} 1 & 0 & 0 & 4 \\ 0 & 1 & 0 & -3 \\ 0 & 0 & 1 & 2 \end{pmatrix}$ **24.** $\begin{pmatrix} 1 & 0 & 0 & 1 & 0 \\ 0 & 1 & 0 & -1 & 0 \\ 0 & 0 & 1 & 0 & 0 \\ 0 & 0 & 0 & 0 & 1 \end{pmatrix}$

25. The system of equations is independent and consistent with solution $x_1 = 4$, $x_2 = 3$, $x_3 = 1$, $x_4 = 2$.

26. The system of equations is dependent and consistent with solution $x_1 = 2 - t$, $x_2 = 3 - t$, $x_3 = t$, $x_4 = 4$.

27. The system of equations is independent and inconsistent and there is no solution.

28. The system of equations is dependent and inconsistent and there is no solution.

29. The Nyack Nursery should raise 48 dahlias, 328 chrysanthemums, and 48 daisies.

30. Each hot dog costs \$1. If the mirror house is \$1.50 and a soda is \$1, then a go-kart ride costs \$4.50.   **31.** $\begin{pmatrix} 4 & -6 \\ 6 & -4 \end{pmatrix}$

32. $\begin{pmatrix} 1 & 2 & 3 & 4 \\ -4 & -3 & -2 & -1 \\ 1 & 2 & 3 & 4 \\ -4 & -3 & -2 & -1 \end{pmatrix}$ **33.** $\begin{pmatrix} 15 & 25 \\ 5 & 2 \end{pmatrix}$ **34.** $\begin{pmatrix} 10 & 3 & -2 \\ -2 & 10 & 11 \\ 9 & 3 & -1 \end{pmatrix}$ **35.** $\begin{pmatrix} 1 & 4 & 1 \\ 2 & 8 & 3 \\ 1 & 5 & 2 \end{pmatrix} \begin{pmatrix} x_1 \\ x_2 \\ x_3 \end{pmatrix} = \begin{pmatrix} 15 \\ 26 \\ 17 \end{pmatrix}$

36. $\begin{pmatrix} 2 & 3 & -1 & 1 \\ 5 & 4 & 1 & 2 \\ 2 & 1 & 1 & 1 \end{pmatrix} \begin{pmatrix} x_1 \\ x_2 \\ x_3 \\ x_4 \end{pmatrix} = \begin{pmatrix} 20 \\ 35 \\ 12 \end{pmatrix}$ **37.** $\begin{pmatrix} 1 & -1 & 0 \\ 0 & 1 & 0 \\ 0 & 0 & 1 \end{pmatrix}$ **38.** $\begin{pmatrix} 0 & 1 & 0 \\ 1 & 0 & -1 \\ 0 & -2 & 1 \end{pmatrix}$

39. Let T be the "this year's" matrix and L be the "last year's" matrix so that the "growth" matrix G is $G = T - L$. Choosing T to be a 3×2 matrix with the rows representing the grandchildren Thomas, Richard, and Harriet in that order and the columns representing their heights and weights in that order,

$$T = \begin{pmatrix} 61 & 90 \\ 54 & 75 \\ 47 & 60 \end{pmatrix} \text{ and } L = \begin{pmatrix} 58 & 80 \\ 52 & 70 \\ 46 & 55 \end{pmatrix} \text{ so then } G = \begin{pmatrix} 3 & 10 \\ 2 & 5 \\ 1 & 5 \end{pmatrix}.$$

40. Let N be the "number of items needed" column matrix and P be the "price" matrix so that the "cost of her order" matrix C is $C = P \cdot N$. Choosing the rows of N to represent the items "jacket," "blouse," "skirt," and "slacks" in that

order, N is the 4×1 matrix $N = \begin{pmatrix} 200 \\ 300 \\ 250 \\ 175 \end{pmatrix}$. Then the columns of P must also represent the items "jacket," "blouse,"

"skirt," and "slacks" in that order, so P is the 2×4 matrix $P = \begin{pmatrix} 195 & 85 & 145 & 130 \\ 190 & 90 & 150 & 125 \end{pmatrix}$ and the rows represent the

prices of the East Coast designer and the Italian team in that order. Then $C = \begin{pmatrix} 123{,}500 \\ 124{,}375 \end{pmatrix}$. The cost of her order from

the East Coast designer is \$123,500 while from the Italian team it is \$124,375.

41. $\begin{pmatrix} 1 & 2 & 3 \\ 1 & 1 & 1 \\ 0 & 1 & 3 \end{pmatrix} \begin{pmatrix} -2 & 3 & 1 \\ 3 & -3 & -2 \\ -1 & 1 & 1 \end{pmatrix} = \begin{pmatrix} 1 & 0 & 0 \\ 0 & 1 & 0 \\ 0 & 0 & 1 \end{pmatrix}$ so this pair of matrices is a matrix and its inverse.

42. $\begin{pmatrix} -3 & 0 & 1 \\ 1 & 3 & 1 \\ -3 & 2 & 2 \end{pmatrix} \begin{pmatrix} -4 & -2 & 3 \\ 5 & 3 & -4 \\ -11 & -6 & 8 \end{pmatrix} = \begin{pmatrix} 1 & 0 & -1 \\ 0 & 1 & -1 \\ 0 & 0 & -1 \end{pmatrix}$ so this pair of matrices is not a matrix and its inverse.

43. $\begin{pmatrix} 3 & 0 & -1 \\ -2 & 0 & 1 \\ -6 & 1 & 3 \end{pmatrix}$ **44.** Singular matrix   **45.** $\begin{pmatrix} x_1 \\ x_2 \end{pmatrix} = \begin{pmatrix} 1 & -4 \\ -3 & 13 \end{pmatrix} \begin{pmatrix} 33 \\ 8 \end{pmatrix} = \begin{pmatrix} 1 \\ 5 \end{pmatrix}$ so $\begin{cases} x_1 = 1 \\ x_2 = 5 \end{cases}$

46. $\begin{pmatrix} x_1 \\ x_2 \end{pmatrix} = \begin{pmatrix} -5 & 8 \\ 2 & -3 \end{pmatrix} \begin{pmatrix} 25 \\ 16 \end{pmatrix} = \begin{pmatrix} 3 \\ 2 \end{pmatrix}$ so $\begin{cases} x_1 = 3 \\ x_2 = 2 \end{cases}$ **47.** $\begin{pmatrix} x_1 \\ x_2 \\ x_3 \end{pmatrix} = \begin{pmatrix} 1 & 1 & -3 \\ 0 & 1 & -1 \\ -2 & -3 & 8 \end{pmatrix} \begin{pmatrix} 11 \\ 7 \\ 5 \end{pmatrix} = \begin{pmatrix} 3 \\ 2 \\ -3 \end{pmatrix}$ so $\begin{cases} x_1 = 3 \\ x_2 = 2 \\ x_3 = -3 \end{cases}$

48. $\begin{pmatrix} x_1 \\ x_2 \\ x_3 \\ x_4 \end{pmatrix} = \begin{pmatrix} 1 & 0 & 0 & -1 \\ 0 & 1 & -2 & 0 \\ 0 & -2 & 6 & -1 \\ -1 & 0 & -1 & 2 \end{pmatrix} \begin{pmatrix} 7 \\ 10 \\ 6 \\ 4 \end{pmatrix} = \begin{pmatrix} 1 \\ 2 \\ -2 \\ 1 \end{pmatrix}$ so $\begin{cases} x_1 = 1 \\ x_2 = 2 \\ x_3 = -2 \\ x_4 = 1 \end{cases}$

49. The Kingman store can display 7 living room suites (2 in the window and 5 more only on the showroom floor) and 8 bedroom suites (3 in the window and 5 more only on the showroom floor); the Prescott store can display 11 living room suites (3 in the window and 8 more only on the showroom floor) and 10 bedroom suites (3 in the window and

7 more only on the showroom floor); and the Holbrook store can display 9 living room suites (3 in the window and 6 more only on the showroom floor) and 12 bedroom suites (2 in the window and 10 more only on the showroom floor).

50. Mr. Dahlman's taxes are $17,100 (federal), $7600 (state), and $3600 (city); Mrs. Farrell's taxes are $8550 (federal), $3800 (state), and $1800 (city); Ms. Mazlin's taxes are $13,680 (federal), $6080 (state), and $2880 (city); and Mr. Seidner's taxes are $25,650 (federal), $11,400 (state), and $5400 (city).

51. $\begin{pmatrix} 0.15 & 0.25 \\ 0.30 & 0.20 \end{pmatrix}$ for sectors A and B   **52.** $\begin{pmatrix} 0.10 & 0.20 & 0.30 & 0 \\ 0.20 & 0.10 & 0 & 0.20 \\ 0.30 & 0 & 0.10 & 0.30 \\ 0 & 0.20 & 0.30 & 0.10 \end{pmatrix}$ for sectors A, B, C, and D

53. 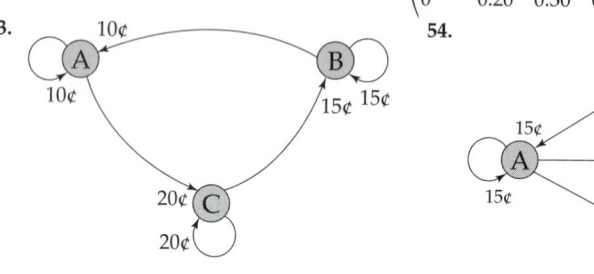   **54.**

55. $Y = \begin{pmatrix} 371 \\ 449 \\ 419 \end{pmatrix}$ **56.** $Y = \begin{pmatrix} 170 \\ 115 \\ 160 \\ 143 \end{pmatrix}$ **57.** $X = \begin{pmatrix} 364 \\ 436 \end{pmatrix}$ **58.** $X = \begin{pmatrix} 457 \\ 605 \\ 529 \end{pmatrix}$

59. Each division must produce $4 million.

60. The other sectors of the economy can use $841 million of domestic oil and $1948 million of foreign oil but all the military protection budget is consumed in the production of this output.

61. $\begin{pmatrix} 0.40 & 0.60 \\ 0.80 & 0.20 \end{pmatrix}$ **62.** $\begin{pmatrix} 0.10 & 0.90 & 0 \\ 0.40 & 0.30 & 0.30 \\ 0.40 & 0.40 & 0.20 \end{pmatrix}$ **63.** Ergodic   **64.** Neither   **65.** Regular

66. Ergodic   **67.** $X = (\frac{29}{130} \quad \frac{21}{65} \quad \frac{59}{130})$ **68.** $X = (\frac{1}{4} \quad \frac{1}{4} \quad \frac{1}{4} \quad \frac{1}{4})$

69. Of the manufacturer's 2655 dealers, 1305 have service departments rated "excellent" by their customers.

70. Out of the next 18 years, 7 will have terrific weather for growing corn, if the old timers are right.

71. $x = 4, y = 3$ **72.** $x = -3, y = 8$ **73.** Inconsistent, yet "almost" satisfied by $x = 5, y = 3$

74. Inconsistent, yet "almost" satisfied by $x = 2, y = -1$ **75.** $y = 7x + 9$ **76.** $y = 73x + 14$ **77.** $y = 13x + 13$

78. $y = 19x - 52$ **79.** They can expect sales of $4405.   **80.** She can expect 12 minor accidents.

EXERCISES 3.1 page 295

1. b   **3.** c   **5.** a   **7.** d   **9.** b

11. The vertices are (0, 0), (40, 0), and (0, 20). The region is bounded.

13. The vertices are (0, 0), (10, 0), (10, 30), and (0, 10). The region is bounded.

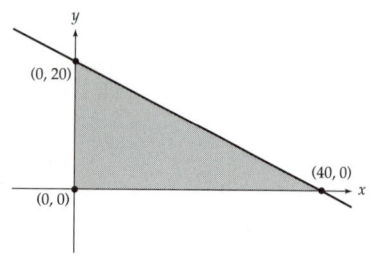

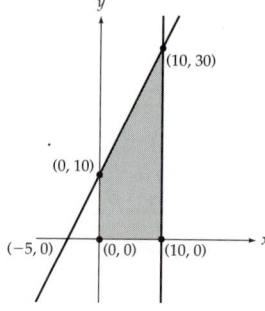

15. The vertices are (0, 0), (6, 0), (4, 2), and (0, 4). The region is bounded.

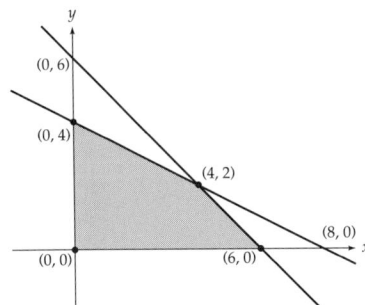

17. The vertices are (4, 0) and (0, 10). The region is unbounded.

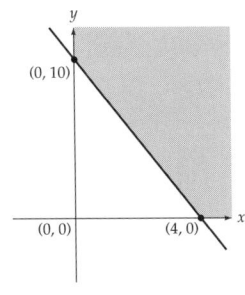

19. The vertices are (3, 4) and (0, 8). The region is unbounded.

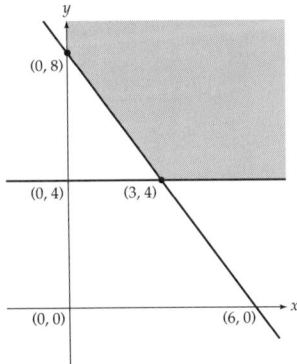

21. The vertices are (8, 0), (2, 6), and (0, 12). The region is unbounded.

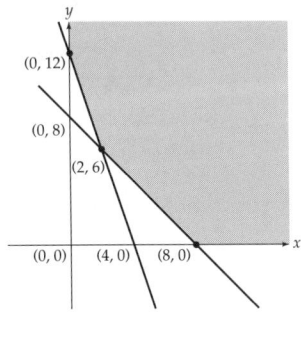

23. The vertices are (0, 0), (15, 0), (15, 5), (10, 10), and (0, 10). The region is bounded.

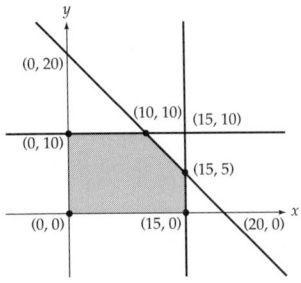

25. The vertices are (30, 0), (40, 0), (0, 80), and (0, 10). The region is bounded.

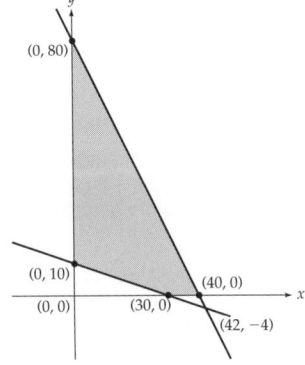

27. The vertices are (15, 0) and (0, 30). The region is unbounded.

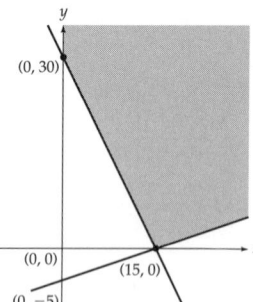

29. The vertices are (0, 0), (9, 0), (8, 2), (3, 7), and (0, 8). The region is bounded.

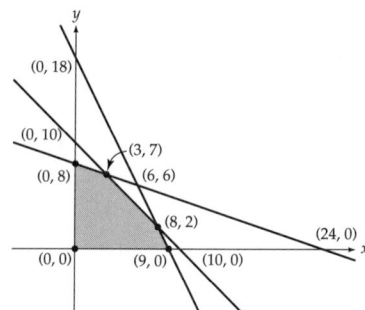

31. Let x be the number of goats and let y be the number of llamas.

$$\begin{cases} 2x + 5y \le 400 & \text{(Land)} \\ 100x + 80y \le 13{,}200 & \text{(Money)} \\ x \ge 0, y \ge 0 & \text{(Nonnegativity)} \end{cases}$$

The vertices are (0, 0), (132, 0), (100, 40), and (0, 80).

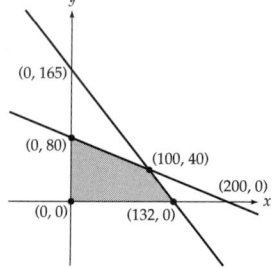

33. Let x be the number of dinghies and let y be the number of rowboats.

$$\begin{cases} 2x + 3y \le 120 & \text{(Metal work)} \\ 2x + 2y \le 100 & \text{(Painting)} \\ x \ge 0, y \ge 0 & \text{(Nonnegativity)} \end{cases}$$

The vertices are (0, 0), (50, 0), (30, 20), and (0, 40).

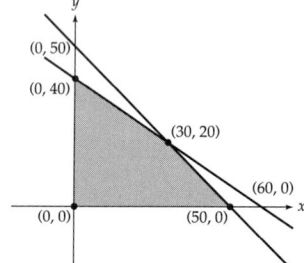

35. Let x be the number of servings of SugarSnaks and let y be the number of bags of Gobbl'Ems.

$$\begin{cases} 5x + 8y \le 80 & \text{(Fat)} \\ 125x + 250y \le 2250 & \text{(Calories)} \\ x \ge 0, y \ge 0 & \text{(Nonnegativity)} \end{cases}$$

The vertices are (0, 0), (16, 0), (8, 5), and (0, 9).

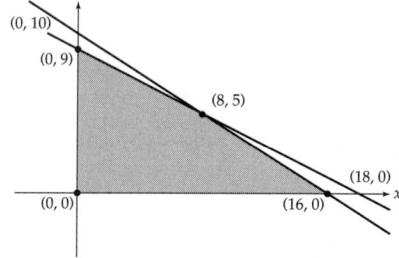

37. Let x be the number of hours the Ohio factory operates and let y be the number of hours the Pennsylvania factory operates.

$$\begin{cases} 4x + 4y \leq 64 & \text{(Sulfur dioxide)} \\ 5x + 3y \leq 60 & \text{(Particulates)} \\ x \geq 0, y \geq 0 & \text{(Nonnegativity)} \end{cases}$$

The vertices are $(0, 0)$, $(12, 0)$, $(6, 10)$, and $(0, 16)$.

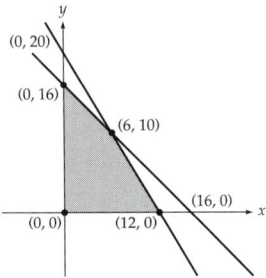

39. Let x be the amount of money (in millions of dollars) invested in stock funds and let y be the amount of money (in millions of dollars) invested in bond funds.

$$\begin{cases} x + y \leq 8 & \text{(Money)} \\ x \leq y & \text{(Limit risk)} \\ x \geq 0, y \geq 0 & \text{(Nonnegativity)} \end{cases}$$

The vertices are $(0, 0)$, $(4, 4)$, and $(0, 8)$.

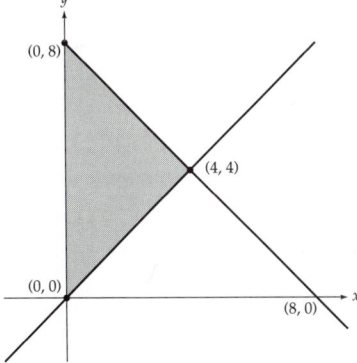

EXERCISES 3.2 page 311

1. The maximum is 45 (when $x = 15$, $y = 15$).   **3.** The minimum is 30 (when $x = 10$, $y = 0$).
5. The maximum is 15 (when $x = 15$, $y = 0$).   **7.** The minimum is 60 (when $x = 0$, $y = 60$).
9. There is no maximum because the region is unbounded in the positive y direction and P increases for increasing y values.
11. The maximum is 400 (when $x = 0$, $y = 10$).   **13.** The minimum is 150 (when $x = 10$, $y = 0$).
15. The maximum is 160 (when $x = 15$, $y = 20$).   **17.** The minimum is 232 (when $x = 6$, $y = 16$).
19. The maximum is 140 (when $x = 0$, $y = 20$).   **21.** The minimum is 180 (when $x = 30$, $y = 0$).
23. The maximum is 230 (when $x = 40$, $y = 10$).   **25.** The maximum is 1080 (when $x = 0$, $y = 90$).
27. The minimum is 885 (when $x = 33$, $y = 9$).   **29.** The minimum is 1545 (when $x = 45$, $y = 220$).
31. The rancher should raise 100 goats and 40 llamas to obtain the greatest possible profit of \$9600.
33. The company should manufacture 32 prams and 9 yawls to obtain the greatest possible profit of \$6420.
35. The county should operate the Norton incinerator 4 hours each day and the Wiseburg incinerator 6 hours each day to obtain the least cost of \$620.
37. Each bunny should receive 30 handfuls of greens and 16 drops of supplement each week to minimize the owner's costs at 76¢.
39. No grassland and 2400 acres of forest should be reclaimed this year to raise the greatest possible amount of \$360,000 in long-term leases for use in next year's reclamation efforts.

EXERCISES 3.3 page 334

1.

	x_1	x_2	x_3	s_1	s_2	
s_1	3	2	4	1	0	12
s_2	6	1	5	0	1	15
P	-8	-9	-7	0	0	0

3.

	x_1	x_2	s_1	s_2	s_3	
s_1	4	3	1	0	0	12
s_2	5	2	0	1	0	20
s_3	1	6	0	0	1	12
P	-13	-7	0	0	0	0

5.

	x_1	x_2	x_3	x_4	s_1	s_2	
s_1	2	1	1	3	1	0	6
s_2	1	4	-2	1	0	1	8
P	-5	2	-10	5	0	0	0

7.

	x_1	x_2	x_3	s_1	s_2	s_3	
s_1	8	1	4	1	0	0	32
s_2	3	5	7	0	1	0	30
s_3	6	2	9	0	0	1	28
P	-10	-20	-15	0	0	0	0

9.

	x_1	x_2	x_3	s_1	s_2	s_3	s_4	
s_1	1	2	3	1	0	0	0	45
s_2	6	5	4	0	1	0	0	40
s_3	7	8	9	0	0	1	0	63
s_4	12	11	10	0	0	0	1	60
P	-90	-80	-100	0	0	0	0	0

11. The smallest negative entry in the bottom row is -9, so the pivot column is column 3. The ratios are $\frac{4}{1} = 4$, $\frac{5}{1} = 5$, and $\frac{6}{3} = 2$, so the pivot row is row 3. The pivot element is the 3 in column 3 and row 3 of the simplex tableau.

	x_1	x_2	x_3	s_1	s_2	s_3	
s_1	5/3	1/3	0	1	0	$-1/3$	2
s_2	2/3	4/3	0	0	1	$-1/3$	3
x_3	1/3	$-1/3$	1	0	0	1/3	2
P	-4	-11	0	0	0	3	18

$R_1^{new} = R_1 - R_{pivot}^{new}$
$R_2^{new} = R_2 - R_{pivot}^{new}$
$R_{pivot}^{new} = R_{pivot}/3$
$R_4^{new} = R_4 + 9R_{pivot}^{new}$

13. The smallest negative entry in the bottom row is -8, so the pivot column is column 3. The ratios are $\frac{3}{1} = 3$, $\frac{4}{1} = 4$, and (omit), so the pivot row is row 1. The pivot element is the 1 in column 3 and row 1 of the simplex tableau.

	x_1	x_2	x_3	s_1	s_2	s_3	
x_3	1	0	1	1	0	0	3
s_2	-1	1	0	-1	1	0	1
s_3	1	1	0	0	0	1	5
P	2	-7	0	8	0	0	24

$R_{pivot}^{new} = R_{pivot}$
$R_2^{new} = R_2 - R_{pivot}^{new}$
$R_3^{new} = R_3$
$R_4^{new} = R_4 + 8R_{pivot}^{new}$

15. The smallest negative entry in the bottom row is -5, so the pivot column is column 2. The ratios are $\frac{12}{2} = 6$ and $\frac{8}{1} = 8$, so the pivot row is row 1. The pivot element is the 2 in column 2 and row 1 of the simplex tableau.

	x_1	x_2	x_3	x_4	s_1	s_2	
x_1	2	1	3	1	1/2	0	6
s_2	1	0	-1	0	$-1/2$	1	2
P	6	0	21	2	5/2	0	30

$R_{pivot}^{new} = R_{pivot}/2$
$R_2^{new} = R_2 - R_{pivot}^{new}$
$R_3^{new} = R_3 + 5R_{pivot}^{new}$

17. No pivot column because there are no negative entries in the bottom row. The solution has been found: the maximum is 90 when $x_1 = 10$, $x_2 = 15$, $x_3 = 0$, and $x_4 = 0$ (and $s_1 = 10$, $s_2 = 0$, and $s_3 = 0$).

19. The smallest negative entry in the bottom row is -10 so the pivot column is column 5. Since all of the entries in the pivot column are zero or negative, there is no pivot row. There is no maximum value.

21. The maximum is 28 when $x_1 = 0$, $x_2 = 14$.   **23.** The maximum is 30 when $x_1 = 0$, $x_2 = 0$, $x_3 = 6$, $x_4 = 0$.

25. The maximum is 2000 when $x_1 = 0$, $x_2 = 50$, $x_3 = 0$.

27. There is no maximum (the second pivot column is column 2 but there is no pivot row).

29. The maximum is 300 when $x_1 = 0$, $x_2 = 20$, $x_3 = 0$, $x_4 = 40$.   **31.** The maximum is 16 when $x_1 = 3$, $x_2 = 2$.

33. The maximum is 12,600 when $x_1 = 0$, $x_2 = 160$, $x_3 = 60$.   **35.** The maximum is 30 when $x_1 = 5$, $x_2 = 5$, $x_3 = 5$.

37. The maximum is 2000 when $x_1 = 0$, $x_2 = 24$, $x_3 = 32$.   **39.** The maximum is 460 when $x_1 = 15$, $x_2 = 10$, $x_3 = 20$.

41. The shop should rebuild no carburetors, 35 fuel pumps, and 20 alternators for a greatest possible profit of $690.

43. The recycling center should accept 300 crates of paper products and 500 crates of glass bottles to raise the greatest possible amount of $59 each week.

45. The company should process no agates, 49 trays of onyxes, and 7 trays of garnets each day to obtain the greatest possible profit of $581.

47. The politician should run 3 daytime ads, 7 prime time ads, and no late night ads to reach 47,000 voters, the most possible.

49. The farmer should plant 30 acres of corn, 90 acres of peanuts, and 120 acres of soybeans for the greatest possible profit of $43,500.

51. a. $5x + 2y = 70$ means $x = 14 - \frac{2}{5}y$. Substitution in the second equation gives $4(14 - \frac{2}{5}y) + 3y = 84$ so $\frac{7}{5}y = 28$. Then $y = 20$ and $x = 14 - \frac{2}{5}(20) = 6$.

b. Pivoting gives

x	y	
1	2/5	14
0	7/5	28

. The first row represents the equation $x + \frac{2}{5}y = 14$, which is equivalent to

$x = 14 - \frac{2}{5}y$. The second row represents the equation $\frac{7}{5}y = 28$, which we found before on the way to obtaining $y = 20$.

53. Yes.

55. For Exercise 19: Pivoting at [column 3, row 1], [column 4, row 2], [column 5, row 3], and [column 6, row 4], the simplex tableau becomes:

	x_1	x_2	s_1	s_2	s_3	s_4	
s_1	0	1	1	0	0	0	30
s_2	0	1	0	1	0	0	20
s_3	−1	1	0	0	1	0	10
s_4	0	1	0	0	0	1	25
P	−10	−15	0	0	0	0	0

That is,

Maximize $P = 10x_1 + 15x_2$

Subject to $\begin{cases} x_2 \le 30 \\ x_2 \le 20 \\ -x_1 + x_2 \le 10 \\ x_2 \le 25 \\ x_1 \ge 0, x_2 \ge 0 \end{cases}$

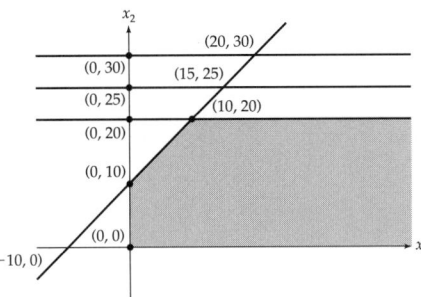

The region is unbounded to the right and the objective function increases in this direction. There is no maximum value.

For Exercise 20: Pivoting at [column 3, row 1], [column 4, row 2], [column 5, row 3], and [column 6, row 4], the simplex tableau becomes:

	x_1	x_2	s_1	s_2	s_3	s_4	
s_1	1	−1	1	0	0	0	15
s_2	1	0	0	1	0	0	20
s_3	1	−1	0	0	1	0	10
s_4	1	0	0	0	0	1	25
P	−20	−15	0	0	0	0	0

That is:

Maximize $P = 20x_1 + 15x_2$

Subject to $\begin{cases} x_1 - x_2 \le 15 \\ x_1 \le 20 \\ x_1 - x_2 \le 10 \\ x_1 \le 25 \\ x_1 \ge 0, x_2 \ge 0 \end{cases}$

The region is unbounded moving upward and the objective function increases in this direction. There is no maximum value.

57. a. The final tableau is

	x_1	x_2	x_3	x_4	s_1	s_2	s_3	
x_3	0	0	1	0	0	0	1	1
s_1	0	-8	0	30	1	-2	3	3
x_1	1	-24	0	6	0	2	1	1
P	0	2	0	21/2	0	3/2	5/4	5/4

. The maximum is $\frac{5}{4}$ when $x_1 = 1$, $x_2 = 0$, $x_3 = 1$, $x_4 = 0$.

b. The final tableau is

	x_1	x_2	x_3	x_4	s_1	s_2	s_3	
x_1	1	-6	0	6	0	2	1	1
x_3	0	0	1	0	0	0	1	1
s_1	0	$-1/2$	0	15/2	1	$-1/2$	3/4	3/4
P	0	1/2	0	21/2	0	3/2	5/4	5/4

. The maximum is $\frac{5}{4}$ when $x_1 = 1$, $x_2 = 0$, $x_3 = 1$, $x_4 = 0$.

59. No.   **61.** The maximum is 100 when $x_1 = 100$, $x_2 = 10$.

63. Maximize $P = \begin{pmatrix} 1 & 10 & 100 & 1000 \end{pmatrix} \begin{pmatrix} x_1 \\ x_2 \\ x_3 \\ x_4 \end{pmatrix}$

Subject to $\begin{pmatrix} 0 & 0 & 0 & 1 \\ 0 & 0 & 1 & 20 \\ 0 & 1 & 20 & 200 \\ 1 & 20 & 200 & 2000 \end{pmatrix} \begin{pmatrix} x_1 \\ x_2 \\ x_3 \\ x_4 \end{pmatrix} \le \begin{pmatrix} 1 \\ 100 \\ 10,000 \\ 1,000,000 \end{pmatrix}$ and $\begin{pmatrix} x_1 \\ x_2 \\ x_3 \\ x_4 \end{pmatrix} \ge 0.$

The maximum is 1,000,000 when $x_1 = 1,000,000$, $x_2 = 0$, $x_3 = 0$, $x_4 = 0$.

EXERCISES 3.4 page 355

1. Minimize $C = \begin{pmatrix} 60 & 100 & 300 \end{pmatrix} \begin{pmatrix} y_1 \\ y_2 \\ y_3 \end{pmatrix}$ subject to $\begin{pmatrix} 1 & 2 & 3 \\ 4 & 5 & 6 \end{pmatrix} \begin{pmatrix} y_1 \\ y_2 \\ y_3 \end{pmatrix} \ge \begin{pmatrix} 180 \\ 120 \end{pmatrix}$ and $\begin{pmatrix} y_1 \\ y_2 \\ y_3 \end{pmatrix} \ge 0$. The dual problem is maxi-

mize $P = \begin{pmatrix} 180 & 120 \end{pmatrix} \begin{pmatrix} x_1 \\ x_2 \end{pmatrix}$ subject to $\begin{pmatrix} 1 & 4 \\ 2 & 5 \\ 3 & 6 \end{pmatrix} \begin{pmatrix} x_1 \\ x_2 \end{pmatrix} \le \begin{pmatrix} 60 \\ 100 \\ 300 \end{pmatrix}$ and $\begin{pmatrix} x_1 \\ x_2 \end{pmatrix} \ge 0$ with initial simplex tableau

	x_1	x_2	s_1	s_2	s_3	
s_1	1	4	1	0	0	60
s_2	2	5	0	1	0	100
s_3	3	6	0	0	1	300
P	-180	-120	0	0	0	0

3. Minimize $C = (3 \quad 20)\begin{pmatrix} y_1 \\ y_2 \end{pmatrix}$ subject to $\begin{pmatrix} 3 & 2 \\ 1 & 4 \\ 3 & 4 \end{pmatrix}\begin{pmatrix} y_1 \\ y_2 \end{pmatrix} \geq \begin{pmatrix} 150 \\ 100 \\ 228 \end{pmatrix}$ and $\begin{pmatrix} y_1 \\ y_2 \end{pmatrix} \geq 0$. The dual problem is maximize

$P = (150 \quad 100 \quad 228)\begin{pmatrix} x_1 \\ x_2 \\ x_3 \end{pmatrix}$ subject to $\begin{pmatrix} 3 & 1 & 3 \\ 2 & 4 & 4 \end{pmatrix}\begin{pmatrix} x_1 \\ x_2 \\ x_3 \end{pmatrix} \leq \begin{pmatrix} 3 \\ 20 \end{pmatrix}$ and $\begin{pmatrix} x_1 \\ x_2 \\ x_3 \end{pmatrix} \geq 0$ with initial simplex tableau

	x_1	x_2	x_3	s_1	s_2	
s_1	3	1	3	1	0	3
s_2	2	4	4	0	1	20˙
P	-150	-100	-228	0	0	0

5. Minimize $C = (84 \quad 21)\begin{pmatrix} y_1 \\ y_2 \end{pmatrix}$ subject to $\begin{pmatrix} 3 & 1 \\ 4 & -1 \end{pmatrix}\begin{pmatrix} y_1 \\ y_2 \end{pmatrix} \geq \begin{pmatrix} 21 \\ 0 \end{pmatrix}$ and $\begin{pmatrix} y_1 \\ y_2 \end{pmatrix} \geq 0$. The dual problem is maximize

$P = (21 \quad 0)\begin{pmatrix} x_1 \\ x_2 \end{pmatrix}$ subject to $\begin{pmatrix} 3 & 4 \\ 1 & -1 \end{pmatrix}\begin{pmatrix} x_1 \\ x_2 \end{pmatrix} \leq \begin{pmatrix} 84 \\ 21 \end{pmatrix}$ and $\begin{pmatrix} x_1 \\ x_2 \end{pmatrix} \geq 0$ with initial simplex tableau

	x_1	x_2	s_1	s_2	
s_1	3	4	1	0	84
s_2	1	-1	0	1	21˙
P	-21	0	0	0	0

7. Minimize $C = (15 \quad 20 \quad 5)\begin{pmatrix} y_1 \\ y_2 \\ y_3 \end{pmatrix}$ subject to $\begin{pmatrix} -1 & 1 & 2 \\ 1 & 2 & 1 \end{pmatrix}\begin{pmatrix} y_1 \\ y_2 \\ y_3 \end{pmatrix} \geq \begin{pmatrix} -30 \\ 30 \end{pmatrix}$ and $\begin{pmatrix} y_1 \\ y_2 \\ y_3 \end{pmatrix} \geq 0$. The dual problem is maxi-

mize $P = (-30 \quad 30)\begin{pmatrix} x_1 \\ x_2 \end{pmatrix}$ subject to $\begin{pmatrix} -1 & 1 \\ 1 & 2 \\ 2 & 1 \end{pmatrix}\begin{pmatrix} x_1 \\ x_2 \end{pmatrix} \leq \begin{pmatrix} 15 \\ 20 \\ 5 \end{pmatrix}$ and $\begin{pmatrix} x_1 \\ x_2 \end{pmatrix} \geq 0$ with initial simplex tableau

	x_1	x_2	s_1	s_2	s_3	
s_1	-1	1	1	0	0	15
s_2	1	2	0	1	0	20˙
s_3	2	1	0	0	1	5
P	30	-30	0	0	0	0

9. Minimize $C = (105 \quad 40)\begin{pmatrix} y_1 \\ y_2 \end{pmatrix}$ subject to $\begin{pmatrix} 7 & 5 \\ 3 & 1 \\ 5 & 2 \end{pmatrix}\begin{pmatrix} y_1 \\ y_2 \end{pmatrix} \geq \begin{pmatrix} 70 \\ 45 \\ 80 \end{pmatrix}$ and $\begin{pmatrix} y_1 \\ y_2 \end{pmatrix} \geq 0$. The dual problem is maximize

$P = (70 \quad 45 \quad 80)\begin{pmatrix} x_1 \\ x_2 \\ x_3 \end{pmatrix}$ subject to $\begin{pmatrix} 7 & 3 & 5 \\ 5 & 1 & 2 \end{pmatrix}\begin{pmatrix} x_1 \\ x_2 \\ x_3 \end{pmatrix} \leq \begin{pmatrix} 105 \\ 40 \end{pmatrix}$ and $\begin{pmatrix} x_1 \\ x_2 \\ x_3 \end{pmatrix} \geq 0$ with initial simplex tableau

	x_1	x_2	x_3	s_1	s_2	
s_1	7	3	5	1	0	105
s_2	5	1	2	0	1	40
P	-70	-45	-80	0	0	0

11. The minimum is 900 when $y_1 = 0$, $y_2 = 40$, $y_3 = 10$.   **13.** The minimum is 43 when $y_1 = 7$, $y_2 = 3$.
15. The minimum is 300 when $y_1 = 20$, $y_2 = 0$.   **17.** The minimum is 45 when $y_1 = 0$, $y_2 = 15$.
19. The minimum is 10 when $y_1 = 0$, $y_2 = 10$.   **21.** The minimum is 4950 when $y_1 = 30$, $y_2 = 45$.
23. The minimum is 98 when $y_1 = 2$, $y_2 = 2$, $y_3 = 0$.   **25.** There is no minimum.
27. The minimum is 90 when $y_1 = 0$, $y_2 = 10$, $y_3 = 30$, $y_4 = 0$.   **29.** The minimum is 2376 when $y_1 = 6$, $y_2 = 12$, $y_3 = 6$.
31. The athlete should use 8 Bulk-Up Bars and 4 cans of Power Drink to receive the needed fat and protein at the least possible cost of $5.64.

33. The office manager should buy 45 packages from Jack's Office Supplies and 30 packages from John's Discount to restock the store room at the least possible cost of $1230.

35. The project engineer should use 350 heavy-duty dump truck loads of dirt, 300 heavy-duty dump truck loads of crushed rock, no regular dump truck loads of dirt, and 150 regular dump truck loads of crushed rock to get the project finished on time at the least possible cost of $66,000.

37. The farmer should buy 8000 pounds of Miracle Mix and 18,000 pounds of the store brand to meet the fertilizer needs of his field at the least possible cost of $2640.

39. The Kentucky warehouse should ship 200 cartons to Kansas, 200 cartons to Texas, and none to Oregon, and the Utah warehouse should ship none to Kansas, 100 cartons to Texas, and 100 cartons to Oregon to incur the smallest possible shipping cost of $1600.

45. The final tableau shows that the solutions of the minimum and maximum problems are both 3390 with the variables taking the values:

Minimum problem:	$y_1 = 7$	$y_2 = 3$	$y_3 = 1$	$y_4 = 0$	$t_1 = 0$	$t_2 = 0$	$t_3 = 0$
Maximum problem:	$s_1 = 0$	$s_2 = 0$	$s_3 = 0$	$s_4 = 75$	$x_1 = 60$	$x_2 = 45$	$x_3 = 30$

Thus one of each pair x_1 and t_1, x_2 and t_2, x_3 and t_3, s_1 and y_1, s_2 and y_2, s_3 and y_3, and s_4 and y_4 is zero.

EXERCISES 3.5 page 370

1. Maximize $P = (15 \quad 20 \quad 18) \begin{pmatrix} x_1 \\ x_2 \\ x_3 \end{pmatrix}$ subject to $\begin{pmatrix} 3 & 2 & 8 \\ -5 & -1 & -6 \end{pmatrix} \begin{pmatrix} x_1 \\ x_2 \\ x_3 \end{pmatrix} \le \begin{pmatrix} 96 \\ -30 \end{pmatrix}$ and $\begin{pmatrix} x_1 \\ x_2 \\ x_3 \end{pmatrix} \ge 0$ with initial simplex

tableau

	x_1	x_2	x_3	s_1	s_2	
s_1	3	2	8	1	0	96
s_2	-5	-1	-6	0	1	-30
P	-15	-20	-18	0	0	0

. The dual pivot element is the -1 in row 2 and column 2 of the

simplex tableau.

3. Maximize $P = (6 \quad 4 \quad 6) \begin{pmatrix} x_1 \\ x_2 \\ x_3 \end{pmatrix}$ subject to $\begin{pmatrix} -2 & -1 & -3 \\ -1 & -1 & -2 \end{pmatrix} \begin{pmatrix} x_1 \\ x_2 \\ x_3 \end{pmatrix} \le \begin{pmatrix} -30 \\ -20 \end{pmatrix}$ and $\begin{pmatrix} x_1 \\ x_2 \\ x_3 \end{pmatrix} \ge 0$ with initial simplex

tableau

	x_1	x_2	x_3	s_1	s_2	
s_1	-2	-1	-3	1	0	-30
s_2	-1	-1	-2	0	1	-20
P	-6	-4	-6	0	0	0

. The dual pivot element is the -1 in row 1 and column 2 of the simplex

tableau.

5. Maximize $P = (20 \quad 30 \quad 10) \begin{pmatrix} x_1 \\ x_2 \\ x_3 \end{pmatrix}$ subject to $\begin{pmatrix} -1 & -1 & -1 \\ 1 & 2 & 3 \\ 1 & 2 & 1 \end{pmatrix} \begin{pmatrix} x_1 \\ x_2 \\ x_3 \end{pmatrix} \le \begin{pmatrix} -8 \\ 30 \\ 18 \end{pmatrix}$ and $\begin{pmatrix} x_1 \\ x_2 \\ x_3 \end{pmatrix} \ge 0$ with initial simplex

tableau

	x_1	x_2	x_3	s_1	s_2	s_3	
s_1	-1	-1	-1	1	0	0	-8
s_2	1	2	3	0	1	0	30
s_3	1	2	1	0	0	1	18
P	-20	-30	-10	0	0	0	0

. The dual pivot element is the -1 in row 1 and column 2 of the

simplex tableau.

7. Maximize $P = (3 \ \ 2 \ \ 5 \ \ 4) \begin{pmatrix} x_1 \\ x_2 \\ x_3 \\ x_4 \end{pmatrix}$ subject to $\begin{pmatrix} -1 & -1 & -1 & -1 \\ -2 & -3 & -2 & -1 \\ 4 & 2 & 1 & 2 \end{pmatrix} \begin{pmatrix} x_1 \\ x_2 \\ x_3 \\ x_4 \end{pmatrix} \leq \begin{pmatrix} -30 \\ -20 \\ 80 \end{pmatrix}$ and $\begin{pmatrix} x_1 \\ x_2 \\ x_3 \\ x_4 \end{pmatrix} \geq 0$ with initial sim-

plex tableau

	x_1	x_2	x_3	x_4	s_1	s_2	s_3	
s_1	-1	-1	-1	-1	1	0	0	-30
s_2	-2	-3	-2	-1	0	1	0	-20
s_3	4	2	1	2	0	0	1	80
P	-3	-2	-5	-4	0	0	0	0

. The dual pivot element is the -1 in row 1 and column 3 of

the simplex tableau.

9. Maximize $P = (2 \ \ 2 \ \ 1) \begin{pmatrix} x_1 \\ x_2 \\ x_3 \end{pmatrix}$ subject to $\begin{pmatrix} -1 & -2 & -1 \\ -2 & -1 & -1 \\ 5 & 3 & 2 \\ 3 & 1 & 2 \end{pmatrix} \begin{pmatrix} x_1 \\ x_2 \\ x_3 \end{pmatrix} \leq \begin{pmatrix} -40 \\ -50 \\ 120 \\ 150 \end{pmatrix}$ and $\begin{pmatrix} x_1 \\ x_2 \\ x_3 \end{pmatrix} \geq 0$ with intial simplex

tableau

	x_1	x_2	x_3	s_1	s_2	s_3	s_4	
s_1	-1	-2	-1	1	0	0	0	-40
s_2	-2	-1	-1	0	1	0	0	-50
s_3	5	3	2	0	0	1	0	120
s_4	3	1	2	0	0	0	1	150
P	-2	-2	-1	0	0	0	0	0

. The dual pivot element is the -1 in row 2 and column 2

of the simplex tableau.

11. The maximum is 80 when $x_1 = 10$, $x_2 = 40$.   13. The maximum is 1200 when $x_1 = 0$, $x_2 = 60$, $x_3 = 0$.

15. There is no maximum.   17. The maximum is 270 when $x_1 = 2$, $x_2 = 20$, $x_3 = 0$.

19. The maximum is 13 when $x_1 = 8$, $x_2 = 5$.   21. There is no maximum.

23. The maximum is 720 when $x_1 = 0$, $x_2 = 0$, $x_3 = 60$, $x_4 = 0$.

25. The maximum is 808 when $x_1 = 0$, $x_2 = 14$, $x_3 = 0$, $x_4 = 24$.

27. The maximum is 30 when $x_1 = 10$, $x_2 = 40$, $x_3 = 0$.   29. The maximum is 240 when $x_1 = 0$, $x_2 = 0$, $x_3 = 0$, $x_4 = 30$.

31. The retired couple should invest $15,000 in certificates of deposit and $5000 in treasury bonds.

33. The manager should place 5 newspaper ads, 15 radio commercials, and 5 TV spots to reach 295,000 potential customers, the greatest possible number.

35. The furniture shop should manufacture 20 desks, 10 tables, and 80 chairs to obtain the greatest possible profit of $7620.

37. The farmer should grow 100 acres of wheat, 300 acres of barley, and 100 acres of oats to obtain the greatest possible profit of $21,500.

39. The electric power plant should purchase 40,000 tons of low-sulfur coal and 50,000 tons of high-sulfur coal to obtain the most energy of 2,300,000 million BTUs.

41. **a.** The minimum of C is 12 when $x_1 = 4$, $x_2 = 0$, while the maximum of P is -12, also when $x_1 = 4$, $x_2 = 0$. Yes.

b. The simplex tableaux for the dual of this problem are

	x_1	x_2	s_1	s_2	
s_1	-1	2	1	0	3
s_2	-1	1	0	1	4
P	10	-8	0	0	0

and then

	x_1	x_2	s_1	s_2	
x_2	$-1/2$	1	$1/2$	0	$3/2$
s_2	$-1/2$	0	$-1/2$	1	$5/2$
P	6	0	4	0	12

.

c. The simplex tableaux for this nonstandard problem are

	x_1	x_2	s_1	s_2	
s_1	1	1	1	0	10
s_2	-2	-1	0	1	-8
P	3	4	0	0	0

and then

	x_1	x_2	s_1	s_2	
s_1	0	1/2	1	1/2	6
x_1	1	1/2	0	−1/2	4
P	0	5/2	0	3/2	−12

d. Yes; yes and yes

47. The maximum is 230 when $x_1 = 0$, $x_2 = 20$, $x_3 = 20$, $x_4 = 10$.   **49.** The maximum is 12 when $x_1 = 0$, $x_2 = 4$.

CHAPTER 3 REVIEW EXERCISES page 379

1.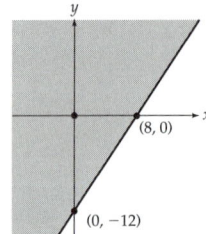

2.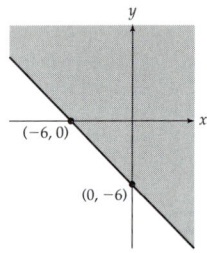

3. The vertices are (0, 0), (20, 0), and (0, 20). The region is bounded.

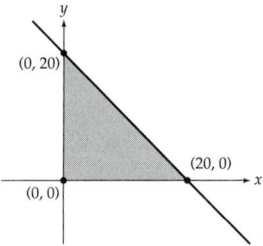

4. The vertices are (10, 0) and (0, 5). The region is unbounded.

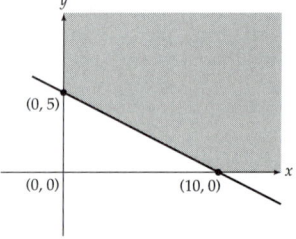

5. The vertices are (0, 0), (10, 0), (5, 10), and (0, 15). The region is bounded.

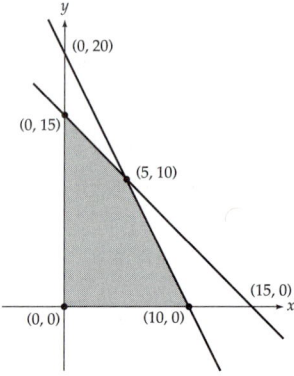

6. The vertices are (8, 0), (4, 4), and (0, 12). The region is unbounded.

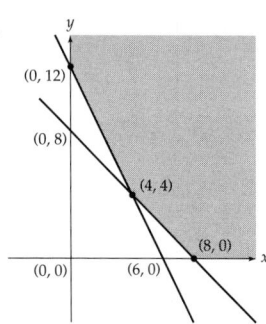

7. The vertices are (2, 0), (8, 0), (4, 6), and (2, 8). The region is bounded.

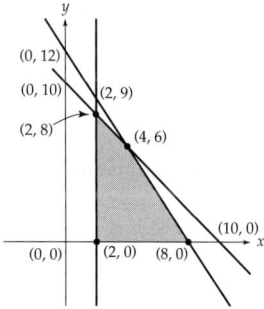

8. The vertices are (0, 0), (9, 0), (9, 1), (3, 7), and (0, 4). The region is bounded.

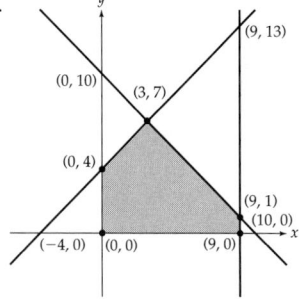

9. Let x be the number of Irish setters and let y be the number of Labrador retrievers.

$$\begin{cases} \frac{1}{2}x + \frac{3}{4}y \leq 6 & \text{(Time in hours)} \\ 8x + 8y \leq 80 & \text{(Dog treats)} \\ x \geq 0, y \geq 0 & \text{(Nonnegativity)} \end{cases}$$

The vertices are (0, 0), (10, 0), (6, 4), and (0, 8).

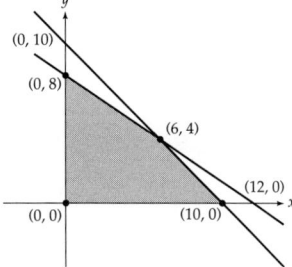

10. Let x be the number of pounds of SongBird brand bird seed and let y be the number of pounds of MeadowMix brand bird seed.

$$\begin{cases} 2x + 4y \geq 104 & \text{(Sunflower hearts)} \\ 3x + 2y \geq 84 & \text{(Crushed peanuts)} \\ x \geq 0, y \geq 0 & \text{(Nonnegativity)} \end{cases}$$

The vertices are $(52, 0)$, $(16, 18)$, and $(0, 42)$.

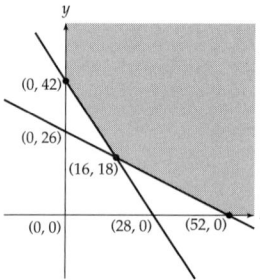

11. The maximum is 40 when $x = 0$, $y = 10$.   **12.** The minimum is 10 when $x = 0$, $y = 5$.
13. There is no maximum because the region is unbounded to the right and upward, and P increases in these directions.
14. The minimum is 60 when $x = 30$, $y = 0$.   **15.** The maximum is 240 when $x = 12$, $y = 0$.
16. The minimum is 2400 when $x = 48$, $y = 24$.   **17.** The maximum is 350 when $x = 4$, $y = 9$.
18. The minimum is 140 when $x = 20$, $y = 0$.
19. The shop should make 24 wall clock cases and 15 mantle clock cases to obtain the greatest possible profit of $9000.
20. The cat's nutritional needs can be met by 30 ounces of canned food and 4 ounces of dry food at a least cost of $1.82.

21.

	x_1	x_2	x_3	s_1	s_2	
s_1	5	2	3	1	0	30
s_2	2	3	4	0	1	24
P	-4	-10	-9	0	0	0

22.

	x_1	x_2	s_1	s_2	s_3	
s_1	1	1	1	0	0	10
s_2	2	1	0	1	0	14
s_3	1	-1	0	0	1	4
P	-5	-7	0	0	0	0

23. The smallest negative entry in the bottom row is -8, so the pivot column is column 3. The ratios are $\frac{6}{1} = 6$, $\frac{8}{2} = 4$, and (omit), so the pivot row is row 2. The pivot element is the 2 in column 3 and row 2 of the simplex tableau.

	x_1	x_2	x_3	s_1	s_2	s_3		
s_1	$-1/2$	$1/2$	0	1	$-1/2$	0	2	$R_1^{new} = R_1 - R_{pivot}^{new}$
x_3	$3/2$	$1/2$	1	0	$1/2$	0	4	$R_{pivot}^{new} = R_{pivot}/2$
s_3	$11/2$	$5/2$	0	0	$3/2$	1	18	$R_3^{new} = R_3 + 3R_{pivot}^{new}$
P	6	8	0	0	4	0	32	$R_4^{new} = R_4 + 8R_{pivot}^{new}$

24. No pivot column because there are no negative entries in the bottom row. The solution has been found: the maximum is 150 when $x_1 = 0$ and $x_2 = 10$ (and $s_1 = 8$, $s_2 = 42$, and $s_3 = 0$).
25. The maximum is 96 when $x_1 = 0$, $x_2 = 12$, $x_3 = 0$.   **26.** The maximum is 90 when $x_1 = 0$, $x_2 = 9$, $x_3 = 0$.
27. The maximum is 37 when $x_1 = 2$, $x_2 = 3$, $x_3 = 0$, $x_4 = 0$.   **28.** The maximum is 61 when $x_1 = 8$, $x_2 = 0$, $x_3 = 7$.
29. The publisher should print no trade edition copies, no book club edition copies, and 250,000 paperback edition copies for the greatest possible profit of $1,000,000.
30. The plastics factory should produce 20 batches of toy racing cars, no toy jet airplanes, and 10 batches of toy speed boats for the greatest possible profit of $3500.

31. Minimize $C = (10 \quad 15 \quad 20) \begin{pmatrix} y_1 \\ y_2 \\ y_3 \end{pmatrix}$ subject to $\begin{pmatrix} 2 & -1 & 1 \\ 1 & 1 & 3 \end{pmatrix} \begin{pmatrix} y_1 \\ y_2 \\ y_3 \end{pmatrix} \geq \begin{pmatrix} 40 \\ 30 \end{pmatrix}$ and $\begin{pmatrix} y_1 \\ y_2 \\ y_3 \end{pmatrix} \geq 0$ is a standard problem because $\begin{pmatrix} 10 \\ 15 \\ 20 \end{pmatrix} \geq 0$. The dual problem is maximize $P = (40 \quad 30)\begin{pmatrix} x_1 \\ x_2 \end{pmatrix}$ subject to $\begin{pmatrix} 2 & 1 \\ -1 & 1 \\ 1 & 3 \end{pmatrix}\begin{pmatrix} x_1 \\ x_2 \end{pmatrix} \leq \begin{pmatrix} 10 \\ 15 \\ 20 \end{pmatrix}$ and

$\begin{pmatrix} x_1 \\ x_2 \end{pmatrix} \geq 0$ with initial simplex tableau

	x_1	x_2	s_1	s_2	s_3	
s_1	2	1	1	0	0	10
s_2	-1	1	0	1	0	15
s_3	1	3	0	0	1	20
P	-40	-30	0	0	0	0

32. Minimize $C = (30 \quad 20)\begin{pmatrix} y_1 \\ y_2 \end{pmatrix}$ subject to $\begin{pmatrix} 5 & 2 \\ 3 & 4 \\ -5 & -4 \end{pmatrix}\begin{pmatrix} y_1 \\ y_2 \end{pmatrix} \geq \begin{pmatrix} 210 \\ 252 \\ -380 \end{pmatrix}$ and $\begin{pmatrix} y_1 \\ y_2 \end{pmatrix} \geq 0$ is a standard problem because

$\begin{pmatrix} 30 \\ 20 \end{pmatrix} \geq 0$. The dual problem is maximize $P = (210 \quad 252 \quad -380)\begin{pmatrix} x_1 \\ x_2 \\ x_3 \end{pmatrix}$ subject to $\begin{pmatrix} 5 & 3 & -5 \\ 2 & 4 & -4 \end{pmatrix}\begin{pmatrix} x_1 \\ x_2 \\ x_3 \end{pmatrix} \leq \begin{pmatrix} 30 \\ 20 \end{pmatrix}$ and

$\begin{pmatrix} x_1 \\ x_2 \\ x_3 \end{pmatrix} \geq 0$ with initial simplex tableau

	x_1	x_2	x_3	s_1	s_2	
s_1	5	3	-5	1	0	30
s_2	2	4	-4	0	1	20
P	-210	-252	380	0	0	0

33. Minimize $C = (42 \quad 36)\begin{pmatrix} y_1 \\ y_2 \end{pmatrix}$ subject to $\begin{pmatrix} 7 & 4 \\ 1 & 1 \end{pmatrix}\begin{pmatrix} y_1 \\ y_2 \end{pmatrix} \geq \begin{pmatrix} 84 \\ 18 \end{pmatrix}$ and $\begin{pmatrix} y_1 \\ y_2 \end{pmatrix} \geq 0$ is a standard problem because

$\begin{pmatrix} 42 \\ 36 \end{pmatrix} \geq 0$. The dual problem is maximize $P = (84 \quad 18)\begin{pmatrix} x_1 \\ x_2 \end{pmatrix}$ subject to $\begin{pmatrix} 7 & 1 \\ 4 & 1 \end{pmatrix}\begin{pmatrix} x_1 \\ x_2 \end{pmatrix} \leq \begin{pmatrix} 42 \\ 36 \end{pmatrix}$ and $\begin{pmatrix} x_1 \\ x_2 \end{pmatrix} \geq 0$

with initial simplex tableau

	x_1	x_2	s_1	s_2	
s_1	7	1	1	0	42
s_2	4	1	0	1	36
P	-84	-18	0	0	0

34. Minimize $C = (130 \quad 40 \quad 98)\begin{pmatrix} y_1 \\ y_2 \\ y_3 \end{pmatrix}$ subject to $\begin{pmatrix} 3 & 1 & 2 \\ 4 & 1 & 2 \\ 3 & 1 & 3 \end{pmatrix}\begin{pmatrix} y_1 \\ y_2 \\ y_3 \end{pmatrix} \geq \begin{pmatrix} 51 \\ 60 \\ 57 \end{pmatrix}$ and $\begin{pmatrix} y_1 \\ y_2 \\ y_3 \end{pmatrix} \geq 0$ is a standard problem because

$\begin{pmatrix} 130 \\ 40 \\ 98 \end{pmatrix} \geq 0$. The dual problem is maximize $P = (51 \quad 60 \quad 57)\begin{pmatrix} x_1 \\ x_2 \\ x_3 \end{pmatrix}$ subject to $\begin{pmatrix} 3 & 4 & 3 \\ 1 & 1 & 1 \\ 2 & 2 & 3 \end{pmatrix}\begin{pmatrix} x_1 \\ x_2 \\ x_3 \end{pmatrix} \leq \begin{pmatrix} 130 \\ 40 \\ 98 \end{pmatrix}$ and

$\begin{pmatrix} x_1 \\ x_2 \\ x_3 \end{pmatrix} \geq 0$ with initial simplex tableau

	x_1	x_2	x_3	s_1	s_2	s_3	
s_1	3	4	3	1	0	0	130
s_2	1	1	1	0	1	0	40
s_3	2	2	3	0	0	1	98
P	-51	-60	-57	0	0	0	0

35. The minimum is 192 when $y_1 = 8$, $y_2 = 5$, $y_3 = 0$.   **36.** The minimum is 510 when $y_1 = 11$, $y_2 = 5$.
37. There is no solution.   **38.** The minimum is 60 when $y_1 = 0$, $y_2 = 12$, $y_3 = 0$.
39. The student should buy no single pens, 3 packages of ink cartridges, and 3 "writer's combo" packages to spend the least possible amount of $6.24.
40. The pasta company should purchase 18 small-capacity machines and 8 large-capacity machines to expand its linguini production at the least possible cost of $138,000.

41. Maximize $P = (80 \quad 30)\begin{pmatrix} x_1 \\ x_2 \end{pmatrix}$ subject to $\begin{pmatrix} 4 & 1 \\ 2 & 3 \\ -1 & -1 \end{pmatrix}\begin{pmatrix} x_1 \\ x_2 \end{pmatrix} \leq \begin{pmatrix} 40 \\ 60 \\ -10 \end{pmatrix}$ and $\begin{pmatrix} x_1 \\ x_2 \end{pmatrix} \geq 0$ with initial simplex tableau

	x_1	x_2	s_1	s_2	s_3	
s_1	4	1	1	0	0	40
s_2	2	3	0	1	0	60
s_3	-1	-1	0	0	1	-10
P	-80	-30	0	0	0	0

60. The dual pivot element is the -1 in row 3 and column 1 of the simplex tableau.

42. Maximize $P = (4 \quad 8 \quad 6 \quad 10)\begin{pmatrix} x_1 \\ x_2 \\ x_3 \\ x_4 \end{pmatrix}$ subject to $\begin{pmatrix} 2 & 3 & -3 & 1 \\ -7 & 5 & -1 & -5 \end{pmatrix}\begin{pmatrix} x_1 \\ x_2 \\ x_3 \\ x_4 \end{pmatrix} \leq \begin{pmatrix} 30 \\ -35 \end{pmatrix}$ and $\begin{pmatrix} x_1 \\ x_2 \\ x_3 \\ x_4 \end{pmatrix} \geq 0$ with initial

simplex tableau

	x_1	x_2	x_3	x_4	s_1	s_2	
s_1	2	3	-3	1	1	0	30
s_2	-7	5	-1	-5	0	1	-35
P	-4	-8	-6	-10	0	0	0

The dual pivot element is the -1 in row 2 and column 3 of

the simplex tableau.
43. The maximum is 40 when $x_1 = 0$, $x_2 = 10$.   **44.** The maximum is 280 when $x_1 = 35$, $x_2 = 0$, $x_3 = 0$.
45. The maximum is 48 when $x_1 = 8$, $x_2 = 0$.   **46.** The maximum is 300 when $x_1 = 25$, $x_2 = 0$, $x_3 = 0$.
47. The maximum is 59 when $x_1 = 4$, $x_2 = 5$.   **48.** The maximum is 300 when $x_1 = 0$, $x_2 = 0$, $x_3 = 30$.
49. The sawmill should produce 6 thousand board-feet of rough-cut lumber and 4 thousand board-feet of finished-grade boards each day to obtain the greatest possible profit of $1120.
50. The money manager should invest $100,000 in U.S. bonds and $50,000 in Canadian stocks (and nothing in U.S. stocks and in Canadian bonds) to obtain the greatest possible return of $10,500.

EXERCISES 4.1 page 396

1. 27   **3.** 60   **5.** 44   **7.** 31   **9.** 17   **11.** $A \cap B^c \cap C^c$ **13.** $A^c \cap B^c \cap C$ **15.** $A \cap B \cap C^c$ **17.** $A \cap B \cap C$
19. 60   **21.** 8   **23.** 17,160; 154,440; 1,235,520   **25.** 10; 30,240; 3,628,800   **27.** 15   **29.** 35
31. 330; 462; 462; 330   **33.** 12; 66; 924; 12; 1   **35.** 17   **37.** 200   **39.** $25 \cdot 9 \cdot 9 = 2025$
41. $15^8 = 2,562,890,625$ **43.** $10^4 = 10,000$ **45.** $_{12}P_4 = 11,880$ **47.** $_8P_8 = 40,320$ **49.** $_{12}C_5 = 792$
51. $_{15}C_9 = 5005$ **53.** $5^35^4 = 78,125$; $5 \cdot 4 \cdot 3 \cdot 5 \cdot 4 \cdot 3 \cdot 2 = 7200$ **55.** $(_{10}C_4)(_{12}C_4) = 103,950$
57.

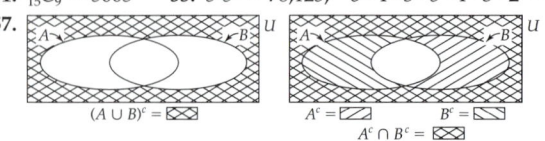

$(A \cup B)^c = $ ▨ $A^c = $ ▨ $B^c = $ ▨
$A^c \cap B^c = $ ▨
Note that the ▨ shadings agree.

59. $n((W \cup F)^c) = 90$, $n(W^c \cap F^c) = 90$ **61.** $_nP_n = n(n-1) \cdots (n-n+1) = n(n-1) \cdots 1 = n!$
63. $_nP_r = n(n-1) \cdots (n-r+1) = \dfrac{n(n-1) \cdots (n-r+1)}{1} \cdot \dfrac{(n-r) \cdots 1}{(n-r) \cdots 1} = \dfrac{n(n-1) \cdots 1}{(n-r) \cdots 1} = \dfrac{n!}{(n-r)!}$
65. $70! = 70 \cdot 69! \approx 70 \cdot 1.7 \cdot 10^{98} = 119 \cdot 10^{98} > 10^2 \cdot 10^{98} = 10^{100}$
so $70! > 10^{100}$, exceeding the maximum number size for this calculator.
67.   **69.**

A   B   C   Windbreaker   Ski Jackets   Overcoats

1 2 3 4   1 2 3 4   1 2 3 4   Red Blue   Red Blue   Red Blue
Therefore, 12 permits.   Therefore, 6 kinds.

71. No: 26 lines each leading to 10 more is impractical for a tree.

EXERCISES 4.2 page 410

1. $\{(A, B), (A, C), (A, D), (B, C), (B, D), (C, D)\}$; $\{(A, B), (A, D), (B, C), (C, D)\}$
3. $\{(S_1, J_1), (S_1, J_2), (S_2, J_1), (S_2, J_2), (S_3, J_1), (S_3, J_2), (S_4, J_1), (S_4, J_2)\}$
5. $\{(R, R), (R, G), (R, B), (G, R), (G, G), (G, B), (B, R), (B, G), (B, B)\}$; $\{(R, G), (R, B), (G, R), (G, B), (B, R), (B, G)\}$
7. a. $\{3\}$ **b.** $\{2, 3, 4, 6\}$ **9. a.** $\{(6, 2), (6, 4), (6, 6), (2, 6), (4, 6)\}$ **b.** $\varnothing$ **11. a.** 0.80 **b.** 1
13. a. 0.20 **b.** 0.15 **15.** $1/(_{12}C_3) = \frac{1}{220}$, $1/(_{12}P_3) = \frac{1}{1320}$ **17.** $P(R) = \frac{6}{10} = \frac{3}{5}$, $P(B) = \frac{4}{10} = \frac{2}{5}$
19. $P(6) = \frac{1}{2}$, $P(8) = \frac{1}{4}$, $P(12) = \frac{1}{4}$ **21. a.** $\frac{1}{20}$ **b.** $\frac{19}{20}$ **23.** 32%

25. $63\% + 48\% - 15\% + 10\% = 106\%$ when it should add to 100%.   **27. a.** $\dfrac{_4C_3}{_{12}C_3} = \dfrac{1}{55}$ **b.** $\dfrac{_4C_3}{_{12}C_3} + \dfrac{_8C_3}{_{12}C_3} = \dfrac{3}{11}$

29. $1 - \dfrac{_9C_3}{_{10}C_3} = \dfrac{3}{10}$ **31.** $\dfrac{_{48}C_1}{_{52}C_5} \approx 0.000018$ **33.** $\dfrac{7 \cdot 6 \cdot 5 \cdot 4 \cdot 3}{7^5} \approx 0.15$ **35.** $\dfrac{_{90}C_{10}}{_{100}C_{10}} \approx 0.33$ **37.** $1 - \dfrac{_{98}C_2}{_{100}C_2} \approx 0.04$

39. $\dfrac{_4C_1}{_6C_1} = \dfrac{2}{3}$ **41.** $P(H) = \frac{1}{2}$ and $P(T) = \frac{1}{2}$ give $\frac{1}{2} : \frac{1}{2}$ or $1 : 1$.
43. $P(E^c) = \frac{m-n}{m}$ so the odds are $\frac{n}{m} : \frac{m-n}{m}$ or $n : (m - n)$.
45. Since the probabilities must add to 1, we divide the odds by $n + m$, so $P(E) = \frac{n}{n+m}$.   **47.** $\frac{7}{11}$ **49.** $\frac{7}{9}$

EXERCISES 4.3 page 424

1. a. $\frac{0.2}{0.4} = \frac{1}{2}$ **b.** $\frac{0.2}{0.6} = \frac{1}{3}$ **3. a.** $\frac{3}{5}$ **b.** $\frac{3}{4}$ (both require first finding $P(A \cap B) = 0.3$)   **5.** $\frac{2}{3}$ **7.** $\frac{1}{3}$ **9.** $\frac{1}{3}$
11. $\frac{0.38}{0.68} \approx 0.56$ **13.** $0.95 \cdot 0.20 + 0.70 \cdot 0.80 = 0.75$ **15.** 0.16 **17.** $\frac{0.55 \cdot 0.60}{0.55 \cdot 0.60 + 0.65 \cdot 0.40} \approx 0.56$
19. 0.019 (approx.)   **21.** 0.26 (approx.)   **23.** Yes $(P(A) \cdot P(B) = \frac{1}{2} \frac{1}{2} = \frac{1}{4}$ and $P(A \cap B) = \frac{1}{4})$
25. No $(P(A) \cdot P(B) = \frac{1}{2} \frac{3}{36} = \frac{1}{24}$ and $P(A \cap B) = \frac{2}{36} = \frac{1}{18})$ **27.** $(\frac{1}{6})^3 = \frac{1}{216}$ **29. a.** $(\frac{1}{6})^3 = \frac{1}{216}$ **b.** $\frac{6}{216} = \frac{1}{36}$ **c.** $\frac{35}{36}$
31. $P(A \cap B^c) = P(A) - P(A \cap B) = P(A) - P(A) \cdot P(B) = P(A)[1 - P(B)] = P(A) \cdot P(B^c)$
33. $P(A \cap B) = \frac{1}{4} = \frac{1}{2} \cdot \frac{1}{2} = P(A) \cdot P(B)$
$P(A \cap C) = \frac{1}{4} = \frac{1}{2} \cdot \frac{1}{2} = P(A) \cdot P(C)$
$P(B \cap C) = \frac{1}{4} = \frac{1}{2} \cdot \frac{1}{2} = P(B) \cdot P(C)$
$P(A \cap B \cap C) = \frac{1}{4}$ but $P(A) \cdot P(B) \cdot P(C) = \frac{1}{2} \cdot \frac{1}{2} \cdot \frac{1}{2} = \frac{1}{8}$

EXERCISES 4.4 page 439

1.

x	0	1	2	3
$P(X = x)$	$\frac{1}{8}$	$\frac{3}{8}$	$\frac{3}{8}$	$\frac{1}{8}$

3.

x	-1	11
$P(X = x)$	$\frac{3}{4}$	$\frac{1}{4}$

5.

x	-12	2	3
$P(X = x)$	$\frac{1}{6}$	$\frac{1}{2}$	$\frac{1}{3}$

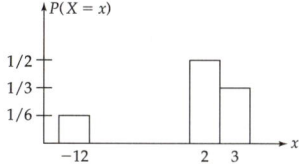

7.

x	1	2	3	4	5	6
$P(X=x)$	$\frac{1}{36}$	$\frac{1}{12}$	$\frac{5}{36}$	$\frac{7}{36}$	$\frac{1}{4}$	$\frac{11}{36}$

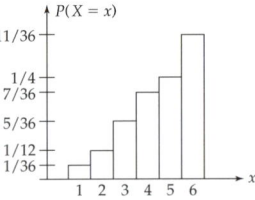

9. $\mu = 0 \cdot \frac{1}{8} + 1 \cdot \frac{3}{8} + 2 \cdot \frac{3}{8} + 3 \cdot \frac{1}{8} = \frac{3}{2}$ **11.** $\mu = 2$, $\sigma = \sqrt{(11-2)^2 \frac{1}{4} + (-1-2)^2 \frac{3}{4}} = \sqrt{27} \approx 5.20$

13. $\mu = 0$, $\sigma = \sqrt{29} \approx 5.39$ **15.** $\mu = \frac{161}{36} \approx 4.5$ **17.** $\mu = 3\frac{1}{4}$ **19.** $\sigma = 1$, $\sigma = 10$

21. $\mu = 20 \cdot \frac{1}{2} = 10$
$\sigma = \sqrt{10 \cdot \frac{1}{2} \cdot \frac{1}{2}} = \sqrt{5} \approx 2.24$

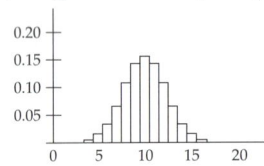

23. $\mu = 16$
$\sigma = \sqrt{3.2} \approx 1.79$

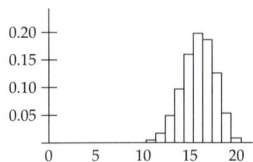

25. 0.78125 or about 78%   **27.** $_8C_4(\frac{1}{2})^8 = \frac{70}{256} \approx 0.273$ or about 27%

29. Most likely: 2 heads, with probability 0.329 (approx.)
Least likely: 6 heads, with probability 0.0014 (approx.)

31. Most likely: 1 six, with probability 0.372 (approx.), $\mu = \frac{4}{3}$ **33.** $14,000   **35.** $\mu = 0.2$, $\sigma = \sqrt{10 \cdot 0.02 \cdot 0.98} \approx 0.443$

37. $P(7 \le X \le 10) \approx 0.172$ **39.** 0.972 (compared to 0.90 for transmitting a single digit)

41. $(_{48}C_{13}/_{52}C_{13})^3 \approx 0.028$, or about 3%   **43.** $_3C_2(\frac{3}{5})^2(\frac{2}{5})^1 + _3C_3(\frac{3}{5})^3(\frac{2}{5})^0 \approx 0.65$, or about 65%

45. Definition of expected value   **47.** $E(X+Y) = E(X) + E(Y)$ and the formula for the expected value of a binomial

49. **(1)** Definition of variance   **(2)** $E(X+Y) = E(X) + E(Y)$, algebra, and independence   **(3)–(5)** Algebra
(6) Definition of variance, probabilities sum to 1, definition of expected value
(7) $E(X - E(X)) = E(X) - E(Y) = 0$, and similarly for Y

51. **(1)** Definition of variance   **(2)** Separating the sum into two parts   **(3)** Dropping a nonnegative quantity
(4) Since $(x - \mu) \ge (k\sigma)^2$

53. Reversing sides and dividing by σ^2 and k^2 **55.** Algebra   **57.** $P(\mu - 5\sigma < X < \mu + 5\sigma) > 1 - \frac{1}{5^2} = 0.96$

CHAPTER 4 REVIEW EXERCISES page 445

1. a. 17   **b.** 32   **c.** 21   **d.** 15   **2. a.** 120   **b.** 20   **3.** 110
4. a. $26^3 = 17,576$ **b.** $26 \cdot 25 \cdot 24 = 15,600$ **5.** $_{20}C_4 = 4845$, $_{20}P_4 = 116,280$ **6.** $_{13}C_5 = 1287$
7. $\{(C_1, S_1, H_1), (C_1, S_1, H_2), (C_1, S_2, H_1), (C_1, S_2, H_2), (C_2, S_1, H_1), (C_2, S_1, H_2), (C_2, S_2, H_1), (C_2, S_2, H_2)\}$
8. a. $\{(B, B), (B, Y), (B, R), (Y, B), (Y, Y), (Y, R), (R, B), (R, Y), (R, R)\}$
b. $\{(B, Y), (B, R), (Y, B), (Y, R), (R, B), (R, Y)\}$
9. a. $\{(H, H), (H, T), (T, H)\}$ **b.** $\{(H, T), (T, H), (T, T)\}$ **c.** $\{(H, T), (T, H)\}$
10. a. $\frac{1}{4}$ **b.** $\frac{3}{4}$ **c.** $\frac{1}{2}$ **11.** $1/_{15}C_2 = \frac{1}{105}$, $1/_{15}P_2 = \frac{1}{210}$ **12.** $P(1) = P(3) = P(5) = P(7) = \frac{1}{4}$
13. $P(1) = \frac{1}{2}$, $P(2) = P(4) = P(6) = \frac{1}{6}$ **14.** $P(R) = \frac{1}{6}$, $P(G) = \frac{1}{3}$, $P(B) = \frac{1}{2}$ **15.** 0.45
16. a. $\frac{1}{3}$ **b.** $\frac{1}{2}$ **17.** $_5C_4/_{40}C_5 \approx 0.00000760$
18. a. $_{12}C_5/_{52}C_5 \approx 0.000305$ or about 0.03%   **b.** $_{40}C_5/_{52}C_5 \approx 0.253$ or about 25%
19. $_{39}C_{13}/_{52}C_{13} \approx 0.0128$ or about 1%   **b.** $_{40}C_{13}/_{52}C_{13} \approx 0.0189$ or about 2%
20. $_{28}C_5/_{30}C_5 \approx 0.690$ or about 69%   **21.** $_{48}C_2/_{50}C_4 \approx 0.00490$ or about 0.5%
22. $_{28}C_1/_{30}C_3 = \frac{1}{145} \approx 0.00690$ or about 0.7%   **23.** $_6C_2/_{10}C_2 = \frac{1}{3}$
24. a. $P(A \text{ given } B) = \frac{3}{4}$ **b.** $P(B \text{ given } A) = \frac{3}{5}$ **25.** $\frac{1}{6}$ **26.** $\frac{1}{7}$
27. 0.00235 or about 0.2%   **28.** $\frac{3}{16}$ **29.** 0.32   **30.** $0.8 \cdot 0.6 + 0.9 \cdot 0.4 = 0.84$ **31.** 0.983 (approx.)
32. $\frac{2}{3}$ **33.** $\frac{0.04 \cdot 0.25}{0.04 \cdot 0.25 + 0.03 \cdot 0.35 + 0.02 \cdot 0.40} \approx 0.351$ or about 35%   **34.** 0.247 or about 25%
35. a. No $(\frac{3}{4} \cdot \frac{3}{4} \ne \frac{1}{2})$ **b.** Yes $(\frac{1}{2} \cdot \frac{1}{2} = \frac{1}{4})$ **36.** 0.9999   **37. a.** $\frac{1}{32}$ **b.** $\frac{1}{16}$ **c.** $\frac{1}{16}$

38.

Value of X	Outcomes
$X = 5$	(H, H, H, H, H)
$X = 1$	(H, T, T, T, T), (T, H, T, T, T), (T, T, H, T, T), (T, T, T, H, T), (T, T, T, T, H)
$X = 0$	(T, T, T, T, T)

39.

x	34	-2
$P(X = x)$	$\frac{1}{4}$	$\frac{3}{4}$

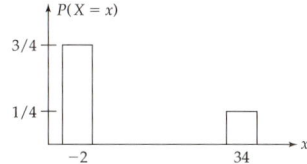

40. $\mu = 7$, $\sigma = \sqrt{243} \approx 15.6$

41.

x	0	1	2
$P(X = x)$	$\frac{5}{18}$	$\frac{5}{9}$	$\frac{1}{6}$

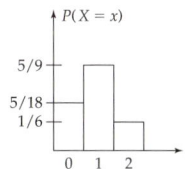

42. $\mu = \frac{8}{9}$

43. $\mu = \$3.10$

44.

x	0	1	2	3	4
$P(X = x)$	$\frac{1}{625}$	$\frac{16}{625}$	$\frac{96}{625}$	$\frac{256}{625}$	$\frac{256}{625}$

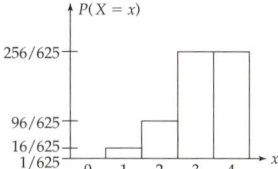

$\mu = \frac{16}{5} = 3\frac{1}{5}$
$\sigma = \frac{4}{5}$

45. 0.711 (approx.) or about 71%   **46.** \$117   **47.** 2
48. $\mu = 12 \cdot 0.01 = 0.12$, $\sigma \approx \sqrt{12 \cdot 0.01 \cdot 0.99} \approx 0.345$
49. 0.859 or about 86%   **50.** 0.554 or about 55%

EXERCISES 5.1 page 459

1. Nominal   **3.** Ordinal   **5.** Ordinal   **7.** Interval   **9.** Ratio

11.

1 = Never married
2 = Married
3 = Widowed
4 = Divorced

13.

1 = Business executives
2 = Real estate developers
3 = Lawyers
4 = Doctors
5 = Certified public accountants
6 = Retirees

15.

Stem	Leaf
0	0, 9, 0, 0
1	8, 9, 2
2	2, 9, 3, 7, 1, 9, 9, 5, 7, 5, 1, 1, 5, 4
3	4, 8, 5, 2, 2
4	8, 2, 3
5	0

17.

Stem	Leaf
2	2, 8, 6, 6, 4
3	1, 4, 3, 7, 8, 4, 3, 6, 4, 5, 2, 7, 5, 7
4	8, 1, 2, 6
5	4

19.

Stem	Leaf
4	6, 0, 9, 2
5	2, 9, 2, 3, 5, 6, 6, 9, 4, 2, 8, 9
6	8, 6, 5, 0, 6, 1, 3
7	6, 4, 0, 6, 4, 8, 7, 1, 5

21. Using 6 classes of width 8 beginning at 12.5:

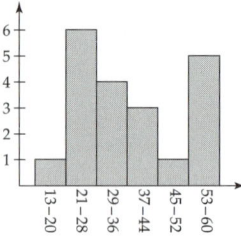

23. Using 8 classes of width 10 beginning at 10.5:

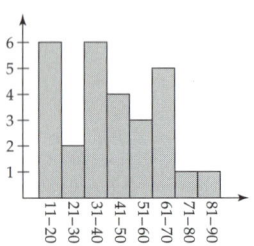

25. Using 11 classes of width 32 beginning at 218.5:

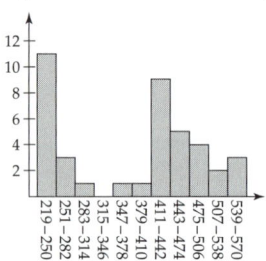

27. Using 5 classes of width 5 beginning at 35.5:

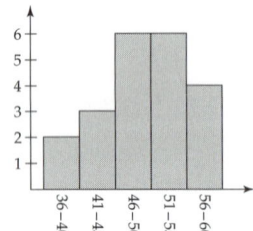

29. Using 7 classes of width 3 beginning at 8.5:

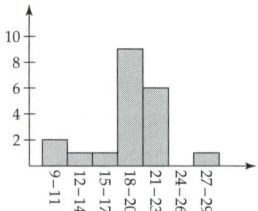

EXERCISES 5.2 page 467

1. Mode = 9, median = 10, $\bar{x}$ = 11   **3.** Mode = 7 and 14 (bimodal), median = 9, $\bar{x}$ = 10
5. Mode = 19, median = 15, $\bar{x}$ = 15.2   **7.** Mode = 14, median = 13, $\bar{x}$ = 12   **9.** Mode = 19, median = 18, $\bar{x}$ = 16
11. Mode = 12 and 20 (bimodal), median = 15, $\bar{x}$ = 15.75, each in minutes
13. Mode = 10, median = 10, $\bar{x}$ = 11.1, each in hours   **15.** Mode = 9, median = 9, $\bar{x}$ = 10, each in weeks
19. $\bar{x} \approx 44.95$ dollars

EXERCISES 5.3 page 475

1. 62   **3.** 20   **5.** 56   **7.** Minimum = 3
First quartile = 10
Median = 14
Third quartile = 21
Maximum = 23

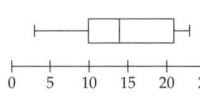

9. Minimum = 8
First quartile = 10
Median = 13
Third quartile = 17
Maximum = 26

11. Minimum = 8
First quartile = 12
Median = 17
Third quartile = 24
Maximum = 25

13. Minimum = 6
First quartile = 8
Median = 16
Third quartile = 18
Maximum = 24

15. Minimum = 5
First quartile = 8.5
Median = 15.5
Third quartile = 17.5
Maximum = 20

17. $s \approx 5.79$ (from $\bar{x} = 11$)   **19.** $s \approx 8.63$ (from $\bar{x} = 13$)

21. Range = 20
Minimum = 5
First quartile = 11
Median = 13   } All in dollars
Third quartile = 17
Maximum = 25
$s \approx 4.64$

23. Range = 11
Minimum = 2
First quartile = 6   } All in
Median = 8   } numbers
Third quartile = 10   } of people
Maximum = 13
$s \approx 2.55$

25. Range = 33
Minimum = 2
First quartile = 15.5
Median = 20.5   } All in weeks
Third quartile = 24.5
Maximum = 35
$s \approx 8.19$

29. $s = \sqrt{\dfrac{739 - \frac{1}{5}(55)^2}{5 - 1}} = \sqrt{33.5} \approx 5.79$ (same answer as Exercise 17)

EXERCISES 5.4 page 487

(*Note:* Answers may differ depending on rounding or use of graphing calculators or tables.)
1. 0.6827   **3.** 0.9973   **5.** 0.3413   **7.** 0.0674   **9.** 0.3361   **11.** $z = 1$ **13.** $z = 0$ **15.** $z = -1$ **17.** $z = -5$
19. $z = 2$ **21.** 0.0298 [from $\mu = 14$, $\sigma \approx 2.05$, after checking that $np > 5$ and $n(1 - p) > 5$]
23. 0.1883 [from $\mu = 14$, $\sigma \approx 2.05$, after checking that $np > 5$ and $n(1 - p) > 5$]
25. 0.0036 [from $\mu = 14$, $\sigma \approx 2.05$, after checking that $np > 5$ and $n(1 - p) > 5$]
27. 0.4029 [from $\mu = 14$, $\sigma \approx 2.05$, after checking that $np > 5$ and $n(1 - p) > 5$]
29. 0.9992 [from $\mu = 14$, $\sigma \approx 2.05$, after checking that $np > 5$ and $n(1 - p) > 5$]   **31.** About 0.89 or 89%
33. About 0.25 or 25% using $P(x \geq 101)$. [Alt. answers: 28% for $P(x > 100.5)$ or 31% for $P(x > 100)$]
35. About 0.31 or 31%   **37.** About 0.24 or 24% [from $\mu = 564$, $\sigma \approx 15$, after checking that $np > 5$ and $n(1 - p) > 5$]
39. About 0.26 or 26% [from $\mu = 80$, $\sigma \approx 6.93$, after checking that $np > 5$ and $n(1 - p) > 5$]

CHAPTER 5 EXERCISES REVIEW page 490

1. Nominal   **2.** Ordinal   **3.** Interval   **4.** Ratio

5.

1 = Army
2 = Navy
3 = Marines
4 = Air Force

6.

1 = Handguns
2 = Other firearms
3 = Knives
4 = Blunt instruments
5 = Hands

7.

Stem	Leaf
2	9
3	6, 8, 3
4	7, 3, 5, 0, 6
5	9, 9, 5, 6, 7, 0, 8, 4, 5
6	0
7	
8	7

8.

Stem	Leaf
0	4, 9, 3
1	4, 8, 3, 9, 7
2	5, 1, 7, 6
3	6, 9, 1, 6, 0, 8, 2, 7
4	8, 5, 9, 1
5	7, 2, 8, 1, 3
6	0

9. Using 8 classes of width 4 beginning at 33.5:

10. Using 10 classes of width 79 beginning at 697.5:

11. Mode = 11, median = 10, $\bar{x} = 9$ **12.** Mode = 6 and 13 (bimodal), median = 11, $\bar{x} = 10$

13. Mode = 13, median = 12.5, $\bar{x} = 11$ **14.** Mode = 8, median = 8, $\bar{x} = 8.50$ (all in dollars)

15. Mode = 4, median = 20, $\bar{x} = 17.4$ (all in numbers of calls)   **16.** 18

17. Minimum = 4
First quartile = 6
Median = 10
Third quartile = 14.5
Maximum = 20

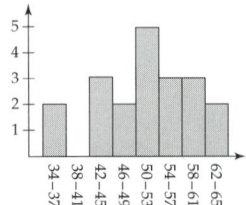

18. $s = 4$

19. Range = 6
Minimum = 3
First quartile = 5
Median = 6 $\Big\rangle$All in minutes
Third quartile = 8
Maximum = 9
$s \approx 1.96$

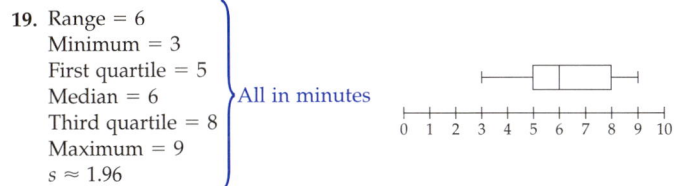

20. Range = 40
Minimum = 114
First quartile = 124 $\Big\rangle$All in
Median = 131 thousands
Third quartile = 134 of dollars
Maximum = 154
$s \approx 9.74$

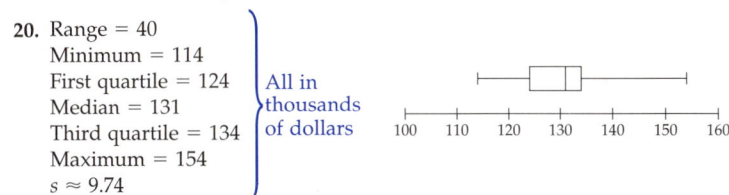

21. 0.8186 $\Big\rangle$Answers may
22. 0.3023 $\Big\rfloor$differ slightly   **23.** $z = 2$ **24.** $z = -2$
25. 0.0151 [from $\mu = 8.75$, $\sigma \approx 2.38$, after checking that $np > 5$ and $n(1 - p) > 5$]
26. 0.6829 [from $\mu = 8.75$, $\sigma \approx 2.38$, after checking that $np > 5$ and $n(1 - p) > 5$]   **27.** About 0.66 or 66%
28. About 0.157 or 16%   **29.** About 0.634 or 63% [from $\mu = 6.9$, $\sigma \approx 2.61$, after checking that $np > 5$ and $n(1 - p) > 5$]
30. About 0.382 or 38% [from $\mu = 48$, $\sigma \approx 5.71$, after checking that $np > 5$ and $n(1 - p) > 5$]

EXERCISES 6.1 page 508

1.

x	$5x - 7$	x	$5x - 7$
1.9	2.5	2.1	3.5
1.99	2.95	2.01	3.05
1.999	2.995	2.001	3.005

$\lim\limits_{x \to 2} (5x - 7) = 3$

3.

x	$\dfrac{1}{x - 5}$	x	$\dfrac{1}{x - 5}$
4.9	-10	5.1	10
4.99	-100	5.01	100
4.999	-1000	5.001	1000

$\lim\limits_{x \to 5} \dfrac{1}{x - 5}$ does not exist

5.

x	$\dfrac{x^3 - 1}{x - 1}$	x	$\dfrac{x^3 - 1}{x - 1}$
0.9	2.71	1.1	3.31
0.99	2.97	1.01	3.03
0.999	2.997	1.001	3.003

$\lim\limits_{x \to 1} \dfrac{x^3 - 1}{x - 1} = 3$

7. Does not exist   **9.** -0.25 **11.** -1 **13.** 2   **15.** 8   **17.** 2   **19.** $\sqrt{2}$ **21.** 6   **23.** $5x^3$ **25.** 4   **27.** -9
29. $2x$ **31.** $4x^2$ **33.** Continuous   **35.** Discontinuous, (3) is violated   **37.** Discontinuous, (1) is violated
39. Discontinuous, (2) is violated   **41. a.**   **b.** Yes

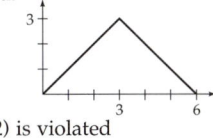

43. a.   **b.** No, (2) is violated

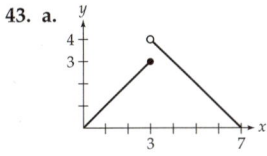

45. Continuous   **47.** Discontinuous at $x = 1$ **49.** Continuous   **51.** Discontinuous at $x = 0$, $x = 4$, and $x = -1$
53. Discontinuous at $x = 0$, ± 1, ± 2, ± 3, . . .
55. The two functions are *not* equal to each other, since at $x = 1$ one is defined and the other is not (see page 53).
57. a. 1.11 (dollars)   **b.** 1.11 (dollars)

EXERCISES 6.2 page 522

1. At P_1: positive slope   **3.** At P_1: positive slope   **5.** At P_1: slope is 3   **7.** Your graph should look roughly
   At P_2: negative slope   At P_2: negative slope   At P_2: slope is $-\frac{1}{2}$ like the following:
   At P_3: zero slope   At P_3: zero slope

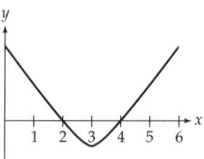

9. $f'(x) = 2x - 3$ **11.** $f'(x) = -2x$ **13.** $f'(x) = 9$ **15.** $f'(x) = \frac{1}{2}$ **17.** $f'(x) = 0$ **19.** $f'(x) = 2ax + b$
21. $f'(x) = 3x^2$ **23.** $f'(x) = \dfrac{-2}{x^2}$ **25.** $f'(x) = \dfrac{1}{2\sqrt{x}}$ **27.** $f'(x) = 3x^2 + 2x$ **29. a.** $y = x + 1$ **b.**

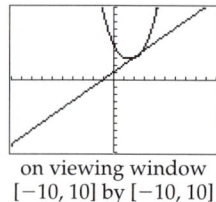

on viewing window
$[-10, 10]$ by $[-10, 10]$

31. a. Intermediate step for tangent line: $y - c^2 = 2c(x - c)$.
   b. Your display will depend on which values of c you choose.
33. a. $f'(x) = 3$ **b.** The graph of $f(x) = 3x - 4$ is a straight line with slope 3.
35. a. $f'(x) = 0$ **b.** The graph of $f(x) = 5$ is a horizontal straight line with slope 0.
37. a. $f'(x) = m$ **b.** The graph of $f(x) = mx + b$ is a straight line with slope m.
39. a. The formula comes from substituting the given function and x-value into the definition of the derivative
   $f'(x) = \lim\limits_{h \to 0} \dfrac{f(x + h) - f(x)}{h}$.   **b.** 0.2411, 0.2491, and 0.2499 (rounded) seem to be approaching 0.25.
   c. 0.2600, 0.2509, and 0.2501 (rounded) also seem to be approaching 0.25, so $f'(1) = 0.25$.
41. a. $f'(x) = 2x - 8$ **b.** Decreasing at the rate of 4 degrees per minute [since $f'(2) = -4$]
   c. Increasing at the rate of $2°/\text{min}$ [since $f'(5) = 2$]
43. a. $f'(x) = 4x - 1$ **b.** When 5 words have been memorized, the memorization time is increasing at the rate of 19
   seconds per word.
45. a. $T'(x) = -2x + 5$ **b.** Increasing at the rate of 1 degree per day
   c. Decreasing at the rate of 1 degree per day   **d.** Deteriorating on day 2, improving on day 3

EXERCISES 6.3 page 538

1. $4x^3$ **3.** $500x^{499}$ **5.** $(\frac{1}{2})x^{-1/2}$ **7.** $2x^3$ **9.** $2w^{-2/3}$ **11.** $-6x^{-3}$ **13.** $8x - 3$ **15.** $(\frac{1}{2})x^{-1/2} + x^{-2}$
17. $4x^{-1/3} + 4x^{-4/3}$ **19.** $-5x^{-3/2} - 15x^{2/3}$ **21. a.** $f'(x) = 0$ **b.** The graph of the constant function $f(x) = 2$ is a
   horizontal line and therefore has slope 0.   **c.** Since $f(x)$ is a constant function, its rate of change is zero.
23. 80   **25.** 3   **27.** 27   **29.** 1   **31.** For $y_1 = 5$ and viewing rectangle $[-10, 10]$ by $[-10, 10]$, your calculator
   screen should look like the following:

33. a. $MP(x) = 0.03x^{1/2}$ **b.** $MP(10,000) = 3$ *Interpretation:* After 10,000 chips, the profit on each additional chip is about $3.

35. $P(10,001) - P(10,000) \approx 3.00007$ **37. a.** $P'(x) = -12,000 + 1200x + 300x^2$
$P(10,000) - P(9999) \approx 2.99992$ **b.** Decreasing by about 10,500 per year
Both are close to $3.   **c.** Increasing by about 30,000 per year

39. Increasing by about 8000 people per additional day   **41.** Increasing by about 0.08 square centimeter per hour

43. Increasing by about six phrases per hour   **45. a.** $MU(x) = 50x^{-1/2}$ **b.** 50   **c.** 0.05

47. a. It becomes steeper, and so profit grows more rapidly as more books are produced.   **b.** About 5800 books
 c. 22.80, so each additional book results in about $22.80 profit (when 6000 books are produced)
 d. Marginal profit increases. It agrees with it: Marginal profit increasing means that profit is growing more rapidly.

49. c. $19,255   **e.** Tuition increasing at the rate of $949 per year

EXERCISES 6.4 page 553

1. $10x^9$ **3.** $9x^8 + 4x^3$ **5.** $5x^4 + 2x$ **7.** $15x^2 - 1$ **9.** $4x^3$ **11.** $9x^2 + 8x + 1$ **13.** 1

15. $36t + 8t^{1/3}$ (after simplification)   **17.** $7z^6 - 1$ (after simplification)   **19.** $6x^5$ **21.** $-\dfrac{3}{x^4}$ or $-3x^{-4}$ **23.** $\dfrac{x^4 - 3}{x^4}$

25. $-\dfrac{2}{(x-1)^2}$ **27.** $\dfrac{4t}{(t^2+1)^2}$ **29.** $\dfrac{2s^3 + 3s^2 + 1}{(s+1)^2}$ (after simplification)   **31.** $\dfrac{2x^5 + 4x^3}{(x^2+1)^2}$ (after simplification)   **33.** 0

35. $\frac{1}{4}$ **37.** $\dfrac{d}{dx}(fgh) = f'[gh] + f[gh]' = f'gh + fg'h + fgh'$ **39.** $2f(x)f'(x)$ **41.** $3x^2\dfrac{x^2+1}{x+1} + (x^3+2)\dfrac{x^2+2x-1}{(x+1)^2}$

43. $\dfrac{3x^6 + 13x^4 + 18x^2 - 2x}{(x^2+2)^2}$ **45.** $\dfrac{x^{-1/2}}{(x^{1/2}+1)^2} = \dfrac{1}{\sqrt{x}\,(\sqrt{x}+1)^2}$ **47.** $MAR(x) = \dfrac{xR'(x) - R(x)}{x^2}$

49. a. $C'(x) = \dfrac{100}{(100-x)^2}$ **b.** Increasing by 4¢ per additional percentage of purity
 c. Increasing by 25¢ per additional percentage of purity

51. b. Rates of change are 4 and 25   **53. a.** $AP(x) = \dfrac{12x - 1800}{x} = 12 - \dfrac{1800}{x} = 12 - 1800x^{-1}$

 b. $MAP(x) = \dfrac{1800}{x^2}$ or $1800x^{-2}$ **c.** $MAP(300) = \dfrac{2}{100}$, so average profit is increasing by 2¢ per additional unit.

55. Increasing at the rate of 7°/hr   **57. b.** 7   **c.** About 104.5°

59. b.   **d.**   **f.**

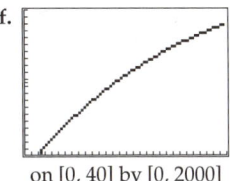

on [0, 27] by [0, 6,000,000]   on [0, 40] by [0, 45,000]   on [0, 40] by [0, 2000]

 f. (continued) $y_3(40) \approx 43,045$, so in 2010 the per capita national debt will be $43,045.
  $y_4(40) \approx 1849$, so in 2010 the per capita national debt will be growing by $1849 per year

61. a. $y' = -2x^{-3} = -\dfrac{2}{x^3}$ (which is undefined at $x = 0$)
 b. Your calculator should give "Error" but may, incorrectly, give "0."

EXERCISES 6.5 page 568

1. a. $4x^3 - 6x^2 - 6x + 5$ **b.** $12x^2 - 12x - 6$ **c.** $24x - 12$ **d.** 24   **3. a.** $1 + x + \dfrac{1}{2}x^2 + \dfrac{1}{6}x^3 + \dfrac{1}{24}x^4$

 b. $1 + x + \dfrac{1}{2}x^2 + \dfrac{1}{6}x^3$ **c.** $1 + x + \dfrac{1}{2}x^2$ **d.** $1 + x$ **5. a.** $\dfrac{5}{2}x^{3/2}$ **b.** $\dfrac{15}{4}x^{1/2}$ **c.** $\dfrac{15}{8}x^{-1/2}$ **d.** $-\dfrac{15}{16}x^{-3/2}$

7. a. $-\dfrac{2}{x^3}$ or $-2x^{-3}$ **b.** $-\dfrac{2}{27}$ **9. a.** $\dfrac{1}{x^3} = x^{-3}$ **b.** $\dfrac{1}{27}$ **11. a.** x^{-4} **b.** $\dfrac{1}{81}$ **13.** $12x^2 + 2$ **15.** $12x^{-7/3}$

17. $\dfrac{2x - 2}{(x^2 - 2x + 1)^2} = \dfrac{2}{(x - 1)^3}$ **19.** 2π **21.** 90 **23.** -720 **25.** 3 **27. a.** iii **b.** i **c.** ii **29.** 0

31. $\dfrac{d^2}{dx^2}(fg) = \dfrac{d}{dx}(f'g + fg') = f''g + f'g' + f'g' + fg'' = f''g + 2f'g' + fg''$

33. a. 54 mph **b.** -42 mph or 42 mph south **c.** 24 mi/hr²
35. 310 ft/sec, 61 ft/sec² **37. a.** 160 ft/sec **b.** 32 ft/sec² **39. a.** $-32t + 1280$ **b.** 40 seconds **c.** 25,600 feet
41. $D'(8) = 24$: After 8 years the debt is growing at \$24 billion per year.
   $D''(8) = 1$: After 8 years the debt will be growing increasingly rapidly, with the rate of growth growing by about
   \$1 billion per year per year.
43. $L'(4) = \dfrac{1}{4}$: After 4 years the sea level will be rising by $\dfrac{1}{4}$ foot per year.

   $L''(4) = -\dfrac{3}{32}$: After 4 years the rate of growth will be slowing by about $\dfrac{3}{32}$ foot per year per year.
45. $P(3) \approx 4.87$, $P'(3) \approx -0.51$, $P''(3) \approx -0.39$
   Interpretation: In 3 years the profit will be about \$4.87 million, decreasing at the rate of \$0.51 million per year, and
   the decline of profit will be accelerating.
47. a. Approximately 9° and $-2°$
   b. Each successive 1 mph increase in wind speed lowers the windchill index, but less so as wind speed rises.
   c. $y_2(15) \approx -1.1°$ and $y_2(30) \approx -0.4°$. *Interpretation:* At a wind speed of 15 mph, each additional mph decreases the
      windchill index by about 1.1°, whereas at a wind speed of 30 mph, each additional mph of wind decreases the
      windchill index by only about 0.4°.
49. $20x^3 - 12x^2 + 6x - 2$ **51.** $\dfrac{2x^5 - 4x^3 - 6x}{(x^2 + 1)^4} = \dfrac{2x^3 - 6x}{(x^2 + 1)^3}$ **53.** $\dfrac{-32x - 16}{(4x^2 + 4x + 1)^2} = \dfrac{-16}{(2x + 1)^3}$

EXERCISES 6.6 page 581

Note: For Exercises 1 through 9 there are other possible correct answers.

1. $f(x) = \sqrt{x}$, $g(x) = x^2 - 3x + 1$ **3.** $f(x) = x^{-3}$, $g(x) = x^2 - x$ **5.** $f(x) = \dfrac{x + 1}{x - 1}$, $g(x) = x^3$

7. $f(x) = x^4$, $g(x) = \dfrac{x + 1}{x - 1}$ **9.** $f(x) = \sqrt{x} + 5$, $g(x) = x^2 - 9$ **11.** $6x(x^2 + 1)^2$ **13.** $4(3z^2 - 5z + 2)^3(6z - 5)$

15. $\dfrac{1}{2}(x^4 - 5x + 1)^{-1/2}(4x^3 - 5)$ **17.** $\dfrac{1}{3}(9z - 1)^{-2/3}(9) = 3(9z - 1)^{-2/3}$ **19.** $-8x(4 - x^2)^3$ **21.** $-12w^2(w^3 - 1)^{-5}$

23. $4x^3 - 4(1 - x)^3$ **25.** $-\dfrac{2}{3}(9x + 1)^{-5/3}(9) = -6(9x + 1)^{-5/3}$ **27.** $3[(x^2 + 1)^3 + x]^2[6x(x^2 + 1)^2 + 1]$
29. $6x(2x + 1)^5 + 30x^2(2x + 1)^4 = 6x(2x + 1)^4(7x + 1)$
31. $6(2x + 1)^2(2x - 1)^4 + 8(2x + 1)^3(2x - 1)^3 = 2(2x + 1)^2(2x - 1)^3(14x + 1)$
33. $-6\dfrac{(x + 1)^2}{(x - 1)^4}$ **35.** $2x(1 + x^2)^{1/2} + x^3(1 + x^2)^{-1/2}$ **37.** $\dfrac{1}{4}x^{-1/2}(1 + x^{1/2})^{-1/2}$ **39. a.** $4x(x^2 + 1)$ **b.** $4x^3 + 4x$
41. a. $-\dfrac{3}{(3x + 1)^2}$ **b.** $-3(3x + 1)^{-2}$ **43.** $\dfrac{d}{dx}L(g(x)) = L'(g(x))\,g'(x) = \dfrac{1}{g(x)}g'(x) = \dfrac{g'(x)}{g(x)}$
45. $20(x^2 + 1)^9 + 360x^2(x^2 + 1)^8$ **47.** $MC(x) = 4x(4x^2 + 900)^{-1/2}$ $MC(20) = \dfrac{8}{5} = 1.60$ **49.** $x = 27$

51. $S'(25) \approx 2.1$; at income level \$25,000, status increases by about 2.1 units for each additional \$1000 of income.
53. $R'(50) = 32\frac{1}{3}$ **55.** 26 mg **57.** $P'(2) = 0.24$; pollution is increasing by about 0.24 ppm per year.

59. b. $y_1 = 1.484x + 324.2$ 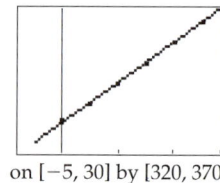 **e.** Slope ≈ 0.036 **f.** $\dfrac{1.8}{0.036} \approx 50$ years

on $[-5, 30]$ by $[320, 370]$

EXERCISES 6.7 page 588

1. $-2, 0,$ and 2 **3.** -3 and 3

5. $\lim\limits_{h\to 0} \dfrac{f(x+h)-f(x)}{h}$ simplifies to $\lim\limits_{h\to 0} \dfrac{|2h|}{h}$, which gives $\begin{cases} 2 & \text{for } h > 0 \\ -2 & \text{for } h < 0 \end{cases}$ so the limit (and therefore the derivative at $x=0$) does not exist.

7. $\lim\limits_{h\to 0} \dfrac{f(x+h)-f(x)}{h}$ simplifies to $\lim\limits_{h\to 0} \dfrac{h^{2/5}}{h} = \lim\limits_{h\to 0} \dfrac{1}{h^{3/5}}$, which does not exist. Therefore the derivative at $x=0$ does not exist.

9. If you got a numerical answer, it is wrong, since the function is undefined at $x=0$, so the derivative at $x=0$ does not exist. (For an explanation, see the Graphing Calculator Exploration on page 586.)

11. a. The formula comes from substituting the given function and x-value into the definition of the derivative and simplifying.   **b.** 3.16, 31.6, 316 (rounded)   **c.** No, No   **d.**

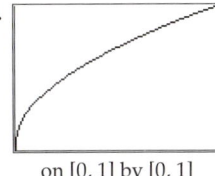

on $[0, 1]$ by $[0, 1]$

CHAPTER 6 REVIEW EXERCISES page 591

1.

x	$4x + 2$	x	$4x + 2$
1.9	9.6	2.1	10.4
1.99	9.96	2.01	10.04
1.999	9.996	2.001	10.004

$\lim\limits_{x\to 2} (4x + 2) = 10$

2.

x	$\dfrac{1}{x-8}$	x	$\dfrac{1}{x-8}$
7.9	-10	8.1	10
7.99	-100	8.01	100
7.999	-1000	8.001	1000

The limit does not exist.

3.

x	$\dfrac{x}{|x|}$	x	$\dfrac{x}{|x|}$
-0.1	-1	0.1	1
-0.01	-1	0.01	1
-0.001	-1	0.001	1

The limit does not exist.

4.

x	$\dfrac{\sqrt{x+1}-1}{x}$	x	$\dfrac{\sqrt{x+1}-1}{x}$
-0.1	0.513	0.1	0.488
-0.01	0.501	0.01	0.499
-0.001	0.500	0.001	0.500

$\lim\limits_{x\to 0} \dfrac{\sqrt{x+1}-1}{x} = 0.5$

5. 5   **6.** π **7.** 4   **8.** 2   **9.** $\frac{1}{2}$ **10.** -3 **11.** $2x^2$ **12.** $-x^2$ **13.** Continuous   **14.** Continuous
15. Continuous   **16.** Continuous   **17.** Discontinuous at $x=0$ and at $x=-1$

18. Discontinuous at $x = 3$ and at $x = -3$ **19.**   Continuous

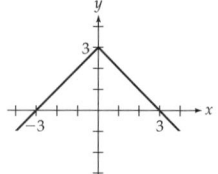

20.

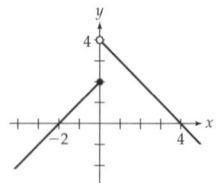

Discontinuous at $x = 0$ **21.** $4x + 3$ **22.** $6x + 2$ **23.** $-\dfrac{3}{x^2} = -3x^{-2}$ **24.** $\dfrac{2}{\sqrt{x}}$

25. $10x^{2/3} + 2x^{-3/2}$ **26.** $10x^{3/2} + 2x^{-4/3}$ **27.** -16 **28.** -9 **29.** 1 **30.** $\frac{1}{2}$ **31 a.** $C'(x) = 3 - 27x^{-3/2}$
b. 2; costs are increasing by about \$2 per additional unit.

32. $f'(10) = -2.3$ (thousand hours); after 10 planes, the construction time is decreasing by about 2300 hours for each additional plane built.

33. a. $\dfrac{dA}{dr} = 2\pi r$ **b.** As the radius increases, the area "grows by a circumference."

34. a. $V' = (\frac{4}{3})\pi r^2 \cdot 3 = 4\pi r^2$ **b.** As radius increases, volume "grows by a surface area."

35. $40x^3 + 6$ (after simplification)   **36.** $15x^4 - 2x$ **37.** $2x(x^2 - 5) + (x^2 + 5)2x = 4x^3$ **38.** $4x^3$

39. $(4x^3 + 2x)(x^5 - x^3 + x) + (x^4 + x^2 + 1)(5x^4 - 3x^2 + 1) = 9x^8 + 5x^4 + 1$

40. $(5x^4 + 3x^2 + 1)(x^4 - x^2 + 1) + (x^5 + x^3 + x)(4x^3 - 2x) = 9x^8 + 5x^4 + 1$ **41.** $\dfrac{2}{(x + 1)^2}$ **42.** $\dfrac{-2}{(x - 1)^2}$

43. $\dfrac{-10x^4}{(x^5 - 1)^2}$ **44.** $\dfrac{12x^5}{(x^6 + 1)^2}$ **45 a.** $-\dfrac{1}{x^2}$ **b.** $-x^{-2}$ (after simplification)   **c.** By simplifying to $f(x) = 2 + x^{-1}$, the derivative is $-x^{-2}$.

46. $S'(6) = -10$, so at a price of \$6 each, sales will decrease by about 10 for each dollar increase in price.

47. a. $\text{AP}(x) = \dfrac{6x - 200}{x}$ **b.** $\text{MAP}(x) = \dfrac{200}{x^2}$
c. $\text{MAP}(10) = 2$, so average profit is increasing by about \$2 per additional unit.

48. a. $\text{AC}(x) = \dfrac{5x + 100}{x}$ **b.** $\text{MAC}(x) = \dfrac{-100}{x^2}$
c. $\text{MAC}(20) = -0.25$, so average cost is decreasing by about 25¢ per additional unit.

49. $9x^{-1/2} + 2x^{-5/3}$ **50.** $-4x^{-4/3} - 3x^{-1/2}$ **51.** $2x^{-4}$ **52.** $6x^{-5}$ **53.** -24 **54.** 60   **55.** 480   **56.** $\frac{3}{8}$ **57.** 15

58. 70   **59.** $P(10) = 200$, $P'(10) = 20$, $P''(10) = 9$. *Interpretation:* Ten years from now the population will be 200 thousand, growing at the rate of 20 thousand per year, and the growth will be accelerating.

60. Velocity 2500 ft/sec, acceleration 150 ft/sec²   **61.** 347.25 feet

62. b.

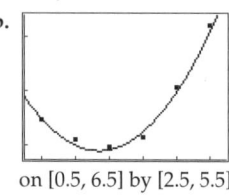

c. $y_1(7) = 6.76$, at the end of year 7 the annual profit will be about \$6.76 million.

on [0.5, 6.5] by [2.5, 5.5]

d. $y_2(7) \approx 1.77$; at the end of the seventh year, the annual profit will be growing by about $1.77 million per year.
e. $y_3(7) \approx 0.41$; profit will be growing increasingly fast, with the rate of growth increasing by about $0.41 million per year per year.

f.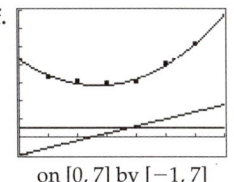

y_1 is a quadratic function, and so its first derivative (y_2) is linear, and its second derivative (y_3) is constant, and the graph of a constant is a horizontal line.

on $[0, 7]$ by $[-1, 7]$

63. $3(4z^2 - 3z + 1)^2(8z - 3)$ **64.** $4(3z^2 - 5z - 1)^3(6z - 5)$ **65.** $-5(100 - x)^4$ **66.** $-4(1000 - x)^3$

67. $\frac{1}{2}(x^2 - x + 2)^{-1/2}(2x - 1)$ **68.** $\frac{1}{2}(x^2 - 5x - 1)^{-1/2}(2x - 5)$ **69.** $2(6z - 1)^{-2/3}$ **70.** $(3z + 1)^{-2/3}$

71. $-2(5x + 1)^{-7/5}$ **72.** $-6(10x + 1)^{-8/5}$ **73.** $2x(2x - 1)^4 + 8x^2(2x - 1)^3 = 2x(2x - 1)^3(6x - 1)$

74. $5(x^3 - 2)^4 + 60x^3(x^3 - 2)^3 = 5(x^3 - 2)^3(13x^3 - 2)$ **75.** $3x^2(x^3 + 1)^{1/3} + x^5(x^3 + 1)^{-2/3}$

76. $4x^3(x^2 + 1)^{1/2} + x^5(x^2 + 1)^{-1/2}$ **77.** $3[(2x^2 + 1)^4 + x^4]^2[16x(2x^2 + 1)^3 + 4x^3]$

78. $2[(3x^2 - 1)^3 + x^3][18x(3x^2 - 1)^2 + 3x^2]$ **79.** $\frac{1}{2}[(x^2 + 1)^4 - x^4]^{-1/2}[8x(x^2 + 1)^3 - 4x^3]$

80. $[(x^3 + 1)^2 + x^2]^{-1/2}[3x^2(x^3 + 1) + x]$ **81.** $12(3x + 1)^3(4x + 1)^3 + 12(3x + 1)^4(4x + 1)^2$

82. $6x(x^2 + 1)^2(x^2 - 1)^4 + 8x(x^2 + 1)^3(x^2 - 1)^3$ **83.** $\dfrac{-20}{x^2}\left(\dfrac{x + 5}{x}\right)^3 = \dfrac{-20(x + 5)^3}{x^5}$

84. $-20\left(\dfrac{x + 4}{x}\right)^4 \dfrac{1}{x^2} = -20\dfrac{(x + 4)^4}{x^6}$ **85.** $20(2w^2 - 4)^4 + 320w^2(2w^2 - 4)^3$ **86.** $24(3w^2 + 1)^3 + 432w^2(3w^2 + 1)^2$

87. $6z(z + 1)^3 + 18z^2(z + 1)^2 + 6z^3(z + 1)$ **88.** $12z^2(z + 1)^4 + 32z^3(z + 1)^3 + 12z^4(z + 1)^2$

89. a. $6x^2(x^3 - 1)$ **b.** $6x^5 - 6x^2$ **90. a.** $\dfrac{-3x^2}{(x^3 + 1)^2}$ **b.** $-(x^3 + 1)^{-2}(3x^2)$

91. $P'(5) = 3$. *Interpretation:* When producing 5 tons, profit is increasing at about 3 thousand dollars for each additional ton.

92. $V'(8) = 17.496$. *Interpretation:* Value increased by about \$17.50 for each additional percentage of interest.

93. a. $P(5) - P(4) \approx 2.73$, $P(6) - P(5) \approx 3.23$, both of which are near 3.   **b.** At about $x = 7.6$ **94.** $x \approx 16$

95. 0.08   **96.** $N'(96) = -250$. At age 96, the number of survivors is decreasing by about 250 people per year.

97. $x = -3, x = 1, x = 3$ **98.** $x = 2, x = -2$ **99.** $x = 0, x = 3.5$ **100.** $x = 0, x = 3$

101. $\displaystyle\lim_{h \to 0} \dfrac{f(x + h) - f(x)}{h}$ simplifies to $\displaystyle\lim_{h \to 0} \dfrac{|5h|}{h}$, which gives: $\begin{cases} 5 & \text{for } h > 0 \\ -5 & \text{for } h < 0 \end{cases}$ and so the limit (and therefore the derivative at $x = 0$) does not exist.

102. $\displaystyle\lim_{h \to 0} \dfrac{f(x + h) - f(x)}{h}$ simplifies to $\displaystyle\lim_{h \to 0} \dfrac{h^{3/5}}{h} = \lim_{h \to 0} \dfrac{1}{h^{2/5}}$, which does not exist. Therefore the derivative at $x = 0$ does not exist.

EXERCISES 7.1 page 608

1. a. $(-\infty, -2)$ and $(0, \infty)$ **b.** $(-2, 0)$ **3.** All but 3 (where the function is undefined)   **5.** 4 and -4
7. $0, -4$, and 1   **9.** 3   **11.** No CVs   **13.** 1 and $\frac{1}{3}$

15.

$f' < 0$	$f' = 0$	$f' > 0$	$f' = 0$	$f' < 0$	$f' = 0$	$f' > 0$
	$x = -4$		$x = 0$		$x = 1$	
↘	→	↗	→	↘	→	↗
	rel min $(-4, -64)$		rel max $(0, 64)$		rel min $(1, 61)$	

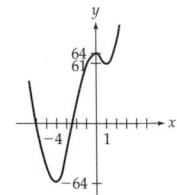

17. $f' > 0$ $f' = 0$ $f' < 0$ $f' = 0$ $f' > 0$ $f' = 0$ $f' < 0$

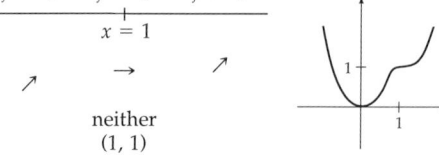

    $x = 0$ $x = 1$ $x = 2$

↗   →   ↘   →   ↗   →   ↘

     rel max   rel min   rel max
     (0, 1)   (1, 0)   (2, 1)

19. $f' < 0$ $f' = 0$ $f' > 0$ $f' = 0$ $f' > 0$

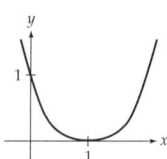

    $x = 0$ $x = 1$

↘   →   ↗   →   ↗

     rel min   neither
     (0, 0)   (1, 1)

21. $f' < 0$ $f' = 0$ $f' > 0$

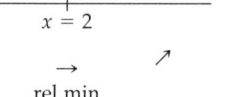

    $x = 1$

↘   →   ↗

     rel min
     (1, 0)

23. $f' < 0$ $f' = 0$ $f' > 0$ $f' = 0$ $f' < 0$ $f' = 0$ $f' > 0$

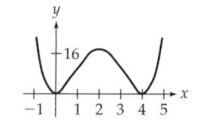

    $x = -2$ $x = 0$ $x = 2$

↘   →   ↗   →   ↘   →   ↗

     rel min   rel max   rel min
     (−2, 0)   (0, 16)   (2, 0)

25. $f' < 0$ $f' = 0$ $f' > 0$ $f' = 0$ $f' < 0$ $f' = 0$ $f' > 0$

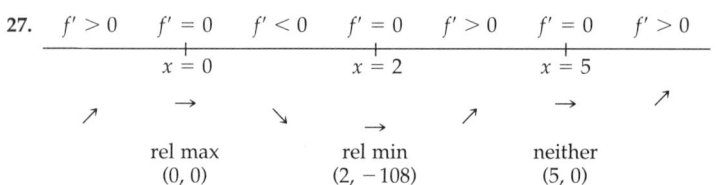

    $x = 0$ $x = 2$ $x = 4$

↘   →   ↗   →   ↘   →   ↗

     rel min   rel max   rel min
     (0, 0)   (2, 16)   (4, 0)

27. $f' > 0$ $f' = 0$ $f' < 0$ $f' = 0$ $f' > 0$ $f' = 0$ $f' > 0$

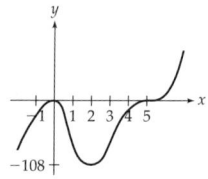

    $x = 0$ $x = 2$ $x = 5$

↗   →   ↘   →   ↗   →   ↗

     rel max   rel min   neither
     (0, 0)   (2, −108)   (5, 0)

29. $f' > 0$ $f' = 0$ $f' < 0$ $f' = 0$ $f' > 0$

    $x = -10$ $x = 10$

↗   →   ↘   →   ↗

     rel max   rel min
     (−10, 2000)   (10, −2000)

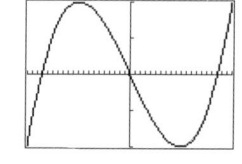

on [−20, 20] by [−2000, 2000]

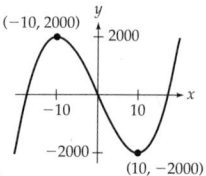

31.

$f' < 0$	$f' = 0$	$f' > 0$	$f' = 0$	$f' < 0$	$f' = 0$	$f' > 0$

$x = -5$ $\qquad$ $x = 0$ $\qquad$ $x = 5$

$\searrow$ $\quad$ $\rightarrow$ $\quad$ $\nearrow$ $\quad$ $\searrow$ $\quad$ $\rightarrow$ $\quad$ $\nearrow$

rel min $\qquad$ rel max $\qquad$ rel min
$(-5, -650)$ $\quad$ $(0, -25)$ $\quad$ $(5, -650)$

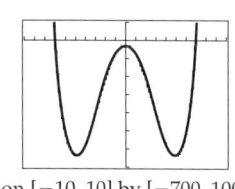 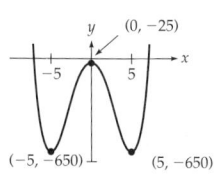

on $[-10, 10]$ by $[-700, 100]$

33.

$f' > 0$	$f' = 0$	$f' > 0$	$f' = 0$	$f' < 0$	$f' = 0$	$f' > 0$

$x = 0$ $\qquad$ $x = 1$ $\qquad$ $x = 3$

$\nearrow$ $\quad$ $\rightarrow$ $\quad$ $\nearrow$ $\quad$ $\rightarrow$ $\quad$ $\searrow$ $\quad$ $\rightarrow$ $\quad$ $\nearrow$

neither $\qquad$ rel max $\qquad$ rel min
$(0, -23)$ $\quad$ $(1, -22)$ $\quad$ $(3, -50)$

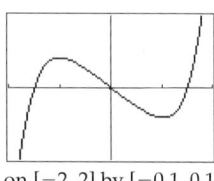

on $[-2, 4]$ by $[-50, 20]$

35.

$f' > 0$	$f' = 0$	$f' < 0$	$f' = 0$	$f' > 0$

$x = -1$ $\qquad$ $x = 1$

$\nearrow$ $\quad$ $\rightarrow$ $\quad$ $\searrow$ $\quad$ $\rightarrow$ $\quad$ $\nearrow$

rel max $\qquad$ rel min
$(-1, 0.04)$ $\quad$ $(1, -0.04)$

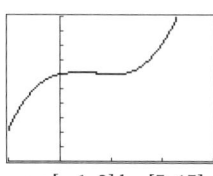

on $[-2, 2]$ by $[-0.1, 0.1]$

37.

$f' > 0$	$f' = 0$	$f' < 0$	$f' = 0$	$f' > 0$

$x = \frac{1}{3}$ $\qquad$ $x = 1$

$\nearrow$ $\quad$ $\rightarrow$ $\quad$ $\searrow$ $\quad$ $\rightarrow$ $\quad$ $\nearrow$

rel max $\qquad$ rel min
$(\frac{1}{3}, 11.15)$ $\quad$ $(1, 11)$

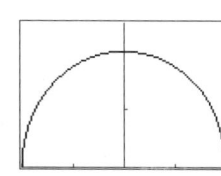

on $[-1, 3]$ by $[5, 15]$

39.

f' und	$f' > 0$	$f' = 0$	$f' < 0$	f' und

$x = -20$ $\qquad$ $x = 0$ $\qquad$ $x = 20$

$\nearrow$ $\quad$ $\rightarrow$ $\quad$ $\searrow$

rel max
$(0, 20)$

on $[-20, 20]$ by $[0, 25]$

41. Critical point: $(1, -\frac{1}{9})$

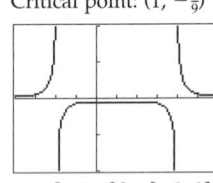

on $[-4, 6]$ by $[-2, 2]$

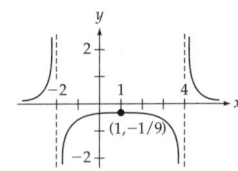

43. Critical point: $(0, 2)$

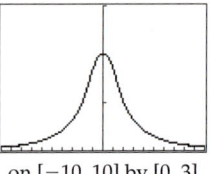

on $[-10, 10]$ by $[0, 3]$

45. Critical point: $(0, 0)$

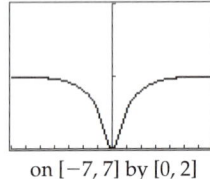

on $[-7, 7]$ by $[0, 2]$

47. Critical points: $(0, 0)$ and $(6, 12)$

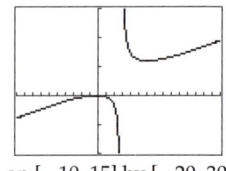

on $[-10, 15]$ by $[-20, 30]$

49. Critical point: $(0, 0)$

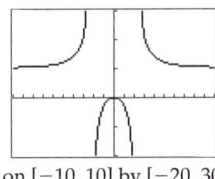

on $[-10, 10]$ by $[-20, 30]$

51. Critical points: $(1, 1), (-1, 1),$ and $(0, 0)$

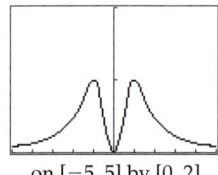

on $[-5, 5]$ by $[0, 2]$

53.

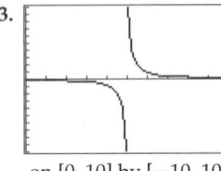

on $[0, 10]$ by $[-10, 10]$

55.

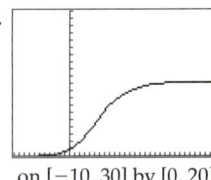

on $[-10, 30]$ by $[0, 20]$

57. $f'(x) = 2ax + b = 0$ at $x = \dfrac{-b}{2a}$

59.

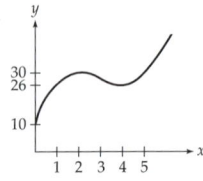

61. d.

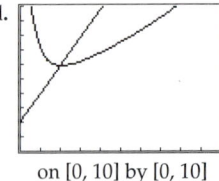

on $[0, 10]$ by $[0, 10]$

63. a.

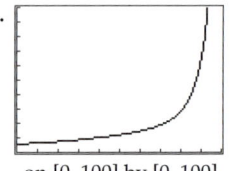

on $[0, 100]$ by $[0, 100]$

EXERCISES 7.2 page 624

1. Point 2   **3.** Points 3 and 5   **5.** Points 4 and 6

7.

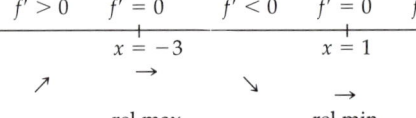

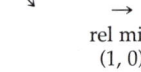

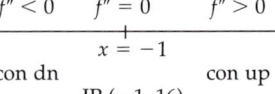

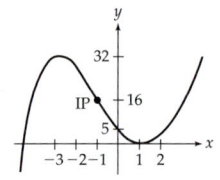

9. $f' > 0$ $f' = 0$ $f' > 0$

———————————+———————————
       $x = 1$

   ↗   →   ↗
         neither
         (1, 5)

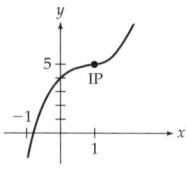

$f'' < 0$ $f'' = 0$ $f'' > 0$

———————————+———————————
       $x = 1$

 con dn   con up
       IP (1, 5)

11. $f' < 0$ $f' = 0$ $f' > 0$ $f' = 0$ $f' > 0$

———————————+—————————————+———————————
       $x = 0$ $x = 3$

   ↘   ↗   →   ↗
         rel min   neither
         (0, 2)   (3, 29)

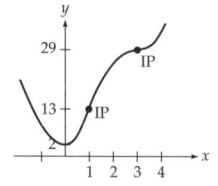

$f'' > 0$ $f'' = 0$ $f'' < 0$ $f'' = 0$ $f'' > 0$

———————————+—————————————+———————————
       $x = 1$ $x = 3$

 con up   con dn   con up
       IP (1, 13)   IP (3, 29)

13. $f' < 0$ $f' = 0$ $f' > 0$ $f' = 0$ $f' < 0$

———————————+—————————————+———————————
       $x = 0$ $x = 4$

   ↘   →   ↗   →   ↘
         rel min   rel max
         (0, 0)   (4, 256)

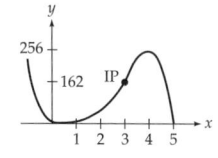

$f'' > 0$ $f'' = 0$ $f'' > 0$ $f'' = 0$ $f'' < 0$

———————————+—————————————+———————————
       $x = 0$ $x = 3$

 con up   con up   con dn
              IP (3, 162)

15. $f' > 0$ $f' = 0$ $f' > 0$

———————————+———————————
       $x = -2$

   ↗   →   ↗
         neither
         (-2, 0)

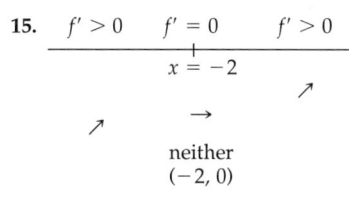

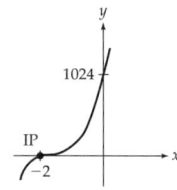

$f'' < 0$ $f'' = 0$ $f'' > 0$

———————————+———————————
       $x = -2$

 con dn   con up
       IP (-2, 0)

17.

$f' > 0$	$f' = 0$	$f' < 0$	$f' = 0$	$f' > 0$

$x = 1$ $x = 3$

↗   →   ↘   →   ↗

rel max   rel min
(1, 4)   (3, 0)

$f'' < 0$	$f'' = 0$	$f'' > 0$

$x = 2$

con dn   con up
IP (2, 2)

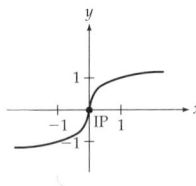

19.

$f' > 0$	f' und	$f' > 0$

$x = 0$

↗   ↗

neither
(0, 0)

$f'' > 0$	f'' und	$f'' < 0$

$x = 0$

con up   con dn
IP (0, 0)

21.

$f' < 0$	f' und	$f' > 0$

$x = 0$

↘   ↗

rel min
(0, 2)

$f'' < 0$	f'' und	$f'' < 0$

$x = 0$

con dn   con dn

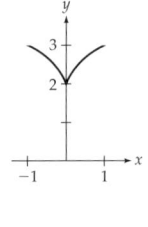

23.

f' und	$f' > 0$

$x = 0$

↗

f'' und	$f'' < 0$

$x = 0$

con dn

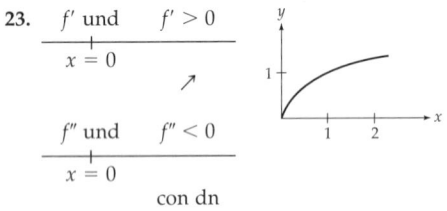

25.

$f' < 0$	f' und	$f' > 0$

$x = 1$

↘   ↗

rel min
(1, 0)

$f'' < 0$	f'' und	$f'' < 0$

$x = 1$

con dn   con dn

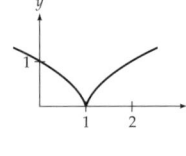

27.

$f' > 0$	$f' = 0$	$f' < 0$	$f' = 0$	$f' > 0$

$x = 2$ $x = 10$

↗   →   ↘   →   ↗

rel max   rel min
(2, 76)   (10, −180)

$f'' < 0$	$f'' = 0$	$f'' > 0$

$x = 6$

con dn   con up
IP (6, −52)

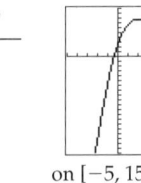

on [−5, 15] by [−200, 100]

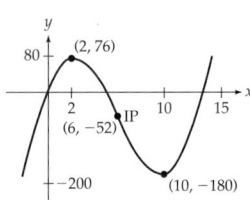

29. $f' < 0$ $f' = 0$ $f' < 0$ $f' = 0$ $f' > 0$

$x = 0$ $x = 12$

↘   →   ↘   →   ↗

neither   rel min
(0, 0)   (12, −6912)

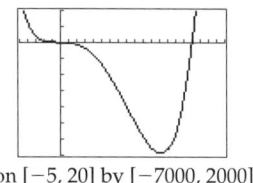

$f'' > 0$ $f'' = 0$ $f'' < 0$ $f'' = 0$ $f'' > 0$

$x = 0$ $x = 8$

con up   con dn   con up

IP (0, 0)   IP (8, −4096)

on [−5, 20] by [−7000, 2000]

31. $f' > 0$ $f' = 0$ $f' < 0$ $f' = 0$ $f' > 0$

$x = -2$ $x = 8$

↗   →   ↘   →   ↗

rel max   rel min
(−2, 100)   (8, −400)

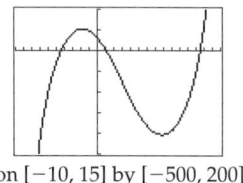

$f'' < 0$ $f'' = 0$ $f'' > 0$

$x = 3$

con dn   con up

IP (3, −150)

on [−10, 15] by [−500, 200]

33. $f' > 0$ $f' = 0$ $f' < 0$ $f' = 0$ $f' > 0$

$x = \dfrac{1}{3}$ $x = 1$

↗   →   ↘   →   ↗

rel max
$\left(\dfrac{1}{3}, 5.15\right)$ rel min
(1, 5)

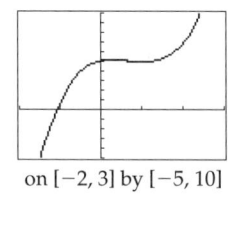

on [−2, 3] by [−5, 10]

$f'' < 0$ $f'' = 0$ $f'' > 0$

$x = \dfrac{2}{3}$

con dn   con up

IP $\left(\dfrac{2}{3}, 5.07\right)$

35. $f' > 0$ f' und $f' > 0$

$x = 1$

↗   ↗

neither
(1, 0)

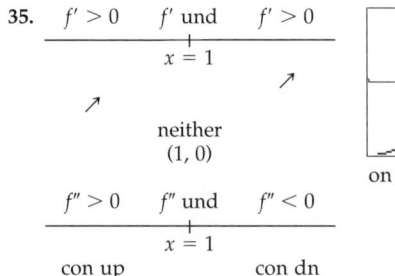

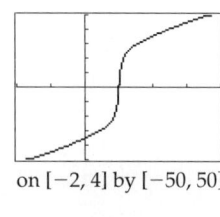

on [−2, 4] by [−50, 50]

$f'' > 0$ f'' und $f'' < 0$

$x = 1$

con up   con dn

IP (1, 0)

37.

$$
\begin{array}{ccc}
f' \text{ und} & f' > 0 \\
\hline
+ \\
x = 0 & \nearrow
\end{array}
$$

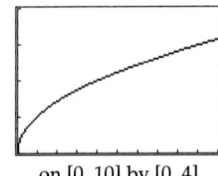

on [0, 10] by [0, 4]

39.

$$
\begin{array}{ccc}
f' \text{ und} & f' < 0 \\
\hline
+ \\
x = 0 & \searrow
\end{array}
$$

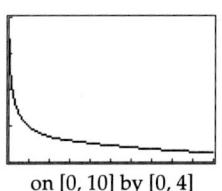

on [0, 10] by [0, 4]

41.

$$
\begin{array}{ccccc}
f' < 0 & f' \text{ und} & f' > 0 & f' = 0 & f' < 0 \\
\hline
& + & & + \\
& x = 0 & & x = 1 \\
\searrow & & \nearrow & \rightarrow & \searrow \\
& \text{rel min} & & \text{rel max} \\
& (0, 0) & & (1, 3)
\end{array}
$$

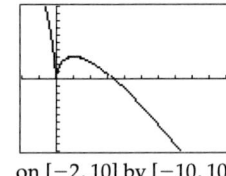

on [−2, 10] by [−10, 10]

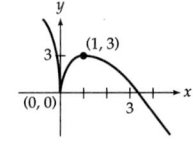

43.

$$
\begin{array}{ccccc}
f' > 0 & f' \text{ und} & f' < 0 & f' = 0 & f' > 0 \\
\hline
& + & & + \\
& x = 0 & & x = 1 \\
\nearrow & & \searrow & \rightarrow & \nearrow \\
& \text{rel max} & & \text{rel min} \\
& (0, 0) & & (1, -2)
\end{array}
$$

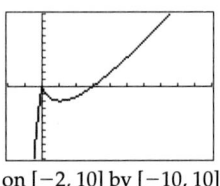

on [−2, 10] by [−10, 10]

45.

$$
\begin{array}{ccccccc}
f' > 0 & f' = 0 & f' < 0 & f' \text{ und} & f' > 0 & f' = 0 & f' < 0 \\
\hline
& + & & + & & + \\
& x = -1 & & x = 0 & & x = 1 \\
\nearrow & \rightarrow & \searrow & & \nearrow & \rightarrow & \searrow \\
& \text{rel max} & & \text{rel min} & & \text{rel max} \\
& (-1, 2) & & (0, 0) & & (1, 2)
\end{array}
$$

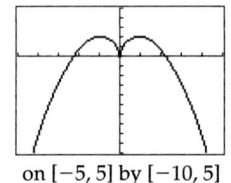

on [−5, 5] by [−10, 5]

47. $f''(x) = 2a$, therefore: For $a > 0$, $f'' > 0$, and so f is concave up.
For $a < 0$, $f'' < 0$, and so f is concave down.

49. Where the curve is concave *up*, it lies *above* its tangent line, and where it is concave *down*, it lies *below* its tangent line, and so *at* an inflection point it must cross its tangent line.

51. −0.77, 0, and 0.77

53. a.

$$
\begin{array}{ccccc}
f' > 0 & f' = 0 & f' < 0 & f' = 0 & f' > 0 \\
\hline
& + & & + \\
& x = 1 & & x = 5 \\
\nearrow & \rightarrow & \searrow & \rightarrow & \nearrow \\
& \text{rel max} & & \text{rel min} \\
& (1, 32) & & (5, 0)
\end{array}
$$

$$
\begin{array}{ccc}
f'' < 0 & f'' = 0 & f'' > 0 \\
\hline
& + \\
& x = 3 \\
\text{con dn} & & \text{con up} \\
& \text{IP (3, 16)}
\end{array}
$$

b.

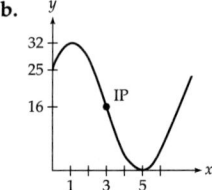

55.

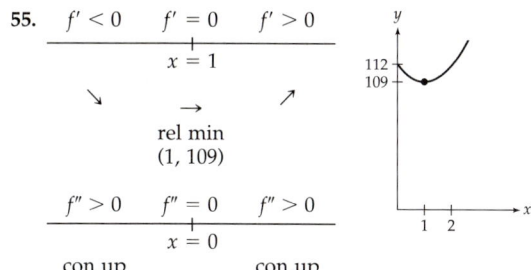

$f' < 0 \quad f' = 0 \quad f' > 0$

$x = 1$

rel min
(1, 109)

$f'' > 0 \quad f'' = 0 \quad f'' > 0$

$x = 0$

con up   con up

57.

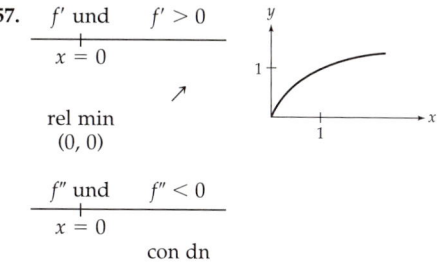

f' und $\quad f' > 0$

$x = 0$

rel min
(0, 0)

f'' und $\quad f'' < 0$

$x = 0$

con dn

59. a.

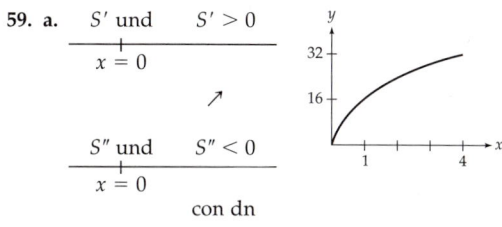

S' und $\quad S' > 0$

$x = 0$

S'' und $\quad S'' < 0$

$x = 0$

con dn

b. Concave down. Status increases more slowly at higher income levels.

61. b. $y = 1.96x^{0.66}$ **c.** Concave down   **d.** About 16   **e.** About 8.5

63. (50, 2.5). The curve is concave up (slope increasing) before $x = 50$ and concave down (slope decreasing) after $x = 50$. Therefore, the slope is maximized at $x = 50$.

EXERCISES 7.3 page 639

1. Max f is 12 (at $x = 1$), min f is -8 (at $x = -1$).   **3.** Max f is 16 (at $x = -2$), min f is -16 (at $x = 2$).

5. Max f is 9 (at $x = 1$), min f is 0 (at $x = 0$ and at $x = -2$).   **7.** Max f is 5 (at $x = 0$), min f is 0 (at $x = 5$).

9. Max f is 1 (at $x = 0$), min f is 0 (at $x = -1$ and at $x = 1$).   **11.** Max f is $\frac{1}{2}$ (at $x = 1$), min f is $-\frac{1}{2}$ (at $x = -1$).

13. a. The number is $\frac{1}{2}$.   **b.** The number is 3.   **15. a.** Both at endpoints   **b.** One at a critical value (the maximum) and one at an endpoint (the minimum)   **c.** Both at critical values   **d.** Yes; for example, [2, 10]

17. On the 20th day   **19.** 31 mph   **21.** 36 years   **23.** 12 miles from A toward B

25. Sell 40 per day, price = $400, max profit = $6500

27. 400 feet along the building and 200 feet perpendicular to the building

29. Each is 200 yards parallel to the river and 150 yards perpendicular to the river

31. 3 inches high with a base 12 inches by 12 inches, volume: 432 cubic inches   **33.** The numbers are 25 and 25.

35. $r = 2$ cm   **37.** $r = 110/\pi \approx 35$ yards, $x = 110$ yards   **39.** $x \approx 1.125$ inches, $y \approx 1.25$ megabytes

41. About 11.9 years; value will be about $254,000 (from $y \approx 254$)

43. Remove a square of side $x \approx 0.96$ inch; volume ≈ 15 square inches

EXERCISES 7.4 page 650

1. Price $14,400, sell 16 cars per day (from $x = 2$ price reductions)

3. Ticket price: $150, number sold: 450 (from $x = 5$ price reductions)

5. Rent the cars for $90, and expect to rent 54 cars (from $x = 2$ price increases)

7. 25 trees per acre (from $x = 5$ extra trees per acre)   **9.** Base: 2 feet by 2 feet, height: 1 foot

11. Base 14 inches by 14 inches, height: 28 inches, volume: 5488 cubic inches

13. 50 feet along the driveway and 100 feet perpendicular to the driveway; cost: $800   **15.** 6.4%   **17.** 16 years

19. (*Hint:* If area is A (a constant) and one side is x, then show that the perimeter is $P = 2x + 2\dfrac{A}{x}$, which is minimized at $x = \sqrt{A}$. Then show that this means the rectangle is a square.)

21. The page should be 8 inches wide and 12 inches tall.

23. $R' = 2cpx - 3cx^2 = xc(2p - 3x)$, which is zero when $x = \dfrac{2}{3} p$. The second derivative test will show that R is maximized.

25. e. 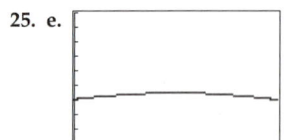　　**f.** Price: \$325; quantity: 35 bicycles　　**g.** Price: \$350; quantity: 40 bicycles

EXERCISES 7.5 page 660

1. Lot size: 400 boxes, with 10 orders during the year　　**3.** Lot size: 500 bottles, with 20 orders during the year

5. Lot size: 40 cars per order, with 20 orders during the year

7. Produce 1000 games per run, with 2 production runs during the year

9. Produce 40,000 tapes per run, with 25 runs for the year　　**11.** Population: 20,000, yield: 40,000 (from $p = 200$)

13. Population: 75,000, yield: 2250 (from $p = 75$)　　**15.** Population: 625,000, yield: 625,000 (from $p = 625$)

17. Population size: 3717, yield: 16,109 (from $x = 3.717$). (The yield exceeds the population because of reproduction later in the year.)

EXERCISES 7.6 page 671

1. $\dfrac{dy}{dx} = \dfrac{2x}{3y^2}$　　**3.** $\dfrac{dy}{dx} = \dfrac{3x^2}{2y}$　　**5.** $\dfrac{dy}{dx} = \dfrac{3x^2 + 2}{4y^3}$　　**7.** $\dfrac{dy}{dx} = -\dfrac{x + 1}{y + 1}$　　**9.** $\dfrac{dy}{dx} = -\dfrac{2y}{x}$ (after simplification)

11. $\dfrac{dy}{dx} = \dfrac{-y + 1}{x}$　　**13.** $\dfrac{dy}{dx} = -\dfrac{y - 1}{2x}$ (after simplification)　　**15.** $\dfrac{dy}{dx} = \dfrac{1}{3y^2 - 2y + 1}$

17. $\dfrac{dy}{dx} = -\dfrac{y^2}{x^2}$ (after simplification)　　**19.** $\dfrac{dy}{dx} = \dfrac{3x^2}{2(y - 2)}$　　**21.** $\dfrac{dy}{dx} = 2$　　**23.** $\dfrac{dy}{dx} = -3$　　**25.** $\dfrac{dy}{dx} = -1$　　**27.** $\dfrac{dy}{dx} = -4$

29. $p' = -\dfrac{2}{2p + 1}$　　**31.** $p' = \dfrac{1}{24p + 4}$　　**33.** $p' = -\dfrac{p}{3x}$ (after simplification)　　**35.** $p' = -\dfrac{p + 5}{x + 2}$

37. $\dfrac{dp}{dx} = -4$ *Interpretation:* The rate of change of price with respect to quantity is -4, so price decreases by about \$4 increases by 1.

39. a. $s' = \dfrac{3r^2}{2s} = 8$　　**b.** $r' = \dfrac{2s}{3r^2} = \dfrac{1}{8}$　　**c.** $\dfrac{ds}{dr} = 8$ means that the rate of change of sales with respect to research expenditures is 8, so that increasing research by \$1 million will increase sales by about \$8 million (at these levels of r and s).

$\dfrac{dr}{ds} = \dfrac{1}{8}$ means that the rate of change of research expenditures with respect to sales is $\dfrac{1}{8}$, so that increasing sales by \$1 million will increase research by about $\dfrac{1}{8}$ million dollars (at these levels of r and s).

41. $3x^2x' + 2yy' = 0$　　**43.** $2xx'y + x^2y' = 0$　　**45.** $6xx' - 7x'y - 7xy' = 0$　　**47.** $2xx' + x'y + xy' = 2yy'$

49. Decreasing by $72\pi \approx 226$ in^3/hr　　**51.** $32\pi \approx 101$ cm^3/week　　**53.** Increasing by \$16,000 per day

55. Increasing by 400 cases per year　　**57.** Slowing by $\dfrac{1}{2}$ mm/sec per year　　**59.** Yes (65.8 mph)

CHAPTER 7 REVIEW EXERCISES page 675

1.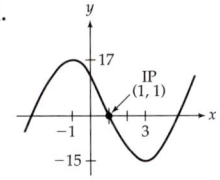

2.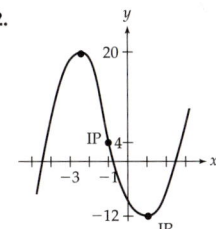

3.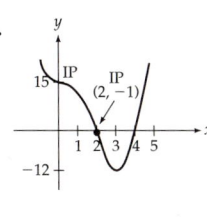

4.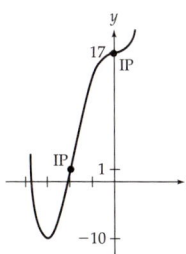

5.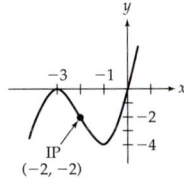

6.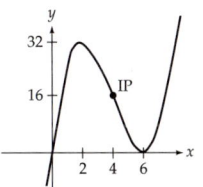

7.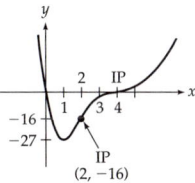

8.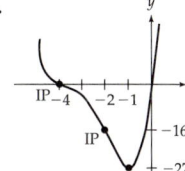

9.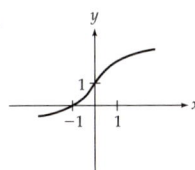

10.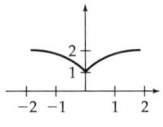

11.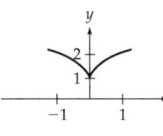

12.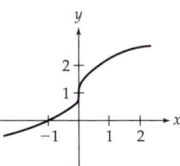

13.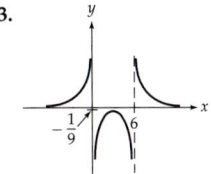

14.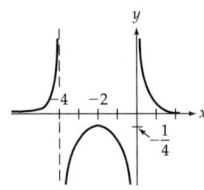

15.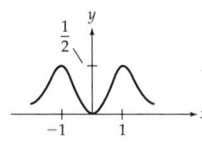

16.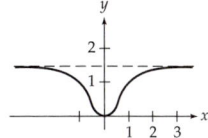

17.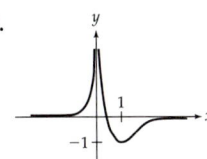

18.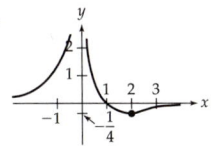

19. Max f is 220 (at $x = 5$), min f is -4 (at $x = 1$).   **20.** Max f is 130 (at $x = 5$), min f is -32 (at $x = 2$).
21. Max f is 64 (at $x = 0$), min f is -64 (at $x = 4$).   **22.** Max f is 6401 (at $x = 10$), min f is 1 (at $x = 0$ and $x = 2$).
23. Max h is 4 (at $x = 9$), min h is 0 (at $x = 1$).   **24.** Max f is 10 (at $x = 0$), min f is 0 (at $x = 10$ and at $x = -10$).
25. Max g is 25 (at $w = 3$ and $w = -3$), min g is 0 (at $w = 2$ and at $w = -2$).

26. Max g is 16 (at $x = 4$), min g is 0 (at $x = 0$ and at $x = 8$).   **27.** Max f is $\dfrac{1}{2}$ (at $x = 1$), min f is $-\dfrac{1}{2}$ (at $x = -1$).

28. Max f is $\dfrac{1}{4}$ (at $x = 2$), min f is $-\dfrac{1}{4}$ (at $x = -2$).   **29.**

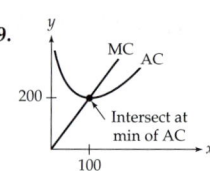

30. ① At the minimum point of AC(x), the derivative of AC(x) must be zero.

② Quotient rule applied to $AC(x) = \dfrac{C(x)}{x}$

③ Factoring out $\dfrac{1}{x}$

④ Simplifying inside the square bracket

⑤ Recognizing $C'(x)$ as the marginal cost function and $\dfrac{C(x)}{x}$ as the average cost function

⑥ $0 = \dfrac{1}{x}[MC(x) - AC(x)]$ means that the quantity in the square bracket must equal zero, and so the marginal cost must equal average cost at this x-value, where average cost is minimized.

31. $v = 2c$, which means that the tugboat should travel through the water at twice the speed of the current.

32. 3600 square feet   **33.** 1800 square feet   **34.** 15 cubits (gilded side) by 135 cubits

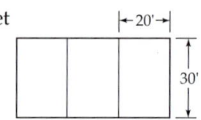

35. Base: 10 inches by 10 inches, height: 5 inches   **36.** Base radius = 2 inches, height = 4 inches

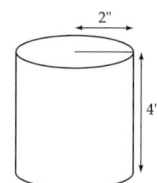

37. Radius = 2 inches, height = 2 inches   **38.** $v = \sqrt[4]{\dfrac{aw^2}{3b}}$

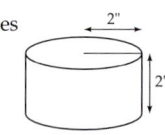

39. Price: \$2400 each, quantity: 9 per week   **40.** 5 weeks   **41.** $x = \frac{3}{4}$ mile
42. a. $t \approx 0.624 = 62.4\%$ **b.** \$1.26   **43.** Radius ≈ 1.2 inches, height ≈ 4.8 inches
44. $x \approx 1.59$ inches, volume ≈ 33.07 cubic inches   **45.** $x \approx 1.13$ inches, volume ≈ 12.13 cubic inches
46. 600 per run, with $1\frac{1}{2}$ runs per year (or 3 runs in 2 years)   **47.** Lot size: 50 per order, requiring 10 orders
48. Population: 150,000 ($p = 150$), yield: 450,000   **49.** Population: $p = 900$ (thousand), yield: 900 (thousand)

50. $y' = \dfrac{-6x - 4y}{4x + y}$ **51.** $y' = \dfrac{-y^2}{2xy - 1}$ (after simplification)   **52.** $y' = \dfrac{-2y^2 + 6xy}{4xy - 3x^2}$

53. $y' = \dfrac{y^{1/2}}{x^{1/2}}$ (after simplification)   **54.** $y' = -1$ **55.** $y' = \dfrac{1}{7}$ **56.** $y' = -\dfrac{1}{6}$ **57.** $y' = 1$ **58.** 600 in³/hr

59. Increasing by \$4200 per day   **60.** Increasing by \$45,000 per month

ANSWERS TO CUMULATIVE REVIEW EXERCISES FOR CHAPTERS 6–7 page 680

1. 20.085   **2.** $f'(x) = 4x - 5$ **3.** $12x^{1/2} + 6x^{-3}$ **4.** $f'(x) = 9x^8 + 10x^4 - 8x^3$ **5.** $f'(x) = \dfrac{11}{(3x - 2)^2}$

6. $P'(8) = 1200$, so in 8 years the population will be increasing by 1200 million people per year.
$P''(8) = -50$, so in 8 years the rate of growth will be slowing (by 50 million people per year per year).

7. $\dfrac{2x}{\sqrt{2x^2 - 5}}$ **8.** $12(3x + 1)^3(4x + 1)^3 + 12(3x + 1)^4(4x + 1)^2 = 12(3x + 1)^3(4x + 1)^2(7x + 2)$ **9.** $\dfrac{12(x - 2)^2}{(x + 2)^4}$

10.

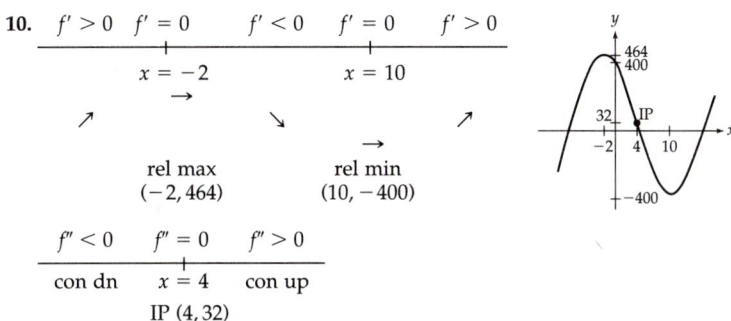

| $f' > 0$ | $f' = 0$ | $f' < 0$ | $f' = 0$ | $f' > 0$ |

$x = -2$ $x = 10$

rel max   rel min
$(-2, 464)$ $(10, -400)$

| $f'' < 0$ | $f'' = 0$ | $f'' > 0$ |

con dn $x = 4$ con up
IP $(4, 32)$

11.

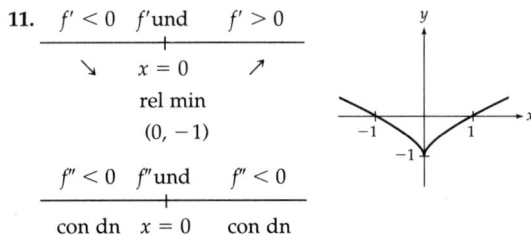

| $f' < 0$ | f'und | $f' > 0$ |

$x = 0$
rel min
$(0, -1)$

| $f'' < 0$ | f''und | $f'' < 0$ |

con dn $x = 0$ con dn

12. 15,000 square feet   **13.** Price: $170, quantity: 18 per day

14. $\dfrac{dy}{dx} = \dfrac{-x^2 - 3y^2}{6xy + 1}$ (after simplification). Evaluating at $(1, 2)$ gives $\dfrac{dy}{dx} = -1$.   **15.** $\dfrac{2}{\pi} \approx 0.64$ ft/min

EXERCISES 8.1 page 694

1.   **3.**

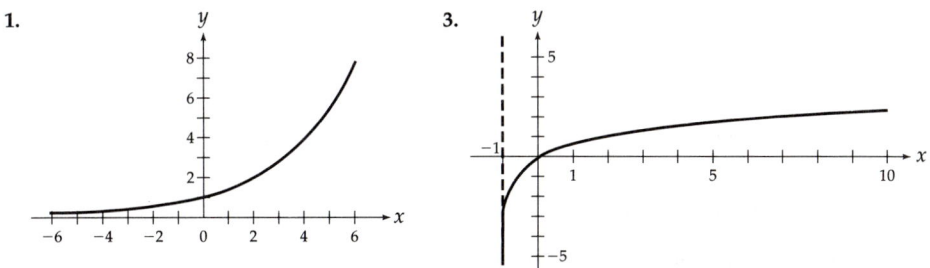

5. a. 99   **b.** $-\frac{1}{2}$ **c.** 0   **7.** $\ln 2.75 \approx 1.011600912$, $e^{1.011600912} \approx 2.75$ **9.** $\ln x$
11. 1   **13.** $2x$ **15. a.** $2143.59   **b.** $2203.76   **c.** $2218.18   **d.** $2225.54

17. First City Trust Co. (paying $11,043.36) is better by $26.34.   **19. a.** $2195.97   **b.** $2095.97
21. a. $1.53 million   **b.** $2.23 million (both approx.)   **23.** 5.8 years (approx.)
25. a. $10,555   **b.** $18,762 (both approx.)   **27.** About 18 years   **29.** About 13.5 years

31. a. About 17 days   **b.** About 58 days   **c.**

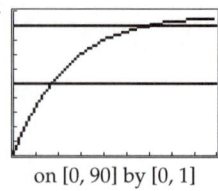

on [0, 90] by [0, 1]

33. a. About 67%   **b.** About 20 minutes   **c.**

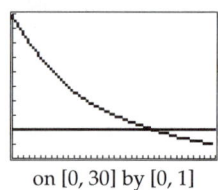

on [0, 30] by [0, 1]

35. a. About 29,670 people   **b.** About 5.4 hours   **c.**

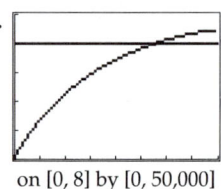

on [0, 8] by [0, 50,000]

37. a. About every 4 hours (from $x \approx 4.2$)   **b.**

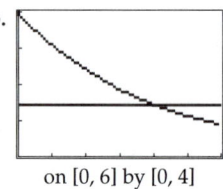

on [0, 6] by [0, 4]

39. About 138 days   **41. a.** $e^{-kt} = \dfrac{1}{2}$ Taking natural logs gives $-kt = \ln \frac{1}{2} = -\ln 2$ so $t = \dfrac{\ln 2}{k}$

b. (Half-life) $= \dfrac{\ln 2}{0.018} \approx 38.5$ hours

EXERCISES 8.2 page 709

1. $2x \ln x + x$ **3.** $\dfrac{2}{x}$ **5.** $\dfrac{1}{2}x^{-1}$ **7.** $\dfrac{6x}{x^2 + 1}$ **9.** $\dfrac{1}{x}$ **11.** $\dfrac{xe^x - 2e^x}{x^3}$ or $\dfrac{e^x(x - 2)}{x^3}$ **13.** $(3x^2 + 2)e^{x^3 + 2x}$

15. $x^2 e^{x^3/3}$ **17.** $1 + e^{-x}$ **19.** 2   **21.** $e^{1 + e^x}e^x$ or $e^{1 + x + e^x}$ **23.** exe^{-1} **25.** 0   **27.** $\dfrac{4x^3}{x^4 + 1} - 2e^{(1/2)x} - 1$

29. $2x \ln x + 2xe^{x^2}$ **31. a.** $\dfrac{1 - 5 \ln x}{x^6}$ **b.** 1   **33. a.** $\dfrac{4x^3}{x^4 + 48}$ **b.** $\dfrac{1}{2}$ **35. a.** $\dfrac{e^x - 3}{e^x - 3x}$ **b.** -2

37. a. $5 \ln x + 5$ **b.** $5 \ln 2 + 5 \approx 8.466$ **39. a.** $\dfrac{xe^x - e^x}{x^2}$ **b.** $\dfrac{2e^3}{9} \approx 4.463$

41. $-4x^3e^{-x^5/5} + x^8e^{-x^5/5}$ or $x^3e^{-x^5/5}(x^5 - 4)$ **43.** $f^{(n)}(x) = k^ne^{kx}$ **45.**

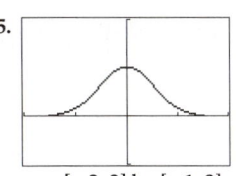

on $[-2, 2]$ by $[-1, 2]$

rel max: $(0, 1)$
IP: $(\frac{1}{2}, 0.61)$
$(-\frac{1}{2}, 0.61)$

47.

on $[-5, 5]$ by $[-1, 4]$

rel min: $(0, 0)$
IP: $(1, 0.69)$
$(-1, 0.69)$

49.

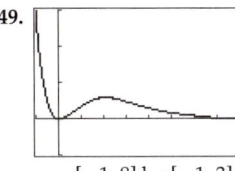

on $[-1, 8]$ by $[-1, 3]$

rel min: $(0, 0)$
rel max: $(2, 0.54)$
IP: $(0.59, 0.19)$
$(3.4, 0.38)$

51.

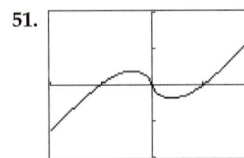

on $[-2, 2]$ by $[-2, 2]$

rel max $(-0.37, 0.37)$
rel min $(0.37, -0.37)$

53. a. Increasing by about \$50 per year   **b.** Increasing by about \$82.44 per year
55. Increasing by about 0.12 billion ($=120$ million) people per year
57. a. Decreasing by 0.06 cc/hr   **b.** Decreasing by 0.054 cc/hr
59. a. Increasing by about 81.4 (thousand) sales per week   **b.** Increasing by about 33 (thousand) sales per week
61. $p = \$100$ **63. a.** $R(x) = 400xe^{-0.20x}$ **b.** Quantity: $x = 5$ (thousand), price: $p = \$147.15$ **65.** $r = a/b$
67. a. After 15 minutes the temperature of the beer is 57.5° and increasing at the rate of 43.8°/hour.
b. After 1 hour the temperature of the beer is 69.1° and increasing at the rate of 3.2°/hour.   **69.** 2.08 seconds
71. $p = \$500$ **73.** After about 2.75 hours
75. a. $(\ln 10)10^x$ **b.** $(\ln 3)(2x)3^{x^2+1}$ **c.** $(\ln 2)3 \cdot 2^{3x}$ **d.** $(\ln 5)6x \cdot 5^{3x^2}$ **e.** $-(\ln 2)2^{4-x}$
77. a. $\dfrac{1}{(\ln 2)x}$ **b.** $\dfrac{2x}{(\ln 10)(x^2 - 1)}$ **c.** $\dfrac{4x^3 - 2}{(\ln 3)(x^4 - 2x)}$

EXERCISES 8.3 page 723

1. a. $\dfrac{2}{t}$ **b.** 2 and 0.2   **3. a.** 0.2   **b.** 0.2   **5. a.** $2t$ **b.** 20   **7. a.** $-2t$ **b.** -20 **9. a.** $\dfrac{1}{2(t - 1)}$ **b.** $\dfrac{1}{10}$
11. 0.0071 or 0.71%   **13. a.** 0.012 or 1.2%   **b.** Yes, in about 15.3 years
15. a. $E(p) = \dfrac{5p}{200 - 5p}$ **b.** Inelastic ($E = \frac{1}{3}$)   **17. a.** $E(p) = \dfrac{2p^2}{300 - p^2}$ **b.** Unitary elastic ($E = 1$)
19. a. $E(p) = 1$ **b.** Unitary elastic ($E = 1$)   **21. a.** $E(p) = \dfrac{3p}{2(175 - 3p)}$ **b.** Elastic ($E = 3$)
23. a. $E(p) = 2$ **b.** Elastic ($E = 2$)   **25. a.** $E(p) = 0.01p$ **b.** Elastic ($E = 2$)   **27.** Lower the price ($E = 8$)
29. No ($E = 1.25$)   **31.** Yes ($E = \frac{3}{8} = 0.375$)   **33.** Lower its price ($E = 1.2$)   **35.** $E = 0.112$
37. $E(p) = \dfrac{-pa(-c)e^{-cp}}{ae^{-cp}} = cp$ **39.** $E_s(p) = n$ **41. a.** $E \approx 0.35$ **b.** Raise the price   **c.** $p \approx \$20,400$

CHAPTER 8 REVIEW EXERCISES page 727

1. a. $18,845.41   **b.** $18,964.81   **2.** East Side Credit ($5718.01) by $18.99
3. a. $V(t) = 800,000(0.8)^t$ **b.** $327,680   **4.** Drug B
5. In about 2060 (from $x \approx 65$ years after 1995)

6. a. About 14 per hour   **b.** About 15 per hour   **c.**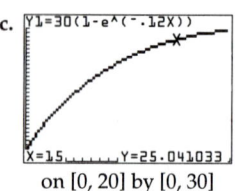
on [0, 20] by [0, 30]

7. About 10 years (from $x \approx 9.7$)   **8.** About 9.2 years   **9.** $f(x) = \ln x + 1$
10. a. $\ln 9.3 \approx 2.2300144$ **b.** $e^{2.2300144} \approx 9.3$ **11.** 2.3 hours   **12.** 13.7 years   **13. a.** 12 years   **b.** 11.9 years
14. a. 72 years   **b.** 69.7 years   **15.** $\dfrac{\ln k}{\ln (1 + r)}$ **16.** $\dfrac{\ln k}{r}$
17. a. In about $6\frac{1}{4}$ years (from $x \approx 25$ quarters)   **b.** In about 6.2 years   **18. a.** About 11 days   **b.** About 16 days
19. $\dfrac{1}{x}$ **20.** $\dfrac{4x}{x^2 - 1}$ **21.** $\dfrac{-1}{1 - x}$ or $\dfrac{1}{x - 1}$ **22.** $\dfrac{x}{x^2 + 1}$ **23.** $\dfrac{1}{3x}$ **24.** 1   **25.** $\dfrac{2}{x}$ **26.** $\ln x$ **27.** $-2xe^{-x^2}$
28. $-e^{1-x}$ **29.** $2x$ **30.** $2x(\ln x)e^{x^2\ln x - x^2/2}$ **31.** $10x + 2 \ln x + 2$ **32.** $6x^2 + 3 \ln x + 3$ **33.** $6x^2 - 3e^{2x} - 6xe^{2x}$

34. $4 - 4xe^{2x} - 4x^2e^{2x}$ **35.** 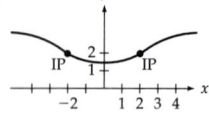   Rel min: $(0, \ln 4) \approx (0, 1.4)$ IPs: $(2, \ln 8) \approx (2, 2.1)$, $(-2, \ln 8) \approx (-2, 2.1)$

36. Rel max: $(0, 16)$ IPs: $(2, 16e^{-1/2}) \approx (2, 9.7)$, $(-2, 16e^{-1/2}) \approx (-2, 9.7)$

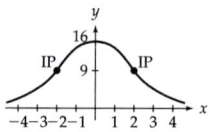

37. a. Increasing by 136 thousand per week   **b.** Increasing by 55 thousand per week
38. a. Decreasing by 0.12 cc per hour   **b.** Decreasing by 0.08 cc per hour   **39.** Decreasing by 33.3% per second
40. a. Increasing by 3.5°/hour   **b.** Increasing by 2.1°/hour
41. a. Increasing by 6667 per hour   **b.** Increasing by 816 per hour   **42.** 25 years
43. a. $R(x) = 200xe^{-0.25x}$ **b.** Quantity $x = 4$ (thousand), price $p = 73.58
44. a. $R(x) = 5x - x \ln x$ **b.** Quantity $x = e^4 \approx 54.60$ Price = $1   **45.** Price = $50

46. 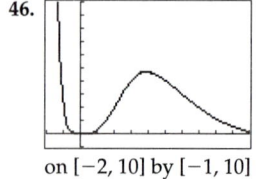   rel min: (0, 0)
rel max: (4, 4.69)
IP: (2, 2.17)
   (6, 3.21)

on [−2, 10] by [−1, 10]

47. 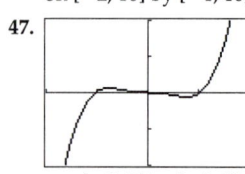   rel max: (−0.72, 0.12)   **48.** $p \approx 769 **49.** $x \approx 130 **50.** 0.0033 or 0.33%
rel min: (0.72, −0.12)
IP: (0, 0)
   (−0.43, 0.07)
   (0.43, −0.07)

on [−2, 2] by [−2, 2]

51. 0.0031 or 0.31%   **52.** Raise prices ($E = 0.8$)   **53.** Raise prices ($E = 0.7$)   **54.** $E = 0.44$ **55.** 0.104 or 10.4%
56. a. $E \approx 1.29$ **b.** Lower the price   **c.** About \$8700 (from $p \approx 8.7$)

EXERCISES 9.1 page 744

1. $\dfrac{1}{5}x^5 + C$ **3.** $\dfrac{3}{5}x^{5/3} + C$ **5.** $\dfrac{2}{3}u^{3/2} + C$ **7.** $-\dfrac{1}{3}w^{-3} + C$ **9.** $2\sqrt{z} + C$ **11.** $x^6 + C$

13. $2x^4 - x^3 + 2x + C$ **15.** $4x^{3/2} + \dfrac{3}{2}x^{2/3} + C$ **17.** $6x^{8/3} + 24x^{-2/3} + C$ **19.** $6t^{5/3} + 3t^{1/3} + C$

21. $\dfrac{1}{3}x^3 - x^2 + x + C$ **23.** $\dfrac{2}{3}w^{3/2} + 4w^{5/2} + C$ **25.** $2x^3 - 3x^2 + x + C$ **27.** $\dfrac{1}{3}x^3 + x^2 - 8x + C$

29. $\dfrac{1}{3}r^3 - r + C$ **31.** $\dfrac{1}{2}x^2 - x + C$ **33.** $\dfrac{1}{4}t^4 + t^3 + \dfrac{3}{2}t^2 + t + C$ **35. a.** $-\dfrac{1}{2}x^{-2} + C$ **b.** $\dfrac{x + C}{\frac{1}{4}x^4 + C_1}$ (where C_1 is

another arbitrary constant)   **37. b.**

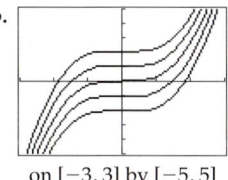

on $[-3, 3]$ by $[-5, 5]$

39. $C(x) = 8x^{5/2} - 9x^{5/3} + x + 4000$

41. $R(x) = 9x^{4/3} + 2x^{3/2}$ **43. a.** $D(t) = -0.03t^3 + 4t^2$ **b.** 370 feet   **45. a.** $6t^{1/2}$ **b.** 30 words
47. a. $P(t) = 16t^{5/2}$ **b.** 512 tons   **c.** No   **49.** $\dfrac{1}{x}$

EXERCISES 9.2 page 755

1. $\dfrac{1}{3}e^{3x} + C$ **3.** $4e^{\frac{1}{4}x} + C$ **5.** $20e^{0.05x} + C$ **7.** $-\dfrac{1}{2}e^{-2y} + C$ **9.** $-2e^{-0.5x} + C$ **11.** $9e^{\frac{2}{3}x} + C$

13. $-5 \ln |x| + C$ **15.** $3 \ln |x| + C$ **17.** $\dfrac{3}{2} \ln |v| + C$ **19.** $\dfrac{1}{3}e^{3x} - 3 \ln |x| + C$ **21.** $6e^{0.5t} - 2 \ln |t| + C$

23. $\dfrac{1}{3}x^3 + \dfrac{1}{2}x^2 + x + \ln |x| - x^{-1} + C$ **25.** $250e^{0.02t} - 200e^{0.01t} + C$ **27. a.** $360e^{0.05t} - 355$ **b.** About 624
29. a. $50 \ln t$ (since $t > 1$, absolute value bars are not needed)   **b.** No (about 170 sold)
31. a. $800e^{0.02t} - 800$ **b.** In about 2016 (21 years from 1995)   **33. a.** $500e^{0.4x} - 500$ **b.** About \$3195
35. a. $60e^{-0.2t} + 10$ **b.** In about 5 hours   **37. a.** $-4000e^{-0.2t} + 4000$ **b.** About $3\frac{1}{2}$ years
39. $\dfrac{1}{2}x^2 + 2x + \ln |x| + C$ **41.** $t + 2 \ln |t| + 3t^{-1} + C$ **43.** $\dfrac{1}{3}x^3 - 3x^2 + 12x - 8 \ln |x| + C$
45. a. $530e^{0.01t} - 530$ **b.** In about 2017 (22 years after 1995)

EXERCISES 9.3 page 773

1. i. 2.8 square units   **ii.** 3 square units
3. i. 4.411 square units (or 4.412, depending on rounding)   **ii.** $\frac{14}{3} \approx 4.667$ square units
5. i. 0.719 square units   **ii.** $\ln 2 \approx 0.693$ square units

7. i.

n	Area
10	2.9
100	2.99
1000	2.999

ii. 3 square units   **9. i.**

n	Area
10	4.515
100	4.652
1000	4.665

ii. $\frac{14}{3} \approx 4.667$ square units

11. i.

n	Area
10	0.719
100	0.696
1,000	0.693

ii. $\ln 2 \approx 0.6931$ square units

13.   9 square units

$y = x^2$

15.   8 square units

$y = 4 - x$

17. $\ln 2$ square units   **19.** 160 square units   **21.** 19 square units

$y = \dfrac{1}{x}$

23. 2 square units   **25.** 16 square units   **27.** $\ln 5$ square units   **29.** $\ln 2 - \frac{7}{3}$ square units

31. 4 square units (from $2e^{\ln 3} - 2$)   **33.** $2e - 2$ square units   **35.** 5 square units   **37.** $\frac{15}{32}$ square unit

39. $4 + \ln 2$ square units   **41.** $\frac{111}{100}$ **43.** 13   **45.** $\frac{3}{4}$ **47.** 1 (from $\ln e - \ln 1$)   **49.** $78 + \ln 3$ **51.** $-3 \ln 2$

53. $4e^3 - 4$ **55.** $5e - 5e^{-1}$ **57.** 1 (from $e^{\ln 3} - e^{\ln 2}$)   **59.** $\frac{7}{2} + \ln 2$ **61.** 1.107   **63.** 2.925   **65.** 92.744

67. a. 9   **b.** Continuing the calculation: $= \left(\dfrac{1}{3} 3^3 + C \right) - \left(\dfrac{1}{3} 0^3 + C \right) = \dfrac{1}{3} 3^3 + C - \dfrac{1}{3} 0^3 - C = \dfrac{1}{3} \cdot 27 = 9$

c. The C always cancels because it is both added and subtracted in the evaluation step.

69. $\displaystyle\int_1^a \dfrac{1}{x}\,dx = \ln |x| \Big|_1^a = \ln a - \ln 1 = \ln a$ **71.** $3\frac{1}{2} + \frac{\pi}{4}$ **73.** 132 units   **75.** 411 checks

77. $-300e^{-2} + 300 \approx \$259.40$ **79.** $162.5e^{0.8} - 162.5 \approx 199$, so about $2   **81.** $23e^{0.1} - 23 \approx 2.4$ million tons

83. $-30e^{-2} + 30 \approx 26$ words   **85.** $\dfrac{a}{-b+1} B^{-b+1} - \dfrac{a}{-b+1} A^{-b+1} = \dfrac{a}{1-b} (B^{1-b} - A^{1-b})$

87. $-4{,}000{,}000e^{-0.6} + 4{,}000{,}000 \approx \$1{,}804{,}753$ **89. a.** $\frac{1}{2}, \frac{1}{3}, \frac{1}{4}, \frac{1}{5}$ **b.** Area $= \dfrac{1}{(n+1)}$ **91.** 586 cars   **93.** 4023 people

EXERCISES 9.4 page 788

1. 3   **3.** 4   **5.** $\dfrac{1}{5}$ **7.** 5   **9.** $\dfrac{104}{3}$ or $34\frac{2}{3}$ **11.** 3   **13.** $e - 1$ **15.** $\ln 2$ **17.** $\dfrac{1}{n+1}$ **19.** $a + b$ **21.** 0.845

23. 318   **25.** 70   **27.** About 25.6 tons   **29.** $3194.53   **31.** $24.04 million   **33.** 6 square units

35. $\dfrac{1}{2} e^4 - e^2 + \dfrac{1}{2}$ square units   **37. a.**

$x^2 + 4$
$2x + 1$

b. 9 square units

39. a. 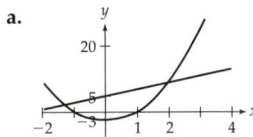   **b.** 18 square units   **41.** 4 square units   **43.** 32 square units   **45.** $\frac{1}{12}$ square unit

47. 1250 square units   **49.** 5.694 square units (rounded)   **51.** About 123 million   **53. a.** 10   **b.** $6000

55. About $529 billion   **57.** About $139 thousand   **59. e.**   **f.** 53,167 lives

on [0, 10] by [0, 55,000]

61. $(2x + 5)e^{x^2 + 5x}$ **63.** $\dfrac{2x + 5}{x^2 + 5x}$

EXERCISES 9.5 page 801

1. $60,000   **3.** $10,000   **5.** $7500   **7.** $5,285 (rounded)   **9.** $1000   **11.** $160,000
13. a. $x = 500$ **b.** $50,000   **c.** $25,000   **15. a.** $x = 50$ **b.** $2500   **c.** $7500
17. a. $x \approx 119.48$ **b.** $10,065   **c.** $3,446 (all rounded)   **19.** 0.52   **21.** 0.35   **23.** 0.2   **25.** $1 - \dfrac{2}{n+1} = \dfrac{n-1}{n+1}$
27. 0.46   **29.** 0.28   **31.** $L(x) = x^{2.13}$, Gini index ≈ 0.36 **33.** $4(x^5 - 3x^3 + x - 1)^3(5x^4 - 9x^2 + 1)$ **35.** $\dfrac{4x^3}{x^4 + 1}$
37. $3x^2 e^{x^3}$

EXERCISES 9.6 page 813

1. $\dfrac{1}{10}(x^2 + 1)^{10} + C$ **3.** $\dfrac{1}{20}(x^2 + 1)^{10} + C$ **5.** $\dfrac{1}{5}e^{x^5} + C$ **7.** $\dfrac{1}{6}\ln(x^6 + 1) + C$
9. $u = x^3 + 1$, $du = 3x^2\,dx$; the powers in the integral and the du do not match.
11. $u = x^4$, $du = 4x^3\,dx$; the powers in the integral and the du do not match.   **13** $\dfrac{1}{24}(x^4 - 16)^6 + C$ **15** $-\dfrac{1}{2}e^{-x^2} + C$
17. $\dfrac{1}{3}e^{3x} + C$ **19.** Cannot be found by our substitution formulas   **21.** $\dfrac{1}{5}\ln |1 + 5x| + C$ **23.** $\dfrac{1}{4}(x^2 + 1)^{10} + C$
25. $\dfrac{1}{5}(z^4 + 16)^{5/4} + C$ **27.** Cannot be found by our substitution formulas   **29.** $\dfrac{1}{24}(2y^2 + 4y)^6 + C$
31. $\dfrac{1}{2}e^{x^2 + 2x + 5} + C$ **33.** $\dfrac{1}{12}\ln |3x^4 + 4x^3| + C$ **35.** $-\dfrac{1}{12}(3x^4 + 4x^3)^{-1} + C$ **37.** $-\dfrac{1}{2}\ln |1 - x^2| + C$
39. $\dfrac{1}{16}(2x - 3)^8 + C$ **41.** $\dfrac{1}{2}\ln(e^{2x} + 1) + C$ **43.** $\dfrac{1}{2}(\ln x)^2 + C$ **45.** $2e^{x^{1/2}} + C$ **47.** $\dfrac{1}{4}x^4 + \dfrac{1}{3}x^3 + C$
49. $\dfrac{1}{6}x^6 + \dfrac{2}{5}x^5 + \dfrac{1}{4}x^4 + C$ **51.** $\dfrac{1}{2}e^9 - \dfrac{1}{2}$ **53.** $\dfrac{1}{2}\ln 2$ **55.** $32\dfrac{2}{3}$ **57.** $-\ln 2$ **59.** $3e^2 - 3e$
61. a. $u^n u'$ **b.** $\int u^n u'\,dx$ **63. a.** $\dfrac{u'}{u}$ **b.** $\int \dfrac{u'}{u}\,dx$ **65.** $\dfrac{1}{2}\ln(2x + 1) + 50$ **67.** $\dfrac{1}{2}$ million
        ⌐—Agree—⌐   ⌐—Agree—⌐
69. $\dfrac{1}{3}\ln 5 - \dfrac{1}{3}\ln 2 \approx 0.305$ million   **71.** $20\dfrac{1}{3}$ units   **73.** About 346   **75.** $\dfrac{1}{2}\ln 10 \approx 1.15$ tons

CHAPTER 9 REVIEW EXERCISES page 817

1. $8x^3 - 4x^2 + x + C$ **2.** $3x^4 + 3x^2 - 3x + C$ **3.** $4x^{3/2} - 5x + C$ **4.** $6x^{4/3} - 2x + C$ **5.** $6x^{5/3} - 2x^2 + C$

6. $2x^{5/2} - 3x^2 + C$ **7.** $\frac{1}{3}x^3 - 16x + C$ **8.** $x^3 + x^2 + 4x + C$ **9.** $C(x) = 2x^{1/2} + 4x + 20,000$

10. a. $P(t) = 200t^{3/2} + 40,000$ **b.** 52,800 people   **11.** $2e^{1/2x} + C$ **12.** $-\frac{1}{2}e^{-2x} + C$ **13.** $4\ln|x| + C$

14. $2\ln|x| + C$ **15.** $2e^{3x} - 6\ln|x| + C$ **16.** $\frac{1}{2}x^2 - \ln|x| + C$ **17.** $3x^3 + 2\ln|x| + 2e^{3x} + C$

18. $-x^{-1} + \ln|x| - e^{-x} + C$ **19. a.** $800e^{0.05t} - 800$ **b.** About 2042 (from $t \approx 47$)

20. a. $2000e^{0.1t} - 2000$ **b.** About 5.6 years

21. a. $400e^{0.02t} - 400$ **b.** In about 15 years from 1995, or in about the year 2010

22. a. $200\ln x$ **b.** In about 20 months   **23.** 36   **24.** 90   **25.** 4   **26.** $\ln 5$ **27.** $1 - e^{-2}$ **28.** $2e - 2$

29. $20e^5 - 100e + 80$ **30.** $25e^{0.4} - 50e^{0.2} + 25$ **31.** 13 square units   **32.** 36 square units

33. $6e^6 - 6$ square units   **34.** $2e^2 - 2$ square units   **35.** $\ln 100$ square units   **36.** $\ln 1000$ square units

37.
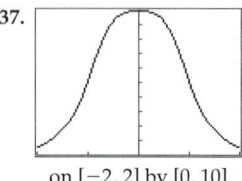
on $[-2, 2]$ by $[0, 10]$
Area ≈ 21.4 square units

38.
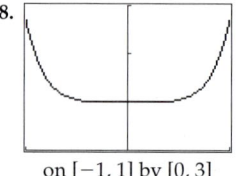
on $[-1, 1]$ by $[0, 3]$
Area ≈ 2.54 square units

39. About 29.5 kilograms   **40.** 12 words   **41.** $1.5e - 1.5 \approx 2.6°$ **42.** \$1640   **43.** \$653.39

44. About 143 pages   **45. a.** 2.28 square units   **b.** $\frac{8}{3} \approx 2.667$ square units

46. a. 4.884 square units   **b.** $\frac{16}{3} \approx 5.333$ square units

47. a.

n	Area
10	5.899
100	7.110
1000	7.239
10000	7.252

b. $e^2 - e^{-2} \approx 7.254$ square units

48. a.

n	Area
10	1.506
100	1.398
1000	1.387
10000	1.386

b. $\ln 4 \approx 1.386$ square units

49. $\frac{4}{3}$ square units   **50.** 108 square units   **51.** $\frac{1}{6}$ square unit   **52.** $\frac{3}{10}$ square unit   **53.** 2 square units

54. $\frac{1}{2}$ square unit   **55.** About 17.13 square units   **56.** About 0.496 square unit   **57.** 0.25   **58.** $\frac{28}{3}$ or $9\frac{1}{3}$

59. About 4.72   **60.** About 2.77   **61.** 10.3 billion   **62.** About \$5800

63. About \$629,000 (from 6.29 hundred thousand dollars)   **64.** About \$49.95   **65.** 27 square meters

66. About \$934 billion   **67.** About \$372 million

68. b.
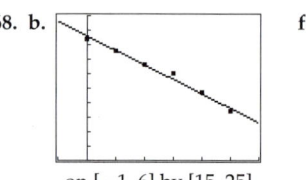
on $[-1, 6]$ by $[15, 25]$
f.
on $[0, 13]$ by $[0, 40]$
g. About 96.8 million people   **69.** \$480,000

70. \$160,000   **71.** About \$5623   **72.** About \$611   **73.** GI ≈ 0.56   **74.** GI ≈ 0.43

75. About 0.23   **76.** About 0.68   **77.** $\frac{1}{4}(x^3 - 1)^{4/3} + C$ **78.** $\frac{1}{6}(x^4 - 1)^{3/2} + C$

79. Cannot be integrated by our substitution formulas   **80.** Cannot be integrated by our substitution formulas

81. $-\dfrac{1}{3}\ln|9-3x|+C$ **82.** $-\dfrac{1}{2}\ln|1-2x|+C$ **83.** $\dfrac{1}{3}(9-3x)^{-1}+C$ **84.** $\dfrac{1}{2}(1-2x)^{-1}+C$

85. $\dfrac{1}{2}(8+x^3)^{2/3}+C$ **86.** $(9+x^2)^{1/2}+C$ **87.** $-\dfrac{1}{2}(w^2+6w-1)^{-1}+C$ **88.** $-\dfrac{1}{2}(t^2-4t+1)^{-1}+C$

89. $\dfrac{2}{3}(1+\sqrt{x})^3+C$ **90.** $(1+x^{1/3})^3+C$ **91.** $\ln|e^x-1|+C$ **92.** $\ln|\ln x|+C$ **93.** $\dfrac{61}{3}$ or $20\dfrac{1}{3}$ **94.** 2

95. 2   **96.** $\dfrac{5}{12}$ **97.** $-\ln 7$ **98.** $-\ln 2$ **99.** $\dfrac{1}{4}e-\dfrac{1}{4}$ **100.** $\dfrac{1}{5}e-\dfrac{1}{5}$ **101.** 8 square units   **102.** 5 square units

103. $\dfrac{1}{4}-\dfrac{1}{4}e^{-4}=\dfrac{1}{4}(1-e^{-4})\approx 0.25$ **104.** $\dfrac{1}{4}\ln 13\approx 0.64$ **105.** $C(x)=(2x+9)^{1/2}+97$ **106.** $\ln 28\approx 3.33$ degrees

EXERCISES 10.1 page 831

1. $\dfrac{1}{2}e^{2x}+C$ **3.** $\dfrac{1}{2}x^2+2x+C$ **5.** $\dfrac{2}{3}x^{3/2}+C$ **7.** $\dfrac{1}{5}(x+3)^5+C$ **9.** $\dfrac{1}{2}xe^{2x}-\dfrac{1}{4}e^{2x}+C$

11. $\dfrac{1}{6}x^6\ln x-\dfrac{1}{36}x^6+C$ **13.** $(x+2)e^x-e^x+C$ **15.** $\dfrac{2}{3}x^{3/2}\ln x-\dfrac{4}{9}x^{3/2}+C$

17. $\dfrac{1}{6}(x-3)(x+4)^6-\dfrac{1}{42}(x+4)^7+C$ **19.** $-2te^{-0.5t}-4e^{-0.5t}+C$ **21.** $-t^{-1}\ln t-t^{-1}+C$

23. $\dfrac{1}{10}s(2s+1)^5-\dfrac{1}{120}(2s+1)^6+C$ **25.** $-\dfrac{1}{2}xe^{-2x}-\dfrac{1}{4}e^{-2x}+C$ **27.** $2x(x+1)^{1/2}-\dfrac{4}{3}(x+1)^{3/2}+C$

29. $\dfrac{1}{a}xe^{ax}-\dfrac{1}{a^2}e^{ax}+C$ **31.** $\dfrac{1}{n+1}x^{n+1}\ln ax-\dfrac{1}{(n+1)^2}x^{n+1}+C$ **33.** $x\ln x-x+C$ **35.** $\dfrac{1}{2}x^2e^{x^2}-\dfrac{1}{2}e^{x^2}+C$

37. a. $\dfrac{1}{2}e^{x^2}+C$ (by substitution)   **b.** $\dfrac{1}{4}(\ln x)^4+C$ (by substitution)   **c.** $\dfrac{1}{3}x^3\ln 2x-\dfrac{1}{9}x^3+C$ (by parts)

d. $\ln(e^x+4)+C$ (by substitution)

39. e^2-1 **41.** $9\ln 3-3+\dfrac{1}{9}$ **43.** $\dfrac{2^6}{30}=\dfrac{32}{15}$ **45.** $4\ln 4-3$

47. a. $\dfrac{1}{6}x(x-2)^6-\dfrac{1}{42}(x-2)^7+C$ **b.** $\dfrac{1}{7}(x-2)^7+\dfrac{1}{3}(x-2)^6+C$
49. Using $u=x^n$ and $dv=e^x\,dx$ the result follows immediately.   **51.** $x^2e^x-2xe^x+2e^x+C$
53. a. The result follows immediately   **b.** (*Hint:* Think of the C.)   **55.** $R(x)=4xe^{x/4}-16e^{x/4}+16$

57. \$105.7 million   **59.** $-14e^{-2.5}+4\approx 2.85$ mg   **61.** $2\ln 2-1+\dfrac{1}{4}\approx 0.64$ square unit

63. $72\ln 6-24+\dfrac{1}{9}\approx 105$ thousand customers   **65.** $-x^2e^{-x}-2xe^{-x}-2e^{-x}+C$

67. $(x+1)^2e^x-2(x+1)e^x+2e^x+C$ **69.** $\dfrac{1}{3}x^3(\ln x)^2-\dfrac{2}{9}x^3\ln x+\dfrac{2}{27}x^3+C$ **71.** $2e^2-2\approx 12.78$

73. $-x^2e^{-x}-2xe^{-x}-2e^{-x}+C=-e^{-x}(x^2+2x+2)+C$ **75.** $\dfrac{1}{2}x^3e^{2x}-\dfrac{3}{4}x^2e^{2x}+\dfrac{3}{4}xe^{2x}-\dfrac{3}{8}e^{2x}+C$

77. $\dfrac{1}{3}(x-1)^3e^{3x}-\dfrac{1}{3}(x-1)^2e^{3x}+\dfrac{2}{9}(x-1)e^{3x}-\dfrac{2}{27}e^{3x}+C$

EXERCISES 10.2 page 842

1. Formula 12 $a=5,b=-1$ **3.** Formula 14 $a=-1,b=7$ **5.** Formula 9 $a=-1,b=1$

7. $\dfrac{1}{6}\ln\left|\dfrac{3+x}{3-x}\right|+C$ **9.** $-\dfrac{1}{x}-2\ln\left|\dfrac{x}{2x+1}\right|+C$ **11.** $-x-\ln|1-x|+C$ **13.** $\ln\left|\dfrac{2x+1}{x+1}\right|+C$

15. $\dfrac{x}{2}\sqrt{x^2-4}-2\ln|x+\sqrt{x^2-4}|+C$ **17.** $-\ln\left|\dfrac{1+\sqrt{1-z^2}}{z}\right|+C$ **19.** $\dfrac{1}{2}x^3e^{2x}-\dfrac{3}{4}x^2e^{2x}+\dfrac{3}{4}xe^{2x}-\dfrac{3}{8}e^{2x}+C$

21. $-\dfrac{1}{100} x^{-100} \ln x - \dfrac{1}{10,000} x^{-100} + C$ **23.** $\dfrac{1}{3} \ln \left| \dfrac{x}{x+3} \right| + C$ **25.** $\dfrac{1}{8} \ln \left| \dfrac{z^2-2}{z^2+2} \right| + C$

27. $\dfrac{x}{2} \sqrt{9x^2+16} + \dfrac{8}{3} \ln |3x + \sqrt{9x^2+16}| + C$ **29.** $-\dfrac{1}{2} \ln \left| \dfrac{2+\sqrt{4-e^{2t}}}{e^t} \right| + C$ **31.** $\dfrac{1}{2} \ln \left| \dfrac{e^t-1}{e^t+1} \right| + C$

33. $\dfrac{1}{4} \ln |x^4 + \sqrt{x^8-1}| + C$ **35.** $\dfrac{1}{3} \ln \left| \dfrac{\sqrt{x^3+1}-1}{\sqrt{x^3+1}+1} \right| + C$ **37.** $\dfrac{1}{2} \ln \left| \dfrac{e^t-1}{e^t+1} \right| + C$ **39.** $2xe^{\frac{1}{2}x} - 4e^{\frac{1}{2}x} + C$

41. $\dfrac{1}{4} \ln \left| \dfrac{e^{-x}+4}{e^{-x}} \right| + C = \dfrac{1}{4} \ln(1+4e^x) + C$ **43.** $\dfrac{15}{2} - 8\ln 8 + 8\ln 4 \approx 1.95$ **45.** $\dfrac{1}{2} \ln \dfrac{1}{2} - \dfrac{1}{2} \ln \dfrac{1}{3} \approx 0.203$

47. $-4 + 5\ln 3 \approx 1.49$ **49.** $\dfrac{1}{2} \ln |2x+6| + C$ **51.** $\dfrac{x}{2} - \dfrac{3}{2} \ln |2x+6| + C$ **53.** $-\dfrac{1}{3}(1-x^2)^{3/2} + C$

55. $\sqrt{1-x^2} - \ln \left| \dfrac{1+\sqrt{1-x^2}}{x} \right| + C$ **57.** $\dfrac{1}{2} \left(\ln |x+1| - \dfrac{1}{3} \ln |3x+1| \right) - \dfrac{1}{2} \ln \left| \dfrac{3x+1}{x+1} \right| + C$

59. $\ln |x + \sqrt{x^2+1}| - \ln \left| \dfrac{1+\sqrt{x^2+1}}{x} \right| + C$ **61.** $x + 2\ln |x-1| + C$

63. $-x^2 e^{-x} - 2xe^{-x} - 2e^{-x} + 2$ million sales   **65.** 24 generations   **67.** $C(x) = \ln(x + \sqrt{x^2+1}) + 2000$

EXERCISES 10.3 page 855

1. 0   **3.** 1   **5.** Does not exist   **7.** Does not exist   **9.** $\dfrac{1}{2}$ **11.** $\dfrac{1}{8}$ **13.** Divergent   **15.** 100   **17.** 20   **19.** $\dfrac{1}{2}$

21. Divergent   **23.** $\dfrac{1}{3}$ **25.** $\dfrac{1}{3}$ **27.** Divergent   **29.** 1   **31.** Divergent

33. $\displaystyle\int_0^\infty e^{\sqrt{x}}\, dx$ diverges and $\displaystyle\int_0^\infty e^{-x^2}\, dx$ converges to 0   **35.** \$200,000   **37. a.** \$10,000   **b.** \$9999.55

39. About \$3,963,000 (from 3963 thousand)   **41.** 1,000,000 barrels (from 1000 thousand)   **43.** 2 square units

45. $\dfrac{1}{a}$ square units   **47.** 0.61 or 61%   **49.** 0.30 or 30%   **51.** 20,000   **53.** \$40,992

EXERCISES 10.4 page 868

Some answers may differ depending on rounding.
1. a. 8.75   **b.** 8.667   **c.** 0.083   **d.** 1%   **3. a.** 0.697   **b.** 0.693   **c.** 0.004   **d.** 0.6%   **5.** 1.154   **7.** 0.743
9. 0.59   **11.** 8.697   **13.** 2.93   **15.** 0.4772   **17.** \$17,300 (from 17.30 thousand)   **19.** 8.667   **21.** 0.693
23. 1.148   **25.** 0.747   **27.** 0.593   **29.** 8.6968   **31.** 2.9253   **33.** 821 feet
35. The justification follows from carrying out the indicated steps.

EXERCISES 10.5 page 884

1. Check that $(4e^{2x} - 3e^x) - 3(2e^{2x} - 3e^x) + 2(e^{2x} - 3e^x + 2) \overset{?}{=} 4$ **3.** Check that $kae^{ax} \overset{?}{=} a\left(ke^{ax} - \dfrac{b}{a}\right) + b$

5. $y = \sqrt[3]{6x^2 + c}$ **7.** Not separable   **9.** $y = ce^{2x^3}$ Check that $c6x^2 e^{2x^3} \overset{?}{=} 6x^2(ce^{2x^3})$

11. $y = cx$ (since $e^{\ln x} = x$) Check that $c \overset{?}{=} \dfrac{cx}{x}$ **13.** $y = \sqrt{4x^2+c}$ and $y = -\sqrt{4x^2+c}$ **15.** Not separable

17. $y = 3x^3 + C$ **19.** $y = \dfrac{1}{2} \ln(x^2+1) + C$ **21.** $y = ce^{x^3/3}$ **23.** $y = \left(\dfrac{1-n}{m+1} x^{m+1} + c\right)^{1/(1-n)}$ **25.** $y = (x+c)^2$

27. Not separable   **29.** $y = ce^{\frac{1}{2}x^2} - 1$ **31.** $y = ce^{e^x} + 1$ **33.** $y = \dfrac{1}{c-ax}$ **35.** $y = ce^{ax} - \dfrac{b}{a}$

37. $y = \sqrt[3]{3x^2+8}$ **39.** $y = -e^{\frac{1}{2}x^2}$ Check that $-xe^{\frac{1}{2}x^2} \overset{?}{=} x(-e^{\frac{1}{2}x^2})$ and $y(0) = -e^0 = -1$
41. $y = (1-x^2)^{-1}$ Check that $-(1-x^2)^{-2}(-2x) \overset{?}{=} 2x[(1-x^2)^{-1}]^2$ and $y(0) = (1-0)^{-1} = 1$

43. $y = 3x$ (using $e^{\ln x} = x$) Check that $3 \overset{?}{=} \dfrac{3x}{x}$ **45.** $y = (x + 1)^2$ **47.** $y = \dfrac{1}{2 - e^x - x}$ **49.** $y = 2e^{\frac{1}{3}ax^3}$

51. $D(p) = cp^{-k}$ (for any constant c)   **53.** $y = 20{,}000e^{0.05t} - 20{,}000$
55. a. $y = 28.6e^{-0.32t} + 70$ **b.** About 3.28 hours earlier   **57.** $y = 150 - 150e^{-0.2t}$
59. a. $y' = 3 + 0.10y$ **b.** $y(0) = 6$ **c.** $y(t) = 36e^{0.1t} - 30$
    d. $y(25) = 408.570$ thousand dollars, or \$408,570 (rounded)
61. a. $y' = 4y^{1/2}$ **b.** $y(0) = 10{,}000$ **c.** $y(t) = (2t + 100)^2$ **d.** 15,376
63. a. $y' = 8y^{3/4}$ **b.** $y(0) = 10{,}000$ **c.** $y(t) = (2t + 10)^4$ **d.** 234,256

65. 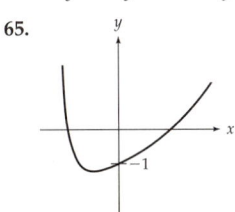   **67.** 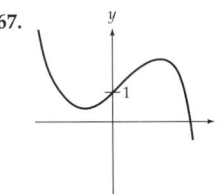   **69.**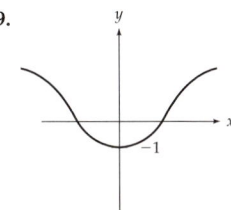

EXERCISES 10.6 page 900

1. $y' = cae^{at} = a(ce^{at}) = ay$ **3.** Unlimited   **5.** Limited   **7.** Neither   **9.** Logistic   **11.** Logistic   **13.** $y = 1.5e^{6t}$
    $y(0) = ce^0 = c$
15. $y = 100e^{-t}$ **17.** $y = -e^{-0.45t}$ **19.** $y = 100(1 - e^{-2t})$ **21.** $y = 0.25(1 - e^{-0.05t})$ **23.** $y = 40(1 - e^{-2t})$

25. $y = 200(1 - e^{-0.01t})$ **27.** $y = \dfrac{100}{1 + 9e^{-500t}}$ **29.** $y = \dfrac{0.5}{1 + 4e^{-0.125t}}$ **31.** $y = \dfrac{10}{1 - \frac{1}{2}e^{-30t}} = \dfrac{20}{2 - e^{-30t}}$

33. $y = \dfrac{3}{1 + 2e^{-6t}}$ **35.** $y' = 0.08y$ **37.** $y' = a(100{,}000 - y)$ **39.** $y' = a(5000 - y)$
                         $y = 1500e^{0.08t}$ $y = 100{,}000(1 - e^{-0.021t})$ $y = 5000(1 - e^{-0.223t})$
                                         about 22,276   about 7.2 weeks
41. $y' = ay(10{,}000 - y)$ **43.** $y' = ay(800 - y)$ **45.** $y' = ay(800 - y)$ **47.** $y = 5e^{-0.15t}$
    $y = \dfrac{10{,}000}{1 + 99e^{-0.535t}}$ $y = \dfrac{800}{1 + 799e^{-0.558t}}$ $y = \dfrac{800}{1 + 7e^{-0.280t}}$ about 3.7
    about 8612   about 675 people   about 6.9 years
49. a. $y' = 0.1(200 - y)$ with $M = 200$ **b.** $y = 200(1 - e^{-0.1t})$ **c.** 30 years [(from solving $200(1 - e^{-0.1t}) = 0.95 \cdot 200$]

51. $y = \dfrac{1}{c - at}$ **53.** $y = ce^{a\ln x} = cx^a$ **55.** The solution follows from the indicated steps.

57. a. About 16.6 ft/sec   **b.** About 0.6 ft/sec   **c.** About 0.006 ft/sec
    d. About $\dfrac{1}{0.006} \approx 167$ seconds, or about 2.7 minutes

CHAPTER 10 REVIEW EXERCISES page 905

1. $\dfrac{1}{2} xe^{2x} - \dfrac{1}{4} e^{2x} + C$ **2.** $-xe^{-x} - e^{-x} + C$ **3.** $\dfrac{1}{9} x^9 \ln x - \dfrac{1}{81} x^9 + C$ **4.** $\dfrac{4}{5} x^{5/4} \ln x - \dfrac{16}{25} x^{5/4} + C$

5. $\dfrac{1}{6} (x - 2)(x + 1)^6 - \dfrac{1}{42} (x + 1)^7 + C$ **6.** $\dfrac{1}{5} (x + 3)(x - 1)^5 - \dfrac{1}{30} (x - 1)^6 + C$ **7.** $2t^{1/2} \ln t - 4t^{1/2} + C$

8. $\dfrac{1}{4} x^4 e^{x^4} - \dfrac{1}{4} e^{x^4} + C$ **9.** $x^2 e^x - 2xe^x + 2e^x + C$ **10.** $x(\ln x)^2 - 2x \ln x + 2x + C$

11. $\dfrac{1}{n + 1} x(x + a)^{n+1} - \dfrac{1}{(n + 1)(n + 2)} (x + a)^{n+2} + C$ **12.** $-\dfrac{1}{n + 1} x(1 - x)^{n+1} - \dfrac{1}{(n + 1)(n + 2)} (1 - x)^{n+2} + C$

13. $4e^5 + 1$ **14.** $\dfrac{1}{4} e^2 + \dfrac{1}{4}$ **15.** $-\ln |1 - x| + C$ **16.** $-\dfrac{1}{2} e^{-x^2} + C$ **17.** $\dfrac{1}{4} x^4 \ln 2x - \dfrac{1}{16} x^4 + C$

18. $(1 - x)^{-1} + C$ **19.** $\dfrac{1}{2} (\ln x)^2 + C$ **20.** $\dfrac{1}{2} \ln (e^{2x} + 1) + C$ **21.** $2e^{\sqrt{x}} + C$ **22.** $\dfrac{1}{8} (e^{2x} + 1)^4 + C$

23. a. $-15,000e^{-0.5} + 10,000 \approx 902$ million dollars   **b.** $902 million   **24.** 6.78 hundred gallons (from $25 - 10e^{0.6}$)

25. $\dfrac{1}{10} \ln \left| \dfrac{5 + x}{5 - x} \right| + C$ **26.** $\dfrac{1}{4} \ln \left| \dfrac{x - 2}{x + 2} \right| + C$ **27.** $2 \ln |x - 2| - \ln |x - 1| + C$

28. $-\ln \left| \dfrac{x - 1}{x - 2} \right| + C$ or $\ln \left| \dfrac{x - 2}{x - 1} \right| + C$ **29.** $\ln \left| \dfrac{\sqrt{x + 1} - 1}{\sqrt{x + 1} + 1} \right| + C$ **30.** $\dfrac{2x - 4}{3} \sqrt{x + 1} + C$

31. $\ln |x + \sqrt{x^2 + 9}| + C$ **32.** $\ln |x + \sqrt{x^2 + 16}| + C$ **33.** $\dfrac{z^2 - 2}{3} \sqrt{z^2 + 1} + C$ (from formula 13)

34. $e^t - 2 \ln (e^t + 2) + C$ **35.** $\dfrac{1}{2} x^2 e^{2x} - \dfrac{1}{2} x e^{2x} + \dfrac{1}{4} e^{2x} + C$ **36.** $x(\ln x)^4 - 4x(\ln x)^3 + 12x(\ln x)^2 - 24x \ln x + 24x + C$

37. $\ln \left| \dfrac{2x + 1}{x + 1} \right| + 1000$ **38.** 1305 (from $750 + 800 \ln 80 - 800 \ln 40$)   **39.** $\dfrac{1}{4}$ **40.** $\dfrac{1}{5}$ **41.** Divergent

42. Divergent   **43.** $\dfrac{1}{2}$ **44.** $2e^{-2}$ **45.** Divergent   **46.** Divergent   **47.** 5   **48.** $10e^{-10}$ **49.** $\dfrac{1}{4}$ **50.** $\dfrac{1}{5}$

51. $\dfrac{1}{2}$ **52.** $\dfrac{1}{4}$ **53.** 1   **54.** 1   **55.** $\dfrac{1}{3}$ **56.** $\dfrac{1}{2}$ **57.** $60,000   **58.** 0.35 or 35%   **59.** 240 thousand

60. 7500 million tons   **61.** $\displaystyle\int_1^\infty \dfrac{1}{x^3} \, dx$ converges to 0.5   **62.** $\displaystyle\int_1^\infty \dfrac{1}{\sqrt[3]{x}} \, dx$ diverges   **63.** 1.102   **64.** 1.09   **65.** 1.204

66. 0.852   **67.** 0.57   **68.** 1.313   **69.** 1.089   **70.** 1.075   **71.** 1.195   **72.** 0.856   **73.** 0.528   **74.** 1.348
75. 1.0894   **76.** 1.0747   **77.** 1.1951   **78.** 0.8556   **79.** 0.5285   **80.** 1.3357   **81.** 1.089429   **82.** 1.074669

83. 1.194958   **84.** 0.855624   **85.** 0.527887   **86.** 1.347855   **87. a.** $-\displaystyle\int_1^0 \dfrac{1}{1 + t^2} \, dt = \int_0^1 \dfrac{1}{1 + t^2} \, dt$ **b.** 0.783

88. a. $-\displaystyle\int_1^0 \dfrac{1}{1 + t^4} \, dt = \int_0^1 \dfrac{1}{1 + t^4} \, dt$ **b.** 0.862   **89.** $y = \sqrt[3]{x^3 + c}$ **90.** $y = ce^{\frac{1}{3}x^3}$ **91.** $y = \dfrac{1}{4} \ln (x^4 + 1) + C$

92. $y = -\dfrac{1}{2} e^{-x^2} + C$ **93.** $y = \dfrac{1}{c - x}$ **94.** $y = \dfrac{1}{\sqrt{c - 2x}}$ and $y = -\dfrac{1}{\sqrt{c - 2x}}$ **95.** $y = 1 + ce^{-x}$ or $y = 1 - e^{c-x}$

96. $y = \sqrt{2x + c}$ and $y = -\sqrt{2x + c}$ **97.** $y = ce^{\frac{1}{2}x^2 - x}$ **98.** $y = ce^{\frac{1}{3}x^3} - 1$ **99.** $y = \sqrt[3]{3x^3 + 1}$

100. $y = e^{1 - x^{-1}}$ (from $ee^{-x^{-1}}$)   **101.** $y = e^{\frac{1}{2} - \frac{1}{2}x^{-2}}$ (from $e^{\frac{1}{2}} e^{-\frac{1}{2}x^{-2}}$)   **102.** $y = \left(\dfrac{2}{3} x - \dfrac{2}{3} \right)^{3/2}$ or $y = -\left(\dfrac{2}{3} x - \dfrac{2}{3} \right)^{3/2}$

103. a. $y' = 4 + 0.05y$ and $y(0) = 10$ **b.** $y = 90e^{0.05t} - 80$ **c.** 68.385 thousand or $68,385
104. a. $y' = 4 - 0.25y$, $y(0) = 0$ **b.** $y = 16 - 16e^{-0.25t}$ **105.** $y = 106 - 36e^{-2.3t}$ **106.** $y(1) \approx 102.4$
$y(2) \approx 105.6$
$y(3) \approx 105.96$

107.
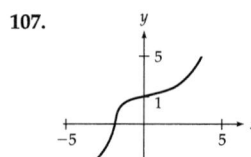
y-intercept is 1   **108.**   **109.** $y' = 0.07y$
$y = 32e^{0.07t}$
about 64¢

110. $y' = 0.22y$
$y = 10.9e^{0.22t}$
about $98.4 billion

111. $y' = ay(8000 - y)$
$y = \dfrac{8000}{1 + 799e^{-2.73t}}$ (t in weeks)
about 1819 cases

112. $y' = ay(500 - y)$
$y = \dfrac{500}{1 + 249e^{-3.78t}}$
about 443

113. $y' = a(10,000 - y)$
$y = 10,000(1 - e^{-0.051t})$
about 4577

114. $y' = a(60 - y)$
$y = 60(1 - e^{-0.269t})$
about 6.7 weeks

115. $y' = a(500,000 - y)$
$y = 500,000(1 - e^{-0.255t})$ (t in weeks)
about 6.3 weeks

EXERCISES 11.1 page 921

1. $\{(x, y) \mid x \neq 0, y \neq 0\}$ **3.** $\{(x, y) \mid x \neq y\}$ **5.** $\{(x, y) \mid x > 0, y \neq 0\}$ **7.** $\{(x, y, z) \mid x \neq 0, y \neq 0, z > 0\}$ **9.** 3
11. 4   **13.** -2 **15.** 1   **17.** $e^{-1} + e$ **19.** e^{-1} **21.** 0   **23.** 0.0157   **25.** 45 minutes   **27.** 472.7
29. $P(2L, 2K) = a(2L)^b(2K)^{1-b} = a2^bL^b\, 2^{1-b}K^{1-b} = 2aL^bK^{1-b} = 2P(L, K)$ **31.** 1548 calls

$$\overset{\displaystyle\searset{}{\ \diagdown\ \diagup\ }}{2}$$

33. $C(x, y) = 210x + 180y + 4000$ **35. a.** $V = xyz$ **b.** $M = xy + 2xz + 2yz$ **37. a.**

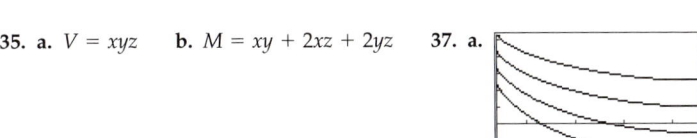

b. A given wind speed will lower the windchill further on a colder day than on a warmer day.
c. For the lowest curve: $dy/dx \approx -2$, meaning that at 20° and 10 mph of wind, windchill drops by about 2° for each additional 1 mph of wind. For the highest curve: $dy/dx \approx -1.1$, meaning that at 50° and 10 mph of wind, windchill drops by only about 1.1° for each additional 1 mph of wind.
d. Yes—the effect of wind on the windchill index is greater on a colder day.

EXERCISES 11.2 page 933

1. a. $3x^2 + 6xy^2 - 1$ **b.** $6x^2y - 6y^2 + 1$ **3. a.** $6x^{-1/2}y^{1/3}$ **b.** $4x^{1/2}y^{-2/3}$ **5. a.** $5x^{-0.95}y^{0.02}$ **b.** $2x^{0.05}y^{-0.98}$

7. a. $-(x + y)^{-2}$ **b.** $-(x + y)^{-2}$ **9. a.** $\dfrac{3x^2}{x^3 + y^3}$ **b.** $\dfrac{3y^2}{x^3 + y^3}$ **11. a.** $6x^2e^{-5y}$ **b.** $-10x^3e^{-5y}$

13. a. ye^{xy} **b.** xe^{xy} **15. a.** $\dfrac{x}{x^2 + y^2}$ or $x(x^2 + y^2)^{-1}$ **b.** $\dfrac{y}{x^2 + y^2}$ or $y(x^2 + y^2)^{-1}$ **17. a.** $3v(uv - 1)^2$

b. $3u(uv - 1)^2$ **19. a.** $ue^{\frac{1}{2}(u^2 - v^2)}$ **b.** $-ve^{\frac{1}{2}(u^2 - v^2)}$ **21.** $18, -10$ **23.** $0, 2e$ **25.** $1\frac{1}{2}$

27. a. $30x - 4y^3$ **b. and c.** $-12xy^2$ **d.** $-12x^2y + 36y^2$ **29. a.** $-2x^{-\frac{5}{3}}y^{\frac{2}{3}}$ **b. and c.** $2x^{-\frac{2}{3}}y^{-\frac{1}{3}} - 12y^2$

d. $-2x^{1/3}y^{-4/3} - 24xy$ **31. a.** ye^x **b. and c.** $e^x - \dfrac{1}{y}$ **d.** xy^{-2} **33.** All three are $36x^2y^2$.

35. a. y^2z^3 **b.** $2xyz^3$ **c.** $3xy^2z^2$ **37. a.** $8x(x^2 + y^2 + z^2)^3$ **b.** $8y(x^2 + y^2 + z^2)^3$ **c.** $8z(x^2 + y^2 + z^2)^3$
39. a. $2xe^{x^2 + y^2 + z^2}$ **b.** $2ye^{x^2 + y^2 + z^2}$ **c.** $2ze^{x^2 + y^2 + z^2}$ **41.** -14 **43.** $4e^6$ **45. a.** $P_x = 4x - 3y + 150$
  b. \$50 (profit per additional tape deck)   **c.** $P_y = -3x + 6y + 75$ **d.** \$75 (profit per additional CD player)
47. a. 250 (the marginal productivity of labor is 250, so production increases by about 250 for each additional unit of labor)
  b. 108 (the marginal productivity of capital is 108, so production increases by about 108 for each additional unit of capital)
  c. Labor
49. $S_x = -0.1$ (sales fall by 0.1 for each dollar price increase)
  $S_y = 0.4y$ (sales rise by 0.4y for each additional advertising dollar above the level y)
51. a. 0.52 (status increases by 0.52 unit for each additional \$1000 of income)
  b. 5.25 (status increases by 5.25 units for each additional year of education)
53. a. 97 (skid distance increases by about 97 feet for each additional ton)
  b. 12.96 (skid distance increases by about 13 feet for each additional mph)
55. a. Rate at which butter sales change as butter prices rise
  b. Negative: as prices rise, sales will fall.
  c. Rate at which butter sales change as margarine prices rise
  d. Positive: as margarine prices rise, people will switch to butter, so butter sales will rise.

EXERCISES 11.3 page 945

1. Rel min value: $f = 5$ at $x = 0, y = -1$ **3.** Rel min value: $f = -12$ at $x = -2, y = 2$
5. Rel max value: $f = 23$ at $x = 5, y = 2$ **7.** No rel extreme values [saddle point at $(2, -4)$]

9. No rel extreme values   **11.** Rel min value: $f = 1$ at $x = 0$, $y = 0$ **13.** Rel min value: $f = 0$ at $x = 0$, $y = 0$
15. Rel max value: $f = 3$ at $x = 1$, $y = -1$ [saddle point at $(-1, -1)$]
17. Rel max value: $f = 17$ at $x = -1$, $y = -2$ [saddle point at $(-1, 2)$]
19. No rel extreme values [saddle point at $(2, 6)$]
21. 10 units of product A, sell for $7000 each; 7 units of product B, sell for $13,000 each. Maximum profit: $22,000
23. a. $P = -0.2x^2 + 16x - 0.1y^2 + 12y - 20$ **b.** 40 cars in America, sell for $12,000; 60 cars in Europe, sell for $10,000
25. 6 hours of practice and 1 hour of rest   **27. a.** $x = 1200$, $p = \$6$, $R = \$7200$
    b. $x = 800$, $y = 800$, $p = \$4$, revenue = $3200 for each   **c.** Duopoly (1600 versus 1200)   **d.** Duopoly
29. a. $P = -0.2x^2 + 16x - 0.1y^2 + 12y - 0.1z^2 + 8z - 22$ **b.** 40 in America, 60 in Europe, 40 in Asia
31. Rel min value: $f = -1$ at $x = 1$, $y = 1$ [saddle point at $(0, 0)$]
33. Rel max value: $f = 32$ at $x = 4$, $y = 4$ [saddle point at $(0, 0)$]
35. Rel min value: $f = -162$ at $x = 3$, $y = 18$ and at $x = -3$, $y = -18$ [saddle point at $(0, 0)$]

EXERCISES 11.4 page 957

Note: Your answers may differ slightly depending on the stage at which you do the rounding.

1. $y = 3.5x - 1.67$ **3.** $y = -0.79x + 6.6$ **5.** $y = 2.4x + 6.9$ **7.** $y = -2.1x + 7.6$
9. $y = 2.2x + 5$, prediction: 16 million   **11.** $y = -8x + 125$, prediction: 85   **13.** $y = -7.7x + 90.5$, prediction: 52
15. $y = -3.7x + 40.5$, prediction: 18.3%   **17.** $y = -0.16x + 71.6$ **19.** $y = 1.09e^{0.63x}$ **21.** $y = 17.45e^{-0.47x}$
23. $y = 0.98e^{0.78x}$ **25.** $y = 16.95e^{-0.52x}$ **27.** $y = 367e^{0.145x}$, prediction: $1,013,000
29. $y = 7.78e^{0.422x}$, prediction: $64,000 per second (from 64 thousand)
31. $y = 13.2e^{0.0519x}$, prediction: $28,800 (from 28.8 thousand)

EXERCISES 11.5 page 974

1. Max $f = 36$ at $x = 6$, $y = 2$ **3.** Max $f = 144$ at $x = 6$, $y = 4$ **5.** Max $f = -28$ at $x = 3$, $y = 5$
7. Max $f = 6$ at $x = 2$, $y = -1$ **9.** Max $f = 2$ (from $\ln e^2$) at $x = e$, $y = e$ **11.** Min $f = 45$ at $x = 6$, $y = 3$
13. Min $f = -16$ at $x = -4$, $y = 4$ **15.** Min $f = 52$ at $x = 4$, $y = 6$ **17.** Min $f = \ln 125$ at $x = 10$, $y = 5$
19. Min $f = e^{20}$ at $x = 2$, $y = 4$ **21.** Max $f = 8$ at $x = 2$, $y = 2$ and at $x = -2$, $y = -2$; Min $f = -8$ at $x = 2$, $y = -2$
    and at $x = -2$, $y = 2$ **23.** Max $f = 18$ at $x = 2$, $y = 8$; Min $f = -18$ at $x = -2$, $y = -8$
25. a. 1000 feet perpendicular to building, 3000 feet parallel to building   **b.** $|\lambda| = 1000$; each additional foot of fence
    adds about 1000 square feet of area   **27.** $r \approx 3.7$ feet, $h \approx 3.7$ feet
29. End: 14 inches by 14 inches, length = 28 inches, volume = 5488 cubic inches   **31. a.** $L = 120$, $K = 20$
    b. $|\lambda| \approx 1.9$, output increases by about 1.9 for each additional dollar
33. Base: 3 inches by 3 inches, height: 5 inches   **35.** Min $f = 24$ at $x = 4$, $y = 2$, $z = -2$
37. Max $f = 6$ at $x = 2$, $y = 2$, $z = 2$ **39.** Base: 50 feet by 50 feet; height: 100 feet

EXERCISES 11.6 page 985

1. $df = 2xy^3 \cdot dx + 3x^2y^2 \cdot dy$ **3.** $df = 3x^{-1/2}y^{1/3} \cdot dx + 2x^{1/2}y^{-2/3} \cdot dy$ **5.** $dg = \dfrac{1}{y} \cdot dx - \dfrac{x}{y^2} \cdot dy$

7. $dg = -(x - y)^{-2} \cdot dx + (x - y)^{-2} \cdot dy$ **9.** $dz = \dfrac{3x^2}{x^3 - y^2} \cdot dx - \dfrac{2y}{x^3 - y^2} \cdot dy$ **11.** $dz = e^{2y} \cdot dx + 2xe^{2y} \cdot dy$

13. $dw = (6x^2 + y) \cdot dx + (x + 2y) \cdot dy$ **15.** $df = 4xy^3z^4 \cdot dx + 6x^2y^2z^4 \cdot dy + 8x^2y^3z^3 \cdot dz$ **17.** $df = \dfrac{1}{x} dx + \dfrac{1}{y} dy + \dfrac{1}{z} dz$

19. $df = yze^{xyz} \cdot dx + xze^{xyz} \cdot dy + xye^{xyz} \cdot dz$ **21. a.** $\Delta f = 0.479$ **b.** $df = 0.4$ **23. a.** $\Delta f = 0.112$ **b.** $df = 0.11$
    $= e^{xyz}(yz \cdot dx + xz \cdot dy + xy \cdot dz)$
25. a. $\Delta f = 0.1407$ **b.** $df = 0.14$ **27.** 125 square feet
29. $4300   **31.** About 113 feet   **33.** 2%   **35.** 0.5 liter per minute
37. a. f is being evaluated at two points along the curve; each of these points give $f = c$, and $c - c = 0$.
    b. Subtracting and adding $F(x)$
    c. Approximating the change $\Delta f = f(x + \Delta x, F + \Delta F) - f(x, F)$ by the total differential $df = f_x \Delta x + f_y \Delta F$
    d. Subtracting $f_y \Delta F$ and dividing by f_y and Δx

    e. Taking the limit as $\Delta x \to 0$ causes $\dfrac{\Delta F}{\Delta x}$ to approach $\dfrac{dF}{dx}$ and the approximation to become exact

EXERCISES 11.7 page 998

1. $2x^9 - 2x$ **3.** $6y^4$ **5.** $2x^2$ **7.** 4   **9.** 2   **11.** $\dfrac{1}{2}$ **13.** 12   **15.** 14   **17.** $-9e^{-3} + 9e^3$ **19.** 0   **21.** -12

23. 0   **25.** 72   **27.** $\dfrac{1}{2}$ **29. a.** $\displaystyle\int_1^3 \int_0^2 3xy^2\, dx\, dy$ and $\displaystyle\int_0^2 \int_1^3 3xy^2\, dy\, dx$ **b.** Both equal 52

31. a. $\displaystyle\int_0^2 \int_{-1}^1 ye^x dx\, dy$ and $\displaystyle\int_{-1}^1 \int_0^2 ye^x\, dy\, dx$ **b.** Both equal $2e - 2e^{-1}$ **33.** 8 cubic units   **35.** $\dfrac{4}{3}$ cubic units

37. $\dfrac{1}{6}$ cubic unit   **39.** $\dfrac{1}{2}e^2 - e + \dfrac{1}{2}$ cubic units   **41.** $45°\left(\text{from } \dfrac{540}{12}\right)$ **43.** About 180,200 people

45. 900,000 cubic feet   **47.** 14   **49.** 10

CHAPTER 11 REVIEW EXERCISES page 1003

1. $\{(x, y) \mid x \geq 0,\, y \neq 0\}$ **2.** $\{(x, y) \mid y > 0\}$ **3.** $\{(x, y) \mid x \neq 0,\, y > 0\}$ **4.** $\{(x, y) \mid x \neq 0,\, y > 0\}$
5. a. $10x^4 - 6xy^3 - 3$ **b.** $-9x^2y^2 + 4y^3 + 2$ **c. and d.** $-18xy^2$
6. a. $12x^3 + 15x^2y^2 - 6$ **b.** $10x^3y - 6y^5 + 1$ **c. and d.** $30x^2y$
7. a. $12x^{-1/3}y^{1/3}$ **b.** $6x^{2/3}y^{-2/3}$ **c. and d.** $4x^{-1/3}y^{-2/3}$
8. a. $\dfrac{2x}{(x^2 + y^3)}$ **b.** $\dfrac{3y^2}{(x^2 + y^3)}$ **c. and d.** $\dfrac{-6xy^2}{(x^2 + y^3)^2}$
9. a. $3x^2e^{x^3 - 2y^3}$ **b.** $-6y^2e^{x^3 - 2y^3}$ **c. and d.** $-18x^2y^2e^{x^3 - 2y^3}$
10. a. $6xe^{-5y}$ **b.** $-15x^2e^{-5y}$ **c. and d.** $-30xe^{-5y}$
11. a. $-ye^{-x} - \ln y$ **b.** $e^{-x} - \dfrac{x}{y}$ **c. and d.** $-e^{-x} - \dfrac{1}{y}$ **12. a.** $2xe^y + yx^{-1}$ **b.** $x^2e^y + \ln x$ **c. and d.** $2xe^y + x^{-1}$
13. a. $\dfrac{1}{2}$ **b.** $\dfrac{1}{2}$ **14. a.** 0   **b.** $\dfrac{1}{2}$ **15. a.** 36   **b.** -24 **16. a.** 216   **b.** -216
17. a. 80: rate at which production increases for each additional unit of labor
b. 135: rate at which production increases for each additional unit of capital   **c.** Capital
18. $S_x = 222$: rate at which sales increase for each additional \$1000 in TV ads. $S_y = 528$: rate at which sales increase for each additional \$1000 in print ads.
19. Min f is -13 (at $x = -1,\, y = -4$).   **20.** Min f is -8 (at $x = 4,\, y = 1$).   **21.** Max f is 8 (at $x = 0,\, y = -1$).
22. Max f is 6 (at $x = 1,\, y = 0$).   **23.** No rel extreme values (saddle point at $x = \dfrac{1}{2},\, y = -3$)

24. No rel extreme values (saddle point at $x = -\dfrac{1}{2},\, y = 1$)

25. Max f is 1 (at $x = 0,\, y = 0$).   **26.** Min f is 1 (at $x = 0,\, y = 0$).   **27.** Min f is 0 (at $x = 0,\, y = 0$).
28. Min f is $\ln 10$ (at $x = 0,\, y = 0$).   **29.** Max f is 25 (at $x = -2,\, y = -3$) (saddle point at $x = 2,\, y = -3$).
30. Min f is -20 at $x = -2,\, y = 2$ (saddle point at $x = 2,\, y = 2$).
31. a. $C(x, y) = 3000x + 5000y + 6000$ **b.** $R(x, y) = 7000x - 20x^2 + 8000y - 30y^2$
c. $P(x, y) = -20x^2 + 4000x - 30y^2 + 3000y - 6000$
d. Make 100 18-foot boats, sell for \$5000 each, and 50 22-foot boats, sell for \$6500 each. Max profit: \$269,000
32. a. $P = -0.2x^2 + 68x - 0.1y^2 + 52y - 100$
b. America: sell 170 for \$46,000 each, Europe: sell 260 for \$38,000 each (since prices are in thousands).
33. $y = 2.6x - 3.2$ **34.** $y = -1.8x + 8.4$ **35.** $y = 4.64x + 7.26,\ 39.7$ million
36. $y = -0.47x + 7.95$, prediction: 4.7% (from 4.66)   **37.** Max f is 292 (at $x = 12,\, y = -24$).
38. Max f is 156 (at $x = 10,\, y = 8$).   **39.** Min f is 90 (at $x = 7,\, y = 4$).   **40.** Min f is -109 (at $x = -3,\, y = 3$).
41. Min f is e^{45} (at $x = 3,\, y = 6$).   **42.** Max f is e^{-5} (at $x = 2,\, y = 1$).
43. Max f is 120 (at $x = 2,\, y = -6$), min f is -120 (at $x = -2,\, y = 6$).
44. Max f is 64 (at $x = 4,\, y = 4$ and at $x = -4,\, y = -4$), min f is -64 (at $x = 4,\, y = -4$ and at $x = -4,\, y = 4$).
45. a. \$40,000 for production, \$20,000 for advertising
b. $|\lambda| \approx 159$, production increases by about 159 units for each additional dollar.
46. a. $\dfrac{1}{2}$ ounces of the first and 7 ounces of the second
b. $|\lambda| = 9$, each additional dollar results in about 9 additional nutritional units.

47. a. $L = 64$, $K = 8$ **b.** $40L^{-1/3}K^{1/3}$, $20L^{2/3}K^{-2/3}$

c. $\dfrac{40L^{-1/3}K^{1/3}}{20L^{2/3}K^{-2/3}} \stackrel{?}{=} \dfrac{25}{100}$ (and now simplify and substitute $L = 64$, $K = 8$)

48. Base: 12 inches by 12 inches, height: 4 inches   **49.** $df = (6x + 2y) \cdot dx + (2x + 2y) \cdot dy$

50. $df = (2x + y) \cdot dx + (x - 6y) \cdot dy$ **51.** $dg = \dfrac{1}{x} dx + \dfrac{1}{y} dy = \dfrac{dx}{x} + \dfrac{dy}{y}$ **52.** $dg = \dfrac{3x^2}{x^3 + y^3} dx + \dfrac{3y^2}{x^3 + y^3} dy$

53. $dz = e^{x-y} dx - e^{x-y} dy$ **54.** $dz = ye^{xy} dx + xe^{xy} dy$

55. Sales would decrease by about $153,000 (from $dS = -153$).   **56.** 2%   **57.** $8e^2 - 8e^{-2}$ **58.** 10   **59.** $\dfrac{4}{3}$

60. $\dfrac{32}{3}$ or $10\dfrac{2}{3}$ **61.** 40 cubic units   **62.** 24 cubic units   **63.** $\dfrac{5}{6}$ cubic unit   **64.** $\dfrac{8}{9}$ cubic unit

65. 12,000 $\left(\text{from } \dfrac{192,000}{16}\right)$ **66.** $640 hundred thousand, or $640,000,000

CUMULATIVE REVIEW EXERCISES FOR CHAPTERS 0 and 6–11 page 1007

1.

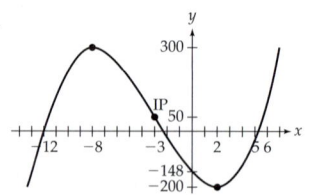

2. 4   **3.** $\dfrac{-1}{x^2}$ (but found using the *definition*)   **4.** 4

5. $S'(12) = -2$. Each $1 price increase (above $12) decreases sales by 2 per week.

6. $3[x^2 + (2x + 1)^4]^2[2x + 8(2x + 1)^3]$ **7.**   **8.**

9. 40 feet parallel to wall, 20 feet perpendicular to wall   **10.** Base: 6 feet by 6 feet, height: 3 feet

11. $\dfrac{2}{\pi} \approx 0.64$ foot per minute   **12. a.** $1268.24   **b.** $1271.25   **13.** In about 14.4 years   **14.** About 6 years

15.   **16.** $4x^3 - 2x^2 + x + C$ **17.** $900e^{0.02t} - 900$ million gallons

18. $85\dfrac{1}{3}$ square units   **19.** 16   **20. a.** $\dfrac{1}{3} \ln |x^3 + 1| + C$ **b.** $2e^{x^{1/2}} + C$ **21.** $\dfrac{1}{4} xe^{4x} - \dfrac{1}{16} e^{4x} + C$

22. $\sqrt{4 - x^2} - 2 \ln \left| \dfrac{2 + \sqrt{4 - x^2}}{x} \right| + C$ **23.** $\dfrac{1}{2}$ **24.** 1.15148 [compared to actual (rounded) value of 1.14779]

25. 1.14778 [compared to actual (rounded) value of 1.14779]   **26. a.** $y = Ce^{\frac{1}{4}x^4}$ **b.** $y = 2e^{\frac{1}{4}x^4}$

27. $f_x = \ln y + 2ye^{2x}$, $f_y = \dfrac{x}{y} + e^{2x}$ **28.** Min $f = 2$ (at $x = 1$, $y = 4$). No relative max.   **29.** $y = 1.75x - 4.75$

30. Min $f = 90$ (at $x = 4$, $y = 7$)   **31.** $df = (4x + y) dx + (x - 6y) dy$ **32.** 48 cubic units

INDEX

FINANCE

Simple Interest: $\quad I = P\,r\,t$ $\qquad\qquad A = P(1 + rt)$ $\qquad\qquad r_s = \dfrac{r}{1 - rt}$

$\qquad\qquad\qquad\qquad A = P(1 + r)^t$

Compound Interest: $\quad A = P(1 + r/m)^{mt}$ $\qquad r_e = (1 + r/m)^m - 1$ $\qquad$ Rule of 72

$\qquad\qquad\qquad\qquad A = Pe^{rt}$ $\qquad\qquad r_e = e^r - 1$ $\qquad\qquad \left(\dfrac{\text{Doubling}}{\text{Time}}\right) \approx \dfrac{72}{r \times 100}$

Annuities and Amortization:

$$A = P\,\frac{(1 + r/m)^{mt} - 1}{r/m} \qquad mt = \frac{\log\!\left(\dfrac{A\,r}{P\,m} + 1\right)}{\log(1 + r/m)} \qquad PV = P\,\frac{1 - (1 + r/m)^{-mt}}{r/m} \qquad P = D\,\frac{r/m}{1 - (1 + r/m)^{-mt}}$$

MATRICES

$$\begin{cases} ax + by = c \\ Ax + By = C \end{cases} \xrightarrow{\text{augmented matrix}} \begin{pmatrix} a & b & c \\ A & B & C \end{pmatrix} \xrightarrow{\text{row reduction may give}} \begin{pmatrix} 1 & 0 & P \\ 0 & 1 & Q \end{pmatrix}$$

$$A \cdot A^{-1} = I = A^{-1} \cdot A \text{ and } (A \mid I) \xrightarrow{\text{row reduction}} (I \mid A^{-1})$$

$$A^{-1} \text{ solves } A \cdot X = B \text{ as } X = A^{-1} \cdot B$$

LINEAR PROGRAMMING

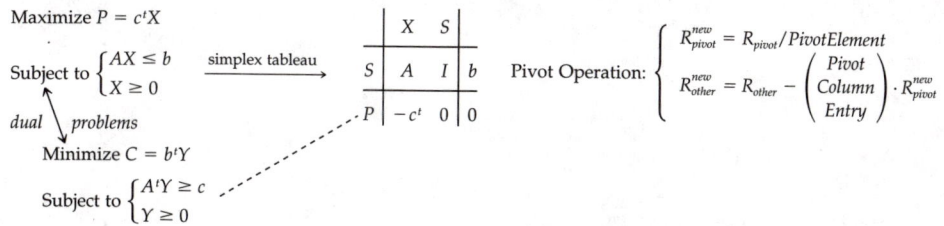

Maximize $P = c^t X$

Subject to $\begin{cases} AX \le b \\ X \ge 0 \end{cases} \xrightarrow{\text{simplex tableau}}$

dual problems

Minimize $C = b^t Y$

Subject to $\begin{cases} A^t Y \ge c \\ Y \ge 0 \end{cases}$

	X	S	
S	A	I	b
P	$-c^t$	0	0

Pivot Operation: $\begin{cases} R^{new}_{pivot} = R_{pivot}/PivotElement \\ R^{new}_{other} = R_{other} - \left(\begin{array}{c}Pivot\\Column\\Entry\end{array}\right) \cdot R^{new}_{pivot} \end{cases}$

PROBABILITY

$$n(A^c) = n(U) - n(A)$$
$$n(A \cup B) = n(A) + n(B) - n(A \cap B)$$

$$_nP_r = n \cdot (n - 1) \cdots (n - r + 1)$$
$$_nC_r = \frac{n \cdot (n - 1) \cdots (n - r + 1)}{r \cdot (r - 1) \cdots (1)}$$

$$P(\varnothing) = 0$$
$$P(S) = 1$$

$$P(E) = \sum_{\text{All } e_i \text{ in } E} P(e_i)$$
$$P(E^c) = 1 - P(E)$$

$$P(A \cup B) = P(A) + P(B) - P(A \cap B)$$

$$P(A \text{ given } B) = \frac{P(A \text{ and } B)}{P(B)} \qquad \text{Independent Events:} \quad P(A \text{ and } B) = P(A) \cdot P(B)$$

Bayes' Formula: $\quad P(U_1 \text{ given } A) = \dfrac{P(A \text{ given } U_1) \cdot P(U_1)}{\displaystyle\sum_{\substack{\text{All } U_i \text{ in} \\ \text{the partition}}} P(A \text{ given } U_1) \cdot P(U_1)}$

$$\mu = E(X) = \sum_{\text{All } x\text{-values}} x \cdot P(X = x) \qquad \sigma = \sqrt{\sum_{\text{All } x\text{-values}} (x - \mu)^2 \cdot P(X = x)}$$

Binomial Distribution: $\quad P(X = x) = {_nC_x}\,p^x(1 - p)^{n-x} \qquad$ Chebyshev's Inequality: $\quad P(|X - \mu| < k\sigma) > 1 - \frac{1}{k^2}$

$$\mu = np \text{ and } \sigma = \sqrt{np(1 - p)}$$

STATISTICS

$$\bar{x} = \frac{1}{n} \sum_{\text{All } x_k\text{-values}} x_k \qquad s = \sqrt{\frac{1}{n - 1} \sum_{\text{All } x_k\text{-values}} (x_k - \bar{x})^2} \qquad z = \frac{x - \mu}{\sigma}$$

DEFINITION OF THE DERIVATIVE

$$f'(x) = \lim_{h \to 0} \frac{f(x + h) - f(x)}{h}$$

DIFFERENTIATION FORMULAS

Power Rule:
$$\frac{d}{dx} x^n = nx^{n-1}$$

Constant Multiple Rule:
$$\frac{d}{dx} [cf(x)] = cf'(x)$$

Sum-Difference Rule:
$$\frac{d}{dx} [f(x) \pm g(x)] = [f'(x) \pm g'(x)]$$

Product Rule:
$$\frac{d}{dx} [f(x)g(x)] = f'(x)g(x) + f(x)g'(x)$$

Quotient Rule:
$$\frac{d}{dx} \left[\frac{f(x)}{g(x)} \right] = \frac{g(x)f'(x) - g'(x)f(x)}{[g(x)]^2}$$

Generalized Power Rule:
$$\frac{d}{dx} [f(x)]^n = n[f(x)]^{n-1}f'(x)$$

Chain Rule:
$$\frac{d}{dx} [f(g(x))] = f'(g(x))g'(x)$$

$$\frac{dy}{dx} = \frac{dy}{du} \frac{du}{dx} \qquad \text{with } y = f(u) \text{ and } u = g(x)$$

Logarithmic Formulas:
$$\frac{d}{dx} \ln x = \frac{1}{x} \qquad \frac{d}{dx} \ln f(x) = \frac{f'(x)}{f(x)}$$

Exponential Formulas:
$$\frac{d}{dx} e^x = e^x \qquad \frac{d}{dx} e^{f(x)} = e^{f(x)}f'(x)$$

AREA AND VOLUME FORMULAS

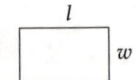

Rectangle

Area $= l \cdot w$

Perimeter $= 2l + 2w$

Circle

Area $= \pi r^2$

Circumference $= 2\pi r$

Rectangular solid

Volume $= l \cdot w \cdot h$

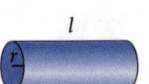

Cylinder

Volume $= \pi r^2 l$

Sphere

Volume $= \frac{4}{3} \pi r^3$

Surface area $= 4\pi r^2$

PROPERTIES OF NATURAL LOGARITHMS

1. $\ln 1 = 0$
2. $\ln e = 1$
3. $\ln e^x = x$
4. $e^{\ln x} = x$
5. $\ln (MN) = \ln M + \ln N$

6. $\ln \left(\dfrac{1}{N} \right) = -\ln N$

7. $\ln \left(\dfrac{M}{N} \right) = \ln M - \ln N$

8. $\ln (M^N) = N \ln M$